Artificial Intelligence in Cancer Diagnosis and Therapy

Artificial Intelligence in Cancer Diagnosis and Therapy

Editors

Hamid Khayyam
Ali Madani
Rahele Kafieh
Ali Hekmatnia

MDPI • Basel • Beijing • Wuhan • Barcelona • Belgrade • Manchester • Tokyo • Cluj • Tianjin

Editors
Hamid Khayyam
School of Engineering,
RMIT University
Australia

Ali Madani
Cyclica Inc.
Canada

Rahele Kafieh
Department of Engineering,
Durham University
UK

Ali Hekmatnia
Radiology Department,
School of Medicine,
Isfahan University of
Medical Sciences
Iran

Editorial Office
MDPI
St. Alban-Anlage 66
4052 Basel, Switzerland

This is a reprint of articles from the Topic published online in the open access journals *AI* (ISSN 2673-2688), *Cancers* (ISSN 2072-6694), *Current Oncology* (ISSN 1718-7729), *Diagnostics* (ISSN 2075-4418), and *Onco* (ISSN 2673-7523) (available at: https://www.mdpi.com/topics/AI_Cancer).

For citation purposes, cite each article independently as indicated on the article page online and as indicated below:

LastName, A.A.; LastName, B.B.; LastName, C.C. Article Title. *Journal Name* **Year**, *Volume Number*, Page Range.

ISBN 978-3-0365-6672-6 (Hbk)
ISBN 978-3-0365-6673-3 (PDF)

Contents

About the Editors

Hamid Khayyam

Hamid Khayyam received a B.Sc. degree (Hons.) from the University of Isfahan, a M.Sc. degree from the Iran University of Science and Technology, and a Ph.D. degree in Mechanical Engineering with specialization in intelligent systems from Deakin University, Australia. Dr. Khayyam has worked in automation and energy productivity in various industrial companies for more than ten years. In his previous position, he was leading the efforts on modeling the control and optimization of energy systems at the Carbon Nexus production line, Deakin University utilizing artificial intelligence and machine learning technologies. Dr. Khayyam is currently a Senior Lecturer in the Department of Mechanical Engineering, School of Engineering at the RMIT University, Australia. He has contributed more than 600 journal articles to professional journals as an editor and reviewer. Additionally, he has published more than 100 high-quality journal articles, 1 (sole) book, 9 book chapters, and currently serves on several Editorial Boards of ISI journals. Dr. Khayyam's research is focused on instituting new technologies to develop distinctive approaches for the integration of artificial intelligence and machine learning to solve complex systems. He is a keynote and invited speaker, in academic committees in international conferences, webinars, and workshops about technology development, artificial intelligence and machine learning. Dr. Khayyam is a senior academic member of the Intelligent Automation Research Group (IARG) at RMIT in Australia and The Materials and Manufacturing Research Institute (MMRI) at The University of British Columbia in Canada. Dr. Khayyam is a Senior Member of IEEE.

Ali Madani

Ali Madani is currently the Director of Machine Learning at Cyclica Inc, a leading Canadian biotechnology company focused on AI-based drug discovery. Dr. Madani is a Ph.D. graduate of the University of Toronto; alumnus of University of Waterloo School of Engineering; and attained a Master of Mathematics from the University of Waterloo. He also received his B.Sc. degree from the University of Tehran. He is an active member of the machine learning community in Toronto and speaks at world- and Canada-wide conferences, webinars, and workshops about technology development, artificial intelligence, drug discovery, and cancer therapeutics. He has published several scientific articles in high-impact factor journals and contributed to technological patents on these subjects. As a believer in industry-oriented education and the pro-democratization of knowledge, Dr. Madani has educated and mentored thousands of students and professionals on topics such as artificial intelligence, programming, and data science.

Rahele Kafieh

Rahele Kafieh received a BSc (2005) followed by an MSc (2008) and a Ph.D. (2014) in Biomedical Engineering from the Sahand University of Technology and Isfahan University of Medical Sciences in Iran. She obtained her first tenure post as an assistant professor in the Department of Advanced Technologies in Medicine at the Isfahan University of Medical Sciences in 2014 and remained there for 7 years, during which she obtained valuable experience in teaching and supervising projects at undergraduate and graduate levels. She has led many successful previously unexplored projects on medical images, including but not limited to classification and segmentation of retinal images with AI; applications of AI in CT-scan and X-ray images; and multi-modality MRI data analysis. Her two stays at the Charité university hospital, Germany and one stay at the Sabanci University in Turkey were awarded by competitive research scholarships from the Einstein forum and TUBITAK, respectively. Then she moved to Newcastle University, where she worked as a Research Associate in an interdisciplinary team at Newcastle University, a position that provided her the great opportunity to conduct in-depth research on the role of artificial intelligence (AI) in the detection of neurological diseases from the eye. Since July 2022, she has been with the Department of Engineering at Durham University as an Assistant Professor. Her current research is on medical image processing in different organs (eye, chest, teeth, brain, heart, and breast), from different modalities (OCT, Fundoscopy, CT-scan, X-ray, cone-beam CT, MRI, fMRI, and infrared), facing challenges such as high dimensionality, noisiness, and imbalance.

Ali Hekmatnia

Ali Hekmatnia is a Professor of Radiology at the Isfahan University of Medical Sciences, Iran. He received his medical degree from the Shahid Beheshti University of Medical Sciences, Tehran in 1989, with an internship Radiology residency from the Isfahan University of Medical Sciences, Iran in 1995. He has had a visiting fellowship in pediatric radiology at the Great Ormond Street Hospital for Sick Children, London, UK in 2000. He had a visiting fellowship in neuroradiology at the National Hospital for Neurology and Neurosurgery, London, UK in 2000 too.He received a M.Sc. degree in Medical Education at the Isfahan University of Medical Sciences, Iran in 2007. Dr Hekmatnia has contributed in more than 120 peer-reviewed articles to professional journals. His research interests are in CT scan, MRI, pediatric radiology, and neuroradiology.

Preface to "Artificial Intelligence in Cancer Diagnosis and Therapy"

Cancer is the second leading cause of death worldwide. According to the World Health Organization (WHO), around 10 million people died from cancer globally in 2020. The early detection of cancer is of utmost importance for the effective treatment and prevention of the spread of cancer cells to other parts of the body (metastasis). However, this task and assigning effective therapies in clinical cancer settings are of great complexity due to inter- and intra- tumor heterogeneities. The detection, diagnosis, and therapy of cancer are challenged by a hidden pattern of seemingly irregular, chaotic medical events requiring methodologies to capture the complexity of cancer to design effective diagnostic systems and therapies.

Artificial Intelligence (AI) and machine learning have been revolutionizing discovery, diagnosis, and treatment designs. It can aid not only in cancer detection but also in cancer therapy design, identification of new therapeutic targets with accelerating drug discovery, and improving cancer surveillance when analyzing patient and cancer statistics. AI-guided cancer care could also be effective in clinical screening and management with better health outcomes. The Machine Learning (ML) algorithms developed based on biological and computer sciences can significantly help scientists in facilitating the discovery process of biological systems behind cancer initiation, growth, and metastasis. They can also be used by physicians and surgeons in the effective diagnosis and treatment design for different types of cancer and for biotechnology and pharmaceutical industries in carrying out more efficient drug discovery. AI and machine learning may be defined as the branch of computer science that is concerned with intelligent behavior. Artificial intelligence techniques learn about the data they are trained on and, subsequently, learning algorithms are designed to generalize from those data.

This book covers some significant impacts in the recent research of AI and machine learning in both the private and public sectors of cancer diagnosis and therapy. The book is divided in five groups:

The first group is AI in prognosis, grading, and prediction. AI is a powerful tool for prognosis, a branch of medicine that aims in predicting the future health of patients. It performs well in assisting cancer prognosis because of its unprecedented accuracy level.

The second group is AI in clinical image analysis. Image-based methods are among the most powerful applications of AI and recent deep learning methods. AI provides real-time and highly accurate image analytics to increase the quality and localize the anatomical features (pre/post processing), facilitate powerful augmented reality and virtual reality applications in the medical domain, and develop the classification and diagnosis of diseases using the medical images.

The third group is AI models for pathological diagnosis. With the impressive growth in the application of AI in health and in pathology, the specific role of AI in supporting routine diagnosis, particularly for patients with cancer, is evident from many published works. AI can handle the enormous quantity of data created throughout the patient care lifecycle to improve pathologic diagnosis.

The Fourth group is ML and statistical models for molecular cancer diagnostics and genetics. Molecular diagnosis involves processing samples of tissue, blood, or other body fluid to look for the presence of certain genes, proteins, or other molecules. They might be a sign of a disease or condition, such as cancer. AI methods provide lots of opportunities for the analysis of such detailed and gigantic data with high accuracy and lead time.

The fifth group is AI in triage, risk stratification, and screening cancer. Due to the complex and expensive procedure needed for the treatment of cancer, triage and risk stratification provide the procedure of assigning levels of priority to patients to determine the most effective order in which to be treated. The health providers can then identify the right level of care and services for distinct subgroups of patients. AI enables this prediction to occur rapidly, immediately, and accurately.

This book is aimed at serving researchers, physicians, biomedical engineers, scientists, engineering graduates, and Ph.D. students of medical, biomedical engineering, and physical science together with interested individuals in medical, engineering, and general science. This book focuses on the application of artificial intelligence and machine learning methods in cancer diagnosis and therapy including prognosis, grading and prediction, clinical image analysis, pathological diagnosis, molecular cancer diagnostics and genetics, and traige, risk stratification, and screening cancer with approaches representing a wide variety of disciplines including medical, engineering, and general science. Throughout the book, great emphasis is placed on medical applications of cancer diagnosis and therapy, as well as methodologies using artificial intelligence and machine learning. The significant impact of the recent research that has been selected is of high interest in cancer diagnosis and therapy as complex systems. An attempt has been made to expose the reading audience of physicians, engineers, and researchers to a broad range of theoretical and practical topics. The topics contained in the present book are of specific interest to physicians and engineers who are seeking expertise in cancer diagnosis and therapy via artificial intelligence methods and machine learning. The primary audience of this book is researchers, graduate students, and engineers in applications of AI in CT-scan and X-ray images, computer engineering, and science and medicine disciplines. In particular, the book can be used for training graduate students as well as senior undergraduate students to enhance their knowledge by undergoing a graduate or advanced undergraduate course in the areas of cancer diagnosis and therapy and engineering applications. The covered research topics are also of interest to researchers in medicine, biomedical engineering, and academia who are seeking to expand their expertise in these areas.

Acknowledgments

This book has been made possible through the effective collaborations of all the enthusiastic chapter author contributors, who have expertise and experience in various disciplines in computer science, clinical cancer settings, and related fields. They deserve the sincerest gratitude for their motivation in creating such a book, for the encouragement in completing the book, for the scientific and professional attitude in constructing each of the chapters of the book, and for the continuous efforts toward improving the quality of the book. Without the collaboration and consistent efforts of the chapter contributors including authors and anonymous reviewers, the completion of this book would have not been possible.

Hamid Khayyam , Ali Madani, Rahele Kafieh, and Ali Hekmatnia

Editors

diagnostics

MDPI

Review

Current Value of Biparametric Prostate MRI with Machine-Learning or Deep-Learning in the Detection, Grading, and Characterization of Prostate Cancer: A Systematic Review

Henrik J. Michaely [1,*], Giacomo Aringhieri [2,3], Dania Cioni [2,3] and Emanuele Neri [2,3]

1 Medical Faculty Mannheim, University of Heidelberg, 69120 Heidelberg, Germany
2 Academic Radiology, Department of Translational Research, University of Pisa, 56126 Pisa, Italy; giacomo.aringhieri@unipi.it (G.A.); dania.cioni@unipi.it (D.C.); emanuele.neri@unipi.it (E.N.)
3 Italian Society of Medical and Interventional Radiology, SIRM Foundation, Via della Signora 2, 20122 Milano, Italy
* Correspondence: michaely@radiologie-karlsruhe.de; Tel.: +49-721-932-480

Abstract: Prostate cancer detection with magnetic resonance imaging is based on a standardized MRI-protocol according to the PI-RADS guidelines including morphologic imaging, diffusion weighted imaging, and perfusion. To facilitate data acquisition and analysis the contrast-enhanced perfusion is often omitted resulting in a biparametric prostate MRI protocol. The intention of this review is to analyze the current value of biparametric prostate MRI in combination with methods of machine-learning and deep learning in the detection, grading, and characterization of prostate cancer; if available a direct comparison with human radiologist performance was performed. PubMed was systematically queried and 29 appropriate studies were identified and retrieved. The data show that detection of clinically significant prostate cancer and differentiation of prostate cancer from non-cancerous tissue using machine-learning and deep learning is feasible with promising results. Some techniques of machine-learning and deep-learning currently seem to be equally good as human radiologists in terms of classification of single lesion according to the PIRADS score.

Keywords: prostate cancer; multiparametric prostate MRI; biparametric prostate MRI; deep-learning; radiomics; artificial intelligence; cancer detection; PIRADS

Citation: Michaely, H.J.; Aringhieri, G.; Cioni, D.; Neri, E. Current Value of Biparametric Prostate MRI with Machine-Learning or Deep-Learning in the Detection, Grading, and Characterization of Prostate Cancer: A Systematic Review. *Diagnostics* **2022**, *12*, 799. https://doi.org/10.3390/diagnostics12040799

Academic Editors: Hamid Khayyam, Ali Madani, Rahele Kafieh and Ali Hekmatnia

Received: 22 February 2022
Accepted: 23 March 2022
Published: 24 March 2022

1. Introduction

1.1. Prostate Cancer

Prostate cancer (PCA) is the second most common cancer in men worldwide and it accounts for up to 25% of all malignancies in Europe [1]. It is the third leading cause of cancer-related death in the United States and Europe [2,3]. The incidence of prostate cancer increases with rising age of patients, and prostate cancer and its management are becoming a major public health challenge. PCA aggressiveness can be linked to specific genes such as BRCA, and behavior such as smoking [4,5]. Accurate and early detection of prostate cancer is therefore paramount to achieve good overall patient outcomes. The tools available for assessing and detecting prostate cancer are digital rectal examination (DRE), PSA screening, transrectal ultrasound, and MRI whereby the latter received the highest amount of attention in the past decade due to its unprecedented capabilities in accuracy [6–8].

In contrast to ultrasound and digital rectal examination, MRI offers an operator-independent tool for objectively assessing the entire prostate gland from base to apex and from the posterior peripheral zone (PZ) to the anterior fibromuscular stroma (AFMS) that are barely assessable with DRE [6,9].

Magnetic resonance imaging of the prostate has a long history going back more than 20 years. In the initial phase, high resolution T2-weighted (T2w) imaging and spectroscopy were mainly used as tools for detecting prostate cancer. Yet, spectroscopy is slow and susceptible to artefacts and was not well perceived. In the recent decade, further developments have taken over including diffusion weighted imaging (DWI), dynamic contrast enhanced imaging (DCE). The entire prostate exam has been standardized worldwide and its reporting has been harmonized by the PIRADS (Prostate Imaging Reporting and Data System) system [10]. This classification system allows to objectively assess the prostate and potential cancerous zones and standardizes reporting over separate sites so that the overall performance of MRI is increased and is more reproducible compared to previous periods. With this development MRI of the prostate follows the trend to standardize the entire radiological procedure from image acquisition to data reporting to achieve a higher reliability, enhanced reproducibility, and a direct implication for radiology-based treatments as it has previously successfully demonstrated in breast imaging with BIRADS (Breast Imaging Reporting and Data System) [11].

The report structuring provided by PIRADS is already a condensation of the imaging information and standardizes reporting and its output. This is one major step toward a more automated and operator-independent radiology. Moreover, the image acquisition parameters, slice orientations, and sequences with its specific sequence characteristics are governed by PIRADS [12]. This automatically sets the stage for a potential automated image analysis. In the past decade, artificial intelligence (AI) with its subdivisions of machine learning (ML), radiomics, and deep learning (DL) has become more prevalent. At this point in time, ML and DL are still no clinical standards. Radiomics, for example, use quantitative imaging features that are often unrecognizable to the human eye. Therefore, it is increasing the number of potential parameters to the multi-parametric approach of prostate MRI and with potential benefits for PCA detection and grading and beyond. DL techniques such as convoluted neural networks (CNN) are currently considered gold standard in computer vision and pattern recognition and hence have potential benefits for PCA detection and grading. With larger data sets as basis, they have the potential to automatically learn and deduct conclusions so that PCA recognition based on unperceivable features to the human eye might be possible. Despite numerous experimental studies which will be discussed further in this study, there is no standardized approach on how to use and implement DL and ML for prostate imaging now.

The aim of this study is to elucidate the status of artificial intelligence in prostate imaging with a focus on the so-called bi-parametric (bp) approach of prostate MRI (bpMRI).

1.2. Prostate Imaging Reporting and Data System

PIRADS was established by key global experts in the field of prostate imaging from America and Europe (European Society of Urogenital Radiology (ESUR), American College of Radiology (ACR)) to facilitate and standardize prostate MRI with the aim of assessing the risk of clinically significant prostate cancer (csPCA). The first version of the PIRADS recommendations was published in December 2011, the latest and current update was published in 2019 (PIRADS v2.1) [10,12,13].

Various studies have compared the predictive performance of PI-RADS v1 for the detection of csPCA compared to image-guided biopsy or radical prostatectomy (RP) specimens as standard of reference. In a 2015 study, Thompson reported multi-parametric MRI detection of csPCA had sensitivity of 96%, specificity of 36%, negative predictive value and positive predictive values of 92% and 52%; when PI-RADS was incorporated into a multivariate analysis (PSA, digital rectal exam, prostate volume, patient age) the area under the curve (AUC) improved from 0.776 to 0.879, $p < 0.001$ [14]. A similar paper showed that PI-RADS v2 correctly identified 94–95% of prostate cancer foci ≥ 0.5 mL but was limited for the assessment of Gleason Score (GS) $\geq 4 + 3$ csPCA ≤ 0.5 mL [15]. An experienced radiologist using PIRADS v2 is reported to achieve an AUC of 0.83 with 77% sensitivity and 81% specificity [16].

1.3. Sequences for Prostate MRI

The initial protocol for MRI of the prostate as provided by PIRADS included high-resolution multiplanar T2w-imaging, DWI, and DCE after the intravenous administration of paramagnetic gadolinium chelate contrast agent. This so-called multi parametric prostate MRI (mpMRI) is considered as the gold standard. T2w-imaging is used to demonstrate zonal anatomy of the prostate. Tumors can be well delineated, and their relation to the prostate capsule can be examined. Benign changes such as benign prostate hyperplasia, post-prostatic changes of the peripheral zone or scars can be identified. T2w-imaging is considered the gold standard for the transitional zone (TZ) of the prostate gland. In addition, T2w-imaging can be used to measure the volume of the prostate. The high anatomic information content of T2w-imaging makes this sequence the perfect roadmap for image-guided biopsy [12,17].

DWI serves as an indirect measure of cellular density. In case of a malignant tumor with high cellular density, the ability of water to freely move in the interstitial compartment is decreased hence the diffusion is impaired. The images with high b-values and even those with more and more common-interpolated calculated b-values allow quick and easy depiction of these suspicious areas in the prostate. The calculated ADC maps give a quantitative measure of cellular density and can be considered as a molecular imaging tool for tumor aggressiveness. DWI imaging is considered as the reference sequence for the peripheral zone (PZ) of the prostate [12,17].

Dynamic contrast enhancement (DCE) is considered as the weakest of the three used approaches for prostate imaging. In contrast to T2w-imaging and DWI, DCE is not being considered as a dominant sequence for any of the prostate zones. It only serves as a tiebreaker in very specific questions in the PIRADS system. In addition, it requires the intravenous administration of contrast agent with the risk of side-effects such as allergies, nephrogenic systemic fibrosis, or Gadolinium deposition in the body [18–21]. While the risk of nephrogenic systemic fibrosis is controllable by using little amounts of macrocyclic Gd-chelates, no harmful consequence for Gd-chelate depositions in the body has been found [22,23]. Nevertheless, patients often try to avoid contrast agent if feasible. Moreover, physicians embrace the idea of non-enhanced exams equally, as it speeds up the acquisition and reduces the number of potential complications. In addition, omitting contrast agent permits to save money.

1.4. Multiparametric and Biparametric MRI of the Prostate

With this in mind and the knowledge that the performance of DCE often yielded limited added value to T2w-imaging and DWI in mpMRI of the prostate bi-parametric MRI (bpMRI) of the prostate is gaining considerable support [15]. Meanwhile, there are several high-ranked studies such as the PROMIS trial and meta-analyses comparing mpMRI and bpMRI of the prostate [24–26]. Current data underline the high negative value of bpMRI in biopsy-naïve patients with a negative predictive value of up to 97% [27,28]. Whether bpMRI might be slightly less accurate in less-experience readers is not yet clearly proven [29,30]. A currently accepted position is that bpMRI of the prostate seems to be equally good as mpMRI of the prostate for patients with low and high risk for csPCA but DCE might be of worth in patients with intermediate risk and PIRADS 3 lesions [25,26,31–35] (Figure 1). bpMRI of the prostate is also commonly used for computer-based postprocessing using artificial intelligence. This is due to the fact that DCE contains a fourth dimension (time) which make those images harder to algin and match with two-dimensional anatomical images such as T2w-imaging and DWI. Another drawback of DCE is that image information is not obvious. The image information on contrast media arrival and distribution which is seen as a surrogate marker for microvascular density have to be extracted using semiquantitative or quantitative pharmacokinetic models which adds another layer of complexity on postprocessing, along with the increase of time necessary to report the exams.

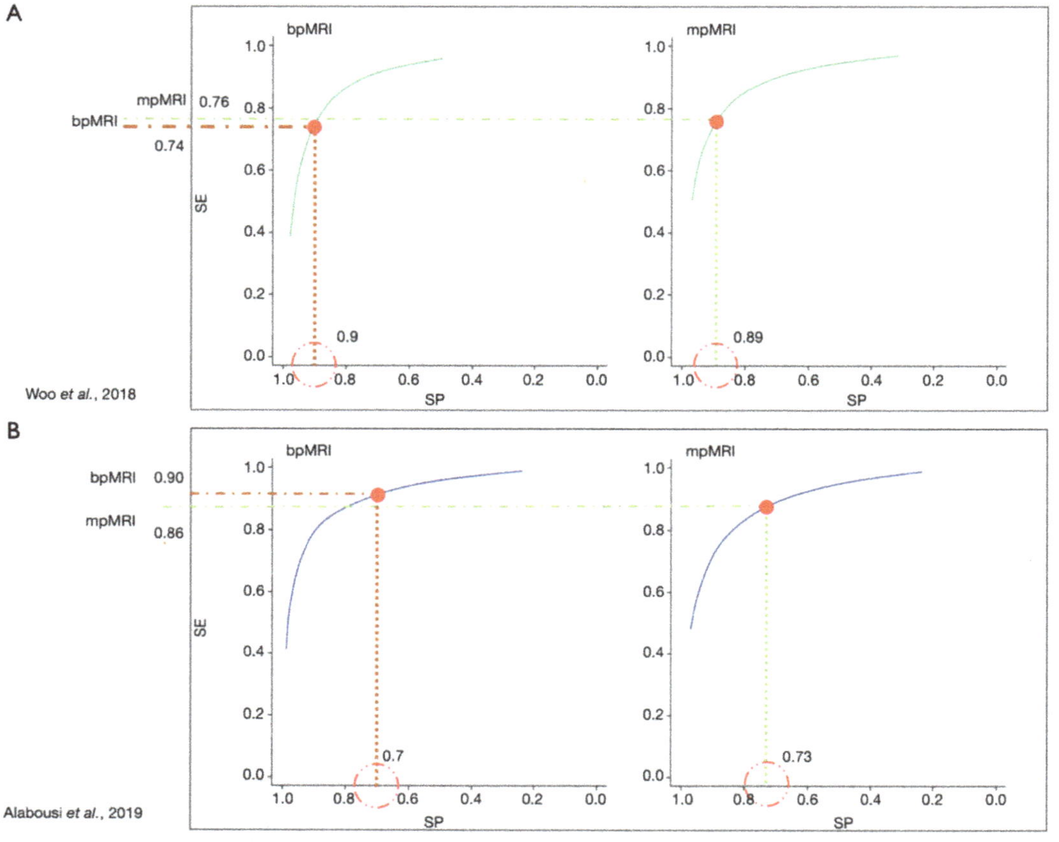

Figure 1. Overview of the performance of mpMRI and bpMRI based on data from Woo et al. [33] and Alabousi et al. [25] demonstrating the near equal performance of bpMRI to mpMRI (reprinted with permission from [17], Copyright 2020 Gland Surgery).

1.5. Artificial Intelligence (AI) for Image Postprocessing

The availability of cheap and high computing power with the additional advent of postprocessing technologies and artificial intelligence such as machine learning techniques and deep neural networks has fostered the application of those techniques for radiology tasks such as tumor detection. The current hierarchical concept of AI is depicted in Figure 2.

Machine-learning (ML) is a subfield of AI in which algorithms are trained to perform tasks by learning rules from data rather than explicit programming. *Radiomics* is seen as a method that extracts large numbers of features from radiological images using data characterization algorithms such as first order statistics, shape-based, histogram-based analyses, Gray Level Co-occurrence Matrix, Gray Level Run Length Matrix, Gray Level Size Zone Matrix, Gray Level Dependence Matrix, Neighboring Gray Tone Difference Matrix to name a few [36–39]. These features are said to have the potential to uncover disease characteristics that are hard to be appreciated by the naked eye. The hypothesis of radiomics is that distinctive imaging features between disease forms may be useful for detecting changes and potentially predicting prognosis and therapeutic response for various conditions such as e.g., detection of csPCA. These radiomic features are then often further analyzed using ML-techniques. An example of a radiomics ML-workflow is shown in Figure 3. An issue concerning ML-techniques is that it often requires the manual placement of a region of interest hence hereby introducing a potential source for errors and biases.

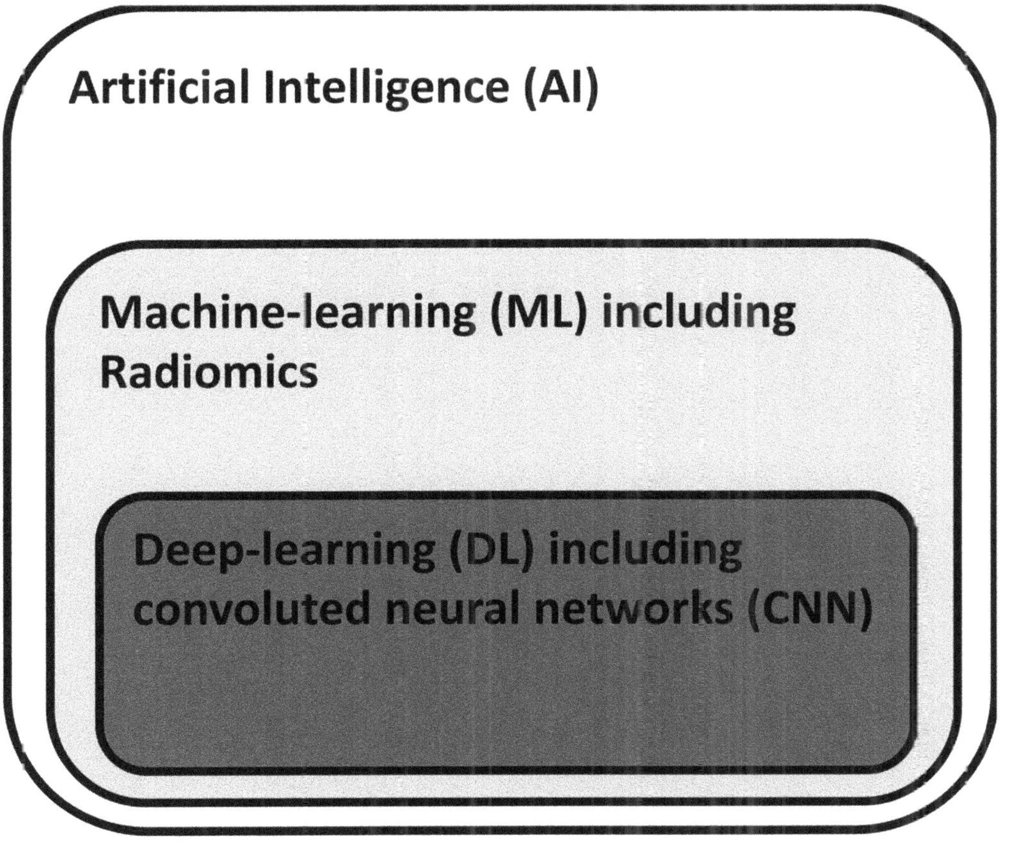

Figure 2. Hierarchical structure of AI-techniques. Whereas ML requires human feature engineering as guidance for learning, DL is based on self-learning algorithms that can detect and process simple and complex image features.

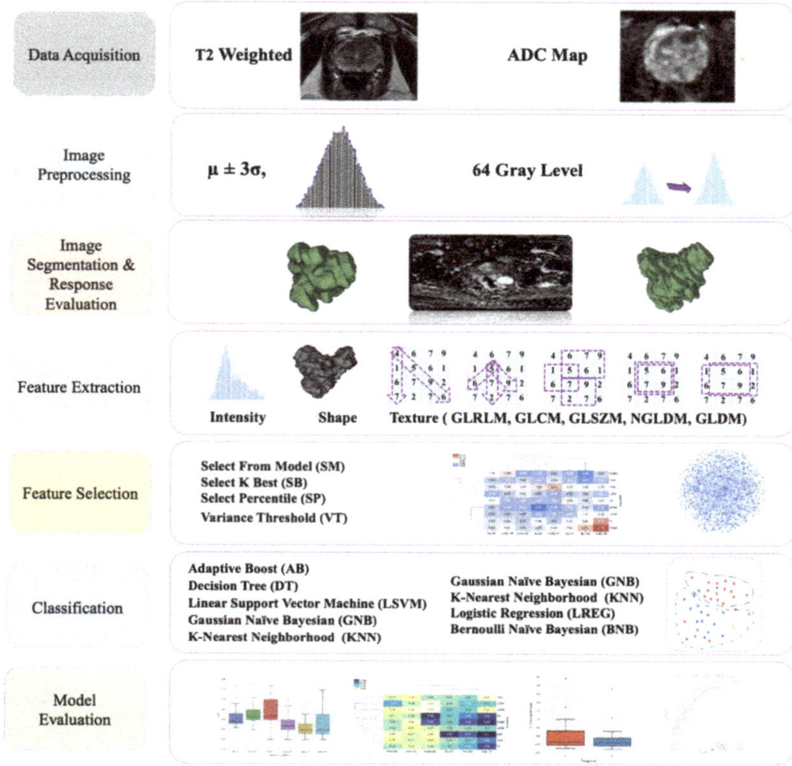

Figure 3. Sample radiomics workflow (reprinted with permission from [40], Copyright 2019 Springer Nature).

Deep learning (DL) is a subfield of AI in which algorithms are trained to perform tasks by learning patterns from data rather than explicit programming. The key factors for the increasing attention that DL attracted in the past years are the availability of large quantities of labelled data, the inexpensive and powerful computing hardware particularly graphic-processing units and improvements in training techniques and architectures. DL is a type of representation learning in which the algorithms learn a composition of features that reflect the hierarchy of structures in the data. Current state-of-the-art for medical image recognition using DL techniques are so called convoluted neural networks (CNN). These networks are characterized by an architecture of connected non-linear functions that learn multiple levels of representations of the input data thereby extracting possibly millions of features [41]. Especially CNNs in which a series of convolution of filter layers are exploited are suitable for image processing [42]. Newer techniques such as transfer learning and data augmentation, or the application of generative methods help in mitigating existing limitations of CNN [43]. The entire process of data processing within the multiple layers of a CNN with convolution filters, pooling, and maximum filtering is beyond the scope of this study. Largely simplified, one might say that bottom layers of the CNN act as a feature extractor while the top layers of the CNN act as a classifier. An overview is given in Figure 4 in which the DL workflow is compared to radiomics or the standard radiology reading process [44]. The reason that CNN-based approaches are considered superior to radiomics is that radiomics depend on hand-crafted features which is limited, whereas CNN can generate features that are most appropriate to the problem itself [45].

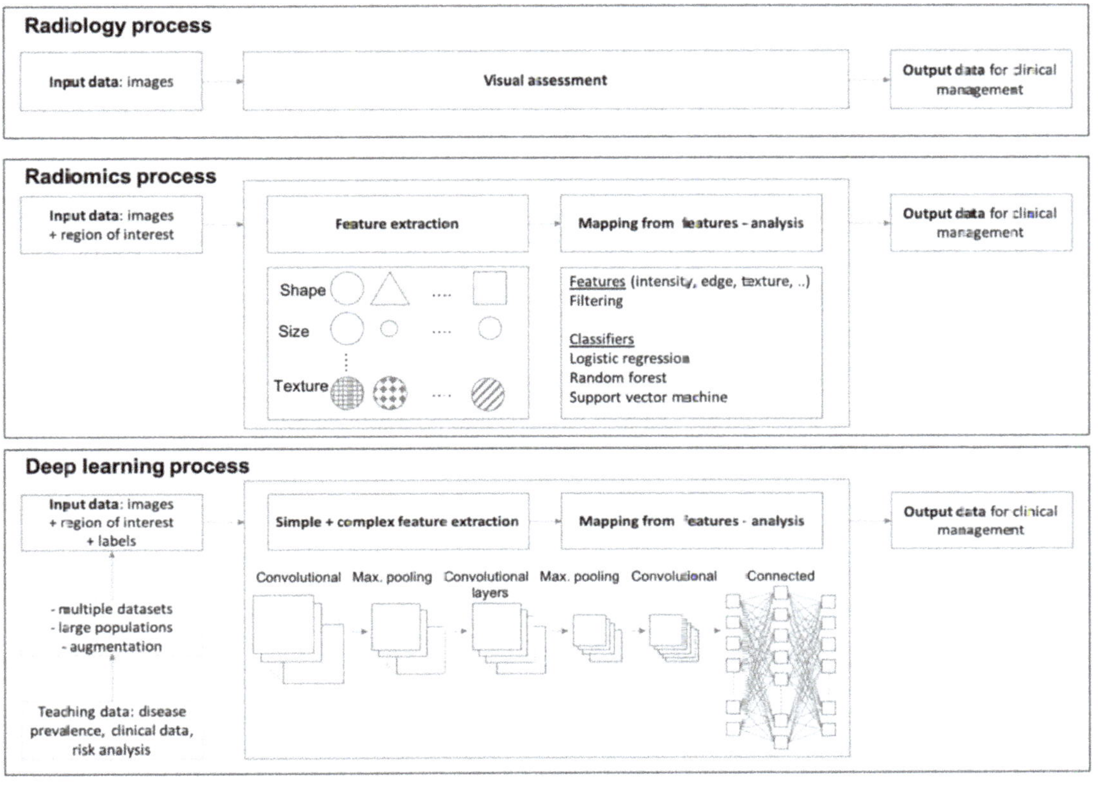

Figure 4. Workflow of standard radiology reporting compared to AI-based methods of radiomic and DL. The entire complexity of deep learning is only schematically shown. There is an abundance of different network architectures or CNN which are beyond the scope of this study. This figure only demonstrates a schematic CNN (reprinted under common creative license 4.0 from [44], Copyright 2021 Springer Nature).

2. Materials and Methods

Literature research for this study took place in August 2021. A PubMed query with the search terms "prostate" and "magnetic" and "deep learning" or "machine learning" or "radiomics" was performed. The aim was to retrieve those studies which made use of ML or DL techniques to facilitate tumor detection and grading. To make sure that only current techniques were included in the analysis only publications from the year 2019 to 2021 were included. Particularly in the field of CNN the technical improvement is rapidly evolving so that elder publications might not represent the current state-of-the-art. Total of 95 publications were initially retrieved. Of these, 66 were omitted for several reasons so that 29 publications were available for analysis (see Figure 5). Clinical data (question to be answered, number of patients, age, AI-technique, lesion segmentation, MRI-technique, sensitivity, specificity, accuracy, AUC) were then manually extracted and transferred to a Microsoft Excel 365 spreadsheet (Microsoft, Redmond, WA, USA). PRISMA guidelines were followed [46]. An overview of the study according to the PRISMA guidelines can be found in the Appendix A.

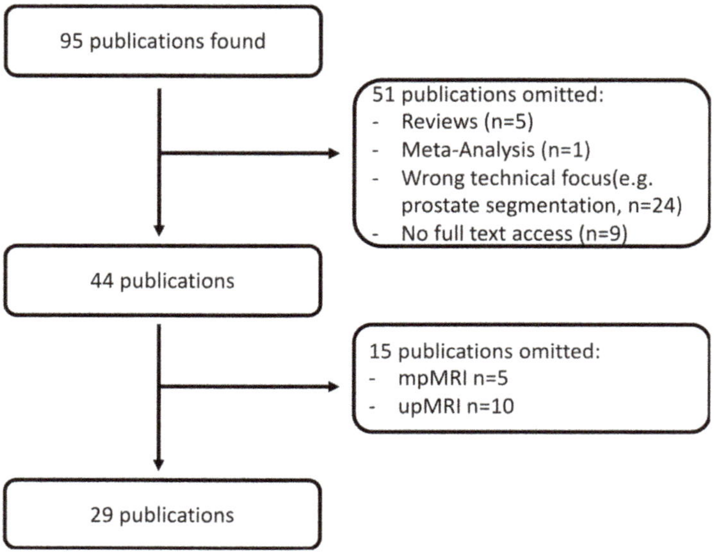

Figure 5. Literature selection work-flow. ML–machine-learning. DL–deep learning. up–uniarametric. bp–biparametric. mp–multiparametric.

This paper focuses on bpMRI. The current PIRADS guidelines state: "Given the limited role of DCE, there is growing interest in performing prostate MRI without DCE, a procedure termed "biparametric MRI" (bpMRI). A number of studies have reported data that supports the value of bpMRI for detection of csPCA in biopsy-naïve men and those with a prior negative biopsy". The potential benefits of bpMRI include: (1) elimination of adverse events and gadolinium, (2) faster MRI-exam times, and (3) overall reduced costs [47]. These factors will potentially make bpMRI easily accessible. Remaining concerns are that the DCE sequence may serve as backup in case of image degradation of the DWI or T2w sequence. It seems as if DCE may be of less value for assessment of treatment of naïve prostate patients but remains essential in assessment for local recurrence following prior treatment, which however is a setting in which current PI-RADS assessment criteria do not apply. The conclusion of the PIRADS steering committee therefore advocates the use of mpMRI particularly in (1) patients with prior negative biopsies with unexplained raised PSA values, (2) those in active surveillance who are being evaluated for fast PSA doubling times or changing clinical/pathologic status, (3) men who previously had undergone a bpMRI exam that did not show findings suspicious for csPCA, and who remain at persistent suspicion of harboring disease, (4) biopsy-naïve men with strong family history, known genetic predispositions, elevated urinary genomic scores, and higher than average risk calculator scores for csPCA, and (5) men with a hip implant or other consideration that will likely degrade DWI [47].

For this paper bpMRI was selected as most studies dealing with ML or DL techniques solely relay on T2w-imaging and DWI. DCE data were rarely included. In contrast to T2w-imaging and DWI the DCE-data must be postprocessed first to generate parameter maps. This process is not yet standardized as several pharmacokinetic models and hereof derived software implementations for postprocessing exist. Without generation of parameter maps a huge number of images would have to be fed into the ML/DL algorithms—a step that most research groups obviously did not want to undertake.

3. Results

All included studies are listed with an abbreviated overview in Table 1.

Table 1. List of include studies and relevant key information.

Reference	Year	ML	DL	Field Strength	Target	Number of Patients	Age	SS/SP/Accuracy	AUC	Sequences Used
Abdollahi H. et al. [40]	2019	1	0	1.5 T	Gleason score prediction	33	73 (51–82)		0.739	T2, ADC
Wu M. et al. [48]	2019	1	0	3 T	TZ PCA detection	44	68 ± 7	93.2%/98.4%	0.989 (LR)	T2, ADC
Varghese B. et al. [49]	2019	1	0	3 T	Grading prediction	68 53		86%/72%	0.71	T2, ADC,
Min X. et al. [50]	2019	1	0	3 T	ci/csPCA discrimination TZ and PZ	280		84.1%/72.7%	0.823	T2, ADC, b1500
Toivonen J. et al. [51]	2019	1	0	3 T	Gleason prediction TZ and PZ	62	65 (45–73)		0.88	T2, b0-b2000, T2mapping
Chen T. et al. [52]	2019	1	0	3 T	Tumor detection aggressiveness prediction TZ and PZ	182 199	73 (55–90)	98.6/99.2%/98.9% (noPCA vs. PCA) 100/98.25 8/99.1% (ci vs. csPCA)	0.999 (noPCA vs. PCA) 0.933 (ciPCA vs. csPCA)	T2, ADC
Xu M. et al. [53]	2019	1	0	3 T	Tumor detection	331	71 (46–94)		0.92 (Radiomics) 0.993 (R + clinical data)	T2, ADC, DWI
Zhong X. et al. [54]	2019	0	1	3 T	ci/cs PCA discrimination DL vs. PIRADS exp. radiologists	140		63.6%/80.6%/72.3% 86.4%/48.0%/86.4%	0.726 (DL) 0.711 (PIRADS v2)	
Yuan Y. et al. [55]	2019	0	1	3 T	ci/cs PCS discrimination (GS > 7)	132 112		−/−/86.9%		T2 ax and sag, ADC
Xu H. et al. [56]	2019	0	1	3 T	Detection of PIRADS ≥ 3 lesions	346		−/−/93.0%	0.950	T2, ADC, high b-value
Schelb P. et al. [57]	2019	0	1	3 T	DL and radiologist for lesion (PIRADS ≥ 3 and 4) detection and segmentation	250 62	64 (58–71) 64 (60–69)	98/17% Rad, PIRADS ≥ 3 84/48% Rad, PRIADS ≥ 4 99/25%, DL, PIRADS ≥ 3 83/55%, DL, PIRADS ≥ 4		T2, ADC, DWI
Montoya Perez I. et al. [58]	2020	1	0	3 T	Detection of csPCA with bpMRI, RNA and clinical data	80	65 ± 7.1		0.92	T2, DWI
Hou Y. et al. [59].	2020	1	0	3 T	csPCA in PIRADS 3 identification in TZ and PZ	263	66.8 ± 11.4		0.89	T2, ADC, b1500
Mehralivand S. et al. [60]	2020	1	0	3 T	Detection csPCA in TZ and PZ	236		50.8%/−/− (TZ, MRI) 61.8%/−/− (TZ, DL)	0.749 (MRI) 0.775 (DL)	T2, b1500
Gong L et al. [61]	2020	1	0	3 T	ci/cs PCA discrimination	326 163		73.8%/65.8%/69.9%	0.788	T2, ADC, b800
Bleker J. et al. [62]	2020	1	0	3 T	ci/cs PCA discrimination in PZ	206	66 (48–83)		0.870 (mpMRI) 0.816 (bpMRI)	T2, ADC, DWI, (DCE)
Zong W. et al. [63]	2020	0	1	3 T	CNN optimization	367		100/92%	0.840	T2, ADC, b0

Table 1. *Cont.*

Reference	Year	ML	DL	Field Strength	Target	Number of Patients	Age	SS/SP/Accuracy	AUC	Sequences Used
Sanford T. et al. [64]	2020	0	1	3 T	Automated PIRADS classification compared to radiologist	687	67 (46–89)			T2, ADC, high b-value
Brunese L. et al. [65]	2020	1	1	1.5 T	Gleason score prediction	52		−/−/98%		T2, DCE
Chen Y. et al. [66]	2020	0	1	3 T	Prostate and cancer segmentation	136	68 (49–62)	75.1/99.9%		T2, ADC, b1200
Winkel D.J. et al. [67]	2020	0	1	3 T	bpMRI PCA Screening	49	58 (45–75)	87/50%		T2, ADC, b2000
Arif M. et al. [68]	2020	0	1	3 T	Detection of csPCA in AS	292	68 (62–72)	92/76%	0.89	T2, ADC, b800
He D. et al. [69]	2021	1	0	3 T	Tumor detection Prediction ECE Prediction PSM	459	65 (30–89)		0.863 0.905 (integrated model)	T2, ADC
Vente C. et al. [70]	2021	0	1	3 T	csPCA detection and grading	99 63				T2, ADC
Chen J. et al. [71]	2021	0	1	3 T	csPCA detection and grading	25		89.6/90.2%/92.1%	0.964	T2, T1
Cao R. et al. [72]	2021	0	1	3 T	PCA detection and grading	126 427	62.4 ± 6.4 61.1 ± 7.1	98/17% PIRADS, ≥3 85/58% PIRADS, ≥4 100/17% Unet ≥ 3 83/58% Unet ≥ 4		T2, ADC
Hou Y. et al. [73]	2021	0	1	3 T	ECE prediction	590 150 103	69.2 (42–86) 69.2 (48–83) 70.2 (52–87)		0.857 0.728	T2, ADC, b1500
Yan Y. et al. [74]	2021	1	1	3 T	BCR prediction	485	69.8		0.802 (C-index)	T2
Schelb P. et al. [75]	2021	0	1	3 T	csPCA detection and grading	284	64 (IQR 61–72)	98/17% PIRADS, ≥3 85/55% PIRADS, ≥4 99/24% Unet ≥ 3 83/55% Unet ≥ 4		T2, ADC, b1500

Total of 29 studies were included in this study. Thirteen of them used ML (44.8%), 14 of them used DL-techniques (48.2%), and 2 of them used a combination of ML and DL (6.9%). The data for 27 of the studies were acquired at 3T (93.1%), 2 of them were acquired at 1.5 T (6.9%). A total of 7466 patients were analyzed within this data set. Hereby, the ProstatEx-data set from the Radbound University, The Netherlands was used seven times. The smallest study had a sample size of 25 patients, the largest study had a sample size of 834 patients. The MRI-technique used for AI-postprocessing most often was T2w-imaging in combination with ADC map and DWI (15 studies/53.6%). Runner-up were T2w-imaging and ADC map (8 studies, 28.6%) and T2w-imaging and DWI (2 studies, 7.1%).

3.1. Tumor Detection and Grading

As seen in Table 1, the results (AUC, sensitivities and specificities) were comparable and no trend clearly favoring ML or DL-approaches in terms of superiority could be detected. Most studies required manual interaction in which a radiologist had to segment the region of interest.

Overall, the rate of detection and correct tumor creating using AI-techniques was comparable to the performance of trained radiologists in most studies. Studies were often hard to compare as they differed in terms of standard of reference (e.g., Gleason score (GS) vs. PIRADS vs. National Comprehensive Cancer Network Guidelines vs. ISUP Guidelines) and different cut-off values within the same grading system (e.g., GS 7 was in one study considered intermediate grade, in most studies considered high-grade tumor). Some studies focused on the PZ only, while others accepted the entire gland as target tissue.

In a small study with 33 patients to predict IMRT response, GS prediction and PCA stage, GS prediction using T2w-radiomic models was found more predictive (mean AUC 0.739) rather than ADC models (mean AUC 0.70), while for stage prediction, ADC models had higher prediction performance (mean AUC 0.675). For T2w-radiomic models, mean AUC was obtained as 0.625 [40].

Using T2w-imaging and 12 b-values from diffusion along with Kurtosis analysis and T2 mapping for differentiation GS $\leq 3 + 3$ vs. GS $> 3 + 3$ an AUC of 0.88 (95% CI 0.82–0.95) could be reached. This study with 72 patients was the only one to employ T2 mapping which, after all, was deemed as of little worth [51].

In a stringent ML-Radiomics study, an equally high AUC for tumor grading according to National Comprehensive Cancer Network guidelines in low-risk vs. high-risk (i.e., GS ≥ 8) was found for the PIRADS assessment as well as for the ML-approach (0.73 vs. 0.71, $p > 0.05$) [49]. Interestingly, the precision and recall were higher with the ML-approach compared to the PIRADS assessment (0.57 and 0.36 vs. 0.45 and 0.61). Similar results were found for the discrimination of ciPCA and csPCA of the PZA using a ML-Radiomics approach with extreme gradient boosting [62]. In this study performed on the ProstaTEx dataset, an AUC of 0.816 for the detection of csPCA using bpMRI was found. Adding DCE slightly increased AUC to 0.870, though this was not statistically significant. Based on the same data set but using optimized CNNs Zong et al. [63] concluded that adding ktrans from DCE deteriorated sensitivity and specificity when compared to bpMRI alone from 100%/83% to 71%/88%. The optimal reported AUC of this study was 0.84

Extremely good ML-radiomics results for differentiation ciPCA vs. csPCA with an AUC of 0.999 were found in a study by Chen et al. They could also show that ML-radiomics exhibited a higher efficacy in differentiation ciPCA from csPCA than PIRADS. A potential explanation for this, compared to the other studies, is that outstanding result might be the study inclusion/exclusion criteria: small lesions <5 mm and lesion not well delineable on MRI were excluded [52].

Somewhat poorer results were presented in a study by Gong et al. [61]. Their ML-radiomics approach that was built on T2w-imaging and b800-DWI images yielded an AUC of 0.787 and an accuracy of 69.9% for the discrimination between ciPCA and csPCA. Adding clinical data to the MRI-based data slightly degraded the results with an AUC of

0.780 and an accuracy of 68.1%. A potential reason for this poorer outcome might be a different set of inclusion parameters.

Zhong et al. compared the performance of DL and Deep Transfer Learning (DTL) with experienced radiologists. They found that DTL further improves DL. The DTL results were comparable to radiologist's performance using PIRADS v2. They concluded that DTL might serve as an adjunct technique to support non-experienced radiologists [54]. Similar results found a study using a CNN-trained algorithm to automatically attribute PIRADS scores to suspicions lesions. A performance comparable to a human radiologist was described [64]. The lowest agreement was found with low PIRADS score, getting better with higher PIRADS scores. There was no statistically significant difference between the radiologist-assigned PIRADS score and the AI-assigned PIRADS score with regards to the presence of csPCA for PIRADS 3–5.

In contrast, for Gleason score prediction one study found better results for AI-based approaches than radiologists for PZ and TZ [76]. This could be particularly useful in the context of active surveillance.

A different study looking into aggressiveness prediction (GS > 8) found equal AUCs for AI and radiologists but higher precision and recall rates for AI than PIRADS mitigating the problem of inter-reader variability [49].

An uncommon approach was presented in [65]. The authors hereby combined Radiomics and DL-based on bpMRI with DCE and T2w-imaging. No ADC/DWI-images were used. In few patients they included, promising results with an AUC of 0.96–0.98 for Gleason score prediction were found. No further study used this subset of DCE and T2w-imaging.

The prospective IMPROD trial also examined if the addition of clinical data and RNA expression profiles of genes associated with prostate cancer increased the accuracy for detection of csPCA [58]. In this study the bpMRI based data yielded the highest AUC 0.92. Adding RNA-based data or clinical data neither improved the results nor yielded better results by itself.

Cao et al. developed an FocalNet to automatically detect and grade PCA (Figure 6) [72]. A similar work was presented by Schelb et al. [75] where a U-Net was trained to detect, segment, and grade PCA. In comparison with radiologists' PIRADS assessment, the U-Net sensitivities and specificities for detection of PCA at different sensitivity levels (PIRADS $\geq$ 3 and PIRADS $\geq$ 4) were comparable.

Positive results for DL-based techniques with a larger number of patients (n = 312) were found in a DL-Study by Schelb et al. using a U-Net [57]. They reported a sensitivity/specificity for radiologists using PIRADS for detection of PIRADS lesions $\geq$ 3 and 4 respectively of 96%/88% and 22%/50% while the U-Net approach yielded 96%/92% and 31%/47% ($p > 0.05$). In their study the U-Net also autocontoured the prostate and the lesion with dice-coefficient of 0.89 (very good) and 0.35 (moderate) respectively.

A ML-approach to generate "attention boxes" for the detection of csPCA was published by Mehralivand et al. [60]. Their multicentric approach with data from five institutions showed an AUC of 0.749 for PIRADS assessment of csPCA and a statistically non-significant AUC of 0.775 for the ML-based approach. For the TZ only, the ML-approach yielded a higher sensitivity for detection of csPCA than PIRADS (61.8% vs. 50.8%, $p = 0.001$). Interestingly, the reading time for the ML-approach was on average 40s longer.

An uncommon approach for CNNs was published by Chen et al. [66]. They used U-Net CNNs to segment the prostate and intraprostatic lesions hereby segmenting the PZ, TZ, CZ, and AFMS. Their approach demonstrated impressive results: a Dice coefficient of 63% and a sensitivity and specificity of 74.1% and 99.9% respectively for correctly segmenting the prostatic zones and the suspicious lesion. Yet, in contrast to most other studies, no grading or discrimination of the suspected PCA lesion was performed. As a segmentation study this study was included in this review as it included segmentation of the prostate and detection of the tumor within the prostate.

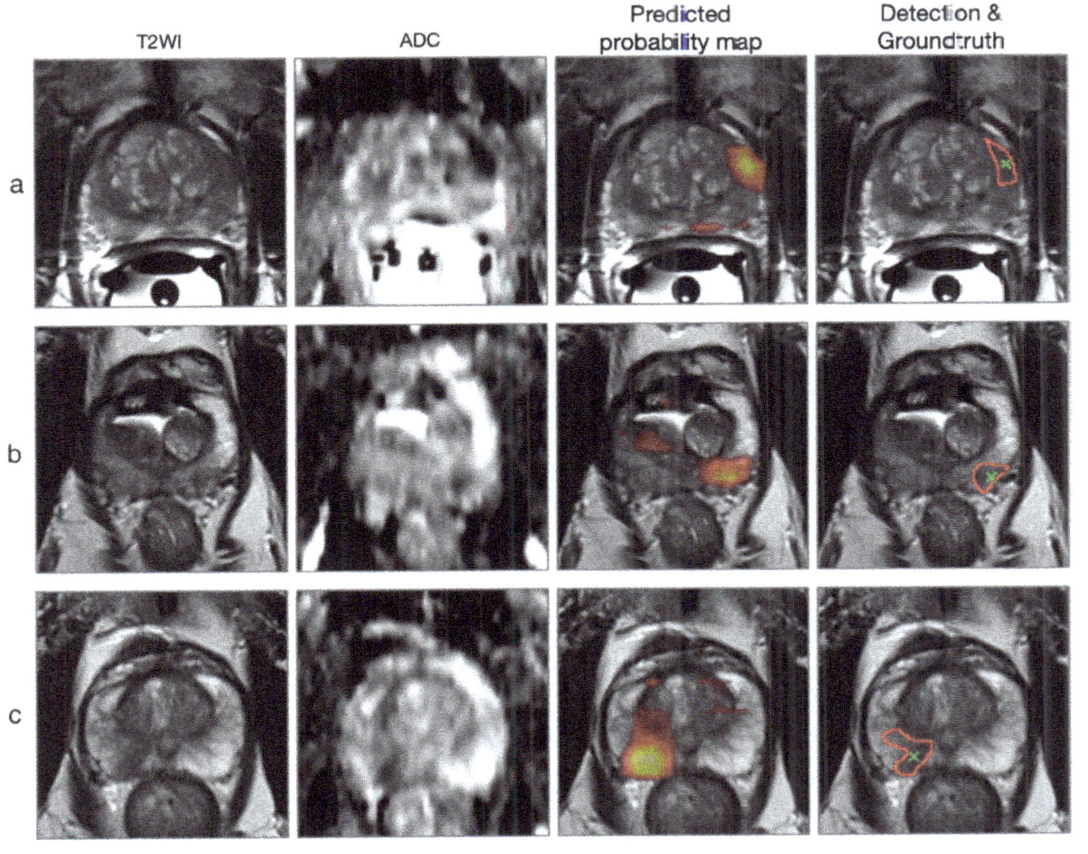

Figure 6. "Examples of lesion detection. The left two columns show the input T2WI and ADC map, respectively. The right two columns show the FocalNet-predicted lesion probability map and detection points (green crosses) with reference lesion annotation (red contours), respectively. (**a**) Patient at age 66, with a prostate cancer (PCa) lesion at left anterior peripheral zone with Gleason Group 5 (Gleason Score 4 + 5). (**b**) Patient at age 68, with a PCa lesion at left posterolateral peripheral zone with Gleason Group 2 (Gleason Score 3 + 4). (**c**) Patient at age 69, with a PCa lesion at right posterolateral peripheral zone with Gleason Group 3 (Gleason Score 4 + 3). ADC = apparent diffusion coefficient; T2WI = T2-weighted imaging"(reprinted with permission from [72], Copyright 2021 John Wiley and Sons).

In a screening study with 3T-bpMRI, Winkel et al. [67] could include and analyze 48 patients, all above 45 years. In a biopsy-correlated reading two human readers and a commercial prototype DL-algorithm were compared in terms of detection of tumor-suspicious lesions and grading according to PIRADS. The DL-approach had a sensitivity and specificity of 87% and 50%. Noteworthy, the DL-analysis required just 14 s.

Different ML-based models were tested and found to be highly accurate for the diagnosis of TZ PCA (sensitivity/specificity/AUC): 93.2%/98.4%/0.989) and their discrimination from BPH-nodules. Reproducibility of segmentation was excellent (DSC 0.84 tumors and 0.87 BPH). Subgroup analyses of TZ PCA vs. stromal BPH (AUC = 0.976) and in <15 mm lesions (AUC = 0.990) remained highly accurate [48].

DL-approach for detection of csPCA in patients under active surveillance was brought up by Arif et al. [68]. Initially 366 patients with low risk were included of which 292 were included in the final study. Sensitivities and specificities for csPCA segmentation rose with increasing tumor volume: tumor volumes > 0.03 cc sensitivity 82% 7 specificity of 43%, AUC 0.65; tumor volume > 0.1 cc sensitivity 85%, specificity of 52%, AUC 0.73. Tumor volumes > 0.5 sensitivity 94%, specificity 74%, AUC 0.89.

A total of six studies among the included studies compared DL/ML-approach to human radiologists [52,57,60,64,72,75]. Overall, due to the small number of studies and because of the different approaches the results cannot be analyzed together. What these studies had in common however was the finding, that at this point AI-based methods revealed a performance similar to that of the radiologists'. No study could either show an advantage of AI-methods of the radiologists or vice versa. An overview about the results can be seen in Table 2.

Table 2. Display of study results comparing human and AI-based performance.

Reference	Year	ML	DL	Metric	Human Radiologist	AI-Approach
Chen T. et al. [52]	2019	1	0	AUC	0.867	0.999
Schelb P. et al. [57]	2019	0	1	Sensitivity/Specificity	98/17% PIRADS $\geq$ 3 84/48% PRIADS $\geq$ 4	99/25% PIRADS $\geq$ 3 83/55% PIRADS $\geq$ 4
Mehralivand S. et al. [60]	2020	1	0	AUC Sensitivity	0.816 89.6%	0.780 87.9%
Sanford T. et al. [64]	2020	0	1	Cancer detection rates	53% PIRADS 3 61% PRIADS 4 92% PIRADS 5	57%, PIRADS 3 60%, PIRADS 4 89% PIRADS 5
Cao R. et al. [72]	2021	0	1	Sensitivity/Specificity	98/17% PIRADS, $\geq$3 85/58% PIRADS, $\geq$4	100/17% PIRADS, $\geq$3 83/58% PIRADS, $\geq$4
Schelb P. et al. [75]	2021	0	1	Sensitivity/Specificity	98/17% PIRADS, $\geq$3 85/55% PIRADS, $\geq$4	99/24% PIRADS, $\geq$3 83/55% PIRADS, $\geq$4

3.2. PIRADS 3 Lesions

Radiomics can detect with high accuracy csPCA in PI-RADS 3 lesions [59,77]. Hou et al. examined in a ML-Radiomics approach the ability of bpMRI to identify csPCA in PIRADS 3 lesions in a group of 253 patients with PIRADS 3 lesions in the TZ and PZ of whom 59 (22.4%) had csPCA [59]. The ML-Radiomics approach including T2w imaging, DWI and ADC had an AUC of 0.89 (95% CI 0.88–0.90) for predicting the presence of csPCA in a PIRADS 3 lesion.

3.3. Extracapsular Extension and Biochemical Recurrence

He et al. set up a large study including 459 patients who underwent 3T bpMRI before prostate biopsy and/or prostatectomy [69]. The aim of the study was first to differentiate between benign and malignant tissue second to predict extracapsular extension (ECE) of prostate tumor and third to predict positive surgical margins (PSM) after RP. Using Radiomics they developed and tested an algorithm that was able to achieve an AUC of 0.905 for the determination of benign and malignant tissue, 0.728 for the prediction of ECE, and a 0.766 for the prediction of PSM. Similarly, Hout et al. found an identical AUC of 0.728 for the prediction of ECE in a DL-based approach using different CNN-architectures [73]. Hence one can infer from the information derived from prostate imaging not only the current situation in the gland but can also predict future developments that might take place under therapy.

Biochemical recurrence (BCR) prediction based on radiomics features was examined in T2w-images only with higher prediction of BCR (C-index 0.802) than conventional scores, particularly also higher than the Gleason scoring system (C-index 0.583) [74]. This work is of particular interest as it first, was one of the few multicentric studies (three centers) with a relatively large number of patients (485) and second, demonstrated the ability of DL-based CNN to look beyond the prostate and infer predictions on the future course of the disease/patient.

4. Discussion

Prostate cancer is a growing medical condition already now being the second most common cancer in men in the western world. The detection and grading of prostate cancer are shifting more toward MRI and is demanding a higher number of MRI-studies to be performed and read. Currently, prostate MRI is considered a specialized exam and requires a highly specific experience to be performed and reported with high quality. A first step toward facilitation of mpMRI prostate acquisition, reading, and reporting was PIRADS, but surely not the last step [10,12,13]. To put it in a nutshell: prostate MRI is developing from the holy grail, and only a few radiologists were being able to read it competently to a commodity in radiology. This is one of the key drivers behind the growing demand for computer-assisted diagnostic tools, such as tumor detection and grading, to facilitate the diagnostic interpretation of prostate MRI also for less-trained radiologists. As the prostate is a densely packed organ with much more information for example as the sparsely packed lung, simple machine learning tools based on e.g., density differences cannot be successfully employed. To distinguish the different prostatic tissues, such as normal transitional and peripheral zones and malignant tissue, higher-developed machine learning tools are required, often based on radiomics or even deep learning techniques. In the papers included in this review, most approaches using either ML or DL were similar to radiologists in their performance [49,54,57,64,75]. For some specific applications, such as tumor detection in the TZ or detection of clinically significant cancers in PIRADS 3 lesions, AI-based methods might even be superior to radiologists' performance [48,59].

These AI-based approaches should enable less well-trained radiologists to read and report prostate-MRI reports with good quality [57,75]. The literature review showed that different approaches to tumor grading and characterization either via ML or DL are capable of differentiating between cancerous and non-cancerous tissue. New approaches are even able to autonomously segment the prostate and the tumor within the gland overcoming a limitation of the elder approaches, where radiologists often had to manually segment the lesions, resulting in a highly time-consuming task [72,75]. Apart from many site-specific implementations of radiomics, ML and DL, another sign of maturation of AI-based approaches is that a first commercial tool was already presented [67]. Compared to the other algorithms, this commercial tool was trained by big data sets for the initial training. This development underlines again the trend in imaging toward commoditization of imaging and democratization of information technology enabling every radiologist to perform on a high-quality imaging.

Yet, there are some obstacles still to overcome. First, MRI is a tricky imaging tool. A major drawback of MRI is the lack of standard quantification of image intensities. Within the same image, the intensities for the same material vary as they are affected by bias field distortions and imaging acquisition parameters, not always perfectly standardized. In addition, not only do MR images taken on different scanner vary in image intensities, but the images for the same patient on the same scanner at different times may appear differently from each other due to a variety of scanner- and patient-dependent variables [45]. Therefore, the initial step in ML/DL image postprocessing is to normalize the MR intensity [45]. This process could induce errors, however. At last, also the reproducibility of CNNs varies resulting in interscan differences, though with less impact [78]. Second, most studies rely on single site source data. Multicentric studies are very rare hence making it harder to compare results of AI-based algorithms across different vendors and sequence parameters. Third, the choice of imaging sequences and their specific parameters is variable. This work focused on bpMRI of the prostate. Even though for a radiologist imaging with T2w-imaging and DWI imaging would be seen as biparametric, things look different in the world of AI-based post-postprocessing: sometimes T2w and ADC, sometimes T2w and a single high b-value, T2w and multiple b-values or T2, ADC and b-values were used (hereby neglecting uncommon outlier studies using DCE and T2 or T1 and T2). Even though DWI source date and ADC are based on the same acquisition, their information content seems different. It was observed in one study that the use of CHB-DWI led to higher specificity while the use of ADC led to highest sensitivity, making the choice of sensing modality useful for different clinical scenarios [79]. For example, maximizing specificity is important for surgery for removal of prostate where minimizing false positive rates to avoid unnecessary surgeries is required. On the other hand, for cancer screening, maximizing sensitivity may be useful to avoid missing cancerous patients [79]. A clear definition what would be considered as truly bpMRI or standards for AI-postprocessing has not been set up. Yet, there is a first European initiative on the development and standardization of AI-based tools for prostate MRI [44]. Fourth, DL-based CNNs are notorious for being a "black box" in terms of the how the decision was achieved. While this may not be entirely true—CNNs can be monitored at any level at some expense—they might never be as transparent as ML-based approaches hence scaring some physicians from using them on real patients outside studies. Moreover, here, commercialization of the techniques might be helpful as larger companies have the means and money to certify algorithms with the FDA or the EU and thus make them broadly (commercially) available.

As seven studies made use of the ProstatEx data, it is worth looking at the overall conclusions the creators of the dataset and initiators of the contest published [80]: the majority of the 71 methods submitted to the challenge (classifying prostate lesions as clinically significant or not) the majority of those methods outperformed random guessing. They conclude that automated classification of clinically significant cancer seems feasible. While in the second contest (computationally assigning lesions to a Gleason grade group) only two out 43 methods did marginally better than random guessing. The creators also conclude that more images and larger data sets with better annotations might be necessary to draw significant conclusions, which brings up again the question of means and money. Another conclusion that can be drawn when looking at the included studies is that 3 T imaging seems to be the standard. This is partly because there is substantial overlap in the source data (ProstatEx) and that, of course, studies are being conducted at University Medical Centers which most often have state-of-the-art equipment. For radiology departments in smaller hospitals or private practices having a 3 T system is less likely. Regarding how far the results of 3T e.g., DWI can be transferred to 1.5 T and how the technological improvement of 1.5 T in the field of signal reception and processing is supportive remain unclear. One might speculate that a state-of-the-art 1.5 T will yield comparable image quality to an elder 3 T system. Looking at the source data of the different studies one can roughly estimate that 30% of these were acquired on elder (>14 a) 3 T systems.

There are some unexpected studies with novel approaches to patient care that should be to highlighted. One was therapy assessment with pre- and post-IMRT T2w-imaging [40] for "delta radiomics", using radiomic features extracted from MR images for predicting response in prostate cancer patients. While there was only one study with this specific design, extrapolating ECE or BCR has roughly the same line of thought: could not it be possible to predict for changes in the future with imaging features measured today [69,73,74]. The AUC values of these studies were unexpectedly high (0.801–0.905) as well as the number of included patients.

Limitations

This review has several limitations that need to be mentioned. First, ML and DL are extremely fast evolving techniques. Data provided in this review simply display a snapshot of the ongoing development. With the ever more powerful hardware and algorithms, future improvements seem likely. Most results are based on small feasibility studies, and larger applications of ML and DL in prostate imaging are not yet available. Whether their results match the promising initial studies remains unclear. Second, the inclusion criteria were narrow so that only 29 studies could be included. With the small sample size, different targets, and the different foci of the studies no wholistic analysis could be performed. Opening up the time window for the included studies would have led to inclusion of elder techniques potentially biasing the results.

5. Conclusions

In summary, this study investigated the current status of bpMRI of the prostate with postprocessing using ML and DL with a focus and tumor detection and grading. The presented results are very promising in terms of detection of csPCA and differentiation of prostate cancer from non-cancerous tissue. ML and DL seem to be equally good in terms of classification of single lesion according to the PIRADS score. Most approaches however rely on human interference and contouring the lesions. Only a few newer approaches automatically segment the entire gland and lesions, along with lesion grading according to PIRADS. There still exist a large variability and methods and just a few multicentric studies. No AI-postprocessing technique is considered gold standard at this time while there seems to be a trend toward CNNs. Regarding the actual MRI-sequences, most studies used T2w-imaging and either b-values from DWI or the ADC maps from DWI. The application of ML and DL to bpMRI postprocessing and the assistance in the reading process surely represent a step into the future of radiology. Currently however, these techniques remain at an experimental level and are not yet ready or available for a broader clinical application.

Author Contributions: Conceptualization, H.J.M., D.C., E.N. and G.A.; methodology, H.J.M., D.C., E.N. and G.A.; formal analysis, H.J.M. and G.A.; data curation, H.J.M. and G.A.; writing—original draft preparation, H.J.M. and G.A.; writing—review and editing, H.J.M., D.C., E.N. and G.A.; visualization, H.J.M.; supervision, H.J.M., D.C., E.N. and G.A. All authors have read and agreed to the published version of the manuscript.

Funding: The research for this paper was done within the University of Pisa Master Course in Oncologic Imaging as part of a research trial of the NAVIGATOR project (funded by Bando Salute 2018, Tuscany Region, Italy).

Institutional Review Board Statement: Not applicable.

Informed Consent Statement: Not applicable as only literature research was performed.

Data Availability Statement: All data can be found in the original publications as listed above.

Acknowledgments: The research for this paper was done within the University of Pisa Master Course in Oncologic Imaging.

Conflicts of Interest: The authors declare no conflict of interest.

Abbreviations

ACR	American College of Radiology
ADC	Apparent Diffusion Coefficient
AFMS	Anterior Fibromuscular Stroma (of the Prostate)
AI	Artificial Intelligence
AS	Anterior stroma (of the Prostate)
AUC	Area under the Curve
BCR	Biochemical Recurrence
bp	bi-parametric
BPH	Benign Prostate Hyperplasia
ciPCA	Clinically Insignificant Prostate Cancer
CNN	Convoluted Neural Network
csPCA	Clinically Significant Prostate Cancer
CZ	Central Zone (of the Prostate)

DCE	Dynamic Contrast-Enhanced Imaging
DL	Deep-Learning
DRE	Digital Rectal Examination
DWI	Diffusion-Weighted Imaging
ECE	Extracapsular Extension
ESUR	European Society of Urologic Radiology
GS	Gleason Score
HBV	High b-Value (of DWI)
IMRT	Intensity-Modulated Radiation Therapy
ML	Machine-Learning
mp	multi-parametric
MR	Magnetic Resonance
MRI	Magnetic Resonance Imaging
nsPCA	Non-Significant Prostate Cancer
PZ	Peripheral Zone (of the Prostate)
PCA	Prostate Cancer
PIRADS	Prostate Imaging Reporting and Data System
PSM	Positive Surgical Margins
RP	Radical Prostatectomy
T2w	T2-weighted Imaging
TSE	TurboSpinEcho
TZ	Transitional Zone (of the Prostate)
up	uni-parametric

Appendix A

Table A1. Display of PRISMA items.

PRISMA Item	Description
Title	Current Value of Biparametric Prostate MRI with Machine-Learning or Deep-Learning in the Detection, Grading and Characterization of Prostate Cancer: a systematic review.
Main objective	Assessing the current value of deep-learning and machine-learning applied to biparametric MRI of the prostate
Inclusion and exclusion criteria	Inclusion criteria: - Study listed in Pubmed - Search terms: "prostate" and "magnetic" and either "deep learning" or "machine learning" or "radiomics" - Full text access available through University of Heidelberg - Paper type: original investigation/research - Focus: Detection or grading of prostate cancer with biparametric prostate MRI - Language: English or German - Year of publication 2019–2021 Exclusion criteria: - No full text access - Wrong paper type: reviews, meta-analysis - Wrong focus (e.g., prostate segmentation, radiation therapy planning) - Wrong technique (uniparametric or multiparametric prostate MRI)
Information source and access time	PubMed query in August 2021
Methods to assess risk of bias in included studies	No structured program was used to assess bias in study selection. Internal review by the authors and critical appraisal of the data was performed.
Methods to present and synthesize results	Descriptive statistics, listing in tabular form
Number of studies and participants included	29 publications included 7466 participants included

Main outcomes	Very heterogenous data did not allow for a general interpretation of all studies. Tumor detection and grading with machine-learning and deep-learning techniques is feasible in trials and shows promising results. Reported values for AUC ranging from 0.71 to 0.999. In studies comparing human radiologists to deep-learning algorithms comparable, statistically not different results for tumor detection were found.
Limitations	- No overall statistical analysis feasible due to the heterogeneity of methods and inclusion criteria reported - 7 out of 29 studies based on the same dataset (ProstatEx, Radbound Nijmwegen, The Netherlands) - Heterogenous studies with different inclusion criteria and ground truth (i.e., if Gleason Grade constitutes high-grade cancer or not) - Often lacking demographic and statistical data
General interpretation	Detection of clinically significant prostate cancer and differentiation of prostate cancer from non-cancerous tissue using machine-learning and deep learning is feasible with promising results. Some techniques of machine-learning and deep-learning currently seem to be equally good as human radiologists in terms of classification of single lesions according to the PIRADS score.
Primary source for funding	No general funding. Publication costs are covered by the Universtiy of Pisa, Pisa, Italy.
Register name and registration number	No registration

References

1. Siegel, R.; Naishadham, D.; Jemal, A. Cancer statistics, 2013. *CA Cancer J. Clin.* **2013**, *63*, 11–30. [CrossRef] [PubMed]
2. Jemal, A.; Center, M.M.; DeSantis, C.; Ward, E.M. Global patterns of cancer incidence and mortality rates and trends. *Cancer Epidemiol. Biomark. Prev.* **2010**, *19*, 1893–1907. [CrossRef] [PubMed]
3. Ferlay, J.; Autier, P.; Boniol, M.; Heanue, M.; Colombet, M.; Boyle, P. Estimates of the cancer incidence and mortality in Europe in 2006. *Ann. Oncol.* **2007**, *18*, 581–592. [CrossRef] [PubMed]
4. Crocetto, F.; Barone, B.; Caputo, V.F.; Fontana, M.; de Cobelli, O.; Ferro, M. BRCA Germline Mutations in Prostate Cancer: The Future Is Tailored. *Diagnostics* **2021**, *11*, 908. [CrossRef] [PubMed]
5. Brookman-May, S.D.; Campi, R.; Henríquez, J.D.S.; Klatte, T.; Langenhuijsen, J.F.; Brausi, M.; Linares-Espinós, E.; Volpe, A.; Marszalek, M.; Akdogan, B.; et al. Latest Evidence on the Impact of Smoking, Sports, and Sexual Activity as Modifiable Lifestyle Risk Factors for Prostate Cancer Incidence, Recurrence, and Progression: A Systematic Review of the Literature by the European Association of Urology Section of Oncological Urology (ESOU). *Eur. Urol. Focus.* **2019**, *5*, 756–787. [CrossRef] [PubMed]
6. Drudi, F.M.; Cantisani, V.; Angelini, F.; Ciccariello, M.; Messineo, D.; Ettorre, E.; Liberatore, M.; Scialpi, M. Multiparametric MRI Versus Multiparametric US in the Detection of Prostate Cancer. *Anticancer Res.* **2019**, *39*, 3101–3110. [CrossRef] [PubMed]
7. Jones, D.; Friend, C.; Dreher, A.; Allgar, V.; Macleod, U. The diagnostic test accuracy of rectal examination for prostate cancer diagnosis in symptomatic patients: A systematic review. *BMC Fam. Pract.* **2018**, *19*, 79. [CrossRef] [PubMed]
8. Naji, L.; Randhawa, H.; Sohani, Z.; Dennis, B.; Lautenbach, D.; Kavanagh, O.; Bawor, M.; Banfield, L.; Profetto, J. Digital Rectal Examination for Prostate Cancer Screening in Primary Care: A Systematic Review and Meta-Analysis. *Ann. Fam. Med.* **2018**, *16*, 149–154. [CrossRef] [PubMed]
9. Pokorny, M.R.; de Rooij, M.; Duncan, E.; Schröder, F.H.; Parkinson, R.; Barentsz, J.O.; Thompson, L.C. Prospective study of diagnostic accuracy comparing prostate cancer detection by transrectal ultrasound-guided biopsy versus magnetic resonance (MR) imaging with subsequent MR-guided biopsy in men without previous prostate biopsies. *Eur. Urol.* **2014**, *66*, 22–29. [CrossRef]
10. Barentsz, J.O.; Richenberg, J.; Clements, R.; Choyke, P.; Verma, S.; Villeirs, G.; Rouviere, O.; Logager, V.; Fütterer, J.J. ESUR prostate MR guidelines 2012. *Eur. Radiol.* **2012**, *22*, 746–757. [CrossRef]
11. Spak, D.A.; Plaxco, J.S.; Santiago, L.; Dryden, M.J.; Dogan, B.E. BI-RADS® fifth edition: A summary of changes. *Diagn. Interv. Imaging* **2017**, *98*, 179–190. [CrossRef]
12. Turkbey, B.; Rosenkrantz, A.B.; Haider, M.A.; Padhani, A.R.; Villeirs, G.; Macura, K.J.; Tempany, C.M.; Choyke, P.L.; Cornud, F.; Margolis, D.J.; et al. Prostate Imaging Reporting and Data System Version 2.1: 2019 Update of Prostate Imaging Reporting and Data System Version 2. *Eur. Urol.* **2019**, *76*, 340–351. [CrossRef]
13. Weinreb, J.C.; Barentsz, J.O.; Choyke, P.L.; Cornud, F.; Haider, M.A.; Macura, K.J.; Margolis, D.; Schnall, M.D.; Shtern, F.; Tempany, C.M.; et al. PI-RADS Prostate Imaging-Reporting and Data System: 2015, Version 2. *Eur. Urol.* **2016**, *69*, 16–40. [CrossRef]
14. Thompson, J.E.; van Leeuwen, P.J.; Moses, D.; Shnier, R.; Brenner, P.; Delprado, W.; Pulbrook, M.; Böhm, M.; Haynes, A.M.; Hayen, A.; et al. The Diagnostic Performance of Multiparametric Magnetic Resonance Imaging to Detect Significant Prostate Cancer. *J. Urol.* **2016**, *195*, 1428–1435. [CrossRef]

15. Vargas, H.A.; Hötker, A.M.; Goldman, D.A.; Moskowitz, C.S.; Gondo, T.; Matsumoto, K.; Ehdaie, B.; Woo, S.; Fine, S.W.; Reuter, V.E.; et al. Updated prostate imaging reporting and data system (PIRADS v2) recommendations for the detection of clinically significant prostate cancer using multiparametric MRI: Critical evaluation using whole-mount pathology as standard of reference. *Eur. Radiol.* **2016**, *26*, 1606–1612. [CrossRef]
16. Kasel-Seibert, M.; Lehmann, T.; Aschenbach, R.; Guettler, F.V.; Abubrig, M.; Grimm, M.O.; Teichgraeber, U.; Franiel, T. Assessment of PI-RADS v2 for the Detection of Prostate Cancer. *Eur. J. Radiol.* **2016**, *85*, 726–731. [CrossRef]
17. Palumbo, P.; Manetta, R.; Izzo, A.; Bruno, F.; Arrigoni, F.; de Filippo, M.; Splendiani, A.; di Cesare, E.; Masciocchi, C.; Barile, A. Biparametric (bp) and multiparametric (mp) magnetic resonance imaging (MRI) approach to prostate cancer disease: A narrative review of current debate on dynamic contrast enhancement. *Gland. Surg.* **2020**, *9*, 2235–2247. [CrossRef]
18. Michaely, H.J.; Thomsen, H.S.; Reiser, M.F.; Schoenberg, S.O. Nephrogenic systemic fibrosis (NSF)—implications for radiology. *Radiologe* **2007**, *47*, 785–793. [CrossRef]
19. Grobner, T. Gadolinium–a specific trigger for the development of nephrogenic fibrosing dermopathy and nephrogenic systemic fibrosis? *Nephrol. Dial. Transplant.* **2006**, *21*, 1104–1108. [CrossRef]
20. Alkhunizi, S.M.; Fakhoury, M.; Abou-Kheir, W.; Lawand, N. Gadolinium Retention in the Central and Peripheral Nervous System: Implications for Pain, Cognition, and Neurogenesis. *Radiology* **2020**, *297*, 407–416. [CrossRef]
21. Radbruch, A.; Weberling, L.D.; Kieslich, P.J.; Eidel, O.; Burth, S.; Kickingereder, P.; Heiland, S.; Wick, W.; Schlemmer, H.P.; Bendszus, M. Gadolinium retention in the dentate nucleus and globus pallidus is dependent on the class of contrast agent. *Radiology* **2015**, *275*, 783–791. [CrossRef]
22. Strickler, S.E.; Clark, K.R. Gadolinium Deposition: A Study Review. *Radiol. Technol.* **2021**, *92*, 249–258.
23. Semelka, R.C.; Ramalho, J.; Vakharia, A.; AlObaidy, M.; Burke, L.M.; Jay, M.; Ramalho, M. Gadolinium deposition disease: Initial description of a disease that has been around for a while. *Magn. Reson. Imaging* **2016**, *34*, 1383–1390. [CrossRef]
24. Ahmed, H.U.; El-Shater Bosaily, A.; Brown, L.C.; Gabe, R.; Kaplan, R.; Parmar, M.K.; Collaco-Moraes, Y.; Ward, K.; Hindley, R.G.; Freeman, A.; et al. Diagnostic accuracy of multi-parametric MRI and TRUS biopsy in prostate cancer (PROMIS): A paired validating confirmatory study. *Lancet* **2017**, *389*, 815–822. [CrossRef]
25. Alabousi, M.; Salameh, J.P.; Gusenbauer, K.; Samoilov, L.; Jafri, A.; Yu, H.; Alabousi, A. Biparametric vs multiparametric prostate magnetic resonance imaging for the detection of prostate cancer in treatment-naïve patients: A diagnostic test accuracy systematic review and meta-analysis. *BJU Int.* **2019**, *124*, 209–220. [CrossRef] [PubMed]
26. Bass, E.J.; Pantovic, A.; Connor, M.; Gabe, R.; Padhani, A.R.; Rockall, A.; Sokhi, H.; Tam, H.; Winkler, M.; Ahmed, H.U. A systematic review and meta-analysis of the diagnostic accuracy of biparametric prostate MRI for prostate cancer in men at risk. *Prostate Cancer Prostatic Dis.* **2020**, *24*, 596–611. [CrossRef] [PubMed]
27. Boesen, L.; Nørgaard, N.; Løgager, V.; Balslev, I.; Bisbjerg, R.; Thestrup, K.C.; Winther, M.D.; Jakobsen, H.; Thomsen, H.S. Assessment of the Diagnostic Accuracy of Biparametric Magnetic Resonance Imaging for Prostate Cancer in Biopsy-Naive Men: The Biparametric MRI for Detection of Prostate Cancer (BIDOC) Study. *JAMA Netw. Open* **2018**, *1*, e180219. [CrossRef]
28. Jambor, I.; Verho, J.; Ettala, O.; Knaapila, J.; Taimen, P.; Syvänen, K.T.; Kiviniemi, A.; Kähkönen, E.; Perez, I.M.; Seppänen, M.; et al. Validation of IMPROD biparametric MRI in men with clinically suspected prostate cancer: A prospective multi-institutional trial. *PLoS Med.* **2019**, *16*, e1002813. [CrossRef]
29. Di Campli, E.; Delli Pizzi, A.; Seccia, B.; Cianci, R.; d'Annibale, M.; Colasante, A.; Cinalli, S.; Castellan, P.; Navarra, R.; Iantorno, R.; et al. Diagnostic accuracy of biparametric vs multiparametric MRI in clinically significant prostate cancer: Comparison between readers with different experience. *Eur. J. Radiol.* **2018**, *101*, 17–23. [CrossRef] [PubMed]
30. Gatti, M.; Faletti, R.; Calleris, G.; Giglio, J.; Berzovini, C.; Gentile, F.; Marra, G.; Misischi, F.; Molinaro, L.; Bergamasco, L.; et al. Prostate cancer detection with biparametric magnetic resonance imaging (bpMRI) by readers with different experience: Performance and comparison with multiparametric (mpMRI). *Abdom. Radiol.* **2019**, *44*, 1883–1893. [CrossRef]
31. Cho, J.; Ahn, H.; Hwang, S.I.; Lee, H.J.; Choe, G.; Byun, S.S.; Hong, S.K. Biparametric versus multiparametric magnetic resonance imaging of the prostate: Detection of clinically significant cancer in a perfect match group. *Prostate Int.* **2020**, *8*, 146–151. [CrossRef]
32. Lee, D.H.; Nam, J.K.; Lee, S.S.; Han, J.Y.; Lee, J.W.; Chung, M.K.; Park, S.W. Comparison of Multiparametric and Biparametric MRI in First Round Cognitive Targeted Prostate Biopsy in Patients with PSA Levels under 10 ng/mL. *Yonsei Med. J.* **2017**, *58*, 994–999. [CrossRef]
33. Woo, S.; Suh, C.H.; Kim, S.Y.; Cho, J.Y.; Kim, S.H.; Moon, M.H. Head-to-Head Comparison Between Biparametric and Multiparametric MRI for the Diagnosis of Prostate Cancer: A Systematic Review and Meta-Analysis. *AJR Am. J. Roentgenol.* **2018**, *211*, W226–W241. [CrossRef]
34. Scialpi, M.; D'Andrea, A.; Martorana, E.; Malaspina, C.M.; Aisa, M.C.; Napoletano, M.; Orlandi, E.; Rondoni, V.; Scialpi, P.; Pacchiarini, D.; et al. Biparametric MRI of the prostate. *Turk. J. Urol.* **2017**, *43*, 401–409. [CrossRef]
35. Scialpi, M.; Prosperi, E.; D'Andrea, A.; Martorana, E.; Malaspina, C.; Palumbo, B.; Orlandi, A.; Falcone, G.; Milizia, M.; Mearini, L.; et al. Biparametric versus Multiparametric MRI with Non-endorectal Coil at 3T in the Detection and Localization of Prostate Cancer. *Anticancer Res.* **2017**, *37*, 1263–1271. [CrossRef]
36. Bernatz, S.; Ackermann, J.; Mandel, P.; Kaltenbach, B.; Zhdanovich, Y.; Harter, P.N.; Döring, C.; Hammerstingl, R.; Bodelle, B.; Smith, K.; et al. Comparison of machine learning algorithms to predict clinically significant prostate cancer of the peripheral zone with multiparametric MRI using clinical assessment categories and radiomic features. *Eur. Radiol.* **2020**, *30*, 6757–6769. [CrossRef]

37. Kumar, V.; Gu, Y.; Basu, S.; Berglund, A.; Eschrich, S.A.; Schabath, M.B.; Forster, K.; Aerts, H.J.; Dekker, A.; Fenstermacher, D.; et al. Radiomics: The process and the challenges. *Magn. Reson. Imaging* **2012**, *30*, 1234–1248. [CrossRef]
38. Lambin, P.; Rios-Velazquez, E.; Leijenaar, R.; Carvalho, S.; van Stiphout, R.G.; Granton, P.; Zegers, C.M.; Gillies, R.; Boellard, R.; Dekker, A.; et al. Radiomics: Extracting more information from medical images using advanced feature analysis. *Eur. J. Cancer* **2012**, *48*, 441–446. [CrossRef]
39. Gillies, R.J.; Kinahan, P.E.; Hricak, H. Radiomics: Images Are More than Pictures, They Are Data. *Radiology* **2016**, *278*, 563–577. [CrossRef]
40. Abdollahi, H.; Mofid, B.; Shiri, I.; Razzaghdoust, A.; Saadipoor, A.; Mahdavi, A.; Galandooz, H.M.; Mahdavi, S.R. Machine learning-based radiomic models to predict intensity-modulated radiation therapy response, Gleason score and stage in prostate cancer. *Radiol. Med.* **2019**, *124*, 555–567. [CrossRef]
41. Wildeboer, R.R.; van Sloun, R.J.G.; Wijkstra, H.; Mischi, M. Artificial intelligence in multiparametric prostate cancer imaging with focus on deep-learning methods. *Comput. Methods Programs Biomed.* **2020**, *189*, 105316. [CrossRef]
42. Litjens, G.; Kooi, T.; Bejnordi, B.E.; Setio, A.A.A.; Ciompi, F.; Ghafoorian, M.; van der Laak, J.; van Ginneken, B.; Sánchez, C.I. A survey on deep learning in medical image analysis. *Med. Image Anal.* **2017**, *42*, 60–88. [CrossRef]
43. Ching, T.; Himmelstein, D.S.; Beaulieu-Jones, B.K.; Kalinin, A.A.; Do, B.T.; Way, G.P.; Ferrero, E.; Agapow, P.M.; Zietz, M.; Hoffman, M.M.; et al. Opportunities and obstacles for deep learning in biology and medicine. *J. R. Soc. Interface* **2018**, *15*, 20170387. [CrossRef]
44. Penzkofer, T.; Padhani, A.R.; Turkbey, B.; Haider, M.A.; Huisman, H.; Walz, J.; Salomon, G.; Schoots, I.G.; Richenberg, J.; Villeirs, G.; et al. ESUR/ESUI position paper: Developing artificial intelligence for precision diagnosis of prostate cancer using magnetic resonance imaging. *Eur. Radiol.* **2021**, *31*, 9567–9578. [CrossRef]
45. Chen, Q.; Hu, S.; Long, P.; Lu, F.; Shi, Y.; Li, Y. A Transfer Learning Approach for Malignant Prostate Lesion Detection on Multiparametric MRI. *Technol. Cancer Res. Treat.* **2019**, *18*, 1–9. [CrossRef]
46. Page, M.J.; Moher, D.; Bossuyt, P.M.; Boutron, I.; Hoffmann, T.C.; Mulrow, C.D.; Shamseer, L.; Tetzlaff, J.M.; Akl, E.A.; Brennan, S.E.; et al. PRISMA 2020 explanation and elaboration: Updated guidance and exemplars for reporting systematic reviews. *BMJ* **2021**, *372*, n160. [CrossRef]
47. American College of Radiology (ACR). PI-RADS. Available online: https://www.acr.org/-/media/ACR/Files/RADS/Pi-RADS/PIRADS-V2-1.pdf (accessed on 4 September 2021).
48. Wu, M.; Krishna, S.; Thornhill, R.E.; Flood, T.A.; McInnes, M.D.F.; Schieda, N. Transition zone prostate cancer: Logistic regression and machine-learning models of quantitative ADC, shape and texture features are highly accurate for diagnosis. *J. Magn. Reson. Imaging* **2019**, *50*, 940–950. [CrossRef]
49. Varghese, B.; Chen, F.; Hwang, D.; Palmer, S.L.; de Castro Abreu, A.L.; Ukimura, O.; Aron, M.; Aron, M.; Gill, I.; Duddalwar, V.; et al. Objective risk stratification of prostate cancer using machine learning and radiomics applied to multiparametric magnetic resonance images. *Sci. Rep.* **2019**, *9*, 1570. [CrossRef] [PubMed]
50. Min, X.; Li, M.; Dong, D.; Feng, Z.; Zhang, P.; Ke, Z.; You, H.; Han, F.; Ma, H.; Tian, J.; et al. Multi-parametric MRI-based radiomics signature for discriminating between clinically significant and insignificant prostate cancer: Cross-validation of a machine learning method. *Eur. J. Radiol.* **2019**, *115*, 16–21. [CrossRef] [PubMed]
51. Toivonen, J.; Montoya Perez, I.; Movahedi, P.; Merisaari, H.; Pesola, M.; Taimen, P.; Boström, P.J.; Pohjankukka, J.; Kiviniemi, A.; Pahikkala, T.; et al. Radiomics and machine learning of multisequence multiparametric prostate MRI: Towards improved non-invasive prostate cancer characterization. *PLoS ONE* **2019**, *14*, e0217702. [CrossRef] [PubMed]
52. Chen, T.; Li, M.; Gu, Y.; Zhang, Y.; Yang, S.; Wei, C.; Wu, J.; Li, X.; Zhao, W.; Shen, J. Prostate Cancer Differentiation and Aggressiveness: Assessment With a Radiomic-Based Model vs. PI-RADS v2. *J. Magn. Reson. Imaging* **2019**, *49*, 875–884. [CrossRef] [PubMed]
53. Xu, M.; Fang, M.; Zou, J.; Yang, S.; Yu, D.; Zhong, L.; Hu, C.; Zang, Y.; Dong, D.; Tian, J.; et al. Using biparametric MRI radiomics signature to differentiate between benign and malignant prostate lesions. *Eur. J. Radiol.* **2019**, *114*, 38–44. [CrossRef] [PubMed]
54. Zhong, X.; Cao, R.; Shakeri, S.; Scalzo, F.; Lee, Y.; Enzmann, D.R.; Wu, H.H.; Raman, S.S.; Sung, K. Deep transfer learning-based prostate cancer classification using 3 Tesla multi-parametric MRI. *Abdom. Radiol.* **2019**, *44*, 2030–2039. [CrossRef] [PubMed]
55. Yuan, Y.; Qin, W.; Buyyounouski, M.; Ibragimov, B.; Hancock, S.; Han, B.; Xing, L. Prostate cancer classification with multiparametric MRI transfer learning model. *Med. Phys.* **2019**, *46*, 756–765. [CrossRef] [PubMed]
56. Xu, H.; Baxter, J.S.H.; Akin, O.; Cantor-Rivera, D. Prostate cancer detection using residual networks. *Int. J. Comput. Assist. Radiol. Surg.* **2019**, *14*, 1647–1650. [CrossRef]
57. Schelb, P.; Kohl, S.; Radtke, J.P.; Wiesenfarth, M.; Kickingereder, P.; Bickelhaupt, S.; Kuder, T.A.; Stenzinger, A.; Hohenfellner, M.; Schlemmer, H.P.; et al. Classification of Cancer at Prostate MRI: Deep Learning versus Clinical PI-RADS Assessment. *Radiology* **2019**, *293*, 607–617. [CrossRef]
58. Montoya Perez, I.; Jambor, I.; Pahikkala, T.; Airola, A.; Merisaari, H.; Saunavaara, J.; Alinezhad, S.; Väänänen, R.M.; Tallgrén, T.; Verho, J.; et al. Prostate Cancer Risk Stratification in Men With a Clinical Suspicion of Prostate Cancer Using a Unique Biparametric MRI and Expression of 11 Genes in Apparently Benign Tissue: Evaluation Using Machine-Learning Techniques. *J. Magn. Reson. Imaging* **2020**, *51*, 1540–1553. [CrossRef]
59. Hou, Y.; Bao, M.L.; Wu, C.J.; Zhang, J.; Zhang, Y.D.; Shi, H.B. A radiomics machine learning-based redefining score robustly identifies clinically significant prostate cancer in equivocal PI-RADS score 3 lesions. *Abdom. Radiol.* **2020**, *45*, 4223–4234. [CrossRef]

60. Mehralivand, S.; Harmon, S.A.; Shih, J.H.; Smith, C.P.; Lay, N.; Argun, B.; Bednarova, S.; Baroni, R.H.; Canda, A.E.; Ercan, K.; et al. Multicenter Multireader Evaluation of an Artificial Intelligence-Based Attention Mapping System for the Detection of Prostate Cancer With Multiparametric MRI. *AJR Am. J. Roentgenol.* **2020**, *215*, 903–912. [CrossRef]
61. Gong, L.; Xu, M.; Fang, M.; Zou, J.; Yang, S.; Yu, X.; Xu, D.; Zhou, L.; Li, H.; He, B.; et al. Noninvasive Prediction of High-Grade Prostate Cancer via Biparametric MRI Radiomics. *J. Magn. Reson. Imaging* **2020**, *52*, 1102–1109. [CrossRef]
62. Bleker, J.; Kwee, T.C.; Dierckx, R.; de Jong, I.J.; Huisman, H.; Yakar, D. Multiparametric MRI and auto-fixed volume of interest-based radiomics signature for clinically significant peripheral zone prostate cancer. *Eur. Radiol.* **2020**, *30*, 1313–1324. [CrossRef]
63. Zong, W.; Lee, J.K.; Liu, C.; Carver, E.N.; Feldman, A.M.; Janic, B.; Elshaikh, M.A.; Pantelic, M.V.; Hearshen, D.; Chetty, I.J.; et al. A deep dive into understanding tumor foci classification using multiparametric MRI based on convolutional neural network. *Med. Phys.* **2020**, *47*, 4077–4086. [CrossRef]
64. Sanford, T.; Harmon, S.A.; Turkbey, E.B.; Kesani, D.; Tuncer, S.; Madariaga, M.; Yang, C.; Sackett, J.; Mehralivand, S.; Yan, P.; et al. Deep-Learning-Based Artificial Intelligence for PI-RADS Classification to Assist Multiparametric Prostate MRI Interpretation: A Development Study. *J. Magn. Reson. Imaging* **2020**, *52*, 1499–1507. [CrossRef]
65. Brunese, L.; Mercaldo, F.; Reginelli, A.; Santone, A. Radiomics for Gleason Score Detection through Deep Learning. *Sensors* **2020**, *20*, 5411. [CrossRef]
66. Chen, Y.; Xing, L.; Yu, L.; Bagshaw, H.P.; Buyyounouski, M.K.; Han, B. Automatic intraprostatic lesion segmentation in multiparametric magnetic resonance images with proposed multiple branch UNet. *Med. Phys.* **2020**, *47*, 6421–6429. [CrossRef]
67. Winkel, D.J.; Wetterauer, C.; Matthias, M.O.; Lou, B.; Shi, B.; Kamen, A.; Comaniciu, D.; Seifert, H.H.; Rentsch, C.A.; Boll, D.T. Autonomous Detection and Classification of PI-RADS Lesions in an MRI Screening Population Incorporating Multicenter-Labeled Deep Learning and Biparametric Imaging: Proof of Concept. *Diagnostics* **2020**, *10*, 951. [CrossRef]
68. Arif, M.; Schoots, I.G.; Castillo Tovar, J.; Bangma, C.H.; Krestin, G.P.; Roobol, M.J.; Niessen, W.; Veenland, J.F. Clinically significant prostate cancer detection and segmentation in low-risk patients using a convolutional neural network on multi-parametric MRI. *Eur. Radiol.* **2020**, *30*, 6582–6592. [CrossRef]
69. He, D.; Wang, X.; Fu, C.; Wei, X.; Bao, J.; Ji, X.; Bai, H.; Xia, W.; Gao, X.; Huang, Y.; et al. MRI-based radiomics models to assess prostate cancer, extracapsular extension and positive surgical margins. *Cancer Imaging* **2021**, *21*, 46. [CrossRef]
70. Vente, C.; Vos, P.; Hosseinzadeh, M.; Pluim, J.; Veta, M. Deep Learning Regression for Prostate Cancer Detection and Grading in Bi-Parametric MRI. *IEEE Trans. Biomed. Eng.* **2021**, *68*, 374–383. [CrossRef]
71. Chen, J.; Wan, Z.; Zhang, J.; Li, W.; Chen, Y.; Li, Y.; Duan, Y. Medical image segmentation and reconstruction of prostate tumor based on 3D AlexNet. *Comput Methods Programs Biomed.* **2021**, *200*, 105878. [CrossRef]
72. Cao, R.; Zhong, X.; Afshari, S.; Felker, E.; Suvannarerg, V.; Tubtawee, T.; Vangala, S.; Scalzo, F.; Raman, S.; Sung, K. Performance of Deep Learning and Genitourinary Radiologists in Detection of Prostate Cancer Using 3-T Multiparametric Magnetic Resonance Imaging. *J. Magn. Reson. Imaging* **2021**, *54*, 474–483. [CrossRef]
73. Hou, Y.; Zhang, Y.H.; Bao, J.; Bao, M.L.; Yang, G.; Shi, H.B.; Song, Y.; Zhang, Y.D. Artificial intelligence is a promising prospect for the detection of prostate cancer extracapsular extension with mpMRI: A two-center comparative study. *Eur. J. Nucl. Med. Mol. Imaging* **2021**, *48*, 3805–3816. [CrossRef]
74. Yan, Y.; Shao, L.; Liu, Z.; He, W.; Yang, G.; Liu, J.; Xia, H.; Zhang, Y.; Chen, H.; Liu, C.; et al. Deep Learning with Quantitative Features of Magnetic Resonance Images to Predict Biochemical Recurrence of Radical Prostatectomy: A Multi-Center Study. *Cancers* **2021**, *13*, 3098. [CrossRef]
75. Schelb, P.; Wang, X.; Radtke, J.P.; Wiesenfarth, M.; Kickingereder, P.; Stenzinger, A.; Hohenfellner, M.; Schlemmer, H.P.; Maier-Hein, K.H.; Bonekamp, D. Simulated clinical deployment of fully automatic deep learning for clinical prostate MRI assessment. *Eur. Radiol.* **2021**, *31*, 302–313. [CrossRef]
76. Antonelli, M.; Johnston, E.W.; Dikaios, N.; Cheung, K.K.; Sidhu, H.S.; Appayya, M.B.; Giganti, F.; Simmons, L.A.M.; Freeman, A.; Allen, C.; et al. Machine learning classifiers can predict Gleason pattern 4 prostate cancer with greater accuracy than experienced radiologists. *Eur. Radiol.* **2019**, *29*, 4754–4764. [CrossRef]
77. Kan, Y.; Zhang, Q.; Hao, J.; Wang, W.; Zhuang, J.; Gao, J.; Huang, H.; Liang, J.; Marra, G.; Calleris, G.; et al. Clinico-radiological characteristic-based machine learning in reducing unnecessary prostate biopsies of PI-RADS 3 lesions with dual validation. *Eur. Radiol.* **2020**, *30*, 6274–6284. [CrossRef]
78. Hiremath, A.; Shiradkar, R.; Merisaari, H.; Prasanna, P.; Ettala, O.; Taimen, P.; Aronen, H.J.; Boström, P.J.; Jambor, I.; Madabhushi, A. Test-retest repeatability of a deep learning architecture in detecting and segmenting clinically significant prostate cancer on apparent diffusion coefficient (ADC) maps. *Eur. Radiol.* **2021**, *31*, 379–391. [CrossRef]
79. Dulhanty, C.; Wang, L.; Cheng, M.; Gunraj, H.; Khalvati, F.; Haider, M.A.; Wong, A. Radiomics Driven Diffusion Weighted Imaging Sensing Strategies for Zone-Level Prostate Cancer Sensing. *Sensors* **2020**, *20*, 1539. [CrossRef] [PubMed]
80. The International Society for Optics and Photonics. Solve for X: Lessons Learned from PROSTATEx. Available online: https://spie.org/news/spie-professional-magazine-archive/2019-january/solve-for-x?SSO=1 (accessed on 10 October 2021).

cancers

MDPI

Article

Discovery of Pre-Treatment FDG PET/CT-Derived Radiomics-Based Models for Predicting Outcome in Diffuse Large B-Cell Lymphoma

Russell Frood [1,2,3,*], Matthew Clark [2], Cathy Burton [4], Charalampos Tsoumpas [5,6], Alejandro F. Frangi [6,7,8], Fergus Gleeson [7,9], Chirag Patel [1,2] and Andrew F. Scarsbrook [1,2,3]

1 Department of Nuclear Medicine, Leeds Teaching Hospitals NHS Trust, Leeds LS2 9JT, UK; chirag.patel13@nhs.net (C.P.); a.scarsbrook@nhs.net (A.F.S.)
2 Department of Radiology, Leeds Teaching Hospitals NHS Trust, Leeds LS2 9JT, UK; matt.clark1@nhs.net
3 Leeds Institute of Health Research, University of Leeds, Leeds LS9 7TF, UK
4 Department of Haematology, Leeds Teaching Hospitals NHS Trust, Leeds LS2 9JT, UK; cathy.burton1@nhs.net
5 Department of Nuclear Medicine and Molecular Imaging, University Medical Center of Groningen, University of Groningen, 9713 AV Groningen, The Netherlands; c.tsoumpas@rug.nl
6 Leeds Institute of Cardiovascular and Metabolic Medicine, University of Leeds, Leeds LS2 9JT, UK; a.frangi@leeds.ac.uk
7 Centre for Computational Imaging and Simulation Technologies in Biomedicine (CISTIB), School of Computing and School of Medicine, University of Leeds, Leeds LS2 9JT, UK; fergus.gleeson@ouh.nhs.uk
8 Medical Imaging Research Center (MIRC), University Hospital Gasthuisberg, Katholieke Universiteit Leuven, 3000 Leuven, Belgium
9 Department of Radiology, Oxford University Hospitals NHS Foundation Trust, Oxford OX3 9DU, UK
* Correspondence: russellfrood@nhs.net; Tel.: +44-113-20-68212

Citation: Frood, R.; Clark, M.; Burton, C.; Tsoumpas, C.; Frangi, A.F.; Gleeson, F.; Patel, C.; Scarsbrook, A.F. Discovery of Pre-Treatment FDG PET/CT-Derived Radiomics-Based Models for Predicting Outcome in Diffuse Large B-Cell Lymphoma. *Cancers* **2022**, *14*, 1711. https://doi.org/10.3390/cancers14071711

Academic Editors: Ali Hekmatnia, Hamid Khayyam, Ali Madani and Rahele Kafieh

Received: 7 March 2022
Accepted: 25 March 2022
Published: 28 March 2022

Simple Summary: Diffuse large B-cell lymphoma (DLBCL) is the most common type of lymphoma. Even with the improvements in the treatment of DLBCL, around a quarter of patients will experience recurrence. The aim of this single centre retrospective study was to predict which patients would have recurrence within 2 years of their treatment using machine learning techniques based on radiomics extracted from the staging PET/CT images. Our study demonstrated that in our dataset of 229 patients (training data = 183, test data = 46) that a combined radiomic and clinical based model performed better than a simple model based on metabolic tumour volume, and that it had a good predictive ability which was maintained when tested on an unseen test set.

Abstract: Background: Approximately 30% of patients with diffuse large B-cell lymphoma (DLBCL) will have recurrence. The aim of this study was to develop a radiomic based model derived from baseline PET/CT to predict 2-year event free survival (2-EFS). Methods: Patients with DLBCL treated with R-CHOP chemotherapy undergoing pre-treatment PET/CT between January 2008 and January 2018 were included. The dataset was split into training and internal unseen test sets (ratio 80:20). A logistic regression model using metabolic tumour volume (MTV) and six different machine learning classifiers created from clinical and radiomic features derived from the baseline PET/CT were trained and tuned using four-fold cross validation. The model with the highest mean validation receiver operator characteristic (ROC) curve area under the curve (AUC) was tested on the unseen test set. Results: 229 DLBCL patients met the inclusion criteria with 62 (27%) having 2-EFS events. The training cohort had 183 patients with 46 patients in the unseen test cohort. The model with the highest mean validation AUC combined clinical and radiomic features in a ridge regression model with a mean validation AUC of 0.75 ± 0.06 and a test AUC of 0.73. Conclusions: Radiomics based models demonstrate promise in predicting outcomes in DLBCL patients.

Keywords: diffuse large B-cell lymphoma; lymphoma; predictive modelling; radiomics; machine learning

1. Introduction

Diffuse large B-cell lymphoma (DLBCL) is the commonest subtype of non-Hodgkin lymphoma (NHL), accounting for approximately 30–40% of adult cases [1]. The gold standard treatment is immunochemotherapy with rituximab, cyclophosphamide, doxorubicin hydrochloride, vincristine (Oncovin) and prednisolone (RCHOP) [2]. Radiotherapy can be added if there is bulky or residual disease. Prophylactic intrathecal methotrexate or intravenous treatment with chemotherapy that crosses the blood-brain barrier may be included if there is high risk for central nervous system (CNS) involvement [3]. Even with current therapy regimes, approximately 20–30% of patients will recur following treatment [4,5]. Staging and response assessment is performed using 2-deoxy-2-[fluorine18]-fluoro-D-glucose (FDG) positron emission tomography/computed tomography (PET/CT). Treatment stratification based on mid-treatment (interim) PET/CT is commonly used in the management of patients with Hodgkin lymphoma but is less established in DLBCL due to the reduced ability to accurately predict treatment outcome in this lymphoma subtype mid-treatment [6,7]. There is increasing interest in the use of PET/CT derived metrics for treatment stratification at baseline in lymphoma to improve patient outcome. A number of groups have explored the potential utility of baseline metabolic tumour volume (MTV) for predicting event free survival (EFS) with promising results, but this has yet to be adopted clinically [8–17]. Others have explored the potential utility of radiomic features extracted from PET/CT for modelling purposes [8,18]. Initial results are promising, however, the published studies with relatively small numbers of patients are heterogenous

This aim of this study was to develop and test models combining baseline clinical information and radiomic features extracted from PET/CT imaging in DLBCL patients to predict 2-year EFS (2-EFS) using data from our tertiary centre. The secondary aim was to compare model performance to the predictive ability of baseline MTV.

2. Materials and Methods

The transparent reporting of a multivariable prediction model for individual prognosis or diagnosis (TRIPOD) guidelines were adhered to as part of this study (Supplementary Material).

2.1. Patient Selection

Radiological and clinical databases were retrospectively reviewed to identify patients who underwent baseline PET/CT for DLBCL at our institution between January 2008 and January 2018. A cut-off of January 2018 was chosen to allow a minimum of 2 years follow up without interference or confounding factors introduced by the COVID-19 pandemic. Patients were excluded if they did not have DLBCL, were under 16 years of age, had no measurable disease on PET/CT, had hepatic involvement, had a concurrent malignancy, were not treated with R-CHOP or if the images were degraded or incomplete. A 2-EFS event was defined as disease progression, recurrence or death from any cause within the 2-year follow up period.

2.2. PET/CT Acquisition

All imaging was performed as part of routine clinical practice. Patients fasted for 6 h prior to administration of intravenous Fluorine-18 FDG (4 MBq/kg). PET acquisition and reconstruction parameters for the four scanners used at our institution are detailed in Table 1. Attenuation correction was performed using a low-dose unenhanced diagnostic CT component acquired using the following settings: 3.75 mm slice thickness; pitch 6; 140 kV; 80 mAs.

Table 1. Reconstruction parameters for the different scanners used.

Scanner	Voxel Size in mm (x, y, z)	Matrix	Reconstruction	Scatter Correction	Randoms Correction
Philips Gemini TF64	4 × 4 × 4	144 or 169	BLOB-OS-TF	SS-Simul	DLYD
GE Healthcare Discovery 690	3.65 × 3.65 × 3.27	192	VPFX	Model based	Singles
GE Healthcare Discovery 710	3.65 × 3.65 × 3.27	192	VPFX	Model based	Singles
GE Healthcare STE	4.6875 × 4.6875 × 3.27	128	OSEM	Convolution subtraction	Singles

BLOB-OS-TF = an ordered subset iterative TOF reconstruction algorithm using blobs instead of voxels; DLYD = delayed event subtraction; OSEM = ordered subsets expectation maximisation; SS-Simul = single-scatter simulation; VPFX = Vue Point FX (OSEM including point spread function and time of flight).

2.3. Image Segmentation

All PET/CT images were reviewed and contoured using a specialised multimodality imaging software package (RTx v1.8.2, Mirada Medical, Oxford, UK). FDG-positive disease segmentation was performed by either a clinical radiologist with six years' experience or a research radiographer with two years' experience. Contours were then reviewed by dual-certified Radiology and Nuclear Medicine Physicians with >15 years' experience of oncological PET/CT interpretation. Any discrepancies were agreed by consensus.

Two different semi-automated segmentation techniques were used. The first applied a fixed standardised uptake value (SUV) threshold of 4.0, and the second used a threshold derived from 1.5 times mean liver SUV. The 4.0 SUV threshold was selected based on previous work assessing different segmentation techniques in a cohort of DLBCL patients by Burggraaff et al. which found it had a higher interobserver reliability [19] and requires less adaption than techniques such as 41% SUVmax. The 1.5 times mean liver SUV threshold was chosen as an adaptive threshold technique which has been used in different cancer types [20,21], and allows for adaptive thresholding which takes into consideration background SUV uptake which can vary between individuals. Mean liver SUV was calculated by placing a 110 cm^3 spherical region of interest (ROI) in the right lobe of the liver. The PET image contour was translated to the CT component of the study with the contours matched to soft tissue with a value of −10 to 100 Hounsfield units (HU). Contours were saved and exported as digital imaging and communications in medicine (DICOM) radiotherapy (RT) structures. Both the images and contours were converted to Neuroimaging Informatics Technology Initiative (NIfTI) files using the python library Simple ITK (v2.0.2) (https://simpleitk.org/, accessed on 1 December 2021).

2.4. Feature Extraction

Feature extraction was performed using PyRadiomics (v2.2.0) (https://pyradiomics.readthedocs.io/en/latest/index.html, accessed on 1 December 2021). Both the CT and PET images were resampled to a uniform voxel size of 2 mm^3. Radiomic features were extracted from the entire segmented disease for each patient. A fixed bin width of 2.5 HU was used for the CT component. Two different bin-widths were used when extracting the radiomic features from the PET component. The first being derived by finding the contour with the maximum range of SUVs and dividing this by 130, the second being derived by dividing the maximum range by 64. This methodology was based on previous work by Orlhac et al. and on PyRadiomics documentation [22]. The first and second order features were extracted from both the PET and CT components. Further higher order features were explored by extracting the first and second order features following application of wavelet, log-sigma, square, square root, logarithm, exponential, gradient and local binary pattern (lbp)-3D filters to the images. All the features extracted and the filters applied are detailed in Table S1. The mathematical definition of each of the radiomic features can be found within the PyRadiomics documentation [23]. PyRadiomics deviates from the image biomarker standardisation initiative (IBSI) by applying a fixed bin width from 0 and not the minimum segmentation value, and the calculation of first order kurtosis being +3 [24,25]. Otherwise, PyRadiomics adheres to IBSI guidelines. Patient age, disease stage and sex were also included as clinical features in the models. Disease stage and sex were dummy encoded using Pandas (v1.2.4) (https://pandas.pydata.org/pandas-docs/stable/whatsnew/v1.2.4.html, accessed on 1 December 2021). This resulted in a total of 3935 features extracted per patient. ComBat harmonisation was applied to account for the different scanners used within the study (https://github.com/Jfortin1/ComBatHarmonization, accessed on 1 December 2021) [26].

2.5. Machine Learning

The dataset was split into a training and test set stratified around 2-EFS, disease stage, age and sex with an 80:20 split using scikit-learn (v0.24.2) (https://scikit-learn.org/stable/whats_new/v0.24.html, accessed on 1 December 2021). Concordance between the demographics of the training and test groups was assessed using a *t*-test for continuous data and a χ^2 test for categorical data. A *p*-value of <0.05 was regarded as significant. Continuous data was normalised using a standard scaler (scikit-learn v0.24.2) which was trained and fit on the training set and subsequently applied to the test set. Highly correlated features were removed from the training and test sets if they had a Pearson coefficient over 0.8. This reduced the number of features from 3935 down to 130 for each patient.

Six different machine learning (ML) classifiers were used: logistic regression with lasso, ridge and elasticnet penalties, support vector machine (SVM), random forest and k-nearest neighbour. A maximum number of five features were included within each model, apart from in the lasso and elasticnet models where these classifiers determined the optimum number of features. To avoid false discoveries (Type 1 errors), a maximum number of five features was chosen guided by the rule of 1 feature per 10 events within the training set. Feature selection for the remaining models was performed using three different methods: a forward wrapper method (mlxtend 0.18.0), a univariate analysis method (scikit-learn v0.24.2), and a recursive feature extraction method (where applicable) (scikitlearn v0.24.2). Each method was used to create a list of features from two to the maximum five features which were to be explored in the training phase. The features selected were based on the highest mean receiver operating characteristic (ROC) curve area under the curve (AUC) in a four-fold stratified cross validation with 25 repeats.

Training of the ML models and the tuning of hyperparameters was performed using a grid search with a stratified four-fold cross validation stratified around 2-EFS with 25 repeats. The list of hyperparameters explored within the grid search are detailed in Table S2. Features and hyperparameters with the highest mean validation AUC which was within 0.05 of the mean training AUC were selected. A 0.05 cut-off was chosen to try and minimise selection of an overfitted model. The model which had the highest mean validation AUC overall was tested once on the unseen test set. The Youden index was used to discover the optimum cut-off value from the ROC curve and the accuracy, sensitivity, specificity, negative predictive value (NPV) and positive predictive value (PPV) were calculated from this for the unseen test set. The pipeline for patient inclusion, feature selection and predictive model creation and testing is depicted in Figure 1.

Given the growing evidence surrounding MTV as a predictor of outcome, two further logistic regression models were derived from the MTVs using the different segmentation. A comparison between results from the different cross validation splits between the radiomic model with the mean highest AUC and the MTV model with the mean higher AUC was performed using a Wilcoxon signed ranked test.

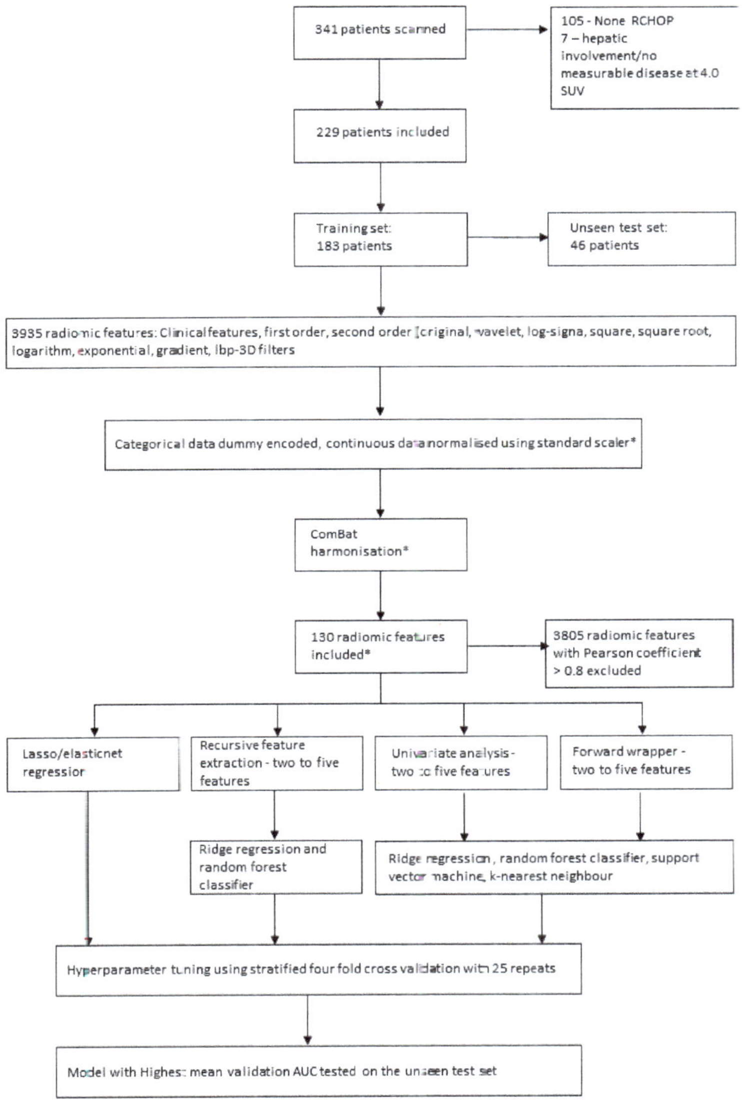

Figure 1. Pathway for patient inclusion, feature selection and model creation. * = initially applied to the training data and then to the test data.

3. Results

A total of 229 DLBCL patients met the inclusion criteria (136 male, 93 female) with 62 2-EFS events. There were 183 patients within the training cohort and 46 patients in the unseen test cohort. No statistically significant differences were identified between the training and test sets (Table 2).

None of the machine learning models created using elasticnet regression, lasso regression or k-nearest neighbour algorithms had a mean validation AUC within 0.05 of the mean training AUC. The remaining model results are presented in Tables 3 and 4.

Table 2. Demographics of the training and testing groups.

Demographic	Training Cohort	Test Cohort	p-Value
Age	67 (IQR = 17)	65 (IQR = 22.5)	0.35
Sex			
Male	107	29	0.69
Female	76	36	
Radiotherapy			
Yes	78	20	0.95
No	105	26	
Stage			
One	42	17	0.26
Two	46	6	
Three	31	6	
Four	64	17	
2-EFS Event			
Yes	50	12	0.98
No	133	34	

2-EFS = 2-year event free survival. The p-values were calculated using a t-test for age and a χ^2 test for the remaining demographic features.

Table 3. Mean training and validation scores for the best performing machine learning models using the 4.0 SUV threshold segmentation technique.

Machine Learning Model	Hyperparameters	Features	AUC Mean Training	AUC Mean Validation
SUVmax/130				
Ridge Regression	C: 1×10^{-5}, penalty: l2, solver: liblinear	Stage One, PET wavelet-LLH GLSZM Large Area Emphasis, PET wavelet-HHH GLSZM Grey Level Non-Uniformity Normalised, PET square 10th Percentile, PET square GLDM Grey Level Non-Uniformity	0.75 (0.02)	0.74 (0.07)
Support Vector Machine	C: 1, gamma: 0.008915428868611115, kernel: sigmoid	PET wavelet-HHH GLSZM Grey Level Non-Uniformity Normalised, PET square 10th Percentile, PET lbp-3D-m1 Interquartile Range, PET lbp-3D-m1 GLDM Large Dependence Low Grey Level Emphasis, PET lbp-3D-k 90th Percentile	0.74 (0.02)	0.73 (0.07)
Random Forest	bootstrap: False, max depth: 1, max features: log2, min samples leaf: 50, min samples split: 50, n estimators: 10	PET original shape Maximum 2D Diameter Column, MTV, PET original first order Kurtosis, PET original GLSZM Large Area Emphasis, PET wavelet-LHL GLCM Correlation, PET wavelet-LHL GLCM Imc2	0.76 (0.02)	0.71 (0.08)
SUVmax/64				
Ridge Regression	C: 0.001, penalty: l2, solver: newton-cg	Stage Four, PET original GLSZM Large Area Emphasis, PET wavelet-HHL GLSZM Small Area Emphasis, PET wavelet-HHH GLSZM Grey Level Non-Uniformity Normalised, PET square 10th Percentile	0.77 (0.02)	0.75 (0.06)
Support Vector Machine	C: 0.1, gamma: 0.07938667031015477, kernel: rbf	PET original GLDM Large Dependence Low Grey Level Emphasis, PET wavelet-HHH GLSZM Grey Level Non-Uniformity Normalised, PET square 10th Percentile, PET lbp-3D-k 90 Percentile, PET lbp-3D-k GLSZM Size Zone Non-Uniformity Normalised	0.75 (0.02)	0.72 (0.06)
Random Forest	bootstrap: True, max depth: 1, max features: log2, min samples leaf: 44, min samples split: 6, n estimators: 243	PET original shape Maximum 2D Diameter Column, PET original shape Surface Volume Ratio, PET original 10th Percentile	0.71 (0.02)	0.69 (0.08)

l2 = Ridge regression penalty, liblinear = A library for large linear classification, GLSZM = grey level size zone matrix, GLDM = grey level dependence matrix, lbp-3D-m1 = local binary pattern filtered image at level 1, lbp-3D-k = local binary pattern kurtosis image, GLCM = grey level co-occurrence matrix, rbf = radial basis function.

Table 4. Mean training and validation scores for the best performing machine learning models using the 1.5 times mean liver SUV thresholding segmentation technique.

Machine Learning Model	Hyperparameters	Features	AUC Mean Training	AUC Mean Validation
SUVmax/130				
Ridge Regression	C: 1×10^{-5}, penalty: l2, solver: saga	Stage Four, Age, PET original GLDM Large Dependence Low Grey Level Emphasis, PET original GLSZM Large Area High Grey Level Emphasis	0.74 (0.03)	0.71 (0.09)
Support Vector Machine	C: 1, gamma: 0.43727367418726576, kernel: rbf	PET square 10th Percentile, PET square first order Energy	0.78 (0.02)	0.73 (0.07)
Random Forest	bootstrap: True, max depth: 10, max features: sqrt, min samples leaf: 33, min samples split: 5, n estimators: 90	Age, PET original shape Elongation, PET original shape Least Axis Length, PET original shape Major Axis Length, PET original shape Maximum 2D Diameter Column, PET original shape Mesh Volume		
SUVmax/64				
Ridge Regression	C: 1.0, penalty: l2, solver: liblinear	Stage Three, Age, PET wavelet-LHL GLCM Imc1, PET square GLDM Dependence Variance, PET square GLSZM Small Area Low Grey Level Emphasis	0.76 (0.02)	0.73 (0.07)
Support Vector Machine	C: 1, gamma: 0.43727367418726576, kernel: rbf	PET square first order 10 Percentile, PET square first order Energy	0.78 (0.02)	0.73 (0.07)
Random Forest	bootstrap: True, max depth: 10, max features: log2, min samples leaf: 42, min samples split: 6, n estimators: 237	PET original shape Sphericity, PET original GLSZM Large Area Emphasis	0.70 (0.02)	0.69 (0.07)

l2 = Ridge regression penalty, liblinear = A library for large linear classification, GLSZM = grey level size zone matrix, GLDM = grey level dependence matrix, lbp-3D-m1 = local binary pattern filtered image at level 1, lbp-3D-k = local binary pattern kurtosis image, GLCM = grey level co-occurrence matrix, rbf = radial basis function.

The model within the highest mean validation ROC AUC was the ridge regression model created using radiomic features extracted from a fixed threshold of 4.0 SUV segmentation using a bin width of the maximum range of SUVs divided by 64. The mean training AUC was 0.77 ± 0.02, the mean validation AUC was 0.75 ± 0.06 and the AUC when tested on the unseen dataset was 0.73 (Figure 2). The features selected with their coefficients and intercept are presented in Table 5. A threshold of 0.5 was chosen and led to an accuracy of 0.70, sensitivity of 0.44, specificity of 0.86, positive predictive value of 0.67, and a negative predictive value of 0.71. The confusion matrix is presented in Table 6.

The logistic regression model created solely from MTV using the 4.0 SUV fixed threshold segmentation technique had a mean training AUC of 0.66 ± 0.03 and a mean validation AUC of 0.66 ± 0.08. The logistic regression model derived from MTV using the 1.5 times mean liver SUV segmentation technique had a mean training AUC of 0.67 ± 0.03 and a mean validation AUC of 0.67 ± 0.08. There was a statistically significant difference when comparing the cross validation AUCs for the 100 splits between the highest performing MTV-based model and the radiomic-based ridge regression model, $p < 0.001$ (Figure 3).

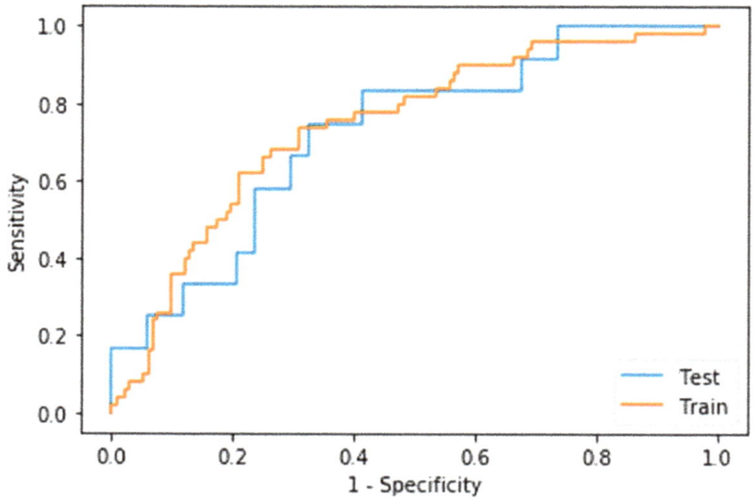

Figure 2. ROC Curve of the training and unseen test data AUCs for the model derived using a 4.0 SUV thresholding segmentation technique with a bin width derived from SUVmax/64.

Table 5. The features selected and their associated coefficients and intercept in the ridge regression model tested on the unseen test dataset.

Feature	Coefficient
Stage Four	0.01153414
PET original GLSZM Large Area Emphasis	0.0161316
PET wavelet-HHL GLSZM Small Area Emphasis	0.01482446
PET wavelet-HHH GLSZM Grey Level Non-Uniformity Normalised	−0.01923886
PET square 10 Percentile	−0.01923886
Intercept	−0.01166859

Table 6. Confusion matrix for the threshold of 0.5.

Prediction	Negative	Positive
Predicted Negative	24	10
Predicted Positive	4	8

Positive = recorded 2-EFS event, Negative = no recorded 2-EFS event, Predicted Positive = predicted to have had a 2-EFS event, Predicted Negative = predicted to not have had a 2-EFS event.

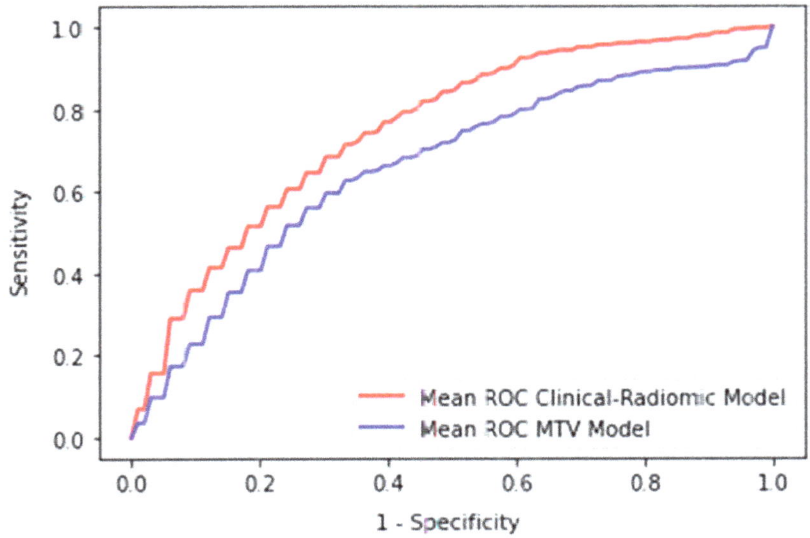

Figure 3. Mean ROC Curve of the MTV-based logistic regression model and the radiomic-based logistic regression model.

4. Discussion

Our study found that a prediction model combining clinical and radiomic features derived from pretreatment PET/CT using a ridge regression model had the highest mean validation AUC when predicting 2-EFS in DLBCL patients. This model had significantly higher validation AUCs than those achieved by a model solely derived from MTV and achieved an AUC of 0.73 on the unseen test set. The radiomic features used within the model that led to the highest mean validation AUC were extracted from a segmentation derived from a fixed threshold of 4.0 SUV using a bin-width calculated from the maximum range of SUVs divided by 64. The model was formed using five features (Stage Four, PET original GLSZM large area emphasis, PET wavelet-HHL GLSZM Small Area Emphasis, PET wavelet-HHH GLSZM Grey Level Non-Uniformity normalised, PET square 10th percentile).

The biological correlate of radiomic features and how these relate to the lesion or disease process can often be overlooked, and can become more complex when image filtering is involved [27]. Three of the radiomic features included in the best model were derived from GLSZM which is a matrix formed by the number of connected voxels with the same grey level intensity. The first was the PET GLSZM Large Area Emphasis, which is a measure of distribution of large area size zones, and was extracted from the PET data without any filter applied. This feature is higher in lesions which have a coarser texture based on the original image. The other two GLZMs are calculated after applying a wavelet filter. Wavelet filters highlight or suppress certain spatial frequencies within an image. In PyRadiomics a combination of high and low filters is passed in each of the different dimensions, which results in eight different decompositions. PET wavelet-HHL GLSZM Small Area Emphasis is a measure of the distribution of small size zones, which are higher in lesions with fine textures following the application of the wavelet filter. PET wavelet-HHH GLSZM Grey Level Non-Uniformity is a measure of the variability of the grey level intensity within the image. A lower value indicates a higher number of similar SUVs on the wavelet filtered image. The last radiomic feature included was PET square 10th percentile which is the 10th percentile value of the SUV after a square of the image SUVs has been taken and normalised to the original SUV range. Interestingly, none of the CT-derived radiomic features were selected as part of the best performing radiomic models. This is likely due to the transposition of the segmentations from the PET on to the unenhanced CT including more areas of non-lymphomatous tissue.

Other studies which have explored the use of radiomic features in outcome prediction in DLBCL are not always directly comparable [12,28–32]. This is mainly due to differences in segmentation methodology, modelling techniques and outcome measures between groups. Aide et al. studied the use of radiomic features in predicting 2-EFS in 132 patients (training = 105, validation = 27) and found that MTV as well as four second-order metrics and five third-order metrics were selected from ROC analyses. However, long-zone high-grey level emphasis was the only independent predictor when analysed with the international prognostic index (IPI) and MTV [29]. In our study long-zone high-grey level emphasis was discarded when checking for multicollinearity. This highlights a potential issue of radiomic model development when applying a methodology on different datasets. It may be that the same features would be chosen between the different datasets, but each method removes the alternate correlated feature and, therefore, appears to create an entirely new model. Both Zhang et al. and Ceriani et al. used lasso in their cox regression models to select the most appropriate features [31,32]. Zhang et al. in a study of 152 patients (training = 100, validation = 52) treated with R-CHOP or R-EPOCH (rituximab, etoposide, prednisone, vincristine, cyclophosphamide, and doxorubicin) found that a survival model created with radiomic features and MTV had a validation time dependent ROC AUC of 0.748 (95% CI 0.596–0.886). A model created with radiomic features and metabolic bulk volume had a validation time dependent ROC AUC 0.759 (95% CI 0.595–0.888). Ceriani et al. reported that a radiomic score derived from a training set of 133 patients and tested on an external dataset of 107 patients had an AUC of 0.71 in both the test and validation datasets. The features selected within their cox regression model were GLCM sum squares, maximum 3D diameter and GLDM grey level variance, GLSZM grey level non-uniformity normalised.

In our study both lasso and elasticnet methods failed to produce a model that achieved mean training and validation scores within 0.05 of each other. Even when allowing for a more generous difference between the training and validation scores, mean validation scores remained below 0.65. This 0.05 cut-off is arbitrary and was applied to try and reduce the impact of overfitting on the dataset and allow selection of a potentially more generalisable model. Despite this, there is still a risk that both training and validation datasets are overfitted and the model would need external validation on an external dataset.

One of the largest published studies by Decazes et al. in 215 DLBCL patients, explored use of tumour volume surface ratio and total tumour surface as outcome predictors for 5-year progression free survival (PFS), but found that MTV outperformed both features with MTV having an AUC of 0.67 [12]. This AUC for MTV is similar to the findings in our study, with the mean validation AUC for MTV prediction of 2-EFS being 0.66 for the 4.0 SUV threshold and 0.67 for the 1.5 times liver threshold segmentation techniques, respectively. Although, there is growing interest in the use of MTV as an imaging biomarker, Adams et al. reported, in a study of 73 DLBCL patients, that the prognostic ability of MTV does not add anything to the prognostic ability of the clinical scoring system National Comprehensive Cancer Network-International Prognostic Index (NCCN-IPI) [33]. Unfortunately, due to missing clinical data it was not possible to compare IPI performance in our patient cohort. However, this does highlight the potential impact of confounders on the generalisability of predictive models. Although, causality is not generally considered in predictive modelling, its use in future models could allow for greater transparency of a model. The issues of generalisability may be compounded by learnt biases towards groups of patients in the training process.

The TRIPOD checklist was completed to increase transparency of model development [34,35]. However, there are limitations to our study including its retrospective nature and uncertainty surrounding the exact timing and recording of recurrence. Use of 2-EFS partially mitigates against this by allowing a wider window for the relapse to be recorded, however, it does mean that data which could have been included in a time to survival type model is lost. 2-EFS was chosen as the majority of patients relapse within the first 2 years. Time to event ML models could be used in future studies to reduce the need to exclude data. The lesions were not re-segmented as part of the study, and therefore, calculations of inter or intra-reliability, as well as robustness of the features have not been performed. ComBat harmonization was used to help mitigate against scanner variation in the extracted feature extraction. However, this limits the ability to apply this model prospectively to patients not scanned using a protocol used to train the model. Lack of clinical data surrounding the IPI and cell of origin (COO) information, meant that these could not be used as direct comparators to radiomic models created.

5. Conclusions

A combined clinical and PET/CT derived radiomics model using ridge regression demonstrated the highest mean AUC validation (AUC = 0.75) when predicting 2-EFS in DLBCL patients treated with R-CHOP, which outperformed a model derived solely from MTV (AUC = 0.67).

Supplementary Materials: The following supporting information can be downloaded at: https://www.mdpi.com/article/10.3390/cancers14071711/s1, TRIPOD Checklist: Prediction Model Development, Table S1: Radiomic features extracted for both the PET and CT components, Table S2: The hyperparameters explored within the grid search.

Author Contributions: Conceptualization, R.F., M.C., C.B., C.T., A.F.F., F.G., C.P. and A.F.S.; methodology, R.F., M.C., C.F. and A.F.S.; formal analysis, R.F., M.C., C.P. and A.F.S.; investigation, R.F. and A.F.S.; data curation, R.F., M.C. and C.B.; writing—original draft preparation, R.F. and A.F.S.; writing—review and editing, R.F., M.C., C.B., C.T., A.F.F., F.G., C.P. and A.F.S. All authors have read and agreed to the published version of the manuscript.

Funding: R.F., F.G., C.P. and A.F.S. are partly funded by Innovate UK via the National Consortium of Intelligent Medical Imaging (NCIMI) (104688) and A.F.F. is funded by the Royal Academy of Engineering (CiET1819/19). The funders had no role in study design, data collection and analysis, decision to publish, or preparation of the manuscript.

Institutional Review Board Statement: Following discussion with the Research and Innovation Department at LTHT it was agreed that this represented a service improvement project. The study was approved by the University of Leeds School of Medicine Research Ethics Committee (SoMREC) (MREC 19-043).

Informed Consent Statement: Informed written consent was obtained prospectively from all patients at the time of imaging for use of their anonymised FDG PET/CT images in research and service development projects.

Data Availability Statement: The data are not publicly available due to institutional data sharing restrictions.

Conflicts of Interest: The authors declare no conflict of interest.

References

1. Armitage, J.O.; Gascoyne, R.D.; Lunning, M.A.; Cavalli, F. Non-Hodgkin lymphoma. *Lancet* **2017**, *390*, 298–310. [CrossRef]
2. Coiffier, B.; Thieblemont, C.; Van Den Neste, E.; Lepeu, G.; Plantier, I.; Castaigne, S.; Lefort, S.; Marit, G.; Macro, M.; Sebban, C.; et al. Long-term outcome of patients in the LNH-98.5 trial, the first randomized study comparing rituximab-CHOP to standard CHOP chemotherapy in DLBCL patients: A study by the Groupe d'Etudes des Lymphomes de l'Adulte. *Blood* **2010**, *116*, 2040–2045. [CrossRef] [PubMed]
3. Kansara, R. Central Nervous System Prophylaxis Strategies in Diffuse Large B Cell Lymphoma. *Curr. Treat. Options Oncol.* **2018**, *19*, 52. [CrossRef] [PubMed]
4. SEER. *SEER Cancer Statistics Review*; SEER: Bethesda, MD, USA, 2021.
5. Cheson, B.D.; Fisher, R.I.; Barrington, S.F.; Cavalli, F.; Schwartz, L.H.; Zucca, E.; Lister, T.A. Recommendations for initial evaluation, staging, and response assessment of hodgkin and non-hodgkin lymphoma: The lugano classification. *J. Clin. Oncol.* **2014**, *32*, 3059–3067. [CrossRef] [PubMed]
6. Yoo, C.; Lee, D.H.; Kim, J.E.; Jo, J.; Yoon, D.H.; Sohn, B.S.; Kim, S.W.; Lee, J.S.; Suh, C. Limited role of interim PET/CT in patients with diffuse large B-cell lymphoma treated with R-CHOP. *Ann. Hematol.* **2011**, *90*, 797–802. [CrossRef] [PubMed]

7. Mikhaeel, N.G.; Cunningham, D.; Counsell, N.; McMillan, A.; Radford, J.A.; Ardeshna, K.M.; Lawrie, A.; Smith, P.; Clifton-Hadley, L.; O'Doherty, M.J.; et al. FDG-PET/CT after two cycles of R-CHOP in DLBCL predicts complete remission but has limited value in identifying patients with poor outcome—Final result of a UK National Cancer Research Institute prospective study. *Br. J. Haematol.* **2021**, *192*, 504–513. [CrossRef]
8. Frood, R.; Burton, C.; Tsoumpas, C.; Frangi, A.F.; Gleeson, F.; Patel, C.; Scarsbrook, A. Baseline PET/CT imaging parameters for prediction of treatment outcome in Hodgkin and diffuse large B cell lymphoma: A systematic review. *Eur. J. Nucl. Med. Mol. Imaging* **2021**, *48*, 3198–3220. [CrossRef]
9. Song, M.K.; Chung, J.S.; Shin, H.J.; Lee, S.M.; Lee, S.E.; Lee, H.S.; Lee, G.W.; Kim, S.J.; Lee, S.M.; Chung, D.S. Clinical significance of metabolic tumor volume by PET/CT in stages II and III of diffuse large B cell lymphoma without extranodal site involvement. *Ann. Hematol.* **2012**, *91*, 697–703. [CrossRef]
10. Song, M.K.; Yang, D.H.; Lee, G.W.; Lim, S.N.; Shin, S.; Pak, K.J.; Kwon, S.Y.; Shim, H.K.; Choi, B.H.; Kim, I.S.; et al. High total metabolic tumor volume in PET/CT predicts worse prognosis in diffuse large B cell lymphoma patients with bone marrow involvement in rituximab era. *Leuk. Res.* **2016**, *42*, 1–6. [CrossRef]
11. Cottereau, A.S.; Nioche, C.; Dirand, A.S.; Clerc, J.; Morschhauser, F.; Casasnovas, O.; Meignan, M.; Buvat, I. 18F-FDG PET dissemination features in diffuse large B-cell lymphoma are predictive of outcome. *J. Nucl. Med.* **2020**, *61*, 40–45. [CrossRef]
12. Decazes, P.; Becker, S.; Toledano, M.N.; Vera, P.; Desbordes, P.; Jardin, F.; Tilly, H.; Gardin, I. Tumor fragmentation estimated by volume surface ratio of tumors measured on 18F-FDG PET/CT is an independent prognostic factor of diffuse large B-cell lymphoma. *Eur. J. Nucl. Med. Mol. Imaging* **2018**, *45*, 1672–1679. [CrossRef] [PubMed]
13. Ilyas, H.; Mikhaeel, N.G.; Dunn, J.T.; Rahman, F.; Möller, H.; Smith, D.; Barrington, S.F. Is there an optimal method for measuring baseline metabolic tumor volume in diffuse large B cell lymphoma? *Eur. J. Nucl. Med. Mol. Imaging* **2019**, *46*, 520–521. [CrossRef] [PubMed]
14. Toledano, M.N.; Desbordes, P.; Banjar, A.; Gardin, I.; Vera, P.; Ruminy, P.; Jardin, F.; Tilly, H.; Becker, S. Combination of baseline FDG PET/CT total metabolic tumour volume and gene expression profile have a robust predictive value in patients with diffuse large B-cell lymphoma. *Eur. J. Nucl. Med. Mol. Imaging* **2018**, *45*, 680–688. [CrossRef] [PubMed]
15. Chang, C.-C.; Cho, S.-F.; Chuang, Y.-W.; Lin, C.-Y.; Chang, S.-M.; Hsu, W.-L.; Huang, Y.-F. Prognostic significance of total metabolic tumor volume on 18F-fluorodeoxyglucose positron emission tomography/ computed tomography in patients with diffuse large B-cell lymphoma receiving rituximab-containing chemotherapy. *Oncotarget* **2017**, *8*, 99587–99600. [CrossRef] [PubMed]
16. Cottereau, A.S.; Lanic, H.; Mareschal, S.; Meignan, M.; Vera, P.; Tilly, H.; Jardin, F.; Becker, S. Molecular profile and FDG-PET/CT Total metabolic tumor volume improve risk classification at diagnosis for patients with diffuse large B-Cell lymphoma. *Clin. Cancer Res.* **2016**, *22*, 3801–3809. [CrossRef]
17. Mikhaeel, N.G.; Smith, D.; Dunn, J.T.; Phillips, M.; Møller, H.; Fields, P.A.; Wrench, D.; Barrington, S.F. Combination of baseline metabolic tumour volume and early response on PET/CT improves progression-free survival prediction in DLBCL. *Eur. J. Nucl. Med. Mol. Imaging* **2016**, *43*, 1209–1219. [CrossRef]
18. Lambin, P.; Leijenaar, R.T.H.; Deist, T.M.; Peerlings, J.; De Jong, E.E.C.; Van Timmeren, J.; Sanduleanu, S.; Larue, R.T.H.M.; Even, A.J.G.; Jochems, A.; et al. Radiomics: The bridge between medical imaging and personalized medicine. *Nat. Rev. Clin. Oncol.* **2017**, *14*, 749–762. [CrossRef]
19. Burggraaff, C.N.; Rahman, F.; Kaßner, I.; Pieplenbosch, S.; Barrington, S.F.; Jauw, Y.W.S.; Zwezerijnen, G.J.C.; Müller, S.; Hoekstra, O.S.; Zijlstra, J.M.; et al. Optimizing Workflows for Fast and Reliable Metabolic Tumor Volume Measurements in Diffuse Large B Cell Lymphoma. *Mol. Imaging Biol.* **2020**, *22*, 1102–1110. [CrossRef]
20. Brown, P.J.; Zhong, J.; Frood, R.; Currie, S.; Gilbert, A.; Appelt, A.L.; Sebag-Montefiore, D.; Scarsbrook, A. Prediction of outcome in anal squamous cell carcinoma using radiomic feature analysis of pre-treatment FDG PET-CT. *Eur. J. Nucl. Med. Mol. Imaging* **2019**, *46*, 2790–2799. [CrossRef]
21. Zhong, J.; Frood, R.; Brown, P.; Nelstrop, H.; Prestwich, R.; McDermott, G.; Currie, S.; Vaidyanathan, S.; Scarsbrook, A.F. Machine learning-based FDG PET-CT radiomics for outcome prediction in larynx and hypopharynx squamous cell carcinoma. *Clin. Radiol.* **2021**, *76*, 78.e9–78.e17. [CrossRef]
22. Orlhac, F.; Soussan, M.; Chouahnia, K.; Martinod, E.; Buvat, I. 18F-FDG PET-derived textural indices reflect tissue-specific uptake pattern in non-small cell lung cancer. *PLoS ONE* **2015**, *10*, e0145063. [CrossRef] [PubMed]
23. PyRadiomics Radiomic Features. Available online: https://pyradiomics.readthedocs.io/en/latest/features.html (accessed on 1 December 2021).
24. Zwanenburg, A.; Leger, S.; Vallières, M.; Löck, S. Image biomarker standardisation initiative. *arXiv* **2016**, arXiv:1612.07003. [CrossRef]
25. PyRadiomics Frequently Asked Questions. Available online: https://pyradiomics.readthedocs.io/en/latest/faq.html (accessed on 1 December 2021).
26. Fortin, J.-P.; Cullen, N.; Sheline, Y.I.; Taylor, W.D.; Aselcioglu, I.; Cook, P.A.; Adams, P.; Cooper, C.; Fava, M.; McGrath, P.J.; et al. Harmonization of cortical thickness measurements across scanners and sites. *Neuroimage* **2018**, *167*, 104–120. [CrossRef] [PubMed]
27. Tomaszewski, M.R.; Gillies, R.J. The biological meaning of radiomic features. *Radiology* **2021**, *298*, 505–516. [CrossRef]
28. Senjo, H.; Hirata, K.; Izumiyama, K.; Minauchi, K.; Tsukamoto, E.; Itoh, K.; Kanaya, M.; Mori, A.; Ota, S.; Hashimoto, D.; et al. High metabolic heterogeneity on baseline 18FDG-PET/CT scan as a poor prognostic factor for newly diagnosed diffuse large B-cell lymphoma. *Blood Adv.* **2020**, *4*, 2286–2296. [CrossRef]

29. Aide, N.; Fruchart, C.; Nganoa, C.; Gac, A.; Lasnon, C. Baseline 18 F-FDG PET radiomic features as predictors of 2-year event-free survival in diffuse large B cell lymphomas treated with immunochemotherapy. *Eur. Radiol.* **2020**, *30*, 4623–4632. [CrossRef]
30. Bera, K.; Braman, N.; Gupta, A.; Velcheti, V.; Madabhushi, A. Predicting cancer outcomes with radiomics and artificial intelligence in radiology. *Nat. Rev. Clin. Oncol.* **2021**, *19*, 132–146. [CrossRef]
31. Ceriani, L.; Milan, L.; Cascione, L.; Gritti, G.; Dalmasso, F.; Esposito, F.; Pirosa, M.C.; Schär, S.; Bruno, A.; Dirnhofer, S.; et al. Generation and validation of a PET radiomics model that predicts survival in diffuse large B cell lymphoma treated with R-CHOP14: A SAKK 38/07 trial post-hoc analysis. *Hematol. Oncol.* **2021**, *40*, 12–22. [CrossRef]
32. Zhang, X.; Chen, L.; Jiang, H.; He, X.; Feng, L.; Ni, M.; Ma, M.; Wang, J.; Zhang, T.; Wu, S.; et al. A novel analytic approach for outcome prediction in diffuse large B-cell lymphoma by [18F]FDG PET/CT. *Eur. J. Nucl. Med. Mol. Imaging* **2021**, *49*, 1298–1310. [CrossRef]
33. Adams, H.J.A.; de Klerk, J.M.H.; Fijnheer, R.; Heggelman, B.G.F.; Dubois, S.V.; Nievelstein, R.A.J.; Kwee, T.C. Prognostic superiority of the National Comprehensive Cancer Network International Prognostic Index over pretreatment whole-body volumetric-metabolic FDG-PET/CT metrics in diffuse large B-cell lymphoma. *Eur. J. Haematol.* **2015**, *94*, 532–539. [CrossRef]
34. Park, J.E.; Kim, D.; Kim, H.S.; Park, S.Y.; Kim, J.Y.; Cho, S.J.; Shin, J.H.; Kim, J.H. Quality of science and reporting of radiomics in oncologic studies: Room for improvement according to radiomics quality score and TRIPOD statement. *Eur. Radiol.* **2020**, *30*, 523–536. [CrossRef] [PubMed]
35. Pinto dos Santos, D.; Dietzel, M.; Baessler, B. A decade of radiomics research: Are images really data or just patterns in the noise? *Eur. Radiol.* **2021**, *31*, 2–5. [CrossRef] [PubMed]

cancers

MDPI

Article

Development of an Image Analysis-Based Prognosis Score Using Google's Teachable Machine in Melanoma

Stephan Forchhammer [1,*], Amar Abu-Ghazaleh [1], Gisela Metzler [2], Claus Garbe [1] and Thomas Eigentler [3]

[1] Eberhardt Karls Universität, Universitäts-Hautklinik, 72076 Tübingen, Germany; amar.abu-ghazaleh@student.uni-tuebingen.de (A.A.-G.); claus.garbe@med.uni-tuebingen.de (C.G.)
[2] Zentrum für Dermatohistologie und Oralpathologie Tübingen/Würzburg, 72072 Tübingen, Germany; metzler@zentrum-dermatohistologie.de
[3] Department of Dermatology, Venereology and Allergology, Charité—Universitätsmedizin Berlin, Corporate Member of Freie Universität Berlin and Humboldt-Universität zu Berlin, Luisenstrasse 2, 10177 Berlin, Germany; thomas.eigentler@charite.de
* Correspondence: stephan.forchhammer@med.uni-tuebingen.de

Simple Summary: The increase in adjuvant treatment of melanoma patients makes it necessary to provide the most accurate prognostic assessment possible, even at early stages of the disease. Although conventional risk stratification correctly identifies most patients in need of adjuvant treatment, there are some patients who, despite having a low tumor stage, have poor prognosis and could therefore benefit from early therapy. To close this gap in prognosis estimation, deep learning-based image analyses of histological sections could play a central role in the future. The aim of this study was to investigate whether such an analysis is possible only using basic image analysis of 831 H&E-stained melanoma sections using Google's Teachable Machine. Although the classification obtained does not provide an additional prognostic estimate to conventional melanoma classification, this study shows that prognostic prediction is possible at the mere cellular image level.

Abstract: Background: The increasing number of melanoma patients makes it necessary to establish new strategies for prognosis assessment to ensure follow-up care. Deep-learning-based image analysis of primary melanoma could be a future component of risk stratification. Objectives: To develop a risk score for overall survival based on image analysis through artificial intelligence (AI) and validate it in a test cohort. Methods: Hematoxylin and eosin (H&E) stained sections of 831 melanomas, diagnosed from 2012–2015 were photographed and used to perform deep-learning-based group classification. For this purpose, the freely available software of Google's teachable machine was used. Five hundred patient sections were used as the training cohort, and 331 sections served as the test cohort. Results: Using Google's Teachable Machine, a prognosis score for overall survival could be developed that achieved a statistically significant prognosis estimate with an AUC of 0.694 in a ROC analysis based solely on image sections of approximately 250 × 250 µm. The prognosis group "low-risk" (n = 230) showed an overall survival rate of 93%, whereas the prognosis group "high-risk" (n = 101) showed an overall survival rate of 77.2%. Conclusions: The study supports the possibility of using deep learning-based classification systems for risk stratification in melanoma. The AI assessment used in this study provides a significant risk estimate in melanoma, but it does not considerably improve the existing risk classification based on the TNM classification.

Keywords: melanoma; prognosis; risk score; deep learning; artificial intelligence; Google's teachable machines

Citation: Forchhammer, S.; Abu-Ghazaleh, A.; Metzler, G.; Garbe, C.; Eigentler, T. Development of an Image Analysis-Based Prognosis Score Using Google's Teachable Machine in Melanoma. *Cancers* **2022**, *14*, 2243. https://doi.org/10.3390/cancers14092243

Academic Editors: Hamid Khayyam, Ali Madani, Rahele Kafieh and Ali Hekmatnia

Received: 4 April 2022
Accepted: 28 April 2022
Published: 29 April 2022

1. Introduction

Over the past years and decades, there has been a significant increase in the incidence of malignant melanoma [1]. Despite major advances in the treatment of metastatic melanoma, including targeted therapy with BRAF inhibitors or immune checkpoint blockade, malignant melanoma remains the skin tumor responsible for the highest number of skin tumor-associated deaths worldwide, with approximately 55,500 cases [2]. Histologically, different subtypes of malignant melanoma can be distinguished. According to the currently valid World Health Organization Classification published in 2018, a distinction is made between melanomas that typically occur in chronically sun-damaged (CSD) skin and those that typically do not occur in chronically sun-damaged skin. These differ in the underlying genetic pathways. The most common subtypes, superficial spreading melanoma (low-CSD) and lentigo maligna melanoma (high-CSD), but also desmoplastic melanoma, are found in association with sun-damaged skin. Representatives of melanomas that do not occur in chronically sun-damaged skin (no-CSD) are acral-, mucosal- and uveal-melanomas, Spitz melanomas, melanomas originating from congenital nevi or blue nevi. Nodular melanomas, on the other hand, can be found in both groups with different underlying genetic pathways [3]. Prognosis prediction and staging of melanoma are mainly based on histologic diagnosis in primary tumors. In this context, tumor thickness (according to Breslow) and ulceration are included in the 8th AJCC classification [4]. In the case of additional histological features, such as regression and mitotic rate, an impact on the further prognosis is assured [5–7]. Other prognostic factors result from the primary staging diagnosis, which includes sonography, CT section imaging and sentinel node biopsy depending on the stage [4,8]. With the advent of adjuvant therapy options for patients with high-risk tumors, the most accurate prognostic prediction possible is already necessary for the primary tumor. Since adjuvant immune checkpoint therapy has a non-negligible side effect profile, it is crucial to identify patients who may particularly benefit from such therapy. Various gene-expression-based assays are in development to distinguish high-risk patients from those with only low risk of metastasis [9–12]. However, these studies are cost intensive and therefore cannot yet be widely used. In addition, these examinations consume tissue that may be needed for further diagnostic workup. The morphology of melanoma already shows a very high diversity in the H&E section, which goes far beyond the detection of tumor thickness, ulceration, mitotic rate and regression. A grading, which is common for most other tumor types, such as cutaneous squamous cell carcinoma, does not exist for melanoma. With the onset of digitalization in pathology, artificial intelligence (AI)-based image analysis has created new opportunities in the evaluation of histological sections. AI is a set of technologies that enables computer systems to acquire intelligent capabilities. One branch of AI is the concept of machine learning, which gives computers the ability to learn without being explicitly programmed [13]. Deep learning, which is popular today, is characterized by greater network depth in terms of multiple layers of neurons; therefore, it makes it possible to learn and solve even complex tasks. Remarkably, artificial neural networks make this possible without having to deposit specific rules or instructions beforehand [14]. It has been shown that programs based on artificial intelligence are able to achieve high diagnostic accuracy in diagnosing melanoma from dermatoscopic images [15–18]. Diagnosis by artificial neuronal networks also seems to lead to very reliable results for histological sections. For epithelial skin tumors, but especially for melanomas, programs have been developed that enable robust diagnostics [19–24]. More exciting, though, is the question of whether image analysis with artificial neural networks can not only confirm a diagnosis but whether it is conceivable that subvisual structures or patterns in histological sections can be detected, leading to improved prognosis assessment. First studies in melanoma have shown that it may be possible to achieve prognosis prediction, prediction of sentinel positivity and prediction of response to immunotherapy by using artificial intelligence-assisted image analysis [25–27]. In particular, the work of Kulkarni et al. was able to make an impressive prognosis prediction based on image analysis; however, here a complex algorithm was used which, in addition to the mere morphological tumor cell information, evaluates in part the distribution of inflammatory cells [26]. Since a clear impact on melanoma prognosis has been well studied, especially for tumor-infiltrating lymphocytes, it remains unclear whether a purely morphological image analysis of tumor cells allows melanoma prognosis [28–30].

The aim of our study was to develop a prognosis score based purely on histological photographs to predict survival in melanoma. Since our score should be made publicly available, easy to use and based solely on morphological image information, Google's Teachable Machine was used as a deep learning program. This is a pre-trained neural network for image analysis that allows the classification of images into certain groups after previous training [31]. Google's Teachable Machine uses the basic framework of TensorFlow. This is a platform released in 2015 that was created to make artificial intelligence and its training accessible to the public. The use of this program has already been investigated in the first studies for image analysis of medical questions [32].

2. Materials and Methods

2.1. Study Population

All 2223 patients diagnosed with primary melanoma at the University Dermatological Clinic Tübingen between 1 January 2012 and 31 December 2015 who provided written informed consent to the nationwide melanoma registry were included in the study. All 831 patients with follow-up data of at least 2 years and histological sections in our archive were included in the further analysis. The group "dead" consists of all patients that died due to melanoma during the observation period up to 114 months. The group "alive" consists of all patients that were alive, lost to follow up or died of another reason. Alive patients with follow-up for less than 2 years were excluded from the study. The diagnosis of melanoma was made by at least two experienced, board-certified dermatopathologists (SF, GM).

2.2. Digitization of HE Sections and AI-Based Evaluation

All H&E sections of primary melanoma were photographed at the site of the highest tumor thickness according to Breslow using 100× magnification (Figure 1). Pictures were taken using a Nikon Eclipse 80i microscope mounted with a Nikon Digital Sight DS-FI2 camera. The program Nikon NIS Elements D Version 4.13.04. was used, and the exposure time was set to 3 ms. The data were saved in JPG format. Images were analyzed using Google's Teachable Machine, a pre-trained neural network [31]. Sixty percent of the 831 images served as the training cohort, and 40% of the images were subsequently evaluated as the test cohort. The allocation of the 500 images to the training cohort or the 331 images to the test cohort was random. The training dataset contains images that were only used for training Google's Teachable Machine. An analysis of these data was not performed later. Of these 500 patients, 429 were alive; thus, these images were used for the training of the "alive" group. Of the 500 patients, 71 were deceased; these were used for the training of the group "dead". The training was carried out twice and separately for the groups "whole images" and "area of interest". The training curves for accuracy and loss were obtained for both training groups and are shown in Figures S1 and S2. The model that emerged from the initial training was used for further assessment. As Google's Teachable Machine does not provide a verification function, a separate set of verification data was not assigned. The remaining 331 patients were used as the test cohort. The images of these patients were not previously seen by Google's Teachable Machine. These 331 patient images were then classified by the program into the categories "dead" and "alive". Patients who were classified as "dead" were given the label "high-risk" in the further study, and patients who were classified as "alive" were given the label "low-risk".

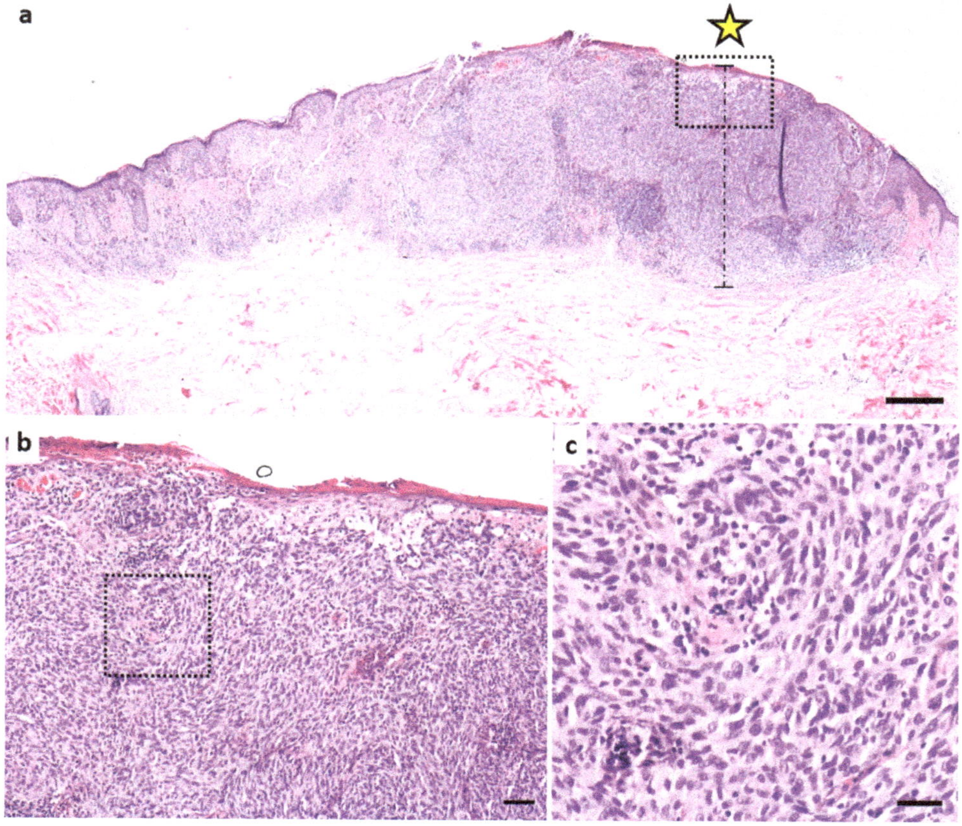

Figure 1. H&E section of a malignant melanoma. (**a**) overview with annotation (star) of the highest tumor thickness (Breslow). The scale is 500 µm. (**b**) Magnification of (**a**) (see square in (**a**)). The image represents one picture of the category "whole image". The scale is 100 µm. (**c**) Magnification of (**b**) (see square in (**b**)). This image represents one picture of the category "area of interest". The scale is 30 µm.

The evaluations of the whole images or the "area of interest" images were performed separately. During the evaluation of whole images, the uploaded images in landscape format 4:3 were cut by Google's Teachable Machine into a square format. To balance the training groups "alive" and "dead", the images of the group "dead" were used 6 times.

For the "area of interest" evaluation, representative image sections of about 250 × 250 µm were selected from the images by a dermatopathologist showing representative tumor areas (file size from 103 kB to 622 kB). Whenever possible, we selected representative areas from the dermal tumor compartment. Only in cases with a very small tumor thickness were areas with an epidermal component included (see Figure 1). To balance the training groups "alive" and "dead" the images of the group "dead" were cut into 6 representative tumor areas. In the advanced settings of the "Teachable Machine" the epochs were set to 1000, the batch size to 16 and the learning rate to 0.001. The 334 images of the test cohort were uploaded individually, and the group allocation of Google's Teachable Machine and the indicated percentage were collected.

2.3. Statistics

Statistical calculations were performed using IBM SPSS Statistics Version 23.0 (IBM SPSS, Chicago, IL, USA). Numerical variables were described by mean value and standard deviation or median values and interquartile range (IQR). Receiver operating characteristic (ROC) curve analyses and corresponding *p*-value calculations were performed using the ROC-Analysis tool in SPSS. *p*-values in Kaplan–Meier curves were calculated using the log-rank (Mantel-Cox) test. Throughout the analysis, p values < 0.05 were considered statistically significant.

3. Results

To create a prognosis score for melanoma, 60% ($n = 500$) of the images were used as a training cohort. For this purpose, the images were categorized as "alive" and "dead", according to the actual survival of the patients. Google's Teachable Machine was used to create an algorithm from these training groups, which was then applied to the test cohort. The training curves of the models showed an overfitting (see Figure S1); therefore, the training was repeated with a new randomized training set to avoid possible bias caused by the grouping (Figure S2). Since the repetition also showed comparable overfitting, the evaluation was continued with the initial trained model. Subsequently, the remaining 40% of the images ($n = 331$) were used as a test of the previously created score. The overall cohort had a median age of 62 years at diagnosis, a preponderance of 55.6% men versus 44.4% women, and a median tumor thickness of 1.05 mm. Ulceration was detectable in 21.3% of the patients. The most common histological subtype was superficial spreading melanoma with 59.3%, followed by nodular melanoma with 16.1%, lentigo maligna melanoma with 9.1%, acrolentiginous melanoma with 6.0%, other melanomas (5.7%) and melanomas of an unknown subtype (3.5%). Most melanomas were found on the trunk (41.4%), followed by melanomas of the lower extremity (26.4%), head and neck (17.7%), and upper extremity (14.1%). At initial diagnosis, 64.3% of patients were classified as stage I, 21% as stage II, 13.4% as stage III, and 1.3% as stage IV. The staging, subtype classification, and epidemiologic data showed comparable values in the training and test cohorts, confirming the randomization of the groups (see Table 1).

Table 1. Demographics, tumor parameters, stage of disease (AJCC 2017), tumor subtype and survival of the cohort.

Demographics and Tumor Parameters	All (n = 831)	Training Cohort (n = 500)	Test Cohort (n = 331)
Age at Diagnosis (years)			
Min./Max.	7/93	9/93	7/91
Median (+IQR)	62 (49/72)	63 (50/73)	59 (48/71)
Mean value (±SD)	59.88 (±15.3)	61.06 (±15.0)	58.11 (±15.7)
Sex (n, %)			
Male (n, %)	462 (55.6%)	285 (57%)	177 (53.5%)
Female (n, %)	369 (44.4%)	215 (43%)	154 (46.5%)
Primary tumor			
Tumor thickness (Breslow, mm), Median (+IQR)	1.05 (0.5/2.4)	1.00 (0.45/2.2)	1.10 (0.55/2.5)
Ulceration (n, %)	177 (21.3%)	103 (20.6%)	74 (22.4%)
Histologic subtype			
Superficially spreading melanoma (SSM) (n, %)	493 (59.3%)	303 (60.6%)	190 (57.4%)
Nodular melanoma (NM) (n, %)	134 (16.1%)	75 (15.0%)	59 (17.8%)
Lentigo Maligna melanoma (LMM) (n, %)	76 (9.1%)	52 (10.4%)	24 (7.3%)
Acrolentiginous melanoma (ALM) (n, %)	50 (6.0%)	27 (5.4%)	23 (6.9%)
Others (n, %)	47 (5.7%)	27 (5.4%)	20 (6.0%)
Unknown (n, %)	29 (3.5%)	15 (3.0%)	14 (4.2%)

Table 1. *Cont.*

Demographics and Tumor Parameters	All (n = 831)	Training Cohort (n = 500)	Test Cohort (n = 331)
Localisation			
Head/neck (n, %)	147 (17.7%)	91 (18.2%)	56 (16.9%)
Trunk (n, %)	344 (41.4%)	224 (44.8%)	120 (36.3%)
Upper Extremities (n, %)	117 (14.1%)	67 (13.4%)	50 (15.1%)
Lower Extremities (n, %)	219 (26.4%)	116 (23.2%)	103 (31.1%)
Others/unknown (n, %)	4 (0.4%)	2 (0.4%)	2 (0.6%)
Stage (AJCC 2017)			
IA (n, %)	401 (48.3%)	248 (49.6%)	153 (46.2%)
IB (n, %)	133 (16.0%)	79 (15.8%)	54 (16.3%)
IIA (n, %)	80 (9.6%)	45 (9%)	35 (10.6%)
IIB (n, %)	60 (7.2%)	37 (7.4%)	23 (6.9%)
IIC (n, %)	35 (4.2%)	19 (3.8%)	16 (4.8%)
IIIA (n, %)	24 (2.9%)	14 (2.8%)	10 (3%)
IIIB (n, %)	23 (2.8%)	15 (3%)	8 (2.4%)
IIIC (n, %)	62 (7.5%)	37 (7.4%)	25 (7.6%)
IIID (n, %)	2 (0.2%)	2 (0.4%)	0
IV (n, %)	11 (1.3%)	4 (0.8%)	7 (2.1%)

Figure 1 shows the procedure for photographing the melanoma sections. In many melanomas, tumors were present in numerous blocks and slides. The H&E section with the highest tumor thickness according to Breslow was selected (see Figure 1a). Here, an image was taken at 100× magnification at the site of the highest tumor thickness. This image was used for the "whole image" analysis. From these "whole images", small image sections (about 250 × 250 µm) were selected that showed representative parts of the tumor. The generation of a prognosis score was initially performed on both groups. These were compared by ROC analysis (see Figure 2a). We investigated how reliably a prognostic prediction of overall survival could be made based only on the AI classifier. When analyzing the "whole images", no significant result (p = 0.101) could be obtained in the prediction of overall survival. The classifier showed an AUC of 0.581, which was only slightly better than a random classification (AUC of 0.5). In contrast, however, a significant prediction estimate with an AUC of 0.694 ($p < 0.001$) could be obtained with the analysis of the AOI images. Therefore, further evaluation was performed using the classifier generated by the analysis of the area of interest images.

If one only uses the classifier, generated solely by image analysis of a H&E-stained melanoma section, this already allows a good prognosis estimate of the overall survival. Of the 331 patients in the test cohort, 230 patients were assigned the AI-classifier "low-risk" and 101 patients were given the AI-classifier "high-risk". Malignant melanoma-related overall survival was 88.2% in the test cohort, with 39 deaths in the observation period up to 114 months. The AI-classifier "low-risk" group showed a statistically significant better overall survival of 93% with 16 deaths, compared to a survival of 77.2% and 23 deaths in the AI-classifier "high-risk" group ($p < 0.001$). Figure 3a shows the Kaplan–Meyer survival curves of the total test cohort, related to melanoma-specific overall survival. Considering recurrence-free survival, there is also a statistically significant distinction by grouping into AI-classifier "low-risk" and "high-risk" ($p < 0.001$). Of the 230 patients in the "low-risk" group, an event such as recurrence, metastasis or death from the disease was recorded in 43 cases. This leads to a recurrence-free survival rate of 81.3%. In contrast, 37 events were recorded in the AI-classifier "high-risk" group out of 101 patients, resulting in a recurrence-free survival of only 63.4% (Figure 3b).

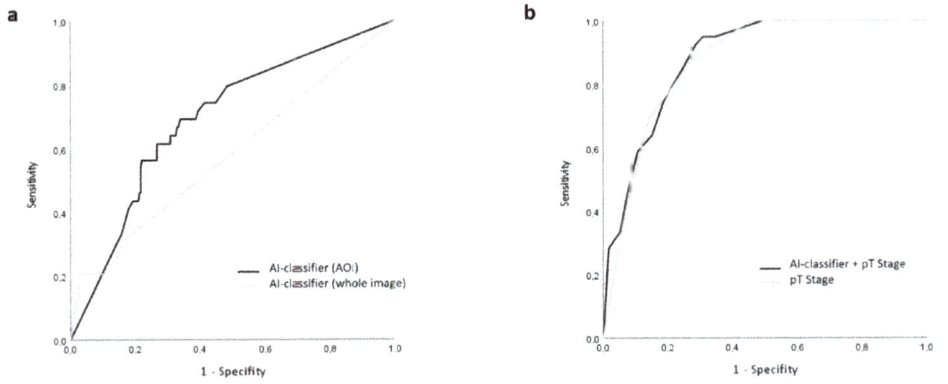

	AUC	Std. error	Significance
AI-classifier (AOI)	0.694	0.044	0.000086
AI-classifier (whole image)	0.581	0.053	0.101

	AUC	Std. error	Significance
AI-classifier + pT Stage	0.881	0.022	1.1529 e^-15
pT Stage	0.872	0.023	4.3077 e^-15

Figure 2. Average receiver operating characteristic (ROC) curves of overall survival prognosis. (**a**) Black line = AI-classifier with "area of interest" analysis. Gray line = AI-classifier with "whole image" analysis. (**b**) Black line = pT stage combined with AI-classifier (AOI). Gray line = pT stage (tumor thickness and presence of ulceration).

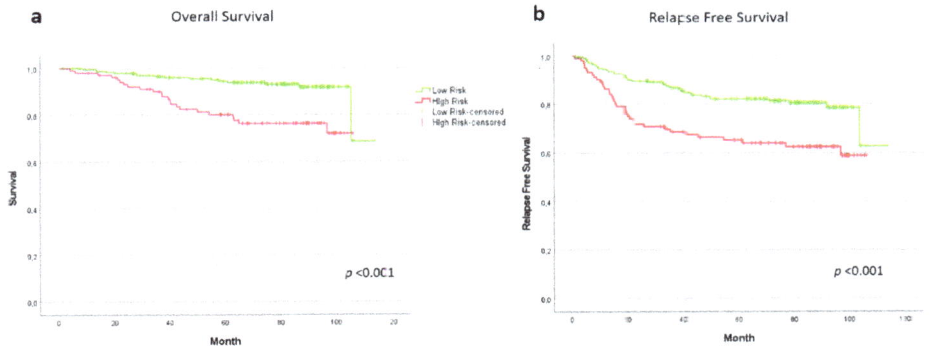

	N	Events OS	OS (%)	Events RFS	RFS (%)
Complete Cohort	331	39	88.2%	80	75.8%
AI-classifier „low-risk"	230	16	93%	43	81.3%
AI-classifier „high-risk"	101	23	77.2%	37	63.4%

Figure 3. Kaplan–Meyer curve of overall survival (**a**) and relapse-free survival (**b**). Green line = AI-classifier "low risk". Red line = AI-classifier "high risk".

Next, we questioned whether the AI classifier could complement the existing forecast prediction with the AJCC 2017 classification. Here, we first performed a ROC analysis. Comparing the prognosis estimate resulting from the existing T-classification (according to AJCC 2017) of the primary tumor (tumor thickness according to Breslow and the presence of an ulceration) (AUC = 0.872) with the prognosis estimate resulting from the addition of the AI-classifier (AUC = 0.881), only a slightly improved risk stratification was shown (see Figure 2b). This was also evident in the analysis of the Kaplan–Meyer curves of overall survival for the individual stages of the AJCC 2017 classification. Looking at AI-based risk classification in stage I, the following picture emerges: of the 207 patients in

Stage I of the test cohort, 163 (79%) received the label AI-classifier "low-risk". Of these 163 patients, 2 died during the observation period, corresponding to an overall survival rate of 98.8%. Forty-four patients (21%) were classified as "high-risk". In this group, there were also two deaths, which corresponds to an overall survival rate of 95.5%. However, with a *p*-value of 0.154, this does not reach statistical significance (Figure 4a). Regarding stage II, of 74 patients, 39 (53%) were classified as "low-risk," and 35 patients (47%) were marked as "high-risk." There were 8 deaths in the "low-risk" group resulting in an overall survival of 79.5%. The "high-risk" group had 10 deaths, resulting in an overall survival of 71.4%. However, this difference did not reach statistical significance with a *p*-value of 0.378 (Figure 3b). Stage III demonstrated the clearest differences in prognosis estimation. In our test cohort, there were 43 patients in stage III, of which 11 patients died during the observation period, resulting in an overall survival of 74.4%. Twenty-five of these patients (58%) were considered "low-risk", and in fact, only 4 deaths occurred in this group, resulting in an overall survival of 84%. Of the 18 patients (42%) designated as "high-risk" by the AI-classifier, 7 patients died, resulting in an overall survival of only 61.1%. Although an early and quite clear separation of the Kaplan–Meyer curves is seen in stage III, no statistically significant difference (p = 0.159) results due to the rather small number of cases in this group (Figure 4c). Seven patients were found to be stage IV at initial diagnosis. Four of these were identified as AI-classifier "high-risk" and 3 were classified as "low-risk". All patients in the "high-risk" group died during the observation period, resulting in an overall survival of 0%. In the "low-risk" group, 2 melanoma-specific deaths were recorded, resulting in a melanoma-specific survival of 33.3%. Patients in the "high-risk" group, in contrast, survived longer than those in the "low-risk" group. This leads to a statistically significant difference in the group classification at this stage (p = 0.018) (Figure 4d).

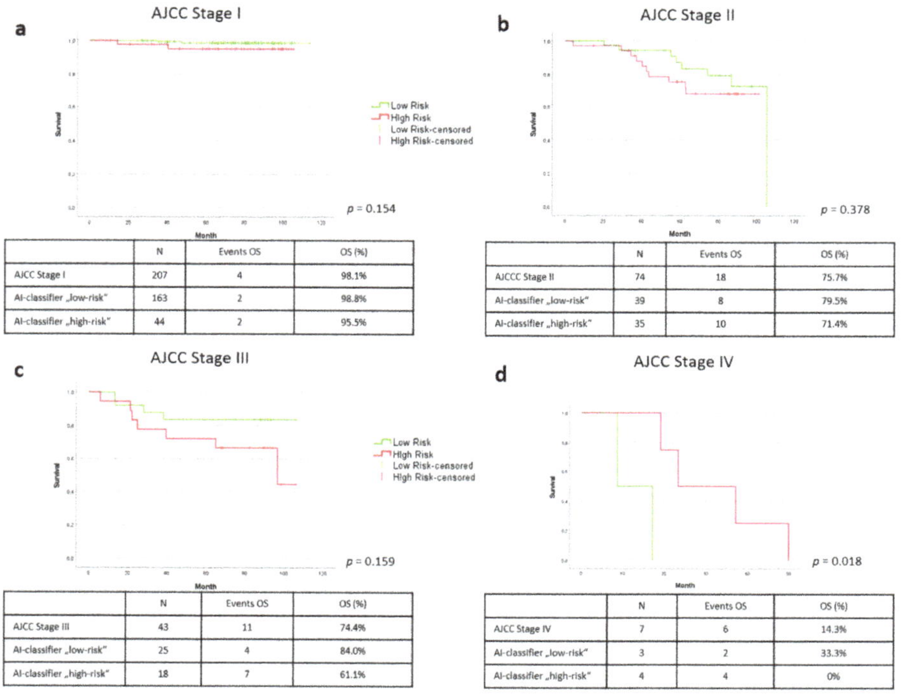

	N	Events OS	OS (%)
AJCC Stage I	207	4	98.1%
AI-classifier „low-risk"	163	2	98.8%
AI-classifier „high-risk"	44	2	95.5%

	N	Events OS	OS (%)
AJCCC Stage II	74	18	75.7%
AI-classifier „low-risk"	39	8	79.5%
AI-classifier „high-risk"	35	10	71.4%

	N	Events OS	OS (%)
AJCC Stage III	43	11	74.4%
AI-classifier „low-risk"	25	4	84.0%
AI-classifier „high-risk"	18	7	61.1%

	N	Events OS	OS (%)
AJCC Stage IV	7	6	14.3%
AI-classifier „low-risk"	3	2	33.3%
AI-classifier „high-risk"	4	4	0%

Figure 4. Kaplan–Meyer curves of overall survival in AJCC (2017) substages I (**a**), II (**b**), III (**c**) and IV (**d**). Green line = AI-classifier "low-risk". Red line = AI-classifier "high-risk".

4. Discussion

4.1. Results

The present study demonstrates the possibilities offered using deep learning-based image analysis in the risk stratification of melanoma. Although the program for risk assessment merely has a tiny image of about 250 × 250 µm at its disposal and no further information is available, a quite reliable and statistically significant risk stratification can be achieved. However, the AI classifier used here does not significantly improve the existing risk classification based on the TNM classification. Nevertheless, it seems possible that such a classifier may add prognostic value to conventional prognostic factors. In particular, our survival data in stage III show a tendency toward improved prognosis with the addition of the AI-classifier, even if this does not reach statistical significance. Further studies with a larger cohort from this advanced tumor stage are needed to confirm this.

The first published studies have investigated the use of AI-based neural networks in melanoma. It has been shown that such image analysis can reliably detect melanomas and differentiate them from benign melanocytic nevi [19–22]. Predicting prognosis, though, is much more complex than mere diagnostic classification of nevus and melanoma. Hence, a study by Brinker et al., published in 2021, failed to predict sentinel lymph node status in malignant melanoma to a clinically meaningful extent using deep learning-based image analysis [25]. In a 2020 study, Kulkarni et al. created a risk classifier that was significantly associated with the occurrence of recurrence in melanoma [26]. However, this score includes other factors for calculation, such as density and distribution of the immune cell infiltrate and nucleus morphology. Therefore, the impressive AUC values of 0.905 and 0.880 achieved in this study are not comparable to the results obtained here. Since other information besides the RGB image had to be included, tumor areas containing lymphocytes in addition to the tumor cells had to be available and the sections should not be too pigmented to allow detection of cellular components [26]. Another unique feature of our risk classifier is that it is a score that can be calculated with an image of only 103 kB to 622 kB in size. There is still a low availability of so-called whole-slide scanners, which can scan and digitize entire histological slides in high resolution in just a few minutes. Although this technology has been established for years, only a few pathological institutes have switched their routine settings to digital reporting, especially because of the high investment costs. Possibly in the coming years, the amount of memory and access to whole-slide scanners will no longer be limiting factors. Currently, a freely available, easy-to-use classifier operating on small data offers massive advantages when it comes to the question of validating that classifier in a large multicenter setting.

4.2. Limitations

The present study has several limitations. One potential point of criticism is the choice of deep learning tool. It is conceivable that an even better prediction of the prognosis could be made with different programs, although this was not investigated in this study. The focus of this research lies in the proof-of-concept, which shows that it is possible to make a prognosis prediction on the histological section with an as simple as possible AI application and as small as possible amount of data. Due to its straightforward transferability as well as its user-friendly interface, the publicly available Google's Teachable Machine was chosen as a deep learning tool. Overfitting describes learning by memorization of the correct answers by the AI model instead of the establishment of a generally applicable assignment rule in the sense of generalization. Such an overfitting was evident in our trained models, even when repeated with reassigned image groups. It is possible that this overfitting could be minimized by various fine adjustments in the AI model, especially by adjusting the number of epochs. However, this was not further investigated in the present study. It is also conceivable that the pre-trained algorithm of the Teachable Machine is not suitable for this complex histological challenge and thus represents the limiting factor in model performance. Further limitations are that the training and test cohorts are retrospective evaluations and that the number of cases in the groups and especially the number of events included (39 deaths in the test cohort) is quite small. Another point of criticism is that all used sections originate from one and the same pathological institute. Possibly, the results show only limited transferability to other institutes, as a slightly different staining pattern in H&E staining may be evident here. In addition, the manual selection of the areas of interest by the pathologist offers the possibility of an influence. A trade-off must be made between large datasets and automated selection and manual selection and small data sets. Additionally, the use of similar images in the "dead" group of the training cohort may have restricted the learning curve of artificial intelligence. The melanoma treatment of the patients in our study was not examined. It is possible that changes in treatment regimens during the study period may have limited the predictive accuracy of AI prognosis. To obtain more meaningful results, a larger, prospectively designed, multicenter study would be necessary. One possibility for such studies in the future could be the use of so-called "swarm learning". This newly described approach uses blockchain-based peer-to-peer networking to decentralize the use of machine learning [33].

Another problem with the method used here is the lack of explainability. A program that offers an explanatory approach implemented in the program would be desirable, so the black box of the AI could be illuminated. A study by Courtoil et al. from 2019 shows such a program that not only forecasts the prognosis of mesothelioma but can also show via a heat map analysis that the decision basis of the AI is to be found in the area of the tumor stroma [34].

5. Conclusions

Finally, the study presented here must be understood as proof-of-concept. It could be shown that prognostic information is contained in tiny image sections of a melanoma, which allows prognosis estimation. To establish a prognosis score that can be used in clinical practice, it must be clearly shown that such a score complements the current classification systems and may in the future be an alternative to invasive diagnostic methods, such as sentinel node biopsy or expensive gene-expression-based prognosis scores.

Supplementary Materials: The following supporting information can be downloaded at: https://www.mdpi.com/article/10.3390/cancers14092243/s1, Figure S1: Training curves for initial training; Figure S2: Training curves for repeated training.

Author Contributions: Conceptualization, S.F., C.G. and T.E.; formal analysis, S.F. and A.A.-G.; resources, S.F., A.A-G., G.M. and T.E.; writing—original draft preparation, S.F.; writing—review and editing, S.F., A.A-G., G.M., C.G. and T.E.; visualization, S.F.; supervision, T.E. and C.G.; project administration, S.F. All authors have read and agreed to the published version of the manuscript.

Funding: This research received no external funding.

Institutional Review Board Statement: The retrospective study on "Prediction of recurrence-free survival in patients with malignant melanoma using image analysis of histological sections by artificial neural networks" was approved by the Ethics Committee of the Medical Faculty of the University of Tübingen (No. 874/2019/BO2).

Informed Consent Statement: Informed consent was obtained from all subjects from the nationwide melanoma registry Tübingen involved in the study.

Data Availability Statement: The trained model for the "whole image" analysis used in this study is available at https://teachablemachine.withgoogle.com/models/q3W4kP4zk/ (accessed on 3 January 2022). The trained model for the "area of interest" analysis used in this study is available at https://teachablemachine.withgoogle.com/models/EWFL98pti/ (accessed on 3 January 2022).

Acknowledgments: The authors thank Peter Martus from the Institute for Clinical Epidemiology and Applied Biometry, Eberhardt Karls Universität, Tübingen, for statistical consulting.

Conflicts of Interest: S.F. received personal fees from Kyowa Kirin and Takeda Pharamceuticals (speaker's honoraria), as well as institutional grants from NeraCare, SkylineDX and BioNTech outside the submitted work. A.A-G. has no conflict of interest to declare. G.M. has no conflict of interest to declare. C.G. reports personal fees from Amgen, personal fees from MSD, grants and personal fees from Novartis, grants and personal fees from NeraCare, grants and personal fees from BMS, personal fees from Philogen, grants and personal fees from Roche, grants and personal fees from Sanofi, outside the submitted work. T.E. acted as a consultant for Almiral Hermal, Bristol-Myers Squibb, MSD, Novartis, Pierre Fabre and Sanofi outside the submitted work.

References

1. Sacchetto, L.; Zanetti, R.; Comber, H.; Bouchardy, C.; Brewster, D.; Broganelli, P.; Chirlaque, M.; Coza, D.; Galceran, J.; Gavin, A. Trends in incidence of thick, thin and in situ melanoma in Europe. *Eur. J. Cancer* **2018**, *92*, 108–118. [CrossRef] [PubMed]
2. Schadendorf, D.; van Akkooi, A.C.; Berking, C.; Griewank, K.G.; Gutzmer, R.; Hauschild, A.; Stang, A.; Roesch, A.; Ugurel, S. Melanoma. *Lancet* **2018**, *392*, 971–984. [CrossRef]
3. Elder, D.E.; Bastian, B.C.; Cree, I.A.; Massi, D.; Scolyer, R.A. The 2018 World Health Organization classification of cutaneous, mucosal, and uveal melanoma: Detailed analysis of 9 distinct subtypes defined by their evolutionary pathway. *Arch. Pathol. Lab. Med.* **2020**, *144*, 500–522. [CrossRef] [PubMed]
4. Gershenwald, J.E.; Scolyer, R.A.; Hess, K.R.; Sondak, V.K.; Long, G.V.; Ross, M.I.; Lazar, A.J.; Faries, M.B.; Kirkwood, J.M.; McArthur, G.A. Melanoma staging: Evidence-based changes in the American Joint Committee on Cancer eighth edition cancer staging manual. *Cancer J. Clin.* **2017**, *67*, 472–492. [CrossRef] [PubMed]
5. Evans, J.L.; Vidri, R.J.; MacGillivray, D.C.; Fitzgerald, T.L. Tumor mitotic rate is an independent predictor of survival for nonmetastatic melanoma. *Surgery* **2018**, *164*, 589–593. [CrossRef]
6. Hale, C.S.; Qian, M.; Ma, M.W.; Scanlon, P.; Berman, R.S.; Shapiro, R.L.; Pavlick, A.C.; Shao, Y.; Polsky, D.; Osman, I. Mitotic rate in melanoma: Prognostic value of immunostaining and computer-assisted image analysis. *Am. J. Surg. Pathol.* **2013**, *37*, 882. [CrossRef]
7. Ribero, S.; Moscarella, E.; Ferrara, G.; Piana, S.; Argenziano, G.; Longo, C. Regression in cutaneous melanoma: A comprehensive review from diagnosis to prognosis. *J. Eur. Acad. Dermatol. Venereol.* **2016**, *30*, 2030–2037. [CrossRef]
8. Garbe, C.; Amaral, T.; Peris, K.; Hauschild, A.; Arenberger, P.; Bastholt, L.; Bataille, V.; Del Marmol, V.; Dreno, B.; Fargnoli, M.C.; et al. European consensus-based interdisciplinary guideline for melanoma. Part 1: Diagnostics—Update 2019. *Eur. J. Cancer* **2020**, *126*, 141–158. [CrossRef]
9. Gambichler, T.; Tsagoudis, K.; Kiecker, F.; Reinhold, U.; Stockfleth, E.; Hamscho, R.; Egberts, F.; Hauschild, A.; Amaral, T.; Garbe, C. Prognostic significance of an 11-gene RNA assay in archival tissue of cutaneous melanoma stage I–III patients. *Eur. J. Cancer* **2021**, *143*, 11–18. [CrossRef]
10. Amaral, T.M.S.; Hoffmann, M.C.; Sinnberg, T.; Niessner, H.; Sulberg, H.; Eigentler, T.K.; Garbe, C. Clinical validation of a prognostic 11-gene expression profiling score in prospectively collected FFPE tissue of patients with AJCC v8 stage II cutaneous melanoma. *Eur. J. Cancer* **2020**, *125*, 38–45. [CrossRef]
11. Gerami, P.; Cook, R.W.; Wilkinson, J.; Russell, M.C.; Dhillon, N.; Amaria, R.N.; Gonzalez, R.; Lyle, S.; Johnson, C.E.; Oelschlager, K.M.; et al. Development of a prognostic genetic signature to predict the metastatic risk associated with cutaneous melanoma. *Clin. Cancer Res.* **2015**, *21*, 175–183. [CrossRef] [PubMed]
12. Bellomo, D.; Arias-Mejias, S.M.; Ramana, C.; Heim, J.B.; Quattrocchi, E.; Sominidi-Damodaran, S.; Bridges, A.G.; Lehman, J.S.; Hieken, T.J.; Jakub, J.W.; et al. Model Combining Tumor Molecular and Clinicopathologic Risk Factors Predicts Sentinel Lymph Node Metastasis in Primary Cutaneous Melanoma. *JCO Precis. Oncol.* **2020**, *4*, 319–334. [CrossRef] [PubMed]
13. Samuel, A.L. Some studies in machine learning using the game of checkers. *IBM J. Res. Dev.* **1959**, *3*, 210–229. [CrossRef]
14. Forchhammer, S.; Hartmann, T. Digitale Dermatopathologie: Vorteile für Befundung, Forschung und Ausbildung. *Dtsch. Dermatol.* **2021**, *69*, 810–813. [CrossRef]
15. Brinker, T.J.; Hekler, A.; Enk, A.H.; Berking, C.; Haferkamp, S.; Hauschild, A.; Weichenthal, M.; Klode, J.; Schadendorf, D.; Holland-Letz, T. Deep neural networks are superior to dermatologists in melanoma image classification. *Eur. J. Cancer* **2019**, *119*, 11–17. [CrossRef]
16. Brinker, T.J.; Hekler, A.; Enk, A.H.; Klode, J.; Hauschild, A.; Berking, C.; Schilling, B.; Haferkamp, S.; Schadendorf, D.; Holland-Letz, T. Deep learning outperformed 136 of 157 dermatologists in a head-to-head dermoscopic melanoma image classification task. *Eur. J. Cancer* **2019**, *113*, 47–54. [CrossRef] [PubMed]
17. Hekler, A.; Utikal, J.S.; Enk, A.H.; Hauschild, A.; Weichenthal, M.; Maron, R.C.; Berking, C.; Haferkamp, S.; Klode, J.; Schadendorf, D. Superior skin cancer classification by the combination of human and artificial intelligence. *Eur. J. Cancer* **2019**, *120*, 114–121. [CrossRef]
18. Phillips, M.; Marsden, H.; Jaffe, W.; Matin, R.N.; Wali, G.N.; Greenhalgh, J.; McGrath, E.; James, R.; Ladoyanni, E.; Bewley, A. Assessment of accuracy of an artificial intelligence algorithm to detect melanoma in images of skin lesions. *JAMA Netw. Open* **2019**, *2*, e1913436. [CrossRef]
19. Brinker, T.J.; Schmitt, M.; Krieghoff-Henning, E.I.; Barnhill, R.; Beltraminelli, H.; Braun, S.A.; Carr, R.; Fernandez-Figueras, M.-T.; Ferrara, G.; Fraitag, S. Diagnostic performance of artificial intelligence for histologic melanoma recognition compared to 18 international expert pathologists. *J. Am. Acad. Dermatol.* **2022**, *86*, 640–642. [CrossRef]
20. Hekler, A.; Utikal, J.S.; Enk, A.H.; Berking, C.; Klode, J.; Schadendorf, D.; Jansen, P.; Franklin, C.; Holland-Letz, T.; Krahl, D.; et al. Pathologist-level classification of histopathological melanoma images with deep neural networks. *Eur. J. Cancer* **2019**, *115*, 79–83. [CrossRef]

21. Hekler, A.; Utikal, J.S.; Enk, A.H.; Solass, W.; Schmitt, M.; Klode, J.; Schadendorf, D.; Sondermann, W.; Franklin, C.; Bestvater, F.; et al. Deep learning outperformed 11 pathologists in the classification of histopathological melanoma images. *Eur. J. Cancer* **2019**, *118*, 91–96. [CrossRef] [PubMed]
22. Hart, S.N.; Flotte, W.; Norgan, A.P.; Shah, K.K.; Buchan, Z.R.; Mounajjed, T.; Flotte, T.J. Classification of melanocytic lesions in selected and whole-slide images via convolutional neural networks. *J. Pathol. Inform.* **2019**, *10*, 5. [CrossRef] [PubMed]
23. Ianni, J.D.; Soans, R.E.; Sankarapandian, S.; Chamarthi, R.V.; Ayyagari, D.; Olsen, T.G.; Bonham, M.J.; Stavish, C.C.; Motaparthi, K.; Cockerell, C.J. Tailored for real-world: A whole slide image classification system validated on uncurated multi-site data emulating the prospective pathology workload. *Sci. Rep.* **2020**, *10*, 3217. [CrossRef] [PubMed]
24. Jiang, Y.; Xiong, J.; Li, H.; Yang, X.; Yu, W.; Gao, M.; Zhao, X.; Ma, Y.; Zhang, W.; Guan, Y. Recognizing basal cell carcinoma on smartphone-captured digital histopathology images with a deep neural network. *Br. J. Dermatol.* **2020**, *182*, 754–762. [CrossRef]
25. Brinker, T.J.; Kiehl, L.; Schmitt, M.; Jutzi, T.B.; Krieghoff-Henning, E.I.; Krahl, D.; Kutzner, H.; Gholam, P.; Haferkamp, S.; Klode, J. Deep learning approach to predict sentinel lymph node status directly from routine histology of primary melanoma tumours. *Eur. J. Cancer* **2021**, *154*, 227–234. [CrossRef]
26. Kulkarni, P.M.; Robinson, E.J.; Sarin Pradhan, J.; Gartrell-Corrado, R.D.; Rohr, B.R.; Trager, M.H.; Geskin, L.J.; Kluger, H.M.; Wong, P.F.; Acs, B.; et al. Deep Learning Based on Standard H&E Images of Primary Melanoma Tumors Identifies Patients at Risk for Visceral Recurrence and Death. *Clin. Cancer Res.* **2020**, *26*, 1126–1134.
27. Johannet, P.; Coudray, N.; Donnelly, D.M.; Jour, G.; Illa-Bochaca, I.; Xia, Y.; Johnson, D.B.; Wheless, L.; Patrinely, J.R.; Nomikou, S. Using machine learning algorithms to predict immunotherapy response in patients with advanced melanoma. *Clin. Cancer Res.* **2021**, *27*, 131–140. [CrossRef]
28. Moore, M.R.; Friesner, I.D.; Rizk, E.M.; Fullerton, B.T.; Mondal, M.; Trager, M.H.; Mendelson, K.; Chikeka, I.; Kurc, T.; Gupta, R. Automated digital TIL analysis (ADTA) adds prognostic value to standard assessment of depth and ulceration in primary melanoma. *Sci. Rep.* **2021**, *11*, 2809. [CrossRef]
29. Taylor, R.C.; Patel, A.; Panageas, K.S.; Busam, K.J.; Brady, M.S. Tumor-infiltrating lymphocytes predict sentinel lymph node positivity in patients with cutaneous melanoma. *J. Clin. Oncol.* **2007**, *25*, 869–875. [CrossRef]
30. Yang, J.; Lian, J.W.; Chin, Y.-P.H.; Wang, L.; Lian, A.; Murphy, G.F.; Zhou, L. Assessing the Prognostic Significance of Tumor-Infiltrating Lymphocytes in Patients with Melanoma Using Pathologic Features Identified by Natural Language Processing. *JAMA Netw. Open* **2021**, *4*, e2126337. [CrossRef]
31. Google.com. Teachable Machine: Train a Computer to Recognize Your Own Images, Sounds, Poses. Available online: https://teachablemachine.withgoogle.com (accessed on 3 January 2022).
32. Jeong, H. Feasibility Study of Google's Teachable Machine in Diagnosis of Tooth-Marked Tongue. *J. Dent. Hyg. Sci.* **2020**, *20*, 206–212.
33. Warnat-Herresthal, S.; Schultze, H.; Shastry, K.L.; Manamohan, S.; Mukherjee, S.; Garg, V.; Sarveswara, R.; Handler, K.; Pickkers, P.; Aziz, N.A.; et al. Swarm Learning for decentralized and confidential clinical machine learning. *Nature* **2021**, *594*, 265–270. [CrossRef] [PubMed]
34. Courtiol, P.; Maussion, C.; Moarii, M.; Pronier, E.; Pilcer, S.; Sefta, M.; Manceron, P.; Toldo, S.; Zaslavskiy, M.; Le Stang, N. Deep learning-based classification of mesothelioma improves prediction of patient outcome. *Nat. Med.* **2019**, *25*, 1519–1525. [CrossRef] [PubMed]

cancers

MDPI

Article

Deep Learning Using CT Images to Grade Clear Cell Renal Cell Carcinoma: Development and Validation of a Prediction Model

Lifeng Xu [1,2,†], Chun Yang [2,3,†], Feng Zhang [1], Xuan Cheng [2,3], Yi Wei [4], Shixiao Fan [2,3], Minghui Liu [2,3], Xiaopeng He [4,5,*], Jiali Deng [2,3], Tianshu Xie [2,3], Xiaomin Wang [2,3], Ming Liu [2,3] and Bin Song [4,*]

1. The Quzhou Affiliated Hospital of Wenzhou Medical University, Quzhou People's Hospital, Quzhou 324000, China; qz1109@wmu.edu.cn (L.X.); fengzhang@wmu.edu.cn (F.Z.)
2. Yangtze Delta Region Institute (Quzhou), University of Electronic Science and Technology of China, Quzhou 324000, China; chunyang@std.uestc.edu.cn (C.Y.); cs_xuancheng@std.uestc.edu.cn (X.C.); shixiaofan@std.uestc.edu.cn (S.F.); minghuiliu@std.uestc.edu.cn (M.L.); dengjiali@std.uestc.edu.cn (J.D.); tianshuxie@std.uestc.edu.cn (T.X.); xmwang@uestc.edu.cn (X.W.); csmliu@uestc.edu.cn (M.L.)
3. University of Electronic Science and Technology of China, Chengdu 610000, China
4. West China Hospital, Sichuan University, Chengdu 610000, China; drweiyi057@scu.edu.cn
5. Affiliated Hospital of Southwest Medical University, Luzhou 646000, China

* Correspondence: xiaopenghe@protonmail.com (X.H.); songbin@wchscu.cn (B.S.)

† These authors contributed equally to this work.

Simple Summary: Clear cell renal cell carcinoma (ccRCC) pathologic grade identification is essential to both monitoring patients' conditions and constructing individualized subsequent treatment strategies. However, biopsies are typically used to obtain the pathological grade, entailing tremendous physical and mental suffering as well as heavy economic burden, not to mention the increased risk of complications. Our study explores a new way to provide grade assessment of ccRCC on the basis of the individual's appearance on CT images. A deep learning (DL) method that includes self-supervised learning is constructed to identify patients with high grade for ccRCC. We confirmed that our grading network can accurately differentiate between different grades of CT scans of ccRCC patients using a cohort of 706 patients from West China Hospital. The promising diagnostic performance indicates that our DL framework is an effective, non-invasive and labor-saving method for decoding CT images, offering a valuable means for ccRCC grade stratification and individualized patient treatment.

Abstract: This retrospective study aimed to develop and validate deep-learning-based models for grading clear cell renal cell carcinoma (ccRCC) patients. A cohort enrolling 706 patients (n = 706) with pathologically verified ccRCC was used in this study. A temporal split was applied to verify our models: the first 83.9% of the cases (years 2010–2017) for development and the last 16.1% (year 2018–2019) for validation (development cohort: n = 592; validation cohort: n = 114). Here, we demonstrated a deep learning(DL) framework initialized by a self-supervised pre-training method, developed with the addition of mixed loss strategy and sample reweighting to identify patients with high grade for ccRCC. Four types of DL networks were developed separately and further combined with different weights for better prediction. The single DL model achieved up to an area under curve (AUC) of 0.864 in the validation cohort, while the ensembled model yielded the best predictive performance with an AUC of 0.882. These findings confirms that our DL approach performs either favorably or comparably in terms of grade assessment of ccRCC with biopsies whilst enjoying the non-invasive and labor-saving property.

Keywords: clear cell renal cell carcinoma; deep learning; tumor grading; self-supervised learning; label noise; class imbalance

Citation: Xu, L.; Yang, C.; Zhang, F.; Cheng, X.; Wei, Y.; Fan, S.; Liu, M.; He, X.; Deng, J.; Xie, T.; et al. Deep Learning Using CT Images to Grade Clear Cell Renal Cell Carcinoma: Development and Validation of a Prediction Model. *Cancers* **2022**, *14*, 2574. https://doi.org/10.3390/cancers14112574

Academic Editors: Hamid Khayyam, Ali Madani, Rahele Kafieh and Ali Hekmatnia

Received: 1 April 2022
Accepted: 29 April 2022
Published: 24 May 2022

Publisher's Note: MDPI stays neutral with regard to jurisdictional claims in published maps and institutional affiliations.

1. Introduction

Renal cell carcinoma (RCC) is one of the most common deadly tumors in the urinary system, originating from the renal parenchymal urinary tubule epithelial system, accounting for 4% of human malignant tumors [1]. Clear cell renal cell carcinoma (ccRCC) is the most common subtype of RCC, accounting for about 75% of all RCC cases [2]. The Fuhrman grading system is highly recognized in the clinical oncology community, and it is widely used for diagnosing the pathological grade of ccRCC. In the Fuhrman grading system, the tumor is classified into one of four different grades (I, II, III, and IV) [3], with higher grades indicating a more serious patient condition. However, to obtain the pathological grade, the biopsy is most often carried out using a sharp tool to remove a small amount of tissue. Inevitably, this invasive procedure may entail great pain physically and mentally, whilst imposing a heavy economic burden on patients' families and society. Recent study [4] also demonstrated that biopsy may increase the risk of complications, including hemorrhage, infection, even tumor rupture. Furthermore, considering the shortage of specialized doctors and conceivable poor conditions of equipment in some rural areas, patients in these areas may be unable to receive timely and appropriate treatment.

In recent years, deep learning (DL) has defined state-of-the-art performance in many computer vision tasks, such as image classification [5], object detection [6,7], and segmentation [7]. DL models will perform satisfactorily once they have learned enough and high-quality data [8]. Thus, given sufficient data, the accuracy of a deep-learning-enabled diagnosis system often matches or even surpasses the level of expert physicians [9,10]. A myriad of studies have validated the utility of DL in various clinical settings through various experiments, including the reduction of false-positive findings in the interpretation of breast ultrasound exams [11], the detection of intensive care unit patient mobilization activities [12], and the improvement of medical technology [13]. In the same way, DL enables the ability to non-invasively and automatically assess the pathological grade for ccRCC, monitor patients' conditions and construct personalized subsequent treatment strategies.

However, to better apply the DL model, there are a few problematic issues that should not be lightly dismissed. First, the domain shift problem. In most deep-learning-enabled medical system, transfer learning is a common practice [14], where researchers use models pretrained on some other dataset, such as ImageNet [15]. Although ImageNet contains a large variety of images, they are all based on real-life situations and do not overlap with medical images in terms of content. The shifts between two datasets represent that the pattern-recognition abilities acquired from large datasets may not apply well to our medical task. Second is the noisy label problem [16]. Inevitably, there are always some cancerous lesions that come from high-grade patients but do not exhibit characteristics sufficient to discriminate them from low-grade patients, resulting in the mismatch between the manual labels and the actual labels. Third, the imbalance dataset problem. In most medical tasks, images for the abnormal class might be challenging to find. Developing on such an unbalanced dataset can wreak havoc on the utility of the DL model. To combat these issues, our study explores a new DL framework initialized by a self-supervised pre-training method, developed with the addition of mixed loss strategy and sample reweighting to identify patients with high grade for ccRCC.

There are also several studies related to that of ours. Zhu and collaborators [17] proposed a system that can accurately discriminate between five related classes, including clear cell RCC, papillary RCC, chromophobe RCC, renal oncocytoma, and normal, based on digitized surgical resection slides and biopsy slides. Different from this, we only focus on the ccRCC and try to explore a non-invasive tool to replace biopsy whilst providing grade assessment. Zheng [18], Cui [19], and Gao [20] had the same intention with us but their works are mainly based on radiomics, which requires using a high-throughput feature extraction method and a series of data-mining algorithms [21,22]. By contrast, our work does not need to use additional procedures, such as feature extraction, which could save labor to some extent. Most relates to our work is that of [14] which also attempted to use the deep learning model to predict the Fuhrman grade of ccRCC patients. However, it is worth nothing that this study still used ImageNet pretraining and did not pay attention to the noise and imbalance problem that may induce performance degradation in most of cases, while our framework provides a new solution to these issues with the addition of the proposed mixed loss strategy and sample reweighting, providing increased power to the common practice. To the best of our knowledge, our study is the first attempt to identify the pathological grades of patients with ccRCC in the context of a large population whilst dealing with the domain shift problem and the noisy label problem, as well as the imbalance dataset problem, simultaneously.

The specific objective of this study was to develop and validate a new DL framework to identify patients with a high grade for ccRCC based on CT images, and the results indicate that it is feasible. In addition to the application of deep learning to ccRCC pathology grading [14], we focused on the solution of these three problems. To improve the network's capabilities, we proposed an innovative self-supervised pre-training methodology, as well as mixed loss strategy and sample reweighting to address label noise and class imbalance problems. To develop and validate our framework, we applied a temporal split to teledermatology cases: the first 83.9% of the cases (years 2010–2017) for development and the last 16.1% (years 2018–2019) for validation as done in [23]. Putting patients with different years into different groups could help avoid the bias that possibly stems from the machines and radiologic technologists, thereby being also a good practice to demonstrate the generalization ability of our method. In addition, to improve the model generalization ability, we combined several excellent single models, which achieved more reliable results. This project provides a convenient, harmless and accurate opportunity for Fuhrman grading, which will not only relieve patients from suffering from biopsies, but also assist radiologists in making diagnostic decisions in routine clinical practice, even for some rural areas.

2. Materials and Methods

The institution's research ethics board approved our study. The ethics board waived informed consent because the data were obtained from preexisting institutional or public databases.

2.1. Patient Cohort

The patient cases covered in this study are all from West China Hospital, with a total case load of 759. We excluded 53 patients for the following reasons: (1) the CT images were incomplete or had poor image quality (n = 24); (2) patients with incomplete indicators (n = 29). Therefore, 706 patients were finally enrolled in this study. All 706 patients were admitted to the hospital from April 2010 to January 2019. From the perspective of the time domain, we assigned a total of 592 patients before year 2018 as the development cohort and a total of 114 patients after year 2018 (including 2018) as the validation cohort according to the acquisition date of the CT images. The characteristics of the included patients are shown in Table 1.

All of the pathological ccRCC patients' grades were reconfirmed by three independent pathologists with extensive pathology experience. The labels of CT images in the validation cohort were verified by professional pathologists. This study employed the Fuhrman grading system as the benchmark. Grades I and II were assessed as low-grade, and grades III and IV were assessed as high grade. Usually, low grade has a better prognosis than high grade [24].

Table 1. Patient characteristics.

Patient Characteristic	Development Cohort	Validation Cohort
Number	592	114
CT Images	9978	2491
Male	374 (63.2%)	71 (62.3%)
Female	218 (36.8%)	43 (37.7%)
Average Age	54.9 (±12.1)	55.8 (±12.1)
Acquisition Date	2010–2017	2018–2019
Low-grade	354 (59.8%)	76 (66.7%)
High-grade	238 (40.2%)	38 (33.3%)

2.2. Image Acquisition

All CT scans used in this study were obtained by one of the six different CT scanners. The PCP, CMP, and NP of the MDCT (multidetector CT) examination were acquired for each ccRCC patient with strict rules. A total of 70–100 mL contrast agents were injected into the antecubital vein using a high-pressure injector at a rate of 3.5 mL/s. The PCP is the precontrast phase. The CMP means that the corticomedullary phase contrast-enhanced scan starting 30 s after injection. The NP means that the nephrographic phase contrast-enhanced scan starting 90 s after the injection. Spiral scanning and thinslice reconstruction were used for all three phases. The CT scanning parameters for the three phases were as follows: the voltage in the tube was 120 kV; the reconstruction thickness was 1 mm to 5 mm, and the matrix was 512 × 512. Only the CMP CT images were used as experimental data most of the time because the CT images are the clearest and most conducive to the analysis of the patient's condition. The selection of only CMP CT images as experimental data somewhat reduces the times of model developing, which may impair the generalization of the model, but since our dataset includes a large enough number of cases, this operation does not have any impact.

2.3. Image Preprocessing

The original CT image contains interference information, of which only the tumor area is really valid for grading, so for each image, the region of interest (ROI) needs to be delineated. With 706 patients containing more than 12,000 CT images, it is clearly not desirable to have a radiologist process every image.

We utilized the DL models in target detection and segmentation to segment tumor regions in the renal CT images. In the detection and segmentation part of the tumor, we used VGG-16 [25] pre-trained on ImageNet [26] as the backbone for extracting features. A small number of images for detection and segmentation training were annotated by experienced doctors. The network was trained for 6000 epochs until its output converged. We used the trained network to detect and segment the tumors in the overall CT images, and the results were tested by an experienced radiologist, largely meeting the criteria. Figure 1 shows the tumor segmentation process. The segmented CT pictures eliminate interference from other bodily regions, allowing the content to be focused on the tumor area on the renal. The CT images involved in subsequent experiments (including pre-training process and developing process) refer to those after detection and segmentation processing. Since the size of the tumor area varies, the sizes of the CT images obtained by partitioning are different. We performed Resize or Padding operations before the data were entered into the network to make the image size uniform to the 224 × 224 × 3.

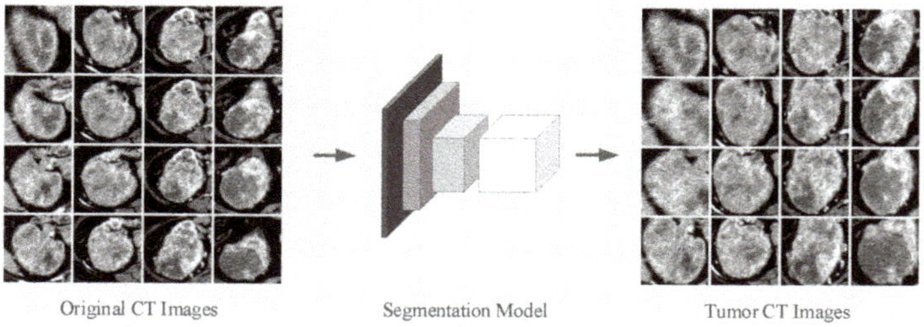

Figure 1. Segmentation model concentrates the CT image's content on the tumor.

2.4. Self-Supervised Learning

We used a self-supervised learning (pre-training) method to equip the network with better awareness of the CT images before developing. In the pre-training and developing process, we used the RegNetY400MF, RegNetY800MF [27], SE-ResNet50 [28] and ResNet-101 [29]. Traditional pre-training models are often obtained by developing on ImageNet [15] and then using transfer learning to satisfy specific classification tasks. Such an approach suffers from the problem that there is segmentation between the pre-training and the actual classification task, with little correlation between the image contents. We used a simpler and more efficient approach to pre-train the network. The images we used in the pre-training are the same as those used in the developing, with the difference that during pre-training, we rotate the input image data clockwise in space in one of four ways (0°, 90°, 180°, 270°), and the images are labeled with the number of 90° of image rotation (0, 1, 2, 3), while during developing, CT images are labeled with the ccRCC grade of the relevant patient (0 for low-grade, 1 for high-grade). Such a pre-training method allows the network to develop feature extraction capability based on the developing images without revealing the original semantics of the developing images. We pre-trained different deep learning models using the stochastic gradient descent (SGD) algorithm and the common cross-entropy loss function. The DL models were finally trained for 60 epochs. The overall structure of the pre-training network is shown in the top half of Figure 2

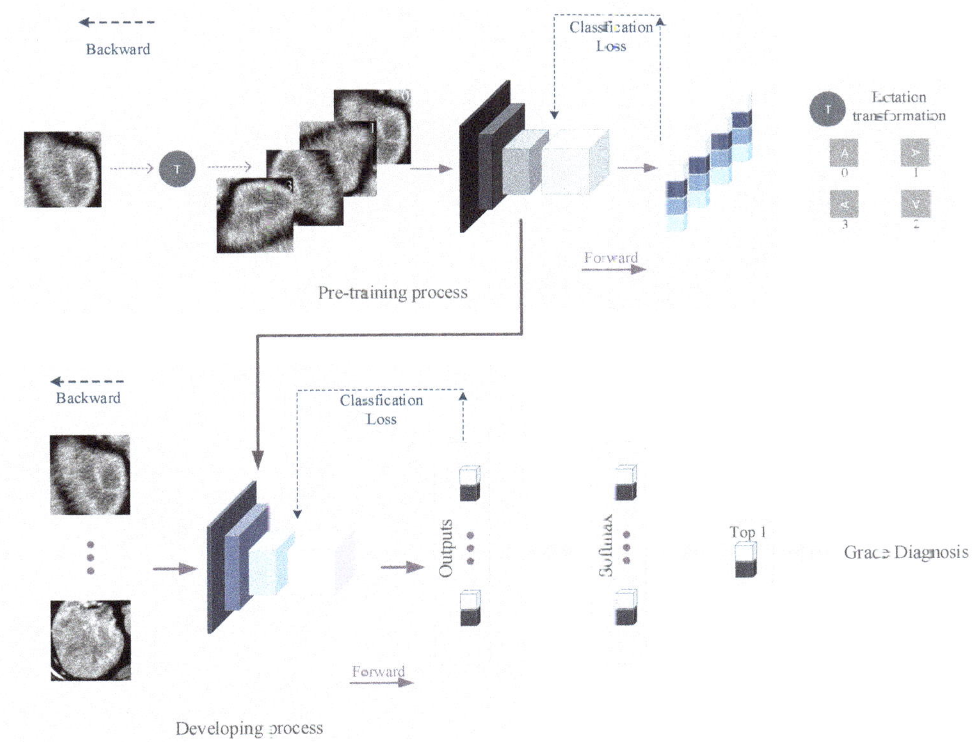

Figure 2. The overall flow of pre-training and developing. The top part of the figure shows the pre-training process. In the pre-training process, the original images are expanded into four images after rotation transformation, and their labels are 0, 1, 2, and 3, representing that they are obtained by quarter-turning the original image 0, 1, 2, and 3 times, clockwise. The bottom part shows the developing process. The developing process network is initialized from the pre-training process network.

2.5. *Mixed Loss Strategy*

There are two pervasive problems in image classification (including medical image classification) tasks: one is the presence of label noise, and the other is the imbalanced data distribution. Both of these problems can be found in the data of our study.

Some malignant lesions that come from higher-grade patients do not exhibit enough characteristics to distinguish them from lower-grade patients, resulting in a mismatch between manual labeling and actual labeling. In simple terms, there are errors in the labels of CT images of some high-grade patients. To tackle the noise problem, we applied the mixed loss strategy similar to that in [30]. Suppose the labeled CT images dataset is $D = (x_i, y_i)_i^N$. During developing, the ordinary cross-entropy loss is as follows:

$$L_{CE} = -\frac{1}{N}\sum_{i=1}^{N}\sum_{j\in\{0,1\}} l_{ij} log p_{ij} \tag{1}$$

where $l_{ij} = 1$ if $y_i = j$, and 0 otherwise. p_{ij} is the network output probability that the ith sample belongs to category j. Since the true labels of some high-grade CT images were supposed to be low-grade., we add loss L_{CE_2} to alleviate the effect of noise in the developing process. Specifically, in the developing phase, under the assumption that the noise rate is $\alpha(0 \le \alpha \le 1)$, the loss is as follows:

$$L_{total} = \alpha L_{CE\_1} + (1-\alpha)L_{CE\_2} \tag{2}$$

$$L_{CE\_1} = -\frac{1}{N}\sum_{i=1}^{N}\sum_{j\in\{0,1\}} l_{ij} log p_{ij} \tag{3}$$

$$L_{CE\_2} = -\frac{1}{N}\sum_{i=1}^{N} l_{i0} log p_{i0} \tag{4}$$

where $l_{i0} = 1$ if $y_i = 0$, and 0 otherwise. p_{i0} is the network output probability that the i-th sample belongs to category 0 (low-grade). The larger the noise rate α, the higher the noise level. In the experiment, the noise rate was set at 0.4 for the best results, which is probably closest to the real noise rate of the data. Through the mixed loss strategy, we made the network learn from the modified data according to a certain probability in the developing process so as to achieve the effect of countering label noise.

2.6. Sample Reweighting

In terms of class imbalance, it is inevitable. For example, the proportion of mild patients in the cases of cancer detection is small, because cancer patients usually feel physical abnormalities in the middle or even late stage of the disease. The sample reweighting method is used to tackle this problem. In order to account for class imbalance when calculating cross-entropy loss, each class was weighed according to its frequency, with rare samples contributing more to the loss function [23]. Specifically, we assigned lower weights to the categories with a larger proportion of sample size. Since we have a bias toward the low-grade patient sample when dealing with the noise problem, we need to take this information into account when calculating the percentage of the number of low-grade and high-grade CT images. Suppose the weight of the low-grade patient sample is λ_0, and the weight of the high-grade patient sample is λ_1; the new weighted cross-entropy is

$$L_{CE\_weight} = -\frac{1}{N}\sum_{i=1}^{N}\sum_{j\in\{0,1\}} \lambda_j l_{ij} log p_{ij} \tag{5}$$

By Equation (5), we made the network learn more from categories with smaller sample sizes. Finally, in order to comprehensively solve the problem of label noise and class imbalance, the overall optimization objective L_{total_weight} is

$$L_{total\_weight} = -\alpha\frac{1}{N}\sum_{i=1}^{N}\sum_{j\in\{0,1\}} \lambda_j l_{ij} log p_{ij} - (1-\alpha)\frac{1}{N}\sum_{i=1}^{N} \lambda_0 l_{i0} log p_{i0} \tag{6}$$

2.7. Developing

After pre-training, we obtained the DL model with feature extraction capability. Then, all models were developed iteratively and used to grade CT images of ccRCC patients.

It is worth noting that during the pre-training process, the classifiers of the models of the four networks are linear, i.e., one fully connected layer (with an avgpooling). During the developing process, we converted the classifier of the original network into nonlinear projection, which can perform more complex mapping and make the dimension reduction of the feature map smoother.

The weights of DL models were initialized from the networks that had been developed to classify four kinds of picture rotation angles (0°, 90°, 180°, 270°), except the projection part. The weights of the projection part are initialized in a common and efficient way [31]. To match the number of classes in our study, the output unit was modified to two (low-grade and high-grade). The developing process is shown in the bottom half of Figure 2.

After five epochs of warm up, the learning rate was set to 0.1 at the beginning and it varied as a cosine function. It is worth noting that the pre-trained backbone already has some feature extraction capability, unlike the untrained projection. Therefore, in the process of network developing, these two parts of the network should adopt different learning rates, i.e., a small learning rate for the backbone and a relatively larger learning rate for the projection. Specifically, we set the learning rate of the backbone to 0.1 times that of the projection. In addition, a weight decay rate of 0.0001 was set to inhibit overfitting, which can keep the weights of the neural network from becoming too large. Data augmentation, including random rotation and horizontal flipping, was performed on the development cohort to avoid overfitting, which can emulate the diversity of data observed in the real world. Four NVIDIA Tesla M40 graphics cards with 24 GB of memory were used in the development process. We used the SGD algorithm and cross-entropy loss defined in Equation (6) to develop the network. The DL model was finally developed with 100 epochs. Pytorch (1.0.1) and Python (3.5.7) were the main tools used in our experiments.

2.8. Validation and Statistics Analysis

After the developing phase, we used a validation cohort to check the generalizability of the developing effect of the model. Since each patient in the experiment contains multiple images, each image is calculated to obtain a probability vector, so for each patient there is a set of probability vectors. We statistically computed the group probability vector for each patient and finally obtained the grade judgment about the patient. When analyzing a patient's condition, the focus is usually on the most severe part of the CT images, which is reasonable because it can accurately identify the patient's condition. Therefore, in the statistical calculation for each patient, we used the highest probability of network output in each patient's CT image as the judgment basis for grading. Suppose the i-th patient has M CT images, and the output of the model for each CT image is $g_j (j = 1, 2, \ldots, M)$. The grading judgment G_i is

$$G_i = max(g_1, g_2, \ldots, g_M) \tag{7}$$

During validation process, the accuracy (ACC), sensitivity (SEN) and specificity (SPC) were calculated to assess the capability of the DL model. In addition, we used the area under the receiver operating characteristic (ROC) curve (AUC) to show the diagnostic ability of the DL model in grading ccRCC patients.

2.9. Model Ensemble

Following the developing method described in Section 2.7, we developed a total of four classes of DL models with different structures in the development cohort. To improve the reliability of DL models, we combined models with different weights according to their performance in order to obtain a prediction that works best. During the experiment, we found that the single model performed close to each other. In order to increase the diversity of weights of different models in the process of model ensemble, we proposed an innovative weight calculation method. We used the model's AUC as a reference for its ensemble weight specifically, as all four types of models have the same decile of AUC, and their ensemble weight is the value of their AUC after decile is removed. Then, for each patient, we weighted the four models' outputs by different weights and summed them to obtain the patient's final grading judgment. Our weight calculation method can make the models with relatively good performance occupy a larger weight in the ensemble process, increasing the difference between the weights of different models and achieving better ensemble results. Assume the weights of the four models are $\gamma_1, \gamma_2, \gamma_3, \gamma_4$, and the i-th patient's predictions are $G_{i1}, G_{i2}, G_{i3}, G_{i4}$. The composite prediction F_i is

$$F_i = \frac{\sum_{k=1}^{4} \gamma_k G_{ik}}{\sum_{k=1}^{4} \gamma_k} \quad (8)$$

3. Results

We divided the CT images of 706 patients into a development cohort and validation cohort according to the acquisition date, where the development cohort contains 592 patients and the validation cohort contains 114 patients.

Four different kinds of networks (including ensemble model) were validated after developing according to our method, and the relevant metrics were calculated statistically; the validation results are shown in Table 2. The results show that our developing method exhibits satisfactory results on different networks, which illustrates the effectiveness of our method, and in contrast to the subsequent ablation experiments, it can also be seen that our method can effectively mitigate the label noise and class imbalance problems in the data. In addition, our ensemble method can effectively improve the prediction accuracy and enhance the reliability of DL model prediction results. This is like combining the opinions of multiple specialists in the patient's diagnosis process to arrive at a more accurate and reliable judgment about the patient. We selected a model with good performance from each of the four types of models and recorded their receiver operating characteristic curves (ROC), as shown in Figure 3. We also recorded the DL model output probability of each patient in the validation cohort (0 for low-grade, 1 for high-grade), and the results are shown in Figure 4. For most high-grade patients, they have larger lesion areas and a more severe condition based on CT images, and are more likely to have a greater probability of network output. The CT images of some high-grade and low-grade patients are similar, and the probability of a corresponding network output is not significantly different. For low-grade patients, they are more likely to have a relatively smaller network output probability, and their CT images reflect a better condition. The percentage of patients who were graded as low grade or high grade by the ensemble model based on their Fuhrman grades (I, II, III, IV) is displayed in Figure 5. Figure 5 shows that the ensemble model can accurately classify patients in grades I and II as low grade and patients in grades III and IV as high grade, which is pathologically justified by treating grades I and II as low grade and grades III and IV as high grade because grades I and II have relatively more similar characteristics than grades III and IV, thus allowing the network to distinguish between low-grade and high-grade patients.

Table 2. Results of different network models and ensemble models in the validation cohort.

Model	Sen (%)	Spec (%)	ACC (%)	AUC (%)
SE_RESNET50	85.5 ± 6.6	76.3 ± 1.3	82.5 ± 4.0	86.4 ± 0.2
RESNET101	77.6 ± 3.9	76.3 ± 4.0	77.1 ± 1.3	82.2 ± 0.3
REGNET400	82.9 ± 4.0	72.4 ± 1.3	79.4 ± 3.1	83.0 ± 0.1
REGNET800	84.2 ± 7.9	74.3 ± 4.6	81.0 ± 3.7	85.9 ± 0.3
ENSEMBLE	85.5 ± 1.3	75.0 ± 2.6	82.0 ± 0.1	88.2 ± 0.6

ACC = Accuracy; SEN = Sensitivity; SPC = Specificity; AUC = Area under the receiver operating characteristic curve.

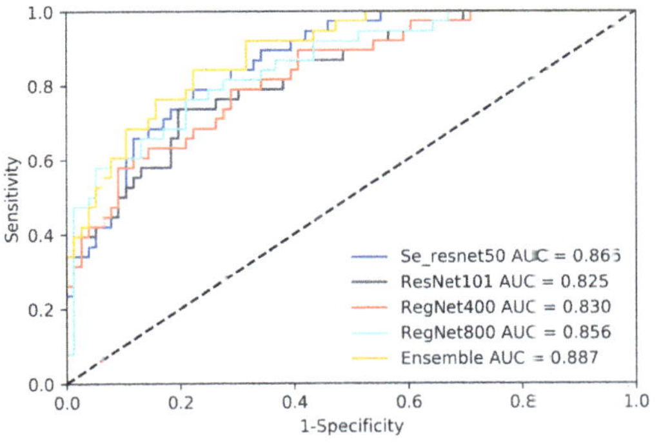

Figure 3. Receiver operating characteristic (ROC) curve of the four different models and the ensemble model.

We also performed a series of ablation experiments to illustrate the effectiveness and necessity of each part of our proposed method. First, we conducted the baseline experiments, i.e., base model experiments without self-supervised pre-training, mixed loss strategy and sample reweighting, and the results are shown in Table 3. From Table 3, we can see that the overall performance of the base model is poor and biased toward the low-grade patients. The overall poor performance is mainly due to the lack of our self-supervised pre-training method. The feature extraction ability of the network is insufficient to accurately identify low-grade and high-grade patients, while the base models are biased toward low-grade patients because they do not solve the label noise and class imbalance problems.

Without using the mixed loss strategy and sample reweighting approaches, we performed experiments with self-supervised pretraining, and the results are shown in Table 4. Compared with the baseline, the self-supervised pre-training method effectively improves the performance of the models, but there is also the problem of excessive bias. Because of the lack of mixed loss strategy and sample reweighting approaches, the network will be more influenced by low-grade patients in the development process, i.e., the number of CT images of low-grade patients is larger than that of high-grade patients, which will make the network biased to low grade in the development process.

We conducted experiments with the addition of the mixed loss strategy and sample reweighting methods without the self-supervised pre-training, and the experimental results are shown in Table 5. From Table 5, we can see that the mixed loss strategy and sample reweighting can effectively solve the bias problem and improve the performance of the model, which is consistent with the fact that they can effectively solve the label noise and class imbalance problems. However, due to the lack of the self-supervised pre-training method, different networks exhibit a large gap in the integrated level relative to Table 2, which once again proves that our self-supervised pre-training method can effectively improve the network feature extraction capability, thus improving the overall network performance.

To validate the effect of different pre-training methods, we pre-trained the SE-ResNet50 model on ImageNet with other settings consistent with the experiments in Table 2. The experimental results are shown in Table 6. Compared with the ImageNet-based pre-training method, our proposed self-supervised pre-training method achieves better experimental results because the ImageNet dataset contains life-like images that have minimal association with the CT images during the developing process, and our proposed pre-training method allows the network to use the same images in the developing process as in the pre-training process and does not reveal the original semantics of the images, which makes the pre-training process and the developing process more relevant and thus allows the pre-training process to better assist the developing process.

We also conducted experiments to compare our method with different traditional machine learning methods [32] including support vector machine (SVM) [33–35],

K-nearest neighbor (KNN), tecision tree [35], random forest [35], and gradient boosting [35]. The degree and tolerance of the SVM were 3 and 0.001. We set the number of neighbors in KNN to 5. For the decision tree, the minimum numbers of samples required to split an internal node and be at a leaf node are 2 and 1. The number of trees in random forest was set to 10. The learning rate of gradient boosting was 0.1, and the number of boosting stages to perform was 100. The experimental results are shown in Table 7. As we can see, our method clearly outperforms all the ML methods. It is worth noting that in our experiments, we did not introduce additional feature extraction methods for the ML methods, saving labor to a great extent while having reliable accuracy. The poor effect of ML methods may be due to the inability to deal with the potential noisy and imbalanced problem intrinsically existing in the data. By contrast, our framework explores a new way to deal with these issues with the help of the proposed mixed loss strategy and sample reweighting, providing increased power to the common practice.

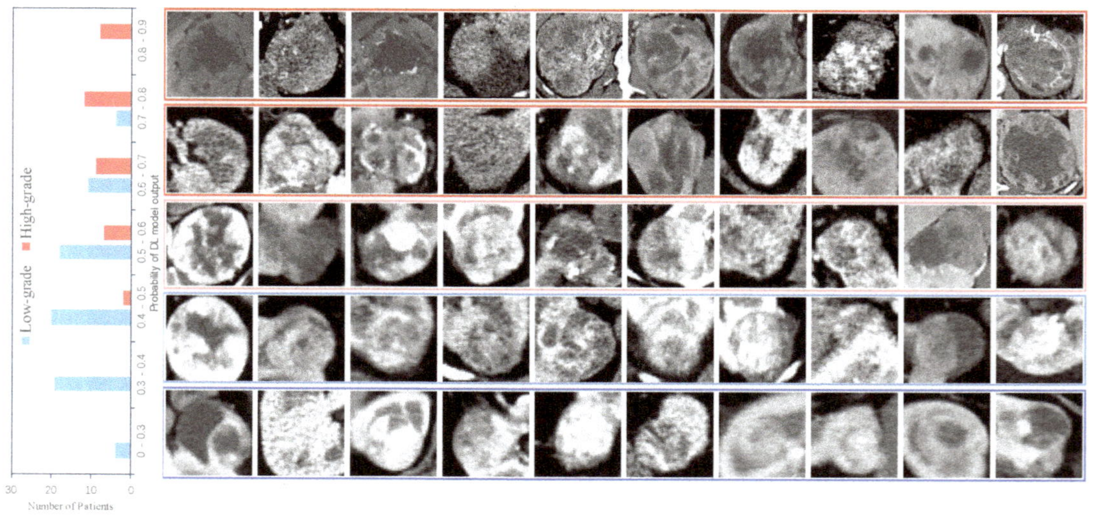

Figure 4. Network output probabilities for low-grade and high-grade patients. The left subplot is the network output probability distribution of low-grade and high-grade patients. The right subplot is the CT images of low-grade and high-grade patients with different network output probabilities.

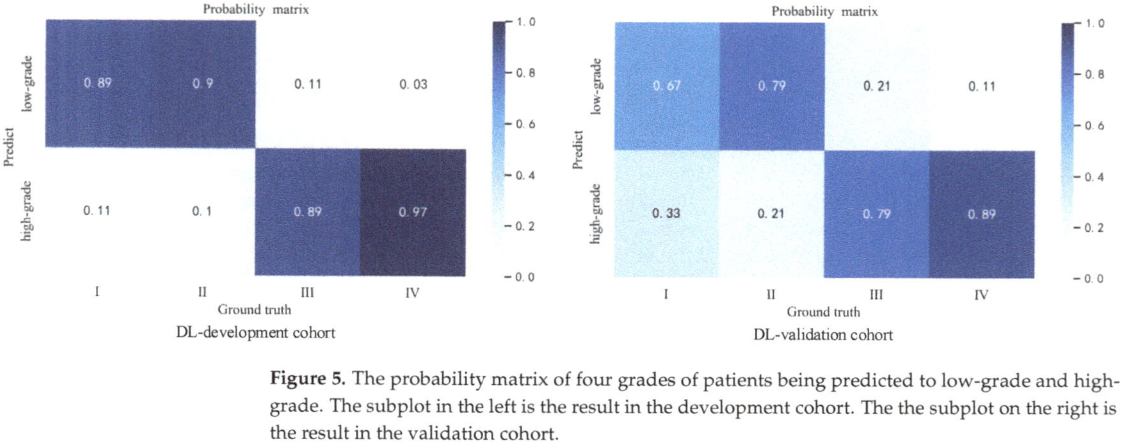

Figure 5. The probability matrix of four grades of patients being predicted to low-grade and high-grade. The subplot in the left is the result in the development cohort. The the subplot on the right is the result in the validation cohort.

Table 3. Performance of the four basic models in the validation cohort.

Model	Sen (%)	Spec (%)	ACC (%)	AUC (%)
SE_RESNET50	65.8 ± 3.7	86.3 ± 3.4	72.6 ± 2.0	78.0 ± 2.3
RESNET101	54.4 ± 13.0	85.5 ± 12.4	64.8 ± 4.7	72.5 ± 0.6
REGNET400	65.7 ± 6.4	85.1 ± 4.5	72.2 ± 2.9	76.6 ± 0.5
REGNET800	66.4 ± 4.7	79.6 ± 2.4	70.8 ± 2.5	75.8 ± 1.4

Table 4. Performance of four types of self-supervised pre-trained models without mixed loss strategy and sample reweighting methods in the validation cohort.

Model	Sen (%)	Spec (%)	ACC (%)	AUC (%)
SE_RESNET50	63.1 ± 2.1	90.3 ± 2.7	72.2 ± 0.5	81.8 ± 0.8
RESNET101	68.4 ± 2.6	80.3 ± 1.3	73.4 ± 0.3	81.2 ± 0.6
REGNET400	69.3 ± 2.5	79.8 ± 1.6	72.8 ± 1.3	80.3 ± 0.2
REGNET800	62.3 ± 2.5	93.0 ± 2.5	72.5 ± 0.8	82.7 ± 0.2

Table 5. Performance of four types of basic models with mixed loss and sample reweighting methods in the validation cohort.

Model	Sen (%)	Spec (%)	ACC (%)	AUC (%)
SE_RESNET50	76.2 ± 3.6	75.0 ± 1.1	75.9 ± 2.2	79.2 ± 1.1
RESNET101	73.7 ± 2.1	76.8 ± 3.3	74.7 ± 1.1	80.4 ± 0.3
REGNET400	72.8 ± 8.9	73.2 ± 10.6	72.9 ± 2.5	79.4 ± 1.1
REGNET800	75.0 ± 2.3	75.3 ± 3.0	75.1 ± 0.6	80.0 ± 0.7

Table 6. Comparison of the SE-ResNet50 model performance based on different pre-training methods in the validation cohort.

Model	Sen (%)	Spec (%)	ACC (%)	AUC (%)
ImageNet	75.0 ± 1.3	77.3 ± 3.3	75.7 ± 1.2	80.3 ± 0.8
Ours	85.5 ± 6.6	76.3 ± 1.3	82.5 ± 4.0	86.4 ± 0.2

Table 7. Performance of machine learning methods in the validation cohort.

Model	Sen (%)	Spec (%)	ACC (%)	AUC (%)
SVM	63.2 ± 18.9	63.2 ± 17.8	63.2 ± 6.7	62.5 ± 7.1
KNN	71.2 ± 16.0	54.6 ± 17.9	60.3 ± 7.1	65.2 ± 2.9
DecisionTree	96.1 ± 2.9	12.8 ± 3.4	40.1 ± 1.6	54.4 ± 1.0
RandomForest	61.8 ± 7.8	68.8 ± 5.7	66.4 ± 1.9	68.4 ± 3.1
GradientBoosting	63.8 ± 11.7	75.7 ± 13.0	71.7 ± 5.9	68.7 ± 4.1
Ours-Ensemble	85.5 ± 1.3	75.0 ± 2.6	82.0 ± 0.1	88.2 ± 0.6

4. Discussion

In this work, we proposed a radiologist-level diagnostic model based on DL approach that is capable of automatically grading ccRCC patients based on CT images. We improved the network's capabilities using innovative self-supervised pre-training approaches. Based on the data in our research, we also proposed solutions to the label noise and class imbalance problems that exist in real world datasets, and the experimental results demonstrate the effectiveness and necessity of our work.

Our best-performing DL model has a high reliability with an accuracy of 88.2% AUC, 82.0% ACC, 85.5% SEN, and 75.0% SPEC. These results confirm that our DL method performs well or equivalent to biopsy in the grade evaluation ccRCC, with the characteristics of noninvasive and labor-saving, which can offer a valuable means for ccRCC grade stratification and individualized patient treatment.

There are four major advantages to our research. Above all, we pre-train the model with the same images (but different labels) as the developing process, in order to provide the network with a better knowledge of the images before developing. Compared with [36–38] using pre-trained models based on ImageNet, our method does not suffer from the problem of small correlation of image contents between the pre-training and developing process, and it allows the network to develop the same images during pre-training and developing without revealing the original semantics of the images.

Furthermore, label noise is the common problem in medical image datasets. The label noise problem degrades the label quality of medical images [39,40], which will make the medical image mismatch with its real label, and have a negative effect in the development of DL. Manually filtering all the samples undoubtedly raises labor costs, and it is inefficient when dealing with large datasets. We have taken the mixed loss strategy for the label noise, with no labor cost overhead but good results. The satisfactory experimental results verify that our method can make the DL model biased toward the correct samples in the development process. Obviously, the actual problem cannot be exactly the same for different datasets; for example, the noise rate differs in size from one dataset to another. Different real situations require different approaches, and we believe that our approach to the two challenges will aid future study in this area.

In addition, class imbalance also occurs frequently in medical image datasets. The class imbalance problem may negatively affect the performance of ML models [41] and DL models [42,43], as most classification methods assume an equal occurrence of different classes. To address this problem, we used the sample reweighting method, which yielded promising benefits. As can be seen from the experimental results, the sample reweighting method effectively prevents the DL model from favoring a certain category in the development process, that is, balance the contribution of samples with different quantity proportions to the loss function. We also expect that our approach of the topic of class imbalance will aid future study in this area.

Last but not leastFinally, DL models with different structures have different independent parameters and are developed to form different perceptions of the dataset. We combined the developed network models with various architectures and obtained more accurate prediction. The model ensemble approach can make up for the shortcomings of individual models in prediction, enhance the network generalization ability, and improve the reliability of results.

In terms of practical significance, our design can help patients in remote areas to further understand their individual conditions, assist doctors to make more accurate clinical judgments on patients' conditions, and to a certain extent compensate for the lack of professional doctors and promote the treatment of patients. With sufficient and noise-free data and reliable developing, our method can reduce or even replace patient biopsy tests, giving patients a safer and more convenient way to be tested.

Despite the contributions of our study in grading ccRCC, it has some potential limitations. The one, although we used model ensemble to improve the generalization ability of the network. For the development of DL models, there are other more DL network architectures that can be utilized, such as VGGNet [25] and GoogleNet [44], but our experiments demonstrated the effectiveness of applying DL to the pathology grading of ccRCC patients. Next, although all cases included in our data are confirmed by professional doctors, there is still a certain human factor, so if our system is to be applied in practice, a large amount of quality data is needed to improve the model in order to make the results more reliable. The WHO/ISUP grading system has superseded the Fuhrman grading system in terms of prognosis assessment and interpretability [45]. Lastly, we take a uniform size operation ($224 \times 224 \times 3$) for tumor images of different sizes, which is necessary for network developing and validation, however, when such an operation is taken for images of small sizes, it may affect the original semantics of the images, which is one of the common problems in the image processing field.However, the intention of using cropped tumor is to exclude the interference of irrelevant information entailed by other normal region. Such normal regions do not contribute positively to the grading of ccRCC. On the contrary, the redundant information may also include a bias or shortcut that would otherwise enforce the model solving a problem differently than intended. For example, there is an observation that the network has learned to detect a metal token that radiology technicians place on the patient in the corner of the image field of view at the time they capture the image in [46].

For the clinical validation of our method, we also look forward to applying our algorithm to real world practice to protect patients from suffering of biopsies as many as possible. However, unfortunately, such a method needs special approval from corresponding authorities, which cannot be easily acquired within short notice. We will positively try this in our future work. In addition, we hope to research a better algorithm to solve the semantic loss problem caused by fixing all images to a uniform size in DL.

5. Conclusions

In this paper, we proposed a DL model that can effectively discriminate different grades of ccRCC patients. Based on the innovative self-supervised pre-training method, different semantics are assigned to the images so that the same images can be used in the pre-training and development tasks, which allows the network to have certain feature extraction capabilities before developing and does not make the pre-training task fragmented from the development task. In addition, we improved the accuracy of the model based on our proposed self-supervised pre-training method and alleviated the effects of label noise and class imbalance problems commonly found in the dataset and the necessity and effectiveness of the proposed method are proved by ablation experiments. With richer and cleaner samples and sufficient developing, the model may become a routine clinical tool to reduce the emotional and physical toll of biopsy on patients.

Author Contributions: Conceptualization, B.S., M.L. (Ming Liu) and X.H.; Data curation, B.S., Y.W. and X.H.; Formal analysis, M.L. (Minghui Liu), J.D. and Y.W.; Funding acquisition, B.S., M.L. (Ming Liu) and X.W.; Investigation, L.X., C.Y. and M.L. (Minghui Liu); Methodology, C.Y., X.C., S.F. and F.Z.; Project administration, L.X., F.Z. and X.C.; Resources, B.S., L.X. and M.L. (Ming Liu); Software, C.Y., S.F., T.X. and J.D.; Supervision, B.S., L.X. and M.L. (Ming Liu); Validation, X.W., M.L. (Ming Liu) and X.C.; Visualization, C.Y., Y.W., S.F. and X.W.; Writing—original draft, C.Y., X.C. and T.X.; Writing—review & editing, B.S., L.X. and X.H. All authors have read and agreed to the published version of the manuscript.

Funding: This work was supported by the Science and Technology Program of Quzhou under Grant 2021D007, Grant 2021D008, Grant 2021D015, and Grant 2021D018, as well as the project LGF22G010009.

Institutional Review Board Statement: The study was conducted according to the guidelines of the Declaration of Helsinki, and approved by the Institutional Review Board of The Quzhou Affiliated Hospital of Wenzhou Medical University (ethical code number 2020-03-002).

Informed Consent Statement: The ethics board waived informed consent because the data were obtained from preexisting institutional or public databases.

Data Availability Statement: The de-identified CT images data used in this study are not publicly available due to restrictions in the data-sharing agreement.

Conflicts of Interest: The funder had no role in the study design, data collection, data analysis, data interpretation, or writing of the report.

References

1. Siegel, R.L.; Miller, K.D.; Fuchs, H.E.; Jemal, A. Cancer Statistics, 2021. *CA Cancer J. Clin.* **2021**, *71*, 7–33. [CrossRef] [PubMed]
2. Hsieh, J.J.; Purdue, M.P.; Signoretti, S.; Swanton, C.; Albiges, L.; Schmidinger, M.; Heng, D.Y.; Larkin, J.; Ficarra, V. Renal cell carcinoma. *Nat. Rev. Dis. Prim.* **2017**, *3*, 1–19. [CrossRef] [PubMed]
3. Fuhrman, S.A.; Lasky, L.C.; Limas, C. Prognostic significance of morphologic parameters in renal cell carcinoma. *Am. J. Surg. Pathol.* **1982**, *6*, 655–663. [CrossRef] [PubMed]
4. Marconi, L.; Dabestani, S.; Lam, T.B.; Hofmann, F.; Stewart, F.; Norrie, J.; Bex, A.; Bensalah, K.; Canfield, S.E.; Hora, M.; et al. Systematic review and meta-analysis of diagnostic accuracy of percutaneous renal tumour biopsy. *Eur. Urol.* **2016**, *69*, 660–673. [CrossRef]

5. Krizhevsky, A.; Sutskever, I.; Hinton, G.E. Imagenet classification with deep convolutional neural networks. *Adv. Neur. Inf.* **2012**, *25*, 1097–1105. [CrossRef]
6. Sermanet, P.; Eigen, D.; Zhang, X.; Mathieu, M.; Fergus, R.; LeCun, Y. Overfeat: Integrated recognition, localization and detection using convolutional networks. *arXiv* **2013**, arXiv:1312.6229.
7. Girshick, R.; Donahue, J.; Darrell, T.; Malik, J. Rich feature hierarchies for accurate object detection and semantic segmentation. In Proceedings of the IEEE Conference on Computer Vision and Pattern Recognition, Columbus, OH, USA, 20–23 June 2014; pp. 580–587
8. Cho, J.; Lee, K.; Shin, E.; Choy, G.; Do, S. How much data is needed to train a medical image deep learning system to achieve necessary high accuracy? *arXiv* **2015**, arXiv:1511.06348.
9. Esteva, A.; Kuprel, B.; Novoa, R. A.; Ko, J.; Swetter, S. M.; Blau, H. M.; Thrun, S. Dermatologist-level classification of skin cancer with deep neural networks. *Nature* **2017**, *542*, 115–118. [CrossRef]
10. Gulshan, V.; Rajan, R.P.; Widner, K.; Wu, D.; Wubbels, P.; Rhodes, T.; Whitehouse, K.; Coram, M.; Peng, L.; Webster, D.R.; et al. Performance of a Deep-Learning Algorithm vs Manual Grading for Detecting Diabetic Retinopathy in India. *JAMA Ophthalmol.* **2019**, *137*, 987–993. [CrossRef]
11. Shen, Y.; Shamout, F.E.; Oliver, J.R.; Witowski, J.; Kannan, K.; Park, J.; Wu, N.; Huddleston, C.; Wolfson, S.; Geras, K.J.; et al. Artificial intelligence system reduces false-positive findings in the interpretation of breast ultrasound exams. *Nat. Commun.* **2021**, *12*, 5645. [CrossRef]
12. Yeung, S.; Rinaldo, F.; Jopling, J.; Liu, B.; Mehra, R.; Downing, N.L.; Guo, M.; Bianconi, G.M.; Fei-Fei, L.; Milstein, A.; et al. A computer vision system for deep learning-based detection of patient mobilization activities in the ICU. *NPJ Digit. Med.* **2019**, *2*, 11. [CrossRef] [PubMed]
13. Topol, E.J. High-performance medicine: The convergence of human and artificial intelligence. *Nat. Med.* **2019**, *25*, 44-56. [CrossRef] [PubMed]
14. Lin, F.; Ma, C.; Xu, J.; Lei, Y.; Li, Q.; Lan, Y.; Sun, M.; Long, W; Cui, E. A CT-based deep learning model for predicting the nuclear grade of clear cell renal cell carcinoma. *Eur. J. Radiol.* **2020**, *129*, 109079. [CrossRef] [PubMed]
15. Deng, J.; Dong, W.; Socher, R.; Li, L. J.; Li, K.; Fei, L. Imagenet: A large-scale hierarchical image database. In Proceedings of the 2009 IEEE Conference on Computer Vision and Pattern Recognition, Miami, FL, USA, 20–25 June 2009; pp. 248–255.
16. Hekler, A.; Kather, J.N.; Krieghoff-Henning, E.; Utikal, J.S.; Meier, F.; Gellrich, F.F.; Upmeier Zu Belzen, J.; French, L.; Schlager, J.G.; Brinker, T.J.; et al. Effects of Label Noise on Deep Learning-Based Skin Cancer Classification. *Front. Med.* **2020**, *7*, 177. [CrossRef] [PubMed]
17. Zhu, M.; Ren, B.; Richards, R.; Suriawinata, M.; Tomita, N.; Hassanpour, S. Development and evaluation of a deep neural network for histologic classification of renal cell carcinoma on biopsy and surgical resection slides. *Sci. Rep.* **2021**, *11*, 7080. [CrossRef] [PubMed]
18. Zheng, Z.; Chen, Z.; Xie, Y.; Zhong, Q.; Xie, W. Development and validation of a CT-based nomogram for preoperative prediction of clear cell renal cell carcinoma grades. *Eur. Radiol.* **2021**, *31*, 6078–6086. [CrossRef]
19. Cui, E.; Li, Z.; Ma, C.; Li, Q.; Lei, Y.; Lan, Y.; Yu, J.; Zhou, Z.; Li, R.; Lin, F.; et al. Predicting the ISUP grade of clear cell renal cell carcinoma with multiparametric MR and multiphase CT radiomics. *Eur. Radiol.* **2020**, *30*, 2912–2921. [CrossRef]
20. Gao, R.; Qin, H.; Lin, P.; Ma, C.; Li, C.; Wen, R.; Huang, J.; Wan, D.; Wen, D.; Yang, H.; et al. Development and Validation of a Radiomic Nomogram for Predicting the Prognosis of Kidney Renal Clear Cell Carcinoma. *Front. Oncol.* **2021**, *11*, 613668. [CrossRef]
21. Ferro, M.; de Cobelli, O.; Vartolomei, M.D.; Lucarelli, G.; Crocetto, F.; Barone, B.; Sciarra, A.; Giudice, F.A.; Muto, M.; Tataru, O.S.; et al. Prostate Cancer Radiogenomics—From Imaging to Molecular Characterization. *Int. J. Mol. Sci.* **2021**, *22*, 9971. [CrossRef]
22. Choi, S.J.; Park, K.J.; Heo, C.; Park, B.W.; Kim, M.; Kim, J.K. Radiomics-based model for predicting pathological complete response to neoadjuvant chemotherapy in muscle-invasive bladder cancer. *Clin. Radiol.* **2021**, *76*, 627.e13–627.e21. [CrossRef]
23. Liu, Y.; Jain, A.; Eng, C.; Way, D.H.; Lee, K.; Bui, P.; Kanada, K.; Marinho, G.O.; Gallegos, J.; Coz, D.; et al. A deep learning system for differential diagnosis of skin diseases. *Nat. Med.* **2020**, *26*, 900–908. [CrossRef] [PubMed]
24. Delahunt, B. Advances and controversies in grading and staging of renal cell carcinoma. *Mod. Pathol.* **2009**, *22*, S24–S36. [CrossRef] [PubMed]
25. Simonyan, K.; Zisserman, A. Very deep convolutional networks for large-scale image recognition. *arXiv* **2014**, arXiv:1409.1556.
26. Russakovsky, O.; Deng, J.; Su, H.; Krause, J.; Satheesh, S.; Ma, S.; Huang, Z.; Karpathy, A.; Khosla, A.; Fei-Fei, L.; et al. Imagenet large scale visual recognition challenge. *Int. J. Comput. Vis.* **2015**, *115*, 211–252. [CrossRef]
27. Radosavovic, I.; Kosaraju, R.P.; Girshick, R.; He, K.; Dollár, P. Designing network design spaces. In Proceedings of the IEEE/CVF Conference on Computer Vision and Pattern Recognition, Seattle, WA, USA, 14–19 June 2020; pp. 10428–10436.
28. Hu, J.; Shen, L.; Sun, G. Squeeze-and-excitation networks. In Proceedings of the IEEE Conference on Computer Vision and Pattern Recognition, Salt Lake City, UT, USA, 18–22 June 2018; pp. 7132–7141.
29. He, K.; Zhang, X.; Ren, S.; Sun, J. Deep residual learning for image recognition. In Proceedings of the IEEE Conference on Computer Vision and Pattern Recognition, Las Vegas, NV, USA, 27–30 June 2016; pp. 770–778.
30. Zhang, H.; Cisse, M.; Dauphin, Y. N.; Lopez-Paz, D. mixup: Beyond empirical risk minimization. *J. Abbr.* **2017**, *10*, 142–149.

31. He, K.; Zhang, X.; Ren, S.; Sun, J. Delving deep into rectifiers: Surpassing human-level performance on imagenet classification. In Proceedings of the IEEE International Conference on Computer Vision, Santiago, Chile, 7–13 December 2015; pp. 1026–1034.
32. Alpaydin, E. *Introduction to Machine Learning*; MIT Press: Cambridge, MA, USA, 2020.
33. Huang, M.W.; Chen, C.W.; Lin, W.C.; Ke, S.W.; Tsai, C.F. SVM and SVM Ensembles in Breast Cancer Prediction. *PLoS ONE* **2017**, *12*, e0161501. [CrossRef]
34. Murugan, A.; Nair, S.A.; Kumar, K.P. Detection of Skin Cancer Using SVM, Random Forest and kNN Classifiers. *J. Med. Syst.* **2019**, *43*, 269. [CrossRef]
35. Tong, J.; Liu, P.; Ji, M.; Wang, Y.; Xue, Q.; Yang, J.T.; Zhou, C.M. Machine Learning Can Predict Total Death After Radiofrequency Ablation in Liver Cancer Patients. *Clin. Med. Insights Oncol.* **2021**, *15*, 11795549211000017. [CrossRef]
36. Zhou, L.; Zhang, Z.; Chen, Y.C.; Zhao, Z.Y.; Yin, X.D.; Jiang, H.B. A deep learning-based radiomics model for differentiating benign and malignant renal tumors. *Transl. Oncol.* **2019**, *12*, 292–300. [CrossRef]
37. Coy, H.; Hsieh, K.; Wu, W.; Nagarajan, M.B.; Young, J.R.; Douek, M.L.; Brown, M.S.; Scalzo, F.; Raman, S.S. Deep learning and radiomics: The utility of Google TensorFlow^{TM} Inception in classifying clear cell renal cell carcinoma and oncocytoma on multiphasic CT. *Abdom. Radiol.* **2019**, *44*, 2009–2020. [CrossRef]
38. Lee, H.; Hong, H.; Kim, J.; Jung, D.C. Deep feature classification of angiomyolipoma without visible fat and renal cell carcinoma in abdominal contrast-enhanced CT images with texture image patches and hand-crafted feature concatenation. *Med. Phys.* **2018**, *45*, 1550–1561. [CrossRef] [PubMed]
39. HosseinKhani, Z.; Hajabdollahi, M.; Karimi, N.; Soroushmehr, R.; Shirani, S.; Najarian, K.; Samavi, S. Adaptive Real-Time Removal of Impulse Noise in Medical Images. *J. Med. Syst.* **2018**, *42*, 216. [CrossRef] [PubMed]
40. Zhang, T.; Cheng, J.; Fu, H.; Gu, Z.; Xiao, Y.; Zhou, K.; Gao, S.; Zheng, R.; Liu, J. Noise Adaptation Generative Adversarial Network for Medical Image Analysis. *IEEE Trans. Med. Imaging* **2020**, *39*, 1149–1159. [CrossRef] [PubMed]
41. Teh, K.; Armitage, P.; Tesfaye, S.; Selvarajah, D.; Wilkinson, I.D. Imbalanced learning: Improving classification of diabetic neuropathy from magnetic resonance imaging. *PLoS ONE* **2020**, *15*, e0243907. [CrossRef]
42. Bria, A.; Marrocco, C.; Tortorella, F. Addressing class imbalance in deep learning for small lesion detection on medical images. *Comput. Biol. Med.* **2020**, *120*, 103735. [CrossRef]
43. Gao, L.; Zhang, L.; Liu, C.; Wu, S. Handling imbalanced medical image data: A deep-learning-based one-class classification approach. *Artif. Intell. Med.* **2020**, *108*, 101935. [CrossRef]
44. Szegedy, C.; Liu, W.; Jia, Y.; Sermanet, P.; Reed, S.; Anguelov, D.; Erhan, D.; Vanhoucke, V.; Rabinovich, A. Going deeper with convolutions. In Proceedings of the IEEE Conference on Computer Vision and Pattern Recognition, Columbus, OH, USA, 20–23 June 2014; pp. 1–9.
45. Dagher, J.; Delahunt, B.; Rioux-Leclercq, N.; Egevad, L.; Srigley, J.R.; Coughlin, G.; Dunglinson, N.; Gianduzzo, T.; Kua, B.; Samaratunga, H.; et al. Clear cell renal cell carcinoma: Validation of World Health Organization/International Society of Urological Pathology grading. *Histopathology* **2017**, *71*, 918–925. [CrossRef]
46. Zech, J.R.; Badgeley, M.A.; Liu, M.; Costa, A.B.; Titano, J.J.; Oermann, E.K. Variable generalization performance of a deep learning model to detect pneumonia in chest radiographs: A cross-sectional study. *PLoS Med.* **2018**, *15*, e1002683. [CrossRef]

cancers

MDPI

Review

Machine Learning Tools for Image-Based Glioma Grading and the Quality of Their Reporting: Challenges and Opportunities

Sara Merkaj [1,2,†], Ryan C. Bahar [1,†], Tal Zeevi [1], MingDe Lin [1,3], Ichiro Ikuta [1], Khaled Bousabarah [4], Gabriel I. Cassinelli Petersen [1], Lawrence Staib [1], Seyedmehdi Payabvash [1], John T. Mongan [5], Soonmee Cha [5] and Mariam S. Aboian [1,*]

1. Department of Radiology and Biomedical Imaging, Yale School of Medicine, 333 Cedar Street, P.O. Box 208042, New Haven, CT 06520, USA; sara.merkaj@uni-ulm.de (S.M.); ryan.bahar@yale.edu (R.C.B.); tal.zeevi@yale.edu (T.Z.); mingde.lin@yale.edu (M.L.); ichiro.ikuta@yale.edu (I.I.); gabriel.cassinellipetersen@yale.edu (G.I.C.P.); lawrence.staib@yale.edu (L.S.); sam.payabvash@yale.edu (S.P.)
2. Department of Neurosurgery, University of Ulm, Albert-Einstein-Allee 23, 89081 Ulm, Germany
3. Visage Imaging, Inc., 12625 High Bluff Dr, Suite 205, San Diego, CA 92130, USA
4. Visage Imaging, GmbH., Lepsiusstraße 70, 12163 Berlin, Germany; kbousabarah@visageimaging.com
5. Department of Radiology and Biomedical Imaging, University of California San Francisco, 505 Parnassus Ave., San Francisco, CA 94143, USA; john.mongan@ucsf.edu (J.T.M.); soonmee.cha@ucsf.edu (S.C.)

* Correspondence: mariam.aboian@yale.edu; Tel.: +650-285-7577

† These authors contributed equally to this work.

Citation: Merkaj, S.; Bahar, R.C.; Zeevi, T.; Lin, M.; Ikuta, I.; Bousabarah, K.; Cassinelli Petersen, G.I.; Staib, L.; Payabvash, S.; Mongan, J.T.; et al. Machine Learning Tools for Image-Based Glioma Grading and the Quality of Their Reporting: Challenges and Opportunities. *Cancers* **2022**, *14*, 2623. https://doi.org/10.3390/cancers14112623

Academic Editor: Axel Pagenstecher

Received: 26 March 2022
Accepted: 23 May 2022
Published: 25 May 2022

Simple Summary: Despite their prevalence in research, ML tools that can predict glioma grade from medical images have yet to be incorporated clinically. The reporting quality of ML glioma grade prediction studies is below 50% according to TRIPOD—limiting model reproducibility and, thus, clinical translation—however, current efforts to create ML-specific reporting guidelines and risk of bias tools may help address this. Several additional deficiencies in the areas of ML model data and glioma classification hamper widespread clinical use, but promising efforts to overcome current challenges and encourage implementation are on the horizon.

Abstract: Technological innovation has enabled the development of machine learning (ML) tools that aim to improve the practice of radiologists. In the last decade, ML applications to neuro-oncology have expanded significantly, with the pre-operative prediction of glioma grade using medical imaging as a specific area of interest. We introduce the subject of ML models for glioma grade prediction by remarking upon the models reported in the literature as well as by describing their characteristic developmental workflow and widely used classifier algorithms. The challenges facing these models—including data sources, external validation, and glioma grade classification methods —are highlighted. We also discuss the quality of how these models are reported, explore the present and future of reporting guidelines and risk of bias tools, and provide suggestions for the reporting of prospective works. Finally, this review offers insights into next steps that the field of ML glioma grade prediction can take to facilitate clinical implementation.

Keywords: artificial intelligence; glioma; machine learning; deep learning; reporting quality

1. Introduction

1.1. Artificial Intelligence, Machine Learning, and Radiomics

Innovations in computation and imaging have rapidly enhanced the potential for artificial intelligence (AI) to impact diagnostic neuroradiology. Emerging areas of implementation include AI in stroke (e.g., early diagnosis, detection of large vessel occlusion, and outcome prediction) [1], AI in spine (fracture detection, and vertebrae segmentation) and detection of intracranial aneurysms and hemorrhage [2], among other disciplines. Machine learning (ML) and its subfield, deep learning (DL), are branches of AI that have received particular attention. ML algorithms, including DL, decipher patterns in input data and independently learn to make predictions [3]. The advent of radiomics—which mines data from images by transforming them into features quantifying tumor phenotypes—has fueled the application of ML methods to imaging, including radiomics-based ML analysis of brain tumors [4–6]. Commonly extracted radiomic features include shape and size, texture, first-order, second-order, higher-order features, etc. (Table 1).

1.2. Machine Learning Applications in Neuro-Oncology

As the most common primary brain tumors, gliomas constitute a major focus of ML applications to neuro-oncology [7,8]. Prominent domains of glioma ML research include the image-based classification of tumor grade and prediction of molecular and genetic characteristics. Genetic information is not only instrumental to tumor diagnosis in the 2021 World Health Organization classification, but also significantly affects survival and underpins sensitivity to therapeutic interventions [9,10]. ML-based models for predicting tumor genotype can therefore guide earlier diagnosis, estimation of prognosis, and treatment-related decision-making [11,12]. Other significant areas of glioma ML research relevant to neuroradiologists include automated tumor segmentation on MRI, detection and prediction of tumor progression, differentiation of pseudo-progression from true progression, glioma survival prediction and treatment response, distinction of gliomas from other tumors and non-neoplastic lesions, heterogeneity assessment based on imaging features, and clinical incorporation of volumetrics [13–15]. Furthermore, ML tools may optimize neuroradiology workflow by expediting the time to read studies from image review to report generation [16]. As an image interpretation support tool, ML importantly may improve diagnostic performance [17,18]. Prior works demonstrate that AI alone can approach the diagnostic accuracy of neuroradiologists and other sub-specialty radiologists [19–21].

1.3. Image-Based Machine Learning Models for Glioma Grading

This review is concerned with the growing body of studies developing predictive ML models for image-based glioma grading, a fundamentally heterogeneous area of literature. While numerous ML models exist to predict high-grade gliomas and low-grade gliomas, they vary in their definitions of high- and low-grade [22–24]. Other models predict individual glioma grades (e.g., 2 vs. 3, 3 vs. 4), but few have combined glioma grading with molecular classification despite the incorporation of both grade and molecular subtype in 2016 World Health Organization central nervous system tumor classification [25,26]. While studies focus on MRI, they are diverse in the sequences used for prediction, with earlier publications relying on conventional imaging and increasing incorporation of advanced MRI sequences throughout the years [27–30]. Finally, studies vary considerably in their feature extraction and selection methods, datasets, validation techniques, and classification algorithms [31].

It is our belief that the ML models with potential to support one of the most fundamental tasks of the neuroradiologist—glioma diagnosis—present obstacles and opportunities relevant to the radiology community, especially as radiologists endeavor to bring ML models into clinical practice. In this article, we aim to introduce the subject of developing ML models for glioma grade prediction, highlight challenges facing these models and their reporting within the literature, and offer insights into next steps the field can take to facilitate clinical implementation.

2. Workflow for Developing Prediction Models

Despite their heterogeneity, ML glioma grade prediction studies follow similar steps in developing their models. The development workflow starts with acquisition, registration, and pre-processing (if necessary) of multi-modal MR images. Common pre-processing tasks include data cleaning, normalization, transformation, and dealing with incomplete data, among other tasks [32]. An in-depth exploration of pre-processing is beyond the scope of this review and readers should refer to Kotsiantis et al. for further explanation. Next, tumors undergo segmentation—the delineation of tumor, necrosis, and edema borders—which can be a manual, semi-automatic, or fully automatic process. Manual segmentations rely on an expert delineating and annotating Regions of Interest (ROIs) by hand. Semi-automated segmentations generate automated ROIs that need to be checked and modified by experts. Fully automatic segmentations, on the other hand, are DL-generated (most frequently by convolutional neural networks (CNNs)), which automatically delineate ROIs and omit the need for manual labor [33]. In general, semi-automated segmentations are considered to be more reliable and transparent than fully automatic segmentations. However, they are less time-efficient than automatic segmentations and always require manual input from experts in the field. Whereas manual segmentation is laborious, time-consuming, and subject to inter-reader variability, fully automatic deep-learning generated segmentations may potentially overcome these challenges [34].

Feature extraction is then performed to extract qualitative and quantitative information from imaging. Commonly extracted data include radiomic features (shape, first-order, second-order, higher-order features, etc.), clinical features (age, sex, etc.), and tumor-specific Visually AcceSAble Rembrandt Images (VASARI) features. Feature types and their explanations are presented in Table 1.

Table 1. Overview of commonly extracted feature types in studies developing ML prediction models.

Feature Type	Explanation
Clinical	Describe patient demographics, e.g., gender and age.
Deep learning extracted	Derived from pre-trained deep neural networks.
First-order	Create a three-dimensional (3D) histogram out of tumor volume characteristics, from which mean, median, range, skewness, kurtosis, etc., can be calculated [35].
Higher-order	Identify repetitiveness in image patterns, suppress noise, or highlight details [35].
Qualitative	Describe visible tumor characteristics on imaging using controlled vocabulary, e.g., VASARI features (tumor location, side of lesion center, enhancement quality, etc.).
Second-order	Classify texture characteristics, e.g., contrast, correlation, dissimilarity, maximum probability, grey level run length features, etc. [35]
Shape and size	Describe the statistical inter-relationships between neighboring voxels, e.g., total volume or surface area, surface-to-volume ratio, tumor compactness, sphericity, etc. [35]

Open-source packages such as PyRadiomics have been developed as a reference standard for radiomic feature extraction [36]. Clinical features are known to be important markers for predicting glioma grades and molecular subtypes [37]. VASARI features, developed by The Cancer Imaging Archive (TCIA), are frequently found in studies that qualitatively describe tumor morphology using visual features and controlled vocabulary/standardized semantics [38].

Current technology permits extraction of over 1000 features per image. As a high number of features may lead to model overfitting, model developers commonly reduce the number of features used through feature selection. Feature selection methods, including Filter, Wrapper, and Embedded methods, remove non-informative features that reduce the model's overall performance [39].

The final set of features is fed into a glioma grade classification algorithm(s)—for example, support vector machine (SVM) and CNN—during the training process. The classifier performance is then measured through performance metrics such as accuracy, area under the curve receiver operating characteristic, sensitivity, specificity, positive predictive value, negative predictive value, and F1 score. The model is validated internally, usually through hold-out or cross-validation techniques. Ideally, the model is externally validated as a final step to ensure reproducibility, generalizability, and reliability in a different setting (Figure 1).

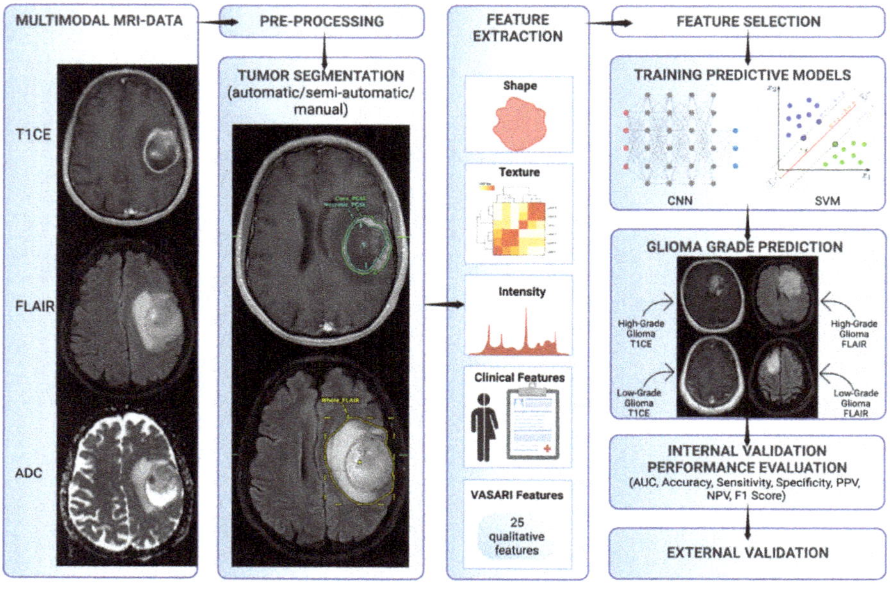

Figure 1. Characteristic workflow for developing ML glioma grade prediction models. VASARI = Visually AcceSAble Rembrandt Images, AUC = area under the curve receiver operating characteristic, CNN = convolutional neural network, ML = machine learning, NPV = negative predictive value, PPV = positive predictive value, and SVM = support vector machine.

3. Algorithms for Glioma Grade Classification

The most common high-performing ML classifiers for glioma grading in the literature are SVM and CNN [13]. SVM is a classical ML algorithm that represents objects as points in an n-dimensional space, with features serving as coordinates. SVMs use a hyperplane, or an n-1 dimensional subspace, to divide the space into disconnected areas [40]. These distinct areas represent the different classes that the model can classify. Unlike CNNs, SVMs require hand-engineered features, such as from radiomics, to serve as inputs. This requirement may be advantageous for veteran diagnostic imagers, whose knowledge of brain tumor appearance may enhance feature design and selection. Hand-engineered features also can undergo feature reduction to mitigate the risks of overfitting, and prior works demonstrate better performance for glioma grading models using a smaller number of quantitative features [41]. However, hand-engineered features are limited since they cannot be adjusted during model training, and it is uncertain if they are optimal features for classification. Moreover, hand-engineered features may not generalize well beyond the training set and should be tested extensively prior to usage [42,43].

CNNs are a form of deep learning based on image convolution. Images are the direct inputs to the neural network, rather than the manually engineered features of classical ML. Numerous interconnected layers each compute feature representations and pass them on to subsequent layers [43,44]. Near the network output, features are flattened into a vector that performs the classification task. CNNs appeared for glioma grading in 2018 and have risen quickly in prevalence while exhibiting excellent predictive accuracies [45–48].

To a greater extent than classical ML, they are suited for working with large amounts of data, and their architecture can be modified to optimize efficiency and performance [46]. Disadvantages include the opaque "black box" nature of deep learning and associated difficulty with interpreting model parameters, along with problems that variably apply to classical ML as well (e.g., high amount of time and data required for training, hardware costs, and necessary user expertise) [49,50].

In our systematic review of 85 published ML studies developing models for image-based glioma grading, we found SVM and CNN to have mean accuracies of 90% and 91%, respectively [51]. Mean accuracies for these algorithms were similar across classification tasks regardless of whether the classification was binary or multi-class (e.g., 90% for the 24 studies whose best models performed binary classification of grades 1/2 vs. 3/4 compared to 86% for the 5 studies classifying grade 2 vs. 3 vs. 4). No consensus has been reached regarding the optimal ML algorithm for image-based glioma classification.

4. Challenges in Image-Based ML Glioma Grading

4.1. Data Sources

Since 2011, a significant number of ML glioma grade prediction studies have used open-source multi-center datasets to develop their models. BraTS [52] and TCIA [53] are two prominent public datasets that contain multi-modal MRI images of high- and low-grade gliomas and patient demographics. BraTS was first made available in 2012, with the 2021 dataset containing 8000 multi-institutional, multi-parametric MR images of gliomas [52]. TCIA first went online in 2011 and contains MR images of gliomas collected across 23 institutions [53]. These datasets were developed with the aim of providing a unified multi-center resource for glioma research. A variety of predictive models have been trained and tested on these large datasets since their 2011 release [54]. Despite their value as public datasets for model development, several limitations should be considered. Images are collected across multiple institutions with variable protocols and image quality. Co-registration and imaging pre-processing integrate these images into a single system. Although these techniques are necessary, they may reduce heterogeneity within the datasets [52]. Models developed on these datasets may perform well in training and testing. Nevertheless, the results may not be reproducible in the real-world clinical setting, where images and tumor presentations are heterogeneous. We strongly support large multi-center datasets in order to demonstrate model performance across distinct hospital settings. We, however, recommend such initiatives incorporate images of various diagnostic qualities into their training datasets, which more closely resemble what is seen in daily practice.

4.2. External Validation

Publications have reported predictive models for glioma grading throughout the last 20 years with the majority relying on internal validation techniques, of which cross-validation is the most popular. While internal validation is a well-established method for measuring how well a model will perform on new cases from the initial dataset, additional evaluation on a separate dataset (i.e., external validation) is critical to demonstrate model generalizability. External validation mitigates site bias (differences amongst centers in protocols, techniques, scanner variability, level of experience, etc.) and sampling/selection bias (performance only applicable to the specific training set population/demographics) [55]. Not controlling for these two major biases undermines model generalizability, yet few publications externally validate their models [13]. Therefore, normalizing external validation is a crucial step in developing glioma grade prediction models that are suitable for clinical implementation.

4.3. Glioma Grade Classification Systems

The classification of glioma subtypes into high- and low-grade gliomas is continuously evolving. In 2016, an integrated histological–molecular classification replaced the previous purely histopathological classification [56]. In 2021, the Consortium to Inform Molecular and Practical Approaches to CNS Tumor Taxonomy (cIMPACT NOW) once more accentuated the diagnostic value of molecular markers, such as the isocitrate dehydrogenase mutation, for glioma classification [57]. As a result of the evolving glioma classification system, definitions for high- and low-grade gliomas vary across ML glioma grade prediction studies and publication years. This reduces the comparability of models themselves and grade-labeled datasets used for model development. We recommend future glioma grade prediction studies focus on both glioma grade and molecular subtypes for more comprehensive and reliable results over time. Neuropathologic diagnostic emphasis has shifted from purely based on microscopic histology to one that combines morphologic and molecular genetic features of tumor including gene mutations, chromosomal copy number alterations, and gene rearrangements to yield integrated diagnosis. Rapid developments in next generation sequencing techniques, multimodal molecular analysis, large scale genomic and epigenomic analyses, and DNA methylation methods promise to fundamentally transform the pathologic CNS tumor diagnostics including glioma diagnosis and grading to whole another level of precision and complexity.

Current and future ML methods must keep abreast of the rapid progress in tissue based integrated diagnostics in order to contribute to and make an impact on the clinical care of glioma patients (Figure 2).

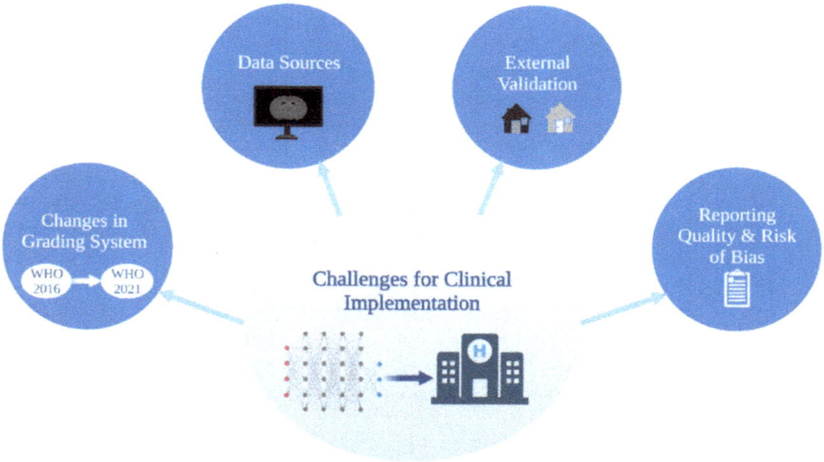

Figure 2. Challenges for clinical implementation of ML glioma grade prediction models. ML = machine learning. WHO = World Health Organization.

4.4. *Reporting Quality and Risk of Bias*

4.4.1. Overview of Current Guidelines and Tools for Assessment

It is critical that studies detailing prediction models, such as those for glioma grading, exhibit a high caliber of scientific reporting in accordance with consensus standards. Clear and thorough reporting enables more complete understanding by the reader and unambiguous assessment of study generalizability, quality, and reproducibility, encouraging future researchers to replicate and use models in clinical contexts. Several instruments have been designed to improve the reporting quality (defined here as the transparency and thoroughness with which authors share key details of their study to enable proper interpretation and evaluation) of studies developing models. The Transparent Reporting of a multivariable prediction model for Individual Prognosis or Diagnosis (TRIPOD) Statement was created in 2015 as a set of recommendations for studies developing, validating, or updating diagnostic or prognostic models [58]. The TRIPOD Statement is a checklist of 22 items considered essential for transparent reporting of a prediction model study. In 2017, with a concurrent rise in radiomics-based model studies, the radiomics quality score (RQS) emerged [59]. RQS is an adaptation of the TRIPOD approach geared toward a radiomics-specific context. The tool has been used throughout the literature for evaluating the methodological quality of radiomics studies, including applications to medical imaging [60]. Radiomics-based approaches for interpreting medical images have evolved to encompass the AI techniques of classical ML and, most recently, deep learning models. Most recently, in recognition of the growing need for an evaluation tool specific to AI applications in medical imaging, the Checklist for AI in Medical Imaging (CLAIM) was published in 2020 [61]. The 42 elements of CLAIM aim to be a best practice guide for authors presenting their research on applications of AI in medical imaging, ranging from classification and image reconstruction to text analysis and workflow optimization. Other tools—the Quality Assessment of Diagnostic Accuracy Studies (QUADAS-2) tool [62] and Prediction model Risk Of Bias ASsessment Tool (PROBAST) [63]—importantly evaluate the risk of bias in studies based on what is reported about their models (Table 2). Bias relates to systematic limitations or flaws in study design, methods, execution, or analysis that distort estimates of model performance [62]. High risk of bias discourages adaptation of the reported model outside of its original research context, and, at a systemic level, undermines model reproducibility and translation into clinical practice.

Table 2. Overview of major reporting guidelines and bias assessment tools for diagnostic and prognostic studies.

Guideline/Tool	Full Name	Year Published	Articles Targeted	Purpose	Specific to ML?
QUADAS-2 [4]	Quality Assessment of Diagnostic Accuracy Studies	2011 (original QUADAS [4]: 2003)	Diagnostic accuracy studies	Evaluates study risk of bias and applicability	No; QUADAS-AI [4] is in development
TRIPOD [6]	Transparent Reporting of a multivariable prediction model for Individual Prognosis Or Diagnosis	2015	Studies developing, validating, or updating a diagnostic or prognostic prediction model	Provides a set of recommendations for study reporting	No; TRIPOD-AI [6] is in development
RQS [5]	Radiomics quality score	2017	Radiomic studies	Assesses study quality (emulating TRIPOD [6])	No
PROBAST [3]	Prediction model Risk Of Bias ASsessment Tool	2019	Studies developing, validating, or updating a diagnostic or prognostic prediction model	Evaluates study risk of bias and applicability	No; PROBAST-AI [3] is in development
CLAIM [2]	Checklist for AI [1] in Medical Imaging	2020	AI [1] studies in medical imaging	Guides authors in presenting (and aids reviewers in evaluating) their research	Yes

[1] AI = artificial intelligence, [2] CLAIM = Checklist for AI in Medical Imaging, [3] PROBAST = Prediction model Risk Of Bias ASsessment Tool, [4] QUADAS-2 = Quality Assessment of Diagnostic Accuracy Studies, [5] RQS = radiomics quality score, and [6] TRIPOD = Transparent Reporting of a multivariable prediction model for Individual Prognosis or Diagnosis.

4.4.2. Reporting Quality and Risk of Bias in Image-Based Glioma Grade Prediction

Assessments of ML-based prediction model studies have demonstrated that risk of bias is high and reporting quality is inadequate. In their systematic review of prediction models developed using supervised ML techniques, Navarro et al. found that the high risk of study bias, as assessed using PROBAST, stems from small study size, poor handling of missing data, and failure to deal with model overfitting [64]. Similar findings have been reported for glioma grade prediction literature. In our prior study conducting a TRIPOD analysis of more than 80 such model development studies, we report a mean adherence rate to TRIPOD of 44%, indicating poor quality of reporting [51]. Areas for improvement included reporting of titles and abstracts, justification of sample size, full model specification and performance, and participant demographics, and missing data. Sohn et al.'s meta-analysis of radiomics studies differentiating high- and low-grade gliomas estimated a high risk of bias according to QUADAS-2, attributing this to the fact that all their analyzed studies were retrospective (and have the potential for bias because patient outcomes are already known), the lack of control over acquisition factors in the studies using public imaging data, and unclear study flow and timing due to poor reporting [41]. Readers should refer directly to Navarro et al., Bahar et al. and Sohn et al. for more detailed discussion of shortcomings in study reporting and risk of bias.

4.4.3. Future of Reporting Guidelines and Risk of Bias Tools for ML Studies

Efforts by authors to refine how they report their studies depend upon existing reporting guidelines. In their systematic review, Yao et al. identified substantial limitations to neuroradiology deep learning reporting standardization and reproducibility [65]. They recommended that future researchers propose a reporting framework specific to deep learning studies. This call for an AI-targeted framework parallels contemporary movements to produce AI extensions of established reporting guidelines. TRIPOD creators have discussed the challenges with ML not captured in the TRIPOD Statement [66]. The introduction of more relevant terminology and movement away from regression-based model approaches will be a part of the forthcoming extension of TRIPOD for studies reporting ML-based diagnostic or prognostic models (TRIPOD-AI) [66,67]. QUADAS-2 creators also announced a plan for an AI-extension (QUADAS-AI), noting that their tool similarly does not accommodate AI-specific terminology and further documenting sources of AI study bias that are not signaled by the tool [68]. PROBAST-AI is in development too [66].

4.4.4. Recommendations

Systematic reviews and meta-analyses in the field [41,51,64] reveal various aspects of reporting and bias risk that need to be addressed in order to promote complete understanding, rigorous assessment, and reproducibility of image-based ML glioma grading studies. Based on the problems identified in this literature (discussed in 4.4.2), we encourage future works to closely adhere to the reporting and risk of bias tools and guidelines most relevant to them, with particular attention to:

- Clearly signifying the development of a prediction model in their titles;
- Increasing the number of participants included in training/testing/validation sets;
- Justifying their choice of sample/sample size (whether that be on practical or logistical grounds) and approach to handling missing data (e.g., imputation);
- Specifying all components of model development (including data pre-processing and model calibration) and a full slate of performance metrics (accuracy, area under the receiver operating characteristic curve (AUC), sensitivity, specificity, positive predictive value, negative predictive value, and F1 score as well as associated confidence intervals) for training/testing/validation. While accuracy is the most comprehensive measure of model performance, AUC is more sensitive to performance differences between classes (e.g., within imbalanced datasets) and should always be reported [69];
- Providing open access to the source code of their algorithms.

For prediction model studies that involve applications of AI to medical imaging, CLAIM is the only framework that is specific to AI and able to capture the nuances of their model reporting—including data preprocessing steps, model layers/connections, software libraries and packages, initialization of model parameters, performance metrics of models on all data partitions, and public access to full study protocols. We, therefore, recommend future studies developing ML models for the prediction of glioma grade from imaging use CLAIM to guide how they present their work. The authors should remain vigilant regarding the release of other AI-specific frameworks that may best suit their studies and seek out AI-specific risk of bias tools to supplement CLAIM once available.

5. Future Directions

ML models present an attractive solution towards overcoming current barriers and accelerating the transition to patient-tailored treatments and precision medicine. Novel algorithms combine information derived from multimodal imaging to molecular markers and clinical information, with the aim of bringing personalized predictions on a patient level into routine clinical care. Relatedly, multi-omic approaches that integrate a variety of advanced techniques such as proteomics, transcriptomics, epigenomics, etc., are increasingly gaining importance in understanding cancer biology and will play a key role in the facilitation of precision medicine [70,71]. The growing presence of ML models in research settings is indisputable, yet several strategies should be considered to facilitate clinical implementation: PACS-based image annotation tools, data-sharing and federated learning, ML fairness, ML transparency, and FDA clearance and real-world use (Figure 3).

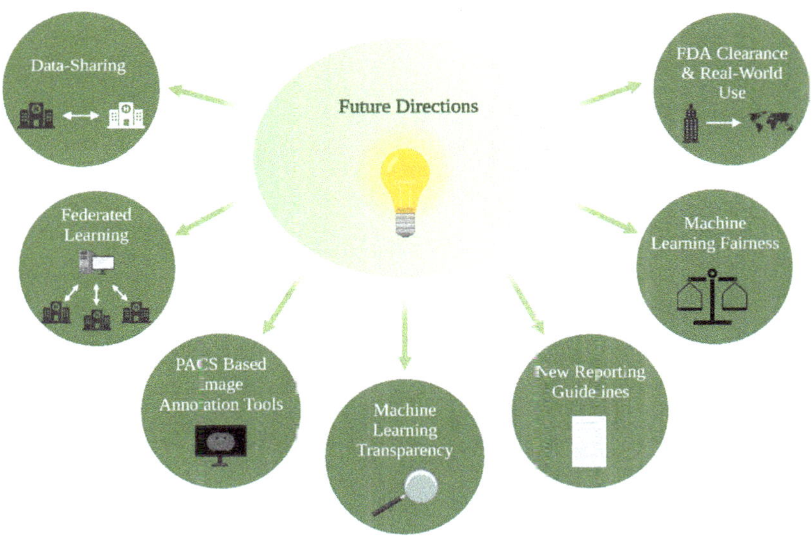

Figure 3. Future directions for clinical implementation of ML glioma grade prediction models, ML = machine learning.

5.1. PACS-Based Image Annotation Tools

Large, annotated datasets that are tailored to the patient populations of individual hospitals and practices are key to training clinically applicable prediction algorithms. An end-to-end solution for generation of these datasets, in which all steps of the ML workflow are performed automatically in clinical picture archiving and communication system (PACS) as the neuroradiologist reads a study, is considered the "holy grail" of AI workflow in radiology [72]. A mechanism for achieving this is through automated/semi-automated segmentation, feature extraction, and prediction algorithms embedded into clinical PACS that provide reports in real-time. The accumulation of saved segmentations through this workflow could accelerate the generation of large, annotated datasets, in addition to providing a decision-support tool for neuroradiologists in daily practice. Under these circumstances, establishing strong academic-industry partnerships for the development of clinically useful image annotation tools is fundamental.

5.2. Data-Sharing and Federated Learning

Multi-institutional academic partnerships are also critical for maximizing clinical applications of ML. Data-sharing efforts are under way in order to accelerate the pace of research [73]. Cross-institutional collaborations not only enrich the quality of the input that goes into training the model, but also provide datasets for externally validating other institutions' models. However, data-sharing across institutions is often hindered by technical, regulatory, and privacy concerns [74]. A promising solution for this is federated learning, an up-and-coming collaborative algorithm training effort that does not require cross-institutional data-sharing. In federated learning, models are trained locally inside an institution's firewalls and learned weights or gradients are transferred from participating institutions for aggregation into a more robust model [75]. This overcomes the barriers of data-sharing and has been shown to be superior to algorithms trained on single-center datasets [76]. Federated learning is not without drawbacks, however; it depends on existing standards for data quality, protocols, and heterogeneity of data distribution. Researchers do not have access to model training data and may face difficulty interpreting unexpected results.

5.3. ML Fairness

A common misconception about AI algorithms is that they are not vulnerable to biases during decision-making. In reality, algorithm unfairness—defined as prejudice or discrimination that skews decisions toward individuals or groups based on their characteristics—has been extensively documented across AI applications. A well-known example is the Correctional Offender Management Profiling for Alternative Sanctions score, which was a tool that assisted judges with their decision to release an offender or keep them in prison. The software was found to be biased towards African Americans, judging them to be at higher risk for recommitting crimes compared to Caucasian individuals [77]. Additional examples of bias have been demonstrated across widely deployed biobanks [78], clinical trial accrual populations [79] and ICU mortality and 30-day psychiatric readmission prediction algorithms [80] among other medical domains. Publicly available tools, including Fairlearn and AI Fairness 360, assess and correct for algorithm unfairness ranging from allocation harms and quality of service harms to feature and racial bias [81,82]. These tools have yet to be applied widely in medical contexts despite their promising utility. Future works on AI in neuro-oncology should consider implementing evidence-based bias detection and mitigation tools tailored to their algorithm development setting and target population prior to clinical integration.

5.4. ML Transparency

The opaqueness of ML models—DL in particular—poses a barrier to their acceptance and usage. In addition, traditional measures such as software validation are insufficient for fulfilling legal, compliance, and/or other requirements for ML tool clarification [83,84]. Explainable artificial intelligence (xAI) approaches may address these concerns by explaining particular prediction outputs and overall model behavior in human-understandable terms [85]. A recent study demonstrates the successful use of state-of-the-art xAI libraries incorporating visual analytics for glioma classification [83]. Other approaches such as Grad-CAM generate visual explanations of DL model decisions and, therefore, enhance algorithm transparency [86]. These tools can support the interpretability of ML model outputs for future research as well as prime ML for dissemination and acceptance in clinical neuroradiology. Guidelines for authors, along with reporting quality assessment and risk of bias tools, should consider encouraging such approaches to further the transparency of literature in the field.

Of relevance to ML model transparency are the concepts of usability and causability. Usability can be defined as the ease of use of a computer system for users, or in other words, the extent to which a user and a system may communicate through an interface without misunderstanding [87,88]. Highly usable tools are associated with positive user satisfaction and performance in the field of human–computer interaction [89]. Causability is a parallel concept to usability and foundational for human–AI interaction. Causability reflects the understandability of an AI model (e.g., CNN) to a human as communicated by an explanation interface [89]. Causability, furthermore, determines relative importance and justifies what should be explained and how [90]. Embracing causability in the development of human–AI interfaces will help people understand the decision-making process of ML algorithms and improve trust. We believe this will lower the threshold for clinical ML utilization.

5.5. FDA Clearance and Real-World Use

Thousands of studies pertaining to applications of AI and ML in medical imaging have been published [15,82]. Yet, few imaging AI/ML algorithms have been cleared by the FDA as medical products [91], perhaps due in part to the lack of standardization and transparency in the FDA clearance process [92]. Bridging the gap between AI/ML research and FDA clearance—as well as FDA clearance and real-world algorithm use—will streamline the adoption of ML models for glioma grading into clinical settings. To this end, Lin presents several suggestions [93]. Partnering of the FDA with professional societies could facilitate the standardization of algorithm development and evaluation. A key focus would be resolving the split between how results are communicated in the literature (e.g., performance metrics) and what is relevant for AI product assessment (e.g., return on investment, integration and flexibility with PACS, ease of use, etc.). Moreover, reporting of post-marketing surveillance could help real-world use and algorithm performance drift.

6. Conclusions

ML glioma grade prediction tools are increasingly prevalent in research but have yet to be incorporated clinically. The reporting quality of ML glioma grade prediction studies is low, limiting model reproducibility and thus preventing reliable clinical translation. However, current efforts to create ML-specific reporting guidelines and risk of bias tools may help address these issues. Future directions for supporting clinical implementation of ML prediction models include data-sharing, federated learning, and development of PACS-based image annotation tools for the generation of large image databases, among other opportunities.

Author Contributions: Conceptualization, S.M., R.C.B. and M.S.A.; methodology, S.M., R.C.B. and M.S.A.; investigation, J.T.M., S.C. and M.S.A.; resources, M.S.A.; writing—original draft preparation, S.M., R.C.B., T.Z. and M.S.A.; writing—review and editing, M.L., I.I., K.B., G.I.C.P., L.S., S.P., J.T.M. and S.C.; visualization, T.Z., I.I. and S.C.; supervision, G.I.C.P., S.C. and M.S.A.; project administration, M.S.A.; funding acquisition, M.S.A. All authors have read and agreed to the published version of the manuscript.

Funding: Sara Merkaj receives funding in part from the Biomedical Education Program (BMEP). Ryan Bahar receives funding in part from the National Institute of Diabetes and Digestive and Kidney Disease of the National Institutes of Health under Award Number T35DK104689. The content is solely the responsibility of the authors and does not necessarily represent the official views of the National Institutes of Health. Mariam Aboian received funding from American Society of Neuroradiology Fellow Award 2018. This publication was made possible by KL2 TR001862 (Mariam Aboian) from the National Center for Advancing Translational Science (NCATS), components of the National Institutes of Health (NIH), and the NIH roadmap for Medical Research. Its contents are solely the responsibility of the authors and do not necessarily represent the official view of NIH. MingDe Lin is an employee and stockholder of Visage Imaging, Inc. and unrelated to this work, receives funding from NIH/NCI

R01 CA206180 and is a board member of Tau Beta Pi engineering honor society. Khaled Bousabarah is an employee of Visage Imaging, GmbH. Seyedmehdi Payabvash has grant support from NIH/NINDS K23NS118056, foundation of the American Society of Neuroradiology (ASNR) #1861150721, the Doris Duke Charitable Foundation (DDCF) #2020097, and NVIDIA.

Institutional Review Board Statement: Not applicable.

Informed Consent Statement: Not applicable.

Data Availability Statement: Not applicable.

Acknowledgments: We thank Yale Department of Radiology for their research resources and space.

Conflicts of Interest: The authors declare no conflict of interest. The funders had no role in the design of the study; in the collection, analyses, or interpretation of data; in the writing of the manuscript; or in the decision to publish the results.

References

1. Yeo, M.; Kok, H.K.; Kutaiba, N.; Maingard, J.; Thijs, V.; Tahayori, B.; Russell, J.; Jhamb, A.; Chandra, R.V.; Brooks, M.; et al. Artificial intelligence in clinical decision support and outcome prediction–Applications in stroke. *J. Med. Imaging Radiat. Oncol.* **2021**, *65*, 518–528. [CrossRef] [PubMed]
2. Kaka, H.; Zhang, E.; Khan, N. Artificial Intelligence and Deep Learning in Neuroradiology: Exploring the New Frontier. *Can. Assoc. Radiol. J.* **2021**, *72*, 35–44. [CrossRef] [PubMed]
3. Sidey-Gibbons, J.A.M.; Sidey-Gibbons, C.J. Machine learning in medicine: A practical introduction. *BMC Med. Res. Methodol.* **2019**, *19*, 64. [CrossRef] [PubMed]
4. Gillies, R.J.; Kinahan, P.E.; Hricak, H. Radiomics: Images Are More than Pictures, They Are Data. *Radiology* **2016**, *278*, 563–577. [CrossRef] [PubMed]
5. Aerts, H.J.W.L.; Velazquez, E.R.; Leijenaar, R.T.H.; Parmar, C.; Grossmann, P.; Carvalho, S.; Bussink, J.; Monshouwer, R.; Haibe-Kains, B.; Rietveld, D.; et al. Decoding tumour phenotype by noninvasive imaging using a quantitative radiomics approach. *Nat. Commun.* **2014**, *5*, 4006. [CrossRef] [PubMed]
6. Giger, M.L. Machine Learning in Medical Imaging. *J. Am. Coll. Radiol.* **2018**, *15*, 512–520. [CrossRef]
7. Ostrom, Q.T.; Cioffi, G.; Gittleman, H.; Patil, N.; Waite, K.; Kruchko, C.; Barnholtz-Sloan, J.S. CBTRUS Statistical Report: Primary Brain and Other Central Nervous System Tumors Diagnosed in the United States in 2012–2016. *Neuro-Oncology* **2019**, *21* Suppl. S5, v1–v100. [CrossRef]
8. Abdel Razek, A.A.K.; Alksas, A.; Shehata, M.; AbdelKhalek, A.; Abdel Baky, K.; El-Baz, A.; Helmy, E. Clinical applications of artificial intelligence and radiomics in neuro-oncology imaging. *Insights Imaging* **2021**, *12*, 152. [CrossRef]
9. Thon, N.; Tonn, J.-C.; Kreth, F.-W. The surgical perspective in precision treatment of diffuse gliomas. *OncoTargets Ther.* **2019**, *12*, 1497–1508. [CrossRef]
10. Hu, L.S.; Hawkins-Daarud, A.; Wang, L.; Li, J.; Swanson, K.R. Imaging of intratumoral heterogeneity in high-grade glioma. *Cancer Lett.* **2020**, *477*, 97–106. [CrossRef]
11. Seow, P.; Wong, J.H.D.; Annuar, A.A.; Mahajan, A.; Abdullah, N.A.; Ramli, N. Quantitative magnetic resonance imaging and radiogenomic biomarkers for glioma characterisation: A systematic review. *Br. J. Radiol.* **2018**, *91*, 20170930. [CrossRef] [PubMed]
12. Lambin, P.; Rios-Velazquez, E.; Leijenaar, R.; Carvalho, S.; van Stiphout, R.G.P.M.; Granton, P.; Zegers, C.M.L.; Gillies, R.; Boellard, R.; Dekker, A.; et al. Radiomics: Extracting more information from medical images using advanced feature analysis. *Eur. J. Cancer* **2012**, *48*, 441–446. [CrossRef] [PubMed]
13. Buchlak, Q.D.; Esmaili, N.; Leveque, J.-C.; Bennett, C.; Farrokhi, F.; Piccardi, M. Machine learning applications to neuroimaging for glioma detection and classification: An artificial intelligence augmented systematic review. *J. Clin. Neurosci.* **2021**, *89*, 177–198. [CrossRef] [PubMed]
14. Chow, D.; Chang, P.; Weinberg, B.D.; Bota, D.A.; Grinband, J.; Filippi, C.G. Imaging Genetic Heterogeneity in Glioblastoma and Other Glial Tumors: Review of Current Methods and Future Directions. *Am. J. Roentgenol.* **2018**, *210*, 30–38. [CrossRef] [PubMed]
15. Pesapane, F.; Codari, M.; Sardanelli, F. Artificial intelligence in medical imaging: Threat or opportunity? Radiologists again at the forefront of innovation in medicine. *Eur. Radiol. Exp.* **2018**, *2*, 35. [CrossRef] [PubMed]
16. Pemberton, H.G.; Zaki, L.A.M.; Goodkin, O.; Das, R.K.; Steketee, R.M.E.; Barkhof, F.; Vernooij, M.W. Technical and clinical validation of commercial automated volumetric MRI tools for dementia diagnosis-a systematic review. *Neuroradiology* **2021**, *63*, 1773–1789. [CrossRef] [PubMed]
17. Richens, J.G.; Lee, C.M.; Johri, S. Improving the accuracy of medical diagnosis with causal machine learning. *Nat. Commun.* **2020**, *11*, 3923. [CrossRef]
18. Rubin, D.L. Artificial Intelligence in Imaging: The Radiologist's Role. *J. Am. Coll. Radiol.* **2019**, *16*, 1309–1317. [CrossRef]
19. Wu, J.T.; Wong, K.C.L.; Gur, Y.; Ansari, N.; Karargyris, A.; Sharma, A.; Morris, M.; Saboury, B.; Ahmad, H.; Boyko, O.; et al. Comparison of Chest Radiograph Interpretations by Artificial Intelligence Algorithm vs Radiology Residents. *JAMA Netw. Open* **2020**, *3*, e2022779. [CrossRef]

20. Cassinelli Petersen, G.I.; Shatalov, J.; Verma, T.; Brim, W.R.; Subramanian, H.; Brackett, A.; Bahar, R.C.; Merkaj, S.; Zeevi, T.; Staib, L.H.; et al. Machine Learning in Differentiating Gliomas from Primary CNS Lymphomas: A Systematic Review, Reporting Quality, and Risk of Bias Assessment. *AJNR Am. J. Neuroradiol.* **2022**, *43*, 526–533. [CrossRef]
21. Rauschecker, A.M.; Rudie, J.D.; Xie, L.; Wang, J.; Duong, M.T.; Botzolakis, E.J.; Kovalovich, A.M.; Egan, J.; Cook, T.C.; Bryan, R.N.; et al. Artificial Intelligence System Approaching Neuroradiologist-level Differential Diagnosis Accuracy at Brain MRI. *Radiology* **2020**, *295*, 626–637. [CrossRef] [PubMed]
22. Decuyper, M.; Bonte, S.; Van Holen, R. Binary Glioma Grading: Radiomics versus Pre-trained CNN Features. *Med. Image Comput. Comput. Assist. Interv.* **2018**, *11072*, 498–505. [CrossRef]
23. Gao, M.; Huang, S.; Pan, X.; Liao, X.; Yang, R.; Liu, J. Machine Learning-Based Radiomics Predicting Tumor Grades and Expression of Multiple Pathologic Biomarkers in Gliomas. *Front. Oncol.* **2020**, *10*, 1676. [CrossRef] [PubMed]
24. Haubold, J.; Demircioglu, A.; Gratz, M.; Glas, M.; Wrede, K.; Sure, U.; Antoch, G.; Keyvani, K.; Nittka, M.; Kannengiesser, S.; et al. Non-invasive tumor decoding and phenotyping of cerebral gliomas utilizing multiparametric 18F-FET PET-MRI and MR Fingerprinting. *Eur. J. Pediatr.* **2020**, *47*, 1435–1445. [CrossRef]
25. Sengupta, A.; Ramaniharan, A.K.; Gupta, R.K.; Agarwal, S.; Singh, A. Glioma grading using a machine-learning framework based on optimized features obtained from T $_1$ perfusion MRI and volumes of tumor components. *J. Magn. Reson. Imaging* **2019**, *50*, 1295–1306. [CrossRef]
26. Tian, Q.; Yan, L.-F.; Zhang, X.; Hu, Y.-C.; Han, Y.; Liu, Z.-C.; Nan, H.-Y.; Sun, Q.; Sun, Y.-Z.; Yang, Y.; et al. Radiomics strategy for glioma grading using texture features from multiparametric MRI. *J. Magn. Reson. Imaging* **2018**, *48*, 1518–1528. [CrossRef]
27. Abdolmaleki, P.; Mihara, F.; Masuda, K.; Buadu, L.D. Neural networks analysis of astrocytic gliomas from MRI appearances. *Cancer Lett.* **1997**, *118*, 69–78. [CrossRef]
28. Christy, P.S.; Tervonen, O.; Scheithauer, B.W.; Forbes, G.S. Use of a Neural-Network and a Multiple-Regression Model to Predict Histologic Grade of Astrocytoma from Mri Appearances. *Neuroradiology* **1995**, *37*, 89–93. [CrossRef]
29. Dandil, E.; Biçer, A. Automatic grading of brain tumours using LSTM neural networks on magnetic resonance spectroscopy signals. *IET Image Process* **2020**, *14*, 1967–1979. [CrossRef]
30. Ji, B.; Wang, S.; Liu, Z.; Weinberg, B.D.; Yang, X.; Liu, T.; Wang, L.; Mao, H. Revealing hemodynamic heterogeneity of gliomas based on signal profile features of dynamic susceptibility contrast-enhanced MRI. *NeuroImage Clin.* **2019**, *23*, 101864. [CrossRef]
31. Tabatabaei, M.; Razaei, A.; Sarrami, A.H.; Saadatpour, Z.; Singhal, A.; Sotoudeh, H. Current Status and Quality of Machine Learning-Based Radiomics Studies for Glioma Grading: A Systematic Review. *Oncology* **2021**, *99*, 433–443. [CrossRef] [PubMed]
32. Kotsiantis, S.B.; Kanellopoulos, D.; Pintelas, P.E. Data preprocessing for supervised leaning. *Int. J. Comput. Sci.* **2006**, *1*, 111–117.
33. Tillmanns, N.; Lum, A.; Brim, W.R.; Subramanian, H.; Lin, M.; Bousabarah, K.; Malhotra, A.; Cui, J.; Brackett, A.; Payabvash, S.; et al. NIMG-71. Identifying clinically applicable machine learning algorithms for glioma segmentation using a systematic literature review. *Neuro-Oncology* **2021**, *23*, vi145. [CrossRef]
34. Shaver, M.M.; Kohanteb, P.A.; Chiou, C.; Bardis, M.D.; Chantaduly, C.; Bota, D.; Filippi, C.G.; Weinberg, B.; Grinband, J.; Chow, D.S.; et al. Optimizing Neuro-Oncology Imaging: A Review of Deep Learning Approaches for Glioma Imaging. *Cancers* **2019**, *11*, 829. [CrossRef]
35. Kumar, V.; Gu, Y.; Basu, S.; Berglund, A.; Eschrich, S.A.; Schabath, M.B.; Forster, K.; Aerts, H.J.W.L.; Dekker, A.; Fenstermacher, D.; et al. Radiomics: The process and the challenges. *Magn. Reson. Imaging* **2012**, *30*, 1234–1248. [CrossRef] [PubMed]
36. Van Griethuysen, J.J.; Fedorov, A.; Parmar, C.; Hosny, A.; Aucoin, N.; Narayan, V.; Aerts, H.J. Computational Radiomics System to Decode the Radiographic Phenotype. *Cancer Res* **2017**, *77*, e104–e107. [CrossRef]
37. Omuro, A.; DeAngelis, L.M. Glioblastoma and other malignant gliomas: A clinical review. *JAMA* **2013**, *310*, 1842–1850. [CrossRef]
38. Rios Velazquez, E.; Meier, R.; Dunn, W.D., Jr.; Alexander, B.; Wiest, R.; Bauer, S.; Aerts, H.J. Fully automatic GBM segmentation in the TCGA-GBM dataset: Prognosis and correlation with VASARI features. *Sci. Rep.* **2015**, *5*, 16822. [CrossRef]
39. Chandrashekar, G.; Sahin, F. A survey on feature selection methods. *Comput. Electr. Eng.* **2014**, *40*, 16–28. [CrossRef]
40. Brereton, R.G.; Lloyd, G.R. Support Vector Machines for classification and regression. *Analyst* **2010**, *135*, 230–267. [CrossRef]
41. Sohn, C.K.; Bisdas, S. Diagnostic Accuracy of Machine Learning-Based Radiomics in Grading Gliomas: Systematic Review and Meta-Analysis. *Contrast Media Mol. Imaging* **2020**, *2020*, 2127062. [CrossRef] [PubMed]
42. Gordillo, N.; Montseny, E.; Sobrevilla, P. State of the art survey on MRI brain tumor segmentation. *Magn. Reson. Imaging* **2013**, *31*, 1426–1438. [CrossRef] [PubMed]
43. LeCun, Y.; Bengio, Y.; Hinton, G. Deep learning. *Nature* **2015**, *521*, 436–444. [CrossRef] [PubMed]
44. Chartrand, G.; Cheng, P.M.; Vorontsov, E.; Drozdzal, M.; Turcotte, S.; Pal, C.J.; Kadoury, S.; Tang, A. Deep Learning: A Primer for Radiologists. *RadioGraphics* **2017**, *37*, 2113–2131. [CrossRef]
45. Ge, C.; Gu, I.Y.-H.; Jakola, A.S.; Yang, J. Deep Learning and Multi-Sensor Fusion for Glioma Classification Using Multistream 2D Convolutional Networks. In Proceedings of the 2018 40th Annual International Conference of the IEEE Engineering in Medicine and Biology Society (EMBC), Honolulu, HI, USA, 17–22 July 2018; pp. 5894–5897. [CrossRef]
46. Kabir-Anaraki, A.; Ayati, M.; Kazemi, F. Magnetic resonance imaging-based brain tumor grades classification and grading via convolutional neural networks and genetic algorithms. *Biocybern. Biomed. Eng.* **2019**, *39*, 63–74. [CrossRef]
47. Ge, C.; Gu, I.Y.-H.; Jakola, A.S.; Yang, J. Deep semi-supervised learning for brain tumor classification. *BMC Med. Imaging* **2020**, *20*, 1–11. [CrossRef]

48. Sharif, M.I.; Li, J.P.; Khan, M.A.; Saleem, M.A. Active deep neural network features selection for segmentation and recognition of brain tumors using MRI images. *Pattern Recognit. Lett.* **2020**, *129*, 181–189. [CrossRef]
49. Hayashi, Y. Toward the transparency of deep learning in radiological imaging: Beyond quantitative to qualitative artificial intelligence. *J. Med. Artif. Intell.* **2019**, *2*, 19. [CrossRef]
50. Topol, E.J. High-performance medicine: The convergence of human and artificial intelligence. *Nat. Med.* **2019**, *25*, 44–56. [CrossRef]
51. Bahar, R.C.; Merkaj, S.; Cassinelli Petersen, G.I.; Tillmanns, N.; Subramanian, H.; Brim, W.R.; Zeevi, T.; Staib, L.; Kazarian, E.; Lin, M.; et al. NIMG-35. Machine Learning Models for Classifying High- and Low-Grade Gliomas: A Systematic Review and Quality of Reporting Analysis. *Front. Oncol.* **2022**, *12*, 856231. [CrossRef] [PubMed]
52. Menze, B.H.; Jakab, A.; Bauer, S.; Kalpathy-Cramer, J.; Farahani, K.; Kirby, J.; Burren, Y.; Porz, N.; Slotboom, J.; Wiest, R.; et al. The Multimodal Brain Tumor Image Segmentation Benchmark (BRATS). *IEEE Trans. Med. Imaging* **2015**, *34*, 1993–2024. [CrossRef] [PubMed]
53. Clark, K.; Vendt, B.; Smith, K.; Freymann, J.; Kirby, J.; Koppel, P.; Prior, F. The Cancer Imaging Archive (TCIA): Maintaining and operating a public information repository. *J. Digit. Imaging* **2013**, *26*, 1045–1057. [CrossRef] [PubMed]
54. Madhavan, S.; Zenklusen, J.-C.; Kotliarov, Y.; Sahni, H.; Fine, H.A.; Buetow, K. Rembrandt: Helping Personalized Medicine Become a Reality through Integrative Translational Research. *Mol. Cancer Res.* **2009**, *7*, 157–167. [CrossRef] [PubMed]
55. Rajan, P.V.; Karnuta, J.M.; Haeberle, H.S.; Spitzer, A.I.; Schaffer, J.L.; Ramkumar, P.N. Response to letter to the editor on "Significance of external validation in clinical machine learning: Let loose too early?". *Spine J.* **2020**, *20*, 1161–1162. [CrossRef]
56. Wesseling, P.; Capper, D. WHO 2016 Classification of gliomas. *Neuropathol. Appl. Neurobiol.* **2018**, *44*, 139–150. [CrossRef]
57. Brat, D.J.; Aldape, K.; Colman, H.; Figrarella-Branger, D.; Fuller, G.N.; Giannini, C.; Weller, M. cIMPACT-NOW update 5: Recommended grading criteria and terminologies for IDH-mutant astrocytomas. *Acta Neuropathol.* **2020**, *139*, 603–608. [CrossRef]
58. Collins, G.S.; Reitsma, J.B.; Altman, D.G.; Moons, K.G. Transparent Reporting of a multivariable prediction model for Individual Prognosis Or Diagnosis (TRIPOD): The TRIPOD Statement. *Br. J. Surg.* **2015**, *102*, 148–158. [CrossRef]
59. Lambin, P.; Leijenaar, R.T.H.; Deist, T.M.; Peerlings, J.; de Jong, E.E.C.; van Timmeren, J.; Sanduleanu, S.; Larue, R.T.H.M.; Even, A.J.G.; Jochems, A.; et al. Radiomics: The bridge between medical imaging and personalized medicine. *Nat. Rev. Clin. Oncol.* **2017**, *14*, 749–762. [CrossRef]
60. Park, J.E.; Kim, H.S.; Kim, D.; Park, S.Y.; Kim, J.Y.; Cho, S.J.; Kim, J.H. A systematic review reporting quality of radiomics research in neuro-oncology: Toward clinical utility and quality improvement using high-dimensional imaging features. *BMC Cancer* **2020**, *20*, 29. [CrossRef]
61. Mongan, J.; Moy, L.; Kahn, C.E. Checklist for Artificial Intelligence in Medical Imaging (CLAIM): A Guide for Authors and Reviewers. *Radiol. Artif. Intell.* **2020**, *2*, e200029. [CrossRef]
62. Whiting, P.F.; Rutjes, A.W.S.; Westwood, M.E.; Mallett, S.; Deeks, J.J.; Reitsma, J.B.; Leeflang, M.M.; Sterne, J.A.; Bossuyt, P.M.; QUADAS-2 Group. QUADAS-2: A Revised Tool for the Quality Assessment of Diagnostic Accuracy Studies. *Ann. Intern. Med.* **2011**, *155*, 529–536. [CrossRef] [PubMed]
63. Wolff, R.F.; Moons, K.G.; Riley, R.D.; Whiting, P.F.; Westwood, M.; Collins, G.S. PROBAST: A Tool to Assess the Risk of Bias and Applicability of Prediction Model Studies. *Ann. Intern. Med.* **2019**, *170*, 51–58. [CrossRef] [PubMed]
64. Andaur Navarro, C.L.; Damen, J.A.A.; Takada, T.; Nijman, S.W.J.; Dhiman, P.; Ma, J.; Collins, G.S.; Bajpai, R.; Riley, R.D.; Moons, K.G.M.; et al. Risk of bias in studies on prediction models developed using supervised machine learning techniques: Systematic review. *BMJ* **2021**, *20*, n2281.
65. Yao, A.D.; Cheng, D.L.; Pan, I.; Kitamura, F. Deep Learning in Neuroradiology: A Systematic Review of Current Algorithms and Approaches for the New Wave of Imaging Technology. *Radiol. Artif. Intell.* **2020**, *2*, e190026. [CrossRef]
66. Collins, G.S.; Dhiman, P.; Navarro, C.L.A.; Ma, J.; Hooft, L.; Reitsma, J.B.; Logullo, P.; Beam, A.L.; Peng, L.; Van Calster, B.; et al. Protocol for development of a reporting guideline (TRIPOD-AI) and risk of bias tool (PROBAST-AI) for diagnostic and prognostic prediction model studies based on artificial intelligence. *BMJ Open* **2021**, *11*, e048008. [CrossRef]
67. Collins, G.S.; Moons, K.G.M. Reporting of artificial intelligence prediction models. *Lancet* **2019**, *393*, 1577–1579. [CrossRef]
68. Sounderajah, V.; Ashrafian, H.; Rose, S.; Shah, N.H.; Ghassemi, M.; Golub, R.; Kahn, C.E.; Esteva, A.; Karthikesalingam, A.; Mateen, B.; et al. A quality assessment tool for artificial intelligence-centered diagnostic test accuracy studies: QUADAS-AI. *Nat. Med.* **2021**, *27*, 1663–1665. [CrossRef]
69. Ling, C.X.; Huang, J.; Zhang, H. AUC: A Better Measure than Accuracy in Comparing Learning Algorithms. *Lect. Notes Artif. Int.* **2003**, *2671*, 329–341. [CrossRef]
70. Olivier, M.; Asmis, R.; Hawkins, G.A.; Howard, T.D.; Cox, L.A. The Need for Multi-Omics Biomarker Signatures in Precision Medicine. *Int. J. Mol. Sci.* **2019**, *20*, 4781. [CrossRef]
71. Subramanian, I.; Verma, S.; Kumar, S.; Jere, A.; Anamika, K. Multi-omics Data Integration, Interpretation, and Its Application. *Bioinform. Biol. Insights* **2020**, *14*, 1177932219899051. [CrossRef]
72. Zaharchuk, G.; Gong, E.; Wintermark, M.; Rubin, D.; Langlotz, C. Deep Learning in Neuroradiology. *Am. J. Neuroradiol.* **2018**, *39*, 1776–1784. [CrossRef] [PubMed]
73. Warren, E. Strengthening Research through Data Sharing. *N. Engl. J. Med.* **2016**, *375*, 401–403. [CrossRef] [PubMed]
74. He, J.; Baxter, S.L.; Xu, J.; Xu, J.; Zhou, X.; Zhang, K. The practical implementation of artificial intelligence technologies in medicine. *Nat. Med.* **2019**, *25*, 30–36. [CrossRef] [PubMed]

75. Rieke, N.; Hancox, J.; Li, W.; Milletarì, F.; Roth, H.R.; Albarqouni, S.; Bakas, S.; Galtier, M.N.; Landman, B.A.; Maier-Hein, K.; et al. The future of digital health with federated learning. *NPJ Digit. Med.* **2020**, *3*, 119. [CrossRef]
76. Sheller, M.J.; Reina, G.A.; Edwards, B.; Martin, J.; Bakas, S. Multi-Institutional Deep Learning Modeling Without Sharing Patient Data: A Feasibility Study on Brain Tumor Segmentation. *Brainlesion* **2019**, *11383*, 92–104.
77. Mehrabi, N.; Morstatter, F.; Saxena, N.; Lerman, K.; Galstyan, A. A Survey on Bias and Fairness in Machine Learning. *ACM Comput. Surv.* **2021**, *54*, 115. [CrossRef]
78. Fry, A.; Littlejohns, T.J.; Sudlow, C.; Doherty, N.; Adamska, L.; Sprosen, T.; Collins, R.; Allen, N.E. Comparison of Sociodemographic and Health-Related Characteristics of UK Biobank Participants with Those of the General Population. *Am. J. Epidemiol.* **2017**, *186*, 1026–1034. [CrossRef]
79. Vickers, S.M.; Fouad, M.N. An overview of EMPaCT and fundamental issues affecting minority participation in cancer clinical trials: Enhancing minority participation in clinical trials (EMPaCT): Laying the groundwork for improving minority clinical trial accrual. *Cancer* **2014**, *120* Suppl. S7, 1087–1090. [CrossRef]
80. Chen, I.Y.; Szolovits, P.; Ghassemi, M. Can AI Help Reduce Disparities in General Medical and Mental Health Care? *AMA J. Ethics* **2019**, *21*, E167–E179.
81. Bird, S.; Dudík, M.; Edgar, R.; Horn, B.; Lutz, R.; Milan, V.; Sameki, M.; Wallach, H.; Walker, K. Fairlearn: A toolkit for assessing and improving fairness in ai. *Tech. Rep.* 2020, pp. 1–7. Available online: https://www.microsoft.com/en-us/research/publication/fairlearn-a-toolkit-for-assessing-and-improving-fairness-in-ai/ (accessed on 24 May 2022).
82. Bellamy, R.K.E.; Dey, K.; Hind, M.; Hoffman, S.C.; Houde, S.; Kannan, K.; Lohia, P.; Martino, J.; Mehta, S.; Mojsilovic, A.; et al. AI Fairness 360: An extensible toolkit for detecting and mitigating algorithmic bias. *IBM J. Res. Dev.* **2019**, *63*, 4:1–4:15. [CrossRef]
83. Gashi, M.; Vuković, M.; Jekic, N.; Thalmann, S.; Holzinger, A.; Jean-Quartier, C.; Jeanquartier, F. State-of-the-Art Explainability Methods with Focus on Visual Analytics Showcased by Glioma Classification. *BioMedInformatics* **2022**, *2*, 139–158. [CrossRef]
84. Königstorfer, F.; Thalmann, S. Software documentation is not enough! Requirements for the documentation of AI. *Digit. Policy Regul. Gov.* **2021**, *23*, 475–488. [CrossRef]
85. Castelvecchi, D. Can we open the black box of AI? *Nat. News* **2016**, *538*, 20. [CrossRef] [PubMed]
86. Selvaraju, R.R.; Cogswell, M.; Das, A.; Vedantam, R.; Parikh, D.; Batra, D. Grad-cam: Visual explanations from deep networks via gradient-based localization. In Proceedings of the IEEE International Conference on Computer Vision, Venice, Italy, 22–29 October 2017; pp. 618–626.
87. Holzinger, A. Usability engineering methods for software developers. *Commun. ACM* **2005**, *48*, 71–74. [CrossRef]
88. Chou, J.R.; Hsiao, S.W. A usability study of human-computer interface for middle-aged learners. *Comput. Hum. Behav.* **2007**, *23*, 2040–2063. [CrossRef]
89. Holzinger, A.; Langs, G.; Denk, H.; Zatloukal, K.; Müller, H. Causability and explainability of artificial intelligence in medicine. *WIREs Data Min. Knowl. Discov.* **2019**, *9*, e1312. [CrossRef]
90. Shin, D. The effects of explainability and causability on perception, trust, and acceptance: Implications for explainable AI. *Int. J. Hum. Comput. Stud.* **2021**, *146*, 102551. [CrossRef]
91. Institute ACoRDS. FDA Cleared AI Algorithms. Available online: https://models.acrdsi.org/ (accessed on 3 December 2021).
92. Ebrahimian, S.; Kalra, M.K.; Agarwal, S.; Bizzo, B.C.; Elkholy, M.; Wald, C.; Allen, B.; Dreyer, K.J. FDA-regulated AI Algorithms: Trends, Strengths, and Gaps of Validation Studies. *Acad. Radiol.* **2021**, *29*, 559–566. [CrossRef]
93. Lin, M. What's Needed to Bridge the Gap between US FDA Clearance and Real-world Use of AI Algorithms. *Acad. Radiol.* **2021**, *29*, 567–568. [CrossRef]

cancers MDPI

Article

Machine Learning Based on MRI DWI Radiomics Features for Prognostic Prediction in Nasopharyngeal Carcinoma

Qiyi Hu [1,†], Guojie Wang [2,†], Xiaoyi Song [3], Jingjing Wan [1], Man Li [3], Fan Zhang [4], Qingling Chen [1], Xiaoling Cao [1], Shaolin Li [2,*] and Ying Wang [1,*]

1 Department of Nuclear Medicine, The Fifth Affiliated Hospital of Sun Yat-sen University, Zhuhai 519099, China; huqy28@mail2.sysu.edu.cn (Q.H.); polaris0817@163.com (J.W.); chenqling23@mail2.sysu.edu.cn (Q.C.); caoxling@mail2.sysu.edu.cn (X.C.)

2 Department of Radiology, The Fifth Affiliated Hospital of Sun Yat-sen University, Zhuhai 519099, China; wanggj5@mail.sysu.edu.cn

3 Guangdong Provincial Key Laboratory of Biomedical Imaging, The Fifth Affiliated Hospital of Sun Yat-sen University, Zhuhai 519099, China; songxy29@mail2.sysu.edu.cn (X.S.); liman26@mail.sysu.edu.cn (M.L.)

4 Department of Head and Neck Oncology, The Cancer Center of the Fifth Affiliated Hospital of Sun Yat-sen University, Zhuhai 519099, China; zhangfan26@mail.sysu.edu.cn

* Correspondence: lishlin5@mail.sysu.edu.cn (S.L.); wangy9@mail.sysu.edu.cn (Y.W.)

† These authors contributed equally to this work.

Simple Summary: In the past, radiomics studies of nasopharyngeal carcinoma (NPC) were only based on basic MR sequences. Previous studies have shown that radiomics methods based on T2-weighted imaging and contrast-enhanced T1-weighted imaging have been successfully used to improve the prognosis of patients with nasopharyngeal carcinoma. The purpose of this study was to explore the predictive efficacy of radiomics analyses based on readout-segmented echo-planar diffusion-weighted imaging (RESOLVE-DWI) which quantitatively reflects the diffusion motion of water molecules for prognosis evaluation in nasopharyngeal carcinoma. Several prognostic radiomics models were established by using diffusion-weighted imaging, apparent diffusion coefficient maps, T2-weighted and contrast-enhanced T1-weighted imaging to predict the risk of recurrence or metastasis of nasopharyngeal carcinoma, and the predictive effects of different models were compared. The results show that the model based on MRI DWI can successfully predict the prognosis of patients with nasopharyngeal carcinoma and has higher predictive efficiency than the model based on the conventional sequence, which suggests MRI DWI-radiomics can provide a useful and alternative approach for survival estimation.

Abstract: Purpose: This study aimed to explore the predictive efficacy of radiomics analyses based on readout-segmented echo-planar diffusion-weighted imaging (RESOLVE-DWI) for prognosis evaluation in nasopharyngeal carcinoma in order to provide further information for clinical decision making and intervention. Methods: A total of 154 patients with untreated NPC confirmed by pathological examination were enrolled, and the pretreatment magnetic resonance image (MRI)—including diffusion-weighted imaging (DWI), apparent diffusion coefficient (ADC) maps, T2-weighted imaging (T2WI), and contrast-enhanced T1-weighted imaging (CE-T1WI)—was collected. The Random Forest (RF) algorithm selected radiomics features and established the machine-learning models. Five models, namely model 1 (DWI + ADC), model 2 (T2WI + CE-T1WI), model 3 (DWI + ADC + T2WI), model 4 (DWI + ADC + CE-T1WI), and model 5 (DWI + ADC + T2WI + CE-T1WI), were constructed. The average area under the curve (AUC) of the validation set was determined in order to compare the predictive efficacy for prognosis evaluation. Results: After adjusting the parameters, the RF machine learning models based on extracted imaging features from different sequence combinations were obtained. The invalidation sets of model 1 (DWI + ADC) yielded the highest average AUC of 0.80 (95% CI: 0.79–0.81). The average AUCs of the model 2, 3, 4, and 5 invalidation sets were 0.72 (95% CI: 0.71–0.74), 0.66 (95% CI: 0.64–0.68), 0.74 (95% CI: 0.73–0.75), and 0.75 (95% CI: 0.74–0.76), respectively. Conclusion: A radiomics model derived from the MRI DWI of patients with nasopharyngeal carcinoma was generated in order to evaluate the risk of recurrence and metastasis. The model

Citation: Hu, Q.; Wang, G.; Song, X.; Wan, J.; Li, M.; Zhang, F.; Chen, Q.; Cao, X.; Li, S.; Wang, Y. Machine Learning Based on MRI DWI Radiomics Features for Prognostic Prediction in Nasopharyngeal Carcinoma. *Cancers* **2022**, *14*, 3201. https://doi.org/10.3390/cancers14133201

Academic Editors: Hamid Khayyam, Ali Madani, Rahele Kafieh and Ali Hekmatnia

Received: 23 May 2022
Accepted: 23 June 2022
Published: 30 June 2022

based on MRI DWI can provide an alternative approach for survival estimation, and can reveal more information for clinical decision-making and intervention.

Keywords: radiomics; nasopharyngeal carcinoma; diffusion-weighted imaging; prognostic prediction; heterogeneity

1. Introduction

Nasopharyngeal carcinoma (NPC) is an epithelial malignancy with distinctive geographic distribution [1]. Over 130,000 patients were newly diagnosed with NPC in 2020, among which more than 70% were located in East and South East Asia [1,2]. Even with advancements in screening and treatments, approximately 5–15% of patients exhibit local recurrence, and 15–30% of NPC patients experience metastatic spread after standard treatment [3]. Therefore, identifying the reliable predictive factors associated with prognosis is necessary. In the last few decades, tumor heterogeneity has continued to be a crucial factor influencing prognosis [4]. At present, the clinical formulation of treatment primarily depends on the TNM staging system. However, similar clinical treatment can result in distinct clinical outcomes for NPC patients with the same TNM stage [5], indicating that the system merely reflects the anatomic invasion and fails to adequately unmask tumor heterogeneity.

Moreover, some specific blood metabolites or cellular and genetic parameters are used to predict the prognosis of nasopharyngeal carcinoma patients, such as EBV-DNA, LDH, ALP, HOPX, miRNAs, and gene expression, etc. [6–10]. Importantly, EBV-DNA and several pretreatment inflammatory biomarkers have been considered as independent prognostic factors for patients with NPC, including lymphocyte and neutrophil counts, and the neutrophil-to-lymphocyte ratio (NLR), etc. [11]. Nevertheless, the former biomarkers present instability and non-specificity, whereas the routine application of the latter parameter modality is restricted by the expensive cost. Therefore, a low-cost, convenient, and accurate approach that can evaluate heterogeneity and prognosis is urgently needed.

The radiomics technique has emerged as a promising approach to the conversion of images into high-dimensional and quantitative features [12]. Radiomics analysis based on clinical images can provide additional information about tumor heterogeneity steadily and accurately, and can thus offer clinical support for decision making, thereby improving tumor treatment with an economic and non-invasive approach [13]. The radiomics model based on MRI to predict the prognosis of patients with nasopharyngeal carcinoma has been observed, and has exported great value in risk stratification and prognosis evaluation [14–16]. However, related studies only extract image features from basic MRI sequences. As a functional imaging technique, DWI can quantitatively demonstrate the diffusion motion of water molecules in the tissue microenvironment, and can detect tissue cellularity, microstructures, and microvasculature at the sub-voxel level, thereby revealing additional internal features of the tumor in order to uncover vital prognostic information [17]. It has been frequently used in clinical trials to report on differential diagnosis, staging, therapeutic evaluation, and prognostic prediction in oncology [18].

In the past, DWI images suffered from insufficient image quality, including obvious artifacts, limited resolution, and blurred images, which may hinder their routine application in radiomics in the head and neck [19]. However, readout-segmented imaging (RS-EPI) approaches have now been introduced to perform high-resolution diffusion-weighted MRI (HR-DWI), and have greatly improved image quality with a higher resolution and fewer artifacts than the extensively adopted single-shot imaging (SS-EPI) DWI [20]. This improvement is achieved by shortening the data-acquisition time and dividing the k-space into multiple interleaved acquisitions in order to diminish the accumulation of phase errors in the phase-encoding direction. Previous studies have shown that a radiomics model based on DWI MRI can accurately reveal the individual prognosis in several cancers, such as bladder, hepatocellular, and prostate cancers [21–23].

According to the literature searched, whether radiomics based on a DWI sequence can extract the tumor heterogeneity of nasopharyngeal carcinoma and evaluate the risk of recurrence and metastasis remains uncertain. Accordingly, we performed the present study to visualize the heterogeneity and disclose the prognosis of nasopharyngeal carcinoma through radiomics analyses based on the RESOLVE-DWI sequence. Furthermore, we sought to compare and combine the radiomics model based on the RESOLVE-DWI sequence and conventional sequence (T2WI and CE-T1WI) in order to provide more clinical decision-making and intervention information.

2. Materials and Methods

2.1. Patients

Approval for this retrospective study was obtained from the Ethics Review Committee of the Fifth Affiliated Hospital of Sun Yat-sen University (ClinicalTrials.gov Identifier: NCT05112510). The Committee exempted the informed consent concurrently. A total of 154 patients with untreated NPC confirmed by pathological examination between March 2014 and June 2018 were enrolled, including 15 patients with local or regional tumor recurrence and 28 patients with distant metastasis (1 of the patients had local recurrence and metastases simultaneously).

The collected clinical features included age, gender, tumor size (T), nodal status (N), metastases (M), TNM staging, and histological subtypes. The staging was based on the Eighth American Joint Committee on Cancer TNM staging manual [24]. According to the criteria from the World Health Organization (WHO), the histological subtypes were classified into three patterns: keratinizing squamous cell carcinoma (type I), nonkeratinizing differentiated carcinoma (type II), and nonkeratinizing undifferentiated carcinoma (type III) [25].

2.2. Inclusion and Exclusion Criteria

The eligibility criteria for patient enrollment were as follows: (1) patients with NPC confirmed by pathological examination; (2) patients with complete MR images and clinical data; (3) patients who did not receive chemotherapy, radiotherapy, or surgery before their MRI scans. Patients were removed by applying the following exclusion criteria: (1) the periodical follow-up data were incomplete; (2) poor image quality; and (3) patients with a concomitant or previous history of cancer.

2.3. Endpoints

Failure-free survival (FFS) was defined as the primary endpoint in this study, and it was considered from the first date of the MR scan, and ended with the progression. Local recurrence was diagnosed through pathological examinations. If any medical report indicated distant metastasis, the suspected site of involvement was subjected to extra histological confirmation. In the case of failed biopsy or no biopsy, regular follow-up was attempted. Distant metastasis was diagnosed when the enlargement of the lesions was observed.

2.4. MRI Acquisition

All 154 patients underwent a series of MRI scans. The sequences included axial T2-weighted imaging (T2WI), contrast-enhanced T1-weighted imaging (CE-T1WI), axial DWI ($b = 800\ \text{s/mm}^2$), and ADC mapping. The MRI scanning was performed on a Magnetom Trio 3.0T MRI scanner (Siemens Medical, Erlangen, Germany). An eight-channel head and neck coil was adopted in order to collect the signals. The scanning range was from the skull base to the subclavian region. The conventional MRI sequence included axial T2WI and CE-T1WI. The contrast agent was a Gadobutrol injection.

The following parameters were set for the axial T2WI: TR/TE, 3760 ms/72 ms; field of view (FOV), 230; matrix size, 320 × 224; layer thickness, 5 mm; interlayer spacing, 1 mm; bandwidth, 340 Hz; acquisition time, 3 min and 23 s; number of excitations (NEX), 2; and resolution, 0.7 × 0.7.

The following parameters were set for CE-T1WI: TR/TE, 4660 ms/10 ms; FOV, 230; matrix size, 320 × 224; layer thickness, 5 mm; interlayer spacing, 1 mm; bandwidth, 347 Hz; acquisition time, 2 min 49 s; NEX, 3; and resolution, 0.7 × 0.7.

The following parameters were set for RESOLVE-DWI: RS-EPI, TR/TE, 3800 ms/65 ms; matrix size, 192 × 192; layer thickness, 4 mm; interlayer spacing, 0.6 mm; bandwidth, 521 Hz; acquisition time, 2 min 55 s; segmented readout times, 9; and b = 0, 800 s/mm^2. The ADC maps were automatically generated from the MRI system.

2.5. Segmentation and Feature Extraction

All of the regions of interest (ROIs) of the images were manually segmented in all of the slices by two radiologists: one with 5 years of clinical experience and the other with 15 years. A total of 5636 features were extracted. Manual segmentation and relative feature extraction were both conducted in the Radcloud platform (https://mics.radcloud.cn, accessed: 23 May 2022). The intraclass correlation coefficient (ICC) in 20 patients was calculated in order to assess the intra- and inter-observer variability for consistency. Features with an ICC below 0.75 were excluded.

2.6. Radiomics Feature and Model Selection

All of the feature columns with the same numerical values were eliminated, and normalization processing at the order of magnitude was performed on all of the features. The extracted features were screened by Random Forest (RF), which creates a decision tree such that the suboptimal segmentation is performed by introducing randomness; this has been adopted extensively in radiomics based on its excellent performance in classification tasks [26]. The workflow for feature selection by Random Forest can be summarized as follows. First, the differential clinical characteristics were added and set as dummy variables. The top 100 features were screened according to importance. Then, the top 10 features in terms of improving the model's predictive power were retained after the cyclical inclusion of each feature with a forward stepwise approach by the RF method. Finally, the features of each model were limited to 10. The training set was randomly split with the k-fold cross-validation method: the training set was divided into five subsets, and one of the K-fold sample sizes was N = 26 (two-folds: N = 27).

The differences in clinical factors between the two groups were investigated by one-way analysis of variance in SPSS (version 25.0, IBM Corp, Armonk, NY, USA). The Chi-square test was used for categorical variables, and the Mann–Whitney U test was used for continuous variables. Hierarchical variables used the Wilcoxon symbol order and test. Python software was performed to screen, choose, and build the machine learning models based on the screened features.

Five of the existing mainstream algorithms (Logistic Regression, kNN, Naive Bayes, Random Forest, and XGB Classifier) were chosen for training and validation. In order to obtain a more robust model, we applied five-fold cross-validation to calculate the average AUC of the training sets and the average AUC of the validation sets. The obtained results were presented as the average AUC of cross-training set and the average AUC of the cross-validation set. The major parameters of the corresponding models were adjusted using GridSearchCV. The model was chosen according to the average AUC of the cross-validation set [27].

2.7. Prediction Model Building

The selected models mentioned above were trained and validated based on the screened features from different sequence combinations, and the parameters were adjusted. All of the major parameters, such as criterion, max_depth, min_samples_leaf, min_samples_split, max_features, and min_impurity decrease, were adjusted within a large range. The OOB_score was chosen as the evaluation criterion, resulting in the parameters of all of the final models.

Models after the parameter adjustment were used for five-fold cross-validation, and were compared in order to obtain the optimal AUC of different sequence combinations. Accordingly, the optimal machine learning models based on the extracted imaging features from different sequence combinations were built, including model 1 (DWI + ADC), model 2 (T2WI + CE-T1WI), model 3 (DWI + ADC + T2WI), model 4 (DWI + ADC + CE-T1WI), and model 5 (DWI + ADC + T2WI + CE-T1WI). The study workflow is briefly displayed in Figure 1.

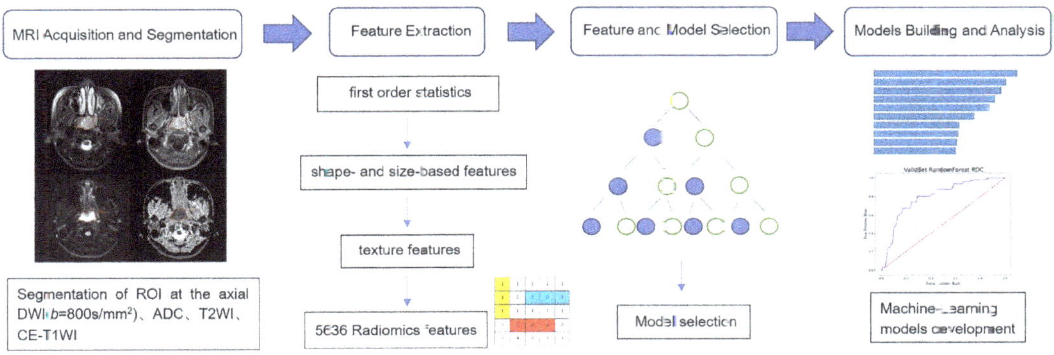

Figure 1. Workflows: (1) MRI acquisition and segmentation; (2) quantitative feature extraction; (3) radiomic feature and model selection; (4) prediction models built based on the extracted imaging features from different sequence combinations.

3. Results

3.1. Clinical Characteristics Analysis

In the present study, 154 patients were included, including 43 females (29%) and 111 males (71%), with a median age of 47 years (19–68). The most common histopathological subtype refers to undifferentiated nonkeratinizing carcinoma (SCC, 80.6%). The relapsed or metastatic group and the non-relapsed or metastatic group presented significant differences in the N, M, and TNM stages ($p < 0.05$). The patient characteristics are presented in Table 1.

Table 1. Clinical characteristics of the patients with NPC in the relapsed or metastatic group and the non-relapsed or metastatic group.

Characteristics	Type	Positive (%) N = 42	Negative (%) N = 112	p-Value
Gender	Male	34	77	0.516
	Female	8	35	
Age (years)	Range	19–68	23–63	0.810
Overall stage	I	0	2	0.026
	II	3	20	
	III	17	56	
	IVa	17	34	
	IVb	5	0	
T stage	I	2	25	0.915
	II	12	22	
	III	13	37	
	IV	15	28	
N stage	0	1	9	0.034
	1	11	48	
	2	21	45	
	3	9	10	
M stage	0	42	107	0.085
	1	0	5	
Histology	WHO type I	0	1	
	WHO type II–III	42	111	0.540

3.2. Machine Learning Model Selection

Five-fold cross-validation was carried out using Logistic Regression, kNN, Naive Bayes, Random Forest, and XGB Classifier, and the results show that the AUC obtained using the RF method is the highest among the different sequence combinations. The results are shown in Figure 2. Therefore, the RF machine learning model was chosen to compare the predictive performances of the different sequence combination models.

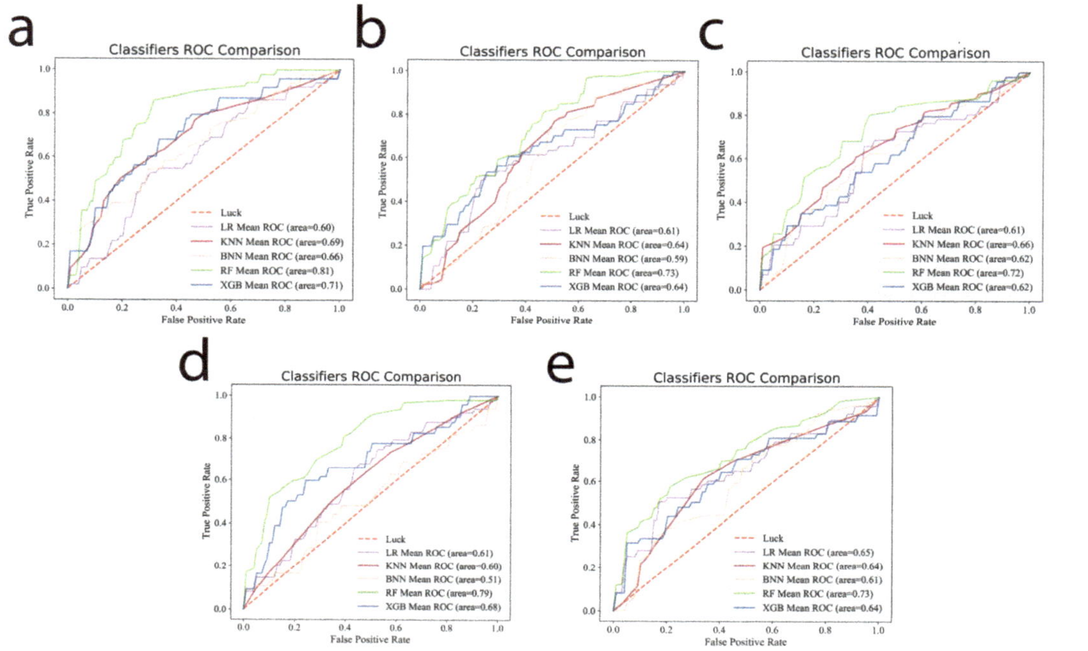

Figure 2. Five existing mainstream algorithms (Logistic Regression, kNN, Naive Bayes, Random Forest, and XGB Classifier) were chosen for the training and validation, which showed that AUC values obtained using the RF method are the highest among all of the models of different sequence combinations: (**a**) DWI + ADC; (**b**) T2WI + CE-T1WI; (**c**) DWI + ADC + T2WI; (**d**) DWI + ADC + CE-T1WI; (**e**) DWI + ADC + T2WI + CE-T1WI.

3.3. Prediction Performance of the Models

Concerning the construction and results of different sequence-combination models, the N and M stages were added according to the dissimilarity tests of the clinical variables, and they were set as dummy variables. The top 100 features were screened by the importance of the RF method. Then, the top 10 features in terms of improving the model's predictive power were retained after the cyclical inclusion of each feature with a forward stepwise approach. The selected features and importances are shown in Figure 3. The selected features were used to construct the RF machine learning prediction model. In order to obtain a more robust outcome, we applied five-fold cross-validation, and the AUC of the validation set in the machine learning model was obtained based on different sequence combinations using the RF method.

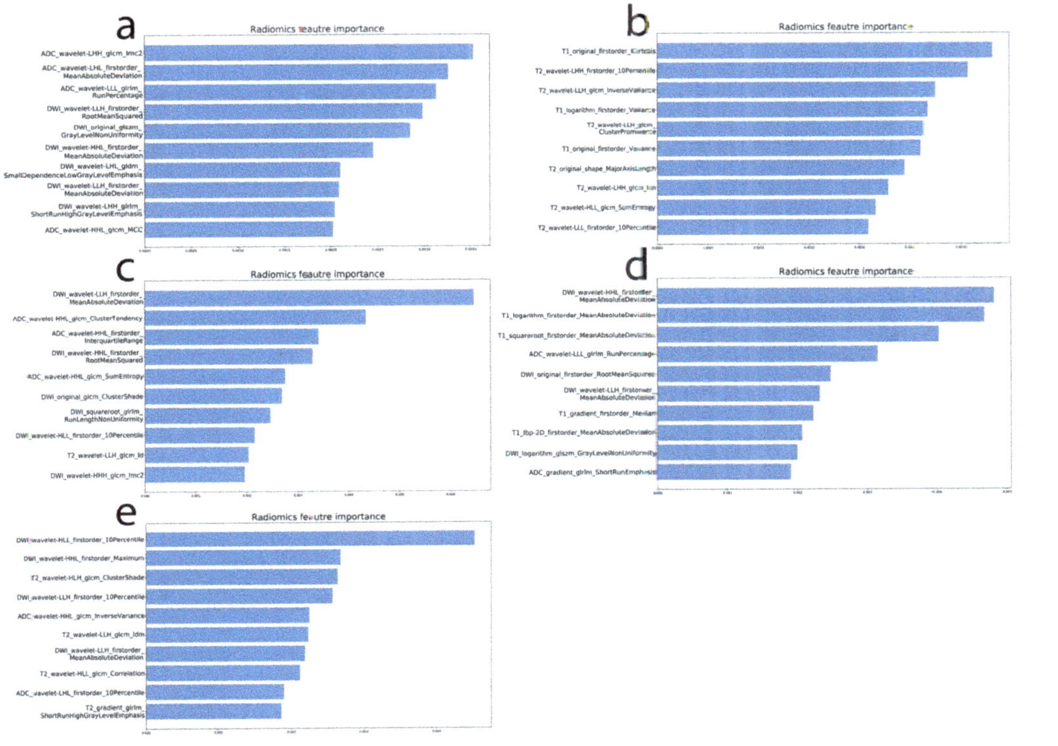

Figure 3. The importance of selected features derived from different sequence combinations: (**a**) DWI + ADC; (**b**) T2WI + CE-T1WI; (**c**) DWI + ADC + T2WI; (**d**) DWI + ADC + CE-T1WI; (**e**) DWI + ADC + T2WI + CE-T1WI.

In order to obtain a more robust outcome, we applied five-fold cross-validation to train and validate the RF machine learning model. After adjusting the parameters, the average AUC of the validation set in the RF machine learning model was obtained based on the extracted imaging features from different sequence combinations. The mean AUCs of the five-fold cross-validation sets of model 1 (DWI + ADC), model 2 (T2WI + CE-T1WI), model 3 (DWI + ADC + T2WI), model 4 (DWI + ADC + CE-T1WI), and model 5 (DWI + ADC + T2WI + CE-T1WI) were 0.80 (95% CI: 0.79–0.81), 0.72 (95% CI: 0.71–0.74), 0.66 (95% CI: 0.64–0.68), 0.74(95% CI: 0.73–0.75), and 0.75 (95% CI: 0.74–0.76), respectively. The average AUC of each model in validation set is shown in Figure 4. The performances of the radiomics models in the validation set are shown in Table 2.

Table 2. The performance metrics for five models in the validation set.

Models	AUC	Accuracy	Specificity	Precision
DWI + ADC	0.80 (95% CI: 0.79–0.81)	0.766	0.926	0.620
T2WI + CE-T1WI	0.72 (95% CI: 0.71–0.74)	0.752	0.930	0.520
DWI + ADC + T2WI	0.66 (95% CI: 0.64–0.68)	0.779	0.925	0.689
DWI + ADC + CE-T1WI	0.74(95% CI: 0.73–0.76)	0.766	0.918	0.548
DWI + ADC + T2WI + CE-T1WI	0.75 (95% CI: 0.74–0.76)	0.766	0.923	0.811

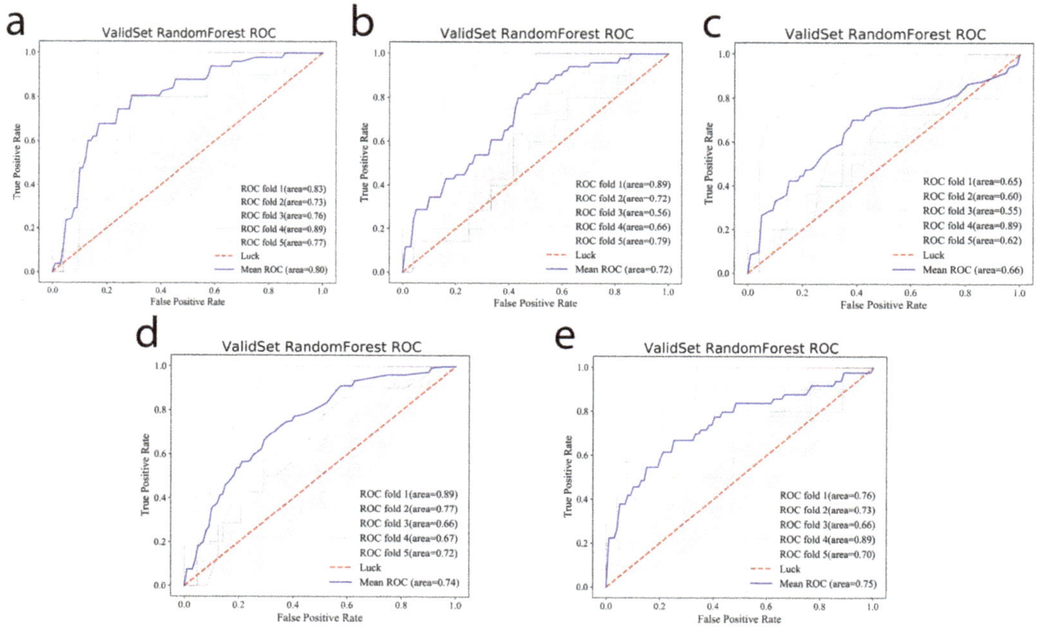

Figure 4. Average AUC values in the validation set of the RF machine learning model based on selected features of model 1 (**a**), model 2 (**b**), model 3 (**c**), model 4 (**d**), and model 5 (**e**).

Based on the results, the RF model based on the extracted features from the DWI and ADC images has higher prognostic prediction efficacy than the RF model based on T2WI and T1WI images. Moreover, the RF model based on the extracted features from DWI, ADC, and T2WI presents better predictive performance for prognosis than the RF model based on DWI, ADC, and CE-T1WI. Finally, the results indicated that the RF model based on the extracted features from the multiple-sequence combination of DWI, ADC, T2WI, and CE-T1WI did not display optimal effects in the prediction of the recurrence and metastasis of nasopharyngeal carcinoma.

4. Discussion

Radiomics models based on MRI features in nasopharyngeal carcinoma (NPC) can predict the prognosis and therapeutic responses [28], but these models were constructed based on basic MR sequences (e.g., T2WI, T1WI, and CE-T1WI). Studies with a radiomics approach based on DWI images in nasopharyngeal carcinoma remain to be explored. Considering that the foregoing radiomics research focuses on tumor heterogeneity and the prognosis of NPC mainly based on T2WI and CE-T1WI [15,16,29–32], we attempted to compare and combine the radiomics model based on RESOLVE-DWI simultaneously with T2WI and CE-T1WI. This process aims to determine the optimal machine learning model for the prognostic prediction of NPC.

Extracted features in various MR sequence combinations were adopted in order to predict the recurrence and metastasis risks of NPC patients in the present study. The results show that the average cross-validated AUC of the RF model based on radiomics features extracted from DWI and ADC sequences reached 0.80, and the AUC of RF models based on conventional MR sequences was 0.72. The AUC of model 2 (T2WI + CE-T1WI) of this study in the validation set closely resembled that of Kim et al.'s study [16], which suggests that the AUC of the radiomics model combining T2WI and CE-T1WI sequences was 0.71 for the prediction of progression-free survival in patients with NPC. At the same time, no data from previous studies were comparable to the results of the radiomics model based on the DWI sequence of the present study on account of the rare usage of DWI in radiomics. However, the radiomics features extracted from the DWI and ADC sequences have higher prediction efficacy in terms of the recurrence and metastasis risks of patients. This finding was potentially obtained due to the quantitative features of models extracted from the image, and the DWI can provide more sub-voxel image information about tumor heterogeneity, which reflects the limited Brownian motion and microarchitecture in tumors [17,33]. Moreover, the machine learning model based on features extracted from the DWI, ADC, and CE-T1WI sequences presents a higher forecast performance than the models based on DWI, ADC, and T2WI sequences. This finding was potentially due to CE-T1WI sequences being able to reflect the blood supply and angiogenesis of tumors [34], and to unmask the proliferation state of tumors better than T2WI, making CE-T1WI sequences more relevant for tumor heterogeneity. Finally, we combined DWI, ADC, T2WI, and CE-T1WI sequences in NPC and extracted the relative features from this combination in order to establish RF machine learning models. The average cross-validated AUC of this model was 0.73 for the prediction of the prognosis of NPC, and this value is not higher than that of the RF model based on DWI and ADC sequences. This finding can be attributed to the increase in mixing factors with the increase in sequences.

Notably, high-resolution DWI was applied to extract related features and build the machine learning model for the prediction of the recurrence and metastasis of nasopharyngeal carcinoma. DWI is a proven non-contrast imaging technology that has become a mature quantitative measurement approach for the identification of benign and malignant lesions in routine clinical work [19,35,36]. In malignant tumors, the diffusion of water molecules is often restricted or limited by the high cell density, which exhibited high signals on DWI and a low value on ADC maps. DWI technology can provide quantitative interpretations as well as qualitative interpretations, thereby increasing the specificity of disease diagnosis [17]. The application in radiomics of a single-shot (SS) EPI-DWI technology extensively used to collect DWI images is easily restricted by magnetic susceptibility artifacts, chemical displacement and geometric distortion, limited spatial resolution, and relatively thick sections, especially in head and neck tumors, such as nasopharyngeal carcinoma with artifacts of the skull base [19]. With the improvement in readout-segmented imaging (RS-EPI) technologies, high-resolution DWI (HR-DWI) was applied to clinical work. It remarkably improved the abovementioned problems by using the same diffusion preparation as SS EPI but dividing the K space into several segments in the phase-encoding direction in order to decrease the echo time [20]. Therefore, readout-segmented imaging (RS-EPI) has obvious advantages and is irreplaceable for the diagnosis of tumors at the head and neck compared with (SS) EPI-DWI [19], and the machine learning model based on DWI collected by RS-EPI is more reliable and robust, providing a good foundation to promote its clinical applications.

Additionally, the acquisition of HR-DWI does not require a contrast agent, making it safer than the CE-T1W in daily clinical work. In present practical applications, it has realized technological advantages of increased speed and decreased artifacts, supporting its extensive use in clinical practice. Based on the above analysis, the radiomics method based on RESOLVE-DWI has higher prediction efficacy than the conventional MR sequence regarding the recurrence and metastasis of NPC. The applications of high-resolution DWI in radiomics might be complementary to—and might even replace—the currently used sequences (T2WI, T1WI, and CE-T1WI) in order to provide more high-specificity information and support for clinical decisions.

The present study has some limitations. First, this study involved a few cases, and it was carried out in one hospital. Therefore, a prospective study should be carried out to support the conclusions. Moreover, minor details are hard to depict, which might influence the extraction of features. Finally, the relationship between radiomics features and prognostic outcomes was not explored further in the present study. Relevant data were collected, and the next step for our research is to discover this relationship and further perform survival analysis according to the radiomics model based on DWI sequences.

5. Conclusions

The results confirmed that the machine learning model based on features extracted by RESOLVE-DWI and corresponding ADC maps could be used as a prognosis detection tool. These features can help to quantify the heterogeneity of patients with NPC and evaluate the risk of recurrence and metastasis in order to quickly provide supporting evidence and thus aid in making a sound clinical decision in clinical practice.

Author Contributions: Segmentation and Feature Extraction, Q.H., G.W. and J.W.; radiomics Feature and Model Selection, Q.H. and X.S.; prediction Model Building, X.S. and M.L.; imaging data collection, Q.H., G.W. and Y.W.; clinical data collection, Q.C. and X.C.; manuscript writing and revision, Q.H., G.W., F.Z. and Y.W.; study design, G.W., F.Z. and Y.W.; supervision, Y.W. and S.L. All authors have read and agreed to the published version of the manuscript.

Funding: This research was funded by the National Natural Science Foundation of China (8210072058), Medical Scientific Research Foundation of Guangdong Province (A2021449), Key Project of Guangdong Province (2018B030335001), the Medical Science and Technology Research Fund of Guangdong province (2018B030322006) and Investigator-Initiated Clinical Trial of the Fifth Affiliated Hospital of Sun Yat-sen University (YNZZ2020-04).

Institutional Review Board Statement: The study was conducted in accordance with the Declaration of Helsinki, and approved by the Ethics Committee of the Fifth Affiliated Hospital of Sun Yat-sen University (protocol code No. ZDWY(2021) Lunzi No. (39-1), approved on 15 July 2021).

Informed Consent Statement: Patient consent was waived due to the retrospective nature of this study.

Data Availability Statement: Not applicable.

Conflicts of Interest: The authors declare no conflict of interest.

References

1. Chen, Y.P.; Chan, A.T.; Le, Q.T.; Blanchard, P.; Sun, Y.; Ma, J. Nasopharyngeal carcinoma. *Lancet* **2019**, *394*, 64–80. [CrossRef]
2. Jozkowiak, M.; Dyszkiewicz-Konwinska, M.; Ramlau, P.; Kranc, W.; Spaczynska, J.; Wierzchowski, M.; Kaczmarek, M.; Jodynis-Liebert, J.; Piotrowska-Kempisty, H. Individual and Combined Treatments with Methylated Resveratrol Analogue DMU-214 and Gefitinib Inhibit Tongue Cancer Cells Growth via Apoptosis Induction and EGFR Inhibition. *Int. J. Mol. Sci.* **2021**, *22*, 6180. [CrossRef] [PubMed]
3. Lee, A.W.; Ma, B.B.; Ng, W.T.; Chan, A.T. Management of Nasopharyngeal Carcinoma: Current Practice and Future Perspective. *J. Clin. Oncol.* **2015**, *33*, 3356–3364. [CrossRef]
4. Dagogo-Jack, I.; Shaw, T.A. Tumour heterogeneity and resistance to cancer therapies. *Nat. Rev. Clin. Oncol.* **2018**, *15*, 81–94. [CrossRef]
5. Li, X.H.; Chang, H.; Xu, B.Q.; Tao, Y.L.; Gao, J.; Chen, C.; Qu, C.; Zhou, S.; Liu, S.R.; Wang, X.H.; et al. An inflammatory biomarker-based nomogram to predict prognosis of patients with nasopharyngeal carcinoma: An analysis of a prospective study. *Cancer Med.* **2017**, *6*, 310–319. [CrossRef]
6. Nilsson, J.S.; Forslund, O.; Andersson, F.C.; Lindstedt, M.; Greiff, L. Intralesional EBV-DNA load as marker of prognosis for nasopharyngeal cancer. *Sci. Rep.* **2019**, *9*, 15432. [CrossRef]
7. Li, G.; Gao, J.; Tao, Y.L.; Xu, B.Q.; Tu, Z.W.; Liu, Z.G.; Zeng, M.S. Increased pretreatment levels of serum LDH and ALP as poor prognostic factors for nasopharyngeal carcinoma. *Chin. J. Cancer Res.* **2019**, *31*, 197–206. [CrossRef]
8. Liu, N.; Chen, N.Y.; Cui, R.X.; Li, W.F.; Li, Y.; Wei, R.R.; Zhang, M.Y.; Sun, Y.; Huang, B.J.; Chen, M.; et al. Prognostic value of a microRNA signature in nasopharyngeal carcinoma a microRNA expression analysis. *Lancet Oncol.* **2012**, *13*, 633–641. [CrossRef]
9. Ren, X.Y.; Yang, X.J.; Cheng, B.; Chen, X.Z.; Zhang, T.p.; He, Q.m.; Li, B.; Li, Y.; Tang, X.; Wen, X.; et al. HOPX hypermethylation promotes metastasis via activating SNAIL transcription in nasopharyngeal carcinoma. *Nat. Commun.* **2017**, *8*, 14053. [CrossRef]
10. Engku, N.S.; Ahmad, A.I.; Chan, Y.Y. Diagnostic and Prognostic Indications of Nasopharyngeal Carcinoma. *Diagnostics* **2020**, *10*, 611. [CrossRef]
11. Su, L.; Zhang, M.W.; Zhang, W.J.; Cai, C.; Hong, J. Pretreatment hematologic markers as prognostic factors in patients with nasopharyngeal carcinoma: A systematic review and meta-analysis. *Medicine* **2017**, *96*, e6364. [CrossRef] [PubMed]
12. Lambin, P.; Leijenaar, R.T.; Deist, T.M.; Peerlings, J.; de Jong, E.E.C.; Timmeren, J.V.; Sanduleanu, S.; Larue, R.; Even, A.J.G.; Jochems, A.; et al. Radiomics: The bridge between medical imaging and personalized medicine. *Nat. Rev. Clin. Oncol.* **2017**, *14*, 749–762. [CrossRef] [PubMed]

13. Limkin, E.J.; Sun, R.; Dercle, L.; Zacharaki, E.I.; Robert, C.; Reuz, S.; Schernberg, A.; Paragios, N.; Deutsch, E.; Ferté, C. Promises and challenges for the implementation of computational medical imaging (radiomics) in oncology. *Ann. Oncol.* **2017**, *28*, 1191–1206. [CrossRef] [PubMed]
14. Zhao, L.; Gong, J.; Xi, Y.B.; Xu, M.; Li, C.; Kang, X.W.; Yin, Y.T.; Qin, W.; Yin, H.; Shi, M. MRI-based radiomics nomogram may predict the response to induction chemotherapy and survival in locally advanced nasopharyngeal carcinoma. *Eur. Radiol.* **2020**, *30*, 537–546. [CrossRef]
15. Zhang, B.; Ouyang, F.S.; Gu, D.S.; Dong, Y.H.; Zhang, L.; Mo, X.K.; Huang, W.H.; Xing, Z.S. Advanced nasopharyngeal carcinoma: Pre-treatment prediction of progression based on multi-parametric MRI radiomics. *Oncotarget* **2017**, *8*, 72457–72465. [CrossRef]
16. Kim, M.J.; Choi, Y.; Sung, Y.E.; Lee, Y.S.; Kim, Y.S.; Ahn, K.J.; Kim, M.S. Early risk-assessment of patients with nasopharyngeal carcinoma: The added prognostic value of MR-based radiomics. *Transl. Oncol.* **2021**, *14*, 101180. [CrossRef]
17. Tang, L.; Zhou, X.J. Diffusion MRI of cancer: From low to high b-values. *J. Magn. Reson. Imaging* **2019**, *49*, 23–40. [CrossRef]
18. Messina, C.; Bignone, R.; Bruno, A.; Bruno, A.; Bruno, F.; Calandri, M.; Caruso, D.; Coppolino, P.; Robertis, R.D.; Gentili, F.; et al. Diffusion-Weighted Imaging in Oncology: An Update. *Cancers* **2020**, *12*, 1493. [CrossRef]
19. Norris, C.D.; Quick, S.E.; Parker, J.G.; Koontz, N.A. Diffusion MR Imaging in the Head and Neck: Principles and Applications. *Neuroimaging Clin. N. Am.* **2020**, *30*, 261–282. [CrossRef]
20. Widmann, G.; Henninger, B.; Kremser, C.; Jaschke, W. MRI Sequences in Head & Neck Radiology—State of the Art. *Rofo* **2017**, *189*, 413–422. [CrossRef]
21. Zhang, S.H.; Song, M.F.; Zhao, Y.S.; Xu, S.S.; Sun, Q.C.; Zhai, G.; Liang, D.; Wu, G.; Li, Z.C. Radiomics nomogram for preoperative prediction of progression-free survival using diffusion-weighted imaging in patients with muscle-invasive bladder cancer. *Eur. J. Radiol.* **2020**, *131*, 109–219. [CrossRef] [PubMed]
22. Zhao, Y.; Wu, J.J.; Zhang, Q.H.; Hua, Z.Y.; Qi, W.J.; Wang, N.; Lin, T.; Sheng, L.J.; Cui, D.H.; Liu, J.H.; et al. Radiomics Analysis Based on Multiparametric MRI for Predicting Early Recurrence in Hepatocellular Carcinoma After Partial Hepatectomy. *J. Magn. Reson. Imaging* **2021**, *53*, 1066–1079. [CrossRef] [PubMed]
23. Zhong, Q.Z.; Long, L.H.; Liu, A.; Li, C.M.; Xiu, X.; Hou, X.Y.; Wu, Q.H.; Gao, H.; Xu, Y.G.; Zhao, T.; et al. Radiomics of Multiparametric MRI to Predict Biochemical Recurrence of Localized Prostate Cancer After Radiation Therapy. *Front. Oncol.* **2020**, *10*, 731. [CrossRef]
24. Amin, M. *AJCC Cancer Staging Manual*, 8th ed.; Springer: Berlin/Heidelberg, Germany, 2018.
25. Ou, S.I.; Zell, J.A.; Ziogas, A.; Anton-Culver, H. Epidemiology of nasopharyngeal carcinoma in the United States: Improved survival of Chinese patients within the keratinizing squamous cell carcinoma histology. *Ann. Oncol.* **2007**, *18*, 29–35. [CrossRef]
26. Breiman, L. Random Forests. *Mach. Learn.* **2001**, *45*, 5–32. [CrossRef]
27. Efron, B.; Tibshirani, R. Improvements on Cross-Validation: The 632+ Bootstrap Method. *J. Am. Stat. Assoc.* **1997**, *92*, 548–560. [CrossRef]
28. Li, S.; Deng, Y.Q.; Zhu, Z.L.; Hua, H.L.; Tao, Z.Z. A Comprehensive Review on Radiomics and Deep Learning for Nasopharyngeal Carcinoma Imaging. *Diagnostics* **2021**, *11*, 1523. [CrossRef]
29. Gao, Y.; Mao, Y.T.; Lu, S.H.; Tan, L.; Li, G.; Chen, J.; Huang, D.H.; Zhang, X.; Qiu, Y.Z.; Liu, Y. Magnetic resonance imaging-based radiogenomics analysis for predicting prognosis and gene expression profile in advanced nasopharyngeal carcinoma. *Head Neck* **2021**, *43*, 3730–3742. [CrossRef]
30. Zhang, B.; Tian, J.; Dong, D.; Gu, D.S.; Dong, Y.H.; Zhang, L.; Lian, Z.Y.; Liu, J.; Luo, X.N.; Pei, S.F.; et al. Radiomics Features of Multiparametric MRI as Novel Prognostic Factors in Advanced Nasopharyngeal Carcinoma. *Clin. Cancer Res.* **2017**, *23*, 4259–4269. [CrossRef]
31. Kang, L.; Niu, Y.L.; Huang, R.; Lin, S.; Tang, Q.L.; Chen, A.L.; Fan, Y.X.; Lang, J.Y; Yin, G.; Zhang, P. Predictive Value of a Combined Model Based on Pre-Treatment and Mid-Treatment MRI-Radiomics for Disease Progression or Death in Locally Advanced Nasopharyngeal Carcinoma. *Front. Oncol.* **2021**, *11*, 774455. [CrossRef]
32. Yang, K.X.; Tian, J.F.; Zhang, B.; Li, M.; Xie, W.J.; Zou, Y.T.; Tan, Q.Y.; Liu, L.H.; Zhu, J.B.; Shou, A.; et al. A multidimensional nomogram combining overall stage, dose volume histogram parameters and radiomics to predict progression-free survival in patients with locoregionally advanced nasopharyngeal carcinoma. *Oral Oncol.* **2019**, *98*, 85–91. [CrossRef]
33. Gatenby, R.A.; Grove, O.; Gillies, R.J. Quantitative imaging in Cancer evolution and ecology. *Radiology* **2013**, *269*, 8–15. [CrossRef] [PubMed]
34. Oostendorp, M.; Post, M.J.; Backes, W.H. Vessel Growth and Function: Depiction with Contrast-enhanced MR Imaging. *Radiology* **2009**, *251*, 217–335. [CrossRef] [PubMed]
35. Song, C.R.; Cheng, P.; Cheng, J.L.; Zhang, Y.; Sun, M.T.; Xie, S.S.; Zhang, X.N. Differential diagnosis of nasopharyngeal carcinoma and nasopharyngeal lymphoma based on DCE-MRI and RESOLVE-DWI. *Eur. Radiol.* **2020**, *30*, 110–118. [CrossRef] [PubMed]
36. Baxter, G.C.; Graves, M.J.; Gilbert, F.J.; Patterson, A.J. A Meta-analysis of the Diagnostic Performance of Diffusion MRI for Breast Lesion Characterization. *Radiology* **2019**, *291*, 632–641. [CrossRef]

cancers

MDPI

Article

Prediction of Nodal Metastasis in Lung Cancer Using Deep Learning of Endobronchial Ultrasound Images

Yuki Ito [1], Takahiro Nakajima [2,*], Terunaga Inage [1], Takeshi Otsuka [3], Yuki Sata [1], Kazuhisa Tanaka [1], Yuichi Sakairi [1], Hidemi Suzuki [1] and Ichiro Yoshino [1]

1 Department of General Thoracic Surgery, Graduate School of Medicine, Chiba University, Chiba 260-8670, Japan; yuki_ito_1989@yahoo.co.jp (Y.I.); potatolunch@yahoo.co.jp (T.I.); y.sata.0506@gmail.com (Y.S.); kazutanaka1118@yahoo.co.jp (K.T.); y_sakairi1@chiba-u.jp (Y.S.); hidemisuzukidesu@yahoo.co.jp (H.S.); iyoshino@faculty.chiba-u.jp (I.Y.)

2 Department of General Thoracic Surgery, Dokkyo Medical University, Tochigi 321-0207, Japan

3 Advanced Image Processing Technology 4, Electrical Engineering, Olympus Medical Systems Corporation, Tokyo 192-8507, Japan; takeshi.otsuka@olympus.com

* Correspondence: t-nakajima@dokkyomed.ac.jp; Tel.: +81-282-861111

Simple Summary: Endobronchial ultrasound-guided transbronchial aspiration is a minimally invasive and highly accurate modality for the diagnosis of lymph node metastasis and is useful for pre-treatment biomarker test sampling in patients with lung cancer. Endobronchial ultrasound image analysis is useful for predicting nodal metastasis; however, it can only be used as a supplemental method to tissue sampling. In recent years, deep learning-based computer-aided diagnosis using artificial intelligence technology has been introduced in research and clinical medicine. This study investigated the feasibility of computer-aided diagnosis for the prediction of nodal metastasis in lung cancer using endobronchial ultrasound images. The outcome of this study may help improve diagnostic efficiency and reduce invasiveness of the procedure.

Abstract: Endobronchial ultrasound-guided transbronchial needle aspiration (EBUS-TBNA) is a valid modality for nodal lung cancer staging. The sonographic features of EBUS helps determine suspicious lymph nodes (LNs). To facilitate this use of this method, machine-learning-based computer-aided diagnosis (CAD) of medical imaging has been introduced in clinical practice. This study investigated the feasibility of CAD for the prediction of nodal metastasis in lung cancer using endobronchial ultrasound images. Image data of patients who underwent EBUS-TBNA were collected from a video clip. Xception was used as a convolutional neural network to predict the nodal metastasis of lung cancer. The prediction accuracy of nodal metastasis through deep learning (DL) was evaluated using both the five-fold cross-validation and hold-out methods. Eighty percent of the collected images were used in five-fold cross-validation, and all the images were used for the hold-out method. Ninety-one patients (156 LNs) were enrolled in this study. A total of 5255 and 6444 extracted images from the video clip were analyzed using the five-fold cross-validation and hold-out methods, respectively. The prediction of LN metastasis by CAD using EBUS images showed high diagnostic accuracy with high specificity. CAD during EBUS-TBNA may help improve the diagnostic efficiency and reduce invasiveness of the procedure.

Keywords: EBUS-TBNA; echo B-mode imaging; deep learning-based computer-aided diagnosis; nodal staging

Citation: Ito, Y.; Nakajima, T.; Inage, T.; Otsuka, T. Sata, Y.; Tanaka, K.; Sakairi, Y.; Suzuki, H.; Yoshino, I. Prediction of Nodal Metastasis in Lung Cancer Using Deep Learning of Endobronchial Ultrasound Images. *Cancers* **2022**, *14*, 3334. https://doi.org/10.3390/cancers14143334

Academic Editor: Samuel C. Mok

Received: 23 May 2022
Accepted: 5 July 2022
Published: 8 July 2022

1. Introduction

Endobronchial ultrasound-guided transbronchial aspiration (EBUS-TBNA) is a minimally invasive and highly accurate modality for the diagnosis of lymph node (LN) metastasis and is useful for pre-treatment biomarker test sampling in patients with lung cancer [1].

According to the current guidelines for lung cancer staging, EBUS-TBNA is recommended as the best first test for nodal staging prior to considering surgical procedures [2].

During EBUS-TBNA, multiple LNs are often encountered within the same nodal station. In this process, selecting the most suspicious LN for sampling is important, considering the difficulty of sampling all LNs using EBUS-TBNA under conscious sedation. Thus, EBUS image analysis is useful for predicting nodal metastasis; however, it can only be used as a supplemental method to tissue sampling [3]. We have previously reported the utility of six distinctive ultrasound and Doppler features on EBUS ultrasound images for predicting nodal metastasis [4,5]. However, categorization of image characteristics was not reliable owing to the fact it was subjective and varied significantly with the operator. Therefore, we sought an objective method to predict nodal metastasis. Elastography is a potential solution since it can visualize the relative stiffness of targeted tissues within the region of interest and helps to predict LN metastases. Moreover, it uses objective parameters such as a stiff area ratio [6,7]. However, elastography requires additional operations during the procedure, and its parameters do not reflect real-time values.

In recent years, deep learning (DL)-based computer-aided diagnosis (CAD) using artificial intelligence (AI) technology has been introduced in research and clinical medicine. CAD has been used for radiology, primarily in the areas of computed tomography (CT), positron emission tomography-CT (PET-CT), and ultrasound images, and for the diagnosis of several tumors, such as breast cancer and gastrointestinal tumors [8–11].

If real-time CAD-based prediction of nodal metastasis during EBUS-TBNA is made possible, the operator can easily identify the most suspicious node for diagnosis, thereby reducing the procedure time of EBUS-TBNA. The well-experienced EBUS operator could predict benign lymph nodes with approximately 90% accuracy by subjective categorization of EBUS ultrasound characters. The AI-CAD technology might make "the expert level prediction of nodal diagnosis" possible even for non-experts. The purpose of this study is to investigate the feasibility of CAD for the prediction of LN metastasis in lung cancer using endobronchial ultrasound images and DL technology.

2. Materials and Methods

2.1. Participants

Patients with lung cancer or those suspected of suffering from lung cancer who underwent EBUS-TBNA for the diagnosis of LN metastasis were enrolled in this study. We prospectively collected clinical information and images related to bronchoscopy since April 2017 (registry ID: UMIN000026942), and the ethical committee allowed prospective case accumulation with written consent (ethical committee approval ID: No. 2563, Chiba University Graduate School of Medicine). The EBUS-TBNA video clips from April 2017 to December 2020 were retrospectively reviewed, and the patient's clinical information was obtained from electronic medical records (ethical committee approval ID: No. 3538, Chiba University Graduate School of Medicine). This was a collaborative study between the Chiba University Graduate School of Medicine and Olympus Medical Systems Corp. (Tokyo, Japan). All patient identifiers were deleted, and the image data were sent to the Olympus Medical Systems Corp.'s laboratory and analyzed using DL technology (ethical committee approval ID: OLET-2019-008, Olympus Medical Systems Corp.). This study was conducted in accordance with the principles of the Declaration of Helsinki.

2.2. EBUS-TBNA Procedure

The patients underwent EBUS-TBNA under local anesthesia with moderate conscious sedation using midazolam and pethidine hydrochloride. OLYMPUS BF-UC290F and EU-ME1 and EU-ME2 PREMIER PLUS were used to observe LNs. Systematic nodal observation starting from the N1, N2, and N3 stations using B-mode, Doppler mode, and elastography was first performed. The size of each LN was measured, and EBUS-TBNA was performed for LNs > 3 mm along the short axis on the EBUS image. TBNA was initiated at N3, N2, and N1 stations to avoid overstating. For TBNA, a dedicated 22-gauge or 21-gauge needle (NA-

201SX-4022, NA-201SX-4021, Olympus Medical Systems Corp., Tokyo, Japan) was used, and rapid on-site evaluation was performed during the procedure. All EBUS procedures were performed by skilled operators (T.N. and Y.Sakairi.) or under their supervision.

2.3. *Confirmation Diagnosis of EBUS-TBNA*

Rapid on-site evaluation by DiffQick staining and conventional cytology by Papanicolaou staining were performed and diagnosed by a cytopathologist. The histological core was collected in CytoLyt solution and fixed in 10% neutral buffered formalin. The formalin-fixed paraffin-embedded specimens were stained with hematoxylin and eosin (H&E) and subjected to immunohistochemistry. Cytology as well as histology was evaluated by independent pathologists who provided pathological diagnosis [12]. The referenced final diagnoses were as follows: (1) malignant cells were proven by EBUS-TBNA, (2) histological diagnosis was made for surgically resected samples after EBUS-TBNA, (3) clinical follow up by radiology after 6 months.

2.4. *EBUS Image Extraction and Image Data Sets*

Ultrasound images were recorded as video clips in the MP4 format; divided into shorter clips featuring each LN using video editing software, XMedia Recode 3.4.3.0 (Sebastian Dörfler, Eschenbergen, Germany); and subsequently anonymized using the dedicated software VideoRectFill (Olympus Medical Systems Corp.). All patient information was manually masked on the software. An anonymized video clip was provided to Olympus Medical Systems Corp. with diagnostic information linked to each LN.

In this study, we retrospectively and prospectively collected cases and investigated the detection of LN metastasis in each LN. The evaluation methods are illustrated in Figure 1. We retrospectively and prospectively collected LNs. We attempted both five-fold cross-validation and hold-out methods for evaluation. Because the images from the video clips included different ultrasound processors (EU-ME1 and EU-ME2 PREMIER PLUS) and different image sizes, these images were allocated equally to each training, validation, and testing group (Figure S1).

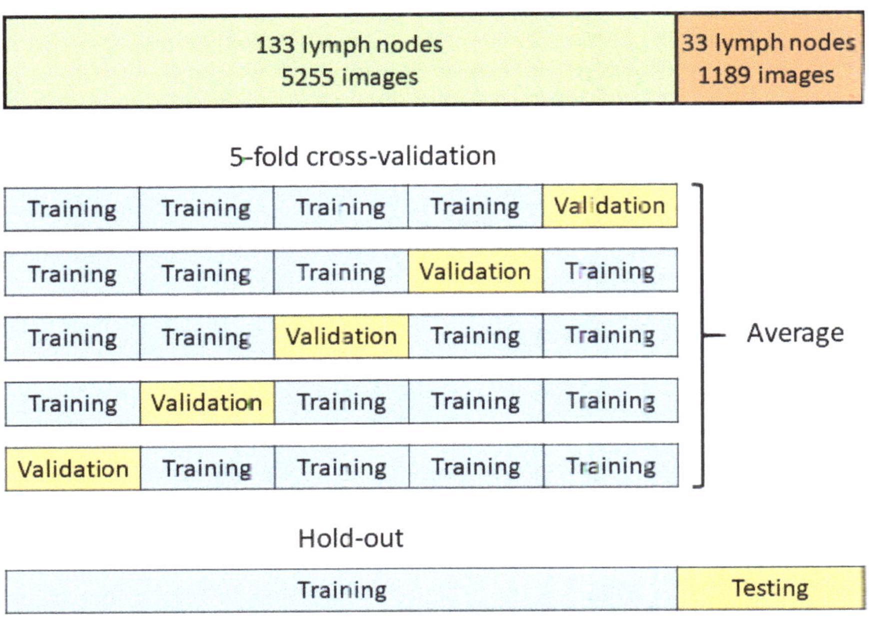

Figure 1. The concept of deep learning algorithm.

2.5. Adjustment of Images for DL

Prior to image analysis, the videos were decomposed into time-series images, from which images of different scenes were extracted. The areas in which the B-mode was drawn were cropped from the images and the cropped images were resized to the same size. To increase the generalizability of the DL algorithm, data augmentation was applied only to the training images, and the number of training images was increased. Scaling and horizontal flipping were used in the data augmentation process.

2.6. DL Algorithm Design

The Convolutional Neural Network (CNN) structure used in this study for LN metastasis detection is shown in Figure S2. The metastasis detection CNN comprises a feature extraction CNN and detection CNN. The feature extraction CNN comprises multiple stages with each stage having multiple blocks and one downsampling layer. The final stage did not include a downsampling layer. We used the Xception block for each block [13]. The downsampling layer comprises two or more strides of the convolution layer. The detection CNN comprises two convolution layers: one for classification and another for positioning.

Initially, the ultrasound image was input to the feature CNN, and local features, such as edges and textures, were extracted from the input image in the first block. As it progressed through the network, its features were integrated and finally converted into features useful for detection.

Subsequently, the features useful for metastasis detection were input into the detection CNN. The detection CNN outputs the probability and bounding box coordinates and sizes for both metastasis and nonmetastasis. The bounding box with the highest probability was selected from among all the metastatic and non-metastatic bounding boxes in the sequence. Finally, the metastasis or non-metastasis parameters, coordinates and size of the bounding box were obtained as the detection result.

2.7. Five-Fold Cross-Validation Method and the Hold-Out Method

For the five-fold cross-validation method, 80% of all the images were used for training and validation. The images were divided into five sections: four sections were used for training, and the last section was used for validation. By changing the validation section, the training and validation were repeated five times. The prediction yield was calculated as the average of the results of each validation.

In the hold-out method, all images were used for training and testing. All of the images comprising the 80% used for the five-fold cross-validation method were used for training. The remaining 20% of the images that were not used for the five-fold cross-validation method were used for testing, following which the prediction yield was calculated.

The images of different sizes from the two ultrasound scanners (EU-ME1 and EU-ME2 PREMIER PLUS) were allocated proportionately in each section to avoid selection bias.

2.8. Statistical Analysis

The "Image" represents "per image" basis analysis and the "Lymph node" represents "per lymph node" basis analysis. The "per image" analysis was based on the accuracy of nodal metastasis prediction for each image. Due to limited number of still images, we used the video clips for analysis. However, in this case, multiple images with varying ultrasound features were included for each targeted lymph node, resulting in variation in the judgement of the AI-CAD system. Therefore, in addition to "per image" analysis, we included "per lymph node" analysis in which multiple images were evaluated for each lymph node. The "per lymph node" analysis included (1) calculation of the ratio between the number of images judged benign and malignant, (2) predicting as benign or malignant based on the ratio >50%, (3) analysis of the accuracy of nodal metastasis prediction for each lymph node.

Sensitivity, specificity, positive predictive value, negative predictive value, and diagnostic accuracy were calculated using standard definitions. Statistical analysis was performed using Fisher's exact test and chi-square test for categorical outcomes, and Student's t-test for continuous variables. Data were analyzed using the JMP Pro 15 software (SAS Institute Inc., Cary, NC, USA). Statistical significance was set at $p < 0.05$.

3. Results

Ninety-five cases with a total of 170 LNs were enrolled in the study. Two cases (two LNs) were excluded because of a history of malignant lymphoma. Cases of large-cell carcinoma and large-cell neuroendocrine carcinoma (one LN each) were also excluded because they could not be assigned to both the training and testing sets. Finally, 91 cases and 166 LNs were analyzed in this study (Figure 2). In this cohort, 64 LNs (38.5%) were diagnosed as metastatic and 102 LNs (61.5%) as non-metastatic by pathology. The characteristics of the enrolled patients and LNs are listed in Table 1.

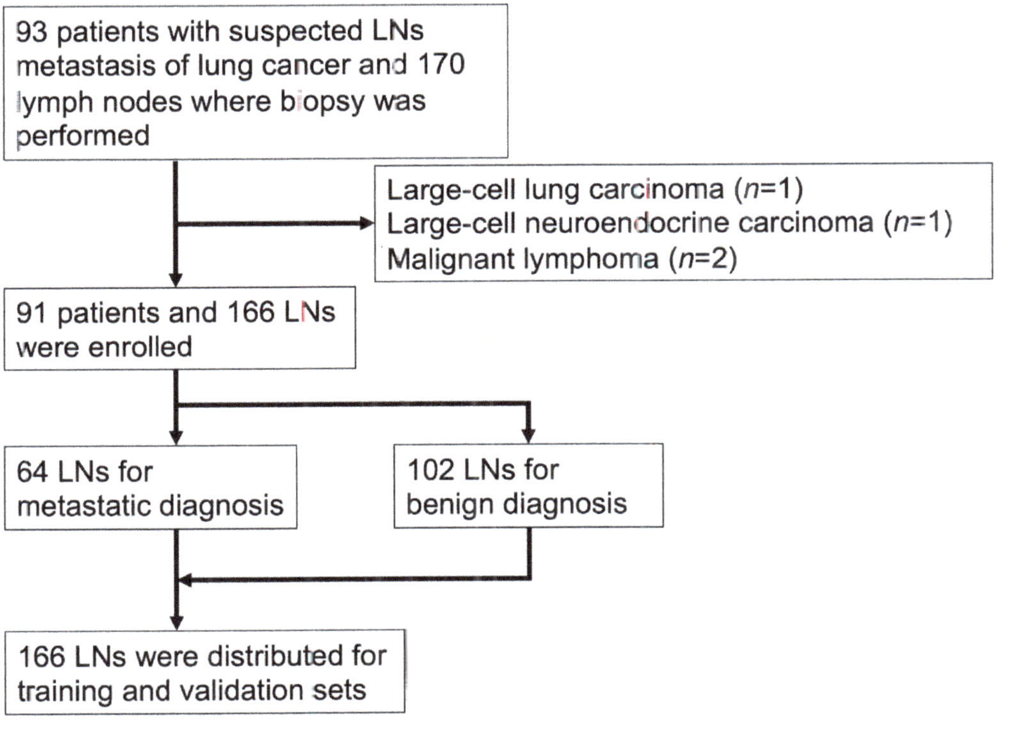

Figure 2. Study cohort flow chart. One hundred sixty-six lymph nodes and 6444 images from 91 patients were enrolled in the final analysis.

Pathological diagnosis including cytology and histology were performed for all lymph nodes. The success rate of each diagnosis was shown in Table 2. For adenocarcinoma cases, molecular biomarker testing was performed for selected cases. For non-small cell lung cancer cases, evaluation for PD-L1 (22C3) immunohistochemistry was done for selected cases. Each success rate, detection rate, and testing rate was shown in Table 2.

Table 1. Patients' and nodal characteristics.

No. of patients	91
Age (y) (median, range)	74 (12–86)
Gender	
male female	61 (67.0%) 30 (33.0%)
No. of lymph nodes	166
Diagnosis	
Metastatic lymph nodes Adenocarcinoma Squamous cell carcinoma Small cell carcinoma Benign lymph nodes	64 (38.5%) 40 (24.0%) 15 (9.0%) 9 (5.4%) 102 (61.5%)
Lymph node station 1 2R 3p 4R/4L 7 8 10R/10L 11s/11i/11(Lt.) 12 13	 1 13 2 41/25 43 1 5 15/6/4 9 1
Lymph node size of long axis	Average (range), mm
All lymph nodes Metastatic lymph nodes Benign lymph nodes	12.9 (3.0–29.2) 15.5 (3.0–29.2) 11.3 (3.5–21.8)

Table 2. Detailed results of pathological diagnosis and biomarker testing in this study.

Metastatic Lymph Node (*n* = 64)	Diagnosed by Cytology	Diagnosed by Histology	Success Rate of Molecular Testing	Detection Rate of Driver Gene Mutations	Testing for PD-L1 Immunohistochemistry
Adenocarcinoma (*n* = 40)	37/40 (92.5%)	37/40 (92.5%)	22/24 (91.7%)	13/22 (59.0%)	22/40 (55.0%)
Squamous cell carcinoma (*n* = 15)	13/15 (86.7%)	14/15 (93.3%)	N/A	N/A	7/15 (46.7%)
Small cell carcinoma (*n* = 9)	9/9 (100%)	9/9 (100%)	N/A	N/A	N/A

First, we evaluated the ability of AI-CAD to detect LN metastasis using endobronchial ultrasound images. A total of 5255 and 6444 extracted images from the video clip were analyzed using the five-fold cross-validation and the hold-out methods, respectively (Figure 1). The representative EBUS images judged by AI-CAD in this study are shown in Figure S3.

Using the five-fold cross-validation method, the LN-based diagnostic accuracy, sensitivity, specificity, positive predictive value, and negative predictive value of the AI-CAD were measured to be 69.9% (95% CI, 32.4–75.2%), 37.3% (95% CI, 27.8–49.1%), 90.2% (95% CI, 82.9–92.3%), 70.4%, and 69.8%, respectively (Figure 3). However, although the specificity was high, the sensitivity of this method was low.

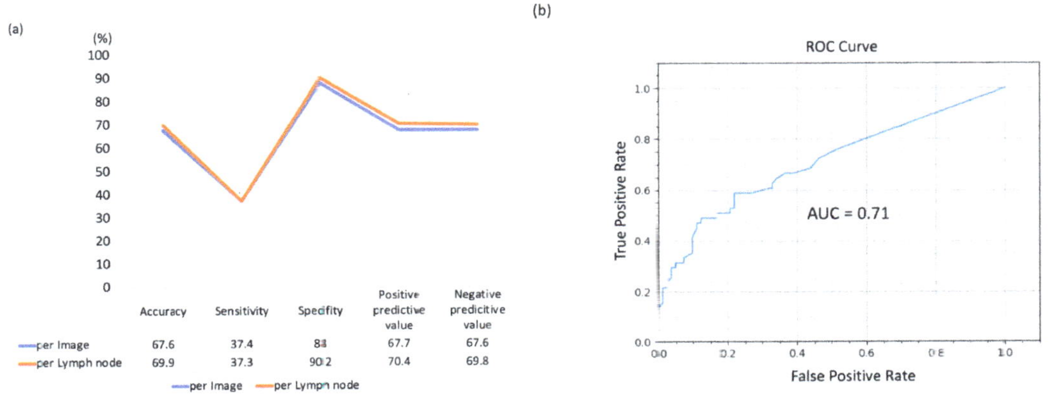

Figure 3. The result of AI-CAD lung cancer lymph node diagnosis accuracy analysis using echo images by five-fold cross validation method. (**a**) Diagnostic yield by per image basis and per lymph node basis. (**b**) ROC curve.

Using the hold-out method, the LN-based diagnostic accuracy, sensitivity, specificity, positive predictive value, and negative predictive value of the AI-CAD were measured to be 87.9% (95% CI, 75.4–94.1%), 76.9% (95% CI, 58.9–92.9%), 95.0% (95% CI, 79.3–100%), and 90.9% and 86.4%, respectively (Figure 4).

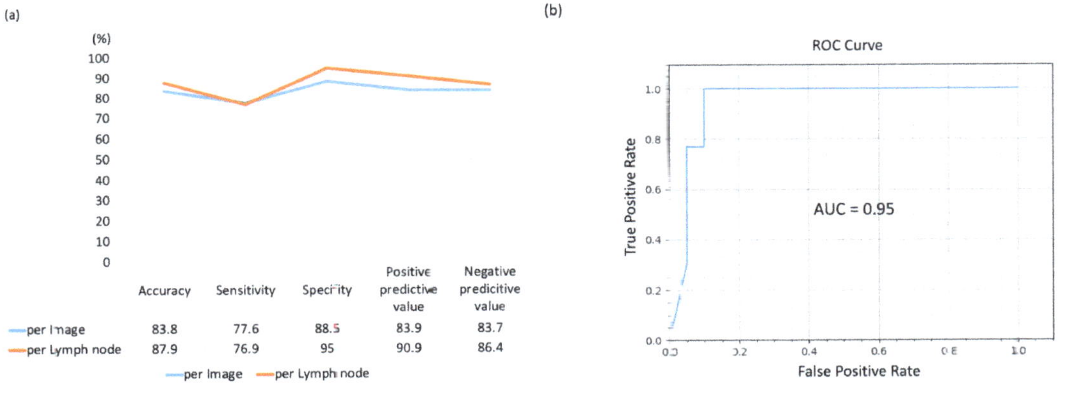

Figure 4. The result of AI-CAD lung cancer lymph node diagnosis accuracy analysis using echo images by hold-out method. (**a**) Diagnostic yield by per image basis and per lymph node basis. (**b**) ROC Curve.

Regarding the diagnostic yield by lung cancer subtypes, the diagnostic accuracy rates were 90.5% for no malignancy, 76.9% for adenocarcinoma, 61.1% for squamous cell carcinoma, and 93.9% for small cell lung cancer (Figure 5).

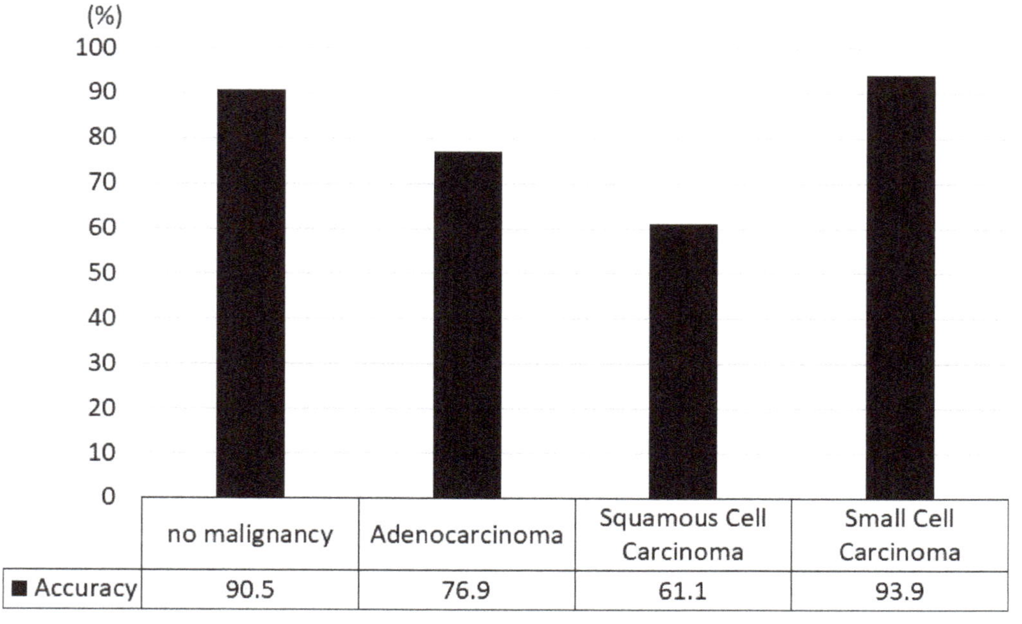

Figure 5. The accuracy rates of the hold-out method by lung cancer subtype.

4. Discussion

The potential applications of AI technology are rapidly growing in the medical field and are expected to facilitate the demanding work of medical staff. The concept of AI, including such systems as machine learning and DL, has been growing in popularity since the evolution of graphics processing units. AI-CAD is one of the AI applications that has been actively developed in radiology. Significant work has been done in the area of combining radiomics and AI-CAD technology, which helps support the diagnosis of benign and malignant tumors, prediction of histology, stage, genetic mutations, and prediction of treatment response and recurrence using CT and PET-CT images [14–18]. AI-CAD is highly useful in analyzing huge amounts of extracted information that includes information invisible to humans. AI-CAD produces objective indicators based on the judgment, knowledge, and experience of experts. During EBUS-TBNA, a highly skilled operator can select the most suspicious LN to sample, based on a subjective categorization of ultrasound image characteristics. In contrast, by applying AI-CAD technology in EBUS, even a trainee can easily select the target LN for sampling, in addition to the dual advantages of a more efficient and less invasive procedure. In this study, we used the CNN algorithm with Xception to predict nodal metastasis based on the ultrasound images of LNs. Using the hold-out method, AI-CAD exhibited a feasible diagnostic accuracy of 84.7%, on average, per LN basis. In this study, the combination of Xception and the hold-out method resulted in the highest diagnostic yield.

The comparison between the five-fold cross-validation and the hold-out methods, demonstrated that the hold-out method exhibited a superior diagnostic yield in this study setting. First, we evaluated using five-fold cross-validation, and then used hold-out method as the standard for developing AI-CAD technology. The number of evaluated images was increased by 20% for the hold-out method compared to five-fold cross-validation. The increased number of images helped with comprehensive covering of image variation and contributed toward better AI-CAD accuracy. The images used in this study were obtained using two different ultrasound image processors (EU-ME1 and EU-ME2 PREMIER PLUS).

In addition, a certain amount of collected images (approximately 10% of all images) were of different sizes owing to the different screen sizes of the various video clips. These variations might affect the diagnostic yield of the five-fold cross-validation and the hold-out methods. Thus, for the analysis of different-size images the image had to be resized and then analyzed, which resulted in an adversarial example (AE). An AE is an event in which AI misrecognizes an image as completely different data owing to the addition of insignificant noises that are imperceptible to humans. [19] Therefore, in this study, these problems were solved by allocating images of different sizes in equal proportions for AI-CAD analysis.

The final diagnostic accuracy and specificity for the prediction of LN metastasis using AI-CAD in this study were 87.9% and 95.0%, respectively. Previous studies have reported comparable but lower values. For instance, Ozcelik et al. reported an accuracy rate of 82% and specificity of 72% for the diagnosis of lung cancer LN metastasis in 345 LNs by CNN using MATLAB [20]. Churchill et al. reported an accuracy rate of 72.8% and a specificity of 90.7% for the diagnosis of lung cancer LN metastasis in 406 LNs by CNN using NeuralSeg [21]. It is noteworthy, however, that the specificity of CNN-based diagnosis for the prediction of nodal metastasis was found to be high, and this might help avoid futile biopsies and reduce examination time as well as the risk of co-morbidities.

Furthermore, we examined the diagnostic yield of lung cancer subtypes (Figure 5). The diagnostic yield was highest for small cell lung cancer, while the accuracy rate was relatively low for squamous cell carcinoma. Squamous cell carcinoma is often accompanied by signs of coagulation necrosis at the center of the LN, which might affect diagnostic accuracy.

In this study, the prediction rate for squamous cell carcinoma was relatively lower than other histology. One of the possible reasons of this phenomenon was that the squamous cell carcinoma often shows various histological characters, such as necrosis and fibrosis, and it reflects the characters on an EBUS ultrasound image such as necrosis sign and heterogeneity of echogram. These various ultrasound image features might cause difficulties for learning and validation by AI-CAD, resulting in a lower prediction rate. Although better AI-CAD analysis required more numbers of squamous carcinoma cases for comprehensive coverage of the image variation of squamous cell carcinoma, the number of actual squamous cell carcinoma cases were relatively low in this study. If we could increase the number of squamous cell carcinoma cases, the diagnostic yield could be better in the future.

This study has several limitations. First, the study population was limited, and we used video clips to overcome the limitations of the small sample size. Some cases underwent multiple LN assessments, and multiple LN images were obtained from a single case, which might show similar image characteristics. Second, we used only B-mode images in this study. Several reports have demonstrated the utility of other imaging modalities such as Doppler mode imaging and elastography [5,6]. Finally, Xception was used for the CNN in this study, although there is currently no consensus as to which algorithm should be used to analyze echo images. To develop the optimal method of AI-CAD for EBUS imaging, a larger prospective cohort study is required in the future. In addition, AI-CAD diagnosis using other imaging modalities such as Doppler mode and elastography should be examined to improve the diagnostic yield of AI-CAD for EBUS imaging.

In this study cohort, the prevalence of nodal metastasis was 38.5%, which was relatively low in comparison with the previous report. Most of the enrolled patients were referred to the surgical department as resectable lung cancer patients. In real clinical setting, the AI-CAD technology will be useful if the operator cannot decide which one to be sampled during EBUS-TBNA. The operator would not need the image analysis support for selecting the target when the lymph node is obviously enlarged. Thus, this study demonstrated that the AI-CAD can be used to support the nodal staging for surgically treatable patients.

5. Conclusions

In this study, we found that AI-CAD (a combination of Xception and the hold-out method) for the prediction of LN metastasis using endobronchial ultrasound images is feasible and exhibits high diagnostic accuracy and specificity. AI-CAD for EBUS may reduce futile biopsies of LNs, shorten examination time, and make EBUS-TBNA a less invasive procedure, regardless of operator experience.

Supplementary Materials: The following supporting information can be downloaded at: https://www.mdpi.com/article/10.3390/cancers14143334/s1.

Author Contributions: Conceptualization, Y.I., T.N., T.O. and I.Y.; methodology, Y.I., T.N.,T.I., Y.S. (Yuki Sata) and T.O.; software, T.O.; validation, Y.I., T.N., T.I., T.O., K.T., Y.S. (Yuichi Sakairi) and H.S.; formal analysis, Y.I.; investigation, Y.I., T.N. and T.O.; resources, Y.I., T.N. and T.O.; data curation, Y.I., T.N. and T.O.; writing—original draft preparation, Y.I.; writing—review and editing, T.N., T.O. and Y.I.; visualization, Y.I.; supervision, T.N. and I.Y.; project administration, Y.I. and T.N.; funding acquisition, T.N. All authors have read and agreed to the published version of the manuscript.

Funding: This is research is a collaboration between Chiba University Graduate School of Medicine and Olympus Medical Systems Corp. (Tokyo, Japan) based on the contract. Funding support was neither provided by Olympus Medical Systems Corp. nor did Olympus Medical Systems Corp. influence the study design or interpretation of the results in this study. This work was supported by JSPS KAKENHI, grant number 21K08880 (PI: Takahiro Nakajima). There was no funding support from Olympus Medical Systems Corp., nor did they influence the study design or interpretation of the results in this study.

Institutional Review Board Statement: We prospectively collected clinical information and images related to bronchoscopy since April 2017 (registry ID: UMIN000026942), and the ethical committee allowed prospective case accumulation with written consent (ethical committee approval ID: No. 2563, Chiba University Graduate School of Medicine). The EBUS-TBNA video clips from April 2017 to December 2020 were retrospectively reviewed, and the patient's clinical information was obtained from electronic medical records (ethical committee approval ID: No. 3538, Chiba University Graduate School of Medicine). This was a collaborative study between the Chiba University Graduate School of Medicine and Olympus Medical Systems Corp. (Tokyo, Japan). All patient identifiers were deleted, and the image data were sent to the Olympus Medical Systems Corp.'s laboratory and analyzed using DL technology (ethical committee approval ID: OLET-2019-008, Olympus Medical Systems Corp.). This study was conducted in accordance with the principles of the Declaration of Helsinki.

Informed Consent Statement: Informed consent was obtained from all subjects involved in the study. Written informed consent has been obtained from the patients to publish this paper.

Data Availability Statement: Not applicable.

Acknowledgments: We would like to thank Junichi Ichikawa, Hironaka Miyaki, Fumiyuki Shiratani, and Jun Ando (Olympus Medical Systems Corporation) for their support with deep learning analysis. We thank the Department of Diagnostic Pathology, Chiba University Hospital, for their support in pathological diagnosis.

Conflicts of Interest: This is a collaborative research project between the Chiba University Graduate School of Medicine and Olympus Medical Systems Corp. (Tokyo, Japan) based on the contract. T.N. received honoraria and lecture fees from Olympus Medical Systems Corp. and AstraZeneca for CME activities related to EBUS-TBNA. The authors other than T.N. have no conflicts of interest related to this study.

References

1. Sakairi, Y.; Nakajima, T.; Yoshino, I. Role of endobronchial ultrasound-guided transbronchial needle aspiration in lung cancer management. *Expert Rev. Respir. Med.* **2019**, *13*, 863–870. [CrossRef] [PubMed]
2. Silvestri, G.A.; Gonzalez, A.V.; Jantz, M.A.; Margolis, M.L.; Gould, M.K.; Tanoue, L.T.; Harris, L.J.; Detterbeck, F.C. Methods for staging non-small cell lung cancer. Diagnosis and management of lung cancer, 3rd ed: American College of Chest Physicians evidence-based clinical practice guidelines. *Chest* **2013**, *143*, e211S–e250S. [CrossRef] [PubMed]

3. Agrawal, S.; Goel, A.D.; Gupta, N.; Lohiya, A.; Gonuguntla, H.K. Diagnostic utility of endobronchial ultrasound (EBUS) features in differentiating malignant and benign lymph nodes—A systematic review and meta-analysis. *Respir. Med.* **2020**, *171*, 106097. [CrossRef] [PubMed]
4. Fujiwara, T.; Yasufuku, K.; Nakajima, T.; Chiyo, M.; Yoshida, S.; Suzuki, M.; Shibuya, K.; Hiroshima, K.; Nakatani, Y.; Yoshino, I. The utility of sonographic features during endobronchial ultrasound-guided transbronchial needle aspiration for lymph node staging in patients with lung cancer: A standard endobronchial ultrasound image classification system. *Chest* **2010**, *138*, 641–647. [CrossRef] [PubMed]
5. Nakajima, T.; Anayama, T.; Shingyoji, M.; Kimura, H.; Yoshino, I.; Yasufuku, K. Vascular image patterns of lymph nodes for the prediction of metastatic disease during EBUS-TBNA for mediastinal staging of lung cancer. *J. Thorac. Oncol.* **2012**, *7*, 1009–1014. [CrossRef] [PubMed]
6. Nakajima, T.; Inage, T.; Sata, Y.; Morimoto, J.; Tagawa, T.; Suzuki, H.; Iwata, T.; Yoshida, S.; Nakatani, Y.; Yoshino, I. Elastography for predicting and localizing nodal metastasis during endobronchial ultrasound. *Respiration* **2015**, *90*, 499–506. [CrossRef] [PubMed]
7. Fujiwara, T.; Nakajima, T.; Inage, T.; Sata, Y.; Sakairi, Y.; Tamura, H.; Wada, H.; Suzuki, H.; Chiyo, M.; Yoshino, I. The combination of endobronchial elastography and sonographic findings during endobronchial ultrasound-guided transbronchial needle aspiration for predicting nodal metastasis. *Thorac. Cancer* **2019**, *10*, 2000–2005. [CrossRef] [PubMed]
8. Ardila, D.; Kiraly, A.P.; Bharadwaj, S.; Choi, B.; Reicher, J.J.; Peng, L.; Tse, D.; Etemadi, M.; Ye, W.; Corrado, G.; et al. End-to-end lung cancer screening with three-dimensional deep learning on low-dose chest computed tomography. *Nat. Med.* **2019**, *25*, 954–961. [CrossRef] [PubMed]
9. Michele, A.; Joseph, S.; Giovanni, P.; Giovanna, S. Radiomics and deep learning in lung cancer. *Strahlenther Onkol.* **2020**, *196*, 879–887.
10. Becker, A.S.; Mueller, M.; Stoffel, E.; Marcon, M.; Ghafoor, S.; Boss, A. Classification of breast cancer in ultrasound imaging using a generic deep learning analysis software: A pilot study. *Br. J. Radiol.* **2018**, *91*, 20170576. [CrossRef] [PubMed]
11. Kim, Y.H.; Kim, G.H.; Kim, K.B.; Lee, W.M.; Lee, E.B.; Baek, H.D.; Kim, H.D.; Park, C.J. Application of A Convolutional Neural Network in The Diagnosis of Gastric Mesenchymal Tumors on Endoscopic Ultrasonography Images. *J. Clin. Med.* **2020**, *9*, 3162. [CrossRef] [PubMed]
12. Nakajima, T.; Yasufuku, K.; Saegusa, F.; Fujiwara, T.; Sakairi, Y.; Hiroshima, K.; Nakatani, Y.; Yoshino, I. Rapid on-site cytologic evaluation during endobronchial ultrasound-guided transbronchial needle aspiration for nodal staging in patients with lung cancer. *Ann. Thorac. Surg.* **2013**, *95*, 1695–1699. [CrossRef] [PubMed]
13. Chollet, F. Xception: Deep Learning with Depthwise Separable Convolutions. In Proceedings of the 2017 IEEE Conference on Computer Vision and Pattern Recognition (CVPR), Honolulu, HI, USA, 21–26 July 2017.
14. Avanzo, M.; Stancanello, J.; Pirrone, G.; Sartor, G. Radiomics and deep learning in lung cancer. *Strahlenther Onkol.* **2020**, *196*, 879–887. [CrossRef] [PubMed]
15. Hawkins, S.; Wang, H.; Liu, Y.; Garcia, A.; Stringfield, O.; Krewer, H.; Li, Q.; Cherezov, D.; Gatenby, R.A.; Balagurunathan, Y. Predicting malignant nodules from screening CT scans. *J. Thorac. Oncol.* **2016**, *11*, 2120–2128. [CrossRef] [PubMed]
16. Onozato, Y.; Nakajima, T.; Yokota, H.; Morimoto, Y.; Nishiyama, A.; Toyoda, T.; Inage, T.; Tanaka, K.; Sakairi, Y.; Suzuki, H.; et al. Radiomics is feasible for prediction of spread through air spaces in patients with nonsmall cell lung cancer. *Sci. Rep.* **2021**, *11*, 13526. [CrossRef] [PubMed]
17. Seijo, L.M.; Peled, N.; Ajona, D.; Boeri, M.; Field, J.K.; Sozzi, G.; Pio, R.; Zulueta, J.J.; Spira, A.; Massion, P.P.; et al. Biomarkers in Lung Cancer Screening: Achievements, Promises, and Challenges. *J. Thorac. Oncol.* **2019**, *14*, 343–357. [CrossRef] [PubMed]
18. Xu, Y.; Hosny, A.; Zeleznik, R.; Parmar, C.; Coroller, T.; Franco, I.; Mak, R.H.; Aerts, H.J.W.L. Deep Learning Predicts Lung Cancer Treatment Response from Serial Medical Imaging. *Clin. Cancer Res.* **2019**, *25*, 3266–3275. [CrossRef] [PubMed]
19. Kurakin, A.; Goodfellow, I.J.; Bengio, S. Adversarial examples in the physical world. *arXiv* **2017**, arXiv:1607.02533.
20. Ozcelik, N.; Ozcelik, A.E.; Bulbul, Y.; Oztuna, F.; Ozlu, T. Can artificial intelligence distinguish between malignant and benign mediastinal lymph nodes using sonographic features on EBUS images? *Curr. Med. Res. Opin.* **2020**, *36*, 2019–2024. [CrossRef] [PubMed]
21. Churchill, I.F.; Gatti, A.A.; Hylton, D.A.; Sullivan, K.A.; Patel, Y.S.; Leontiadis, G.I.; Phil, F.F.M.; Hanna, W.C. An Artificial Intelligence Algorithm to Predict Nodal Metastasis in Lung Cancer. *Ann. Thorac. Surg.* **2021**, *114*, 248–256. [CrossRef] [PubMed]

Current Oncology

MDPI

Article

Novel Harmonization Method for Multi-Centric Radiomic Studies in Non-Small Cell Lung Cancer

Marco Bertolini [1], Valeria Trojani [1,*], Andrea Botti [1], Noemi Cucurachi [1], Marco Galaverni [2], Salvatore Cozzi [3], Paolo Borghetti [4], Salvatore La Mattina [4], Edoardo Pastorello [4], Michele Avanzo [5], Alberto Revelant [6], Matteo Sepulcri [7], Chiara Paronetto [7], Stefano Ursino [8], Giulia Malfatti [8], Niccolò Giaj-Levra [9], Lorenzo Falcinelli [10], Cinzia Iotti [3], Mauro Iori [1] and Patrizia Ciammella [3]

1 S.C. Fisica Medica, Azienda USL-IRCCS di Reggio Emilia, 42124 Reggio Emilia, Italy; marco.bertolini@ausl.re.it (M.B.); andrea.botti@ausl.re.it (A.B.); noemi.cucurachi@ausl.re.it (N.C.); mauro.iori@ausl.re.it (M.I.)
2 S.C. Radioterapia, Azienda Ospedaliero-Universitaria Maggiore, 43126 Parma, Italy; mgalaverni@ao.pr.it
3 S.C. Radioterapia, Azienda USL-IRCCS di Reggio Emilia, 42124 Reggio Emilia, Italy; salvatore.cozzi@ausl.re.it (S.C.); cinzia.iotti@ausl.re.it (C.I.); patrizia.ciammella@ausl.re.it (P.C.)
4 Department of Radiation Oncology, University and Spedali Civili Hospital, 25123 Brescia, Italy; paolo.borghetti@asst-spedalicivili.it (P.B.); s.lamattina@unibs.it (S.L.M.); e.pastorello@unibs.it (E.P.)
5 Medical Physics Department, Centro di Riferimento Oncologico di Aviano (CRO) IRCCS, 33081 Aviano, Italy; mavanzo@cro.it
6 Radiation Oncology Department, Centro di Riferimento Oncologico di Aviano (CRO) IRCCS, 33081 Aviano, Italy; alberto.revelant@cro.it
7 Radiotherapy, Veneto Institute of Oncology IOV—IRCCS, 35128 Padova, Italy; matteo.sepulcri@iov.veneto.it (M.S.); chiara.paronetto@iov.veneto.it (C.P.)
8 Department of Radiation Oncology, Santa Chiara University Hospital, 56100 Pisa, Italy; stefano.ursino@med.unipi.it (S.U.); giulia.malfatti@med.unipi.it (G.M.)
9 Department of Radiation Oncology, IRCCS Sacro Cuore Don Calabria Hospital Negrar, 37024 Verona, Italy; niccolo.giajlevra@sacrocuore.it
10 Radiation Oncology Section, S. Maria della Misericordia Hospital, 06129 Perugia, Italy; lorenzo.falcinelli@ospedale.perugia.it
* Correspondence: valeria.trojani@ausl.re.it

Citation: Bertolini, M.; Trojani, V.; Botti, A.; Cucurachi, N.; Galaverni, M.; Cozzi, S.; Borghetti, P.; La Mattina, S.; Pastorello, E.; Avanzo, M.; et al. Novel Harmonization Method for Multi-Centric Radiomic Studies in Non-Small Cell Lung Cancer. *Curr. Oncol.* **2022**, *29*, 5179–5194. https://doi.org/10.3390/curroncol29080410

Received: 2 May 2022
Accepted: 14 July 2022
Published: 22 July 2022

Abstract: The purpose of this multi-centric work was to investigate the relationship between radiomic features extracted from pre-treatment computed tomography (CT), positron emission tomography (PET) imaging, and clinical outcomes for stereotactic body radiation therapy (SBRT) in early-stage non-small cell lung cancer (NSCLC). One-hundred and seventeen patients who received SBRT for early-stage NSCLC were retrospectively identified from seven Italian centers. The tumor was identified on pre-treatment free-breathing CT and PET images, from which we extracted 3004 quantitative radiomic features. The primary outcome was 24-month progression-free-survival (PFS) based on cancer recurrence (local/non-local) following SBRT. A harmonization technique was proposed for CT features considering lesion and contralateral healthy lung tissues using the LASSO algorithm as a feature selector. Models with harmonized CT features (B models) demonstrated better performances compared to the ones using only original CT features (C models). A linear support vector machine (SVM) with harmonized CT and PET features (A1 model) showed an area under the curve (AUC) of 0.77 (0.63–0.85) for predicting the primary outcome in an external validation cohort. The addition of clinical features did not enhance the model performance. This study provided the basis for validating our novel CT data harmonization strategy, involving delta radiomics. The harmonized radiomic models demonstrated the capability to properly predict patient prognosis.

Keywords: imaging biomarkers and radiomics; quantitative imaging/analysis; computed tomography (ct); multi-modality ct-positron emission tomography (pet); machine learning; non-small-cell lung cancer; stereotactic body radiation therapy (sbrt)

1. Introduction

Non-small-cell lung cancer (NSCLC) is, overall, the second-most-common cancer and a leading cause of cancer-related death worldwide, despite recent therapeutic advances [1]. Stage I disease represents approximately 25% of the patients receiving diagnoses of NSCLC and accounts for the most curable cohort of the population [2]. Surgery is the gold standard for these patients: lobectomy with hilar and mediastinal lymph node dissection is the preferred approach, given the Lung Cancer Study Group (LCSG) trial results [3]. Instead, sublobar resection has shown inferior local control and a trend toward decreased survival. However, evaluation of sublobar resection in selected patients is currently underway. The historical standard therapy for unresectable early-stage NSCLC was conventionally fractionated radiation therapy (RT) (e.g., 2 Gy per fraction, for a total dose of 54–60 Gy). However, the reported long-term local control (LC; 30–70%) and overall survival (OS;15–30%) rates with this approach are suboptimal [4–6].

Advances in imaging and radiation treatment planning and delivery (e.g., with image guidance and motion management) made the delivery of "ablative doses" of radiation to small targets possible with better results in terms of local control [7–11].

Stereotactic Body radiation therapy (SBRT) has proved to be the first therapeutic option in inoperable stage I NSCLC patients or for those who refuse surgical treatment, with similar rates of local tumor control and overall clinical outcomes [12,13]. Recently, a meta-analysis by Li et al. reported a significant superiority in the local control rate and in 3-year and 5-year OS (54.73% and 29.30 % vs. 39.5 and 27.47) in the SBRT group compared with conventionally fractionated RT [14].

SBRT was reported to have a local control rate in excess of 85% at 3 years [14,15]. Despite consistent clinical outcomes, it is well known that dose fractionation heterogeneity and technical expertise may influence the outcome with SBRT [16–18]. A recent study reported that the factors affecting outcomes after SBRT for early-stage NSCLC are Biological Effective Dose (BED) and tumor size [19].

Radiomics is a recent technique introduced in medicine to describe characteristics of medical images quantitatively. Radiomics belongs to artificial intelligence (AI) applications, but it is based on the calculation of features using well-defined mathematical formulas applied directly to the image pixel values (or to a filtered version of the original images). The mathematical definitions of radiomic features are based on the distribution and the relationship between pixels and voxels in the images' region of interest. The concept behind this method lies in the fact that the human eye cannot appreciate all the characteristics of a medical image. Haralick et al. [20] described how the textural features, highlighting the behavior of gray levels' dependencies, can identify different areas in an image. Later, textural information was proposed as an application in medical imaging [21,22]. The improvement in hardware calculation power made these techniques able to compute a high number of medical imaging biomarkers in an acceptable span of time; those indices should help the physician during the treatment decision task, allowing a personalized care pathway for different patients. However, these biomarkers are not yet ready to be used in oncology without a robust validation or a demonstration of their reliability [23]. Among them, radiomic indices and feature signatures are increasingly present in the panorama of modern scientific literature [24,25]. The main issue and challenge up to date are to understand how to overcome the limits of this approach [26,27].

To date, in the literature, several studies have investigated the ability of radiomics features in the tumor-healthy tissue differentiation task, both for computed tomography (CT) and positron emission tomography (PET) datasets, as described by Chu et al. [28] that used feature values in a random forest classifier for diagnostic purposes. In another study [29], healthy tissues' features were used as additional information for an automatic segmentation algorithm. More recently, feature-extracted CT images were combined with BED values to predict tumor response to SBRT [30].

Despite the great work done to date, to our knowledge, there is still no characterization of the radiomic features' ability to give specific information about healthy tissue compared to the sick one when machine learning models for prognosis are involved. One of the main challenges in the field of radiomics, which makes its clinical application difficult, is the harmonization of the features to be analyzed.

Our present multicentric work aims to propose a novel concept of harmonizing the CT radiomic signal using a combination derived from both the tumor and the healthy contralateral tissue, to overcome the variability typical in each patient in different conditions (i.e., manufacturer/technical characteristics, acquisition, reconstruction protocol, and different anatomy).

2. Materials and Methods

2.1. Study Design

This study was a retrospective multicentric work. It was approved by the Area Vasta Emilia Nord (AVEN) Ethics Committee (ID: 817/2018/OSS*/IRCCSRE). The study was also approved by the ethics committees of all the participating institutions; it was performed in accordance with the principles of Good Clinical Practice (GCP) in respect of the ICH GCP guidelines, the ethical principles contained in the Helsinki declaration and its subsequent updates. Each patient gave informed consent for joining the study.

2.1.1. Patient Cohort

Patients who underwent SBRT for histologically proven diagnosis of primary early-stage NSCLC were retrospectively collected from January 2010 to December 2019. A multicenter research project named "TEXture Analysis of PET/CT in lung cancer patients treated with Stereotactic body radiation therapy (TEXAS)" was designed to involve seven Italian Centers.

Inclusion criteria were: (1) histologically proven diagnosis of NSCLC; (2) early-stage T1–T3N0M0 (TNM 7th edition); (3) patients who underwent SBRT, with treatment biological effective dose $BED_{10} \geq 100$ Gy; and (4) age > 18 years.

Exclusion criteria were: (1) lung tumor greater than 7 cm; (2) histologically proven diagnosis of small cell lung cancer or metastasis; (3) previous thoracic irradiation; (4) presence of bone, lymph node, or visceral metastatic lesions; (5) patients with secondary pulmonary nodules from non-NSCLC or NSCLC; (6) past non-NSCLC tumors with evidence of active disease at the time of SBRT and synchronous non-NSCLC tumors (arising within six months of SBRT diagnosis of NSCLC) with the exception in both cases of non-melanomatous skin tumors.

The patient cohort was divided into training (76 patients from three centers) and external validation (41 patients from the other four centers) datasets. This strategy for the distribution of centers among datasets was made to balance the two groups according to the patients' outcomes as described in the following sections. The external validation step was a fundamental part of the study in order to confirm the performances obtained in the training phase.

2.1.2. SBRT Details

Conventional computed tomography (CT) simulation scans were obtained. The radiation oncologist contoured gross tumor volume (GTV) on the CT, as part of the therapeutic pathway. A 5–10 mm isotropic margin was added to GTV to generate the planning target volume (PTV). Intensity-modulated radiation therapy (IMRT) was delivered to all patients. The dose normalization ensured that at least 95% of PTV receives 100% of the prescribed dose with a homogeneous distribution. For all patients, ipsilateral and contralateral lung, heart, chest wall, esophagus, spinal cord, and bronchial trees were contoured as organs at risk (OARs).

2.2. Image Acquisition

All patients included in the study had PET/CT images, previously acquired as part of their care pathways, and a pre-treatment CT used for planning of SBRT. The planning

CT acquisition protocols and scanning devices differed among institutions, as reported in Table 1. PET image sets, corrected for attenuation, were acquired no more than three months before the start of the treatment. Patients fasted at least 6 h before the injection 18F-FDG tracer and the serum glucose level measured at the injection time was below 160 mg/mL in all patients. PET examinations were performed 60 min after the intravenous administration of the radiotracer using a specific protocol for each institution shown in Table 1.

Table 1. Protocol acquisition parameters for simulation CT and PET examinations stratified for centers. Whenever two scanners were used, a " | " indicated the different configurations.

	CT						
	Center	**kV**	**mAs (Min–Max)**	**Slice Thickness (mm)**	**Manufacturer (s)**	**Convolution Kernel**	**Recon Diameter**
TRAIN	BS	120	191–401	3.0	PHILIPS	B	500
	RE	120	83–355	3.0	GE	STD +	500
	PD	120	70–363	2.5	GE	BODY FILTER	500
EXT VAL	AV	120	108–138	2.5	PHILIPS	B	500
	NE	120	40–73	3.0	SIEMENS	B30f	500
	PI	120	27–236	2.0	SIEMENS	B30f–B31s	500
	PG	120	80–200	2.5–3	GE	STD +	500

	PET				
	Center	Slice thickness (mm)	Manufacturer (s)	Recon diameter	Recon method
TRAIN	BS	3.27	GE	700–815	3D IR/VPFXS
	RE	3.27	GE	700–700	3D IR/VPFXS
	PD	2–4	PHILIPS | SIEMENS	576–815	3D-RAMLA/BLOB-OS-TF(PHILIPS) | PSF 3i21s/(SIEMENS)
EXT VAL	AV	4	PHILIPS | GE	500–700	BLOB-OS-TF/VPFXS
	NE	2–5	SIEMENS	576–700	PSF+TOF 3i21s
	PI	3.27	GE | PHILIPS	576–700	3D IR (GE) | BLOB-OS-TF(PHILIPS)
	PG	3.27	GE | SIEMENS	600–700	OSEM | OSEM 2i8s

2.3. *Image Segmentation*

Computed tomography and PET image sets were exported in DICOM format into a dedicated research computer for radiomics analysis. For the present study, gross tumor volume contouring was separately performed on the CT (manually, referred to as GTVCT) and PET (automatically, hereinafter named GTVPET) images of the pre-treatment PET/CT studies.

Two radiation oncologists with experience in lung cancer contoured each lesion on every sequential slice of the planning CT using standardized window settings for parenchyma (W = 1600 and L = −600), according to EORTC guidelines [31] for all patients. Regarding GTVPET delineation, the radiation oncologists placed a region of interest (ROI) on the area of tumor FDG uptake on PET images and an automatic contour—consisting of the region encompassed by a given fixed percent intensity level relative to the maximum registered tumor activity (40% of SUV max)—was generated. We decided to use this approach as a previous study showed that GTVPET delineation using this fixed threshold was better correlated with the gross tumor (based on pathologic examination) instead of using as basis the manually delineated GTVCT [32].

In order to perform radiomic feature harmonization, we used an ROI from the healthy tissue. This was obtained by copying the GTVCT into a healthy lung region, i.e., the contralateral lung at the same level of the GTV (named Contra_Lung). The Contra Lung initial volume was also shifted by 0.6 and 0.3 cm in six directions for a total of 12 shifts, avoiding the inclusion of surrounding tissues of the healthy lung. These shifts had the aim of simulating the uncertainty in the positioning of the healthy ROI (for future reproducibility of the harmonization method). We chose the shifts in accordance with PET image resolution to account for a likely uncertainty in ROI positioning since PET imaging can be used to localize the GTV before the treatment.

An example picture showing the location of the ROIs mentioned above is shown in Figure 1.

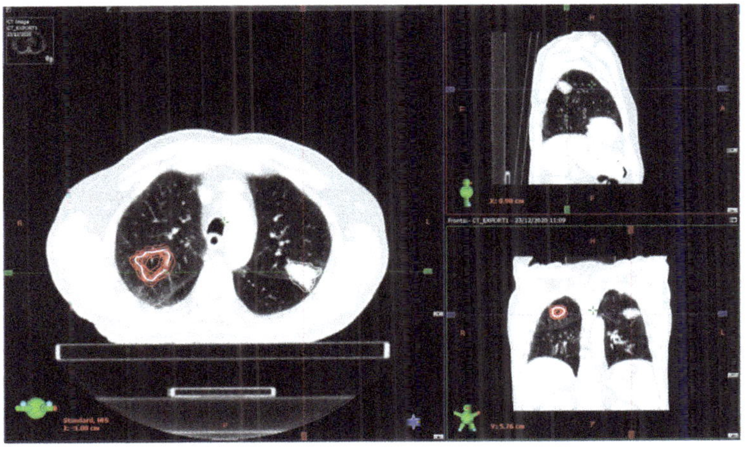

Figure 1. Visualization of the CT ROIs in a patient. The contralateral ROI was shifted in 12 different positions (shown in red).

2.4. Outcome

In this study, PFS was considered as the primary endpoint and was converted into a binary outcome, which was set to 1 for patients who were alive and without disease progression at 24 months, 0 otherwise. PFS was defined as the time from the start of the SBRT to documented relapse or death. The use of the 2-year threshold was chosen because it could properly describe the treatment effectiveness. In fact, a preliminary analysis of the Kaplan–Meier curves of PFS after SBRT for our cohort showed that the majority of the progressions occurred in a period ranging from 2 to 3 years.

2.5. Radiomics Analysis

Our analysis followed the steps defined for our radiomic study (Figure 2), which included image preprocessing. The first phase consisted of spatial resampling to an isotropic voxel size to obtain reproducible and rotationally invariant features. Then, image range re-segmentation updated the ROI voxels according to a chosen intensity range to remove all voxels for which intensity values fall outside the selected intensity range. Finally, the images were discretized by intensity, grouping the original intensity values (256) into specific ranges (bins). The aim was to reduce image noise and computational burden. The intensity discretization process fixed the width of the re-segmentation interval and the bin width, defining a new bin for each intensity interval. Selecting the bin width allowed direct control of the absolute range represented on each bin. The image preprocessing of intensity and spatial discretization is described in Supplementary Material Table S1. Intensity discretization parameters were chosen accordingly to the guidelines proposed by Orlhac et al. [33,34].

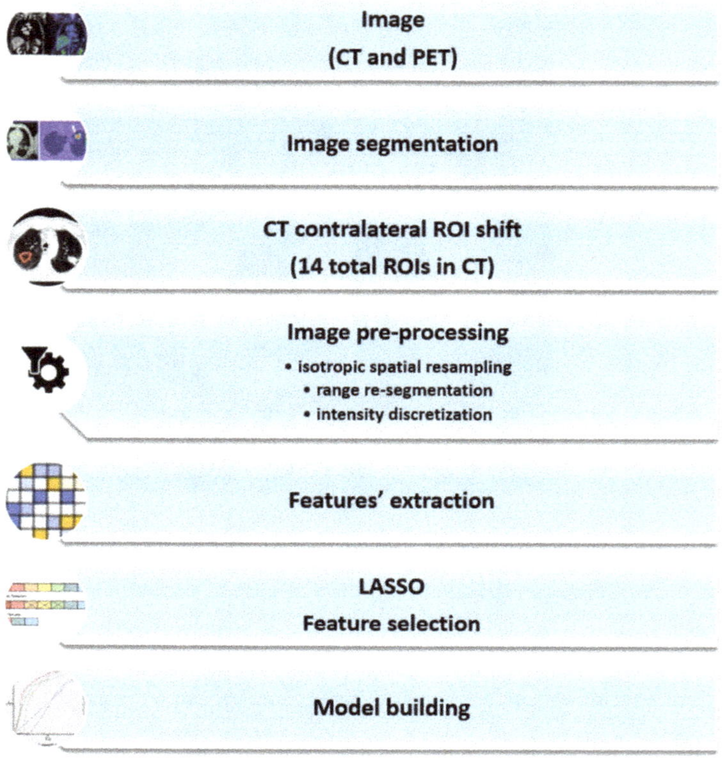

Figure 2. Radiomic pipeline description of the implemented steps in our evaluation process.

2.5.1. Feature Extraction

After the image preprocessing steps, feature calculation and their extraction was performed. Features (intensity-based, shape-based, and second-order) were extracted from original images and filtered images (using wavelets, Laplacian of Gaussian (LoG), and gamma modifier filters) [35]. Radiomic features were calculated using a homemade software employing the widely used pyRadiomics library in order to apply pre-determined filters to the original images and compute features from the edited results. The list of the extracted radiomic features can be found at https://pyradiomics.readthedocs.io/en/latest/features.html, (accessed on 15 July 2022).

2.5.2. Harmonization Process

The harmonization process consisted of, for our two available image modalities, calculating features for 14 different ROIs. One of them coincided with GTVCT, the other 13 with the duplicated GTVCT positioned in the healthy region and its shifts, as described in Section 2.3. This allows us to consider operators' variability in the positioning of the healthy ROI. The general idea of employing this harmonization formula was inspired by another work [36], and it is shown in Equation (1):

$$f_{HARM}(i) = \frac{f_{GTV}(i) - f_{HEALTHY}(i)}{\sigma(i)} \tag{1}$$

where: $f_{HARM}(i)$ is the ith harmonized feature, $f_{HEALTHY}(i)$ is the median ith feature value calculated on the 13 healthy tissue samples for each patient and modality, and $\sigma(i)$ is the difference between the 75th and the 25th percentile of the ith feature distribution. Shape features were not harmonized. We applied this harmonization only to CT data because of its intrinsic dependence on the protocol acquisition parameters. Furthermore, CT is used for a morphologic and anatomical characterization and pixel values are related to a physical characteristic of the tissue (the linear attenuation coefficient). On the other hand, PET, being a functional imaging modality, is less sensitive to low signal changes in spatial coordinates. Especially in this case, for the healthy lung, in PET pixel values, there is no useful physical information regarding a region where we do not register a signal from the radiotracer absorption.

2.5.3. Feature Selection

LASSO feature selection was applied, in which the following function is minimized (Equation (2))

$$\sum_{i=1}^{n}\left(y_i - \beta_0 - \sum_{j=1}^{p}\beta_j x_{ij}\right)^2 + \lambda\sum_{j=1}^{p}|\beta_j| \quad (2)$$

where: y_i is the observed value, β_j is the LASSO coefficients, and $\lambda \sum_{j=1}^{p}|\beta_j|$ is the shrinkage penalty [37].

The parameter λ was chosen using 10-fold cross-validation (CV) computing its error. LASSO penalty brings to zero the weight coefficients (β_j) of irrelevant features not predictive of the chosen outcome. In addition, LASSO handles sets of collinear features by increasing the weight of one of them while setting the other weights to zero. Because the considered outcome was binary, we used a binomial function for LASSO regression. In Table S2 we show the shrinkage penalties for our trained models.

2.5.4. Model Building

The original and harmonized features were used to develop a supervised machine learning binary classifier. A linear support vector machine (SVM, Model 1) [38] and an Ensemble Subspace Discriminant (ESD, Model 2) [39] were trained by optimizing their performance in 10-fold cross-validation in the training dataset.

Linear SVM classifiers provide low generalization error, even with small learning sample datasets. ESD classifiers are used to decide an explicit discriminant subspace of low dimension.

The two described model types were applied to five different combinations of input features: (A) harmonized CT + PET, (B) harmonized CT, (C) original CT, (D) only original PET, and (E) harmonized CT + PET + selected clinical variables in order to assess the effect of harmonization on the performance of the predictive models. The interested reader can find more information in Text S1 in Supplementary Materials.

The clinical variables in method (E) were chosen among the available ones by using Kaplan–Meier survival curves as described in Section 2.5.5.

2.5.5. Statistical Analysis

For each model, a 95% confidence interval (CI) of the AUCs was calculated for the training and external validation sets. Furthermore, accuracy (95% CI are reported), precision, and recall were calculated.

Subsequently, Kaplan–Meier survival curves were computed using the PFS to select the clinical features. A clinical feature exhibiting a p-value from a log-rank test less than 0.05 was considered significant and included in model E. Matlab R2021b (Mathworks, Natick, MA) and R (Vienna, Austria), available at https://www.R-project.org (accessed on 15 July 2022), were used to perform the statistical analysis.

The p-values related to statistical differences among the AUC values of each model were calculated using two-sided DeLong test.

3. Results

3.1. Clinical Results

One-hundred and seventeen early-stage NSCLC patients met the inclusion criteria. The baseline characteristics of the patients are summarized in Table 2. The median age was 78 years and there were more male (72.6%) than female patients. With a median follow-up of 29.8 months, the median PFS was 24.2 months, and 2-year PFS percentage was 51.2%. Median OS and 2-year OS percentage were 28.5 months and 64%, respectively. The clinical characteristics, including age, gender, Charlson comorbidity index (CCI), diffusing capacity of carbon monoxide (DLCO), tumor size, Eastern Cooperative Oncology Group (ECOG) performance status, and biological equivalent dose to PTV, showed no significant differences between the training and external validation cohorts (Table 2).

Table 2. Statical analysis of clinical variables. Abbreviations: PS: Performance status according to ECOG scale, BPCO: chronic obstructive pulmonary disease; ADK: Adenocarcinoma, SCC: squamous cell carcinoma, Fr: fraction; RT: radiotherapy VMAT: volumetric arc-therapy; IMRT: intensity modulated radiotherapy, TOMO: Tomotherapy, PTV: planning target volume. *p*-values in bold mean the statistical significance.

Characteristics	Training Cohort (N = 76)	External Validation Co#Hort (N = 41)	*p*
Gender			
Male	61	24	**0.04**
Female	15	17	
Age (years)	78 [51–87]	79 [57–88]	0.72
Smoking Status			
Yes	50	27	0.22
No	26	14	
Performance Status			
0	37	18	
1	35	15	0.75
2	4	7	
BMI	25.2 [16.4–37.1]	24.8 [18.3–44.7]	0.17
Diabetes mellitus			
Yes	16	12	0.58
No	60	29	
BPCO			
Yes	43	17	0.54
No	19	24	
Charlson Comorbidity Index (CCI)			
Median	6.5	6	0.55
Range	[3–13]	[4–10]	
T diameter			
Median	2.35	2.3	0.58
Range	[0.6–5.5]	[0.72–27]	
Lesion type			
Subsolid	5	4	0.42
Solid	71	37	
Lung Side			
Lung right	42	22	**0.006**
Lung left	34	19	

Table 2. *Cont.*

Characteristics	Training Cohort (N = 76)	External Validation Cohort (N = 41)	p
Lobe Site			
Upper Lobe	44	23	
Lower Lobe	30	15	0.89
Middle Lobe	2	1	
Lesion Site			
Peripheral	55	34	0.92
Central	21	7	
BED_{10}			
Median	115.5	100	0.64
Range	[100–180]	[100–132]	

No clinical or treatment-related features were shown to be significantly related to PFS in the univariate analysis of the whole population, except for gender ($p = 0.04$ in favor of female) and lung site (right vs. left in favor of the right one, $p = 0.006$).

3.2. PFS Models

The PFS predictive performance of the linear SVM and ESD models using radiomic features and clinical features are reported in Table 3. In Figure 3, all the models are graphically compared considering their confidential intervals. Models using harmonized features and PET (A,E) achieved AUCs greater than 0.70, both in training and validation. The performances of models using CT-only harmonized features (B models) are not confirmed on the validation dataset (AUC training > 0.75, AUC validation <0.60), while adding PET features leads to better stability between the training and validation sets. Only C-type models (original CT-only features) showed a low-mean AUC (< 0.62), both in training and validation.

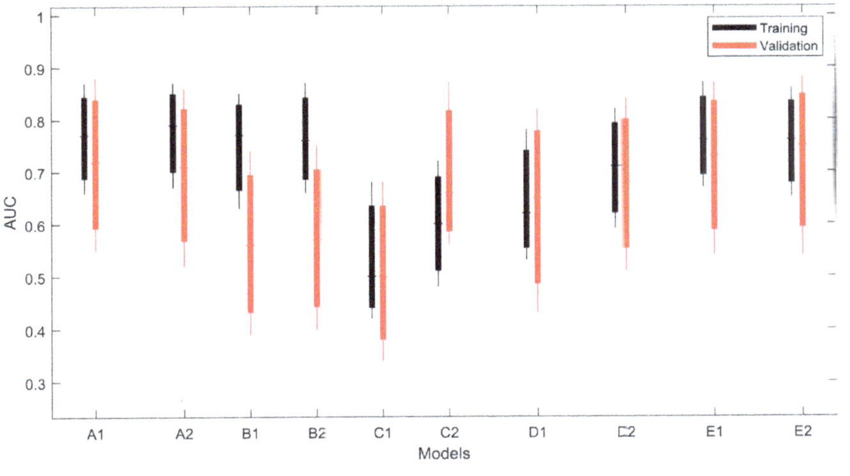

Figure 3. Performances (AUC) of the studied models. The boxplot shows the minimum, maximum, and average values of the bootstrapped 95% CIs.

Table 3. Models' results in terms of AUC, accuracy, precision, and recall.

	AUC	Accuracy	Precision	Recall	p
Harmo CT + Original PET Features (A)					
Linear SVM (A1)					
	AUC *	Accuracy	Precision **	Recall **	p ***
Training dataset	0.77 [0.66–0.87]	0.72 ± 0.02	0.67	0.83	**1.0×10^{-4}**
External validation dataset	0.75 [0.55–0.88]	0.66 ± 0.01	0.68	0.65	**0.01**
Subspace Discriminant (A2)					
	AUC *	Accuracy	Precision **	Recall **	p ***
Training dataset	0.79 [0.67–0.87]	0.71 ± 0.01	0.69	0.83	**0.02**
External validation dataset	0.71 [0.52–0.86]	0.63 ± 0.02	0.68	0.65	**0.046**
Harmo CT features (B)					
Linear SVM (B1)					
	AUC	Accuracy	Precision **	Recall **	p ***
Training dataset	0.77 [0.63–0.85]	0.67 ± 0.02	0.74	0.58	**1.0×10^{-4}**
External validation dataset	0.56 [0.39–0.74]	0.58 ± 0.01	0.67	0.52	0.5
Subspace Discriminant (B2)					
	AUC	Accuracy	Precision **	Recall **	p ***
Training dataset	0.76 [0.66–0.87]	0.71 ± 0.02	0.73	0.6	**0.01**
External validation dataset	0.57 [0.4–0.75]	0.58 ± 0.01	0.67	0.52	0.50
Original CT features (C)					
Linear SVM (C1)					
	AUC	Accuracy	Precision **	Recall **	
Training dataset	0.56 [0.42–0.68]	0.52 ± 0.03	0.49	0.45	
External validation dataset	0.50 [0.34–0.68]	0.43 ± 0.02	0.54	0.65	
Subspace Discriminant (C2)					
	AUC	Accuracy	Precision **	Recall **	
Training dataset	0.63 [0.48–0.72]	0.56 ± 0.03	0.58	0.56	
External validation dataset	0.51 [0.39–0.74]	0.54 ± 0.01	0.58	0.65	
PET features only (D)					
Linear SVM (D1)					
	AUC	Accuracy	Precision **	Recall **	p ***
Training dataset	0.68 [0.53-0.78]	0.64 ± 0.03	0.64	0.80	0.09
External validation dataset	0.65 [0.43-0.82]	0.64 ± 0.01	0.67	0.78	0.18
Subspace Discriminant (D2)					
	AUC	Accuracy	Precision **	Recall **	p ***
Training dataset	0.71 [0.59–0.82]	0.69 ± 0.01	0.67	0.8	0.10
External validation dataset	0.68 [0.51–0.84]	0.60 ± 0.01	0.67	0.61	0.08
Harmo CT + Original PET + Clinical features (E)					
Linear SVM (E1)					
	AUC *	Accuracy	Precision **	Recall **	p ***
Training dataset	0.79 [0.67–0.87]	0.73 ± 0.02	0.72	0.83	**6.0×10^{-5}**
External validation dataset	0.73 [0.54–0.87]	0.73 ± 0.01	0.77	0.74	**0.02**
Subspace Discriminant (E2)					
	AUC *	Accuracy	Precision **	Recall **	p ***
Training dataset	0.76 [0.65–0.86]	0.74 ± 0.01	0.72	0.83	0.01
External validation dataset	0.75 [0.54–0.88]	0.68 ± 0.02	0.73	0.70	0.02

* AUCs in square brackets are their bootstrapped 95% CIs. ** Precision and recall are presented for class 1. *** p-values are calculated with respect to the conditions C1 and C2 for linear SVM and ESD models, respectively. Values in bold mean the statistical significance.

It is worth noting that both A1 and A2 models significantly outperformed C1 and C2 models, both in the training and external validation datasets ($p = 0.0001$, $p = 0.01$ and $p = 0.02$, $p = 0.046$, respectively, for linear SVM and subspace discriminant models), likewise for E and C models (E1: $p < 0.0001$, and $p = 0.02$, and E2: $p = 0.01$, and $p = 0.02$, respectively, for training and external validation datasets). C models outperformed B models, but only in the training dataset ($p < 0.0001$ and $p = 0.01$, respectively, for linear SVM and subspace discriminant models) and D and C models had the same performance metrics. In summary, models using clinical information (E models) do not add a significant improvement to A models. This effect is also appreciable in Figure 4, where all the p-values among the models are represented.

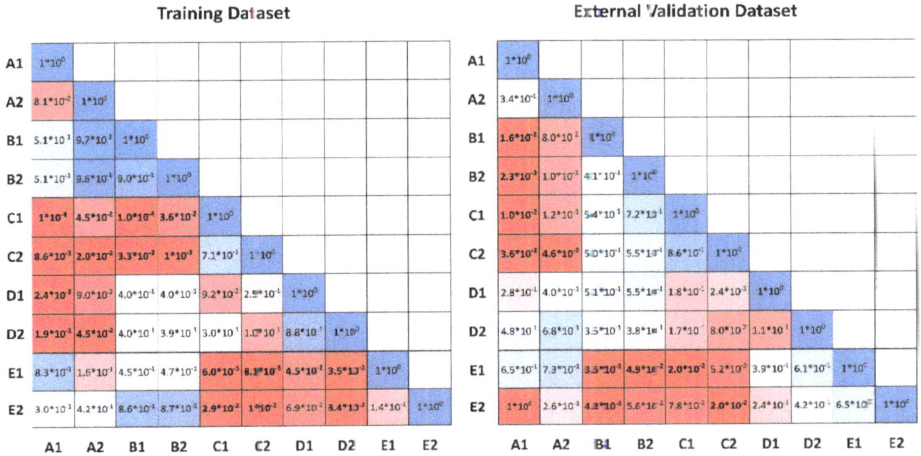

Figure 4. p-values calculated using the two-sided DeLong test. Numbers in bold mean the statistical significance.

4. Discussion

In this work, a multi-centric cohort of early-stage NSCLC patients treated with SBRT was used to build and validate predictive models of PFS greater than 24 months using radiomic features from CT and PET exams and clinical information. Several existing works in the literature [40] describe acquisition protocol variabilities in multi-centric studies, which could affect the performance of radiomic models. Since radiomics computes features from the pixel values in the images, differences in acquisition protocol can lead to biased results. The rationale behind our harmonization method lies in the fact that retrospective multicenter radiomic studies are challenging but necessary, as gathering data from several centers for a centralized analysis is complex for legal, ethical, administrative, and technical reasons. Most of the time, the different centers involved do not follow standardized acquisition and reconstruction protocols; therefore, the collected data suffer from intra-, and inter-variability, making radiomic features sensitive to multicenter variability. Our novel harmonization technique aims to reduce the bias caused by the absence of standardized protocols. Generally, feature analysis is performed by calculating them inside an ROI that coincides with the lesion target. Our study aims to tackle this issue by attempting to reduce this effect using the healthy region of the patient as the baseline from which to harmonize the radiomic data computed from the lesion. From our results, the harmonization improves models' performance when it is used on CT image sets. On the other hand, we expected that for PET-only images, the harmonization method is not easily applicable due to the functional aim of this imaging modality. In such a modality, several healthy tissues (i.e., lungs) are not 18-FDG-avid, while a harmonization based on healthy radiotracer accumulation, to our knowledge, has yet to be studied. Our work investigated and evaluated the feasibility of this technique, which could be employed and better analyzed in future studies.

When using original feature values, PET features were preferred over CT during feature selection, resulting in an only PET-based model. Furthermore, in A models, two features from CT were included in the final prediction score (Log_Sigma30mm_GLDM_Small-Dependence-High-Gray-Level-Emphasis (SDHGLE) and Wavelet_LHH_NGTDM_Busyness), which were also selected in B models. In the same way, a subset of selected features in the A method (Log_sigma10mm_GLSZM_Size-Zone-Low-Gray-Level-Zone-Emphasis (LGLZE), Exponential_FIRST ORDER_Median, Square_GLSZM_ZoneEntropy (ZE)) is also present among the selected features in the PET-only-based method. This could mean that our approach, also given the higher performance metric of the A method, was able to merge the information hidden in the CT and PET image sets in a multi-centric cohort of patients, highlighting the importance of properly handling hybrid imaging in radiomic models.

Other harmonization techniques were previously described in the literature [41]. Among these, the ComBat harmonization method, which removes batch effects, mostly based on an empirical Bayes framework, is one of the most used. Conversely, the ComBat method has some limitations: for instance, the dimension of the homogeneous group cannot be too few (in our study, it would have not been applicable as most of the centers provided less than 15 patients). Indeed, recently, some methods have been proposed to overcome these limitations, e.g., using bootstrap and Monte Carlo technique to improve robustness in the estimation [42].

Even if Monte Carlo and bootstrap strategies aimed at overcoming the cohort size limitations, the objective of this method still focuses on removing differences in radiomic feature distributions among different labels (corresponding to different centers). ComBat, thus, relies on the individual distributions and changes made to feature values are dependent on a group of patients. While we know that there are data supporting the effectiveness of this method (especially in making the feature distributions uniform), we wanted to tackle the multicenter studies issue from a different angle, which is to account for the individual patients' differences (caused both by the scanner/institution protocols and their anatomy) taken directly from the lesion imaging. This renders the method easier, both computationally and for cohort eligibility reasons (which, in the ComBat method, is needed for representing the single center in terms of homogeneity being an assumption of the method). In fact, if validated further, our method can be applied even in heterogenous cohorts since it uses only the single image set of the patient.

Our approach aimed to use all the information present in the CT data, both from cancer lesions and the contralateral healthy tissue, simulating the radiologists' skill in subjectively evaluating a lesion and adding this information in quantifiable and statistical terms (through the radiomic features).

In our work, the well-known and studied concept of delta radiomics was implemented not in a temporal sense but spatially (cancer vs. healthy tissue), which is an approach that, to our knowledge, was not applied in other prognostic oncological works. Traditional radiomics uses absolute values extracted from regions of interest to predict a clinical outcome. On the other hand, delta radiomics predicts a clinical outcome through the combination of radiomic values computed from image sets acquired at different time points (i.e., radiographs to monitor follow-ups or differences between basal PET and interim PET), which is a rationale also used in clinical practice to assess lesion progression (i.e., PERCIST). In our manuscript, we decided to apply delta radiomics not between different time points but between different anatomic locations (healthy vs. tumor tissues). The assumption behind this use of delta radiomics is that each patient can have an intrinsic "baseline" value for a certain radiomic feature (caused by individual anatomy and institution protocols) that needs to be accounted for when building predictive models. Comparison between normal and tumor tissue behavior (even in terms of pixel values) is also common in clinical practice (i.e., SUV values typical of physiological metabolism or HU/density values of healthy tissue). Some authors [43–45] explain that delta radiomics—which is the use of textural indices associated with different time points or anatomical regions—is more successful than traditional radiomics. Our work aims to provide the basis for a framework where the study of simple absolute feature values can make room for the analysis of their relationship to a reference, used as a threshold or as a comparison.

There are several limitations to the current work. Our study suffered a restricted number of patients selected retrospectively. Nonetheless, the patients' number seemed reasonable at the current phase of our study. It assures the homogeneity in terms of pathologies, as including only NSCLC lesions prevented possible biases created by evaluating different diseases, even in the lung anatomical district. Our study highlighted the necessity to monitor and carefully use features related to pixel values and their relationships. In this case, we can assume that the importance of the radiomic features is not only held in their numerical value but also in the intrinsic relationship among those values. The textural indices' ability to perform more complex clinical tasks (i.e., predicting toxicity and its grade) could be further examined in the next phase of our work or in a prospective study design, which could also assess the robustness of our method. We believe that a prospective study will be able to validate these models within a cohort gathered with a better strategy.

Interestingly enough, we found that the improved performance in models employing harmonized features in the training phase was also confirmed in the validation dataset; the use of an external dataset is becoming more and more crucial to radiomics studies to assure and facilitate their introduction in clinical practice.

In our study, no clinical or treatment-related features were shown to be significantly related to PFS, except for gender and lung site. It is well known in the literature that gender is a prognostic factor for PFS [46–48]. Due to the size of our cohort, we did not find significant correlations between PFS and other studied clinical prognostic variables, such as age or histology (also due to the inclusion criteria). To our knowledge, we did not find any other study reporting a significant correlation between PFS and tumor laterality; thus, we will investigate this finding together with our model generalization power in a future prospective study. Indeed, in the literature, many studies showed a significant correlation between some clinical or treatment-related characteristics and outcomes (PFS and OS) and some created predictive models. Among the various statistical prediction models, nomograms can be accurate and feasible prognostic instruments with high utility in estimating individual patient risk and may, thus, help guide treatment decisions in clinical practice. At present, there are some nomograms, based on clinical features, developed for early-stage NSCLC treated with SBRT [49–51], but there is still need for validation of the clinical variables found in those studies and their experimental results in more robust cohorts, such as prospective ones. Therefore, a need exists for a robust recurrence-related prediction model to help select high-risk candidates who may benefit from additional systemic therapies.

In this scenario, a predictive model based on radiomics and clinical and treatment-related characteristics can improve the prediction of clinical outcomes, as already demonstrated by other works. We also found out that clinical variables did not improve the radiomics models, but only the proposed harmonization process statistically significantly improves the model's performance.

As previously stated, our future aim is to apply our method to a prospective multicentric cohort to further validate the framework's stability. In addition, other anatomical regions should be explored to generalize the harmonization, even when the healthy area is not so easily defined as in the lung case. Regarding the employment of this method also in PET datasets, it could be interesting to explore the feasibility of applying our harmonization to 18F-FDG-avid anatomical regions, such as the liver or the brain, which are, however, not related to a pathologic response. Such a method could be especially useful where CT-PET is the only exam included in the care pathway of the patient.

5. Conclusions

A novel strategy of CT data harmonization involving delta radiomics, considering both cancer and healthy tissue in the contralateral lung, was tested and externally validated in a multi-centric study for NSCLC patients, to initially assess its feasibility.

The radiomics models with harmonized features can predict better the selected patient outcome in our cohort, providing valuable additional information to the clinician.

Supplementary Materials: The following are available online at https://www.mdpi.com/article/10.3390/curroncol29080410/s1, Table S1: The image preprocessing of intensity and spatial discretization, Table S2: Lasso feature selection, Text S1: Models' description.

Author Contributions: M.B.: study, design, literature search, data analysis/interpretation, writing original draft, supervision, V.T.: study, design, literature search, data analysis/interpretation, writing original draft, A.B.: Data analysis/interpretation, data validation, reviewing draft, N.C.: literature search, data analysis/interpretation, data validation, figures, writing, M.G.: study design, data collection, S.C.: literature search, reviewing draft, P.B.: resources, data collection, S.L.M.: resources, data collection, E.P.: resources, data collection, M.A.: resources, data collection, reviewing draft, A.R.: resources, data collection, M.S.: resources, data collection, C.P.: resources, data collection, S.U.: resources, data collection, G.M.: resources, data collection, N.G.-L.: resources, data collection, L.F.: resources, data collection, C.I.: resources, data collection, M.I.: resources, data collection, P.C.: resources, data collection, supervision. All authors have read and agreed to the published version of the manuscript.

Funding: This study was partially supported by the Italian Ministry of Health—Ricerca Corrente.

Institutional Review Board Statement: The study was conducted in accordance with the Declaration of Helsinki, and approved by the Ethics Committee of by the Area Vasta Emilia Nord (AVEN) (protocol code 817/2018/OSS*/IRCCSRE approved on 16 July 2019).

Informed Consent Statement: Informed consent was obtained from all subjects involved in the study.

Data Availability Statement: The training weights of the models proposed in this work (namely, models E and A described in Section 2.5.4) are available in a GitHub repository at the link: https://github.com/ausl-re/TEXAS, accessed on 15 July 2022. Further instructions on how to perform the harmonization and to use the model will be added.

Acknowledgments: A heartfelt thanks to all the people involved in the various centers in collecting and managing the data, especially Simona Marani, Maria Paola Ruggieri, Giulia Mascari, and Cinthia Aristei. A special thanks to the AIRO Lung Group, especially Vieri Scotti, Stefano Vagge, and Alessio Bruni.

Conflicts of Interest: The authors declare no conflict of interest.

References

1. Siegel, R.; Naishadham, D.; Jemal, A. Cancer statistics, 2013. *CA Cancer J. Clin.* **2013**, *63*, 11–30. [CrossRef]
2. Edge, S.B.; Byrd, D.R.; Compton, C.C.; Fritz, A.G.; Greene, F.L.; Trotti, A. *AJCC Cancer Staging Manual*, 7th ed.; Springer: New York, NY, USA, 2010; pp. 253–264.
3. Ginsberg, R.J.; Rubinstein, L.V. Randomized trial of lobectomy versus limited resection for T1 N0 non-small cell lung cancer. Lung Cancer Study Group. *Ann. Thorac. Surg.* **1995**, *60*, 615–622. [CrossRef]
4. Bogart, J.A. Fractionated radiotherapy for high-risk patients with early-stage non-small cell lung cancer. *Semin. Thorac. Cardiovasc. Surg.* **2010**, *22*, 44–52. [CrossRef]
5. Hayman, J.A.; Martel, M.K.; Ten Haken, R.K.; Normolle, D.P.; Todd, R.F., III; Littles, J.F.; Sullivan, M.A.; Possert, P.W.; Turrisi, A.T.; Lichter, A.S. Dose escalation in non-small-cell lung cancer using three-dimensional conformal radiation therapy: Update of a phase I trial. *J. Clin. Oncol.* **2001**, *19*, 127–136. [CrossRef]
6. Qiao, X.; Tullgren, O.; Lax, I.; Sirzén, F.; Lewensohn, R. The role of radiotherapy in the treatment of stage I non-small cell lung cancer. *Lung Cancer* **2003**, *41*, 1–11. [CrossRef]
7. Papiez, L.; Timmerman, R.; DesRosiers, C.; Randall, M. Extracranial stereotactic radioablation: Physical principles. *Acta Oncol.* **2003**, *42*, 882–894. [CrossRef]
8. Timmerman, R.; Papiez, L.; McGarry, R.; Likes, L.; DesRosiers, C.; Frost, S.; Williams, M. Extracranial stereotactic radioablation: Results of a phase I study in medically inoperable stage I non-small cell lung cancer. *Chest* **2003**, *124*, 1946–1955. [CrossRef]
9. Timmerman, R.; Paulus, R.; Galvin, J.; Michalski, J.; Straube, W.; Bradley, J.; Fakiris, A.; Bezjak, A.; Videtic, G.; Johnstone, D.; et al. Stereotactic body radiation therapy for inoperable early-stage lung cancer. *JAMA* **2010**, *303*, 1070–1076. [CrossRef]
10. Timmerman, R.D.; Herman, J.; Cho, L.C. Emergence of stereotactic body radiation therapy and its impact on current and future clinical practice. *J. Clin. Oncol.* **2014**, *32*, 2847–2854. [CrossRef]
11. Videtic, G.M.; Stephans, K.L.; Woody, N.M.; Reddy, C.A.; Zhuang, T.; Magnelli, A.; Djemil, T. 30 Gy or 34 Gy? Comparing 2 single-fraction SBRT dose schedules for stage I medically inoperable non-small cell lung cancer. *Int. J. Radiat. Oncol. Biol. Phys.* **2014**, *90*, 203–208. [CrossRef]
12. Simone, C.B., 2nd; Wildt, B.; Haas, A.R.; Pope, G.; Rengan, R.; Hahn, S.M. Stereotactic body radiation therapy for lung cancer. *Chest* **2013**, *143*, 1784–1790. [CrossRef]

13. Ackerson, B.G.; Tong, B.C.; Hong, J.C.; Gu, L.; Chino, J.; Trotter, J.W.; D'Amico, T.A.; Torok, J.A.; Lafata, K.; Chang, C.; et al. Stereotactic body radiation therapy versus sublobar resection for stage I NSCLC. *Lung Cancer* **2018**, *125*, 185–191. [CrossRef]
14. Li, C.; Wang, L.; Wu, Q.; Zhao, J.; Yi, F.; Xu, J.; Wei, Y.; Zhang, W. A meta-analysis comparing stereotactic body radiotherapy vs. conventional radiotherapy in inoperable stage I non-small cell lung cancer. *Medicine* **2020**, *99*, e21715. [CrossRef]
15. Fakiris, A.J.; McGarry, R.C.; Yiannoutsos, C.T.; Papiez, L.; Williams, M.; Henderson, M.A.; Timmerman, R. Stereotactic body radiation therapy for early-stage non-small-cell lung carcinoma: Four-year results of a prospective phase II study. *Int. J. Radiat. Oncol. Biol. Phys.* **2009**, *75*, 677. [CrossRef]
16. Koshy, M.; Malik, R.; Mahmood, U.; Husain, Z.; Sher, D.J. Stereotactic body radiotherapy and treatment at a high volume facility is associated with improved survival in patients with inoperable stage I non-small cell lung cancer. *Radiother Oncol.* **2015**, *114*, 148–154. [CrossRef]
17. Olsen, J.R.; Robinson, C.G.; El Naqa, I.; Creach, K.M.; Drzymala, R.E.; Bloch, C.; Parikh, P.J.; Bradley, J.D. Dose Response for stereotactic body radiotherapy in early-stage nonsmall-cell lung cancer. *Int. J. Radiat. Oncol. Biol. Phys.* **2013**, *81*, e299–e303. [CrossRef]
18. Onishi, H.; Shirato, H.; Nagata, Y.; Hiraoka, M.; Fujino, M.; Gomi, K.; Niibe, Y.; Karasawa, K.; Hayakawa, K.; Takai, Y.; et al. Hypofractionated Stereotactic Radiotherapy (HypoFXSRT) for stage I non-small cell lung cancer: Updated results of 257 patients in a Japanese multi-institutional study. *J. Thorac. Oncol.* **2011**, *2*, S94–S100. [CrossRef]
19. Lee, P.; Loo, B.W., Jr.; Biswas, T.; Ding, G.X.; El Naqa, I.M.; Jackson, A.; Kong, F.M.; LaCouture, T.; Miften, M.; Solberg, T.; et al. Local Control After Stereotactic Body Radiation Therapy for Stage I Non-Small Cell Lung Cancer. *Int. J. Radiat. Oncol. Biol. Phys.* **2021**, *110*, 160–171. [CrossRef]
20. Haralick, R.M.; Shanmugam, K.; Dinstein, I. Textural Features for Image Classification. *IEEE Trans. Syst. Man. Cybern.* **1973**, *SMC-3*, 610–621. [CrossRef]
21. Castellano, G.; Bonilha, L.; Li, L.M.; Cendes, F. Texture analysis of medical images. *Clin. Radiol.* **2004**, *59*, 1061–1069. [CrossRef]
22. Tourassi, G.D. Journey toward computer-aided diagnosis: Role of image texture analysis. *Radiology* **1999**, *213*, 317–320. [CrossRef] [PubMed]
23. Yip, S.S.; Aerts, H.J. Applications and limitations of radiomics. *Phys. Med. Biol.* **2016**, *61*, R150–R166. [CrossRef] [PubMed]
24. Lambin, P.; Rios-Velazquez, E.; Leijenaar, R.; Carvalho, S.; van Stiphout, R.G.; Granton, P.; Zegers, C.M.; Gillies, R.; Boellard, R.; Dekker, A.; et al. Radiomics: Extracting more information from medical images using advanced feature analysis. *Eur. J. Cancer* **2012**, *48*, 441–446. [CrossRef] [PubMed]
25. Lambin, P.; Leijenaar, R.T.-H.; Deist, T.M.; Peerlings, J.; de Jong, E.E.C.; van Timmeren, J.; Sanduleanu, S.; Larue, R.T.H.M.; Even, A.J.G.; Jochems, A.; et al. Radiomics: The bridge between medical imaging and personalized medicine. *Nat. Rev. Clin. Oncol.* **2017**, *14*, 749–762. [CrossRef]
26. Drucker, E.; Krapfenbauer, K. Pitfalls and limitations in translation from biomarker discovery to clinical utility in predictive and personalised medicine. *EPMA J.* **2013**, *4*, 7. [CrossRef]
27. Limkin, E.J.; Sun, R.; Dercle, L.; Zacharaki, E.I.; Robert, C.; Reuzé, S.; Schernberg, A.; Paragios, N.; Deutsch, E.; Ferté, C. Promises and challenges for the implementation of computational medical imaging (radiomics) in oncology. *Ann. Oncol.* **2017**, *28*, 1191–1206. [CrossRef]
28. Chu, L.C.; Park, S.; Kawamoto, S.; Fouladi, D.F.; Shayesteh, S.; Zinreich, E.S.; Graves, J.S.; Horton, K.M.; Hruban, R.H.; Yuille, A.L.; et al. Utility of CT Radiomics Features in Differentiation of Pancreatic Ductal Adenocarcinoma from Normal Pancreatic Tissue. *Am. J. Roentgenol.* **2019**, *213*, 349–357. [CrossRef]
29. Markel, D.; Caldwell, C.; Alasti, H.; Soliman, H.; Ung, Y.; Lee, J.; Sun, A. Automatic Segmentation of Lung Carcinoma Using 3D Texture Features in 18-FDG PET/CT. *Int. J. Mol. Imaging.* **2013**, *2013*, 980769. [CrossRef]
30. Avanzo, M.; Gagliardi, V.; Stancanello, J.; Blanck, O.; Pirrone, G.; El Naqa, I.; Revelant, A.; Sartor, G. Combining computed tomography and biologically effective dose in radiomics and deep learning improves prediction of tumor response to robotic lung stereotactic body radiation therapy. *Med. Phys.* **2021**, *48*, 6257–6269. [CrossRef]
31. De Ruysscher, D.; Faivre-Finn, C.; Moeller, D.; Nestle, U.; Hurkmans, C.W.; le Péchoux, C.; Belderbos, J.; Guckenberger, M.; Senan, S. Lung Group and the Radiation Oncology Group of the European Organization for Research and Treatment of Cancer (EORTC). European Organization for Research and Treatment of Cancer (EORTC) recommendations for planning and delivery of high-dose, high precision radiotherapy for lung cancer. *Radiother. Oncol.* **2017**, *124*, 1–10.
32. Van Loon, J.; Siedschlag, C.; Stroom, J.; Blauwgeers, H.; van Suylen, R.J.; Knegjens, J.; Rossi, M.; van Baardwijk, A.; Boersma, L.; Klomp, H.; et al. Microscopic disease extension in three dimensions for non–small-cell lung cancer: Development of a prediction model using pathology-validated positron emission tomography and computed tomography features. *Int. J. Radiat. Oncol. Biol. Phys.* **2012**, *82*, 448–456. [CrossRef] [PubMed]
33. Reuzé, S.; Orlhac, F.; Chargari, C.; Nioche, C.; Limkin, E.; Riet, F.; Escande, A.; Haie-Meder, C.; Dercle, L.; Gouy, S.; et al. Prediction of cervical cancer recurrence using textural features extracted from 18F-FDG PET images acquired with different scanners. *Oncotarget* **2017**, *8*, 43169–43179. [CrossRef] [PubMed]
34. Orlhac, F.; Soussan, M.; Chouahnia, K.; Martinod, E.; Buvat, I. 18F-FDG PET-Derived Textural Indices Reflect Tissue-Specific Uptake Pattern in Non-Small Cell Lung Cancer. *PLoS ONE* **2015**, *10*, e0145063. [CrossRef]

35. Van Griethuysen, J.J.M.; Fedorov, A.; Parmar, C.; Hosny, A.; Aucoin, N.; Narayan, V.; Beets-Tan, R.G.H.; Fillon-Robin, J.C.; Pieper, S.; Aerts, H.J.W.L. Computational Radiomics System to Decode the Radiographic Phenotype. *Cancer Res.* **2017**, *77*, e104–e107. [CrossRef] [PubMed]
36. Haga, A.; Takahashi, W.; Aoki, S.; Nawa, K.; Yamashita, H.; Abe, O.; Nakagawa, K. Standardization of imaging features for radiomics analysis. *J. Med. Investig.* **2019**, *66*, 35–37. [CrossRef]
37. James, G.; Witten, D.; Hastie, T.; Tibshirani, R. Support vector machines. In *An Introduction to Statistical Learning: Applications in R*; Springer: New York, NY, USA, 2013; p. 359.
38. Cortes, C.; Vapnik, V. Support-vector networks. *Mach. Learn.* **1995**, *20*, 273–297. [CrossRef]
39. Polikar, R. Ensemble based systems in decision making. *IEEE Circuits Syst. Mag.* **2006**, *6*, 21–45. [CrossRef]
40. Kothari, G.; Korte, J.; Lehrer, E.J.; Zaorsky, N.G.; Lazarakis, S.; Kron, T.; Hardcastle, N.; Siva, S. A systematic review and meta-analysis of the prognostic value of radiomics based models in non-small cell lung cancer treated with curative radiotherapy. *Radiother. Oncol.* **2021**, *155*, 188–203. [CrossRef]
41. Mali, S.A.; Ibrahim, A.; Woodruff, H.C.; Andrearczyk, V.; Müller, H.; Primakov, S.; Salahuddin, Z.; Chatterjee, A.; Lambin, P. Making Radiomics More Reproducible across Scanner and Imaging Protocol Variations: A Review of Harmonization Methods. *J. Pers. Med.* **2021**, *11*, 842. [CrossRef]
42. Da-ano, R.; Masson, I.; Lucia, F.; Doré, M.; Robin, P.; Alfieri, J.; Rousseau, C.; Mervoyer, A.; Reinhold, C.; Castelli, J.; et al. Performance comparison of modified ComBat for harmonization of radiomic features for multicenter studies. *Sci. Rep.* **2020**, *10*, 10248. [CrossRef]
43. Fave, X.; Zhang, L.; Yang, J.; Mackin, D.; Balter, P.; Gomez, D.; Followill, D.; Jones, A.K.; Stingo, F.; Liao, Z.; et al. Delta-radiomics features for the prediction of patient outcomes in non-small cell lung cancer. *Sci. Rep.* **2017**, *7*, 588. [CrossRef] [PubMed]
44. Alahmari, S.S.; Cherezov, D.; Goldgof, D.B.; Hall, L.O.; Gillies, R.J.; Schabath, M.B. Delta Radiomics Improves Pulmonary Nodule Malignancy Prediction in Lung Cancer Screening. *IEEE Access* **2018**, *6*, 77796–77806. [CrossRef] [PubMed]
45. Nardone, V.; Reginelli, A.; Guida, C.; Belfiore, M.P.; Biondi, M.; Mormile, M.; Banci Buonamici, F.; di Giorgio, E.; Spadafora, M.; Tini, P.; et al. Delta-radiomics increases multicentre reproducibility: A phantom study. *Med. Oncol.* **2020**, *37*, 38. [CrossRef] [PubMed]
46. Pinto, J.A.; Vallejos, C.S.; Raez, L.E.; Mas, L.A.; Ruiz, R.; Torres-Roman, J.S.; Morante, Z.; Araujo, J.M.; Gómez, H.L.; Aguilar, A.; et al. Gender and outcomes in non-small cell lung cancer: An old prognostic variable comes back for targeted therapy and immunotherapy? *ESMO Open* **2018**, *3*, e000344. [CrossRef]
47. De Perrot, M.; Licker, M.; Bouchardy, C.; Usel, M.; Robert, J.; Spiliopoulos, A. Sex differences in presentation, management, and prognosis of patients with non-small cell lung carcinoma. *J. Thorac. Cardiovasc. Surg.* **2000**, *119*, 21–26. [CrossRef]
48. Hsu, L.H.; Chu, N.M.; Liu, C.C.; Tsai, S.Y.; You, D.L.; Ko, J.S.; Lu, M.C.; Feng, A.C. Sex-associated differences in non-small cell lung cancer in the new era: Is gender an independent prognostic factor? *Lung Cancer* **2009**, *66*, 262–267. [CrossRef]
49. Louie, A.V.; Haasbeek, C.J.; Mokhles, S.; Rodrigues, G.B.; Stephans, K.L.; Lagerwaard, F.J.; Palma, D.A.; Videtic, G.M.; Warner, A.; Takkenberg, J.J.; et al. Predicting Overall Survival After Stereotactic Ablative Radiation Therapy in Early-Stage Lung Cancer: Development and External Validation of the Amsterdam Prognostic Model. *Int. J. Radiat. Oncol. Biol. Phys.* **2015**, *93*, 82–90. [CrossRef]
50. Ye, L.; Shi, S.; Zeng, Z.; Huang, Y.; Hu, Y.; He, J. Nomograms for predicting disease progression in patients of Stage I non-small cell lung cancer treated with stereotactic body radiotherapy. *Jpn. J. Clin. Oncol.* **2018**, *48*, 160–166. [CrossRef]
51. Kang, J.; Ning, M.S.; Feng, H.; Li, H.; Bahig, H.; Brooks, E.D.; Welsh, J.W.; Ye, R.; Miao, H.; Chang, J.Y. Predicting 5-Year Progression and Survival Outcomes for Early Stage Non-small Cell Lung Cancer Treated with Stereotactic Ablative Radiation Therapy: Development and Validation of Robust Prognostic Nomograms. *Int. J. Radiat. Oncol. Biol. Phys.* **2020**, *106*, 90–99. [CrossRef]

cancers

MDPI

Article

ViSTA: A Novel Network Improving Lung Adenocarcinoma Invasiveness Prediction from Follow-Up CT Series

Wei Zhao [1,†], Yingli Sun [2,†], Kaiming Kuang [3], Jiancheng Yang [3,4], Ge Li [5], Bingbing Ni [4], Yingjia Jiang [1], Bo Jiang [1], Jun Liu [1,6,*] and Ming Li [2,7,*]

1 Department of Radiology, The Second Xiangya Hospital, Central South University, Changsha 410011, China; wei.zhao@csu.edu.cn (W.Z.); moshangqingcheng@163.com (Y.J.); jiangbo@csu.edu.cn (B.J.)
2 Department of Radiology, Huadong Hospital, Fudan University, Shanghai 200040, China; sunyingli208ok@126.com
3 Dianei Technology, Shanghai 200051, China; kaiming.kuang@dianei-ai.com (K.K.); jekyll4168@sjtu.edu.cn (J.Y.)
4 Department of Electronic Engineering, Shanghai Jiao Tong University, Shanghai 200240, China; nibingbing@sjtu.edu.cn
5 Department of Radiology, The Xiangya Hospital, Central South University, Changsha 410008, China; ligeanyi@126.com
6 Radiology Quality Control Center, Changsha 410011, China
7 Institute of Functional and Molecular Medical Imaging, Fudan University, Shanghai 200437, China
* Correspondence: junliu123@csu.edu.cn (J.L.); ming_li@fudan.edu.cn (M.L.); Tel.: +86-137-8708-5002 (J.L.); +86-138-1662-0371 (M.L.); Fax: +86-0731-85292116 (J.L.); +86-21-57643271 (M.L.)
† These authors contributed equally to this work.

Simple Summary: Assessing follow-up computed tomography(CT) series is of great importance in clinical practice for lung nodule diagnosis. Deep learning is a thriving data mining method in medical imaging and has obtained surprising results. However, previous studies mostly focused on the analysis of single static time points instead of the entire follow-up series and required regular intervals between CT examinations. In the current study, we propose a new deep learning framework, named ViSTA, that can better evaluate tumor invasiveness using irregularly serial follow-up CT images to avoid aggressive procedures or delay diagnosis in clinical practice. ViSTA provides a new solution for irregularly sampled data. ViSTA delivers superior performance compared with other static or serial deep learning models. The proposed ViSTA framework is capable of improving performance close to the human level in the prediction of invasiveness of lung adenocarcinoma while being transferrable to other tasks analyzing serial medical data.

Abstract: To investigate the value of the deep learning method in predicting the invasiveness of early lung adenocarcinoma based on irregularly sampled follow-up computed tomography (CT) scans. In total, 351 nodules were enrolled in the study. A new deep learning network based on temporal attention, named Visual Simple Temporal Attention (ViSTA), was proposed to process irregularly sampled follow-up CT scans. We conducted substantial experiments to investigate the supplemental value in predicting the invasiveness using serial CTs. A test set composed of 69 lung nodules was reviewed by three radiologists. The performance of the model and radiologists were compared and analyzed. We also performed a visual investigation to explore the inherent growth pattern of the early adenocarcinomas. Among counterpart models, ViSTA showed the best performance (AUC: 86.4% vs. 60.6%, 75.9%, 66.9%, 73.9%, 76.5%, 78.3%). ViSTA also outperformed the model based on Volume Doubling Time (AUC: 60.6%). ViSTA scored higher than two junior radiologists (accuracy of 81.2% vs. 75.4% and 71.0%) and came close to the senior radiologist (85.5%). Our proposed model using irregularly sampled follow-up CT scans achieved promising accuracy in evaluating the invasiveness of the early stage lung adenocarcinoma. Its performance is comparable with senior experts and better than junior experts and traditional deep learning models. With further validation, it can potentially be applied in clinical practice.

Citation: Zhao, W.; Sun, Y.; Kuang, K.; Yang, J.; Li, G.; Ni, B.; Jiang, Y.; Jiang, B.; Liu, J.; Li, M. ViSTA: A Novel Network Improving Lung Adenocarcinoma Invasiveness Prediction from Follow-Up CT Series. *Cancers* **2022**, *14*, 3675. https://doi.org/10.3390/cancers14153675

Academic Editors: Hamid Khayyam, Ali Madani, Rahele Kafieh and Ali Hekmatnia

Received: 7 June 2022
Accepted: 20 July 2022
Published: 28 July 2022

Keywords: adenocarcinoma; invasiveness; X-ray computed tomography; deep learning; temporal attention

1. Introduction

Low-dose computed tomography (LDCT) is recommended for lung cancer screening in high-risk populations based on the National Lung Cancer Screening Trial (NLST) report, which is now included in US screening guidelines [1]. Owing to LDCT, more and more early stage lung adenocarcinomas are diagnosed and treated. In clinical practice, most people require follow-up CT scans due to the indeterminate diagnosis or low probability of malignancy on baseline CT. Assessing the changes in size, CT value, and other imaging features can substantially help the diagnosis and invasiveness evaluation of early stage lung adenocarcinomas. However, the evaluation process is tedious and lacks objectivity, which means that radiologists could be overwhelmed by numerous serial CT image evaluations. Moreover, the features indicating malignancy may not be present in the early stages of lung adenocarcinoma. As we know, biological changes may precede morphological changes. Therefore, an efficient tool for objectively evaluating the changes and mining the internal patterns of lung nodules on serial CTs is of great importance.

Deep learning is a thriving data mining method in medical imaging and has obtained surprising results [2–4]. It can efficiently and automatically process medical images and has achieved promising performances on par with clinicians on various clinical tasks, including disease classification, medical image registration, and organ segmentation [5–9]. Previous studies have shown that deep learning could aid clinical decision-making for early lung cancer in disease management and invasiveness prediction [10–14]. However, most prior studies only included single-time CT scan images, while serial CT scan images were not fully investigated. Several powerful deep learning methods have been invented to process serial data, e.g., Long Short-term Memory, Gated Recurrent Unit Network, and Transformer [15–17]. Equipped with the aforementioned tools, a deep learning system can include serial images, better evaluate the biological behavior and changes, and then better predict different clinical events, such as prognosis, therapeutic effect, and subsequent growth patterns [18].

Serial deep learning models have achieved great success in serial data domains, including natural language processing, video classification, and speech recognition [15,19,20]. Nonetheless, it is important to notice that medical serial data such as electronic health records [21] or medical examinations are almost always sampled irregularly in time, separating them from the aforementioned modalities. Since the progression of the disease is strongly correlated with the time intervals between two time points, the asynchronous (sampled irregularly) nature of medical data requires special treatment. For example, by limiting sampling time intervals to 1, 3, and 6 months, deep learning methods proved effective in integrating multiple time points and improving the prediction of lung cancer treatment response [22]. However, this restriction on time intervals still limits the usage of the deep learning method in processing clinical serial data, epically for irregularly serial data.

In this article, we propose ViSTA (Visual Simple Temporal Attention), a deep learning framework capable of predicting the tumor invasiveness of pulmonary adenocarcinomas from Follow-up CT Series. The main contributions are three-fold: First, by introducing a simple temporal attention mechanism, we propose a new deep learning network, named ViSTA, to evaluate the invasiveness of early stage lung adenocarcinoma using irregularly serial CT scans images. ViSTA is able to gather information throughout the entire series and improve the prediction performance. Compared with serial analysis using traditional recurrent neural networks [22], ViSTA is not limited by different time intervals and can process completely irregularly sampled serial data. ViSTA was trained and validated on a dataset of 1121 CT scans from 282 follow-up series and evaluated on a hold-out test set of 113 CT scans from 69 follow-up series. Second, ViSTA delivers superior performances compared with other static or serial deep learning models. ViSTA also outperforms size-based predictive methods (Volume Doubling Time [23]) by a large margin. Third, ViSTA was proven to achieve higher scores than two junior radiologists and came close to one senior radiologist in the observer study. Our results prove ViSTA's superiority in terms of processing irregularly sampled series and its great potential of being put into clinical practice in reality. Additionally, ViSTA is completely transferrable to other medical imaging tasks where analyzing serial data should yield better performances.

2. Materials and Methods

2.1. Data Collection

From January 2011 to October 2017, a search of the electronic medical records and the radiology information systems of the hospital was performed by one author (Yingli Sun). The inclusion criteria are as follows: (1) two or more available CT examinations with thin-slice (≤1.5 mm) images before resection. If there were only two CT examinations, the interval between two scans should be over 30 or more days. (2) Complete pathologic reports. The exclusion criteria for this analysis were: (1) prior treatment before surgery; (2) poor quality CT images; (3) lesions that were difficult to clearly delineate. Finally, a total of 351 nodules from 347 patients (mean age 58.41 years ±11.79 (SD); range, 22–84 years) were enrolled in the study. Among the 351 lung nodules, 191 nodules were pathologically identified as preinvasive lesions, including 1 atypical adenomatous hyperplasia (AAH), 39 adenocarcinomas in situ (AIS), and 151 minimally invasive adenocarcinoma (MIA); whereas 160 nodules were identified as invasive adenocarcinoma (IA). In total, 1234 serials CT scans of the 351 nodules were enrolled in this study. The median interval between the first and the last CT examinations was 366 ± 500 days (range, 30–2813 days; interquartile range, 165–852 days). The 351 nodules were randomly separated into a training set (245 nodules), validation set (37 nodules), and test set (69 nodules) (see Table 1).

Table 1. Number of CT scans/nodules in training, validation, and test set.

Pathological Type		No. CT Scans/Nodules			
		Training	Validation	Test	Total
Non-IA	AAH	5/1	0/0	0/0	5/1
	AIS	98/29	9/4	19/6	126/39
	MIA	383/104	40/16	114/31	537/151
	Total	486/134	49/20	133/37	668/191
IA		398/111	64/17	104/32	566/160
Total		884/245	113/37	237/69	1234/351

2.2. CT Scanning Parameters

Preoperative chest CT in our department was performed using the following four scanners: GE Discovery CT750 HD, 64-slice LightSpeed VCT (GE Medical Systems, Chicago, IL, USA); Somatom Definition flash, Somatom Sensation-16 (Siemens Medical Solutions, Erlangen, Germany) with the following parameters: 120 kVp; 100–200 mAs; pitch, 0.75–1.5; and collimation, 1–1.5 mm, respectively. All imaging data were reconstructed using a medium sharp reconstruction algorithm with a thickness of 1–1.5 mm.

2.3. Nodule Labeling, Segmentation and Imaging Preprocessing

A medical image processing and navigation software 3D Slicer (v4.8.0, Brigham and Women's Hospital, Boston, MA, USA) was used to manually delineate the volume of interest (VOI) of the included nodules at the voxel level by one radiologist (Yingli Sun, with 5 years of experience in chest CT interpretation), then the VOI was confirmed by another radiologist (Ming Li, with 12 years of experience in chest CT interpretation). Large vessels and bronchioles were excluded as much as possible from the volume of the nodule. The lung CT DICOM (Digital Imaging and Communications in Medicine) format images were imported into the software for delineation, and then the images with VOI information were extracted with NII format for next step analysis. Each segmented nodule was attributed a specific pathological label (AAH, AIS, MIA, IA), according to the detailed pathological report. Two steps were performed to preprocess CT images before path extraction. First, the whole-volume CT image was resampled to the spacing of 1 mm in all three dimensions to guarantee isotropy. Second, HU values were clipped to the range of (−1000, 400) and normalized to (0, 1) using minimum–maximum normalization. Normalization can accelerate the convergence in the training of the deep learning model and improve its generalization ability.

2.4. Development of the Deep Learning Model

We developed a deep learning model named ViSTA to classify IA/non-IA lung nodules. The overall architecture of ViSTA is presented in Figure 1. ViSTA first extracts features from CT image patches using a CNN backbone and then integrates information from time series using a lightweight attention module named SimTA [24], which is designed specifically for analyzing asynchronous time series. Details regarding the architecture of ViSTA are provided in Supplementary Section S1, and a single SimTA layer was shown in Figure S1. To avoid overoptimization, we did not heavily tune the hyperparameters of our deep learning model and simply adopted common settings. ViSTA and all its counterparts are trained end-to-end for 100 epochs using the AdamW optimizer [25]. We used a cosine decay learning schedule from 10^{-3} to 10^{-6}. The batch size of each update was 32. The drop-out probability and weight decay were set at 0.2 and 0.01 to avoid overfitting.

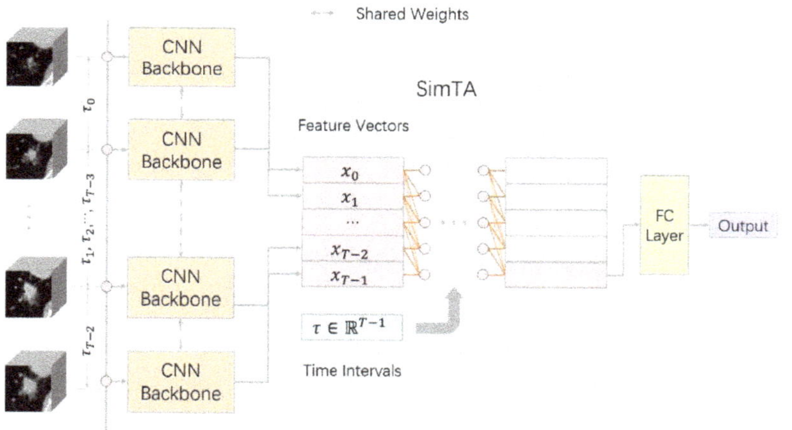

Figure 1. The model overview of the proposed ViSTA. It consists of a CNN backbone followed by the SimTA module made up of several SimTA layers.

2.5. Counterpart Methods

For comparison with ViSTA, we conducted experiments on a few of its counterparts:

- VDT (Volume Doubling Time). VDT is an important volumetric indicator used in follow-up examinations. It represents the time it takes for a nodule to double its volume. The formula of VDT is provided in Supplementary Section S1. Nodules with VDT < 400 days are considered fast-growing and are more likely to be malignant [23]. In this research, we evaluated VDT's metrics under two different thresholds: 400 days and the cutoff that provides the best Youden index on the validation set. Youden index's formula is presented in Supplementary Section S1;
- CNN (Convolutional Neural Network): To compare ViSTA against static models, we introduced CNN as a counterpart. We conducted the following experiments to further investigate the source of performance difference between ViSTA and CNN;

- CNN-last: This experiment was conducted to train and validate CNN only on the last time point of each follow-up series. It is obvious that the last time point is most relevant to the final diagnosis;
- CNN-first: This experiment was conducted to train and validate CNN only on the first time point of each follow-up series. In this experiment, the first steps were treated as if they had the same label as the last one. This setting was used to confirm that earlier time points convey less information than later ones;
- CNN-all: This experiment was conducted to train CNN on all time points of each follow-up series and validate it on the first and last time point separately (named CNN-all-first and CNN-all-last, respectively). This was used is to investigate if ViSTA's superior performance only comes from the larger data size it enjoys;
- CNN+LSTM (Long Short-term Memory) [16]: LSTM is a subtype of RNN (Recurrent Neural Network) designed to analyze serial data and capture long-term relations. This setting is quite similar to previous research which combined CNN and RNN to predict lung cancer treatment response [22]. However, we did not limit time intervals to specific values so that we could fairly compare ViSTA and RNN-based methods. One major difference between CNN+LSTM and ViSTA is that CNN+LSTM treats all time points as if they had the same interval (synchronous). By comparing the previous two methods, we would like to see if ViSTA is more suitable for analyzing irregularly sampled time series.

2.6. Evaluation and Statistical Analysis

We evaluated the proposed ViSTA model both quantitatively and qualitatively. To evaluate each method's performance, we used a variety of metrics, including accuracy, precision, sensitivity, F1 score, and AUC. Formulas of evaluation metrics are presented in Supplementary Section S1.

To explore the visual representation and interpretability of ViSTA, we followed Simonyan, K. et al. [26] and plotted our model's saliency maps through backpropagation, and investigated the mechanism under ViSTA and where it directed its attention.

2.7. Observer Study

To further evaluate the performance of ViSTA, we conducted an observer study to compare the performance of radiologists in the same task against other models. In the observer study, all 69 CT series in the test set were evaluated by three radiologists. One is a senior radiologist with 22 years of experience, and the other two are junior radiologists with 5 and 3 years of experience, respectively. Radiologists gave the results based on the evaluation of all available serials CTs. The reviewed results were analyzed and compared with the performance of our proposed model. Radiologists' performances were evaluated using accuracy, sensitivity, precision, and F1 score.

3. Results

3.1. Performance of Deep Learning Models in Predicting the Invasiveness of Early Lung Adenocarcinoma

To validate the effectiveness of ViSTA in predicting IA/non-IA nodules, we evaluated its performance using a variety of metrics against its counterparts: VDT (cutoff value set at best Youden index or 400 days), CNN (including CNN-last, CNN-first, CNN-all-first and CNN-all-last), and CNN+LSTM.

Tables S1 and S2 show their performances on the training dataset and validation dataset. Figure 2 provide the ROC curves of all models on the test dataset. Our proposed model outperformed all deep learning models and VDT-based methods in every metric by considerable margins (best among models are highlighted with an underscore). It is worth noting that VDT is far from effective in terms of invasiveness classification. It underperformed almost all deep learning models in terms of AUC, accuracy, and F1 score. Secondly, sequential models (ViSTA and CNN+LSTM) delivered better performances than CNN models that utilize static data points. ViSTA outperformed CNN+LSTM by considerable margins in all metrics. This performance gap can be attributed to ViSTA's suitability to analyze asynchronous time series. Unlike CNN+LSTM, which treats all time points as if they were regularly sampled, ViSTA takes time intervals into account and is better at processing follow-up series. Furthermore, we trained CNN on all time points (CNN all-first and CNN all-last) to investigate if sequential models gain superiority over larger training datasets. It turned out that ViSTA and CNN+LSTM still outperformed CNN even when it was trained on all data.

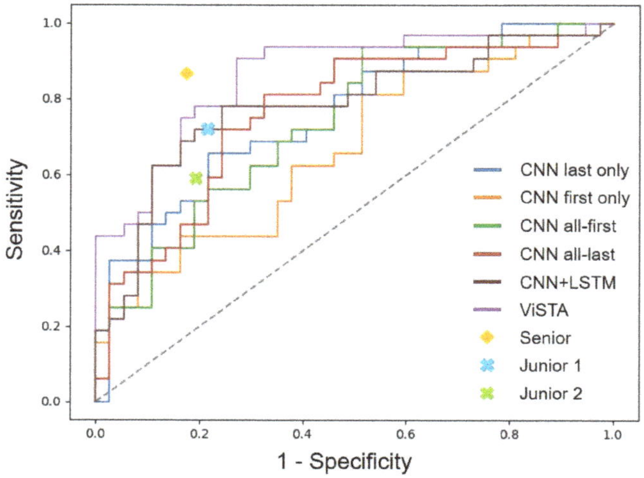

Figure 2. ROC curves of different models compared with performances of radiologists. The gray dotted line indicates the performance of a random classifier with no predictive ability.

3.2. *Performance Comparison against Radiologists*

In the observer study, we compared the performances of ViSTA and its counterparts against three radiologists (Table 2). One is a senior radiologist with 22 years of experience, and the other two are junior radiologists with 5 and 3 years of experience, respectively. All 69 follow-up series from the test set were included in the observer study. We evaluated radiologists' performances using accuracy, sensitivity and precision, and F1 score and compared them against the proposed model. In terms of metrics that require specifying threshold, we chose the threshold that delivers the best Youden Index on the validation set as the cutoff value. Figure 2 plot deep learning models' ROC curves against radiologists' metrics. In terms of accuracy and F1 score, ViSTA scored higher than the two junior radiologists (accuracy of 81.2% vs. 75.4% and 71.0%; F1 score of 81.7% vs. 73.0% and 65.5%) and came close to the senior radiologist (accuracy of 81.2% vs. 85.5%; F1 score of 81.7% vs. 84.8%).

3.3. *Visual Presentation Investigation*

To investigate the mechanism of ViSTA, we used a neural network visualization technique [26] to visualize the attention heatmap of the model, which was mostly attributed to the predicted results and potentially correlated to the biological behavior (Figure 3). We took the absolute value of the raw heatmap and clipped it to the range of (0, 0.01) for better visualization and interpretation. In view of the created heatmaps, we can see that the "attention" of the deep learning system was mostly focused on the nodule. Areas surrounding the nodule draw the attention of ViSTA as well, meaning that they also carry valuable information as the nodule does (Figure 3A,B). Figure 3A show a long follow-up series of 11 time points. We observed that heatmaps stay blank in the first half of the series, during which both nodule volume and IA probability remain relatively stable. In the latter half, heatmaps begin to show along with significant increases in nodule volume and IA probability. Heatmaps are sometimes only lit up at the last time point (Figure 3B). We contribute this to the sudden increase of nodule volume between the third and the fourth time point, which provides sufficient information for the model. This argument is supported by the spike of IA probability at the fourth time point. In some rare cases, heatmaps on all time points are close to invisible (Figure 3C). We conjecture that this is because the lung nodule had almost no progression, which was proven by the fact that both nodule volume and IA probability stayed almost unchanged throughout the entire series.

Table 2. The performance of different models and radiologists on the test dataset. The highest among all is highlighted in bold, and the highest among models and VDT (Volume Doubling Time)-based methods is highlighted with an underscore.

	AUC	Acc.	Prec.	Sens.	F1
Senior	-	**85.5%**	**82.4%**	87.5%	**84.8%**
Junior 1	-	75.4%	74.2%	71.9%	73.0%
Junior 2	-	71.0%	73.1%	59.4%	65.5%
1/VDT (best Youden index)	60.6%	62.3%	56.3%	84.4%	67.5%
1/VDT (400 days)	60.6%	58.0%	71.4%	15.6%	25.6%
CNN last only	75.9%	72.5%	72.4%	65.6%	68.9%
CNN first only	66.9%	65.2%	70.0%	43.8%	64.3%
CNN all-first	73.9%	65.2%	60.5%	71.9%	65.7%
CNN all-last	76.5%	73.9%	71.9%	71.9%	71.9%
CNN+LSTM	78.3%	76.8%	73.5%	78.1%	75.8%
ViSTA	**<u>86.4%</u>**	<u>81.2%</u>	<u>74.4%</u>	**<u>90.6%</u>**	<u>81.7%</u>

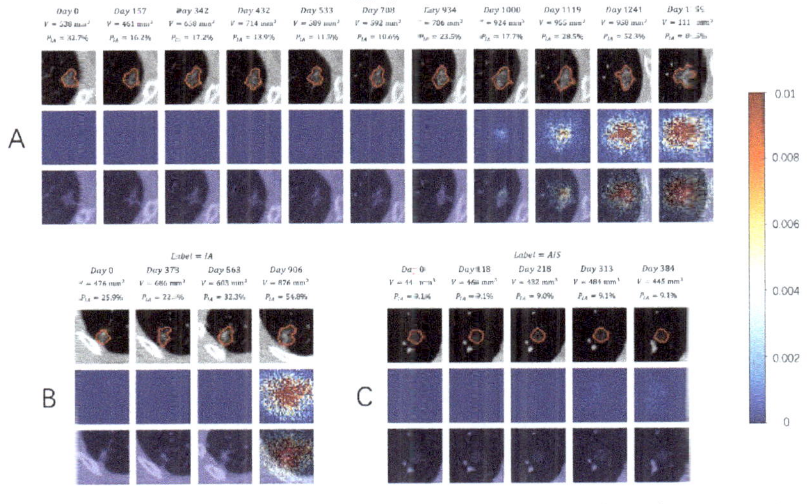

Figure 3. Visualization investigation of ViSTA. The top row shows CT slices of each time point in the follow-up series. The middle row shows attention heatmaps extracted using the technique proposed by Simonyan, K. et al. [26]. The bottom row masks heatmaps on top of CT slices. **(A)** Attention gradually grew along with the nodule volume and IA probability as the nodule progressed to the end of the series. **(B)** The heatmap only lit up at the last time point as it is considered the one carrying valuable information. **(C)** All time points are allocated with little to no attention, which may be caused by the slow progress of the nodule.

4. Discussion

In the current study, we proposed a deep learning framework named ViSTA to predict the invasiveness of lung adenocarcinomas using serial CT images. Our results showed that models fed with serial CT images substantially and consistently outperformed models fed with single CT images. Moreover, our proposed model can effectively process asynchronous time series and outperform the traditional serial network, i.e., LSTM. Our models achieved an AUC of 86.4% and an F1 score of 81.7% in the test dataset, which were higher than those of all its counterparts. In the observer study, ViSTA achieved higher accuracy and F1 scores than two junior radiologists (accuracy of 81.2% vs. 75.4% and 71.0%, F1 score of 81.7% vs. 73.0% and 65.5%). When compared with the senior radiologist, our proposed model delivered close performance (accuracy of 81.2% vs. 85.5%, F1 score of 81.7% vs. 84.8%).

Timely and accurately assessing the biological behavior of early stage lung adenocarcinomas has been a continuous focus of attention in clinical practice. In contrast to traditional radiographic features and handcraft features, deeper and higher dimension level features mined by the deep learning method present promising advantages in many tasks, including predicting the invasiveness of the early lung adenocarcinoma. Kim et al. performed a comparison study and revealed that the predictive accuracy of the deep learning method was superior to those of the size-based logistic model [11]. We also analyzed the predictive value of VDT [27], a size-based key parameter in the differentiation of aggressive tumors from slow-growing tumors in clinical practice [24]. Not surprisingly, the performance of our proposed model substantially exceeded that of the VDT-based methods. It indirectly verified the conjecture that a deep learning system could extract and learn deeper and more valuable features, then better discover the biological behavior of the tumors and predict the invasiveness of early stage lung adenocarcinoma.

Although the deep learning method can obtain better performance, most previous studies only used single CT scan data prior to the surgery for training and extracting features, which cannot reveal and learn the internal growth pattern of the nodules. In clinical scenarios, internal growth is a vital component of Lung-RADS, a guideline to standardize image interpretation by radiologists and dictate management recommendations. Including serial CTs can facilitate medical tasks, such as differentiating benign tumors from malignant ones [28] and monitoring and predicting treatment response [22,29]. The discovery of our study supports this. By modelling serial CTs, the predictive performance of ViSTA substantially surpassed its counterparts in analyzing static data. In clinical practice, sequential medical data is generally sampled irregularly, i.e., with different follow-up time intervals. To address the irregular sampling issue, we adopted SimTA in our proposed model to process irregularly sampled time series. This lightweight module enables modeling sequential information in an efficient way. It turned out that the proposed ViSTA significantly outperformed the standard serial framework, i.e., CNN+LSTM, with considerably fewer parameters and less computation and memory footprint. ViSTA can better take advantage of the complete information of all time point CTs by modelling simple yet effective exponentially decay attention in time series. This was proved by our experiments comparing ViSTA, CNN+LSTM, and pure CNN models trained with all time point CTs (CNN-all). ViSTA's superiority over CNN-all proved that its performance gain does not come from a larger training dataset.

In the visualization analysis, we found that ViSTA can drive its attention on the nodule and the surrounding tissue and drop more attention when the probability of invasiveness increases. It can partly explain the mechanism of the deep learning system. We also found some cases where the model appeared to use features close to the nodule, such as the vasculature and parenchyma surrounding the nodule. In fact, peritumoral tissue may possess valuable information, such as tumor-infiltrating status. Features extracted from the peritumoral tissues can improve the efficiency of intramodular radiomic analysis [30,31]. However, we still cannot fully interpret whether the model incorporates other abnormalities such as background emphysema in its predictions. Further investigation using more comprehensive model attribution techniques may allow clinicians to take advantage of the same visual features used by the model to assess the biological status of tumors. It is worth noting that some of the heatmaps in the time series are completely blue, meaning that the deep learning model allocated close to zero attention to these time points. Though this phenomenon is not completely interpretable, we argue that it can be attributed to these two facts: these time points are too far from the current one, and they lack findings informative for the deep learning model.

Even though the proposed ViSTA proved effective in processing irregularly sampled CT series in our experiments, there are several limitations left untouched. First, due to the difficulty of collecting complete lung nodule follow-up series, we only included data from a single center in this study. In clinical practice, it is preferable if the proposed method generalizes to multiple data domains. Furthermore, it is possible that a single follow-up series contains CT scans from different centers, which would be an important challenge to solve if the proposed model were to be put into clinical usage. In future studies, we will include CT series from external centers to validate the generalization performance of ViSTA. Second, the SimTA module in ViSTA models a simple temporal attention mechanism that monotonically increases weights as the time point gets closer to the current time. However, it is viable to model more complicated attention relations using deep learning models such as Transformeror Informer [15,32]. These temporal models enable capturing non-monotonic and dynamic temporal attention that could be useful in predicting invasiveness. Last but not least, even though we conducted a visual investigation on ViSTA, the interpretation of deep learning model predictions still remains a major challenge. Additionally, the final clinical decision is still up to clinicians to date. In our future research, we will further investigate the underlying mechanism of ViSTA or other similar attention mechanisms.

5. Conclusions

To summarize, we designed a deep learning model processing irregularly sampled CT series to predict the invasiveness of early stage lung adenocarcinoma from follow-up CT scans. The model achieved promising accuracy comparable with senior experts and better than junior experts and its counterparts. With further validation, the proposed model could better evaluate the invasiveness of early stage lung adenocarcinoma, avoiding aggressive procedures or delayed diagnosis and helping precise management in clinical practice.

Supplementary Materials: The following supporting information can be downloaded at: https://www.mdpi.com/article/10.3390/cancers14153675/s1. References [24,33–36]. Section S1: Supplementary Methods. Figure S1: The illustration of a single SimTA layer. The colormap of the attention matrix indicates the magnitude of attention allocated in a particular timepoint. Gray represents zero attention. The darker the orange color is, the more attention the timepoint gets; Table S1: Different models' performances in predicting invasiveness on the training dataset. The highest among all is highlighted in bold; Table S2: Different models' performances in predicting invasiveness on the validation dataset. The highest among all is highlighted in bold.

Author Contributions: Conceptualization, K.K. and J.L.; Data curation, W.Z., Y.S. and M.L.; Formal analysis, W.Z., Y.S. and J.Y.; Funding acquisition, M.L.; Investigation, W.Z., K.K., Y.S., G.L., B.N., Y.J. and B.J.; Methodology, W.Z., K.K., Y.S., J.Y., Y.J. and B.J.; Project administration, M.L.; Resources, Y.S.; Software, K.K., B.N. and M.L.; Supervision, J.Y., J.L. and M.L.; Validation, G.L.; Visualization, B.N.; Writing—original draft, W.Z. and Y.S.; Writing—review and editing, W.Z., J.L. and M.L. All authors have read and agreed to the published version of the manuscript.

Funding: This study was supported by the National Natural Science Foundation of China 82102157 (Wei Zhao), 61976238 (Ming Li), Hunan Provincial Natural Science Foundation for Excellent Young Scholars 2022JJ20089 (Wei Zhao), Hunan Provincial Natural Science Foundation of China 2021JJ40895 (Wei Zhao), the Research Project of Postgraduate Education and Teaching Reform of Central South University 2021JGB147 (Jun Liu), 2022JGB117 (Wei Zhao), the Clinical Research Center For Medical Imaging In Hunan Province 2020SK4001 (Jun Liu), the science and technology innovation program of Hunan Province 2021RC4016 (Jun Liu), and the Clinical Medical Technology Innovation Guidance Project in Hunan Province 2020SK53423 (Wei Zhao), Science and Technology Planning Project of Shanghai Science and Technology Commission 22Y11910700 (Ming Li), Science and Technology

Planning Project of Shanghai Science and Technology Commission 20Y11902900 (Ming Li), Shanghai "Rising Stars of Medical Talent" Youth Development Program "Outstanding Youth Medical Talents" SHWJRS [2021]-99 (Ming Li).

Institutional Review Board Statement: This retrospective study was approved by the institutional review board of Huadong Hospital, affiliated with Fudan University (No. 2019K134).

Informed Consent Statement: The institutional review board waived the requirement for patients' written informed consent.

Data Availability Statement: Data are available for bona fide researchers who request it from the authors.

Conflicts of Interest: The authors declare no conflict of interest.

References

1. Ettinger, D.S.; Wood, D.E.; Aisner, D.L.; Akerley, W.; Bauman, J.R.; Bharat, A.; Bruno, D.S.; Chang, J.Y.; Chirieac, L.R.; D'Amico, T.A.; et al. Non-Small Cell Lung Cancer, Version 3.2022, NCCN Clinical Practice Guidelines in Oncology. *J. Natl. Compr. Cancer Netw.* **2022**, *20*, 497–530. [CrossRef] [PubMed]
2. Shen, D.; Wu, G.; Suk, H.I. Deep Learning in Medical Image Analysis. *Annu. Rev. Biomed. Eng.* **2017**, *19*, 221–248. [CrossRef] [PubMed]
3. Krittanawong, C.; Johnson, K.W.; Rosenson, R.S.; Wang, Z.; Aydar, M.; Baber, U.; Min, J.K.; Tang, W.H.W.; Halperin, J.L.; Narayan, S.M. Deep learning for cardiovascular medicine: A practical primer. *Eur. Heart J.* **2019**, *40*, 2058–2073. [CrossRef] [PubMed]
4. Bi, W.L.; Hosny, A.; Schabath, M.B.; Giger, M.L.; Birkbak, N.; Mehrtash, A.; Allison, T.; Arnaout, O.; Abbosh, C.; Dunn, I.F.; et al. Artificial intelligence in cancer imaging: Clinical challenges and applications. *CA Cancer J. Clin.* **2019**, *69*, 127–157. [CrossRef]
5. Bulten, W.; Pinckaers, H.; van Boven, H.; Vink, R.; De Bel, T.; Van Ginneken, B.; Van der Laak, J.; Hulsbergen-van de Kaa, C.; Litjens, G. Automated deep-learning system for Gleason grading of prostate cancer using biopsies: A diagnostic study. *Lancet Oncol.* **2020**, *21*, 233–241. [CrossRef]
6. Balakrishnan, G.; Zhao, A.; Sabuncu, M.R.; Guttag, J.; Dalca, A.V. VoxelMorph: A Learning Framework for Deformable Medical Image Registration. *IEEE Trans. Med. Imaging* **2019**, *38*, 103–111. [CrossRef]
7. Isensee, F.; Jaeger, P.F.; Kohl, S.A.A.; Petersen, J.; Maier-Hein, K.H. nnU-Net: A self-configuring method for deep learning-based biomedical image segmentation. *Nat. Methods* **2021**, *18*, 203–211. [CrossRef]
8. Ahmad, M.; Qadri, S.F.; Ashraf, M.U.; Subhi, K.; Khan, S.; Zareen, S.S.; Qadri, S. Efficient Liver Segmentation from Computed Tomography Images Using Deep Learning. *Comput. Intell. Neurosci.* **2022**, *2022*, 2665283. [CrossRef]
9. Qadri, S.F.; Shen, L.; Ahmad, M.; Qadri, S.; Zareen, S.S.; Akbar, M.A. SVseg: Stacked Sparse Autoencoder-Based Patch Classification Modeling for Vertebrae Segmentation. *Mathematics* **2022**, *10*, 796. [CrossRef]
10. Massion, P.P.; Antic, S.; Ather, S.; Arteta, C.; Brabec, J.; Chen, H.; Declerck, J.; Dufek, D.; Hickes, W.; Kadir, T.; et al. Assessing the Accuracy of a Deep Learning Method to Risk Stratify Indeterminate Pulmonary Nodules. *Am. J. Respir. Crit. Care Med.* **2020**, *202*, 241–249. [CrossRef]
11. Kim, H.; Lee, D.; Cho, W.S.; Lee, J.C.; Goo, J.M.; Kim, H.C.; Park, C.M. CT-based deep learning model to differentiate invasive pulmonary adenocarcinomas appearing as subsolid nodules among surgical candidates: Comparison of the diagnostic performance with a size-based logistic model and radiologists. *Eur. Radiol.* **2020**, *30*, 3295–3305. [CrossRef]
12. Zhao, W.; Yang, J.; Sun, Y.; Li, C.; Wu, W.; Jin, L.; Yang, Z.; Ni, B.; Gao, P.; Wang, P.; et al. 3D Deep Learning from CT Scans Predicts Tumor Invasiveness of Subcentimeter Pulmonary Adenocarcinomas. *Cancer Res.* **2018**, *78*, 6881–6889. [CrossRef]
13. De Margerie-Mellon, C.; Gill, R.R.; Salazar, P.; Oikonomou, A.; Nguyen, E.T.; Heidinger, B.H.; Medina, M.A.; VanderLaan, P.A.; Bankier, A.A. Assessing invasiveness of subsolid lung adenocarcinomas with combined attenuation and geometric feature models. *Sci. Rep.* **2020**, *10*, 14585. [CrossRef]
14. Wang, X.; Li, Q.; Cai, J.; Wang, W.; Xu, P.; Zhang, Y.; Fang, Q.; Fu, C.; Fan, L.; Xiao, Y.; et al. Predicting the invasiveness of lung adenocarcinomas appearing as ground-glass nodule on CT scan using multi-task learning and deep radiomics. *Transl. Lung Cancer Res.* **2020**, *9*, 1397–1406. [CrossRef]
15. Vaswani, A.; Shazeer, N.; Parmar, N.; Uszkoreit, J.; Jones, L.; Gomez, A.N.; Kaiser, Ł.; Polosukhin, I. Attention Is All You Need. In Proceedings of the Conference on Neural Information Processing Systems (NIPS 2017), Long Beach, CA, USA, 4–9 December 2017; pp. 5998–6008.
16. Hochreiter, S.; Schmidhuber, J. Long Short-Term Memory. *Neural Comput.* **1997**, *9*, 1735–1780. [CrossRef]
17. Cho, K.; Van Merrienboer, B.; Bahdanau, D.; Bengio, Y. On the Properties of Neural Machine Translation: Encoder-Decoder Approaches. In Proceedings of the SSST-8, Eighth Workshop on Syntax, Semantics and Structure in Statistical Translation, Doha, Qatar, 25 October 2014; pp. 103–111. [CrossRef]
18. Li, Y.; Yang, J.; Xu, Y.; Xu, J.; Ye, X.; Tao, G.; Xie, X.; Liu, G. Learning Tumor Growth via Follow-Up Volume Prediction for Lung Nodules. *arXiv* **2020**, arXiv:2006.13890. [CrossRef]

19. Wu, C.-Y.; Girshick, R.; He, K.; Feichtenhofer, C.; Krahenbuhl, P. A Multigrid Method for Efficiently Training Video Models. In Proceedings of the 2020 IEEE/CVF Conference on Computer Vision and Pattern Recognition (CVPR), Seattle, WA, USA, 13–19 June 2020.
20. Saon, G.; Kurata, G.; Sercu, T.; Audhkhasi, K.; Thomas, S.; Dimitriadis, D.; Cui, X.; Ramabhadran, B.; Picheny, M.; Lim, L.-L.; et al. English Conversational Telephone Speech Recognition by Humans and Machines. In Proceedings of the INTERSPEECH 2017: Conference of the International Speech Communication Association, Stockholm, Sweden, 20–24 August 2017.
21. Rajkomar, A.; Oren, E.; Chen, K.; Dai, A.M.; Hajaj, N.; Hardt, M.; Liu, P.J.; Liu, X.; Marcus, J.; Sun, M.; et al. Scalable and accurate deep learning with electronic health records. *NPJ Digit. Med.* **2018**, *1*, 18. [CrossRef]
22. Xu, Y.; Hosny, A.; Zeleznik, R.; Parmar, C.; Coroller, T.; Franco, I.; Mak, R.H.; Aerts, H.J. Deep Learning Predicts Lung Cancer Treatment Response from Serial Medical Imaging. *Clin. Cancer Res.* **2019**, *25*, 3266–3275. [CrossRef]
23. Heuvelmans, M.A.; Oudkerk, M.; De Bock, G.H.; De Koning, H.J.; Xie, X.; Van Ooijen, P.M.A.; Greuter, M.; De Jong, P.A.; Groen, H.J.M.; Vliegenthart, R. Optimisation of volume-doubling time cutoff for fast-growing lung nodules in CT lung cancer screening reduces false-positive referrals. *Eur. Radiol.* **2013**, *23*, 1836–1845. [CrossRef]
24. Yang, J.; Chen, J.; Kuang, K.; Lin, T.; He, J.; Ni, B. MIA-Prognosis: A Deep Learning Framework to Predict Therapy Response. In Proceedings of the International Conference on Medical Image Computing and Computer-Assisted Intervention, Lima, Peru, 4–8 October 2020.
25. Loshchilov, I.; Hutter, F. Decoupled Weight Decay Regularization. *arXiv* **2017**, arXiv:1711.05101.
26. Simonyan, K.; Vedaldi, A.; Zisserman, A. Deep Inside Convolutional Networks: Visualising Image Classification Models and Saliency Maps. *arXiv* **2013**, arXiv:1312.6034. [CrossRef]
27. Park, S.; Lee, S.M.; Kim, S.; Lee, J.-G.; Choi, S.; Do, K.-H.; Seo, J.B. Volume Doubling Times of Lung Adenocarcinomas: Correlation with Predominant Histologic Subtypes and Prognosis. *Radiology* **2020**, *295*, 703–712. [CrossRef] [PubMed]
28. Ardila, D.; Kiraly, A.P.; Bharadwaj, S.; Choi, B.; Reicher, J.J.; Peng, L.; Tse, D.; Etemadi, M.; Ye, W.; Corrado, G.; et al. End-to-end lung cancer screening with three-dimensional deep learning on low-dose chest computed tomography. *Nat. Med.* **2019**, *25*, 954–961. [CrossRef]
29. Trebeschi, S.; Bodalal, Z.; Boellaard, T.N.; Bucho, T.M.T.; Drago, S.G.; Kurilova, I.; Calin-Vainak, A.M.; Pizzi, A.D.; Muller, M.; Hummelink, K.; et al. Prognostic Value of Deep Learning-Mediated Treatment Monitoring in Lung Cancer Patients Receiving Immunotherapy. *Front. Oncol.* **2021**, *11*, 609054. [CrossRef]
30. Choi, Y.; Ahn, K.-J.; Nam, Y.; Jang, J.; Shin, N.-Y.; Choi, H.S.; Jung, S.-L.; Kim, B.-S. Analysis of heterogeneity of peritumoral T2 hyperintensity in patients with pretreatment glioblastoma: Prognostic value of MRI-based radiomics. *Eur. J. Radiol.* **2019**, *120*, 108642. [CrossRef]
31. Beig, N.; Khorrami, M.; Alilou, M.; Prasanna, P.; Braman, N.; Orooji, M.; Rakshit, S.; Bera, K.; Rajiah, P.; Ginsberg, J.; et al. Perinodular and Intranodular Radiomic Features on Lung CT Images Distinguish Adenocarcinomas from Granulomas. *Radiology* **2019**, *290*, 783–792. [CrossRef]
32. Zhou, H.; Zhang, S.; Peng, J.; Zhang, S.; Zhang, W. Informer: Beyond Efficient Transformer for Long Sequence Time-Series Forecasting. *Proc. AAAI Conf. Artif. Intell.* **2020**, *35*, 11106–11115.
33. Krizhevsky, A.; Sutskever, I.; Hinton, G. ImageNet Classification with Deep Convolutional Neural Networks. *Commun. ACM* **2017**, *60*, 84–90. [CrossRef]
34. Clevert, D.-A.; Unterthiner, T.; Hochreiter, S. Fast and Accurate Deep Network Learning by Exponential Linear Units (ELUs). *arXiv* **2015**, arXiv:1511.07289. [CrossRef]
35. Ba, J.L.; Kiros, J.R.; Hinton, G.E. Layer Normalization. *arXiv* **2016**, arXiv:1607.06450.
36. Xie, S.; Girshick, R.; Dollár, P.; Tu, Z.; He, K. Aggregated Residual Transformations for Deep Neural Networks. In Proceedings of the IEEE Conference on Computer Vision and Pattern Recognition (CVPR), Honolulu, HI, USA, 21–26 July 2017.

cancers

MDPI

Article

Development and Validation of Novel Deep-Learning Models Using Multiple Data Types for Lung Cancer Survival

Jason C. Hsu [1,2,3,4], Phung-Anh Nguyen [1,2,3], Phan Thanh Phuc [4], Tsai-Chih Lo [5], Min-Huei Hsu [6,7], Min-Shu Hsieh [8,9], Nguyen Quoc Khanh Le [10,11], Chi-Tsun Cheng [3], Tzu-Hao Chang [2,5,*] and Cheng-Yu Chen [11,12,*]

1 Clinical Data Center Office of Data Science, Taipei Medical University, Taipei 110, Taiwan
2 Clinical Big Data Research Center, Taipei Medical University Hospital, Taipei Medical University, Taipei 110, Taiwan
3 Research Center of Health Care Industry Data Science, College of Management, Taipei Medical University, Taipei 110, Taiwan
4 International Ph.D. Program in Biotech and Healthcare Management, College of Management, Taipei Medical University, Taipei 110, Taiwan
5 Graduate Institute of Biomedical Informatics, College of Medical Science and Technology, Taipei Medical University, 250 Wu-Hsing Str., Xinyi Dist., Taipei 110, Taiwan
6 Office of Data Science, Taipei Medical University, Taipei 110 Taiwan
7 Graduate Institute of Data Science, College of Management, Taipei Medical University, Taipei 110, Taiwan
8 Department of Pathology, National Taiwan University Hospital, Taipei 100, Taiwan
9 Graduate Institute of Pathology, College of Medicine, National Taiwan University, Taipei 100, Taiwan
10 Professional Master Program in Artificial Intelligence in Medicine, College of Medicine, Taipei Medical University, Taipei 110, Taiwan
11 Research Center for Artificial Intelligence in Medicine, Taipei Medical University, Taipei 110, Taiwan
12 Department of Radiology, College of Medicine, Taipei Medical University, 250 Wu-Hsing Str., Xinyi Dist., Taipei 110, Taiwan
* Correspondence: kevinchang@tmu.edu.tw (T.-H.C.); sandychen@tmu.edu.tw (C.-Y.C.); Tel.: +886-02-66382736 (ext.1508) (T.-H.C.); +886-02-27361661 (ext. 2018) (C.-Y.C.)

Citation: Hsu, J.C.; Nguyen, P.-A.; Phuc, P.T.; Lo, T.-C.; Hsu, M.-H.; Hsieh, M.-S.; Le, N.Q.K.; Cheng, C.-T.; Chang, T.-H.; Chen, C.-Y. Development and Validation of Novel Deep-Learning Models Using Multiple Data Types for Lung Cancer Survival. *Cancers* **2022**, *14*, 5562. https://doi.org/10.3390/cancers14225562

Academic Editors: Hamid Khayyam, Ali Madani, Rahele Kafieh and Ali Hekmatnia

Received: 21 September 2022
Accepted: 10 November 2022
Published: 12 November 2022

Simple Summary: Previous survival-prediction studies have had several limitations, such as a lack of comprehensive clinical data types, testing only in limited machine-learning algorithms, or a lack of a sufficient external testing set. This lung-cancer-survival-prediction model is based on multiple data types, multiple novel machine-learning algorithms, and external testing. This predicted model demonstrated a higher performance (ANN, AUC, 0.89; accuracy, 0.82; precision, 0.91) than previous similar studies.

Abstract: A well-established lung-cancer-survival-prediction model that relies on multiple data types, multiple novel machine-learning algorithms, and external testing is absent in the literature. This study aims to address this gap and determine the critical factors of lung cancer survival. We selected non-small-cell lung cancer patients from a retrospective dataset of the Taipei Medical University Clinical Research Database and Taiwan Cancer Registry between January 2008 and December 2018. All patients were monitored from the index date of cancer diagnosis until the event of death. Variables, including demographics, comorbidities, medications, laboratories, and patient gene tests, were used. Nine machine-learning algorithms with various modes were used. The performance of the algorithms was measured by the area under the receiver operating characteristic curve (AUC). In total, 3714 patients were included. The best performance of the artificial neural network (ANN) model was achieved when integrating all variables with the AUC, accuracy, precision, recall, and F1-score of 0.89, 0.82, 0.91, 0.75, and 0.65, respectively. The most important features were cancer stage, cancer size, age of diagnosis, smoking, drinking status, EGFR gene, and body mass index. Overall, the ANN model improved predictive performance when integrating different data types.

Keywords: lung cancer; survival; prediction models; real-world data; artificial intelligence; machine learning

1. Introduction

Lung cancer is the leading cause of cancer deaths worldwide [1]. Globally, there were around 2.21 million new cases of lung cancer and 1.80 million fatalities in 2020 [2]. One study reported that lung cancer incidence and mortality rates were 22.2 and 18.0 per 100,000 people in 2020, respectively [3,4]. Lung cancer can be divided clinically into two types based on histological features: non-small-cell lung cancer (NSCLC) and small-cell lung cancer (SCLC). NSCLC is the most common among them, accounting for 80–90% of lung cancers [5]. Cell deterioration and metastasis are slower in NSCLC than in SCLC. Around 70% of patients are diagnosed at an advanced stage, making surgical resection and complete treatment challenging [6,7].

Artificial intelligence (AI) has been increasingly used in medical research and clinical practice [8,9]. The accurate prediction of disease prognosis and the outcome of drug treatment, which may serve as a reference for treatment decision-making and drug selection, has become an essential topic in the clinical medicine [9,10]. Developing disease-risk and prognosis-prediction models using machine-learning or deep-learning algorithms with big data is a major area of AI-based academic research in the medical field [10,11]. Studies have used machine-learning and/or deep-learning algorithms to develop lung cancer risk and prognosis-prediction models [12–15]. Among them, Lai et al. [16] used 15 biomarkers with clinical data (including gene expression) from 614 patients to develop a deep neural network to predict the five-year overall survival of NSCLC patients.

This study aimed to develop survival-prediction models for lung cancer patients using a large number of samples, different data types, various machine-learning algorithms, and external testing. In addition to the basic clinical data (including demographic information, disease condition, comorbidity, and current medication), we examined the role of laboratory and genomic test results, which are generally not easy to obtain in predicting lung cancer survival. Moreover, we also explored the important predictors for developing prediction models.

2. Methods

2.1. Study Design and Data Source

We conducted a retrospective study in which we obtained data from the Taiwan Cancer Registry (TCR) database and the Taipei Medical University Clinical Research Database (TMUCRD). The TCR database was established in 1979 and is managed by Taiwan's Health Promotion Administration, Ministry of Health and Welfare. It covers 98% of Taiwanese cancer patients and includes diagnosis and other related information. The TMUCRD retrieved data from various electronic medical records (EHR) of three hospitals, Taipei Medical University Hospital (TMUH), Wan-Fang Hospital (WFH), and Shuang-Ho Hospital (SHH). The database contains the electronic medical record data of 3.8 million people from 1998 to 2020, including structured data (e.g., basic information of patients, medical information, test reports, diagnosis results, treatment process, surgery, and medication history) and unstructured data (e.g., progress notes, pathology reports, and medical imaging reports) [17]. This study has been approved by the Joint Institute Review Board of Taipei Medical University (TMU-JIRB), Taipei, Taiwan (approval number N202101080). All the data were anonymous before conducting analysis.

2.2. Cohort Selection

This study selected patients with lung cancer (ICD-O-3 code: C33, C34) from 2008 to 2018 in the TCR database. Exclusion criteria included individuals under 20 years old, SCLC patients, and patients who did not have any medical history in the three hospitals (TMUH, WFH, SHH). Thus, a total of 3714 patients were included in this study, including 960 patients from TMUH, 1320 from WFH, and 1434 from SHH (Figure S1 in the Supplementary Materials).

2.3. Outcome Measurement

We ascertained the study outcomes using TMUCRD EHR and vital status data from the Taiwan Death Registry (TDR) [18]. We used the diagnosis date of NSCLC as the index date, and the outcome of this study was death within two years following diagnosis. Data were censored at the date of death or loss to follow-up, insurance termination, or the study's end on 31 December 2018.

2.4. Feature Selection

Based on a literature review and consultation with clinicians, we selected features that may lead to the mortality of NSCLC patients to build prediction models. These features consisted of:

1. Demographic information: age, gender, body mass index (BMI), smoking, drinking;
2. Cancer conditions: tumor size and cancer stage;
3. Comorbidities: cardiovascular problems (i.e., myocardial infarction (MI), congestive heart failure (CHF), peripheral vascular disease (PVD), and cardiovascular disease (CVD)), dementia, chronic obstructive pulmonary disease (COPD), rheumatic disease, peptic ulcer disease (PUD), renal disease, liver disease, diabetes, anemia, depression, hyperlipidemia, hypertension, Parkinson's disease, and Charlson Comorbidity Index (CCI) score. These conditions were considered if they were diagnosed in at least two outpatient claims or one hospitalization over a year before the cancer diagnosis date.
4. Medications: alimentary tract and metabolism, blood and blood-forming organs, cardiovascular system, genitourinary system and hormones, musculoskeletal system, nervous system, and respiratory system. We measured patients who had used medications by receiving them for more than a month (i.e., 30 days) during a year (i.e., 360 days) before the index date.
5. Laboratory tests: basophil, blood urea nitrogen (BUN), calcium, cholesterol, chloride, creatinine, eosinophil, ferritin, glucose AC, HbA1c, HCT, HGB, potassium, lymphocyte, MCH, MCHC, MCV, monocyte, sodium, neutrophil, platelet (PLT), RBC, triglyceride, and WBC. We only selected laboratory tests with a missing rate of less than 70% values a year before or a month after the index date.
6. Genomic tests: ALK, EGFR, KRAS, PDL1, and ROS1. We collected genomic tests if patients had ever taken one a month after the cancer diagnosis date.

2.5. Development of the Algorithms

This study established prediction models based on four modes and different algorithms:

- The primary mode (e.g., Mode 1) included demographic information, cancer conditions, comorbidities, and medications.
- The second mode (Mode 2) included the data from Mode 1 and the laboratory tests.
- The third mode (Mode 3) included the data from Mode 1 and genomic tests.
- The fourth mode (Mode 4) considered all the above features.

This study aims to predict the survival of lung cancer patients; therefore, the problem can be formulated as a classification model as it could occur in the same patients. We used possible machine-learning techniques such as logistic regression (LR), linear discriminant analysis (LDA), light gradient-boosting machine (LGBM), gradient-boosting machine (GBM), extreme gradient boosting (XGBoost), random forest (RF), AdaBoost, support vector machine (SVC), and artificial neural network (ANN). These methods are briefly introduced below.

Logistic Regression (LR): This is a discrete choice model that models the relationship between a response and multiple explanatory variables and is based on the concept of probability [19]. It is widely used and more practical in fields such as biostatistics, clinical medicine, and quantitative psychology. Its Equation (1) is:

$$y = \frac{e^{(b_0+b_1X)}}{1+e^{(b_0+b_1X)}} \tag{1}$$

where x is the input value, y is the predicted output, b_0 is the bias or intercept term, and b_1 is the coefficient for input (x). In this study, we used the LR function with the parameter C (inverse of regularization strength) of 0.0001 to reduce the model's overfitting.

Linear Discriminant Analysis (LDA): This is generally used to classify patterns between two classes; however, it can be extended to multiple patterns. LDA assumes that all classes are linearly separable, and according to the multiple linear discrimination functions representing several hyperplanes in the feature space are created to distinguish the classes [20]. In this study, we set the parameters' *shrinkage* to '0' and the *solver* to 'lsqr' to improve estimation and classification accuracy.

Light Gradient-Boosting Machine (LGBM): This is a gradient-boosting framework that uses tree-based learning algorithms. It is designed to be distributed and efficient with the following advantages: faster training speed and higher efficiency; lower memory usage; better accuracy; support of parallel, distributed, and GPU learning; and capability to handle large-scale data [21]. The model's *class_weight* parameter was set as 'balanced', which uses the output's value to automatically adjust weights inversely proportional to class frequencies in the input data. The *learning_rate*, l1 regularization—*reg_alpha*, and l2 regularization—*reg_lambda* parameters were set as 0.05, 0.1, and 0.1, respectively.

Gradient-Boosting Machine (GBM): Gradient-boosting regression trees produce competitive, highly robust, and interpretable procedures for regression and classification. The ability of TreeBoost procedures to give a quick indication of potential predictability, coupled with their extreme robustness, makes them a useful preprocessing tool that can be applied to imperfect data [22]. The default parameters were used in this model.

Extreme Gradient Boosting (XGBoost): XGBoost, an efficient and scalable implementation of the gradient-boosting framework, is a machine-learning system for tree boosting. The scalability of XGBoost is attributed to several critical systems and algorithmic optimizations. These innovations include a novel tree-learning algorithm for handling sparse data; a theoretically justified weighted quantile sketch procedure allows the handling of instance weights in approximate tree learning [23]. The default parameters were used in this model.

Random Forest (RF): RF is an ensemble-learning method that operates by constructing many small scales of classification modules (most often decision trees) at the training time. The model outputs the class that combines the result of the individual modules based on some voting algorithms [24]. In this study, we set the parameters as follows: *n_estimators* (the number of trees) of 500, *max_depth* of 10, *min_samples_split* of 400, and *class_weight* of 0.5 for each class.

AdaBoost: The AdaBoost algorithm is an iterative procedure that combines several weak classifiers to approximate the Bayes classifier $C*(x)$. AdaBoost builds a classifier, e.g., a classification tree that produces class labels, starting with the unweighted training sample. If a training data point is misclassified, the weight of that data point is increased (boosted). A second classifier is built using the new weights, which are no longer equal. Again, misclassified training data have their weights boosted, and the procedure is repeated [25]. The number of estimators (*n_estimators*) used was 100.

Support Vector Machine (SVC): This is a machine-learning algorithm that can be applied to linear and nonlinear data. SVC transforms the original data to a higher dimension, from which it can use the super vectors in the training data set to find the hyperplane for categorizing the data. An SVC mainly identifies the hyperplane with the most significant margin, e.g., the maximum marginal hyperplane, to achieve higher accuracy [26]. The SVC can be represented by the following Equation (2):

$$f(x) = \sum_{i=1}^{N} (\alpha_i^* - \alpha_i) K(x, x_i) + \mathrm{B} \tag{2}$$

where $K(x, x_i)$ is the kernel function, $\alpha_i, \alpha_i^* \geq 0$ are the Lagrange multipliers, and B is a bias term. In this study, we used a *linear* kernel for computations.

Artificial Neural Network (ANN): This is a learning algorithm vaguely inspired by biological neural networks. Computations are structured in terms of an interconnected group of artificial neurons, and these neutrons process information using a connectionist approach to computation. They are usually used to model complex relationships between inputs and outputs, find patterns in data, or capture the statistical structure [27]. The number of hidden layers with the number of neurons in each layer was set at 3 and 16, respectively. Additionally, for each layer, the *l2 regularization* of 0.01 and the 'relu' *activation* were used in the study. We set the 'softmax' activation for the output layer. We also used the 'Adam' *optimizer*, a highly performant stochastic gradient descent algorithm, and 'binary_crossentropy' as the binary classification outcome for the *loss* function.

2.6. Evaluating the Algorithms

The training dataset contained the data of patients from TMUH and WFH. The stratified 5-fold cross-validation was applied in the training set to assess the different machine-learning models' performance and general errors. In other words, patients in the training set were divided into five groups, each used repeatedly as the internal validation set. We recruited data from SHH and used it for the external testing dataset to generalize the model.

The performance of the algorithms was measured by the area under the receiver operating characteristic curve (AUC), accuracy, sensitivity (recall), specificity, positive predictive value (PPV, precision), negative predictive value (NPV), and F1-score. We defined the best model using the highest AUC by comparing various models based on the external testing set Furthermore, we analyzed the feature's contribution (i.e., the feature's importance) of the best model using SHAP values (SHapley Additive exPlanations) [28].

All the data processing was performed using MSSQL server 2017 (Redmond, WA, USA), and the model training and testing were performed using Python version 3.8 (Wilmington, DE, USA) with scikit-learn version 1.1 (Paris, France) [29].

3. Results

3.1. Baseline Characteristics of Patients

We identified 3714 eligible lung cancer patients diagnosed for the first time and registered at the TCR. Among those patients, 2280 patients were included in the training dataset, whereas 1434 were in the testing dataset. Demographic characteristics, comorbidities, tumor size, tumor stage, genomic tests, medication uses, and laboratory tests are presented in Table 1. The mean (standard deviation, SD) ages and BMI of cohort patients were 68 (13.7) and 23.4 (4.33), respectively. Most of the patients were male (57.5%) with late-stage lung cancer (i.e., stage IV, 54.8%), and patients were less likely to smoke (26.7%) or drink (11%). The cohort of patients had comorbidities related to hypertension (19.8%), hyperlipidemia (13.9%). COPD (16.1%), and CVD problems (11.6%). The follow-up durations for the cohort patents were a mean (SD) of 2.25 (2.47) years and a median (interquartile range (IQR)) of 1.41 [0.46–3.04] years. Detailed information is shown in Table S1 in the Supplementary Materials.

Table 1. Basic Characteristics of the Study Cohort.

Features	Overall n = 3714	Training Set [a] n = 2280	Testing Set [b] n = 1434
Male, N (%)	2136 (57.5)	1258 (55.2)	878 (61.2)
Age, Mean (SD), yrs.	68.0 (13.7)	67.9 (13.8)	68.0 (13.4)
BMI, Mean (SD), kg/m^2	23.4 (4.33)	23.4 (3.93)	23.4 (4.81)
Smoking, N (%)			
No	1170 (31.5)	710 (31.1)	460 (32.1)
Yes	993 (26.7)	523 (22.9)	470 (32.8)
Unknown	1551 (41.8)	1047 (45.9)	504 (35.1)

Table 1. *Cont.*

Features	Overall n = 3714	Training Set [a] n = 2280	Testing Set [b] n = 1434
Drinking, N (%)			
No	1750 (47.1)	983 (43.1)	767 (53.5)
Yes	408 (11.0)	247 (10.8)	161 (11.2)
Unknown	1556 (41.9)	1050 (46.1)	506 (35.3)
Tumor size, cm			
Mean (SD)	4.23 (2.45)	4.11 (2.39)	4.46 (2.55)
Median [IQR]	3.8 [2.4–5.5]	3.6 [2.3–5.5]	4.0 [2.5–5.7]
Cancer stage, N (%)			
0	11 (0.3)	10 (0.4)	1 (0.1)
I	533 (14.4)	348 (15.3)	185 (12.9)
II	139 (3.7)	88 (3.9)	51 (3.6)
III	527 (14.1)	330 (14.5)	197 (13.7)
IV	2034 (54.8)	1207 (52.9)	827 (57.7)
Missing	470 (12.7)	297 (13.0)	173 (12.1)
Genomic Test			
ALK, N (%)			
Negative	681 (18.3)	457 (20.0)	224 (15.6)
Positive	39 (1.1)	21 (0.9)	18 (1.3)
Unknown	2994 (80.6)	1802 (79.0)	1192 (83.1)
EGFR, N (%)			
Negative	842 (22.7)	473 (20.7)	369 (25.7)
Positive	787 (21.2)	467 (20.5)	320 (22.3)
Unknown	2085 (56.1)	1340 (58.8)	745 (52.0)
KRAS, N (%)			
Negative	45 (1.2)	32 (1.4)	13 (0.9)
Positive	5 (0.1)	2 (0.1)	3 (0.2)
Unknown	3664 (98.7)	2246 (98.5)	1418 (98.9)
PDL1, N (%)			
Negative	269 (7.2)	149 (6.5)	120 (8.4)
Positive	66 (1.8)	42 (1.8)	24 (1.7)
Unknown	3379 (91.0)	2089 (91.6)	1290 (90.0)
ROS1, N (%)			
Negative	288 (7.8)	287 (12.6)	1 (0.1)
Positive	29 (0.8)	27 (1.2)	2 (0.1)
Unknown	3397 (91.4)	1966 (86.2)	1431 (99.8)
Comorbidity, N (%)			
CVD problems	432 (11.6)	296 (13.0)	136 (9.5)
Dementia	124 (3.3)	71 (3.1)	53 (3.7)
COPD	599 (16.1)	391 (17.1)	208 (14.5)
Rheumatic disease	28 (0.75)	16 (0.7)	12 (0.8)
PUD	365 (9.8)	246 (10.8)	119 (8.3)
Renal disease	128 (3.4)	92 (4.0)	31 (2.2)
Liver disease	211 (5.7)	147 (6.4)	64 (4.5)
DM	372 (10.0)	248 (10.9)	124 (8.6)
Anemia	107 (2.9)	76 (3.3)	31 (2.2)
Depression	245 (6.6)	175 (7.7)	70 (4.9)
Hyperlipidemia	516 (13.9)	385 (16.9)	131 (9.1)
Hypertension	736 (19.8)	503 (22.1)	233 (16.2)
Parkinson's disease	50 (1.3)	29 (1.3)	21 (1.5)
Charlson Comorbidity Index (CCI)			
Mean (SD)	3.08 (2.07)	3.13 (2.19)	2.97 (1.86)
Median [IQR]	3.0 [2.0–4.0]	3.0 [2.0–4.0]	3.0 [2.0–4.0]
Follow-up, yrs.			
Mean (SD)	2.25 (2.47)	2.44 (2.61)	1.96 (2.19)
Median [IQR]	1.41 [0.46–3.04]	1.51 [0.53–3.36]	1.24 [0.38–2.64]

Table 1. *Cont.*

Features	Overall *n* = 3714	Training Set [a] *n* = 2280	Testing Set [b] *n* = 1434
Medications, N (%)			
Alimentary tract and metabolism	591 (15.9)	394 (17.3)	197 (14.7)
Blood and blood-forming organs	446 (12.0)	293 (12.9)	153 (11.3)
Cardiovascular system	675 (18.2)	448 (19.6)	227 (16.9)
Genitourinary system and hormones	132 (3.6)	74 (3.2)	53 (4.3)
Musculoskeletal system	252 (6.8)	141 (6.2)	111 (8.3)
Nervous system	391 (10.5)	254 (11.1)	137 (10.2)
Respiratory system	319 (8.6)	226 (9.9)	93 (6.9)
Laboratory Test, Mean (SD)			
Basophil	0.50 (0.40)	0.53 (0.42)	0.48 (0.39)
BUN	19.4 (14.9)	18.8 (13.1)	20.5 (17.6)
Creatinine	1.05 (0.98)	1.02 (0.90)	1.10 (1.07)
Eosinophil	1.89 (2.31)	2.03 (2.59)	1.76 (1.97)
HCT	38.3 (5.69)	38.5 (5.61)	37.9 (5.30)
HGB	12.9 (1.97)	13.0 (1.91)	12.7 (2.05)
K	3.99 (0.56)	4.02 (0.53)	3.95 (0.50)
Lymphocyte	18.7 (9.98)	19.6 (9.55)	17.8 (10.3)
MCH	29.9 (3.02)	29.9 (3.03)	29.8 (3.00)
MCHC	33.6 (0.95)	33.7 (0.96)	33.6 (0.94)
MCV	88.6 (7.61)	88.5 (7.64)	88.7 (7.57)
Monocyte	7.45 (2.90)	7.42 (2.93)	7.48 (2.87)
Na	137 (4.46)	137 (4.39)	137 (4.53)
Neutrophil	71.3 (11.9)	70.2 (11.4)	72.3 (12.2)
PLT	263 (109)	258 (100)	259 (121)
RBC	4.35 (0.68)	4.38 (0.67)	4.29 (0.69)
WBC	9.72 (5.38)	9.16 (4.16)	10.6 (6.80)

Note: SD, Standard deviation; yrs., Years; IQR, Interquartile Range; BMI, Body mass index; COPD, Chronic obstructive pulmonary disease; PUD, Peptic ulcer disease; CVD, Cardiovascular; DM, Diabetes; BUN, Blood urea nitrogen; HCT, Hematocrit; HGB, Hemoglobin; K, Potassium; MCH, Mean corpuscular hemoglobin; MCHC, Mean corpuscular hemoglobin concentration; MCV, Mean corpuscular volume; Na, Sodium; PLT, Platelet; RBC, Red blood count; WBC, White blood count; [a] The training set included the data from Taipei Medical University and Wan-Fang hospitals; [b] The testing set included the data from Shuang Ho hospital.

3.2. The Performances of Different Prediction Models

The performances of different prediction models are shown in Table 2. In Mode 1, the highest AUC of 0.88 was observed for the ANN model (i.e., accuracy, 0.82; precision, 0.90; recall, 0.75; and F1-score, 0.64), followed by the GBM and RF models with an AUC of 0.83 and 0.82, respectively. In Mode 3, the best performance was found with an AUC of 0.89 for the ANN model (i.e., accuracy, 0.83; precision, 0.89; recall, 0.81; and F1-score, 0.64). The following AUCs were observed 0.85 for LGBM, GBM, and 0.84 for RF models. Moreover, when considering all features in Mode 4, we found that the best model was the ANN model with an AUC of 0.89 (i.e., accuracy, 0.82; precision, 0.91; recall, 0.75; and F1-score, 0.65). Figures 1 and 2 show the ROC curves of different prediction models in four modes. Detailed information on the various models' measurements (i.e., sensitivity, specificity, PPV, NPV, accuracy, and F1-score) is shown in Table S2 in the Supplementary Materials.

Table 2. Performance of various Prediction Models by Modes.

Modes	Models	AUC Training	AUC Testing	Accuracy	Precision	Recall	F1-score
Mode 1	LR	0.70	0.72	0.65	0.88	0.64	0.75
	LDA	0.78	0.78	0.71	0.90	0.70	0.80
	LGBM	0.98	0.81	0.73	0.92	0.72	0.81
	GBM	0.96	0.83	0.75	0.91	0.76	0.84
	XGBoost	0.99	0.80	0.75	0.90	0.77	0.84
	RF	0.90	0.82	0.72	0.92	0.70	0.80
	AdaBoost	0.94	0.81	0.73	0.91	0.72	0.81
	SVC	0.78	0.78	0.71	0.89	0.72	0.79
	ANN *	**0.89**	**0.88**	**0.82**	**0.90**	**0.75**	**0.64**
Mode 2	LR	0.74	0.75	0.60	0.93	0.53	0.67
	LDA	0.81	0.79	0.71	0.90	0.70	0.80
	LGBM	0.99	0.83	0.78	0.91	0.79	0.86
	GBM	0.96	0.84	0.78	0.91	0.80	0.87
	XGBoost	1.00	0.81	0.78	0.90	0.81	0.86
	RF	0.92	0.83	0.69	0.94	0.64	0.76
	AdaBoost	0.95	0.80	0.74	0.90	0.76	0.83
	SVC	0.81	0.79	0.70	0.91	0.68	0.78
	ANN *	**0.89**	**0.89**	**0.80**	**0.91**	**0.75**	**0.64**
Mode 3	LR	0.70	0.73	0.65	0.88	0.63	0.74
	LDA	0.80	0.81	0.75	0.91	0.76	0.83
	LGBM	0.98	0.85	0.80	0.92	0.81	0.87
	GBM	0.96	0.85	0.79	0.92	0.79	0.86
	XGBoost	1.00	0.83	0.79	0.91	0.80	0.86
	RF	0.91	0.84	0.72	0.93	0.69	0.80
	AdaBoost	0.95	0.83	0.79	0.91	0.80	0.86
	SVC	0.80	0.81	0.75	0.90	0.75	0.83
	ANN *	**0.89**	**0.89**	**0.83**	**0.89**	**0.81**	**0.64**
Mode 4	LR	0.74	0.75	0.61	0.93	0.53	0.67
	LDA	0.83	0.82	0.76	0.90	0.77	0.84
	LGBM	0.99	0.86	0.81	0.92	0.83	0.88
	GBM	0.97	0.85	0.79	0.92	0.81	0.87
	XGBoost	1.00	0.84	0.77	0.92	0.77	0.85
	RF	0.93	0.85	0.75	0.93	0.73	0.82
	AdaBoost	0.96	0.83	0.76	0.92	0.75	0.83
	SVC	0.83	0.81	0.75	0.90	0.76	0.84
	ANN *	**0.89**	**0.89**	**0.82**	**0.91**	**0.75**	**0.65**

Note: LR, Logistic Regression; LDA, Linear Discriminant Analysis; LGBM, Light Gradient Boosting Machine; GBM, Gradient Boosting Machine; XGBoost, Extreme Gradient Boosting; RF, Random Forest; SVC, Support Vector Machine; ANN, Artificial Neural Network; *, Best model based on AUC values.

Figure 3 shows the top 20 important features of the ANN model in Mode 4. The most important features were the cancer stage, size, age of diagnosis, smoking, and EGFR gene. In other words, patients with advanced cancer stage, large cancer size, older age, and smoking behavior had a higher risk of death within two years. The SHAP value presented the important features of the GBM model in Mode 4 and was consistent with the ANN model, such as cancer stage, age at diagnosis, cancer size, and smoking status (Figure S2 in the Supplementary Materials).

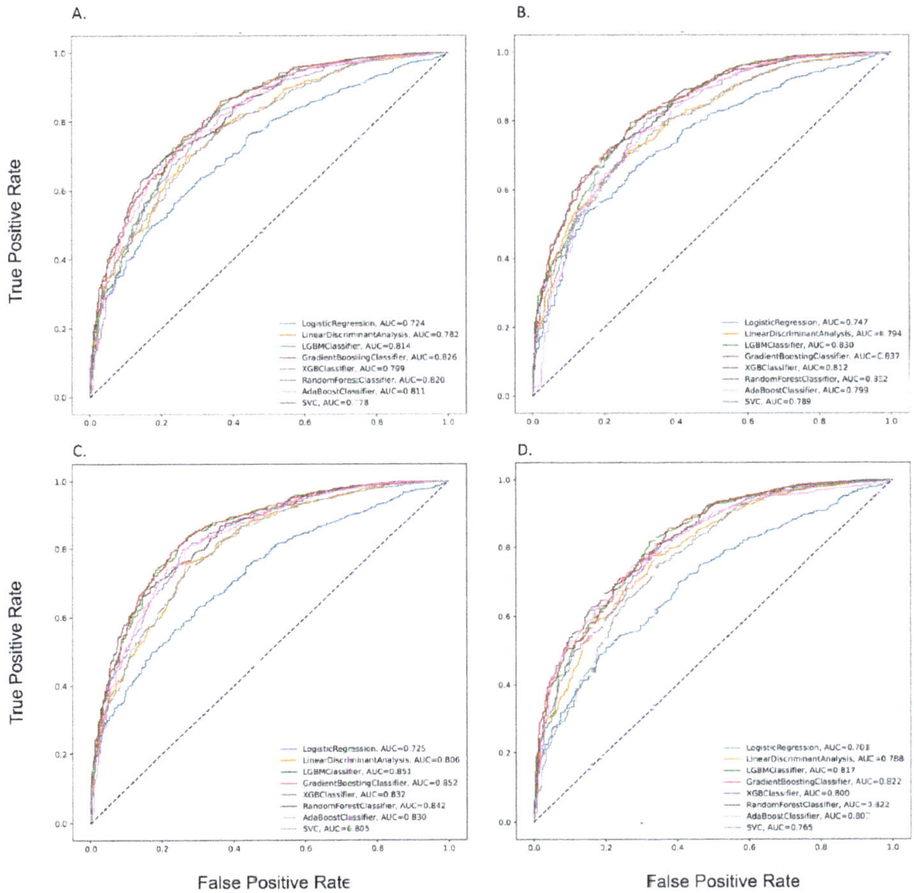

Figure 1 The Performance of the Prediction Models in the Testing dataset by different Modes. **Note**: (**A**), Mode 1; (**B**), Mode 2; (**C**), Mode 3; (**D**), Mode 4.

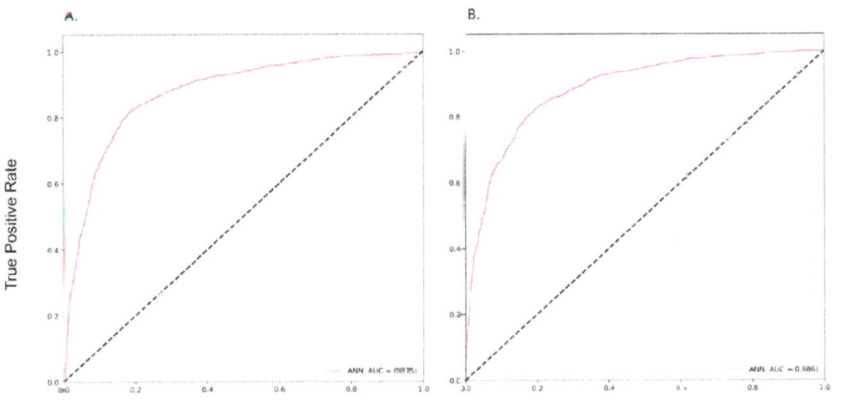

Figure 2. *Cont.*

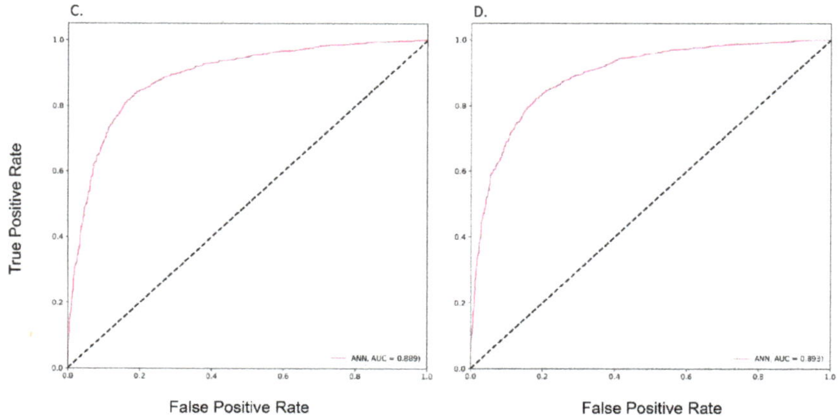

Figure 2. The Performance of the ANN Prediction Models in the Testing dataset by different Modes. **Note**: (**A**), Mode 1; (**B**), Mode 2; (**C**), Mode 3; (**D**), Mode 4.

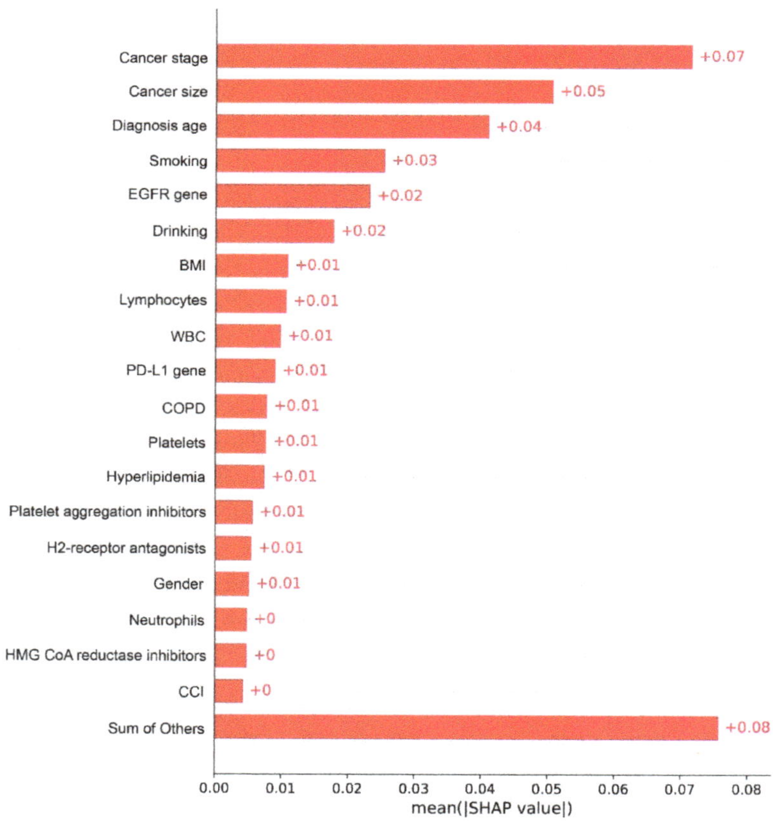

Figure 3. Feature Importance of the ANN Prediction Model in Mode 4. **Note**: BMI, Body mass index; EGFR, Epidermal growth factor receptor; WBC, White blood cell; PD-L1, Programmed death-ligand 1; COPD, Chronic obstructive pulmonary disease; CCI, Charlson comorbidity index.

4. Discussion

In recent years, the prediction of cancer patients' survival has attracted the medical community's attention in various countries because it can facilitate medical decision making, strengthen the relationship between doctors and patients, and improve the quality of medical care. Rapid progress in the development of AI based on machine learning has led to more diversified applications of AI in the field of precision medicine. Based on previously published studies on machine-learning algorithms to build prediction models for the survival of lung cancer patients [12,14–16], this study further compared the performance of various novel machine-learning algorithms. In addition, we also analyzed the relationship between the diversity of features and the accuracy of prediction results and determined the most important features affecting lung cancer survival.

Studies using multiple data types and multiple novel machine-learning algorithms simultaneously are limited. In previous studies on lung cancer prediction, most of them used a single machine-learning (e.g., RF [30]) or deep-learning (e.g., NN [14–16]) algorithm or a few basic machine-learning algorithms (e.g., LR, SVM, decision tree, RF, GBM [12,31]) to develop prediction models. Our results showed that the ANN model had the highest AUC value (it was the most suitable tool for survival prediction). In contrast, the AUC value of the traditional LR algorithm exhibited the lowest performance (it had the lowest predictive ability). Lai Y.H. et al. [16] presented a deep neural network to predict the overall survival of NSCLC patients. They obtained a good predictive performance (AUC = 0.82, accuracy = 75.4%) by integrating microarray and clinical data. While only using basic clinical data (demographics, comorbidities, and medications), our predicted model demonstrated a higher performance (ANN, AUC, 0.88; accuracy, 0.82; precision, 0.90, recall, 0.75, and F1-score, 0.64). Furthermore, when combining other variables, such as laboratory and genomic tests, the AUC values of the predicted model were better (based on the external testing, the AUCs of the ANN model in Mode 1 and Mode 4 were 0.88 and 0.89, respectively; the AUCs of LGBM model in Mode 1 and Mode 4 were 0.81 and 0.86, respectively; the AUCs of the RF model in Mode 1 and Mode 4 were 0.82 and 0.85, respectively).

In this study, we explored the variables that might affect the predictive performance of the survival model. As expected, these variables were highly correlated to the mortality of lung cancer patients, such as advanced cancer stage, tumor size, age at diagnosis, and smoking and drinking status [32]. Our findings also showed that lymphocytes, platelets, and neutrophils tests were associated with the likelihood of lung cancer survival [33]. Thus, lymphocytes play an essential role in producing cytokines, inhibiting the proliferation of cancer cells, and provoking cytotoxic cell death [34]. In words, a decrease in lymphocyte count may predict worse survival in cancer patients. Neutrophils are recruited with cytokines released by the tumor microenvironment, enhancing carcinogenesis and cancer progression [35]. Platelets modulate the tumor microenvironment by releasing factors contributing to tumor growth, invasion, and angiogenesis [36]. Another study by Wang J. et al. [37] reported that lung cancer patients with a higher BMI have prolonged survival compared to those with a lower BMI. The same was true for our study's results, which may be due to the poor nutrition and weight loss caused by respiratory diseases [38], such as COPD.

There are limitations to this study. First, although the study used data from various clinical settings (e.g., TMUH and WFH for establishing the prediction model and SHH for conducting an external test) located in the north of Taiwan, the results may not directly apply to lung cancer patients in other regions. Future studies may need to consider validating the model using data from other areas. Second, this study used retrospective data for development and validation. Further experiments with a prospective study design in clinical settings are needed. Third, to obtain a highly accurate prediction, we developed the machine-learning algorithms with binary outcomes (i.e., survival and death) rather than expected continuous outcomes (i.e., length of survival) for the NSCLC patients. Further studies should be conducted with larger sample sizes to deal with continuous outcomes for lung cancer survival.

5. Conclusions

In summary, to observe the expected survival of NSCLC patients during a two-year period, we designed an artificial neural network model with high AUC, precision, and recall. Moreover, integrating different data types (especially laboratory and genomic data) led to better predictive performance. Further research is necessary to determine the feasibility of applying the algorithm in the clinical setting and explore whether this tool could improve care and outcomes.

Supplementary Materials: The following supporting information can be downloaded at: https://www.mdpi.com/article/10.3390/cancers14225562/s1, Figure S1: Cohort Selection Process; Figure S2: Feature Importance of the GBM Prediction Model of Mode 4; Table S1: Detailed Demographic Characteristics of Cohort Patients; Table S2: Detailed Performance of various Prediction Models by Modes.

Author Contributions: T.-H.C., P.-A.N. and J.C.H. conceptualized and designed the study. P.-A.N., P.T.P. and T.-C.L. collected the data, performed the analysis, and drafted the manuscript. C.-Y.C. and T.-H.C. provided suggestions for the research design and article content. M.-H.H., M.-S.H., N.Q.K.L., C.-T.C. and J.C.H. reviewed all data and revised the manuscript critically for intellectual content. All authors have read and agreed to the published version of the manuscript.

Funding: This study was supported by Taiwan Ministry of Science and Technology grants (grant numbers: MOST109-2321-B-038-004; MOST110-2321-B-038-004). The funders had no role in the study design, data collection and analysis, publication decision, or manuscript preparation.

Institutional Review Board Statement: This study has been approved by the TMU-Joint Institutional Review Board (Project number: TMU-JIRB N202101080).

Informed Consent Statement: Not applicable.

Data Availability Statement: The authors obtained data from the Taiwan Cancer Registry (TCR) database and the Taipei Medical University Clinical Research Database (TMUCRD).

Conflicts of Interest: The authors declare no conflict of interest.

Abbreviations

NSCLC	Non-small cell lung cancer
SCLC	Small cell lung cancer
AI	Artificial intelligence
TCR	Taiwan Cancer Registry
TDR	Taiwan Death Registry
TMUCRD	Taipei Medical University Clinical Research Database
TMUH	Taipei Medical University Hospital
WFH	Wan-Fang Hospital
SHH	Shuang-Ho Hospital
BMI	Body mass index
MI	Myocardial infarction
CHF	Congestive heart failure
PVD	Peripheral vascular disease
CVD	Cardiovascular disease
COPD	Chronic obstructive pulmonary disease
PUD	Peptic ulcer disease
CCI	Charlson Comorbidity Index
BUN	Blood urea nitrogen
PLT	Platelet
LR	Logistic regression
LDA	Linear discriminant analysis

LGBM	Light gradient boosting machine
GBM	Gradient boosting machine
XGBoost	Extreme gradient boosting
RF	Random forest
SVC	Support vector machine
ANN	Artificial neural network
AUC	The area under the receiver operating characteristic curve
PPV	Positive predictive value
NPV	Negative predictive value
SHAP	Shapley additive explanations

References

1. Bray, F.; Ferlay, J.; Soerjomataram, I.; Siegel, R.L.; Torre, L.A.; Jemal, A. Global cancer statistics 2018: GLOBOCAN estimates of incidence and mortality worldwide for 36 cancers in 185 countries. *CA Cancer J. Clin.* **2018**, *68*, 394–424. [CrossRef] [PubMed]
2. World Health Organization. Cancer Fact Sheets. Available online: https://www.who.int/news-room/fact-sheets/detail/cancer (accessed on 1 November 2022).
3. Sung, H.; Ferlay, J.; Siegel, R.L.; Laversanne, M.; Soerjomataram, I.; Jemal, A.; Bray, F. Global Cancer Statistics 2020: GLOBOCAN Estimates of Incidence and Mortality Worldwide for 36 Cancers in 185 Countries. *CA Cancer J. Clin.* **2021**, *71*, 209–249. [CrossRef]
4. World Health Organization. Lung Cancer Statistics. Available online: https://www.wcrf.org/cancer-trends/lung-cancer-statistics/ (accessed on 1 November 2022).
5. Siddiqui, F.; Vaqar, S.; Siddiqui, A.H. Lung Cancer. In *StatPearls*; StatPearls Publishing LLC.: Treasure Island, FL, USA, 2022.
6. Testa, U.; Castelli, G.; Pelosi, E. Lung Cancers: Molecular Characterization, Clonal Heterogeneity and Evolution, and Cancer Stem Cells. *Cancers* **2018**, *10*, 248. [CrossRef] [PubMed]
7. Ryan, C.; Burke, L. Pathology of lung tumours. *Surgery* **2017**, *35*, 234–242. [CrossRef]
8. Liang, C.-W.; Yang, H.-C.; Islam, M.M.; Nguyen, P.A.A.; Feng, Y.-T.; Hou, Z.Y.; Huang, C.-W.; Poly, T.N.; Li, Y.-C.J. Predicting Hepatocellular Carcinoma With Minimal Features From Electronic Health Records: Development of a Deep Learning Model. *JMIR Cancer* **2021**, *7*, e19812. [CrossRef] [PubMed]
9. Poly, T.N.; Islam, M.M.; Muhtar, M.S.; Yang, H.-C.; Nguyen, P.A.; Li, Y.-C. Machine Learning Approach to Reduce Alert Fatigue Using a Disease Medication–Related Clinical Decision Support System: Model Development and Validation. *JMIR Med Inform.* **2020**, *8*, e19489. [CrossRef] [PubMed]
10. Le, N.Q.K.; Ho, Q.-T. Deep transformers and convolutional neural network in identifying DNA N6-methyladenine sites in cross-species genomes. *Methods* **2022**, *204*, 199–206. [CrossRef] [PubMed]
11. Dang, H.H.; Ta, H.D.K.; Nguyen, T.T.T.; Anuraga, G.; Wang, C.-Y.; Lee, K.-H.; Le, N.Q.K. Prospective role and immunotherapeutic targets of sideroflexin protein family in lung adenocarcinoma: Evidence from bioinformatics validation. *Funct. Integr. Genom.* **2022**, *22*, 1057–1072. [CrossRef] [PubMed]
12. Lynch, C.M.; Abdollahi, B.; Fuqua, J.D.; de Carlo, A.R.; Bartholomai, J.A.; Balgemann, R.N.; van Berkel, V.H.; Frieboes, H.B. Prediction of lung cancer patient survival via supervised machine learning classification techniques. *Int. J. Med Inform.* **2017**, *108*, 1–8. [CrossRef]
13. Siah, K.W.; Khozin, S.; Wong, C.H.; Lo, A.W. Machine-Learning and Stochastic Tumor Growth Models for Predicting Outcomes in Patients With Advanced Non-Small-Cell Lung Cancer. *JCO Clin. Cancer Inform.* **2019**, *3*, 1–11. [CrossRef]
14. Cui, L.; Li, H.; Hui, W.; Chen, S.; Yang, L.; Kang, Y.; Bo, Q.; Feng, J. A deep learning-based framework for lung cancer survival analysis with biomarker interpretation. *BMC Bioinform.* **2020**, *21*, 112. [CrossRef]
15. She, Y.; Jin, Z.; Wu, J.; Deng, J.; Zhang, L.; Su, H.; Jiang, G.; Liu, H.; Xie, D.; Cao, N.; et al. Development and Validation of a Deep Learning Model for Non–Small Cell Lung Cancer Survival. *JAMA Netw. Open* **2020**, *3*, e205842. [CrossRef] [PubMed]
16. Lai, Y.-H.; Chen, W.-N.; Hsu, T.-C.; Lin, C.; Tsao, Y.; Wu, S. Overall survival prediction of non-small cell lung cancer by integrating microarray and clinical data with deep learning. *Sci. Rep.* **2020**, *10*, 4679. [CrossRef] [PubMed]
17. Lu, Y.; Van Zandt, M.; Liu, Y.; Li, J.; Wang, X.; Chen, Y.; Chen, Z.; Cho, J.; Dorajoo, S.R.; Feng, M.; et al. Analysis of Dual Combination Therapies Used in Treatment of Hypertension in a Multinational Cohort. *JAMA Netw. Open* **2022**, *5*, e223877. [CrossRef] [PubMed]
18. Nguyen, P.-A.; Chang, C.-C.; Galvin, C.J.; Wang, Y.-C.; An, S.Y.; Huang, C.-W.; Wang, Y.-H.; Hsu, M.-H.; Li, Y.-C.; Yang, H.-C. Statins use and its impact in EGFR-TKIs resistance to prolong the survival of lung cancer patients: A Cancer registry cohort study in Taiwan. *Cancer Sci.* **2020**, *111*, 2965–2973. [CrossRef]
19. Dreiseitl, S.; Ohno-Machado, L. Logistic regression and artificial neural network classification models: A methodology review. *J. Biomed. Inform.* **2002**, *35*, 352–359. [CrossRef]
20. Izenman, A.J. Linear discriminant analysis. In *Modern Multivariate Statistical Techniques*; Springer: Berlin/Heidelberg, Germany, 2013; pp. 237–280.
21. Ke, G.; Meng, Q.; Finley, T.; Wang, T.; Chen, W.; Ma, W.; Ye, Q.; Liu, T.-Y. Lightgbm: A highly efficient gradient boosting decision tree. *Adv. Neural Inf. Process. Syst.* **2017**, *30*, 3149–3157.
22. Friedman, J.H. Greedy function approximation: A gradient boosting machine. *Ann. Stat.* **2001**, *29*, 1189–1232. [CrossRef]

23. Chen, T.; He, T.; Benesty, M.; Khotilovich, V.; Tang, Y.; Cho, H.; Chen, K. Xgboost: Extreme gradient boosting. *R Package Version 0.4-2* **2015**, *1*, 1–4.
24. Ho, T.K. Random decision forests. In Proceedings of the 3rd International Conference on Document Analysis and Recognition, IEEE, Montreal, QC, Canada, 14–16 August 1995; pp. 278–282.
25. Hastie, T.; Rosset, S.; Zhu, J.; Zou, H. Multi-class adaboost. *Stat. Its Interface* **2009**, *2*, 349–360. [CrossRef]
26. Gunn, S.R. Support vector machines for classification and regression. *ISIS Tech. Rep.* **1998**, *14*, 5–16.
27. Agatonovic-Kustrin, S.; Beresford, R. Basic concepts of artificial neural network (ANN) modeling and its application in pharmaceutical research. *J. Pharm. Biomed. Anal.* **2000**, *22*, 717–727. [CrossRef]
28. Lundberg, S.M.; Lee, S.-I. A unified approach to interpreting model predictions. *Adv. Neural Inf. Process. Syst.* **2017**, *30*, 4768–4777.
29. Pedregosa, F.; Varoquaux, G.; Gramfort, A.; Michel, V.; Thirion, B.; Grisel, O.; Blondel, M.; Prettenhofer, P.; Weiss, R.; Dubourg, V. Scikit-learn: Machine learning in Python. *J. Mach. Learn. Res.* **2011**, *12*, 2825–2830.
30. He, J.; Zhang, J.X.; Chen, C.T.; Ma, Y.; De Guzman, R.; Meng, J.; Pu, Y. The Relative Importance of Clinical and Socio-demographic Variables in Prognostic Prediction in Non-Small Cell Lung Cancer: A Variable Importance Approach. *Med Care* **2020**, *58*, 461–467. [CrossRef]
31. Bartholomai, J.A.; Frieboes, H.B. Lung Cancer Survival Prediction via Machine Learning Regression, Classification, and Statistical Techniques. In Proceedings of the 2018 IEEE International Symposium on Signal Processing and Information Technology (ISSPIT), Louisville, KY, USA, 6–8 December 2018; Volume 2018, pp. 632–637. [CrossRef]
32. Goussault, H.; Gendarme, S.; Assié, J.B.; Bylicki, O.; Chouaïd, C. Factors associated with early lung cancer mortality: A systematic review. *Expert Rev. Anticancer Ther.* **2021**, *21*, 1125–1133. [CrossRef]
33. Kang, J.; Chang, Y.; Ahn, J.; Oh, S.; Koo, D.H.; Lee, Y.G.; Shin, H.; Ryu, S. Neutrophil-to-lymphocyte ratio and risk of lung cancer mortality in a low-risk population: A cohort study. *Int. J. Cancer* **2019**, *145*, 3267–3275. [CrossRef]
34. Spicer, J.D.; McDonald, B.; Cools-Lartigue, J.J.; Chow, S.C.; Giannias, B.; Kubes, P.; Ferri, L.E. Neutrophils promote liver metastasis via Mac-1-mediated interactions with circulating tumor cells. *Cancer Res.* **2012**, *72*, 3919–3927. [CrossRef]
35. Powell, D.R.; Huttenlocher, A. Neutrophils in the Tumor Microenvironment. *Trends Immunol.* **2016**, *37*, 41–52. [CrossRef]
36. Contursi, A.; Grande, R.; Dovizio, M.; Bruno, A.; Fullone, R.; Patrignani, P. Platelets in cancer development and diagnosis. *Biochem. Soc. Trans.* **2018**, *46*, 1517–1527. [CrossRef]
37. Wang, J.; Xu, H.; Zhou, S.; Wang, D.; Zhu, L.; Hou, J.; Tang, J.; Zhao, J.; Zhong, S. Body mass index and mortality in lung cancer patients: A systematic review and meta-analysis. *Eur. J. Clin. Nutr.* **2018**, *72*, 4–17. [CrossRef] [PubMed]
38. Nakagawa, T.; Toyazaki, T.; Chiba, N.; Ueda, Y.; Gotoh, M. Prognostic value of body mass index and change in body weight in postoperative outcomes of lung cancer surgery. *Interact. Cardiovasc. Thorac. Surg.* **2016**, *23*, 560–566. [CrossRef] [PubMed]

cancers

MDPI

Article

Prognostication in Advanced Cancer by Combining Actigraphy-Derived Rest-Activity and Sleep Parameters with Routine Clinical Data: An Exploratory Machine Learning Study

Shuchita Dhwiren Patel [1,*], Andrew Davies [2], Emma Laing [3], Huihai Wu [3], Jeewaka Mendis [4] and Derk-Jan Dijk [5,6]

1 Department of Clinical and Experimental Medicine, Faculty of Health and Medical Sciences, University of Surrey, Guildford GU2 7XP, UK
2 Trinity College Dublin, University College Dublin and Our Lady's Hospice, DRW RY72 Dublin, Ireland
3 School of Biosciences, Faculty of Health and Medical Sciences, University of Surrey, Guildford GU2 7XP, UK
4 Surrey Clinical Trials Unit, University of Surrey, Guildford GU2 7XP, UK
5 Surrey Sleep Research Centre, Department of Clinical and Experimental Medicine, Faculty of Health and Medical Sciences, University of Surrey, Guildford GU2 7XP, UK
6 Care Research and Technology Centre, UK Dementia Research Institute, Imperial College London and University of Surrey, Guildford GU2 7XP, UK
* Correspondence: suchipatel_2501@hotmail.com

Simple Summary: Survival prediction is an important aspect of oncology and palliative care. Measures of night-time relative to daytime activity, derived from a motion sensor, have shown promise in patients receiving chemotherapy. Measuring rest-activity and sleep may, therefore, result in improved prognostication in advanced cancer patients. Fifty adult outpatients with advanced cancer were recruited, and rest-activity, sleep, and routine clinical variables were collected just over a one week period, and used in machine learning models. Our findings confirmed the importance of some well-established survival predictors and identified new ones. We found that sleep-wake parameters may be useful in prognostication in advanced cancer patients when combined with routinely collected data.

Abstract: Survival prediction is integral to oncology and palliative care, yet robust prognostic models remain elusive. We assessed the feasibility of combining actigraphy, sleep diary data, and routine clinical parameters to prognosticate. Fifty adult outpatients with advanced cancer and estimated prognosis of <1 year were recruited. Patients were required to wear an Actiwatch® (wrist actigraph) for 8 days, and complete a sleep diary. Univariate and regularised multivariate regression methods were used to identify predictors from 66 variables and construct predictive models of survival. A total of 49 patients completed the study, and 34 patients died within 1 year. Forty-two patients had disrupted rest-activity rhythms (dichotomy index (I < O ≤ 97.5%) but I < O did not have prognostic value in univariate analyses. The Lasso regularised derived algorithm was optimal and able to differentiate participants with shorter/longer survival (log rank $p < 0.0001$). Predictors associated with increased survival time were: time of awakening sleep efficiency, subjective sleep quality, clinician's estimate of survival and global health status score, and haemoglobin. A shorter survival time was associated with self-reported sleep disturbance, neutrophil count, serum urea, creatinine, and C-reactive protein. Applying machine learning to actigraphy and sleep data combined with routine clinical data is a promising approach for the development of prognostic tools.

Keywords: biomarkers; circadian; machine learning; palliative care; prognosis; survival

Citation: Patel, S.D.; Davies, A.; Laing, E.; Wu, H.; Mendis, J.; Dijk, D.-J. Prognostication in Advanced Cancer by Combining Actigraphy-Derived Rest-Activity and Sleep Parameters with Routine Clinical Data: An Exploratory Machine Learning Study. *Cancers* **2023**, *15*, 503. https://doi.org/10.3390/cancers15020503

Academic Editors: Hamid Khayyam, Ali Madani, Rahele Kafieh and Ali Hekmatnia

Received: 10 August 2022
Revised: 23 December 2022
Accepted: 6 January 2023
Published: 13 January 2023

1. Introduction

Prognostication (i.e., estimation of survival) is an important aspect of the management of patients with cancer. It is of particular importance in advanced cancer where it has immediate implications for clinicians' decisions about the treatment of the cancer, treatment of co-morbidities, so-called "ceilings of care", and referral to palliative care services [1,2]. Furthermore, it has implications for patients (and families) in terms of current decision-making, advance care planning, and "getting one's affairs in order".

Healthcare professionals are inaccurate prognosticators, often overestimating survival [3], and the accuracy of estimates is inversely related to survival [2]. Healthcare professionals are relatively good at predicting if patients will die within a couple of days, but not so good at predicting if patients will live for a couple of months or longer.

Various prognostic tools/algorithms have been developed to improve prognostication in patients with cancer [2,4]: these tools vary in their content (e.g., objective items only; subjective items only; objective and subjective items). However, none of these tools have been shown to be consistently better than clinicians' predictions of survival [2]. Current prognostication tools often include measures such as performance status, symptoms, venous blood sample data, and clinician-predicted survival [2,5]. The integration of other physiological and behavioural parameters, such as rest-activity rhythms ("diurnal or circadian") and sleep parameters are yet to be considered in prognostic models. (The term 'circadian' is meant to refer to rhythms that persist in constant conditions. Rhythms assessed in the presence of environmental rhythms, as in the present study, are referred to as diurnal or 24 h rhythms, although increasingly these rhythms are also referred to as 'circadian')

Sleep-wake cycles and circadian rhythms have a key role in sustaining normal body function and homeostasis [6]. Deterioration of rest-activity rhythmicity (loss of rhythmicity) and fragmentation of the sleep-wake cycle may be a marker of deterioration of health and, indeed, a predictor of illness including cancer, as well as cancer survival [7–9].

Several studies in cancer patients have incorporated actigraphy to objectively assess daytime activity, 24 h variation in rest-activity, as well as nocturnal and daytime sleep [7]. A number of actigraphy-derived parameters have been used to quantify rest-activity rhythms in this population including acrophase (time of peak activity), amplitude (peak to nadir difference, i.e., height of activity rhythm peak), mesor (average activity over a 24 h period), and the "dichotomy index" ($I < O$). Of these parameters, the $I < O$ is one of the most commonly studied rest-activity measures in cancer studies. The $I < O$ has been identified as an independent prognostic biomarker for overall survival, particularly in patients with metastatic colorectal cancer [10,11]. The $I < O$ is defined as the percentage of the activity counts measured when the patient is in bed that are inferior to the median of the activity counts measured when the patient is out of bed [12]. An $I < O$ of $\leq 97.5\%$ is indicative of a disrupted rest-activity circadian rhythm (i.e., increased fragmented sleep and reduced daytime activity patterns) [7]. However, the $I < O$ has not been used to prognosticate per se, either alone or in combination with other items. Furthermore, few studies have explored the potential of actigraphy-derived sleep parameters as prognostic markers in advanced cancer patients [13].

The first aim of this study was to investigate the feasibility of using $I < O$ and other actigraphy-derived parameters as stand-alone items, to prognosticate in patients with advanced cancer. The second aim of the study was to determine whether the $I < O$ and other actigraphy and sleep parameters should be combined with established prognostic indicators, e.g., Eastern Cooperative Oncology Group performance status (ECOG-PS), modified version of the Glasgow Prognostic Score (mGPS), Prognosis in Palliative Care Study (PiPS) –B, as well as putative prognostic variables from routine clinical data derived from blood samples, to improve prognostic accuracy. To achieve this second aim we deployed regularised regression, a supervised machine learning approach which overcomes some of the limitations of classical multiple regression, to identify effective prognostic indicators and develop more robust prognostic algorithms [14].

2. Materials and Methods

2.1. Study Design and Setting

The study was a prospective observational study conducted in a medium-sized district general hospital/cancer centre in the United Kingdom. The study was sponsored by the Royal Surrey County Hospital and received ethical approval from the London–Bromley REC (reference number—16/LO/0243). The study was registered on the CancerTrials.gov registry (reference number—NCT03283683). The study was funded by the Palliative Care Research Fund (Prof. Davies—Royal Surrey County Hospital), including an unrestricted donation from the family of Mr. John Spencer.

2.2. Study Participants

Participants were recruited from outpatients at the study site. All patients that met the criteria for the study were eligible for entry into the study (convenience sampling, consecutive recruitment). The inclusion criteria were: (a) age ≥ 18 years; (b) diagnosis of locally advanced/metastatic cancer; (c) clinician estimated prognosis of more than 2 weeks but less than 1 year; and (d) known to a specialist palliative care team. The exclusion criteria were: (a) cognitive impairment; (b) physical disability that affected general activity; and (c) physical disability that affected non-dominant arm movement.

Patients were diagnosed with locally advanced/metastatic cancer according to NHS guidelines, which consider TNM staging. All patients who met the inclusion criteria were deemed eligible for entry into the study. Potentially eligible patients were identified by the clinical team and approached by a member of the research team and invited to participate in the study. Any patient referred to the specialist palliative care team was expected to die within the next twelve months (as per the General Medical Council definition for end-of-life care [14]).

2.3. Routine Data Collection

Written informed consent was obtained from participants prior to entry into the study. The initial review (day 0) involved a collection of routine clinical data: patient demographics, information about cancer diagnosis/treatment, information about co-morbidities/ medication, assessment of Eastern Cooperative Oncology Group performance status (ECOG-PS) (by clinician and patient) [15], and completion of the Abbreviated Mental Test Score [16], the Memorial Symptom Assessment Scale—Short Form (MSAS-SF) [17], and the Global Health Status question from the PiPS-B algorithm [18]. The participant's pulse was measured (as part of the PiPS), and a venous blood sample was taken to measure haemoglobin, white blood cell count (WBC), neutrophil count, lymphocyte count, platelet count, sodium, potassium, urea, creatinine, albumin, alanine aminotransferase (ALT), alkaline phosphatase (ALP), and C-reactive protein (CRP). The final review (day 8) involved further assessment of ECOG-PS (by clinician and patient), completion of the MSAS-SF, the Pittsburgh Sleep Quality Index (PSQI) [19], and a patient acceptability questionnaire. The blood test results were used to complete the PiPS-B scoring algorithm, and serum CRP and albumin were used to calculate the mGPS [20].

2.4. Wrist Actigraphy and Consensus Sleep Diary

Wrist actigraphy was used to measure physical activity and standard sleep measures. Participants were fitted with the Actiwatch Spectrum Plus® (Philips Respironics, Bend, OR, USA) on the non-dominant arm after the initial review (day 0) and were instructed to wear the device for eight consecutive 24 h periods. The Actiwatch Spectrum Plus® is a CE-marked device with an accelerometer (i.e., motion sensor) that samples movement at 32 Hz [21] with a sensitivity of 0.025 G (at 2 count level). Participants were also given a Consensus Sleep Dairy in order to provide confirmatory information about specific sleep parameters (e.g., number of awakenings, time of final awakening) [22]: the "diary" was completed for eight consecutive sleep periods. The Actiwatches were configured and data were retrieved using device-specific software (Actiware version 6.0.9: Philips Respironics,

Bend, OR, USA). The Actiwatches were adjusted to provide an epoch length (sampling interval) of one minute, which is the most common epoch length used in studies of cancer patients [23]. The Consensus Sleep Diary was used in conjunction with the Actiwatch to assist in actigraphy data interpretation (i.e., determine the major sleep/wake periods) [24].

The data from the Actiwatches was downloaded into an Excel spreadsheet, and the following rest-activity parameters were calculated using a study specific SAS programme (SAS® Version 9.4 Statistical Analysis Software, SAS Institute, Cary, NC, USA): I < O, r24 (an autocorrelation coefficient at 24 h, that is "a measure of the regularity and reproducibility of the activity pattern over a 24 h period from one day to the next") [25], mean daily activity (MDA), and mean activity during daytime wakefulness. MDA was calculated as the average number of wrist movements per minute throughout the recording time [25], and the mean duration of activity during wakefulness was calculated as the mean activity score (counts/minute) during the time period between two major sleep period intervals [26]. In addition, the following sleep parameters were calculated both automatically from the Actiwatches (using the Actiware sleep scoring algorithm) and manually from the sleep diary [27]: bedtime (BT), get-up time (GUT), time in bed (TIB), sleep onset latency (SOL), total sleep time (TST), sleep efficiency (SE), wake after sleep onset (WASO), and number of awake episodes (NA). The sleep parameters derived manually solely from the sleep diary were: time tried to sleep, time of final awakening and terminal awakening (TWAK) [22]. See Table 1 for definitions of the sleep parameters.

Table 1. Definitions of actigraphy-derived sleep/consensus sleep diary parameters [22,26,28].

Sleep Parameter	Definition
Actigraphy and sleep diary	
Bed-time (BT) (hh:mm)	Clock time attempted to fall asleep based on actigraphy event marker or sleep diary
Get-up time (GUT) (hh:mm)	Clock time attempted to rise from bed for the final time based on actigraphy event marker or sleep diary
Time in bed (TIB) (hh:mm)	Duration between reported BT and GUT (reported in hours and minutes) or as self-reported in sleep diary
Sleep onset latency (SOL) (min)	Duration between reported BT and actigraph scored sleep onset time or as self-reported in sleep diary
Total sleep time (TST) (hh:mm)	Duration of sleep during the major sleep period calculated by Actiware;
	Sleep diary manual calculation: TIB minus (SOL plus WASO plus TWAK)
Sleep efficiency (SE) (%)	Proportion of time the patient is asleep out of the total time in bed (reported as a percentage) calculated by Actiware;
	Sleep diary manual calculation: TST divided by TIB × 100
Wake after sleep onset (WASO) (min)	Sum of wake times from sleep onset to the final awakening calculated by Actiware or as self-reported in sleep diary
Number of awake episodes (NA)	Number of continuous blocks of wake during the major sleep period calculated by Actiware or as self-reported in sleep diary
Sleep Diary	
Time tried to sleep (hh:mm)	Self-reported time participant began 'trying' to fall asleep
Time of final awakening (hh:mm)	Self-reported time participant last woke up in the morning
Terminal awakening (TWAK) (hh:mm)	GUT minus time of final awakening

2.5. Follow-Up

During the study period (from time of first patient recruited to six months after last patient recruited), participants' survival status (and date of death, if applicable) was determined every three months by reviewing the hospital clinical records, and/or contacting the general practitioner.

3. Statistical Analyses

The sample size for the study (n = 50) was derived from guidance on sample sizes for feasibility studies (and represents the upper range) [29]. Statistical support was provided by statisticians, within the Research Design Service South-East (based in the Clinical Trials Unit at the University of Surrey). Descriptive statistics were used to explain much of the data (e.g., mean and standard error; median and range). The Intraclass Correlation Coefficient (ICC) was used to assess the robustness of I < O as a marker of the rest-activity rhythm, and its stability throughout the actigraphy recording. The Spearman's Rank correlation coefficient was used to measure the association between I < O and other actigraphy-derived parameters. The Spearman's rank correlation 'r' values were defined as follows: $0 \leq r < 0.3$ indicated a negligible correlation, $0.3 \leq r < 0.5$ a low correlation, $0.5 \leq r < 0.7$ a moderate correlation, $0.7 \leq r < 0.9$ a high correlation, and $0.9 \leq r \leq 1$ a very high correlation [30]. Kaplan–Meier plots, a non-parametric statistical method, were used to estimate the probability of survival past a given time point along with the log rank test to compare the survival distribution of two groups. Statistical significance was evaluated at 5%.

The "per protocol set" refers to participants that wore the Actiwatch for the eight consecutive 24 h periods with the corresponding sleep diary, whilst the "full analysis set" refers to participants that wore the Actiwatch for at least three consecutive 24 h periods (i.e., 72 h) and completed the corresponding sleep diary for the actigraphy rest-activity and sleep analysis, or for at least three consecutive or non-consecutive nights in the sleep diary for the subjective sleep analysis (i.e., calculation of the sleep diary parameters).

4. Machine Learning Methods and Data Analysis

Cox regression has been the standard approach to survival analysis in oncology. However, Cox regression has a number of limitations. In particular, it is not an adequate approach for situations in which the number of predictors is high relative to the number of observations, as is the case in this feasibility study. We therefore opted to use simple alternative methods that can (1) adequately deal with situations in which the number of predictors is large relative to the number of observations and (2) yield models that are interpretable, i.e., are not 'black box models'. Penalised (Regularised) regression models represent such an approach.

A supervised machine learning algorithm was used to develop a predictive model, where the collated subjective and objective parameters (i.e., routine clinical data and actigraphy-derived rest-activity and sleep parameters) were individual predictor variables and survival was the 'response' variable [31]. Sixty-six predictor variables were tested for potential predictive value (Appendix A, see Table A1 for descriptive statistics of the numerical predictor variables). Overall survival was defined as the time from initial review (day 0) to death or until 14 May 2020 for patients that remained alive until the end of the study.

4.1. Machine Learning Dataset

All patients recruited into the study (n = 50) were used for the machine learning analysis. The predictor variables were classified into the relevant variable type (e.g., binary, categorical_nominal, etc.) and entered into a .csvfile in Excel. Binary variables, such as 'use of opioid analgesia' were transformed into dummy variables (0 or 1). Categorical_ordinal variables with a numerical ranking, such as ECOG-PS were labelled using the 'LabelEncoder' approach, where the output integer value from the LabelEncoder function was used to reflect the ordering of the original integer. Categorical_ordinal variables with non-numerical values, such as PSQI sleep disturbance, were assigned a numerical ranking. Numerical_continuous variables involving sleep/wake times were entered in the 24 h format. Missing data values were imputed with the average of the group or with the corresponding subjective/objective data from the same participant. Missing data accounted for <4% of the dataset.

4.2. Regularised Regression Methods

Regularised regression was used to reduce "overfitting" and aid the generalisability of the model. 'Regularisation' corresponds to a penalty that limits the overall weight that can be assigned across all predictor variables in the model, which reduces model complexity (compared to traditional multivariate regression). For some regularised regression approaches, the penalty can drive the weight of a variable to zero, effectively selecting the optimal combination of predictor variables that can be used to predict the given outcome.

Here, three regularised multivariate regression methods were applied and compared: ridge regression, least absolute shrinkage and selection operator (Lasso) and elastic net. The ridge regression algorithm includes all the predictor variables, shrinking the coefficients towards (but not set at) zero in a continuous manner [32]. The Lasso-derived algorithm combines the method of shrinkage with the sub-selection of predictor variables, using a penalty 'L_1 norm' [32,33], creating a 'sparse' model (i.e., selecting only a few variables from the dataset) [32]. The elastic net algorithm is broadly a combination of the ridge and Lasso [34]. This method simultaneously performs continuous shrinkage and feature selection, selecting groups of correlated variables, using a penalty of 'L_1 norm' and 'L_2 norm' [34]. Highly correlated predictor variables are averaged and entered into the model to remove any deviances caused by extreme correlations [35]. Since survival data are censored, i.e., at the end of the observation period some participants may still be alive, we applied regularised Cox regression using the glmnet package in R.

4.3. Model Development

The models were validated using a k-fold (10 folds used) cross-validation approach [32]. For each of the 50 individuals, the predicted survival was based on a model which was constructed on '$k-1$' subjects, i.e., the model was blind to the participant and the participant did not contribute to the estimation of the prediction. All analyses were carried out within the statistical computing environment R (version 3.6.2). For machine learning, ridge, Lasso and elastic net (alpha = 0.5) regression the package glmnet (version 2.0) was used. Here, an exhaustive search for lambda able to produce the minimum Mean Cross-Validated Error (CVM) was performed. All subjects were used as the training set to build a final model, then k-fold cross-validation for performances (CVM) was performed. Analyses were performed with different settings of elastic net mixing parameter (alpha), which were elastic net (alpha = 0.5), Lasso (alpha = 0.99) and ridge (alpha = 0.01). The models generated a predicted hazard, which was compared to the actual survival in days using Pearson's correlation coefficient. To estimate the intra-variable variation in their contribution to the predictor, we computed the mean cross-validated error of the weights of each of the variables that were consistently identified in all 50 participants.

5. Results

A total of 50 patients were recruited to the study, and 49 participants completed the study (Figure 1): the full analysis set consisted of 44 participants, whilst the per protocol set consisted of 37 participants. See Table 2 for characteristics of the participants. A total of 46 participants were followed up for 12 months (40 in the full analysis set, 33 in the per protocol set), and 34 died within this time period (28 in the full analysis set, 22 in the per protocol set). Unless otherwise stated, the following results relate to the full analysis set.

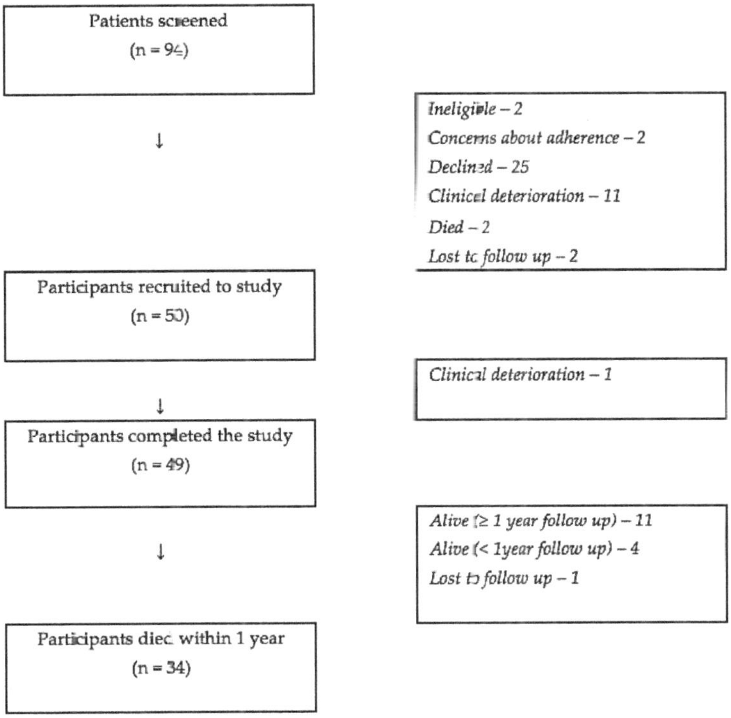

Figure 1. Study flow chart.

Table 2. Participant characteristics.

Characteristic	All Participants (n = 50)	"Full Analysis Set" (n = 40)
Age	Median—63 yr (range 40–81 yr)	Median—66 yr (range 43–81 yr)
Sex	Female—21 (42%)	Female–17 (39%)
Male—29 (58%)	Male—27 (61%)	
Cancer diagnosis	Breast—6 (12%) Endocrine—1 (2%) Gastrointestinal—16 (32%) Gynaecological—6 (12%) Haematological—2 (4%) Head and Neck—3 (6%) Lung—6 (12%) Skin—2 (4%) Urological—8 (16%)	Breast—6 (14%) Endocrine—1 (2%) Gastrointestinal—14 (32%) Gynaecological—4 (9%) Haematological—2 (5%) Head and Neck—2 (5%) Lung—6 (14%) Skin—2 (5%) Urological—7 (16%)
ECOG-PS (Physician-assessed at baseline)	0–0 (0%) 1–26 (52%) 2–13 (26%) 3–11 (22%) 4–0 (0%)	0–0 (0%) 1–24 (55%) 2–10 (23%) 3–10 (23%) 4–0 (0%)

Note: Percentages may not sum to 100 due to rounding.

5.1. Acceptability of Actigraphy and Sleep Diary Acceptability

Actigraphy data were missing from one participant due to a technical problem. Forty-two (84%) participants reported that the Actiwatch was "comfortable to wear", and only four (8%) reported that the Actiwatch interfered with their normal activities. No adverse effects were reported from using the Actiwatch. Fourteen (28%) participants reported that the Consensus Sleep Diary was difficult to complete, and two (4%) subjects reported that the diary interfered with their normal activities.

5.2. Univariate Analyses of Actigraphy Parameters

5.2.1. Characteristics of the Dichotomy Index (I < O) and Correlation with Other Actigraphy and Sleep Parameters

Table 3 shows the results for the I < O. Forty-two (95%) participants had an I < O of ≤97.5%, indicating a disrupted rest-activity circadian rhythm [7]. The I < O can be considered a stable variable since the intraclass correlation coefficient for values obtained over eight days using the per protocol set, was 0.93 (95% CI: 0.88–1.00; $p < 0.0005$), which is considered an "excellent" correlation [36]. In fact, there was a "high" positive correlation between the I < O for the first three days (72 h) and for the full eight days (Spearman's correlation: $r = 0.82$; $p < 0.0005$) [31]. Moreover, there was a "high" positive correlation between the I < O on weekdays and on the weekend (Spearman's correlation: $r = 0.76$; $p < 0.0005$). Additionally, there was a "very high" positive correlation between the I < O calculated using 24 h of data, and the I < O calculated using 20 h of data, i.e., excluding the one-hour periods before/after going to bed, and the one-hour periods before/after getting out of bed (Spearman's correlation: $r = 0.98$; $p < 0.0005$).

Table 3. Dichotomy Index (I < O) data.

I < O Parameter	Full Analysis Set ($n = 44$)	Per Protocol Set ($n = 37$)
Mean (+/− standard error)	88.90% (+/− 1.04)	89.90% (+/− 0.97)
Minimum	70.90%	70.90%
25th Centile	86.90%	87.40%
Median	90.40%	90.80%
75th Centile	93.60%	93.60%
Maximum	98.10%	97.60%
Distribution	Non-normal (Shapiro-Wilk test: $p = 0.001$)	Non-normal (Shapiro-Wilk test: $p = 0.001$)

There was a "moderate" positive correlation between the I < O and the r24 (Spearman's correlation: $r = 0.66$; $p < 0.0005$), and the mean activity during wakefulness (Spearman's correlation: $r = 0.51$; $p < 0.0005$). However, there was only a "low" positive correlation between the I < O and the mean daily activity (Spearman's correlation: $r = 0.43$; $p = 0.003$). Other standard actigraphy parameters correlated with the I < O were SE, i.e., number of minutes of sleep divided by total number of minutes in bed (Spearman's correlation: $r = 0.47$, "low" correlation; $p = 0.001$), and WASO, i.e., number of minutes awake after sleep onset during sleep period (Spearman's correlation: $r = -0.51$, "moderate" correlation; $p < 0.0005$).

5.2.2. I < O: Predictor of Survival and Correlation with ECOG-PS

Amongst participants that completed one year of follow-up ($n = 40$), there was no significant difference in overall survival between those separated into two groups (based on the median I < O; log rank test, $p = 0.917$), or four groups (based on the quartiles of the I < O; log rank test, $p = 0.838$). However, I < O had a "moderate" negative correlation with the physician assessed ECOG-PS (Spearman rank correlation: $r = -0.63$; $p < 0.0005$). The ECOG-PS was an independent prognostic indicator in this cohort of patients (log rank test, $p < 0.0005$). The median survival for participants with an ECOG-PS of 1 (end of study) was 141 days, ECOG-PS of 2 was 135 days, ECOG-PS of 3 was 57 days, and ECOG-PS of 4 was 17 days.

5.2.3. Autocorrelation Coefficient at 24 h (r24)

The median r24 was 0.16 (range 0.04–0.37). Amongst participants that completed one year of follow-up ($n = 40$), there was no significant difference in overall survival between those separated into two groups (based on the median r24; log rank test, $p = 0.318$), or four groups (based on the quartiles of the r24; log rank test, $p = 0.800$).

5.2.4. Other Actigraphy Parameters

None of the other actigraphy-derived sleep parameters were associated with a decreased overall survival: (a) TIB (log rank, $p = 0.574$: based on group median of 9 h 29 min); (b) TST (log rank, $p = 0.147$: based on normative cut-off value of ≥6.5 h [28]; (c) SOL (log rank, $p = 0.283$: based on normative cut-off value of ≤30 min [28]; (d) SE log rank, $p = 0.224$: based on normative cut-off value of ≥85% [28]; (e) WASO (log rank, $p = 0.549$: based on normative cut-off value of >30 min [28]; and (f) NA (log rank, $p = 0.972$: based on group median of 23 episodes).

5.3. Multivariate Predictors of Survival: Machine Learning Results

In the machine learning dataset, 46 participants had died within the specified time period of follow-up (i.e., by 14 May 2020). The Lasso model selected 22 predictor variables, with 14 variables consistently selected in all 50 participants during the process of validation (Figure 2). These involved eight predictor variables associated with greater survival time and six predictor variables, associated with a reduced survival time. The predictor variables associated with increased survival time, i.e., smaller hazard (in order of the coefficient associated with the predictor variable) were: later sleep diary time of final awakening, later actigraphy get up time, longer PiPS-E clinician's estimate of survival, better PSQI subjective sleep quality, greater PiPS-B global health status score (indicating better health), better actigraphy sleep efficiency, and higher haemoglobin values. The variables associated with reduced survival time were more frequent PSQI sleep disturbance wake middle of the night/early morning, higher neutrophil count, higher serum urea, serum creatinine, and serum C-reactive protein. On the contrary, a larger MSAS-SF total symptom distress was associated with a lower risk of death and a higher I < O was associated with a worse prognosis. The predicted median hazard was 0.00052, and the model was able to successfully differentiate between participants with a shorter/longer overall survival (log rank $p < 0.0001$) (Figure 3). Figure A1 shows the correlation between the actual survival and predicted hazard (Pearson's correlation coefficient $r = -0.5$; $p = 0.0002$).

The ridge model consistently identified 28 predictor variables in all 50 participants (Figure 4). During the process of validation, the top 10 variables consistently selected involved seven predictors associated with longer survival time and three predictors associated with shorter survival time. The seven predictor variables associated with longer survival time (in order of the coefficient associated with the predictor variable) were: actigraphy get-up time, sleep diary time of final awakening, sleep diary get-up time and PSQI usual get-up time; PiPS-B clinician's estimate of survival, PSQI subjective sleep quality and PiPS-B global health status score. The 3 predictor variables associated with shorter survival time were: use of opioid analgesia, modified Glasgow Prognostic Score and physician-assessed ECOG-PS day 8. The predicted median hazard was 0.44; however, there was no significant difference in overall survival when a median split was applied (log rank, $p = 0.0914$) (Figure 5). Figure A2 shows the correlation between the actual survival and predicted hazard (Pearson's correlation coefficient $r = -0.5$; $p = 0.0002$).

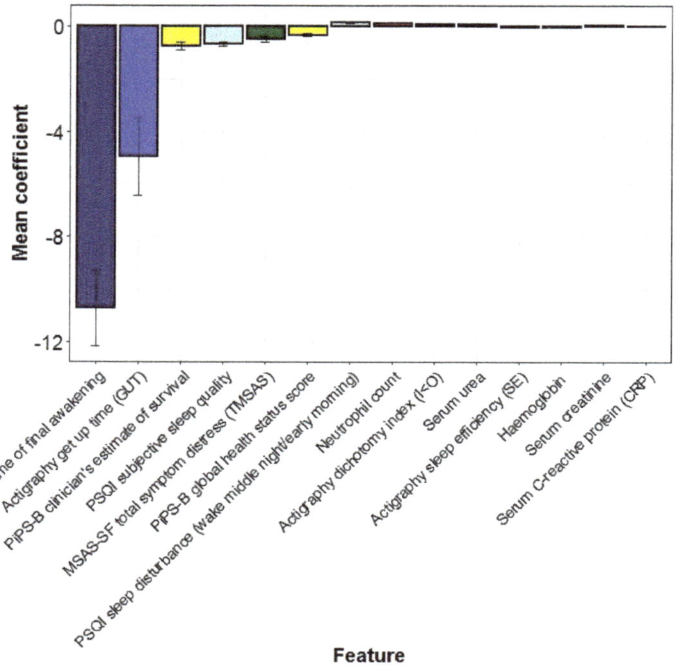

Figure 2. The mean cross—validated error (CVM) of predictor variables for hazard selected by the Lasso model.

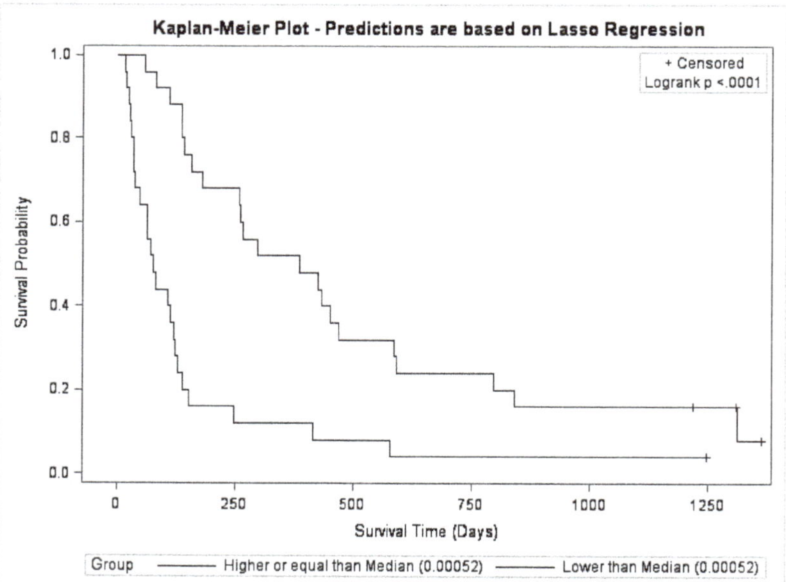

Figure 3. Kaplan–Meier curve comparing survival probability predicted by the Lasso-derived algorithm (log rank, $p < 0.0001$).

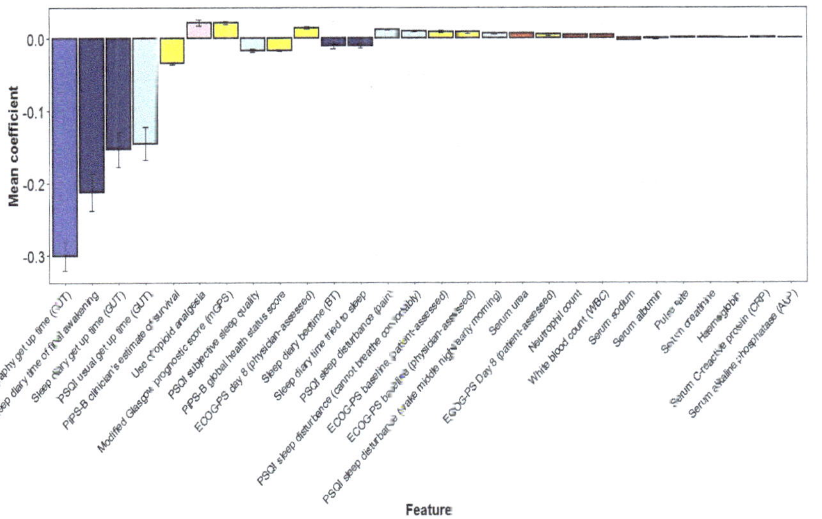

Figure 4. The mean cross—validated error (CVM) of predictor variables for hazard selected by the ridge model.

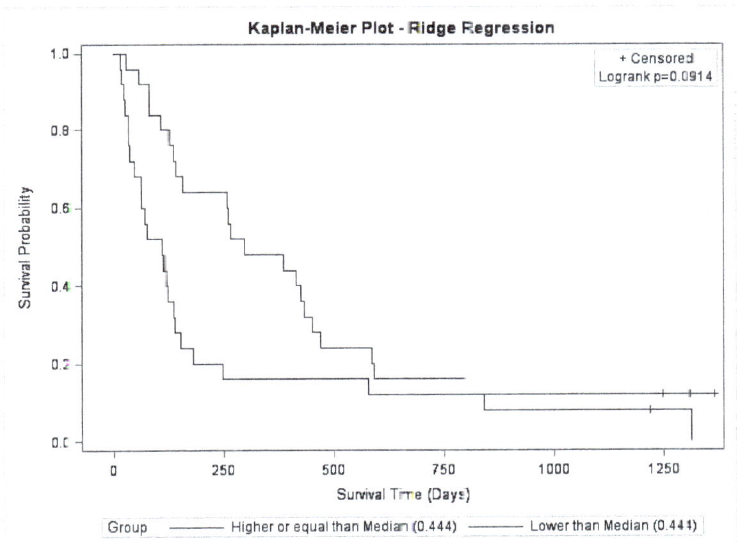

Figure 5. Kaplan–Meier curve comparing survival probability predicted by the ridge-derived algorithm (log rank, $p = 0.0914$).

The elastic net model selected 10 predictor variables, with 6 variables being consistently selected during the process of validation: the two consistently selected predictor variables associated with longer survival time were (in order of the coefficient associated with the predictor variable): later actigraphy get-up time and greater PiPS-B global health status score; the 4 consistently selected predictor variables associated with shorter survival time were: higher serum urea, neutrophil count, serum C-reactive protein, and serum creatinine (Figure A3). The predicted median hazard was 0.408, but there was no significant difference in overall survival (log rank, $p = 0.9877$) (Figure A4). Figure A5 shows the correlation between the actual survival and predicted hazard (Pearson's correlation coefficient $r = -0.08$; $p = 0.5808$).

6. Discussion

The results of this study show that univariate approaches to survival prediction, based on, for example, the I < O, are not very powerful; whereas, multivariate approaches appear to hold promise. To the best of our knowledge, this is the first study describing the application of supervised machine learning methods, involving a combination of actigraphy-derived rest-activity and sleep parameters, and data collected in routine clinical practice (i.e., simple questionnaires such as the MSAS-SF, ECOG-PS, PSQI, venous blood sampling) to prognosticate patients with advanced cancer, receiving supportive and palliative care [37]. Our study confirmed certain established novel predictors and identified some for survival in this group of patients and points to the importance of sleep characteristics for prognostication. The results of the study also confirm that clinicians are inaccurate prognosticators [3], since 11 (24%) participants were still alive at 1 year (despite the inclusion criteria of clinician estimated prognosis of more than 2 weeks but less than 1 year).

The literature had suggested that actigraphy-derived parameters, and the I < O index in particular, could be used as predictors because a low I < O is associated with increased morbidity (worse symptoms, worse quality of life), and with decreased survival [7]. At the outset of this study, we therefore focused on the I < O and other parameters describing the robustness of the rest-activity. We indeed observed a very high prevalence (i.e., 95%) of disrupted rest-activity rhythms in these advanced cancer patients, which is much higher than the reported prevalence of 19.1–54.9% [7]. This disparity undoubtedly reflects different populations, with our population having more advanced disease (and worse performance status) than previous studies [11,38]. However, in the univariate analyses of the data in our study there was no direct association between I < O and survival. Furthermore, other actigraphy-derived parameters, when used in isolation, are also not very accurate in the population.

However, the results of the study suggest that novel models developed through machine learning can facilitate improvements in prognostication. Penalised regression methods implement a feature selection strategy, providing a combination of subjective and objective predictor variables of survival that are ranked based on their contribution to the model. The models manage collinearity within the dataset, which is particularly useful in datasets involving terminal cancer patients, where often the number of features exceeds the relative sample size. The best performing method was Lasso regression which reduces the coefficients of variables with a minor contribution to zero and thereby creates a simple 'model' with only a few variables. Sleep parameters were amongst the most important variables, not only in the Lasso model but also in the more complex elastic net and ridge models. These measures primarily represented positive predictors of survival. Sleep diary final awakening (lasso and ridge) and actigraphy-derived GUT (all models) were found to have particular prognostic relevance in our study, suggesting that a later sleep diary determined 'time of final awakening' and a later actigraphy-derived 'get-up time' are associated with a lower risk of death and improved survival. Furthermore, actigraphy-derived SE, which may be considered an objective measure of sleep quality, was selected as a positive predictor of survival in the lasso model (i.e., greater sleep efficiency was associated with enhanced survival) for our population. Whilst actigraphy-derived sleep quality, as opposed to sleep quantity, has been reported to have prognostic significance in advanced breast cancer patients [13], we identified quantitative sleep measures as important contributors to survival prediction.

Studies have reported actigraphy-derived circadian disruption [10,12,39] and fragmented sleep [13] to have prognostic implications in cancer patients, yet little is known about the prognostic impact of subjective sleep measures. A recent study identified the PSQI sleep duration component as a prognostic indicator in a cohort of advanced hepatobiliary/pancreatic cancer patients [40], yet a novel finding in our study was the selection of other sleep parameters from the PSQI: (1) usual get up time and (2) subjective sleep quality, where a later get up time and very good sleep quality are associated with longer overall survival, and (3) PSQI sleep disturbance components—pain, cannot breathe comfortably and wake up in the middle of the night or early morning—were associated with poorer survival. Furthermore, subjective sleep parameters, as opposed to actigraphy-derived sleep parameters, were more commonly identified in all participants in the ridge model.

Venous blood sample measurements were also significant contributors to predicting survival in our study. Previous studies have reported moderate evidence for the prognostic significance of an elevated C-reactive protein (CRP) and leucocytosis being associated with a shorter survival [1,5,18,41]. Whilst our study was able to echo these findings, we were able to further identify novel biomarkers, such as an elevated urea and serum creatinine, that may also be associated with a poorer survival, and raised haemoglobin that may be associated with a lower risk of death. Blood sampling is generally deemed 'inappropriate' when patients are in their last days/weeks of life [42], regardless only one of the 94 patients screened for our study, declined participation. Our findings endorse further evaluation of biological parameters from venous blood sample data, as they may be beneficial to improving prognostication in these patients.

Although our multivariate findings controversially imply that a higher I < O is associated with a shorter predicted survival time, all participants in our population had poor health, i.e., an I < O of <99%, which has recently been identified as an optimal cut-off for distinguishing between healthy controls and patients with advanced cancer [43]. Further inspection of our data identified that all our participants, whether they had shorter or longer survival had disrupted rest-activity rhythms, equally both groups had moderate symptom distress as measured by TMSAS, inevitably expected in an advanced cancer population. Therefore, whilst it may be a simple way of quantifying rest-activity rhythms, I < O may be a more meaningful prognostic indicator during the earlier trajectories of cancer, as opposed to the progressive stages.

In summary, our data suggests that subjective sleep parameters, measured using the consensus sleep diary and the PSQI, and actigraphy-derived sleep parameters may be especially useful when combined with routine clinical data using machine learning approaches, with no substantial additional costs or burden to the health service. Thus, further investigation of these parameters as prognostic indicators is warranted. Indeed, we plan to undertake a larger (definitive) study in the near future. Sleep-wake disturbances and circadian dysregulation are deemed to have a reciprocal relationship [43,44] and our findings are suggestive of sleep/circadian rhythm parameters as potential prognostic indicators. Whether improving the patient's sleep disturbance may improve overall survival remains an open question. Rehabilitation of the circadian system by means of behavioural and pharmacological strategies, to re-synchronise the circadian system, may ultimately improve circadian function and sleep, as well as overall survival [44,45].

The Lasso model was the only model able to successfully differentiate between long and short survival in our study, and the correlation between observed and predicted hazard was only significant for the Lasso and ridge models. The Lasso model is 'sparse' (i.e., only a few variables from the dataset are selected) [32] and therefore may be favourable if a consolidated model were needed to aid prognostication. However, the Lasso selects at most '*n*' variables before it saturates; therefore, the number of predictors is restricted by the number of observations [32]. The ridge model, therefore, may be beneficial due to the greater inclusivity of variables, at the expense of an increased risk of overfitting. Indeed, the absence of significant results cannot be overlooked with the small sample size in this feasibility study. In the definitive study, all three supervised machine learning methods would be deployed after the recruitment of a larger sample size as well as the inclusion of additional variables that may be clinically relevant (e.g., stage of disease, number of comorbidities, nutritional status, presence of specific symptoms/problems) [1,2], More data would enable robustness of the predictive ability of the models to be assessed as well as enable generalisability of our findings with further confidence in our observations.

Interestingly, a recent systematic review described the prediction of survival to be a process as opposed to an event, and that predictors of survival may develop as the disease progresses [5]. Therefore, there may be added value in predicting the trajectory of death, as opposed to the time of death in future studies. Machine learning approaches would be particularly valuable in such cases, where relevant predictor variables may be identified as the disease trajectory evolves only to ultimately enhance our true understanding of prognostication.

A few limitations need consideration. Firstly, our small sample size is unlikely to capture the true variance of the population. Secondly, the Lasso and elastic net models involve only a subset of predictors and the value of the coefficient associated with each of these predictor variables is dependent on the presence of the other (non-zero) predictor variables in the model. Our results are essentially correlational and demonstrate that the relevant predictor variables (above non-zero coefficient value) may be associated in a positive or negative way with the risk of death. Thirdly, imputation of missing data values with the sample population average may not have been a true reflection of the individual sample's actual score nor using subjective data to impute objective values, particularly if the tools were measuring different timeframes, i.e., actigraphy (over a one-week duration) versus the PSQI questionnaire (measures on average over the previous one month). The *K*-fold cross-validation approach also has some limitations. As it is executed '*k*' times (where '*k*' is the number of subsets of observations), this approach may not be resourceful in a small dataset. Furthermore, *K*-fold cross-validation is likely to have a high variance as well as a higher bias, given the small size of the training set from a small dataset. Therefore, the number '*k*' highly influences the estimation of the prediction error, and the presence of outliers can lead to a higher variation. Indeed, it can be a challenge to find the appropriate '*k*' number to reach a good 'bias-variance' trade-off. In future studies, it will be essential to include an independent validation set.

7. Conclusions

This study suggests that subjective sleep parameters, measured using the consensus sleep diary and the PSQI, and actigraphy-derived sleep parameters may be useful for prognostication in patients with advanced cancer, and that it may be especially useful when combined with routine clinical data and machine learning approaches.

Author Contributions: Conceptualization, A.D. and D.-J.D.; methodology, S.D.P., A.D. and D.-J.D.; data collection, S.D.P.; data interpretation, S.D.P.; supervision, A.D., D.-J.D., E.L., J.M. and H.W.; dichotomy index software and analysis, J.M. and S.D.P.; machine learning software and data analysis, H.W.; data curation, S.D.P. and H.W.; writing—original draft preparation, S.D.P. and A.D.; writing—review and editing, S.D.P., A.D., D.-J.D., E.L., J.M. and H.W.; graphical figures—S.D.P., H.W., J.M. and D.-J.D.; supervision, A.D. and D.-J.D.; project administration, S.D.P.; funding acquisition, A.D. All authors have read and agreed to the published version of the manuscript.

Funding: This research was funded by the Palliative Care Research Fund (Davies—Royal Surrey County Hospital), including an unrestricted donation from the family of John Spencer.

Institutional Review Board Statement: The study was sponsored by the Royal Surrey County Hospital and received ethical approval from the London-Bromley REC (reference number—16/LO/0243 on 9 May 16). The study was registered on the CancerTrials.gov registry (reference number—NCT03283683).

Informed Consent Statement: Informed consent was obtained from all subjects involved in the study. Written informed consent has been obtained from the patients to publish this paper.

Data Availability Statement: The data presented in this study are available on request from the corresponding author. The data are not publicly available due to the nature of the consent obtained from participants.

Acknowledgments: The authors would like to thank Simon Skene for his statistical expertise and Victoria Robinson for her invaluable administrative support.

Conflicts of Interest: The authors declare no conflict of interest.

Appendix A. Prognostic Parameters for Machine Learning

1. Medication: Use of opioid analgesia
2. ECOG-PS at baseline: Physician-assessed
3. ECOG-PS at baseline: Patient-assessed
4. ECOG-PS Day 8: Physician-assessed
5. ECOG-PS Day 8: Patient-assessed
6. MSAS-SF: Number of symptoms
7. MSAS-SF: Physical symptom subscale score (MSASPHYS)
8. MSAS-SF: Psychological symptom subscale score (MSASPSYCH)
9. MSAS-SF: Total symptom distress score (TMSAS)
10. MSAS-SF: Global Distress Index (GDI)
11. PSQI: Usual Bedtime (BT)
12. PSQI: Time to fall asleep (SOL)
13. PSQI: Usual getting up time (GUT)
14. PSQI: Hours of sleep per night (TST)
15. PSQI: Sleep disturbance—Cannot get to sleep within 30 min
16. PSQI: Sleep disturbance—Wake up in the middle of the night or early morning
17. PSQI: Sleep disturbance—Have to get up to use bathroom
18. PSQI Sleep disturbance—Cannot breathe comfortably
19. PSQI Sleep disturbance—Cough or snore loudly
20. PSQI Sleep disturbance—Feel too cold
21. PSQI Sleep disturbance—Feel too hot
22. PSQI: Sleep disturbance—Had bad dreams
23. PSQI: Sleep disturbance—Have pain
24. PSQI: Subjective sleep quality
25. PSQI: Use of medication for sleep
26. PSQI: Daytime dysfunction: Trouble staying awake
27. PSQI: Keep up enough enthusiasm to get things done
28. PSQI: Presence of bed partner or roommate
29. PiPS-B algorithm: Abbreviated Mental Test Score (out of 10)
30. PiPS-B algorithm: Patient's pulse rate
31. PiPS-B algorithm: Global Health Status Score (1 = extremely poor health; 7 = normal health)
32. PiPS-B algorithm: Clinician's estimate of survival (Days/Weeks/Months+)
33. Modified Glasgow Prognostic Score (mGPS)
34. Bloods: Haemoglobin (g/L) (130–180)
35. Bloods: White Blood Count (10^9/L) (4–11)
36. Bloods: Neutrophils (10^9/L) (2.0–7.5)
37. Bloods: Lymphocytes (10^9/L) (1.0–4.0)
38. Bloods: Platelets (10^9/L) (150–450)
39. Bloods: Sodium (mmol/L) (133–146)
40. Bloods: Potassium (mmol/L) (3.5–5.3)
41. Bloods: Urea (mmol/L) (2.5–7.8)
42. Bloods: Creatinine (μmol/L) (64–104)
43. Bloods: ALP (IU/L) (30–130)
44. Bloods: ALT (IU/L) (<50)
45. Bloods: Albumin (g/L) (35–50)
46. Bloods: C-reactive protein (CRP) (mg/L) (<10)
47. Wrist actigraphy: Rest-activity parameter—Dichotomy Index (I < O) at least 72 h
48. Wrist actigraphy: Rest-activity parameter—r24 (autocorrelation coefficient) at least 72 h
49. Wrist actigraphy: Activity parameters—Mean activity during wakefulness at least 72 h
50. Wrist actigraphy: Activity parameters—Mean daily activity (MDA) at least 72 h
51. Wrist actigraphy: Sleep parameter—Bedtime (BT)
52. Wrist actigraphy: Sleep parameter—Get up time (GUT)

53. Wrist actigraphy: Sleep parameter—Time in bed (TIB)
54. Wrist actigraphy: Sleep parameter—Total sleep time (TST)
55. Wrist actigraphy: Sleep parameter—Sleep onset latency (SOL)
56. Wrist actigraphy: Sleep parameter—Sleep Efficiency (%)
57. Wrist actigraphy: Sleep parameter—Wake after sleep onset (WASO)
58. Wrist actigraphy: Sleep parameter—Number of awake episodes (NA)
59. Consensus Sleep Diary: Time in bed (BT)
60. Consensus Sleep Diary: Time of final awakening
61. Consensus Sleep Diary: Time out of bed (GUT)
62. Consensus Sleep Diary: Time tried to go to sleep
63. Consensus Sleep Diary: Time to fall asleep (SOL)
64. Consensus Sleep Diary: Quality of Sleep
65. Consensus Sleep Diary: Total amount of time awakenings lasted (WASO)
66. Consensus Sleep Diary: Number of times awakened in the night (NA)

Table A1. Mean and standard deviation for prognostic parameters for machine learning.

Numerical Prognostic Parameter (n = 42)	Mean	Standard Deviation
MSAS-SF: Number of symptoms	11.9	5.2
MSAS-SF: Physical Symptom Subscale Score (MSASPHYS)	2.3	0.7
MSAS-SF: Psychological Symptom Subscale Score (MSASPSYCH)	1.9	0.8
MSAS-SF: Total symptom distress score (TMSAS)	2.2	0.6
MSAS-SF: Global Distress Index (GDI)	2.3	0.6
PSQI: Usual Bedtime (BT) (hh:mm)	22:28	1:13
PSQI: Time to fall asleep (SOL) (min)	28.3	38.3
PSQI: Usual getting up time (GUT) (hh:mm)	07:51	1:11
PSQI: Hours of sleep per night (TST) (h)	6.7	1.8
PiPS-B algorithm: Patient's pulse rate (beats per min)	84	16
Bloods: Haemoglobin (g/L)	111.5	20.8
Bloods: White Blood Count (10^9/L)	7.8	4.3
Bloods: Neutrophils (10^9/L)	5.7	4.1
Bloods: Lymphocytes (10^9/L)	1.3	1.2
Bloods: Platelets (10^9/L)	315.3	166.7
Bloods: Sodium (mmol/L)	138.5	4.0
Bloods: Potassium (mmol/L)	4.3	0.6
Bloods: Urea (mmol/L)	6.2	2.7
Bloods: Creatinine (µmol/L)	71.4	26.1
Bloods: ALP (IU/L)	284.9	436.0
Bloods: ALT (IU/L)	59.4	149.3
Bloods: Albumin (g/L)	37.4	4.3
Bloods: C-reactive protein (CRP) (mg/L)	45.2	48.5
Wrist actigraphy: (I < O) at least 72 h (%)	89.0	6.5
Wrist actigraphy: r24 at least 72 h (autocorrelation coefficient)	0.17	0.1
Wrist actigraphy: Mean activity during wakefulness at least 72 h (number of accelerations per min)	143.7	62.1
Wrist actigraphy: Mean daily activity (MDA) at least 72 h (number of accelerations per min)	96.8	39.8
Wrist actigraphy: Bedtime (BT) (hh:mm)	22:41	1:07
Wrist actigraphy: Get up time (GUT) (hh:mm)	08:03	1:01
Wrist actigraphy: Time in bed (TIB) (hh:mm)	09:22	1:33
Wrist actigraphy: Total sleep time (TST) (hh:mm)	7:18	1:39
Wrist actigraphy: Sleep onset latency (SOL) (min)	21.7	21.6
Wrist actigraphy: Sleep efficiency (SE) (%)	78.2	12.0
Wrist actigraphy: Wake after sleep onset (WASO) (min)	68.4	31.6
Wrist actigraphy: Number of awake episodes (NA)	22.4	10.1

Table A1. *Cont.*

Numerical Prognostic Parameter (n = 42)	Mean	Standard Deviation
Consensus Sleep Diary: Time in bed (BT) (hh:mm)	22:35	1:06
Consensus Sleep Diary: Time of final awakening (hh:mm)	07:08	1:05
Consensus Sleep Diary: Time out of bed (GUT) (hh:mm)	08:03	1:01
Consensus Sleep Diary: Time tried to go to sleep (hh:mm)	22:58	1:02
Consensus Sleep Diary: Time to fall asleep (SOL) (min)	32.4	32.7
Consensus Sleep Diary: Total amount of time awakenings lasted (WASO) (min)	37.7	37.6
Consensus Sleep Diary: Number of times awakened in the night (NA)	2.5	1.3

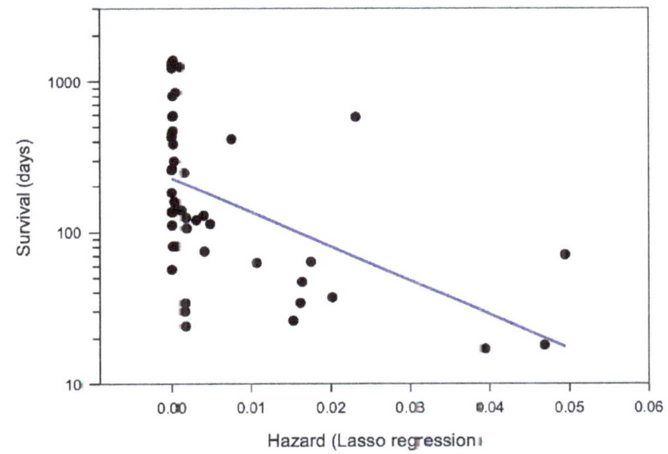

Figure A1. Scatterplot showing correlation between actual survival and predicted hazard using the lasso model.

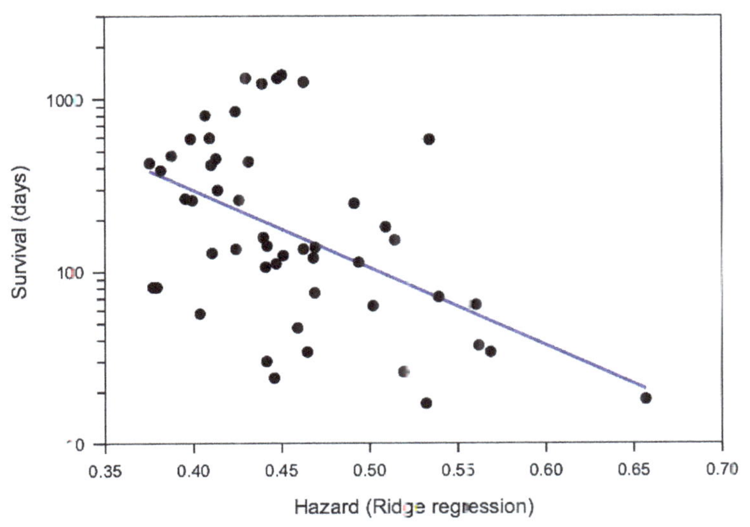

Figure A2. Scatterplot showing correlation between actual survival and predicted hazard using the ridge regression model.

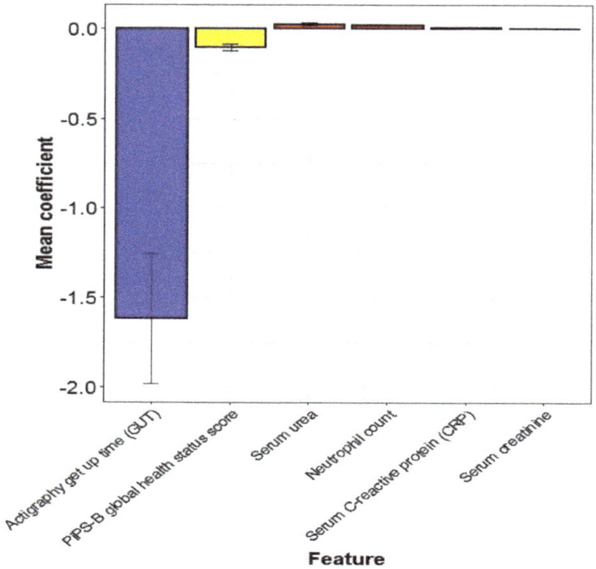

Figure A3. The mean cross—validated error (CVM) of predictor variables for hazard selected by the elastic net model.

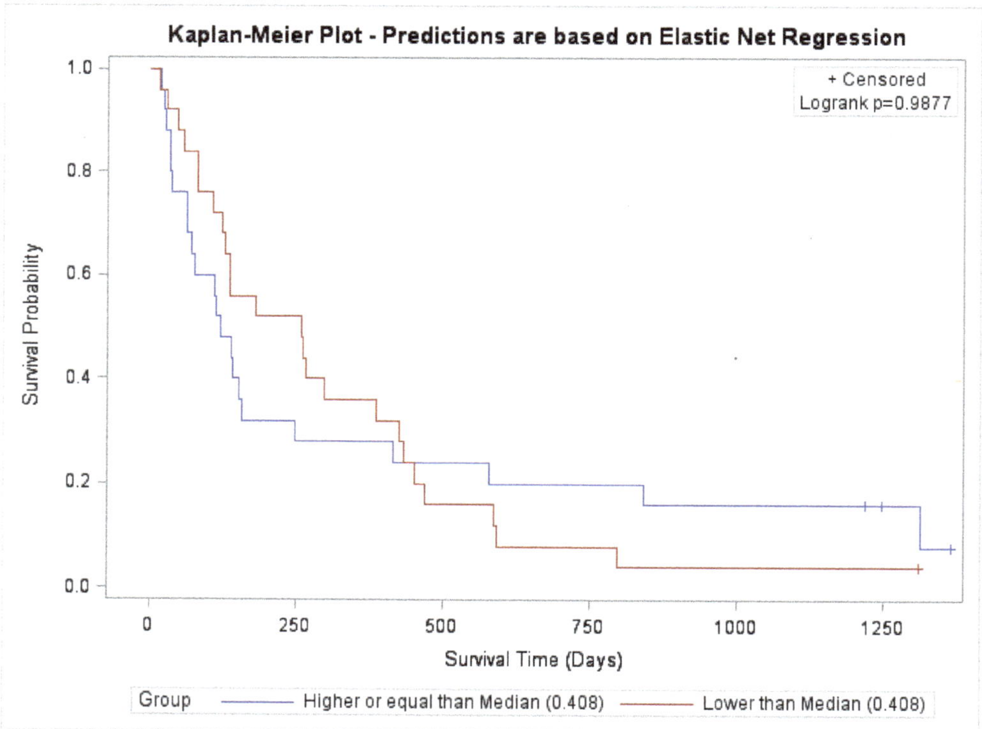

Figure A4. Kaplan–Meier curve comparing survival probability predicted by the elastic net-derived algorithm (log rank, $p = 0.9877$).

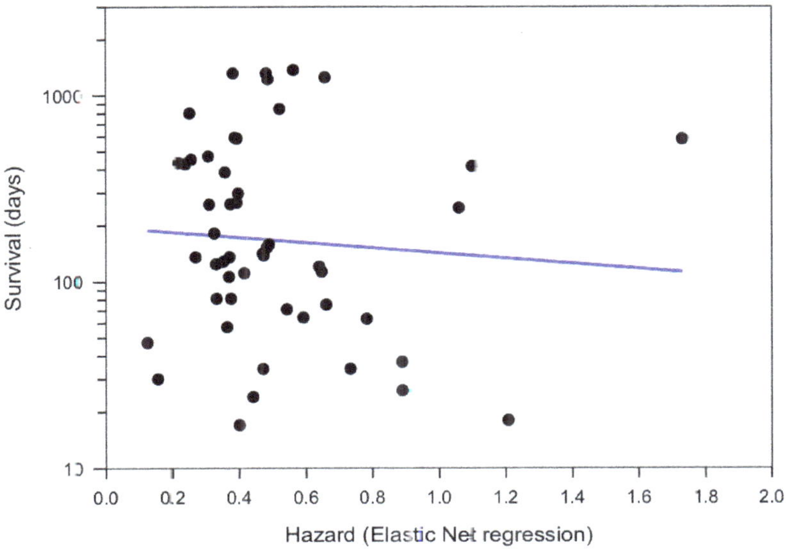

Figure A5. Scatterplot showing correlation between actual survival and predicted hazard using the elastic net model.

References

1. Maltoni, M.; Caraceni, A.; Brunelli, C.; Broeckaert, B.; Christakis, N.; Eychmueller, S.; Glare, P.; Nabal, M.; Vigano, A.; Larkin, P.; et al. Prognostic factors in advanced cancer patients: Evidence-based clinical recommendations–a study by the steering committee of the european association for palliative care. *J. Clin. Oncol.* **2005**, *23*, 6240–6248. [CrossRef] [PubMed]
2. Chu, C.; White, N.; Stone, P. Prognostication in palliative care. *Clin. Med.* **2019**, *19*, 306–310. [CrossRef] [PubMed]
3. White, N.; Reid, F.; Harris, A.; Harries, P.; Stone, P. A systematic review of predictions of survival in palliative care: How accurate are clinicians and who are the experts? *PLoS ONE* **2016**, *11*, e0161407. [CrossRef]
4. Hui, D.; Paiva, C.E.; Del Fabbro, E.G.; Steer, C.; Naberhuis, J.; van de Wetering, M.; Fernández-Ortega, P.; Morita, T.; Suh, S.Y.; Bruera, E.; et al. Prognostication in advanced cancer: Update and directions for future research. *Support. Care Cancer* **2019**, *27*, 1973–1984. [CrossRef] [PubMed]
5. Hui, D. Prognostication of survival in patients with advanced cancer: Predicting the unpredictable? *Cancer Control* **2015**, *22*, 489–497. [CrossRef]
6. Allada, R.; Bass, J. Circadian mechanisms in medicine. *N. Engl. J. Med.* **2021**, *384*, 550–561. [CrossRef]
7. Milanti, A.; Chan, D.N.S.; Li, C.; So, W.K.W. Actigraphy-measured rest-activity circadian rhythm disruption in patients with advanced cancer: A scoping review. *Support. Care Cancer* **2021**, *29*, 7145–7169. [CrossRef]
8. Balachandran, D.D.; Miller, M.A.; Faiz, S.A.; Yennurajalingam, S.; Innominato, P.F. Evaluation and management of sleep and circadian rhythm disturbance in cancer. *Curr. Treat. Options Oncol.* **2021**, *22*, 81. [CrossRef]
9. Medic, G.; Wille, M.; Hemels, M.E. Short- and long-term health consequences of sleep disruption. *Nat. Sci. Sleep* **2017**, *9*, 151–161. [CrossRef]
10. Levi, F.; Dugue, P.A.; Innominato, P.; Karaboue, A.; Dispersyn, G.; Parganiha, A.; Giacchetti, S.; Moreau, T.; Focan, C.; Waterhouse, J.; et al. Wrist actimetry circadian rhythm as a robust predictor of colorectal cancer patients survival. *Chronobiol. Int.* **2014**, *31*, 891–900. [CrossRef]
11. Innominato, P.F.; Komarzynski, S.; Palesh, O.G.; Dallmann, R.; Bjarnason, G.A.; Giacchetti, S.; Ulusakarya, A.; Bouchahda, M.; Haydar, M.; Ballesta, A.; et al. Circadian rest-activity rhythm as an objective biomarker of patient-reported outcomes in patients with advanced cancer. *Cancer Med.* **2018**, *7*, 4396–4405. [CrossRef] [PubMed]
12. Mormont, M.C.; Waterhouse, J.; Bleuzen, P.; Giacchetti, S.; Jami, A.; Bogdan, A.; Lellouch, J.; Misset, J.L.; Touitou, Y.; Levi, F. Marked 24-h rest/activity rhythms are associated with better quality of life, better response, and longer survival in patients with metastatic colorectal cancer and good performance status. *Clin. Cancer Res.* **2000**, *6*, 3038–3045. [PubMed]
13. Palesh, O.; Aldridge-Gerry, A.; Zeitzer, J.M.; Koopman, C.; Neri, E.; Giese-Davis, J.; Jo, B.; Kraemer, H.; Nouriani, B.; Spiegel, D. Actigraphy-measured sleep disruption as a predictor of survival among women with advanced breast cancer. *Sleep* **2014**, *37*, 837–842. [CrossRef] [PubMed]
14. *Treatment and Care towards the End of Life: Good Practice in Decision-Making*; General Medical Council: London, UK, 2010.

15. Oken, M.M.; Creech, R.H.; Tormey, D.C.; Horton, J.; Davis, T.E.; McFadden, E.T.; Carbone, P.P. Toxicity and response criteria of the eastern cooperative oncology group. *Am. J. Clin. Oncol.* **1982**, *5*, 649–655. [CrossRef] [PubMed]
16. Hodkinson, H.M. Evaluation of a mental test score for assessment of mental impairment in the elderly. *Age Ageing* **1972**, *1*, 233–238. [CrossRef] [PubMed]
17. Chang, V.T.; Hwang, S.S.; Feuerman, M.; Kasimis, B.S.; Thaler, H.T. The memorial symptom assessment scale short form (msas-sf). *Cancer* **2000**, *89*, 1162–1171. [CrossRef] [PubMed]
18. Gwilliam, B.; Keeley, V.; Todd, C.; Gittins, M.; Roberts, C.; Kelly, L.; Barclay, S.; Stone, P.C. Development of prognosis in palliative care study (pips) predictor models to improve prognostication in advanced cancer: Prospective cohort study. *BMJ* **2011**, *343*, d4920. [CrossRef]
19. Buysse, D.J.; Reynolds, C.F., 3rd; Monk, T.H.; Berman, S.R.; Kupfer, D.J. The pittsburgh sleep quality index: A new instrument for psychiatric practice and research. *Psychiatry Res.* **1989**, *28*, 193–213. [CrossRef]
20. McMillan, D.C. The systemic inflammation-based glasgow prognostic score: A decade of experience in patients with cancer. *Cancer Treat. Rev.* **2013**, *39*, 534–540. [CrossRef]
21. Philips Respironics. *Professional Sleep and Activity Monitoring Solutions: Specifications for Actiwatch 2, Actiwatch Spectrum Plus, and Actiwatch Spectrum Pro*; Koninklijke Philips Electronics N.V.: Amsterdam, The Netherlands, 2013.
22. Carney, C.E.; Buysse, D.J.; Ancoli-Israel, S.; Edinger, J.D.; Krystal, A.D.; Lichstein, K.L.; Morin, C.M. The consensus sleep diary: Standardizing prospective sleep self-monitoring. *Sleep* **2012**, *35*, 287–302. [CrossRef]
23. Berger, A.M.; Wielgus, K.K.; Young-McCaughan, S.; Fischer, P.; Farr, L.; Lee, K.A. Methodological challenges when using actigraphy in research. *J. Pain Symptom Manag.* **2008**, *36*, 191–199. [CrossRef]
24. Berger, A.M. Update on the state of the science: Sleep-wake disturbances in adult patients with cancer. *Oncol. Nurs. Forum* **2009**, *36*, E165–E177. [CrossRef] [PubMed]
25. Innominato, P.F.; Focan, C.; Gorlia, T.; Moreau, T.; Garufi, C.; Waterhouse, J.; Giacchetti, S.; Coudert, B.; Iacobelli, S.; Genet, D.; et al. Circadian rhythm in rest and activity: A biological correlate of quality of life and a predictor of survival in patients with metastatic colorectal cancer. *Cancer Res.* **2009**, *69*, 4700–4707. [CrossRef]
26. Ancoli-Israel, S.; Martin, J.L.; Blackwell, T.; Buenaver, L.; Liu, L.; Meltzer, L.J.; Sadeh, A.; Spira, A.P.; Taylor, D.J. The sbsm guide to actigraphy monitoring: Clinical and research applications. *Behav. Sleep Med.* **2015**, *13*, S4–S38. [CrossRef]
27. Sadeh, A.; Hauri, P.J.; Kripke, D.F.; Lavie, P. The role of actigraphy in the evaluation of sleep disorders. *Sleep* **1995**, *18*, 288–302. [CrossRef] [PubMed]
28. Schutte-Rodin, S.; Broch, L.; Buysse, D.; Dorsey, C.; Sateia, M. Clinical guideline for the evaluation and management of chronic insomnia in adults. *J. Clin. Sleep Med.* **2008**, *4*, 487–504. [CrossRef] [PubMed]
29. Research Design Service London. Justify Sample Size for a Feasibility Study. Available online: https://www.rds-london.nihr.ac.uk/resources/justify-sample-size-for-a-feasibility-study/ (accessed on 14 June 2022).
30. Mukaka, M.M. Statistics corner: A guide to appropriate use of correlation coefficient in medical research. *Malawi Med. J.* **2012**, *24*, 69–71.
31. Hastie, T.; Tibshirani, R.; Friedman, J.H. *The Elements of Statistical Learning: Data Mining, Inference and Prediction*, 2nd ed.; Springer: New York, NY, USA, 2009.
32. James, G.; Witten, D.; Hastie, T.; Tibshirani, R. *An Introduction to Statistical Learning: With Applications in R*; Springer: New York, NY, USA, 2013.
33. Tibshirani, R. Regression shrinkage and selection via the lasso. *J. R. Stat. Soc. Ser. B* **1996**, *58*, 267–288. [CrossRef]
34. Zou, H.; Hastie, T. Regularization and variable selection via the elastic net. *J. R. Stat. Soc. Ser. B* **2005**, *67*, 301–320. [CrossRef]
35. Friedman, J.; Hastie, T.; Tibshirani, R. Regularization paths for generalized linear models via coordinate descent. *J. Stat. Softw.* **2010**, *33*, 1–22. [CrossRef]
36. Koo, T.K.; Li, M.Y. A guideline of selecting and reporting intraclass correlation coefficients for reliability research. *J. Chiropr. Med.* **2016**, *15*, 155–163. [CrossRef] [PubMed]
37. Storick, V.; O'Herlihy, A.; Abdelhafeez, S.; Ahmed, R.; May, P. Improving palliative and end-of-life care with machine learning and routine data: A rapid review. *HRB Open Res.* **2019**, *2*, 13. [CrossRef] [PubMed]
38. Innominato, P.F.; Lim, A.S.; Palesh, O.; Clemons, M.; Trudeau, M.; Eisen, A.; Wang, C.; Kiss, A.; Pritchard, K.I.; Bjarnason, G.A. The effect of melatonin on sleep and quality of life in patients with advanced breast cancer. *Support. Care Cancer* **2016**, *24*, 1097–1105. [CrossRef]
39. Innominato, P.F.; Giacchetti, S.; Bjarnason, G.A.; Focan, C.; Garufi, C.; Coudert, B.; Iacobelli, S.; Tampellini, M.; Durando, X.; Mormont, M.C.; et al. Prediction of overall survival through circadian rest-activity monitoring during chemotherapy for metastatic colorectal cancer. *Int. J. Cancer* **2012**, *131*, 2684–2692. [CrossRef]
40. Collins, K.P.; Geller, D.A.; Antoni, M.; Donnell, D.M.; Tsung, A.; Marsh, J.W.; Burke, L.; Penedo, F.; Terhorst, L.; Kamarck, T.W.; et al. Sleep duration is associated with survival in advanced cancer patients. *Sleep Med.* **2017**, *32*, 208–212. [CrossRef] [PubMed]
41. Laird, B.J.; Kaasa, S.; McMillan, D.C.; Fallon, M.T.; Hjermstad, M.J.; Fayers, P.; Klepstad, P. Prognostic factors in patients with advanced cancer: A comparison of clinicopathological factors and the development of an inflammation-based prognostic system. *Clin. Cancer Res.* **2013**, *19*, 5456–5464. [CrossRef]
42. Ellershaw, J.; Ward, C. Care of the dying patient: The last hours or days of life. *BMJ* **2003**, *326*, 30–34. [CrossRef]

43. Natale, V.; Innominato, P.F.; Boreggiani, M.; Tonetti, L.; Filardi, M.; Parganiha, A.; Fabbri, M.; Martoni, M.; Lévi, F. The difference between in bed and out of bed activity as a behavioral marker of cancer patients: A comparative actigraphic study. *Chronobiol. Int.* **2015**, *32*, 925–933. [CrossRef]
44. Palesh, O.; Haitz, K.; Levi, F.; Bjarnason, G.A.; Deguzman, C.; Alizeh, I.; Ulusakarya, A.; Packer, M.M.; Innominato, P.F. Relationship between subjective and actigraphy-measured sleep in 237 patients with metastatic colorectal cancer. *Qual. Life Res.* **2017**, *26*, 2783–2791. [CrossRef]
45. Innominato, P.F.; Roche, V.P.; Palesh, O.G.; Ulusakarya, A.; Spiegel, D.; Levi, F.A. The circadian timing system in clinical oncology. *Ann. Med.* **2014**, *46*, 191–207. [CrossRef]

diagnostics

MDPI

Article

3D Convolutional Neural Network-Based Denoising of Low-Count Whole-Body ^{18}F-Fluorodeoxyglucose and ^{89}Zr-Rituximab PET Scans

Bart M. de Vries [1,*], Sandeep S. V. Golla [1], Gerben J. C. Zwezerijnen [1], Otto S. Hoekstra [1], Yvonne W. S. Jauw [1,2], Marc C. Huisman [1], Guus A. M. S. van Dongen [1], Willemien C. Menke-van der Houven van Oordt [3], Josée J. M. Zijlstra-Baalbergen [1,2], Liesbet Mesotten [4,5], Ronald Boellaard [1] and Maqsood Yaqub [1]

1 Cancer Center Amsterdam, Department of Radiology and Nuclear Medicine, Vrije Universiteit Amsterdam, Amsterdam UMC, De Boelelaan 1117, 1081 HV Amsterdam, The Netherlands; s.golla@amsterdamumc.nl (S.S.V.G.); g.zwezerijnen@amsterdamumc.nl (G.J.C.Z.); os.hoekstra@amsterdamumc.nl (O.S.H.); yws.jauw@amsterdamumc.nl (Y.W.S.J.); m.huisman@amsterdamumc.nl (M.C.H.); gams.vandongen@amsterdamumc.nl (G.A.M.S.v.D.); j.zijlstra@amsterdamumc.nl (J.J.M.Z.-B.); r.boellaard@amsterdamumc.nl (R.B.); maqsood.yaqub@amsterdamumc.nl (M.Y.)

2 Cancer Center Amsterdam, Department of Hematology, Vrije Universiteit Amsterdam, Amsterdam UMC, De Boelelaan 1117, 1081 HV Amsterdam, The Netherlands

3 Cancer Center Amsterdam, Department of Medical Oncology, Vrije Universiteit Amsterdam, Amsterdam UMC, De Boelelaan 1117, 1081 HV Amsterdam, The Netherlands; c.menke@amsterdamumc.nl

4 Faculty of Medicine and Life Sciences, Hasselt University, Agoralaan Building D, B-3590 Diepenbeek, Belgium; liesbet.mesotten@zol.be

5 Department of Nuclear Medicine, Ziekenhuis Oost Limburg, Schiepse Bos 6, B-3600 Genk, Belgium

* Correspondence: b.devries1@amsterdamumc.nl; Tel.: +31-643628806

Citation: de Vries, B.M.; Golla, S.S.V.; Zwezerijnen, G.J.C.; Hoekstra, O.S.; Jauw, Y.W.S.; Huisman, M.C.; van Dongen, G.A.M.S.; Menke-van der Houven van Oordt, W.C.; Zijlstra-Baalbergen, J.J.M.; Mesotten, L.; et al. 3D Convolutional Neural Network-Based Denoising of Low-Count Whole-Body ^{18}F-Fluorodeoxyglucose and ^{89}Zr-Rituximab PET Scans. *Diagnostics* **2022**, *12*, 596. https://doi.org/10.3390/diagnostics12030596

Academic Editor: Ayman El-Baz

Received: 20 January 2022
Accepted: 24 February 2022
Published: 25 February 2022

Abstract: Acquisition time and injected activity of ^{18}F-fluorodeoxyglucose (^{18}F-FDG) PET should ideally be reduced. However, this decreases the signal-to-noise ratio (SNR), which impairs the diagnostic value of these PET scans. In addition, ^{89}Zr-antibody PET is known to have a low SNR. To improve the diagnostic value of these scans, a Convolutional Neural Network (CNN) denoising method is proposed. The aim of this study was therefore to develop CNNs to increase SNR for low-count ^{18}F-FDG and ^{89}Zr-antibody PET. Super-low-count, low-count and full-count ^{18}F-FDG PET scans from 60 primary lung cancer patients and full-count ^{89}Zr-rituximab PET scans from five patients with non-Hodgkin lymphoma were acquired. CNNs were built to capture the features and to denoise the PET scans. Additionally, Gaussian smoothing (GS) and Bilateral filtering (BF) were evaluated. The performance of the denoising approaches was assessed based on the tumour recovery coefficient (TRC), coefficient of variance (COV; level of noise), and a qualitative assessment by two nuclear medicine physicians. The CNNs had a higher TRC and comparable or lower COV to GS and BF and was also the preferred method of the two observers for both ^{18}F-FDG and ^{89}Zr-rituximab PET. The CNNs improved the SNR of low-count ^{18}F-FDG and ^{89}Zr-rituximab PET, with almost similar or better clinical performance than the full-count PET, respectively. Additionally, the CNNs showed better performance than GS and BF.

Keywords: low-count; CNN; denoising; ^{18}F-FDG; ^{89}Zr-antibody

1. Introduction

^{18}F-fluorodeoxyglucose (^{18}F-FDG) positron emission tomography (PET) is essential in staging of a broad spectrum of malignancies [1–3]. Currently, a whole-body ^{18}F-FDG PET scan is acquired using a scan duration of 2 min per bed position and an injected activity of 3.7 MBq/kg. A shorter scan duration per bed position could ideally decrease the total scan duration, and therefore, minimize movement artefacts and increase patient comfort and throughput. A reduction of injected activity would decrease the radiation burden for the patient, and therefore makes it possible to perform more frequent ^{18}F-FDG PET scans per patient for restaging and therapy-response assessments, in case of scanning children and/or for non-oncological cases. However, a shorter scan duration and lower injected activity would result in a lower signal-to-noise ratio (SNR). Poor scan quality due to low-count (LC) is also observed for ^{89}Zr-antibody PET scans, which are obtained after relatively low injected activity imposed by the radiation burden of ^{89}Zr [4]. Therefore, denoising LC whole-body ^{18}F-FDG and ^{89}Zr-antibody PET scans is of interest for improving image quality.

Traditionally, Gaussian smoothing (GS) has been used to denoise PET images [5]. However, GS reduces the spatial resolution of the images, and therefore, could impair detectability and quantification of small (tumour) lesions [6]. Bilateral filtering (BF) exhibits superior properties in comparison to the more commonly used GS for noise reduction in PET [7]. BF reduces the noise of PET scans, while preserving spatial information (e.g., edges). However, BF parameters are difficult to optimize in a generic way because both an optimized intensity and spatial parameter need to be determined, which depend on both the tracer and site of interest. Therefore, another adaptive/data-driven denoising method with high accuracy is warranted.

Convolutional Neural Networks (CNN) are a specialized type of Neural Networks that use convolution to extract features from the PET scan. This is done by convolution filters which assign importance/weights to (learnable) features present in the PET scan. It can therefore learn and detect features such as PET intensities, edges, shapes, etc. Therefore, CNNs are highly beneficial in various medical image processing/segmentation tasks [8–11]. A CNN-based deep-learning algorithm may also be superior in denoising tasks since it can learn non-linear latent/hidden (not observable by humans) features (which you want to preserve) from LC PET scans and increase the SNR [12–18]. Therefore, denoising LC whole-body ^{18}F-FDG and ^{89}Zr-antibody PET scans using a CNN may be performed to improve the SNR and thus their diagnostic value. In previous studies [12–14,17,19] a successful application of CNNs for improving ^{18}F-FDG PET scans has been presented. However, these studies were performed in a small (oncology) patient cohort, were based on unsupervised deep learning networks and on improving full-count (FC) ^{18}F-FDG PET scans or a longer scan duration per bed position.

Therefore, the aim of this study was to develop, train, and extensively evaluate the performance of CNNs to denoise LC whole-body ^{18}F-FDG and ^{89}Zr-antibody PET scans. A secondary aim was to compare the diagnostic value of the CNNs to that of GS and BF.

2. Materials and Methods

2.1. Participants

We included PET scans of 60 patients with stage I–IV non-small-cell lung carcinoma (NSCLC) (40 patients from Limburg PET-Center Hasselt Belgium (LPC) [20], and 20 patients from Amsterdam UMC, location VUmc), of which five patients with diffuse large B cell lymphoma (DLBCL) non-Hodgkin lymphoma (Amsterdam UMC, location VUmc) [21] (Table 1). The study at LPC was registered at clinical trials.gov, NCT02024113. The data from the patients with lung cancer at Amsterdam UMC were retrospectively obtained from medical records, with a waiver for informed consent from the Medical Ethics Review Committee of Amsterdam UMC, location VUmc. This study was registered as IRB2018.029. The patients with non-Hodgkin lymphoma were included as part of studies performed by Jauw et al. These patients provided written informed consent, and the studies were approved by the Medical Ethics Review Committee of Amsterdam UMC, location VUmc. This study was registered at Dutch Trial Register http://www.trialregister.nl (accessed on 19 January 2022), NTR3392.

Table 1. The patients characteristics included in this study from Amsterdam UMC.

	^{18}F-FDG (n = 20) NSCLC–Amsterdam UMC	^{18}F-FDG (n = 40) NSCLC–LPC	^{89}Zr-Rituximab (n = 5) Non-Hodgkin Lymphoma
Male/female (n)	11/9	25/15	3/2
Injected dose (MBq)	259.4 ± 43.8	298.2 ± 49.4	73.7 ± 0.3
Tumour volume (cubic centimeter (cc))–Test data	29.2 ± 54.3	-	25.6 ± 51.3

2.2. Data Acquisition

Whole-body ^{18}F-FDG PET scans in LPC were acquired with a Gemini Big Bore TF PET/CT scanner, and in Amsterdam UMC with an Ingenuity TF PET/CT and Vereos Digital PET/CT scanners (Philips Medical Systems, Best, The Netherlands). ^{89}Zr-rituximab PET scans (patients with non-Hodgkin lymphoma) were acquired with an Ingenuity TF PET/CT. For ^{18}F-FDG PET scans, 60 min after 259.4 ± 43.8 MBq tracer injection, a low-dose computed tomography (LDCT) scan was performed for attenuation correction and anatomical localisation, and subsequently a 20 min static (exact time depends on patient length) whole-body ^{18}F-FDG PET scan was acquired (2 min per bed position). Six days after the injection of 73.7 ± 0.3 MBq ^{89}Zr-rituximab, an LDCT scan was obtained, directly followed by a 60 min static whole-body PET scan (5 min per bed position). Corrections for decay, dead time, normalization (detector sensitivities), attenuation, random coincidences and scatter were applied.

Amsterdam UMC ^{18}F-FDG PET data were reconstructed with a 10 s (super-low-count (SLC), 92% scan time reduction), 30 s (low-count (LC), 75% scan time reduction) and 2 min (full-count (FC)) scan duration per bed-position. The (S)LC PET scans were reconstructed using multiple time points/delays, which was later used for data augmentation during training. These scans were reconstructed using the blob-basis function ordered-subsets time of flight (BLOB-OS-TF) for the Ingenuity TF PET/CT scanner and the novel ordered subset expectation maximization (OSEM 3i15s, 1i6r-PSF, 4 mm FWHM GAUSS, OSEM 3i15s, 3 mm FWHM GAUSS) for the Vereos Digital PET/CT scanner. The ^{89}Zr-rituximab scans, and the ^{18}F-FDG PET data from LPC were reconstructed with a FC 5 min and 2 min scan duration per bed position only using BLOB-OS-TF, respectively.

The ^{18}F-FDG and ^{89}Zr-rituximab PET scans from Amsterdam UMC were reconstructed according to current European Association of Nuclear Medicine Research Ltd. (Vienna, Austria) EARL1 standards and settings associated with EARL accreditation [22], respectively. Matrix and voxels sizes were 144 × 144 and 4 mm in all directions, respectively. ^{18}F-FDG PET scans from LPC were reconstructed according to EARL1 standards, with matrix and voxel sizes of 169 × 169 and 4 mm, respectively.

2.3. Image Processing

For each FC whole-body ^{18}F-FDG PET scan from LPC, SLC PET, scans were simulated using the SiMulAtion and ReconsTruction (SMART)-PET package [23]. Simulation-reconstruction settings were chosen so that the simulated noisy ^{18}F-FDG PET scans from LPC showed an almost similar coefficient of variation as the SLC-reconstructed ^{18}F-FDG PET scans from Amsterdam UMC. The simulated PET images were used to initially train the CNN, while parts of the actual reconstructed images were used for further fine-training of the CNN, details are explained later.

2.4. Model Architecture

A supervised U-Net based [11] 3D-CNN (Figure 1 and Appendix B) was used to denoise the (S)LC 18F-FDG PET scans while maintaining their diagnostic value. However, instead of the max-pooling layer that is traditionally used, in this study the down sampling layers consisted of convolution layers with a stride of two [24]. Although the convolution layer compresses the feature image just as is the case for a max-pooling layer, it does not exclude voxels by only looking at the maximum values. It therefore, not only reduces computation time (although less than max-pooling), but most importantly increases the model its ability to learn [24]. Additionally, in contrast to conventional CNNs, a kernel size of 6 × 6 × 6 instead of 3 × 3 × 3 was applied to learn inter-slice morphological features [25].

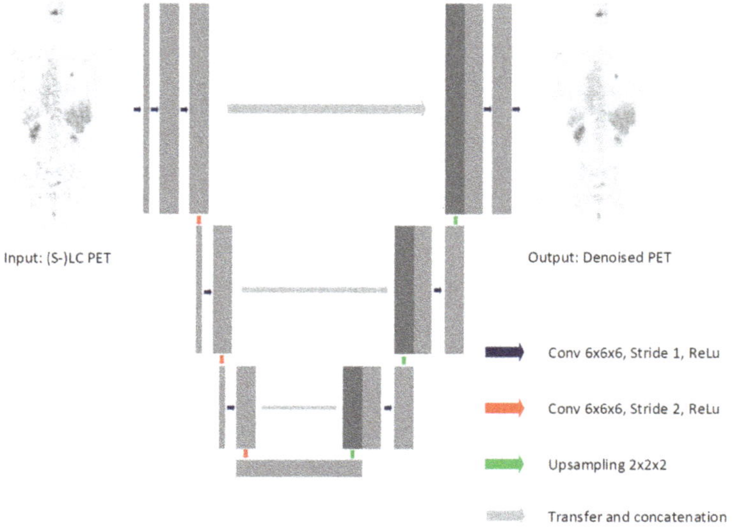

Figure 1. Architecture of the U-net shaped 3D-CNN used in the study. It consists of an encoding and decoding path, which are connected with concatenation layers at each resolution block.

2.5. Model Performance

2.5.1. Quantitative Performance

The simulated SLC whole-body ^{18}F-FDG PET scans from LPC were used to pre-train a 3D-CNN. Next, the reconstructed SLC and LC ^{18}F-FDG PET data from Amsterdam UMC were used for fine-training (transfer-learning) the pre-trained model, which generated two additional models (SLC-CNN and LC-CNN) that are tailored to manage low or super low count/quality images. These two models were subsequently used for further evaluation. Training of the CNN model on the simulated noisy LPC ^{18}F-FDG data was performed to avoid overfitting due to the small dataset. Noise characteristics of ^{89}Zr-rituximab and LC ^{18}F-FDG PET scans were almost similar. However, we used the SLC-CNN to denoise the ^{89}Zr-rituximab PET scans instead of the LC-CNN, because of the higher level of noise reduction.

The ^{18}F-FDG PET data from LPC was split into a training (80%, n = 32) and a validation (20%, n = 8) set. Thereafter, for further refinement, validation and testing, the two CNN models, ^{18}F-FDG and ^{89}Zr-rituximab data from Amsterdam UMC, were used. During this training, the data were split into a training (32%, n = 8), a validation (8%, n = 2) and an independent test (60%, n = 15) set. The training and validation set from Amsterdam UMC consisted of only ^{18}F-FDG PET scans from the Ingenuity TF PET/CT scanner. The test set, however, consisted both of ^{18}F-FDG PET scans from the Ingenuity TF PET/CT scanner, the Vereos Digital PET/CT scanner, and ^{89}Zr-rituximab PET scans from the Ingenuity TF PET/CT scanner. PET data augmentation was applied during each training epoch (train-data only) by randomly sampling the different (time points/delays) (S)LC ^{18}F-FDG PET scans for each patient during training. In other words, instead of traditional augmentation (shifts, zoom, translation, rotation, etc.), in each training epoch, minor differences in noise characteristics were present.

To compare the performance of the CNNs with other denoising methods, the (S)LC test PET scans were also denoised using traditional GS (^{18}F-FDG) and more advanced BF [17] (^{18}F-FDG and ^{89}Zr-rituximab) denoising methods (Table 2). A Mann–Whitney U test ($p < 0.05$) was used to compare tumour recovery coefficients (TRC) and levels of noise in the images after applying the denoising methods.

Table 2. Gaussian smoothing (GS) and Bilateral Filtering (BF) settings evaluated for the denoising of the low-count whole-body PET scans.

	GS (FWHM)	BF (FWHM; SUV)
^{18}F-FDG PET		
SLC	8 mm, 10 mm and 12 mm	4 mm and 5 mm; SUV2.5
LC	4 mm, 6 mm and 8 mm	3 mm and 4 mm; SUV2.5
^{89}Zr-rituximab PET	6 mm, 8 mm and 10 mm	2 mm, 3 mm and 4 mm; SUV2.5

We calculated TRC for the ^{18}F-FDG and ^{89}Zr-rituximab PET scans (test data) using Equation (1). TRC was computed for the test data post-processed with a 3D-CNN, GS, or BF denoising method and compared this to the FC data. PET uptake features from the tumour volumes were extracted (U_X, X = average, maximum and 3Dpeak) for both the denoised (U_X *denoised*) as the FC (U_X *FC*) PET scans, using the in-house built and open-access ACCURATE tool (quAntitative onCology moleCUlaR Analyses SuiTE) [26,27]. From the ^{18}F-FDG scans, only the primary lung tumour was extracted using a 50% SUV3Dpeak isocontour (Table 1 and Figure A1). For the ^{89}Zr-rituximab PET scans, tumours were extracted using manual delineation (Table 1 and Figure A1). For patients with non-Hodgkin lymphoma with more than three tumours, bootstrapping was applied to randomly choose three tumours for analysis.

$$\text{TRC} = \frac{U_X\ denoised}{U_X\ FC} \quad (1)$$

The level of noise was presented as the coefficient of variance (COV; Equation (2)). Four spherical volume of interest (VOIs) were drawn in the liver (because the liver showed homogeneous tracer uptake in this cohort, and therefore, could be used to reliably assess the level of noise). Average standard deviation $\left(\overline{\sigma\ liver}\right)$ and average uptake $\left(\overline{U_{avg}\ liver}\right)$ were extracted using these four VOIs.

$$COV = \frac{\overline{\sigma\ liver}}{\overline{U_{avg}\ liver}} \quad (2)$$

2.5.2. Qualitative Performance

For a qualitative assessment of the denoising methods (CNN and BF), the images after denoising were independently evaluated by two experienced nuclear medicine physicians (BZ and OH). GS was not included in this assessment due to a mostly significant ($p < 0.05$) lower quantitative performance in comparison to the CNNs and BF. The questionnaire was drafted to assess the reliability and effectiveness of the denoising methods. The assessment was blinded, i.e., the scans presented to the physicians were a random combination (without labels) of the FC, SLC, LC (with and without denoising) PET scans per patient. The ^{18}F-FDG and ^{89}Zr-rituximab PET scans were scored per patient (1–5: low to high) based on the level of noise, tumour detectability, overall scan quality, clinical acceptability (yes/no), and overall best performance (1st/2nd/3rd/4th/(5th)). A Mann–Whitney U test ($p < 0.05$) was used to compare the performance of the denoising methods.

3. Results

3.1. Quantitative Assessment

3.1.1. ^{18}F-FDG

The BF and the proposed CNN (SLC- and LC-CNN) denoised PET scans have an overall higher TRC and more similar COV to the FC PET scans than GS (Figure 2 and Table A1). In contrast with BF, the SLC-CNN denoised PET scans showed a higher average uptake TRC, a higher 3Dpeak uptake TRC, but a lower maximum uptake TRC. With regard to the LC scans, the LC-CNN denoised PET scans showed a trend ($0.05 < p < 0.1$) of a higher TRC than the BF denoised PET scans for the average uptake, and 3Dpeak uptake. Additionally, the LC-CNN showed a higher but not significant maximum uptake TRC than the BF. In addition, the LC-CNN denoised PET scans showed a similar COV as the FC PET scans. The SLC-CNN had the second closest COV to the FC PET scans.

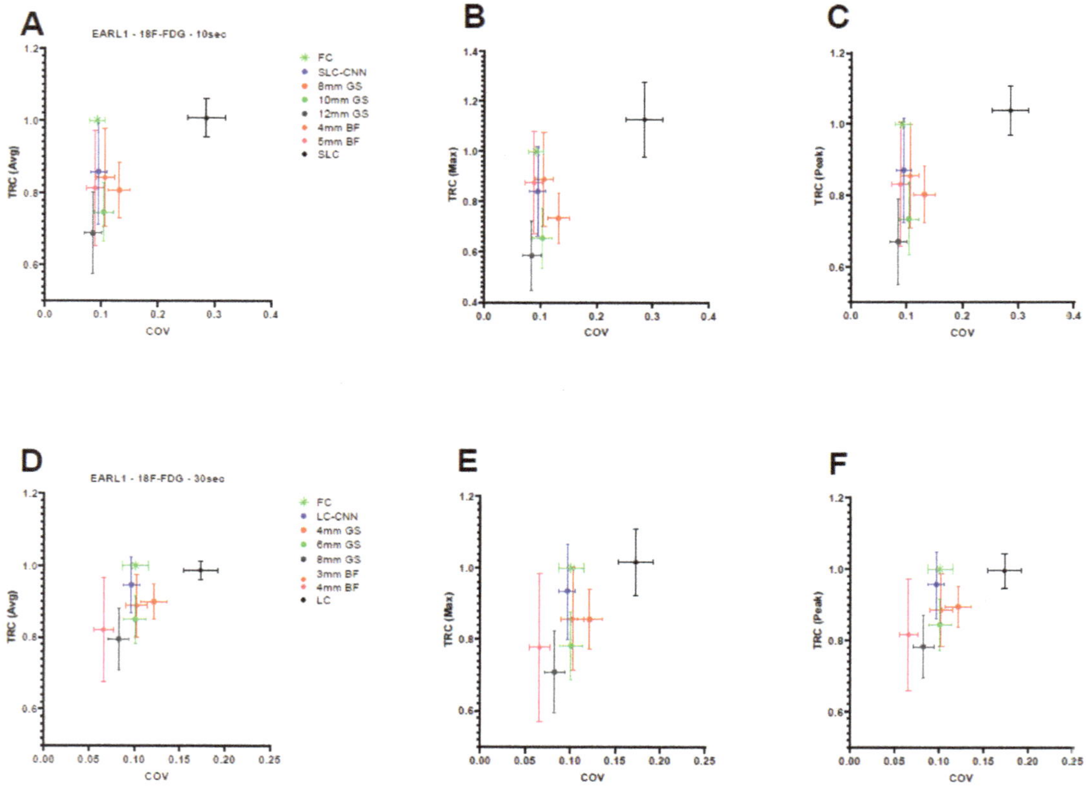

Figure 2. The performance of the denoising methods for low-count ^{18}F-FDG PET. The (**A**) average, (**B**) maximum, (**C**) 3Dpeak TRC of the SLC-CNN, GS (8 mm, 10 mm and 12 mm) and BF (4 mm and 5 mm) denoising methods of the SLC ^{18}F-FDG PET from the Ingenuity TF PET/CT and the Vereos Digital PET/CT scanner. The (**D**) average, (**E**) maximum, (**F**) 3Dpeak TRC of the LC-CNN, GS (4 mm, 6 mm and 8 mm) and BF (3 mm and 4 mm) denoising methods of the LC ^{18}F-FDG PET from the Ingenuity TF PET/CT and the Vereos Digital PET/CT scanner.

3.1.2. ^{89}Zr-Rituximab

The SLC-CNN denoised PET scans showed a predominant trend of a TRC higher than the 3 mm and 4 mm BF (Figure 3 and Table A2). The SLC-CNN even showed a significantly ($p < 0.05$) higher average uptake TRC than the 3 mm and 4 mm BF. The 3 mm and 4 mm BF presented a comparable COV as the SLC-CNN. The 2 mm BF had a significantly ($p < 0.05$) higher COV than the 3 mm and 4 mm BF and the SLC-CNN.

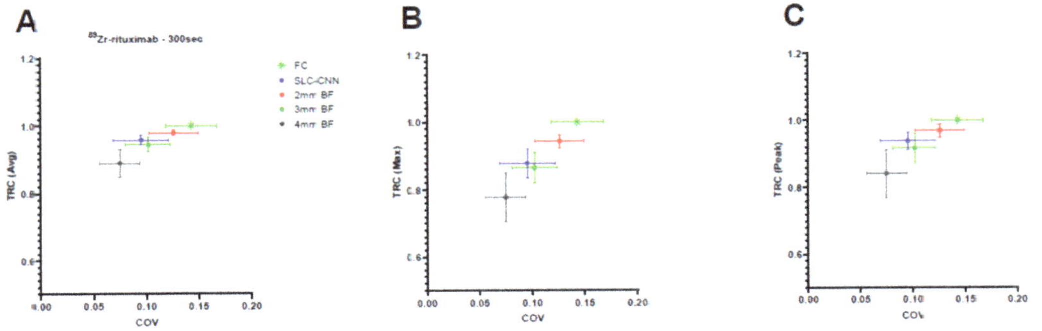

Figure 3. The performance of the denoising methods for low-count ^{89}Zr-rituximab PET. The (**A**) average, (**B**) maximum, (**C**) 3Dpeak TRC of the SLC-CNN and BF (2 mm, 3 mm and 4 mm) denoising methods of the FC ^{89}Zr-rituximab PET from the Ingenuity TF PET/CT scanner.

3.2. *Qualitative Assessment*

For the ^{18}F-FDG scans, the observers found lower levels of noise, better tumour detectability, better overall scan quality and higher clinical acceptability for all the CNN models in comparison to the BF denoising methods (Figures 4 and A2, Table 3), with the only exception being SLC-CNN in terms of noise levels and tumour detectability.

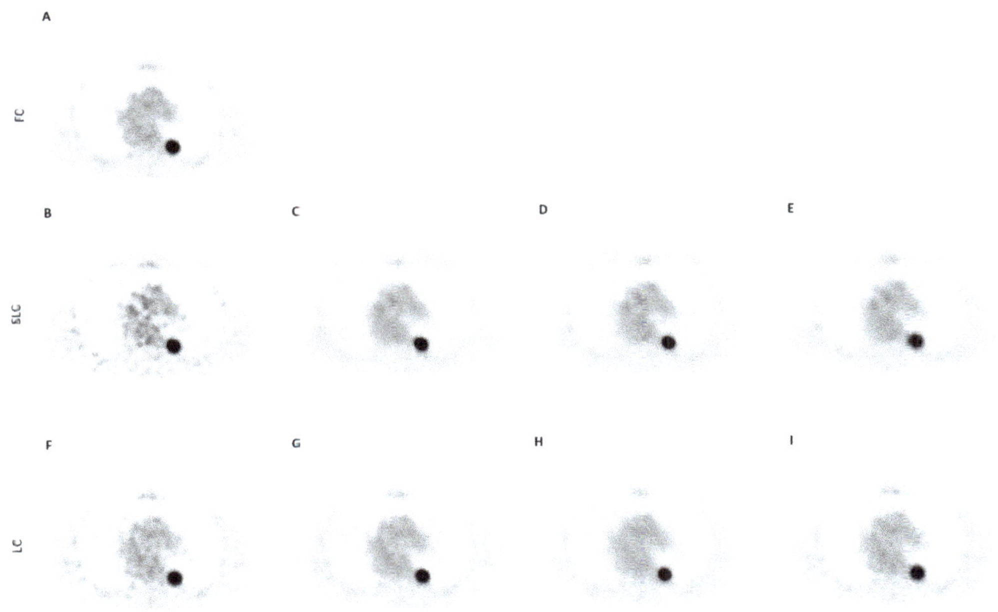

Figure 4. Illustration of a (**A**) FC, (**B**) SLC, (**C**) SLC-CNN, (**D**) BF 4 mm, (**E**) GS 10 mm denoised ^{18}F-FDG PET scan (axial orientation) from the Ingenuity TF PET/CT scanner. Illustration of a (**F**) LC (**G**) LC-CNN, (**H**) BF 4 mm, (**I**) GS 6 mm denoised ^{18}F-FDG PET scan (axial orientation) from the Ingenuity TF PET/CT scanner.

Table 3. Scores provided by the Nuclear Medicine Physicians as part of qualitative assessment of the denoised ^{18}F-FDG PET scans. The best performing method is indicated in bold based on the average score of both physicians.

	SLC			LC			FC
Metrics [1–5: Low–High]	**SLC-CNN**	**BF-4 mm**	**BF-5 mm**	**LC-CNN**	**BF-3 mm**	**BF-4 mm**	
Level of noise	**3.0–3.2**	4.0–4.0 *	3.6–4.0 *	**1.8–2.0**	2.6–2.0 **	2.2–3.8 *	1–1 *
Tumour detectability	2.0–3.0	**2.2–3.0**	2.0–3.0	**4.0–4.0**	3.6–4.0	2.8–3.0 *	5–5 *
Overall scan quality	**2.4–2.6**	1.6–2.0 *	1.8–2.0 *	**4.4–4.0**	3.8–4.0	2.8–2.2 *	5–5 *
Clinically acceptable? [%]	**0–80**	0–0 *	0–0 *	**100–100**	80–100	20–0 *	100–100
Best scan (1/2/3/4)	**2–2**	3–3	4–4	**2–3**	**3–2**	4–4	1–1

* significant ($p < 0.05$) higher/lower than (S)LC-CNN. ** trend ($p < 0.1$).

For the ^{89}Zr-rituximab scans, the observers found a comparable level of noise, but similar/better tumour detectability, better overall scan quality and higher clinical acceptability for the SLC-CNN in comparison to the BF denoising methods (Figure 5 and Table 4).

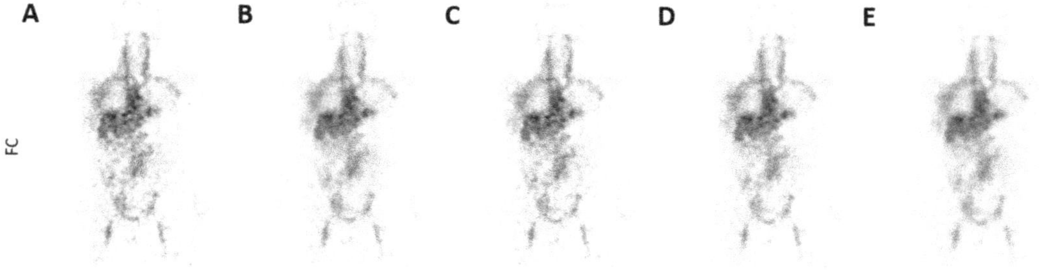

Figure 5. Illustration of a (**A**) FC, (**B**) SLC-CNN, (**C**) BF 3 mm, (**D**) BF 4 mm, (**E**) BF 5 mm denoised ^{89}Zr-rituximab PET scan (coronal orientation) from the Ingenuity TF PET/CT scanner.

Table 4. Qualitative assessment of the ^{89}Zr-rituximab PET scans. Scores were given for the PET scans with (SLC-CNN and BF) and without (FC) denoising by both Nuclear Medicine Physicians. In bold the best performing method (or scan) is indicated based on the average score of both physicians.

Metrics [1–5: Low–High]	SLC-CNN	BF-2 mm	BF-3 mm	BF-4 mm	FC
Level of noise	2.4–2.6	3.8–4.6 *	2.8–2.6	**1.4–1.2** *	4.6–4.8 *
Tumour detectability	3.4–3.8	4.4–4.0 *	2.4–2.4 *	1.4–1.4 *	**4.6–4.2** *
Overall scan quality	**3.8–3.8**	3.6–3.8	3.4–3.0 *	2.0–1.4 *	3.4–3.0 *
Clinical acceptable? [%]	**100–100**	**100–100**	80–80	**0–0** *	80–80
Best scan (1/2/3/4/5)	**1–1**	3–2	2–4	4–5	3–3

* significant ($p < 0.05$) higher/lower than SLC-CNN.

4. Discussion

CNN models to denoise (S)LC ^{18}F-FDG and ^{89}Zr-rituximab PET scans were trained and extensively evaluated. Overall, the CNN models performed better than the conventional GS and the more advanced BF denoising methods for both the ^{18}F-FDG and ^{89}Zr-rituximab PET scans. As such, the CNN models show promise for reducing the acquisition time and injected activity of ^{18}F-FDG PET scans and increasing the image quality of ^{89}Zr-rituximab PET scans.

In this study, we trained noise-specific CNN models, to address the difference in noise levels seen for different scan acquisition times and injected activity in ^{18}F-FDG PET scans. However, in case of PET tracers such as ^{89}Zr-antibody PET, training a noise specific CNN model was not feasible. Due to the dose limits of ^{89}Zr, the overall image quality was impaired (low SNR), and therefore, no high quality ^{89}Zr-antibody PET images were available for training a CNN. So, the only possible solution was to directly apply the SLC-CNN (trained using SLC ^{18}F-FDG PET scans) to the ^{89}Zr-rituximab PET scans and test its performance. The main advantage of this approach is that this validation is the ultimate way of externally testing the CNN on data that are obtained with a different tracer. Although the SLC-CNN is not trained on ^{89}Zr-rituximab PET scans, it obtained a higher TRC than the 3 mm and 4 mm BF denoising methods (Figure 3 and Table A2).

With regard to ^{18}F-FDG PET scans with a low injected tracer activity, such as scans with shorter scan duration, a lower SNR will be observed, which impairs both the quantitative and qualitative value of these scans. The CNNs could therefore also be useful to maintain a good image quality when reducing injected ^{18}F-FDG activity in whole-body ^{18}F-FDG PET studies, and therefore, reduce radiation burden for the patient, but maintain diagnostic value. However, further assessment is necessary to evaluate the performance of CNNs when used for a reduction in the injected activity for whole-body ^{18}F-FDG PET acquisitions.

The qualitative assessment also showed that the proposed CNNs were preferred over BF. However, the CNN denoised (S)LC ^{18}F-FDG PET scans did show an overall lower qualitative performance than the FC ^{18}F-FDG PET scans. Yet, the LC-CNN denoised LC ^{18}F-FDG PET scans obtained a similar clinical acceptability score as the FC ^{18}F-FDG PET scans (Table 3), while for ^{89}Zr-rituximab PET scans, the SLC-CNN increased the overall image quality of the FC ^{89}Zr-rituximab PET scans (Table 4). The observers preferred the SLC-CNN denoised ^{89}Zr-rituximab PET scans over both the BF denoised and FC ^{89}Zr-rituximab PET scans. This can be explained by a higher ratio between tumour signal and background signal present in the SLC-CNN denoised ^{89}Zr-rituximab PET scans (Figure 5). This indicates that the SLC-CNN shows promise for establishing an optimal denoising setup for ^{89}Zr-antibody PET scans.

In this study, several strategies to prevent overfitting were applied. First, data augmentation was applied by randomly sampling the different (S)LC ^{18}F-FDG PET scans for each patient. By using traditional augmentation, interpolation may be different between the training data ((S)LC) and training labels (FC), and therefore, this was not applied in this study. Another method by which overfitting was reduced is by using the symmetric connections in the U-Net based 3D-CNN [28]. As shown in previous studies [12,13], training a model using a small dataset could result in overfitting. Since acquiring sufficient real (S)LC ^{18}F-FDG PET data were not feasible, SLC ^{18}F-FDG PET data were generated using the already available LPC data. SLC ^{18}F-FDG PET data from LPC were simulated using SMART, which facilitated the development of a pre-trained model familiar with morphological features. This resulted in a shorter learning time, lower probability of overfitting, and a more accurate and robust model. Even though pre-training of the model was only performed on SLC ^{18}F-FDG PET data from LPC, the fine-trained LC-CNN showed a higher performance than a LC-CNN without a pre-trained model.

The main limitation of this study is the size of the patient cohort. Small-sized tumours in the (S)LC PET scans are more prone to being underestimated by the proposed CNNs This is because the training data were devoid of small tumours. Therefore, it could be that the model specifies this signal as noise rather than a tumour-specific signal [29]. However, the proposed CNNs showed better correspondence with the FC PET scans than GS and BF So, although small tumours were present in a small number in the training data, by using the proposed CNNs, more quantitative information was retained in comparison to GS and BF. Even though the differences in performance between the proposed CNNs and BF were small, contrary to BF, a CNN still has the ability to learn and improve by incorporating more patients. Thus, further evaluation in a larger and more heterogeneous cohort could further improve CNNs performances. However, although the proposed method showed promising results for denoising low-count ^{18}F-FDG PET scans, obtaining a similar quantitative and qualitative value as the FC ^{18}F-FDG PET scans may not be fully feasible and we therefore foresee that the main applications of the CNNs are denoising and improving image quality of ^{89}Zr-antibody PET studies.

5. Conclusions

The 3D-CNNs used in this study to denoise (S)LC whole body ^{18}F-FDG and ^{89}Zr-rituximab PET scans were constructed and tested. The CNN denoised (S)LC ^{18}F-FDG and ^{89}Zr-rituximab PET scans showed almost similar or better clinical performance than the FC scans, respectively. Therefore, the proposed CNNs show promise for reducing PET scan duration or lowering injected activity of whole-body ^{18}F-FDG PET scans but are particularly useful to increase the quantitative and qualitative image quality of ^{89}Zr-rituximab PET scans.

Author Contributions: Conceptualization, B.M.d.V., M.Y., S.S.V.G. and R.B.; methodology, B.M.d.V., M.Y., S.S.V.G. and R.B.; software, B.M.d.V.; validation, B.M.d.V., M.Y., S.S.V.G. and R.B.; formal analysis, B.M.d.V., M.Y., S.S.V.G. and R.B.; investigation, B.M.d.V., M.Y., S.S.V.G. and R.B.; resources, R.B.; data curation, B.M.d.V.; writing—original draft preparation, B.M.d.V.; writing—review and editing, B.M.d.V., M.Y., S.S.V.G., R.B., G.J.C.Z., O.S.H., Y.W.S.J., M.C.H., G.A.M.S.v.D., W.C.M.-v.d.H.v.O., J.J.M.Z.-B. and L.M.; visualization, B.M.d.V.; supervision, S.S.V.G., M.Y. and R.B.; project administration, B.M.d.V.; funding acquisition, R.B. All authors have read and agreed to the published version of the manuscript.

Funding: This research received no external funding.

Institutional Review Board Statement: The study was conducted in accordance with the Declaration of Helsinki, and approved by the Institutional Review Board (or Ethics Committee) of Amsterdam UMC, location VUmc.

Informed Consent Statement: Informed consent was obtained from all subjects involved in the study.

Data Availability Statement: Data can make available on reasonable request.

Conflicts of Interest: The authors declare no conflict of interest.

Appendix A

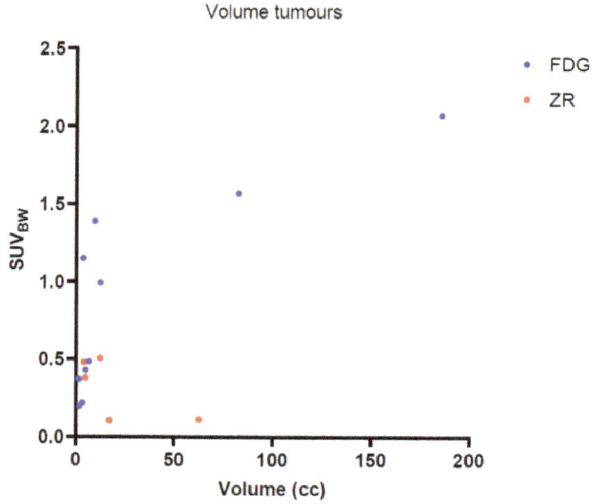

Figure A1. Volume (cc) and SUV_{BW} distribution of the tumours in the ^{18}F-FDG PET and the ^{89}Zr-rituximab PET scans (test data).

Figure A2. Illustration of a (**A**) FC, (**B**) SLC, (**C**) SLC-CNN, (**D**) BF 4 mm, (**E**) BF 5 mm, (**F**) GS 10 mm denoised ^{18}F-FDG PET scan (coronal orientation) from the Ingenuity TF PET/CT scanner. Illustration of a (**G**) LC, (**H**) LC-CNN, (**I**) BF 3 mm, (**J**) BF 4 mm, (**K**) GS 6 mm denoised ^{18}F-FDG PET scan (coronal orientation) from the Ingenuity TF PET/CT scanner.

Table A1. Overview of the performance of the GS, the BF and the proposed CNNs denoising methods using the external test ^{18}F-FDG PET scans from the Ingenuity TF PET/CT and Vereos Digital PET/CT scanners. For the FC (green), SLC and LC (with (blue) and without (orange) post-processing) PET scans, the TRC and COV values are shown in each column.

	TRC–Average	TRC–Maximum	TRC–3DPeak	COV
FC				0.09
SLC	0.99	1.11	1.03	0.29
GS (12; 10; 8 mm)	[0.69 *; 0.74 *; 0.81 **]	[0.59 *; 0.66 *; 0.74]	[0.68 *; 0.74 *; 0.80]	[0.09; 0.11; 0.14]
BF (5; 4 mm)	[0.79; 0.82]	[0.85; 0.87]	[0.82; 0.84]	[0.01; 0.11]
SLC-CNN	0.86	0.85	0.87	0.10
LC	0.98	1.02	0.99	0.17
GS (8; 6; 4 mm)	[0.80 *; 0.85 **; 0.90]	[0.71 *; 0.79 *; 0.86 *]	[0.79 *; 0.84 *; 0.90 *]	[0.09; 0.11; 0.13]
BF (4; 3 mm)	[0.81 **; 0.88]	[0.78; 0.87]	[0.82 **; 0.89]	[0.07; 0.11]
LC-CNN	0.95	0.94	0.96	0.10

* significant ($p < 0.05$) higher/lower than (S)LC-CNN. ** trend ($0.05 < p < 0.1$).

Table A2. Overview of the performance of the BF and the SLC-CNN on the external test ^{89}Zr-rituximab PET scans from the Ingenuity TF PET/CT scanner. For the FC (orange), and post-processed (blue) PET scans, the TRC and COV values are shown in each column.

TRC–Average	TRC–Maximum	TRC–3Dpeak	COV	
FC				0.14
BF (4; 3; 2 mm)	[0.88 *; 0.94 *; 0.98]	[0.78 **; 0.86; 0.94 *]	[0.84 **; 0.92 **; 0.97 *]	[0.07; 0.10; 0.13]
SLC-CNN	0.96	0.88	0.94	0.10

* significant ($p < 0.05$) higher/lower than SLC-CNN. ** trend ($0.05 < p < 0.1$).

Appendix B

Appendix B.1. Image Processing

Matrix dimension of the PET scans varied between centre, scanner, and patients. Therefore, EARL1 scans from the Ingenuity TF PET/CT, the Vereos Digital PET/CT and the Gemini Big Bore TF PET/CT scanner were zero-padded to a uniform matrix size of

192 × 192 × 320. The EARL2 scans from the Vereos Digital PET/CT scanner were zero-padded to a uniform matrix size of 384 × 384 × 640. Next to overcome capacity limitations of the computer system, all the PET scans were rebinned to a matrix size of 192 × 192 × 80 with a voxel size of 4 mm for EARL1 and 2 mm for EARL2 scans in all directions.

Appendix B.2. CNN Architecture

Noise reduction inevitably degrades some of the quantitative features of the PET image. To repress this, the network uses symmetric connections (concatenate two layers) while decoding (upsampling) to alleviate the loss of details during encoding (downsampling). So, the decoding layers use the features from the previous layer (encoded scans) but also the retained details from the downsampling layers (uncompressed scans). This results in a network that increases the SNR of the PET scans, and simultaneously retain the quantitative and qualitative features of the scan.

The proposed model was implemented with the Keras library (v2.2) in Python (v3.6), which is based on Tensorflow (v.1.13.1) as backend. The model was trained, validated, and tested on two NV-linked Nvidia 11GB RTX 2080Ti GPUs. For optimisation of the weights, an Adam optimizer was used with a low learning rate of 1×10^{-5} with a decay of 1×10^{-6}. The batch size for training the CNNs was set to 2.

Box A1. Python code of the architecture of the U-Net

```
kernel_size = (6,6,6)
inputShape = (192,192,80,1)
inputs = Input(inputShape)
model = Convolution3D(16,kernel_size,strides = 1,padding = 'same')(inputs)
model = Activation('relu')(model)
model = Convolution3D(16,kernel_size,strides = 1,padding = 'same')(model)
model_1 = Activation('relu')(model)
model = Convolution3D(32,kernel_size,strides = 2,padding = 'same')(model_1)
model = Activation('relu')(model)
model = Convolution3D(32,kernel_size,strides = 1,padding = 'same')(model)
model_2 = Activation('relu')(model)
model = Convolution3D(64,kernel_size,strides = 2,padding = 'same')(model_2)
model = Activation('relu')(model)
model = Convolution3D(64,kernel_size,strides = 1,padding = 'same')(model)
model_3 = Activation('relu')(model)
model = Convolution3D(128,kernel_size,strides = 2,padding = 'same')(model_3)
model = Activation('relu')(model)
model = Convolution3D(128,kernel_size,strides = 1,padding = 'same')(model)
model = Activation('relu')(model)
model = UpSampling3D((2,2,2))(model)
model = concatenate([model_3,model])
model = Convolution3D(64,kernel_size,strides = 1,padding = 'same')(model)
model = Activation('relu')(model)
model = UpSampling3D((2,2,2))(model)
model = concatenate([model_2,model])
model = Convolution3D(32,kernel_size,strides = 1,padding = 'same')(model)
model = Activation('relu')(model)
model = UpSampling3D((2,2,2))(model)
model = concatenate([model_1,model])
model = Convolution3D(16,kernel_size,strides = 1,padding = 'same')(model)
model = Activation('relu')(model)
model = Convolution3D(1,kernel_size,strides = 1,padding = 'same')(model)
model = Activation('relu')(model)
```

Appendix B.3. Model Performance

For training and validation, the model performance of the CNNs was measured using structural similarity (SSIM) and peak-signal-to-noise ratio (PSNR) [30]. The PSNR represent the peak signal error, whereas the SSIM is a measure of the similarity between two scans, which have been proven to be consistent with human-eye perception. Based on these two metrics the optimal (highest PSNR and SSIM) trained SLC-CNN and LC-CNN weights were chosen for further assessment using the test PET scans.

References

1. Almuhaideb, A.; Papathanasiou, N.; Bomanji, J. 18F-FDG PET/CT imaging in oncology. *Ann. Saudi Med.* **2011**, *31*, 3–13. (In English) [CrossRef]
2. Budak, E.; Çok, G.; Akgün, A. The Contribution of Fluorine (18)F-FDG PET/CT to Lung Cancer Diagnosis, Staging and Treatment Planning. *Mol. Imaging Radionucl. Ther.* **2018**, *27*, 73–80. (In English) [CrossRef]
3. Verhagen, A.; Bootsma, G.; Tjan-Heijnen, V.; Van Der Wilt, G.; Cox, A.; Brouwer, M.; Corstens, F.; Oyen, W. FDG-PET in staging lung cancer: How does it change the algorithm? *Lung Cancer* **2004**, *44*, 175–181. (In English) [CrossRef] [PubMed]
4. Jauw, Y.W.; O'Donoghue, J.A.; Zijlstra, J.M.; Hoekstra, O.S.; Menke-Van Der Houven, C.W.; Morschhauser, F.; Carrasquillo, J.A.; Zweegman, S.; Pandit-Taskar, N.; Lammertsma, A.A.; et al. (89)Zr-Immuno-PET: Toward a Noninvasive Clinical Tool to Measure Target Engagement of Therapeutic Antibodies In Vivo. *J. Nucl. Med. Off. Publ. Soc. Nucl. Med.* **2019**, *60*, 1825–1832. (In English) [CrossRef]
5. Tsutsui, Y.; Awamoto, S.; Himuro, K.; Umezu, Y.; Baba, S.; Sasaki, M. Characteristics of Smoothing Filters to Achieve the Guideline Recommended Positron Emission Tomography Image without Harmonization. *Asia Ocean J. Nucl. Med. Biol.* **2018**, *6*, 15–23. (In English) [CrossRef] [PubMed]
6. Soret, M.; Bacharach, S.L.; Buvat, I. Partial-Volume Effect in PET Tumor Imaging. *J. Nucl. Med.* **2007**, *48*, 932–945. [CrossRef]
7. Hofheinz, F.; Langner, J.; Beuthien-Baumann, B.; Oehme, L.; Steinbach, J.; Kotzerke, J.; Van den Hoff, J. Suitability of bilateral filtering for edge-preserving noise reduction in PET. *EJNMMI Res.* **2011**, *1*, 23. (In English) [CrossRef]
8. de Vries, B.M.; Golla, S.S.V.; Ebenau, J.; Verfaillie, S.C.J.; Timmers, T.; Heeman, F.; Cysouw, M.C.F.; van Berckel, B.N.M.; van der Flier, W.M.; Yaqub, M.; et al. Classification of negative and positive 18F-florbetapir brain PET studies in subjective cognitive decline patients using a convolutional neural network. *Eur. J. Nucl. Med. Mol. Imaging* **2021**, *48*, 721–728. [CrossRef]
9. Liu, M.; Cheng, D.; Yan, W. Classification of Alzheimer's Disease by Combination of Convolutional and Recurrent Neural Networks Using FDG-PET Images. *Front. Neuroinformatics* **2018**, *12*, 35. (In English) [CrossRef]
10. Milletari, F.; Navab, N.; Ahmadi, S.-A. V-Net: Fully Convolutional Neural Networks for Volumetric Medical Image Segmentation. In Proceedings of the 2016 Fourth International Conference on 3D Vision (3DV), Stanford, CA, USA, 25–28 October 2016; pp. 565–571.
11. Ronneberger, O.; Fischer, P.; Brox, T. U-net: Convolutional networks for biomedical image segmentation. In *Proceedings of the International Conference on Medical Image Computing and Computer-Assisted Intervention, Munich, Germany, 5–9 October 2015*; Springer: Berlin/Heidelberg, Germany, 2015; pp. 234–241.
12. Gong, K.; Guan, J.; Liu, C.-C.; Qi, J. PET Image Denoising Using a Deep Neural Network Through Fine Tuning. *IEEE Trans. Radiat. Plasma Med. Sci.* **2018**, *3*, 153–161. [CrossRef]
13. Hashimoto, F.; Ohba, H.; Ote, K.; Teramoto, A.; Tsukada, H. Dynamic PET Image Denoising Using Deep Convolutional Neural Networks Without Prior Training Datasets. *IEEE Access* **2019**, *7*, 96594–96603. [CrossRef]
14. Cui, J.; Gong, K.; Guo, N.; Wu, C.; Meng, X.; Kim, K.; Zheng, K.; Wu, Z.; Fu, L.; Xu, B.; et al. PET image denoising using unsupervised deep learning. *Eur. J. Nucl. Med. Mol. Imaging* **2019**, *46*, 2780–2789. [CrossRef] [PubMed]
15. Matsubara, K.; Ibaraki, M.; Nemoto, M.; Watabe, H.; Kimura, Y. A review on AI in PET imaging. *Ann. Nucl. Med.* **2022**, *36*, 133–143. [CrossRef]
16. Schaefferkoetter, J.; Yan, J.; Ortega, C.; Sertic, A.; Lechtman, E.; Eshet, Y.; Metser, U.; Veit-Haibach, P. Convolutional neural networks for improving image quality with noisy PET data. *EJNMMI Res.* **2020**, *10*, 105. [CrossRef] [PubMed]
17. Spuhler, K.; Serrano-Sosa, M.; Cattell, R.; DeLorenzo, C.; Huang, C. Full-count PET recovery from low-count image using a dilated convolutional neural network. *Med. Phys.* **2020**, *47*, 4928–4938. [CrossRef]
18. Chen, K.T.; Gong, E.; Macruz, F.B.D.C.; Xu, J.; Boumis, A.; Khalighi, M.; Poston, K.L.; Sha, S.J.; Greicius, M.D.; Mormino, E.; et al. Ultra-Low-Dose 18F-Florbetaben Amyloid PET Imaging Using Deep Learning with Multi-Contrast MRI Inputs. *Radiology* **2019**, *290*, 649–656. [CrossRef]
19. Lu, W.; A Onofrey, J.; Lu, Y.; Shi, L.; Ma, T.; Liu, Y.; Liu, C. An investigation of quantitative accuracy for deep learning based denoising in oncological PET. *Phys. Med. Biol.* **2019**, *64*, 165019. [CrossRef]
20. Pfaehler, E.; Mesotten, L.; Zhovannik, I.; Pieplenbosch, S.; Thomeer, M.; Vanhove, K.; Adriaensens, P.; Boellaard, R. Plausibility and redundancy analysis to select FDG-PET textural features in non-small cell lung cancer. *Med. Phys.* **2021**, *48*, 1226–1238. (In English) [CrossRef]
21. Jauw, Y.W.S.; Zijlstra, J.M.; De Jong, D.; Vugts, D.J.; Zweegman, S.; Hoekstra, O.S.; Van Dongen, G.A.M.S.; Huisman, M.C. Performance of 89Zr-Labeled-Rituximab-PET as an Imaging Biomarker to Assess CD20 Targeting: A Pilot Study in Patients with Relapsed/Refractory Diffuse Large B Cell Lymphoma. *PLoS ONE* **2017**, *12*, e0169828. (In English) [CrossRef]
22. Makris, N.E.; Boellaard, R.; Visser, E.P.; de Jong, J.R.; Vanderlinden, B.; Wierts, R.; van der Veen, B.J.; Greuter, H.J.; Vugts, D.J.; van Dongen, G.A.; et al. Multicenter harmonization of 89Zr PET/CT performance. *J. Nucl. Med. Off. Publ. Soc. Nucl. Med.* **2014**, *55*, 264–267. [CrossRef]

23. Pfaehler, E.; Jong, J.; Dierckx, R.; van Velden, F.; Boellaard, R. SMART (SiMulAtion and ReconsTruction) PET: An efficient PET simulation-reconstruction tool. *EJNMMI Phys.* **2018**, *5*, 16. [CrossRef] [PubMed]
24. Brownlee, J. A Gentle Introduction to Pooling Layers for Convolutional Neural Networks. Deep Learning for Computer Vision. Available online: https://machinelearningmastery.com/pooling-layers-for-convolutional-neural-networks/ (accessed on 18 December 2021).
25. Chen, J.; Shen, Y. The effect of kernel size of CNNs for lung nodule classification. In Proceedings of the 2017 9th International Conference on Advanced Infocomm Technology (ICAIT), Chengdu, China, 22–24 November 2017; pp. 340–344. [CrossRef]
26. Boellaard, R. Quantitative oncology molecular analysis suite: ACCURATE. *J. Nucl. Med.* **2018**, *59*, 1753.
27. Boellaard, R. Accurate: An Oncology PET/CT Quantitative Analysis Tool. Available online: https://doi.org/10.5281/zenodo.3908203 (accessed on 19 January 2022).
28. Hou, W.; Wang, W.; Liu, R.; Lu, T. Cropout: A General Mechanism for Reducing Overfitting on Convolutional Neural Networks. In Proceedings of the 2019 International Joint Conference on Neural Networks (IJCNN), Budapest, Hungary, 14–19 July 2019; pp. 1–8. [CrossRef]
29. Johnson, J.M.; Khoshgoftaar, T.M. Survey on deep learning with class imbalance. *J. Big Data* **2019**, *6*, 27. [CrossRef]
30. Horé, A.; Ziou, D. Image Quality Metrics: PSNR vs. SSIM. In Proceedings of the 2010 20th International Conference on Pattern Recognition, Istanbul, Turkey, 23–26 August 2010; pp. 2366–2369. [CrossRef]

cancers

MDPI

Article

Thermal Ablation of Liver Tumors Guided by Augmented Reality: An Initial Clinical Experience

Marco Solbiati [1], Tiziana Ierace [2], Riccardo Muglia [3], Vittorio Pedicini [2], Roberto Iezzi [4], Katia M. Passera [1], Alessandro C. Rotilio [1], S. Nahum Goldberg [5] and Luigi A. Solbiati [2,6,*]

1 R&D Unit, R.A.W. Srl, 20127 Milano, Italy; m.solbiati@endo-sight.it (M.S.); k.passera@endo-sight.it (K.M.P.); a.rotilio@endo-sight.it (A.C.R.)

2 Department of Radiology, IRCCS Humanitas Research Hospital, Rozzano, 20089 Milan, Italy; t.ierace@tin.it (T.I.); vittorio.pedicini@humanitas.it (V.P.)

3 Department of Radiology, ASST Papa Giovanni XXIII, 24127 Bergamo, Italy; rmuglia@asst-pg23.it

4 Department of Diagnostic Imaging, Oncologic Radiotherapy and Hematology, IRCCS Policlinico Universitario A. Gemelli, Università Cattolica del Sacro Cuore, 00168 Rome, Italy; roberto.iezzi.md@gmail.com

5 Department of Radiology, Hadassah Hebrew University Medical Centre, Jerusalem 90221, Israel; sgoldber@bidmc.harvard.edu

6 Department of Biomedical Sciences, Humanitas University Pieve Emanuele, 20089 Milan, Italy

* Correspondence: lusolbia@gmail.com; Tel.: +39-3358134341

Simple Summary: We report the first clinical use of Endosight, a new guidance system for percutaneous interventional procedures based on augmented reality, to guide percutaneous thermal ablations. The new system was demonstrated to be precise and reliable, with a targeting accuracy of 3.4 mm. Clinically acceptable, rapid setup and procedural times can be achieved.

Abstract: Background: Over the last two decades, augmented reality (AR) has been used as a visualization tool in many medical fields in order to increase precision, limit the radiation dose, and decrease the variability among operators. Here, we report the first in vivo study of a novel AR system for the guidance of percutaneous interventional oncology procedures. Methods: Eight patients with 15 liver tumors (0.7–3.0 cm, mean 1.56 + 0.55) underwent percutaneous thermal ablations using AR guidance (i.e., the Endosight system). Prior to the intervention, the patients were evaluated with US and CT. The targeted nodules were segmented and three-dimensionally (3D) reconstructed from CT images, and the probe trajectory to the target was defined. The procedures were guided solely by AR, with the position of the probe tip was subsequently confirmed by conventional imaging. The primary endpoints were the targeting accuracy, the system setup time, and targeting time (i.e., from the target visualization to the correct needle insertion). The technical success was also evaluated and validated by co-registration software. Upon completion, the operators were assessed for cybersickness or other symptoms related to the use of AR. Results: Rapid system setup and procedural targeting times were noted (mean 14.3 min; 12.0–17.2 min; 4.3 min, 3.2–5.7 min, mean, respectively). The high targeting accuracy (3.4 mm; 2.6–4.2 mm, mean) was accompanied by technical success in all 15 lesions (i.e., the complete ablation of the tumor and 13/15 lesions with a >90% 5-mm periablational margin). No intra/periprocedural complications or operator cybersickness were observed. Conclusions: AR guidance is highly accurate, and allows for the confident performance of percutaneous thermal ablations.

Keywords: augmented reality; three-dimensional (3D) reconstruction; interventional oncology; computed tomography; liver

Citation: Solbiati, M.; Ierace, T.; Muglia, R.; Pedicini, V.; Iezzi, R.; Passera, K.M.; Rotilio, A.C.; Goldberg, S.N.; Solbiati, L.A. Thermal Ablation of Liver Tumors Guided by Augmented Reality: An Initial Clinical Experience. *Cancers* **2022**, *14*, 1312. https://doi.org/10.3390/cancers14051312

Academic Editor: David Wong

Received: 21 January 2022
Accepted: 27 February 2022
Published: 3 March 2022

1. Introduction

Precision and targeting accuracy are key for the success of all image-guided interventional procedures. Over the last 20 years, several new navigational tools have been added to conventional imaging modalities (ultrasound, CT, MRI) with the purpose of increasing precision, favouring dose reduction, decreasing variability among operators, and thus promoting the diffusion of diagnostic and therapeutic interventional procedures based on ever-increasing reliability. Image fusion platforms based on electromagnetic or optical devices [1–4], CT with laser marker systems [5], CT fluoroscopy [6], cone-beam CT [7,8], CT with electromagnetic tracking [9], and robotic systems [10] have been incorporated into clinical practice in many centers. However, these tools still have some limitations, such as the inability to provide a real, live, 3D visualization of the target and the surrounding structures, the need for the operator to alternate their gaze between the interventional field and the instrumentation screen(s), a steep learning curve, and, for CT-guided procedures, potentially substantial radiation doses to patients and operators [11]. Recently, spatial computing technology has allowed the development of simulated reality environments, virtual reality (VR) and augmented reality (AR), which enable real-time interaction by the user. VR completely immerses the user in an artificial, digitally created 3D world through head-mounted displays (HMDs), with the user having no direct interaction with the real world. Therefore, in the medical field, VR can be used for surgical planning and simulation, but not for the direct guidance of interventional procedures [11,12]. To the contrary, AR overlays digital content onto the visualized real world through an external device [12–14], enhancing reality with superimposed information, using optical see-through head-mounted displays (HMDs or "goggles"), screens, smartphones, tablets and videoprojectors, such that digital and physical objects are visualized simultaneously. This permits their interaction with each other, thus allowing guidance of interventional procedures. The capability for computers to enhance visibility and navigate through 3D coordinates during minimally invasive interventional procedures was first noted in 1997 [15]. Since then, AR has been clinically applied as a visualization tool to augment anatomical [16] and pathological structures in neurosurgery [17–19] and vascular [20,21], orthopedic [22,23], urologic [24–26], plastic [27], and abdominal surgery [28,29]. This was achieved by creating 3D anatomic volumes from cross-sectional scans or angiographic images, and manually overlapping them over patients positioned in the real operating field [3] through electromagnetic or optical tracking systems and computer vision algorithms. In Interventional Oncology, AR was initially tested on phantoms to assist with percutaneous biopsies [30,31], and subsequently for the assessment of its potential role for the augmentation of minimally invasive surgery for the accurate localization of organ, or the guidance of radiofrequency ablation (RFA) or irreversible electroporation (IRE) electrodes on phantoms [32,33], but not for the direct guidance of interventional procedures in humans. To our knowledge, this is the first report of the targeting and ablation of small hepatic malignancies in human patients using AR as the sole modality of guidance.

2. Materials and Methods

This study was performed at two tertiary referral centres for liver diseases (Humanitas Research Hospital and IRCCS Policlinico Universitario A. Gemelli), with the approval of the local Institutional Ethics Committees. Written informed consent was obtained from all of the subjects involved in the study.

2.1. Patient Population

Fifteen hepatic malignancies (9 hepatocellular carcinomas (HCCs), 3 metastases from breast carcinoma, and 3 from pancreatic adenocarcinoma) in eight patients (5 males and 3 females, median age 72.5 years, range 56–83) underwent AR-guided percutaneous thermal ablation. The treated nodule size ranged from 0.7 to 3.0 cm (mean 1.56 + 0.55).

For all of the cases, the treatment decision was determined by the consensus of an Institutional Multidisciplinary Liver Team. According to the BCLC classification, the nine HCCs in five patients were either very early (8/9 cases) or early stage (1/9), in a subset of HCV-related early stage cirrhosis (Child-Pugh A, ECOG PS 0) [34]. These were located in segments VIII (n = 4), V (n = 2), II (n = 2) and VI (n = 1); the sizes ranged from 1.2 to 3.0 cm (mean 1.69 + 0.53). One patient had four HCCs, and one two HCCs. All of the nodules were treated in the same session. The other three patients had only one HCC. All of the HCCs were diagnosed through a non-invasive radiological work-up, following the European Association of the Study of the Liver (EASL) 2018 clinical practice guidelines [35].

The six metastases in the three patients ranged from 0.7 to 2.1 cm (mean 1.35 cm + 0.56) in size, and were diagnosed by percutaneous US-guided biopsies using 20 G Menghini-modified needles (Sterylab, Milan, Italy).

2.2. Pre-Treatment Diagnostic Assessment

All of the patients were initially evaluated with a baseline ultrasound of the liver which included contrast enhanced ultrasound (CEUS) after the intravenous administration of 2.4 to 4.8 mL second-generation contrast agent (SonoVue, Bracco, Milan, Italy) (Figure 1A), and an abdominal contrast enhanced computed tomography (CECT) in the arterial, portal, and late phases (Figure 1B). In order to achieve registration for the orientation reference of the AR display, twenty radiopaque markers with no repetitive pattern were applied to the abdominal skin in the right hypochondrium surrounding the area of interest (Figure 1C) immediately prior to the treatment. A new CECT in the arterial and portal phases was acquired during free breathing (i.e., normal respiration), paying particular attention to include all of the markers within the scanning area. In 14 of the 15 patients, CT scans were acquired with two different machines (Ingenuity, Philips Healthcare, Cleveland, OH, USA for 4 patients, and Revolution EVO, General Electric, Boston, MA, USA for 3 patients) following the injection of Iopamidol (Iopamiro 370, Bracco, Milan, Italy) at 4 mL/s, using a 2-mm slice thickness, a matrix of 512 × 512 pixels, an in-plane pixel size of 0.48–0.78 mm, 1:1 pitch, 120 kVp and 180 mA. In the last patient, 70 mL Iomeprol (Iomeron 400 mg/mL, Bracco, Milan, Italy) was injected at 3 mL/s using Lightspeed VCT 64 (General Electric, Boston, MA, USA) using a 2.5-mm slice thickness, a matrix of 512 × 512 pixels, an in-plane pixel size of 0.48–0.78 mm, 1:1 pitch, 120 kVp and 180 mA.

2.3. Augmented Reality Settings

The AR set-up comprised a proprietary augmented reality system (Endosight, R.A.W. Srl, Milan, Italy) that features a 27″ medical display (ACL, Leipzig, Germany), a laptop (Dell Technologies, Round Rock, TX, USA) with installed proprietary image processing and augmented reality software, and a commercially available head-mounted display (HMD) (Oculus Rift-S, Facebook Technologies, Menlo Park, CA, USA) paired with a binocular camera (Zed Mini, Stereolabs, San Francisco, CA, USA) (Figure 2).

The binocular camera viewed the patient from two different angles in order to register the patient model in the camera frame using the markers visible in both video images while tracking the ablation applicator. The software enabled the 3D reconstruction (from CT scans to 3D volumes), co-registration, and AR intervention. Specifically, after uploading the CECT scans into the system, followed by the automatic segmentation and 3D reconstruction of the liver, spleen, bones, liver blood vessels and radiopaque markers, the semi-automatic segmentation of the target lesions occurred using proprietary reconstruction algorithms. In addition, the most suitable trajectory path from the skin to the target was defined. Subsequently, by moving the HMD around the patient, the system software co-registered (matched) all of the radiopaque markers segmented on the CT scans with all of the real markers applied to the patient's skin. This allowed for the simultaneous visualization of the patient's surface and internal anatomy, the target lesion, and the trajectory path to the target in 3D, by superimposing—in real-time—virtual images on the operator's real field of sight (Figure 1D). Next, in order to allow the visualization of the probe position during the procedure, a clip with five markers with no repetitive pattern was attached to either a 14 G (for 14 ablations) or a 11 G (for one ablation) coaxial needle, 7.8 cm in length (Bard Inc., Murray Hill, New York, NY, USA), that was used as a coaxial ablation device introducer (Figure 1E).

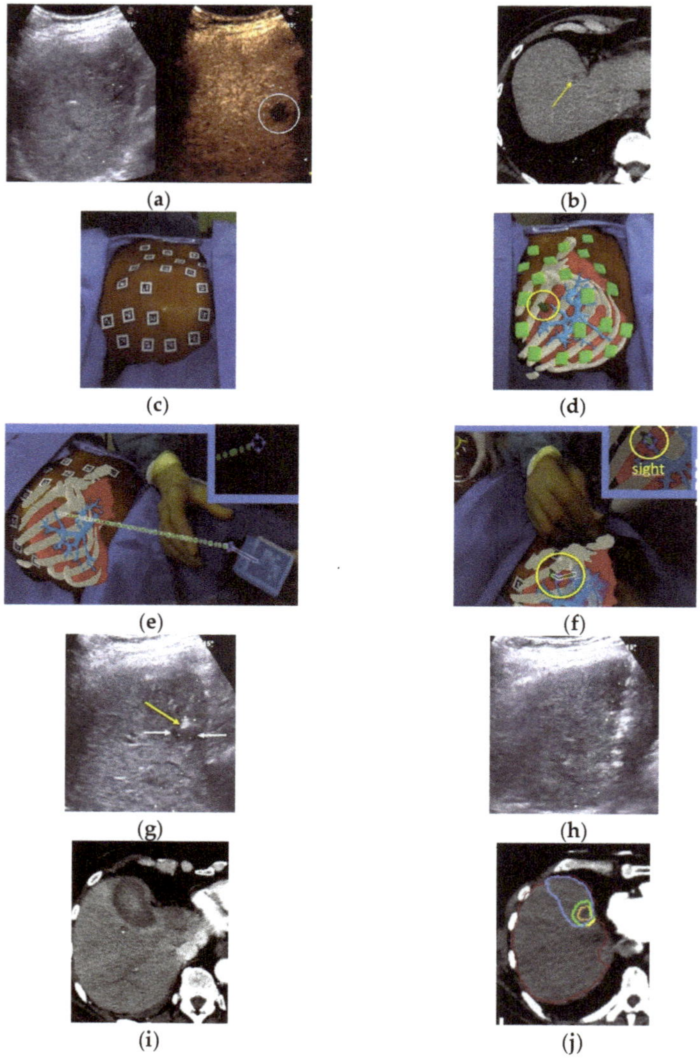

Figure 1. Augmented reality guided ablation: a 1.5-cm pancreatic carcinoma metastasis at segment VIII, poorly visible on B-mode US and clearly seen by CEUS (**a**), and seen on pre-ablation CT scan (arrow) (**b**). Radiopaque markers with no repetitive pattern applied to the patient's skin (**c**). View through the operator's HMD: ribs (in white), major hepatic blood vessels (light blue), liver (red), and target lesion (green, in a yellow circle) (**d**). View through the HMD, showing that the operator can see the virtual needle (blue line) and the line that connects the tip of the needle to the center of the target (in green) (**e**). Following the trajectory line permits successful tumor targeting with AR guidance alone (**f**). The 5.4-mm distance between the tip of the coaxial needle and the target center by US (**g**). Subsequently, the microwave antenna is inserted into the coaxial needle (**h**). On a post-ablation CT scan, a large ablation volume completely surrounds the metastasis (**i**). Using ablation confirmation software (Ablation-fit^{TM}), the technical success achieved was precisely demonstrated. The margins of the target tumor are shown in orange, the 5-mm ablation margin is shown in green, and the margins of the necrosis volume are shown in blue. Complete tumor ablation with only 5.4% of the safety margin out of the necrosis volume was achieved (**j**).

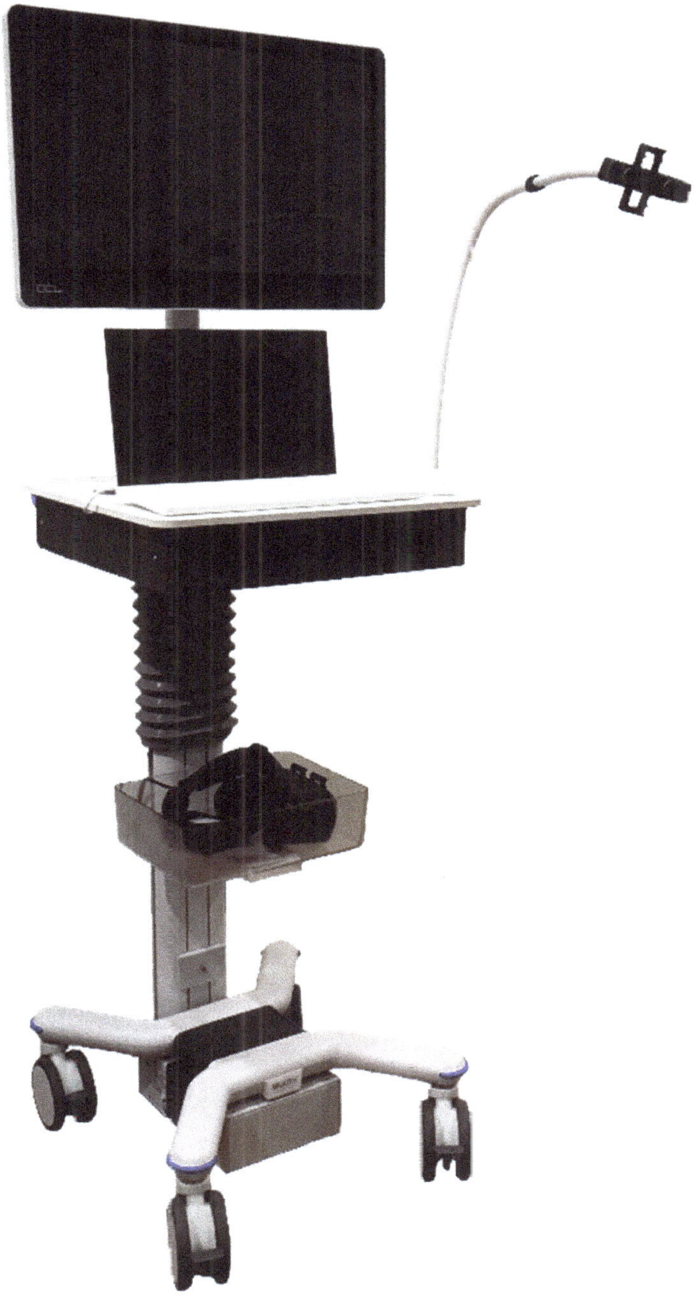

Figure 2. Endosight system overview: cart, medical display, laptop, and Oculus Rift-S paired with a Zed Mini camera.

2.4. Treatment Procedure

All of the procedures were performed by three interventional radiologists with more than 15 years of experience in percutaneous thermal ablations. In 14 of the 15 patients, the ablations were performed in the CT room coupled with real-time ultrasound, under assisted ventilation, during short-acting anaesthesia using propofol (AstraZeneca, Cambridge, UK) (10 mg/mL) and alfentanil (Hameln Pharma, Gloucester, UK) (0.5 mg/mL), with continuous hemodynamic monitoring throughout the procedure. In the remaining patient, the ablation was performed under direct CT control (Lightspeed VCT 64) after local anesthesia and deep sedation with 0.2 mg Fentanyl (Janssen-Cilag, Beerse, Belgium) without additional ultrasound guidance. Using AR guidance alone, the coaxial needle was inserted following the predefined trajectory line planned during the setup (Figure 1F). This was facilitated by color coding, in that when the predefined trajectory line overlapped the virtual needle line, this path turned from blue to green in the AR visual field, highlighting and denoting the correct alignment. The insertion was conducted during the patient's free breathing (as in the pre-ablation acquisition of the CT scans) in order to minimize the organ displacement caused by breathing. The depth from the entry point (i.e., the skin) to the target centre was measured in real-time by the software, and was visualized on the operator's HMD. Before the introduction of the ablation device into the coaxial needle, the position of the coaxial needle and its correspondence with the real location of the target nodule was verified using real-time US when the target nodule was visible with US, or with CT when the target was invisible on US. In order to assess the precision of the AR, the distance from the real target centre visualized on the US or CT and the virtual target centre shown by the trajectory line starting from the tip of the coaxial needle was measured (Figure 1G). The ablation probe was then inserted, positioning its tip 5–7 mm beyond the deep margin of the target in order to achieve sufficient ablative margins (Figure 1H). Then, the coaxial needle was partly retracted while maintaining the positioning of the ablation device in order to achieve the complete exposition of the active tip. Microwave ablations (MWA) were performed with 13 G, 15 cm-long antennae (Medtronic, Dublin, Ireland) for three malignancies of three patients, and 14 G, 15 cm-long antennae (HS Hospital Service, Aprilia, Italy) for eleven nodules of five patients. The remaining patient recieved RFA performed with a 14 G, 15 cm-long electrode with a 3-cm exposed tip (RF Medical, Seoul, Korea). The treatment power and duration, and the total amount of energy delivered were selected based upon the size and location of each nodule, according to the device manufacturer's technical recommendation and operator experience. Figure 3 shows the complete treatment procedure workflow.

2.5. Post-Procedural Assessment

The CECT was performed immediately after withdrawing the ablation device (Figure 1I). A proprietary ablation-confirmation software (Ablation-fit^{TM}, R.A.W. Srl, Milan, Italy) [36]—whichl enables the automatic segmentation of the liver and intrahepatic blood vessels, and semi-automatically co-registers the target nodules on pre-ablation CT scans with the volumes of necrosis achieved on post-ablation scans using a non-rigid registration tool—was used in order to assess the precision and completeness of the ablation volume achieved (Figure 1J). Using a 3D model, the software verified whether the volume of ablative necrosis included entirely or partially the tumor and a pre-defined ablative margin (5-mm thick, in these cases), as well as quantifying, as a percentage, the amount of tumor and ablative margin (if any) external to the ablation volume, thus allowing us to assess the technical success of the procedure [37,38]. After the ablation, all of the operators were interviewed regarding the need for manual adjustments of the HMDs and the occurrence of eye fatigue, dizziness, or cybersickness.

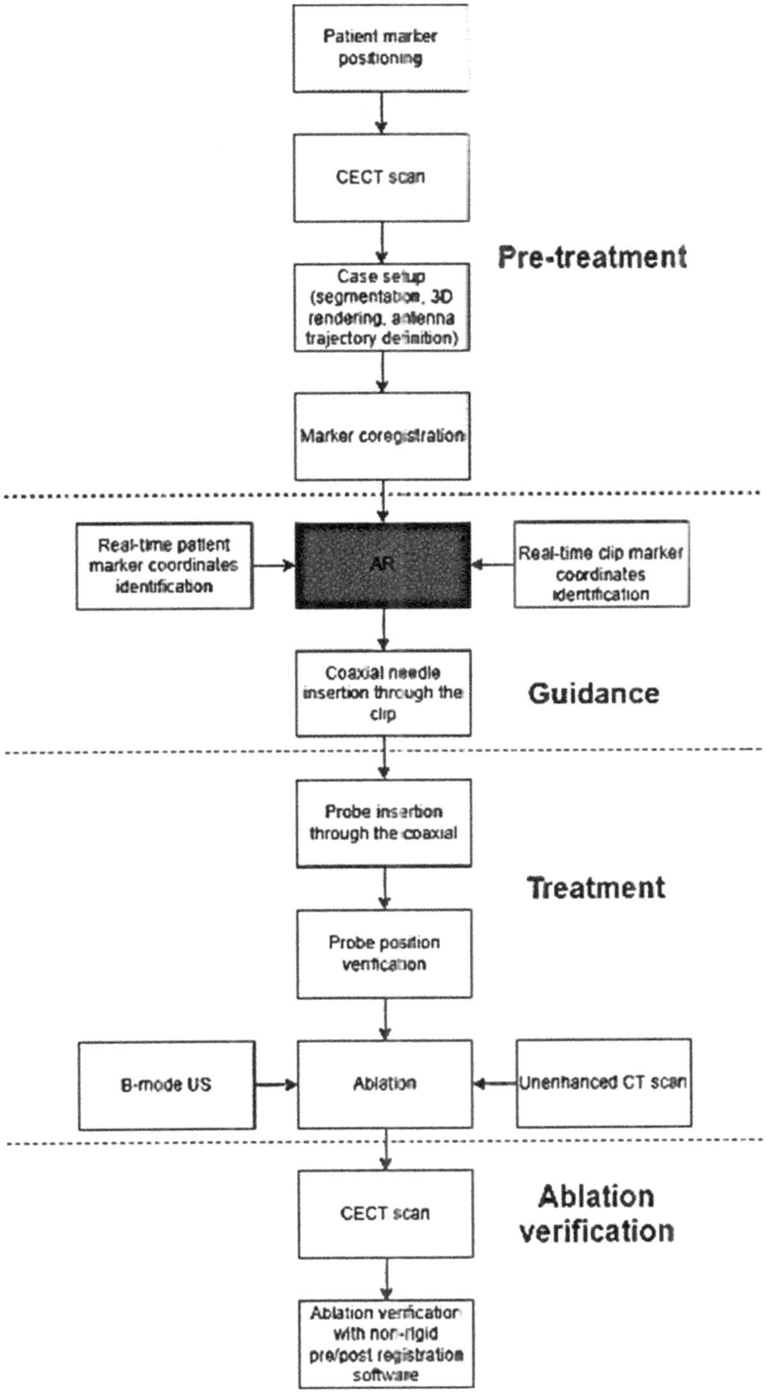

Figure 3. Workflow of the AR-guided thermal ablations.

2.6. Statistical Analysis

The primary endpoints evaluated included the time required to set up the system and to position the antenna tip inside the nodule, the mean depth of the target centre from the needle entry point on the skin, and the deployment accuracy, defined as the mean distance between the geometric center of the target and the ablation device tip measured on unenhanced CT or US. Secondary endpoints included the technical success, i.e., the complete ablation of the entire tumor and the achievement of an >90% 5-mm periablational margin ablation [36], complications, and operator sensations regarding the procedure. The data were analyzed with statistical software (SPSS, version 17.0), and were reported as the mean ± standard deviation (SD), or as the mean and range.

3. Results

The time required to set up the system ranged from 12.0 to 17.2 min (12.3 ± 2.1 min), and the time required to perform each insertion and tumor targeting ranged from 3.2 to 5.7 min (4.3 ± 0.9 min). In 7 of the 15 (46.7%) cases, the target nodule was visible on the US, and the real location of the target nodule and the position of the coaxial needle tip in respect to the target centre were verified using real-time US. In the remaining 8 of the 15 (53.3%) cases, unenhanced CT was employed for verification. The mean depth of the target centre from the needle entry point on the skin was 76.0 ± 28.2 mm. The distance between the geometric center of the target and the ablation device tip measured on unenhanced CT or US ranged from 2.1 to 4.5 mm (3.2 ± 0.7 mm). Table 1 shows—for each target—the size, the distance of the interventional device tip from the tumor center, the time taken to reach the target, and the modality used for the verification.

Table 1. Sizes of the targets, the distance of the interventional device tip from the center of each target tumor, the time needed to reach the target, and the modality used for the distance measurement.

	Size [mm]	Distance from Target Center [mm]	Time to Reach Target [min]	Modality Used for Measurement
Patient 1—Target 1	1.8	3.1	3.3	US
Patient2—Target 1	1.8	3.8	4.1	US
Patient 3—Target 1	1.5	2.1	5.7	CT
Patient 3—Target 2	1.7	2.4	3.2	CT
Patient 3—Target 3	1.4	3.6	4.9	CT
Patient 3—Target 4	1.2	2.7	4.2	CT
Patient 4—Target 1	1.4	3.9	5.3	US
Patient 4—Target 2	1.4	2.9	3.4	US
Patient 5—Target 1	2.1	3.6	5.3	CT
Patient 6—Target 1	1.8	2.4	4.0	CT
Patient 6—Target 2	0.8	2,2	4.2	CT
Patient 7—Target 1	3.0	4.5	5.2	CT
Patient 8—Target 1	1.5	4.1	3.3	US
Patient 8—Target 2	1.2	3.1	3.6	US
Patient 8—Target 3	0.7	3.4	4.9	US
Overall:	1.56 ± 0.55 mm	3.2 ± 0.7 mm	4.3 ± 0.9	

For the MWA, the power delivered ranged from 50 to 60 W, with a treatment duration of 5 min in four HCCs, and 6 min in the remaining five HCCs and the five metastases. For the case of radiofrequency ablation (RFA), the power delivered was 1500 mA for 12 min. A single ablation device insertion was performed for each target tumor. Technical success was achieved in each case. After the automatic coregistration of the 3D volumes of the pre-ablation tumors and post-ablation necrotic changes, achieved with the Ablation-fit™ software, the complete ablation of the tumors (i.e., no residual unablated portion of the target tumors) was found. The residual 5-mm ablative margin percentage ranged from 0 to 14.1% (5.5 ± 4.3%), with 13 of the 15 (86.7%) patients showing >90% ablation of this margin. Table 2 shows the residual 5-mm safety margin (in percentage) of each target lesion.

Table 2. Residual 5-mm safety margin (as a percentage) of each target tumor, calculated by the Ablation-fit™ software.

	Residual 5 mm Safety Margin [%]
Patient 1—Target 1	5.4
Patient 2—Target 1	2.8
Patient 3—Target 1	3.1
Patient 3—Target 2	9.2
Patient 3—Target 3	12.1
Patient 3—Target 4	1.9
Patient 4—Target 1	0
Patient 4—Target 2	4.9
Patient 5—Target 1	8.1
Patient 6—Target 1	14.1
Patient 6—Target 2	10.1
Patient 7—Target 1	4.1
Patient 8—Target 1	3.3
Patient 8—Target 2	3.1
Patient 8—Target 3	0

No intra- or periprocedural adverse events occurred. No user-dependent calibration and adjustment for the HMD was needed, and no significant eye fatigue or "cybersickness" was reported by any of the users.

4. Discussion

Modern imaging modalities enable the visualization of increasingly small target lesions, often in difficult-to-target locations, which is particularly suitable for local, image-guided treatments (IGTs). Consequently, the requests for image-guided therapy, accompanied by expectations of favorable outcomes, are constantly increasing. However, some problems still remain unsolved. First of all, the learning curve for the use of these technologies is often long, and this limits the diffusion of interventional procedures, particularly among young operators and/or in low-referral centers. The lack of the real, live, 3D visualization of targets, and the poor working ergonomics (the need to check many screens simultaneously, restricted line-of-sight to screens, and the need to alternate the operator's gaze between the interventional field and the instrumentation screens) are additional important limitations. The mental registration of the target position seen in the reference image (US, CT, MRI) with the corresponding position in patients is often challenging, particularly for liver dome lesions requiring non-orthogonal or out-of-plane approaches, even when CT guidance is used. The difficulty and subjectivity of this process may also increase the risks for patients. Thus, the need for a technically easy combination of "real-world" visualization with virtual objects precisely superimposed upon the scene is increasingly desired. This can be achieved with AR technology in the actual interventional field, where the operator can visualize and interact simultaneously with the real world (patient, interventional instrumentation) and virtual objects (hidden organs and targets, surrounding structures seen on CT and MRI, etc.) based on the superimposition of the "two worlds", as displayed on

HMD, smartphones, tablets, screens or videoprojectors. Moreover, HMDs can be relatively advantageous compared to the direct line of sight through the lens display [39].

The most critical issue for the use of AR in medical applications is the superimposition precision, i.e., the registration accuracy. Multiple studies on phantoms, animal models, and human cadavers have primarily focused upon the assessment of registration accuracy, either for AR navigation [40] or the AR guidance of needles [19,31–33,41,42]. Hecht et al. [41] reported a smartphone-based AR system for needle trajectory planning and real-time guidance on phantoms. In their first experiment, the mean error of the needle insertion was 2.69 + 2.61 mm, which was 78% lower than the CT-guided freehand procedure. In their second experiment, the operators successfully navigated the needle tip within 5 mm on each first attempt under the guidance of the AR system, which eliminated the need for further needle adjustments. In addition, the procedural time was 66% lower than the CT-guided freehand procedure. Long et al. [42] compared the accuracy and the placement time needed by five interventional radiologists and a resident with a range of clinical experience (3–25 years) to place biopsy needles on millimetric targets positioned in an anthropomorphic abdominal phantom at different depths, using cone-beam CT (CBCT)-guided fluoroscopy, and smartphone- and smartglasses-based AR navigation platforms. The placement error was extremely small and virtually identical for all of the three modalities (4–5 mm), and the placement time was significantly shorter for smartphones and HMDs (38% and 55% respectively) than for CBCT. Additionally, the results were achieved by AR without intra-procedural radiation, and with a learning curve of only 15 min.

Using the same system employed for the present study, Solbiati et al. recently published a proof-of-concept study on phantoms, animal models, and human cadavers targeted with AR guidance. In the rigid phantom, sub–5-mm accuracy (2.0 + 1.5 mm) (mean + standard deviation) was achieved. In a porcine model with small (2 × 1 mm) metallic targets embedded, the accuracy was 3.9 + 0.4 mm when the targeting was performed with respiration suspended at maximum expiration, as in the initial CT scan, and 8.0 + 0.5 mm when the procedure was performed without breathing control. In a human cadaver attached to a ventilator to induce simulated respirations, two liver metastases (1.8 cm and 3.0 cm) were targeted with an accuracy of 2.5 mm and 2.8 mm, respectively [43].

Here, we note the similar accuracy of 3.4 mm in living, breathing patients. Regarding AR-guided needle insertions in human patients, De Paolis et al. [32] reported their preliminary experience in locating a focal liver lesion in the operating room just before open surgery. The surgeon was able to determine the correct position of the real tumor by touching it and applying the ablation applicator to it in order to verify the correct overlap between the virtual and the real tumor. Although an excellent accuracy of 2 mm was reported, problems of depth perception and instrument visibility occurred whenever the surgeon's body was located between the tracker and the instrument, both of which related to the use of the optical tracker.

The AR system used for our current report is specifically designed to guide percutaneous biopsies and ablation procedures. It is based on disposable markers with no repetitive pattern affixed to the patient's abdominal skin before performing the CT scans. The associated software enables us to visualize and segment the markers on the patient (virtual objects) and the target tumor, to automatically register and superimpose virtual and real images in real-time, to define the safe and accurate trajectory line to the target center, to depict the guided movements of the interventional device without the need for additional imaging, and to show the whole procedure on a display, HMD, or screen [43,44]. The main advantage of HMD is the 3D visualization, which tops the 2D visualization of smartphones and tablets. The results achieved were very promising: the accuracy of the antenna tip with respect to the center of the target was well below the 5-mm threshold (with a mean of 3.2 + 0.7 mm), and technical success of the ablation was achieved in all cases. The mean times required to set up the system and to perform each insertion were 14.3 min and 4.3 min, respectively, and were independent of the type of ablation system used. This is not substantially different from the time usually required to perform CT-guided procedures by expert radiologists, even after a long learning curve. Moreover, the software for the assessment and quantification of the tumor ablation margins in 3D was integrated into the AR system, enabling the immediate and accurate evaluation of the technical success [36].

In recent years, two issues have been raised regarding the technology of HMDs used for AR, i.e., the field of view (FOV) and the need for calibration [39]. The binocular FOV of human eyes is naturally about 200 o in the horizontal plane and 135 o in the vertical plane, while commercially available HMDs had initial FOVs ranging around 30–40 o. This limitation has recently increased to 90 o both horizontally and vertically. Nevertheless, the calibration of HMDs is needed to tailor projections to the user's interpupillary distance. Given that most HMDs have fixed focal planes, when the calibration is inaccurate, the eyes can focus and converge at separate distances, causing distorted depth perception, eye fatigue and "cybersickness" due to discrepancies between the visual and vestibular senses. Nowadays, commercially available HMDs are provided with two videocameras, which has eliminated the need for user-dependent calibration and adjustment. This has limited the common occurrence of the cybersickness which was reported previously, as noted in our study.

The patient's respiratory movement and motion remain one of the largest technical and practical hurdles, as AR guidance systems are currently unable to follow respiratory excursions in mobile organs with real-time corrections, bearing a risk of the shifting of the intended target relative to the expected location. Other target-related limitations arise from the abilities of lesions to warp or move within their environments. Respiratory motion tracking and the monitoring of respiration during deep sedation or general anesthesia seem to offer the best solutions to date. The guiding information is provided regularly at the point of the breathing that matches the respiratory phase during which the preoperative CT image was acquired (the middle respiratory or expiratory phase). In this time interval window, the operator can move the needle toward the target as rapidly as possible. In our study, the insertion was conducted during the patient's free breathing (as in the pre-ablation acquisition of the CT scans) in order to minimize the organ displacement caused by breathing. Given that this was the initial study of AR-guided thermal ablation, we selected only tumors which were visible on US or on CT in order to be able to check the position of the device tip after its insertion, before starting the ablation. Probe repositioning was never required, as the position achieved with AR guidance was always sufficiently accurate. Nevertheless, we acknowledge that this will not always be invariable, and note that—should minor placement corrections be needed—the virtual system will potentially save a substantial amount of radiation exposure compared to fully CT-guided procedures, be they CT-guided freehand, cone-beam CT, or CT fluoroscopy guidance [20]. Indeed, in the experimental study conducted by Park et al. [39] comparing a HoloLens-based 3D AR-assisted navigation system with CT-guided simulations, the AR system reduced the radiation dose by 41%.

An additional potential challenge of AR-guided interventional procedures is needle bending during the insertion, exacerbated by increased applied pressure or the use of thinner needles. The solution we successfully utilized was the use of a rigid coaxial needle to maintain the interventional device fixed in 3D space during its advancement, minimizing the bending of the ablation device inserted into the coaxial needle. We further demonstrated that the attachment of a clip with markers with no repetitive pattern to the coaxial needle permits precise monitoring by AR of the probe advancement towards the target, and the interventional device subsequently inserted into the coaxial needle can easily hit the target center. Coaxial needles have been used for interventional procedures for decades, and do not appreciably increase the risk of bleeding because their construction is engineered to result in an ultimate size only 1–2 G larger than that of ablation devices or biopsy needles.

With respect to other navigation systems, AR guidance offers an ergonomic advantage that the overlay of treatment information (anatomy, target, trajectory line, etc.) is shown directly in the procedural environment, and not on a display screen away from the patient on a monitor, as occurs with CT- or CBCT-guided fluoroscopy. Additional advantages of

AR guidance are the ease of use, the reduced procedural time compared to more traditional guidance systems, and the short learning curve (compared to that of CT-guided procedures), which is particularly useful for young operators with limited experience, who perform equally or even better than senior operators with long experience. Furthermore, AR guidance systems are significantly less expensive than all of the other needle guidance systems. This may favour the diffusion of AR, and consequently of image-guided procedures in small centers, and in developing countries that cannot afford to buy complex and expensive guidance technologies (the so called "democratization" of interventional procedures).

With AR, the same images seen by the operator can also be visualized on monitors inside and outside the interventional room, and can be broadcast on a larger scale, allowing interventional radiologists to visualize live or recorded procedures performed by experts. AR can provide not only an excellent opportunity for physician training and education but also a very useful tool to exchange collaborative experiences among various centers for remote real-time instruction or expert assistance [12,45].

We acknowledge that this study has some limitations, most notably the small number of patients within the cohort, and the non-randomized type of lesions treated, all of which visible on both US and CT despite their small size. Nonetheless, we believe that it will encourage new prospective studies, and will work as the basis for the development of AR technology in the clinical field.

5. Conclusions

In this retrospective study, we obtained high standards of targeting accuracy, technical efficacy, procedural time, and radiation dose reduction using AR as the sole guidance method for percutaneous thermal ablation, without encountering any complications. In spite of the small cohort analyzed, we propose that our preliminary data demonstrate the potential for AR, with further validation, to become a leading and low-cost modality for the guidance of interventional procedures worldwide.

Author Contributions: Conceptualization, M.S. and L.A.S.; methodology, L.A.S. and M.S.; software, M.S., K.M.P. and A.C.R.; validation, M.S., A.C.R. and R.M.; formal analysis, S.N.G. and L.A.S.; investigation, L.A.S., T.I., V.P., R.M. and R.I.; resources, A.C.R.; data curation, L.A.S.; writing—original draft preparation, L.A.S., K.M.P. and M.S.; writing—review and editing, R.M., R.I. and S.N.G.; visualization, M.S. and A.C.R.; supervision, S.N.G.; project administration, M.S. and A.C.R. All authors have read and agreed to the published version of the manuscript.

Funding: This research received no external funding.

Institutional Review Board Statement: The study was conducted in accordance with the Declaration of Helsinki, and approved by the Institutional Ethics Committee of Humanitas Research Hospital and IRCCS Policlinico Universitario A. Gemelli (protocol code Rad-HCC Eudra CT # 2016-004004-60).

Informed Consent Statement: Informed consent was obtained from all subjects involved in the study.

Data Availability Statement: The data presented in this study are available on request from the corresponding author.

Conflicts of Interest: M.S., K.M.P and A.C.R. are employees of R.A.W. Srl, and were the developers of the described technique. T.I., L.A.S., R.M., V.P. and R.I. declare that they have no conflict of interest to disclose. S.N.G. performs unrelated consulting for Angiodynamics and Cosman Instruments.

References

1. Chehab, M.A.; Brinjikji, W.; Copelan, A.; Venkatesan, A.M. Navigational Tools for Interventional Radiology and Interventional Oncology Applications. *Semin. Interv. Radiol.* **2015**, *32*, 416–427. [CrossRef] [PubMed]
2. Mauri, G.; Cova, L.; De Beni, S.; Ierace, T.; Tondolo, T.; Cerri, A.; Goldberg, S.N.; Solbiati, L. Real-Time US-CT/MRI Image Fusion for Guidance of Thermal Ablation of Liver Tumors Undetectable with US: Results in 295 Cases. *Cardiovasc. Interv. Radiol.* **2014**, *38*, 143–151. [CrossRef] [PubMed]
3. Mauri, G.; De Beni, S.; Forzoni, L.; D'Onofrio, S.; Kolev, V.; Laganà, M.M.; Solbiati, L. Virtual Navigator Automatic Registration Technology in Abdominal Application. *Annu. Int. Conf. IEEE Eng. Med. Biol. Soc.* **2014**, *2014*, 5570–5574. [CrossRef] [PubMed]

4. Rajagopal, M.; Venkatesan, A.M. Image fusion and navigation platforms for percutaneous image-guided interventions. *Abdom. Radiol.* **2016**, *41*, 620–628. [CrossRef]
5. Kloeppel, R.; Weisse, T.; Deckert, F.; Wilke, W.; Pecher, S. CT-guided intervention using a patient laser marker system. *Eur. Radiol.* **2000**, *10*, 1010–1014. [CrossRef] [PubMed]
6. Schweiger, G.D.; Brown, B.P.; Pelsang, R.E.; Dhadha, R.S.; Barloon, T.J.; Wang, G. CT fluoroscopy: Technique and utility in guiding biopsies of transiently enhancing hepatic masses. *Gastrointest. Radiol.* **2000**, *25*. 81–85. [CrossRef] [PubMed]
7. Braak, S.J.; Van Strijen, M.J.L.; Van Leersum, M.; Van Es, H.W.; Van Heesewijk, J.P.M. Real-Time 3D Fluoroscopy Guidance during Needle Interventions: Technique, Accuracy, and Feasibility. *Am. J. Roentgenol.* **2010**, *194*, W445–W451. [CrossRef] [PubMed]
8. Wallace, M.J.; Kuo, M.D.; Glaiberman, C.; Binkert, C.A.; Orth, R.; Soulez, G. Three-dimensional C-arm Cone-beam CT: Applications in the Interventional Suite. *J. Vasc. Interv. Radiol.* **2009**, *20*, S523–S537. [CrossRef] [PubMed]
9. Kim, E.; Ward, T.J.; Patel, R.S.; Fischman, A.M.; Nowakowski, S.; Lookstein, R.A. CT-Guided Liver Biopsy with Electromagnetic Tracking: Results From a Single-Center Prospective Randomized Controlled Trial. *Am. J. Roentgenol.* **2014**, *203*, W715–W723. [CrossRef] [PubMed]
10. Kettenbach, J.; Kronreif, G. Robotic systems for percutaneous needle-guided interventions. *Minim. Invasive Ther. Allied Technol.* **2015**, *24*, 45–53. [CrossRef] [PubMed]
11. de Ribaupierre, S.; Eagleson, R. Editorial: Challenges for the usability of AR and VR for clinical neurosurgical procedures. *Healthc. Technol. Lett.* **2017**, *4*, 151. [CrossRef] [PubMed]
12. Uppot, R.N.; Laguna, B.; McCarthy, C.J.; De Novi, G.; Phelps, A.; Siegel, E.; Courtier, J. Implementing Virtual and Augmented Reality Tools for Radiology Education and Training, Communication, and Clinical Care. *Radiology* **2019**, *291*, 570–580. [CrossRef] [PubMed]
13. Auloge, P.; Cazzato, R.L.; Ramamurthy, N.; DE Marini, P.; Rousseau, C.; Garnon, J.; Charles, Y.P.; Steib, J.-P.; Gangi, A. Augmented reality and artificial intelligence-based navigation during percutaneous vertebroplasty: A pilot randomised clinical trial. *Eur. Spine J.* **2019**, *29*, 1580–1589. [CrossRef]
14. Elsayed, M.; Kadom, N.; Ghobadi, C.; Strauss, B.; Al Dandan, O.; Aggarwal, A.; Anzai, Y.; Griffith, B.; Lazarow, F.; Straus, C.M.; et al. Virtual and augmented reality: Potential applications in radiology. *Acta Radiol.* **2020**, *61*, 1258–1265. [CrossRef]
15. Jolesz, F.A. 1996 RSNA Eugene P. Pendergrass New Horizons Lecture. Image-guided procedures and the operating room of the future. *Radiology* **1997**, *204*, 601–612. [CrossRef]
16. Rolland, J.P.; Wright, D.L.; Kancherla, A.R. Towards a Novel Augmented-Reality Tool to Visualize Dynamic 3-D Anatomy. *Stud. Health Technol. Inform.* **1997**, *39*, 337–348. [PubMed]
17. Léger, É.; Reyes, J.; Drouin, S.; Popa, T.; Hall, J.A.; Collins, D.L.; Kersten-Oertel, M. MARIN: An open-source mobile augmented reality interactive neuronavigation system. *Int. J. Comput. Assist. Radiol. Surg.* **2020**, *15*, 1013–1021. [CrossRef]
18. Maruyama, K.; Watanabe, E.; Kin, T.; Saito, K.; Kumakiri, A.; Noguchi, A.; Nagane, M.; Shiokawa, Y. Smart Glasses for Neurosurgical Navigation by Augmented Reality. *Oper. Neurosurg.* **2018**, *15*, 551–556. [CrossRef]
19. Watanabe, E.; Satoh, M.; Konno, T.; Hirai, M.; Yamaguchi, T. The Trans-Visible Navigator: A See-Through Neuronavigation System Using Augmented Reality. *World Neurosurg.* **2016**, *87*, 399–405. [CrossRef]
20. Kuhlemann, I.; Kleemann, M.; Jauer, P.; Schweikard, A.; Ernst, F. Towards X-ray free endovascular interventions—using HoloLens for on-line holographic visualisation. *Healthc. Technol. Lett.* **2017**, *4*, 184–187. [CrossRef]
21. Mohammed, M.A.A.; Khalaf, M.H.; Kesselman, A.; Wang, D.S.; Kothary, N. A Role for Virtual Reality in Planning Endovascular Procedures. *J. Vasc. Interv. Radiol.* **2018**, *29*, 971–974. [CrossRef] [PubMed]
22. El-Hariri, H.; Pandey, P.; Hodgson, A.J.; Garbi, R. Augmented reality visualisation for orthopaedic surgical guidance with pre- and intra-operative multimodal image data fusion. *Healthc. Technol. Lett.* **2018**, *5*, 189–193. [CrossRef]
23. Gregory, T.M.; Gregory, J.; Sledge, J.; Allard, R.; Mir, O. Surgery guided by mixed reality: Presentation of a proof of concept. *Acta Orthop.* **2018**, *89*, 480–483. [CrossRef] [PubMed]
24. Detmer, F.J.; Hettig, J.; Schindele, D.; Schostak, M.; Hansen, C. Virtual and Augmented Reality Systems for Renal Interventions: A Systematic Review. *IEEE Rev. Biomed. Eng.* **2017**, *10*, 78–94. [CrossRef] [PubMed]
25. Samei, G.; Tsang, K.; Kesch, C.; Lobo, J.; Hor, S.; Mohareri, O.; Chang, S.; Goldenberg, S.L.; Black, P.C.; Salcudean, S. A partial augmented reality system with live ultrasound and registered preoperative MRI for guiding robot-assisted radical prostatectomy. *Med. Image Anal.* **2020**, *60*, 101588. [CrossRef] [PubMed]
26. Wake, N.; Bjurlin, M.A.; Rostami, P.; Chandarana, H.; Huang, W. Three-dimensional Printing and Augmented Reality: Enhanced Precision for Robotic Assisted Partial Nephrectomy. *Urology* **2018**, *116*, 227–228. [CrossRef] [PubMed]
27. Tepper, O.M.; Rudy, H.L.; Lefkowitz, A.; Weimer, K.A.; Marks, S.M.; Stern, C.S.; Garfein, E.S. Mixed Reality with HoloLens. *Plast. Reconstr. Surg.* **2017**, *140*, 1066–1070. [CrossRef] [PubMed]
28. Nicolau, S.; Soler, L.; Mutter, D.; Marescaux, J. Augmented reality in laparoscopic surgical oncology. *Surg. Oncol.* **2011**, *20*, 189–201. [CrossRef] [PubMed]
29. Tang, R.; Ma, L.-F.; Rong, Z.-X.; Li, M.-D.; Zeng, J.-P.; Wang, X.-D.; Liao, H.-E.; Dong, J.-H. Augmented reality technology for preoperative planning and intraoperative navigation during hepatobiliary surgery: A review of current methods. *Hepatobiliary Pancreat Dis. Int.* **2018**, *17*, 101–112. [CrossRef] [PubMed]
30. Racadio, J.M.; Nachabe, R.; Homan, R.; Schierling, R.; Racadio, J.M.; Babić, D. Augmented Reality on a C-Arm System: A Preclinical Assessment for Percutaneous Needle Localization. *Radiology* **2016**, *281*, 249–255. [CrossRef] [PubMed]

31. Rosenthal, M.; State, A.; Lee, J.; Hirota, G.; Ackerman, J.; Keller, K.; Pisano, E.D.; Jiroutek, M.; Muller, K.; Fuchs, H. Augmented reality guidance for needle biopsies: An initial randomized, controlled trial in phantoms. *Med Image Anal.* **2002**, *6*, 313–320. [CrossRef]
32. De Paolis, L.T.; De Luca, V. Augmented visualization with depth perception cues to improve the surgeon's performance in minimally invasive surgery. *Med Biol. Eng. Comput.* **2019**, *57*, 995–1013. [CrossRef] [PubMed]
33. Kuzhagaliyev, T.; Janatka, M.; Vasconcelos, F.; Clancy, N.T.; Clarkson, M.J.; Hawkes, D.J.; Gurusamy, K.; Davidson, B.; Stoyanov, D.; Tchaka, K. Augmented reality needle ablation guidance tool for irreversible electroporation in the pancreas. In *Medical Imaging 2018: Image-Guided Procedures, Robotic Interventions, and Modeling*; International Society for Optics and Photonics: Houston, TX, USA, 2018. [CrossRef]
34. Reig, M.; Forner, A.; Rimola, J.; Ferrer-Fàbrega, J.; Burrel, M.; Garcia-Criado, Á.; Kelley, R.K.; Galle, P.R.; Mazzaferro, V.; Salem, R.; et al. BCLC strategy for prognosis prediction and treatment recommendation: The 2022 update. *J. Hepatol.* **2021**, *76*, 681–693. [CrossRef] [PubMed]
35. European Association for the Study of the Liver. EASL Clinical Practice Guidelines: Management of hepatocellular carcinoma. *J. Hepatol.* **2018**, *69*, 182–236. [CrossRef] [PubMed]
36. Solbiati, M.; Muglia, R.; Goldberg, S.N.; Ierace, T.; Rotilio, A.; Passera, K.M.; Marre, I.; Solbiati, L. A novel software platform for volumetric assessment of ablation completeness. *Int. J. Hyperth.* **2019**, *36*, 336–342. [CrossRef] [PubMed]
37. Ahmed, M.; Solbiati, L.; Brace, C.L.; Breen, D.J.; Callstrom, M.R.; Charboneau, J.W.; Chen, M.-H.; Choi, B.I.; De Baère, T.; Dodd, G.D.; et al. Image-guided Tumor Ablation: Standardization of Terminology and Reporting Criteria—A 10-Year Update. *Radiology* **2014**, *273*, 241–260. [CrossRef] [PubMed]
38. Puijk, R.S.; Ahmed, M.; Adam, A.; Arai, Y.; Arellano, R.; de Baère, T.; Bale, R.; Bellera, C.; Binkert, C.A.; Brace, C.L.; et al. Consensus Guidelines for the Definition of Time-to-Event End Points in Image-guided Tumor Ablation: Results of the SIO and DATECAN Initiative. *Radiology* **2021**, *301*, 533–540. [CrossRef] [PubMed]
39. Park, B.J.; Hunt, S.J.; Martin, C.; Nadolski, G.J.; Wood, B.; Gade, T.P. Augmented and Mixed Reality: Technologies for Enhancing the Future of IR. *J. Vasc. Interv. Radiol.* **2020**, *31*, 1074–1082. [CrossRef] [PubMed]
40. Pratt, P.; Ives, M.; Lawton, G.; Simmons, J.; Radev, N.; Spyropoulou, L.; Amiras, D. Through the HoloLens™ looking glass: Augmented reality for extremity reconstruction surgery using 3D vascular models with perforating vessels. *Eur. Radiol. Exp.* **2018**, *2*, 1–7. [CrossRef]
41. Hecht, R.; Li, M.; De Ruiter, Q.M.B.; Pritchard, W.F.; Li, X.; Krishnasamy, V.; Saad, W.; Karanian, J.W.; Wood, B. Smartphone Augmented Reality CT-Based Platform for Needle Insertion Guidance: A Phantom Study. *Cardiovasc. Interv. Radiol.* **2020**, *43*, 756–764. [CrossRef] [PubMed]
42. Long, D.J.; Li, M.; De Ruiter, Q.M.B.; Hecht, R.; Li, X.; Varble, N.; Blain, M.; Kassin, M.T.; Sharma, K.V.; Sarin, S.; et al. Comparison of Smartphone Augmented Reality, Smartglasses Augmented Reality, and 3D CBCT-guided Fluoroscopy Navigation for Percutaneous Needle Insertion: A Phantom Study. *Cardiovasc. Interv. Radiol.* **2021**, *44*, 774–781. [CrossRef] [PubMed]
43. Solbiati, M.; Passera, K.M.; Rotilio, A.; Oliva, F.; Marre, I.; Goldberg, S.N.; Ierace, T.; Solbiati, L. Augmented reality for interventional oncology: Proof-of-concept study of a novel high-end guidance system platform. *Eur. Radiol. Exp.* **2018**, *2*, 18. [CrossRef] [PubMed]
44. Solbiati, L.; Gennaro, N.; Muglia, R. Augmented Reality: From Video Games to Medical Clinical Practice. *Cardiovasc. Interv. Radiol.* **2020**, *43*, 1427–1429. [CrossRef]
45. Wang, S.; Parsons, M.; Stone-McLean, J.; Rogers, P.; Boyd, S.; Hoover, K.; Meruvia-Pastor, O.; Gong, M.; Smith, A. Augmented Reality as a Telemedicine Platform for Remote Procedural Training. *Sensors* **2017**, *17*, 2294. [CrossRef] [PubMed]

cancers

MDPI

Article

System for the Recognizing of Pigmented Skin Lesions with Fusion and Analysis of Heterogeneous Data Based on a Multimodal Neural Network

Pavel Alekseevich Lyakhov [1,2], Ulyana Alekseevna Lyakhova [3,*] and Nikolay Nikolaevich Nagornov [2]

1 North-Caucasus Center for Mathematical Research, North-Caucasus Federal University, 355017 Stavropol, Russia; ljahov@mail.ru
2 Department of Automation and Control Processes, Saint Petersburg Electrotechnical University "LETI", 197376 Saint Petersburg, Russia; sparta1392@mail.ru
3 Department of Mathematical Modeling, North-Caucasus Federal University, 355017 Stavropol, Russia
* Correspondence: uljahovs@mail.ru; Tel.: +7-(906)-474-76-57

Simple Summary: Skin cancer is one of the most common cancers in humans. This study aims to create a system for recognizing pigmented skin lesions by analyzing heterogeneous data based on a multimodal neural network. Fusing patient statistics and multidimensional visual data allows for finding additional links between dermoscopic images and medical diagnostic results, significantly improving neural network classification accuracy. The use by specialists of the proposed system of neural network recognition of pigmented skin lesions will enhance the efficiency of diagnosis compared to visual diagnostic methods.

Abstract: Today, skin cancer is one of the most common malignant neoplasms in the human body. Diagnosis of pigmented lesions is challenging even for experienced dermatologists due to the wide range of morphological manifestations. Artificial intelligence technologies are capable of equaling and even surpassing the capabilities of a dermatologist in terms of efficiency. The main problem of implementing intellectual analysis systems is low accuracy. One of the possible ways to increase this indicator is using stages of preliminary processing of visual data and the use of heterogeneous data. The article proposes a multimodal neural network system for identifying pigmented skin lesions with a preliminary identification, and removing hair from dermatoscopic images. The novelty of the proposed system lies in the joint use of the stage of preliminary cleaning of hair structures and a multimodal neural network system for the analysis of heterogeneous data. The accuracy of pigmented skin lesions recognition in 10 diagnostically significant categories in the proposed system was 83.6%. The use of the proposed system by dermatologists as an auxiliary diagnostic method will minimize the impact of the human factor, assist in making medical decisions, and expand the possibilities of early detection of skin cancer.

Keywords: digital image processing; pattern recognition; convolutional neural networks; multimodal neural networks; heterogeneous data; metadata; dermatoscopic images; pigmented skin lesions; hair removal; melanoma

Citation: Lyakhov, P.A.; Lyakhova, U.A.; Nagornov, N.N. System for the Recognizing of Pigmented Skin Lesions with Fusion and Analysis of Heterogeneous Data Based on a Multimodal Neural Network. *Cancers* **2022**, *14*, 1819. https://doi.org/10.3390/cancers14071819

Academic Editor: Christine Decaestecker

Received: 9 February 2022
Accepted: 30 March 2022
Published: 3 April 2022

Publisher's Note: MDPI stays neutral with regard to jurisdictional claims in published maps and institutional affiliations.

1. Introduction

According to World Health Organization statistics, non-melanoma and melanoma skin cancer incidence has significantly increased over the past decade [1]. Up to three million cases of non-melanoma skin cancer [2] and about 140,000 cases of melanoma skin cancer are recorded annually [3]. According to the Skin Cancer Foundation Statistics [4], every third case of cancer diagnostics is caused by skin cancer, making it one of the most common types of malignant lesions in the body [5]. This is because the bulk of the population of the countries of the Northern Hemisphere of the Earth are owners of I and II skin phototypes according to Fitzpatrick's classification [6]. A feature of these phototypes is the genetic inability to increase the level of Ultraviolet radiation (UV) [7] and the greatest tendency to develop melanoma [8]. In modern conditions of decreasing the thickness of the atmosphere's ozone layer, UV directly affects the skin, a factor in the activation of oncogenes. It is estimated that a 10% decrease in the ozone layer will lead to an additional 300,000 non-melanoma and 4500 melanoma skin cancers [9]. In regions with high sun exposure, skin cancer is preceded by solar keratosis, the diagnosis of which can help prevent the transformation of pigmented skin lesions into a cancer-positive form [10].

Rapid and highly accurate early diagnosis of skin cancer can reduce patients' risk of death [11]. When detected early, the 5-year survival rate for patients with melanoma is 99%. In the later stages of diagnosis, when the disease reaches the lymph nodes and metastasizes to distant organs, the survival rate in patients is only 27% [3]. Dermatoscopy is the most common method for diagnosing pigmented skin lesions visually [12]. This method is based on the visual acuity and experience of the practitioner and can only be effectively used by qualified professionals [13]. With the help of dermatoscopy, an experienced dermatologist can achieve an average accuracy in the classification of pigmented skin lesions that ranges from 65% to 75% [14]. The early manifestations of malignant and benign neoplasms are visually indistinguishable [15].

Today medicine is considered one of the strategic and promising areas for the effective implementation of systems based on artificial intelligence [16]. There is an improvement in mathematical models and methods, as well as an increase in the amount of digital information in various fields of medicine due to the accumulation of data from electronic medical records, the results of laboratory and instrumental studies, mobile devices for monitoring human physiological functions, etc. [17]. The development of artificial intelligence technologies allowed algorithms for computer analysis of data to be equal to inefficiency, and some tasks surpass human capabilities [18]. A comparison of the classification accuracy of pigmented skin lesions in dermatologists with different levels of experience and a computer program using an artificial intelligence algorithm is presented in articles such as [19–21]. Studies show that artificial intelligence can outperform 136 out of 157 dermatologists and achieve higher accuracy in recognizing pigmented lesions. Despite the higher quality of recognition in artificial intelligence systems than visual diagnostics in physicians, the problem of low accuracy in general in neural network classification systems remains relevant. One of the possible ways to improve recognition accuracy is using the image pre-processing stage [22].

There are many methods for pre-processing dermoscopic images to improve and visually highlight diagnostically significant features. One of these methods is segmentation to highlight pigmented skin lesions' contours. Segmentation can be performed using a biorthogonal two-dimensional wavelet transform and the Otsu algorithm [23]. Edge extraction can be done using Gaussian contrast enhancement and edge extraction using the saliency map construction [24]. Saliency maps use inner and outer non-overlapping windows, making the foreground and background distinct. A significant disadvantage of segmentation methods using filters is the lack of versatility in selecting contours in images of different quality. Illumination, skin color, and sharpness of the contours of a pigmented skin lesion significantly reduce the accuracy of these algorithms. Another way to highlight contours on dermoscopic images is contrast stretching with further detection using Faster Region-Based Convolutional Neural Network (Faster R-CNN) [25,26]. Segmentation based on neural network algorithms makes it possible to accurately identify the contours of pigmented skin lesions, separate a pigmented neoplasm from a skin area, and exclude the influence of skin color type when recognized by artificial intelligence. At the same time, the problem of the presence of hair structures remains, which can be perceived by both neural network algorithms and filter-based algorithms as part of a pigmented skin lesion.

The presence of hair in dermatoscopic images can drastically change the size, shape, color, and texture of the lesion, which significantly affects the automatic analysis of the neural network [27]. Removing hair from images during digital pre-processing is an important step in improving the accuracy of automated diagnostic systems [28]. Today, several methods are designed for pre-processing dermatoscopic images of pigmented skin lesions to remove hair or other noise elements [29]. For example, the essence of the DullRazor process [30] is to use the morphological operation of closing. A significant drawback of DullRazor is the distortion of the dark areas of pigmented lesions, which can change diagnostic signs and have a substantial impact on the quality of recognition. In [31], another hair removal method on dermatoscopic images is presented based on non-linear Partial Differential Equation diffusion (PDE-diffusion). The algorithm is designed to fill linear hair structures by diffusion. This method is also used in [32,33].

Another way to improve the accuracy of intelligent classification systems is to combine heterogeneous data and further analyze them to find additional relationships. In database dermatology, heterogeneous data mining makes it possible to combine patient statistical metadata and dermoscopic images, greatly improving the recognition of pigmented skin lesions. The use of multimodal neural network systems [34–37], as well as methods for combining metadata and multidimensional visual data [38], has significantly improved the accuracy in recognizing pigmented skin lesions.

Despite significant progress in implementing artificial intelligence technologies to analyze dermatological data, developing neural network systems of varying complexity is relevant to achieving higher recognition accuracy. The main hypothesis of the manuscript is a potential increase in the quality of neural network systems for analyzing medical data due to the emerging synergy when using various methods to improve recognition accuracy together. This study aims to develop and model a multimodal neural network system for analyzing dermatological data through the preliminary cleaning of hair structures from images. The proposed system makes it possible to achieve higher recognition accuracy levels than similar neural network systems due to the preliminary cleaning of hair structures from dermoscopic images. The use of the proposed system by dermatologists as an auxiliary diagnostic method will minimize the impact of the human factor in making medical decisions.

The rest of the work is structured as follows. Section 2 is divided into several subsection. In Section 2.1 a description of a method for identifying and cleaning hair structures as pre-processing dermatoscopic images of pigmented skin lesions is proposed. In Section 2.2 a description of the method for pre-processing statistical metadata about patients has been made. In Section 2.3 the definition of a multimodal neural network system for processing statistical data and dermatoscopic images of pigmented skin lesions is presented. Section 3 presents practical modeling of the proposed multimodal neural network system to classify pigmentary neoplasms with a preliminary stage of hair removal on dermatoscopic images. Section 4 discusses the results obtained and their comparison with known works in neural network classification of dermatoscopic skin images. In conclusion, the results of the work are summed up.

2. Materials and Methods

The paper proposes a multimodal neural network system for recognizing pigmented skin lesions with a stage of preliminary processing of dermatoscopic images. The proposed multimodal neural network system for analysis and classification combines heterogeneous diagnostic data represented by multivariate visual data and patient statistics. The scheme of a multimodal neural network system for the classification of dermatoscopic images of pigmented skin lesions with preliminary processing of heterogeneous data is shown in Figure 1.

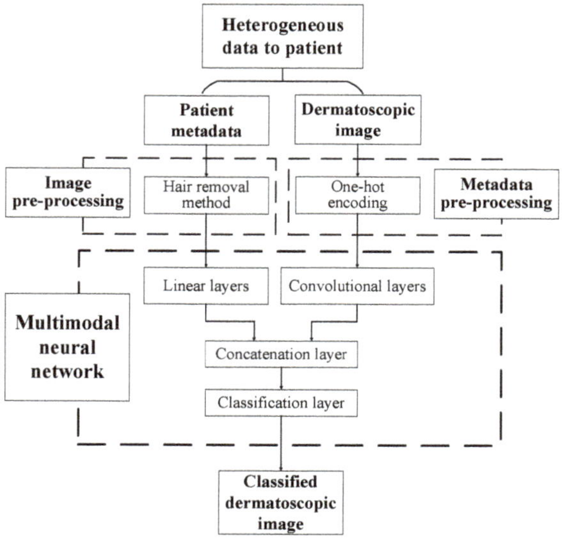

Figure 1. Multimodal neural network system for the classification of dermatoscopic images of pigmented skin lesions with preliminary heterogeneous data processing.

The multidimensional visual data undergoes a pre-processing stage, which identifies and cleans hair structures from dermatoscopic images of pigmented skin lesions. Patient statistics also undergo a one-hot encoding process to generate a feature vector. The multimodal neural network system for recognizing pigmented lesions in the skin consists of two neural network architectures. Dermatoscopic images are processed using the specified Convolutional Neural Network (CNN) architecture. Statistical metadata is processed using a linear multilayer neural network. The resulting feature vector at the CNN output and the output signal of the linear neural network are combined on the concatenation layer. The combined signal is fed to the layer for classification. The output signal from the proposed multimodal neural network system for recognizing pigmented skin lesions is the percentage of 10 diagnostically significant categories, including a recognized dermatoscopic image.

2.1. Hair Removal

The main diagnostic method in the field of dermatology is visual analysis. Today, many imaging approaches have been developed to help dermatologists overcome the problems caused by the apperception of tiny skin lesions. The most widely used imaging technique in dermatology is dermatoscopy, a non-invasive technique for imaging the skin surface using a light magnifying device and immersion fluid [39]. Statistics show that dermatoscopy has increased the efficiency of diagnosing malignant neoplasms by 50% [40]. A significant problem when working with this method is the possible presence of hair on the area of the pigmented lesion, which causes occlusion.

The presence of such noisy structures as hair significantly complicates the work of dermatologists and specialists. It can also cause errors in recognizing pigmented skin lesions in automatic analysis systems. Hair violates the geometric properties of the pigmented lesion areas, which negatively affects the diagnostic accuracy [41]. Figure 2 shows dermatoscopic images of pigmented skin lesions with hair structures present that cause occlusion by altering the size, shape of the lesion, and texture of the image.

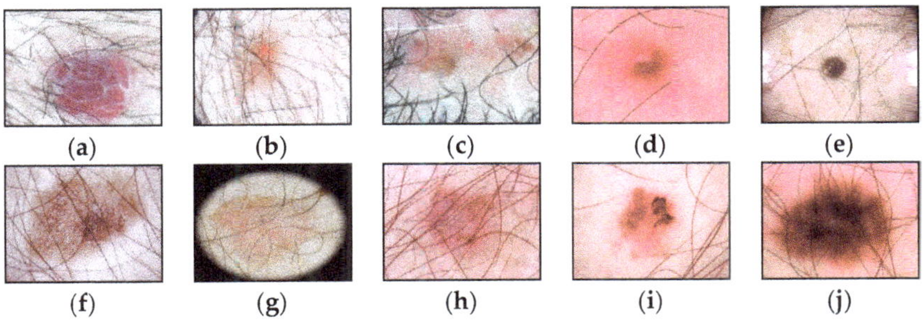

Figure 2. Examples of pigmented skin lesions images with hairy structures: (**a**) vascular lesions; (**b**) nevus; (**c**) solar lentigo; (**d**) dermatofibroma; (**e**) seborrheic keratosis; (**f**) benign keratosis; (**g**) actinic keratosis; (**h**) basal cell carcinoma; (**i**) squamous cell carcinoma; (**j**) melanoma.

The most common way to solve the occlusion problem of pigmented skin lesions is to remove the visible part of the hair with a cutting instrument before performing a dermatoscopic examination. However, this approach leads to skin irritation. Also, it causes diffuse changes in the color of the entire pigmented lesion, which distorts diagnostically significant signs to a greater extent than the presence of hair itself. An alternative solution is digitalizing dermatoscopic visual data to remove hair structures. The essence of the hair pre-cleaning methods is to identify each pixel of the image as a pixel-hair or pixel-skin and then replace the pixels of the hair structures with skin pixels [42]. Preliminary digital processing of dermatoscopic images using morphological operations is one of the possible methods for identifying and replacing pixels of hair structures.

This paper proposes a method for digital pre-processing dermoscopic images using morphological operations on multidimensional visual data. A step-by-step scheme of the proposed method is shown in Figure 3.

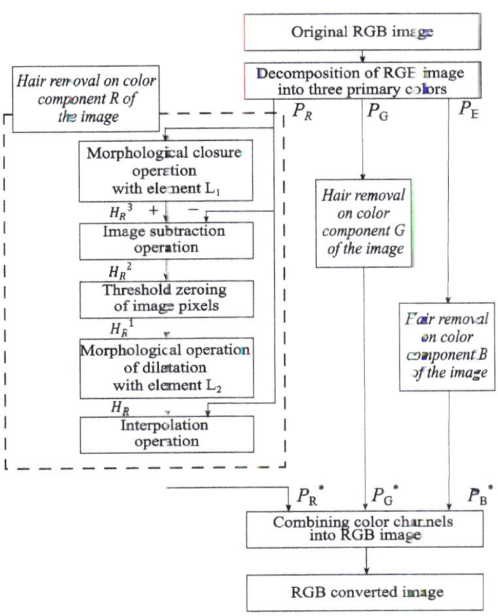

Figure 3. Scheme of the proposed method of identification and hair removal from dermatoscopic images of pigmented skin lesions.

Image processing of pigmented skin lesions consists of four main stages. At the first stage, the RGB image is decomposed into color components. The second step is to locate the locations of the hair structures. At the third stage, the hair pixels are replaced with neighboring pixels. The fourth step is to reverse engineer an RGB color dermatoscopic image.

The input of the proposed method is RGB dermatoscopic images of pigmented neoplasms of the skin $P_{(x,y)}$. The color components P_R, P_G, and P_B are extracted from the image. The following processing steps are performed separately for each color component. The variables L_1 and L_2 are defined as follows:

$$L_{1,2} = \{(x,y) : \rho(T,(x,y)) \le r\} \quad (1)$$

where ρ is the distance from the center T of the set $L_{1,2}$ by the chosen metric, and r is the radius of the set specified by the user. The next stage is a morphological closure operation using the L_1 element to determine the location of hair structures on dermatoscopic images:

$$H_{CC}{}^3 = P_{CC} \cdot L_1 = (P_{CC} \oplus L_1) \ominus L_1 \quad (2)$$

where CC stands for the color channel, $CC \in \{R, G, B\}$, $\oplus$ is the operation of dilatation of the set P along L_1 and $\ominus$ is the operation erosion by element L_1. The closure operation smooths out the contours of the hair structures in dermatoscopic images, eliminates voids, and fills in narrow gaps and long small-width depressions.

At the next stage, the original image P_{CC} is subtracted from the image obtained as a result of the $H_{CC}{}^3$ close operation:

$$H_{CC}{}^2 = H_{CC}{}^3 - P_{CC} \quad (3)$$

The operator of zeroing the pixels δ of the image $P_{(x,y)}$ for further operations is defined as follows:

$$\delta\left(P_{(x,y)}\right) = \begin{cases} P_{(x,y)}, \text{ if } P_{(x,y)} > K \\ 0, \text{if } P_{(x,y)} \le K \end{cases} \quad (4)$$

where K is the user-defined threshold of pixel intensity values. The next stage is the threshold zeroing of the pixels of the detected hair structures. For this, the entered zeroing operator δ is applied to the resulting dermatoscopic image $H_{CC}{}^2$:

$$H_{CC}{}^1 = \delta(H_{CC}{}^2) \quad (5)$$

After the operation of threshold zeroing of pixels, a morphological operation of dilatation with the L_2 element is performed to expand the boundaries of the hair structures:

$$H_{CC} = H_{CC}{}^1 \oplus L_2 \quad (6)$$

The next step is to replace the pixels of the hair structure with neighboring pixels. Using the Laplace equation, pixels are interpolated from the area's border of the selected hair structures. In this case, the pixels from the border of the hair structures cannot be changed. The last step is the reverse construction of the RGB color image from the extracted color components. For this, the color channels $P_R{}^*$, $P_G{}^*$, and $P_B{}^*$ are combined.

An example of the step-by-step work of the proposed method for identifying and cleaning hair structures from dermatoscopic images of pigmented skin lesions is shown in Figure 4. To improve the visual perception of the intermediate results of each method stage, Figure 4d–f were inverted.

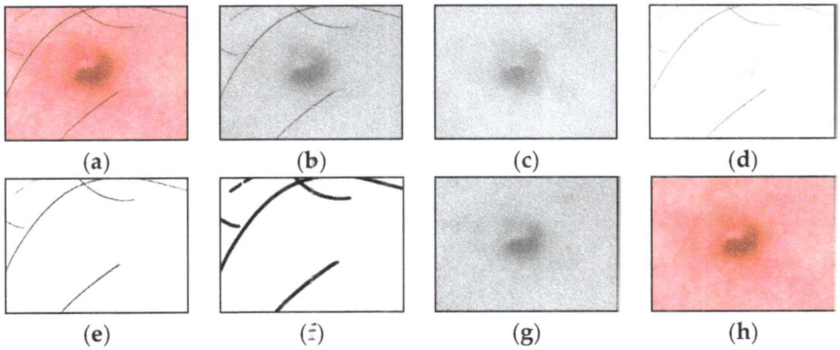

Figure 4. Images obtained as a result of passing each stage of the method of identification and hair removal: (**a**) input RGB image P_{RGB}; (**b**) the color component P_R, presented in shades of gray; (**c**) the result of the $H_R{}^3$ closing operation; (**d**) the result of the subtraction operation $H_R{}^2$ (inverted image); (**e**) the result of zeroing pixels $H_R{}^1$ (inverted image); (**f**) the result of the H_R dilatation operation (inverted image); (**g**) pixel interpolation result $P_R{}^*$; (**h**) output RGB image $P_{RGB}{}^*$. Scale bar or magnification.

2.2. Metadata Pre-Processing

Today, in medicine, there is an increase in the volume of digital information due to the accumulation of data from electronic medical records, the results of laboratory and instrumental studies, mobile devices for monitoring human physiological functions, and others [17]. Patient biomedical statistics are structured data that describe the characteristics of research subjects. Statistical data includes gender, age, race, predisposition to various diseases, bad habits, etc. Such information facilitates the search for connections between research objects and the analysis result.

Metadata pre-processing is converting statistical data into the format required by the selected data mining method. Since the proposed multimodal system for recognizing pigmented skin lesions is a fully connected neural network, it must encode the data as a vector of features. A corresponding metadata information vector is generated for each image in the dataset, which depends on the amount and type of statistical information. One-hot encoding can sometimes outperform complex encoding systems [43]. All multi-categorical variables (discrete variables with more than two categories) are converted to a new set of binary variables for one-hot encoding. For example, the categorical variable to denote a pigmented lesion on the patient's body will be replaced by 8 dummy variables indicating whether the pigmented lesion is located on the anterior torso, head/neck, lateral torso, lower extremity, oral/genital, palms/soles, posterior torso, or upper extremity.

Suppose the M metadata includes various statistics $M = \{M_1, M_2, \ldots, M_n\}$ with $M_n \in m_n$, where m_n is a pointer to a specific patient parameter. If m_n is a pointer to the gender of the patient, then $M_1 = \{male, female\}$. For each set M_n, which is one of the patient's indicators, its power $\mu_n = |M_n|$ is calculated. For metadata pre-processing, an $\vec{m}$ feature vector of $\sum_n \mu_n$ the dimension is generated. The first coordinate of the $\vec{m}$ metadata vector of the μ_1 the dimension will encode the statistical data m_1. The next coordinate of the μ_2 the dimension will encode the m_2 statistical data, and so on.

One-hot encoding is used to encode the statistic $m_n \in M_n$ as follows. For the set of M_n, the ordering is performed in an arbitrary fixed way for all considered cases. After that, the binary code $\underbrace{1000\ldots0}_{\mu_n}$ is reserved for the first element of the set M_n. For the second element of the set M_n, the binary code $\underbrace{0100\ldots0}_{\mu_n}$ is reserved, and so on. The statistical metadata pre-processing scheme is shown in Figure 5.

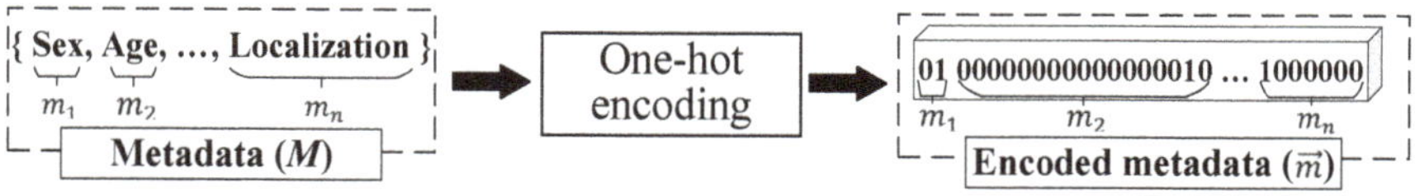

Figure 5. Metadata pre-processing scheme.

2.3. Multimodal Neural Network

In deep learning, multimodal fusion or heterogeneous synthesis combines different data types obtained from various sources [44]. In the field of diagnosis of pigmented skin lesions, the most common types of data are dermatoscopic images and patient statistics such as age, sex, and location of the pigmented lesion on the patient's body. Combining visual data, signals, and multidimensional statistical data about patients allows you to create heterogeneous medical information databases that can be used to build intelligent systems for diagnostics and decision support for specialists, doctors, and clinicians [45]. The rationale for using heterogeneous databases is that the fusion of heterogeneous data can provide additional information and increase the efficiency of neural network analysis and classification systems [46]. The use of heterogeneous data in training multimodal neural network systems will improve the accuracy of diagnostics by searching for connections between visual objects of research and statistical metadata [47].

For the recognition of multidimensional visual data, the most optimal neural network architecture is CNN [48]. The input of the proposed multimodal system for neural network classification of pigmented skin lesions is supplied with dermatoscopic images of $P_{(img)}$, pre-processed metadata in the vector form of $\vec{m} = (m_1, m_2, \ldots, m_n)$ and tags with a diagnosis of $l \in \{1, \ldots, N_{lab}\}$, where N_{lab} is the number of diagnostic categories.

The dermatoscopic image includes R rows, C columns, and D color components. In this case, for the RGB format $= 3$, the color components are represented by the levels of red, green, and blue colors of the image pixels. The input of the convolutional layer receives a dermatoscopic $P_{(img)}$ image, while the input is a three-dimensional function $P(x, y, z)$, where $0 \le x < R$, $0 \le y < C$ and $0 \le z < D$ are spatial coordinates, and the amplitude P at any point with coordinates $(x,\ y,\ z)$ is the intensity of the pixels at a given point. Then the procedure for obtaining feature maps in the convolutional layer is as follows:

$$P_f(x, y) = g + \sum_{i=\frac{-w-1}{2}}^{\frac{w-1}{2}} \sum_{j=\frac{-w-1}{2}}^{\frac{w-1}{2}} \sum_{k=0}^{D-1} w_{ijk}^{(1)} P(x+i,\ y+j, k), \tag{7}$$

where P_f is a feature map; $w_{ijk}^{(1)}$ is the coefficient of a filter of size $w \times w$ for processing D arrays; g is offset.

The concatenation layer at the input receives the feature map, which was obtained on the last layer intended for processing dermatoscopic images P_f, and the metadata vector $\vec{m}$. The P_f feature map contains a set of x_{ijk}, where i is the height coordinate, j is the width coordinate, k is the number of the map obtained on the last layer from the set of layers that were intended for processing dermatoscopic images. The operation of combining heterogeneous data on the concatenation layer can be represented as follows:

$$f_l = \sum_i \sum_j \sum_k x_{ijk} w_{ijkl}^{(2)} + \sum_{i=1}^{n} m_i w_{il}^{(3)}, \tag{8}$$

where $w_{ijkl}^{(2)}$ is a set of weights for processing feature maps of dermatoscopic images; $w_{il}^{(3)}$ is a set of weights for processing metadata vectors.

The activation of the last layer of the multimodal neural network is displayed through the $softmax$ function with the distribution $P(y|x, \theta)$ and has the form:

$$P(y|x, \theta) = softmax(x;\theta) = \frac{\exp\left(w_l^n\right)^T x_l^n + g_l^n}{\sum_{k=1}^{K} \exp\left(w_l^n\right)^T x_l^n + g_l^n}, \quad (9)$$

where w_l^n is the weight vector leading to the output node that is associated with class l. The proposed multimodal system for recognizing pigmented skin lesions based on CNN AlexNet is shown in Figure 6.

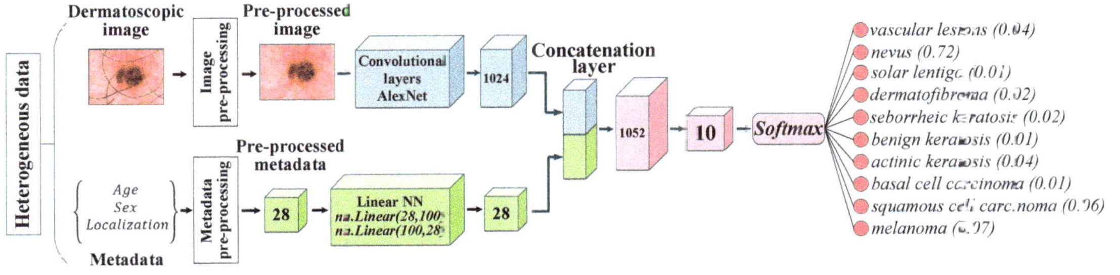

Figure 6. Neural network architecture for multimodal classification of pigmented skin lesions based on CNN AlexNet. Scale bar or magnification.

3. Results

Data from the open archive of The International Skin Imaging Collaboration (ISIC), which is the largest available set of confidential data in dermatology, was used for the simulations [49]. The main clinical goal of the ISIC project is to support efforts to reduce mortality associated with melanoma and reduce biopsies by improving the accuracy and efficiency of early detection of melanoma. ISIC develops proposed digital imaging standards and engages the dermatological and bioinformatics communities to improve diagnostic accuracy using artificial intelligence. While the initial focus in the ISIC collaboration is on melanoma, diagnosing non-melanoma skin cancer and inflammatory dermatoses is equally important. ISIC has developed an open-source platform for hosting images of skin lesions under Creative Commons licenses. Dermatoscopic photos are associated with reliable diagnoses and other clinical metadata and are available for public use The ISIC archive contains 41,725 dermatoscopic photographs of various sizes, representing a database of digital representative images of the 10 most important diagnostic categories. Most of the photographs are digitized transparencies of the Roffendal Skin Cancer Clinic in Queensland, Australia, and the Department of Dermatology at the Medical University of Vienna, Austria [50]. The dataset also contains statistical meta-information about the patient's age group (in five-year increments), anatomical site (eight possible sites), and gender (male/female). Figure 7 shows a diagram of the distribution of dermatoscopic images for 10 diagnostically significant categories. Diagnostically significant categories are divided into groups "benign" and "malignant", and are also arranged in order of increasing risk and severity of the course of the disease. Since actinic keratosis can be considered as intraepithelial dysplasia of keratinocytes and, therefore, as a "precancerous" skin lesion, or as in situ squamous cell carcinoma, this category was therefore assigned to the group of "malignant" pigmented neoplasms [51–53]. The diagram shows how unbalanced the available images of pigmented skin lesions are towards the "nevus" category. Figure 8 shows diagrams of the distribution of the base of dermatoscopic images according to the statistical data of patients. The database is dominated by male patients and patients aged 15 to 20 years. At the same time, in patients, pigmented skin lesions were most often found on the back (posterior torso).

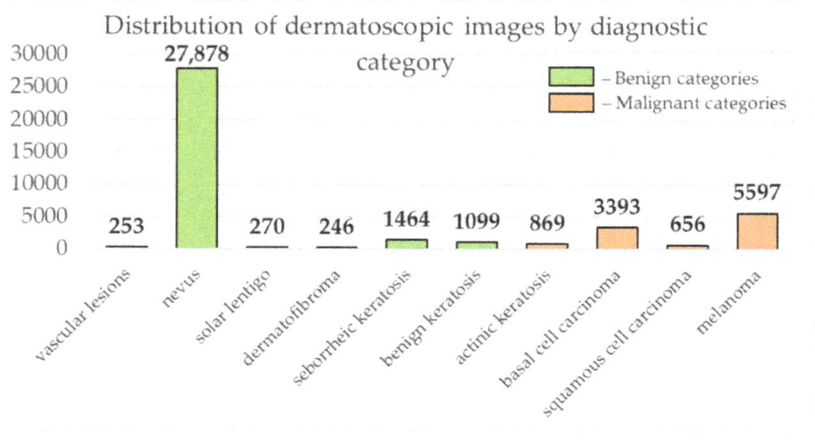

Figure 7. Diagram of the distribution of the number of dermatoscopic images in 10 diagnostically significant categories.

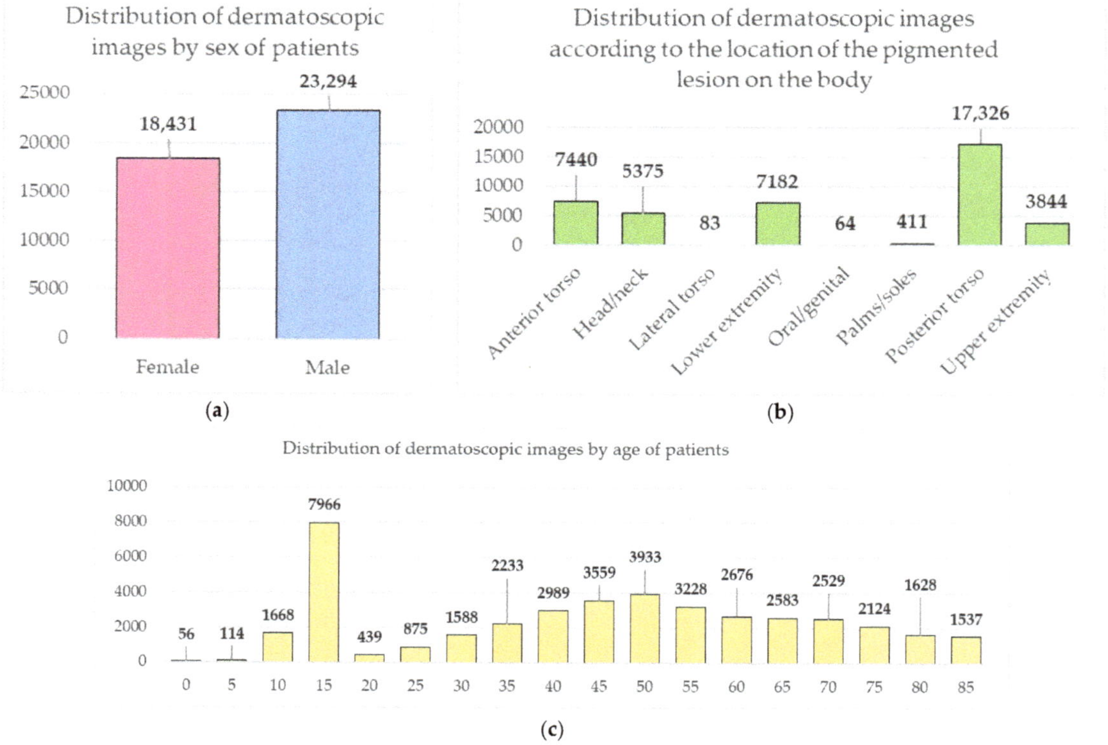

Figure 8. Diagrams of the distribution of the base of dermatoscopic images according to the statistical data of patients: (**a**) by gender; (**b**) by age; (**c**) by the location of the pigmented lesion on the body.

The modeling was performed using the high-level programming language Python 3.8.8. All calculations were performed on a PC with an Intel (R) Core (TM) i5-8500 CPU

@ 3.00 GHz 3.00 GHz with 16 GB of RAM and a 64-bit Windows 10 operating system. Multimodal CNN training was carried out using a graphics processing unit (GPU) based on an NVIDIA video chipset GeForce GTX 1050TI.

Preliminary heterogeneous data processing was carried out at the first stage of the proposed multimodal classification system. Dermatoscopic image pre-processing consisted of stepwise hair removal and image resizing. The removal of hair structures was carried out using the developed method based on morphological operations, presented in Section 2.1. An empirical analysis of the application of Formula (1) showed that the best result of identification and cleaning of hair structures is achieved at $r = 5$ for the element L_1 and at $r = 3$ for the element L_2. In the calculations, the Euclidean norm (L2) was used as a metric. It was also empirically found that the optimal threshold value in Formula (4) is $K = 40$. Examples of pre-cleaning dermatoscopic images are shown in Figure 9. Figure 9b was inverted to improve the visual perception of the results of the stage of hair extraction in the pictures.

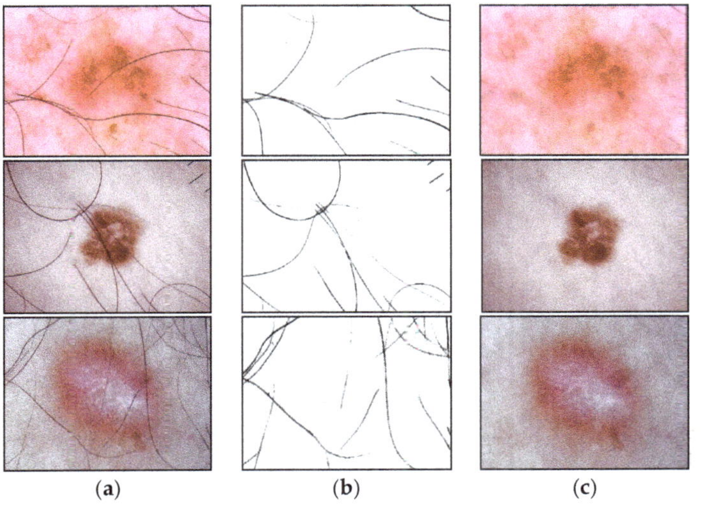

Figure 9. Examples of identification and cleaning of hair structures from dermatoscopic images of pigmented skin lesions using the proposed method: (**a**) original dermatoscopic image; (**b**) the result of extracting hair in the image (inverted image); (**c**) dermatoscopic image cleared of hair structures. Scale bar or magnification.

The pre-processing of patient metadata consisted of one-hot encoding to convert the vector format required for further mining. The coding tables for each patient metadata index are presented in Tables 1–3. An example of pre-processing statistical patient metadata using one-hot encoding is shown in Figure 10.

Table 1. A coding table for patient gender metadata.

Patient Gender (Sex)	One-Hot Code	
male	0	1
female	1	0

Table 2. A coding table for localization of pigmented lesion on the patient body.

Localization of Pigmented Lesion on the Patient Body (Anatomloc)	One-Hot Code							
anterior torso	1	0	0	0	0	0	0	0
head/neck	0	1	0	0	0	0	0	0
lateral torso	0	0	1	0	0	0	0	0
lower extremity	0	0	0	1	0	0	0	0
oral/genital	0	0	0	0	1	0	0	0
palms/soles	0	0	0	0	0	1	0	0
posterior torso	0	0	0	0	0	0	1	0
upper extremity	0	0	0	0	0	0	0	1

Table 3. A coding table for patient age metadata.

The Age of the Patient (Age)	One-Hot Code																	
0	1	0	0	0	0	0	0	0	0	0	0	0	0	0	0	0	0	0
5	0	1	0	0	0	0	0	0	0	0	0	0	0	0	0	0	0	0
10	0	0	1	0	0	0	0	0	0	0	0	0	0	0	0	0	0	0
15	0	0	0	1	0	0	0	0	0	0	0	0	0	0	0	0	0	0
20	0	0	0	0	1	0	0	0	0	0	0	0	0	0	0	0	0	0
25	0	0	0	0	0	1	0	0	0	0	0	0	0	0	0	0	0	0
30	0	0	0	0	0	0	1	0	0	0	0	0	0	0	0	0	0	0
35	0	0	0	0	0	0	0	1	0	0	0	0	0	0	0	0	0	0
40	0	0	0	0	0	0	0	0	1	0	0	0	0	0	0	0	0	0
45	0	0	0	0	0	0	0	0	0	1	0	0	0	0	0	0	0	0
50	0	0	0	0	0	0	0	0	0	0	1	0	0	0	0	0	0	0
55	0	0	0	0	0	0	0	0	0	0	0	1	0	0	0	0	0	0
60	0	0	0	0	0	0	0	0	0	0	0	0	1	0	0	0	0	0
65	0	0	0	0	0	0	0	0	0	0	0	0	0	1	0	0	0	0
70	0	0	0	0	0	0	0	0	0	0	0	0	0	0	1	0	0	0
75	0	0	0	0	0	0	0	0	0	0	0	0	0	0	0	1	0	0
80	0	0	0	0	0	0	0	0	0	0	0	0	0	0	0	0	1	0
85	0	0	0	0	0	0	0	0	0	0	0	0	0	0	0	0	0	1

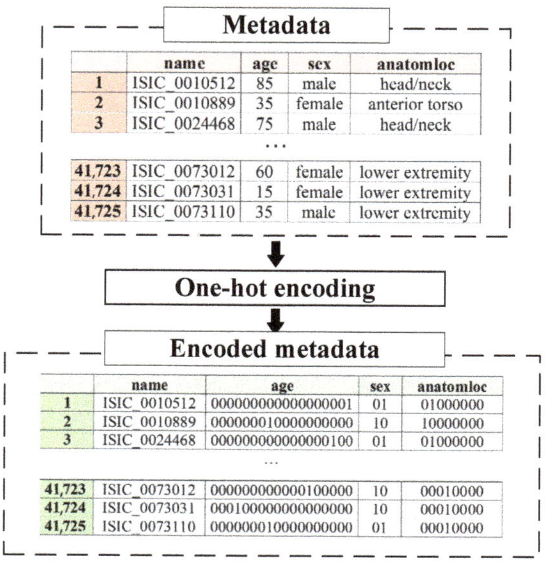

Figure 10. An example of pre-processing statistical patient metadata using one-hot encoding.

CNN AlexNet [54], SqueezeNet [55], and ResNet-101 [56] were selected to simulate a multimodal neural network system for recognizing pigmented skin lesions, which were pre-trained on a set on a set of natural images ImageNet. The most common size of dermatoscopic images in the ISIC database is 450 × 600 × 3, where 3 is the color channels. For neural network architectures AlexNet and SqueezeNet, the images were transformed to a size of 227 × 227 × 3. For CNN ResNet-101, the images were converted to 224 × 224 × 3. For further modeling, the base of dermatoscopic photographs was divided into images for training and images for validation in a percentage ratio of 80 to 20. Since the ISIC dermatoscopic image base is strongly unbalanced towards the "nevus" category, the training images were augmented using affine transformations.

Large volumes of training data make it possible to increase the classification accuracy of automated systems for neural network recognition of dermatoscopic images of pigmented skin lesions. Creating large-scale medical imaging datasets is costly and time-consuming because diagnosis and further labeling require specialized equipment and trained practitioners. It also requires the consent of patients to process and provides personal data. Existing training datasets for the intelligent analysis of pigmented skin lesions, including the ISIC open archive, are imbalanced across benign lesion classes. All of this leads to inaccurate classification results due to CNN overfitting.

Affine transformations are one of the main methods for increasing and balancing the amount of multidimensional visual data in each class. The possible affine transformations are rotation, displacement, reflection, scaling, etc. The selected dermatoscopic images of pigmented skin lesions include multidimensional visual data of various sizes. Different CNN architectures require input images of a certain size. Scaling using affine transformations transforms visual data into a set of images of the same size. Scaling is usually combined with cropping to achieve the desired image size.

Augmentation of dermatoscopic images of pigmented skin lesions included all of the above methods of affinity transformations, examples of which are shown in Figure 11.

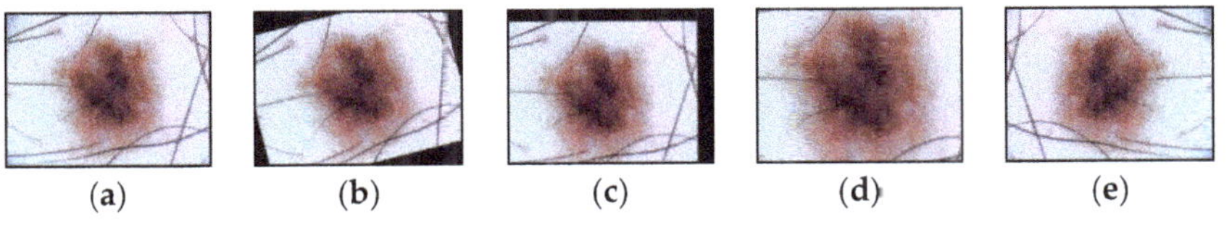

Figure 11. Images obtained as a result of affine transformations: (**a**) original image; (**b**) image after the operation of rotation by a given angle; (**c**) image after shift operation; (**d**) image after the scaling operation; (**e**) image after the reflection operation. Scale bar or magnification.

New multidimensional visual data were created from existing ones using augmentation for more effective training. This allowed us to increase the number of training images. Training data augmentation has proven effective enough to improve accuracy in neural network recognition systems for medical data [57]. When trained, neural network classifiers tend to lean towards classes containing the largest number of images [58]. The use of data augmentation made it possible to minimize the imbalance and achieve uniform learning across all diagnostically significant classes presented. An example of transformed dermatoscopic images from the database for training a multimodal neural network for recognizing pigmented skin lesions is shown in Figure 12.

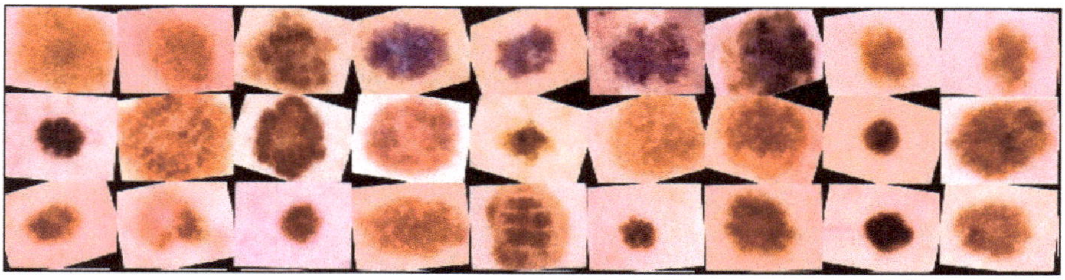

Figure 12. Examples of dermatoscopic training images that have been previously cleaned and enlarged using affinity transformations. Scale bar or magnification.

Pre-processed images of pigmented skin lesions were fed into CNN architectures. The vector of pre-processed metadata was provided to the input of a linear neural network, which consisted of several linear layers and ReLu activation layers. After passing the different input signals through the CNN and the linear neural network, the heterogeneous data passed fusion on the concatenation layer. The combined data was fed to the softmax layer for classification. Figures A1–A3 from Appendix A show graphs of the learning outcomes of a multimodal neural network system for recognizing pigmented skin lesions based on various CNNs.

Table 4 presents the results of assessing the recognition accuracy of dermatoscopic images of pigmented skin lesions. The highest indicator of the accuracy of recognition of pigmented skin lesions was achieved using a multimodal neural network system for recognizing pigmented skin lesions with a stage of preliminary hair cleaning with a pre-trained AlexNet architecture [54] and amounted to 83.56%. When training each multimodal neural network architecture using the method of preliminary identification and cleaning of hair structures, the obtained percentage of recognition accuracy was higher than when training original CNNs without a preliminary processing stage. The increase in recognition accuracy during training of multimodal neural network recognition systems for pigmented skin lesions with a stage of preliminary hair cleaning was 4.93–6.28%, depending on the CNN architecture. The best indicator of improving the recognition accuracy was obtained when training a multimodal neural network classification system with a preliminary hair cleaning stage with a pre-trained ResNet-101 [56] architecture amounted to 6.28%. The smallest result of an increase in recognition accuracy of 4.93% was shown by a multimodal system based on AlexNet [54]. Adding each of the components to the system improves the accuracy by 2.18–4.11%. As a result of modeling the original CNN architecture with the stage of preliminary cleaning of hair structures based on SqueezeNet, the increase in recognition accuracy was 2.13%. At the same time, adding the stage of neural network analysis of statistical data made it possible to increase the accuracy by another 4.11%. For the AlexNet neural network architecture, this increase was 2.18% and 2.75%, respectively. For the ResNet-101 neural network architecture, recognition accuracy increased by 3.17% and 3.11%, respectively. The results obtained indicate that the combined use of various methods for improving the accuracy of recognition can significantly increase the accuracy of neural network data analysis.

Table 4. Results of modeling a multimodal neural network classification system for dermatoscopic images of pigmented skin lesions. Bold font indicates the best result in each column of the table.

CNN Architecture	Results of Recognition			
	Original CNN Architecture, %	Original CNN Architecture with a Stage of Preliminary Hair Removal, %	Proposed Multimodal Neural Network System with a Stage of Preliminary Hair Removal, %	Different in Recognition Accuracy between Original and Proposed Neural Network Systems, %
AlexNet [54]	**78.63**	**80.81**	**83.56**	4.93
SqueezeNet [55]	71.63	73.76	77.87	6.24
ResNet-101 [56]	76.75	79.92	83.03	**6.28**

The results predicted by the multimodal neural network from the test sample were converted to a binary form to construct the Receiver Operating Characteristic curve (ROC curve). Each predicted class label consisted of a combination of two characters with a length of 10 characters. The ROC curve represents the number of correctly classified positive values on incorrectly classified negative values.

$$TPR = \frac{TP}{TP + FN} \times 100\% \quad (10)$$

$$FPR = \frac{FP}{TN + FP} \times 100\%, \quad (11)$$

where TP is true positive cases; TN is true negative cases; FN is false-negative cases; FP is false positives cases. The ROC curve is plotted so that the x-axis is the proportion of false positives FPR, and the y-axis is the proportion of true positive TPR cases. The AUC is the area under the ROC curve and is calculated as follows:

$$AUC = \int_0^1 TPR \, d(FPR). \quad (12)$$

Table 5 shows the results of testing the proposed multimodal neural network system for recognizing pigmented lesions with a stage of preliminary cleaning from hair structures. Figures 13–15 show confusion matrices resulting from testing multimodal neural network systems for identifying pigmented skin lesions based on various CNNs.

Table 5. Testing results of the proposed multimodal neural network system to recognize pigmented lesions. Bold font indicates the best result in each column of the table.

CNN Architecture	Recognition Accuracy, %	Loss Function	AUC
AlexNet [54]	**83.56**	**0.47**	0.90
SqueezeNet [55]	77.87	0.67	0.88
ResNet-101 [56]	83.03	0.66	**0.93**

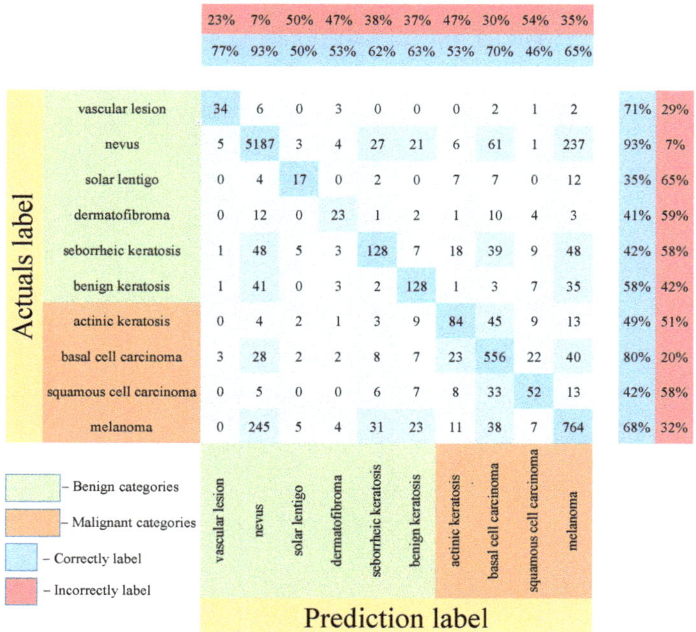

Figure 13. Confusion matrix in the testing results in a multimodal neural network system for recognizing pigmented skin lesions based on CNN AlexNet.

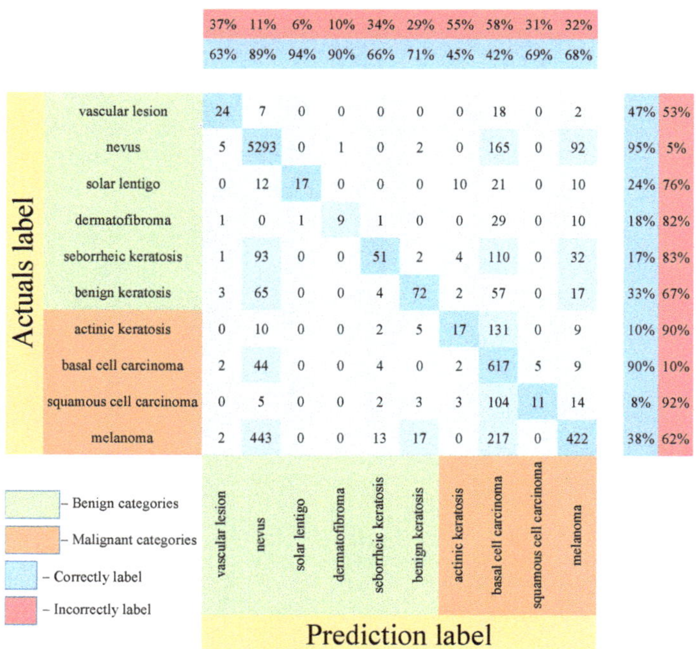

Figure 14. Confusion matrix in the testing results in a multimodal neural network system for recognizing pigmented skin lesions based on CNN SqueezeNet.

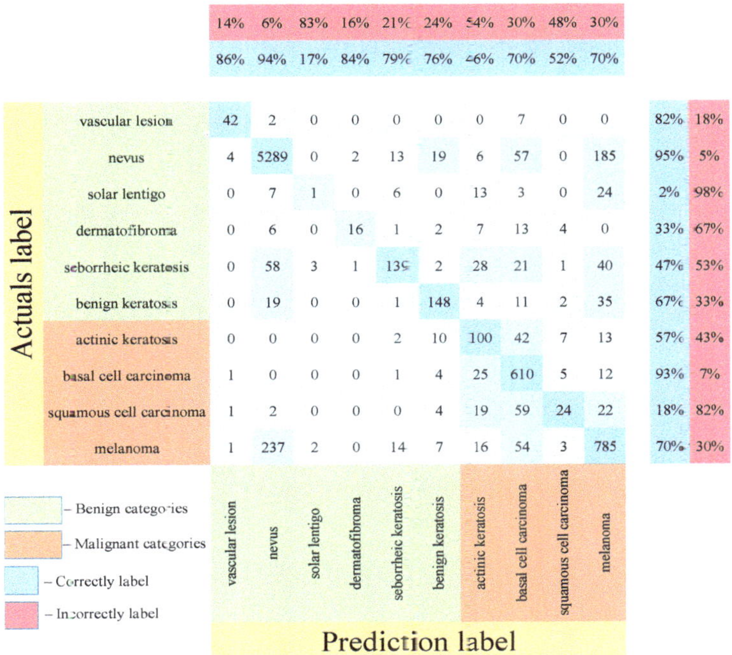

Figure 15. Confusion matrix in the testing results in a multimodal neural network system for recognizing pigmented skin lesions based on CNN ResNet-101.

Following the analysis of the confusion matrices in Figures 13–15, it can be concluded that the most frequently erroneous prediction results concern the different categories of malignant skin neoplasms (see percentages at the top of the columns). As summarized in Figure 16, part of these errors are benign lesions predicted as malignant (i.e. false positives). In addition, the malignant categories of "basal cell carcinoma" and "melanoma" are often predicted as pigmented neoplasms of benign categories. Based on the lines of the confusion matrices in Figure 16, malignant pigmented neoplasms are falsely recognized as benign in an average of 19.6% of cases.

The χ^2 McNemar statistic was calculated as follows:

$$\chi^2 = \frac{(b-c)^2}{b+c} \quad (13)$$

where b is the value when the proposed multimodal system incorrectly predicted the images and the results of the original CNN were correct; c is the value when the results of the original CNN were incorrect and the results of the multimodal system were correct.

The results of the analysis of the McNemar test from Figure 17 show that the proposed multimodal neural network system made it possible to correctly recognize pigmented neoplasms in 825–1238 images that were incorrectly classified by the original CNN with a pre-cleaning step for oatmeal structures; in 86–181 the image was misclassified, in contrast to the results of the original CNN with a pre-cleaning step for oat structures. Based on the results of the McNemar test, the proposed multimodal neural network system correctly classifies images of pigmented neoplasms on average, 12% of the time, compared to the original convolutional neural network architectures with a hair pre-clearing step.

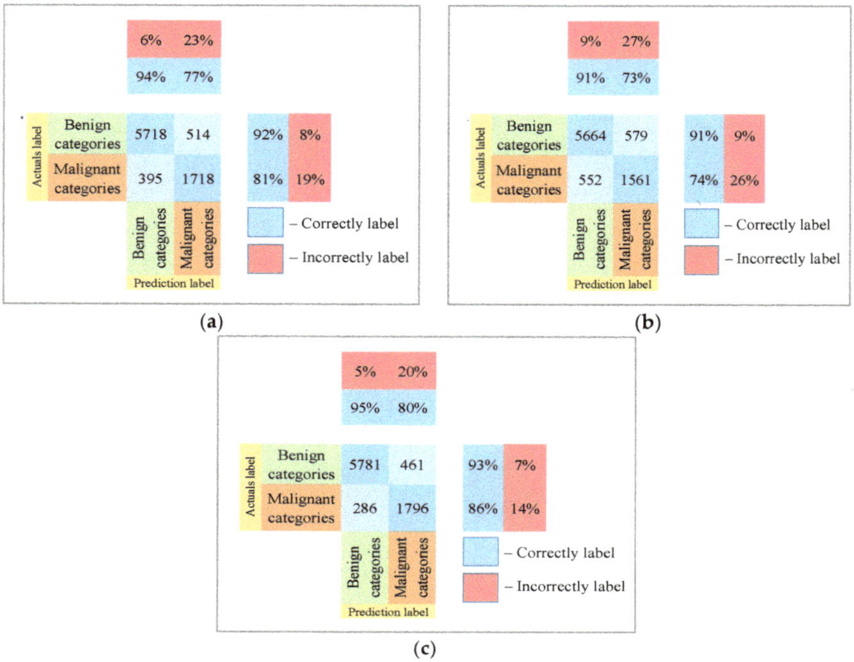

Figure 16. The confusion matrix of the test results of the proposed multimodal neural network system based on CNN is divided into two groups: (**a**) AlexNet; (**b**) SqueezeNet; (**c**) ResNet-101.

	Proposed multi-modal neural network system with a stage of preliminary hair removal. **Correct**	Proposed multi-modal neural network system with a stage of preliminary hair removal. **Incorrect**	Total:
CNN architecture ...age of preliminary ...ir removal. **Correct**	6553	181	6734
CNN architecture ...age of preliminary ...ir removal. **Incorrect**	825	920	1745
Total:	7378	1101	
χ^2 McNemar	412.262		
...cal Significance:	0.001		

(**a**)

	Proposed multi-modal neural network system with a stage of preliminary hair removal. **Correct**	Proposed multi-modal neural network system with a stage of preliminary hair removal. **Incorrect**	Total:
Original CNN architecture with a stage of preliminary hair removal. **Correct**	6293	104	6397
Original CNN architecture with a stage of preliminary hair removal. **Incorrect**	975	831	1806
Total:	7268	935	
χ^2 McNemar	703.096		
Statistical Significance:	0.001		

(**b**)

	Proposed multi-modal neural network system with a stage of preliminary hair removal. **Correct**	Proposed multi-modal neural network system with a stage of preliminary hair removal. **Incorrect**	Total:
Original CNN architecture with a stage of preliminary hair removal. **Correct**	5982	86	6068
Original CNN architecture with a stage of preliminary hair removal. **Incorrect**	1238	959	2197
Total:	7220	1045	
χ^2 McNemar	1002.344		
Statistical Significance:	0.001		

(**c**)

Figure 17. Classification table neural network systems for recognizing pigmented skin lesions for analysis McNemar based on CNN: (**a**) AlexNet; (**b**) SqueezeNet; (**c**) ResNet-101.

Even though the proposed multimodal neural network system with the stage of preliminary cleaning of hair structures shows higher results in recognition accuracy compared to existing similar systems, as well as compared to visual diagnostic methods for physicians in the field of dermatology, the use of the proposed system as an independent diagnostic tool is impossible due to the presence of a false-negative response in cases of malignant neoplasms. This system can only be used as a high-precision auxiliary tool for physicians and specialists.

Figure 18 shows the ROC curve when testing a multimodal neural network system to identify pigmented skin lesions based on various CNNs.

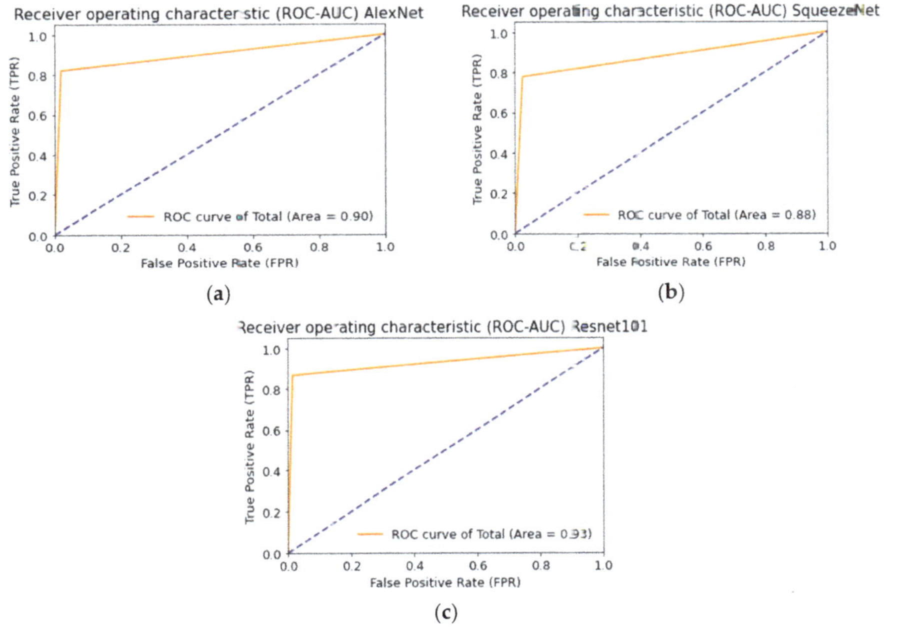

Figure 18. Receiver operative characteristics (ROC) curve when testing a multimodal neural network system for recognizing pigmented lesions and skin based on CNN: (**a**) AlexNet; (**b**) SqueezeNet; (**c**) ResNet-101.

AlexNet deep neural network architecture is superior to other architectures in the following ways: it does not require specialized hardware and works well with limited GPU; learning AlexNet is faster than other deeper architectures; more filters are used on each layer; a pooling layer follows each convolutional layer; ReLU is used as the activation function, which is more biological and reduces the likelihood of the gradient disappearing [59]. The listed characteristics substantiate the best result of training a multimodal neural network to recognize pigmented skin lesions based on the AlexNet neural network architecture.

4. Discussion

As a result of modeling the proposed multimodal neural network system, the best recognition accuracy was 83.6%. The preliminary cleaning of hair structures and the analysis of heterogeneous data made it possible to significantly exceed the classification accuracy compared to simple neural network architectures to recognize dermoscopic images. In [20] CNN GoogleNet Inception v3 was trained based on dermoscopic images, consisting of nine diagnostically significant categories. The recognition accuracy of CNN

GoogleNet Inception v3 was 72.1%, which is 11.46% lower than modeling the multimodal neural network system proposed in this paper; in [21], the authors present CNN ResNet50 training based on benign and malignant pigmented skin lesions. The trained ResNet50 CNN achieved 82.3% accuracy, which is 1.26% lower than the recognition accuracy of the proposed system with the hair pre-cleaning step. The superior recognition accuracy of the multimodal neural network system proposed in this paper compared to the results of pre-trained CNNs is explained by different data processing methods, which, when used together, enter into synergy.

In [60], preliminary hair cleaning is performed using the DullRazor method, and the skin lesion image classification using a neural network classifier. The best result of recognition accuracy was 78.2%. The analysis of heterogeneous data using the proposed multimodal neural network system made it possible to increase the recognition accuracy by 5.4% compared to recognition using a neural network classifier; [61] presents a skin cancer detection system. The preliminary cleaning of dermatoscopic images from hair was performed at the first stage using the DullRazor method. Neural network classification was performed using the K-Nearest Neighbor (KNN). The system's accuracy was 82.3%, which is 1.3% lower than the recognition accuracy of the proposed multimodal neural network system with the stage of preliminary cleaning of hair structures. The authors of [62] proposed a neural network system for classifying benign and malignant pigmented skin lesions with the stage of preliminary hair removal. This approach made it possible to achieve a classification accuracy of 79.1%, which is 4.5% lower than the recognition accuracy of the proposed multimodal neural network system. Combining and analyzing heterogeneous dermatological data allows the multimodal neural network algorithm to find additional links between images and metadata and improve recognition accuracy compared to the classification accuracy of visual data only by neural network algorithms.

A comparison of the recognition accuracy of various multimodal neural network systems for recognizing pigmented lesions and skin with the proposed system is presented in Table 6.

Table 6. Results of recognition accuracy of various multimodal neural network systems for recognizing pigmented lesions and skin.

Multimodal Neural Network Systems for the Classification of Skin Pigmentation Lesions		**Accuracy of Detection of Pigmented Skin Lesions, %**
Known neural network systems	[34]	63.4
	[35]	72.0
	[36]	72.9
	[38]	79.0
Proposed neural network system		83.6

In [34], the authors solved two problems for neural network classification of pigmented skin lesions. The modeling was carried out based on the open archive ISIC 2019, which is currently the most suitable for research in this area since it contains the largest amount of visual and statistical data. The authors selected 25,331 dermatoscopic images for modeling, divided into eight diagnostically significant categories. The authors used various CNNs to classify dermatoscopic images for the first task. For the second task, statistical metadata about patients was also used along with the photos. The multimodal neural network system for the second task consisted of CNN for dermatoscopic imaging and a dense neural network for metadata. In the first step, the authors trained CNN only on visual multivariate data, then fixed the CNN weights and connected a neural network with metadata. The core architecture of CNN was a pre-trained EfficientNets consisting of eight different models. Pre-trained SENet154 and ResNext were also used for modeling variability. The images were cropped to the required size $224 \times 224 \times 3$ and augmented as a pre-processing stage. Metadata pre-processing consisted of simple numeric coding. In this case, the missing values were coded as "-5". Most of the training was done on an NVIDIA GTX 1080TI graphics card. The use of metadata has improved the accuracy by 1–2%. At the same time, the increase was observed mainly on smaller models. On the test set, the accuracy of the neural network recognition system in the first task was 63.4%. For the second task using metadata, the accuracy on the test set was 63.4%. At the same time, the most optimal results were 72.5 ± 1.7 and 74.2 ± 1.1 for the first and second tasks, respectively.

Identical conditions for modeling, hardware resources, image base, and many diagnostic categories used make it possible to compare the results obtained with the proposed multimodal neural network system with the stage of preliminary hair removal with the results from work. The recognition accuracy of the proposed multimodal system with the stage of preliminary hair removal on the test set was 83.6%, which is about 20.2% higher than the results of testing the system from [34]. The main difference between the multimodal neural network system proposed in the work is the use of the hair removal method at the stage of preliminary processing of visual data, which significantly increased the accuracy.

In [35], a multimodal convolutional neural network (IM-CNN) is presented, a model for the multiclass classification of dermatoscopic images and patient metadata as input for diagnosing pigmented skin lesions. The modeling was carried out on the open dataset HAM10.000 ("Human versus machine with 10,000 training images"), part of the ISIC Melanoma Project open database, and consists of seven diagnostic categories. This set includes statistical metadata about patients such as age, gender, location of pigmented lesions, and diagnosis. The pre-trained DenseNet and ResNet architectures were used as CNNs to classify dermatoscopic images. The best test result for the proposed model was 72% recognition accuracy. That is about 11.6% lower than the proposed multimodal system with a stage of preliminary hair removal. The main differences in the operation of the proposed multimodal system for the recognition of pigmented lesions of the skin are, firstly, the stage of preliminary hair removal, and, secondly, the use of a larger number of diagnostically significant recognition classes and a more substantial amount of data for training. These distinctive features made it possible to improve the visual quality of diagnostically significant signs on dermatoscopic images due to the removal of hair structures and improve the correctness and balance of the training of the neural network system.

The authors of [36] presented a method combining visual data and patient metadata to improve the efficiency of automatic diagnosis of pigmented skin lesions. The modeling was carried out on the ISIC Melanoma Project database, which consisted of 2917 dermatoscopic images of five classes (nevi, melanoma, basal cell carcinoma, squamous cell carcinoma, pigmented benign keratoses). For image recognition, a modified CNN architecture, ResNet-50, was used. Simulation results have shown that the combination of dermatoscopic images and metadata can improve the accuracy of the classification of skin lesions. The best average recognition accuracy (mAP) using metadata on the test set was 72.9%. This result is 10.7% lower than the recognition accuracy of the proposed multimodal system for recognizing pigmented skin lesions with a stage of preliminary removal of hair structures. A small variation in the database of dermatoscopic examples for training in [36] can significantly affect the reliability of the neural network classification system.

In [38] proposed two methods for classifying pigmented skin lesions. The first method was to use CNN to recognize dermatoscopic images. The authors selected 1000 images from the International Skin Imaging Collaboration (ISIC) archive, divided into two categories (benign and melanoma). The result of recognition accuracy in two categories on the basis for validation was 82.2%. The second method used 600 images from the ISIC archive and patient metadata. Metadata has been added to the dermatoscopic image pixel matrix in each RGB layer at the bottom. After repeatedly adding metadata, a colored bar appeared on the images. The accuracy of CNN recognition and the metadata on the validation set was 79.0%, which is 4.6% lower than the recognition accuracy of the proposed multimodal neural network system. Although adding metadata directly to the image matrix allowed the authors from [38] to improve the classification accuracy, using a separate full-fledged classifier for statistical data is a more rational solution. Convolutional layers in CNN highlight such features on dermatoscopic images as contour, color, size. The metadata added to the pixel matrix of each dermatoscopic image does not require feature extraction.

The main limitation in using the proposed multimodal neural network system for recognizing pigmented lesions in the skin is that specialists can only use the system as an additional diagnostic tool. The proposed system is not a medical device and cannot independently diagnose patients. Since the major dermatoscopic training databases are biased towards benign image classifications, misclassification is possible. The use of augmentation based on affine transformations makes it possible to minimize this factor but not completely exclude it.

A promising direction for further research is constructing more complex multimodal systems for neural network classification of pigmented skin neoplasms. The use of segmentation and preliminary cleaning of the hair's visual data will help highlight the contour of the pigmented skin lesion. Distortion of the shapes of the skin neoplasm is an important diagnostic sign that may indicate the malignancy of this lesion.

5. Conclusions

The article presents a multimodal neural network system for recognizing pigmented skin lesions with a stage of preliminary cleaning from hair structures. The fusion of dissimilar data made it possible to increase the recognition accuracy by 4.93–6.28%, depending on the CNN architecture. The best recognition accuracy for 10 diagnostically significant categories was 83.56% when using the AlexNet pre-trained CNN architecture. At the same time, the best indicator of improving the accuracy was obtained using the pre-trained ResNet-101 architecture and amounted to 6.28%. The use of the stage of preliminary processing of visual data made it possible to prepare dermatoscopic images for further analysis and improve the quality of diagnostically important visual information. At the same time, the fusion of patient statistics and visual data made it possible to find additional links between dermatoscopic images and the results of medical diagnostics, which significantly increased the accuracy of the classification of neural networks.

Creating systems for automatically recognizing the state of pigmented lesions of patients' skin can be a good incentive for cognitive medical monitoring systems. This can reduce the consumption of financial and labor resources involved in the medical industry. At the same time, the creation of mobile monitoring systems to monitor potentially dangerous skin neoplasms will automatically receive feedback on the condition of patients.

Author Contributions: Conceptualization, P.A.L.; methodology, P.A.L.; software, U.A.L.; validation, U.A.L. and N.N.N.; formal analysis, N.N.N.; investigation, P.A.L.; resources, U.A.L.; data curation, N.N.N.; writing—original draft preparation, U.A.L.; writing—review and editing, P.A.L.; visualization, N.N.N.; supervision, P.A.L.; project administration, U.A.L.; funding acquisition, P.A.L. All authors have read and agreed to the published version of the manuscript.

Funding: The work of Pavel Lyakhov (Section 2.1) is supported by the Ministry of Science and Higher Education of the Russian Federation 'Goszadanie' №075-01024-21-02 from 29 September 2021 (project FSEE-2021-0015). Research in Section 2.2 was supported by the Presidential Council for grants (project no. MK-3918.2021.1.6). Research in Section 2.3 was supported by the Presidential Council for grants (project no. MK-371.2022.4). Research in Section 3 was supported by Russian Science Foundation, project 21-71-00017.

Institutional Review Board Statement: Not applicable.

Informed Consent Statement: Not applicable.

Data Availability Statement: A publicly available dataset was analyzed in this study. This data can be found in https://www.isic-archive.com/#!/topWithHeader/wideContentTop/main, accessed on 7 February 2022. Both the data analyzed during the current study and code are available from the corresponding author upon request.

Acknowledgments: The authors would like to thank the North-Caucasus Federal University for supporting the contest of projects competition of scientific groups and individual scientists of the North-Caucasus Federal University.

Conflicts of Interest: The authors declare no conflict of interest.

Appendix A

Appendix A shows the training and testing graphs of the proposed multimodal neural network systems based on various CNN architectures with preliminary cleaning of hair structures.

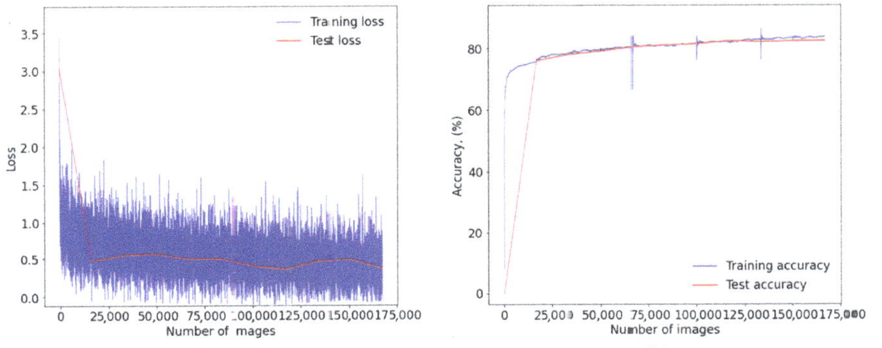

(a) (b)

Figure A1. Graph of learning outcomes of a multimodal neural network system for classifying dermatoscopic images of pigmented skin lesions based on CNN AlexNet: (**a**) loss function; (**b**) recognition accuracy.

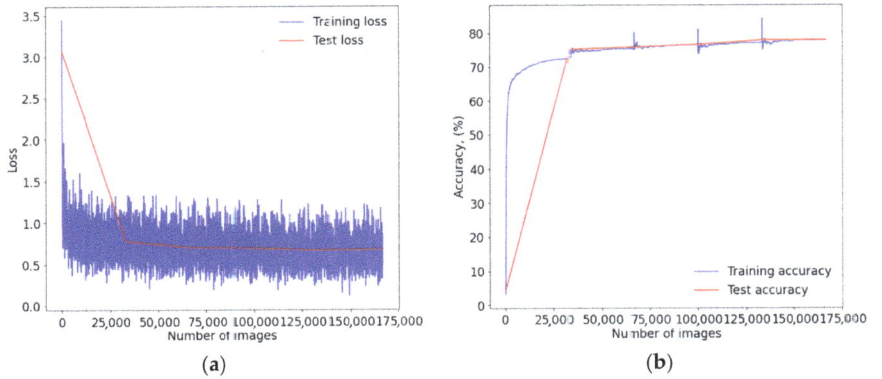

(a) (b)

Figure A2. Graph of learning outcomes of a multimodal neural network system for classifying dermatoscopic images of pigmented skin lesions based on CNN SqueezeNet: (**a**) loss function; (**b**) recognition accuracy.

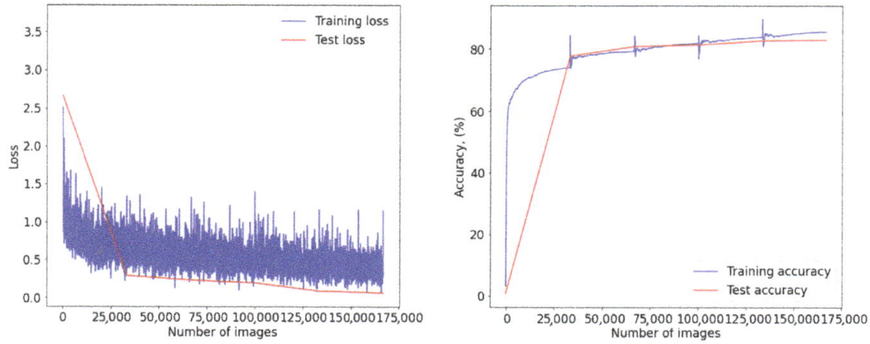

(a) (b)

Figure A3. Graph of learning outcomes of a multimodal neural network system for classifying dermatoscopic images of pigmented skin lesions based on CNN ResNet-101: (**a**) loss function; (**b**) recognition accuracy.

References

1. Health Consequences of Excessive Solar UV Radiation. Available online: https://www.who.int/news/item/25-07-2006-health-consequences-of-excessive-solar-uv-radiation (accessed on 18 October 2021).
2. Rogers, H.W.; Weinstock, M.A.; Harris, A.R.; Hinckley, M.R.; Feldman, S.R.; Fleischer, A.B.; Coldiron, B.M. Incidence Estimate of Nonmelanoma Skin Cancer in the United States, 2006. *Arch. Dermatol.* **2010**, *146*, 283–287. [CrossRef] [PubMed]
3. Madan, V.; Lear, J.T.; Szeimies, R.-M. Non-Melanoma Skin Cancer. *Lancet* **2010**, *375*, 673–685. [CrossRef]
4. The Skin Cancer Foundation. Skin Cancer Facts & Statistics. Available online: https://www.skincancer.org/skin-cancer-information/skin-cancer-facts/ (accessed on 21 October 2021).
5. Stern, R.S. Prevalence of a History of Skin Cancer in 2007: Results of an Incidence-Based Model. *Arch. Dermatol.* **2010**, *146*, 279–282. [CrossRef] [PubMed]
6. Fitzpatrick, T.B. Pathophysiology of Hypermelanoses. *Clin. Drug Investig.* **2012**, *10*, 17–26. [CrossRef]
7. Fitzpatrick, T.B. The Validity and Practicality of Sun-Reactive Skin Types I Through VI. *Arch. Dermatol.* **1988**, *124*, 869–871. [CrossRef] [PubMed]
8. Pathak, M.A.; Jimbow, K.; Szabo, G.; Fitzpatrick, T.B. Sunlight and Melanin Pigmentation. *Photochem. Photobiol. Rev.* **1976**, *1*, 211–239. [CrossRef]
9. Rogers, H.W.; Weinstock, M.A.; Feldman, S.R.; Coldiron, B.M. Incidence Estimate of Nonmelanoma Skin Cancer (Keratinocyte Carcinomas) in the US Population, 2012. *JAMA Dermatol.* **2015**, *151*, 1081–1086. [CrossRef]
10. Hoey, S.E.H.; Devereux, C.E.J.; Murray, L.; Catney, D.; Gavin, A.; Kumar, S.; Donnelly, D.; Dolan, O.M. Skin Cancer Trends in Northern Ireland and Consequences for Provision of Dermatology Services. *Br. J. Dermatol.* **2007**, *156*, 1301–1307. [CrossRef]
11. Diepgen, T.L.; Mahler, V. The Epidemiology of Skin Cancer. *Br. J. Dermatol.* **2002**, *146*, 1–6. [CrossRef]
12. Alzahrani, S.; Al-Bander, B.; Al-Nuaimy, W. A Comprehensive Evaluation and Benchmarking of Convolutional Neural Networks for Melanoma Diagnosis. *Cancers* **2021**, *13*, 4494. [CrossRef]
13. Siegel, R.; Naishadham, D.; Jemal, A. Cancer Statistics for Hispanics/Latinos, 2012. *CA Cancer J. Clin.* **2012**, *62*, 283–298. [CrossRef] [PubMed]
14. Nami, N.; Giannini, E.; Burroni, M.; Fimiani, M.; Rubegni, P. Teledermatology: State-of-the-Art and Future Perspectives. *Expert Rev. Dermatol.* **2012**, *7*, 1–3. [CrossRef]
15. Bratchenko, I.A.; Alonova, M.v.; Myakinin, O.O.; Moryatov, A.A.; Kozlov, S.V.; Zakharov, V.P. Hyperspectral Visualization of Skin Pathologies in Visible Region. *Comput. Opt.* **2016**, *40*, 240–248. [CrossRef]
16. Chen, M.; Zhou, P.; Wu, D.; Hu, L.; Hassan, M.M.; Alamri, A. AI-Skin: Skin Disease Recognition Based on Self-Learning and Wide Data Collection through a Closed-Loop Framework. *Inf. Fusion* **2020**, *54*, 1–9. [CrossRef]
17. Andreu-Perez, J.; Poon, C.C.Y.; Merrifield, R.D.; Wong, S.T.C.; Yang, G.Z. Big Data for Health. *IEEE J. Biomed. Health Inform.* **2015**, *19*, 1193–1208. [CrossRef]
18. Neubauer, C. Evaluation of Convolutional Neural Networks for Visual Recognition. *IEEE Trans. Neural Netw.* **1998**, *9*, 685–696. [CrossRef]
19. Brinker, T.J.; Hekler, A.; Enk, A.H.; Klode, J.; Hauschild, A.; Berking, C.; Schilling, B.; Haferkamp, S.; Schadendorf, D.; Holland-Letz, T.; et al. Deep Learning Outperformed 136 of 157 Dermatologists in a Head-to-Head Dermoscopic Melanoma Image Classification Task. *Eur. J. Cancer* **2019**, *113*, 47–54. [CrossRef]
20. Esteva, A.; Kuprel, B.; Novoa, R.A.; Ko, J.; Swetter, S.M.; Blau, H.M.; Thrun, S. Dermatologist-Level Classification of Skin Cancer with Deep Neural Networks. *Nature* **2017**, *542*, 115–118. [CrossRef]

21. Brinker, T.J.; Hekler, A.; Enk, A.H.; Berking, C.; Haferkamp, S.; Hauschild, A.; Weichenthal, M.; Klode, J.; Schadendorf, D.; Holland-Letz, T.; et al. Deep Neural Networks Are Superior to Dermatologists in Melanoma Image Classification. *Eur. J. Cancer* **2019**, *119*, 11–17. [CrossRef]
22. Zimmermann, M.; Benning, L.; Peintner, A.; Peintner, L. Advances in and the Applicability of Machine Learning-Based Screening and Early Detection Approaches for Cancer: A Primer. *Cancers* **2022**, *14*, 623. [CrossRef]
23. Amin, J.; Sharif, A.; Gul, N.; Anjum, M.A.; Nisar, M.W.; Azam, F.; Bukhari, S.A.C. Integrated Design of Deep Features Fusion for Localization and Classification of Skin Cancer. *Pattern Recognit. Lett.* **2020**, *131*, 63–70. [CrossRef]
24. Khan, M.A.; Akram, T.; Sharif, M.; Javed, K.; Rashid, M.; Bukhari, S.A.C. An Integrated Framework of Skin Lesion Detection and Recognition through Saliency Method and Optimal Deep Neural Network Features Selection. *Neural Comput. Appl.* **2020**, *32*, 15929–15948. [CrossRef]
25. Khan, M.A.; Sharif, M.; Akram, T.; Bukhari, S.A.C.; Nayak, R.S. Developed Newton-Raphson Based Deep Features Selection Framework for Skin Lesion Recognition. *Pattern Recognit. Lett.* **2020**, *129*, 293–303. [CrossRef]
26. Manzoor, K.; Majeed, F.; Siddique, A.; Meraj, T.; Rauf, H.T.; El-Meligy, M.A.; Sharaf, M.; Abd Elgawad, A.E.E. A Lightweight Approach for Skin Lesion Detection through Optimal Features Fusion. *Comput. Mater. Contin.* **2021**, *70*, 1617–1630. [CrossRef]
27. Gomez Garcia, A.M.; McLaren, C.E.; Meyskens, F.L. Melanoma: Is Hair the Root of the Problem? *Pigment Cell Melanoma Res.* **2011**, *24*, 110. [CrossRef] [PubMed]
28. Zaqout, I.S. Image Processing, Pattern Recognition: An Efficient Block-Based Algorithm for Hair Removal in Dermoscopic Images. *Comput. Opt.* **2017**, *41*, 521–527. [CrossRef]
29. Zhou, H.; Chen, M.; Gass, R.; Rehg, J.M.; Ferris, L.; Ho, J.; Drogowski, L. Feature-preserving artifact removal from dermoscopy images. In *Medical Imaging 2008: Image Processing*; International Society for Optics and Photonics: Bellingham, WA, USA, 2008; Volume 6914, p. 69141B. [CrossRef]
30. Lee, T.; Ng, V.; Gallagher, R.; Coldman, A.; McLean, D. Dullrazor®: A Software Approach to Hair Removal from Images. *Comput. Biol. Med.* **1997**, *27*, 533–543. [CrossRef]
31. Abbas, Q.; Celebi, M.E.; Garcia, I.F. Hair Removal Methods: A Comparative Study for Dermoscopy Images. *Biomed. Signal Process. Control* **2011**, *6*, 395–404. [CrossRef]
32. Barcelos, C.A.Z.; Pires, V.B. An Automatic Based Nonlinear Diffusion Equations Scheme for Skin Lesion Segmentation. *Appl. Math. Comput.* **2009**, *215*, 251–261. [CrossRef]
33. Xie, F.Y.; Qin, S.Y.; Jiang, Z.G.; Meng, R.S. PDE-Based Unsupervised Repair of Hair-Occluded Information in Dermoscopy Images of Melanoma. *Comput. Med. Imaging Graph.* **2009**, *33*, 275–282. [CrossRef] [PubMed]
34. Gessert, N.; Nielsen, M.; Shaikh, M.; Werner, R.; Schlaefer, A. Skin Lesion Classification Using Ensembles of Multi-Resolution EfficientNets with Meta Data. *MethodsX* **2020**, *7*, 100864. [CrossRef] [PubMed]
35. Wang, S.; Yin, Y.; Wang, D.; Wang, Y.; Jin, Y. Interpretability-Based Multimodal Convolutional Neural Networks for Skin Lesion Diagnosis. *IEEE Trans. Cybern.* **2021**, 1–15. [CrossRef] [PubMed]
36. Yap, J.; Yolland, W.; Tschandl, P. Multimodal Skin Lesion Classification Using Deep Learning. *Exp. Dermatol.* **2018**, *27*, 1261–1267. [CrossRef]
37. Bi, L.; Feng, D.D.; Fulham, M.; Kim, J. Multi-Label Classification of Multi-Modality Skin Lesion via Hyper-Connected Convolutional Neural Network. *Pattern Recognit.* **2020**, *107*, 107502. [CrossRef]
38. Ruiz-Castilla, J.-S.; Rangel-Cortes, J.-J.; Garcia-Lamont, F.; Trueba-Espinosa, A.; Zumpango, J.; Tejocote, E.; de Mexico, E. CNN and Metadata for Classification of Benign and Malignant Melanomas. In *Lecture Notes in Computer Science*; Springer: Berlin/Heidelberg, Germany, 2019; Volume 11644, pp. 569–579. [CrossRef]
39. Pellacani, G.; Seidenari, S. Comparison between Morphological Parameters in Pigmented Skin Lesion Images Acquired by Means of Epiluminescence Surface Microscopy and Polarized-Light Videomicroscopy. *Clin. Dermatol.* **2002**, *20*, 222–227. [CrossRef]
40. Sinz, C.; Tschandl, P.; Rosendahl, C.; Akay, B.N.; Argenziano, G.; Blum, A.; Braun, R.P.; Cabo, H.; Gourhant, J.Y.; Kreusch, J.; et al. Accuracy of Dermatoscopy for the Diagnosis of Nonpigmented Cancers of the Skin. *J. Am. Acad. Dermatol.* **2017**, *77*, 1100–1109. [CrossRef]
41. Li, W.; Joseph Raj, A.N.; Tjahjadi, T.; Zhuang, Z. Digital Hair Removal by Deep Learning for Skin Lesion Segmentation. *Pattern Recognit.* **2021**, *117*, 107994. [CrossRef]
42. Fiorese, M.; Peserico, E.; Silletti, A. VirtualShave: Automated Hair Removal from Digital Dermatoscopic Images. In Proceedings of the Annual International Conference of the IEEE Engineering in Medicine and Biology Society, Boston, MA, USA, 30 August–3 September 2011; pp. 5145–5148. [CrossRef]
43. Waldmann, P. Approximate Bayesian Neural Networks in Genomic Prediction. *Genet. Sel. Evol.* **2018**, *50*, 1–9. [CrossRef]
44. Atrey, P.K.; Hossain, M.A.; El Saddik, A.; Kankanhalli, M.S. Multimodal Fusion for Multimedia Analysis: A Survey. *Multimedia Syst.* **2010**, *16*, 345–379. [CrossRef]
45. Li, J.; Wang, Q. Multi-Modal Bioelectrical Signal Fusion Analysis Based on Different Acquisition Devices and Scene Settings: Overview, Challenges, and Novel Orientation. *Inf. Fusion* **2022**, *79*, 229–247. [CrossRef]
46. Xu, Z.; Luo, J.; Yan, J.; Pulya, R.; Li, X.; Wells, W.; Jagadeesan, J. Adversarial Uni- and Multi-Modal Stream Networks for Multimodal Image Registration. In *Lecture Notes in Computer Science*; Springer: Berlin/Heidelberg, Germany, 2020; Volume 12263, pp. 222–232. [CrossRef]

47. Huertas-Tato, J.; Martin, A.; Fierrez, J.; Camacho, D. Fusing CNNs and Statistical Indicators to Improve Image Classification. *Inf. Fusion* **2022**, *79*, 174–187. [CrossRef]
48. Li, H.; Pan, Y.; Zhao, J.; Zhang, L. Skin Disease Diagnosis with Deep Learning: A Review. *Neurocomputing* **2021**, *464*, 364–393. [CrossRef]
49. ISIC Archive. Available online: https://www.isic-archive.com/#!/topWithHeader/wideContentTop/main (accessed on 18 November 2021).
50. Tschandl, P.; Rosendahl, C.; Kittler, H. The HAM10000 Dataset, a Large Collection of Multi-Source Dermatoscopic Images of Common Pigmented Skin Lesions. *Sci. Data* **2018**, *5*, 1–9. [CrossRef] [PubMed]
51. Heaphy, M.R.; Ackerman, A.B. The nature of solar keratosis: A critical review in historical perspective. *J. Am. Acad. Dermatol.* **2000**, *43*, 138–150. [CrossRef]
52. Siegel, J.A.; Korgavkar, K.; Weinstock, M.A. Current perspective on actinic keratosis: A review. *Br. J. Dermatol.* **2017**, *177*, 350–358. [CrossRef]
53. Jeffes, E.W.B.; Tang, E.H. Actinic keratosis. *Am. J. Clin. Dermatol.* **2000**, *1*, 167–179. [CrossRef]
54. Krizhevsky, A.; Sutskever, I.; Hinton, G.E. ImageNet Classification with Deep Convolutional Neural Networks. *Commun. ACM* **2017**, *60*, 84–90. [CrossRef]
55. Iandola, F.N.; Han, S.; Moskewicz, M.W.; Ashraf, K.; Dally, W.J.; Keutzer, K. SqueezeNet: AlexNet-Level Accuracy with 50x Fewer Parameters and <0.5 MB Model Size. *arXiv* **2016**, arXiv:1602.07360.
56. He, K.; Zhang, X.; Ren, S.; Sun, J. Deep Residual Learning for Image Recognition. In Proceedings of the IEEE Computer Society Conference on Computer Vision and Pattern Recognition, Las Vegas, NV, USA, 27–30 June 2016; pp. 770–778. [CrossRef]
57. Zhao, A.; Balakrishnan, G.; Durand, F.; Guttag, J.V.; Dalca, A.V. Data Augmentation Using Learned Transformations for One-Shot Medical Image Segmentation. In Proceedings of the IEEE Computer Society Conference on Computer Vision and Pattern Recognition, Long Beach, CA, USA, 15–20 June 2019; pp. 8543–8553. [CrossRef]
58. Bisla, D.; Choromanska, A.; Berman, R.S.; Stein, J.A.; Polsky, D. Towards Automated Melanoma Detection with Deep Learning: Data Purification and Augmentation. In Proceedings of the 2019 IEEE/CVF Conference on Computer Vision and Pattern Recognition Workshops (CVPRW), Long Beach, CA, USA, 16–17 June 2019; IEEE: Piscataway, NJ, USA; pp. 2720–2728.
59. Hosny, K.M.; Kassem, M.A.; Foaud, M.M. Classification of Skin Lesions Using Transfer Learning and Augmentation with Alex-Net. *PLoS ONE* **2019**, *14*, e0217293. [CrossRef]
60. Lynn, N.C.; Kyu, Z.M. Segmentation and Classification of Skin Cancer Melanoma from Skin Lesion Images. In Proceedings of the Parallel and Distributed Computing, Applications and Technologies, PDCAT, Taipei, Taiwan, 18–20 December 2017; pp. 117–122. [CrossRef]
61. Kamboj, A. A Color-Based Approach for Melanoma Skin Cancer Detection. In Proceedings of the 1st International Conference on Secure Cyber Computing and Communications, ICSCCC 2018, Jalandhar, India, 15–17 December 2018; pp. 508–513. [CrossRef]
62. Zagrouba, E.; Barhoumi, W. A Prelimary Approach for the Automated Recognition of Malignant Melanoma. *Image Anal. Stereol.* **2004**, *23*, 121–135. [CrossRef]

cancers

MDPI

Review

Virtual Reality Rehabilitation Systems for Cancer Survivors: A Narrative Review of the Literature

Antonio Melillo [1,2], Andrea Chirico [3], Giuseppe De Pietro [4], Luigi Gallo [4], Giuseppe Caggianese [4], Daniela Barone [5], Michelino De Laurentiis [6,*] and Antonio Giordano [2]

1 Department of Mental and Physical Health and Preventive Medicine, "Luigi Vanvitelli" University of Campania, 80129 Naples, Italy; antonio.melillo2@studenti.unicampania.it

2 Department of Biology, Sbarro Institute for Cancer Research and Molecular Medicine, Center for Biotechnology, College of Science and Technology, Temple University, Philadelphia, PA 19122, USA; giordano@temple.edu

3 Department of Social and Developmental Psychology, "Sapienza" University of Rome, 00185 Rome, Italy; andrea.chirico@uniroma1.it

4 Institute for High Performance Computing and Networking, National Research Council of Italy (ICAR-CNR), 80131 Naples, Italy giuseppe.depietro@icar.cnr.it (G.D.P.); luigi.gallo@icar.cnr.it (L.G.); giuseppe.caggianese@icar.cnr.it (G.C.)

5 Cell Biology and Biotherapy Unit, Istituto Nazionale Tumori-IRCCS-Fondazione G. Pascale, 80131 Naples, Italy; d.barone@istitutotumori.na.it

6 Department of Breast and Thoracic Oncology, National Cancer Institute "Fondazione Pascale", 80131 Naples, Italy

* Correspondence: m.delaurentiis@istitutotumori.na.it; Tel.: +39-0815903512

Simple Summary: To the best of our knowledge, this is the first review aiming to assess the impact of VR on the rehabilitation care of cancer survivors. We conducted a general review of the current evidence on the efficacy of virtual reality rehabilitation (VRR) systems on cancer-related impairments as retrieved through a systematic search of the main research databases. VRR systems may improve adherence to rehabilitation training programs and be better tailored to cancer patients' needs, but more data is needed.

Abstract: Rehabilitation plays a crucial role in cancer care, as the functioning of cancer survivors is frequently compromised by impairments that can result from the disease itself but also from the long-term sequelae of the treatment. Nevertheless, the current literature shows that only a minority of patients receive physical and/or cognitive rehabilitation. This lack of rehabilitative care is a consequence of many factors, one of which includes the transportation issues linked to disability that limit the patient's access to rehabilitation facilities. The recent COVID-19 pandemic has further shown the benefits of improving telemedicine and home-based rehabilitative interventions to facilitate the delivery of rehabilitation programs when attendance at healthcare facilities is an obstacle. In recent years, researchers have been investigating the benefits of the application of virtual reality to rehabilitation. Virtual reality is shown to improve adherence and training intensity through gamification, allow the replication of real-life scenarios, and stimulate patients in a multimodal manner. In our present work, we offer an overview of the present literature on virtual reality-implemented cancer rehabilitation. The existence of wide margins for technological development allows us to expect further improvements, but more randomized controlled trials are needed to confirm the hypothesis that VRR may improve adherence rates and facilitate telerehabilitation.

Keywords: virtual; reality; cancer; rehabilitation; disability; robotics; lymphedema; pain; fatigue; telemedicine

Citation: Melillo, A.; Chirico, A.; De Pietro, G.; Gallo, L.; Caggianese, G.; Barone, D.; De Laurentiis, M.; Giordano, A. Virtual Reality Rehabilitation Systems for Cancer Survivors: A Narrative Review of the Literature. *Cancers* **2022**, *14*, 3163. https://doi.org/10.3390/cancers14133163

Academic Editor: Hamid Khayyam

Received: 25 March 2022
Accepted: 24 June 2022
Published: 28 June 2022

1. Introduction

Cancer ranks as a leading healthcare issue, striking 19.3 million new cases worldwide in just 2020 and with an estimated projection of 28.4 million new cases for 2040 [1]. Contemporarily to this increase in incidence, mainly explainable by the world population's growth and aging, cancer mortality rates have been steadily decreasing by 1% per year, both in high- and low-income countries and for both sexes [2]. Thanks to both diagnostic and therapeutic advancements, the 5-year survival rate of cancer patients has indeed increased from 49% in 1979 to roughly 67% in the US in 2015 [3,4]. As a consequence of these trends, the population of individuals who have received a cancer diagnosis in their life is set to increase rapidly, with the latest projections showing an increase from 16.9 million in the US to 26.1 million people in 2040 [5]. "Cancer survivors" is a term generally used to define anyone living with the physical and or psychological consequences of a recent or past cancer diagnosis and its treatment, with some researchers even advocating for the inclusion of even cancer patients' caregivers and family members under the term [6]. These consequences have a long and significant impact on the physical functioning of this population, as both the disease, the long-term toxicity of chemotherapeutic drugs and radiotherapy, as well as surgical procedures can result in chronic symptoms and long-standing physical and cognitive impairment.

Pain is by far one of the most common chronic symptoms cancer survivors experience, with prevalence rates of 55.0% during anticancer treatment, 39.3% after curative treatment, and 66.4% in advanced, metastatic, or terminal disease [7]. Persistent pain not only significantly undermines quality of life but also causes functional limitations and hence disability. Cancer-related fatigue (CRF) is another extremely common symptom in cancer patients, with a prevalence ranging from 25% to 99% depending on the specific disease, the treatment, and age [8]. Lymphedema is an extremely frequent consequence of cancer treatment, as it can be secondary to the surgical removal of lymph nodes, radiation therapy, chemotherapy, or a combination of such [9]. The condition may severely impact patients' lives, as it causes both pain and function limitations. Its incidence is influenced by both the cancer and the intervention type: rates range from 75% of breast cancer patients after axillary nodes removal to between 14.5 and 41.4% after chest and breast radiation therapy depending on the extension of the area involved, to 50% for melanoma patients and a 16% incidence for genitourinary cancers [10,11]. Many cancer survivors experience not only physical but also cognitive impairment, in particular in areas such as memory, attention span, word-finding, and speed of processing and execution. This impairment is sometimes colloquially referred to as "chemo brain", referring to the well-known neurotoxicity of many chemotherapeutic drugs [12]. However, recent findings on the existence of mild cognitive impairment already existing before chemo treatment pose doubts on the true cause(s) of this condition [13]. Chemotherapy-induced peripheral neuropathy (CIPN) is a severe collateral effect of chemotherapy. Many chemotherapeutic drugs can indeed cause different types of nerve damage depending on the exact chemical compound [14]. Its incidence also varies depending on the treatment, ranging from 19% to 85%. Clinically, CIPN usually manifests itself mainly as a distal sensory deficit, with symptoms of dysesthesia, paresthesia, pain symptoms, or complete anesthesia. Motor symptoms occur less frequently and also usually involve distal limbs, causing balance and gait problems as well. CIPN usually gradually develops months after chemotherapeutic treatment and may affect the patient for years.

These conditions have been shown to benefit from rehabilitation, and in the last years, many systematic reviews and guidelines have contributed to the establishment of specific recommendations for the prescription of specific exercise programs for different cancer types [15–19]. Despite this indication, many studies have shown how just a minority of cancer survivors are referred to rehabilitation programs. Reporting data collected from 163 breast cancer survivors, Cheville et al. found that 91% of women had physical impairments, but only 30% were receiving proper rehabilitative care [20]. Concordantly, a study by Hansen et al. examining a cohort of 3439 cancer survivors reported a total of 60% of patients referring to the unmet need for either physical or psychological rehabilitation [21]. In a more recent 20-year follow-up of pediatric brain cancer survivors in Norway, the percentage rose to as high as 86% [22]. Through a non-systematic review of the previous literature, Cheville attempted to explain the lack of proper rehabilitative care, mentioning as possible causes the insidious and gradual genesis of these impairments as well as the incapability of the cancer care system to deliver the early detection of the impairing symptoms [23]. However, even when the program is initiated, it is often discontinued as early as within the first twelve months, mainly as a result of the difficulty of traditional training programs in motivating the patients' adherence [24]. In addition, the recent pandemic has very well exposed another cause of this underutilization of rehabilitative cancer care, which is the inadequacy of the present rehabilitative care system in delivering home-based interventions [25,26]. Indeed, many cancer survivors suffer from disabilities or transportation issues which may limit their attendance at rehabilitation facilities. Therefore, in the last years, many studies have been investigating the role of telerehabilitation in the rehabilitative care of cancer survivors to improve adherence and as a safe and more accessible alternative to traditional rehabilitation [27–29]. One of the latest technologies proposed to remotely connect patients and rehabilitation professionals is Virtual Reality (VR) [26,30–34]. Virtual Reality Rehabilitation (VRR) has been tested in various clinical conditions, such as stroke-related deficits [35], spinal cord injuries [36], multiple sclerosis [37], Parkinson's disease [32], cerebral palsy [38–40], and cancer rehabilitation. Many studies have argued that VRR may improve both adherence rates and training intensity thanks to its entertaining and game-like nature [41–43].

The purpose of the present narrative review is to contribute to the investigation of whether VR may be a useful implementation in the cancer rehabilitation field and to give an overview of the current evidence on this application. At the moment, the scientific literature registers either attempts to evaluate the advantages of VR implementation in the rehabilitation field in general [41,44] or to review the implementation of VR in palliative care for single cancer symptoms, mainly during acute cancer care, as highlighted by Zeng et al. [45,46]. From our perspective, the former fails to assess the advantages of VR-integrated rehabilitation when applied to the specifics of cancer survivor disabilities, which often result from the slow and insidious accrual of more symptoms and physical impairments [20]. The latter, on the other hand, does not examine the potential application of VR technology to cancer survivors with chronic symptoms and their role in an impairment-driven rehabilitation of disabilities resulting from a cancer history. Hence, to the best of our knowledge, this is the first review aiming to assess the impact of VR on the rehabilitation care of cancer survivors.

2. Methods

Database Search

The main online databases (PubMed, Scopus) were searched from inception until May 2022. The query string was the following: Cancer Survivor*" OR "cancer" OR "cancer patient*" AND "Lymphedema" OR "cancer-related fatigue" OR "Fatigue" OR "Chronic Pain" OR "Cancer Pain" OR "cognitive" OR "motor" OR "symptom management" OR "peripheral neuropathy" AND "Rehabilitation" OR "Telerehabilitation" OR "Exercise" OR "physical therapy" OR "sensorimotor rehabilitation" OR "exercise training" OR "postural balance" OR "sensorimotor" AND "Virtual Reality" OR "body sensors" OR "avatar*". The first author performed the literature search. The first and second authors independently screened titles and abstracts as well as full texts' reference lists against eligibility criteria. The final selection of articles was discussed by the first and second authors. Study eligibility was assessed using the PICOS tool [47]: to be included, studies had to fulfill the following inclusion criteria: (1) population: individuals with a history of cancer; (2) intervention: Virtual Reality-based rehabilitation; (3) comparison for RCCTs: standard physiotherapy; (4) outcomes for clinical trials: functional parameters, pain, lymphedema volume, cancer-related fatigue, program adherence, exercise performance; and (5) study design: RCT with or without control, perspective studies, comparative studies, feasibility studies. Studies published in English, Spanish, or Italian were all considered.

3. Results

The search of the main databases (PubMed, Scopus) produced a total of 7733 results. Duplicate detection led to the elimination of 149 results. After screening through eligibility criteria, a total of nine studies were selected for our review (Figure 1). We will here, therefore, review the design of the included studies, summarized in Table 1.

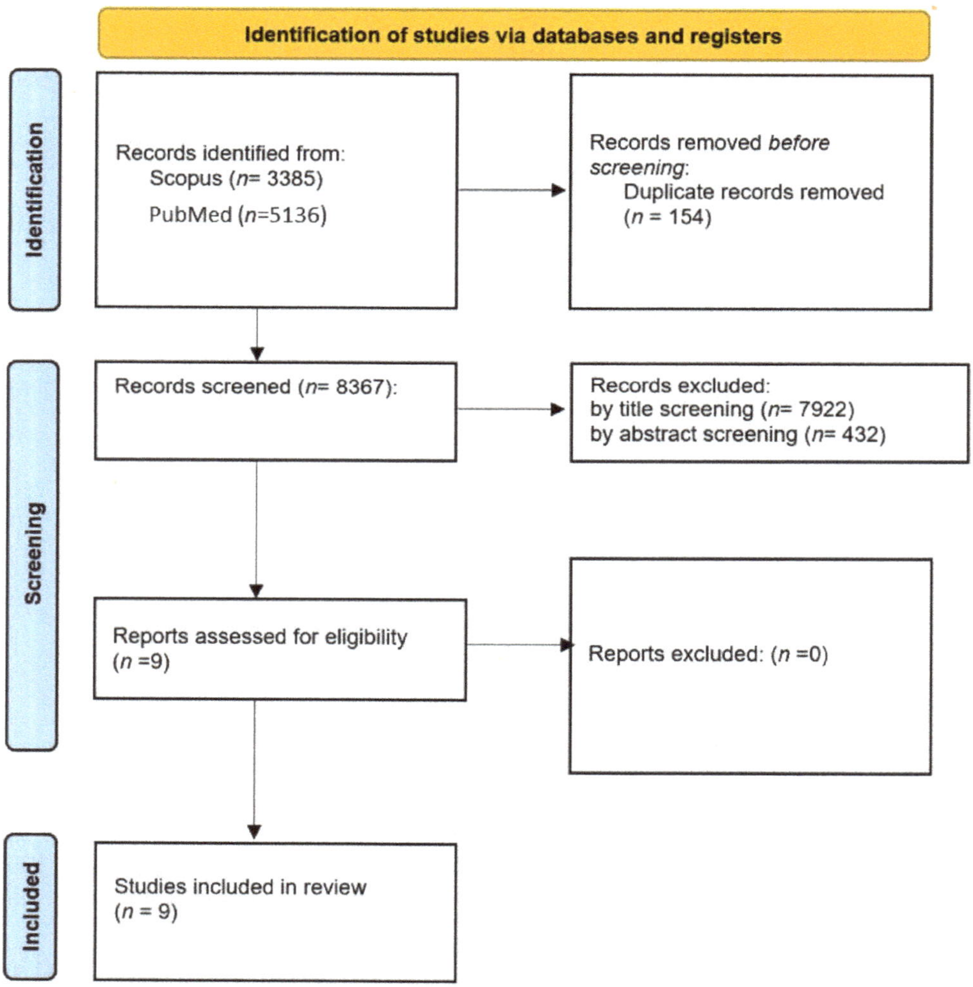

Figure 1. Prisma flowchart of the study selection.

Table 1. Features of the included studies.

Included Study	Study Design	VRR System	Considered Impairment	Outcome	Conclusions
Atef et al., 2020 [48]	Comparative study	Nintendo Wii games	Post-mastectomy lymphedema	Upper limb function (quickDASH); arm volume	VR training was not inferior to regular proprioceptive neuromuscular facilitation in improving functioning and reducing volume.
Axenie et al., 2020 [49]	Perspective study	Virtual reality avatar-based kinematics assessment and sensorimotor training	Chemotherapy-induced polyneuropathy	Not applicable	Virtual reality software allowed for simultaneous kinematics assessment and multimodal sensorimotor stimulation. In addition, it may facilitate motion training through the use of avatars.
Basha et al., 2021 [50]	Comparative study	Xbox Kinect with games involving upper limb movement	Breast cancer-related lymphedema	Pain (VAS), upper limb function (DASH), shoulder and elbow ROM, hand grip strength, quality of life	VR training was superior to resistance exercises for pain, upper limb function, and shoulder ROM outcomes.
Feyzioğlu et al., 2019 [51]	Comparative study	Xbox Kinect	Post-mastectomy arm and shoulder impairment	Pain (VAS), grip strength, functionality (disabilities of the arm, shoulder and hand questionnaire), muscle strength, ROM and fear of movement (TKS)	Both standardized therapy and VRR resulted in significant changes in pain, ROM, muscle strength, grip strength, functionality, and TKS scores, without any significant differences between groups. Fear of movement was significantly improved in the VRR group but the standard physiotherapy group displayed more improvement in functionality.
Hoffman et al., 2014 [52]	Randomized non-controlled trial	Nintendo Wii Fit Plus	Post-thoracotomy cancer-related fatigue	Levels of adherence (days of training), exercise performance, cancer-related fatigue (0–10 scale), perceived self-efficacy for fatigue self-management (0–10 scale), perceived self-efficacy for walking 30 min (%)	Non-immersive virtual reality improved both CRF and perceived self-efficacy.
House et al., 2016 [53]	Feasibility study	BrightArm Duo: robotic rehabilitation table, computerized forearm supports, and display	Post-mastectomy arm impairment, depression in cancer survivors	Pain (NRS); arm function (FMA, upper extremity section); bimanual function (CAHAI-9); hand function (JHFT); upper arm autonomy in ADL (UEFI-20); depression (BDI-II); cognitive function (NAB, HVLT-R, BVM-T, TMT);	VR rehabilitation significantly improved 10/11 cognitive parameters and depression scores. In addition, it improved arm function as well.
Reynolds et al., 2022 [54]	Randomized non-controlled trial	Immersive VR headset (Pico Goblin)	Pain, fatigue, depression, anxiety, and stress in metastatic breast cancer patients	Pain (BPI), quality of life (EQ-5D-5L scale), fatigue (FACIT-Fatigue) depression, anxiety, and stress levels, (DASS-SF)	VRR scenarios had significant effects on all considered outcomes. VRR scenarios did not significantly differ in any outcome
Schwenk et al., 2015 [55]	Randomized controlled trial	Non-immersive Virtual Reality software connected to triaxial accelerometers, gyroscopes, and magnetometers	Chemotherapy-induced polyneuropathy	Balance (sway of hip, sway of ankle, center of mass movement), gait speed, fear of falling (FES-I score)	Virtual reality improved balance through patient-tailored, sensor-based exercise but did not improve gait speed and fear of falling
Tsuda et al., 2016 [56]	Randomized non-controlled trial	Nintendo Wii Fit	Physical performance worsening related to chemotherapy and hematological malignancies	Levels of adherence, physical performance (Barthel index), muscle strength, emotive state (hospital anxiety and depression scale)	Virtual reality exercise programs showed good adherence rates (65.5%) and helped maintain physical performance in hospitalized patients.

* Table 1: Features of the included studies. VR: Virtual reality; VAS: visual analogue scale; DASH: disability of the arm, hand, and shoulder questionnaire; ROM: range of motion; TKS: Tampa Kinesiophobia Scale; CRF: cancer-related fatigue; NRS: numeric rating scale; FMA: Fulg-Meyer assessment; CAHAI-9: Chedokee arm and hand activity inventory; JHFT: Jebsen hand function test; ADL: activities of daily living; UEFI-20: upper extremity function index; BDI-II: Beck Depression Inventory, Second Edition; NAB: Neuropsychological Assessment Battery; HVLT-R: Hopkins Verbal Learning Test; BVMT-R: the Brief Visuospatial Memory Test, Revised; TMT: Trail Making Test; FES-I: Falls efficacy scale—international; pain, measured by BPI: (Brief Pain Inventory scale) (BFI); quality of life, measured through the EQ-5D-5L scale; fatigue, measured through the Functional Assessment of Chronic Illness Therapy Fatigue scale (FACIT-Fatigue); and depression, anxiety, and stress levels, measured through the short version of the Depression, Anxiety, and Stress Scales (DASS-SF).

Atef et al. conducted a quasi-randomized clinical trial comparing the efficacy of VRR and proprioceptive neuromuscular facilitation (PNF) on post-mastectomy lymphedema upper-arm exceeding volume and upper arm function recovery, measured through the QuickDASH-9 scale [48]. The experimental procedure consisted of a 30 min exercise program using a Wii Fit non-immersive VR game. Both the VRR and the PNF procedures were conducted two times per week for a total of 4 weeks. During these sessions, both groups, consisting of 15 women each, also received a procedure of pneumatic compression for the treatment of lymphedema.

Axenie and Kurz conducted a prospective study on the combination of Virtual Reality avatars and Machine Learning to drive patient-tailored CIPN-related motor deficit compensation [49]. They proposed a closed-loop system based on wearable devices designed to precisely assess the kinematics of the sensorimotor deficits. Furthermore, they conceptualized a VR avatar designed to reproduce the patient's movements and to display the discrepancies between the desired movement and the measured/executed one, so as to trigger deficit compensation.

Basha et al. conducted a randomized clinical trial comparing the therapeutic efficiency of non-immersive VR training and resistance exercise training on breast cancer-related lymphedema [50]. The experimental protocol consisted of an exercise program conducted through Xbox Kinect games involving upper arm motion. Both rehabilitation groups, consisting of 30 patients each, received five rehabilitation sessions per week for 8 weeks. The outcome measures included excessive limb volume and pain, measured through the visual analog scale (VAS); the impairment of the upper arm, measured through the Disability of the Arm, Shoulder, and Hand (DASH) questionnaire; shoulder range of motion (ROM); shoulder muscle strength; and hand grip strength.

Feyzioğlu et al., 2019 presented a prospective randomized controlled trial comparing the efficacy of a non-immersive VRR intervention with standard physiotherapy on breast cancer survivors who had undergone surgery with axillary dissection [51]. The experimental and control groups, both consisting of 20 individuals, both received the treatment for 45 min per session and two times a week for 6 weeks. The experimental intervention consisted of playing Xbox Kinect games involving upper arm motion in the presence of a trained physiotherapist. However, the intervention group also received a scar tissue massage for 5 min and passive shoulder joint mobilization for 5 min, performed by the same physiotherapist assisting them. The outcomes considered were pain (VAS), grip strength, functionality (assessed through the DASH questionnaire), muscle strength, ROM, and fear of movement, measured through the Tampa Kinesiophobia Scale (TKS).

Hoffman et al. (2014) conducted a non-controlled trial investigating the feasibility of a home-based VRR intervention on seven lung cancer patients who had received thoracotomy [52]. The home-based rehabilitation program, divided into two phases of 5 and 10 weeks, respectively, consisted of playing Nintendo Wii Fit Plus exergames of gradually increasing intensity and duration 5 days a week. The VRR sessions did not require the presence of rehabilitation professionals. The outcomes considered were the levels of adherence, measured as the days of actual training, exercise performance, cancer-related fatigue (0–10 scale), perceived self-efficacy for fatigue self-management (0–10 scale), and perceived self-efficacy for walking 30 min (%).

House et al. conducted a trial on a sample of six patients to investigate the feasibility of a rehabilitative intervention based on a novel technology, named BrightArm Duo, on breast cancer survivors with post-surgical pain and depression [53]. The novel technological tool tested consisted of a combination of a robotic table for forearm rehabilitation and a computer executing non-immersive VR rehabilitation games. The rehabilitation program consisted of training sessions lasting 20 to 50 min of training twice a week for a period of 8 weeks. The outcomes considered were pain, measured through the Numeric Rating Scale (NRS); arm, hand, and bimanual function measured through the Fulg-Meyer assessment, the Chedokee arm and hand activity inventory, and the Jebsen hand function test; upper arm autonomy in the activities of daily living, measured through the Upper extremity function index (UEFI-20); depression, measured through the Beck Depression Inventory (BDI-II); and cognitive function, measured through the Neuropsychological Assessment Battery (NAB), the Hopkins Verbal Learning Test (HVLT-R), the Brief Visuospatial Memory Test (BVMT-R), and the Trail Making Test (TMT).

Reynolds et al. conducted a pilot study to evaluate the efficacy of two different VRR interventions on pain, CRF, and quality of life [54]. The study involved two groups of 19 and 20 women with metastatic breast cancer who were asked to participate in an immersive home-based VR intervention. The technology involved consisted of a Pico Goblin VR headset playing two different relaxing scenarios. The outcomes considered were pain, measured through the Brief Pain Inventory scale (BPI); quality of life, measured through the EQ-5D-5L scale; fatigue, measured through the Functional Assessment of Chronic Illness Therapy Fatigue scale (FACIT-Fatigue); and depression, anxiety, and stress levels, measured through the short version of the Depression, Anxiety, and Stress Scales (DASS-SF).

Schwenk and colleagues conducted a randomized trial on VR-based balance training [55]. The authors used inertial sensors equipped with gyroscopes and accelerometers on the lower limbs to assess positions and joint angles and a multi-step balance retraining virtual game based on the inputs of the sensors. In particular, the intervention group, consisting of 11 individuals with chemotherapy-induced polyneuropathy, conducted exercises and balance retraining tasks while receiving visual and auditory feedback on their motor errors. The outcomes measured were the sway of the hip, the sway of the ankle, the center of mass movement, gait speed, and fear of falling, measured through the Falls Efficacy Scale (FES-I).

Tsuda et al. conducted a preliminary study on a VR-based exercise program on over 60-year-old hospitalized patients with hematological malignancies receiving chemotherapy [56]. The virtual reality exercise program involved Nintendo Wii Fit games, which were played for 20 min a day, five times a week until hospital discharge. The primary outcomes were adherence rates, physical performance (measured through the Barthel index), muscle strength, and emotive state (hospital anxiety and depression scale).

In summary, eight of the considered studies were clinical trials, with one study conducting a preclinical investigation [49]. Of the clinical trials, four compared VRR to a standard rehabilitation program [48,50,51,55]. One study involved an immersive VR program [54], while the remaining eight studies used non-immersive VR technology. As for the population considered by the clinical trials, five of the included studies involved breast cancer survivors [48,50,51,53,54]. As for the outcomes considered, four of the retrieved studies tested VRR on more than one physical impairment [50,51,53,54]. Overall, we found four studies testing the efficacy of VRR on chronic pain [50,51,53,54], two studies on cancer fatigue [52,54], two studies on lymphedema-related excessive arm volume [48,50], one on cognitive function [53], four on motor performance impairment [48,50,51,53], and two on chemotherapy-induced polyneuropathy [49,55]. Finally, we here report the results of the two included studies considering adherence rates as an outcome [52,56].

3.1. Pain

Feyzioğlu et al. did not find a statistical difference in pain [51]. The study, however, found significant differences in the decreased fear of movement as calculated through the Tampa Kinesiophobia Scale. Moreover, House et al. reported a 20% decrease in pain after treatment ($p = 0.1$) [53]. Basha and colleagues, comparing non-immersive VR exercise with regular resistance exercise in patients with breast cancer-related lymphedema, found significant differences in pain intensity ($p = 0.002$) between groups [50]. Reynolds et al. found that both scenarios significantly reduced pain (mean difference = −6.01, $p = 0.004$) [54]. To summarize, four of the included studies considered pain as their outcome, but only two found a statistically significant effect.

3.2. Fatigue

Hoffman et al. reported statistically significant improvements in both CRF severity and perceived self-efficacy for walking [52]. Reynolds et al. found a statistical difference in pain and at follow-up compared to before the intervention (mean difference −5.00, $p < 0.001$) [54]. To summarize, two of the included studies found statistically significant effects of VR on cancer-related fatigue.

3.3. Lymphedema

Atef et al. found that both VR and PNF exercise reduced edema, with no significant differences ($p = 0.902$) [48]. Basha et al.'s trial showed no significant differences among groups for lymphedema-related excessive shoulder volume (mean difference = −11.1 mL, $p = 0.15$) [50]. In conclusion, none of the included studies found statistically significant evidence in favor of a VRR intervention compared to standard rehabilitation.

3.4. Cognitive Impairment

House et al.'s study on VR rehabilitation found it effective on cognitive function, with 10 out of 11 parameters improved ($p = 0.004$) [53].

3.5. Motor Performance

The Feyzioğlu trial on arm rehabilitation following mastectomy recorded improvements in range of motion, grip strength, and arm muscle strength but did not find any significant differences with the control group [51]. House et al.'s study, also considering arm rehabilitation in breast cancer patients following surgery, reported a significant improvement of the affected shoulder in 17 of 18 range-of-motion metrics ($p < 0.01$), of which five were above the Minimal Clinically Important Difference [53]. The study also reported a recovery in 13 out of 15 strength and function metrics ($p = 0.02$). Basha et al.'s trial also found statistical differences in physical and motility outcomes (shoulder flexion strength, external rotation strength, abduction strength, and handgrip strength) in favor of the control group, who performed regular resistance exercises [50]. The trial also reported that VRR was, however, significantly superior to standard rehabilitation for the range of motion outcome ($p < 0.001$). Lastly, the Atef et al. trial reported statistically significant differences among the VRR group and the control group regarding the functional improvements of the arm following mastectomy ($p = 0.045$) [48]. To summarize, four trials considered motor impairment as their outcome, but only two reported a statistically significant effect of VRR, while one trial found it inferior compared to standard rehabilitation on some of the considered outcomes.

3.6. Chemotherapy-Induced Peripheral Neuropathy

Schwenk et al. reported how the sway of the hip, ankle, and center of mass while standing with eyes opened and in a semi-tandem position was significantly reduced in the intervention group compared to the control ($p = 0.010$–0.022 and $p = 0.008$–0.035, respectively, for the two positions) [55]. No significant effects were found for balance with eyes closed, gait speed, and fear of falling ($p > 0.05$).

3.7. Adherence to Rehabilitation Programs

Tsuda et al. recorded an adherence rate of 66.5% in 88 sessions among 16 hospitalized patients and noted the maintenance of physical performance [56]. The Hoffman et al. study reported a mean adherence rate at the end of Phase I of 96.6% (SD: 3.4%) and of 87.6% (SD: 12.2%) at the end of phase II [52]. To summarize, two studies considered adherence rates as an outcome, but none of the two compared it to standard rehabilitation adherence rates.

In summary, VRR was found to be significantly effective for cancer-related fatigue, cognitive impairment, and CIPN-related balance impairment. VRR was found to be effective for cancer survivors' pain, but only two studies found it significantly superior to standard rehabilitation. The included studies showed mixed results for the motor impairment outcome, with two studies reporting statistically significant data in favor of VRR and one study reporting statistically significant results in favor of the control group for some of the motor performance outcomes. None of the included studies found a statistically significant effect on lymphedema.

4. Discussion

The present review aimed to offer an overview of the present evidence regarding the benefits of the integration of VR for the rehabilitation of the chronic symptoms and impairments of a specific population, cancer survivors. As previously discussed, the impairments and chronic symptoms considered by the present review are indications for and can be treated through rehabilitation programs [15–17]. The studies retrieved by our database search found VRR effective on cancer survivors' pain, accordantly with previous reviews which found VR interventions effective not only for acute but also for chronic pain [57–59]. However, only two of the included studies found VRR significantly superior to standard rehabilitation for cancer survivors, so more studies will need to address this comparison. Two of the included studies found statistically significant effects of VR on cancer-related fatigue. This is consistent with the previous literature, which found VRR effective for the treatment of chronic fatigue in other conditions, such as multiple sclerosis [60]. Regarding specifically cancer-related fatigue, however, the previous studies have focused on testing the effects of VR on acute cancer fatigue, for example during procedures such as chemotherapy infusions. Indeed, a 2020 systematic review concluded that VR had a statistically significant beneficial effect on cancer-related fatigue immediately after VR-assisted chemotherapy infusions [61]. Consequently, it must be concluded that more studies are needed to confirm the efficacy of VRR for the long-term treatment of chronic cancer-related fatigue. One study found VRR effective for the treatment of CIPN-related balance impairment, coherently with the results of previous studies on the use of VRR for the treatment of balance impairment secondary to other conditions such as diabetic neuropathy, stroke, and senility [62–64]. Two of the included studies considered lymphedema-related excessive arm volume as an outcome, but none found statistically significant evidence in favor of a VRR intervention compared to standard rehabilitation. The included studies also showed mixed results for the motor impairment outcome, with two studies reporting statistically significant data in favor of VRR and one study reporting statistically significant results in favor of the control group for some motor performance outcomes. This result is inconsistent with previous studies showing the efficacy of VRR compared to regular exercise for motor performance and strength outcomes in different conditions, such as cerebral palsy, senility, and after stroke [65–67]. One study found VRR effective for the treatment of cognitive impairment in cancer survivors, consistent with the previous literature stating the efficacy of VRR interventions for cognitive impairment [68–72].

Among the included studies, three conducted a home-based intervention [51,52,54]. This area of research is particularly crucial for cancer survivors: as previously discussed, one of the factors contributing to the limited access that cancer patients have to rehabilitative care seems to be represented by the transportation issues resulting from the patients' disability [16,23,73]. For this reason, many studies have been investigating the potential role of telerehabilitation in improving cancer patients' access to rehabilitative care [29]. Furthermore, the previous literature has addressed how virtual reality may more generally improve and facilitate remote-assisted and home-based healthcare interventions [26,33,74,75]. Considering more particularly the studies included in our review, Hoffman et al. employed a Wii Fit device to deliver a rehabilitative program of increasing intensity. The program involved only two home visits by a rehabilitation professional, one of which was before the start of the training program to set up the device, later involving only remote phone assistance. The study showed promising results in terms of adherence rates; however, its single-arm design did not allow the authors to conclude whether the VR-implemented program actually improved adherence rates compared to standard facility-based or home-based training programs. Reynolds and colleagues also reported the results of a VRR home-based intervention that did not require assistance from a rehabilitation professional but did not report adherence rates. However, discussing the acceptability of their intervention, they reported a feedback comment which may be found suggestive, although of course far from acceptable as evidence:

> "With my lack of mobility that's resulted from my illness, I really enjoyed the VR as it made me feel like I'm not house bound ... "

Feyzioğlu et al., on the other hand, conducted a randomized controlled trial, comparing two home-based interventions, an Xbox 360 Kinect-based intervention and a standard physiotherapy intervention. However, the experimental intervention involved a combination of standard physiotherapy and VRR, as it consisted of a phase of active training through a VRR gaming session and passive mobilization and scar tissue massaging, both performed by the trained physiotherapist. As such, this home-based intervention required the constant physical presence of a rehabilitation professional rather than involving remote assistance. So it must be concluded that more studies are needed to examine whether the VR implementation would facilitate remote supervision and whether the implementation of this technology in home-based interventions would improve the cancer survivors' adherence. A possible limitation emerging from the overview of the included studies is, however, the compatibility of some applied VRR systems and especially some of their more complex additional devices with home-based interventions in terms of both costs and usability. However, other included studies did test the application of VR devices currently already commercially available, mainly for entertainment and gaming purposes, and which may even be already present in the patients' houses [48,50,51,54,56]. As previously reported, two of the included trials considered adherence as an outcome [52,56]. However, both consisted of single-arm studies, so more studies are needed to confirm the hypothesis that VRR may actually improve adherence in cancer patients compared to traditional rehabilitation. This result would be consistent with previous studies reporting how VRR may benefit both adherence rates and training intensity [41–43,62,76]. More evidence on this subject would be very significant, as many studies highlighted how cancer survivors often discontinue rehabilitation programs as early as within the first 12 months [24]. One of the contributing factors to these statistics seems to be represented by the patient's lack of confidence and motivation, as standard rehabilitation programs typically require high numbers of repetitions of exercises, which are found to be tiring and boring, when not very frustrating [77]. On this subject, it has been theorized how VRR may increase the patients' enjoyment and excitement about the rehabilitation task administered, which many researchers argue may benefit both adherence rates and training intensity [41–43]. Part of the excitement added by the VR implementation may be explained by the novelty of interacting with a virtual world or even simply wearing an HMD instead of using standard training tools. However, part of its potential in terms of increased engagement seems to derive from the possibility of adding game-like features, rules, and designs to the training tasks, a process named gamification [34,78–80]. Indeed, the virtually unlimited possibilities of the virtual scenario design allow adding positive feedback and an exciting narrative to the training activities through the setting of goals, challenges, and competition elements such as score points and badges [79,81–83]. In addition, VR scenarios can replicate real-life tasks and situations with the result of greater physical and cognitive fidelity of the trained task to the everyday task the patient needs to reacquire. So, it may be argued that VRR may improve motivation by structuring a more goal-oriented training program compared to the execution of physical exercises in the context of a rehabilitation facility.

Another possible advantage of VRR comes from the multisensorial nature of VR experiences, which allow the stimulation of the patient in a multimodal manner [74]. This is particularly important when it comes to cancer-related disabilities, which, as previously discussed, often derive from the sum of more than one impairment. On this subject, we aim to stress how four of the retrieved studies tested VRR on more than one physical impairment [50,51,53,54]. In addition, three of the included studies considered the effects of VRR on both psychological and physical outcomes [53,54,56], with one also considering cognitive outcomes [53]. Furthermore, we would also like to note how two of the included studies tested VRR systems integrating VR with other technologies [53,55]. In particular, House et al. tested a system consisting of a low-friction robotic rehabilitation table, computerized forearm supports, and a display delivering the non-immersive VR scenario. Schwenk et al. used inertial sensors equipped with gyroscopes and accelerometers on the lower limbs connected to the VRR software, to deliver error-based retraining in the motor tasks required. Many previous studies also integrated VR with other technologies, utilizing the VR software to process the data sent live from different digital rehabilitation tools including treadmills [40,84–88], data gloves [89–91], and robotically-assisted orthoses [92–96]. So, regarding this subject, we aim to stress how VR software can represent an integration platform for the function of many devices currently being tested or already clinically used in the rehabilitation field and for cancer survivors.

5. Conclusions

The included studies and the previous literature suggest that VRR may be better tailored to cancer survivors' needs, such as the need for home-based rehabilitation, the need for incentives for adherence and motivation, and the need for a multimodal approach. More randomized controlled trials are needed to produce evidence on the possible advantages of VRR compared to standard rehabilitative care. In particular, it would be crucial to confirm the hypothesis that VRR may improve adherence rates thanks to its more entertaining nature and multimodal stimulation. Lastly, we wish to encourage the development of new VRR systems and VRR training programs structured to support remote connections in order to allow patients to more easily reach the assistance of healthcare and rehabilitation professionals. Nonetheless, the existence of wide margins for technological development allows us to expect further improvements in the clinical efficacy and usability of VRR systems as well as a reduction in their prices.

Author Contributions: Conceptualization, A.C. A.M. and L.G.; methodology, A.C., G.C. and L.G.; investigation, G.C., L.G., D.B. and A.M.; writing—original draft preparation, A.M.; writing—review and editing, D.B., A.C. and A.M.; visualization, G.C. and D.B.; resources, G.D.F., M.D.L. and A.G.; supervision, A.G., G.D.P. and M.D.L.; project administration, A.G. and G.D.P.; funding acquisition, M.D.L. All authors have read and agreed to the published version of the manuscript.

Funding: This research received no external funding.

Data Availability Statement: Not applicable.

Conflicts of Interest: The authors declare no conflict of interest.

References

1. Sung, H.; Ferlay, J.; Siegel, R.L.; Laversanne, M.; Soerjomataram, I.; Jemal, A.; Bray, F. Global Cancer Statistics 2020: GLOBOCAN Estimates of Incidence and Mortality Worldwide for 36 Cancers in 185 Countries. *CA Cancer J. Clin.* **2021**, *71*, 209–249. [CrossRef] [PubMed]
2. Hashim, D.; Boffetta, P.; La Vecchia, C.; Rota, M.; Bertuccio, P.; Malvezzi, M.; Negri, E. The Global Decrease in Cancer Mortality: Trends and Disparities. *Ann. Oncol.* **2016**, *27*, 926–933. [CrossRef] [PubMed]
3. Siegel, R.L.; Miller, K.D.; Jemal, A. Cancer Statistics, 2020. *CA Cancer J. Clin.* **2020**, *70*, 7–30. [CrossRef] [PubMed]
4. Mattiuzzi, C.; Lippi, G. Current Cancer Epidemiology. *J. Epidemiol. Glob. Health* **2019**, *9*, 217. [CrossRef]
5. American Cancer Society. *Cancer Treatment & Survivorship Facts & Figures 2019–2021*; American Cancer Society: New York, NY, USA, 2019.
6. Shapiro, C.L. Cancer Survivorship. *N. Engl. J. Med.* **2018**, *379*, 2438–2450. [CrossRef]
7. Van den Beuken-van Everdingen, M.H.J.; Hochstenbach, L.M.J.; Joosten, E.A.J.; Tjan-Heijnen, V.C.G.; Janssen, D.J.A. Update on Prevalence of Pain in Patients with Cancer: Systematic Review and Meta-Analysis. *J. Pain Symptom Manag.* **2016**, *51*, 1070–1090.e9. [CrossRef]
8. Bernier Carney, K.; Starkweather, A.; Lucas, R.; Ersig, A.L.; Guite, J.W.; Young, E. Deconstructing Pain Disability through Concept Analysis. *Pain Manag. Nurs.* **2019**, *20*, 482–488. [CrossRef]
9. Bernas, M.; Thiadens, S.R.J.; Smoot, B.; Armer, J.M.; Stewart, P.; Granzow, J. Lymphedema Following Cancer Therapy: Overview and Options. *Clin. Exp. Metastasis* **2018**, *35*, 547–551. [CrossRef]

10. Bernas, M.; Thiadens, S.R.J.; Stewart, P.; Granzow, J. Secondary Lymphedema from Cancer Therapy. *Clin. Exp. Metastasis* **2021**, *39*, 239–247. [CrossRef]
11. Shaitelman, S.F.; Cromwell, K.D.; Rasmussen, J.C.; Stout, N.L.; Armer, J.M.; Lasinski, B.B.; Cormier, J.N. Recent Progress in the Treatment and Prevention of Cancer-Related Lymphedema: Lymphedema Treatment and Prevention. *CA Cancer J. Clin.* **2015**, *65*, 55–81. [CrossRef]
12. Eide, S.; Feng, Z.-P. Doxorubicin Chemotherapy-Induced "Chemo-Brain": Meta-Analysis. *Eur. J. Pharmacol.* **2020**, *881*, 173078. [CrossRef] [PubMed]
13. Hermelink, K. Chemotherapy and Cognitive Function in Breast Cancer Patients: The So-Called Chemo Brain. *JNCI Monogr.* **2015**, *2015*, 67–69. [CrossRef] [PubMed]
14. Hu, L.-Y.; Mi, W.-L.; Wu, G.-C.; Wang, Y.-Q.; Mao-Ying, Q.-L. Prevention and Treatment for Chemotherapy-Induced Peripheral Neuropathy: Therapies Based on CIPN Mechanisms. *Curr. Neuropharmacol.* **2019**, *17*, 184–196. [CrossRef]
15. Campbell, K.L.; Winters-Stone, K.M.; Wiskemann, J.; May, A.M.; Schwartz, A.L.; Courneya, K.S.; Zucker, D.S.; Matthews, C.E.; Ligibel, J.A.; Gerber, L.H.; et al. Exercise Guidelines for Cancer Survivors: Consensus Statement from International Multidisciplinary Roundtable. *Med. Sci. Sports Exerc.* **2019**, *51*, 2375–2390. [CrossRef] [PubMed]
16. Silver, J.K.; Baima, J.; Mayer, R.S. Impairment-Driven Cancer Rehabilitation: An Essential Component of Quality Care and Survivorship: Impairment-Driven Cancer Rehabilitation. *CA Cancer J. Clin.* **2013**, *63*, 295–317. [CrossRef] [PubMed]
17. Turner, R.R.; Steed, L.; Quirk, H.; Greasley, R.U.; Saxton, J.M.; Taylor, S.J.; Rosario, D.J.; Thaha, M.A.; Bourke, L. Interventions for Promoting Habitual Exercise in People Living with and beyond Cancer. *Cochrane Database Syst. Rev.* **2018**, *9*, CD010192. [CrossRef]
18. Mitchell, L.J.; Bisdounis, L.; Ballesio, A.; Omlin, X.; Kyle, S.D. The Impact of Cognitive Behavioural Therapy for Insomnia on Objective Sleep Parameters: A Meta-Analysis and Systematic Review. *Sleep Med. Rev.* **2019**, *47*, 90–102. [CrossRef]
19. Dimeo, F.C. Effects of Exercise on Cancer-Related Fatigue. *Cancer* **2001**, *92*, 1689–1693. [CrossRef]
20. Cheville, A.L.; Mustian, K.; Winters-Stone, K.; Zucker, D.S.; Gamble, G.L.; Alfano, C.M. Cancer Rehabilitation. *Phys. Med. Rehabil. Clin. N. Am.* **2017**, *28*, 1–17. [CrossRef]
21. Hansen, D.G.; Larsen, P.V.; Holm, L.V.; Rottmann, N.; Bergholdt, S.H.; Søndergaard, J. Association between Unmet Needs and Quality of Life of Cancer Patients: A Population-Based Study. *Acta Oncol.* **2013**, *52*, 391–399. [CrossRef]
22. Stensvold, E.; Stadskleiv, K.; Myklebust, T.Å.; Wesenberg, F.; Helseth, E.; Bechensteen, A.G.; Brandal, P. Unmet Rehabilitation Needs in 86% of Norwegian Paediatric Embryonal Brain Tumour Survivors. *Acta Paediatr.* **2020**, *109*, 1875–1886. [CrossRef] [PubMed]
23. Cheville, A.L.; Kornblith, A.B.; Basford, J.R. An Examination of the Causes for the Underutilization of Rehabilitation Services Among People with Advanced Cancer. *Am. J. Phys. Med. Rehabil.* **2011**, *90*, S27–S37. [CrossRef] [PubMed]
24. Pudkasam, S.; Polman, R.; Pitcher, M.; Fisher, M.; Chinlumprasert, N.; Stojanovska, L.; Apostolopoulos, V. Physical Activity and Breast Cancer Survivors: Importance of Adherence, Motivational Interviewing and Psychological Health. *Maturitas* **2018**, *116*, 66–72. [CrossRef] [PubMed]
25. Nuara, A.; Fabbri-Destro, M.; Scalona, E.; Lenzi, S.E.; Rizzolatti, G.; Avanzini, P. Telerehabilitation in Response to Constrained Physical Distance: An Opportunity to Rethink Neurorehabilitative Routines. *J. Neurol.* **2021**, *269*, 627–638. [CrossRef]
26. Meyding-Lamadé, U.; Bassa, B.; Tibitanzl, P.; Davtyan, A.; Lamadé, E.K.; Craemer, E.M. Telerehabilitation: Von der virtuellen Welt zur Realität—Medizin im 21. Jahrhundert: Videogestützte Therapie in Zeiten von COVID-19. *Nervenarzt* **2021**, *92*, 127–136. [CrossRef]
27. Lambert, G.; Alos, N.; Bernier, P.; Laverdière, C.; Drummond, K.; Dahan-Oliel, N.; Lemay, M.; Veilleux, L.-N.; Kairy, D. Patient and Parent Experiences with Group Telerehabilitation for Child Survivors of Acute Lymphoblastic Leukemia. *Int. J. Environ. Res. Public Health* **2021**, *18*, 3610. [CrossRef]
28. Galiano-Castillo, N.; Cantarero-Villanueva, I.; Fernández-Lao, C.; Ariza-García, A.; Díaz-Rodríguez, L.; Del-Moral-Ávila, R.; Arroyo-Morales, M. Telehealth System: A Randomized Controlled Trial Evaluating the Impact of an Internet-Based Exercise Intervention on Quality of Life, Pain, Muscle Strength, and Fatigue in Breast Cancer Survivors: Telehealth System in Breast Cancer. *Cancer* **2016**, *122*, 3166–3174. [CrossRef]
29. Dennett, A.; Harding, K.E.; Reimert, J.; Morris, R.; Parente, P.; Taylor, N.F. Telerehabilitation Was Safe, Feasible and Increased Exercise Uptake in Cancer Survivors: A Process Evaluation (Preprint). *JMIR Cancer* **2021**, *7*, e33130. [CrossRef]
30. Navarra-Ventura, G.; Gomà, G.; de Haro, C.; Jodar, M.; Sarlabous, L.; Hernando, D.; Bailón, R.; Ochagavía, A.; Blanch, L.; López-Aguilar, J.; et al. Virtual Reality-Based Early Neurocognitive Stimulation in Critically Ill Patients: A Pilot Randomized Clinical Trial. *J. Pers. Med.* **2021**, *11*, 1260. [CrossRef]
31. Villiger, M.; Liviero, J.; Awai, L.; Stoop, R.; Pyk, P.; Clijsen, R.; Curt, A.; Eng, K.; Bolliger, M. Home-Based Virtual Reality-Augmented Training Improves Lower Limb Muscle Strength, Balance, and Functional Mobility Following Chronic Incomplete Spinal Cord Injury. *Front. Neurol.* **2017**, *8*, 635. [CrossRef]
32. Gandolfi, M.; Geroin, C.; Dimitrova, E.; Boldrini, P.; Waldner, A.; Bonadiman, S.; Picelli, A.; Regazzo, S.; Stirbu, E.; Primon, D.; et al. Virtual Reality Telerehabilitation for Postural Instability in Parkinson's Disease: A Multicenter, Single-Blind, Randomized, Controlled Trial. *BioMed Res. Int.* **2017**, *2017*, 1–11. [CrossRef] [PubMed]

33. Lloréns, R.; Noé, E.; Colomer, C.; Alcañiz, M. Effectiveness, Usability, and Cost-Benefit of a Virtual Reality–Based Telerehabilitation Program for Balance Recovery After Stroke: A Randomized Controlled Trial. *Arch. Phys. Med. Rehabil.* **2015**, *96*, 418–425.e2. [CrossRef] [PubMed]
34. Berton, A.; Longo, U.G.; Candela, V.; Fioravanti, S.; Giannone, L.; Arcangeli, V.; Alciati, V.; Berton, C.; Facchinetti, G.; Marchetti, A.; et al. Virtual Reality, Augmented Reality, Gamification, and Telerehabilitation: Psychological Impact on Orthopedic Patients' Rehabilitation. *J. Clin. Med.* **2020**, *9*, 2567. [CrossRef] [PubMed]
35. Laver, K.E.; George, S.; Thomas, S.; Deutsch, J.E.; Crotty, M. Virtual Reality for Stroke Rehabilitation. *Cochrane Database Syst. Rev.* **2015**, *2*, CD008349. [CrossRef]
36. Chi, B.; Chau, B.; Yeo, E.; Ta, P. Virtual Reality for Spinal Cord Injury-Associated Neuropathic Pain: Systematic Review. *Ann. Phys. Rehabil. Med.* **2019**, *62*, 49–57. [CrossRef]
37. Maggio, M.G.; Russo, M.; Cuzzola, M.F.; Destro, M.; La Rosa, G.; Molonia, F.; Bramanti, P.; Lombardo, G.; De Luca, R.; Calabrò, R.S. Virtual Reality in Multiple Sclerosis Rehabilitation: A Review on Cognitive and Motor Outcomes. *J. Clin. Neurosci.* **2019**, *65*, 106–111. [CrossRef]
38. Amirthalingam, J.; Paidi, G.; Alshowaikh, K.; Iroshani Jayarathna, A.; Salibindla, D.B.A.M.R.; Karpinska-Leydier, K.; Ergin, H.E. Virtual Reality Intervention to Help Improve Motor Function in Patients Undergoing Rehabilitation for Cerebral Palsy, Parkinson's Disease, or Stroke: A Systematic Review of Randomized Controlled Trials. *Cureus* **2021**, *13*, e16763. [CrossRef]
39. Ravi, D.K.; Kumar, N.; Singhi, P. Effectiveness of Virtual Reality Rehabilitation for Children and Adolescents with Cerebral Palsy: An Updated Evidence-Based Systematic Review. *Physiotherapy* **2017**, *103*, 245–258. [CrossRef]
40. Gagliardi, C.; Turconi, A.C.; Biffi, E.; Maghini, C.; Marelli, A.; Cesareo, A.; Diella, E.; Panzeri, D. Immersive Virtual Reality to Improve Walking Abilities in Cerebral Palsy: A Pilot Study. *Ann. Biomed. Eng.* **2018**, *46*, 1376–1384. [CrossRef]
41. Rose, T.; Nam, C.S.; Chen, K.B. Immersion of Virtual Reality for Rehabilitation—Review. *Appl. Ergon.* **2018**, *69*, 153–161. [CrossRef]
42. Perez-Marcos, D. Virtual Reality Experiences, Embodiment, Videogames and Their Dimensions in Neurorehabilitation. *J. Neuroeng. Rehabil.* **2018**, *15*, 113. [CrossRef] [PubMed]
43. Oesch, P.; Kool, J.; Fernandez-Luque, L.; Brox, E.; Evertsen, G.; Civit, A.; Hilfiker, R.; Bachmann, S. Exergames versus Self-Regulated Exercises with Instruction Leaflets to Improve Adherence during Geriatric Rehabilitation: A Randomized Controlled Trial. *BMC Geriatr.* **2017**, *17*, 77. [CrossRef] [PubMed]
44. Asadzadeh, A.; Samad-Soltani, T.; Salahzadeh, Z.; Rezaei-Hachesu, P. Effectiveness of Virtual Reality-Based Exercise Therapy in Rehabilitation: A Scoping Review. *Inform. Med. Unlocked* **2021**, *24*, 100562. [CrossRef]
45. Chirico, A.; Lucidi, F.; De Laurentiis, M.; Milanese, C.; Napoli, A.; Giordano, A. Virtual Reality in Health System: Beyond Entertainment. A Mini-Review on the Efficacy of VR During Cancer Treatment: Efficacy of vr during cancer treatment. *J. Cell. Physiol.* **2016**, *231*, 275–287. [CrossRef]
46. Zeng, Y.; Zhang, J.-E.; Cheng, A.S.K.; Cheng, H.; Wefel, J.S. Meta-Analysis of the Efficacy of Virtual Reality–Based Interventions in Cancer-Related Symptom Management. *Integr. Cancer Ther.* **2019**, *18*, 1534735419871108. [CrossRef]
47. Methley, A.M.; Campbell, S.; Chew-Graham, C.; McNally, R.; Cheraghi-Sohi, S. PICO, PICOS and SPIDER: A Comparison Study of Specificity and Sensitivity in Three Search Tools for Qualitative Systematic Reviews. *BMC Health Serv. Res.* **2014**, *14*, 579. [CrossRef]
48. Atef, D.; Elkeblawy, M.M.; El-Sebaie, A.; Abouelnaga, W.A.I. A Quasi-Randomized Clinical Trial: Virtual Reality versus Proprioceptive Neuromuscular Facilitation for Postmastectomy Lymphedema. *J. Egypt. Natl. Cancer Inst.* **2020**, *32*, 29. [CrossRef]
49. Axenie, C.; Kurz, D. Role of Kinematics Assessment and Multimodal Sensorimotor Training for Motion Deficits in Breast Cancer Chemotherapy-Induced Polyneuropathy: A Perspective on Virtual Reality Avatars. *Front. Oncol.* **2020**, *10*, 1419. [CrossRef]
50. Basha, M.A.; Aboelnour, N.H.; Alsharidah, A.S.; Kamel, F.H. Effect of Exercise Mode on Physical Function and Quality of Life in Breast Cancer–Related Lymphedema: A Randomized Trial. *Support. Care Cancer* **2021**, *3*, 2101–2110. [CrossRef]
51. Feyzioğlu, Ö.; Dinçer, S.; Akan, A.; Algun, Z.C. Is Xbox 360 Kinect-Based Virtual Reality Training as Effective as Standard Physiotherapy in Patients Undergoing Breast Cancer Surgery? *Support. Care Cancer* **2020**, *28*, 4295–4303. [CrossRef]
52. Hoffman, A.J.; Brintnall, R.A.; Brown, J.K.; von Eye, A.; Jones, L.W.; Alderink, G.; Ritz-Holland, D.; Enter, M.; Patzelt, L.H.; VanOtteren, G.M. Virtual Reality Bringing a New Reality to Postthoracotomy Lung Cancer Patients Via a Home-Based Exercise Intervention Targeting Fatigue While Undergoing Adjuvant Treatment. *Cancer Nurs.* **2014**, *37*, 23–33. [CrossRef] [PubMed]
53. House, G.; Burdea, G.; Grampurohit, N.; Polistico, K.; Roll, D.; Damiani, F.; Hundal, J.; Demesmin, D. A Feasibility Study to Determine the Benefits of Upper Extremity Virtual Rehabilitation Therapy for Coping with Chronic Pain Post-Cancer Surgery. *Br. J. Pain* **2016**, *10*, 186–197. [CrossRef] [PubMed]
54. Reynolds, L.M.; Cavadino, A.; Chin, S.; Little, Z.; Akroyd, A.; Tennant, G.; Dobson, R.; Broom, R.; Gautier, A. The Benefits and Acceptability of Virtual Reality Interventions for Women with Metastatic Breast Cancer in Their Homes; a Pilot Randomised Trial. *BMC Cancer* **2022**, *22*, 360. [CrossRef]
55. Schwenk, M.; Grewal, G.S.; Holloway, D.; Muchna, A.; Garland, L.; Najafi, B. Interactive Sensor-Based Balance Training in Older Cancer Patients with Chemotherapy-Induced Peripheral Neuropathy: A Randomized Controlled Trial. *Gerontology* **2016**, *62*, 553–563. [CrossRef]
56. Tsuda, K.; Sudo, K.; Goto, G.; Takai, M.; Itokawa, T.; Isshiki, T.; Takei, N.; Tanimoto, T.; Komatsu, T. A Feasibility Study of Virtual Reality Exercise in Elderly Patients with Hematologic Malignancies Receiving Chemotherapy. *Intern. Med.* **2016**, *55*, 347–352. [CrossRef]

57. Goudman, L.; Jansen, J.; Billot, M.; Vets, N.; De Smedt, A.; Roulaud, M.; Rigoard, P.; Moens, M. Virtual Reality Applications in Chronic Pain Management: Systematic Review and Meta-Analysis. *JMIR Serious Games* **2022**, *10*, e34402. [CrossRef] [PubMed]
58. Alemanno, F.; Houdayer, E.; Emedoli, D.; Locatelli, M.; Mortini, P.; Mandelli, C.; Raggi, A.; Iannaccone, S. Efficacy of Virtual Reality to Reduce Chronic Low Back Pain: Proof-of-Concept of a Non-Pharmacological Approach on Pain, Quality of Life, Neuropsychological and Functional Outcome. *PLoS ONE* **2019**, *14*, e0216858. [CrossRef] [PubMed]
59. Brea-Gómez, B.; Torres-Sánchez, I.; Ortiz-Rubio, A.; Calvache-Mateo, A.; Cabrera-Martos, I.; López-López, L.; Valenza, M.C. Virtual Reality in the Treatment of Adults with Chronic Low Back Pain: A Systematic Review and Meta-Analysis of Randomized Clinical Trials. *Int. J. Environ. Res. Public Health* **2021**, *18*, 11806. [CrossRef]
60. Cortés-Pérez, I.; Sánchez-Alcalá, M.; Nieto-Escámez, F.A.; Castellote-Caballero, Y.; Obrero-Gaitán, E.; Osuna-Pérez, M.C. Virtual Reality-Based Therapy Improves Fatigue, Impact, and Quality of Life in Patients with Multiple Sclerosis. A Systematic Review with a Meta-Analysis. *Sensors* **2021**, *21*, 7389. [CrossRef]
61. Ioannou, A.; Papastavrou, E.; Avraamides, M.N.; Charalambous, A. Virtual Reality and Symptoms Management of Anxiety, Depression, Fatigue, and Pain: A Systematic Review. *SAGE Open Nurs.* **2020**, *6*, 2377960820936163. [CrossRef]
62. Coons, M.J.; Roehrig, M.; Spring, B. The Potential of Virtual Reality Technologies to Improve Adherence to Weight Loss Behaviors. *J. Diabetes Sci. Technol.* **2011**, *5*, 340–344. [CrossRef] [PubMed]
63. Grewal, G.S.; Sayeed, R.; Schwenk, M.; Bharara, M.; Menzies, R.; Talal, T.K.; Armstrong, D.G.; Najafi, B. Balance Rehabilitation: Promoting the Role of Virtual Reality in Patients with Diabetic Peripheral Neuropathy. *J. Am. Podiatr. Med. Assoc.* **2013**, *103*, 498–507. [CrossRef] [PubMed]
64. Iruthayarajah, J.; McIntyre, A.; Cotoi, A.; Macaluso, S.; Teasell, R. The Use of Virtual Reality for Balance among Individuals with Chronic Stroke: A Systematic Review and Meta-Analysis. *Top. Stroke Rehabil.* **2017**, *24*, 68–79. [CrossRef] [PubMed]
65. Donath, L.; Rössler, R.; Faude, O. Effects of Virtual Reality Training (Exergaming) Compared to Alternative Exercise Training and Passive Control on Standing Balance and Functional Mobility in Healthy Community-Dwelling Seniors: A Meta-Analytical Review. *Sports Med.* **2016**, *46*, 1293–1309. [CrossRef] [PubMed]
66. Phu, S.; Vogrin, S.; Al Saedi, A.; Duque, G. Balance Training Using Virtual Reality Improves Balance and Physical Performance in Older Adults at High Risk of Falls. *Clin. Interv. Aging* **2019**, *14*, 1567–1577. [CrossRef]
67. Cho, H.; Sohng, K.-Y. The Effect of a Virtual Reality Exercise Program on Physical Fitness, Body Composition, and Fatigue in Hemodialysis Patients. *J. Phys. Ther. Sci.* **2014**, *26*, 1661–1665. [CrossRef] [PubMed]
68. Tieri, G.; Morone, G.; Paolucci, S.; Iosa, M. Virtual Reality in Cognitive and Motor Rehabilitation: Facts, Fiction and Fallacies. *Expert Rev. Med. Devices* **2018**, *15*, 107–117. [CrossRef]
69. Faria, A.L.; Andrade, A.; Soares, L.; Badia, S.B.I. Benefits of Virtual Reality Based Cognitive Rehabilitation through Simulated Activities of Daily Living: A Randomized Controlled Trial with Stroke Patients. *J. Neuroeng. Rehabil.* **2016**, *13*, 96. [CrossRef] [PubMed]
70. Ahn, S.-N. Combined Effects of Virtual Reality and Computer Game-Based Cognitive Therapy on the Development of Visual-Motor Integration in Children with Intellectual Disabilities: A Pilot Study. *Occup. Ther. Int.* **2021**, *2021*, 6696779. [CrossRef]
71. Aminov, A.; Rogers, J.M.; Middleton, S.; Caeyenberghs, K.; Wilson, P.H. What Do Randomized Controlled Trials Say about Virtual Rehabilitation in Stroke? A Systematic Literature Review and Meta-Analysis of Upper-Limb and Cognitive Outcomes. *J. Neuroeng. Rehabil.* **2018**, *15*, 29. [CrossRef]
72. Carelli, L.; Morganti, F.; Poletti, B.; Corra, B.; Weiss, P.L.T.; Kizony, R.; Silani, V.; Riva, G. A NeuroVR Based Tool for Cognitive Assessment and Rehabilitation of Post-Stroke Patients: Two Case Studies. *Stud. Health Technol. Inform.* **2009**, *144*, 243–247. [PubMed]
73. Sleight, A.G.; Lyons, K.D.; Vigen, C.; Macdonald, H.; Clark, F. The Association of Health-Related Quality of Life with Unmet Supportive Care Needs and Sociodemographic Factors in Low-Income Latina Breast Cancer Survivors: A Single-Centre Pilot Study. *Disabil. Rehabil.* **2019**, *41*, 3151–3156. [CrossRef] [PubMed]
74. Navarro, E.; González, P.; López-Jaquero, V.; Montero, F.; Molina, J.P.; Romero-Ayuso, D. Adaptive, Multisensorial, Physiological and Social: The Next Generation of Telerehabilitation Systems. *Front. Neuroinform.* **2018**, *12*, 43. [CrossRef] [PubMed]
75. Smits, M.; Staal, J.B.; van Goor, H. Could Virtual Reality Play a Role in the Rehabilitation after COVID-19 Infection? *BMJ Open Sport Exerc. Med.* **2020**, *6*, e000943. [CrossRef]
76. Annesi, J.J.; Mazas, J. Effects of Virtual Reality-Enhanced Exercise Equipment on Adherence and Exercise-Induced Feeling States. *Percept. Mot. Skills* **1997**, *85*, 835–844. [CrossRef] [PubMed]
77. Howard, M.C. A Meta-Analysis and Systematic Literature Review of Virtual Reality Rehabilitation Programs. *Comput. Hum. Behav.* **2017**, *70*, 317–327. [CrossRef]
78. Edwards, E.A.; Lumsden, J.; Rivas, C.; Steed, L.; Edwards, L.A.; Thiyagarajan, A.; Sohanpal, R.; Caton, H.; Griffiths, C.J.; Munafò, M.R.; et al. Gamification for Health Promotion: Systematic Review of Behaviour Change Techniques in Smartphone Apps. *BMJ Open* **2016**, *6*, e012447. [CrossRef]
79. Johnson, D.; Deterding, S.; Kuhn, K.-A.; Staneva, A.; Stoyanov, S.; Hides, L. Gamification for Health and Wellbeing: A Systematic Review of the Literature. *Internet Interv.* **2016**, *6*, 89–106. [CrossRef]
80. Mitchell, R.; Schuster, L.; Jin, H.S. Playing Alone: Can Game Design Elements Satisfy User Needs in Gamified MHealth Services? *Health Promot. Int.* **2021**, *37*, daab168. [CrossRef]

81. Goršič, M.; Cikajlo, I.; Novak, D. Competitive and Cooperative Arm Rehabilitation Games Played by a Patient and Unimpaired Person: Effects on Motivation and Exercise Intensity. *J. Neuroeng. Rehabil.* **2017**, *14*, 23. [CrossRef]
82. Vang, M.H.; Fox, J. Race in Virtual Environments: Competitive versus Cooperative Games with Black or White Avatars. *Cyberpsychol. Behav. Soc. Netw.* **2014**, *17*, 235–240. [CrossRef]
83. Navarro, M.D.; Llorens, R.; Borrego, A.; Alcañiz, M.; Noé, E.; Ferri, J. Competition Enhances the Effectiveness and Motivation of Attention Rehabilitation After Stroke. A Randomized Controlled Trial. *Front. Hum. Neurosci.* **2020**, *14*, 575403. [CrossRef] [PubMed]
84. Bang, Y.-S.; Son, K.H.; Kim, H.J. Effects of Virtual Reality Training Using Nintendo Wii and Treadmill Walking Exercise on Balance and Walking for Stroke Patients. *J. Phys. Ther. Sci.* **2016**, *28*, 3112–3115. [CrossRef] [PubMed]
85. Bekkers, E.M.J.; Mirelman, A.; Alcock, L.; Rochester, L.; Nieuwhof, F.; Bloem, B.R.; Pelosin, E.; Avanzino, L.; Cereatti, A.; Della Croce, U.; et al. Do Patients with Parkinson's Disease With Freezing of Gait Respond Differently Than Those Without to Treadmill Training Augmented by Virtual Reality? *Neurorehabil. Neural Repair* **2020**, *34*, 440–449. [CrossRef]
86. Cho, K.H.; Lee, W.H. Effect of Treadmill Training Based Real-World Video Recording on Balance and Gait in Chronic Stroke Patients: A Randomized Controlled Trial. *Gait Posture* **2014**, *39*, 523–528. [CrossRef]
87. Mirelman, A.; Rochester, L.; Reelick, M.; Nieuwhof, F.; Pelosin, E.; Abbruzzese, G.; Dockx, K.; Nieuwboer, A.; Hausdorff, J.M. V-TIME: A Treadmill Training Program Augmented by Virtual Reality to Decrease Fall Risk in Older Adults: Study Design of a Randomized Controlled Trial. *BMC Neurol.* **2013**, *13*, 15. [CrossRef] [PubMed]
88. Mirelman, A.; Rochester, L.; Maidan, I.; Del Din, S.; Alcock, L.; Nieuwhof, F.; Rikkert, M.O.; Bloem, B.R.; Pelosin, E.; Avanzino, L.; et al. Addition of a Non-Immersive Virtual Reality Component to Treadmill Training to Reduce Fall Risk in Older Adults (V-TIME): A Randomised Controlled Trial. *Lancet Lond. Engl.* **2016**, *388*, 1170–1182. [CrossRef]
89. Proulx, C.E.; Beaulac, M.; David, M.; Deguire, C.; Haché, C.; Klug, F.; Kupnik, M.; Higgins, J.; Gagnon, D.H. Review of the Effects of Soft Robotic Gloves for Activity-Based Rehabilitation in Individuals with Reduced Hand Function and Manual Dexterity Following a Neurological Event. *J. Rehabil. Assist. Technol. Eng.* **2020**, *7*, 2055668320918130. [CrossRef]
90. Demolder, C.; Molina, A.; Hammond, F.L.; Yeo, W.-H. Recent Advances in Wearable Biosensing Gloves and Sensory Feedback Biosystems for Enhancing Rehabilitation, Prostheses, Healthcare, and Virtual Reality. *Biosens. Bioelectron.* **2021**, *190*, 113443. [CrossRef]
91. Wang, Q.; Kang, B.; Kristensson, P.O. Supporting Physical and Mental Health Rehabilitation at Home with Virtual Reality Headsets and Force Feedback Gloves. In Proceedings of the 2021 IEEE Conference on Virtual Reality and 3D User Interfaces Abstracts and Workshops (VRW), Lisbon, Portugal, 27 March–1 April 2021; pp. 685–686.
92. Abd El-Kafy, E.M.; Alshehri, M.A.; El-Fiky, A.A.-R.; Guermazi, M.A.; Mahmoud, H.M. The Effect of Robot-Mediated Virtual Reality Gaming on Upper Limb Spasticity Poststroke: A Randomized-Controlled Trial. *Games Health J.* **2022**, *11*, 93–103. [CrossRef]
93. Baur, K.; Schättin, A.; de Bruin, E.D.; Riener, R.; Duarte, J.E.; Wolf, P. Trends in Robot-Assisted and Virtual Reality-Assisted Neuromuscular Therapy: A Systematic Review of Health-Related Multiplayer Games. *J. Neuroeng. Rehabil.* **2018**, *15*, 107. [CrossRef] [PubMed]
94. Burdea, G.C.; Cioi, D.; Kale, A.; Janes, W.E.; Ross, S.A.; Engsberg, J.R. Robotics and Gaming to Improve Ankle Strength, Motor Control, and Function in Children with Cerebral Palsy—A Case Study Series. *IEEE Trans. Neural Syst. Rehabil. Eng.* **2013**, *21*, 165–173. [CrossRef] [PubMed]
95. Calabrò, R.S.; Cacciola, A.; Bertè, F.; Manuli, A.; Leo, A.; Bramanti, A.; Naro, A.; Milardi, D.; Bramanti, P. Robotic Gait Rehabilitation and Substitution Devices in Neurological Disorders: Where Are We Now? *Neurol. Sci.* **2016**, *37*, 503–514. [CrossRef] [PubMed]
96. De Mauro, A.; Carrasco, E.; Oyarzun, D.; Ardanza, A.; Frizera Neto, A.; Torricelli, D.; Pons, J.L.; Gil, A.; Florez, J. Virtual Reality System in Conjunction with Neurorobotics and Neuroprosthetics for Rehabilitation of Motor Disorders. *Stud. Health Technol. Inform.* **2011**, *163*, 163–165. [PubMed]

cancers

MDPI

Review

Diagnostic Strategies for Breast Cancer Detection: From Image Generation to Classification Strategies Using Artificial Intelligence Algorithms

Jesus A. Basurto-Hurtado [1,2], Irving A. Cruz-Albarran [1,2], Manuel Toledano-Ayala [3], Mario Alberto Ibarra-Manzano [4], Luis A. Morales-Hernandez [1,*] and Carlos A. Perez-Ramirez [2,*]

1 C.A. Mecatrónica, Facultad de Ingeniería, Campus San Juan del Río, Universidad Autónoma de Querétaro, Rio Moctezuma 249, San Cayetano, San Juan del Rio 76807, Mexico; jesus.alberto.basurto@uaq.mx (J.A.B.-H.); irving.cruz@uaq.mx (I.A.C.-A.)

2 Laboratorio de Dispositivos Médicos, Facultad de Ingeniería, Universidad Autónoma de Querétaro, Carretera a Chichimequillas S/N, Ejido Bolaños, Santiago de Querétaro 76140, Mexico

3 División de Investigación y Posgrado de la Facultad de Ingeniería (DIPFI), Universidad Autónoma de Querétaro, Cerro de las Campanas S/N Las Campanas, Santiago de Querétaro 76010, Mexico; toledano@uaq.mx

4 Laboratorio de Procesamiento Digital de Señales, Departamento de Ingeniería Electrónica, Division de Ingenierias Campus Irapuato-Salamanca (DICIS), Universidad de Guanajuato, Carretera Salamanca-Valle de Santiago KM. 3.5 + 1.8 Km., Salamanca 36885, Mexico; ibarram@ugto.mx

* Correspondence: luis.morales@uaq.mx (L.A.M.-H.) carlos.perez@uaq.mx (C.A.P.-R.)

Simple Summary: With the recent advances in the field of artificial intelligence, it has been possible to develop robust and accurate methodologies that can deliver noticeable results in different health-related areas, where the oncology is one the hottest research areas nowadays, as it is now possible to fuse information that the images have with the patient medical records in order to offer a more accurate diagnosis. In this sense, understanding the process of how an AI-based methodology is developed can offer a helpful insight to develop such methodologies. In this review, we comprehensively guide the reader on the steps required to develop such methodology, starting from the image formation to its processing and interpretation using a wide variety of methods; further, some techniques that can be used in the next-generation diagnostic strategies are also presented. We believe this helpful insight will provide deeper comprehension to students and researchers in the related areas, of the advantages and disadvantages of every method.

Abstract: Breast cancer is one the main death causes for women worldwide, as 16% of the diagnosed malignant lesions worldwide are its consequence. In this sense, it is of paramount importance to diagnose these lesions in the earliest stage possible, in order to have the highest chances of survival. While there are several works that present selected topics in this area, none of them present a complete panorama, that is, from the image generation to its interpretation. This work presents a comprehensive state-of-the-art review of the image generation and processing techniques to detect Breast Cancer, where potential candidates for the image generation and processing are presented and discussed. Novel methodologies should consider the adroit integration of artificial intelligence-concepts and the categorical data to generate modern alternatives that can have the accuracy, precision and reliability expected to mitigate the misclassifications.

Keywords: breast cancer; mammography; magnetic resonance; ultrasound; thermography; image processing; artificial intelligence

Citation: Basurto-Hurtado, J.A.; Cruz-Albarran, I.A.; Toledano-Ayala, M.; Ibarra-Manzano, M.A.; Morales-Hernandez, L.A.; Perez-Ramirez, C.A. Diagnostic Strategies for Breast Cancer Detection: From Image Generation to Classification Strategies Using Artificial Intelligence Algorithms. *Cancers* **2022**, *14*, 3442. https://doi.org/10.3390/cancers14143442

Academic Editor: David Wong

Received: 21 May 2022
Accepted: 12 July 2022
Published: 15 July 2022

1. Introduction

According to the World Health Organization, Breast Cancer (BC) represents around 16% of the malignant tumors diagnosed worldwide [1]. In Mexico, BC is the leading death cause for cancer in the female population [2]. BC develops when any lump begins an angiogenesis process, that is, the process that causes the development of new blood vessels and capillaries from the existent vasculature [3]. Unfortunately, BC has a mortality rate of 69% in emergent countries, which is greater than the one in developed countries [1]. This increase is explained as the cancer is detected in a later stage, making the treatment a financial obstacle as its price increases, especially if the disease is detected in an advanced stage [4]. Hence, the development of strategies that can perform an early detection of BC is a priority topic for governments, as an early detection increases the survival chances and lowers the financial burden the disease imposes to families and health systems [4].

A methodology for the BC detection can be composed of 4 steps: (1) image acquisition, (2) Segmentation and preprocessing, (3) feature extraction, and (4) classification. An illustration of the abovementioned concepts is described in Figure 1.

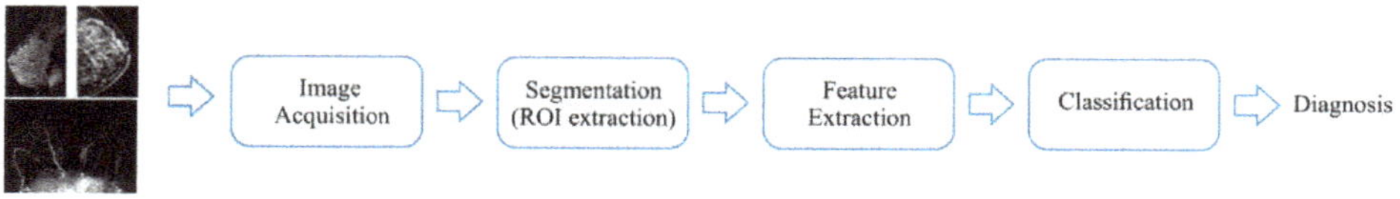

Figure 1. BC detection using image processing strategies.

From this figure, it can be seen that the first step uses the different technologies available to acquire the internal tissue dynamics of the breast, so they can be expressed in an image; the second step is used to execute algorithms that perform basic tasks on the images (for instance, correcting the color scale), so the segmentation, which is the detection of Region-of-interest (ROI), can be done; then, the third step quantifies the differences between images that have abnormalities from the ones that do not have; finally, once the differences are quantified, it is necessary to classify them to provide a diagnosis. With the rapid development of novel technologies that can capture more accurately the dynamics of the breast tissues, numerous advances have been done in all the aforementioned fields; in this sense, the goal of detecting all the abnormalities without generating false alarms is still a highly desirable feature for all the proposals [5,6]. Recently, some articles have reviewed some proposals regarding the feature classification and its interpretation [6–9]; yet, an article that presents the main technologies used to form the breast image as well as the processing stages required to provide a diagnosis is still missing. This article presents a state-of-the-art review of both the technologies used to create the breast image as well as the strategies employed to perform the image processing and classification. The article is organized as follows: Section 2 describes the main technologies used for the image generation; Section 3 describes the methods used to perform the segmentation, feature extraction, as well as the interpretation; next, Sections 4 and 5 present some emerging techniques that can be used to improve the image formation and the algorithms used for the interpretation. The article ends with some concluding remarks.

2. Technologies Used to Obtain Breast Tissue Images

One of the steps require to develop a diagnose system is the representation of the breast tissue dynamics. In this sense, there are several technologies that are commonly used to represent the tissue by means of images. This section presents the most used ones.

2.1. Mammography

Mammography is a study used to screen the breast tissue in order to detect abnormalities that could indicate the prescience of cancer or other breast diseases [10]. This technique has a sensibility of up to 85% in the recommended population. Essentially, mammography uses low doses of X-ray to form a picture of the breast internal tissues [11]. To form the picture, the breasts are compressed by two plates with the aim of mitigating the dispersion of the rays, allowing to obtain a better picture without using an X-ray high-dose [11], where the tissue changes might appear as white zones on a grey contrast [11]. On average, the total radiation dose for a typical mammogram with 2 views for each breast is about 0.4 [11]. Figure 2 illustrates the mammography procedure.

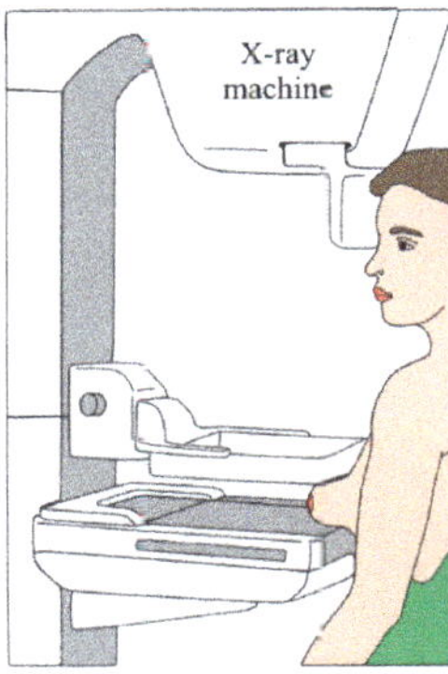

Figure 2. Mammography procedure.

Several works have focused on the processing of the digital mammographies to detect the most common symptoms that could indicate the presence of cancer calcifications or masses [12]. Traditionally, the specialist locks for zones that have a different appearance (size, shape, contrast, edges, or bright spots) than the normal tissue. With the employment of segmentation algorithms [13–15], the automatization of this task has been proposed, where some attempts using neural networks have done [12,16,17], delivering encouraging results.

Recently, the utilization of the Breast tomosynthesis (BT) and the Contrast-Enhanced Mammography (CEM) [10] have been proposed as improvements to the traditional digital mammography. The former is a 3D breast reconstruction that allows to further improve the image resolution whereas the latter improves the image resolution injecting a contrast agent; in this way, the anatomic and vascularity definition of the abnormalities is exposed. In this sense, some improvements when dealing with breast-dense tissue patients are obtained; yet, the detection of clustered micro calcifications is still an issue [10]; on the other hand, additional screening tests are required to determine if the abnormality detected by CEM is cancer or not, besides of requiring more expensive equipment.

2.2. Ultrasound

Ultrasound is a non-invasive and non-irradiating technique that uses sound waves to create images from organs, in this case the breasts, to detect changes in their form. To create the images, a transducer sends high-frequency sound waves (>20 kHz) and measures the reflected ones [10]. The image is formed using the wave sound reflected from the internal tissues. This procedure is depicted in Figure 3.

Ultrasound is used for three purposes: (1) assessing and determining the abnormality condition, that is, to help doctors if the abnormal mass is solid, which might require further examination, is fluid-filled, or has both features; (2) as an auxiliary screen tool, which is used when the patient has dense breasts and the mammography is not the reliable enough, (3) or as a guide to develop a biopsy in the suspected abnormality [10]. Several computer-aided diagnose (CAD) systems that analyze ultrasound images have been proposed [18]. One of the points they note it is necessary to improve is the resolution of the images [19] using specific-designed filters. Another modification proposed is the utilization of micro-bubbles that are injected into the abnormalities detected at first sight [20].

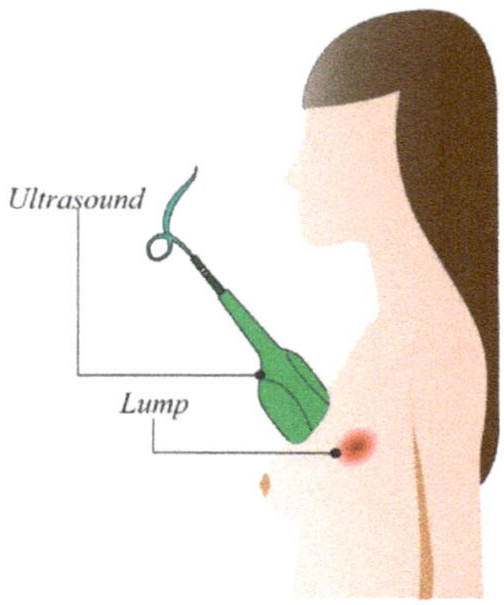

Figure 3. Ultrasound procedure.

It should be noticed that the mass tends to stay in its position when compressed, i.e., they do not displace. Elastography is the technique that is employed to measure the tumor displacement when compressed using a special transducer [21]. These developments have led to discover masses that usually require performing a biopsy to determine the mass nature, which delay the diagnosis confirmation [10,21]; moreover, the image interpretation requires a well-trained specialist, which is not always available to perform all the studies.

2.3. Magnetic Resonance Imagining (MRI)

Breast MRI (BMRI) uses a magnetic field and radio waves to create a detailed image from the breast. Usually, a 1.5 T magnet is used along with a contrast, usually gadolinium, to generate the images of both breasts [22]. To acquire the images, the patient is located in a prone position, in order to minimize the respiration movement and to allow the expansion of the breast tissue [10,22]. When the magnet is turned on, the magnetic field temporary realigns the water molecules; thus, when radio waves are applied, the emitted radiation is captured using specific-designed coils, located at the breast positions, which transforms the captured radiation in electrical signals. The coils position must ensure an appropriate field-of-vision from the clavicle to the infra-mammary fold, including axilla [10]. An illustration of the patient position is depicted in Figure 4.

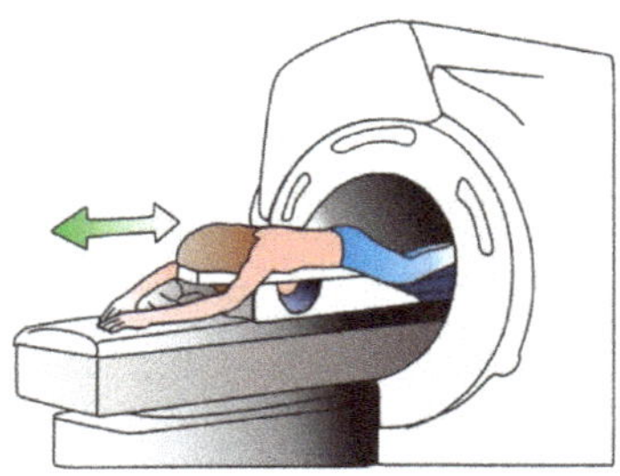

Figure 4. BMRI procedure.

The main objective of getting the images is to assess for the breast symmetry and the possible changes in the parenchymal tissue, since those changes might indicate the presence of lesions that can be malignant. In general, malignant lesions have irregular margins (or asymmetry), whereas the benign ones usually have a round or oval geometrical shape with well-defined margins (symmetry). To deliver the best possible result, it is necessary to remove the homogenous fat around the breast and parenchyma since fat can render images that can be uninterpretable, specially to detect subtle lesions [10,22].

On the other hand, one of the problems that BMRI has is the false-positive (specificity) rate, as the technique can detect low-size masses (lesions whose size is less than 5 mm) that are benign [10,22]. To mitigate the aforementioned issue, nanomaterials have been developed, so they stick to the cancer masses but not to the benign ones [23] as well as contrast agents [24]. Recently, it has been proposed that a multiparametric approach has been suggested as a strategy to improve the specificity rate [10].

2.4. Other Approaches

Recently, microwave radiation has been employed as an alternative to obtain information about the breast tissue. The microwaves, whose frequency range varies from 1 to 20 GHz, are applied to the breast and the reflected waves are measured using specific-designed antennas. To have the best possible results, some works propose that the tissue must be immersed in a liquid [25]. In this sense, some works have proposed acquisition systems that deal with this issue [26–29].

When it is necessary to perform a biopsy to confirm, images from the cells that form the abnormalities are obtained using among other techniques, the fine needle aspiration citology (FNAC), core or excisional biopsy. Once the cell images are captured, an image processing technique is applied in order to detect the differences between normal and malignant cells, which are classified using modern strategies [30–32] such as neural networks, probabilistic-based algorithms and association rules coupled with neural networks.

It should be pointed out that other alternatives for imaging are employed such as Computed Tomography (CT) or Positron Emission Tomography (PET). The former employ X-rays to form images from the chest using different angles; using image processing and reconstruction algorithms, a 3D image of the chest (including the breasts) is obtained [33,34]; on the other hand, the latter uses a small amount of tracer, that is a specific-designed sugar with radioactive properties known as fluorodeoxyglucose-18. The main idea of using this type of sugar is that cancer cells have an increased consume of glucose compared with the normal cells; in this sense, the tracer sticks in the zones where there is an increased glucose consume [35,36]. It is worth noticing that these techniques are recommended to determine the cancer stage rather than first-line diagnosis scheme [10,37]. In this way, they complement the three main techniques to provide more information from the tissues surrounding the breasts [37]. Table 1 presents a table that summarizes the abovementioned methods.

Table 1. Summary of the used breast image generation technologies.

Imagining Technique	Advantages	Disadvantages	Recommended Population	Some Types of Cancer Detected	Sensitivity and/or Specificity
Mammography	1. Equipment is widely available worldwide. 2. Methods, such as tomosynthesis, can improve the specificity and sensibility of the technique with patients that have dense breasts [10]	1. The rate of both false positive and false negatives increases since there is no possibility to determine if the masses are benign 2. The procedure used to obtain the images could be bothersome. 3. Dense breasts or young patients are not indicated to use this imaging technique.	Women whose age is greater than 40 years, have low-dense breast and an average risk of contracting the disease.	1. Ductal Carcinoma in Situ 2. Invasive Breast Cancer.	Sensitivity up to 85%.
Ultrasound	1. Can be used in young patients or have dense breast. 2. The equipment used is available in most of the hospitals	1. Calcifications could not be detected. 2. Sensitivity depends on the operator ability to interpret the images 3. False-positivity rate is an issue.	Women with heterogeneously or extremely dense breast tissue [38,39]. Women that are pregnant or lactating [40].	1. Ductal Carcinoma in Situ. 2. Invasive ductal carcinoma	Sensitivity ranging between 40–75% in younger high-risk women [40].

Table 1. *Cont.*

Imagining Technique	Advantages	Disadvantages	Recommended Population	Some Types of Cancer Detected	Sensitivity and/ or Specificity
Magnetic Resonance Imaging	1. Effective for detecting suspicious masses in high-risk population [10]. 2. The breast tissue density is no longer an issue [38–40]. 3. Multifocal lesions can be detected [10,41]	1. Equipment is only available in specialized hospitals. 2. Expensive 3. False positive findings are an important concern [41]	1. Women that may carry mutation in ATM, BRCA1, BRCA2, CHEK2, PALB2, PTEN, TP53 genes. 2. Women that had radiation therapy in the chest zone during the childhood.	1. Ductal in situ carcinomas 2. Invasive ductal carcinomas. 3. Invasive lobular carcinomas 4. Invasive mammary carcinomas with mixed ductal and lobular features [24]	Sensitivity ranging from 83 to 100% [42–44].

As it is seen in Table 1, numerous advances for imagining techniques have been achieved in the last years; still, there is a necessity of developing strategies that can allow obtaining sharp images, even for dense breast tissues. In this sense, the obtained images can be used to perform a focused surveillance on the patients that have a higher risk for developing the disease, allowing to achieve the cancer detection in the earliest possible stage. On the other hand, these novel imagining techniques should be able to operate without requiring additional requirements, such as specific electrical or mechanical conditions, so they can be easily adopted in hospitals, or in an ambulatory area.

3. Image Processing and Classification Strategies

3.1. ROI Estimation

Once the image is acquired, the next step required is its interpretation. To this purpose, it is necessary to identify the suspicious regions that might contain masses or calcifications, where model, region, or counter-based algorithms for the image segmentation are employed [45]. It should be noticed that these approaches often rely on the manual entries to refine the segmentation zones, which limits the applicability of the proposals on different datasets [45], making necessary to develop novel strategies that can automatically detect all the interest zones. Recently, Sha et al. [46] proposed a convolutional neural network (CNN)-based method for segmentation. The authors develop an optimization scheme to determine the best parameters for the CNN in order to segment the suspicious zones. The results presented show the proposal has a reasonable sensitivity and specificity (89% and 88%, respectively) to determine if a mammograph presents cancerous tumors or not. Wang et al. [47] present a CNN-based strategy. They modify the convolutional layer to increase the detection of multiple suspicious zones. Heidari et al. [48] employ a Gaussian bandpass filter to detect suspicious zones using local properties of the image. On the other hand, Suresh et al. [49] and Sapate et al. [50] employ a fuzzy-based strategy to cluster all the pixels with similar features in order to detect all the zones that have differences. Other strategies involve the utilization of mathematical morphology [51–55], image contrast and intensity [56,57], geometrical features [58,59], correlation and convolution [60,61], non-linear filtering [62,63], texture features [64], deep learning [65–69], entropy [70,71], among other strategies. It is worth noticing that from the diversity of the employed strategies, some of them still require an initial guidance to detect the suspicious zones, either by manually selecting pixels inside of the zone or using the radiologist notes about the localization. An effective approach for the automatic detection should employ a denoising stage in order to remove residual noise generated during the acquisition and equalization, so the intensity pixel disparities associated to the environment light can be mitigated as much as possible.

3.2. Feature Extraction

After the suspicious zones are detected and segmented, it is necessary to extract features from them to generate the necessary information to classify the detected lesions as cancer or benign. To this purpose, Fourier Transform-based methods [48,72], wavelet transform-based strategies [73–76], geometric features [77,78], information theory algorithms [79], co-occurrence matrix features [47,80–82], histogram-based values [46,83–85], morphology [86,87], among others. On the other hand, with the increased capabilities (the number of simultaneous operations that can be done) of the new-generation graphical processor units, it is now possible to execute high-load computational algorithms faster than in a multicore processor [88]; in consequence, novel neural networks algorithms that perform the feature extraction and quantification are now being proposed. For instance, Xu et al. [89], use a CNN to extract and classify ultrasound images with suspicious areas in four categories: skin, glandular tissue, masses, and fat. They modify the convolutional filters to speed up the process. Arora et al. [90] also use an ensemble of CNN architectures to extract directly the suspicious zones. They only modify the final layers to speed up the training process. Gao et al. [91] use a deep neural network to generate the features from mammograms. They employ a modified architecture where the outputs and inputs of the network are used to update the model parameters during its training. Similar approaches are described in [92–95].

It should be pointed out that a reduction of the estimated features is often used to reduce the amount of computational resources used in the training scheme and to mitigate the overfitting problem, which reduce the algorithm efficacy. This step is known as dimensionality reduction [45] and the most employed algorithms are the principal component analysis (PCA) and linear discriminant analysis (LDA). PCA use eigenvalue-based algorithms to determine the features that are unrelated between them, that is, they have the maximum variance between them as this will indicate the maximum variation of the information contained, whereas LDA perform a projection of the samples to find out the distance between the classes' mean. In this sense, the greater the distance between the means, the more unrelated the features are [96]. Nevertheless, these algorithms use global properties of the values which might cause to deliver suboptimal results [96]. For these reasons, hybrid strategies are proposed such as neurofuzzy algorithms [97,98], diffusion maps [99], deep learning [100–102], independent component analysis (ICA) [103], clustering-based approaches [104], multidimensional scaling [105], among other strategies. It should be pointed out that hybrid approaches, as abovementioned ones, are particularly effective when a non-linear relationship between the features exists.

To the best of the authors' knowledge, there are no papers that compare some of the abovementioned techniques using the same database to compare the techniques efficacy. In this sense, it is an interesting research topic, since the results of this comparison can provide some guidelines about the image used (mammogram, ultrasound, or MRI) and the technique that has the best performance.

3.3. *Classifiers*

The last step of this stage is the classification of the extracted features to make a diagnosis. Broadly speaking, a classifier uses the input data to find out relationships that can be used to determine the class where the input data belongs to. The evaluation of the classifier is done using three basic measurements: accuracy, specificity, and sensitivity [106,107]. Accuracy refers to the percentage of images that are correctly classified in their corresponding classes; sensitivity is the percentage of classified images as malignant that truly are specificity is the percentage of classified images as benign that truly are, and the area under the curve is a parameter that allows choosing the optimal models. It takes a value between 0 and 1, being a good classifier the one that has a value close to 1 [108]. In this sense, depending on the training algorithm required by the strategy, classifiers can be divided in unsupervised or supervised [45,106,107].

3.3.1. Unsupervised Classifiers

An unsupervised classifier aims to find the underlying structures that the input data has without making explicit the class the input data belongs to [109]. In this sense, input data that has similar values is assigned to the same class [109]. Dubey et al. [110] studied the effects that the selection scheme for the size of the number of clusters in the K-means algorithm has. To this purpose, the random and foggy methods were employed. They note that foggy initialization method and the Euclidean-type distances produced the best results, as a 92%-accuracy is obtained. K-means and K-nearest neighbor classifiers have been also employed by Singh et al. [58] and Hernandez-Capistran et al. [111]. This family of classifiers is effective when the distance between the clusters is reasonable; but, when the aforementioned concept is not possible, the accuracy rate is highly degraded. For this reason, Onan [112] introduced the concepts of the fuzzy logic to measure the distance between the set of features used as input and the clusters, where the mutual information, an information theory algorithm, is the chosen to measure the aforementioned distance. The author reports an accuracy of 99%, and a specificity and sensitivity of 99% and 100%, respectively. Similar results are achieved using the fuzzy c-means algorithm [113,114], fuzzy-based classifier for time-series [115], fuzzy rule classifier [116,117], among others. Other clustering-based approaches employed for classification are hierarchical clustering [118] and Unsupervised Test Vector Optimization [119]. It should be pointed out that unsupervised classifiers require a careful selection of the features used to train the algorithm, since an incorrect mix of features will degrade the performance of the classifier.

3.3.2. Supervised Classifiers

Supervised classifiers require to know a-priori the class of which the input data belongs to, that is, the input data must be labeled. The Decision Tree (DT) is an algorithm that uses a set of rules to determine the class of the data input. DT has been employed by Mughal et al. [71], where they perform the detection of masses in mammograms using texture features in the region of interest. Using a DT, they obtain an accuracy, specificity, and sensibility of 89%, 89% and 88.5%, respectively. Shan et al. [120] employ geometrical features to classify abnormalities detected in ultrasound images. The obtained results show an accuracy, sensitivity, and specificity of 77.7%, 74.0%, and 82.0%, respectively. An improvement of DT is the Random Forest (RF). During the training stage, RF uses several DT, where the ones that have the lowest error are chosen; in this way, the accuracy is enhanced. RF are considered as ensemble classifiers, where some applications have been reported [121–124]. The accuracy, specificity, and sensitivity reported show an improvement. Another type of ensemble classifier is the Adaptive Boosting (AdaBoost) algorithm. It consists in the utilization of weak classifiers, which are usually features that can generate a classification accuracy greater than 50% by themselves; thus, using them in an ensemble way, the outliers that the features value have are used, improving the classifier accuracy. AdaBoost applications have been reported [125–127], achieving good results (the accuracy, specificity, and sensitivity values are greater than 90%); yet, the authors note that extensive investigation is still required to ensure that these results can be obtained with different types of images (mammograms, ultrasound, and MRI).

Another classification algorithm widely used for BC detection is the support vector machine (SVM). SVM finds the hyperplane that divides the zones where the values of the input features are located. In this regard, Liu et al. [52] use the morphological and edge features combined with a SVM classifier with a linear kernel, to detect benign and malignant masses in ultrasound images. They obtain an accuracy, sensitivity, and specificity of 82.6%, 66.67%, and 93.55%, respectively. It should be noted that most of the revised works use the term malignant to describe masses or lesions that are cancer regardless its type. To improve the aforementioned results, Sharma and Khanna [128] use the Zernike moments as features and a SVM classifier using a non-linear function as a kernel. The authors obtain a specificity and sensitivity of 99%. Similar approaches have been reported [87,129–133]. It is worth noticing that if the features have a strong nonlinear relationship, other classifiers could deliver better results.

3.4. *Artificial Intelligence-Based Classifiers*

Artificial Intelligence (AI) is the section of the computer science that develops algorithms to perform complex tasks that previously are solved with the human knowledge [134]. Evidently, since classification is a task usually solved by the physician, AI can provide automated solutions. In this sense, Artificial Neural Networks (ANN) are a type of AI algorithms employed to perform the classification in different classes. ANN are brain-inspired algorithms that store the knowledge that the input data using a training process [135]. An ANN consists in a three-layer scheme: input, hidden, and output, as depicted in Figure 5.

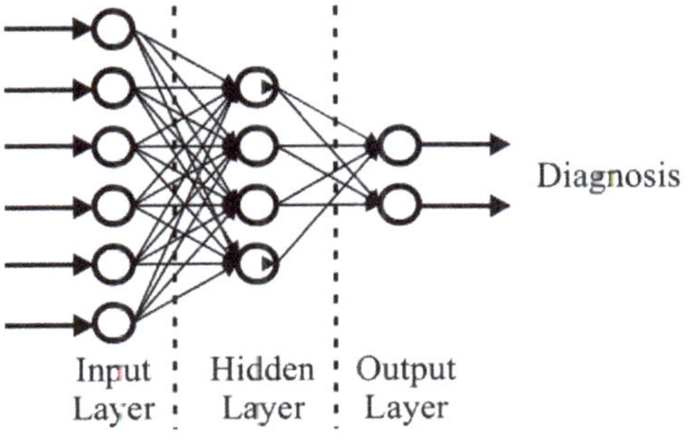

Figure 5. Artificial Neural Network.

The training process takes the information contained in the input variables and adjust the values of the variables (weights) that connect all the layers in order to match the input with its respecting class; in this way, the hidden pattern that share all the input and their corresponding class is detected and stored. Consequently, it is necessary to use a sufficient database, with representative scenarios, to train the ANN. Beura et al. [136] present a methodology that employs mammograms to detect masses (benign and malignant) using the two-dimension discrete wavelet transform (2D-DWT) with normalized gray-level co-occurrence matrices (NGLCM). The images are segmented using a cropping-based strategy to obtain the ROI, which are analyzed with the symmetric biorthogonal 4.4 wavelet mother and a decomposition level of 2. All the frequency bands are processed to obtain the features (NGLCM), where the *t*-test is selected to perform the optimal choice of the most discriminant features. The obtained results show that the proposal achieves an accuracy, sensitivity, and specificity of 94.2%, 100%, and 90% respectively, using the ANN classifier, whereas a RF classifier, using the same database, obtains an 82.4%-accuracy. Mohammed et al. [137] uses fractal dimension values as features to classify ultrasound breast images in benign and malignant. They obtain the ROIs using a cropping-based algorithm, which are processed to obtain multifractal dimension features. They obtain an accuracy, sensitivity, and specificity of 82.04%, 79.4%, and 84.76% respectively using an ANN classifier. They point out that the ROI extraction algorithm must be improved. Gallego-Ortiz and Martel [138] classifies MRI breast images using graph-based features, the Deep Embedded Clustering algorithm to select the most relevant features and an ANN classifier. The ROIs are obtained using a graph model, where they obtain an area under the curve, which is another feature to measure the classifier effectiveness, of 0.80 (the closer to 1, the better). ANN classifiers have been also used in [139–142].

Deep neural networks (DNN) are a specific type of AI algorithms based on the architecture of an ANN [134]. DNN resembles how the brain stores, in multiple layers, the acquired knowledge to solve a specific task [8]. The Convolutional Neural Network (CNN) is a DNN that emulates the visual processing cortex to determine the class that an image belongs to [8,134]. A CNN typical scheme is depicted in Figure 6.

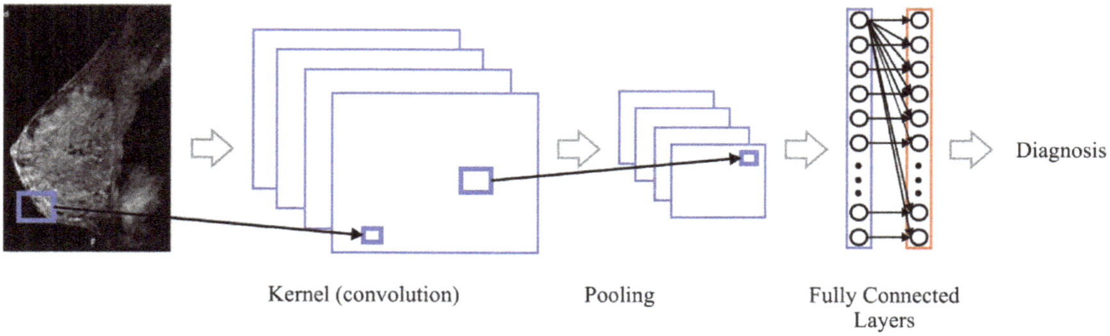

Figure 6. Convolutional Neural Network.

From the figure, it is seen that a CNN consists of a kernel, pooling and fully connected layers. The purpose of the kernel layer is to detect and extract spatial features that the image has, which is usually done with the convolution operator. The output of this layer, known as feature map, might contain negative values that might cause numerical instabilities in the training stage; thus, map is processed using a function to avoid the negative values. Once the feature map is processed, the pooling layer reduces the amount of information contained in order to eliminate redundant information; finally, the output of the pooling layer goes to the fully connected layer to be classified. In this sense, several works [143–148], have been employed CNN to detect benign and malignant tissues in either mammography or MRI images. They note that the depth of the network, i.e., the number of layers, the fine-tuning of some of the kernel or pooling layers, as well as the number of images, affect the classifier performance.

Ribli et al. [149] add an additional layer to implement specific-designed filters for mammograms. The CNN they employ has 16-layers and classifies the detected lesions in benign or malignant, obtaining an area under the curve of 0.85. A similar approach is proposed in [150]. The modification they propose is that a fully connected layer is placed as the first layer of the CNN so when the images are noise-corrupted, the feature extraction process is not degraded. They obtain an accuracy, sensitivity, and specificity of 98.7%, 98.65%, and 99.57% for the detection of benign and malignant lesions in mammograms. Zhang et al. [151] carry out a test to find out the specific-suited process for the pooling layer. They found out that rank-based stochastic process is the best-suited algorithm, obtaining an accuracy, sensibility, and specificity of 94.0%, 93.4%, and 94.6%, respectively, for classifying lesions for normal or abnormal using mammograms. Similar approaches have been proposed [152–155]. Table 2 presents a summary of the classifiers above discussed. It should be noted that a mix of images from mammograms, ultrasound, MRI are usually employed. These images usually came from private databases.

From the data shown in Table 2, it can be seen that it is necessary to standardize the minimum requirements regarding the number of images that the databases must have. In this way, the performance metrics that are employed, i.e., accuracy, specificity, and sensitivity, can be compared in a better way. Moreover, even when the presented approaches show interesting results, one thing they found out is the necessity of having a considerable database that contain significant labeled images to obtain the best possible results, which in many real-life scenarios is not always possible. For these reasons, algorithms that can work with both labeled and unlabeled images are still a necessity.

Table 2. Summary of the used image classification algorithms.

Type of Classifier	Classifier	Advantages	Disadvantages	Number of Images	Performance Metrics
Unsupervised	K-means	• Easiness of implementation. • Fast implementation. • Fast computing (distance to the centroids is only required).	• The initial value of the centroids length influences the performance. • Samples must be presented in an organized and normalized way. • The centroid distance of the classes might induce misclassifications.	• Some works have used the Wisconsin Breast Cancer Dataset with 569 instances [110,112].	• Accuracy: up to 92% [110] • Specificity: up to 99% [112] • Sensitivity: up to 100% [112]
	Hierarchical Clustering	• No distance measurement is required. • Similarity measures could be employed. • Easy to implement.	• Large datasets increase the time complexity to deliver a result. • Outliers degrade the classifier performance. • Normalization of the samples values is required.	• 117 images are analyzed [118].	• Accuracy: 88.0% • Specificity: 89.3% • Sensitivity: 85.7% [118]
Supervised	Decision Trees	• Its construction no imposes any probabilistic distribution to the data. • Can deal with large datasets. • Easy to understand.	• They can be too complex if the training data is not carefully chosen. • Their performance will decrease if several classes exist in the data.	• Some works have analyzed from 283 [120] to 722 images [71].	• Accuracy: up to 89%, • Specificity: up to 89% • Sensitivity: up to 90% [71,120]
	Random Forest	• Non-linear relationships between the features are well processed. • Outliers do not degrade the classifier performance. • Noisy measurements do not affect the accuracy.	• Training time increases due to the number of trees generated. • The classifier complexity is increased as the number of trees needed to be evaluated.	• Several works have used different number of images from 59 [121], 283 [120] to 512 [122]. • On the other hand, some authors have used ten different datasets, the shortest with 155 images and the largest with 569 images [123].	• Accuracy: up to 80%. • Specificity: up to 80%. • Sensitivity: up to 90% [120–123]
	AdaBoost	• Base classifiers only need to have an accuracy greater than 50%. • They can be from different domains (spatial, frequency, among others)	• Noise can degrade the classifier performance, as the weight assigned to each weakly classifier is increased to reduce the error. • Sensitive to the base classifiers employed	• Some works have used from 1062 [126] to 2336 [125] images.	• Accuracy: up to 90%. • Specificity: up to 90%. • Sensitivity: up to 90% [12,125,126]
	Support Vector Machines	• Can deal with high-dimensional data (features). • Robust against outliers. • Overfitting is reduced due to the training process.	• Accuracy is kernel dependent. • Large datasets are not properly handled. • Overlapping and noise degrade the accuracy. • Uncertainty cannot be incorporated.	• Some authors have used different number of images from 207 [87], 240 [132] to 1187 [131].	• Accuracy: up to 90%. • Specificity: up to 90%. • Sensitivity: up to 90% [74,87,131,132]

Table 2. *Cont.*

Type of Classifier	Classifier	Advantages	Disadvantages	Number of Images	Performance Metrics
Supervised	Artificial Neural Networks	• Can deal with highly non-linear relationships. • Can deal with noisy data. • Uncertainty can be incorporated. • Fine-tuning could be done using different activation functions.	• High-dimensional data might cause instabilities to the training algorithms. • Prone to overfitting. • Selection of the number of neurons could be troublesome.	• Other authors have been used 111 [139], 184 [137], and 569 [140] images.	• Accuracy: up to 95%. • Sensitivity: up to 100%. • Specificity: up to 90% [134,137–140]
	Convolutional Neural Networks	• Can process the image without any preprocessing stage. • They can perform feature extraction task automatically. • Moderate noisy images can be properly handled.	• They require a large dataset to avoid overfitting. • They require a high computational load to their training.	• Some authors have used different number of images from 87 [144], 221 [143] to 229,426 digital screening mammography exams [145].	• Accuracy: up to 99%. • Sensitivity: up to 99% • Specificity: up to 99.6% [7,8,143–145,149]

4. Recent Image Generation Techniques

Infrared Thermography (IRT) Applied to Breast Cancer

Temperature has been documented as an indicator of health [156]. Specifically speaking of breast cancer, when a tumor exists, it makes use of nutrients for its growth (angiogenesis), resulting in an increase in metabolism, thus the temperature around the tumor will increase in all directions [157]. To detect the temperature changes, IRT has been used as it measures the intensity of the thermal radiation (in the form of energy) that bodies emit, converting it into temperature [158]. The emitted energy can be visualized in the electromagnetic spectrum, as shown in Figure 7, where it is seen that the infrared (IR) wave ranges from 0.76 to 1000 µm and in turn is divided into Near-IR, Mid-IR and Far-IR. The available technology to measure IR allows performing the aforementioned task using non-invasive, contactless, safe, and painless equipment [159–161], making a suitable proposal for developing scanning technologies.

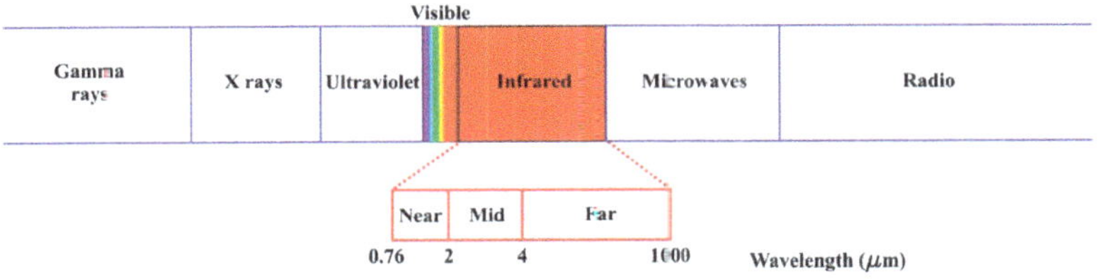

Figure 7. Electromagnetic spectrum.

To obtain the best possible images, there are mainly three factors that influence thermographic imaging in humans [162,163]

1. Individual factors: everything that has to do with the patient's conditions, such as age, sex, height, medical history, among others. As well as the inclusion and exclusion criteria. An aspect of vital importance is the emissivity of humans, which is 0.98 [164].
2. Technical factors: it has to do with everything related to the technology used during the study, such as the thermal imager (considering the distance from the lens to the patient), the protocol, the processing of the medical thermal images obtained, as well such as feature extraction and subsequent analysis.
3. Environmental factors: room position (it should be located in the area of the lowest possible incidence of light), temperature, relative humidity of the space where the thermographic images are to be taken, as well as the patient's air conditioning time.

Considering the all the above discussed aspects, a suitable location for developing a controlled scenario to acquire thermographic images focused on breast cancer is depicted in Figure 8.

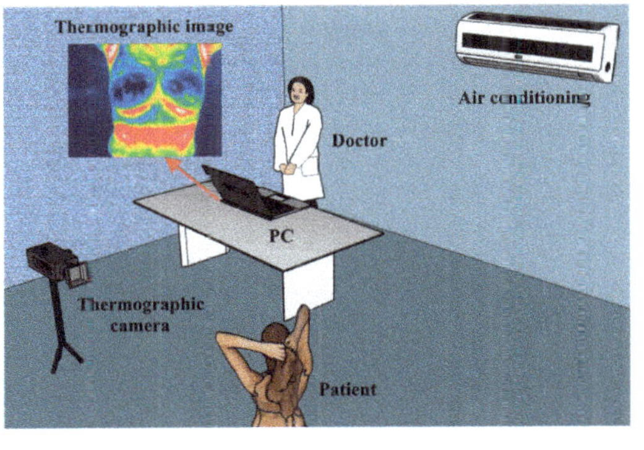

Figure 8. Proposed experimental set up for the breast thermal images acquisition.

Once the room is conditioned for obtaining the thermographic images, the acquisition can be done. The reported results make use of the previously discussed image processing and classification algorithms. Table 3 shows a brief resume of the most recent proposed works.

Table 3. Summary of the breast lesions detection using infrared thermography.

Authors	Number of Patients	IR System	Image Processing and Classification Algorithms		Accuracy (%)	Room Temperature (°C)	Acclimation Time (min)
			Features	Classification			
Ekici and Jawzal [165]	140	FLIR SC-620	Bio-data, image analysis, and image statistics	CNNs optimized by Bayes algorithm	98.95	17–24	15
AlFayez et al. [166]	Public dataset DMR-IR		Geometrical and textural features	Extreme Learning Machine (ELM) and Multilayer Perceptron (MLP)	ELM—100 MLP—82.2	Public dataset DMR-IR	
Rani et al. [167]	60	FLIR T650SC	Temperature and intensity	SVM with Radial basis function kernel	83.22	20–24	15
Saxena et al. [168]	32	FLIR A320	ROI thermal	Cut-off value	88	22 ± 0.5	Not specified
Tello-Mijares [169]	63	FLIR SC-620	Shape, colour, texture, and left and right breast relation	CNN	100	20–22	15
Garduño-Ramón et al. [170]	454	FLIR A300	Temperature	Difference of temperature	79.60	18–22	15
Raghavendra et al. [171]	50	Thermo TVS200	Student's *t*-test based feature selection algorithm	Decision Tree	98	20–22	15
Lashkari et al. [172]	67	Thermoteknix VisIR 640	23 features, including statistical, morphological, frequency domain, histogram and Gray Level Co-occurrence Matrix	Adaboost, SVM, kNN, Naive, PNN	85.33 and 87.42	18–23	ice test: 20 min
Francis et al. [173]	22	med2000™ IRIS	Statistical and texture features are extracted from thermograms in the curvelet domain	SVM	90.91	25	15
Milosevic et al. [174]	40 images	VARIOSCAN 3021 ST	Texture measures derived from the Gray Level Co-occurrence Matrix	K-Nearest Neighbor	92.5	20–23	Few minutes
Araujo et al. [175]	50	FLIR S45	Thermal interval for each breast	Linear discriminant classifier, minimum distance classifier, and Parzen window	-	24–28	At least 10 min

Recently, dynamic infrared thermography (DIT) has been proposed as an alternative to further improve the image quality and sharpness [64]. DIT is a sequence of thermograms captured after stimulating the breasts by means of a cold stressor [176]. The objective of this stressor is to generate a contrast between areas with abnormal vascularity and metabolic activity with areas free of abnormalities. Therefore, it is possible to analyze the sinus response after removing this stimulus. In this way, the image sharpness is enhanced. Silva et al. [177] proposed a technology that analyzes the information from the DIT to indicate patients at risk of breast cancer, where they segment the area of interest (breast) and analyze the changes in temperature through the different thermograms acquired. Saniei et al. [178] proposed a system that segments both breasts to obtain the branching point of the vascular network, which represents the pattern of the veins; finally, these patterns are classified to obtain the diagnosis. As it can be seen, the DIT requires robust systems that allow the analysis of the acquired thermograms over time, which should be considered in order to generate the next generation of equipment that can allow the early detection of the angiogenesis process. By doing this, patients can be properly monitored so the changes in the patterns of the angiogenesis process be detected.

5. Recent Classification Algorithms

As pointed out in the Classifiers subsection, it is necessary to overcome the lack of a large database of images (mammograms, ultrasound or BRMI) that have been diagnosed to generate robust and efficient classifiers. In this sense, semi-supervised methods can be an attractive choice to explore. They usually combine an unsupervised algorithm to cluster the images available, so a representation of the dataset is obtained; then, the supervised classifier assigns the classes that images have [109,179]. The data that is used in the unsupervised algorithm assumes the unlabeled images are close to the labeled ones in their input space, so their labels are the same [109] Some of the most recent developments that could be applied in the breast cancer detection are presented.

5.1. Autoencoders

An autoencoder is a neural network that has one or more hidden layers that is used to reconstruct the input compactly, as the hidden layers have few neurons. The autoencoder is depicted in Figure 9.

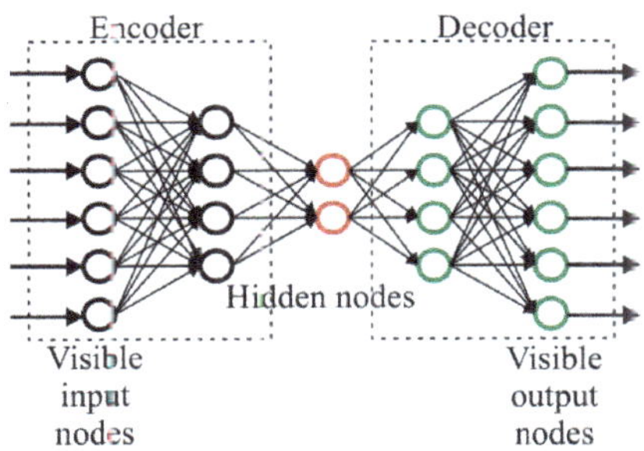

Figure 9. Autoencoder structure.

From the figure, it is seen that it has two parts: the encoder, that represents the input into its compact representation, and the decoder. which performs the inverse operation, that is, use the compact representation to recover the original data. The most common training scheme consists in employing a loss function that aims to reduce the error between the original and reconstructed data. For breast cancer detection, autoencoders can be used feature extraction stages, as the encoder part obtains the compact representation or features of the input image, that are followed by a supervised classifier. Recently, this approach has been explored [79,94,180–183] showing promising results to generate robust methodologies, where accuracies values above 95% are obtained.

5.2. Deep Belief Networks (DBF)

They are based on the usage of restricted Boltzmann machines (RBMs) RBMs only use two layers: input and hidden, to represent, as in the case of the autoencoders, the most important features that can represent the input data but in a stochastic way [99] This ensure that the outliers do not affect the network performance. Detailed information can be found in [184,185]. The main idea in employing DBF is that the image segmentation can be done without external guidance; thus, a totally automated methodology can be proposed. Recent works have been explored this idea to perform the liver segmentation [186], lung lesions detection [187], and fusion of medical images [188]. Its use could deliver promising results to detect BC.

5.3. Ladder Networks

Ladder Neural Network, proposed by Rasmus et al. [189], uses an autoencoder as the first part of a feedforward network to denoise the inputs; further, by determining the minimum features that represent the inputs, the classification can be done using simple algorithms. The network uses a penalization term in the training algorithm to ensure the maximum similarity between the original and reconstructed inputs.

5.4. Deep Neural Network (DNN)-Based Algorithms

Recently, DNN-based classification strategies have been proposed to maximize the accuracy that the classifiers achieve while reducing the computational resources required to perform its training and execution, being the physics-informed neural network or more recently, the Deep Kronecker neural network [190] are one of the most recent algorithms that have been proposed. In particular, these NNs are designed to take full advantage of the adaptive activation functions. Traditional activation functions, such as the unipolar and bipolar sigmoid and the ReLU, might have problem when dealing with low-amplitude features as the training algorithm fails to achieve the lowest point in the error surface, thus generating classifiers prone to have generalization issues [190].

In this sense, by introducing a parameter into the activation function equations that can be modified during the training process, it can be avoided that the gradient function does not stall in a local minimum on the error surface [191]; thus, the highest accuracy can be obtained since the global minimum is reached [192]. The results presented [190–192] suggest that the utilization of this type of activation function might increase the classifier accuracy without increasing the computational burden required to train the network as the geometrical shape that the activation function defines can be adapted during the training time to the boundary decision zone where classification is required. It should be noted that the proposed Rowdy family of activation functions could be an interesting research topic for designing classification algorithms, as the presented results demonstrate that the lowest error is achieved in a prediction task.

6. Concluding Remarks

This paper presents a state-of-the-art review of the technologies used to acquire images from the breast and the algorithms used to detect BC. To the best of the author's knowledge, this is the first review article that deals with all the required steps to propose a reliable methodology for the BC detection. This is important as the earliest detection of the disease can save a considerable amount of money in the required treatments, and the most important, potentially saving numerous lives.

The analyzed papers are focused on the research on the processing of images obtained using non-invasive methods: X-ray, ultrasound, or magnetic resonance, as they are the most accessible technologies in hospitals. The strategy used in most of the papers has 4 steps: image acquisition, ROI estimation, feature extraction, and interpretation. For the ROI estimation, the strategies proposed are based on radiologist annotations or require external help in order to be executed. This is an opportunity area to develop automatic algorithms that can detect the abnormalities. The feature estimation is used to quantify the detected zones in numerical values. In this sense, texture-based and geometrical-based features are by far, the most employed due to its estimation simplicity; still, frequency or spatial features have recently begun to be explored and can lead to detect minimal changes that might increase the sensitivity required to further improve the classification accuracy. It should be noticed that feature reduction strategies are commonly employed in order to reduce the training time or avoid potential misclassifications, where the most popular are LDA and PCA. On the other hand, classification strategies employ either supervised or unsupervised algorithms. The selection of the type of classifier heavily depends on the nature of the features extracted. If they are highly discriminant between them, then an unsupervised classifier is usually selected. On the other hand, when the features used have an overlap zone, then it is necessary to employ a supervised classifier. It should be noticed that AI-based algorithms, especially those based on deep learning, have the edge in terms of the performance they get at the expense of being very expensive in terms of the computational resources employed.

Emerging imaging technologies such as the microwave and thermography are being explored recently. In particular, the latter has recently obtained the attention of researchers as it is easy-to-use, and, with a proper cooling protocol, can reach an interesting level of accuracy to detect, at least, suspected masses that might evolved into malignant ones. With the development of semi-supervised strategies, some of the stages employed can be integrated into one, allowing the development of effective feature extraction, selection and classification strategies that have the same performance of supervised classifier, with lower computational resources employed, even in the presence of limited labeled images, which is a major obstacle to the training of the classifiers.

Modern BC detection strategies should rely using artificial intelligence (AI)-based algorithms that can use both on the information of the images acquired and categorical data [193–195], i.e., information about the daily life of the patients, with the aim of proposing algorithms that can determine if the patient has malignant lesions with a higher certainty and with the lowest false alarm at the earliest stage possible in order to get an effective treatment that can prevent the disease propagation. To achieve this goal, it is necessary to develop a database that contains the aforementioned features and whose size can reflect the main scenarios that can be found in real-life. Further, having algorithms that can deal with the aforementioned information, it can be possible to design personalized surveillance and clinical screening strategies that could offer the best health outcome for every patient.

Author Contributions: Conceptualization: I.A.C.-A., L.A.M.-H. and C.A.P.-R.; methodology: J.A.B.-H. and M.A.I.-M.; investigation: C.A.P.-R. and M.T.-A.; writing—original draft preparation, C.A.P.-R., J.A.B.-H. and I.A.C.-A.; writing—review and editing: M.A.I.-M., M.T.-A. and L.A.M.-H. All authors have read and agreed to the published version of the manuscript.

Funding: This research received no external funding.

Institutional Review Board Statement: Not applicable.

Informed Consent Statement: Not applicable.

Data Availability Statement: Not applicable.

Conflicts of Interest: The authors declare no conflict of interest.

References

1. World Health Organization (WHO). Cáncer de Mama: Prevención y Control. Available online: https://www.who.int/topics/cancer/breastcancer/es/index1.html (accessed on 3 May 2022).
2. Villa-Guillen, D.E.; Avila-Monteverde, E.; Gonzalez-Zepeda, J.H. Breast cancer risk and residential exposure to environmental hazards in Hermosillo, Sonora, Mexico [abstract]. In Proceedings of the 2019 San Antonio Breast Cancer Symposium, San Antonio, TX, USA, 10–14 December 2019; AACR: Philadelphia, PA, USA.
3. Keith, B.; Simon, M.C. *Tumor Angiogenesis. The Molecular Basis of Cancer*, 4th ed.; Mendelsohn, J., Gray, J.W., Howley, P.M., Israel, M.A., Thompson, C.B., Eds.; Elsevier: Philadelphia, PA, USA, 2015; pp. 257–268.
4. Semin, J.N.; Palm, D.; Smith, L.M.; Ruttle, S. Understanding breast cancer survivors' financial burden and distress after financial assistance. *Support. Care Cancer* **2020**, *28*, 4241–4248. [CrossRef]
5. Mann, R.M.; Cho, N.; Moy, L. Breast MRI: State of the Art. *Radiology* **2019**, *292*, 520–536. [CrossRef] [PubMed]
6. Vobugari, N.; Raja, V.; Sethi, U.; Gandhi, K.; Raja, K.; Surani, S.R. Advancements in Oncology with Artificial Intelligence—A Review Article. *Cancers* **2022**, *14*, 1349. [CrossRef]
7. Chougrad, H.; Zouaki, H.; Alheyane, O. Multi-label transfer learning for the early diagnosis of breast cancer. *Neurocomputing* **2020**, *392*, 168–180. [CrossRef]
8. Le, E.P.V.; Wang, Y.; Huang, Y.; Hickman, S.; Gilbert, F. Artificial intelligence in breast imaging. *Clin. Radiol.* **2019**, *74*, 357–366. [CrossRef] [PubMed]
9. Yassin, N.I.R.; Omran, S.; Houby, E.M.F.; Allam, H. Machine learning techniques for breast cancer computer aided diagnosis using different image modalities: A systematic review. *Comput. Methods Programs Biomed.* **2018**, *156*, 25–45. [CrossRef]

10. Jochelson, M. *Advanced Imaging Techniques for the Detection of Breast Cancer*; American Society of Clinical Oncology Educational Book: Alexandria, VA, USA, 2012; pp. 65–69.
11. Yaffe, M.J. AAPM tutorial. Physics of mammography: Image recording process. *RadioGraphics* **1990**, *10*, 341–363. [CrossRef]
12. Pak, F.; Kanan, H.R.; Alikhassi, A. Breast cancer detection and classification in digital mammography based on Non-Subsampled Contourlet Transform (NSCT) and Super Resolution. *Comput. Methods Programs Biomed.* **2015**, *122*, 89–107. [CrossRef]
13. Geweid, G.G.N.; Abdallah, M.A. A Novel Approach for Breast Cancer Investigation and Recognition Using M-Level Set-Based Optimization Functions. *IEEE Access* **2019**, *7*, 136343–136357. [CrossRef]
14. Guzmán-Cabrera, R.; Guzmán-Sepúlveda, J.R.; Torres-Cisneros, M.; May-Arrioja, D.A.; Ruiz-Pinales, J.; Ibarra-Manzano, O.G.; Aviña-Cervantes, G.; Parada, A.G. Digital Image Processing Technique for Breast Cancer Detection. *Int. J. Thermophys.* **2012**, *34*, 1519–1531. [CrossRef]
15. Avuti, S.K.; Bajaj, V.; Kumar, A.; Singh, G.K. A novel pectoral muscle segmentation from scanned mammograms using EMO algorithm. *Biomed. Eng. Lett.* **2019**, *9*, 481–496. [CrossRef]
16. Vijayarajeswari, R.; Parthasarathy, P.; Vivekanandan, S.; Basha, A.A. Classification of mammogram for early detection of breast cancer using SVM classifier and Hough transform. *Measurement* **2019**, *146*, 800–805. [CrossRef]
17. Rodríguez-Álvarez, M.X.; Tahoces, P.G.; Cadarso-Suárez, C.; Lado, M.J. Comparative study of ROC regression techniques—Applications for the computer-aided diagnostic system in breast cancer detection. *Comput. Stat. Data Anal.* **2011**, *55*, 888–902. [CrossRef]
18. Cheng, H.D.; Shan, J.; Ju, W.; Guo, Y.; Zhang, L. Automated breast cancer detection and classification using ultrasound images: A survey. *Pattern Recognit.* **2010**, *43*, 299–317. [CrossRef]
19. Ouyang, Y.; Tsui, P.-H.; Wu, S.; Wu, W.; Zhou, Z. Classification of Benign and Malignant Breast Tumors Using H-Scan Ultrasound Imaging. *Diagnostics* **2019**, *9*, 182. [CrossRef] [PubMed]
20. Ouyang, Y.; Tsui, P.-H.; Wu, S.; Wu, W.; Zhou, Z. Breast cancer detection by B7-H3–targeted ultrasound molecular imaging. *Cancer Res.* **2015**, *75*, 2501–2509. [CrossRef]
21. Athanasiou, A.; Tardivon, A.; Ollivier, L.; Thibault, F.; El Khoury, C.; Neuenschwander, S. How to optimize breast ultrasound. *Eur. J. Radiol.* **2009**, *69*, 6–13. [CrossRef]
22. Mumin MRad, N.A.; Hamid MRad, M.T.R.; Ding Wong, J.H.; Rahmat MRad, K.; Hoong Ng, K. Magnetic Resonance Imaging Phenotypes of Breast Cancer Molecular Subtypes: A Systematic Review. *Acad. Radiol.* **2022**, *29*, S89–S106. [CrossRef]
23. Han, C.; Zhang, A.; Kong, Y.; Yu, N.; Xie, T.; Dou, B.; Li, K.; Wang, Y.; Li, J.; Xu, K. Multifunctional iron oxide-carbon hybrid nanoparticles for targeted fluorescent/MR dual-modal imaging and detection of breast cancer cells. *Anal. Chim. Acta* **2019**, *1067*, 115–128. [CrossRef]
24. Mango, V.L.; Morris, E.A.; Dershaw, D.D.; Abramson, A.; Fry, C.; Moskowitz, C.S.; Hughes, M.; Kaplan, J.; Jochelson, M.S. Abbreviated protocol for breast MRI: Are multiple sequences needed for cancer detection? *Eur. J. Radiol.* **2015**, *84*, 65–70. [CrossRef]
25. Nikolova, N.K. Microwave Imaging for Breast Cancer. *IEEE Microw. Mag.* **2011**, *12*, 78–94. [CrossRef]
26. Xu, M.; Thulasiraman, P.; Noghanian, S. Microwave tomography for breast cancer detection on Cell broadband engine processors. *J. Parallel Distrib. Comput.* **2021**, *72*, 1106–1116. [CrossRef]
27. Grzegorczyk, T.M.; Meaney, P.M.; Kaufman, P.A.; di Florio-Alexander, R.M.; Paulsen, K.D. Fast 3-D Tomographic Microwave Imaging for Breast Cancer Detection. *IEEE Trans. Med. Imaging* **2012**, *31*, 1584–1592. [CrossRef] [PubMed]
28. AlSawaftah, N.; El-Abed, S.; Dhou, S.; Zakaria, A. Microwave Imaging for Early Breast Cancer Detection: Current State, Challenges, and Future Directions. *J. Imaging* **2022**, *8*, 123. [CrossRef] [PubMed]
29. Zerrad, F.-E.; Taouzari, M.; Makroum, E.M.; El Aoufi, J.; Islam, M.T.; Özkaner, V.; Abdulkarim, Y.I.; Karaaslan, M. Multilayered metamaterials array antenna based on artificial magnetic conductor's structure for the application diagnostic breast cancer detection with microwave imaging. *Med. Eng. Phys.* **2022**, *99*, 103737. [CrossRef]
30. Karabatak, M.; Ince, M.C. An expert system for detection of breast cancer based on association rules and neural network. *Expert Syst. Appl.* **2009**, *36*, 3465–3469. [CrossRef]
31. Wang, P.; Hu, X.; Li, Y.; Liu, Q.; Zhu, X. Automatic cell nuclei segmentation and classification of breast cancer histopathology images. *Signal Process.* **2016**, *122*, 1–13. [CrossRef]
32. Wahab, N.; Khan, A.; Lee, Y.S. Two-phase deep convolutional neural network for reducing class skewness in histopathological images based breast cancer detection. *Comput. Biol. Med.* **2017**, *85*, 86–97. [CrossRef]
33. Fan, Y.; Wang, H.; Gemmeke, H.; Hopp, T.; Hesser, J. Model-data-driven image reconstruction with neural networks for ultrasound computed tomography breast imaging. *Neurocomputing* **2022**, *467*, 10–21. [CrossRef]
34. Koh, J.; Yoon, Y.; Kim, S.; Han, K.; Kim, E.-K. Deep Learning for the Detection of Breast Cancers on Chest Computed Tomography. *Clin. Breast Cancer* **2021**, *22*, 26–31. [CrossRef]
35. Zangheri, B.; Messa, C.; Picchio, M.; Gianolli, L.; Landoni, C.; Fazio, F. PET/CT and breast cancer. *Euro. J. Nuclear Med. Mol. Imaging.* **2004**, *31*, S135–S142. [CrossRef] [PubMed]
36. Sollini, M.; Cozzi, L.; Ninatti, G.; Antunovic, L.; Cavinato, L.; Chiti, A.; Kirienko, M. PET/CT radiomics in breast cancer: Mind the step. *Methods* **2020**, *188*, 122–132. [CrossRef] [PubMed]
37. Salaün, P.-Y.; Abgral, R.; Malard, O.; Querellou-Lefranc, S.; Quere, G.; Wartski, M.; Coriat, R.; Hindie, E.; Taieb, D.; Tabarin, A.; et al. Good clinical practice recommendations for the use of PET/CT in oncology. *Eur. J. Nuclear Med. Mol. Imaging* **2020**, *47*, 28–50. [CrossRef] [PubMed]

38. Yi, A.; Jang, M.-J.; Yim, D.; Kwon, B.R.; Shin, S.U.; Chang, J.M. Addition of Screening Breast US to Digital Mammography and Digital Breast Tomosynthesis for Breast Cancer Screening in Women at Average Risk. *Radiology* **2021**, *298*, 568–575. [CrossRef]
39. Spak, D.A.; Le-Petross, H.T. Screening Modalities for Women at Intermediate and High Risk for Breast Cancer. *Curr. Breast Cancer Rep.* **2019**, *11*, 111–116. [CrossRef]
40. Lee, T.C.; Reyna, C.; Shaughnessy, E.; Lewis, J.D. Screening of populations at high risk for breast cancer. *J. Surg. Oncol.* **2019**, *120*, 820–830. [CrossRef]
41. Shah, T.A.; Guraya, S.S. Breast cancer screening programs: Review of merits, demerits, and recent recommendations practiced across the world. *J. Microsc. Ultrastruct.* **2017**, *5*, 59–69. [CrossRef]
42. Nguyen, D.L.; Myers, K.S.; Oluyemi, E.; Mullen, L.A.; Panigrahi, B.; Rossi, J.; Ambinder, E.B. BI-RADS 3 Assessment on MRI: A Lesion-Based Review for Breast Radiologists. *J. Breast Imaging* **2022**, wbac032. [CrossRef]
43. Daimiel Naranjo, I.; Gibbs, P.; Reiner, J.S.; Lo Gullo, R.; Thakur, S.B.; Jochelson, M.S.; Thakur, N.; Baltzer, P.A.T.; Helbich, T.H.; Pinker, K. Breast Lesion Classification with Multiparametric Breast MRI Using Radiomics and Machine Learning: A Comparison with Radiologists' Performance. *Cancers* **2022**, *14*, 1743. [CrossRef]
44. Shimauchi, A.; Jansen, S.A.; Abe, H.; Jaskowiak, N.; Schmidt, R.A.; Newstead, G.M. Breast Cancers Not Detected at MRI: Review of False-Negative Lesions. *Am. J. Roentgenol.* **2010**, *194*, 1674–1679. [CrossRef]
45. Tasdemir, S.B.Y.; Tasdemir, K.; Aydin, Z. A review of mammographic region of interest classification. *WIREs Data Min. Knowl. Discov.* **2020**, *10*, 1357. [CrossRef]
46. Sha, Z.; Hu, L.; Rouyendegh, B.D. Deep learning and optimization algorithms for automatic breast cancer detection. *Int. J. Imaging Syst. Technol.* **2020**, *30*, 495–506. [CrossRef]
47. Wang, C.; Brentnall, A.R.; Mainprize, J.G.; Yaffe, M.; Cuzick, J.; Harvey, J.A. External validation of a mammographic texture marker for breast cancer risk in a case–control study. *J. Med. Imaging* **2020**, *7*, 014003. [CrossRef] [PubMed]
48. Heidari, M.; Mirniaharikandehei, S.; Liu, W.; Hollingsworth, A.B.; Liu, H.; Zheng, B. Development and Assessment of a New Global Mammographic Image Feature Analysis Scheme to Predict Likelihood of Malignant Cases. *IEEE Trans. Med. Imaging* **2020**, *39*, 1235–1244. [CrossRef]
49. Suresh, R.; Rao, A.N.; Reddy, B.E. Detection and classification of normal and abnormal patterns in mammograms using deep neural network. *Concurr. Comput. Pract. Exp.* **2018**, *31*, 5293. [CrossRef]
50. Sapate, S.; Talbar, S.; Mahajan, A.; Sable, N.; Desai, S.; Thakur, M. Breast cancer diagnosis using abnormalities on ipsilateral views of digital mammograms. *Biocybern. Biomed. Eng.* **2020**, *40*, 290–305. [CrossRef]
51. Pezeshki, H. Breast tumor segmentation in digital mammograms using spiculated regions. *Biomed. Signal Process. Control* **2022**, *76*, 103652. [CrossRef]
52. Liu, Y.; Ren, L.; Cao, X.; Tong, Y. Breast tumors recognition based on edge feature extraction using support vector machine. *Biomed. Signal Process. Control* **2020**, *58*, 101825. [CrossRef]
53. Liu, Y.; Ren, L.; Cao, X.; Tong, Y. Diffusion-Weighted MRI of Breast Cancer: Improved Lesion Visibility and Image Quality Using Synthetic b-Values. *J. Magn. Reson. Imaging* **2019**, *50*, 1754–1761.
54. Almalki, Y.E.; Soomro, T.A.; Irfan, M.; Alduraibi, S.K.; Ali, A. Impact of Image Enhancement Module for Analysis of Mammogram Images for Diagnostics of Breast Cancer. *Sensors* **2022**, *22*, 1868. [CrossRef]
55. Rani, V.M.K.; Dhenakaran, S.S. Classification of ultrasound breast cancer tumor images using neural learning and predicting the tumor growth rate. *Multimed. Tools Appl.* **2020**, *79*, 16967–16985. [CrossRef]
56. Bria, A.; Karssemeijer, N.; Tortorella, F. Learning from unbalanced data: A cascade-based approach for detecting clustered microcalcifications. *Med. Image Anal.* **2014**, *18*, 241–252. [CrossRef] [PubMed]
57. Shrivastava, N.; Bharti, J. Breast Tumor Detection in Digital Mammogram Based on Efficient Seed Region Growing Segmentation. *IETE J. Res.* **2020**. [CrossRef]
58. Singh, H.; Sharma, V.; Singh, D. Comparative analysis of proficiencies of various textures and geometric features in breast mass classification using k-nearest neighbor. *Vis. Comput. Ind. Biomed. Art* **2022**, *5*, 1–19. [CrossRef] [PubMed]
59. Sasaki, M.; Tozaki, M.; Rodríguez-Ruiz, A.; Yotsumoto, D.; Ichiki, Y.; Terawaki, A.; Oosako, S.; Sagara, Y.; Sagara, Y. Artificial intelligence for breast cancer detection in mammography: Experience of use of the ScreenPoint Medical Transpara system in 310 Japanese women. *Breast Cancer* **2020**, *27*, 642–651. [CrossRef]
60. Junior, G.B.; da Rocha, S.V.; de Almeida, J.D.S.; de Paiva, A.C.; Silva, A.C.; Gattass, M. Breast cancer detection in mammography using spatial diversity, geostatistics, and concave geometry. *Multimed. Tools Appl.* **2019**, *78*, 13005–13031. [CrossRef]
61. Fanizzi, A.; Basile, T.M.A.; Losurdo, L.; Bellotti, R.; Bottigli, U.; Dentamaro, R.; Didonna, V.; Fausto, A.; Massafra, R.; Moschetta, M.; et al. A machine learning approach on multiscale texture analysis for breast microcalcification diagnosis. *BMC Bioinform.* **2020**, *21*, 1–11. [CrossRef]
62. Green, C.A.; Goodsitt, M.M.; Lau, J.H.; Brock, K.K.; Davis, C.L.; Carson, P.L. Deformable Mapping Method to Relate Lesions in Dedicated Breast CT Images to Those in Automated Breast Ultrasound and Digital Breast Tomosynthesis Images. *Ultrasound Med. Biol.* **2020**, *46*, 750–765. [CrossRef]
63. Padmavathy, T.V.; Vimalkumar, M.N.; Bhargava, D.S. Adaptive clustering based breast cancer detection with ANFIS classifier using mammographic images. *Clust. Comput.* **2019**, *22*, 13975–13984. [CrossRef]
64. Raghavendra, U.; Gudigar, A.; Ciaccio, E.J.; Ng, K.H.; Chan, W.Y.; Rahmat, K.; Acharya, U.R. 2DSM vs FFDM: A computer aided diagnosis based comparative study for the early detection of breast cancer. *Expert Syst.* **2021**, *38*, e12474. [CrossRef]

65. Wang, Z.; Li, M.; Wang, H.; Jiang, H.; Yao, Y.; Zhang, H.; Xin, J. Breast Cancer Detection Using Extreme Learning Machine Based on Feature Fusion with CNN Deep Features. *IEEE Access* **2019**, *7*, 105146–105158. [CrossRef]
66. Yap, M.H.; Pons, G.; Marti, J.; Ganau, S.; Sentis, M.; Zwiggelaar, R.; Davison, A.K.; Marti, R. Automated Breast Ultrasound Lesions Detection Using Convolutional Neural Networks. *IEEE J. Biomed. Health Inform.* **2018**, *22*, 1218–1226. [CrossRef] [PubMed]
67. Teare, P.; Fishman, M.; Benzaquen, O.; Toledano, E.; Elnekave, E. Malignancy Detection on Mammography Using Dual Deep Convolutional Neural Networks and Genetically Discovered False Color Input Enhancement. *J. Digit. Imaging* **2017**, *30*, 499–505. [CrossRef] [PubMed]
68. Shen, L.; Margolies, L.R.; Rothstein, J.H.; Fluder, E.; McBride, R.; Sieh, W. Deep Learning to Improve Breast Cancer Detection on Screening Mammography. *Sci. Rep.* **2019**, *9*, 1–12. [CrossRef] [PubMed]
69. Gamage, T.P.B.; Malcolm, D.T.K.; Talou, G.D.M.; Mîra, A.; Doyle, A.; Nielsen, P.M.F.; Nash, M.P. An automated computational biomechanics workflow for improving breast cancer diagnosis and treatment. *Interface Focus* **2019**, *9*, 20190034. [CrossRef]
70. Bouron, C.; Mathie, C.; Seegers, V.; Morel, O.; Jézéquel, P.; Lasla, H.; Guillerminet, C.; Girault, S.; Lacombe, M.; Sher, A.; et al. Prognostic Value of Metabolic, Volumetric and Textural Parameters of Baseline [^{18}F]FDG PET/CT in Early Triple-Negative Breast Cancer. *Cancers* **2022**, *14*, 637. [CrossRef]
71. Mughal, B.; Sharif, M.; Muhammad, N. Bi-model processing for early detection of breast tumor in CAD system. *Eur. Phys. J. Plus* **2017**, *132*, 266. [CrossRef]
72. Wang, S.; Rao, R.V.; Chen, P.; Zhang, Y.; Liu, A.; Wei, L. Abnormal Breast Detection in Mammogram Images by Feed-forward Neural Network Trained by Jaya Algorithm. *Fundam. Inform.* **2017**, *151*, 191–211. [CrossRef]
73. Muduli, D.; Dash, R.; Majhi, B. Automated breast cancer detection in digital mammograms: A moth flame optimization based ELM approach. *Biomed. Signal Process. Control* **2020**, *59*, 101912. [CrossRef]
74. Shiji, T.P.; Remya, S.; Lakshmanan, R.; Pratab, T.; Thomas, V. Evolutionary intelligence for breast lesion detection in ultrasound images: A wavelet modulus maxima and SVM based approach. *J. Intell. Fuzzy Syst.* **2020**, *38*, 6279–6290. [CrossRef]
75. Chakraborty, J.; Midya, A.; Rabidas, R. Computer-aided detection and diagnosis of mammographic masses using multi-resolution analysis of oriented tissue patterns. *Expert Syst. Appl.* **2018**, *99*, 168–179. [CrossRef]
76. Jara-Maldonado, M.; Alarcon-Aquino, V.; Rosas-Romero, R. A new machine learning model based on the broad learning system and wavelets. *Eng. Appl. Artif. Intell.* **2022**, *112*, 104886. [CrossRef]
77. Hajiabadi, H.; Babaiyan, V.; Zabihzadeh, D.; Hajiabadi, M. Combination of loss functions for robust breast cancer prediction. *Comput. Electr. Eng.* **2020**, *84*, 106624. [CrossRef]
78. Eltrass, A.S.; Salama, M.S. Fully automated scheme for computer-aided detection and breast cancer diagnosis using digitised mammograms. *IET Image Process.* **2020**, *14*, 495–505. [CrossRef]
79. Parekh, V.S.; Jacobs, M.A. Multiparametric radiomics methods for breast cancer tissue characterization using radiological imaging. *Breast Cancer Res. Treat.* **2020**, *180*, 407–421. [CrossRef]
80. Wang, C.; Brentnall, A.R.; Cuzick, J.; Harkness, E.F.; Evans, D.G.; Astley, S. A novel and fully automated mammographic texture analysis for risk prediction: Results from two case-control studies. *Breast Cancer Res.* **2017**, *19*, 114. [CrossRef]
81. Bajaj, V.; Pawar, M.; Meena, V.K.; Kumar, M.; Sengur, A.; Guo, Y. Computer-aided diagnosis of breast cancer using bi-dimensional empirical mode decomposition. *Neural Comput. Appl.* **2019**, *31*, 3307–3315. [CrossRef]
82. Li, Z.; Yu, L.; Wang, X.; Yu, H.; Gao, Y.; Ren, Y.; Wang, G.; Zhou, X. Diagnostic Performance of Mammographic Texture Analysis in the Differential Diagnosis of Benign and Malignant Breast Tumors. *Clin. Breast Cancer* **2018**, *18*, e621–e627. [CrossRef]
83. Huang, Q.; Huang, Y.; Luo, Y.; Yuan, F.; Li, X. Segmentation of breast ultrasound image with semantic classification of superpixels. *Med. Image Anal.* **2020**, *61*, 101657. [CrossRef]
84. Bressan, R.S.; Bugatti, P.H.; Saito, P.T. Breast cancer diagnosis through active learning in content-based image retrieval. *Neurocomputing* **2019**, *357*, 1–10. [CrossRef]
85. Suradi, S.H.; Abdullah, K.A.; Isa, N.A.M. Improvement of image enhancement for mammogram images using Fuzzy Anisotropic Diffusion Histogram Equalisation Contrast Adaptive Limited (FADHECAL). *Comput. Methods Biomech. Biomed. Eng. Imaging Vis.* **2022**, *10*, 67–75. [CrossRef]
86. Al-Antari, M.A.; Al-Masni, M.; Park, S.-U.; Park, J.; Metwally, M.K.; Kadah, Y.M.; Han, S.-M.; Kim, T.-S. An Automatic Computer-Aided Diagnosis System for Breast Cancer in Digital Mammograms via Deep Belief Network. *J. Med. Biol. Eng.* **2018**, *38*, 443–456. [CrossRef]
87. Zhang, Q.; Peng, Y.; Liu, W.; Bai, J.; Zheng, J.; Yang, X.; Zhou, L. Radiomics Based on Multimodal MRI for the Differential Diagnosis of Benign and Malignant Breast Lesions. *J. Magn. Reson. Imaging* **2020**, *52*, 596–607. [CrossRef]
88. Dhouibi, M.; Ben Salem, A.K.; Saidi, A.; Ben Saoud, S. Accelerating Deep Neural Networks implementation: A survey. *IET Comput. Digit. Tech.* **2021**, *15*, 79–96. [CrossRef]
89. Xu, Y.; Wang, Y.; Yuan, J.; Cheng, Q.; Wang, X.; Carson, P.L. Medical breast ultrasound image segmentation by machine learning. *Ultrasonics* **2019**, *91*, 1–9. [CrossRef] [PubMed]
90. Arora, R.; Rai, P.K.; Raman, B. Deep feature–based automatic classification of mammograms. *Med. Biol. Eng. Comput.* **2020**, *58*, 1199–1211. [CrossRef] [PubMed]
91. Gao, F.; Wu, T.; Li, J.; Zheng, B.; Ruan, L.; Shang, D.; Patel, B. SD-CNN: A shallow-deep CNN for improved breast cancer diagnosis. *Comput. Med. Imaging Graph.* **2018**, *70*, 53–62. [CrossRef]

92. Romeo, V.; Clauser, P.; Rasul, S.; Kapetas, P.; Gibbs, P.; Baltzer, P.A.T.; Hacker, M.; Woitek, R.; Helbich, T.H.; Pinker, K. AI-enhanced simultaneous multiparametric 18F-FDG PET/MRI for accurate breast cancer diagnosis. *Eur. J. Pediatr.* **2022**, *49*, 596–608. [CrossRef]
93. Tsochatzidis, L.; Koutla, P.; Costaridou, L.; Pratikakis, I. Integrating segmentation information into CNN for breast cancer diagnosis of mammographic masses. *Compt. Meth. Prog. Biomed.* **2021**, *200*, 105913. [CrossRef]
94. Toğaçar, M.; Ergen, B.; Cömert, Z. Application of breast cancer diagnosis based on a combination of convolutional neural networks, ridge regression and linear discriminant analysis using invasive breast cancer images processed with autoencoders. *Med. Hypotheses* **2020**, *135*, 109503. [CrossRef]
95. Singh, V.K.; Rashwan, H.A.; Romani, S.; Akram, F.; Pandey, N.; Sarker, M.K.; Saleh, A.; Arenas, M.; Arquez, M.; Puig, D.; et al. Breast tumor segmentation and shape classification in mammograms using generative adversarial and convolutional neural network. *Expert Syst. Appl.* **2020**, *139*, 112855. [CrossRef]
96. Khan, S.; Islam, N.; Jan, Z.; Din, I.U.; Rodrigues, J.J.P.C. A novel deep learning based framework for the detection and classification of breast cancer using transfer learning. *Pattern Recognit. Lett.* **2019**, *125*, 1–6. [CrossRef]
97. Li, J.-B.; Yu, Y.; Yang, Z.-M.; Tang, L.-L. Breast Tissue Image Classification Based on Semi-supervised Locality Discriminant Projection with Kernels. *J. Med. Syst.* **2012**, *36*, 2779–2786. [CrossRef] [PubMed]
98. Algehyne, E.A.; Jibril, M.L.; Algehainy, N.A.; Alamri, O.A.; Alzahrani, A.K. Fuzzy Neural Network Expert System with an Improved Gini Index Random Forest-Based Feature Importance Measure Algorithm for Early Diagnosis of Breast Cancer in Saudi Arabia. *Big Data Cogn. Comput.* **2022**, *6*, 13. [CrossRef]
99. Akhbardeh, A.; Jacobs, M.A. Comparative analysis of nonlinear dimensionality reduction techniques for breast MRI segmentation. *Med. Phys.* **2012**, *39*, 2275–2289. [CrossRef] [PubMed]
100. Ragab, M.; Albukhari, A.; Alyami, J.; Mansour, R.F. Ensemble Deep-Learning-Enabled Clinical Decision Support System for Breast Cancer Diagnosis and Classification on Ultrasound Images. *Biology* **2022**, *11*, 439. [CrossRef]
101. Jabeen, K.; Khan, M.A.; Alhaisoni, M.; Tariq, U.; Zhang, Y.-D.; Hamza, A.; Mickus, A.; Damaševičius, R. Breast Cancer Classification from Ultrasound Images Using Probability-Based Optimal Deep Learning Feature Fusion. *Sensors* **2022**, *22*, 807. [CrossRef]
102. Bacha, S.; Taouali, O. A novel machine learning approach for breast cancer diagnosis. *Measurement* **2021**, *187*, 110233. [CrossRef]
103. Mert, A.; Kılıç, N.; Akan, A. An improved hybrid feature reduction for increased breast cancer diagnostic performance. *Biomed. Eng. Lett.* **2015**, *4*, 285–291. [CrossRef]
104. Zheng, B.; Yoon, S.W.; Lam, S.S. Breast cancer diagnosis based on feature extraction using a hybrid of K-means and support vector machine algorithms. *Expert Syst. Appl.* **2014**, *41*, 1476–1482. [CrossRef]
105. Sun, W.; Tseng, T.-L.; Zhang, J.; Qian, W. Enhancing deep convolutional neural network scheme for breast cancer diagnosis with unlabeled data. *Comput. Med. Imaging Graph.* **2017**, *57*, 4–9. [CrossRef] [PubMed]
106. Sharif, M.I.; Li, J.P.; Naz, J.; Rashid, I. A comprehensive review on multi-organs tumor detection based on machine learning. *Pattern Recognit. Lett.* **2020**, *131*, 30–37. [CrossRef]
107. Shehab, M.; Abualigah, L.; Shambour, Q.; Abu-Hashem, M.A.; Shambour, M.K.Y.; Alsalibi, A.I.; Gandomi, A.H. Machine learning in medical applications: A review of state-of-the-art methods. *Comput. Biol. Med.* **2022**, *145*, 105458. [CrossRef] [PubMed]
108. Hicks, S.A.; Strümke, I.; Thambawita, V.; Hammou, M.; Riegler, M.A.; Halvorsen, P.; Parasa, S. On evaluation metrics for medical applications of artificial intelligence. *Sci. Rep.* **2022**, *12*, 5979. [CrossRef]
109. van Engelen, J.E.; Hoos, H.H. A survey on semi-supervised learning. *Mach. Learn.* **2019**, *109*, 373–440. [CrossRef]
110. Dubey, A.K.; Gupta, U.; Jain, S. Analysis of k-means clustering approach on the breast cancer Wisconsin dataset. *Int. J. Comput. Assist. Radiol. Surg.* **2016**, *11*, 2033–2047. [CrossRef] [PubMed]
111. Hernández-Capistrán, J.; Martínez-Carballido, J.F.; Rosas-Romero, R. False Positive Reduction by an Annular Model as a Set of Few Features for Microcalcification Detection to Assist Early Diagnosis of Breast Cancer. *J. Med. Syst.* **2018**, *42*, 134. [CrossRef]
112. Onan, A. A fuzzy-rough nearest neighbor classifier combined with consistency-based subset evaluation and instance selection for automated diagnosis of breast cancer. *Expert Syst. Appl.* **2015**, *42*, 6844–6852. [CrossRef]
113. Hosseinpour, M.; Ghaemi, S.; Khanmohammadi, S.; Daneshvar, S. A hybrid high-order type-2 FCM improved random forest classification method for breast cancer risk assessment. *Appl. Math. Comput.* **2022**, *424*. [CrossRef]
114. Sadad, T.; Munir, A.; Saba, T.; Hussain, A. Fuzzy C-means and region growing based classification of tumor from mammograms using hybrid texture feature. *J. Comput. Sci.* **2018**, *29*, 34–45. [CrossRef]
115. Saberi, H.; Rahai, A.; Hatami, F. A fast and efficient clustering based fuzzy time series algorithm (FEFTS) for regression and classification. *Appl. Soft Comput.* **2017**, *61*, 1088–1097. [CrossRef]
116. Thani, I.; Kasbe, T. Expert system based on fuzzy rules for diagnosing breast cancer. *Health Technol.* **2022**, *12*, 473–489. [CrossRef]
117. Nguyen, T.-L.; Kavuri, S.; Park, S.-Y.; Lee, M. Attentive Hierarchical ANFIS with interpretability for cancer diagnostic. *Expert Syst. Appl.* **2022**, *201*, 117099. [CrossRef]
118. Zhang, Q.; Xiao, Y.; Suo, J.; Shi, J.; Yu, J.; Guo, Y.; Wang, Y.; Zheng, H. Sonoelastomics for Breast Tumor Classification: A Radiomics Approach with Clustering-Based Feature Selection on Sonoelastography. *Ultrasound Med. Biol.* **2017**, *43*, 1058–1069. [CrossRef] [PubMed]
119. Indra, P.; Manikandan, M. Multilevel Tetrolet transform based breast cancer classifier and diagnosis system for healthcare applications. *J. Ambient Intell. Humaniz. Comput.* **2020**, *12*, 3969–3978. [CrossRef]

120. Shan, J.; Alam, S.K.; Garra, B.; Zhang, Y.; Ahmed, T. Computer-Aided Diagnosis for Breast Ultrasound Using Computerized BI-RADS Features and Machine Learning Methods. *Ultrasound Med. Biol.* **2016**, *42*, 980–988. [CrossRef]
121. Abdel-Nasser, M.; Melendez, J.; Moreno, A.; Omer, O.A.; Puig, D. Breast tumor classification in ultrasound images using texture analysis and super-resolution methods. *Eng. Appl. Artif. Intell.* **2017**, *59*, 84–92. [CrossRef]
122. Muramatsu, C.; Hara, T.; Endo, T.; Fujita, H. Breast mass classification on mammograms using radial local ternary patterns. *Comput. Biol. Med.* **2016**, *72*, 43–53. [CrossRef]
123. Alam, Z.; Rahman, M.S. A Random Forest based predictor for medical data classification using feature ranking. *Inform. Med. Unlocked* **2019**, *15*, 100180. [CrossRef]
124. Wu, J.-X.; Chen, P.-Y.; Lin, C.-H.; Chen, S.; Shung, K.K. Breast Benign and Malignant Tumors Rapidly Screening by ARFI-VTI Elastography and Random Decision Forests Based Classifier. *IEEE Access* **2020**, *8*, 54019–54034. [CrossRef]
125. Lu, W.; Li, Z.; Chu, J. A novel computer-aided diagnosis system for breast MRI based on feature selection and ensemble learning. *Comput. Biol. Med.* **2017**, *83*, 157–165. [CrossRef] [PubMed]
126. Huang, Q.; Chen, Y.; Liu, L.; Tao, D.; Li, X. On Combining Biclustering Mining and AdaBoost for Breast Tumor Classification. *IEEE Trans. Knowl. Data Eng.* **2019**, *32*, 728–738. [CrossRef]
127. Vamvakas, A.; Tsivaka, D.; Logothetis, A.; Vassiou, K.; Tsougos, I. Breast Cancer Classification on Multiparametric MRI—Increased Performance of Boosting Ensemble Methods. *Technol. Cancer Res. Treat.* **2022**, *21*. [CrossRef]
128. Sharma, S.; Khanna, P. Computer-Aided Diagnosis of Malignant Mammograms using Zernike Moments and SVM. *J. Digit. Imaging* **2015**, *28*, 77–90. [CrossRef]
129. Agossou, C.; Atchadé, M.N.; Djibril, A.M.; Kurisheva, S.V. Support Vector Machine, Naive Bayes Classification, and Mathematical Modeling for Public Health Decision-Making: A Case Study of Breast Cancer in Benin. *SN Comput. Sci.* **2022**, *3*, 1–19. [CrossRef]
130. Alshutbi, M.; Li, Z.; Alrifaey, M.; Ahmadipour, M.; Murtadha Othman, M. A hybrid classifier based on support vector machine and Jaya algorithm for breast cancer classification. *Neural Compt. App.* **2022**, 1–13. [CrossRef]
131. Samma, H.; Lahasan, B. Optimized Two-Stage Ensemble Model for Mammography Mass Recognition. *IRBM* **2020**, *41*, 195–204. [CrossRef]
132. Wu, W.; Li, B.; Mercan, E.; Mehta, S.; Bartlett, J.; Weaver, D.L.; Elmore, J.G.; Shapiro, L.G. MLCD: A Unified Software Package for Cancer Diagnosis. *JCO Clin. Cancer Inform.* **2020**, *4*, 290–298. [CrossRef]
133. Badr, E.; Almotairi, S.; Salam, M.A.; Ahmed, H. New Sequential and Parallel Support Vector Machine with Grey Wolf Optimizer for Breast Cancer Diagnosis. *Alex. Eng. J.* **2022**, *61*, 2520–2534. [CrossRef]
134. Mendelson, E.B. Artificial Intelligence in Breast Imaging: Potentials and Limitations. *Am. J. Roentgenol.* **2019**, *212*, 293–299. [CrossRef]
135. Amato, F.; López, A.; Mendez, E.P.; Vanhara, P.; Hampl, A.; Havel, J. Artificial neural networks in medical diagnosis. *J. Appl. Biomed.* **2013**, *11*, 47–58. [CrossRef]
136. Beura, S.; Majhi, B.; Dash, R. Mammogram classification using two dimensional discrete wavelet transform and gray-level co-occurrence matrix for detection of breast cancer. *Neurocomputing* **2015**, *154*, 1–14. [CrossRef]
137. Mohammed, M.A.; Al-Khateeb, B.; Rashid, A.N.; Ibrahim, D.A.; Ghani, M.K.A.; Mostafa, S.A. Neural network and multi-fractal dimension features for breast cancer classification from ultrasound images. *Comput. Electr. Eng.* **2018**, *70*, 871–882. [CrossRef]
138. Gallego-Ortiz, C.; Martel, A.L. A graph-based lesion characterization and deep embedding approach for improved computer-aided diagnosis of nonmass breast MRI lesions. *Med. Image Anal.* **2019**, *51*, 116–124. [CrossRef] [PubMed]
139. Danala, G.; Patel, B.; Aghaei, F.; Heidari, M.; Li, J.; Wu, T.; Zheng, B. Classification of Breast Masses Using a Computer-Aided Diagnosis Scheme of Contrast Enhanced Digital Mammograms. *Ann. Biomed. Eng.* **2018**, *46*, 1419–1431. [CrossRef] [PubMed]
140. Punitha, S.; Amuthan, A.; Joseph, K.S. Enhanced Monarchy Butterfly Optimization Technique for effective breast cancer diagnosis. *J. Med. Syst.* **2019**, *43*, 206. [CrossRef]
141. Alshayeji, M.H.; Ellethy, H.; Abed, S.; Gupta, R. Computer-aided detection of breast cancer on the Wisconsin dataset: An artificial neural networks approach. *Biomed. Signal Process. Control* **2022**, *71*, 103141. [CrossRef]
142. Rezaeipanah, A.; Ahmadi, G. Breast Cancer Diagnosis Using Multi-Stage Weight Adjustment In The MLP Neural Network. *Comput. J.* **2022**, *65*, 788–804. [CrossRef]
143. Ting, F.F.; Tan, Y.J.; Sim, K.S. Convolutional neural network improvement for breast cancer classification. *Expert Syst. Appl.* **2019**, *120*, 103–115. [CrossRef]
144. Yousefi, M.; Krzyżak, A.; Suen, C.Y. Mass detection in digital breast tomosynthesis data using convolutional neural networks and multiple instance learning. *Comput. Biol. Med.* **2018**, *96*, 283–293. [CrossRef]
145. Wu, N.; Phang, J.; Park, J.; Shen, Y.; Huang, Z.; Zorin, M.; Jastrzebski, S.; Fevry, T.; Katsnelson, J.; Kim, E.; et al. Deep Neural Networks Improve Radiologists' Performance in Breast Cancer Screening. *IEEE Trans. Med. Imaging* **2019**, *39*, 1184–1194. [CrossRef] [PubMed]
146. AlBalawi, U.; Manimurugan, S.; Varatharajan, R. Classification of breast cancer mammogram images using convolution neural network. *Concurr. Comput. Pract. Exp.* **2022**, *34*, e3803. [CrossRef]
147. Inan, M.S.K.; Alam, F.I.; Hasan, R. Deep integrated pipeline of segmentation guided classification of breast cancer from ultrasound images. *Biomed. Signal Process. Control* **2022**, *75*, 103553. [CrossRef]
148. Feizi, A. A gated convolutional neural network for classification of breast lesions in ultrasound images. *Soft Comput.* **2022**, *26*, 5241–5250. [CrossRef]

149. Ribli, D.; Horváth, A.; Unger, Z.; Pollner, P.; Csabai, I. Detecting and classifying lesions in mammograms with Deep Learning. *Sci. Rep.* **2018**, *8*, 4165–4167. [CrossRef]
150. Liu, K.; Kang, G.; Zhang, N.; Hou, B. Breast Cancer Classification Based on Fully-Connected Layer First Convolutional Neural Networks. *IEEE Access* **2018**, *6*, 23722–23732. [CrossRef]
151. Zhang, Y.-D.; Pan, C.; Chen, X.; Wang, F. Abnormal breast identification by nine-layer convolutional neural network with parametric rectified linear unit and rank-based stochastic pooling. *J. Comput. Sci.* **2018**, *27*, 57–68. [CrossRef]
152. Oyetade, I.S.; Ayeni, J.O.; Ogunde, A.O.; Oguntunde, B.O.; Olowookere, T.A. Hybridized Deep Convolutional Neural Network and Fuzzy Support Vector Machines for Breast Cancer Detection. *SN Comput. Sci.* **2022**, *3*, 58. [CrossRef]
153. Takahashi, K.; Fujioka, T.; Oyama, J.; Mori, M.; Yamaga, E.; Yashima, Y.; Imokawa, T.; Hayashi, A.; Kujiraoka, Y.; Tsuchiya, J.; et al. Deep Learning Using Multiple Degrees of Maximum-Intensity Projection for PET/CT Image Classification in Breast Cancer. *Tomography* **2022**, *8*, 131–141. [CrossRef]
154. Muduli, D.; Dash, R.; Majhi, B. Automated diagnosis of breast cancer using multi-modal datasets: A deep convolution neural network based approach. *Biomed. Signal Process. Control* **2021**, *71*, 102825. [CrossRef]
155. Ayana, G.; Park, J.; Jeong, J.-W.; Choe, S.-W. A Novel Multistage Transfer Learning for Ultrasound Breast Cancer Image Classification. *Diagnostics* **2022**, *12*, 135. [CrossRef] [PubMed]
156. Dey, S.; Roychoudhury, R.; Malakar, S.; Sarkar, R. Screening of breast cancer from thermogram images by edge detection aided deep transfer learning model. *Multimed. Tools Appl.* **2022**, *81*, 9331–9349. [CrossRef] [PubMed]
157. Ring, E. The historical development of temperature measurement in medicine. *Infrared Phys. Technol.* **2007**, *49*, 297–301. [CrossRef]
158. Ng, E.Y.K.; Kee, E.C. Advanced integrated technique in breast cancer thermography. *J. Med. Eng. Technol.* **2008**, *32*, 103–114. [CrossRef] [PubMed]
159. Lahiri, B.; Bagavathiappan, S.; Jayakumar, T.; Philip, J. Medical applications of infrared thermography: A review. *Infrared Phys. Technol.* **2012**, *55*, 221–235. [CrossRef]
160. Singh, D.; Singh, A.K. Role of image thermography in early breast cancer detection- Past, present and future. *Comput. Methods Programs Biomed.* **2020**, *183*, 105074. [CrossRef]
161. Baic, A.; Plaza, D.; Lange, B.; Michalecki Stanek, A.; Kowalczyk, A.; Ślosarek, K.; Cholewka, A. Long-Term Skin Temperature Changes after Breast Cancer Radiotherapy. *Int. J. Environ. Res. Public Health* **2022**, *19*, 6891. [CrossRef]
162. Fernández-Cuevas, I.; Marins, J.C.B.; Lastras, J.A.; Carmona, P.M.G.; Cano, S.P.; García-Concepción, M.Á.; Sillero-Quintana, M. Classification of factors influencing the use of infrared thermography in humans: A review. *Infrared Phys. Technol.* **2015**, *71*, 28–55. [CrossRef]
163. Ioannou, S.; Gallese, V.; Merla, A. Thermal infrared imaging in psychophysiology: Potentialities and limits. *Psychophysiology* **2014**, *51*, 951–963. [CrossRef]
164. Bernard, V.; Staffa, E.; Mornstein, V.; Bourek, A. Infrared camera assessment of skin surface temperature—Effect of emissivity. *Phys. Med.* **2013**, *29*, 583–591. [CrossRef]
165. Ekici, S.; Jawzal, H. Breast cancer diagnosis using thermography and convolutional neural networks. *Med. Hypotheses* **2019**, *137*, 109542. [CrossRef]
166. AlFayez, F.; El-Soud, M.W.A.; Gaber, T. Thermogram Breast Cancer Detection: A Comparative Study of Two Machine Learning Techniques. *Appl. Sci.* **2019**, *10*, 551. [CrossRef]
167. Gogoi, U.R.; Majumdar, G.; Bhowmik, M.K.; Ghosh, A.K. Evaluating the efficiency of infrared breast thermography for early breast cancer risk prediction in asymptomatic population. *Infrared Phys. Technol.* **2019**, *99*, 201–211. [CrossRef]
168. Saxena, A.; Ng, E.; Raman, V.; Hamli, M.S.B.M.; Moderhak, M.; Kolacz, S.; Jankau, J. Infrared (IR) thermography-based quantitative parameters to predict the risk of post-operative cancerous breast resection flap necrosis. *Infrared Phys. Technol.* **2019**, *103*, 103063. [CrossRef]
169. Tello-Mijares, S.; Woo, F.; Flores, F. Breast Cancer Identification via Thermography Image Segmentation with a Gradient Vector Flow and a Convolutional Neural Network. *J. Health Eng.* **2019**, *2019*, 1–13. [CrossRef]
170. Garduño-Ramón, M.A.; Vega-Mancilla, S.G.; Morales-Henández, L.A.; Osornio-Rios, R.A. Supportive Noninvasive Tool for the Diagnosis of Breast Cancer Using a Thermographic Camera as Sensor. *Sensors* **2017**, *17*, 497. [CrossRef]
171. Raghavendra, U.; Acharya, U.R.; Ng, E.Y.K.; Tan, J.-H.; Gudigar, A. An integrated index for breast cancer identification using histogram of oriented gradient and kernel locality preserving projection features extracted from thermograms. *Quant. Infrared Thermogr. J.* **2016**, *13*, 195–209. [CrossRef]
172. Lashkari, A.; Pak, F.; Firouzmand, M. Full Intelligent Cancer Classification of Thermal Breast Images to Assist Physician in Clinical Diagnostic Applications. *J. Med. Signals Sens.* **2016**, *6*, 12–24. [CrossRef]
173. Francis, S.V.; Sasikala, M.; Saranya, S. Detection of Breast Abnormality from Thermograms Using Curvelet Transform Based Feature Extraction. *J. Med. Syst.* **2014**, *38*, 1–9. [CrossRef]
174. Milosevic, M.; Jankovic, D.; Peulic, A. Thermography based breast cancer detection using texture features and minimum variance quantization. *EXCLI J.* **2014**, *13*, 1204. [CrossRef]
175. Araújo, M.C.; Lima, R.C.; de Souza, R.M. Interval symbolic feature extraction for thermography breast cancer detection. *Expert Syst. Appl.* **2014**, *41*, 6728–6737. [CrossRef]
176. Gonzalez-Hernandez, J.-L.; Recinella, A.N.; Kandlikar, S.G.; Dabydeen, D.; Medeiros, L.; Phatak, P. Technology, application and potential of dynamic breast thermography for the detection of breast cancer. *Int. J. Heat Mass Transf.* **2018**, *131*, 558–573. [CrossRef]

177. Silva, L.F.; Santos, A.A.S.; Bravo, R.S.; Silva, A.C.; Muchaluat-Saade, D.C.; Conci, A. Hybrid analysis for indicating patients with breast cancer using temperature time series. *Comput. Methods Programs Biomed.* **2016**, *130*, 142–153. [CrossRef]
178. Saniei, E.; Setayeshi, S.; Akbari, M.E.; Navid, M. A vascular network matching in dynamic thermography for breast cancer detection. *Quant. Infrared Thermogr. J.* **2015**, *12*, 1–13. [CrossRef]
179. Kumar, A.; Bi, L.; Kim, J.; Feng, D.D. Machine learning in medical imaging. In *Biomedical Information Technology*; Feng, D.D., Ed.; Academic Press: Cambridge, MA, USA, 2020; pp. 167–196. [CrossRef]
180. Nayak, D.R.; Dash, R.; Majhi, B.; Pachori, R.B.; Zhang, Y. A deep stacked random vector functional link network autoencoder for diagnosis of brain abnormalities and breast cancer. *Biomed. Signal Process. Control* **2020**, *58*, 101860. [CrossRef]
181. Kadam, V.J.; Jadhav, S.M.; Vijayakumar, K. Breast Cancer Diagnosis Using Feature Ensemble Learning Based on Stacked Sparse Autoencoders and Softmax Regression. *J. Med. Syst.* **2019**, *43*, 263. [CrossRef]
182. Zhang, E.; Seiler, S.; Chen, M.; Lu, W.; Gu, X. BIRADS features-oriented semi-supervised deep learning for breast ultrasound computer-aided diagnosis. *Phys. Med. Biol.* **2020**, *65*, 125005. [CrossRef]
183. Zhang, H.; Guo, W.; Zhang, S.; Lu, H.; Zhao, X. Unsupervised Deep Anomaly Detection for Medical Images Using an Improved Adversarial Autoencoder. *J. Digit. Imaging* **2022**, *35*, 153–161. [CrossRef]
184. Movahedi, F.; Coyle, J.L.; Sejdic, E. Deep Belief Networks for Electroencephalography: A Review of Recent Contributions and Future Outlooks. *IEEE J. Biomed. Health Inform.* **2017**, *22*, 642–652. [CrossRef]
185. Le Roux, N.; Bengio, Y. Representational Power of Restricted Boltzmann Machines and Deep Belief Networks. *Neural Comput.* **2007**, *20*, 1631–1649. [CrossRef]
186. Ahmad, M.; Ai, D.; Xie, G.; Qadri, S.F.; Song, H.; Huang, Y.; Wang, Y.; Yang, J. Deep Belief Network Modeling for Automatic Liver Segmentation. *IEEE Access* **2019**, *7*, 20585–20595. [CrossRef]
187. Kaur, M.; Singh, D. Fusion of medical images using deep belief networks. *Clust. Comput.* **2019**, *23*, 1439–1453. [CrossRef]
188. Zhao, Z.; Zhao, J.; Song, K.; Hussain, A.; Du, Q.; Dong, Y.; Liu, J.; Yang, X. Joint DBN and Fuzzy C-Means unsupervised deep clustering for lung cancer patient stratification. *Eng. Appl. Artif. Intell.* **2020**, *91*, 103571. [CrossRef]
189. Rasmus, A.; Berglund, M.; Honkala, M.; Valpola, H.; Raiko, T. Semi-supervised learning with ladder networks. In *Proceedings of the 28th International Conference on Neural Information Processing Systems—Volume 2 (NIPS'15)*; MIT Press: Cambridge, MA, USA, 2015; pp. 3546–3554.
190. Zahoor, S.; Shoaib, U.; Lali, I.U. Breast Cancer Mammograms Classification Using Deep Neural Network and Entropy-Controlled Whale Optimization Algorithm. *Diagnostics* **2022**, *12*, 557. [CrossRef]
191. Jagtap, A.D.; Kawaguchi, K.; Karniadakis, G.E. Adaptive activation functions accelerate convergence in deep and physics-informed neural networks. *J. Comput. Phys.* **2020**, *404*, 109136. [CrossRef]
192. Jagtap, A.D.; Shin, Y.; Kawaguchi, K.; Karniadakis, G.E. Deep Kronecker neural networks: A general framework for neural networks with adaptive activation functions. *Neurocomputing* **2022**, *468*, 165–180. [CrossRef]
193. Zhang, J.; Wang, G.; Ren, J.; Yang, Z.; Li, D.; Cui, Y.; Yang, X. Multiparametric MRI-based radiomics nomogram for preoperative prediction of lymphovascular invasion and clinical outcomes in patients with breast invasive ductal carcinoma. *Eur. Radiol.* **2022**, *32*, 4079–4089. [CrossRef] [PubMed]
194. Schaffter, T.; Buist, D.S.M.; Lee, C.I.; Nikulin, Y.; Ribli, D.; Guan, Y.; Lotter, W.; Jie, Z.; Du, H.; Wang, S.; et al. Evaluation of Combined Artificial Intelligence and Radiologist Assessment to Interpret Screening Mammograms. *JAMA Netw. Open* **2020**, *3*, e200265. [CrossRef]
195. Grimm, L.J.; Mazurowski, M.A. Breast Cancer Radiogenomics: Current Status and Future Directions. *Acad. Radiol.* **2020**, *27*, 39–46. [CrossRef]

cancers

MDPI

Review

Efficacy of Artificial Intelligence-Assisted Discrimination of Oral Cancerous Lesions from Normal Mucosa Based on the Oral Mucosal Image: A Systematic Review and Meta-Analysis

Ji-Sun Kim [1], Byung Guk Kim [1] and Se Hwan Hwang [2,*]

[1] Department of Otolaryngology-Head and Neck Surgery, Eunpyeong St. Mary's Hospital, College of Medicine, Catholic University of Korea, Seoul 03312, Korea; skswltjs23@hanmail.net (J.-S.K.); entkbg@gmail.com (B.G.K.)

[2] Department of Otolaryngology-Head and Neck Surgery, Bucheon St. Mary's Hospital, College of Medicine, Catholic University of Korea, Bucheon 14647, Korea

* Correspondence: yellobird@catholic.ac.kr; Tel.: +82-32-340-7044

Simple Summary: Early detection of oral cancer is important to increase the survival rate and reduce morbidity. For the past few years, the early detection of oral cancer using artificial intelligence (AI) technology based on autofluorescence imaging, photographic imaging, and optical coherence tomography imaging has been an important research area. In this study, diagnostic values including sensitivity and specificity data were comprehensively confirmed in various studies that performed AI analysis of images. The diagnostic sensitivity of AI-assisted screening was 0.92. In subgroup analysis, there was no statistically significant difference in the diagnostic rate according to each image tool. AI shows good diagnostic performance with high sensitivity for oral cancer. Image analysis using AI is expected to be used as a clinical tool for early detection and evaluation of treatment efficacy for oral cancer.

Abstract: The accuracy of artificial intelligence (AI)-assisted discrimination of oral cancerous lesions from normal mucosa based on mucosal images was evaluated. Two authors independently reviewed the database until June 2022. Oral mucosal disorder, as recorded by photographic images, autofluorescence, and optical coherence tomography (OCT), was compared with the reference results by histology findings. True-positive, true-negative, false-positive, and false-negative data were extracted. Seven studies were included for discriminating oral cancerous lesions from normal mucosa. The diagnostic odds ratio (DOR) of AI-assisted screening was 121.66 (95% confidence interval [CI], 29.60; 500.05). Twelve studies were included for discriminating all oral precancerous lesions from normal mucosa. The DOR of screening was 63.02 (95% CI, 40.32; 98.49). Subgroup analysis showed that OCT was more diagnostically accurate (324.33 vs. 66.81 and 27.63) and more negatively predictive (0.94 vs. 0.93 and 0.84) than photographic images and autofluorescence on the screening for all oral precancerous lesions from normal mucosa. Automated detection of oral cancerous lesions by AI would be a rapid, non-invasive diagnostic tool that could provide immediate results on the diagnostic work-up of oral cancer. This method has the potential to be used as a clinical tool for the early diagnosis of pathological lesions.

Keywords: mouth neoplasms; imaging; optical image; precancerous conditions; artificial intelligence; screening

Citation: Kim, J.-S.; Kim, B.G.; Hwang, S.H. Efficacy of Artificial Intelligence-Assisted Discrimination of Oral Cancerous Lesions from Normal Mucosa Based on the Oral Mucosal Image: A Systematic Review and Meta-Analysis. *Cancers* **2022**, *14*, 3499. https://doi.org/10.3390/cancers14143499

Academic Editor: Rahele Kafieh

Received: 29 June 2022
Accepted: 17 July 2022
Published: 19 July 2022

1. Introduction

Oral cancer accounts for 4% of all malignancies and is the most common type of head and neck cancer [1]. The diagnosis of oral cancer is often delayed, resulting in a poor prognosis. It has been reported that early diagnosis increases the 5-year survival rate to 83%, but if a diagnosis is delayed and metastasis occurs, the survival rate drops to less than

30% [2]. Therefore, there is an urgent need for early and accurate detection of oral lesions and for distinguishing precancerous and cancerous tissues from normal tissues.

The conventional screening method for oral cancer is visual examination and palpation of the oral cavity. However, the accuracy of this method is highly dependent on the subjective judgment of the clinician. Diagnostic methods such as toluidine blue staining, autofluorescence, optical coherence tomography (OCT), and photographic imaging were useful as adjunctive methods for oral cancer screening [3–6].

Over the past decade, studies have increasingly showed that artificial intelligence (AI) technology is consistent with or even superior to human experts in identifying abnormal lesions in additional images of various organs [7–11]. These results give us hope for the potential of AI in the screening of oral cancer. However, large-scale statistical approaches to diagnostic power for using oral imaging with AI are lacking. Therefore, in this study, the sensitivity and specificity were analyzed through meta-analysis to evaluate the accuracy of detecting oral precancerous and cancerous lesions in AI-assisted oral mucosa images. We also performed subgroup analysis to determine whether accuracy differs between imaging tools.

2. Materials and Methods

2.1. Literature Search

Searches were performed in six databases: PubMed, Embase, Web of Science, SCOPUS, Cochrane Central Register of Controlled Trials, and Google Scholar. The search terms were: "artificial intelligence", "photo", "optical image", "dysplasia", "oral precancer", "oral cancer", and "oral carcinoma". The search period was set to June 2022, and data written in English were reviewed. Two independent reviewers reviewed all abstracts and titles of candidate studies. Among studies diagnosing oral cancer using images, studies that did not deal with AI were excluded.

2.2. Selection Criteria

The inclusion criteria were: (1) use of AI; (2) prospective or retrospective study protocol; (3) comparison of AI-assisted screening of oral mucosal lesions with the reference test (histology); and (4) sensitivity and specificity analyses. The exclusion criteria were: (1) case report format; (2) review article format; (3) diagnosis of other tumors (laryngeal cancer or nasal cavity tumors); and (4) lack of diagnostic AI data. The search strategy is summarized in Figure 1.

2.3. Data Extraction and Risk of Bias Assessment

All data were collected using standardized forms. As diagnostic accuracy, diagnostic odds ratio (DOR), areas under the curve (AUC), and summary receiver operating characteristic (SROC) were identified. The diagnostic performance was compared with histological examination results.

A random-effect model was used in this study. DOR represents the effectiveness of a diagnostic test. DOR is mathematically defined as (true positive/false positive)/(false negative/true negative). When DOR is greater than 1, higher values indicate better performance of the diagnostic method. A value of 1 means that the presence or absence of a disease cannot be determined and that the method cannot provide diagnostic information. To obtain an approximately normal distribution, we calculated the logarithm of each DOR and then calculated 95% confidence intervals [12]. SROC is a statistical technique used when performing a meta-analysis of studies that report both sensitivity and specificity. As the diagnostic ability of the test increases, the SROC curve shifts towards the upper-left corner of the ROC space, where both sensitivity and specificity are 1. AUC ranges from 0 to 1, with higher values indicating better diagnostic performance. We collected data on the number of patients, true-positive, true-negative, false-positive, and false-negative values in all included studies, and calculated AUCs and DORs from these values. The methodological quality of the included studies was evaluated using the Quality Assessment of Diagnostic Accuracy Study (QUADAS-2) tool.

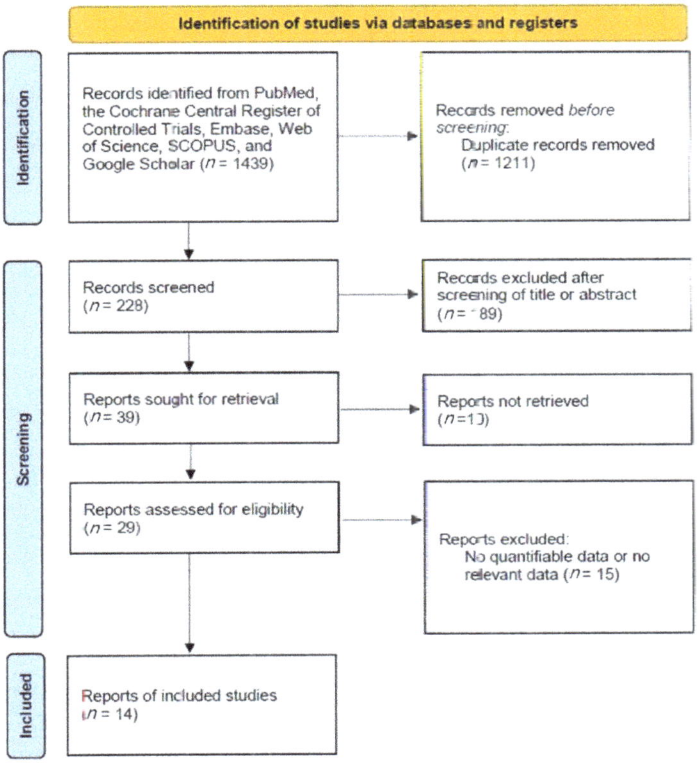

Figure 1. Summary of the search strategy.

2.4. Statistical Analysis and Outcome Measurements

R statistical software (R Foundation for Statistical Computing, Vienna, Austria) was used to conduct a meta-analysis of the studies. Homogeneity analyses were then performed using the Q statistic. Forest plots were drawn for the sensitivity, specificity, and negative predictive values, and for the SROC curves. A meta-regression analysis was performed to determine the potential influence of imaging tools on AI-based diagnostic accuracy for all premalignant lesions.

3. Results

This analysis included 14 studies [6,13–25]. Table 1 presents the assessment of bias. The characteristics of the studies are attached in Table S1.

3.1. Diagnostic Accuracy of AI-Assisted Screening of Oral Mucosal Cancerous Lesions

Seven prospective and retrospective studies were included for discriminating oral cancerous lesions from normal mucosa. The diagnostic odds ratio (DOR) of AI-assisted screening was 121.6609 (95% confidence interval [CI], 29.5996; 500.0534, $I^2 = 93.5\%$) (Figure 2A).

Table 1. Methodological quality of all included studies.

Reference	Risk of Bias				Concerns about Application		
	Patient Selection	Index Test	Reference Standard	Flow and Timing	Patient Selection	Index Test	Reference Standard
Nayak 2006 [13]	Unclear	Low	Unclear	Unclear	Low	Low	Low
Heidari 2018 [14]	Low	Low	Low	Low	Low	Low	Low
Song 2018 [15]	Low	Low	Low	Low	Low	Low	Low
Fu 2020 [6]	high	Low	Low	Low	Low	Low	Low
Duran-Sierra 2021 [16]	Unclear	Low	Unclear	Unclear	Low	Low	Low
James 2021 [17]	Low	Low	Unclear	Low	Low	Low	Low
Jubair 2021 [18]	Unclear	Low	Low	Low	Low	Low	Low
Lin 2021 [19]	Unclear	Low	Unclear	Low	Low	Low	Low
Song 2021 [20]	Low	Low	Low	Low	Low	Low	Low
Tanriver 2021 [21]	Low	Low	Low	Low	Low	Low	Low
Warin 2021 [22]	Low	Low	Low	Low	Low	Low	Low
Yang 2021 [23]	Low	Low	Low	Low	Low	Low	Low
Warin 2022 [24]	Low	Low	Low	Unclear	Low	Low	Low
Yuan 2022 [25]	Low	Low	Low	Low	Low	Low	Low

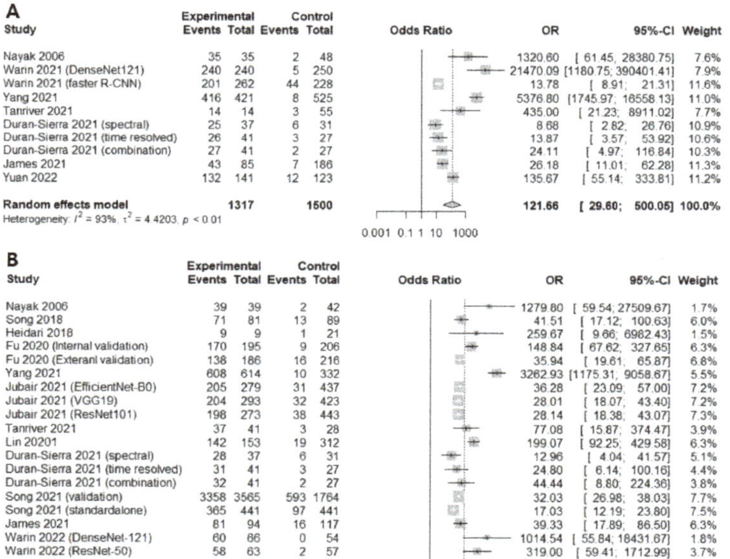

Figure 2. Forest plot of the diagnostic odds ratios for (**A**) screening only oral cancerous lesions [13,16,17,21–23,25] and (**B**) screening all premalignant mucosal lesions [13–21,23,24].

The area under the summary receiver operating characteristic curve was 0.948, suggesting excellent diagnostic accuracy (Figure 3A).

The correlation between the sensitivity and the false-positive rate was 0.437, indicating the absence of heterogeneity. AI-assisted screening exhibited good sensitivity (0.9232 [0.8686; 0.9562]; I^2 = 81.9%), specificity (0.9494 [0.7850; 0.9897], I^2 = 98.3%), and negative predictive value (0.9405 [0.8947; 0.9671]. I^2 = 83.6%) (Figure 4). The Begg's funnel plot (Supplementary Figure S1) shows that a source of bias was not evident in the included studies. The Egger's test result ($p > 0.05$) also shows that the possibility of publication bias is low.

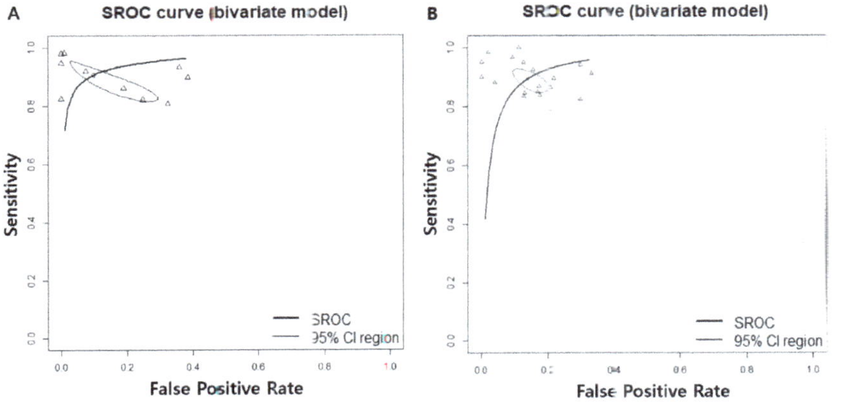

Figure 3. Area under the summary receiver operating characteristic for (**A**) screening only the oral cancerous lesions and (**B**) screening all premalignant mucosal lesions. SROC; summary receiver operating characteristic, CI; confidence interval.

Subgroup analyses were performed to determine which image tool assisted by AI had higher discriminating power between oral cancer lesions and normal mucosa. This analysis showed that that there were no significant differences between the photographic image, autofluorescence, and OCT in AI based on the screening for oral cancer lesion (Table 2).

Table 2. Subgroup analysis regarding image tool in discriminating oral cancerous lesions from normal mucosa.

Subgroup	Study (*n*)	DOR [95% CIs]	Sensitivity [95% CIs]	Specificity [95% CIs]	NPV [95% CIs]	AUC
	7	121.6609 [29.5996; 500.0534]; I^2 = 93.5%	0.9232 [0.8686; 0.9562]; I^2 = 81.9%	0.9494 [0.7850; 0.9897]; I^2 = 98.3%	0.9405 [0.8947; 0.9671]; I^2 = 83.6%	0.948
			Image tool			
Autofluorescence	2	25.9083 [6.3059; 106.4464]; I^2 = 68.0%	0.8972 [0.8262; 0.9413]; I^2 = 63.5%	0.8213 [0.4430; 0.9637]; 94.0%	0.9041 [0.8263; 0.9492]; 23.9%	
Optical coherense tomography	3	261.9981 [14.7102; 4666.3521]; I^2 = 96.3%	0.9419 [0.8544; 0.9781]; I^2 = 84.4%	0.9461 [0.7931; 0.9877]; 94.6%	0.9625 [0.9106; 0.9848]; 81.9%	
Photographic image	2	431.6524 [4.0037; 46537.4743]; I^2 = 93.0%	0.9149 [0.7475; 0.9750]; I^2 = 87.4%	0.9983 [0.2905; 1.0000]; 94.9%	0.9381 [0.8109; 0.9816]; 87.5%	
		0.2332	0.5910	0.2907	0.2291	

DOR; diagnostic odds ratio, AUC; area under the curve, NPV; negative predictive value.

3.2. Diagnostic Accuracy of AI-Assisted Screening of Oral Mucosal Precancerous and Cancerous Lesions

Twelve prospective and retrospective studies were included for discriminating oral precancerous and cancerous lesions from normal mucosa. The diagnostic odds ratio (DOR) of AI-assisted screening was 63.0193 (95% confidence interval [CI], 40.3234; 98.4896, I^2 = 88.2%) (Figure 2B). The area under the summary receiver operating characteristic curve was 0.943, suggesting excellent diagnostic accuracy (Figure 3B). The correlation between the sensitivity and the false-positive rate was 0.337, indicating the absence of heterogeneity. AI-assisted screening exhibited good sensitivity (0.9094 [0.8725; 0.9364]; I^2 = 92.3%), specificity (0.8848 [0.8400; 0.9183], I^2 = 93.8%), and negative predictive value (0.9169 [0.8815; 0.9424], I^2 = 92.8%) (Figure 5).

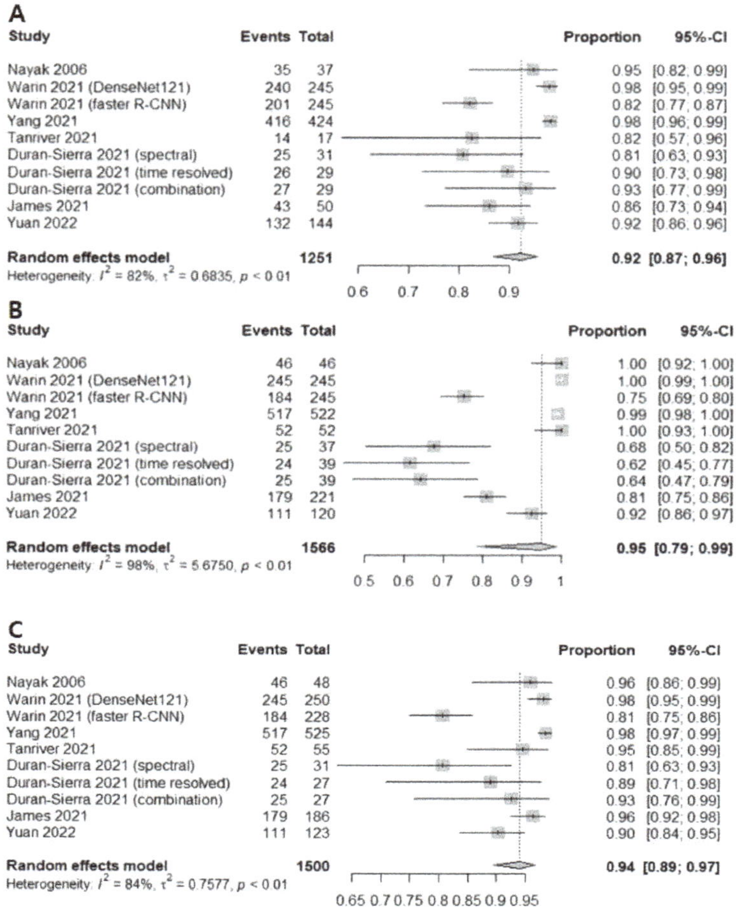

Figure 4. Forest plots of (**A**) sensitivity, (**B**) specificity, and (**C**) negative predictive values for screening oral cancerous lesions [13,16,17,21–23,25].

The Egger's test results of sensitivity ($p = 0.02025$) and negative predictive value ($p < 0.001$) also show that the possibility of publication bias is high. To compensate for the publication bias using statistical methods, trim-and-fill methods (trimfill) were applied to the outcomes. After implementation of trimfill, sensitivity dropped from 0.9094 [0.8725; 0.9364] to 0.8504 [0.7889; 0.8963] and NPV also dropped from 0.9169 [0.8815; 0.9424] to 0.7815 [0.6577; 0.8694]. These results could mean that the diagnostic power of AI-assisted screening of precancerous and cancerous lesions would be overestimated and clinicians would need to be careful when interpreting these outcomes.

Subgroup analyses were performed to determine which image tool assisted by AI had higher discriminating power of oral mucosal cancerous lesions including precancerous lesions. Subgroup analysis showed that OCT was more diagnostically accurate (324.3335 vs. 66.8107 and 27.6313) and more negatively predictive (0.9399 vs. 0.9311 and 0.8405) than photographic images and autofluorescence in AI based on the screening for oral precancerous and cancerous lesions from normal mucosa (Table 3). Meta-regression of AI diagnostic accuracy for oral precancerous and cancerous lesions on the basis of imaging tool revealed the significant correlations ($p = 0.0050$).

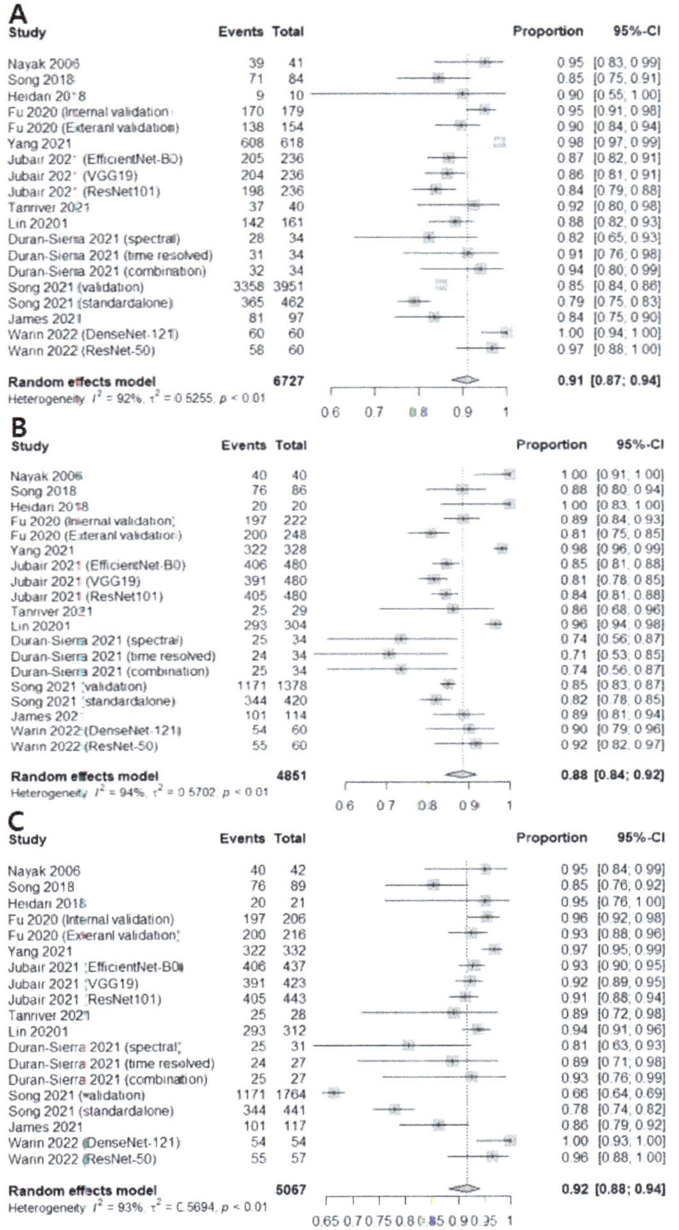

Figure 5. Forest plots of (**A**) sensitivity, (**B**) specificity, and (**C**) negative predictive values for screening all premalignant mucosal lesions [6,13–21,23,24].

Table 3. Subgroup analysis regarding image tool in discriminating oral precancerous and cancerous lesions from normal mucosa.

Subgroup	Study (*n*)	DOR [95% CIs]	Sensitivity [95% CIs]	Specificity [95% CIs]	NPV [95% CIs]	AUC
	12	63.0193 [40.3234; 98.4896]; I^2 = 88.2%	0.9094 [0.8725; 0.9364]; I^2 = 92.3%	0.8848 [0.8400; 0.9183]; I^2 = 93.8%	0.9169 [0.8815; 0.9424]; I^2 = 92.8%	0.943
			Image tool			
Autofluorescence	4	27.6313 [17.2272; 44.3186]; I^2 = 69.3%	0.8562 [0.8002; 0.8985]; I^2 = 69.6%	0.8356 [0.7591; 0.8913]; 86.8%	0.8405 [0.7487; 0.9031]; 91.1%	
Optical coherense tomography	3	324.3335 [10.2511; 10261.6006]; I^2 = 95.6%	0.9424 [0.8000; 0.9853]; I^2 = 88.3%	0.9653 [0.8737; 0.9911]; 79.8%	0.9399 [0.8565; 0.9762]; 75.7%	
Photographic image	5	66.8107 [38.0216; 117.3983]; I^2 = 81.7%	0.9123 [0.8683; 0.9426]; I^2 = 79.5%	0.8779 [0.8322; 0.9125]; 87.4%	0.9311 [0.9196; 0.9410]; 0.0%	
		0.0312	0.1120	0.0659	0.0073	

DOR; diagnostic odds ratio, AUC; area under the curve, NPV; negative predictive value.

4. Discussion

Oral cancer is a malignant disease with high disease-related morbidity and mortality due to its advanced loco-regional status at diagnosis. Early detection of oral cancer is the most effective means to increase the survival rate and reduce morbidity, but a significant number of patients experience delays between noticing the first symptoms and receiving a diagnosis from a clinician [26]. In clinical practice, a conventional visual examination is not a strong predictor of oral cancer diagnosis, and a quantitatively validated diagnostic method is needed [27]. Radiographic imaging, such as magnetic resonance imaging and computed tomography, can help determine the size and extent of oral cancer before treatment, but these techniques are not sensitive enough to distinguish precancerous lesions. Accordingly, various adjunct clinical imaging techniques such as autofluorescence and OCT have been used [28].

AI has been introduced in various industries, including healthcare, to increase efficiency and reduce costs, and the performance of AI models is improving day by day [29]. For the past few years, the early detection of oral cancer using AI technology based on autofluorescence imaging, photographic imaging, and OCT imaging has been an important research area. In this study, diagnostic values including sensitivity and specificity data were comprehensively confirmed in various studies that performed AI analysis of images. The diagnostic sensitivity of oral cancer analyzed by AI was as high as 0.92, and the analysis including precancerous lesions was slightly lower than the diagnostic sensitivity for cancer, but this also exceeded 90%. In subgroup analysis, there was no statistically significant difference in the diagnostic rate according to each image tool. In particular, the sensitivity of OCT to all precancerous lesions was found to be very high at 0.94.

Autofluorescence images are created using the characteristic that autofluorescence naturally occurring from collagen, elastin, and other endogenous fluorophores such as nicotinamide adenine dinucleotide in mucosal tissues by blue light or ultraviolet light is expressed differently in cancerous lesions [30,31]. Although it has been used widely in the dental field for the purpose of screening abnormal lesions in the oral cavity, it has been reported that the accuracy is low, with a sensitivity of only 30–50% [32,33]. It has been noted that autofluorescence images have a low diagnostic rate when used in oral cancer screening. Most of the previous clinical studies on autofluorescence-obtained images used differences in spectral fluorescence signals between normal and diseased tissues. Recently, time-resolved autofluorescence measurements using the characteristics of different fluorescence lifetimes of endogenous fluorophores have been used to solve the problem of broadly overlapping spectra of fluorophores, improving image accuracy [34]. Using various AI algorithms for advanced autofluorescence images, the diagnostic sensitivity of precancerous and cancerous lesions was reported to be as high as 94% [15]. As confirmed in our study, AI diagnosis sensitivity using autofluorescence images was confirmed to be 85% in all precancerous lesions. It showed relatively low diagnostic accuracy when compared to other imaging tools in this study. However, autofluorescence imaging is of sufficient value as an adjunct diagnostic tool. Efforts are also being made to improve the diagnostic accuracy for oral cancer by using AI to analyze images obtained using other tools along with the autofluorescence image [19].

The photographic image is a fast and convenient method with high accessibility compared to other adjunct methods. However, there is a disadvantage in that the image quality varies greatly depending on the camera, lighting, and resolution used while obtaining the image. Unlike external skin lesions, the oral cavity is surrounded by a complex, three-dimensional structure including the lips, teeth, and buccal mucosa, which may decrease the image accuracy [6]. In a recent study introducing a smartphone-based device, it was reported that the problem of the image itself was solved through a probe that can easily access the inside of the mouth and increasing images pixel [35]. Image diagnosis using a smartphone is very accessible in the current era of billions of phone subscribers worldwide, and in particular, it is expected that accurate and efficient screening will be possible by diagnosing a vast number of these images with AI. According to our analysis, AI-aided diagnosis from photographic images was confirmed to have a diagnostic sensitivity of over 91% for precancerous and cancerous lesions.

OCT is a medical technology that images tissues using the difference in physical properties between the reference light path and the sample light path reflected after interaction in the tissue [13]. OCT is non-invasive and uses infrared light, unlike other radiology tests that use X-rays. It is also a good diagnostic method that allows real-time image verification. Since its introduction in 1991 [36], OCT has been developed to provide high-resolution images at a faster speed and has played an important role in the biomedical field. In an AI analysis study of OCT images published by Yang et al., it was reported that the sensitivity and specificity of oral cancer diagnosis was 98% or more [22]. In our study, OCT images were found to be the most accurate diagnostic test, with sensitivity of 94% in AI diagnosis compared to other image tools (sensitivity of autofluorescence and photographic images of 89% and 91%, respectively). Therefore, AI diagnosis using OCT images is considered to be of sufficient value as a screening method for oral lesions. Each image tool included in our study has its own pros and cons to be considered when using it in actual clinical practice. In addition, accessibility of equipment or systems that can be performed on patients in actual outpatient treatment will be an important factor.

Based on our results, AI analysis of images in cancer diagnosis is thought to be helpful in making fast decisions regarding further examination and treatment. The accuracy of discriminating between precancerous lesions and normal tissues showed a high sensitivity of over 90%, showing good accuracy as a screening method. Although the question of whether AI can replace experts still exists, it is expected that oral cancer diagnosis using AI will sufficiently improve mortality and morbidity due to disease in low- and middle-income countries with poor health care systems. Acquisition of large-scale image datasets to improve AI analysis accuracy will be a clinically important key.

Our study has several limitations. First, our results include data from multiple imaging tools analyzed at once. This created heterogeneity in the results. Therefore, the sensitivity of each imaging tool was checked separately. The study is meaningful as it is the first meta-analysis to judge the accuracy of AI-based image analysis. Second, even with the same imaging tool, differences in the quality of the devices used in each study and differences between techniques may affect the accuracy of diagnosis. The images used to train the AI algorithm may not fully represent the diversity of oral lesions. Third, there is a limit to the interpretation of the results due to the absolute lack of prospective studies between the conventional examination and AI imaging diagnosis. It is our task to study this in various clinical fields in order to prepare for a future in which AI-assisted healthcare will be successful

5. Conclusions

AI shows good diagnostic performance with high sensitivity for oral cancer. Through the development of image acquisition devices and the grafting of various AI algorithms, the diagnostic accuracy is expected to increase. As new studies in this field are published frequently, a comprehensive review of the clinical implications of AI in oral cancer will be necessary again in the future.

Supplementary Materials: The following supporting information can be downloaded at: https://www.mdpi.com/article/10.3390/cancers14143499/s1, Figure S1: Begg's funnel plot; Table S1: Study characteristics.

Author Contributions: Conceptualization, J.-S.K., B.G.K. and S.H.H.; methodology, J.-S.K. and S.H.H.; software, S.H.H.; validation, S.H.H.; formal analysis, J.-S.K. and S.H.H.; investigation, J.-S.K. and S.H.H.; data curation, J.-S.K. and S.H.H.; writing—original draft preparation, J.-S.K. and S.H.H.; writing—review and editing, J.-S.K. and S.H.H.; visualization, J.-S.K. and S.H.H.; supervision, J.-S.K., B.G.K. and S.H.H. All authors have read and agreed to the published version of the manuscript.

Funding: This work was supported by the National Research Foundation of Korea (NRF) grant funded by the Korea government (MSIT) (2022R1F1A1066232). The sponsors had no role in the study design, data collection and analysis, decision to publish, or preparation of the manuscript.

Conflicts of Interest: The authors declare no conflict of interest.

References

1. Cunningham, M.J.; Johnson, J.T.; Myers, E.N.; Schramm, V.L., Jr.; Thearle, P.B. Cervical lymph node metastasis after local excision of early squamous cell carcinoma of the oral cavity. *Am. J. Surg.* **1986**, *152*, 361–366. [CrossRef]
2. Messadi, D.V. Diagnostic aids for detection of oral precancerous conditions. *Int. J. Oral Sci.* **2013**, *5*, 59–65. [CrossRef] [PubMed]
3. Kim, D.H.; Song, E.A.; Kim, S.W.; Hwang, S.H. Efficacy of toluidine blue in the diagnosis and screening of oral cancer and pre-cancer: A systematic review and meta-analysis. *Clin. Otolaryngol.* **2021**, *46*, 23–30. [CrossRef] [PubMed]
4. Awan, K.; Morgan, P.; Warnakulasuriya, S. Evaluation of an autofluorescence based imaging system (VELscope™) in the detection of oral potentially malignant disorders and benign keratoses. *Oral Oncol.* **2011**, *47*, 274–277. [CrossRef]
5. Tsai, M.-T.; Lee, H.-C.; Lee, C.-K.; Yu, C.-H.; Chen, H.-M.; Chiang, C.-P.; Chang, C.-C.; Wang, Y.-M.; Yang, C. Effective indicators for diagnosis of oral cancer using optical coherence tomography. *Opt. Express* **2008**, *16*, 15847–15862.
6. Fu, Q.; Chen, Y.; Li, Z.; Jing, Q.; Hu, C.; Liu, H.; Bao, J.; Hong, Y.; Shi, T.; Li, K. A deep learning algorithm for detection of oral cavity squamous cell carcinoma from photographic images: A retrospective study. *eClinicalMedicine* **2020**, *27*, 100558. [CrossRef]
7. LeCun, Y.; Bengio, Y.; Hinton, G. Deep learning. *Nature* **2015**, *521*, 436–444. [CrossRef]
8. Gulshan, V.; Peng, L.; Coram, M.; Stumpe, M.C.; Wu, D.; Narayanaswamy, A.; Venugopalan, S.; Widner, K.; Madams, T.; Cuadros, J.; et al. Development and Validation of a Deep Learning Algorithm for Detection of Diabetic Retinopathy in Retinal Fundus Photographs. *JAMA* **2016**, *316*, 2402–2410. [CrossRef]
9. Esteva, A.; Kuprel, B.; Novoa, R.A.; Ko, J.; Swetter, S.M.; Blau, H.M.; Thrun, S. Dermatologist-level classification of skin cancer with deep neural networks. *Nature* **2017**, *542*, 115–118. [CrossRef]
10. Varshni, D.; Thakral, K.; Agarwal, L.; Nijhawan, R.; Mittal, A. Pneumonia detection using CNN based feature extraction. In Proceedings of the 2019 IEEE International Conference on Electrical, Computer and Communication Technologies (ICECCT), Coimbatore, India, 20–22 February 2019; pp. 1–7.
11. Ilhan, B.; Guneri, P.; Wilder-Smith, P. The contribution of artificial intelligence to reducing the diagnostic delay in oral cancer. *Oral Oncol.* **2021**, *116*, 105254. [CrossRef]
12. Kim, D.H.; Kim, S.W.; Kim, S.H.; Jung, J.H.; Hwang, S.H. Usefulness of imaging studies for diagnosing and localizing cerebrospinal fluid rhinorrhea: A systematic review and meta-analysis. *Int. Forum. Allergy Rhinol.* **2022**, *12*, 828–837. [CrossRef] [PubMed]
13. Nayak, G.; Kamath, S.; Pai, K.M.; Sarkar, A.; Ray, S.; Kurien, J.; D'Almeida, L.; Krishnanand, B.; Santhosh, C.; Kartha, V. Principal component analysis and artificial neural network analysis of oral tissue fluorescence spectra: Classification of normal premalignant and malignant pathological conditions. *Biopolym. Orig. Res. Biomol.* **2006**, *82*, 152–166. [CrossRef] [PubMed]
14. Heidari, A.E.; Sunny, S.P.; James, B.L.; Lam, T.M.; Tran, A.V.; Yu, J.; Ramanjinappa, R.D.; Uma, K.; Birur, P.; Suresh, A. Optical coherence tomography as an oral cancer screening adjunct in a low resource settings. *IEEE J. Sel. Top. Quantum Electron.* **2018**, *25*, 7202008. [CrossRef]
15. Song, B.; Sunny, S.; Uthoff, R.D.; Patrick, S.; Suresh, A.; Kolur, T.; Keerthi, G.; Anbarani, A.; Wilder-Smith, P.; Kuriakose, M.A. Automatic classification of dual-modalilty, smartphone-based oral dysplasia and malignancy images using deep learning. *Biomed. Opt. Express* **2018**, *9*, 5318–5329. [CrossRef] [PubMed]
16. Duran-Sierra, E.; Cheng, S.; Cuenca, R.; Ahmed, B.; Ji, J.; Yakovlev, V.V.; Martinez, M.; Al-Khalil, M.; Al-Enazi, H.; Cheng, Y.-S.L. Machine-Learning Assisted Discrimination of Precancerous and Cancerous from Healthy Oral Tissue Based on Multispectral Autofluorescence Lifetime Imaging Endoscopy. *Cancers* **2021**, *13*, 4751. [CrossRef]
17. James, B.L.; Sunny, S.P.; Heidari, A.E.; Ramanjinappa, R.D.; Lam, T.; Tran, A.V.; Kankanala, S.; Sil, S.; Tiwari, V.; Patrick, S. Validation of a Point-of-Care Optical Coherence Tomography Device with Machine Learning Algorithm for Detection of Oral Potentially Malignant and Malignant Lesions. *Cancers* **2021**, *13*, 3583. [CrossRef]

18. Jubair, F.; Al-karadsheh, O.; Malamos, D.; Al Mahdi, S.; Saad, Y.; Hassona, Y. A novel lightweight deep convolutional neural network for early detection of oral cancer. *Oral Dis.* **2022**, *28*, 1123–1130. [CrossRef]
19. Lin, H.; Chen, H.; Weng, L.; Shao, J.; Lin, J. Automatic detection of oral cancer in smartphone-based images using deep learning for early diagnosis. *J. Biomed. Opt.* **2021**, *26*, 086007. [CrossRef]
20. Song, B.; Sunny, S.; Li, S.; Gurushanth, K.; Mendonca, P.; Mukhia, N.; Patrick, S.; Gurudath, S.; Raghavan, S.; Imchen, T.; et al. Mobile-based oral cancer classification for point-of-care screening. *J. Biomed. Opt.* **2021**, *26*, 065003. [CrossRef]
21. Tanriver, G.; Soluk Tekkesin, M.; Ergen, O. Automated Detection and Classification of Oral Lesions Using Deep Learning to Detect Oral Potentially Malignant Disorders. *Cancers* **2021**, *13*, 2766. [CrossRef]
22. Warin, K.; Limprasert, W.; Suebnukarn, S.; Jinaporntham, S.; Jantana, P. Automatic classification and detection of oral cancer in photographic images using deep learning algorithms. *J. Oral Pathol. Med.* **2021**, *50*, 911–918. [CrossRef] [PubMed]
23. Yang, Z.; Shang, J.; Liu, C.; Zhang, J.; Liang, Y. Identification of oral precancerous and cancerous tissue by swept source optical coherence tomography. *Lasers Surg. Med.* **2022**, *54*, 320–328. [CrossRef] [PubMed]
24. Warin, K.; Limprasert, W.; Suebnukarn, S.; Jinaporntham, S.; Jantana, P. Performance of deep convolutional neural network for classification and detection of oral potentially malignant disorders in photographic images. *Int. J. Oral Maxillofac. Surg.* **2022**, *51*, 699–704. [CrossRef] [PubMed]
25. Yuan, W.; Cheng, L.; Yang, J.; Yin, B.; Fan, X.; Yang, J.; Li, S.; Zhong, J.; Huang, X. Noninvasive oral cancer screening based on local residual adaptation network using optical coherence tomography. *Med. Biol. Eng. Comput.* **2022**, *60*, 1363–1375. [CrossRef]
26. Scott, S.E.; Grunfeld, E.A.; McGurk, M. Patient's delay in oral cancer: A systematic review. *Community Dent. Oral Epidemiol.* **2006**, *34*, 337–343. [CrossRef]
27. Epstein, J.B.; Güneri, P.; Boyacioglu, H.; Abt, E. The limitations of the clinical oral examination in detecting dysplastic oral lesions and oral squamous cell carcinoma. *J. Am. Dent. Assoc.* **2012**, *143*, 1332–1342. [CrossRef]
28. Camalan, S.; Mahmood, H.; Binol, H.; Araujo, A.L.D.; Santos-Silva, A.R.; Vargas, P.A.; Lopes, M.A.; Khurram, S.A.; Gurcan, M.N. Convolutional Neural Network-Based Clinical Predictors of Oral Dysplasia: Class Activation Map Analysis of Deep Learning Results. *Cancers* **2021**, *13*, 1291. [CrossRef]
29. Mintz, Y.; Brodie, R. Introduction to artificial intelligence in medicine. *Minim. Invasive Ther. Allied Technol.* **2019**, *28*, 73–81. [CrossRef]
30. Pavlova, I.; Williams, M.; El-Naggar, A.; Richards-Kortum, R.; Gillenwater, A. Understanding the biological basis of autofluorescence imaging for oral cancer detection: High-resolution fluorescence microscopy in viable tissue. *Clin. Cancer Res.* **2008**, *14*, 2396–2404. [CrossRef]
31. Skala, M.C.; Riching, K.M.; Gendron-Fitzpatrick, A.; Eickhoff, J.; Eliceiri, K.W.; White, J.G.; Ramanujam, N. In vivo multiphoton microscopy of NADH and FAD redox states, fluorescence lifetimes, and cellular morphology in precancerous epithelia. *Proc. Natl. Acad. Sci. USA* **2007**, *104*, 19494–19499. [CrossRef]
32. Mehrotra, R.; Singh, M.; Thomas, S.; Nair, P.; Pandya, S.; Nigam, N.S.; Shukla, P. A cross-sectional study evaluating chemiluminescence and autofluorescence in the detection of clinically innocuous precancerous and cancerous oral lesions. *J. Am. Dent. Assoc.* **2010**, *141*, 151–156. [CrossRef] [PubMed]
33. Farah, C.S.; McIntosh, L.; Georgiou, A.; McCullough, M.J. Efficacy of tissue autofluorescence imaging (VELScope) in the visualization of oral mucosal lesions. *Head Neck* **2012**, *34*, 856–862. [CrossRef] [PubMed]
34. Lagarto, J.L.; Villa, F.; Tisa, S.; Zappa, F.; Shcheslavskiy, V.; Pavone, F.S.; Cicchi, R. Real-time multispectral fluorescence lifetime imaging using Single Photon Avalanche Diode arrays. *Sci. Rep.* **2020**, *10*, 8116. [CrossRef]
35. Uthoff, R.D.; Song, B.; Sunny, S.; Patrick, S.; Suresh, A.; Kolur, T.; Keerthi, G.; Spires, O.; Anbarani, A.; Wilder-Smith, P. Point-of-care, smartphone-based, dual-modality, dual-view, oral cancer screening device with neural network classification for low-resource communities. *PLoS ONE* **2018**, *13*, e0207493. [CrossRef] [PubMed]
36. Huang, D.; Swanson, E.A.; Lin, C.P.; Schuman, J.S.; Stinson, W.G.; Chang, W.; Hee, M.R.; Flotte, T.; Gregory, K.; Puliafito, C.A.; et al. Optical coherence tomography. *Science* **1991**, *254*, 1178–1181. [CrossRef] [PubMed]

cancers

MDPI

Article

Comparative Multicentric Evaluation of Inter-Observer Variability in Manual and Automatic Segmentation of Neuroblastic Tumors in Magnetic Resonance Images

Diana Veiga-Canuto [1,2,*], Leonor Cerdà-Alberich [1], Cinta Sangüesa Nebot [2], Blanca Martínez de las Heras [3], Ulrike Pötschger [4], Michela Gabelloni [5], José Miguel Carot Sierra [6], Sabine Taschner-Mandl [4], Vanessa Düster [4], Adela Cañete [3], Ruth Ladenstein [4], Emanuele Neri [5] and Luis Martí-Bonmatí [1,2]

1 Grupo de Investigación Biomédica en Imagen, Instituto de Investigación Sanitaria La Fe, Avenida Fernando Abril Martorell, 106 Torre A 7planta, 46026 Valencia, Spain; leonor_cerda@iislafe.es (L.C.-A.); marti_lui@gva.es (L.M.-B.)
2 Área Clínica de Imagen Médica, Hospital Universitario y Politécnico La Fe, Avenida Fernando Abril Martorell, 106 Torre A 7planta, 46026 Valencia, Spain; sanguesa_cin@gva.es
3 Unidad de Oncohematología Pediátrica, Hospital Universitario y Politécnico La Fe, Avenida Fernando Abril Martorell, 106 Torre A 7planta, 46026 Valencia, Spain; blanca_martinez@iislafe.es (B.M.d.l.H.); canyete_ade@gva.es (A.C.)
4 St. Anna Children's Cancer Research Institute, Zimmermannplatz 10, 1090 Vienna, Austria; ulrike.poetschger@ccri.at (U.P.); sabine.taschner@ccri.at (S.T.-M.); vanessa.duester@ccri.at (V.D.); ruth.ladenstein@ccri.at (R.L.)
5 Academic Radiology, Department of Translational Research, University of Pisa, Via Roma, 67, 56126 Pisa, Italy; michela.gabelloni@med.unipi.it (M.G.); emanueleneri1@gmail.com (E.N.)
6 Departamento de Estadística e Investigación Operativa Aplicadas y Calidad, Universitat Politècnica de València, Camí de Vera s/n, 46022 Valencia, Spain; jcarot@eio.upv.es
* Correspondence: veiga_dia@gva.es

Citation: Veiga-Canuto, D.; Cerdà-Alberich, L.; Sangüesa Nebot, C.; Martínez de las Heras, B.; Pötschger, U.; Gabelloni, M.; Carot Sierra, J.M.; Taschner-Mandl, S.; Düster, V.; Cañete, A.; et al. Comparative Multicentric Evaluation of Inter-Observer Variability in Manual and Automatic Segmentation of Neuroblastic Tumors in Magnetic Resonance Images. *Cancers* **2022**, *14*, 3648. https://doi.org/10.3390/cancers14153648

Academic Editors: Hamid Khayyam, Ali Madani, Rahele Kafieh and Ali Hekmatnia

Received: 25 May 2022
Accepted: 26 July 2022
Published: 27 July 2022

Simple Summary: Tumor segmentation is a key step in oncologic imaging processing and is a time-consuming process usually performed manually by radiologists. To facilitate it, there is growing interest in applying deep-learning segmentation algorithms. Thus, we explore the variability between two observers performing manual segmentation and use the state-of-the-art deep learning architecture nnU-Net to develop a model to detect and segment neuroblastic tumors on MR images. We were able to show that the variability between nnU-Net and manual segmentation is similar to the inter-observer variability in manual segmentation. Furthermore, we compared the time needed to manually segment the tumors from scratch with the time required for the automatic model to segment the same cases, with posterior human validation with manual adjustment when needed.

Abstract: Tumor segmentation is one of the key steps in imaging processing. The goals of this study were to assess the inter-observer variability in manual segmentation of neuroblastic tumors and to analyze whether the state-of-the-art deep learning architecture nnU-Net can provide a robust solution to detect and segment tumors on MR images. A retrospective multicenter study of 132 patients with neuroblastic tumors was performed. Dice Similarity Coefficient (DSC) and Area Under the Receiver Operating Characteristic Curve (AUC ROC) were used to compare segmentation sets. Two more metrics were elaborated to understand the direction of the errors: the modified version of False Positive (FPRm) and False Negative (FNR) rates. Two radiologists manually segmented 46 tumors and a comparative study was performed. nnU-Net was trained-tuned with 106 cases divided into five balanced folds to perform cross-validation. The five resulting models were used as an ensemble solution to measure training (n = 106) and validation (n = 26) performance, independently. The time needed by the model to automatically segment 20 cases was compared to the time required for manual segmentation. The median DSC for manual segmentation sets was 0.969 (±0.032 IQR). The median DSC for the automatic tool was 0.965 (±0.018 IQR). The automatic segmentation model achieved a better performance regarding the FPRm. MR images segmentation variability is similar between radiologists and nnU-Net. Time leverage when using the automatic model with posterior visual validation and manual adjustment corresponds to 92.8%.

Keywords: tumor segmentation; neuroblastic tumors; deep learning; manual segmentation; automatic segmentation; inter-observer variability

1. Introduction

Neuroblastic tumors are the most frequent extracranial solid cancers in children. They comprise ganglioneuroma, ganglioneuroblastoma and neuroblastoma. Ganglioneuroma is composed of gangliocytes and mature stroma and is the most benign. Ganglioneuroblastoma is formed by mature gangliocytes and immature neuroblasts and has an intermediate malignant potential [1]. The most frequent type is neuroblastoma, which is more immature and undifferentiated. It is a heterogeneous neoplasm that shows different behavior based on biological, clinical and prognostic features, some tumors undergo spontaneous regression, while others progress with fatal outcomes despite therapy [2]. Neuroblastic tumors show a wide range of variability in their position. The most common sites of origin of neuroblastic tumors are the adrenal region (48%), extra-adrenal retroperitoneum (25%) and the chest (16%), followed by the neck (3%) and the pelvis (3%) [3]. Furthermore, they show high variability in their size, shape and boundaries, resulting in a common challenging task to differentiate them from the neighboring structures.

Tumor diagnosis, prognosis and the decision on respective treatment/disease management are mainly based on information obtained from imaging, including magnetic resonance (MR) [4]. Additionally, multiparametric data, radiomic features and imaging biomarkers, can provide the clinician with relevant information for disease diagnosis, characterization, and evaluation of aggressiveness and treatment response [5].

In order to ensure the best usability of imaging, it is essential to develop a robust and reproducible imaging processing pipeline. One of the most relevant steps involves segmentation, which consists of placing a Region of Interest (ROI) on a specific area (e.g., a tumor), with the assignment and labeling of voxels in the image that correspond to the ROI. Tumor segmentation can be performed in three different ways: manual, semiautomatic and automatic. Manual segmentation is usually performed by an experienced radiologist. This is usually done slice-by-slice, but is also possible in 3D, with the expert either encircling the tumor or annotating the voxels of interest. This is a reliable but time-consuming method that hinders the radiologists' workflow, especially in cases of mass data processing. However, manual segmentation is observer-dependent and may show wide inter and intra-observer variability [6,7]. This variability is influenced by some objective factors, such as organ/tumor characteristics or contour, and by some subjective factors related to the observer, such as their expertise or coordination skills [6].

Semiautomatic segmentation tries to solve some of the problems related to manual segmentation [8]. By assisting the segmentation with algorithms, for example, by growing the segmentation over a region or expanding the segmentation to other slices to eliminate the need for a slice-by-slice segmentation, the effort and time required from the user can be reduced. However, inter-observer variability is still present, as the manual part of the segmentation and the settings of the algorithm influence the result.

In the case of neuroblastic tumors, several studies have explored the development of semiautomatic segmentation algorithms. They have been performed on Computed Tomography (CT) or MR images, making use of mathematical morphology, fuzzy connectivity and other imaging processing tools [9–11]. However, they have included a very low number of cases and the findings show little improvement with respect to manual approaches. To the best of our knowledge, a robust and generalizable solution for neuroblastoma segmentation has not been yet devised.

Nowadays, most advanced tools are built to be used as automatic segmentation methods, which, by definition, do not rely on user interaction. These solutions are built with deep-learning segmentation algorithms [12], usually based on convolutional neural networks (CNNs). CNNs use several sequential convolution and pooling operations in which images are processed to extract features and recognize patterns using the image itself to be trained during the learning process [13] One of the most commonly used is the U-Net architecture, consisting of a contracting path to capture context and a symmetric expanding path that enables precise localization, which achieves very good performance in segmentation of different types of cancer [14,15]. Nevertheless, its applicability to specific image analysis and its reproducibility in different structures or lesions has been observed to be limited [16].

Recently, a new solution based on CNNs algorithms called nnU-Net has been proposed. It consists of an automatic deep learning-based segmentation framework that automatically configures itself, including preprocessing, network architecture, training and post-processing, and adapts to any new dataset, surpassing most existing approaches [16,17].

The aim of this study was to assess the inter-observer variability in manual and automatic segmentation of neuroblastic tumors. We hypothesize that the state-of-the-art deep learning framework nnU-Net can be used to automatically detect and segment neuroblastic tumors on MR images, providing a more robust, universal and error-free solution than that obtained by the manual segmentation process. This comparison is performed by evaluating the inter-observer variability between two radiologists. The automatic segmentation model is trained, fine-tuned and validated with cases from different European institutions and then compared to manual segmentation. Previous expert tumor delineation is performed as there does not exist an open-access annotated data set dedicated to this specific tumor.

The automatic segmentation model is then applied to a group of patients from the training set. The time needed for the automatic segmentation (with manual adjustment when necessary) is compared to the time required to manually segment the same cases from scratch.

2. Materials and Methods

2.1. Participants

A retrospective multicenter and international collection of 132 pediatric patients with neuroblastic tumors who had undergone a diagnostic MR examination was conducted.

All patients had received a diagnosis of neuroblastic tumor with pathological confirmation between 2002 and 2021. Patients and MR data were retrospectively obtained from 3 centers in Spain (n = 73, La Fe University and Polytechnic Hospital, including 21 cases from European Low and Intermediate Risk Neuroblastoma Protocol clinical trial (LINES)), Austria (n = 57, Children's Cancer Research Institute from SIOPEN High Risk Neuroblastoma Study (HR-NBL1/SIOPEN) current accrual over 3000 patients from 12 countries), and Italy (n = 4, Pisa University Hospital). The study had the corresponding institutional Ethics Committee approvals from all involved institutions. This data set was collected within the scope of PRIMAGE (PRedictive In-silico Multiscale Analytics to support cancer personalized diaGnosis and prognosis, empowered by imaging biomarkers) project [5]. Age at first diagnosis was 37.6 ± 39.3 months (mean ± standard deviation, range 0 to 252 months, median of 24.5 months ± 54 interquartile range (IQR)), with a slight female predominance (70 females, 62 males).

Histology of the tumor was neuroblastoma (104 cases), ganglioneuroblastoma (18 cases) and ganglioneuroma (10 cases). Tumor location was classified as abdominopelvic (105 cases, 59 of them from the adrenal gland, 32 with abdominal non-adrenal location and 14 with a pelvic location) or cervicothoracic (27 cases, 18 of them thoracic, 2 with an exclusive cervical location and 7 affecting both thoracic and cervical regions).

Imaging data from SIOPEN clinical trials (HR-NBL1 and LINES) were collected and centrally stored on an Image Management Server maintained by the Austrian Institute of Technology (AIT) in order to be properly pseudonymized with the European Unified Patient Identity Management (EUPID) [18] system enabling a privacy-preserving record linkage and a secure data transition to the PRIMAGE context. Other imaging data not coming from a SIOPEN trial received a EUPID pseudonym through the direct upload to the PRIMAGE platform. All collected images have been stored in the PRIMAGE platform to be used for further investigation.

The MR images accounted for a high data acquisition variability, including different scanners, vendors and protocols, from the different institutions. MR images were acquired with either a 1.5 T (n = 116) or 3 T (n = 16) scanner, manufactured by either General Electric Healthcare (Signa Excite HDxt, Signa Explorer) (n = 51), Siemens Medical (Aera, Skyra, Symphony, Avanto) (n = 54) or Philips Healthcare (Intera, Achieva, Ingenia) (n = 27). The MR protocol varied among the institutions. Essentially, MR studies consisted of T1-weighted (T1W), T2- weighted (T2w) and/or T2w with fat suppression (T2w fat-sat), Diffusion-weighted (DW) and Dynamic Contrast-enhanced (CET). Chest images were acquired with respiratory synchronization. Mean FOV size was 410 mm, and median FOV was 440 mm (range of 225 to 500 mm).

2.2. Manual Image Labeling

Tumor segmentation was performed on the transversal T2w fat-sat images as they yield the maximum contrast between the tumor and the surrounding organs (48 cases). T2w images were used when T2w fat-sat images were not available (84 cases). All images were obtained in DICOM format. The open source ITK-SNAP (version 3.8.0) (www.itksnap.org) tool [19] was used for the manual tumor segmentation by two radiologists (with 30 (Radiologist 1) and 5 (Radiologist 2) years of experience in pediatric radiology, respectively) with prior experience with manual segmentation tools. All the tumors (132 cases) were manually segmented by Radiologist 2. For the inter-observer variability study, 46 cases were independently segmented by both radiologists after agreement on the best tumor definition criteria. To increase reproducibility, a restrictive segmentation methodology was established, excluding doubtful peripheral areas. If the tumor contacted or encased a vessel, the vessel was excluded from the segmentation. When the tumor infiltrated neighboring organs with ill-defined margins, the DWI and CE images were reviewed to exclude non-tumoral areas. Lymph nodes separated from the tumor and metastases were also excluded. Each of the readers performed a blinded annotation of all the cases independently. Finally, the obtained segmentation masks were exported in NIfTI format (nii) and were considered the ground truth ROIs.

Tumor volume was obtained from the 132 masks performed by Radiologist 2. The median volume of all the masks was 116,518 mm^3 (±219,084 IQR) and the mean volume was 193,634 mm^3.

2.3. Study Design and Data Partitioning

Our study consisted of two parts (Figure 1). Firstly, the inter-observer variability in manual MR segmentation was analyzed by comparing the performance of two radiologists in 46 cases of neuroblastic tumor. Secondly, the training and validation of the automatic segmentation model based on nnU-Net architecture were performed, dividing the dataset into two cohorts: training-tuning and validation. A balanced and stratified split of the cases from both cohorts was implemented to eliminate sampling bias and to guarantee the heterogeneity of both datasets in order to construct a reproducible and universal model. Stratified sampling with the scikit-learn library [20] was used, considering four variables: manufacturer (Siemens/Philips/GE), magnetic field strength (1.5 T/3 T), tumor location (abdominopelvic/cervicothoracic) and segmented sequence (T2w/T2w fat-sat). (Table 1).

A first cohort (80% of cases, n = 106) was selected to train and fine-tune the model with a 5-fold cross-validation approach. A second cohort (20% of patients, n = 26) was used for validation.

2.4. Convolutional Neural Network Architecture

The automatic segmentation model was developed using the state-of-the-art, self-configuring framework for medical segmentation, *nnU-Net* [16]. All the images were resampled with a new voxel spacing: [z, x, y] = [8, 0.695, 0.695], corresponding to the average values within the training data set. The model training was performed along 1000 epochs with 250 iterations each and a batch size of 2. The loss function to optimize each iteration was based on the Dice Similarity Coefficient (DSC). A z-score normalization was applied to the images.

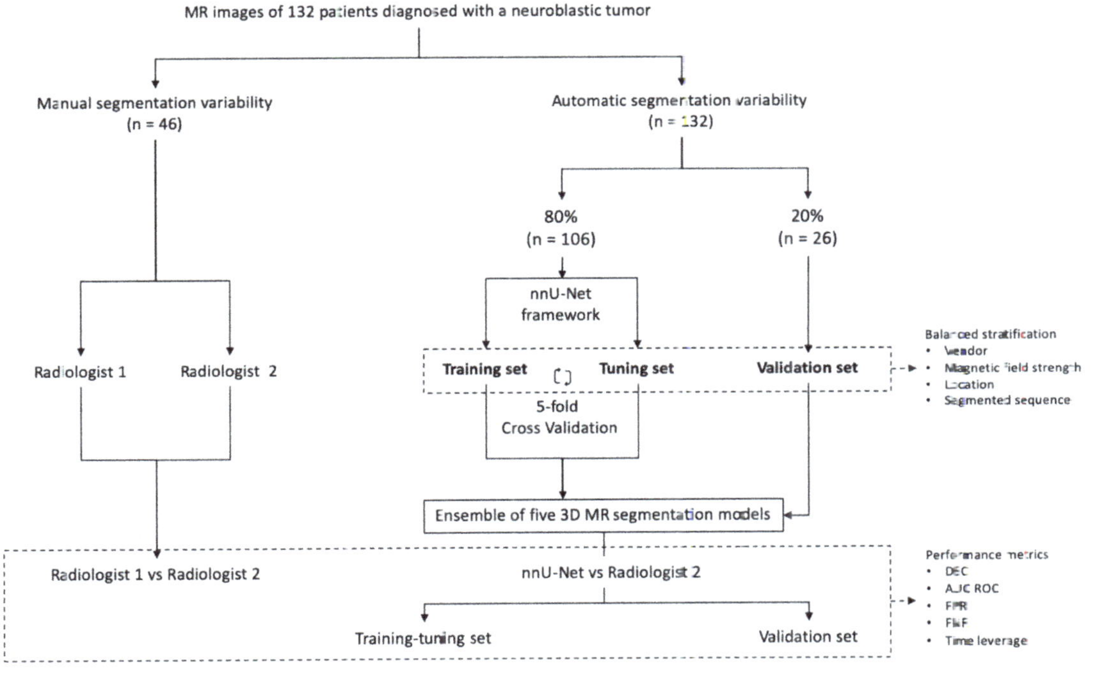

Figure 1. Study design. The first part consisted of manual segmentation variability, comparing the performance of two radiologists (n = 46). The second part included the training and validation of the nnU-Net using 132 cases manually segmented by Radiologist 2. Training-tuning with cross-validation was performed. The 5 resulting segmentation models obtained with the cross-validation method were used as an ensemble solution to test all the cases of the training-tuning (n = 106) and the validation (n = 26) data sets in order to measure training and validation performance independently. The previous split of the cases into balanced groups considering vendor, magnetic field strength, location and segmented sequence was performed.

The model employed a 3D net and was trained with a cross-validation strategy, which is a statistical technique frequently used to estimate the skill of a machine learning model on unseen data [21]. The training-tuning dataset (n = 106) was partitioned into 5 subsets or folds of 21 or 22 non-overlapping cases each. Each of the 5 folds was given an opportunity to be used as a held-back test set, whilst all other folds collectively were used as a training dataset. A total of 5 models were fit and evaluated on the 5 hold-out test sets and performance metrics were reported (median and IQR were reported as the distribution of the results was not normal in all the cases. Confidence interval (CI) was also calculated).

Additionally, the 5 resulting segmentation models obtained using the cross-validation method were used as an ensemble solution to test all the cases of the training-tuning (n = 106) and the validation (n = 26) data sets in order to measure training and validation performance independently.

Table 1. Composition of validation dataset (20% of cases, n = 26), considering four variables for a balanced split: vendor, magnetic field strength, location and segmented sequence.

Validation Set (n = 26)			
Sequence	**Equipment**	**Field Strength**	**Location**
T2	Philips	1.5	Abdominopelvic
T2	Siemens	1.5	Abdominopelvic
T2	Philips	1.5	Abdominopelvic
T2	GE	1.5	Abdominopelvic
T2	GE	1.5	Cervicothoracic
T2	Philips	1.5	Abdominopelvic
T2	Siemens	1.5	Abdominopelvic
T2	Philips	1.5	Cervicothoracic
T2	Siemens	1.5	Abdominopelvic
T2 fat sat	GE	1.5	Abdominopelvic
T2	Siemens	3	Abdominopelvic
T2	GE	1.5	Abdominopelvic
T2	GE	1.5	Cervicothoracic
T2	Philips	1.5	Abdominopelvic
T2 fat sat	Philips	1.5	Abdominopelvic
T2	Siemens	3	Abdominopelvic
T2	Siemens	1.5	Abdominopelvic
T2 fat sat	Siemens	1.5	Abdominopelvic
T2 fat sat	GE	1.5	Cervicothoracic
T2 fat sat	GE	1.5	Abdominopelvic
T2 fat sat	GE	1.5	Abdominopelvic
T2 fat sat	Siemens	1.5	Abdominopelvic
T2	GE	1.5	Cervicothoracic
T2 fat sat	Siemens	1.5	Abdominopelvic
T2 fat sat	Siemens	3	Abdominopelvic
T2 fat sat	GE	1.5	Cervicothoracic

2.5. Analysis and Metrics

To compare segmentation results, different metrics have been described in the literature. The main metric used in this study for the evaluation of results was the DSC [22], a spatial overlap index and a reproducibility validation metric [23]. Its value can range from 0 (meaning no spatial overlap between two sets) to 1 (indicating complete overlap). DSC index has been widely used to calculate the overlap metric between the results of segmentation and ground truth, and is defined as [24,25]:

$$\mathrm{DSC} = \frac{2\mathrm{TP}}{2\mathrm{TP} + \mathrm{FP} + \mathrm{FN}}$$

The ROC AUC metric was also calculated. The ROC curve, as a plot of sensitivity against 1-specificity, normally assumes more than one measurement. For the case where a test segmentation is compared to a ground truth segmentation, we consider a definition of the AUC as [26]:

$$\mathrm{AUC} = 1 - \frac{1}{2}\left(\frac{\mathrm{FP}}{\mathrm{FP} + \mathrm{TN}} + \frac{\mathrm{FN}}{\mathrm{FN} + \mathrm{TP}}\right)$$

Since metrics have different properties, selecting a suitable one is not a trivial task, and therefore a wide range of metrics have been previously developed and implemented to approach 3D image segmentation [26]. For our study, two spatial overlap-based metrics were specifically designed to gain a deeper understanding of the direction of the errors encountered: the false positive (FP) and false negative (FN) rates, independently, with respect to the ground truth, which consisted of the manual segmentation performed by Radiologist 2 (Figure 2).

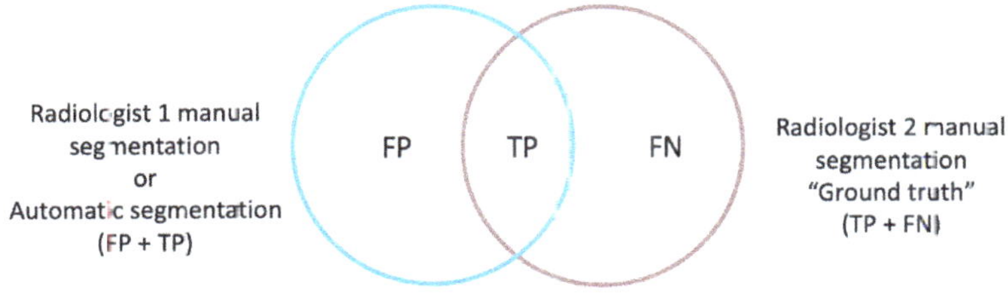

Figure 2. The ground truth (true positive and false negative voxels) corresponds to the manual segmentation performed by Radiologist 2, which was compared firstly to the manual segmentation performed by Radiologist 1 and then to the automatic segmentation obtained by the automatic segmentation model (non-ground truth mask, true positive and false positive voxels). The FPRm considered those voxels that were identified by the model as tumor but corresponded to other structures, divided by the voxels that actually corresponded to the ground truth mask. The FNR measured those voxels belonging to the tumor that the model did not include as such, divided by the ground truth voxels.

The rate of FP of the automatic segmentation to the ground truth (modified version of FPR) considered those voxels that were identified by the net as tumor but corresponded to other structures, divided by the voxels that actually corresponded to the ground truth mask (TP + FN voxels). This definition differs from the FPR used in standard statistical problems in the exclusion of the true negative (TN) term from the mathematical expression, as the TN voxels correspond to the image background in a segmentation task and not to the segmentation masks intended to be compared.

$$\text{FPRm} = \frac{\text{FP}}{\text{TP} + \text{FN}}$$

The rate of FN of the automatic segmentation to the ground truth (FNR) measured voxels belonging to the tumor that the net did not include as such, divided by the ground truth voxels.

$$\text{FNR} = \frac{\text{FN}}{\text{TP} + \text{FN}} = 1 - \text{Sensitivity or Recall}$$

For consistency reasons and to facilitate the understanding of the results, the FPRm and FNR metrics are reported as 1-self, resulting in a maximum of 1 for a complete voxel-wise agreement and a minimum of 0 for a null similitude.

2.6. Time Sparing

For comparing the time leverage, the final automatic segmentation model was applied to 20 cases from the training set, corresponding to 4 cases per fold to account for the heterogeneity of the data set. Cases were independently segmented manually from scratch by Radiologist 2, and the mean time (in minutes) necessary to perform that task was compared to the mean time required to obtain the masks with the automatic model. As some variability may exist in the final automatic masks, a human-based validation by Radiologist 2 was performed, and the mean time required to visually validate and manually edit the resulting automatic masks (when needed) was compared to the time necessary to manually segment them from scratch. To remove a potential software-related bias, the open source ITK-SNAP tool [19] was used for both manual and automatic correction approaches.

3. Results

3.1. Inter-Observer Comparison for Manual Segmentation

The segmentation results obtained by Radiologist 1 were compared to those of Radiologist 2 to measure inter-observer variability (Table 2). The median DSC was found to be 0.969 (±0.032 IQR). The median FPRm was 0.939 (±0.063 IQR), resulting in a high concordance between both radiologists according to the non-tumor included voxels. The median FNR was 0.998 (±0.008 IQR), meaning that Radiologist 1 did not miss tumor during the segmentation. AUC ROC was 0.998 (Figure 3).

Table 2. Inter-observer variability. Performance metrics for inter-observer comparison for manual segmentation, considering DSC, AUC ROC, 1-FPRm and 1-FNR.

	DSC	AUC ROC	1-FPRm	1-FNR
Median	0.969	0.998	0.939	0.998
IQR	0.032	0.004	0.063	0.008
CI	0.042	0.021	0.044	0.042

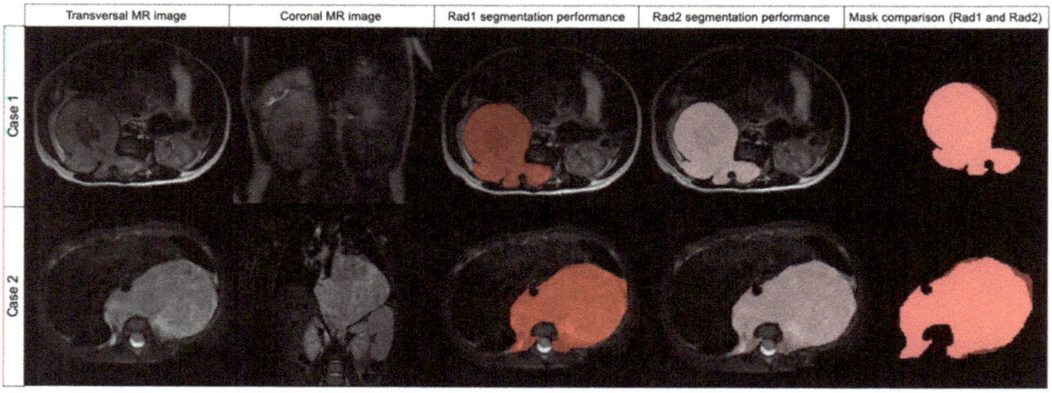

Figure 3. Comparison of two cases segmented by Radiologist 1 (red label) and Radiologist 2 (pink label) and mask superposition and comparison. Case 1 was segmented in T2w while case 2 was segmented in T2w fat-sat. In both cases, DSC was 0.957.

3.2. *Comparison between Radiologist and nnU-Net*

As the 106 cases of the training group were divided into five folds of 21 or 22 cases to perform cross-validation, each fold achieved different DSC results (Table 3) (Figures 4 and 5).

Table 3. Performance metrics for comparison between nnU-Net and Radiologist 2. Cases were divided into 5 folds to perform cross-validation. DSC, AUC ROC, 1-FPRm and 1-FNR for each fold are described.

Fold	Metric	DSC	AUC ROC	1-FPRm	1-FNR
Fold 0	Median	0.895	0.940	0.922	0.882
	IQR	0.121	0.116	0.082	0.233
	CI	0.146	0.117	0.074	0.148
Fold 1	Median	0.873	0.926	0.944	0.856
	IQR	0.110	0.100	0.100	0.100
	CI	0.127	0.066	0.088	0.132
Fold 2	Median	0.899	0.936	0.935	0.875
	IQR	0.131	0.064	0.133	0.133
	CI	0.123	0.062	0.125	0.124
Fold 3	Median	0.901	0.948	0.949	0.897
	IQR	0.122	0.062	0.088	0.124
	CI	0.046	0.030	0.090	0.061
Fold 4	Median	0.874	0.927	0.958	0.856
	IQR	0.134	0.110	0.033	0.221
	CI	0.141	0.071	0.032	0.142

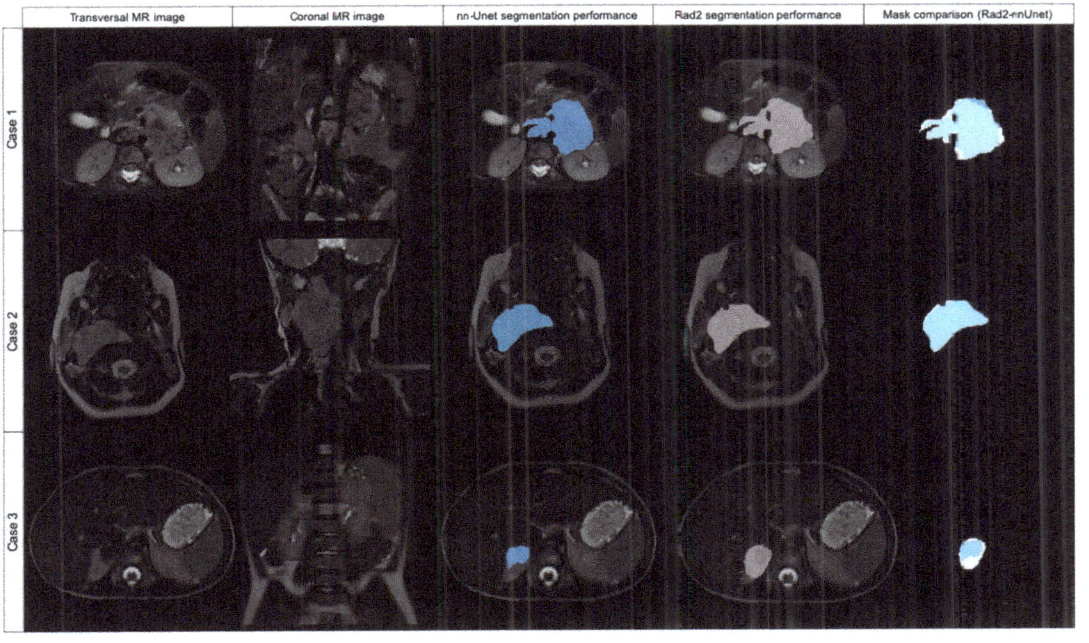

Figure 4. Original transversal and coronal MR images and examples of three cases automatically segmented by nnU-Net (blue labeled) and Radiologist 2 (pink labeled), with mask superposition for comparison. Case 1 was segmented in T2w fat-sat with a DSC of 0.869. Case 2 was segmented on T2w and the DSC obtained was 0.954. Case 3 was segmented with a DSC of 0.617.

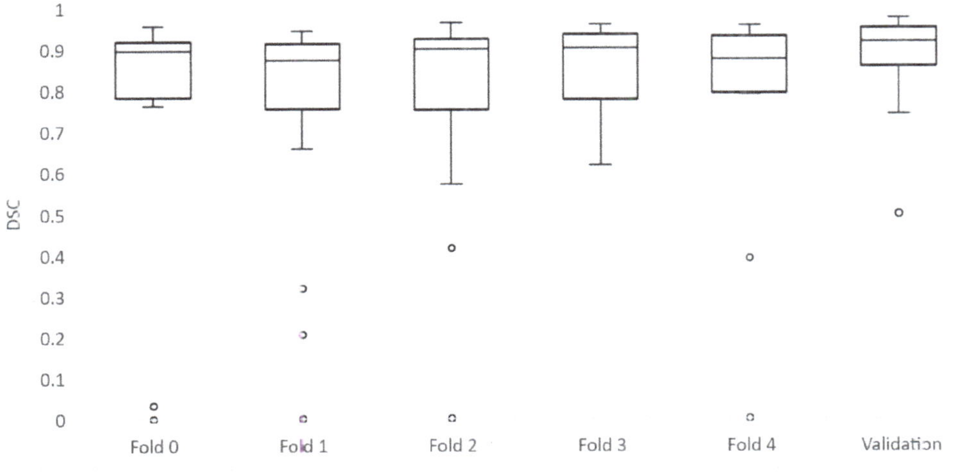

Figure 5. Box plots depicting the whole set of DSC for each fold of the training group and validation set.

Of the 106 cases, 27 had a DSC value <0.8: folds 0, 2 and 4 had 6 cases each; folds 1 and 3 had 5 cases each. The mean age for these cases was 32.7 ± 30.3 months and the median age was 19.8 months. They had a median volume of 75,733 mm^3 (±42,882 IQR).

From these 27 cases, 8 had a DSC from 0 to 0.19, being in all the cases < 0.01; 3 cases had a DSC ≥ 0.2 to 0.39; 1 case had a DSC ≥ 0.4 to 0.59; and 11 cases had a DSC ≥ 0.6 to 0.8. Cases showing high variability (DSC < 0.8) after automatic segmentation were analyzed by Radiologist 2 to identify the reasons for the low level of agreement. Regarding the eight cases with DSC < 0.01, the net had segmented extensive lymph nodes instead of the primary tumor in three cases. In another two cases, the net segmented other structures instead of the tumor (gallbladder or left kidney). In another three cases, the net did not identify any structure from the original DICOM and thus did not perform any mask.

Of the remaining 19 cases with a DSC < 0.8, 18 cases showed differences as the net localized the tumor well but did not completely segment it or presented variability in the borders, especially in cases with surrounding lymph nodes. One case had bilateral tumors and the net only detected one of them.

Posteriorly, the five resulting segmentation models obtained using the cross-validation method were used as an ensemble solution to test all the cases of the training-tuning (n = 106). We obtained a median DSC of 0.965 (±0.018 IQR) and AUC ROC of 0.981. The FPRm for this ensemble solution was 0.968, and FNR was 0.963. For comparing means of DSC attending to the effects of location (abdominopelvic or cervicothoracic) and magnetic field strength (1.5 or 3 T) (Table 4), an ANOVA test was performed, showing that there were no differences in DSC mean values for the location and magnetic field factors, and the results repeated after considering atypical values and removing them.

Table 4. The 5 resulting segmentation models obtained using the cross-validation method were used as an ensemble solution to test all the cases of the training-tuning (n = 106). Performance metrics for the final results are described. Results are detailed according to location (abdominopelvic or cervicothoracic) and magnetic field strength (1.5 T or 3 T).

	DSC	AUC ROC	1-FPRm	1-FNR
Median	0.965	0.981	0.968	0.963
IQR	0.018	0.010	0.015	0.021
CI	0.031	0.015	0.025	0.031
		Cervicothoracic (n = 21)		
Median	0.956	0.975	0.962	0.950
IQR	0.024	0.012	0.015	0.024
CI	0.036	0.018	0.037	0.036
		Abdominopelvic (n = 85)		
Median	0.966	0.982	0.969	0.645
IQR	0.015	0.009	0.014	0.019
CI	0.037	0.018	0.030	0.038
		1.5 T (n = 93)		
Median	0.965	0.981	0.969	0.963
IQR	0.018	0.011	0.016	0.021
CI	0.029	0.014	0.021	0.029
		3 T (n = 13)		
Median	0.964	0.982	0.967	0.964
IQR	0.013	0.005	0.007	0.010
CI	0.145	0.073	0.138	0.145

When introducing age and volume as corrective factors in the evaluation of the effects of location and magnetic field in the DICE, no differences were observed in the results of the analyses. Age and volume have no significant effect and do not show any trend in the DICE (p-value = 0.052 for age and 0.169 for volume). Therefore, the effects of location and magnetic field, as well as their interaction, continue to be insignificant when the correction for age and volume is introduced.

Focusing on the direction of the errors between both sets (ground truth vs. automatic segmentation), the median FPRm is 0.968 (±0.015 IQR), meaning that the mask is including as tumor 3.2% of voxels that are not included in the ground truth. The median FNR is 0.963 (±0.021 IQR), so the automatic tool does not include 3.7% of the voxels included in the ground truth mask.

3.3. Validation

The validation was performed at the end of the model development to test for model overfitting which could result in an overestimation of the model performance. The median DSC result for validation was 0.918 (±0.057 IQR) and AUC ROC was 0.968 (Table 5) (Figure 5).

Table 5. Performance metrics for the validation cohort results (n = 26) considering DSC. AUC ROC, 1-FPRm and 1-FNR. Results for Radiologist 2 vs. automatic model are shown. To compare these results to inter-radiologist agreement, Radiologist 1 segmented the 26 cases from the validation dataset and comparisons with Radiologist 2 and to the automatic model were made.

	DSC	AUC ROC	1-FPRm	1-FNR
		Radiologist 2 vs. automatic model		
Median	0.918	0.968	0.943	0.938
IQR	0.080	0.051	0.088	0.104
CI	0.059	0.473	0.134	0.063
		Radiologist 1 vs. Radiologist 2		
Median	0.920	0.950	0.929	0.930
IQR	0.090	0.192	0.015	0.024
CI	0.038	0.053	0.166	0.058
		Radiologist 1 vs. automatic model		
Median	0.915	0.950	0.915	0.912
IQR	0.443	0.122	0.436	0.189
CI	0.114	0.054	0.161	0.104

Of the 26 cases in the validation dataset, 4 had a DSC value < 0.8: 3 cases had a DSC ≥ 0.4 to 0.59; and 1 case had a DSC ≥ 0.6 to 0.8. These cases were analyzed by Radiologist 2 to identify the reasons for the low level of agreement. Regarding the three cases with DSC <0.6, the net had segmented extensive lymph nodes besides the primary tumor in two cases, and identified only a part of the tumor in one case. In the case with a DSC ≥ 0.6 to 0.8, the net segmented lymph nodes besides the primary tumor.

To compare the validation results to the inter-radiologist agreement, Radiologist 1 manually segmented the cases from the validation dataset. We compared the segmentation of Radiologist 1 to the segmentations of Radiologist 2 and the automatic model (Table 5).

For comparing the time leverage, we performed a comparison of the mean time needed to manually segment 20 cases (418 slices) from scratch with the mean time required by the automatic model to segment them. Cases segmented manually required a mean time of 56 min per case, while the mean time needed to obtain each mask with the automatic model was 10 s (0.167 min), resulting in a time reduction of 99.7%.

As some variability may exist in the final automatic masks, a human-based visual validation of the masks was performed. All the segmentations were visually validated, and manual editing and adjustment of the automatic masks were performed when needed (12 cases were edited, including 92 slices). The mean time to perform these processes was 4.08 min (±2.35 SD) and the median time was 4 min. This was compared to the time necessary to manually segment the masks from scratch, showing a time reduction of 92.8%.

4. Discussion

The inter-observer variability when performing manual segmentation of neuroblastic tumors in T2w MR images indicates that there is a high concordance between observers (median DSC overlap index of 0.969 (±0.032 IQR)). The discrepancies between observers may be due to the heterogeneous nature of the neuroblastic tumors and to the intrinsic variability of the manual segmentation related to individual skills and level of attention to detail. In our study, both radiologists are pediatric radiologists with previous experience in segmentation tasks. Radiologist 2 was considered the ground truth for practical reasons, as the segmentation of the whole dataset had been performed by this observer. Nevertheless, the manual ground truth mask may itself have some errors that are intrinsically associated with the human-based segmentation methodology. Joskowicz et al [6] investigated the variability in manual delineations on CT for liver tumors, lung tumors, kidney contours and brain hematomas between 11 observers, and concluded that inter-observer variability is large and even two or three observers may not be sufficient to establish the full range of inter-observer variability. Montagne et al [27] compared the inter-observer variability for prostate segmentation on MR performed by seven observers and concluded that variability is influenced by changes in prostate morphology. Therefore, delineation volume overlap variability for different structures and observers is large [28].

In our study, expert tumor delineation performed as the best (although not perfect) approach to truth. The evaluation of the voxel-wise similarity between the ground truth and the automatically segmented mask demonstrates that the state-of-the-art deep learning architecture nnU-Net can be used to detect and segment neuroblastic tumors on MR images, with a median DSC of 0.965 (±0.018 IQR), achieving a strong performance and surpassing the methods and results obtained in previous studies that approached the problem of neuroblastoma segmentation [9–11]. However, no previous literature has demonstrated the performance of a CNN-based solution in neuroblastic tumor. nnU-Net sets a new state-of-the-art in various semantic segmentation challenges and displays strong generalization characteristics for other structures [16,17]. Our results suggest that this automatic segmentation tool introduces a variability equivalent to that observed in the manual segmentation process in neuroblastic tumors. Previous studies related to breast tumors showed that segmentation algorithms may improve manual variability [29].

When analyzing the direction of the errors in a tumor segmentation problem, our recommendation is to give more relevance to the FPRm, aiming to minimize the included FP voxels with respect to the ground truth, as this metric represents those voxels that belong to adjacent organs or structures, which could introduce a strong bias in the extraction of quantitative imaging features for the development of radiomics models. The influence of FN in radiomics models seems less important, as it may not have a significant impact if some peripheral tumor voxels are missed. The effect of manual inter-observer segmentation variability on MR-based radiomics feature robustness has been described previously in other tumors such as breast cancer [30].

When assessing the FPRm and FNR between the manual segmentations performed by the two radiologists, the median FPRm is 0.939 (±0.063 IQR), indicating that 6.1% of the voxels were misclassified as tumor, while the median FNR is 0.998 (±0.008 IQR), therefore, the manual segmentation of Radiologist 1 did not include 0.2% of voxels included in the ground truth mask. Regarding manual ground truth vs. automatic segmentation, we observe that the median FPRm is 0.968 (±0.015 IQR). Therefore, the automatic segmentation tool generates masks with an average of 3.2% non-tumoral voxels. The median FNR corresponds to a value of 0.963 (±0.021 IQR), therefore, the automatic tool fails to include 3.7% of tumoral voxels. The results obtained demonstrate that the automatic segmentation model achieves a better performance regarding the FPRm, which is a great advantage in segmentation tasks for the posterior extraction of quantitative imaging features.

With regards to the time required for the segmentation process, an average time reduction of 99.7% was obtained when comparing the automatic model with the manual segmentation methodology. As some errors and variability may exist in the final automatic masks, this result is over-optimistic, as in practice the reader has to visually validate the quality of all the segmentations provided by the automatic tool before introducing some corrections, if needed. A human-based visual validation of the masks is recommended to edit and adjust the automatic masks. In our study, this process of validation and correction of the automatic masks that needed adjustment reduced the time required for segmentation from scratch by 92.8%. Correction time was influenced by the intrinsic difficulty of segmentation of each tumor, as there were tumors easier to segment (e.g., more homogeneous, with sharper margins, without lymph nodes) that did not need corrections or required slight adjustments, while there were more challenging tumors with contrast variations close to organ borders and of similar appearance to surrounding structures that required more time to be corrected. Overall, the application of the automatic model results in a great leverage of the time required to perform the segmentation process, facilitating the workflow for radiologists.

In addition, the human-based analysis of the masks performed by the net is useful to gain insights and correct for potential human mistakes and biases/outliers within the data set, which could be used to retrain the model, increasing the model's overall accuracy and robustness.

There are some limitations to this study. Segmentations were performed only by two observers, so it may only represent a fraction of the full range of inter-observer variability and may not be enough to establish a reference standard. Furthermore, both were experienced radiologists, as previous medical knowledge and expertise are assumed to contour highly heterogeneous neuroblastic tumors. Therefore, manual segmentations performed by other users (less experienced radiologists, other clinical users, non-medical staff) are expected to encounter higher inter-observer variability. Another potential limitation is that tumors may associate extensive lymph nodes or can present contact with them, making their differentiation difficult in some cases, which, as we have proved, can lead to errors in the segmentation performed by the net. Finally, as has been pointed out, the mask that is considered to be the ground truth may itself have some errors that are associated intrinsically with the manual segmentation process, and the comparison of the nnU-Net results was performed with the segmentations done by a single radiologist.

5. Conclusions

MR image segmentation accuracy of neuroblastic tumors is observed to be comparable between radiologists and the state-of-the-art deep learning architecture nnU-Net. The automatic segmentation model achieves a better performance regarding the FPRm, which is a great advantage in segmentation tasks for the posterior extraction of quantitative imaging features. Moreover, the time leverage when using the automatic model corresponds to 99.7%. A human-based validation based on manual editing of the automatic masks is recommended and corresponds to a reduction of time of 92.8% compared to the fully manual approach, reducing the radiologist's involvement in this task.

Author Contributions: L.M.-B. and L.C.-A. conceived the idea for this study and supervised the work. D.V.-C. and C.S.N. performed the manual segmentations. L.C.-A. developed the mathematical method, D.V.-C. and L.C.-A. performed the computations. D.V.-C. took the lead in writing the manuscript in consultation with L.M.-B., L.C.-A. and J.M.C S. U.P. verified the analytical and statistical methods. M.G., C.S.N. and E.N. contributed to the segmentation methodology and to the radiological point of view. B.M.d.l.H., S.T.-M., V.D., R.L. and A.C. supervised the findings of this work from an oncological and clinical point of view. All authors discussed the results and contributed to the preparation of the final manuscript submitted. All authors have read and agreed to the published version of the manuscript.

Funding: This study was funded by PRIMAGE (PRedictive In-silico Multiscale Analytics to support cancer personalized diaGnosis and prognosis, empowered by imaging biomarkers), a Horizon 2020 | RIA project (Topic SC1-DTH-07-2018), grant agreement no: 826494.

Institutional Review Board Statement: The study was conducted in accordance with the Declaration of Helsinki, and approved by the Hospital's Ethics Committee (The Ethics Committee for Investigation with medicinal products of the University and Polytechnic La Fe Hospital, ethic code: 2018/0228, 27 March 2019).

Informed Consent Statement: Informed consent was obtained from all subjects involved in the study.

Conflicts of Interest: All authors certify that there is no actual or potential conflict of interest in relation to this article.

References

1. Lonergan, G.J.; Schwab, C.M.; Suarez, E.S.; Carlson, C.L. From the Archives of the AFIP: Neuroblastoma, Ganglioneuroblastoma, and Ganglioneuroma: Radiologic-Pathologic Correlation. *RadioGraphics* **2002**, *22*, 911–934. [CrossRef] [PubMed]
2. Cohn, S.L.; Pearson, A.D.J.; London, W.B.; Monclair, T.; Ambros, P.F.; Brodeur, G.M.; Faldum, A.; Hero, B.; Iehara, T.; Machin, D.; et al. The International Neuroblastoma Risk Group (INRG) Classification System: An INRG Task Force Report. *J. Clin. Oncol.* **2009**, *27*, 289–297. [CrossRef] [PubMed]
3. Brisse, H.J.; McCarville, M.B.; Granata, C.; Krug, K.B.; Wootton-Gorges, S.L.; Kanegawa, K.; Giammarile, F.; Schmidt, M.; Shulkin, B.; Matthay, K.K.; et al. Guidelines for Imaging and Staging of Neuroblastic Tumors: Consensus Report from the International Neuroblastoma Risk Group Project. *Radiology* **2011**, *261*, 243–257. [CrossRef] [PubMed]
4. Matthay, K.K.; Maris, J.M.; Schleiermacher, G.; Nakagawara, A.; Mackall, C.L.; Diller, L.; Weiss, W.A. Neuroblastoma. *Nat. Rev. Primer* **2016**, *10*, 16078. [CrossRef]
5. Martí-Bonmatí, L.; Alberich-Bayarri, Á.; Ladenstein, R.; Blanquer, I.; Segrelles, J.D.; Cerdá-Alberich, L.; Gkontra, P.; Hero, B.; Garcia-Aznar, J.M.; Keim, D.; et al. PRIMAGE project: Predictive in silico multiscale analytics to support childhood cancer personalised evaluation empowered by imaging biomarkers. *Eur. Radiol. Exp.* **2020**, *4*, 22. [CrossRef]
6. Joskowicz, L.; Cohen, D.; Caplan, N.; Sosna, J. Inter-observer variability of manual contour delineation of structures in CT. *Eur. Radiol.* **2019**, *29*, 1391–1399. [CrossRef]
7. Bø, H.K.; Solheim, O.; Jakola, A.S.; Kvistad, K.-A.; Reinertsen, I.; Berntsen, E.M. Intra-rater variability in low-grade glioma segmentation. *J. Neurooncol.* **2017**, *131*, 393–402. [CrossRef]
8. Yushkevich, P.A.; Gao, Y.; Gerig, G. ITK-SNAP: An interactive tool for semi-automatic segmentation of multi-modality biomedical images. In Proceedings of the 2016 38th Annual International Conference of the IEEE Engineering in Medicine and Biology Society (EMBC), Orlando, FL, USA, 16–20 August 2016; pp. 3342–3345. [CrossRef]
9. Deglint, H.J.; Rangayyan, R.M.; Ayres, F.J.; Boag, G.S.; Zuffo, M.K. Three-Dimensional Segmentation of the Tumor in Computed Tomographic Images of Neuroblastoma. *J. Digit. Imaging* **2007**, *20*, 72–87. [CrossRef]
10. Gassenmaier, S.; Tsiflikas, I.; Fuchs, J.; Grimm, R.; Urla, C.; Esser, M.; Maennlin, S.; Ebinger, M.; Warmann, S.W.; Schäfer, J.F. Feasibility and possible value of quantitative semi-automated diffusion weighted imaging volumetry of neuroblastic tumors. *Cancer Imaging* **2020**, *20*, 89. [CrossRef]
11. Rangayyan, R.M.; Banik, S.; Boag, G.S. Landmarking and segmentation of computed tomographic images of pediatric patients with neuroblastoma. *Int. J. Comput. Assist. Radiol. Surg.* **2009**, *4*, 245–262. [CrossRef]
12. McKinney, S.M.; Sieniek, M.; Godbole, V.; Godwin, J.; Antropova, N.; Ashrafian, H.; Back, T.; Chesus, M.; Corrado, G.S.; Darzi, A.; et al. International evaluation of an AI system for breast cancer screening. *Nature* **2020**, *577*, 89–94. [CrossRef]
13. Yasaka, K.; Akai, H.; Abe, O.; Kiryu, S. Deep Learning with Convolutional Neural Network for Differentiation of Liver Masses at Dynamic Contrast-enhanced CT: A Preliminary Study. *Radiology* **2018**, *286*, 887–896. [CrossRef]
14. Feng, F.; Ashton-Miller, J.A.; DeLancey, J.O.L.; Luo, J. Convolutional neural network-based pelvic floor structure segmentation using magnetic resonance imaging in pelvic organ prolapse. *Med. Phys.* **2020**, *47*, 4281–4293. [CrossRef]
15. Ronneberger, O.; Fischer, P.; Brox, T. U-Net: Convolutional Networks for Biomedical Image Segmentation. In Proceedings of the Medical Image Computing and Computer-Assisted Intervention—MICCAI 2015, Munich, Germany, 5–9 October 2015; pp. 234–241.
16. Isensee, F.; Jaeger, P.F.; Kohl, S.A.A.; Petersen, J.; Maier-Hein, K.H. nnU-Net: A self-configuring method for deep learning-based biomedical image segmentation. *Nat. Methods* **2021**, *18*, 203–211. [CrossRef]
17. Alves, N.; Schuurmans, M.; Litjens, G.; Bosma, J.S.; Hermans, J.; Huisman, H. Fully Automatic Deep Learning Framework for Pancreatic Ductal Adenocarcinoma Detection on Computed Tomography. *Cancers* **2022**, *14*, 376. [CrossRef]
18. Ebner, H.; Hayn, D.; Falgenhauer, M.; Nitzlnader, M.; Schleiermacher, G.; Haupt, R.; Erminio, G.; Defferrari, R.; Mazzocco, K.; Kohler, J.; et al. Piloting the European Unified Patient Identity Management (EUPID) Concept to Facilitate Secondary Use of Neuroblastoma Data from Clinical Trials and Biobanking. *Stud. Health Technol. Inf.* **2016**, *31*, 223. [CrossRef]
19. Yushkevich, P.A.; Piven, J.; Hazlett, H.C.; Smith, R.G.; Ho, S.; Gee, J.C.; Gerig, G. User-guided 3D active contour segmentation of anatomical structures: Significantly improved efficiency and reliability. *NeuroImage* **2006**, *31*, 1116–1128. [CrossRef]
20. Pedregosa, F.; Varoquaux, G.; Gramfort, A.; Michel, V.; Thirion, B.; Grisel, O.; Blondel, M.; Prettenhofer, P.; Weiss, R.; Dubourg, V.; et al. Scikit-learn: Machine Learning in Python. *J. Mach. Learn. Res.* **2011**, *12*, 2825–2830.
21. Bey, R.; Goussault, R.; Grolleau, F.; Benchoufi, M.; Porcher, R. Fold-stratified cross-validation for unbiased and privacy-preserving federated learning. *J. Am. Med. Inform. Assoc.* **2020**, *27*, 1244–1251. [CrossRef]

22. Dice, L.R. Measures of the amount of ecologic association between species. *Ecology* **1945**, *26*, 297–302. [CrossRef]
23. Zou, K.H.; Warfield, S.; Bharatha, A.; Tempany, C.M.; Kaus, M.R.; Haker, S.J.; Wells, W.M.; Jolesz, F.A.; Kikinis, R. Statistical validation of image segmentation quality based on a spatial overlap index1. *Acad. Radiol.* **2004**, *11*, 178–189. [CrossRef]
24. Luo, C.; Shi, C.; Li, X.; Gao, D. Cardiac MR segmentation based on sequence propagation by deep learning. *PLoS ONE* **2020**, *15*, e0230415. [CrossRef]
25. Chlebus, G.; Meine, H.; Thoduka, S.; Abolmaali, N.; Van Ginneken, B.; Hahn, H.K.; Schenk, A. Reducing inter-observer variability and interaction time of MR liver volumetry by combining automatic CNN-based liver segmentation and manual corrections. *PLoS ONE* **2019**, *14*, e0217228. [CrossRef] [PubMed]
26. Taha, A.A.; Hanbury, A. Metrics for evaluating 3D medical image segmentation: Analysis, selection, and tool. *BMC Med. Imaging* **2015**, *15*, 29. [CrossRef]
27. Montagne, S.; Hamzaoui, D.; Allera, A.; Ezziane, M.; Luzurier, A.; Quint, R.; Kalai, M.; Ayache, N.; Delingette, H.; Renard-Penna, R. Challenge of prostate MRI segmentation on T2-weighted images: Inter-observer variability and impact of prostate morphology. *Insights Imaging* **2021**, *12*, 71. [CrossRef]
28. Meyer, C.R.; Johnson, T.D.; McLennan, G.; Aberle, D.R.; Kazerooni, E.A.; MacMahon, H.; Mullan, B.F.; Yankelevitz, D.F.; van Beek, E.J.; Armato, S.G.; et al. Evaluation of Lung MDCT Nodule Annotation Across Radiologists and Methods. *Acad. Radiol.* **2006**, *13*, 1254–1265. [CrossRef]
29. Saha, A.; Grimm, L.J.; Harowicz, M.; Ghate, S.V.; Kim, C.; Walsh, R.; Mazurowski, M.A. Interobserver variability in identification of breast tumors in MRI and its implications for prognostic biomarkers and radiogenomics: MRI breast tumor annotation: Interobserver variability analysis. *Med. Phys.* **2016**, *43 Pt 1*, 4558–4564. [CrossRef]
30. Granzier, R.W.Y.; Verbakel, N.M.H.; Ibrahim, A.; Van Timmeren, J.E.; Van Nijnatten, T.J.A.; Leijenaar, R.T.H.; Lobbes, M.B.I.; Smidt, M.L.; Woodruff, H.C. MRI-based radiomics in breast cancer: Feature robustness with respect to inter-observer segmentation variability. *Sci. Rep.* **2020**, *10*, 14163. [CrossRef]

diagnostics

MDPI

Article

Deep Learning-Based Classification of Uterine Cervical and Endometrial Cancer Subtypes from Whole-Slide Histopathology Images

JaeYen Song [1], Soyoung Im [2], Sung Hak Lee [3,*] and Hyun-Jong Jang [4,*]

1 Department of Obstetrics and Gynecology, Seoul St. Mary's Hospital, College of Medicine, The Catholic University of Korea, Seoul 06591, Korea
2 Department of Hospital Pathology, St. Vincent's Hospital, College of Medicine, The Catholic University of Korea, Seoul 16247, Korea
3 Department of Hospital Pathology, Seoul St. Mary's Hospital, College of Medicine, The Catholic University of Korea, Seoul 06591, Korea
4 Catholic Big Data Integration Center, Department of Physiology, College of Medicine, The Catholic University of Korea, Seoul 06591, Korea
* Correspondence: hakjjang@catholic.ac.kr (S.H.L.); hjjang@catholic.ac.kr (H.-J.J.); Tel.: +82-2-2258-1617 (S.H.L.); +82-2-2258-7274 (H.-J.J.)

Abstract: Uterine cervical and endometrial cancers have different subtypes with different clinical outcomes. Therefore, cancer subtyping is essential for proper treatment decisions. Furthermore, an endometrial and endocervical origin for an adenocarcinoma should also be distinguished. Although the discrimination can be helped with various immunohistochemical markers, there is no definitive marker. Therefore, we tested the feasibility of deep learning (DL)-based classification for the subtypes of cervical and endometrial cancers and the site of origin of adenocarcinomas from whole slide images (WSIs) of tissue slides. WSIs were split into 360 × 360-pixel image patches at 20× magnification for classification. Then, the average of patch classification results was used for the final classification. The area under the receiver operating characteristic curves (AUROCs) for the cervical and endometrial cancer classifiers were 0.977 and 0.944, respectively. The classifier for the origin of an adenocarcinoma yielded an AUROC of 0.939. These results clearly demonstrated the feasibility of DL-based classifiers for the discrimination of cancers from the cervix and uterus. We expect that the performance of the classifiers will be much enhanced with an accumulation of WSI data. Then, the information from the classifiers can be integrated with other data for more precise discrimination of cervical and endometrial cancers.

Keywords: computational pathology; computer-aided diagnosis; convolutional neural network; digital pathology

Citation: Song, J.; Im, S.; Lee, S.H.; Jang, H.-J. Deep Learning-Based Classification of Uterine Cervical and Endometrial Cancer Subtypes from Whole-Slide Histopathology Images. *Diagnostics* **2022**, *12*, 2623. https://doi.org/10.3390/diagnostics12112623

Academic Editors: Hamid Khayyam, Ali Madani, Rahele Kafieh and Ali Hekmatnia

Received: 4 September 2022
Accepted: 26 October 2022
Published: 28 October 2022

1. Introduction

Uterine cervical and endometrial cancers are two major cancer types threatening women's health worldwide [1]. Although they originate from the same organ, i.e., uterus, cervical and endometrial cancers have different subtypes with different clinical outcomes [2–6]. The main histologic subtypes of cervical cancers are squamous cell carcinoma and endocervical adenocarcinoma. The two major histologic subtypes of endometrial cancers are endometrioid adenocarcinoma and serous adenocarcinoma. Because management and prognosis are different between the subtypes, differential diagnosis is crucial for proper treatment decisions. Furthermore, an endometrial and endocervical origin for an adenocarcinoma should be distinguished considering the marked differences in their management [7]. The first step for the discrimination of the subtypes of these cancers is to investigate hematoxylin and eosin (H&E)-stained tissue slides by pathologists. However, the visual discrimination of subtypes is not always clear because some morphologic features are overlapping [7,8]. Furthermore, there is considerable inter- and intra-observer variations in the histological subtyping by pathologists [8]. Although various immunohistochemical markers can help distinguish the subtypes, there is no definitive marker [7,8]. Therefore, ancillary methods for the discrimination of the subtypes of cervical and endometrial cancers, and also the origin of the cancers are necessary to improve treatment decisions.

Because whole-slide images (WSIs) were approved for primary diagnostic purposes, many pathologic laboratories have been adopting digitized diagnosis processes [9]. The digitization enabled computer-aided analysis of pathologic tissues. Computer-aided analysis of H&E-stained WSIs could provide valuable information in a cost- and time-effective manner, considering the wide availability of H&E-stained pathologic tissue slides for most cancer patients. Recently, deep learning (DL) has been widely applied for various analysis tasks on H&E-stained WSIs [10]. DL usually performs better than many previous machine learning methods because it can automatically learn the most discriminative features directly from large datasets [11]. Many studies showed that DL can correctly diagnose various cancers from WSIs [12]. Furthermore, DL can even detect molecular alterations of cancer tissues from H&E-stained WSIs [13]. Therefore, DL has tremendous potential to improve the precision of pathologic diagnosis with minimal additional cost.

In the present study, we applied sequential DL models for the subtyping of cervical and endometrial cancers. First, cervical and endometrial cancer regions were automatically selected with DL models. Then, two separate DL models were trained to discriminate cervical and endometrial cancers into cervical squamous cell carcinoma and endocervical adenocarcinoma, and into endometrioid endometrial adenocarcinoma and serous endometrial adenocarcinoma, respectively. Furthermore, we trained an additional DL model to discriminate whether an adenocarcinoma has an endocervical or endometrial origin. The three models showed excellent performance proving the potential of DL for the discrimination of subtypes in gynecologic tumors.

2. Materials and Methods

2.1. Datasets

Classifiers for the subtypes of cervical and endometrial cancers and the origin of adenocarcinomas were trained with the WSIs provided by The Cancer Genome Atlas (TCGA) program. From the TCGA cervical (TCGA-CESC) and endometrial (TCGA-UCEC) datasets, we collected formalin-fixed paraffin-embedded (FFPE) slides after the basic slide quality reviews. The TCGA-CESC dataset provided slides from 255 patients for cervical squamous cell carcinoma and from 47 patients for endocervical adenocarcinoma. From the TCGA-UCEC dataset, tissue slides of 399 and 109 patients were obtained for endometrioid endometrial adenocarcinoma and serous endometrial adenocarcinoma, respectively. When there are huge differences in the numbers of data between the classes, performance evaluation can be skewed by the majority class. Therefore, we randomly selected 70 and 160 patients for cervical squamous cell carcinoma and endometrioid endometrial adenocarcinoma, respectively, to make the differences between the major and minor classes under 1.5-fold.

The performance of the classifier for the subtypes of endometrial carcinoma was also evaluated on The National Cancer Institute's Clinical Proteomic Tumor Analysis Consortium (CPTAC) endometrial cancer dataset (CPTAC-UCEC). There were 83 patients for endometrioid endometrial adenocarcinoma and 12 patients for serous endometrial adenocarcinoma.

2.2. Deep Learning Model

To fully automate the classification tasks, we sequentially applied different DL-based classifiers to the WSIs (Figure 1). The WSIs were divided into non-overlapping, 360 × 360-pixel image patches at 20× magnification because a WSI is too big to be analyzed by a current DL-system as a whole. In a WSI, various artifacts can exist including air bubbles, blurring, compression artifacts, pen markings, and tissue folding. Patches with these artifacts should be discarded because they can interfere with proper learning of relevant features. In our previous study for gastric cancer subtyping, we trained a simple DL classifier that can discriminate various artifacts and white backgrounds all at once [14]. The DL network consisted of three convolution layers with 12 [5 × 5] filters, 24 [5 × 5] filters and 24 [5 × 5] filters, each followed by a [2 × 2] max-pooling layer. We reused the classifier and only proper tissue image patches were selected for the next steps (Figure 1a).

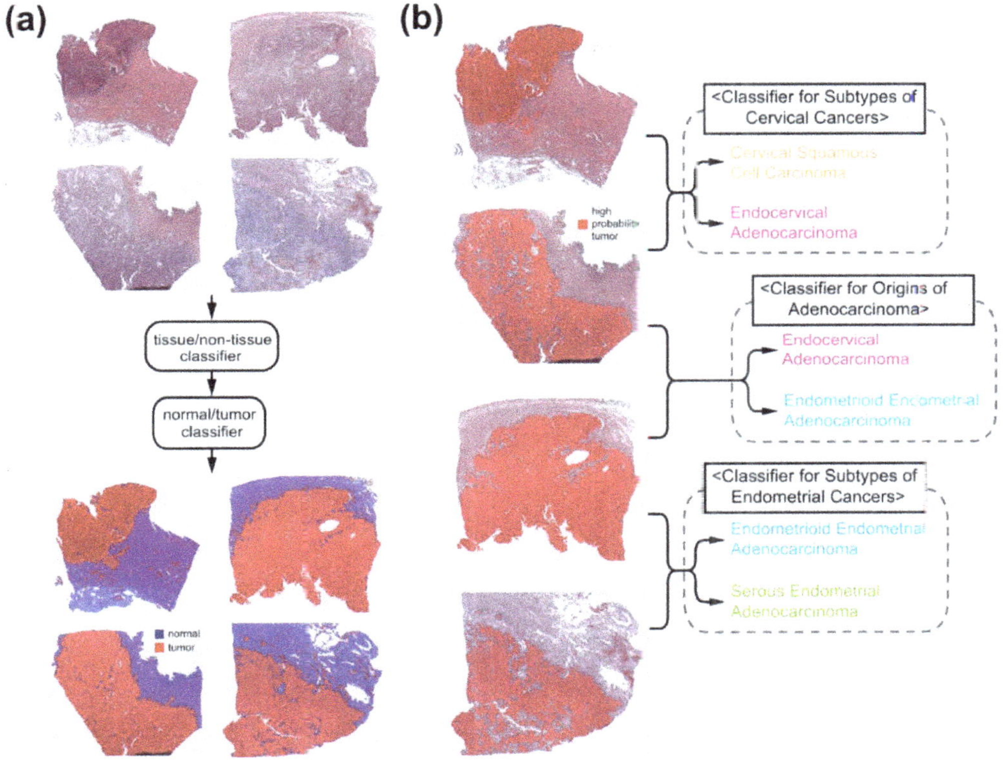

Figure 1. Classification procedure. (**a**) Sequential application of tissue/non-tissue and normal/tumor classifiers can discriminate proper tumor tissues. (**b**) Three separate classifiers for subtypes of cervical cancers, subtypes of endometrial cancers, and site of origin for adenocarcinomas were trained from tumor tissue image patches.

Cancer subtype classifiers should be trained on the cancer tissues. Therefore, normal and tumor tissue classifiers are prerequisites for cancer subtyping. To train the normal/tumor classifiers, two pathologists (S.L and S.H.L.) annotated normal and tumor regions for cervical and endometrial cancer tissue slides (Figure 2 left panels). Then, normal and tumor tissue image patches were collected based on the annotation. From these patches, classifiers to discriminate normal and tumor tissues for cervical and endometrial cancers were trained separately for each cancer type.

Next, we trained classifiers for the subtypes of cervical and endometrial cancers, and the origin of adenocarcinomas on prominent tumor tissue patches selected by the normal/tumor classifiers. To evaluate the general performance of the classifiers for the TCGA-CESC and -UCEC datasets, 5-fold cross validation was adopted. Therefore, the WSIs were split into 5 non-overlapping patient-level subsets and classifiers were trained and evaluated for each subset. As we noted, 70 and 160 patients for cervical squamous cell carcinoma and endometrioid endometrial adenocarcinoma were selected for evaluation.

However, performance can be enhanced when the classifiers were exposed to more various tissue images during training. Therefore, we randomly sampled tumor image patches from all cervical squamous cell carcinoma and endometrioid endometrial adenocarcinoma WSIs other than the test sets to match the 1.5-fold data ratio of major/minor class tissue patches for training, as this strategy could include a greater variety of tissue images. Therefore, we included sampled data from all patients other than the test sets during training and selected patients for the testing to avoid skewed test results. For the selection of the samples, we made a random selection to avoid selection biases from human selectors. The numbers of image patches used for the training of the classifiers were summarized in Supplementary Table S1.

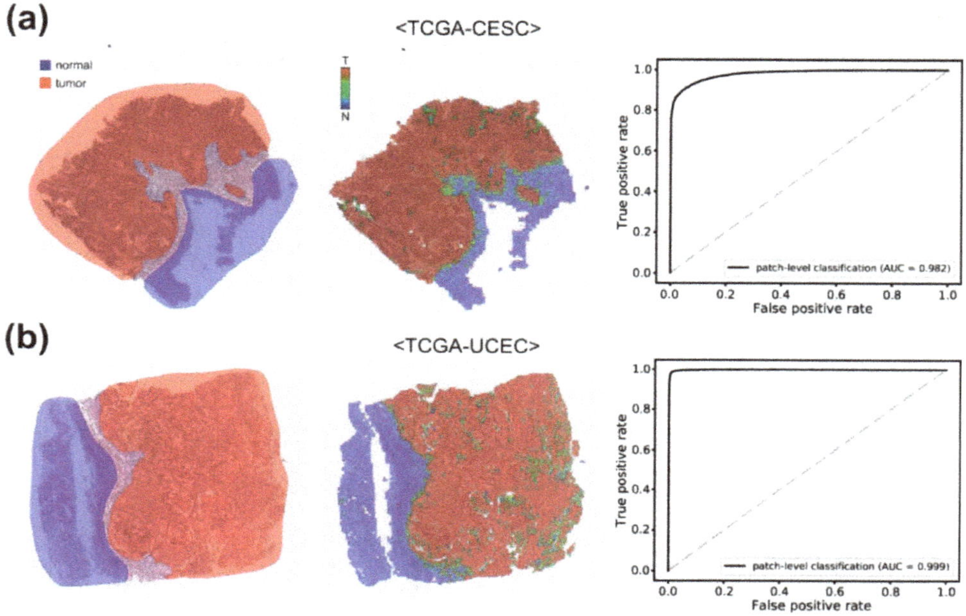

Figure 2. Normal/tumor classification results for (**a**) cervical and (**b**) endometrial cancers. Left panels: annotation made by pathologists. Middle panels: classification results of the normal/tumor classifiers. Right panels: Receiver operating characteristic curves for normal/tumor classification results. AUC: area under the curve.

Inception-v3 model was adopted for the normal/tumor, cancer subtypes, and origin classifiers because the Inception-v3 model yielded good results for normal/tumor classification or tissue subtype classification in our previous studies [14,15]. The models were implemented using the Tensorflow deep learning library version 1.15 (http://tensorflow.org (accessed on 22 January 2022)). The overall structure of the model is presented in Supplementary Figure S1. RMSPropOptimizer was adopted to optimize the model and the hyperparameters were as follows: initial learning rate 0.1, number of epochs per decay 10.0, learning rate decay factor 0.16, RMSPROP decay 0.9, RMSPROP_MOMENTUM 0.9, RMSPROP_EPSILON 1.0. Tissue images were color normalized before the training and testing. During training, data augmentation techniques such as random rotation by 90° and random horizontal/vertical flipping were applied to the tissue patches. Four computer systems equipped with an Intel Core i9-12900K Processor (Intel Corporation, Santa Clara, CA, USA) and dual NVIDIA RTX 3090 GPUs (NVIDIA corporation, Santa Clara, CA, USA) were used for the training and testing of the models.

2.3. Visualization and Statistics

To visualize the distribution of different tissue types, heatmaps of classification results of tissue patches were overlaid on the WSIs with specific colors demonstrated in Figure 1. To obtain the overall classification result of a WSI, patch classification results were averaged to obtain the result for the WSI. Receiver operating characteristic (ROC) curves and area under the curves for the ROC curves (AUROCs) were presented to demonstrate the performance of each classifier. For 5-fold cross validated datasets, ROC curves for the folds with the lowest and highest AUROCs and for the concatenated results of all 5 folds were provided for more precise evaluation of the performance of the classifiers. For the concatenated results of all 5 folds, 95% confidence intervals (CIs) were presented. To obtain accuracy, sensitivity, specificity and F1 score of the classification results, cutoff values yielding maximal Youden index (sensitivity + specificity − 1) were adopted.

When a comparison between the ROC curves is necessary, Venkatraman's permutation test with 1000 iterations was applied [16]. A p-value < 0.05 was considered significant.

2.4. Ethical Statement

Informed consent of patients in the TCGA cohorts was acquired by the TCGA consortium [17]. The Institutional Review Board of the College of Medicine at The Catholic University of Korea approved this study (XC21ENDI0031K).

3. Results

3.1. Normal/Tumor Classification

To classify the subtypes of cancer tissues, proper cancer tissue image patches should be selected (Figure 1). First, we removed image patches containing various artifacts and white background with a tissue/non-tissue classifier from our previous study [14]. Then, normal/tumor classifiers for cervical and endometrial cancers were trained based on pathologists' annotation (Figure 2). Pathologists annotated 100 slides for each cervical and endometrial cancer. The normal/tumor classifiers were trained with 80 slides and tested on the remaining 20 slides. The representative WSIs in Figure 2 are the cervical and endometrial cancer WSIs from the test sets. The classification results of the normal/tumor classifiers matched well with the pathologists' annotation. The AUROCs for the patch-level classification results of the normal/tumor classifiers are 0.982 and 0.999 for cervical and endometrial cancers, respectively.

3.2. Cervical Cancer Subtypes Classification

With the tissue/non-tissue and normal/tumor classifiers, we can collect proper tumor patches for the training of the cancer subtype classifiers. First, we trained classifiers for the cervical cancer subtypes. The patches from a WSI are labeled as either cervical squamous cell carcinoma or endocervical adenocarcinoma based on the information obtained from cBioPortal for Cancer Genomics (https://www.cbioportal.org/ (accessed on 12 March 2022)). Then, separate classifiers were trained to distinguish the subtypes for each 5-fold. For each fold, four classifiers were trained repeatedly and a classifier yielding the best AUROC was used to present the results. The classification results of cervical squamous cell carcinoma and endocervical adenocarcinoma are presented in Figure 3. The upper panels show the representative WSIs of clear cervical squamous cell carcinoma, clear endocervical adenocarcinoma, and confusing case with mixed classification results. The ROC curves of slide-level classification results for folds with the lowest and highest AUROCs and concatenated results of all 5-folds are presented in the lower panels. The AUROCs were 0.979 and 1.000 for the folds with the lowest and highest AUROCs, respectively. The AUROC for the concatenated results was 0.977 (95% CI, 0.957–0.998).

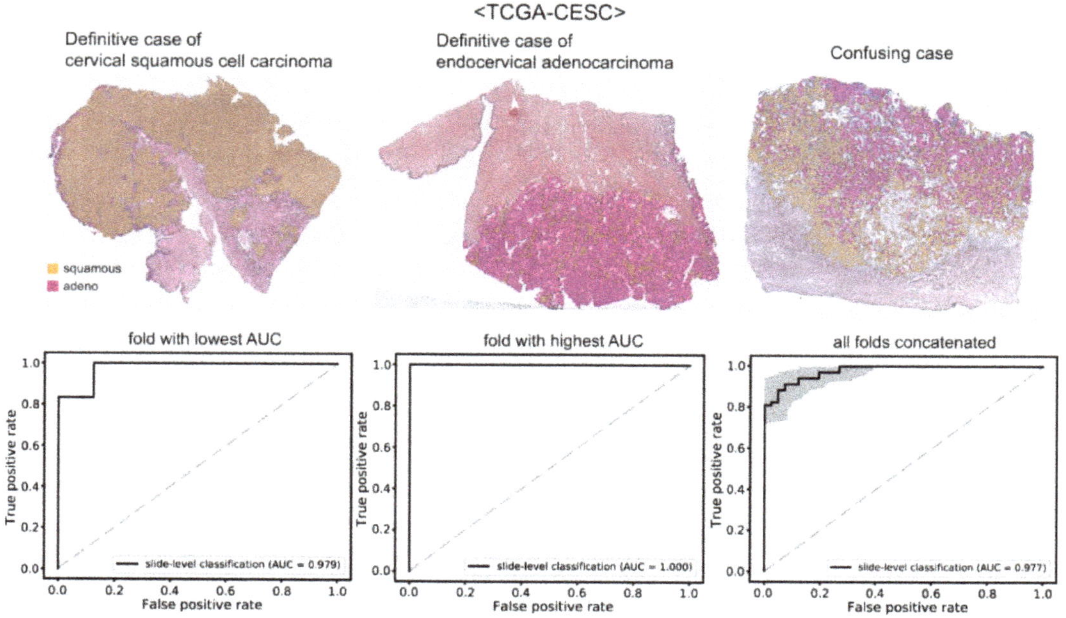

Figure 3. Classification results for cervical cancer subtypes. Upper panels: the representative whole slide images of clear cervical squamous cell carcinoma, clear endocervical adenocarcinoma, and confusing case with mixed classification results. Lower panels: the receiver operating characteristic curves of slide-level classification results for folds with the lowest and highest area under the curve (AUC) and concatenated results of all 5-folds.

3.3. Endometrial Cancer Subtypes Classification

Next, we trained other classifiers for the endometrial cancer subtypes. The patches from a WSI are labeled as either endometrioid endometrial adenocarcinoma or serous endometrial adenocarcinoma based on the information obtained also from the cBioPortal. The classification results are presented in Figure 4a. The representative WSIs of clear endometrioid endometrial adenocarcinoma, clear serous endometrial adenocarcinoma, and confusing case with mixed classification results are presented in the upper panels. The AUROCs were 0.923 and 0.982 for the folds with the lowest and highest AUROCs, respectively. The AUROC for the concatenated results was 0.944 (95% CI, 0.916–0.969).

It is of interest whether the classifiers trained on the TCGA datasets work well or not on other datasets. Therefore, we tested the classifier on the CPTAC-UCEC dataset. CPTAC-UCEC provides multiple WSIs for a patient with pure normal tissue WSIs (Figure 5a). We discarded normal WSIs and selected all WSIs with more than 30% of tumor tissue regions for the testing. The classification results are presented in Figure 4b. The AUROC was 0.826 (95% CI, 0.727–0.925), much poorer compared to the AUROC for the TCGA dataset ($p < 0.05$ between CPTAC and TCGA by Venkatraman's permutation test).

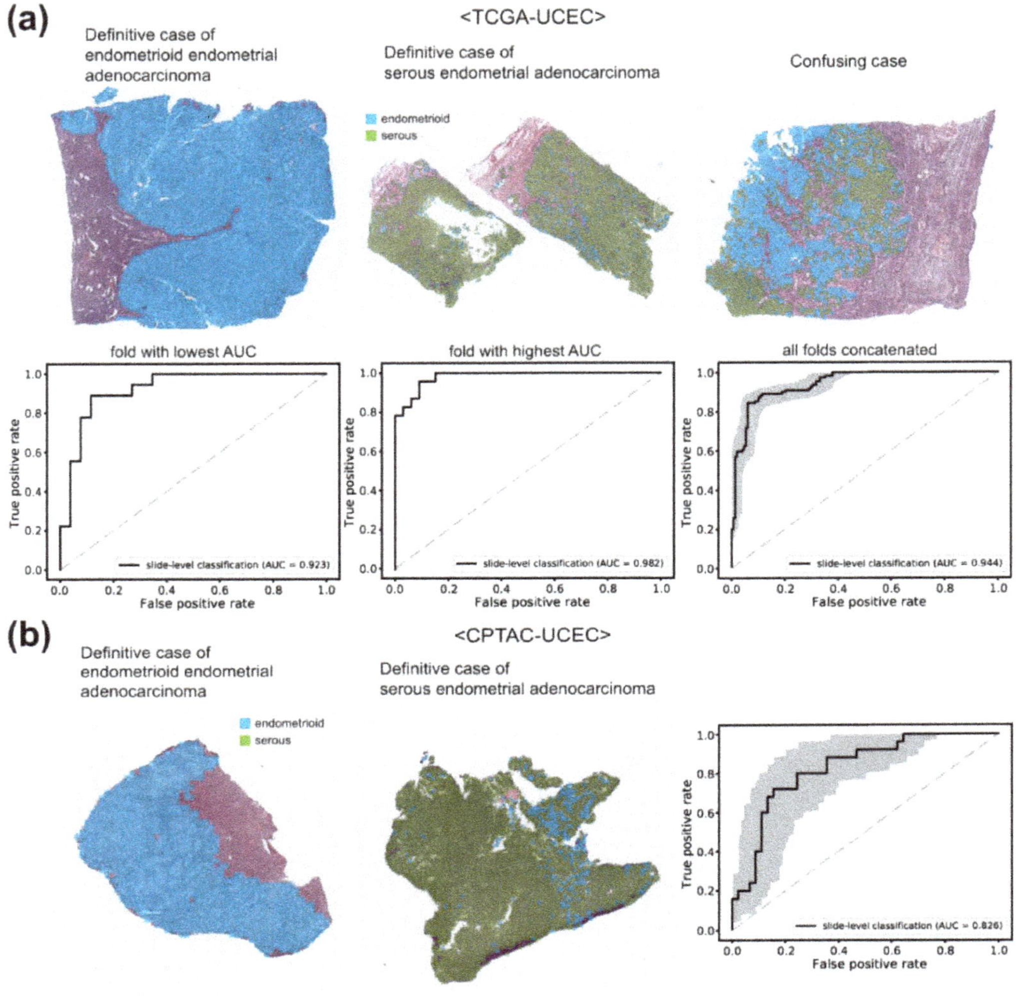

Figure 4. Classification results for endometrial cancer subtypes. (**a**) Results for the TCGA-UCEC dataset. Upper panels: the representative whole slide images (WSIs) of clear endometrioid endometrial adenocarcinoma, clear serous endometrial adenocarcinoma, and confusing case with mixed classification results. Lower panels: the receiver operating characteristic (ROC) curves of slide-level classification results for folds with the lowest and highest area under the curve (AUC) and concatenated results of all 5-folds. (**b**) The classification results of the CPTAC-UCEC dataset by the classifier trained with the TCGA-UCEC dataset. Left two representative WSIs demonstrate clear endometrioid endometrial adenocarcinoma and clear serous endometrial adenocarcinoma. The ROC curve is obtained from all CPTAC-UCEC tissues with more than 30% of tumor tissue regions.

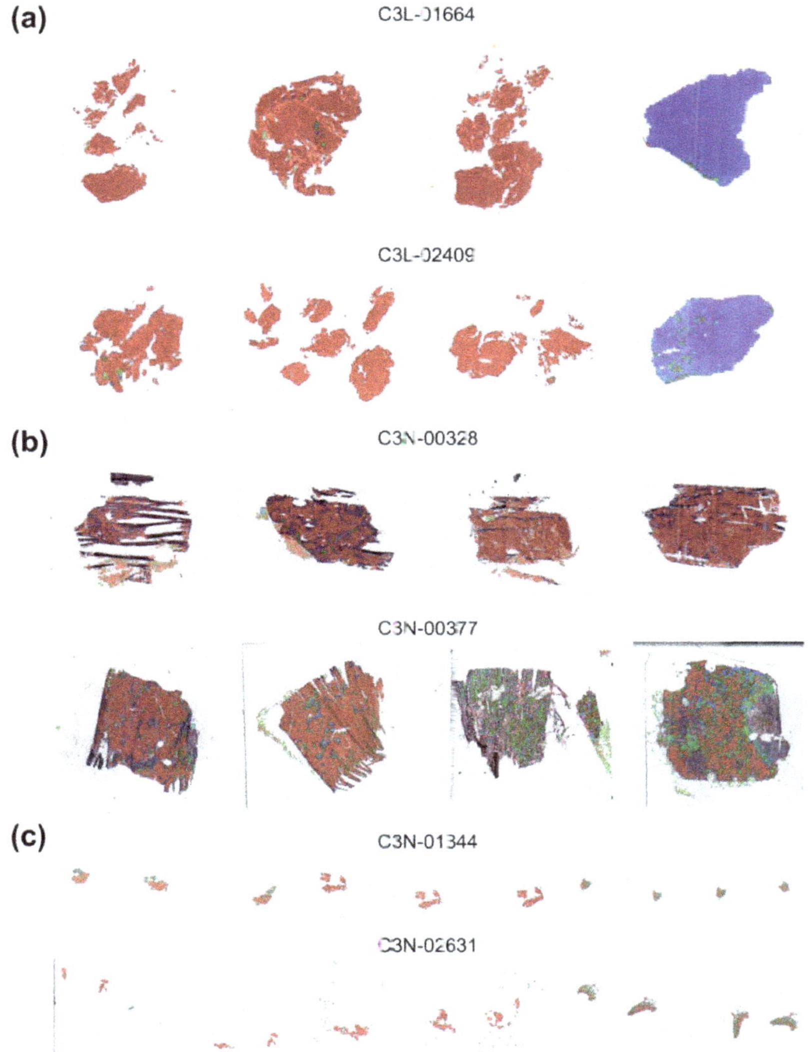

Figure 5. Characteristics of CPTAC-UCEC tissues. Examples of tissues from six patients indicated by IDs. (**a**) Patients with both pure tumor and pure normal tissue samples. (**b**) Patients with frozen tissue samples. (**c**) Patients with small curettage tissue samples.

3.4. Tumor Origin Classification

Lastly, we trained classifiers to distinguish the origin of adenocarcinomas: endocervical adenocarcinoma vs. endometrioid endometrial adenocarcinoma. The classification results are presented in Figure 6. The upper panels show the representative WSIs of clear endocervical adenocarcinoma, clear endometrioid endometrial adenocarcinoma, and confusing case with mixed classification results. The AUROCs were 0.904 and 0.987 for the folds with the lowest and highest AUROCs, respectively. The AUROC for the concatenated results was 0.939 (95% CI, 0.896–0.982).

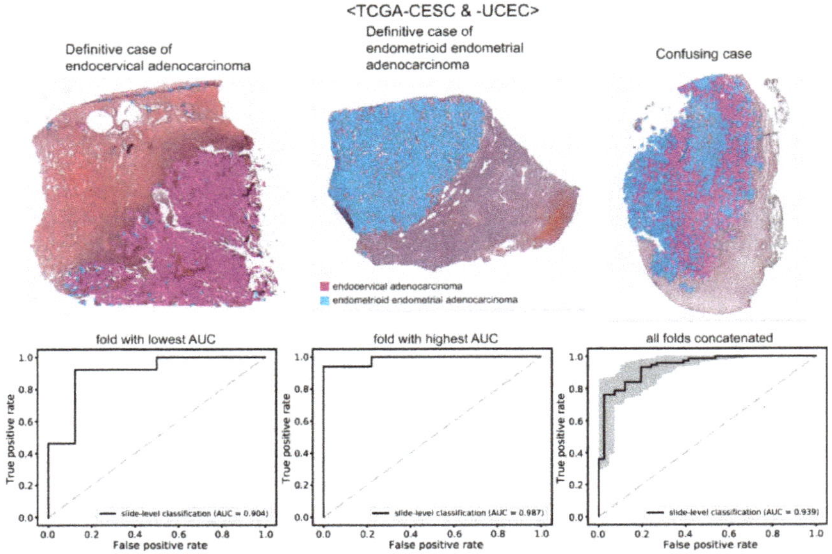

Figure 6. Classification results for the origin of adenocarcinomas. Upper panels: the representative whole slide images of clear endocervical adenocarcinoma, clear endometrioid endometrial adenocarcinoma, and confusing case with mixed classification results. Lower panels: the receiver operating characteristic curves of slide-level classification results for folds with the lowest and highest area under the curve (AUC) and concatenated results of all 5-folds.

In Table 1, accuracy, sensitivity, specificity, and F1 score of the classification results for these classifiers were presented with cutoff values yielding maximal Youden index (sensitivity + specificity − 1).

Table 1. Accuracy, sensitivity, specificity, and F1 score of the classification results. The measures were obtained with cutoff values yielding maximal Youden index (sensitivity + specificity − 1).

	Accuracy	Sensitivity	Specificity	F1 Score
TCGA-CESC cervical squamous cell carcinoma/ endocervical adenocarcinoma	0.917	0.912	0.927	0.932
TCGA-UCEC endometrioid endometrial adenocarcinoma/ serous endometrial adenocarcinoma	0.899	0.846	0.939	0.876
CPTAC-UCEC endometrioid endometrial adenocarcinoma/ serous endometrial adenocarcinoma	0.757	0.8	0.733	0.702
TCGA-CESC/UCEC endocervical adenocarcinoma/ endometrioid endometrial adenocarcinoma	0.888	0.933	0.805	0.915

4. Discussion

In the present study, we investigated the feasibility of DL-based classification for the subtypes of cervical and endometrial cancers and the site of origin of adenocarcinomas. Although the performance of the classifiers was not perfect, high AUROCs of all the classifiers revealed the potential of DL-based classification of H&E-stained tissue slides of

cervical and uterine cancers. The performance can be much enhanced when more WSI data can be collected for the training of the classifiers.

The DL-based classifiers for cervical cancer showed the best performance among the classifiers in the study. Pure adenocarcinoma and squamous cell carcinoma of the cervix can be relatively clearly separable because their morphologies have many differences [5]. However, there are also confusing cases including adenosquamous carcinoma which is defined as a tumor with both glandular and squamous components. This explains why the classifier could not accomplish perfection. In clinical practice, tissue slides with mixed classification results need more careful attention by pathologists when a DL-based assistant system for tissue slides is adopted.

Serous endometrial adenocarcinoma represents only about 10% of endometrial carcinomas. However, it is responsible for almost 40% of cancer deaths [8,18]. The distinction between endometrioid and serous endometrial adenocarcinoma is not very clear. Although serous carcinoma typically shows a predominant papillary growth pattern, which is also found in some endometrioid carcinomas. Antibodies for p53, p16, IMP2, and IMP3 can help to distinguish serous endometrial adenocarcinoma, but the markers are not definitive [19]. Therefore, there is an opportunity for DL-based classifiers to improve the diagnostic accuracy of subtypes of endometrial cancers.

One of the important issues of DL application is the generalizability of trained classifiers for external datasets. The TCGA-trained classifiers did not perform well on the CPTAC dataset in the present study. There can be various reasons for the decreased performance. First, the quality of H&E-stained tissue slides can vary between TCGA and CPTAC datasets because of the differences in tissue processing including tissue cutting, fixation, dye concentration, and staining time [20]. Furthermore, the differences in the settings of the slide scanners can also affect the color features of the WSIs. Although we normalized color, it may not be able to overcome the innate differences in the datasets. In addition, there are many other differences between TCGA and CPTAC datasets. CPTAC dataset contains not only FFPE tissues but also frozen tissue sections (Figure 5b). In our previous study, we clearly demonstrated that the classifiers trained on either frozen or FFPE tissue did not perform well on another tissue type [21]. Therefore, the classifiers trained on the TCGA-UCEC FFPE tissues cannot perform properly on the CPTAC frozen tissues. Furthermore, the CPTAC dataset also contains small tissue samples such as biopsy or small curettage specimens (Figure 5c). The dilatation and curettage may be able to deform tissue morphology. In addition, because biopsy samples have fundamental limitations in reflecting the overall contour of tumor histomorphology, the classifiers trained on resection specimens may not perform well on biopsy or small curettage tissues. Whatever the reason, the limited generalizability suggests that the TCGA dataset is not enough to train a classifier performing generally well on real-world problems. More data from various institutes should be collected to establish high generalizability. Recently, many countries started to construct large datasets of pathologic tissue slides [22,23]. Therefore, the performance and generalizability of DL-based tissue classifiers will be much enhanced with the accumulation of more training data in the near future.

The distinction of the site of origin between cervical adenocarcinomas and endometrial adenocarcinomas is important for clinical decisions especially for tumors involving both the endometrium and the endocervix or for tumors with multiple lesions [7]. The decision can be supported by immunohistochemistry for ER, p16, CEA, and vimentin or HPV in situ hybridization [5]. However, there is no decisive marker and additional methods are necessary to support the distinction. It is strongly recommended that various information including clinicopathologic, immunohistochemical, and molecular data should be integrated for proper differentiation of these cancers. We suggest that information from the DL-based classifier can also be integrated into these data for more accurate decisions.

In the present study, we applied DL to classify H&E-stained tissues of cervical and endometrial cancers. There have been other studies applying DL to assist the analysis of gynecologic tumors. Many studies tried to improve cervical cancer screening results based

on cervical cytology tests [24–26]. In these studies, DL can discriminate normal/cancer cells from conventional Pap smear or liquid-based cytology. Grades of cervical intraepithelial neoplasia can be determined by DL from either colposcopy images [27,28] or histology images [29]. DL can also analyze hysteroscopy images to discriminate different types of endometrial legions [30,31]. Normal endometrium, endometrial polyp, endometrial hyperplasia, and endometrial adenocarcinoma can be discriminated by DL from H&E-stained histopathologic slides [32]. Molecular profiles such as molecular subtypes or microsatellite instability status of endometrial cancers can be predicted by DL directly from H&E-stained WSIs [33]. These studies indicate that DL has tremendous potential to support the assessment of patients with gynecologic tumors.

However, there are also limitations of DL. First, it is almost impossible for human interpreters to understand how DL reaches to the classification results. This "black-box" nature is one of the most important hurdles for the adoption of DL in clinical practice [34]. The effort to enhance the interpretability of DL is actively ongoing [35]. Second, DL cannot perform well in inexperienced settings although the difference is not tremendous. For example, a classifier trained on FFPE tissues has limited performance on frozen tissues although the difference is not limiting to human. Therefore, separate DL models should be trained for slightly different settings. Otherwise, a huge dataset covering every variation should be used to train a widely available model.

In the present study, we demonstrated the feasibility of DL-based classifiers for the subtypes of cervical and endometrial cancers and the site of origin of adenocarcinomas. Although there is still room for improvement, our results showed that DL can capture selective features for the discrimination of cancer tissues. We believe the performance will be much enhanced with an accumulation of training data in the near future. The classification results of DL can be integrated with other clinical information for a more precise analysis of cervical and endometrial cancers.

Supplementary Materials: The following supporting information can be downloaded at: https://www.mdpi.com/article/10.3390/diagnostics12112623/s1, Figure S1: Schematic view of the network structure for the classification of tissue subtypes; Table S1: Numbers of tissue image patches for the training of the classifiers.

Author Contributions: Study design, J.S. and H.-J.J.; methodology, S.H.L. and H.-J.J.; software, H.-J.J.; validation, S.I. and S.H.L.; formal analysis, J.S., S.I., S.H.L. and H.-J.J.; data curation, J.S., S.H.L. and H.-J.J.; writing—original draft preparation, J.S., S.I. and H.-J.J.; writing—review and editing, S.H.L. and H.-J.J.; visualization, H.-J.J.; supervision, S.H.L. and H.-J.J.; funding acquisition, S.H.L. and H.-J.J. All authors have read and agreed to the published version of the manuscript.

Funding: This research was funded by a grant from the National Research Foundation of Korea (NRF-2021R1A4A5028966) and a grant from the Korea Health Technology R&D Project through the Korea Health Industry Development Institute (KHIDI), funded by the Ministry of Health & Welfare, Republic of Korea (grant number: HI21C0940).

Institutional Review Board Statement: The study was conducted in accordance with the Declaration of Helsinki, and approved by the Institutional Review Board of the College of Medicine at The Catholic University of Korea (XC21ENDI0031K, approved on 1 July 2021).

Informed Consent Statement: Informed consent of patients in the TCGA cohorts was acquired by the TCGA consortium.

Data Availability Statement: The TCGA data presented in this study are openly available in the GDC data portal (https://portal.gdc.cancer.gov/ (accessed on 11 April 2022)). The CPTAC data presented in this study are openly available in The Cancer Imaging Archive website (http://www.cancerimagingarchive.net/ (accessed on 29 April 2022)). Further information is available from the corresponding authors upon request.

Conflicts of Interest: The authors declare no conflict of interest. The funders had no role in the design of the study; in the collection, analyses, or interpretation of data; in the writing of the manuscript; or in the decision to publish the results.

References

1. Bray, F.; Ferlay, J.; Soerjomataram, I.; Siegel, R.L.; Torre, L.A.; Jemal, A. Global cancer statistics 2018: GLOBOCAN estimates of incidence and mortality worldwide for 36 cancers in 185 countries. *CA Cancer J. Clin* **2018**, *68*, 394–424. [CrossRef] [PubMed]
2. Feinberg, J.; Albright, B.; Black, J.; Lu, L.; Passarelli, R.; Gysler, S.; Whicker, M.; Altwerger, G.; Menderes, G.; Hui, P.; et al. Ten-Year Comparison Study of Type 1 and 2 Endometrial Cancers: Risk Factors and Outcomes. *Gynecol. Obstet. Investig.* **2019**, *84*, 290–297. [CrossRef] [PubMed]
3. Hu, K.; Wang, W.; Liu, X.; Meng, Q.; Zhang, F. Comparison of treatment outcomes between squamous cell carcinoma and adenocarcinoma of cervix after definitive radiotherapy or concurrent chemoradiotherapy. *Radiat. Oncol.* **2018**, *13*, 249. [CrossRef] [PubMed]
4. Takeuchi, S. Biology and treatment of cervical adenocarcinoma. *Chin. J. Cancer Res.* **2016**, *28*, 254–262. [CrossRef] [PubMed]
5. Lax, S. Histopathology of cervical precursor lesions and cancer. *Acta Dermatovenerol Alpina Pannonica Adriat.* **2011**, *20*, 125–133.
6. Zhou, J.; Zhang, W.W.; Wu, S.G.; He, Z.Y.; Sun, J.Y.; Yang, G.F.; Li, F.Y. The prognostic value of histologic subtype in node-positive early-stage cervical cancer after hysterectomy and adjuvant radiotherapy. *Int. J. Surg.* **2017**, *44*, 1–6. [CrossRef]
7. Stewart, C.J.R.; Crum, C.P.; McCluggage, W.G.; Park, K.J.; Rutgers, J.K.; Oliva, E.; Malpica, A.; Parkash, V.; Matias-Guiu, X.; Ronnett, B.M. Guidelines to Aid in the Distinction of Endometrial and Endocervical Carcinomas, and the Distinction of Independent Primary Carcinomas of the Endometrium and Adnexa From Metastatic Spread Between These and Other Sites. *Int. J. Gynecol. Pathol.* **2019**, *38* (Suppl. 1), S75–S92. [CrossRef]
8. Gatius, S.; Matias-Guiu, X. Practical issues in the diagnosis of serous carcinoma of the endometrium. *Mod. Pathol.* **2016**, *29* (Suppl. 1), S45–S58. [CrossRef]
9. Evans, A.J.; Bauer, T.W.; Bui, M.M.; Cornish, T.C.; Duncan, H.; Glassy, E.F.; Hipp, J.; McGee, R.S.; Murphy, D.; Myers, C.; et al. US Food and Drug Administration Approval of Whole Slide Imaging for Primary Diagnosis: A Key Milestone Is Reached and New Questions Are Raised. *Arch. Pathol. Lab. Med.* **2018**, *142*, 1383–1387. [CrossRef]
10. Cifci, D.; Foersch, S.; Kather, J.N. Artificial intelligence to identify genetic alterations in conventional histopathology. *J. Pathol.* **2022**, *257*, 430–444. [CrossRef]
11. LeCun, Y.; Bengio, Y.; Hinton, G. Deep learning. *Nature* **2015**, *521*, 436–444. [CrossRef]
12. Klein, C.; Zeng, Q.; Arbaretaz, F.; Devevre, E.; Calderaro, J.; Lomenie, N.; Maiuri, M.C. Artificial intelligence for solid tumour diagnosis in digital pathology. *Br. J. Pharmacol.* **2021**, *178*, 4291–4315. [CrossRef]
13. Lee, S.H.; Jang, H.J. Deep learning-based prediction of molecular cancer biomarkers from tissue slides: A new tool for precision oncology. *Clin. Mol. Hepatol.* **2022**, *28*, 754–772. [CrossRef]
14. Jang, H.J.; Song, I.H.; Lee, S.H. Deep Learning for Automatic Subclassification of Gastric Carcinoma Using Whole-Slide Histopathology Images. *Cancers* **2021**, *13*, 3811. [CrossRef]
15. Cho, K.O.; Lee, S.H.; Jang, H.J. Feasibility of fully automated classification of whole slide images based on deep learning. *Korean J. Physiol. Pharmacol.* **2020**, *24*, 89–99. [CrossRef]
16. Venkatraman, E.S. A Permutation Test to Compare Receiver Operating Characteristic Curves. *Biometrics* **2000**, *56*, 1134–1138 [CrossRef]
17. Yu, K.H.; Zhang, C.; Berry, G.J.; Altman, R.B.; Re, C.; Rubin, D.L.; Snyder, M. Predicting non-small cell lung cancer prognosis by fully automated microscopic pathology image features. *Nat. Commun.* **2016**, *7*, 12474. [CrossRef]
18. Kim, H.J.; Kim, T.J.; Lee, Y.Y.; Choi, C.H.; Lee, J.W.; Bae, D.S.; Kim, B.G. A comparison of uterine papillary serous, clear cell carcinomas, and grade 3 endometrioid corpus cancers using 2009 FIGO staging system. *J. Gynecol. Oncol.* **2013**, *24*, 120–127. [CrossRef]
19. Zaidi, A.; Gupta, P.; Gupta, N.; Rajwanshi, A.; Rai, B.; Gainder, S. Role of Immunohistochemistry to Distinguish Grade 3 Endometrioid Carcinoma and Uterine Serous Carcinoma. *Appl. Immunohistochem. Mol. Morphol.* **2020**, *28*, 42–48. [CrossRef]
20. Nam, S.; Chong, Y.; Jung, C.K.; Kwak, T.Y.; Lee, J.Y.; Park, J.; Rho, M.J.; Go, H. Introduction to digital pathology and computer-aided pathology. *J. Pathol. Transl. Med.* **2020**, *54*, 125–134. [CrossRef]
21. Jang, H.J.; Song, I.H.; Lee, S.H. Generalizability of Deep Learning System for the Pathologic Diagnosis of Various Cancers. *Appl. Sci.* **2021**, *11*, 808. [CrossRef]
22. Kang, Y.; Kim, Y.J.; Park, S.; Ro, G.; Hong, C.; Jang, H.; Cho, S.; Hong, W.J.; Kang, D.U.; Chun, J.; et al. Development and operation of a digital platform for sharing pathology image data. *BMC Med. Informatics Decis. Mak.* **2021**, *21*, 114. [CrossRef] [PubMed]
23. Moulin, P.; Grunberg, K.; Barale-Thomas, E.; der Laak, J.V. IMI—Bigpicture: A Central Repository for Digital Pathology. *Toxicol. Pathol.* **2021**, *49*, 711–713. [CrossRef] [PubMed]
24. Zhang, L.; Le, L.; Nogues, I.; Summers, R.M.; Liu, S.; Yao, J. DeepPap: Deep Convolutional Networks for Cervical Cell Classification. *IEEE J. Biomed. Health Inform.* **2017**, *21*, 1633–1643. [CrossRef] [PubMed]
25. Kanavati, F.; Hirose, N.; Ishii, T.; Fukuda, A.; Ichihara, S.; Tsuneki, M. A Deep Learning Model for Cervical Cancer Screening on Liquid-Based Cytology Specimens in Whole Slide Images. *Cancers* **2022**, *14*, 1159. [CrossRef]
26. Cheng, S.; Liu, S.; Yu, J.; Rao, G.; Xiao, Y.; Han, W.; Zhu, W.; Lv, X.; Li, N.; Cai, J.; et al. Robust whole slide image analysis for cervical cancer screening using deep learning. *Nat. Commun.* **2021**, *12*, 5639. [CrossRef]
27. Cho, B.J.; Choi, Y.J.; Lee, M.J.; Kim, J.H.; Son, G.H.; Park, S.H.; Kim, H.B.; Joo, Y.J.; Cho, H.Y.; Kyung, M.S.; et al. Classification of cervical neoplasms on colposcopic photography using deep learning. *Sci. Rep.* **2020**, *10*, 13652. [CrossRef]

28. Miyagi, Y.; Takehara, K.; Nagayasu, Y.; Miyake, T. Application of deep learning to the classification of uterine cervical squamous epithelial lesion from colposcopy images combined with HPV types. *Oncol. Lett.* **2020**, *19*, 1602–1610. [CrossRef]
29. Cho, B.J.; Kim, J.W.; Park, J.; Kwon, G.Y.; Hong, M.; Jang, S.H.; Bang, H.; Kim, G.; Park, S.T. Automated Diagnosis of Cervical Intraepithelial Neoplasia in Histology Images via Deep Learning. *Diagnostics* **2022**, *12*, 548. [CrossRef]
30. Takahashi, Y.; Sone, K.; Noda, K.; Yoshida, K.; Toyohara, Y.; Kato, K.; Inoue, F.; Kukita, A.; Taguchi, A.; Nishida, H.; et al. Automated system for diagnosing endometrial cancer by adopting deep-learning technology in hysteroscopy. *PLoS ONE* **2021**, *16*, e0248526. [CrossRef]
31. Zhang, Y.; Wang, Z.; Zhang, J.; Wang, C.; Wang, Y.; Chen, H.; Shan, L.; Huo, J.; Gu, J.; Ma, X. Deep learning model for classifying endometrial lesions. *J. Transl. Med.* **2021**, *19*, 10. [CrossRef]
32. Sun, H.; Zeng, X.; Xu, T.; Peng, G.; Ma, Y. Computer-Aided Diagnosis in Histopathological Images of the Endometrium Using a Convolutional Neural Network and Attention Mechanisms. *IEEE J. Biomed. Health Inform.* **2020**, *24*, 1664–1676. [CrossRef]
33. Hong, R.; Liu, W.; DeLair, D.; Razavian, N.; Fenyö, D. Predicting endometrial cancer subtypes and molecular features from histopathology images using multi-resolution deep learning models. *Cell Rep. Med.* **2021**, *2*, 100400. [CrossRef]
34. Bera, K.; Schalper, K.A.; Rimm, D.L.; Velcheti, V.; Madabhushi, A. Artificial intelligence in digital pathology—New tools for diagnosis and precision oncology. *Nat. Rev. Clin. Oncol.* **2019**, *16*, 703–715. [CrossRef]
35. Rudin, C. Stop explaining black box machine learning models for high stakes decisions and use interpretable models instead. *Nat. Mach. Intell.* **2019**, *1*, 206–215. [CrossRef]

cancers

MDPI

Review

Artificial Intelligence-Driven Diagnosis of Pancreatic Cancer

Bahrudeen Shahul Hameed [1,2] and Uma Maheswari Krishnan [1,2,3,*]

1 Centre for Nanotechnology & Advanced Biomaterials (CeNTAB), Shanmugha Arts, Science, Technology and Research Academy, Deemed University, Thanjavur 613401, India
2 School of Chemical & Biotechnology (SCBT), Shanmugha Arts, Science, Technology and Research Academy, Deemed University, Thanjavur 613401, India
3 School of Arts, Sciences, Humanities & Education (SASHE), Shanmugha Arts, Science, Technology and Research Academy, Deemed University, Thanjavur 613401, India
* Correspondence: umakrishnan@sastra.edu; Tel.: +91-4362264101 (ext. 2677); Fax: +91-436226412

Simple Summary: Pancreatic cancer poses a grave threat to mankind, due to its poor prognosis and aggressive nature. An accurate diagnosis is critical for implementing a successful treatment plan given the risk of exacerbation. The diagnosis of pancreatic cancer relies on medical imaging, which provides inaccurate information about the prognosis of the patient and makes it difficult for clinicians to select the optimal treatment. Data derived from medical imaging has been integrated with artificial intelligence, an emerging technology, to facilitate clinical decision making. This review explores the implementation of artificial intelligence for various imaging modalities to obtain a precise cancer diagnosis.

Abstract: Pancreatic cancer is among the most challenging forms of cancer to treat, owing to its late diagnosis and aggressive nature that reduces the survival rate drastically. Pancreatic cancer diagnosis has been primarily based on imaging, but the current state-of-the-art imaging provides a poor prognosis, thus limiting clinicians treatment options. The advancement of a cancer diagnosis has been enhanced through the integration of artificial intelligence and imaging modalities to make better clinical decisions. In this review, we examine how AI models can improve the diagnosis of pancreatic cancer using different imaging modalities along with a discussion on the emerging trends in an AI-driven diagnosis, based on cytopathology and serological markers. Ethical concerns regarding the use of these tools have also been discussed.

Keywords: pancreatic cancer; artificial intelligence; deep learning; cancer imaging; risk prediction

Citation: Hameed, B.S.; Krishnan, U.M. Artificial Intelligence-Driven Diagnosis of Pancreatic Cancer. *Cancers* **2022**, *14*, 5382. https://doi.org/10.3390/cancers14215382

Academic Editors: Hamid Khayyam, Ali Madani, Rahele Kafieh and Ali Hekmatnia

Received: 5 October 2022
Accepted: 28 October 2022
Published: 31 October 2022

1. Introduction

Pancreatic cancer (PC) is among the most fatal and invasive tumors of the digestive system [1]. It has been referred to as the king of cancer, due to its aggressiveness, invasiveness and rapid metastasis, poor survival, and poor prognosis [2]. Recent decades have witnessed a surge in the incidence of pancreatic cancer across the globe that has been largely linked to ageing, alcohol consumption, smoking, sedentary lifestyle, obesity, chronic pancreatitis, diabetes, hereditary factors, long-term exposure to air and water pollutants, unhealthy lifestyle, and diet [1,3,4]. Surgery has been the main therapeutic intervention for these patients. However, several factors, including the absence of specific clinical manifestations and molecular markers, have resulted in the detection of the disease only at advanced stages, thereby making surgical options ineffective. Therefore, an early diagnosis and accurate stratification of pancreatic cancer stages are important for improved therapeutic outcomes. Pancreatic cancer diagnosis is challenging because the pancreas is a deep-seated retro-peritoneal organ with complex surrounding structures. The highly vascularized environment surrounding the pancreas facilitates rapid metastasis of the cancer that makes pancreatic cancer highly aggressive. The common symptoms of pancreatic

cancer include abdominal pain, changes in the consistency of faeces, nausea, bloated body, co-morbidities, such as diabetes and jaundice, abnormal liver function parameters, loss of weight, etc. [5]. These symptoms usually become prominent only during the advanced stage of the disease and are often missed during the early stages. Further, serological markers for pancreatic cancer, such as CA-19-9 (Carbohydrate antigen), are not highly specific and indicate only the advanced stage of the disease, thereby increasing the mortality risk of the affected individual. Several imaging tools, including magnetic resonance imaging (MRI), computed tomography (CT), endoscopic ultrasound (EUS), etc., have also been explored for the diagnosis of pancreatic cancer. Due to rapid advances in recent years, imaging technology has emerged in the forefront for the diagnosis, staging, and prognosis of pancreatic cancer [6]. However, distinction of a cancerous lesion from other pancreatic disorders, such as pancreatitis, a chronic inflammation of the pancreas, remains a major roadblock in the accurate and early diagnosis of pancreatic cancer. Despite the existence of advanced imaging equipment, confirmation of pancreatic cancer is confirmed through biopsy after imaging. Not only is this time-consuming, but it also increases the probability of mortality in the affected individual, due to the inordinate delay. A study had reported that nearly 90% of the misdiagnosis of pancreatic cancer was due to the inability to identify the vascular invasion and the difficulty in spotting the underlying tumour mass, due to the inflammation [7]. Table 1 lists some of the common imaging techniques used for the clinical diagnosis of pancreatic cancer, along with their merits and limitations.

Table 1. Major imaging techniques employed for the diagnosis of pancreatic cancer and their limitations.

Technique	Merit(s)	Demerit(s)
Multidetector computed tomography (MDCT)	• High sensitivity and specificity for the detection of the vascular invasion; • Short acquisition time; • 3D image processing aids in the staging of the cancer; • Obtaining thin collimation images with a high spatial and temporal resolution.	• Nephrotoxicity; • Tissue/organ damage due to the radiation exposure; • Lack of an attenuation gradient between the cancer tissue and pancreatic parenchyma, leading to erroneous predictions.
Magnetic resonance imaging (MRI)	• Low risk of ionizing radiation; • Better sensitivity, specificity, and accuracy when compared to CT techniques; • Non-invasive imaging of the pancreato-biliary system by magnetic resonance cholangio-pancreatography (MRCP).	• Expensive; • Limited availability; • Problems associated with individuals having metal implants.
Endoscopic ultrasound (EUS) with or without fine needle aspiration (FNA)	• Can detect small cancerous lesions 2–5 mm in dimension; • Highest diagnostic accuracy; • Highly specific; • Loco-regional staging can be detected.	• Cannot detect extra-abdominal metastasis; • Limited availability; • Requires a trained operator.
Positron emission tomography (PET)	• Useful in detecting metastasis.	• Staging of pancreatic cancer cannot be conclusively determined; • Expensive; • Exposure to radiation.

Several approaches to improve the sensitivity and prediction accuracy of these imaging techniques have been reported in the literature. These include the use of image contrast agents to improve the resolution and sensitivity and the use of image processing software for a better diagnostic accuracy. In recent years, the emergence of artificial intelligence and deep learning has transformed the landscape of an image-driven diagnosis of pancreatic cancer with a dramatic improvement in the prediction accuracy. The vari-

ous attempts to integrate artificial intelligence for the diagnosis of pancreatic cancer are discussed in the following sections.

2. Artificial Intelligence for Diagnostic Applications

The advances in computer technology, witnessed in the recent decades, coupled with the development of effective image processing strategies, have ushered in a new era of digital medicine. As a result, clinical personnel can avoid the laborious medical image analysis performed manually, thus saving time as well as overcome errors in diagnosis arising, due to the differences in expertise and clinical exposure [8]. The 21st century has witnessed the widespread use of artificial intelligence (AI) that employs computer programs to perform tasks associated with human intelligence, such as learning and problem-solving. The phrase artificial intelligence was first coined by John McCarthy in the mid-1950s, and has since evolved from a set of if-then commands to more complex algorithms that mimic the human brain in some aspects [9,10]. The application of AI tools has resulted in the emergence of a new field of clinical diagnosis, namely, precision oncology that uses a large volume of data from genomics, proteomics, and metabolomics [11]. AI-based cancer diagnosis is mainly driven by machine learning (ML) and deep learning (DL) techniques. Machine learning uses computational methods to analyse large volumes of data and identify patterns for prediction [12]. ML can be supervised where it uses data from previous trials/measurements for the identification of patterns or trends for making predictions. Thus, for a pancreatic cancer diagnosis, CT or PET scans, ultrasonographs, and MRI data can be used to train the system to identify abnormalities that can be classified as pancreatic cancer. The prediction accuracy can be better if large numbers of dataset are used for the training. Different mathematical models and algorithms can be iteratively used during the training period to identify the most efficient model, the accuracy of which can be validated using a testing dataset. The advantage of such supervised ML models is that they can extract meaningful features and identify patterns or subtle changes that could be missed by human personnel, due to oversight or exhaustion. Hence, the prediction accuracy of ML for a cancer diagnosis is higher. ML can also be unsupervised where it can discern patterns and trends from unclassified data. However, the accuracy of the prediction is slightly compromised when compared to the supervised models [13]. The 3D reconstruction of images has also been realized by the ML models for a superior diagnostic accuracy [6].

Another type of ML that is yet to be applied for cancer diagnosis, is reinforcement learning where the algorithm uses the data to understand and respond to the environment predominantly by a trial-and-error process [14]. In other words, reinforcement learning is an advanced concept that could also facilitate decision-making, in addition to prediction [15]. Thus, apart from a diagnosis of pancreatic cancer, reinforcement learning could be used to alert clinicians in remote locations or trigger actuators for releasing a therapeutic agent. These concepts, though attractive, are yet to be realized, but could very well represent the diagnostic technology of the future. Deep learning is another sub-type of AI that uses large data sets and complex algorithms that mimic the human brain to enable prediction, forecasting, and decision-making [16,17]. Most of the DL is supervised and uses data for training for the decision-making process, unlike reinforcement learning that is a dynamic process which relies on a trial-and-error method for the same. Both DL and reinforcement learning are advanced concepts that require a longer duration for training and testing [18]. DL employs convolutional neural networks (CNNs) and artificial neural networks (ANNs) extensively for decision-making [19].

A plethora of supervised and unsupervised ML and DL models continue to be developed and explored for improving the accuracy of a pancreatic cancer diagnosis at the early stage which could be invaluable in enhancing the survival of the affected individual [20]. The complexity of the algorithms will reflect the type of functions they can perform ranging from feature extraction, simple clustering or segregation of data, classification of data, prediction, forecasting, and decision-making [21]. Algorithms such as Naive–Bayes, support vector machine, linear regression analysis, ensemble methods, decision tree, K-mode,

hidden Markov model, hierarchical, Gaussian mixture, and neural networks have all been explored with different imaging data sets for distinguishing cancerous tissue from non-cancerous tissues [22]. The work flow in the detection of cancer using ML is depicted in Figure 1.

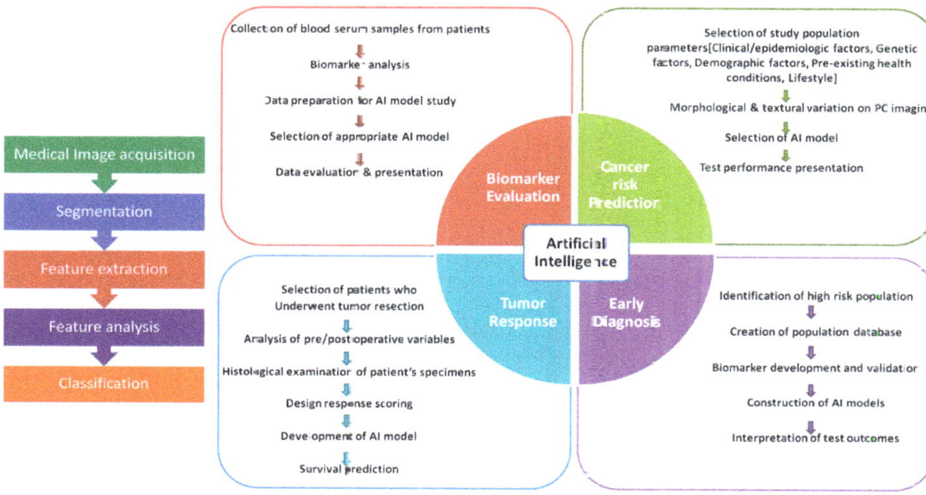

Figure 1. Work-flow of the stages during the training of the ML models for the diagnosis of cancer lesions.

The classification of images for diagnosis using various AI models can be broadly divided into one-stage and two-stage methods. The one-stage method segments the medical image into grids and applies the model for classification while the two-stage method demarcates several candidate zones that are used for classification during the training. Though time-consuming, the two-stage object method identifies and screens regions of interest resulting in more accurate predictions. Region-based convolution network (R-CNN), Fast R-CNN, and Faster R-CNN have been employed in the two-stage method as an integrated network for discriminative feature extraction, segmentation, and classification for an improved cancer detection without compromising the spatial structures [6].

3. AI Models for the Diagnosis of Pancreatic Cancer

Medical imaging has been widely used for locating and diagnosing cancerous tissue in the gastrointestinal tract. Current analysis is largely dependent upon the expertise and experience of the clinician. The quality of the images also influences the diagnosis through conventional methods [23]. The field of digital pathology continues to evolve from the first generation of image processing that involved the use of image processing tools to analyse a single slide, to much more advanced second-generation tools that could scan, analyse, and store records of whole tissue samples. The current paradigm in digital pathology involves the use of AI-based algorithms to analyse images, diagnose the condition with a high accuracy, and even predict the possibility of developing the disease even before the onset of the disease [24]. The development of AI-based tools has enabled the rapid and high precision diagnosis of cancer using different medical images [25] In the context of pancreatic cancer, AI-based diagnostic tools have been employed for risk prediction, survival prediction, and the distinction of cancer masses from other pancreatic lesions as well as for the evaluation of the response post-therapy.

Machine learning tools, such as the K-nearest neighbour (k-NN), ANN, and SVM, have been extensively investigated for their ability to extract unique signatures from med-

ical images that could be used for the identification of abnormalities [26] in different types of digestive system cancers that also includes pancreatic cancer [27]. The k-NN algorithm, first introduced in 1967 by Cover and Hart, calculates and predicts the distance between the values of the specified features in the sample data and training data. Based on the calculated distance, the sample data is grouped with its nearest neighbour class [28]. The k-NN concept was employed by Kilicet al. [29] to identify colonic polyps using region covariance in CT-colonography images as the distinguishing features. In another report employing k-NN [30], the gray level co-occurrence matrix was employed as the classifying feature in medical images of the brain and pancreatic cancers. However, k-NN is limited by issues pertaining to local structure sensitivity and the possibility of over-fitting, leading to errors.

Artificial Neural Networks (ANNs), the concept of which was first proposed in the early 1940s by McCulloch and Pitts, attempt to mimic the human neuronal network. The input layer receives the input signal that is then passed on to each of the inner hidden layers that understands and transforms them and passes it on to the next layers, until it reaches the final output layer [31], as shown in Figure 2. Unlike k-NN models that can only handle limited data, the ANN model is adaptive and can be trained using large volumes of data to become more robust and accurate. The progress in ANNs has been accelerated, due to advances in big data, affordable graphics processing units (GPUs) and the development of novel algorithms [32]. The ANN method used in diagnosing digestive cancers is the back-propagating (BP) network that was first introduced in 1986 by Rumelhart [33]. This strategy enables the error correction as the output is sent back to the inner layers if found erroneous, to refine the output parameters during the training period. This iterative process ensures the minimization of errors and the improved accuracy. In the context of a pancreatic cancer diagnosis, Săftoiu et al. [34] successfully employed ANNs to differentiate chronic pancreatitis and pancreatic adenocarcinoma, using endoscopic ultrasound images with a sensitivity of 94%. The ANN method has advantages of being able to handle large data sets and predict all types of interactions and inter-relationships between dependent and independent variables [35]. However, ANN algorithms are slow when large numbers of inputs are provided during the training period and require a large computational load, apart from adopting a black-box approach that makes it challenging for achieving accuracy in multi-layer networks [36].

To overcome some of the limitations of ANNs, Vapnik et al. [37] developed a supervised learning algorithm, in 1995, known as the support vector machine (SVM) algorithm, that defines the boundaries known as support vectors to construct a hyperplane, which is used to classify data [38]. The negative and positive boundaries and the maximum margin are defined, based upon the training set of data fed as inputs. The SVM is capable of pattern recognition and regression analysis in addition to the classification of data [39]. Zhang et al. [40] had effectively applied the SVM to identify pancreatic cancers from EUS images, by classifying textural features to achieve a detection accuracy of 99.07%. Though SVM models display a high accuracy and can work with remarkable efficiency when there is a clear demarcation of the data classes, its efficiency reduces when the size of the data set increases or when there is extensive overlap of the data. In addition, despite being memory efficient, SVM algorithms are slow, both during the training, as well as the testing phases.

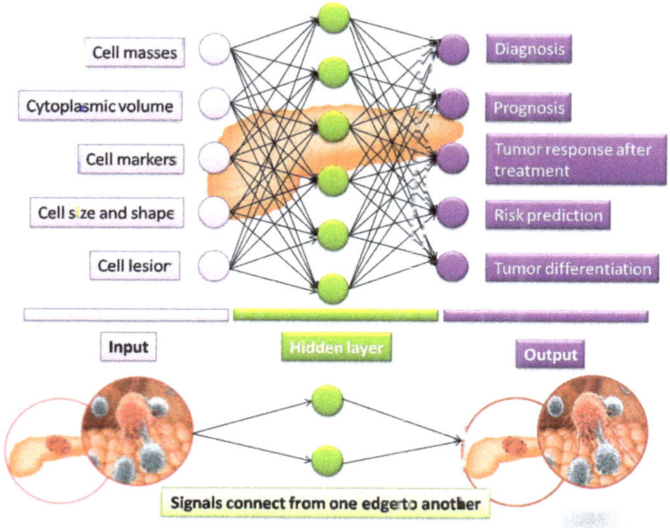

Figure 2. Schematic representation of the process flow in a sample ANN model for the diagnosis of pancreatic cancer.

Deep learning networks exhibit superior diagnostic abilities when compared to ML models as they could extract all features rather than selected ones from the medical images, as in the case of ML. As a result, DL models are preferred for the detection of digestive cancers and image segmentation [41]. Convolutional neural networks (CNNs) are among the most extensively employed supervised DL techniques. These consist of input layers where different clusters of nodes, each for a specific feature, interact with the hidden layers that have the same weightage and bias and perform convolutional operations on these inputs. These are then pooled and transformed to give the final output [42]. A typical CNN network comprises the input, convolutional, activating, pooling, fully connected, and output layers [43]. CNNs are computationally efficient but consume lots of computational power and are slow. CNNs provide a probabilistic depiction of the complete image that can be preferably employed for the image classification, rather than the segmentation [44]. Among the various types of CNNs, U-Net algorithms that use fewer convolutional layers have also been commonly employed for the diagnosis of digestive cancers, including pancreatic cancer, by classifying and segmenting specific features in the medical images [45]. The LeNet, proposed by Lecun et al. [46] in 1989, is considered the basic structure of CNNs. Several other variants, such as AlexNet, VGGNet (visual geometry group), Inception Net, and ResNet, have been introduced, between 2012 and 2015, that vary in the number of convolutional and pooling layers employed [47]. In the context of digestive cancers, Sharma et al. [48] classified and detected necrosis in medical images of gastric carcinoma using the AlexNet architecture with a classification accuracy of 69.9% and a detection accuracy of 81%. Colonic polyps were automatically detected by Shin et al. [49] from colonoscopy images using the Inception-Resnet network. Long et al. [50] proposed a fully convolutional network (FCN) model, in 2015, for the semantic segmentation where each pixel is classified as an image. As the final fully connected layer is substituted by a convolutional layer in the FCN, resulting in the superior segmentation effects, it has been extensively studied for the diagnosis of digestive cancers. Oda et al. [51] employed a three-dimensional FCN model to segment the pancreas automatically using CT images and an average Dice score

of 89.7 ± 3.8, was obtained. The Dice score indicates the precision of the segmentation model employed by eliminating false positives and is computed as follows:

$$Dicescore = 2 \times \frac{areaofoverlapbetweentwoimagesets}{totalnumberofpixelsinbothimages} \quad (1)$$

Generally, a Dice score above 88% is considered highly precise. In another study, Guo et al. [52] employed a Gaussian mixture model and used morphological operations on a three-dimensional U-Net segmentation technique, to achieve an improved segmentation accuracy with a Dice score of 83.2 ± 7.8%. It is also evident from the various reports, that the type of AI tool employed will be different for various imaging techniques. The following sections highlight some recent AI-based strategies for different imaging modalities.

4. Endoscopic Ultrasound (EUS)

Endoscopic ultrasound (EUS) employs high-frequency ultrasound (US) for the visualization of the size and location of the primary tumor in the pancreas. The ultrasound probe can be maneuvered close to the pancreas for acquiring images of the entire pancreas or the specific locations of suspicious masses or lesions [53]. Advances in the transducer design and the advent of colour Doppler techniques, have contributed to an improved diagnosis and staging of pancreatic cancer. Currently, the sensitivity of EUS, for identifying cancerous lesions in the pancreas, lie in the range 85–99%, that is comparatively superior to CT techniques. Specifically, EUS can detect small lesions in the range of 2–3 mm [54]. For instance, the accuracy of diagnosis for pancreatic tumors with a diameter of 3 cm was reported to be 93% for EUS images, which was significantly superior to CT (53%) and MRI (67%) techniques [55]. Though several literature reports have highlighted the effectiveness of EUS over other medical imaging techniques for the diagnosis of pancreatic cancer and its staging, the resectability has been found to be better predicted only using a combination of CT and EUS images [56,57]. The EUS-driven fine needle aspiration (EUS-FNA) technique has enabled tissue sampling and the evaluation of the primary tumour site, as well as the neighbouring lymph nodes with nearly 100% specificity, that otherwise pose a challenge for detection, using other imaging modalities [58]. The EUS-FNA combination achieved diagnostic accuracies of up to 85%, that are a significant improvement over the 50% accuracy obtained using a CT-assisted diagnosis [59]. However, the EUS-FNA combination is not available in many healthcare institutions. Additionally, the combination requires experienced operators for the precise insertion of the needle that has a major bearing on the diagnostic outcomes [60].

One of the major challenges for clinicians is to distinguish cancerous lesions in the presence of chronic pancreatitis (CP), as the neoplastic features are masked by the inflammation [61]. Norton et al. in 2001 [62], employed neural network models to analyse EUS images for differentiating pancreatic ductal adenocarcinoma (PDAC) and CP, using four different image parameters. Though a high sensitivity was achieved, this strategy resulted in a poor specificity of only 50%. In another attempt, Zhu et al. [63] employed a support vector machine model to extract features from EUS images recorded for 262 individuals affected with pancreatic cancer and 126 individuals with CP. The model extracted 105 distinctive features out of which 16 were selected to differentiate pancreatic cancer and CP with a 94% sensitivity. Similarly, the SVM was used by Zhang et al. [40] to differentiate PDAC and normal tissue using29 features identified in EUS images with a sensitivity of 97.98%. In another attempt, Das et al. [64] employed a combination of image analysis and ANNs to demarcate the cancerous zones in EUS images, acquired from individuals affected with pancreatic cancer with a high accuracy of 93%. In another effort employing multilayer perceptron neural networks (MNNs), a type of ANN, Ozkan et al. [65] categorized EUS images of non-malignant and malignant tissues, based upon various age groups of the patients namely, <40 y, 40–60 y, and >60 y. The MNNs employ a visible layer that receives an input that is passed onto inner units that are denoted as hidden layers, as they do not directly receive the input. The final hidden layer turns out the output. The error is

calculated, based on the deviations from the expected output and these are used to modify the layers to reduce the error during the training period. In another study [66], both one and two hidden layers were employed that exhibited a 97% accuracy with the training data set and a 95% accuracy with the testing data set for discriminating the malignant and non-malignant samples in the different age categories. The high accuracy was achieved for the data sets that were initially segregated into different age groups when compared to their uncategorized counterparts.

Yet another independent study employed MNNs for identifying pancreatic cancer from images of cell clusters, obtained from individuals using fine needle aspiration (FNA). Post-training, the MNN model was found to match the accuracy of an experienced cytopathologist. Additionally, the MNN model was able to predict accurately even inconclusive images, with 80% sensitivity, clearly demonstrating the promise of this tool for the screening of FNA specimens for pancreatic cancer with a conclusive diagnosis, especially those that are deemed inconclusive by cytopathologists. In an interesting study, a computer-assisted diagnosis (CAD) system was developed to analyse EUS images, using deep learning models (EUS-CAD) to identify PDAC, CP, and a normal pancreas (NP). The training set used 920 EUS images and the testing set used 470 EUS images. The detection efficiency was 92% and 94% in the validation and testing phases, respectively. Errors in diagnosis were identified only using the multivariate analysis of non-PDAC cases that was attributed to mass formation resulting in an over diagnosis of tumours [67].

EUS images of intraductal papillary mucinous neoplasms (IPMNs), that are precursors of PDAC, were analysed using deep learning algorithms to predict malignancy, using EUS images of patients acquired before a pancreatectomy. A total of 3970 images were used for the study and the malignant probability was calculated. The probability of the deep learning algorithm to diagnose malignant IPMN was 0.98 ($p < 0.001$) with a sensitivity, specificity, and accuracy of calculated to be 95.7%, 92.6%, and 94.0%, respectively. The accuracy was significantly superior to the corresponding human diagnosis (56.0%) [68] A comparison of the literature on pancreatic cancer discrimination from EUS images using AI tools revealed that deep learning and ANN techniques exhibited the greatest accuracy. followed by CNNs and the SVM. However, the literature reports chosen for the study had used images that compared normal and pancreatic cancer while some had tried to differentiate pancreatic cancer with CP. Similarly, the size of the cancerous tissues varied between the studies [69]. Therefore, additional studies are required to address if these differences could reflect in the prediction accuracy of the AI tool employed.

5. MRI

MRI is used to visualise the thinned slices of two-dimensional or three-dimensional soft tissues, due to the presence of water molecules in our body. The shift in the precessional frequency and alignment of the nuclei of the protons in the water molecule, in the presence of an external applied magnetic field and radiofrequency. is used for acquiring the image. The technique measures the relaxation times, T1 and T2 that denote the spin-lattice and spin-spin relaxation, respectively, to reach the original equilibrium position [70]. Relaxivities (r1 and r2), which are the inverse of the respective relaxation times are also measured. Most of the cases employ positive or negative contrast agents, such as gadolinium-based chelates or iron oxide, respectively, to significantly enhance the ratio of the relaxivities for an improved resolution and sensitivity [71].

Early detection of pancreatic cancer is essential to provide the affected individual with a fair chance of survival beyond five years. However, most imaging techniques, including MRI, fail to identify conclusively subtle changes observed in the pre-malignant stages, such as the pancreatic intraepithelial neoplasia, which is commonly associated with the tumorigenesis of PDAC [72]. Even an individual with stage I (localized) pancreatic cancer has only a 39% survival rate over a five-year period [73]. In a typical example of the use of AI for diagnosis using MRI images, a supervised machine learning (ML) algorithm was developed to predict the overall survival rates in PDAC affected patients, using

a cohort of 102 MRI images during training and a further 30 images during the testing period [74]. The algorithm was used to segment the images extract features. The sensitivity of the ML algorithm was 87%, while the specificity was determined to be 80%. The considerable overlap between the clinical histopathological conclusions and the ML-driven predictions indicates the promise of this strategy for classifying pancreatic cancer sub-types and diagnosis.

Another study [75] had investigated the ability of deep learning to distinguish between different pancreatic diseases from magnetic resonance (MR) images that were contrast-enhanced, using the T1 contrast agent gadopentetate dimeglumine. The generative adversarial network (GAN) form of machine learning can generate new sets of data which resemble/mimic the data used for training. GAN was employed to generate synthetic images that augmented the T1 contrast enhanced MRI data of 398 subjects within the age range of 16 and 85 years, acquired before the commencement of any treatment from a single hospital centre. The Inception-V4 network, a type of CNN with multiple hidden layers, was trained on the GAN augmented data set. Following the training, the MRI images acquired from two different hospital centres, comprising 50 images from subjects in the age group 24–85 years, and 56 images from patients aged between 26–80 years, were used for validating the performance of the Inception-V4 network towards the disease classification. The results were compared with the predictions made by the radiologist.

To augment the diagnostic accuracy of MRI on paediatric pancreatic cancer, Zhang et al. [76] used a quantum genetic algorithm to optimize the parameters of a traditional SVM classification model, for the improved prediction accuracy. In addition, this study acquired test samples from real life cases, and assessed the image processing performance of the algorithm for an efficient detection. The results revealed that the model distinguished clearly the cancer features with a high accuracy when compared with the conventional detection algorithm. Another study had employed a robust and intelligent method of ANNs combined with the SVM for the classification of pancreatic cancer to improve the diagnostic process, in terms of both accuracy and time [77]. Here, features of the MR images of the pancreas were extracted using the GLCM (gray-level co-occurrence matrix) method, a second order image texture analysis technique, that defines the spatial relationships among pixels in the region of interest. The best features extracted, using the JAFER algorithm, were analysed using five classification techniques: ANN BP (back propagation ANN), ANN RBF (radial basis function ANN), SVM Linear, SVM Poly (polynomial kernel), and SVM RBF (radial basis function SVM). The two best features selected, using the ANN BP techniques were used for the classification of pancreatic cancer with a 98% accuracy.

Corral et al. [78] employed a deep learning tool to identify neoplasia in intraductal papillary mucinous neoplasia (IPMN), using CNNs for the classification of the MRI scans of the pancreas. The classification was based on the guidelines issued by the American Gastroenterology Association, as well as the Fukuoka guidelines. When tested in 139 MRI scans of individuals, among which 22% were of a normal pancreas, 34% had a low-grade dysplasia while 14% were diagnosed with a high-grade dysplasia and the remaining 29% had adenocarcinoma, the model exhibited a detection sensitivity of 92% and a specificity of 52% for the detection of dysplasia. The deep learning technique exhibited an accuracy of 78%, in comparison to the 76% obtained by the classification using the American Gastroenterology Association guidelines.

For improving the accuracy, reliability, and efficiency of diagnosis, Chen et al. [79] developed an automated deep learning model (ALAMO) for the segmentation of multiple organs-at-risk (OARs) from the clinical MR images of the abdomen. The model had included training procedures, such as Multiview, deep connection, and auxiliary supervision. The model used multislice MR images as the input and generated segmented images as the output. The model was investigated using ten different OARs, such as the pancreas, liver, spleen, stomach, duodenum, small intestine, kidneys, spinal cord, and vertebral bodies. The results from the model correlated well with those obtained using the manual techniques. However, further studies integrating AI-based algorithms with these ALAMO

generated segmented MR images of the pancreas are required for the extraction of features to confirm the onset or progression of PC.

6. Computed Tomography

A computed tomography (CT) scan is a non-invasive clinical imaging technique that employs X-rays to obtain images at different angles. The resultant images are processed using customized software to obtain a reconstructed 3D image, which provides valuable anatomical information [80]. This technique is widely employed in healthcare centres for the diagnosis of tumours or internal injuries [81,82]. Despite its merits, CT scan images pose a challenge to clinicians for the accurate diagnosis of cancers, owing to irregular contours presented by regions with lesions, vasculature, bony structures, and soft tissues that display a mosaic of densities and intensities [83]. Additional challenges involved in the precise prediction of the disease from the CT scans are associated with fuzzy and noisy images that lack adequate contrast [84]. AI-driven methods that enable image segmentation, contour identification, and disease classification, therefore will be invaluable in improving the prediction efficiency for pancreatic diseases from CT images [85]. The currently employed conventional image segmentation models consume considerable computational time and power, as they perform every operation for each pixel in the image [86]. Further, the resultant processed image quality also lacks quality, thereby necessitating the development of more robust tools for AI-driven tools for image segmentation and processing that may provide a better diagnostic accuracy [87]. In an interesting study [88], about 19,500 non-contrast CT scan images, acquired from 469 scans, were segmented using CNNs and the mean pancreatic tissue density, in terms of the Hounsfield unit (HU), as well as the pancreatic volume, were computed using the CNN algorithm. The comparison of the results of the pre-diagnostic scans from individuals who later developed PDAC and those that remained cancer-free, revealed that there was a significant reduction in the mean whole gland pancreatic HU of 0.2 vs. 7.8 in individuals who developed PDAC. This suggests that the attenuation of the HU intensity in the CT images of the pancreas could imply a risk of PDAC. This study has opened new avenues for employing CNNs as a tool for the pre-diagnosis/very early diagnosis of PDAC from CT scan images.

In another attempt to classify PDAC, a regular CNN algorithm with four hidden layers was trained using CT images obtained from 222 affected individuals and 190 noncancerous individuals. Though a diagnostic accuracy of 95% was achieved using CNNs, it was not superior to the predictions made by human experts indicating the need for an appropriate AI architecture for the classification of pancreatic cancer [89]. Zhang et al. [90] employed feature pyramid networks with a recurrent CNN (R-CNN) that could identify the sequential patterns and predict the subsequent patterns of a given data set for classifying PDAC from CT scan images. A dataset of 2890 CT images was employed for training the network to achieve a classification accuracy of about 94.5%. Though this method proved to be superior to the existing methods, it was limited by the input uncertainty that is generally associated with closed-source data. This drawback could be eliminated by using a public data set for training. In a more advanced variant, a 16-layer VGG16 CNN model was employed along with R-CNN to diagnose PDAC from 6084 enhanced CT scans obtained from 338 PDAC-affected individuals. The combination of VGG16 and R-CNN exhibited a high prediction accuracy of about 96%. Each CT image was processed by the R-CNN within 0.2 s that was considerably faster than a clinical imaging expert [6]. Additionally, a deep learning algorithm has been developed by Chen et al. [91] for detecting pancreatic cancer that is smaller than 2 cm on CT scans. The study result showed that the CNN was effective in distinguishing patients with pancreatic cancer from normal pancreatic individuals, achieving an 89.7% sensitivity and a 92.8% specificity. It also showed a higher sensitivity of 74.7% for the identification of pancreatic cancer malignancies, smaller than 2 cm.

An attempt to employ CNN models to distinguish different kinds of pancreatic cysts was made using CT images from 206 patients. Among these individuals, 64 suffered from

intraductal papillary mucinous neoplasms (IPMNs), 66 had been diagnosed with serous cystic neoplasms (SCN), 35 had mucinous cystic neoplasms (MCNs) while 41 individuals suffered from solid pseudopapillary epithelial neoplasms (SPENs). The feature extraction from the CT images and classification of the type of pancreatic cyst, was accomplished using densely connected convolutional network (Dense-Net) architecture that uses dense layers which receives inputs from all neurons/nodes and dense blocks connecting all layers by the shortest route. The Dense-Net algorithm performed better than the conventional CNN model in discriminating between the different types of cysts with the highest accuracy of 81.3% observed for IPMNs followed by 75.8% for the SCNs and 61% for the SPENs [92]. Though the Dense-Net model outperformed the CNNs in all categories, the study lacked information on the tumour size and failed to provide reasons for the positive and negative errors encountered in the identification of the type of pancreatic cysts. The model needs to be tested rigorously with a wider range of cysts to understand its capability for discriminating between different types of pancreatic cysts if it is to be adopted in the clinics.

7. Positron Emission Tomography (PET)

Positron emission tomography (PET) employs short-lived radioisotope tracers that emit positrons. These positrons destructively interact with an electron to generate photons, which are recorded for generating the PET image. The tracer can be differentially localized in various tissues by conjugating with a biomolecule for a better target specificity [93,94]. The PET scans provide additional information about the functioning of an organ. Commonly employed tracers include ^{18}F, ^{15}O, ^{13}N, and ^{11}C, that have half-lives of 109.74 min, 122.24 s, 9.97 min, and 20.38 min, respectively [95]. PET imaging has been also used to diagnose the recurrence of pancreatic cancer as well as to understand the response of the cancer tissue to different therapeutic interventions. Despite several studies that have shown the diagnostic efficiency of PET scans towards pancreatic cancers with a sensitivity in the range of 85% and above [96], several factors, such as the dysregulated glucose metabolism and inflammation interfere with the sensitivity of the diagnosis from PET images, resulting in false positives [97]. PET scans are also ineffective in diagnosing pancreatic cancers when the tumour mass has a diameter below 2 cm [98]. This necessitates the use of advanced AI-aided algorithms for the discrimination and classification of cancerous masses from the PET scan images.

For imaging cancers, ^{18}F substituted glucose or fluorodeoxyglucose (FDG) has been frequently used, due to the high consumption of glucose by cancer cells to meet its metabolic requirements [97]. PET scans have been employed frequently in conjunction with MRI or non-contrast CT, owing to their poor spatial resolution for the diagnosis of cancers, including their staging [99]. To overcome challenges in discriminating cancerous lesions from non-contrast CT images, 18F- FDG PET/CT imaging of pancreatic cancers was used by Li et al. [100], in conjunction with a SVM algorithm. The region of interest (ROI) identified in the CT image of the pancreas was initially segmented, using a simple linear iterative clustering (SLIC) followed by the feature extraction using the dual threshold principal component analysis (DT-PCA). Finally, a hybrid feedback-SVM-random forest algorithm (HFP-SVM-RF) was used to classify the pancreatic cancerous lesions. The random forest model is a type of supervised machine learning model that is widely used for classification and decision making. The hybrid model exhibited an accuracy of 96.5% when tested using the PET/CT images of 40 patients with pancreatic cancer and 40 non-cancer individuals. The hybrid algorithm when tested using 82 public PET/CT scans exhibited a similarity score of 78.9% and 65.4%, when compared with the ground-truth contours using the Dice coefficient and Jaccard index, respectively, suggesting there is scope for further improvement in the diagnostic performance.

Radiomics is a feature extraction method that has been widely used in image processing tools. A combination of radiomics with machine learning was employed for the prognostic prediction of the survival rate from ^{18}F-FDG-PET scans of 138 patients with

pancreatic cancer. A random forest model was used for the classification of 42 features extracted from the PET images. The model revealed that the gray-level zone length matrix (GLZLM), the gray-level non-uniformity (GLNU) in the images as the top factor that influenced the one year survival, while the total lesion glycolysis ranked second. This information was used to stratify individuals into poor prognosis groups with a high risk of mortality [101].

It is thus evident that every imaging technique will require customized robust algorithms to extract the subtle but distinctive features of pancreatic cancer for the accurate identification and stratification. The evolution of new ML algorithms continues to improve the sensitivity and selectivity of the diagnosis of pancreatic cancer at an early stage, thereby improving the survival chances of the affected individual. Table 2 lists some of the major studies, using various AI driven models for the diagnosis of pancreatic cancer.

Table 2. Summary of the AI driven models for the pancreatic cancer diagnosis.

Modality	AI Model	Study Population	Purpose	Sensitivity	Specificity	Accuracy	Reference
CT	CNN	27	Pancreatic cystic neoplasm malignancy prediction	-	-	92.9	Watson et al., 2021 [102]
CT	Naïve Bayer classifier	72	PDAC identification	-	-	86	Ahamed et al., 2022 [103]
CT	CNN	1006	Pancreas segmentation	-	-	-	Lim et al., 2022 [104]
CT	CNN	68	Serum tumor marker analysis	89.31	92.31	-	Qiao et al., 2022 [105]
CT	CNN	513	Pancreatico enteric Anastomotic Fistulas prediction after a pancreatoduodenectomy	86.7	87.3	87.1	Mu et al., 2020 [106]
CT	ANN	62	Acute pancreatitis risk prediction	-	-	-	Keogan et al., 2002 [107]
CT	Support vector machine	56	PDAC histopathological grade discrimination	78	95	86	Qiu et al., 2019 [108]
CT	CNN	370 patients, 320 controls	PC detection	97.3 (Test set 1) 99 (Test set 2)	100 (Test set 1) 98.9 (Test set 2)	98.6(Test set 1) 98.9 (Test set 2)	Liu et al., 2020 [109]
CT	Deep learning	750 patients 575 controls	PDAC detection	-	-	87.8	Chu et al., 2019 [110]
CT	CNN	222 patients 190 controls	PC diagnosis	91.58	98.27	95.47	Ma et al., 2020 [89]
CT	DCNN	2890 CT images	Pancreatic cancer detection	83.76	91.79	94	Zhang et al., 2020 [90]
CT	Deep learning	319	Preoperative pancreatic cancer diagnosis	86.8	69.5	87.1	Si et al., 2021 [111]

Table 2. *Cont.*

Modality	AI Model	Study Population	Purpose	Sensitivity	Specificity	Accuracy	Reference
CT	ANN	898	Cancer risk prediction	80.7	80.7	-	Muhammad et al., 2019 [112]
CT	CNN	669 patients 804 controls	PC differentiation	89.7	92.8	-	Chen et al., 2022 [91]
MRI	CNN	139	Identification of intraductal papillary mucinous neoplasia	75	78	-	Juan et al., 2019 [78]
MRI	CNN	27	Automatic image segmentation	-	-	-	Liang et al., 2020 [113]
MRI	ANN	168	PDAC differentiation	-	-	96	Devi et al., 2018 [114]
EUS	CNN	583	Autoimmune pancreatitis from PDAC	90	85	-	Marya et al., 2021 [115]
EUS	CAD	920 (Validation) +470 (test)	PDAC detection	-	-	-	Tonozuka et al., 2021 [67]
EUS	ANN	202 (cancerous) & 130 (Non-cancerous)	Computer-aided pancreatic cancer diagnosis using image processing	83.3	93.3	87.5	Ozkan et al., 2019 [65]
EUS	ANN	258	Pancreatic lesion characterization	-	-	91	Saftoiu et al., 2012 [34]
EUS	ANN	388	PDAC and CP differentiation	96	93	94	Zhu et al., 2013 [63]
EUS	ANN	167	PDAC and CP differentiation	94	94	-	Saftoiu et al., 2015 [116]
EUS	ANN	56	Normal, CP and PDAC differentiation	-	-	93	Das et al., 2008 [64]
EUS	ANN	21	PDAC and CP differentiation	-	-	89	Norton et al., 2001 [62]
PET/CT	SVM	80	Pancreatic cancer segmentation	95.23	97.51	96.47	Li et al., 2018 [100]

8. Pancreatic Cancer Risk Prediction Using AI

Since pancreatic cancer is a highly aggressive form of cancer that is largely asymptomatic in the early stages and has a tendency to spread rapidly, leading to poor survival duration post-diagnosis, the AI-based prediction of the risk of developing pancreatic cancer could be an immensely useful strategy for improving the prognosis for an individual. Muhammad et al. [112] had successfully employed ANNs from personal health data to predict and stratify the pancreatic cancer risk as a low, medium, or high risk ,with a sensitivity and specificity of 80.7%.This study highlights the ability of the AI-based predictive tools for the effective management of the pancreatic cancer risk even before the manifestation of symptoms. Similarly, Corral et al. [78] had employed an AI algorithm to identify pancreatic cysts that pose a high risk of transforming into cancerous lesions. Such a pre-diagnosis could help clinicians in designing adequate preventive interventions to save

lives. The detection of subtle textural and morphological changes in CT and MRI scans of the pancreas could also be facilitated through customized AI algorithms [117]. Several attempts have also been reported to employ AI tools to predict the risk of developing pancreatic cancer from biomarker measurements, as well as abdominal scans to discern pre-cancerous abnormalities [117].

9. AI-Driven Diagnosis Based on Cancer Biomarkers

The serological detection of PC is based on the quantification of a biomarker whose levels are altered in cancerous conditions. However, a single marker could not accurately diagnose a specific type of cancer as there are several other conditions that could modulate the levels of said biomarker. Hence, multiple biomarkers need to be analysed, to accurately diagnose PC. In an earlier work, protein markers from the serum of 27 normal and 27 individuals diagnosed with pancreatic cancer, were profiled using surface-enhanced laser desorption ionization (SELDI), and were classified using a decision tree algorithm, based on which six serum proteins were identified as pancreatic cancer biomarkers [118]. Carbohydrate antigen 19-9(CA19-9) is the most extensively explored protein biomarker of pancreatic cancer. However, several studies have indicated that CA19-9, by itself, could not be an effective predictor of pancreatic cancer and hence the search for additional diagnostic protein markers in serum are underway [119]. Analysis of datasets from microarray and the next generation sequencing of samples for the gene expression or serum protein expressions using deep learning and machine learning algorithms, could aid in identifying the most promising protein biomarkers that aid in the early detection of pancreatic cancer. For instance, the SVM based algorithm, in combination with the recursive feature elimination (RFE), was employed to screen the gene expression datasets of 78 samples, for additional pancreatic cancer biomarkers. Seven gene targets were short-listed among the genes encoding for the proteins FOS that encodes for the leucine zipper protein, MMP7 (matrix metalloproteinase-7), and A2M (alpha-2-macroglobulin), were predicted to be more accurate diagnostic markers for pancreatic cancer, not only in serum, but also in urine samples [120]. Similarly, ANN-based methods have been employed to analyse the levels of key serum biomarkers implicated in PC, such as CA19-9, CA125, and carcinoembryonic antigen (CEA), from 913 samples obtained from individuals with a normal and a cancerous pancreas. The results showed an improved detection accuracy when compared with a single marker-based prediction, clearly highlighting the benefits of an AI-integrated multi-analyte diagnosis [121]. Exosomes, which are vesicular structures containing miRNA, specific to the source cells, are gaining importance for the disease diagnosis. Several exosome entrapped miRNA have been identified in PC, such as miR-16, miR-20a, miR-21, miR-21-5p, miR-24, miR25, miR99a, miR-133a, miR185, miR191, miR-196a, miR-223, miR-642b-3p, miR-663a, miR-1290, miR-1246, miR-5100, and miR-8073 [122]. In a seminal work, the exosomes obtained from a panel of mouse and human origin PC cell lines, were captured using antibodies against the surface expressed EpCAM (epithelial cell adhesion molecule). The RNA cargo was isolated from the exosomes and the miRNA was identified using qPCR. The cancer miRNA signatures were identified using a custom-developed machine learning algorithm. The system was validated using samples isolated from individuals with a normal pancreas and those with pancreatic cancer, with a good prediction accuracy [123]. In another study, a neural network algorithm was employed to screen 140 datasets of individuals diagnosed with pancreatic cancer, for gene biomarkers in urine samples, namely REG1A/1B, LYVE1, TFF1, and CA19-9. Following the training, the neural network algorithm predicted REG1A/1B as the most important biomarker in the urine samples with an importance ratio exceeding 80% [124]. With the discovery of new circulating markers, such as glycoproteins and genetic markers, such a machine learning-based diagnosis could herald in the rapid and accurate detection of PC.

The histological analysis or tissue biopsies have been conventionally employed for the identification and stratification of cancers. However, this is a time-consuming process. Further, there is a constant increase in the number of samples that are sent for analysis to

the anatomical pathological laboratory and this, coupled with insufficient skilled pathologists, leads to long turn-around-times [125]. Additionally, cytopathology requires the accurate slide preparation and optimal staining of the tissue slices. However, the staining intensity of biopsy slides exhibit analyst-based, sample thickness-based and laboratory protocol-based variations in the intensity [125]. In this context, deep learning algorithms, such as VGG, DenseNet, ResNet etc., and machine learning algorithms, based on SVM and the random forest, can be employed to extract specific tumour features from the tissue slices to improve the speed of detection and reduce the burden on the clinical pathologists. Similarly, the use of algorithms, such as SA-GAN (stain acclimatization generative adversarial network) that employs a generator that imports the input source image and generates a target image that incorporates the features of the input sample and the colour intensity of a training sample. Two discriminators are also incorporated into this deep learning model, which ensure that the colour intensity of the desired training image and textural features of the source image are maintained in the generated image, thus ensuring the stain colour normalization across the different images [126]. Such approaches have been attempted, to identify various types of gastrointestinal and breast cancer, using mammograms and tissue biopsies [127]. Using a similar concept, a deep learning-based spiral algorithm was employed to transform 3D MRI images of the pancreatic tissue into 2D images without compromising then original image texture and edge parameters. The CNN-based models were employed for the feature extraction and the bilinear pooling module was used to improve the prediction accuracy. Parameters, such as size, shape, volume, texture, and intensity, were employed to classify the image as pancreatic cancer with TP53 gene mutation or otherwise. The prediction results agreed well with the actual mutation status. This approach overcomes the drawback of the need for painful biopsies for classifying a tumour as TP53 positive. In addition, this novel method offers a non-invasive approach for predicting gene mutations, using AI-driven cytopathology that may also be extended for other forms of cancer or gene mutations [128]. Similarly, ResNet and DenseNet models have been employed to identify *Helicobacter pylori*, a key causative pathogen in different gastric cancers from stained tissue biopsy specimens [129]. The advantage of using machine learning models in this case over conventional cytopathology, is the ability of the model to identify even small numbers of the bacteria, which is very tedious and time-consuming in the conventional mode. Abnormal goblet cells have been identified with an 86% accuracy in tissue samples of individuals with Barretts esophagus, using VGG algorithms [130]. AI-driven algorithms can be useful in detecting microsatellite instabilities in the biopsy samples, that area hallmark of many forms of cancers [125]. These studies clearly demonstrate that the integration of machine learning in cytopathology can be useful for the faster, efficient, and early diagnosis of pancreatic cancer. This field is slowly gaining prominence and may soon lead to the establishment of a digital cytopathology as a mainstay in the detection and stratification of cancers.

10. Ethics of Using AI for Diagnosis

Though AI offers a plethora of benefits in improving the detection and stratification of PC, there are several ethical concerns that have emerged among a section of the society, on the extensive use of AI-based diagnostics. Since AI tools require large datasets for training and validation, concerns on data privacy and confidentiality have been raised. Additionally, data security and safety issues have also been associated with use of an AI-based diagnosis [131]. There exists a regulatory vacuum in the realm of AI-based tool development and no structured white document is available on the data collection, storage, processing and sharing. Furthermore, frequent comparisons between expert predictions by clinicians and the AI algorithm, have given rise to the theory of inadequate training or de-skilling of clinicians, in future, owing to the over-dependence of AI-based detections. A lack of patient-doctor connect, or dissolution of the trust factor are additional issues that have been associated with the deployment of AI-driven technologies in healthcare [132]. Accountability and professional responsibility issues, in the case of a wrong diagnosis by

AI-based tools, that may result in disastrous consequences, is another facet that is being debated as a negative aspect of all AI-driven cancer diagnoses.

11. Concluding Remarks

The use of AI for cancer detection and biomarker discovery, is expected to be the target of several research studies involving AI over the next decade. Several studies in this direction have clearly demonstrated the benefits of the AI-driven detection of pancreatic cancer, especially those employing imaging tools. However, the widespread clinical deployment of this technology is yet to be realized, owing to lack of large datasets to convincingly train and validate the developed algorithms. Most AI-based models have been developed in a black box mode and as a result, the clinicians are unable to understand or explain the basis of identification or stratification, thereby leading to a reticence in employing this technology. Additional ethical issues concerning data privacy and security further have slowed down the translation of an AI-based diagnosis in clinics. However, the exponential growth, witnessed in computing resources, including open-source tools, has triggered an avalanche of studies focused on developing more robust algorithms for the accurate, rapid, and early diagnosis of PC. As this field continues to grow, new regulatory policies concerning its use and deployment will emerge so that the benefits of this technology can be harnessed to save lives.

Author Contributions: Conceptualization, U.M.K.; resources, B.S.H. and U.M.K.; data curation, B.S.H.; writing—original draft preparation, B.S.H. and U.M.K.; writing—review and editing, B.S.H. and U.M.K.; visualization, U.M.K.; supervision, U.M.K.; project administration, U.M.K.; funding acquisition, U.M.K. All authors have read and agreed to the published version of the manuscript.

Funding: This research was funded by Nano Mission, Department of Science & Technology grant number (SR/NM/NS-1095/2012) and FIST, Department of Science & Technology (SR/FST/LSI-622/2014).

Acknowledgments: The authors gratefully acknowledge the financial support from Nano Mission, Department of Science & Technology and FIST, Department of Science & Technology. The authors also thank SASTRA Deemed University for financial and infrastructural support.

Conflicts of Interest: The authors declare no conflict of interest.

References

1. Rawla, P.; Sunkara, T.; Gaduputi, V. Epidemiology of Pancreatic Cancer: Global Trends, Etiology and Risk Factors. *World J. Oncol.* **2019**, *10*, 10–27. [CrossRef] [PubMed]
2. Hu, C.; Li, M. In Advanced Pancreatic Cancer: The Value and Significance of Interventional Therapy. *J. Interv. Med.* **2020**, *3*, 118–121. [CrossRef] [PubMed]
3. Gordon-Dseagu, V.L.; Devesa, S.S.; Goggins, M.; Stolzenberg-Solomon, R. Pancreatic Cancer Incidence Trends: Evidence from the Surveillance, Epidemiology and End Results (SEER) Population-Based Data. *Int. J. Epidemiol.* **2018**, *47*, 427–439. [CrossRef] [PubMed]
4. Maisonneuve, P.; Lowenfels, A.B. Epidemiology of Pancreatic Cancer: An Update. *Dig. Dis.* **2010**, *28*, 645–656. [CrossRef]
5. Kamisawa, T.; Wood, L.D.; Itoi, T.; Takaori, K. Pancreatic Cancer. *Lancet* **2016**, *388*, 73–85. [CrossRef]
6. Liu, S.-L.; Li, S.; Guo, Y.-T.; Zhou, Y.-P.; Zhang, Z.-D.; Li, S.; Lu, Y. Establishment and Application of an Artificial Intelligence Diagnosis System for Pancreatic Cancer with a Faster Region-Based Convolutional Neural Network. *Chin. Med. J.* **2019**, *132*, 2795–2803. [CrossRef]
7. Kang, J.D.; Clarke, S.E.; Costa, A.F. Factors Associated with Missed and Misinterpreted Cases of Pancreatic Ductal Adenocarcinoma. *Eur. Radiol.* **2021**, *31*, 2422–2432. [CrossRef]
8. Lee, D.; Yoon, S.N. Application of Artificial Intelligence-Based Technologies in the Healthcare Industry: Opportunities and Challenges. *Int. J. Environ. Res. Public Health* **2021**, *18*, 271. [CrossRef]
9. González García, C.; Núñez-Valdez, E.; García-Díaz, V.; Pelayo G-Bustelo, C.; Cueva-Lovelle, J.M. A Review of Artificial Intelligence in the Internet of Things. *Int. J. Interact. Multimed. Artif. Intell.* **2019**, *5*, 9–20. [CrossRef]
10. Cohen, S. The Evolution of Machine Learning: Past, Present, and Future. In *Artificial Intelligence and Deep Learning in Pathology*; Elsevier: Amsterdam, The Netherlands, 2021; pp. 1–12. ISBN 978-0-323-67538-3.
11. Luchini, C.; Pea, A.; Scarpa, A. Artificial Intelligence in Oncology: Current Applications and Future Perspectives. *Br. J. Cancer* **2022**, *126*, 4–9. [CrossRef]

12. Induja, S.N.; Raji, C.G. Computational Methods for Predicting Chronic Disease in Healthcare Communities. In Proceedings of the 2019 International Conference on Data Science and Communication (IconDSC), Bangalore, India, 1–2 March 2019; pp. 1–6.
13. Kumar, U. Applications of Machine Learning in Disease Pre-screening. In *Research Anthology on Artificial Intelligence Applications in Security*; Information Resources Management Association, Ed.; IGI Global: Hershey, PA, USA, 2020; pp. 1052–1084. ISBN 978-1-79987-705-9.
14. Noori, A.; Alfi, A.; Noori, G. An Intelligent Control Strategy for Cancer Cells Reduction in Patients with Chronic Myelogenous Leukaemia Using the Reinforcement Learning and Considering Side Effects of the Drug. *Expert Syst.* **2021**, *38*, e12655. [CrossRef]
15. Zhao, Y.; Kosorok, M.R.; Zeng, D. Reinforcement Learning Design for Cancer Clinical Trials. *Statist. Med.* **2009**, *28*, 3294–3315. [CrossRef] [PubMed]
16. Zhu, W.; Xie, L.; Han, J.; Guo, X. The Application of Deep Learning in Cancer Prognosis Prediction. *Cancers* **2020**, *12*, 603. [CrossRef] [PubMed]
17. Vial, A.; Stirling, D.; Field, M.; Ros, M.; Ritz, C.; Carolan, M.; Holloway, L.; Miller, A.A. The Role of Deep Learning and Radiomic Feature Extraction in Cancer-Specific Predictive Modelling: A Review. *Transl. Cancer Res.* **2018**, *7*, 803–816. [CrossRef]
18. Ghosh, P.; Azam, S.; Hasib, K.M.; Karim, A.; Jonkman, M.; Anwar, A. A Performance Based Study on Deep Learning Algorithms in the Effective Prediction of Breast Cancer. In Proceedings of the 2021 International Joint Conference on Neural Networks (IJCNN), Shenzhen, China, 18–22 July 2021; pp. 1–8.
19. Lin, C.-Y.; Chien, T.-W.; Chen, Y.-H.; Lee, Y.-L.; Su, S.-B. An App to Classify a 5-Year Survival in Patients with Breast Cancer Using the Convolutional Neural Networks (CNN) in Microsoft Excel: Development and Usability Study. *Medicine* **2022**, *101*, e28697. [CrossRef] [PubMed]
20. Bakasa, W.; Viriri, S. Pancreatic Cancer Survival Prediction: A Survey of the State-of-the-Art. *Comput. Math. Methods Med.* **2021**, *2021*, 1188414. [CrossRef] [PubMed]
21. Capobianco, E. High-Dimensional Role of AI and Machine Learning in Cancer Research. *Br. J. Cancer* **2022**, *126*, 523–532. [CrossRef]
22. Hussain, L.; Saeed, S.; Awan, I.A.; Idris, A.; Nadeem, M.S.A.; Chaudhry, Q.-A. Detecting Brain Tumor Using Machines Learning Techniques Based on Different Features Extracting Strategies. *CMIR* **2019**, *15*, 595–606. [CrossRef]
23. Gassenmaier, S.; Afat, S.; Nickel, D.; Mostapha, M.; Herrmann, J.; Othman, A.E. Deep Learning–Accelerated T2-Weighted Imaging of the Prostate: Reduction of Acquisition Time and Improvement of Image Quality. *Eur. J. Radiol.* **2021**, *137*, 109600. [CrossRef]
24. Iqbal, M.J.; Javed, Z.; Sadia, H.; Qureshi, I.A.; Irshad, A.; Ahmed, R.; Malik, K.; Raza, S.; Abbas, A.; Pezzani, R.; et al. Clinical Applications of Artificial Intelligence and Machine Learning in Cancer Diagnosis: Looking into the Future. *Cancer Cell Int.* **2021**, *21*, 270. [CrossRef]
25. Xu, J.; Jing, M.; Wang, S.; Yang, C.; Chen, X. A Review of Medical Image Detection for Cancers in Digestive System Based on Artificial Intelligence. *Expert Rev. Med. Devices* **2019**, *16*, 877–889. [CrossRef] [PubMed]
26. Davatzikos, C.; Sotiras, A.; Fan, Y.; Habes, M.; Erus, G.; Rathore, S.; Bakas, S.; Chitalia, R.; Gastounioti, A.; Kontos, D. Precision Diagnostics Based on Machine Learning-Derived Imaging Signatures. *Magn. Reson. Imaging* **2019**, *64*, 49–61. [CrossRef] [PubMed]
27. Chen, W.; Chen, Q.; Parker, R.A.; Zhou, Y.; Lustigova, E.; Wu, B.U. Risk Prediction of Pancreatic Cancer in Patients with Abnormal Morphologic Findings Related to Chronic Pancreatitis: A Machine Learning Approach. *Gastro Hep Adv.* **2022**, *1*, 1014–1026. [CrossRef]
28. Avuçlu, E.; Elen, A. Evaluation of Train and Test Performance of Machine Learning Algorithms and Parkinson Diagnosis with Statistical Measurements. *Med. Biol. Eng. Comput.* **2020**, *58*, 2775–2788. [CrossRef]
29. Kilic, N.; Kursun, O.; Ucan, O.N. Classification of the Colonic Polyps in CT-Colonography Using Region Covariance as Descriptor Features of Suspicious Regions. *J. Med. Syst.* **2010**, *34*, 101–105. [CrossRef]
30. Reddy, D.J.; Arun Prasath, T.; PallikondaRajasekaran, M.; Vishnuvarthanan, G. Brain and Pancreatic Tumor Classification Based on GLCM—K-NN Approaches. In *International Conference on Intelligent Computing and Applications*; Bhaskar, M.A., Dash, S.S., Das, S., Panigrahi, B.K., Eds.; Advances in Intelligent Systems and Computing; Springer: Singapore, 2019; Volume 846, pp. 293–302. ISBN 9789811321818.
31. Jamshidi, M.; Zilouchian, A. *Intelligent Control Systems Using Soft Computing Methodologies*; CRC Press: Boca Raton, FL, USA, 2001; ISBN 978-0-8493-1875-7.
32. Lee, J.-G.; Jun, S.; Cho, Y.-W.; Lee, H.; Kim, G.B.; Seo, J.B.; Kim, N. Deep Learning in Medical Imaging: General Overview. *Korean J. Radiol.* **2017**, *18*, 570. [CrossRef]
33. Rumelhart, D.E.; Hinton, G.E.; Williams, R.J. Learning Representations by Back-Propagating Errors. *Nature* **1986**, *323*, 533–536. [CrossRef]
34. Săftoiu, A.; Vilmann, P.; Gorunescu, F.; Janssen, J.; Hocke, M.; Larsen, M.; Iglesias–Garcia, J.; Arcidiacono, P.; Will, U.; Giovannini, M.; et al. Efficacy of an Artificial Neural Network–Based Approach to Endoscopic Ultrasound Elastography in Diagnosis of Focal Pancreatic Masses. *Clin. Gastroenterol. Hepatol.* **2012**, *10*, 84–90.e1. [CrossRef]
35. Tu, J.V. Advantages and Disadvantages of Using Artificial Neural Networks versus Logistic Regression for Predicting Medical Outcomes. *J. Clin. Epidemiol.* **1996**, *49*, 1225–1231. [CrossRef]

36. Hakkoum, H.; Idri, A.; Abnane, I. Assessing and Comparing Interpretability Techniques for Artificial Neural Networks Breast Cancer Classification. *Comput. Methods Biomech. Biomed. Eng. Imaging Vis.* **2021**, *9*, 587–599. [CrossRef]
37. Cortes, C.; Vapnik, V. Support-Vector Networks. *Mach. Learn.* **1995**, *20*, 273–297. [CrossRef]
38. Reeves, D.M.; Jacyna, G.M. Support Vector Machine Regularization. *WIREs Comp. Stat.* **2011**, *3*, 204–215. [CrossRef]
39. Huang, C.-H. A Reduced Support Vector Machine Approach for Interval Regression Analysis. *Inf. Sci.* **2012**, *217*, 56–64. [CrossRef]
40. Zhang, M.-M.; Yang, H.; Jin, Z.-D.; Yu J.-G.; Cai, Z.-Y.; Li, Z.-S. Differential Diagnosis of Pancreatic Cancer from Normal Tissue with Digital Imaging Processing and Pattern Recognition Based on a Support Vector Machine of EUS Images. *Gastrointest. Endosc.* **2010**, *72*, 978–985. [CrossRef] [PubMed]
41. Du, W.; Rao, N.; Liu, D.; Jiang, H.; Luo, C.; Li, Z.; Gan, T.; Zeng, B. Review on the Applications of Deep Learning in the Analysis of Gastrointestinal Endoscopy Images. *IEEE Access* **2019**, *7*, 142053–142069. [CrossRef]
42. Kim, P. Convolutional Neural Network. In *MATLAB Deep Learning*; Apress: Berkeley, CA, USA, 2017; pp. 121–147. ISBN 978-1-4842-2844-9.
43. Liu, Y.H. Feature Extraction and Image Recognition with Convolutional Neural Networks. *J. Phys. Conf. Ser.* **2018**, *1087*, 062032. [CrossRef]
44. Kumar, A.; Kim, J.; Lyndon, D.; Fulham, M.; Feng, D. An Ensemble of Fine-Tuned Convolutional Neural Networks for Medical Image Classification. *IEEE J. Biomed. Health Inform.* **2017**, *21*, 31–40. [CrossRef]
45. Siddique, N.; Paheding, S.; Elkin, C.P.; Devabhaktuni, V. U-Net and Its Variants for Medical Image Segmentation: A Review of Theory and Applications. *IEEE Access* **2021**, *9*, 82031–82057. [CrossRef]
46. Lecun, Y.; Bottou, L.; Bengio, Y.; Haffner, P. Gradient-Based Learning Applied to Document Recognition. *Proc. IEEE* **1998**, *86*, 2278–2324. [CrossRef]
47. Özyurt, F. A Fused CNN Model for WBC Detection with MRMR Feature Selection and Extreme Learning Machine. *Soft Comput.* **2020**, *24*, 8163–8172. [CrossRef]
48. Sharma, H.; Zerbe, N.; Klempert, I.; Hellwich, O.; Hufnagl, P. Deep Convolutional Neural Networks for Automatic Classification of Gastric Carcinoma Using Whole Slide Images in Digital Histopathology. *Comput. Med. Imaging Graph.* **2017**, *61*, 2–13. [CrossRef] [PubMed]
49. Shin, Y.; Qadir, H.A.; Aabakken, L.; Bergsland, J.; Balasingham, I. Automatic Color Polyp Detection Using Region Based Deep CNN and Post Learning Approaches. *IEEE Access* **2018**, *6*, 40950–40962. [CrossRef]
50. Long, J.; Shelhamer, E.; Darrell, T. Fully Convolutional Networks for Semantic Segmentation. In Proceedings of the IEEE Conference on Computer Vision and Pattern Recognition (CVPR), Boston, MA, USA, 7–12 June 2015. [CrossRef]
51. Oda, M.; Shimizu, N.; Oda, H.; Hayashi, Y.; Kitasaka, T.; Fujiwara, M.; Misawa, K.; Mori, K.; Roth, H.R. Towards Dense Volumetric Pancreas Segmentation in CT Using 3D Fully Convolutional Networks. In Proceedings of the Medical Imaging 2018: Image Processing, Houston, TX, USA, 2 March 2018; p. 10.
52. Guo, Z.; Zhang, L.; Lu, L.; Bagheri, M.; Summers, R.M.; Sonka, M.; Yao, J. Deep LOGISMOS: Deep Learning Graph-Based 3D Segmentation of Pancreatic Tumors on CT Scans. In Proceedings of the 2018 IEEE 15th International Symposium on Biomedical Imaging (ISBI 2018), Washington, DC, USA, 4–7 April 2018; pp. 1230–1233.
53. Luthra, A.K.; Evans, J.A. Review of Current and Evolving Clinical Indications for Endoscopic Ultrasound. *World J. Gastrointest. Endosc.* **2016**, *8*, 157. [CrossRef] [PubMed]
54. Gonzalo-Marin, J. Role of Endoscopic Ultrasound in the Diagnosis of Pancreatic Cancer. *World J. Gastrointest. Oncol.* **2014**, *6*, 360. [CrossRef] [PubMed]
55. Munroe, C.A.; Fehmi, S.M.A.; Savides, T.J. Endoscopic Ultrasound in the Diagnosis of Pancreatic Cancer. *Expert Opin. Med. Diagn.* **2013**, *7*, 25–35. [CrossRef] [PubMed]
56. Bhutani, M.; Koduru, P.; Joshi, V.; Saxena, P.; Suzuki, R.; Irisawa, A.; Yamao, K. The Role of Endoscopic Ultrasound in Pancreatic Cancer Screening. *Endosc. Ultrasound* **2016**, *5*, 8. [CrossRef] [PubMed]
57. DeWitt, J.; Devereaux, B.M.; Lehman, G.A.; Sherman, S.; Imperiale, T.F. Comparison of Endoscopic Ultrasound and Computed Tomography for the Preoperative Evaluation of Pancreatic Cancer: A Systematic Review. *Clin. Gastroenterol. Hepatol.* **2006**, *4* 717–725. [CrossRef]
58. Pausawasdi, N.; Hongsrisuwan, P.; Chalermwai, W.V.; Butt, A.S.; Maipang, K.; Charatchareonwitthaya, P. The Diagnostic Performance of Combined Conventional Cytology with Smears and Cell Block Preparation Obtained from Endoscopic Ultrasound-Guided Fine Needle Aspiration for Intra-Abdominal Mass Lesions. *PLoS ONE* **2022**, *17*, e0263982. [CrossRef]
59. Hayashi, H.; Uemura, N.; Matsumura, K.; Zhao, L.; Sato, H.; Shiraishi, Y.; Yamashita, Y.; Baba, H. Recent Advances in Artificial Intelligence for Pancreatic Ductal Adenocarcinoma. *World J. Gastroenterol.* **2021**, *27*, 7480–7496. [CrossRef]
60. Herth, F.J.F.; Rabe, K.F.; Gasparini, S.; Annema, J.T. Transbronchial and Transoesophageal (Ultrasound-Guided) Needle Aspirations for the Analysis of Mediastinal Lesions. *Eur. Respir. J.* **2006**, *28*, 1264–1275. [CrossRef]
61. Cazacu, I.; Udristoiu, A.; Gruionu L.; Iacob, A.; Gruionu, G.; Saftoiu, A. Artificial Intelligence in Pancreatic Cancer: Toward Precision Diagnosis. *Endosc. Ultrasound* **2019**, *8*, 357. [CrossRef] [PubMed]
62. Norton, I.D.; Zheng, Y.; Wiersema, M.S.; Greenleaf, J.; Clain, J.E.; DiMagno, E.P. Neural Network Analysis of EUS Images to Differentiate between Pancreatic Malignancy and Pancreatitis. *Gastrointest. Endosc.* **2001**, *54*, 625–629. [CrossRef] [PubMed]

63. Zhu, M.; Xu, C.; Yu, J.; Wu, Y.; Li, C.; Zhang, M.; Jin, Z.; Li, Z. Differentiation of Pancreatic Cancer and Chronic Pancreatitis Using Computer-Aided Diagnosis of Endoscopic Ultrasound (EUS) Images: A Diagnostic Test. *PLoS ONE* **2013**, *8*, e63820. [CrossRef]
64. Das, A.; Nguyen, C.C.; Li, F.; Li, B. Digital Image Analysis of EUS Images Accurately Differentiates Pancreatic Cancer from Chronic Pancreatitis and Normal Tissue. *Gastrointest. Endosc.* **2008**, *67*, 861–867. [CrossRef] [PubMed]
65. Ozkan, M.; Cakiroglu, M.; Kocaman, O.; Kurt, M.; Yilmaz, B.; Can, G.; Korkmaz, U.; Dandil, E.; Eksi, Z. Age-Based Computer-Aided Diagnosis Approach for Pancreatic Cancer on Endoscopic Ultrasound Images. *Endosc. Ultrasound* **2016**, *5*, 101. [CrossRef] [PubMed]
66. Săftoiu, A.; Vilmann, P.; Gorunescu, F.; Gheonea, D.I.; Gorunescu, M.; Ciurea, T.; Popescu, G.L.; Iordache, A.; Hassan, H.; Iordache, S. Neural Network Analysis of Dynamic Sequences of EUS Elastography Used for the Differential Diagnosis of Chronic Pancreatitis and Pancreatic Cancer. *Gastrointest. Endosc.* **2008**, *68*, 1086–1094. [CrossRef]
67. Tonozuka, R.; Itoi, T.; Nagata, N.; Kojima, H.; Sofuni, A.; Tsuchiya, T.; Ishii, K.; Tanaka, R.; Nagakawa, Y.; Mukai, S. Deep Learning Analysis for the Detection of Pancreatic Cancer on Endosonographic Images: A Pilot Study. *J. Hepato-Biliary-Pancreat. Sci.* **2021**, *28*, 95–104. [CrossRef]
68. Kuwahara, T.; Hara, K.; Mizuno, N.; Okuno, N.; Matsumoto, S.; Obata, M.; Kurita, Y.; Koda, H.; Toriyama, K.; Onishi, S.; et al. Usefulness of Deep Learning Analysis for the Diagnosis of Malignancy in Intraductal Papillary Mucinous Neoplasms of the Pancreas. *Clin. Transl. Gastroenterol.* **2019**, *10*, e00045. [CrossRef]
69. Dumitrescu, E.A.; Ungureanu, B.S.; Cazacu, I.M.; Florescu, L.M.; Streba, L.; Croitoru, V.M.; Sur, D.; Croitoru, A.; Turcu-Stiolica, A.; Lungulescu, C.V. Diagnostic Value of Artificial Intelligence-Assisted Endoscopic Ultrasound for Pancreatic Cancer: A Systematic Review and Meta-Analysis. *Diagnostics* **2022**, *12*, 309. [CrossRef]
70. Viard, A.; Eustache, F.; Segobin, S. History of Magnetic Resonance Imaging: A Trip Down Memory Lane. *Neuroscience* **2021**, *474*, 3–13. [CrossRef]
71. Mao, X.; Xu, J.; Cui, H. Functional Nanoparticles for Magnetic Resonance Imaging. *WIREs Nanomed. Nanobiotechnol.* **2016**, *8*, 814–841. [CrossRef] [PubMed]
72. Hanada, K.; Shimizu, A.; Kurihara, K.; Ikeda, M.; Yamamoto, T.; Okuda, Y.; Tazuma, S. Endoscopic Approach in the Diagnosis of High-grade Pancreatic Intraepithelial Neoplasia. *Dig. Endosc.* **2022**, *34*, 927–937. [CrossRef] [PubMed]
73. Enriquez, J.S.; Chu, Y.; Pudakalakatti, S.; Hsieh, K.L.; Salmon, D.; Dutta, P.; Millward, N.Z.; Lurie, E.; Millward, S.; McAllister, F.; et al. Hyperpolarized Magnetic Resonance and Artificial Intelligence: Frontiers of Imaging in Pancreatic Cancer. *JMIR Med. Inform.* **2021**, *9*, e26601. [CrossRef] [PubMed]
74. Kaissis, G.; Ziegelmayer, S.; Lohöfer, F.; Algül, H.; Eiber, M.; Weichert, W.; Schmid, R.; Friess, H.; Rummeny, E.; Ankerst, D.; et al. A Machine Learning Model for the Prediction of Survival and Tumor Subtype in Pancreatic Ductal Adenocarcinoma from Preoperative Diffusion-Weighted Imaging. *Eur. Radiol. Exp.* **2019**, *3*, 41. [CrossRef]
75. Gao, X.; Wang, X. Performance of Deep Learning for Differentiating Pancreatic Diseases on Contrast-Enhanced Magnetic Resonance Imaging: A Preliminary Study. *Diagn. Interv. Imaging* **2020**, *101*, 91–100. [CrossRef]
76. Zhang, Y.; Wang, S.; Qu, S.; Zhang, H. Support Vector Machine Combined with Magnetic Resonance Imaging for Accurate Diagnosis of Paediatric Pancreatic Cancer. *IET Image Process.* **2020**, *14*, 1233–1239. [CrossRef]
77. Balasubramanian, A.D.; Murugan, P.R.; Thiyagarajan, A.P. Analysis and Classification of Malignancy in Pancreatic Magnetic Resonance Images Using Neural Network Techniques. *Int. J. Imaging Syst. Technol.* **2019**, *29*, 399–418. [CrossRef]
78. Corral, J.E.; Hussein, S.; Kandel, P.; Bolan, C.W.; Bagci, U.; Wallace, M.B. Deep Learning to Classify Intraductal Papillary Mucinous Neoplasms Using Magnetic Resonance Imaging. *Pancreas* **2019**, *48*, 805–810. [CrossRef]
79. Chen, Y.; Ruan, D.; Xiao, J.; Wang, L.; Sun, B.; Saouaf, R.; Yang, W.; Li, D.; Fan, Z. Fully Automated Multiorgan Segmentation in Abdominal Magnetic Resonance Imaging with Deep Neural Networks. *Med. Phys.* **2020**, *47*, 4971–4982. [CrossRef]
80. Brooks, S.L. Computed Tomography. *Dent. Clin. N. Am.* **1993**, *37*, 575–590. [CrossRef]
81. Raman, S.P.; Horton, K.M.; Fishman, E.K. Multimodality Imaging of Pancreatic Cancer—Computed Tomography, Magnetic Resonance Imaging, and Positron Emission Tomography. *Cancer J.* **2012**, *18*, 511–522. [CrossRef] [PubMed]
82. Múnera, F.; Cohn, S.; Rivas, L.A. Penetrating Injuries of the Neck: Use of Helical Computed Tomographic Angiography. *J. Trauma Inj. Infect. Crit. Care* **2005**, *58*, 413–418. [CrossRef] [PubMed]
83. Miller, T.T.; Sofka, C.M.; Zhang, P.; Khurana, J.S. Systematic Approach to Tumors and Tumor-Like Conditions of Soft Tissue. In *Diagnostic Imaging of Musculoskeletal Diseases*; Bonakdarpour, A., Reinus, W.R., Khurana, J.S., Eds.; Humana Press: Totowa, NJ, USA, 2009; pp. 313–349. ISBN 978-1-58829-947-5.
84. Willemink, M.J.; Persson, M.; Pourmorteza, A.; Pelc, N.J.; Fleischmann, D. Photon-Counting CT: Technical Principles and Clinical Prospects. *Radiology* **2018**, *289*, 293–312. [CrossRef] [PubMed]
85. Litjens, G.; Kooi, T.; Bejnordi, B.E.; Setio, A.A.A.; Ciompi, F.; Ghafoorian, M.; van der Laak, J.A.W.M.; van Ginneken, B.; Sánchez, C.I. A Survey on Deep Learning in Medical Image Analysis. *Med. Image Anal.* **2017**, *42*, 60–88. [CrossRef] [PubMed]
86. Razzak, M.I.; Naz, S.; Zaib, A. Deep Learning for Medical Image Processing: Overview, Challenges and the Future. In *Classification in BioApps*; Dey, N., Ashour, A.S., Borra, S., Eds.; Lecture Notes in Computational Vision and Biomechanics; Springer International Publishing: Cham, Switzerland, 2018; Volume 26, pp. 323–350. ISBN 978-3-319-65980-0.
87. Dhruv, B.; Mittal, N.; Modi, M. Early and Precise Detection of Pancreatic Tumor by Hybrid Approach with Edge Detection and Artificial Intelligence Techniques. *EAI Endorsed Trans. Pervasive Health Technol.* **2021**, *7*, e1. [CrossRef]

88. Drewes, A.M.; van Veldhuisen, C.L.; Bellin, M.D.; Besselink, M.G.; Bouwense, S.A.; Olesen, S.S.; van Santvoort, H.; Vase, L.; Windsor, J.A. Assessment of Pain Associated with Chronic Pancreatitis: An International Consensus Guideline. *Pancreatology* **2021**, *21*, 1256–1284. [CrossRef]
89. Ma, H.; Liu, Z.-X.; Zhang, J.-J.; Wu, F.-T.; Xu, C.-F.; Shen, Z.; Yu, C.-H.; Li, Y.-M. Construction of a Convolutional Neural Network Classifier Developed by Computed Tomography Images for Pancreatic Cancer Diagnosis. *World J. Gastroenterol.* **2020**, *26*, 5156–5168. [CrossRef]
90. Zhang, Z.; Li, S.; Wang, Z.; Lu, Y. A Novel and Efficient Tumor Detection Framework for Pancreatic Cancer via CT Images. In Proceedings of the 2020 42nd Annual International Conference of the IEEE Engineering in Medicine & Biology Society (EMBC), Montreal, QC, Canada, 20–24 July 2020; pp. 1160–1164.
91. Chen, P.-T.; Wu, T.; Wang, P.; Chang, D.; Liu, K.-L.; Wu, M.-S.; Roth, H.R.; Lee, P.-C.; Liao, W.-C.; Wang, W. Pancreatic Cancer Detection on CT Scans with Deep Learning: A Nationwide Population-Based Study. *Radiology* **2022**, *2022*, 220152. [CrossRef]
92. Barat, M.; Chassagnon, G.; Dohan, A.; Gaujoux, S.; Coriat, R.; Hoeffel, C.; Cassinotto, C.; Soyer, P. Artificial Intelligence: A Critical Review of Current Applications in Pancreatic Imaging. *Jpn. J. Radiol.* **2021**, *39*, 514–523. [CrossRef]
93. Sandland, J.; Malatesti, N.; Boyle, R. Porphyrins and Related Macrocycles: Combining Photosensitization with Radio- or Optical-Imaging for next Generation Theranostic Agents. *Photodiagnosis Photodyn. Ther.* **2018**, *23*, 281–294. [CrossRef]
94. Kumar, K.; Ghosh, A. ^{18}F-AlF Labeled Peptide and Protein Conjugates as Positron Emission Tomography Imaging Pharmaceuticals. *Bioconjug. Chem.* **2018**, *29*, 953–975. [CrossRef] [PubMed]
95. Jacobson, O.; Chen, X. PET Designated Flouride-18 Production and Chemistry. *Curr. Top. Med. Chem.* **2010**, *10*, 1048–1059. [CrossRef] [PubMed]
96. Buchmann, I.; Ganten, T.; Haberkorn, U. [^{18}F]-FDG-PET in der Diagnostik gastrointestinaler Tumoren. *Z. Gastroenterol.* **2008**, *46*, 367–375. [CrossRef] [PubMed]
97. Rosenbaum, S.J.; Lind, T.; Antoch, G.; Bockisch, A. False-Positive FDG PET Uptake—The Role of PET/CT. *Eur. Radiol.* **2006**, *16*, 1054–1065. [CrossRef] [PubMed]
98. Pakzad, F.; Groves, A.M.; Ell, P.J. The Role of Positron Emission Tomography in the Management of Pancreatic Cancer. *Semin. Nucl. Med.* **2006**, *36*, 248–256. [CrossRef] [PubMed]
99. Rankin, S. [^{18}F]2-Fluoro-2-Deoxy-D-Glucose PET/CT in Mediastinal Masses. *Cancer Imaging* **2010**, *10*, S156–S160. [CrossRef]
100. Li, S.; Jiang, H.; Wang, Z.; Zhang, G.; Yao, Y. An Effective Computer Aided Diagnosis Model for Pancreas Cancer on PET/CT Images. *Comput. Methods Programs Biomed.* **2018**, *165*, 205–214. [CrossRef]
101. Toyama, Y.; Hotta, M.; Motoi, F.; Takanami, K.; Minamimoto, R.; Takase, K. Prognostic Value of FDG-PET Radiomics with Machine Learning in Pancreatic Cancer. *Sci. Rep.* **2020**, *10*, 17024. [CrossRef]
102. Watson, M.D.; Lyman, W.B.; Passeri, M.J.; Murphy, K.J.; Sarantou, J.P.; Iannitti, D.A.; Martinie, J.B.; Vrochides, D.; Baker, E.H. Use of Artificial Intelligence Deep Learning to Determine the Malignant Potential of Pancreatic Cystic Neoplasms with Preoperative Computed Tomography Imaging. *Am. Surg.* **2021**, *87*, 602–607. [CrossRef]
103. Qureshi, T.A.; Gaddam, S.; Wachsman, A.M.; Wang, L.; Azab, L.; Asadpour, V.; Chen, W.; Xie, Y.; Wu, B.; Pandol, S.J.; et al. Predicting Pancreatic Ductal Adenocarcinoma Using Artificial Intelligence Analysis of Pre-Diagnostic Computed Tomography Images. *Cancer Biomark.* **2022**, *33*, 211–217. [CrossRef]
104. Lim, S.-H.; Kim, Y.J.; Park, Y.-H.; Kim, D.; Kim, K.G.; Lee, D.-H. Automated Pancreas Segmentation and Volumetry Using Deep Neural Network on Computed Tomography. *Sci. Rep.* **2022**, *12*, 4075. [CrossRef] [PubMed]
105. Qiao, Z.; Ge, J.; He, W.; Xu, X.; He, J. Artificial Intelligence Algorithm-Based Computerized Tomography Image Features Combined with Serum Tumor Markers for Diagnosis of Pancreatic Cancer. *Comput. Math. Methods Med.* **2022**, *2022*, 8979404. [CrossRef] [PubMed]
106. Mu, W.; Liu, C.; Gao, F.; Qi, Y.; Lu, H.; Liu, Z.; Zhang, X.; Cai, X.; Ji, R.Y.; Hou, Y.; et al. Prediction of Clinically Relevant Pancreatico-Enteric Anastomotic Fistulas after Pancreatoduodenectomy Using Deep Learning of Preoperative Computed Tomography. *Theranostics* **2020**, *10*, 9779–9788. [CrossRef] [PubMed]
107. Keogan, M.T.; Lo, J.Y.; Freed, K.S.; Raptopoulos, V.; Blake, S.; Kamel, I.R.; Weisinger, K.; Rosen, M.P.; Nelson, R.C. Outcome Analysis of Patients with Acute Pancreatitis by Using an Artificial Neural Network. *Acad. Radiol.* **2002**, *9*, 410–419. [CrossRef]
108. Qiu, W.; Duan, N.; Chen, X.; Ren, S.; Zhang, Y.; Wang, Z.; Chen, R. Pancreatic Ductal Adenocarcinoma: Machine Learning–Based Quantitative Computed Tomography Texture Analysis for Prediction of Histopathological Grade. *Cancer Manag. Res.* **2019**, *11*, 9253–9264. [CrossRef] [PubMed]
109. Liu, K.-L.; Wu, T.; Chen, P.-T.; Tsai, Y.M.; Roth, H.; Wu, M.-S.; Liao, W.-C.; Wang, W. Deep Learning to Distinguish Pancreatic Cancer Tissue from Non-Cancerous Pancreatic Tissue: A Retrospective Study with Cross-Racial External Validation. *Lancet Digit. Health* **2020**, *2*, e303–e313. [CrossRef]
110. Chu, L.C.; Park, S.; Kawamoto, S.; Wang, Y.; Zhou, Y.; Shen, W.; Zhu, Z.; Xia, Y.; Xie, L.; Liu, F.; et al. Application of Deep Learning to Pancreatic Cancer Detection: Lessons Learned from Our Initial Experience. *J. Am. Coll. Radiol.* **2019**, *16*, 1338–1342. [CrossRef]
111. Si, K.; Xue, Y.; Yu, X.; Zhu, X.; Li, Q.; Gong, W.; Liang, T.; Duan, S. Fully End-to-End Deep-Learning-Based Diagnosis of Pancreatic Tumors. *Theranostics* **2021**, *11*, 1982–1990. [CrossRef]
112. Muhammad, W.; Hart, G.R.; Nartowt, B.; Farrell, J.J.; Johung, K.; Liang, Y.; Deng, J. Pancreatic Cancer Prediction Through an Artificial Neural Network. *Front. Artif. Intell.* **2019**, *2*, 2. [CrossRef]

113. Liang, Y.; Schott, D.; Zhang, Y.; Wang, Z.; Nasief, H.; Paulson, E.; Hall, W.; Knechtges, P.; Erickson, B.; Li, X.A. Auto-Segmentation of Pancreatic Tumor in Multi-Parametric MRI Using Deep Convolutional Neural Networks. *Radiother. Oncol.* **2020**, *145*, 193–200. [CrossRef]
114. Aruna Devi, B.; PallikondaRajasekaran, M. Performance Evaluation of MRI Pancreas Image Classification Using Artificial Neural Network (ANN). In *Smart Intelligent Computing and Applications*; Smart Innovation, Systems and Technologies; Satapathy, S.C., Bhateja, V., Das, S., Eds.; Springer: Singapore, 2019; Volume 104, pp. 671–681. ISBN 9789811319204.
115. Marya, N.B.; Powers, P.D.; Chari, S.T.; Gleeson, F.C.; Leggett, C.L.; Abu Dayyeh, B.K.; Chandrasekhara, V.; Iyer, P.G.; Majumder, S.; Pearson, R.K.; et al. Utilisation of Artificial Intelligence for the Development of an EUS-Convolutional Neural Network Model Trained to Enhance the Diagnosis of Autoimmune Pancreatitis. *Gut* **2021**, *70*, 1335–1344. [CrossRef] [PubMed]
116. Săftoiu, A.; Vilmann, P.; Dietrich, C.F.; Iglesias-Garcia, J.; Hocke, M.; Seicean, A.; Ignee, A.; Hassan, H.; Streba, C.T.; Ioncică, A.M.; et al. Quantitative Contrast-Enhanced Harmonic EUS in Differential Diagnosis of Focal Pancreatic Masses (with Videos). *Gastrointest. Endosc.* **2015**, *82*, 59–69. [CrossRef] [PubMed]
117. Qureshi, T.A.; Javed, S.; Sarmadi, T.; Pandol, S.J.; Li, D. Artificial Intelligence and Imaging for Risk Prediction of Pancreatic Cancer: A Narrative Review. *Chin. Clin. Oncol.* **2022**, *11*, 1. [CrossRef] [PubMed]
118. Yu, Y.; Chen, S.; Wang, L.-S.; Chen, W.-L.; Guo, W.-J.; Yan, H.; Zhang, W.-H.; Peng, C.-H.; Zhang, S.-D.; Li, H.-W.; et al. Prediction of Pancreatic Cancer by Serum Biomarkers Using Surface-Enhanced Laser Desorption/Ionization-Based Decision Tree Classification. *Oncology* **2005**, *68*, 79–86. [CrossRef] [PubMed]
119. Brezgyte, G.; Shah, V.; Jach, D.; Crnogorac-Jurcevic, T. Non-Invasive Biomarkers for Earlier Detection of Pancreatic Cancer—A Comprehensive Review. *Cancers* **2021**, *13*, 2722. [CrossRef] [PubMed]
120. Wang, Y.; Liu, K.; Ma, Q.; Tan, Y.; Du, W.; Lv, Y.; Tian, Y.; Wang, H. Pancreatic Cancer Biomarker Detection by Two Support Vector Strategies for Recursive Feature Elimination. *Biomark. Med.* **2019**, *13*, 105–121. [CrossRef] [PubMed]
121. Wu, H.; Ou, S.; Zhang, H.; Huang, R.; Yu, S.; Zhao, M.; Tai, S. Advances in Biomarkers and Techniques for Pancreatic Cancer Diagnosis. *Cancer Cell Int.* **2022**, *22*, 220. [CrossRef]
122. Yang, J.; Xu, R.; Wang, C.; Qiu, J.; Ren, B.; You, L. Early Screening and Diagnosis Strategies of Pancreatic Cancer: A Comprehensive Review. *Cancer Commun.* **2021**, *41*, 1257–1274. [CrossRef]
123. Ko, J.; Bhagwat, N.; Yee, S.S.; Ortiz, N.; Sahmoud, A.; Black, T.; Aiello, N.M.; McKenzie, L.; OHara, M.; Redlinger, C.; et al. Combining Machine Learning and Nanofluidic Technology to Diagnose Pancreatic Cancer Using Exosomes. *ACS Nano* **2017**, *11*, 11182–11193. [CrossRef]
124. Patel, H.Y.; Mukherjee, I. A Novel Neural Network to Predict Locally Advanced Pancreatic Cancer Using 4 Urinary Biomarkers: REG1A/1B, LYVE1, and TFF1. *J. Am. Coll. Surg.* **2022**, *235*, S144–S145. [CrossRef]
125. Wong, A.N.N.; He, Z.; Leung, K.L.; To, C.C.K.; Wong, C.Y.; Wong, S.C.C.; Yoo, J.S.; Chan, C.K.R.; Chan, A.Z.; Lacambra, M.D.; et al. Current Developments of Artificial Intelligence in Digital Pathology and Its Future Clinical Applications in Gastrointestinal Cancers. *Cancers* **2022**, *14*, 3780. [CrossRef] [PubMed]
126. Kausar, T.; Kausar, A.; Ashraf, M.A.; Siddique, M.F.; Wang, M.; Sajid, M.; Siddique, M.Z.; Haq, A.U.; Riaz, I. SA-GAN: Stain Acclimation Generative Adversarial Network for Histopathology Image Analysis. *Appl. Sci.* **2021**, *12*, 288. [CrossRef]
127. Hamidinekoo, A.; Denton, E.; Rampun, A.; Honnor, K.; Zwiggelaar, R. Deep Learning in Mammography and Breast Histology, an Overview and Future Trends. *Med. Image Anal.* **2018**, *47*, 45–67. [CrossRef] [PubMed]
128. Chen, X.; Lin, X.; Shen, Q.; Qian, X. Combined Spiral Transformation and Model-Driven Multi-Modal Deep Learning Scheme for Automatic Prediction of TP53 Mutation in Pancreatic Cancer. *IEEE Trans. Med. Imaging* **2021**, *40*, 735–747. [CrossRef] [PubMed]
129. Zhou, S.; Marklund, H.; Blaha, O.; Desai, M.; Martin, B.; Bingham, D.; Berry, G.J.; Gomulia, E.; Ng, A.Y.; Shen, J. Deep Learning Assistance for the Histopathologic Diagnosis of Helicobacter Pylori. *Intell. Based Med.* **2020**, *1–2*, 100004. [CrossRef]
130. Gehrung, M.; Crispin-Ortuzar, M.; Berman, A.G.; ODonovan, M.; Fitzgerald, R.C.; Markowetz, F. Triage-Driven Diagnosis of Barretts Esophagus for Early Detection of Esophageal Adenocarcinoma Using Deep Learning. *Nat. Med.* **2021**, *27*, 833–841. [CrossRef]
131. Carter, S.M.; Rogers, W.; Win, K.T.; Frazer, H.; Richards, B.; Houssami, N. The Ethical, Legal and Social Implications of Using Artificial Intelligence Systems in Breast Cancer Care. *Breast* **2020**, *49*, 25–32. [CrossRef]
132. Shreve, J.T.; Khanani, S.A.; Haddad, T.C. Artificial Intelligence in Oncology: Current Capabilities, Future Opportunities, and Ethical Considerations. *Am. Soc. Clin. Oncol. Educ. Book* **2022**, *42*, 842–851. [CrossRef]

cancers

Article

Deep Learning for Automated Elective Lymph Node Level Segmentation for Head and Neck Cancer Radiotherapy

Victor I. J. Strijbis [1,2,*], Max Dahele [1,2], Oliver J. Gurney-Champion [3,4], Gerrit J. Blom [1], Marije R. Vergeer [1], Berend J. Slotman [1,2] and Wilko F. A. R. Verbakel [1,2]

1 Department of Radiation Oncology, Amsterdam UMC Location Vrije Universiteit Amsterdam, De Boelelaan 1117, 1081 HV Amsterdam, The Netherlands
2 Cancer Center Amsterdam, Cancer Treatment and Quality of Life, 1081 HV Amsterdam, The Netherlands
3 Department of Radiology and Nuclear Medicine, Amsterdam UMC Location University of Amsterdam, Meibergdreef 9, 1105 AZ Amsterdam, The Netherlands
4 Cancer Center Amsterdam, Imaging and Biomarkers, 1081 HV Amsterdam, The Netherlands
* Correspondence: v.strijbis@amsterdamumc.nl; Tel.: +31-6-54-32-64-03

Simple Summary: When treating patients with head-and-neck cancer (HNC), in addition to the primary tumour, commonly involved lymph node (LN) levels are often electively irradiated. This requires the definition of the elective LN target volume. Because the LN levels that will be included in the target depend on the clinical situation, and because manual contouring is a laborious task that can also introduce inter- and intra-observer variation, being able to automate the segmentation of individual LN levels would reduce the clinical burden and would allow use of contours regardless of the primary tumor location. We trained and evaluated three patch- and/or voxel-based deep learning frameworks to segment elective LN levels. Our results suggest that accurate segmentations can be obtained using an ensemble of patch-based UNets and that this result can be further refined by sequentially applying a 2.5D, multi-view voxel classification network.

Abstract: Depending on the clinical situation, different combinations of lymph node (LN) levels define the elective LN target volume in head-and-neck cancer (HNC) radiotherapy. The accurate auto-contouring of individual LN levels could reduce the burden and variability of manual segmentation and be used regardless of the primary tumor location. We evaluated three deep learning approaches for the segmenting individual LN levels I–V, which were manually contoured on CT scans from 70 HNC patients. The networks were trained and evaluated using five-fold cross-validation and ensemble learning for 60 patients with (1) 3D patch-based UNets, (2) multi-view (MV) voxel classification networks and (3) sequential UNet+MV. The performances were evaluated using Dice similarity coefficients (DSC) for automated and manual segmentations for individual levels, and the planning target volumes were extrapolated from the combined levels I–V and II–IV, both for the cross-validation and for an independent test set of 10 patients. The median DSC were 0.80, 0.66 and 0.82 for UNet, MV and UNet+MV, respectively. Overall, UNet+MV significantly ($p < 0.0001$) outperformed other arrangements and yielded DSC = 0.87, 0.85, 0.86, 0.82, 0.77, 0.77 for the combined and individual level I–V structures, respectively. Both PTVs were also significantly ($p < 0.0001$) more accurate with UNet+MV, with DSC = 0.91 and 0.90, respectively. The accurate segmentation of individual LN levels I–V can be achieved using an ensemble of UNets. UNet+MV can further refine this result.

Keywords: computed tomography; deep learning; head-and-neck cancer; lymph nodes; radiation oncology; auto-contouring

Citation: Strijbis, V.I.J.; Dahele, M.; Gurney-Champion, O.J.; Blom, G.J.; Vergeer, M.R.; Slotman, B.J.; Verbakel, W.F.A.R. Deep Learning for Automated Elective Lymph Node Level Segmentation for Head and Neck Cancer Radiotherapy. *Cancers* **2022**, *14*, 5501. https://doi.org/10.3390/cancers14225501

Academic Editors: Hamid Khayyam, Ali Madani, Rahele Kafieh and Ali Hekmatnia

Received: 30 September 2022
Accepted: 7 November 2022
Published: 9 November 2022

1. Introduction

Head-and-neck cancer (HNC) radiotherapy (RT) planning frequently includes the contouring of neck lymph nodes (LN) as a part of the elective RT target volume. However,

the manual delineation of the elective target volume is a labour-intensive task that is prone to inter-observer variation [1], despite the availability of delineation guidelines [2], making automated methods attractive, as an alternative to manual segmentation.

Over the last few years, developments in deep learning approaches have shown impressive results for automated segmentation of organs at risk (OAR) by using convolutional neural networks (CNN) [3–6] and for pathology detection [7], including the deep learning-based delineation of elective targets such as the combinations of neck LN levels, which has only more recently been investigated [8]. Most studies that demonstrated automated LN segmentation with deep learning, incorporated all LN levels or all of those levels relevant to the primary HNC location in one structure, rather than focusing on individual LN levels [7–11]. The methods that segment multiple lymph levels in one structure, however, are not generalizable to all primary HNC locations and tumour stages and require separate networks for contouring different combinations of lymph node levels. Therefore, it would be desirable to have a more general and flexible approach that concurrently and accurately contours individual LN levels and hence can be used for all HNC patients regardless of the subtype and the specific lymph levels required for RT treatment planning.

The automated segmentation of the LN levels is a challenging task because of anatomical limitations in the manual reference. The guidelines prescribe delineation based on anatomical markers in axial slices and assume that no voxels of levels II, III and IV can exist in the same axial plane, irrespective of the curvature and pitch of the neck. In addition, the LN target volumes do not encompass anatomical structures, but rather the expansions of groups of LNs.

In this work, three combinations of deep learning networks were investigated to segment individual LN levels I–V as separate structures. To do this, we evaluated the performance of two CNNs, alone and in combination. First, since UNet is a widely established CNN that is used for a variety of imaging-related problems [12] and since it was used in two other studies for combined lymph structure segmentation [9,13], we included a patch-based UNet variant as a baseline model configuration. Other works have suggested the use of voxel-classification methods for individual LN level segmentation using a 3D multi-scale network [14], as well as 2.5D (multi-view; MV) networks for several segmentation challenges (multiple sclerosis [15], ocular structures [16], abdominal lymph structures [17], head-and-neck tumors [18]). Because 2.5D networks may more effectively learn features in the presence of little data [19] and because voxel classification may better resolve local ambiguities near level transitions, a multi-view convolutional neural network (MV-CNN) was included as our second configuration. This method, however, appears limited by a systematic over-estimation of foreground classes [18]. Therefore, as our third configuration, UNet was used for foreground segmentation, and subsequently MV was used for classifying the foreground voxels into individual LN levels. This way, the over-estimation of foreground classes seen in MV models was effectively eliminated.

This work expands the existing literature by demonstrating the feasibility of deep learning for auto-segmentation for the target definition of individual LN levels I–V towards a flexible RT planning for locally advanced HNC. Based on earlier work, we estimate that accurate performance levels are attained for the segmentation of individual LN levels I–V with Dice similarity coefficient (DSC) of at least 0.8 [9,13,14] and we hypothesize that the contours can be obtained with such accuracy levels for the majority of patients, using one or more of the proposed deep learning configurations.

2. Materials and Methods

2.1. Data Acquisition

This retrospective study was exempted from requiring participants' informed consent by the medical ethics committee and was performed using the three-dimensional (3D) planning computed tomography (CT) scans (GE Discovery 590RT, helically scanned) of 70 patients treated between 2019 and 2022 with (chemo-)radiotherapy for locally advanced HNC, of which 60 were used for training and testing, and 10 were retained for an inde-

pendent test set. We used isotropic, in-plane acquisition resolutions of [0.92–1.56] mm and a 2.5 mm slice thickness, except for two cases in which the slice thickness was 1.25 mm. The CT acquisition dimensions were $512 \times 512 \times (147 - 361)$ voxels. Patient-specific radiotherapy head-and-neck moulds and immobilisation masks were used to position the patients in a neutral position. Ground truth contours were created for the specific purpose of this study by manual contouring of individual LN levels I–V according to contouring guidelines [1], by two experienced radiation oncologists (GJB, MRV). During contouring, levels IV and V are regarded as the combinations of IVa, IVb and Va, Vb, respectively. No HNC disease stages or patients with positive LNs were excluded, provided that they had elective LN levels contoured for at least one side. In patients with only one side contoured, the LN level contours of the side that contained no diagnosed disease were added, such that all patients had all individual levels at both sides contoured.

2.2. Pre-Processing

For all patients, planning CTs and structure sets were initially interpolated to the same isotropic 1.25 mm^3 voxel spacing by 3rd-order and nearest-neighbour interpolation, respectively. This spacing was chosen to minimize image interpolations, whilst making sure the network's filters were of equal size in each orthogonal plane for all patients.

2.3. Experimental Outline

We investigated the performance of three model configurations, i.e., UNet (Figure 1), MV (Figure 1) and UNet+MV (Figure 1). As a baseline reference, we investigated a multi-class, patch-based UNet, which concurrently classifies all lymph levels as separate classes in a single step. This was compared to a per-voxel classification approach that uses an MV-CNN, which is a 2.5D network that uses multiple resolutions of orthogonal views to classify the voxel where the planes cross. In the interest of time, this model used a preconstructed mask to provide the network with the information on which voxels it should consider for segmentation (cyan in Figure 1). Lastly, we investigated a two-step approach, which is essentially a combination of UNet and MV: we used a single-class UNet for segmenting the combined structure of LN levels I–V in an initial step, after which MV was applied only to the detected foreground voxels and classified each voxel in the combined structure into individual lymph levels UNet+MV. Schematic representations of the used UNet and MV networks are displayed in the blue and red boxes in Figures 1 and 2, respectively.

2.4. Model Training

Model training, validation and evaluation were performed on four NVIDIA-GeForce GTX 2080 TI graphics processor units (GPUs), a 64 GB RAM system with an Intel® Core™ i9-9900KF CPU @3.6 GHz processor, using the GPU version of TensorFlow (Version 2.2.0) with Cuda 10.1 and Python (Version 3.8.10). The TensorBoard (Version 2.2.2) callback was used for tracking the training and validation scores, whilst only the best model in terms of DSC was saved. The models were trained using the Adam optimizer [20]. All models were trained using standard values in Keras, with an initial learning rate of 0.001, $\beta_1 = 0.9$, $\beta_2 = 0.999$ and $\varepsilon = 1 \times 10^{-7}$ To reduce the divergence of the model weights at later stages of training, an exponential learning rate decay scheduler was used to decrease the learning rate by 5% with every epoch, up to a minimum of 0.0001. Dropout was switched off at test time. All models were trained using 5-fold cross-validation, with a train\test split of 48\12 cases every fold. To minimize the training variation, we used ensemble learning [9,21–23], where the highest cumulated in-class segmentation probability of 5 sequentially trained networks decided the final segmentation map. The training and evaluation times were saved.

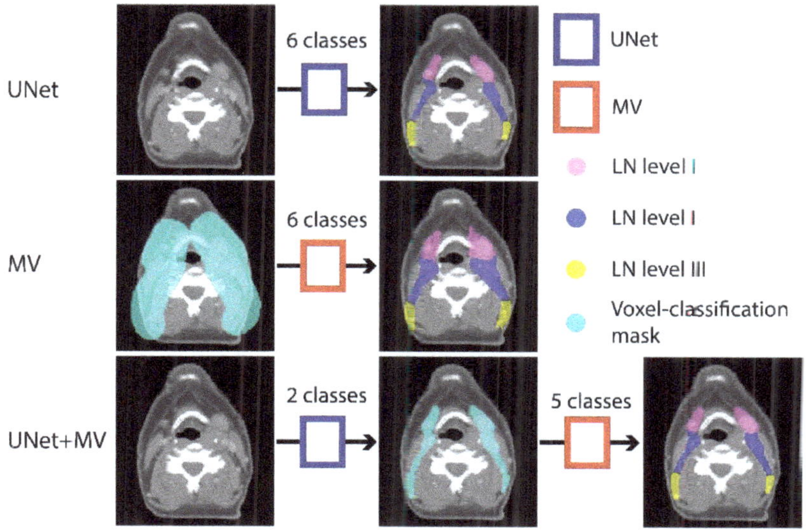

Figure 1. Schematic overview of the experimental outline. UNet (blue boxes) and MV (red boxes) were used to make three model configurations. In the first configuration, a patch-based UNet segments the background and LN levels I–V directly from the planning CT. In the second configuration, MV classifies the background and LN levels I–V voxels from within a preconstructed mask (cyan). In UNet+MV, a patch-based UNet first segments the combined structure of LN levels I–V. This is subsequently used as a mask (cyan) for MV to subsequently classify positive voxels into individual levels I–V. The details of both models are given in Figure 2. Abbreviations: MV: multi-view; CT: computed tomography.(Also shows in Figure S2).

2.4.1. UNet

The network that was used is an adaptation of a vanilla UNet [12], where residual blocks were added to reduce the effect from vanishing gradients in deeper layers of the model [24,25], similar to those used by Milleran et al. [26] Batch normalization was performed after every (3 × 3 × 3) 3D convolution, before the non-linear activation function. We used patch-based training of the 3D UNet to ensure the network fitted on our video card [27,28]. During training, patches of 64 × 64 × 64 voxels were sampled randomly from two pre-defined, unilateral regions of interest (ROI) of 280 × 200 × 280 mm^3 in volume that were known to contain the combined structure of LN levels I–V for every patient on each side. Binary and multi-class dice loss functions were used for optimization. The multi-class DSC loss was defined as the sum of individual foreground class losses (Equation (1)):

$$DSC_{loss} = \sum_{m=1}^{M} W_m \cdot DL_m \quad (1)$$

Here, W_m are the class weights that are calculated using the Python's scikit-learn module [29], m ranges from 1 to M and denotes class indices, where M is the number of classes. DL is the DSC loss, defined as 1 minus the DSC score (Equation (2)):

$$DL_m = 1 - \frac{2 \cdot |A_m \cap B_m|}{|A_m| + |B_m|} \quad (2)$$

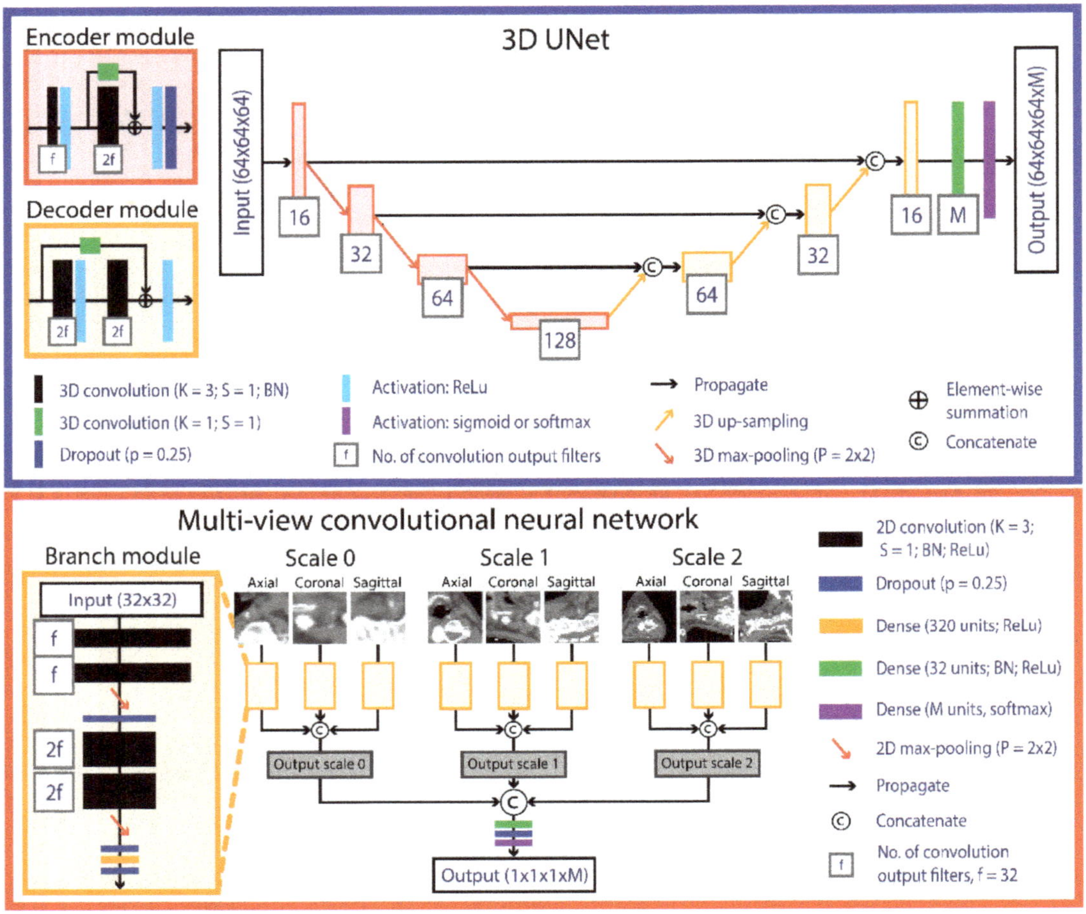

Figure 2. Schematic overview of the UNet (blue box) and MV (red box) networks. UNet consists of an encoder (**left**) and a decoder (**right**) pathway that generates binary segmentation maps from 64 cubed voxel patches sampled from planning CTs. MV uses three multi-view branches that build up to each anatomical plane within a scale block, the output of which is concatenated and used as the input for the multi-scale branched architecture. The thickness of the convolutional blocks corresponded with the number of filters used. The number of output classes (M) was six for UNet in the UNet-only configuration and two for UNet in the UNet+MV configuration. M was six for MV in the MV-only configuration and five in the UNet+MV configuration. Abbreviations: MV: multi-view; ch: number of channels; BN: batch normalization; ReLu: rectified linear unit; f: number of output filters; M: number of output classes; K: convolution kernel size; S: convolution stride; BN: batch normalization; p: dropout fraction: CT: computed tomography.

Here, A_m and B_m denote the predicted and manual reference binary sets of class m, respectively. In the case of binary segmentation, DSC_{loss} is reduced to the latter loss function. For patches that contain a limited amount of foreground voxels, DSC_{loss} becomes ill-defined (the denominator in DL_m is not constrained to values larger than 0). To ameliorate this, we used a Gaussian sampling method, where the mean and standard deviation of the x, y and z coordinates are calculated from the centre of mass of the combined, binary structure of LN levels I–V of all patients. Subsequently, we used a truncated normal distribution to sample

patches, such that they were constrained to be entirely within the region of interest. The weights were initialized using the standard initialization method in Keras (glorot uniform initialization). The models were optimized for 100 epochs. However, it should be noted that the use of an epoch in a patch-based setting is arbitrary, because patch sampling is perfrmed at random, and thus a different sub-set of all data is seen by the network in each epoch. The number of training pairs seen by the network per epoch was set to 4096, which corresponded to roughly 34 training patches per side per patient.

2.4.2. Multi-View

MV-CNN is a voxel-wise classification method, for which we predefined which voxels to classify. For C2, this information was provided by a pre-constructed mask, indicated in cyan in Figure 1, which was constructed by a uniform expansion of the manual reference by a margin of 15 mm. This margin was chosen as a balance such that no foreground voxels would be segmented at the border of this mask, while also minimizing the training and evaluation times. In contrast, for C3, the pre-constructed mask was determined by the foreground segmentation result of UNet. Our multi-view network was adapted from a previous classification study [16]. Batch normalization was applied after every (3 × 3) 2D convolution layer, before the non-linear activation function. Three context pyramid scales, 0, 1 and 2, were included to incorporate multi-view information from 4, 8 and 16 cm around the query voxel, respectively. This was done by sampling every, every other and every fourth voxel for scales 0, 1 and 2, respectively, for each view. Fewer pyramid scales yielded inferior results, and more pyramid scales would cause the field of view to fall far outside the ROI. The loss function used for voxel classification was categorical cross-entropy (CCE; Equation (3)):

$$H(p,q) = -\sum_{m=1}^{M}\sum_{a=1}^{A} p(a,m)\log(q(a,m)) \quad (3)$$

where $p(a,m)$ represents a reference distribution of $a \in A$, given by the manual annotations, $q(a,m)$ is a query distribution, A is a set of observations, m denotes class indices and ranges from 1 to M, and M is the number of classes. The network was optimized for 1000 epochs (batch size = 32). In every epoch, a different random sub-set of at maximum 20% of all training pairs was sampled to allow for varied training and validation. Random over-sampling of minority classes was applied to reduce the effects of class imbalance.

2.4.3. Data Augmentation

Data augmentations were the same for all models and were performed on the fly. Augmentation involved random flipping, rotation and contrast adaptation, with chances of each augmentation occurring being 50%, 40% and 40%, respectively. Flipping was carried out in the left–right direction. Rotation was applied in either the sagittal or the transversal plane, with an angle that was uniformly sampled from [−5, +5 degrees]. Rotated images were acquired by 3rd order spline interpolation for the CT image and by nearest-neighbour interpolation for the corresponding segmentation maps. The default window level center (C_C) [width (C_W)] was 0 [700], as was previously used for lymph structure segmentation [9]. If contrast adaptation was applied, alternative window level center, and width were sampled from normal distributions, with $\mu_C = 0$; $\sigma_C = 3\% \times 700$ and $\mu_W = 700$; $\sigma_W = 3\% \times 700$, respectively.

2.5. Post-Processing

In all segmentation maps, the combined structure of LN levels I–V was post-processed with hole filling and by subsequently removing all but the largest connected components. To investigate the agreement in the resulting planning target volumes (PTV), the resulting segmentations of combined structures of LN levels I–V and II-IV were expanded by a margin of 4 mm and were denoted as PI–PV and PII–PIV. These two PTVs were chosen because they were used for planning the majority of HNC sub-types.

2.6. Evaluation and Statistical Analysis

The evaluation of a full 3D image by UNet was achieved by sliding the 64 × 64 × 64 UNet field of view over the image with stride 32 and subsequently only evaluating the central 32 × 32 × 32 voxels. By doing this, we ensured that the network had sufficient context for reliable inferences, while also making sure that each voxel was classified exactly once. The spatial performance of all models was measured by using DSC, Hausdorff distance (HD) and mean surface distance (MSD) between predictions and manual contours and between the PTVs that resulted from the predictions and manual contours. Because the measures were not normally distributed upon histogram inspection and omnibus test of normality [30], the differences in spatial performance were evaluated by a two-sided Wilcoxon signed-rank test. Bonferroni correction was applied for each model and spatial metric separately to account for multiple comparisons. The volumetric agreement was assessed with intra-class correlation [31] (ICC; two-way mixed effects, single measurement, consistency) coefficients and volume outside of the manual contour. Finally, cases with a median DSC in the lowest quartile of the UNet+MV configuration were qualitatively reviewed by GJB. Cases from each quartile (Q1–Q3), as well as several informative examples, were chosen for display, such as one patient who underwent laryngectomy surgery. This case was included during training to maximize the number of training samples but was omitted from the calculations of the model performance metrics, because the anatomical landmarks normally required for manual contouring were not present in this patient's anatomy.

2.7. Independent Validation

To assess the model generalizability, the two best performing models (UNet and UNet+MV) were tested on the independent test set of 10 patients. These were unique samples that were not seen or used during the model development. For this independent testing, the UNet and UNet+MV models were re-trained using the complete cross-validation dataset (60 patients) as the training data. All training and evaluation settings were identical to the cross-validation setting.

3. Results

In the cross-validation set, the mean age ± standard deviation was 64.0 ± 10.4 (N = 49) and 58.5 ± 4.9 (N = 11) for males and females, respectively. UNet and UNet+MV showed better agreement with the manual reference than MV for the complete LN structure, all individual LN levels and both PTVs (Table 1). UNet+MV typically showed the better segmentation performance of the combined LN structure, individual levels II–IV and both PTV structures (Figures 3–6; Table 1). In addition, UNet+MV showed the highest volumetric agreement with the manual reference for all structures (Figure 4). Overall, UNet+MV significantly ($p < 0.0001$) outperformed the other models, with the DSCs (median [interquartile range (Q1–Q3)]) of all individual LN structures present in the dataset being 0.804 [0.763–0.814], 0.658 [0.616–0.678] and 0.821 [0.769–0.831] for the models UNet, MV and UNet+MV, respectively. Even with some deformation, e.g., patient not aligned straight in the mask, median-level DSC results were attained (e.g., Figure 3, second column). MV often (Figure 3, Ax. 1 and Cor. 1 rows) overestimated the segmented combined LN volume medially.

UNet+MV showed significantly higher DSCs for the complete LN level I–V structure, individual levels II–IV and both PTV structures (Figure 5, p-values in figure). However, UNet showed higher spatial agreement with the manual reference for LN level I. The transitions of LN levels II–III by UNet+MV typically agreed most strongly with the manual reference. All models commonly disagreed with the manual reference on the caudal and cranial ends of LN level V. In addition, there existed a substantial disagreement on the lateral and dorsal ends of this structure in the model predictions. The models benefitted marginally from ensembling all model configurations for all classes, as the results from the model ensembles were more consistent (Table 1). The models were optimized for a median [range] of 10.3 [9.6–10.9] h, except for MV–only, which was optimized for 19.8 [18.7–21.2] h. The inference time for all UNet models was 2.1 [1.8–2.4] minutes, whereas the MV inference

time, which is proportional to the size of the input mask, was 6.0 [5.4–7.0] and 1.0 [0.8–1.3] minutes per patient for the MV–only and UNet+MV configurations, respectively.

Table 1. The reported values denote the range of median DSCs produced by five individual models and ensemble model combinations of UNet, MV and UNet+MV configurations after post-processing. Ensemble results that showed higher spatial agreement than the most accurate individual model are denoted in bold. Ensembles increased result consistency and typically outperformed any of the standalone models for all configurations. Abbreviations: MV: multi-view; Ind. individual; Ens: ensemble; LN: lymph node.

	Cross-Validation						Independent Test	
	UNet		MV		UNet+MV		UNet	UNet+MV
	Ind.	Ens.	Ind.	Ens.	Ind.	Ens.	Ens.	Ens.
LN I–V	[0.850–0.852]	**0.857**	[0.692–0.706]	**0.708**	[0.860–0.862]	**0.867**	0.846	0.865
LN I	[0.849–0.855]	**0.860**	[0.682–0.695]	**0.700**	[0.851–0.856]	**0.857**	0.856	0.852
LN II	[0.827–0.834]	**0.840**	[0.702–0.720]	**0.726**	[0.856–0.858]	**0.862**	0.824	0.850
LN III	[0.771–0.781]	0.781	[0.628–0.653]	**0.656**	[0.802–0.812]	0.810	0.755	0.825
LN IV	[0.714–0.746]	**0.748**	[0.559–0.585]	0.583	[0.757–0.764]	0.764	0.743	0.724
LN V	[0.738–0.751]	**0.754**	[0.572–0.604]	**0.610**	[0.753–0.761]	**0.763**	0.697	0.707
PI–PV	[0.897–0.898]	**0.899**	[0.779–0.788]	**0.798**	[0.899–0.900]	**0.908**	0.892	0.904
PII–PIV	[0.887–0.891]	**0.892**	[0.768–0.782]	**0.788**	[0.899–0.900]	**0.902**	0.893	0.892

By visually comparing the model and manual reference contour pairs in the worst-performing quartile (N = 15), several trends were observed. First, the manual reference was judged to be suboptimal (i.e., not according to the contouring guidelines; Figure 6A–E) for at least one level in 6/15 cases. In these six patients, one, four, two, one and three inaccuracies were found in each respective LN level I–V. Second, the level II–III transitions predicted by UNet+MV were typically more accurate than those obtained from the manual reference, and UNet+MV also often outperformed UNet at this transition (Figure 6F–I). Third, the predictions of LN level II by UNet and UNet+MV were visually more accurate than those of the manual reference at the cranial limit (Figure 6J). Fourth, the automated methods showed a large variation in disagreement with the manual reference for LN level V (Figure 6A,E,H,I,L). In cases where the automated methods showed considerable disagreement with the manual reference, pitch, rotation and/or tilt were often underlying confounders (Figure 6K–M), especially for LN level V (Figure 6M), or there were anatomical variations such as malnourishment (Figure 6F) and laryngectomy (with fewer anatomical landmarks available; Figure 6N–O)). Cases with a coronal tilt showed disagreement in contralateral structures of the same level (Figure 6M). Among cases of the first quartile, there were no particularities in the manual reference.

In the independent test, the mean age ± standard deviation was 66.3 ± 10.1 (N = 7) and 64.3 ± 13.6 (N = 3) years for males and females, respectively. The median [interquartile range (Q1–Q3)] DSCs of all individual LN level structures were 0.769 [0.703–0.834] and 0.809 [0.729–0.852] by the UNet and UNet+MV configurations, respectively, and differed significantly ($p < 0.0001$). UNet+MV showed significantly higher DSCs for the complete I–V structure, LN levels II and III, as well as both extrapolated PTVs (Table 1; Figure 7). For reference, volumetric performances of the independent test set are included in Figure S1.

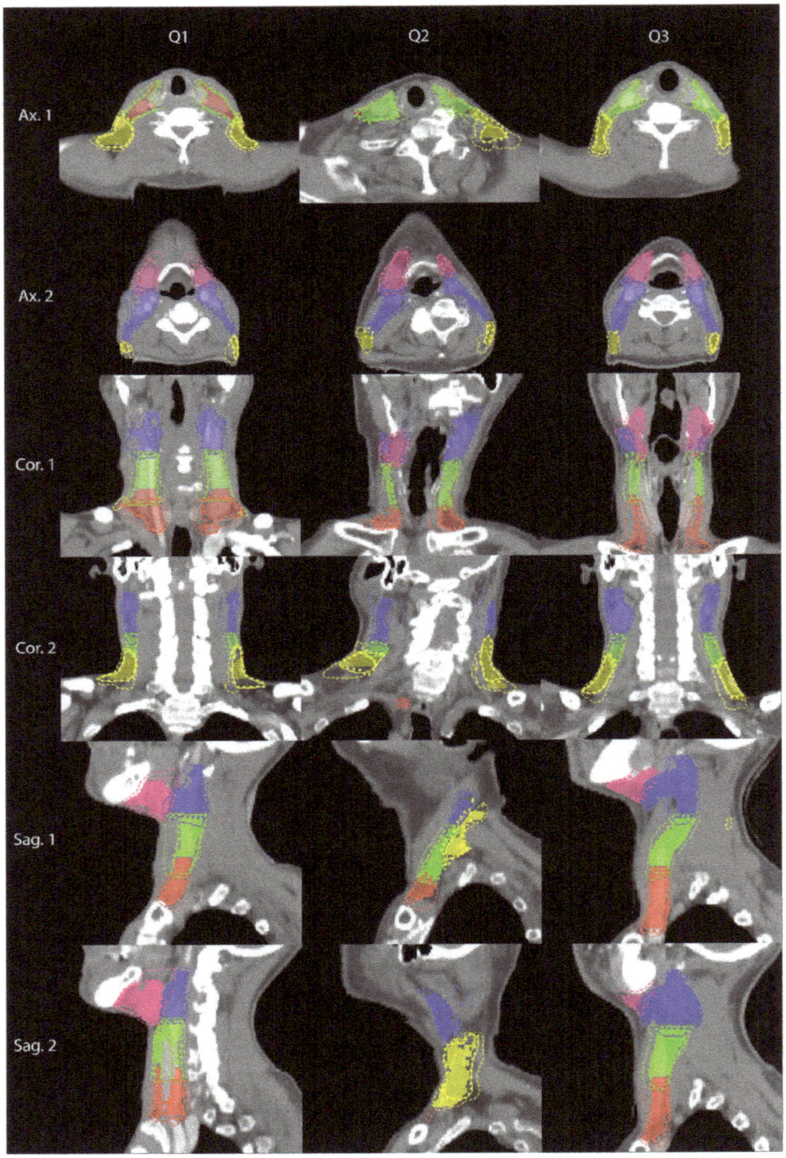

Figure 3. Example segmentations selected from the first (Q1), second (Q2) and third (Q3) quartile in terms of DSC averaged over individual LN levels I–V. The filled region is the manual reference. The solid, dashed and dotted lines correspond to the predictions of the model configurations of UNet, MV and UNet+MV, respectively. LN levels I–V are indicated in pink, blue, green, red and yellow, respectively. The low average DSC in Q1 was in part attributed to an error in the manual reference level III–IV transition. Abbreviations: DSC: dice similarity coefficient; LN: lymph node.

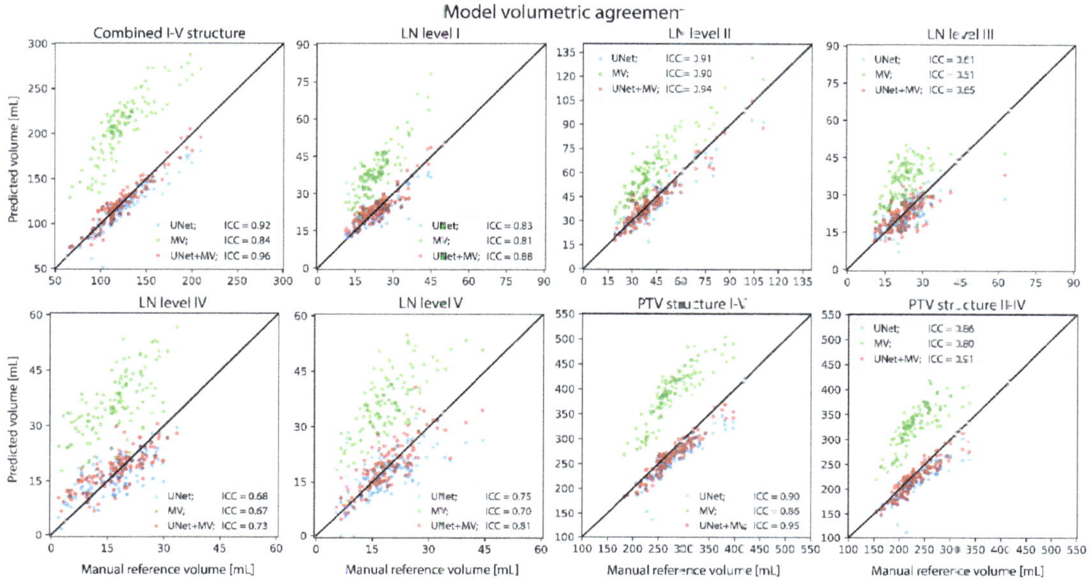

Figure 4. Predicted and manual reference volumes for all structures. Abbreviations: ICC: intra-class correlation (two-way mixed, single measures, consistency).

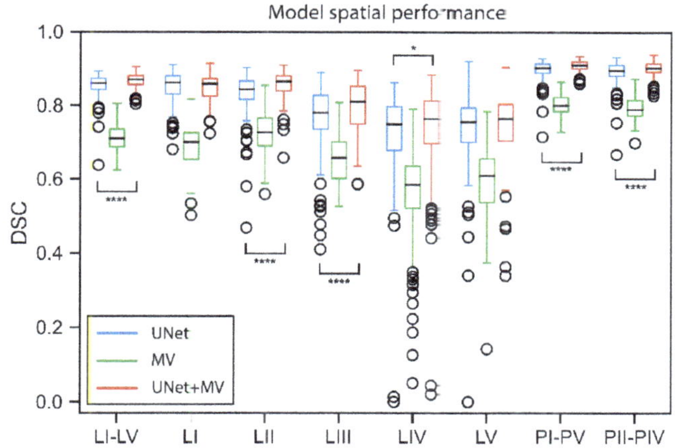

Figure 5. *Cont.*

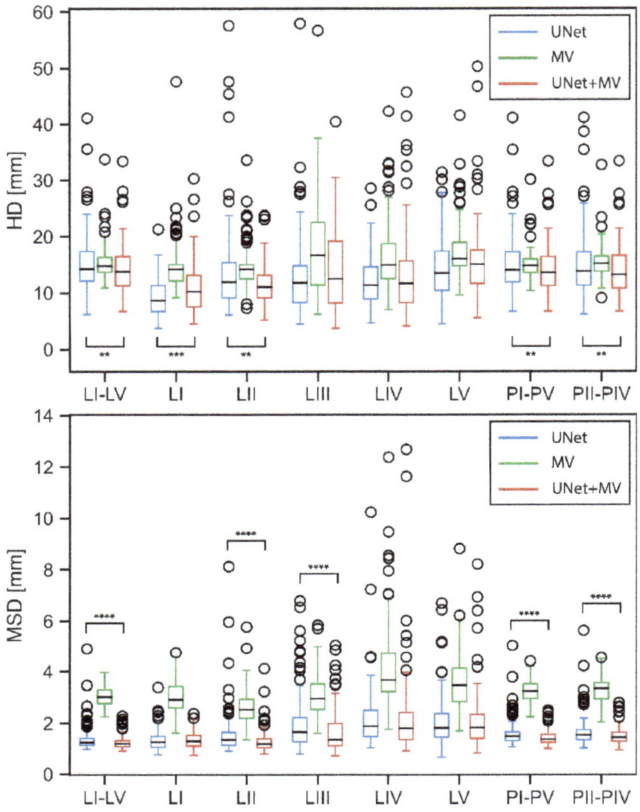

Figure 5. Spatial performances of UNet, MV and UNet+MV model configurations for DSC, HD and MSD measures. Statistical significance marking of the MV configuration was omitted because differences between MV and other model configurations were always significant. Structures for which differences between UNet and UNet+MV were statistically significant are denoted by significance bars. *: $p < 0.05$; **: $p < 0.01$; ***: $p < 0.001$; ****: $p < 0.0001$; Abbreviations: DSC: dice similarity coefficient; MV: multi-view; HD: Hausdorff distance; MSD: mean surface distance.

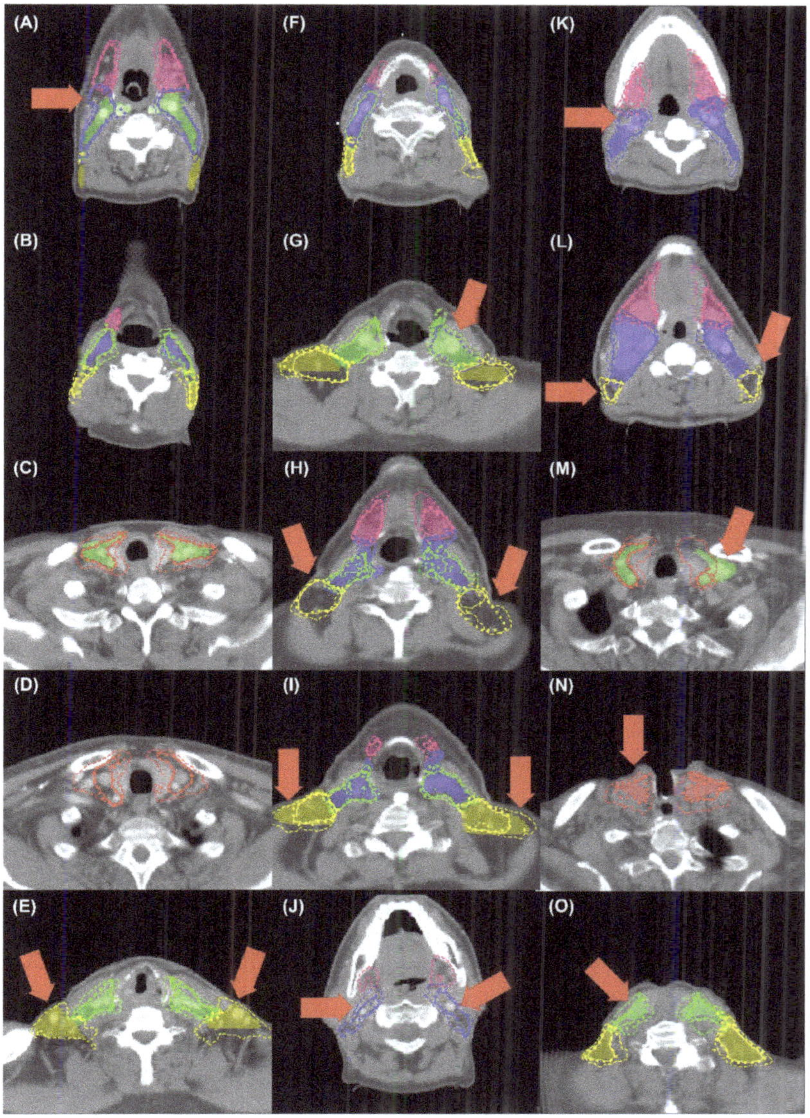

Figure 6. Examples from the worst-performing quartile samples in terms of DSC averaged over individual LN levels I–V. The filled region is the manual reference. The solid, dashed and dotted lines correspond to the predictions of the UNet, MV and UNet+MV model configurations, respectively. LN levels I–V are indicated in pink, blue, green, red and yellow, respectively. Arrows indicate specific locations of interest. Abbreviations: DSC: dice similarity coefficient; LN: lymph node.

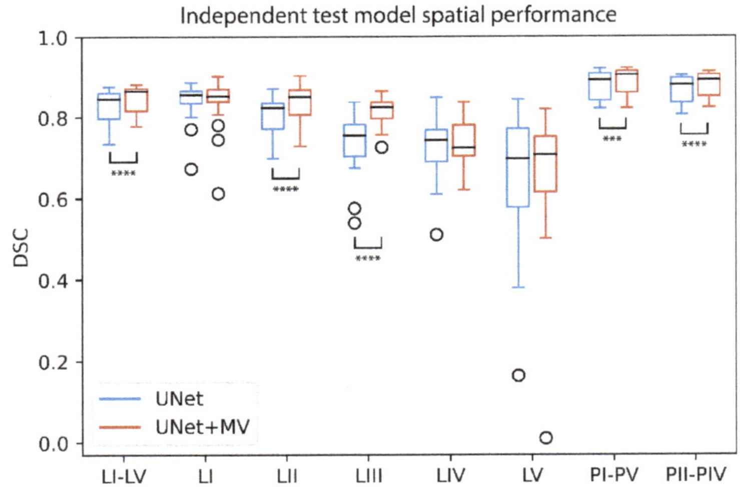

Figure 7. UNet and UNet+MV spatial model performances in the independent test. Structures for which differences between model configurations were statistically significant are denoted by significance bars. ***: $p < 0.001$; ****: $p < 0.0001$; Abbreviations: DSC: Dice similarity coefficient; MV: multi-view.

4. Discussion

Our results suggest that accurate contours of individual LN levels I–V can be obtained using UNet (complete I–V structure median DSC = 0.859; individual structure DSC = 0.804), and that these results can be further refined by using a UNet+MV sequential model (complete I–V structure DSC = 0.866; individual structure DSC = 0.821). Despite a limited gain compared to UNet, UNet+MV exhibited a significantly better spatial performance for the complete I–V structure, individual levels II–IV and both PTV structures, and better volumetric performance for all structures. Comparable results were achieved using an independent test set for the model configurations UNet and UNet+MV, suggesting that the models have the ability to generalize beyond the data used for model training and development.

These results, however, should be interpreted with some care. A review of patients with a median DSC in the lowest quartile ($N = 15$) highlighted cases where the automated methods were factually closer to the truth than the manual reference, due to inconsistencies in the manual reference that arose from patient angulation and anatomical limitations in contouring guidelines (Figure 6A–M). In addition, all models were considerably less accurate for levels IV and V. Several factors may have contributed to this. First, it is known that DSC is dependent on the structure size [32]; therefore, the small volumes of the levels IV–V likely negatively influenced DSC, which was especially true for malnourished patients (Figure 6M). Such a case was observed in the independent test, where LN level V had a manual reference volume below the typical range (5 mL) and was almost completely missed (Figure 7; Figure S1-LN level V). Second, despite the measures that were taken to prevent most patient angulation during scanning, considerable patient angulation was sometimes seen. This could be due to anatomical variations and to some patients' inability to lie with their head down. This may also have contributed to a larger variation in the manual reference and may have led to disagreements between the predictions and the manual reference. This problem has recently been addressed in another study by Weissmann et al. [13]. Because the contouring guidelines do not take into account the curvature of the neck and the patient's pitch, tilt and rotation, it can be argued that the predictions may be more factually "correct" than the manual reference when this is the case. Alternatively, if the goal of DL methods is to emulate the contouring guidelines, the networks could be trained using explicit information of slice orientation. Variations

in slice plane orientation are especially problematic for level V, because, for example, the lower axial end of this structure contour in the manual reference is defined by the "plane just below the transverse cervical vessel" [2,33]. This caused larger inconsistencies in the manual reference for LN-level V highly pitched patients, compared to patients with other levels. The same holds for the contralateral structures from the same level for patients with a coronal tilt. The current guidelines prescribe level contours of both sides starting at the same axial slice clinically, even though the tilt leads to different predictions for either side for the automated methods. Similarly, although predictions generally show disagreement in the caudal end of level IV and both axial ends and dorsal borders of level V, it should not be concluded that predictions are inaccurate for these regions. Rather, the way that the contouring guidelines were set up can cause peculiarities for patients with large pitch and/or tilt when comparing with more standardized, automated methods. Although it could seem like the rational step to take, it is not a given fact that redefining contouring guidelines to be less dependent on anatomical landmarks in a certain slice and patient angulation would be better for the clinical practice. Such guidelines would be more labour-intense for the clinician, which will need to consider more strongly the 3D information of the patient. However, such an approach may result in more accurate data, which in the long run, will be more informative to the network and result in more consistent contours.

To put the results of this study into perspective, we compared our results to others in the relevant literature on automated lymph level segmentation of combined lymph levels, which reported a mean DSC range of 0.64–0.82 [34]. Commercially available contouring software (Limbus Contour build 1.0.22) was evaluated for the neck lymph nodal structures [11], but it was reported that the performance could still be improved (mean DSC = 0.75). Cardenas et al. reported an accurate segmentation performance of the combined LN level I–V and II-IV clinical target volumes (CTV; both DSC = 0.90) [9], but it should be noted that an inspection of example segmentations suggested that these structures more closely resembled PTV structures from our institute. We believe that our finding of PTV overlap of UNet and UNet+MV (PTV I–V and II–IV DSCs = 0.91, 0.90, respectively) is in line with, if not better than, the segmented structures reported by Cardenas et al. To the best of our knowledge, the work of Van der Veen et al. [14] was the first to involve the automated segmentation of individual levels I and V and reported segmentation accuracies (without expert intervention) of DSC = 0.73, 0.61 and 0.79 for levels I and V and the combined II–IV structure, respectively. Interestingly, however, these results seem to more closely resemble the results obtained with our second configuration (level I, V DSCs = 0.70, 0.61, respectively). This is not unexpected, because the MV configuration involves a direct voxel classification method that uses multiple scales, similar to the proposed method by Van der Veen et al., but differs in the 2.5D convolution kernel, whereas Van der Veen et al. used a fully 3D kernel.

The model application times are sufficient for clinical use, but can still be improved. Typical whole-image full segmentation by UNet takes time in the order of seconds, but since this UNet was trained in a patch-based fashion, it required application to all parts of the image, such that each part of the image was seen by the 32 × 32 × 32 center patch exactly once. This procedure was not optimized for speed and could likely still be accelerated considerably. Similarly, the MV models were not optimized for speed. For example, when processing neighbouring voxels, there existed much overlap between the extracted patches, even though each patch was extracted separately in the current implementation.

Our research has some limitations. First, we only indirectly investigated the implications of model predictions for RT treatment planning by investigating the overlap of the two predicted PTVs with the manual reference. Future work may investigate whether the predicted volumes lead to improved dose–volume histograms in OARs and target volumes when using them in a treatment planning system. Second, we did not include LN levels VI and VII because these are less frequently clinically used. Since these are central levels and require a larger region of interest to be considered for learning, deep learning frameworks

aiming to include these structures may focus on patch-based training with sampling from both sides simultaneously or by defining two left/right and one central ROI.

5. Conclusions

We demonstrated that a UNet can accurately (DSC > 0.8) segment individual LN levels I–V for the majority of patients and that this result can be further refined by using a UNet for the segmentation of foreground structures, followed by a sequential voxel classification network. With this generalized approach, any set of lymph levels can be combined to define patient-specific LN level target structures. When dealing with angulated patients, one should be aware that the current contouring guidelines can lead to situations where the LN level contours may become inconsistent, which may be prevented by using more standardized, automated deep learning methods. Future work should investigate whether clinically acceptable RT plans can be obtained using predicted contours.

Supplementary Materials: The following supporting information can be downloaded at: https://www.mdpi.com/article/10.3390/cancers14225501/s1, Figure S1: Predicted and manual reference volumes for all structures resulting from the independent test set. Abbreviations: ICC: intra-class correlation (two-way mixed, single measures, consistency). Figure S2: Graphical representation of the summary of this work.

Author Contributions: Conceptualization, V.I.J.S., M.D. and W.F.A.R.V.; Methodology: V.I.J.S., M.D., O.J.G.-C. and W.F.A.R.V.; Software: V.I.J.S. and O.J.G.-C.; Validation: V.I.J.S.; Formal analysis: V.I.J.S.; Investigation: V.I.J.S., M.D., O.J.G.-C. and W.F.A.R.V.; Resources: G.J.B., B.J.S. and M.R.V.; Data curation: V.I.J.S., G.J.B. and M.R.V.; Writing—original draft preparation: V.I.J.S.; Writing—review and editing: V.I.J.S., M.D., O.J.G.-C., G.J.B., B.J.S. and W.F.A.R.V.; Visualization: V.I.J.S.; Supervision: M.D., O.J.G.-C. and W.F.A.R.V.; Project administration: W.F.A.R.V. and B.J.S.; Funding acquisition: M.D., B.J.S. and W.F.A.R.V. All authors have read and agreed to the published version of the manuscript.

Funding: This work was funded by a research grant from Varian Medical Systems.

Institutional Review Board Statement: Due to the retrospective nature of this study, not including tests or different treatments on human individuals and the use of pseudonamized data, it was concluded by the Medical ethics committee that the Medical Research involving Human Subjects Act (WMO) does not apply to this study and that official approval of this study by the committee is not required.

Informed Consent Statement: This retrospective study was exempted from requiring participants' informed consent by the medical ethics committee.

Data Availability Statement: The data used in this study is not made publically available for reasons of consent of the participants.

Conflicts of Interest: This work was funded by a research grant from Varian Medical Systems. W.F.A.R.V. and M.D. have received speakers honoraria/travel expenses from Varian Medical Systems outside the current work. O.J.G. was funded by the Dutch cancer society (KWF) grant numbers KWF-UVA 2017.10873 and KWF-UVA 2021.13785. KWF funding was outside the scope of this work.

References

1. Van der Veen, J.; Gulyban, A.; Nuyts, S. Interobserver variability in delineation of target volumes in head and neck cancer. *Radiother. Oncol.* **2019**, *137*, 9–15. [CrossRef] [PubMed]
2. Grégoire, V.; Ang, K.; Budach, W.; Grau, C.; Hamoir, M.; Langendijk, J.A.; Lee, A.; Le, Q.-T.; Maingon, P.; Nutting, C.; et al. Delineation of the primary tumour Clinical Target Volumes (CTV-P) in laryngeal, hypopharyngeal, oropharyngeal and oral cavity squamous cell carcinoma: AIRO, CACA, DAHANCA, EORTC, GEORCC, GORTEC, HKNPCSG, HNCIG, IAG-KHT, LPRHHT, NCIC CTG, NCRI, NRG Oncolog. *Int. J. Radiat. Oncol.* **2019**, *104*, 677–684. [CrossRef]
3. Van Rooij, W.; Dahele, M.; Brandao, H.R.; Delaney, A.R.; Slotman, B.J.; Verbakel, W.F.A.R. Deep Learning-Based Delineation of Head and Neck Organs at Risk: Geometric and Dosimetric Evaluation. *Int. J. Radiat. Oncol.* **2019**, *104*, 677–684. [CrossRef] [PubMed]
4. Zhu, W.; Huang, Y.; Zeng, L.; Chen, X.; Liu, Y.; Qian, Z.; Du, N.; Fan, W.; Xie, X. AnatomyNet: Deep learning for fast and fully automated whole-volume segmentation of head and neck anatomy. *Med. Phys.* **2018**, *46*, 576–589. [CrossRef] [PubMed]
5. Wang, W.; Wang, Q.; Jia, M.; Wang, Z.; Yang, C.; Zhang, D.; Wen, S.; Hou, D.; Liu, N.; Wang, P. Deep Learning-Augmented Head and Neck Organs at Risk Segmentation From CT Volumes. *Front. Phys.* **2021**, *9*, 1–11. [CrossRef]

6. Kawahara, D.; Tsuneda, M.; Ozawa, S.; Okamoto, H.; Nakamura, M.; Nishio, T.; Saito, A.; Nagata, Y. Stepwise deep neural network (stepwise-net) for head and neck auto-segmentation on CT images. *Comput. Biol. Med.* **2022**, *143*, 105295. [CrossRef]
7. Ali, R.; Hardie, R.C.; Narayanan, B.N.; Kebede, T.M. IMNets: Deep Learning Using an Incremental Modular Network Synthesis Approach for Medical Imaging Applications. *Appl. Sci.* **2022**, *12*, 5500. [CrossRef]
8. Men, K.; Chen, X.; Zhang, Y.; Zhang, T.; Dai, J.; Yi, J.; Li, Y. Deep deconvolutional neural network for target segmentation of nasopharyngeal cancer in planning computed tomography images. *Front. Oncol.* **2017**, *7*, 315. [CrossRef]
9. Cardenas, C.E.; Beadle, B.M.; Garden, A.S.; Skinner, H.D.; Yang, J.; Rhee, D.J.; McCarroll, R.E.; Netherton, T.J.; Gay, S.S.; Zhang, L. Generating High-Quality Lymph Node Clinical Target Volumes for Head and Neck Cancer Radiation Therapy Using a Fully Automated Deep Learning-Based Approach. *Int. J. Radiat. Oncol.* **2021**, *109*, 801–812. [CrossRef]
10. Chen, A.; Deeley, M.A.; Niermann, K.J.; Moretti, L.; Dawant, B.M. Combining registration and active shape models for the automatic segmentation of the lymph node regions in head and neck CT images. *Med. Phys.* **2010**, *37*, 6338–6346. [CrossRef]
11. Stapleford, L.J.; Lawson, J.D.; Perkins, C.; Edelman, S.; Davis, L.; McDonald, M.W.; Waller, A.; Schreibmann, E.; Fox, T. Evaluation of Automatic Atlas-Based Lymph Node Segmentation for Head-and-Neck Cancer. *Int. J. Radiat. Oncol.* **2010**, *77*, 959–966. [CrossRef]
12. Ronneberger, O.; Fischer, P.; Brox, T. U-net: Convolutional networks for biomedical image segmentation. In *Lecture Notes in Computer Science, Proceedings of the Medical Image Computing and Computer-Assisted Intervention 2015, Munich, Germany, 5–9 October 2015*; Springer: Cham, Switzerland, 2015. [CrossRef]
13. Weissmann, T.; Huang, Y.; Fischer, S.; Roesch, J. Deep Learning for automatic head and neck lymph node level delineation. *Int. J. Radiat. Oncol. Biol. Phys.* **2022**, 1–17. Available online: https://arxiv.org/abs/2208.13224 (accessed on 1 October 2022).
14. Van der Veen, J.; Willems, S.; Bollen, H.; Maes, F.; Nuyts, S. Deep learning for elective neck delineation: More consistent and time efficient. *Radiother. Oncol.* **2020**, *153*, 180–188. [CrossRef]
15. Birenbaum, A.; Greenspan, H. Multi-view longitudinal CNN for multiple sclerosis lesion segmentation. *Eng. Appl. Artif. Intell.* **2017**, *65*, 111–118. [CrossRef]
16. Strijbis, V.I.J.; de Bloeme, C.M.; Jansen, R.W.; Kebiri, H.; Nguyen, H.-G.; de Jong, M.C.; Moll, A.C.; Bach-Cuadra, M.; de Graaf, P.; Steenwijk, M.D. Multi-view convolutional neural networks for automated ocular structure and tumor segmentation in retinoblastoma. *Sci. Rep.* **2021**, *11*, 14590. [CrossRef]
17. Roth, H.R.; Lu, L.; Seff, A.; Cherry, K.M.; Hoffman, J.; Wang, S.; Liu, J.; Turkbey, E.; Summers, R.M. A New 2.5D Representation for Lymph Node Detection Using Random Sets of Deep Convolutional Neural Network Observations. In *Lecture Notes in Computer Science, Proceedings of the Medical Image Computing and Computer-Assisted Intervention (MICCAI), Boston, MA, USA, 14–18 September 2014*; Springer: Cham, Switzerland, 2014; Volume 17, pp. 520–527. [CrossRef]
18. Schouten, J.P.; Noteboom, S.; Martens, R.M.; Mes, S.W.; Leemans, C.R.; de Graaf, P.; Steenwijk, M.D. Automatic segmentation of head and neck primary tumors on MRI using a multi-view CNN. *Cancer Imaging* **2022**, *22*, 8. [CrossRef]
19. Aslani, S.; Dayan, M.; Storelli, L.; Filippi, M.; Murino, V.; Rocca, M.A.; Sona, D. Multi-branch convolutional neural network for multiple sclerosis lesion segmentation. *NeuroImage* **2019**, *196*, 1–15. [CrossRef]
20. Kingma, D.P.; Ba, J. Adam: A Method for Stochastic Optimization. In Proceedings of the 3rd International Conference of Learning Representations (ICLR), San Diego, USA, 7–9 May 2015.
21. Ren, J.; Eriksen, J.G.; Nijkamp, J.; Korreman, S.S. Comparing different CT, PET and MRI multi-modality image combinations for deep learning-based head and neck tumor segmentation. *Acta Oncol.* **2021**, *60*, 1399–1406. [CrossRef]
22. Van Rooij, W.; Dahele, M.; Nijhuis, H.; Slotman, B.J.; Verbakel, W.F.A.R. OC-0346: Strategies to improve deep learning-based salivary gland segmentation. *Radiat. Oncol.* **2020**, *15*, 272. [CrossRef]
23. Van Rooij, W.; Verbakel, W.F.; Slotman, B.J.; Dahele, M. Using Spatial Probability Maps to Highlight Potential Inaccuracies in Deep Learning-Based Contours: Facilitating Online Adaptive Radiation Therapy. *Adv. Radiat. Oncol.* **2021**, *6*, 100658. [CrossRef]
24. He, K.; Zhang, X.; Ren, S.; Sun, J. Identity Mappings in Deep Residual Networks. In *European Conference on Computer Vision*; Springer: Cham, Switzerland, 2016.
25. He, K.; Zhang, X.; Ren, S.; Sun, J. Deep residual learning for image recognition. In Proceedings of the 2016 IEEE Conference on Computer Vision and Pattern Recognition (CVPR), Las Vegas, NV, USA, 27–30 June 2016. [CrossRef]
26. Milletari, F.; Navab, N.; Ahmadi, S.-A. V-Net: Fully convolutional neural networks for volumetric medical image segmentation. In Proceedings of the 2016 Fourth International Conference on 3D Vision (3DV), Stanford, CA, USA, 25–28 October 2016.
27. Wu, D.; Kim, K.; Li, Q. Computationally efficient deep neural network for computed tomography image reconstruction. *Med. Phys.* **2019**, *46*, 4763–4776. [CrossRef] [PubMed]
28. Bouman, P.M.; Strijbis, V.I.J.; Jonkman, L.E.; Hulst, H.E.; Geurts, J.J.G.; Steenwijk, M.D. Artificial double inversion recovery images for (juxta)cortical lesion visualization in multiple sclerosis. *Mult. Scler. J.* **2022**. [CrossRef] [PubMed]
29. Pedregosa, F.; Varoquaux, G.; Gramfort, A.; Michel, V.; Thirion, B.; Grisel, O.; Blondel, M.; Prettenhofer, P.; Weiss, R.; Dubourg, V. Scikit-learn: Machine learning in Python. *J. Mach. Learn. Res.* **2011**, *12*, 2825–2830.
30. D'Agostino, B.B. An omnibus test of normality for moderate and large size samples. *Biometrika* **1971**, *58*, 341–348. [CrossRef]
31. Koo, T.K.; Li, M.Y. A Guideline of Selecting and Reporting Intraclass Correlation Coefficients for Reliability Research. *J. Chiropr. Med.* **2016**, *15*, 155–163. [CrossRef]

32. Wack, D.S.; Dwyer, M.G.; Bergsland, N.; Di Perri, C.; Ranza, L.; Hussein, S.; Ramasamy, D.; Poloni, G.; Zivadinov, R. Improved assessment of multiple sclerosis lesion segmentation agreement via detection and outline error estimates. *BMC Med. Imaging* **2012**, *12*, 17. [CrossRef]
33. Grégoire, V.; Ang, K.; Budach, W.; Grau, C.; Hamoir, M.; Langendijk, J.A.; Lee, A.; Quynh-Thu, L.; Maingon, P.; Nutting, C.; et al. Delineation of the neck node levels for head and neck tumors: A 2013 update. DAHANCA, EORTC, HKNPCSG, NCIC CTG, NCRI, RTOG, TROG consensus guidelines. *Radiother. Oncol.* **2014**, *110*, 172–181. [CrossRef]
34. Nogues, I.; Lu, L.; Wang, X.; Roth, H.; Bertasius, G.; Lay, N.; Shi, J.; Tsehay, Y.; Summers, R.M. Automatic lymph node cluster segmentation using holistically-nested neural networks and structured optimization in CT images. In *Lecture Notes in Computer Science, Proceedings of the Medical Image Computing and Computer-Assisted Intervention–MICCAI 2016, Athens, Greece, 17–21 October 2016*; Springer: Cham, Switzerland, 2016. [CrossRef]

diagnostics

Article

DECT-CLUST: Dual-Energy CT Image Clustering and Application to Head and Neck Squamous Cell Carcinoma Segmentation

Faicel Chamroukhi [1,*], Segolene Brivet [2], Peter Savadjiev [3], Mark Coates [2] and Reza Forghani [3,4]

1 IRT SystemX, 2 Boulevard Thomas Gobert, 91120 Palaiseau, France
2 Electrical and Computer Engineering Department, McGill University, Montreal, QC H3A 0G4, Canada
3 Augmented Intelligence and Precision Health Laboratory (AIPHL), Department of Radiology, McGill University, Montreal, QC H3G 1A4, Canada
4 Radiomics and Augmented Intelligence Laboratory (RAIL), Department of Radiology and the Norman Fixel Institute for Neurological Diseases, University of Florida College of Medicine, Gainesville, FL 32610, USA
* Correspondence: faicel.chamroukhi@irt-systemx.fr

Abstract: Dual-energy computed tomography (DECT) is an advanced CT computed tomography scanning technique enabling material characterization not possible with conventional CT scans. It allows the reconstruction of energy decay curves at each 3D image voxel, representing varied image attenuation at different effective scanning energy levels. In this paper, we develop novel unsupervised learning techniques based on mixture models and functional data analysis models to the clustering of DECT images. We design functional mixture models that integrate spatial image context in mixture weights, with mixture component densities being constructed upon the DECT energy decay curves as functional observations. We develop dedicated expectation–maximization algorithms for the maximum likelihood estimation of the model parameters. To our knowledge, this is the first article to develop statistical functional data analysis and model-based clustering techniques to take advantage of the full spectral information provided by DECT. We evaluate the application of DECT to head and neck squamous cell carcinoma. Current image-based evaluation of these tumors in clinical practice is largely qualitative, based on a visual assessment of tumor anatomic extent and basic one- or two-dimensional tumor size measurements. We evaluate our methods on 91 head and neck cancer DECT scans and compare our unsupervised clustering results to tumor contours traced manually by radiologists, as well as to several baseline algorithms. Given the inter-rater variability even among experts at delineating head and neck tumors, and given the potential importance of tissue reactions surrounding the tumor itself, our proposed methodology has the potential to add value in downstream machine learning applications for clinical outcome prediction based on DECT data in head and neck cancer.

Keywords: spectral image clustering; dual-energy CT imaging; mixture models; functional data analysis; HNSCC cancer

Citation: Chamroukhi, F.; Brivet, S.; Savadjiev, P.; Coates, M.; Forghani, R. DECT-CLUST: Dual-Energy CT Image Clustering and Application to Head and Neck Squamous Cell Carcinoma Segmentation. *Diagnostics* **2022**, *12*, 3072. https://doi.org/10.3390/diagnostics12123072

Academic Editor: Kalevi Kairemo

Received: 28 October 2022
Accepted: 1 December 2022
Published: 6 December 2022

1. Introduction

Computed tomography (CT) has been one of the most common and widespread imaging techniques used in the clinic for the last few decades. There is increasing interest in a more advanced CT technique known as dual-energy CT (DECT) or spectral CT that enables additional material or tissue characterization beyond what is possible with conventional CT. In conventional CT, X-rays are emitted at a certain level of energy, whereas in DECT, they are emitted at two separate energy levels, which brings important benefits as compared to standard CT. First, since different materials can have different attenuation coefficients at different energy levels, DECT allows for the separation of materials with different atomic numbers. In particular, DECT enables the computation of image attenuation levels at

multiple effective energy levels. This results in the association of a decay curve with each reconstructed image voxel, representing energy-dependent changes in attenuation at that body location. Conventional CT imaging is the first-line modality for the clinical evaluation of many different types of known or suspected cancers in adults [1]. However, because of the properties described above, there has been an increased interest in the use of DECT in oncology in recent years, as it provides a new and exciting way of characterizing tumors as well as their surrounding tissues.

Current expert evaluation of CT scans of head and neck cancer patients in clinical practice is largely based on qualitative image evaluation for the delineation of the tumor anatomic extent and basic two- or three-dimensional measurements. However, an increasing body of evidence suggests that quantitative texture or radiomic features extracted from CT images can be used to enhance diagnostic evaluation, including the prediction of tumor molecular phenotypes, prediction of therapy response, and outcome prediction [2–5]. The typical radiomic approach for tumor evaluation can be separated into two steps: (i) identification and segmentation of the tumor in the image; and (ii) prediction of a clinical endpoint of interest based on features extracted from the segmented image region. As such, the ability of the segmentation algorithm to correctly target the tumor region is the first essential step in this process. If performed by an expert, manual processing is prohibitively time-consuming and prone to intra- and inter-observer variability. This step is ideally suited for computerized analysis to make these types of analyses feasible and more reproducible. When conventional single-energy CT (SECT) scans represent a 3D image of a patient, DECT scans may be viewed as a 4D image: a 3D body volume over a range of spectral attenuation levels. The latter dimension provides, for each voxel, a decay curve representing energy-dependent changes in attenuation, enabling tissue characterization beyond what is possible with conventional CT [6,7].

DECT has been shown to improve qualitative image interpretation for the evaluation of head and neck cancer and preliminary results also suggest that the energy-dependent curves associated with each image voxel can be used to improve predictions using radiomic approaches [8–12]. However, there is currently no widely accepted method for the use of spectral data from DECT scans for radiomic type studies. In this study, we propose a clustering method that incorporates the spectral tissue attenuation curves as a fourth dimension of the 3D representation of tissue voxels in head and neck DECT scans. We demonstrate that by combining spectral information from the voxel-associated curves and spatial information from the voxel coordinates, we can create a segmentation map with high concordance with tumoral tissue voxels. The proposed model provides a clustering of high qualitative aspect, that can act as the basis for identifying tumor or peritumoral regions to be used in subsequent radiomic studies on DECT scans.

In this paper, we evaluate the application of DECT to head and neck squamous cell carcinoma (HNSCC). Current image-based evaluation of HNSCC tumors in clinical practice is largely qualitative, based on a visual assessment of tumor anatomic extent and basic one- or two-dimensional tumor size measurements. However, the frequently complex shape of mucosal head and neck cancers and at times poorly defined boundaries and potential adjacent tissue reactions can result in a high inter-observer variability in defining the extent of the tumor [13–16], especially among radiologists without sub-specialty expertise in head and neck imaging. Furthermore, there is strong evidence that alterations of gene expression and protein–protein interactions in the peri-tumoral tissue or normal adjacent tissue may play a critical role in the evolution and risk of recurrence of HNSCC tumors (e.g., [17,18]). Yet these adjacent regions may not be obviously salient upon visual image examination. For all these reasons, the accurate and consistent determination of a predictive region around the tumor is essential, both for conventional staging which determines patient management and for future automated quantitative image-based predictive algorithms based on machine learning. Such a predictive region may include the tumor itself, but also surrounding tissues of biological relevance.

With these considerations in mind, DECT provides new and unexplored opportunities to answer an important question: what can be learned from DECT about tumor heterogeneity and its associations with surrounding tissues? As a first step toward answering this question, in this paper, we adapt functional data analysis (FDA) techniques to DECT data in order to explore underlying patterns of association in and around the tumor. FDA is a classical branch of statistics dedicated to the analysis of functional data, in situations where each data object is considered a function. This is particularly appropriate for DECT image data, as each 3D image voxel is associated with a curve of image intensity decay over multiple reconstructed energy levels (more details are provided below in Section 2.1). Thus, we adapt FDA statistical models for the clustering of 3D image voxels based on the full functional information provided by the decay curves associated with each voxel. More specifically, the architecture underpinning our proposed method is a functional mixture model, where the mixture component densities are built upon functional approximation of the spectral decay curves at each image voxel, and the mixture weights are constructed to integrate spatial constraints. We then derive an expectation–maximization (EM) algorithm to the maximum likelihood estimation (MLE) of the model parameters.

To our knowledge, this is the first article to propose spatial clustering utilizing the full spectral information available in DECT data, based on an appropriate FDA statistical framework. Existing methods for automatic tumor delineation in DECT (reviewed in detail in Section 2.2) are mostly based on deep learning techniques and utilize only a small subset of the available information, due to the sheer amount of 4D (spatial + spectral) data available in a single DECT scan.

We apply the proposed methodology on 91 DECT scans of HNSCC tumors, and we compare our results to manually traced tumor contours performed by an experienced expert radiologist. We also compare to other baseline clustering methods. However, tumor segmentation on its own is not a clinical outcome. A full demonstration of the clinical utility of our method necessitates an analysis of its ability to predict actual clinical outcomes, and how this prediction performance compares to the performance in the case of manually drawn contours, or contours drawn using alternative automatic methods. We leave this prediction analysis for a subsequent paper. As the first article to adapt FDA statistical tools to DECT data, the main focus of the present paper is on the statistical methodology and on algorithm development. As such, we can summarize our contributions as follows:

1. We extend the statistical framework of mixture models to the spatio-spectral heterogeneous DECT data. In particular, DECT energy decay curves observed at each image voxel are modeled as spatially distributed functional observations;
2. We develop unsupervised learning algorithms for clustering by incorporating full spectral information from DECT data;
3. To our knowledge, this is the first time that these models are applied to DECT;
4. The source codes of our algorithms are publicly available https://github.com/fchamroukhi/DECT-CLUST (accessed on 28 October 2022), free of charge.

The rest of this paper is organized as follows; First, as a background, we describe in Section 2 related work on dual-energy CT and dedicated segmentation methods. Then, in Section 3, we introduce the proposed methodology and present the developed algorithms. Section 4 is dedicated to the experimental study and the obtained quantitative results are provided in Section 5. Finally, in Section 6, we discuss the proposed approach and the obtained results.

2. Background and Related Work

2.1. Dual-Energy CT

The use of DECT techniques has very recently attracted interest in different clinical applications, including diagnostics; for example using DECT for the improved detection of portal vein thrombosis via virtual monoenergetic reconstructions [19], for reducing visceral-

motion-related artifacts on the liver by comparing different CT scanner techniques [20], or also using DECT of the heart to the study of coronary artery disease [21].

DECT data may be viewed as a 4D image of a patient: a 3D body volume over a range of energy levels. The dual-energy image acquisition using two X-ray energy peaks at the source provides enough attenuation information to be combined and to be able to reconstruct a curve at multiple "virtual monochromatic" energy levels. These simulate what the attenuation (in Hounsfield units; HU) would be if the study was acquired with a monochromatic X-ray beam at that energy value (in kilo-electron-Volt; keV). The reconstructed curve of attenuation numbers over each energy level translates the energy-dependent changes and is commonly called the spectral Hounsfield unit attenuation curve, or an *energy decay curve* [7]. In our method, we will make use of this spectral information through functional approximations, and thus consider the curves as functional observations. An energy decay curve is calculated for each image voxel, and thus, a DECT scan is represented as a 4D image with 3 dimensions for X, Y and Z spatial coordinates and 1 dimension for energy level coordinates. The virtual monochromatic image (VMI) is the 3D image representation at a given energy level. See Figure 1 (left) for examples of a 2D slice from different VMIs and Figure 1 (right) for examples of decay curves for different tissue characteristics.

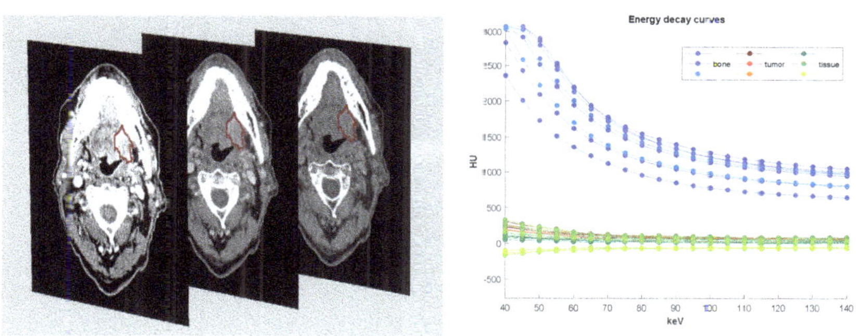

Figure 1. (**Left**) 2D slices of VMIs at 40,65,140 keV with tumor contour in red. At lower energy levels, VMIs are more constrated; at higher levels, VMIs are less noisy. A VMI at 65 keV is similar to a standard CT scan. (**Right**) Examples of decay curves for different body locations. A blue (resp. red, green) curve represents attenuation information stored at one voxel within bone (resp. tumor, tissue).

2.2. *Segmentation of Dual-Energy CT Data*

Segmentation is a process of delineating an image region of interest. For example, radiation oncologists usually manually segment tumors for radiation planning. Automatic tumor segmentation has a long history of developments: from knowledge-driven early techniques to data-driven newer techniques, algorithms aim to extract image features to make a decision on region boundaries [22]. However, this process remains challenging in medical imaging due to the heterogeneity over the image or the acquisition process; most of the current algorithms need manual adjustments on the result [23].

In head and neck CT imaging, the difficulty to contour precisely a tumor region results in large inter-observer variability in the segmentation results, even among trained radiation oncologists. A study among radiologists from 14 different institutions obtained a median Dice similarity score (DSC) ranging from 0.51 to 0.82 [15], depending on the delineation criteria used. Another study assessing the same variability among 3 experienced radiologists over 10 tumors obtained a mean DSC of 0.57 [16].

To the best of our knowledge, only a few studies have focused on DECT segmentation. These employ deep learning approaches [24–26]. The four dimensions of the data required workarounds in order to apply neural networks. For example, using two VMIs sampled from the energy level spectrum, one at a low- and one at a high-energy level, Chen et al. in [24] merged the two VMIs in a layer connected to a U-Net architecture [27].

Wang et al. in [26] learned features from two pyramid networks on the two VMIs independently and combined them through deep attention into a mask-scoring regional convolutional neural network (R-CNN). They achieved good performance in segmenting large-sized organs (DSC larger than 0.8), and performance was less impressive for small-sized organs (DSC between 0.5 and 0.8). Deep learning techniques have indeed revolutionized the world and are very popular in many domains. However, despite what they can provide as good results in practice, we note that as a mixture model-based approach, our proposed approach is not necessarily as complex as a deep learning one can be and is not regarded as a black box. It also enjoys interpretability and relies on the statistical sound background of mixture models and the desirable properties of the EM algorithm, in particular, the fact of monotonically improving the loglikelihood as a loss-function. It is also user-friendly, which can be useful in particular for clinical use, and its implementation is also quite simple.

2.3. Decay Curve Clustering via Functional Data Analysis

FDA aims to represent infinite-dimensional functional data into a finite-dimensional vector of coefficients [28]. To achieve this, FDA consists in expanding functional data into function bases. One approach relies on projection on bases which consist in projecting functional data onto finite dimensional function bases (e.g., splines, B-splines, polynomials, Fourier, and wavelet). It associates a finite vector of projection coefficients. This is what we use in this paper. Analogously another common approach would be to run a functional principal component analysis (fPCA) to obtain a basis of eigenfunctions of the covariance of the process describing our functional data. It associates a (truncated) projection vector of PCA coefficients.

Our objective is to partition our data, modeled with FDA, in different groups of voxels having similar decay curve characteristics. Among the available clustering approaches (e.g., centroid-based clustering, such as k-means; connectivity-based clustering, such as hierarchical clustering; density-based clustering, such as DBSCAN [29]; and distribution-based clustering with model-based methods), since we have a model for each decay curve, a model-based approach is preferred.

Model-based clustering is a thoroughly developed field [30,31], particularly for multivariate analysis. Model-based clustering approaches rely on the finite mixture modeling framework [32] to represent the density of a set of independent multivariate observations and on an optimization algorithm to automatically find a partition into groups of such observations.

To represent different groups of data, mixture models assume each datum to follow a known distribution (e.g., Gaussian), and build a mean representation (i.e., model) for each group of data. The mixture model calculates, for each data point, a value defined by the sum over $k = \{1 \ldots \#\text{groups}\}$, of the probability distribution function (pdf) that this point belongs to group k model, emphasized by a weight giving a higher or lower chance of belonging to this group (derived in Section 3.1). Mixture models have the advantage of being interpretable, parametric, thus well-understood, and flexible, as the pdf modeling the data in each cluster can be chosen explicitly.

The expectation–maximization (EM) algorithm [33] is a popular and adapted tool with desirable properties that can be used to conduct an iterative estimation of the mixture model parameters and thus the cluster membership probabilities.

Mixture models for clustering have been applied and adapted to different kind of data, including time-series data [34], gene expression data [35], 3D noisy medical images [36], or spatio-temporal data (non image data) [37]. They also have been recently investigated for functional data [38], and thus provide an avenue to model the spectral decay curves, but in this context of spectral images, we also need to incorporate the spatial information into the clustering.

A related idea was proposed in [39] to develop a spatio-temporal mixture of hidden process models for fMRI analysis. The authors built a temporal probabilistic model, and

reshaped the prior probability with spatial constraints to determine a "region of influence" for the temporal model. A specification of our model covers this approach, and our model goes further in generalizing it via the construction of more flexible Gaussian-mixture weights around the spatial coordinates. The resulting model enjoys better numerical learning properties with faster convergence due to closed-form updating rules for the spatial weights parameters. Alternative constructions of the proposed mixture model are also presented in order to validate the method and to accommodate potential user specifications.

3. Methodology

3.1. Generative Modeling Framework

We adopt the framework of generative modeling for image clustering using different families of extended mixture distributions. The general form of the generative model for the image assumes that the ith datum (i.e., pixel and voxel) in the image has the general semi-parametric mixture density:

$$f_i(\boldsymbol{\theta}) = \sum_{k=1}^{K} \pi_{ik} f(\boldsymbol{\theta}_k), \quad (1)$$

which is a convex combination of K component densities, $f_i(\boldsymbol{\theta}_k)$, $k \in [K] = \{1, \cdots, K\}$, weighted by non-negative mixture weights π_{ik} that sum to one, that is $\sum_{k=1}^{K} \pi_{ik} = 1$ for all i, $i \in [n]$. The unknown parameter vector $\boldsymbol{\theta}$ of density (1) is composed of the set of component density parameters $\{\boldsymbol{\theta}_k\}$ and their associated weights $\{\pi_{ik}\}$, i.e., $\boldsymbol{\theta} = \{\pi_{ik}, \boldsymbol{\theta}_k\}_{k=1}^{K}$.

From the perspective of model-based clustering of the image, each component density can be associated with a cluster, and hence the clustering problem becomes one of parametric density estimation. Suppose that the image has K segments and let $Z_i \in [K]$ be the random variable representing the unknown segment label of the ith observation in the image. Suppose that the distribution of the data within each segment $k \in [K]$ is $f(\boldsymbol{\theta}_k)$. i.e, $f_{iZ_i=k}(\boldsymbol{\theta}) = f_i(\boldsymbol{\theta}_k)$. Then, from a generative point of view, model (1) is equivalent to *(i)* sampling a segment label according to the discrete distribution with parameters being the mixture weights $\boldsymbol{\pi} = \{\pi_1, \cdots, \pi_K\}$, then *(ii)* sampling an observation Im_i from the conditional distribution $f(\boldsymbol{\theta}_k)$. Given a model of the form (1) represented by $\widehat{\boldsymbol{\theta}}$, typically fitted by maximum likelihood estimation (MLE) from the n observations composing the image Im_n, as

$$\widehat{\boldsymbol{\theta}} \in \arg\max_{\theta \in \Theta} \log L(\boldsymbol{\theta}\mathrm{Im}_n) \quad (2)$$

where $L(\boldsymbol{\theta}\mathrm{Im}_n)$ is the likelihood of $\boldsymbol{\theta}$ given the image data Im_n, then, the segment labels can be determined via the Bayes' allocation rule,

$$\widehat{Z}_i = \arg\max_{k \in [K]} \mathbb{P}(Z_i = k\mathrm{Im}(i); \widehat{\boldsymbol{\theta}}), \quad (3)$$

which consists of maximizing the conditional probabilities

$$\mathbb{P}(Z_i = k\mathrm{Im}(i); \widehat{\boldsymbol{\theta}}) = \frac{\widehat{\pi}_{ik} f_i(\widehat{\boldsymbol{\theta}}_k)}{f_i(\widehat{\boldsymbol{\theta}})}. \quad (4)$$

that the ith observation originates from segment k, $k \in [K]$, given the image data and the fitted model.

Model (1) has many different specifications in the literature, depending on the nature of the data generative process, resulting in a multitude of choices for the mixture weights and for the component densities. Mixtures of multivariate distributions [32] are in particular more popular in model-based clustering of vectorial data using multivariate mixtures. These include multivariate Gaussian mixtures [30,31], where $\pi_{ik} = \pi_k$, $\forall i$ are constant mixture weights, and the component densities $f_i(\boldsymbol{\theta}_k) = \phi_i(\boldsymbol{\mu}_k, \boldsymbol{\Sigma}_k)$ are multivariate Gaussians with

means $\boldsymbol{\mu}_k$ and covariance matrices $\boldsymbol{\Sigma}_k$. Mixtures of regression models, introduced in [40], are common in the modeling and clustering of regression-type data. For example, in the widely used Gaussian regression mixture model [41], we have constant mixing proportions, i.e., $\pi_{ik} = \pi_k$, and the mixture components $f_i(\boldsymbol{\theta}_k)$'s are Gaussian regressors $\phi(\cdot, \boldsymbol{\beta}_k^\top \boldsymbol{x}_i, \sigma_k^2)$ with typically linear means $\boldsymbol{\beta}_k^\top \boldsymbol{x}_i$ and variances σ_k^2 in the case of a univariate response.

In this paper, we consider a more flexible mixture of regressions model in which both the mixture weights and the mixture components are covariate-dependent, and are constructed upon flexible semi-parametric functions. More specifically, in this full conditional mixture model, the mixture weights π_{ik} are constructed upon parametric functions $\pi_{ik} = \pi_k(\cdot, \boldsymbol{x}_i; \boldsymbol{\alpha})$ of some covariates $\boldsymbol{x}_i$ represented by a parameter vector $\boldsymbol{\alpha}$, and the regression functions $f_i(\boldsymbol{\theta}_k)$ are Gaussian regressors $\phi(\cdot, \mu(\boldsymbol{x}_i; \boldsymbol{\beta}_k), \sigma_k^2)$ with semi-parametric (non-)linear mean functions $\mu(\boldsymbol{x}_i; \boldsymbol{\beta}_k)$. This flexible modeling allows us to better capture more non-linear relationships in the functional data via the semi-parametric mean functions. Heterogeneity is accommodated via the mixture distribution, and spatial organization can be captured via spatial-dependent mixture weights.

3.2. Spatial Mixture of Functional Regressions for Dual-Energy CT Images

We propose a spatialized mixture of functional regressions model, adapted to the given type of image data, for the model-based clustering of dual-energy CT scans. The images we analyze include spectral curves for each 3D voxel. Each image, denoted Im, is represented as a sample of n observations, $\mathrm{Im} = \{\boldsymbol{v}_i, \boldsymbol{x}_i, \boldsymbol{y}_i\}_{i=1}^n$ where $\boldsymbol{v}_i = (v_{i1}, v_{i2}, v_{i3})$ is the ith voxel 3D spatial coordinates.The ith voxel is represented by the curve $(\boldsymbol{x}_i, \boldsymbol{y}_i)$ composed of HU attenuation values $\boldsymbol{y}_i = (y_{i1}, \ldots, y_{im})$ measured at energy levels (covariates) $\boldsymbol{x}_i = (x_{i1}, \ldots, x_{im})$, with m being the number of energy levels.

To accommodate the spatial organization of the image together with the functional nature of each of its voxels, we propose spatialized conditional extensions of the general family of model (1), in which we model the ith voxel observation of the image using the conditional density $f(\boldsymbol{y}_i \boldsymbol{x}_i, \boldsymbol{v}_i; \boldsymbol{\theta})$ that relates the attenuation curve levels $\boldsymbol{y}_i$, given the associated energy levels $\boldsymbol{x}_i$, and spatial location $\boldsymbol{v}_i$ via a convex combination of (non-)linear (semi-)parametric functional regressors $f(\boldsymbol{y}_i \boldsymbol{x}_i; \boldsymbol{\theta}_k)$ with spatial weights $\pi_k(\boldsymbol{v}_i; \boldsymbol{\alpha})$, that is,

$$f(\boldsymbol{y}_i \boldsymbol{x}_i, \boldsymbol{v}_i; \boldsymbol{\theta}) = \sum_{k=1}^{K} \pi_k(\boldsymbol{v}_i; \boldsymbol{\alpha}) f(\boldsymbol{y}_i \boldsymbol{x}_i; \boldsymbol{\theta}_k) \,. \tag{5}$$

To this purpose, we consider two different spatial constructions of the mixing weights (gating functions) $\pi_k(\boldsymbol{v}_i; \boldsymbol{\alpha})$: (i) softmax gates; and (ii) normalized Gaussian gates. The latter is an appropriate choice if more approximation quality is needed, and facilitates the computations in the learning process. We also consider different families to model the functional regressors, including spline and B-spline regression functions that enjoy better curve approximation capabilities, compared to linear or polynomial regression functions.

3.2.1. Functional Regression Components

We have a 3D image volume over a range of energy levels that provide, for each voxel i, an attenuation curve $(\boldsymbol{x}_i, \boldsymbol{y}_i)$ which represents energy-dependent changes in attenuation, which enables a better characterization of the tissue at voxel i. We therefore model the component densities $f(\boldsymbol{y}_i \boldsymbol{x}_i; \boldsymbol{\theta}_k)$ as functional regression models constructed upon the attenuation curves as functional observations. This allows us to accommodate the spectral curve nature of the data. More specifically, in the case of univariate energy levels, we use smooth functional approximations to model, for the ith voxel, the mean spectral curve of the kth component $\mu(\boldsymbol{x}_i; \boldsymbol{\beta}_k) = \mathbb{E}_{\boldsymbol{\theta}}[\mathbf{Y}_i Z_i = k, \boldsymbol{x}_i]$, that is, $\mu(\boldsymbol{x}_i; \boldsymbol{\beta}_k) = (\mu(x_{i1}; \boldsymbol{\beta}_k), \ldots, \mu(x_{im}; \boldsymbol{\beta}_k))$ using polynomial or (B)-spline functions, whose coefficients are $\boldsymbol{\beta}_k$.

The conditional density model for each regression is then modeled as a functional Gaussian regressor defined by $f(\boldsymbol{y}_i \boldsymbol{x}_i; \boldsymbol{\theta}_k) = \phi_m(\boldsymbol{y}_i; \mu(\boldsymbol{x}_i; \boldsymbol{\beta}_k), \sigma_k^2 \mathbf{I})$, with $\mu(\boldsymbol{x}_i; \boldsymbol{\beta}_k) = \mathbf{B}(\boldsymbol{x}_i) \boldsymbol{\beta}_k$

being the function approximation onto polynomial or (B-)spline bases $\mathbf{B}(x_i)$, and the matrix form of the functional regression model is then given by

$$f(y_i x_i; \boldsymbol{\theta}_k) = \phi_m(y_i; \mathbf{B}(x_i)\boldsymbol{\beta}_k, \sigma_k^2\mathbf{I}), \tag{6}$$

where $\boldsymbol{\theta}_k = (\boldsymbol{\beta}_k^\top, \sigma_k^2)^\top \in \mathbb{R}^{p+q+2}$ is the unknown parameter vector of regression k.

3.2.2. Spatial Gating Functions

The constructed functional mixture of regressions model (5) specifically integrates the spatial constraints in the mixture weights $\pi_k(v_i; \boldsymbol{\alpha})$ via functions of the spatial locations v_i parametrized by vectors of coefficients $\boldsymbol{\alpha}$. We investigate two choices to this end. The first proposed model is a spatial softmax-gated functional mixture of regression and is defined by (5) with a softmax gating function:

$$\pi_k(v_i; \boldsymbol{\alpha}) = \frac{\exp(\boldsymbol{\alpha}_k^\top v_i)}{1 + \sum_{k'=1}^{K-1} \exp(\boldsymbol{\alpha}_{k'}^\top v_i)}, \tag{7}$$

where $\boldsymbol{\alpha} = (\boldsymbol{\alpha}_1^\top, \ldots, \boldsymbol{\alpha}_K^\top)^\top$ is the unknown parameter vector of the gating functions. We will refer to this model, defined by (5)–(7), as the spatial softmax-gated mixture of functional regressions, abbreviated as **SsMFR**. The softmax modeling of the mixture weights is a standard choice known in the mixtures-of-experts community. However, its optimization performed at the M step of the EM algorithm, is not analytic and requires numerical Newton–Raphson optimization. This can become costly, especially in larger image applications, such as the one we address.

In the second proposed model, we use a spatial Gaussian-gated functional mixture of regressions, defined by (5) with a Gaussian-gated function:

$$\pi_k(v_i; \boldsymbol{\alpha}) = \frac{w_k \phi_3(v_i; \boldsymbol{\mu}_k, \mathbf{R}_k)}{\sum_{\ell=1}^{K} w_\ell \phi_3(v_i; \boldsymbol{\mu}_\ell, \mathbf{R}_\ell)}, \tag{8}$$

in which w_k are non-negative weights that sum to one, $\phi_d(v_i; \boldsymbol{\mu}_k, \mathbf{R}_k)$ is the density function of a multivariate Gaussian vector of dimension d with mean $\boldsymbol{\mu}_k$ and covariance matrix $\boldsymbol{\Sigma}_k$, and $\boldsymbol{\alpha} = (\boldsymbol{\alpha}_1^\top, \ldots, \boldsymbol{\alpha}_K^\top)^\top$ is the parameter vector of the gating functions with $\boldsymbol{\alpha}_k = (w_k, \boldsymbol{\mu}_k^\top, \mathrm{vech}(\mathbf{R}_k)^\top)^\top$.

We will refer to this model, defined by (5), (6) and (8), as the spatial Gaussian-gated mixture of functional regressions, abbreviated as **SgMFR**. This Gaussian gating function was introduced in [42] to bypass the need for an additional numerical optimization in the inner loop of the EM algorithm. We obtain a closed form updating formula at the M-Step, that is detailed in the next section presenting the derived EM algorithm.

3.2.3. MLE of the SgMFR Model via the EM Algorithm

Based on Equations (5), (6) and (8), the SgMFR joint density $f(y_i, x_i, v_i; \boldsymbol{\theta})$ is then derived and the joint log-likelihood we maximize via EM is

$$\log L(\boldsymbol{\theta}) = \sum_{i=1}^{n} \log f(y_i, x_i, v_i; \boldsymbol{\theta}) = \sum_{i=1}^{n} \log \sum_{k=1}^{K} w_k \phi_3(v_i; \boldsymbol{\mu}_k, \mathbf{R}_k) \phi_m(y_i; \mathbf{B}(x_i)\boldsymbol{\beta}_k, \sigma_k^2\mathbf{I}). \tag{9}$$

The complete-data log-likelihood, upon which the EM algorithm is constructed, is

$$\log L_c(\boldsymbol{\theta}) = \sum_{i=1}^{n} \sum_{k=1}^{K} Z_{ik} \log\Big[w_k \phi_3(v_i; \boldsymbol{\mu}_k, \mathbf{R}_k) \phi_m(y_i; \mathbf{B}(x_i)\boldsymbol{\beta}_k, \sigma_k^2\mathbf{I})\Big]. \tag{10}$$

where Z_{ik} is an indicator variable such that $Z_{ik} = 1$ if $Z_i = k$ (i.e., if the ith pair (x_i, y_i) is generated from the kth regression component) and $Z_{ik} = 0$, otherwise. The EM algorithm,

after starting with an initial solution $\boldsymbol{\theta}^{(0)}$, alternates between the E and M steps until convergence (when there is no longer a significant change in the log-likelihood).

The E-step: Compute the conditional expectation of the complete-data log-likelihood (10), given the image Im_n and the current estimate $\boldsymbol{\theta}^{(t)}$:

$$Q(\boldsymbol{\theta};\boldsymbol{\theta}^{(t)}) = \mathbb{E}\Big[L_c(\boldsymbol{\theta})|\mathrm{Im}_n;\boldsymbol{\theta}^{(t)}\Big] = \sum_{i=1}^{n}\sum_{k=1}^{K}\tau_{ik}^{(t)}\log\Big[\alpha_k\phi_3(\boldsymbol{v}_i;\boldsymbol{\mu}_k,\mathbf{R}_k)\phi_m(\boldsymbol{y}_i;\mathbf{B}(\boldsymbol{x}_i)\boldsymbol{\beta}_k,\sigma_k^2\mathbf{I})\Big], \quad (11)$$

where $\tau_{ik}^{(t)} = \mathbb{P}(Z_i = k\boldsymbol{y}_i, \boldsymbol{x}_i, \boldsymbol{v}_i; \boldsymbol{\theta}^{(t)})$ given by

$$\tau_{ik}^{(t)} = \frac{w_k^{(t)}\phi_3(\boldsymbol{v}_i;\boldsymbol{\mu}_k^{(t)},\mathbf{R}_k^{(t)})\phi_m(\boldsymbol{y}_i;\mathbf{B}(\boldsymbol{x}_i)\boldsymbol{\beta}_k^{(t)},\sigma_k^{2(t)}\mathbf{I})}{f(\boldsymbol{v}_i,\boldsymbol{x}_i,\boldsymbol{y}_i;\boldsymbol{\theta}^{(t)})} \quad (12)$$

is the posterior probability that the observed pair $(\boldsymbol{x}_i, \boldsymbol{y}_i)$ is generated by the kth regressor. This step therefore only requires the computation of the posterior component membership probabilities $\tau_{ik}^{(t)}$ $(i = 1,\ldots,n)$, for $k = 1,\ldots,K$.

The M-step: Calculate the parameter vector update $\boldsymbol{\theta}^{(t+1)}$ by maximizing the Q-function (11), i.e., $\boldsymbol{\theta}^{(t+1)} = \arg\max_{\boldsymbol{\theta}} Q(\boldsymbol{\theta};\boldsymbol{\theta}^{(t)})$. By decomposing the Q−function as

$$Q(\boldsymbol{\theta};\boldsymbol{\theta}^{(t)}) = \sum_{k=1}^{K} Q(\boldsymbol{\alpha}_k;\boldsymbol{\theta}^{(t)}) + Q(\boldsymbol{\theta}_k;\boldsymbol{\theta}^{(t)}), \quad (13)$$

with $Q(\boldsymbol{\alpha}_k;\boldsymbol{\theta}^{(t)}) = \sum_{i=1}^{n}\tau_{ik}^{(t)}\log[w_k\phi_3(\boldsymbol{v}_i;\boldsymbol{\mu}_k,\mathbf{R}_k)]$ and $Q(\boldsymbol{\theta}_k;\boldsymbol{\theta}^{(t)}) = \sum_{i=1}^{n}\tau_{ik}^{(t)}\log[\phi_m(\boldsymbol{y}_i;\mathbf{B}(\boldsymbol{x}_i)\boldsymbol{\beta}_k,\sigma_k^2\mathbf{I})]$, the maximization can then be performed by K separate maximizations with respect to the parameters of the gating and the regression functions.

Updating the gating functions parameters: Maximizing (13) with respect to $\boldsymbol{\alpha}_k$'s corresponds to the M step of a Gaussian mixture model [32]. The closed-form expressions for updating the parameters are given by

$$w_k^{(t+1)} = \sum_{i=1}^{n}\tau_{ik}^{(t)}\Big/n, \quad (14)$$

$$\boldsymbol{\mu}_k^{(t+1)} = \sum_{i=1}^{n}\tau_{ik}^{(t)}\boldsymbol{v}_i\Big/\sum_{i=1}^{n}\tau_{ik}^{(t)}, \quad (15)$$

$$\mathbf{R}_k^{(t+1)} = \sum_{i=1}^{n}\tau_{ik}^{(t)}(\boldsymbol{v}_i-\boldsymbol{\mu}_k^{(t+1)})(\boldsymbol{v}_i-\boldsymbol{\mu}_k^{(t+1)})^\top\Big/\sum_{i=1}^{n}\tau_{ik}^{(t)}. \quad (16)$$

Updating the regression functions parameters: Maximizing (13) with respect to $\boldsymbol{\theta}_k$ corresponds to the M step of standard mixtures of experts with univariate Gaussian regressions. The closed-form updating formulas are given by

$$\boldsymbol{\beta}_k^{(t+1)} = \Big[\sum_{i=1}^{n}\tau_{ik}^{(t)}\mathbf{B}^\top(\boldsymbol{x}_i)\mathbf{B}(\boldsymbol{x}_i)\Big]^{-1}\sum_{i=1}^{n}\tau_{ik}^{(t)}\mathbf{B}(\boldsymbol{x}_i)^\top\boldsymbol{y}_i, \quad (17)$$

$$\sigma_k^{2(t+1)} = \sum_{i=1}^{n}\tau_{ik}^{(t)}(\boldsymbol{y}_i-\mathbf{B}(\boldsymbol{x}_i)\boldsymbol{\beta}_k^{(t+1)})^2\Big/\sum_{i=1}^{n}\tau_{ik}^{(t)}m_i. \quad (18)$$

3.3. Alternative Two-Fold Approaches

We also investigate an alternative approach to the one derived before, which consists of a two-fold approach, rather than a simultaneous functional approximation and model estimation for segmentation. We first construct approximations of the functional data onto polynomial or (B-)splines bases $\mathbf{B}(\boldsymbol{x}_i)$ via MLE (ordinary least squares in this case) to obtain

$$\widehat{\boldsymbol{\beta}}_i = \Big[\mathbf{B}(\boldsymbol{x}_i)^\top\mathbf{B}(\boldsymbol{x}_i)\Big]^{-1}\mathbf{B}(\boldsymbol{x}_i)^\top\boldsymbol{y}_i. \quad (19)$$

Then, we model the density of the resulting coefficient vectors $\widehat{\beta}_i$, which is regarded as the ith curve representative, by a mixture density with spatial weights of the form

$$f(\widehat{\beta}_i, v_i, \theta) = \sum_{k=1}^{K} \pi_k(v_i; \alpha)\phi_d(\widehat{\beta}_i; m_k, C_k), \quad (20)$$

where m_k and C_k are the mean and the covariance matrix of each component. The spatial weights $\pi_k(v_i; \alpha)$ are normalized Gaussians as in (8) or softmax as in (7). We will refer to these methods as spatial Gaussian-gated (resp. softmax-gated) mixtures of vectorized functional regressions, **SgMVFR** (resp. **SsMVFR**).

The EM algorithm for fitting this mixture of spatial mixtures, constructed upon pre-computed polynomial or (B-)spline coefficients with its two variants for modeling the spatial weights, takes a similar form to the previously presented algorithm, and is summarized as follows. The conditional memberships of the E step are given for the softmax-gated model by

$$\tau_{ik}^{(t)} = \frac{\pi_k(v_i; \alpha^{(t)})\phi_d(\widehat{\beta}_i; m_k^{(t)}, C_k^{(t)})}{f(\widehat{\beta}_i v_i; \theta^{(t)})}, \quad (21)$$

and for the Gaussian-gated model by

$$\tau_{ik}^{(t)} = \frac{w_k^{(t)}\phi_3(v_i; \mu_k^{(t)}, R_k^{(t)})\phi_d(\widehat{\beta}_i; m_k^{(t)}, C_k^{(t)})}{f(v_i, \widehat{\beta}_i; \theta^{(t)})}. \quad (22)$$

The latter has the same advantage as explained above. In the M step, the gating functions parameter updates are given by (14)–(16) for the Gaussian-gated model, or through a Newton–Raphson optimization algorithm for the softmax-gated model. The component parameter updates are those of classical multivariate Gaussian mixtures

$$m_k^{(t+1)} = \sum_{i=1}^{n} \tau_{ik}^{(t)}\widehat{\beta}_i \Big/ \sum_{i=1}^{n} \tau_{ik}^{(t)}, \quad (23)$$

$$C_k^{(t+1)} = \sum_{i=1}^{n} \tau_{ik}^{(t)}(\widehat{\beta}_i - m_k^{(t+1)})(\widehat{\beta}_i - m_k^{(t+1)})^\top \Big/ \sum_{i=1}^{n} \tau_{ik}^{(t)}. \quad (24)$$

In a nutshell, to compute a clustering of the image, the label of voxel i, given the fitted parameters $\widehat{\theta}$, is calculated by the Bayes' allocation rule (3), in which $\mathrm{Im}(i)$ is the spatial coordinates of voxel i with either its direct spectral curve representative (x_i, y_i) or its pre-calculated functional approximation coefficients $\widehat{\beta}_i$ given by (19).

Appendix A contains the pseudo-codes summarizing the proposed method.

3.4. Time Complexity of the Proposed Algorithms

In this subsection we investigate the time complexity of the proposed algorithms. The time complexity of the E-step of the proposed EM algorithms for the SgMFR and the SsMFR models is of $\mathcal{O}(Kd^2pnm)$, with n being the number of voxels, m the number of energy levels, d is the number of spatial coordinates (2 or 3), p the number of the number of regression coefficients, and K the number of clusters. For the M step, the SgMFR and the SgMVFR have closed-form updates; The SgMFR requires the calculation of the regression coefficients via weighted least squares with a complexity of $\mathcal{O}(Kp^2nm)$ The SgMVFR models require the calculations of the Gaussian means and covariance matrices as in multivariate Gaussian mixtures, and have a complexity of $\mathcal{O}(Kp^2n)$. However, the SsMFR and SsMVFR algorithms require at each iteration inside the M step of the EM algorithm an IRLS loop and the inversion of the Hessian matrix which is of dimension $d(K-1)$. Therefore, the complexity of the IRLS is approximately of $\mathcal{O}(I_{IRLS}d^2K^2)$, where IRLS is the average number of iterations required by the internal IRLS algorithm. The complexity here can therefore be an issue for a large number of clusters, and the SgMFR and SgMVFR algorithms can be preferred.

4. Experimental Study

In this section, we describe the evaluation of different versions of our method: mixtures of B-spline and polynomial functional regressions with spatial Gaussian gates (resp. SgMFR-Bspl and SgMFR-poly), mixture of B-spline regressions with softmax gates (SsMFR-Bspl), and mixture of vectorized B-spline regressions with Gaussian gates (SgMVFR-Bspl).

4.1. Data

In total, 91 head and neck DECT scans were evaluated, consisting of HNSCC tumors of different sizes and stages from different primary sites. In our dataset, 34% of tumors are located in the oral cavity, 26% in the oropharynx, 21% in the larynx, 8% in the hypopharynx and 11% in other locations. The tumors' T-stage [43] ranges from T1 to T4. Of the patients, 75% were coming for a first diagnostic while 25% were recurrent patients. Institutional review board approval was obtained for this study with a waiver of informed consent. Tumors were contoured by an expert head and neck radiologist. All scans were acquired using a fast kVp switching DECT scanner (GE Healthcare) after administration of IV contrast and reconstructed into 1.25 mm sections of axial slices with a resolution of 0.61 mm, as previously described [11]. Multienergy VMIs were reconstructed at energy levels from 40 to 140 keV in 5 keV increments at the GE Advantage workstation (4.6; GE Healthcare).

In each DECT scan, we crop volumes of interest (VOIs) of size 150*150*6 containing a tumor, along with the 21-point-spectral curve associated to each selected voxel, in order to reduce the computational demands for an exploratory study, and to exclude regions containing a majority of air voxels around the body. A pre-processing step is also applied to mask any remaining air voxels in the VOI to focus the clustering on tissues.

4.2. Regularization Parameter

In our study, we augmented the statistical estimator in Equation (16) of the covariance matrix of the spatial coordinates within cluster k, with a regularization parameter $\lambda \in (0,1]$, which controls the amount of spatial dispersion (neighborhood) taken into account in the spatial mixture weights, by

$$\widetilde{\mathbf{R}}_k^{(t+1)} = \lambda \mathbf{R}_k^{(t+1)}. \tag{25}$$

By doing so, we can numerically control the amount of data within cluster k (i.e., its volume). Indeed, if we decompose $\mathbf{C}_k = \lambda \mathbf{D}\mathbf{A}\mathbf{D}^T$ where $\mathbf{A}$ is the $\mathbf{C}_k$ eigenvectors matrix, and $\mathbf{D}$ is a diagonal matrix whose diagonal elements are the eigenvalues in decreasing order, then λ is the volume of cluster k. Since the tumor cluster has in general no strong spatial dispersion, then in practice, we take small values of order 0.1.

4.3. Parameter Initialization

For the sake of reproducibility, we start by initializing the regression mixture and weight parameters of the EM algorithm with a coarse clustering solution given by a Voronoi diagram. We build up Voronoi tiles over the selected voxels in the VOI (a square region where air voxels are deleted) with the k-means algorithm applied only on spatial coordinates. Then to fix the number of clusters K, and to fix the spatio-spectral hyperparameter λ, we assess on a small training set (10 patients) the three metrics described in Section 4.5. In our experiments, K is taken to be large enough, say 20, 30 or 40, so that we do not have to perform a full grid search which could be computationally demanding, and the value λ around 0.075 works very well. The process is run similarly for both methods of mixtures of functional regressions and mixtures of vectorized functional regressions. The range of search values is adapted for each method, and 5 search values are taken in each range. In the end, we use the mean of optimal values over the patients in the train set.

4.4. Baselines

We compare the quantitative and qualitative performance of our methodology with three baseline algorithms. First, we implement a Gaussian mixture model (GMM) with the iterative EM algorithm, using the standard non-reshaped algorithm [32] to cluster the spectral curves, thus leading to not include spatial coordinates. Because several clusters can become empty through the optimization, we fix an initial number of clusters ($K = 150$) that ends up providing, on average, the same resulting number of clusters for our method (i.e., $K = 40$). Second, we implement k-means clustering, using all vector information available, that is, the input vector is built with spectral information (i.e., the energy decay curve points) concatenated to a vector of spatial information (i.e., the 3D coordinates). The number of clusters is picked to be also $K = 40$, the number of clusters being stable throughout the optimization. We use the Matlab k-means implementation for images with a reproducible initialization through the built-in 'imsegkmeans' function. Third and last, we implement selective search, a machine learning graph-based segmentation method for object recognition. Using a region merging hierarchical approach with an SVM classifier to select the hierarchical rank of the resulting regions, the authors published an open-source code [44]. We apply selective search on low-, intermediate- and high-energy levels (resp. 40, 65 and 140 keV). These energy levels are used instead of the three RGB image channels. We note that selective search does not predetermine the number of clusters, but specifies a cluster minimum size or favors smaller or larger cluster sizes.

4.5. Metrics

Our clustering methods, as well as the baseline clustering algorithms, are all evaluated using the following three metrics:

1. A cluster separation index, *Davies–Bouldin index (DB)*, computed on spatial content and on spectral content.
2. A clustering separation index focused only on the relationship between tumor clusters and other clusters, *Davies–Bouldin index on tumor (DB_t)*, an adaptation of DB, computed on spatial and on spectral content.
3. A segmentation score computed on tumor clusters versus ground truth region, *the Dice similarity score*, that can be computed only on spatial content.

To define tumor clusters, we select the cluster(s) which best cover the tumor, i.e., the ones that give the best Dice score when merged together. Several clusters segmenting the tumor area can indeed represent different tumor subparts, but we only know the tumor primary site contour as the ground truth.

When C is the ensemble of clusters, $d(\cdot,\cdot)$ is the Euclidean distance operator, $\overline{c_k}$ is the centroid of cluster c_k, the Davies–Bouldin index is defined as

$$\mathrm{DB}(C) = 1/C \sum_{c_k \in C} \max_{c_l \in C \setminus c_k} (S(c_k) + S(c_l))/d(\overline{c_k}, \overline{c_l}) \tag{26}$$

where $S(c_k) = 1/c_k \sum_{x_i \in c_k} d(x_i, \overline{c_k})$. The 'tumor' Davies–Bouldin index is adapted as

$$\mathrm{DB}_t(C) = \max_{c_l \in C \setminus c_{tum}} (S(c_{tum}) + S(c_l))/d(\overline{c_{tum}}, \overline{c_l})), \tag{27}$$

where c_{tum} is the region of the merged tumor clusters. The Dice score is defined as

$$\mathrm{Dice}(c_{tum}, c_{truth}) = 2 * c_{tum} \cap c_{truth}/(c_{tum} + c_{truth}), \tag{28}$$

where c_{truth} is the ground truth region. We summarize the distribution of these index values across the population by computing the mean, median and interquartile range.

5. Results

Figure 2 shows an overview of our results for one tumor example. We visualize the results on a 2D slice when the model has been run on the 3D VOI containing this slice.

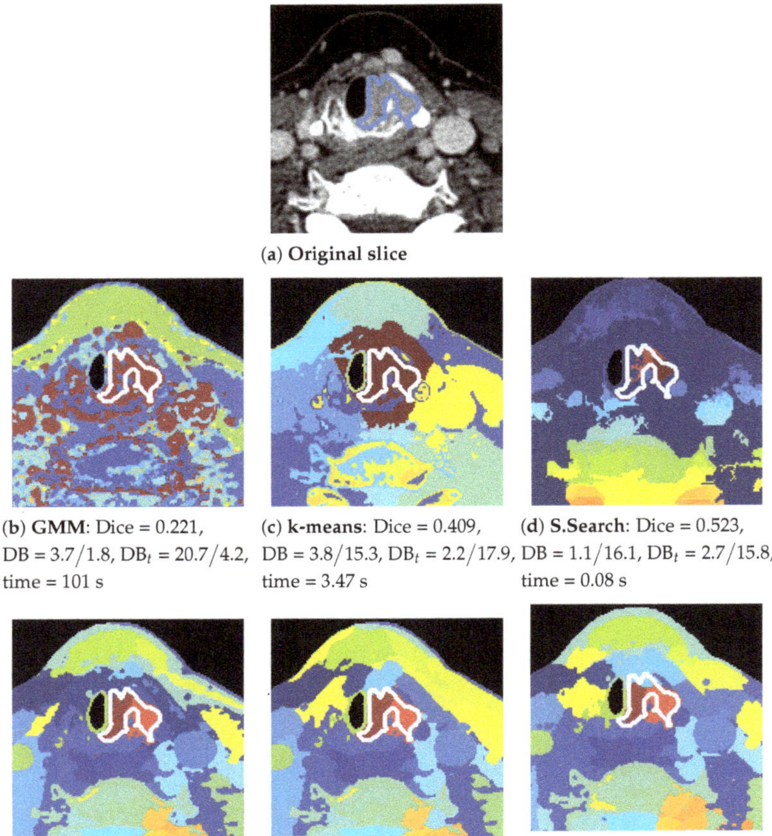

(**a**) **Original slice**

(**b**) **GMM**: Dice = 0.221, DB = 3.7/1.8, DB_t = 20.7/4.2, time = 101 s

(**c**) **k-means**: Dice = 0.409, DB = 3.8/15.3, DB_t = 2.2/17.9, time = 3.47 s

(**d**) **S.Search**: Dice = 0.523, DB = 1.1/16.1, DB_t = 2.7/15.8, time = 0.08 s

(**e**) **SgMFR-Bspl**: Dice = 0.809, DB = 1.4/7.6, DB_t = 4.3/5.6, time = 1521 s

(**f**) **SgMFR-poly**: Dice = 0.838, DB = 1.2/6.5, DB_t = 1.8/3.6, time = 603 s

(**g**) **SgMVFR-Bspl**: Dice = 0.761, DB = 1.3/9.0, DB_t = 2.0/6.2, time = 199 s

Figure 2. Clustering results for each approach in one tumor ($DB_{(t)}$ = spatial/spectral index). Our proposed approaches are on the bottom row. One random color is assigned per cluster, ground truth tumor contour is in blue (**a**) or white (**b**–**g**).

The top row shows the performance of the baseline algorithms, whereas the bottom row shows our proposed methods. While the baseline approaches attribute a high number of clusters to bone regions containing big spectral variations and miss smaller variations in tissue regions, our methods with Gaussian gates in Figure 2e–g are able to adapt to relative variations and split the image with more spatial coherence. The results also demonstrate that our method is able to capture tissue characteristics invisible in Figure 2b–d,h. Note that DB and DB_t scores depend on the number of clusters and SsMFR in Figure 2h has a very low number of clusters (softmax having vanishing clusters in the optimization). GMM and selective search in Figure 2b,d have around 40 clusters (varying number as explained in Section 4.4). The results in Figure 2c,e–g were obtained with 40 clusters.

Table 1 and Figure 3 present the quantitative results obtained with the three clustering metrics defined in Section 4.5. Among the three proposed methods that outperform the baseline methods in terms of Dice score (SgMFR-Bspl, SgMFR-poly, and SgMVFR-Bspl), we compared the Dice score distribution obtained with SgMFR-poly (which has the lowest

median Dice score among the three) to that obtained with the k-means-like baseline (which has the highest median Dice score) with a two-sample t-test, obtaining p=0.0014.

Figure 4 showcases the clustering results obtained when varying the tuning of λ. Here, we understand that a smaller λ gives a higher preference to the spatial information: clusters are compact and define well-separated areas. On the other hand, a larger λ gives a higher priority to the spectral information: clusters more closely match tissue characteristics, but one cluster can be split into tiny voxel groups spread all over the image. The ideal λ choice would be a λ that prioritizes spectral information, but still achieves some cluster spatial compactness. We assess this through metrics calculated on spectral and spatial content as explained in Section 4.5. The general tuning of $\lambda = 0.075$ is determined to be, on average, the optimal hyper-parameter. However, we can see strong improvements in tumor separation, on a case by case basis, with small variation of λ. As shown in Figure 4e,f, the Dice score increases from 0.36 to 0.64. This shows an example out of several results belonging to the lower quartile in the boxplot of Dice scores in Figure 3 that could be highly improved simply with a specific tuning. Some other examples of results in the lower quartile could be due to small tumors (size inferior to 1cm), although half of these small tumors are actually well-separated with our method, reaching a Dice score as high as 0.84 in the best case (see in Figure 5).

The proposed algorithms clearly outperform the investigated standard clustering algorithms in terms of the considered metrics, such as the Dice score. In particular, the two approaches based on Gaussian-gating mixture weights, whenever they are directly built upon a functional mixture model (SgMFR), or used with a prior functional data representation of the energy curves (SgMVFR), enjoy both high segmentation capabilities while being computationally effective. The two proposed alternative approaches based on the use of softmax-gating mixture weights (SsMFR and SsMVFR), can, however, be computationally expensive, given that they use, at each EM iteration, a Newton–Raphson optimization; they may lead in some situations to less precise segmentation, typically due to a numerical convergence issue, as compared to the proposed Gaussian-gating-based approaches. As a result, we can suggest to the interested users prioritize the use of the SgMFR and SgMVF approaches when investigating our proposed techniques for DECT clustering. This being said, in order to investigate the statistical significance of the differences in the results of the proposed family of algorithms, it is interesting to perform a statistical study with appropriate statistical testing.

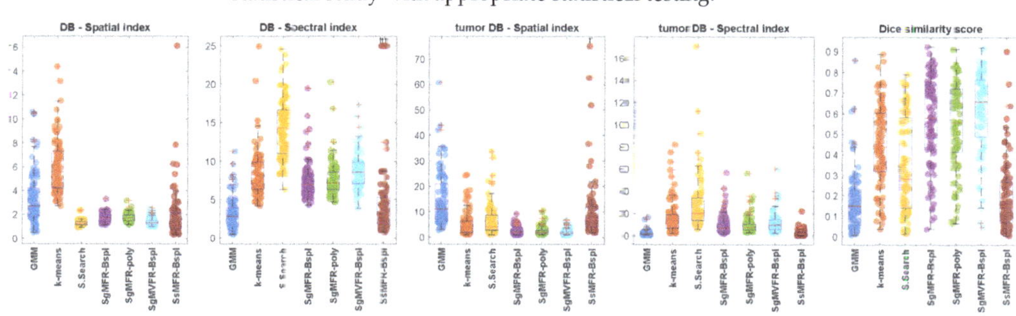

Figure 3. Boxplots for each metric per method. Note that SsMFR-Bspl method gives few outliers out of reach (order of 10^{13}) on the DB spectral index, and one outlier of 135 for tumor DB spatial index. These values are shifted in the displayed range and exhibit a top arrow.

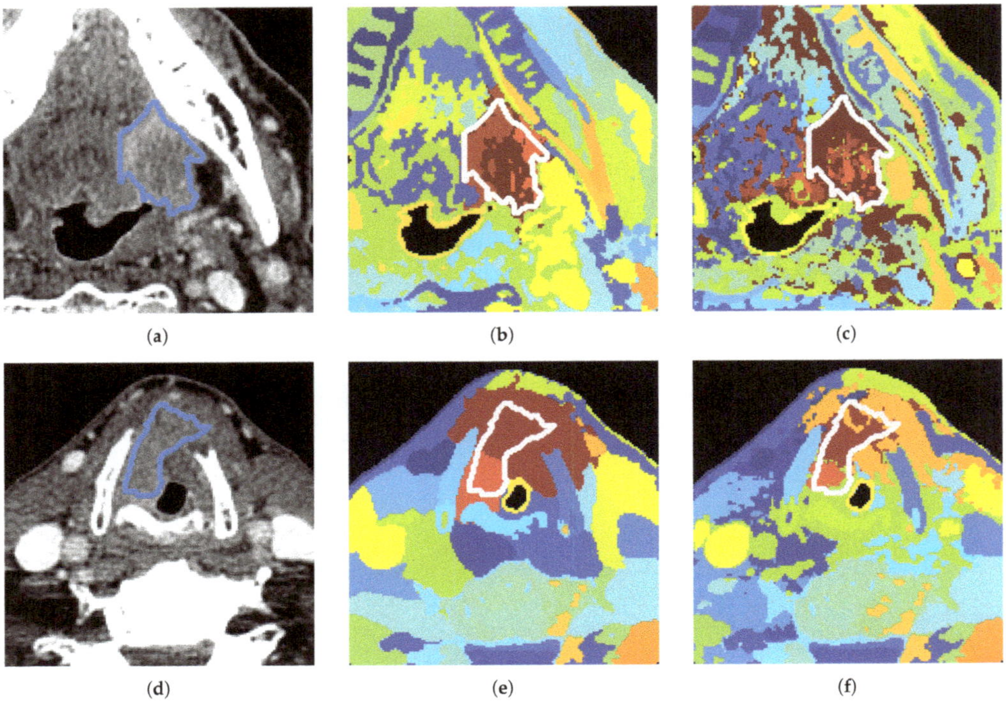

Figure 4. Clustering results for our SgMFR method with different λ tuning. (a) Original slice. (b) $\lambda = \mathbf{0.075}$, Dice = 0.79, DB = 1.68/7.29, DB_t = 1.47/6.79. (c) $\lambda = \mathbf{0.100}$, Dice = 0.39, DB = 2.59/4.27, DB_t = 3.34/3.74. (d) Original slice. (e) $\lambda = \mathbf{0.075}$, Dice = 0.36, DB = 1.75/6.70, DB_t = 1.28/23.71. (f) $\lambda = \mathbf{0.080}$, Dice = 0.64, DB = 1.87/6.57, DB_t = 1.78/2.93. One random color is assigned per cluster, ground truth tumor contour is in blue (a,d) or white (b,c,e,f).

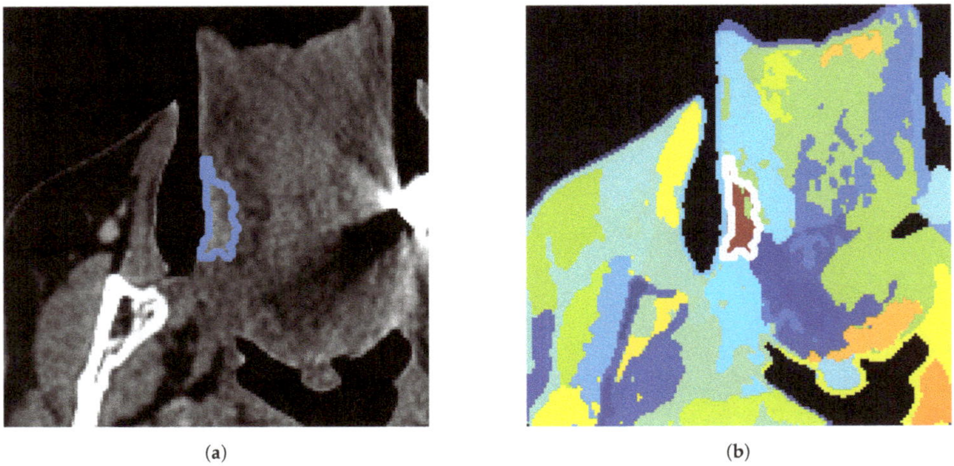

Figure 5. Clustering results with our SgMFR for a small tumor. Note the robustness of the result in the presence of a metallic artifact in the right-hand side of the anatomical image. (a) Original slice. (b) Dice = 0.84, DB = 1.64/6.92, DB_t = 1.98/7.22, $\lambda = 0.075$.

Table 1. Average median (interquartile range) of metrics and runtime over 81 patient scans. DB cluster separation index is computed for spatial content (spat-DB) and spectral content (spec-DB); idem for DB_t indices focused on tumor separability; runtime is given in seconds.

	GMM			**k-Means**			**S.Search**			**SgMFR-Bspl**			**SgMFR-poly**			**SgMVFR-Bspl**			**SsMFR-Bspl**		
	avg	**med**	**itq**	**avg**	**med**	**itq**	**avg**	**med**	**itq**	**avg**	**med**	**itq**	**avg**	**med**	**itq**	**avg**	**med**	**itq**	**avg**	**med**	**itq**
Dice	0.17	0.15	*(0.15)*	0.45	0.44	*(0.28)*	0.35	0.28	*(0.44)*	0.56	0.58	*(0.32)*	0.57	0.56	*(0.31)*	0.61	0.66	*(0.28)*	0.20	0.15	*(0.20)*
spat-DB	3.32	2.69	*(2.38)*	6.00	5.57	*(3.08)*	1.30	1.28	*(0.24)*	1.75	1.75	*(0.45)*	1.74	1.72	*(0.45)*	1.40	1.32	*(0.35)*	1.84	1.22	*(1.31)*
spec-DB	3.17	2.80	*(2.21)*	8.41	7.75	*(3.51)*	14.36	14.36	*(5.66)*	7.52	6.95	*(1.83)*	7.66	7.13	*(2.27)*	8.69	8.58	*(2.81)*	3.78	2.96	*(3.48)*
spat DB_t	11.97	10.95	*(13.94)*	5.07	3.68	*(4.30)*	6.57	4.07	*(6.00)*	2.49	2.05	*(1.27)*	2.59	2.00	*(1.33)*	2.05	1.70	*(0.95)*	11.07	7.12	*(7.89)*
spec-DB_t	1.97	1.11	*(1.71)*	16.52	11.42	*(12.17)*	28.47	19.86	*(20.11)*	9.64	6.78	*(6.90)*	9.46	7.52	*(5.46)*	11.57	9.53	*(8.64)*	2.52	1.36	*(2.20)*
runtime	342	165	*(210)*	3.64	3.78	*(0.78)*	0.062	0.047	*(0.014)*	1423	1321	*(600)*	902	901	*(345)*	437	361	*(277)*	2425	2121	*(2049)*

6. Discussion and Conclusions

In this paper, we developed a statistical methodology to cluster intensity attenuation curves in DECT scans. We applied our proposed methods, together with other alternative clustering algorithms used as baselines, to a set of 91 DECT scans of HNSCC tumors. The classical manner of evaluating algorithms for clustering/segmentation is via measures of overlap (such as the Dice score) with a ground truth segmentation. However, as mentioned in the Introduction above, the manual segmentations of HNSCC tumors that are used as "ground truth" can suffer from large inter-rater variability, and do not incorporate in any systematic manner regions immediately adjacent to the tumor that may be biologically important for determining the course of evolution of the tumor. Because of this inherent uncertainty in the appropriate contours of an HNSCC tumor, the main objective in our paper was to compare our clustering results to the manual contouring, but also to explore associations between voxels within the ground truth tumor contour and voxels in the surrounding tissue areas.

Compared to the baseline algorithms, it is clear both visually and quantitatively that our methods using Gaussian gates (SgM(V)FR) produce results that match better the manual segmentation contours. Our method using softmax gates (SsMFR) is less flexible compared to the one with Gaussian gating functions, and thus sometimes leads to non-satisfactory results. Although in terms of qualitative assessment, clusters of SsMFR-Bspl are indeed more spatially compact, quantitative performance in some situations stays similar to GMM baseline. Thus, this variant of the algorithm does not appear to perform well in practice. Using Gaussian gates, however, Dice score distributions are significantly better than the k-means-like algorithm, the best of our baseline methods.

That being said, it is also clear that with Dice scores ranging from nearly 0 to nearly 1, our proposed methods do not recover the "ground truth" segmentations in a reliable and consistent manner. Several reasons may explain this finding. First, our clinical dataset of DECT scans is not uniform, i.e., it includes tumors of highly variable characteristics, in highly variable sizes, locations and environments, which makes it particularly challenging. Moreover, as seen in Figure 4, changes in parameter tuning can lead to substantial improvement in Dice scores for some tumors. Finally, because of their intricate morphology and often small sizes, HNSCC tumors are inherently difficult to segment. In a recent international challenge, Dice scores of head and neck tumor segmentation ranged across different competition entries between 0.56 and 0.76 [45].

Most importantly, as argued throughout this paper, the clinical value of recovering the manual segmentations of HNSCC tumors as an objective criterion for evaluating the algorithm is also not clear. In fact, it was recently argued in the clinical literature that AI methods in medical imaging would be more meaningful if evaluated against clinical outcomes, as opposed to an evaluation against radiologists' performance, due to inherent subjectivity and variability of the latter [46]. For all the reasons, the objective of this paper was moved away from reproducing the manual contours produced by the radiologist, and was focused instead on developing tools that discover patterns of association in the DECT data.

Our study has several limitations. As discussed above, in our view, the appropriate way of evaluating the methodology's clinical utility is not by computing Dice scores relative to manually drawn contours. Rather, a more clinically informative evaluation would determine the performance of the recovered clusters in predicting clinical outcome in a machine learning setting, compared to the same predictive algorithm applied with the manual tumor segmentations. Such an evaluation is missing from the present paper; it will be part of a subsequent paper in future work. Another limitation stems from the lack of an automated identification of those clusters that are associated with the tumor region. Right now, we choose those clusters that maximize overlap with the manually segmented tumor region. Ideally, however, the abnormal tumor clusters should be identified automatically, by selecting those clusters that have the highest

association with clinical outcomes. In this manner, the automated cluster identification can be naturally made part of a single machine learning pipeline for predicting clinical outcomes. Yet another limitation comes from the very small size of the subset of tumors ($n = 10$) over which we estimated the algorithm parameters (λ and number of clusters), before applying the algorithm to the remaining 81 tumors in our dataset. A larger dataset, together with additional patient-specific tuning will help tune the algorithm's performance.

The need for the improvements described above is clear, and they will be made part of a subsequent publication. In the present article, we chose to focus on the theoretical and algorithmic developments. As mentioned in the Introduction, this is the first time to our knowledge that statistical tools from the functional data analysis field are put into practice with DECT data. As such, the present paper remains an inherently exploratory one in its experimental framework.

Nevertheless, we believe that we provide several important technical and methodological contributions. We constructed a functional regression mixture model that integrates spatial content into the mixture weights, and we developed a dedicated EM algorithm to estimate the optimal model parameters. Our mixture-based model is a highly flexible statistical approach allowing for many choices of the parametric form of the component densities. We proposed two candidate designs for the mixture weights, normalized Gaussian gates and softmax gates. The Gaussian-gate closed-form solution for spatial mixture weight updates considerably reduces the computation time while also providing solutions with better clustering index values, compared to the Newton–Raphson optimization algorithm needed at each update of the softmax-gating parameters.

Author Contributions: Conceptualization, F.C.; methodology, F.C.; software, F.C., S.B.; validation, P.S., M.C. and R.F.; formal analysis, F.C., S.B. and P.S.; investigation, F.C., S.B. and P.S.; resources, R.F.; data curation, S.B. and F.C.; writing—original draft preparation, F.C. and S.B; writing—review and editing, all authors; visualization, F.C. and S.B; supervision, F.C., P.S., M.C. and R.F. All authors have read and agreed to the published version of the manuscript.

Funding: This research was funded by Fonds de Recherche du Québec - Santé (FRQS) and Fondation de l'Association des Radiologistes du Québec (FARQ) (R.F.); by the Natural Sciences and Engineering Research Council of Canada (NSERC), [funding reference 260250] (M.C.) and by IRT SystemX (F.C.).

Institutional Review Board Statement: Not applicable.

Informed Consent Statement: Not applicable.

Data Availability Statement: The source codes for the proposed methods are publicly available at https://github.com/fchamroukhi/DECT-CLUST (accessed on 28 October 2022), free of charge.

Conflicts of Interest: The authors declare no conflict of interest. The funders had no role in the design of the study; in the collection, analyses, or interpretation of data; in the writing of the manuscript; or in the decision to publish the results.

Abbreviations

The following abbreviations are used in this manuscript:

DECT	Dual-energy computed tomography
DSC	Dice similarity score
EM	Expectation–maximization
FDA	Functional data analysis
fPCA	Functional PCA
HNSCC	Head and neck squamous cell carcinoma
MLE	Maximum likelihood estimation

PCA Principal component analysis
R-CNN Regional convolutional neural network
SgMFR Spatial Gaussian-gated mixtures of functional regressions
SgMVFR Spatial Gaussian-gated mixtures of vectorized functional regressions
SsMFR Spatial softmax-gated mixtures of functional regressions
SsMVFR Spatial softmax-gated mixtures of vectorized functional regressions
VOI Volumes of interest
VMI Virtual monochromatic image

Appendix A. Pseudo-Codes for the Proposed Methods

Algorithm A1 Pseudo code of DECT-CLUST with SsMFR and SgMFR models.

Inputs: 4D-image (n curves $(\boldsymbol{v}_i, \boldsymbol{x}_i, \boldsymbol{y}_i)_{i=1}^n$), # clusters K, degree p (and # knots q)

Initialization: $\boldsymbol{\theta}^{(0)} = (\boldsymbol{\alpha}^{(0)}, \boldsymbol{\theta}_1^{(0)}, \ldots, \boldsymbol{\theta}_K^{(0)})$; set $t \leftarrow 0$
while increment in log-likelihood $> \epsilon$ (e.g., $1e^{-6}$) **do**
  E-Step: `% Conditional memberships :`
  for $k = 1, \ldots, K$ **do**
    compute $\tau_{ik}^{(t)}$ for $i = 1, \ldots, n$ using (12) for `SgMFR` or the standard one for `SsMFR`
  end for
  M-Step: `%a. Spatial Mixture Weights`
  if `SgMFR` model is used: **then**
    `%Update Spatial Gaussian-Gating Functions:`
    for $k = 1, \ldots, K$ **do**
      compute $w_k^{(t+1)}$ (14), $\boldsymbol{\mu}_k^{(t+1)}$ (15), and $\mathbf{R}_k^{(t+1)}$ (16)
    end for
  end if
  if `SsMFR` model is used: **then**
    `% Update Spatial Softmax-Gating Functions:`
    IRLS Algorithm:
    Initialize: $\boldsymbol{\alpha}^{(s)} \leftarrow \boldsymbol{\alpha}^{(t)}$ and $s \leftarrow 0$ (IRLS iteration)
    while increment in $Q_{\boldsymbol{\alpha}}(\boldsymbol{\alpha}, \boldsymbol{\theta}^{(t)}) > \delta$ (eg. 1e-6) **do**
      compute $\boldsymbol{\alpha}^{(s+1)}$ using *IRLS*
      $s \leftarrow s + 1$
    end while
    $\boldsymbol{\alpha}^{(t+1)} \leftarrow \boldsymbol{\alpha}^{(s)}$
  end if
  `%b. Update Functional Mixture Components:`
  for $k = 1, \ldots, K$ **do**
    compute $\boldsymbol{\beta}_k^{(t+1)}$ using (17) and $\sigma_k^{2(t+1)}$ using (18)
  end for
  `% Convergence test`
  Compute the joint log-likelihood (9) for `SgMFR` or the standard marginal log-likelihood for `SsMFR`.
  $t \leftarrow t + 1$
end while

Outputs: $\widehat{\boldsymbol{\theta}} = (\boldsymbol{\alpha}^{(t)}, \boldsymbol{\theta}_1^{(t)}, \ldots \boldsymbol{\theta}_K^{(t)})$ the MLE of $\boldsymbol{\theta}$ and the conditional probabilities $\tau_{ik}^{(t)}$

Algorithm A2 Pseudo code of DECT-CLUST with alternative SgMVFR and SsMVFR models.

```
Inputs: 4D-image (n curves (v_i, x_i, y_i)_{i=1}^n), # clusters K, degree p (and # knots q)
for i = 1, ..., n do
  Compute the functional data representations \hat{β}_i by (19)
end for
Initialization: θ^(0) = (α^(0), θ_1^(0), ..., θ_K^(0)); set t ← 0 (EM iteration)
while increment in log-likelihood > ε (eg. 1e-6) do
  E-Step: % Conditional memberships :
  for k = 1, ..., K do
    compute τ_ik^(t) for i = 1, ..., n using (21) for SsMVFR or using (22) for SgMVFR
  end for
  M-Step: %(a) Update Spatial Weights:
  if SgMVFR model is used: then
    %Update Spatial Gaussian-Gating Functions:
    for k = 1, ..., K do
      compute w_k^(t+1) using (14), μ_k^(t+1) using (15), and R_k^(t+1) using (16)
    end for
  end if
  if SsMVFR model is used: then
    % Update Spatial Softmax-Gating Functions:
    α^(t+1) ← is returned by IRLS:
  end if
  %(b) Update Multivariate Mixture Components:
  for k = 1, ..., K do
    compute m_k^(t+1) using (23) and C_k^(t+1) using (24)
  end for
  % Convergence test
  Compute the joint log-likelihood (9) for SsMVFR or the standard marginal one for SgMVFR.
  t ← t + 1
end while
Outputs: \hat{θ} = (α^(t), θ_1^(t), ... θ_K^(t)) the MLE of θ and the conditional probabilities τ_ik^(t)
```

References

1. Forghani, R.; Johnson, J.; Ginsberg, L. *Cancer of the Head and Neck*; Chapter Imaging of Head and Neck Cancer; Wolters Kluwer: Philadelphia, PA, USA, 2017; pp. 92–148.
2. Aerts, H.J.W.L.; Velazquez, E.R.; Leijenaar, R.T.H.; Parmar, C.; Grossmann, P.; Carvalho, S.; Bussink, J.; Monshouwer, R.; Haibe-Kains, B.; Rietveld, D.; et al. Decoding tumour phenotype by noninvasive imaging using a quantitative radiomics approach. *Nat. Commun.* **2014**, *5*, 4006. [CrossRef] [PubMed]
3. Parmar, C.; Leijenaar, R.T.H.; Grossmann, P.; Velazquez, E.R.; Bussink, J.; Rietveld, D.; Rietbergen, M.M.; Haibe-Kains, B.; Lambin, P.; Aerts, H.J.W.L. Radiomic feature clusters and prognostic signatures specific for Lung and Head & Neck cancer. *Sci. Rep.* **2015**, *5*, 11044. [CrossRef] [PubMed]
4. Gillies, R.J.; Kinahan, P.E.; Hricak, H. Radiomics: Images Are More than Pictures, They Are Data. *Radiology* **2016**, *278*, 563–577. [CrossRef] [PubMed]
5. Forghani, R. Precision Digital Oncology: Emerging Role of Radiomics-based Biomarkers and Artificial Intelligence for Advanced Imaging and Characterization of Brain Tumors. *Radiol. Imaging Cancer* **2020**, *2*. [CrossRef] [PubMed]
6. Forghani, R.; Man, B.D.; Gupta, R. Dual-Energy Computed Tomography: Physical Principles, Approaches to Scanning, Usage, and Implementation: Part 1. *Neuroimaging Clin. N. Am.* **2017**, *27*, 371–384. [CrossRef] [PubMed]
7. Forghani, R.; Man, B.D.; Gupta, R. Dual-Energy Computed Tomography: Physical Principles, Approaches to Scanning, Usage, and Implementation: Part 2. *Neuroimaging Clin. N. Am.* **2017**, *27*, 385–400. [CrossRef] [PubMed]
8. Forghani, R.; Levental, M.; Gupta, R.; Lam, S.; Dadfar, N.; Curtin, H. Different spectral hounsfield unit curve and high-energy virtual monochromatic image characteristics of squamous cell carcinoma compared with nonossified thyroid cartilage. *Am. J. Neuroradiol.* **2015**, *36*, 1194–1200. [CrossRef]

9. Albrecht, M.H.; Scholtz, J.E.; Kraft, J.; Bauer, R.W.; Kaup, M.; Dewes, P.; Bucher, A.M.; Burck, I.; Wagenblast, J.; Lehnert, T.; et al. Assessment of an Advanced Monoenergetic Reconstruction Technique in Dual-Energy Computed Tomography of Head and Neck Cancer. *Eur. Radiol.* **2015**, *25*, 2493–2501. [CrossRef]
10. Forghani, R.; Kelly, H.R.; Curtin, H.D. Applications of Dual-Energy Computed Tomography for the Evaluation of Head and Neck Squamous Cell Carcinoma. *Neuroimaging Clin. N. Am.* **2017**, *27*, 445–459. [CrossRef]
11. Forghani, R.; Chatterjee, A.; Reinhold, C.; Pérez-Lara, A.; Romero-Sanchez, G.; Ueno, Y.; Bayat, M.; Alexander, J.W.M.; Kadi, L.; Chankowsky, J.; et al. Head and Neck Squamous Cell Carcinoma: Prediction of Cervical Lymph Node Metastasis by Dual-Energy CT Texture Analysis with Machine Learning. *Eur. Radiol.* **2019**, *29*, 6172–6181. [CrossRef] [PubMed]
12. Forghani, R.; Srinivasan, A.; Forghani, B. Advanced Tissue Characterization and Texture Analysis Using Dual-Energy Computed Tomography: Horizons and Emerging Applications. *Neuroimaging Clin. N. Am.* **2017**, *27*, 533–546. [CrossRef] [PubMed]
13. Vinod, S.K.; Jameson, M.G.; Min, M.; Holloway, L.C. Uncertainties in volume delineation in radiation oncology: A systematic review and recommendations for future studies. *Radiother. Oncol.* **2016**, *121*, 169–179. [CrossRef] [PubMed]
14. Hong, T.; Tome, W.; Chappell, R.; Harari, P. Variations in target delineation for head and neck IMRT: An international multi-institutional study. *Int. J. Radiat. Oncol. Biol. Phys.* **2004**, *60*, S157–S158. [CrossRef]
15. van der Veen, J.; Gulyban, A.; Nuyts, S. Interobserver variability in delineation of target volumes in head and neck cancer. *Radiother. Oncol.* **2019**, *137*, 9–15. [CrossRef]
16. Gudi, S.; Ghosh-Laskar, S.; Agarwal, J.P.; Chaudhari, S.; Rangarajan, V.; Paul, S.N.; Gupta, T. Interobserver Variability in the Delineation of Gross Tumour Volume and Specified Organs-at-risk During IMRT for Head and Neck Cancers and the Impact of FDG-PET/CT on Such Variability at the Primary Site. *J. Med. Imaging. Radiat. Sci.* **2017**, *48*, 184–192. [CrossRef]
17. Ganci, F.; Sacconi, A.; Manciocco, V.; Covello, R.; Benevolo, M.; Rollo, F.; Strano, S.; Valsoni, S.; Bicciato, S.; Spriano, G.; et al. Altered peritumoral microRNA expression predicts head and neck cancer patients with a high risk of recurrence. *Mod. Pathol.* **2017**, *30*, 1387–1401. [CrossRef]
18. Aran, D.; Camarda, R.; Odegaard, J.; Paik, H.; Oskotsky, B.; Krings, G.; Goga, A.; Sirota, M.; Butte, A.J. Comprehensive analysis of normal adjacent to tumor transcriptomes. *Nat. Commun.* **2017**, *8*, 1077. [CrossRef]
19. Martin, S.S.; Kolaneci, J.; Czwikla, R.; Booz, C.; Gruenewald, L.D.; Albrecht, M.H.; Thompson, Z.M.; Lenga, L.; Yel, I.; Vogl, T.J.; et al. Dual-Energy CT for the Detection of Portal Vein Thrombosis: Improved Diagnostic Performance Using Virtual Monoenergetic Reconstructions. *Diagnostics* **2022**, *12*, 1682. [CrossRef]
20. Grosu, S.; Vijittrakarnrung, K.; Wang, Z.J.; Obmann, M.M.; Sun, Y.; Sugi, M.D.; Yeh, B.M. Reducing Visceral-Motion-Related Artifacts on the Liver with Dual-Energy CT: A Comparison of Four Different CT Scanner Techniques. *Diagnostics* **2022**, *12*, 2155. [CrossRef]
21. Dell'Aversana, S.; Ascione, R.; De Giorgi, M.; De Lucia, D.R.; Cuocolo, R.; Boccalatte, M.; Sibilio, G.; Napolitano, G.; Muscogiuri, G.; Sironi, S.; et al. Dual-Energy CT of the Heart: A Review. *J. Imaging* **2022**, *8*, 236. [CrossRef]
22. Savadjiev, P.; Reinhold, C.; Martin, D.; Forghani, R. Knowledge Based Versus Data Based: A Historical Perspective on a Continuum of Methodologies for Medical Image Analysis. *Neuroimaging Clin. N. Am.* **2020**, *30*, 401–415. [CrossRef] [PubMed]
23. Savadjiev, P.; Chong, J.; Dohan, A.; Agnus, V.; Forghani, R.; Reinhold, C.; Gallix, B. Image-based biomarkers for solid tumor quantification. *Eur. Radiol.* **2019**, *29*, 5431—5440. [CrossRef] [PubMed]
24. Chen, S.; Zhong, X.; Hu, S.; Dorn, S.; Kachelrieß, M.; Lell, M.; Maier, A. Automatic multi-organ segmentation in dual-energy CT (DECT) with dedicated 3D fully convolutional DECT networks. *Med. Phys.* **2020**, *47*, 552–562. [CrossRef] [PubMed]
25. Chen, S.; Zhong, X.; Dorn, S.; Ravikumar, N.; Tao, Q.; Huang, X.; Maier, A. Improving Generalization Capability of Multi-Organ Segmentation Models Using Dual-Energy CT. *IEEE Trans. Radiat. Plasma. Med. Sci.* **2021**, *6*, 1. [CrossRef]
26. Wang, T.; Lei, Y.; Roper, J.; Ghavidel, B.; Beitler, J.J.; McDonald, M.; Curran, W.J.; Liu, T.; Yang, X. Head and neck multi-organ segmentation on dual-energy CT using dual pyramid convolutional neural networks. *Phys. Med. Biol.* **2021**, *66*, 115008. [CrossRef]
27. Ronneberger, O.; Fischer, P.; Brox, T. U-Net: Convolutional Networks for Biomedical Image Segmentation. In Proceedings of the Medical Image Computing and Computer-Assisted Intervention—MICCAI 2015, Munich, Germany, 5–9 October 2015; Navab, N., Hornegger, J., Wells, W.M., Frangi, A., Eds.; Springer International Publishing: Cham, Swizerland, 2015; pp. 234–241.
28. Wang, J.L.; Chiou, J.M.; Müller, H.G. Functional Data Analysis. *Annu. Rev. Stat. Appl.* **2016**, *3*, 257–295. [CrossRef]
29. Ester, M.; Kriegel, H.P.; Sander, J.; Xu, X. *A Density-Based Algorithm for Discovering Clusters in Large Spatial Databases with Noise*; AAAI Press: Washington, DC, USA, 1996.
30. McLachlan, G.; Basford, K. *Mixture Models: Inference and Applications to Clustering*; Marcel Dekker: New York, NY, USA, 1988. [CrossRef]
31. Fraley, C.; Raftery, A.E. Model-Based Clustering, Discriminant Analysis, and Density Estimation. *JASA* **2002**, *97*, 611–631. [CrossRef]
32. McLachlan, G.; Peel, D. *Finite Mixture Models*; Wiley: New York, NY, USA, 2000.
33. Dempster, A.P.; Laird, N.M.; Rubin, D.B. Maximum likelihood from incomplete data via the EM algorithm. *J. R Stat. Soc. Ser. Stat. Methodol.* **1977**, *39*, 1–38.
34. Samé, A.; Chamroukhi, F.; Govaert, G.; Aknin, P. Model-based clustering and segmentation of time series with changes in regime. *Adv. Data Anal. Classif.* **2011**, *5*, 301–321. [CrossRef]
35. Yeung, K.Y.; Fraley, C.; Murua, A.; Raftery, A.; Ruzzo, W. Model-Based Clustering and Data Transformations for Gene Expression Data. *Bioinformatics* **2001**, *17*, 977–987. [CrossRef]

36. Balafar, M.A. Spatial based Expectation Maximizing (EM). *Diagn. Pathol.* **2011**, *6*, 103. [CrossRef] [PubMed]
37. Vanhatalo, J.; Foster, S.D.; Hosack, G.R. Spatiotemporal clustering using Gaussian processes embedded in a mixture model. *Environmetrics* **2021**, *32*, e2681. [CrossRef]
38. Chamroukhi, F.; Nguyen, H.D. Model-Based Clustering and Classification of Functional Data. *Wiley Interdiscip. Rev. Data Min. Knowl. Discov.* **2019**, *9*, e1298. [CrossRef]
39. Shen, Y.; Mayhew, S.D.; Kourtzi, Z.; Tiňo, P. Spatial–temporal modelling of fMRI data through spatially regularized mixture of hidden process models. *NeuroImage* **2014**, *84*, 657–671. [CrossRef]
40. Quandt, R.E.; Ramsey, J.B. Estimating Mixtures of Normal Distributions and Switching Regressions. *JASA* **1978**, *73*, 730–738. [CrossRef]
41. Montuelle, L.; Pennec, E.L. Mixture of Gaussian regressions model with logistic weights, a penalized maximum likelihood approach. *Electron. J. Stat.* **2014**, *8*, 1661–1695. [CrossRef]
42. Jordan, M.I.; Xu, L. Convergence results for the EM approach to mixtures of experts architectures. *Neural Netw* **1995**, *8*, 1409–1431. [CrossRef]
43. Amin, M.B.; Edge, S.B.; Greene, F.L.; Byrd, D.R.; Brookland, R.K.; Washington, M.K.; Gershenwald, J.E.; Compton, C.C.; Hess, K.R.; Sullivan, D.C.; et al. *AJCC Cancer Staging Manual*, 8th ed.; Springer International Publishing: New York, NY, USA, 2017.
44. Uijlings, J.; Sande, K.; Gevers, T.; Smeulders, A. Selective Search for Object Recognition. *Int. J. Comput. Vis.* **2013**, *104*, 154–171. [CrossRef]
45. Andrearczyk, V.; Oreiller, V.; Jreige, M.; Vallières, M.; Castelli, J.; Elhalawani, H.; Boughdad, S.; Prior, J.O.; Depeursinge, A. Overview of the HECKTOR Challenge at MICCAI 2020: Automatic Head and Neck Tumor Segmentation in PET/CT. In Proceedings of the Head and Neck Tumor Segmentation, Lima, Peru, 4 October 2020; Springer: Cham, Switzerland, 2021.
46. Savadjiev, P.; Chong, J.; Dohan, A.; Vakalopoulou, M.; Reinhold, C.; Paragios, N.; Gallix, B. Demystification of AI-driven medical image interpretation: Past, present and future. *Eur. Radiol.* **2019**, *29*, 1616–1624. [CrossRef]

diagnostics

MDPI

Article

Development and Validation of an Ultrasound-Based Radiomics Nomogram for Identifying HER2 Status in Patients with Breast Carcinoma

Yinghong Guo †, Jiangfeng Wu *,†, Yunlai Wang and Yun Jin

Department of Ultrasound, Dongyang People's Hospital, No. 60 Wuning West Road, Dongyang 322100, China
* Correspondence: wjfhospital@163.com
† These authors have contributed equally to this work and should be considered as co-first author.

Abstract: (1) Objective: To evaluate the performance of ultrasound-based radiomics in the preoperative prediction of human epidermal growth factor receptor 2-positive (HER2+) and HER2− breast carcinoma. (2) Methods: Ultrasound images from 309 patients (86 HER2+ cases and 223 HER2− cases) were retrospectively analyzed, of which 216 patients belonged to the training set and 93 patients assigned to the time-independent validation set. The region of interest of the tumors was delineated, and the radiomics features were extracted. Radiomics features underwent dimensionality reduction analyses using the intra-class correlation coefficient (ICC), Mann–Whitney U test, and the least absolute shrinkage and selection operator (LASSO) algorithm. The radiomics score (Rad-score) for each patient was calculated through a linear combination of the nonzero coefficient features. The support vector machine (SVM), K nearest neighbors (KNN), logistic regression (LR), decision tree (DT), random forest (RF), naive Bayes (NB) and XGBoost (XGB) machine learning classifiers were trained to establish prediction models based on the Rad-score. A clinical model based on significant clinical features was also established. In addition, the logistic regression method was used to integrate Rad-score and clinical features to generate the nomogram model. The leave-one-out cross validation (LOOCV) method was used to validate the reliability and stability of the model. (3) Results: Among the seven classifier models, the LR achieved the best performance in the validation set, with an area under the receiver operating characteristic curve (AUC) of 0.786, and was obtained as the Rad-score model, while the RF performed the worst. Tumor size showed a statistical difference between the HER2+ and HER2− groups (p = 0.028). The nomogram model had a slightly higher AUC than the Rad-score model (AUC, 0.788 vs. 0.786), but no statistical difference (Delong test, p = 0.919). The LOOCV method yielded a high median AUC of 0.790 in the validation set. (4) Conclusion: The Rad-score model performs best among the seven classifiers. The nomogram model based on Rad-score and tumor size has slightly better predictive performance than the Rad-score model, and it has the potential to be utilized as a routine modality for preoperatively determining HER2 status in BC patients non-invasively.

Keywords: ultrasound; HER2; breast carcinoma; radiomics

Citation: Guo, Y.; Wu, J.; Wang, Y.; Jin, Y. Development and Validation of an Ultrasound-Based Radiomics Nomogram for Identifying HER2 Status in Patients with Breast Carcinoma. *Diagnostics* **2022**, *12*, 3130. https://doi.org/10.3390/diagnostics12123130

Academic Editor: Mauro Giuseppe Mastropasqua

Received: 27 September 2022
Accepted: 9 December 2022
Published: 12 December 2022

Publisher's Note: MDPI stays neutral with regard to jurisdictional claims in published maps and institutional affiliations.

1. Introduction

Breast carcinoma (BC) is the most common malignancy and the most frequent cause of carcinoma mortality in women worldwide [1], and it is a complex and heterogeneous disease [2–4]. Currently, BC is mainly classified into hormone-receptor-positive, human epidermal growth factor receptor 2-positive (HER2+), and triple-negative BC on the basis of histopathological characteristics [5,6].

HER2+ BC, in which the cells do not express estrogen receptors and progesterone receptors, accounts for about 15% of all BC cases and presents a high rate of recurrence and poor prognosis compared with hormone-receptor-positive BC [7–9]. Nevertheless, over the last two decades, as agents that target HER2, including trastuzumab and pertuzumab,

are extensively applied in clinical practice, significant advances have been made in the treatment of HER2+ BC and overall survival has improved [10–12]. Hence, the status of HER2 is one of the most significant and decisive factors in the treatment decision and prognosis for breast carcinoma patients.

So far, the evaluation of HER2 status in breast carcinoma patients mainly relies on immunohistochemistry (IHC) examination after surgical tumor excision or biopsy [13], whereas both biopsy and surgery are invasive procedures and may lead to an increased risk of complications such as seroma, local pain, and infection [14,15]. Moreover, the evaluation results of a few tissue biopsies do not necessarily represent HER2 status of the whole tumor [16]. In addition, in our center, routine histopathological findings are analyzed, but patients still need to spend extra to get results from IHC. Therefore, it is urgent to develop an economical, non-invasive, and precise pretreatment technology to predict HER2 status in breast carcinoma patients.

Radiomics is a new research field on the basis of quantitative imaging methods, which are mainly adopted to extract and analyze a large number of imaging features hardly perceived by radiologists to reflect tissue information [17,18]. Recent studies demonstrate that radiomics features extracted from magnetic resonance imaging (MRI) and computed tomography (CT) images have been widely used in diagnosis, prediction of tumor stage and histological subtype, as well as prognostic evaluation [19–22]. MRI and CT are limited by economic cost and/or equipment availability. Compared with the above imaging technologies, ultrasound, recognized as a radiation-free, convenient, and reasonably priced technology, is universally used for breast carcinoma screening and diagnosis [23]. A number of researchers have extended radiomics to ultrasound imaging [24,25]. Prior ultrasound radiomics studies have shown that molecular subtypes of BC are related to qualitative imaging characteristics and histopathologic features [26,27].

To the best of our knowledge, there are still relatively few studies to predict HER2 status of breast carcinoma using the method of ultrasound-based radiomics. We hypothesized that ultrasound radiomics features might provide guidance for predicting HER2 status in patients with breast carcinoma and would like to develop and validate an ultrasound radiomics model that could predict HER2 status.

2. Materials and Methods

2.1. Patient Cohorts

The institutional review board approved this retrospective study, and the requirement for written informed consent was waived.

In total, 522 female patients confirmed as primary BC based on pathology examination by means of biopsy or surgical excision and examined by ultrasound before treatment at our institution from March 2019 to November 2021 were retrospectively collected.

Exclusion criteria were as follows: (a) ultrasound images not suitable for radiomics study because of poor quality, artifacts, calcifications, or cystic changes (n = 48); (b) tumors larger than 50 mm in diameter (incompletely displayed in a single plane) (n = 27); (c) patients who underwent biopsy, radiotherapy, and/or chemotherapy before ultrasound examination (n = 65); (d) patients with multifocal lesions or non-mass BC (n = 4040); and (e) patients with missing clinical characteristics and/or postoperative histopathology (n = 32). Finally, there were 309 eligible patients with BC, of whom those from March 2019 to November 2020 served as the training set (n = 216), while the remaining patients formed the time-independent validation set (n = 93). The flowchart of patient selection is shown in Figure 1.

2.2. Pathological Assessment

IHC is the leading clinical technology for immunostaining, which can precisely determine the molecular subtypes of BC with high specificity. The estrogen receptor (ER) and progesterone receptor (PR) status was considered positive if ≥1% of tumor cells had positively stained nuclei [28]. For HER2 status identification, an IHC score 3+ of HER2 was

considered as positive, while an IHC score 0 or 1+ of HER2 was considered as negative. An IHC score of 2+ was considered indeterminate, and then fluorescence in situ hybridization (FISH) was carried out to assess gene amplification, and HER2 was classified as positive if the ratio was $\geq$2.0 [6]. For Ki-67 status, tumors with greater than 14% positive nuclei were considered to have high expression, while other cases were considered to have low expression [29].

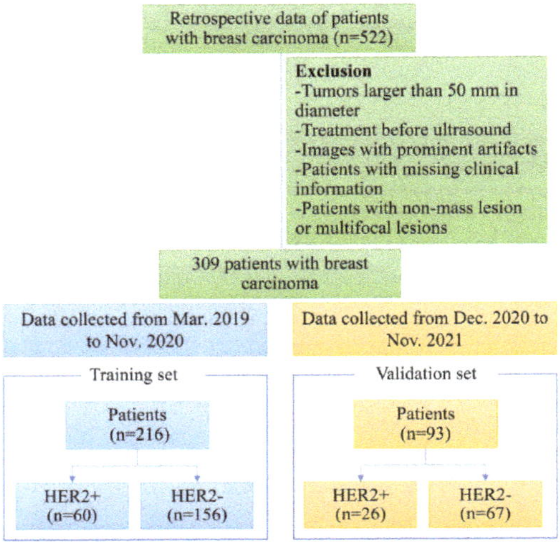

Figure 1. The patient enrollment process for this study.

2.3. Clinical Characteristics

Clinical data such as age, tumor size, and tumor location were obtained from patients' medical records. Status of ER, PR, and HER2, Ki-67 levels, molecular subtype, lymph node metastasis, and histological type of tumor were obtained by reviewing patients' pathology reports.

2.4. Image Acqusition and Segmentation

Breast ultrasound examinations were carried out by sonographers with more than 5 years of experience in breast ultrasound imaging, within 2 weeks before surgical resection. Ultrasound was performed using the LOGIQ E9 ultrasound system with a 6–15 L linear array probe and the Siemens Acuson S2000 with a 6–18 L linear array probe with radial, transverse, and longitudinal scanning on both breasts. The imaging parameters were consistent among patients: gain was about 50%; image depth was about 3.0 cm to 5.0 cm; and focus paralleled the lesion. The ultrasound image was 1164 $\times$ 873 pixels and 1024 $\times$ 768 pixels in size on the LOGIQ E9 and Siemens Acuson S2000 devices, respectively. The image of the largest section of the breast tumor with the clearest imaging was saved in the format of Digital Imaging and Communications in Medicine to maximize the preservation of the image information. Manual segmentation was performed on gray-scale ultrasound images of breast lesions. Sonographer 1 (with more than 5 years of experience in breast ultrasound imaging) with no information about the patient's clinical history selected the largest plane of each breast lesion and drew an outline of the region of interest (ROI) by using ITK-SNAP software (version 3.4.0).

2.5. Radiomic Feature Extraction

A total of 788 radiomics features, consisting of shape, statistics, texture, and wavelet features, were extracted. Radiomics features were extracted using the "pyradiomics" package of Python (version 3.7.11). These ultrasound radiomic features were divided into

four categories, including 14 two-dimension shape-based features, 18 first-order statistics features, 22 gray-level co-occurrence matrix (GLCM) features, 16 gray-level run length matrix (GLRLM) features, 16 gray-level size zone matrix (GLSZM) features, 14 gray-level dependence matrix (GLDM) features, and 688 features derived from first-order GLCM, GLRLM, GLSZM, and GLDM features using wavelet filter images. Supplementary Material Data S1 contains details on the ultrasound radiomics extraction settings.

2.6. Evaluation of Inter- and Intra-Class Correlation Coefficient

The inter- and intra-class correlation coefficients (ICCs) were adopted to test the reproducibility of feature extraction. Sonographers 1 and 2 (both with more than 5 years of experience in breast ultrasound imaging) drew ROIs on the same ultrasound images from the 50 randomly selected patients and extracted the radiomics features. Two weeks later, sonographer 1 repeated ROI segmentation on the same ultrasound images and extracted the radiomics features to assess the intra-observer reproducibility. An ICC greater than 0.75 suggested a good agreement for the feature extraction.

2.7. Radiomics Feature Selection

All the radiomics features were standardized by the z-score algorithm to ensure that the scale of feature value was uniform and improve the comparability between features, which was realized in the proportional scaling of the original data. The features with ICCs less than 0.75 were excluded.

In the training set, the Kolmogorov-Smirnov test was first performed to assess whether variances were normally distributed, and Levene's test was used to assess the equality of variance. An independent sample *t* test was used for variables with a normal distribution and homogeneity of variance. Otherwise, the Mann–Whitney U test was used. The radiomics features that showed no significant differences were excluded. The remaining radiomics features were further screened by using penalized logistic regression with a least absolute shrinkage and selection operator (LASSO) algorithm. An optimal lambda was selected through 10-fold stratified cross-validation, which was tuned to achieve minimum mean square error. Thus, features with a non-zero coefficient in the model were regarded as the most representative features.

2.8. Development and Validation of the Prediction Model

The radiomics score (Rad-score) was calculated for each lesion using LASSO regression and a linear combination of the values of the selected features weighted by their respective non-zero coefficients. Based on the Rad-score, seven machine learning classifiers consisting of decision tree (DT), K nearest neighbors (KNN), random forest (RF), support vector machine (SVM), logistic regression (LR), naive Bayes (NB), and XGBoost were used to construct the prediction model in the training set. The classifier with the highest AUC value in the validation set was obtained as the Rad-score model.

2.9. Clinical Model and Nomogram Model

Clinical features that showed a statistical difference between the HER2− and HER2− BC in the training set were adopted to develop the clinical model by using the logistic regression method. In addition, the nomogram model combining significant clinical factors and the Rad-score was constructed for personalized HER2 status prediction.

We evaluated the performances of all the models in the time-independent validation set in terms of sensitivity, specificity, positive predictive value (PPV), negative predictive value (NPV), accuracy, and the area under the receiver operating characteristic (ROC) curve (AUC). To verify the robustness of the nomogram model, the calibration curve [25] was plotted. Furthermore, decision curve analysis (DCA) [26] was also utilized to select the model that maximized patient benefits. The flowchart of this research is shown in Figure 2.

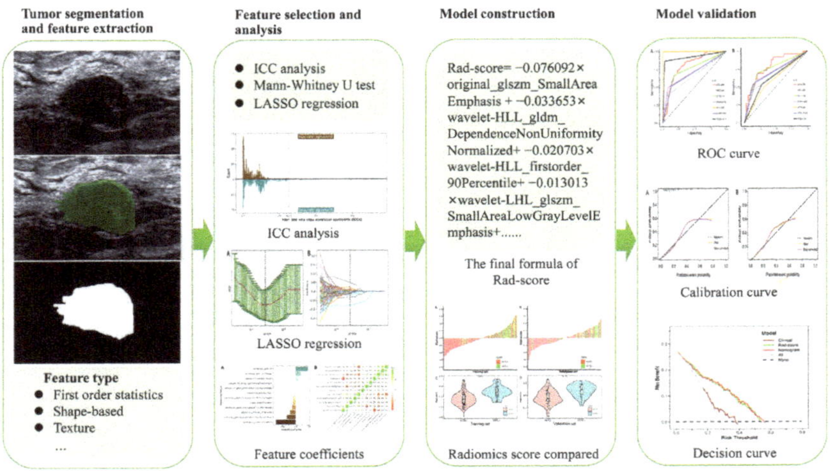

Figure 2. Schematic representation of the radiomics analysis steps.

2.10. Statistical Analysis

R version 3.5.1 software was used for statistical analysis and figure plotting. Radiomics features were extracted from each ROI using the "pyradiomics" package of Python (version 3.7.11). The continuous variables with normal distribution and homogeneity of variance were shown as the mean (standard deviation) and tested by an independent sample *t* test; otherwise, the data were analyzed by the Mann–Whitney U test and expressed as the median (interquartile range). For categorical variables, the chi-square analysis or Fisher's exact tests were applied to compare the results. A two-tailed $p < 0.05$ indicated a significant difference.

3. Results

3.1. Clinical and Pathological Characteristics

The clinical and pathological characteristics of the training and validation sets were compared, and there was no statistically significant difference found ($p > 0.05$) (Table 1). This suggested that the training and validation sets were harmonious in these clinical and pathological characteristics.

Table 1. The baseline characteristics of the enrolled patients in the training and validation sets.

Characteristic	Total Set (n = 309)	Training Set (n = 216)	Validation Set (n = 93)	p-Value
Age (year, mean ± SD)	52.88 ± 10.96	53.61 ± 10.98	51.18 ± 10.76	0.073
Size (mm, mean ± SD)	24.58 ± 11.06	25.25 ± 11.03	23.02 ± 11.03	0.106
Tumor location				0.480
Right lobe	165	112	53	
Left lobe	144	104	40	
BI-RADS				0.297
4A	46	29	17	
4B	116	79	37	
4C	81	63	18	
5	66	45	21	

Table 1. *Cont.*

Characteristic	Total Set (n = 309)	Training Set (n = 216)	Validation Set (n = 93)	p-Value
ER				0.973
Positive	228	160	68	
Negative	91	56	25	
PR				0.597
Positive	188	134	54	
Negative	121	82	39	
HER2				1.000
Positive	86	60	26	
Negative	223	156	67	
Histologic type				0.581
Invasive ductal	259	184	75	
Invasive lobular	14	9	5	
Other	36	23	13	
Ultrasound equipment				0.636
Siemens Acuson S2000	246	174	72	
LOGIQ E9	63	42	21	
US-reported LN				0.875
Metastasis positive	130	92	38	
Metastasis negative	179	124	55	
Pathology-reported LN				0.868
Metastasis positive	170	120	50	
Metastasis negative	139	96	43	
Ki-67 (%, mean ± SD)	28.52 ± 22.16	28.16 ± 21.96	29.38 ± 22.72	0.663
Radiomics score (median, IQR)	−0.0097 (−0.0975, 0.0794)	−0.0099 (−0.1030, 0.0787)	−0.0029 (−0.0883, 0.0808)	0.678

ER, estrogen receptor; PR, progesterone receptor; HER2, human epidermal growth factor receptor 2; SD, standard deviation; IQR, interquartile range; LN, lymph node; US, ultrasound; BI-RADS, Breast Imaging Reporting and Data System.

3.2. Radiomics Feature Extraction and Selection

A total of 788 radiomics features were extracted from the ultrasound images of each patient. The reproducibility of ultrasound radiomics features extraction was assessed. The intra-observer correlation coefficient of sonographer 1 in two extractions was between 0.296 and 0.996, while the inter-observer correlation coefficient of extraction by sonographer 1 and sonographer 2 was between 0.323 and 0.989. Finally, 23 radiomics features (ICC < 0.75) were excluded. The ICC evaluation results are shown in Figure 3. The morphological characteristics of the randomly selected lesions for ICC assessment are provided as Supplementary Material Data S2. All of the following analyses were based on the radiomics features extracted by sonographer 1.

In the training set, after evaluating the differences of radiomics features by the Mann–Whitney U test, 321 radiomics features were used for further analysis. Then, the optimum Lambda (Lambda = 0.027464741148160516) was determined for the LASSO regression, and 12 radiomics features with nonzero coefficients were selected to differentiate HER2+ from HER2− BC (Figure 4).

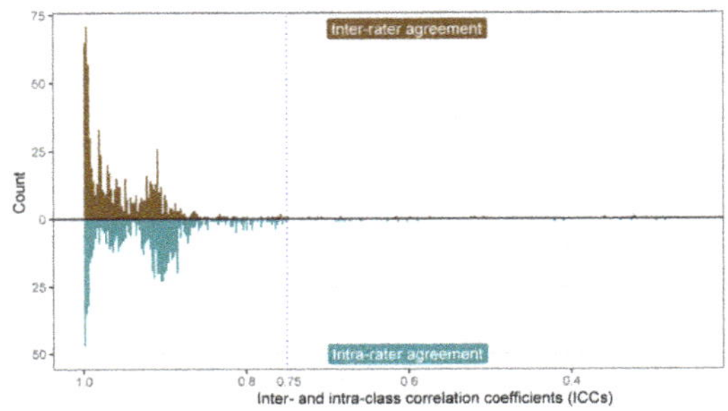

Figure 3. Bar plots of intra- and inter-observer ICC. Upper: inter-rater agreement; Lower: intra-rater agreement. ICC: intra-class correlation coefficient.

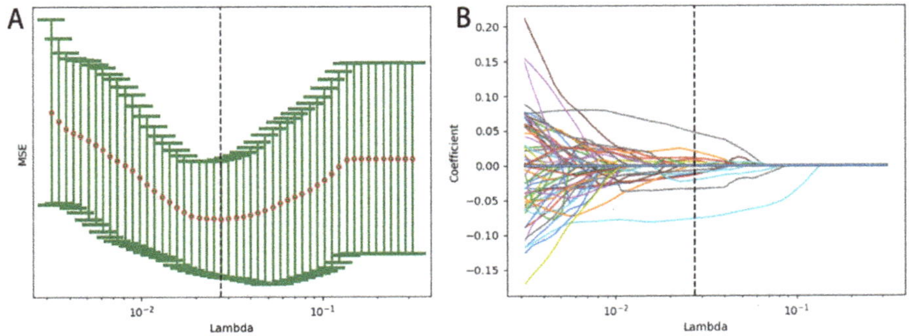

Figure 4. Feature selection and Rad-score building by LASSO. (**A**) A 10-fold cross validation was used to predict mean square error of the Rad-score building by different Lambda values. (**B**) The coefficient profiles of the radiomics features determined by different Lambda values.

Detailed information on the HER2+ BC-related features is shown in Table 2, and the nonzero coefficients of the selected features based on the LASSO regression are shown in Figure 5A. Moreover, the Pearson correlation coefficient between any pair of selected features was computed, and the correlation coefficient matrix heatmap is shown in Figure 5B.

Table 2. List of features with nonzero coefficients.

Image Type	Feature Class	Feature Name	Coefficient
original	shape	Elongation	−0.011322
original	glszm	SmallAreaEmphasis	−0.076092
wavelet-LHL	glcm	Idn	0.047259
wavelet-LHL	glszm	SmallAreaLowGrayLevelEmphasis	−0.013013
wavelet-LHH	glszm	HighGrayLevelZoneEmphasis	0.008385
wavelet-LHH	glszm	SizeZoneNonUniformityNormalized	0.005098
wavelet-HLL	firstorder	90Percentile	−0.020703
wavelet-HLL	glcm	JointEntropy	0.020412
wavelet-HLL	glszm	GrayLevelNonUniformityNormalized	−0.010225
wavelet-HLL	gldm	DependenceNonUniformityNormalized	−0.033653
wavelet-HLH	firstorder	Mean	−0.00703
wavelet-HHH	firstorder	Median	0.010776

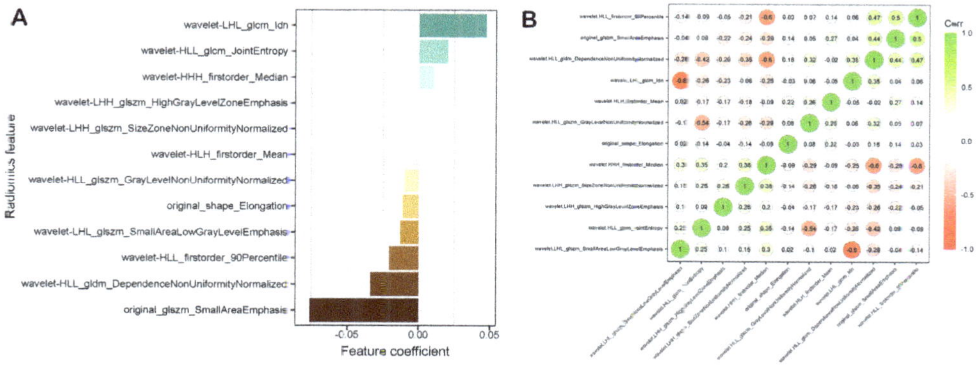

Figure 5. (**A**) The coefficients of radiomics features to construct the Rad-score; (**B**) a Pearson correlation coefficient heatmap of the selected features for predicting HER2 status. Green color denotes a positive correlation, the red color denotes a negative correlation, and the shade of the color indicates the degree of correlation.

3.3. Radiomics Score Calculation

The radiomics score (Rad-score) for each patient in the training and validation sets was calculated through a linear combination of the nonzero coefficient features based on the LASSO regression, as shown in Figure 6A,B. The corresponding fitting formula is listed in Supplementary Material Data S3. In the training set, the medians of Rad-score showed a statistical difference between the HER2+ and HER2− BC (0.0838 vs. −0.0546, $p < 0.001$), and the same results were achieved in the validation set (0.0936 vs. −0.0518, $p < 0.001$) (Figure 6C,D, Table 3).

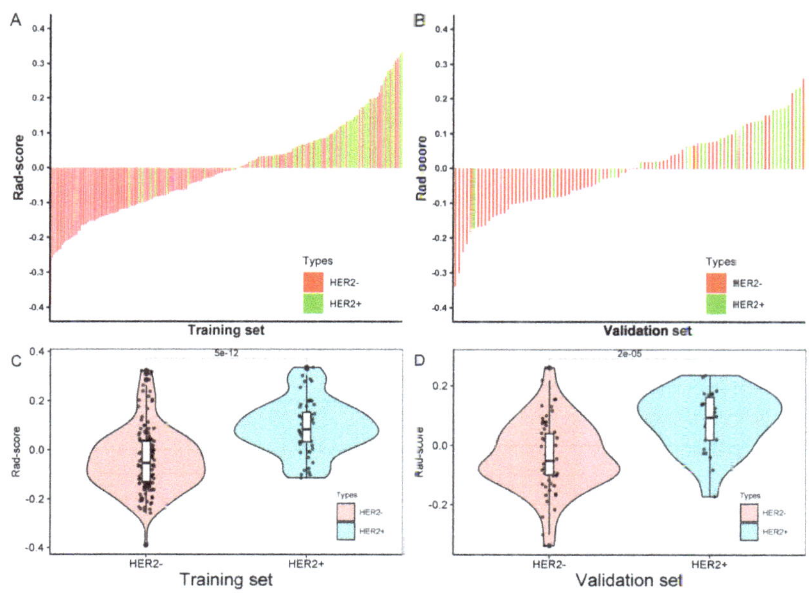

Figure 6. Radiomics score for each breast carcinoma patient in the training (**A**) and validation sets (**B**); Distribution of radiomics score values of the HER2+ and HER2− groups in the training (**C**) and validation sets (**D**).

Table 3. Rad-score for the training and validation sets.

Rad-Score	HER2− (Median, IQR)	HER2+ (Median, IQR)	*p*-Value
Training set	−0.0546 (−0.1303, 0.0338)	0.0838 (0.0336, 0.1523)	<0.001
Validation set	−0.0518 (−0.0985, 0.0394)	0.0936 (0.0185, 0.1623)	<0.001

IQR, interquartile range.

3.4. Construction and Evaluation of Machine Learning Classifier

Seven machine learning classifiers (KNN, DT, RF, SVM, LR, NB, and XGBoost) were then adopted to develop the prediction model based on the Rad-score. The sensitivity, specificity, accuracy, PPV, NPV, and AUC values of the seven machine learning classifiers are shown in Table 4.

Table 4. Diagnostic performance of seven machine learning classifiers in training and validation sets.

	Training Set				Time-Independent Validation Set			
Model	AUC (95%CI)	SEN	SPE	ACC	AUC (95%CI)	SEN	SPE	ACC
LR	0.804 (0.742–0.865)	80.0%	70.5%	73.1%	0.786 (0.683–0.890)	69.2%	79.1%	76.3%
SVM	0.691 (0.622–0.760)	51.7%	86.5%	76.9%	0.702 (0.596–0.808)	53.8%	86.6%	77.4%
KNN	0.708 (0.641–0.776)	50.0%	91.7%	80.1%	0.699 (0.592–0.806)	57.7%	82.1%	75.3%
RF	1.000 (1.000–1.000)	100.0%	100.0%	100.0%	0.593 (0.480–0.706)	50.0%	68.7%	63.4%
DT	0.747 (0.680–0.814)	66.7%	82.7%	78.2%	0.742 (0.639–0.845)	69.2%	79.1%	76.3%
XGB	0.917 (0.872–0.963)	86.7%	96.8%	94.0%	0.627 (0.516–0.739)	53.8%	71.6%	66.7%
NB	0.655 (0.589–0.722)	40.0%	91.0%	76.9%	0.667 (0.564–0.770)	42.3%	91.0%	77.4%

DT, decision tree; RF, random forest; SVM, support vector machine; LR, logistic regression; NB, naive Bayes; KNN, K nearest neighbors; XGB, XGBboost; AUC, area under the curve; SEN, sensitivity; SPE, specificity; ACC, accuracy.

Among the classifiers, the general accuracies of the RF and XGBoost were 100.0% and 94.0% in the training set and 63.4% and 66.7% in the validation set, which suggested overfitting. The accuracy was 63.4% in the RF classifier and 77.4% in the SVM and NB classifiers; the AUC values of the seven machine learning classifiers ranged from 0.593 to 0.786 in the validation set, with the LR classifier performing the best and the RF classifier performing the worst. The LR classifier with the highest AUC value was selected as the Rad-score model. In addition, a comparison of the ROC curves of the seven machine learning classifiers in the training set and validation set is shown in Figure 7. Furthermore, the AUC values between any pair of the classifiers were compared, and the *p* values were obtained by DeLong test, which are shown in Table 5.

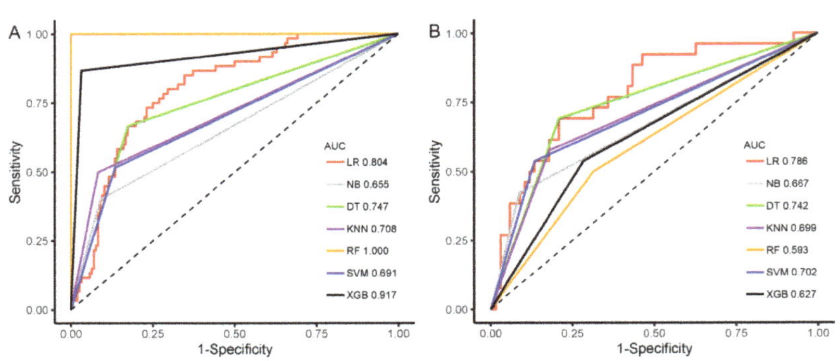

Figure 7. Receiver operating characteristic curves of seven machine learning classifiers predicting HER2+ status in training (**A**) and validation sets (**B**).

Table 5. P values for AUC comparison between any pair of models tested by the DeLong method in the validation set.

Model (AUC Value)	LR (0.786)	SVM (0.702)	KNN (0.699)	RF (0.593)	DT (0.742)	XGB (0.627)	NB (0.667)
LR (0.786)	1	-	-	-	-	-	-
SVM (0.702)	0.023	1	-	-	-	-	-
KNN (0.699)	0.054	0.955	1	-	-	-	-
RF (0.593)	0.004	0.164	0.101	1	-	-	-
DT (0.742)	0.124	0.317	0.225	0.021	1	-	-
XGB (0.627)	0.042	0.344	0.367	0.674	0.142	1	-
NB (0.667)	0.006	0.305	0.574	0.329	0.124	0.512	1

LR, logistic regression; KNN, K nearest neighbors; DT, decision tree; RF, random forest; SVM, support vector machine; NB, naive Bayes; XGB, XGBboost; AUC, area under the curve. The bold numbers (<0.05) mean statistical difference.

3.5. Clinical Model and Nomogram Model

Comparison of the clinical features between the HER2+ and the HER2− BC in the training set was performed. Tumor size ($p = 0.028$) and Rad-score ($p < 0.001$) were the significant factors to distinguish the HER2+ from HER2− BC. Other clinical features such as age, tumor location, ultrasound equipment, and ultrasound-reported lymph node status were not identified as potential factors for predicting the HER2+ type (Table 6). Then, the clinical model based on tumor size was constructed using logistic regression. At the same time, the nomogram model was established by combining the tumor size and Rad-score (Figure 8).

Table 6. Comparison of the clinical features between the HER2+ and HER2− BC groups in the training set.

Clinical Feature	Training Set (n = 216)		p-Value
	HER2− (n = 156)	HER2+ (n = 60)	
Age (year, mean ± SD)	54.04 ± 11.78	52.47 ± 8.55	0.279
Tumor location			0.673
Right	79	33	
Left	77	27	
Tumor size (mm, mean ± SD)	24.21 ± 10.90	27.93 ± 11.02	0.028
US equipment			0.064
Siemens Acuson S2000	131	43	
LOGIQ E9	25	17	
US-reported LN			0.550
Metastasis positive	64	28	
Metastasis negative	92	31	
Rad-score (median, IQR)	−0.0546 (−0.1303, 0.0338)	0.0838 (0.0336, 0.1523)	$p < 0.001$

SD, standard deviation; LN, lymph node; US, ultrasound; IQR, interquartile range.

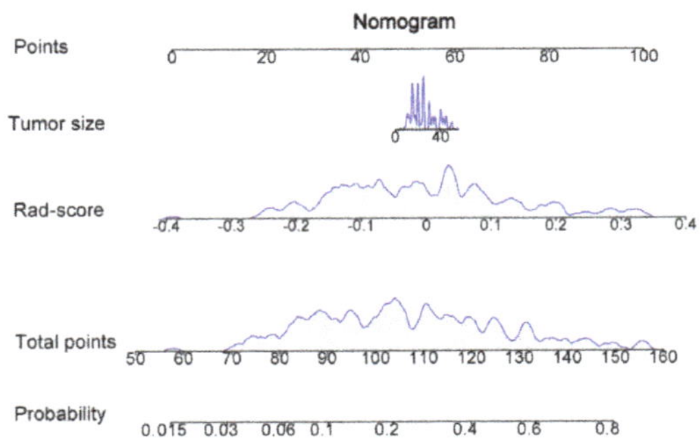

Figure 8. Nomogram based on the combination of the tumor size and Rad-score was developed using logistic regression analysis.

Moreover, the predictive abilities of the clinical, Rad-score and nomogram models were compared. The results for each model are summarized in Table 7. The ROC curves of the three models to predict the HER2+ type are shown in Figure 9. In the time-independent validation set, the AUC value of the nomogram was significantly higher than that of the clinical model (AUC, 0.788 vs. 0.618; DeLong test, $p = 0.016$). Although the nomogram model performed slightly better than the Rad-score model, there was no statistically significant difference between them (AUC, 0.788 vs. 0.786; DeLong test, $p = 0.919$).

Table 7. Predictive performances of the models identifying HER2+ status in patients with BC.

	Training Set				Time-Independent Validation Set			
Model	**AUC (95%CI)**	**SEN**	**SPE**	**ACC**	**AUC (95%CI)**	**SEN**	**SPE**	**ACC**
Clinical	0.594 (0.509–0.679)	48.3%	69.9%	63.9%	0.618 (0.485–0.751)	61.5%	62.7%	62.4%
Rad-score	0.804 (0.742–0.865)	80.0%	70.5%	73.1%	0.786 (0.683–0.890)	69.2%	79.1%	76.3%
Nomogram	0.804 (0.742–0.866)	81.7%	71.8%	74.5%	0.788 (0.685–0.891)	73.1%	80.6%	78.5%

AUC, area under the curve; SEN, sensitivity; SPE, specificity; ACC, accuracy.

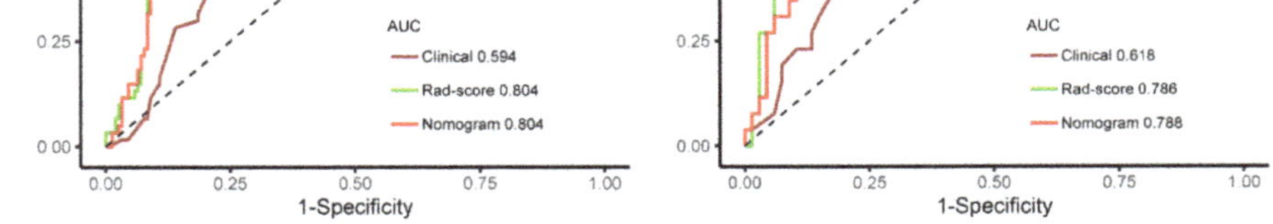

Figure 9. Receiver operating characteristic curves of the three models predicting HER2+ type in the training (**A**) and validation sets (**B**).

The LOOCV algorithm was carried out to validate the reliability and stability of the results, which yielded a high median AUC (0.790 in the validation set), indicating that the predictive performance of the nomogram model was reliable and stable.

3.6. Model Performance Evaluation

The predictive performances of the nine models, including seven machine learning classifiers, a clinical model, and a nomogram model, in the validation set are shown in Figure 10. The nomogram model has the highest AUC value (0.788), sensitivity (73.1%), and accuracy (78.5%), and NB has the highest specificity (91.0%). To sum up, the overall discrimination performance of the nomogram model was better than that of other models.

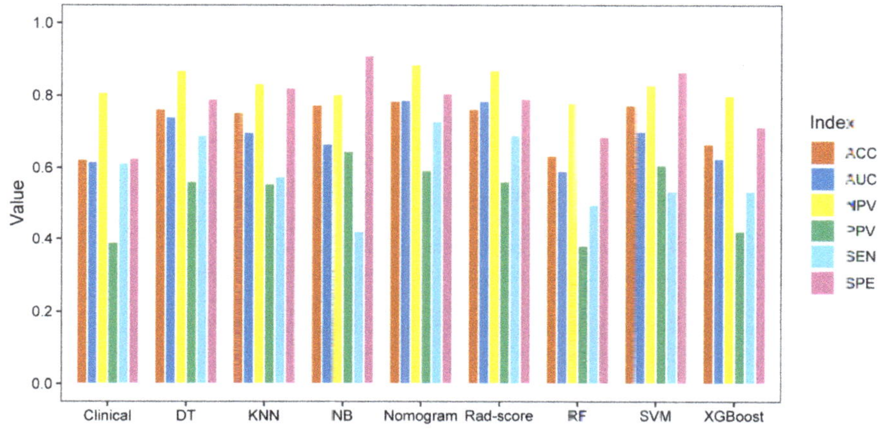

Figure 10. Bar plot of the performances of the nine prediction models in the validation set.

3.7. Clinical Application of the Prediction Models

The calibration curve for the nomogram was tested using the Hosmer-Lemeshow test and yielded nonsignificant results due to both p values > 0.05 in the training and validation sets, showing good agreements between the observed and predicted results (Figure 11).

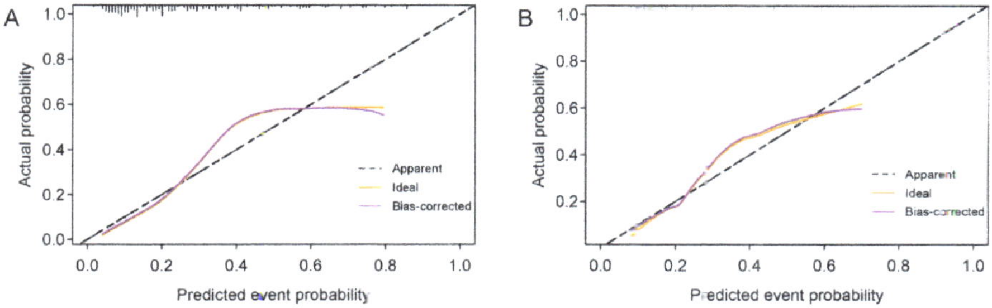

Figure 11. Calibration curves of the nomogram model in the training (**A**) and validation sets (**B**).

Decision curve analysis of the clinical, Rad-score and nomogram models is shown in Figure 12. The gray line represents the assumption that all lesions were HER2+ type. The black line represents the assumption that all lesions were HER2− type. If the threshold probability was less than 56.9%, using the nomogram would add more benefit (red line).

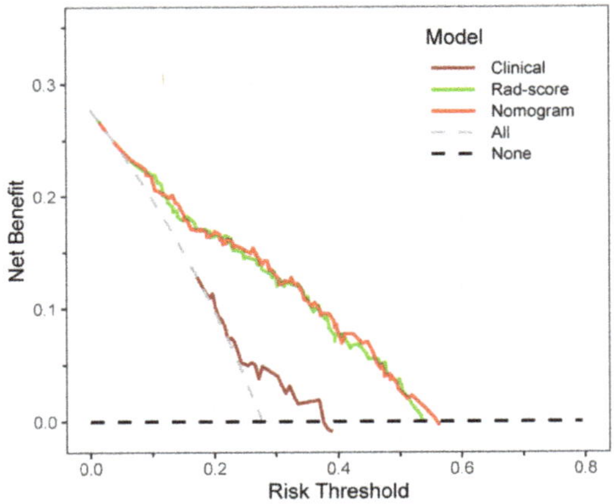

Figure 12. Decision curves of the models. If the risk threshold is less than 56.9%, the nomogram model will obtain more benefit than all treatment (assuming all breast cancer patients were HER2+) or no treatment (assuming all breast cancer patients were HER2−).

4. Discussion

Mineable data can be extracted from digital medical images by radiomics and analyzed to improve detection, diagnosis, staging, and prognosis prediction [20–22,24]. Ultrasound radiomics might be helpful to answer questions like what the molecular subtype of BC is, and this might affect the treatment strategy in patients with BC.

In our study, seven machine learning classifiers, such as KNN, LR, SVM, DT, NB, RF, and XGBoost, were established based on the Rad-score in the training set and tested in the time-independent validation set. Among them, the LR classifier with the AUC value of 0.786 performed the best, which might be that complex classifiers needed more training samples. Then the LR classifier was selected as the Rad-score model. The results indicated that the ultrasound-related Rad-score could predict the HER2+ status of patients with breast carcinoma. In addition, by establishing a nomogram model combining the Rad-score with clinical risk factors, we found that the nomogram model had significantly improved predictive performance compared with the model only involving clinical risk factors (AUC, 0.788 vs. 0.618, in the validation set) and slightly improved the ability compared with the Rad-score model (AUC, 0.788 vs. 0.786, in the validation set). The consistency between the nomogram model's predicted probability of HER2 status and the actual results were evaluated by the calibration curve, and p-values in the training and validation sets were all > 0.05, which suggested that the stability of the model is fine. In addition, patients with BC could obtain a pronounced net benefit from the nomogram model when the threshold probability is less than 56.9%, which is shown in the decision curve analysis, demonstrating the good clinical utility of this model. The nomogram model could be potentially utilized as a routine tool to assist clinicians in preoperatively predicting HER2 status non-invasively.

In recent years, radiomics studies have mainly been carried out based on computer tomography or magnetic resonance imaging [19–22], demonstrating that radiomics features could reflect the heterogeneity of tumors and have become a reliable potential biomarker for improving diagnosis and treatment decisions. In recent radiomics studies on breast ultrasound imaging, researchers have mainly focused on the differential diagnosis of benign and malignant breast tumors [27,30,31], prediction of preoperative axillary lymph node metastasis [26,32,33], and prediction of molecular subtypes [28], with mixed findings that might be due to the heterogeneity of ultrasound machines, algorithms, and extracted features. The results of our study facilitate a possible clinical role for the nomogram model in the identification of HER2 status in BC, in accordance with the mentioned studies above carried out by ultrasound radiomics.

In the present study, the ultrasound images of breast carcinomas were analyzed by radiomics, and finally 12 features were screened out to calculate the radiomics score. A majority of the selected ultrasound radiomics features were wavelet-based features that were supposed to redisplay tumor characteristics hidden behind the speckle and show discriminative ability [32,34]. Among the 12 features, original_glszm_SmallAreaEmphasis revealed the strongest correlation with HER2+, while wavelet-LHL_glcm_Idn and wavelet-HLL_gldm_DependenceNonUniformityNormalized also showed a strong correlation. The relationship between the combinations of gray levels in the image parameters is calculated by glcm texture features, which have been widely used in many texture analysis applications and can reflect the internal spatial heterogeneity of the tumor lesions [35,36]. In the present study, glcm features extracted from an ultrasound image of BC were correlated with HER2 status. Radiomics features extracted from ultrasound image of BC could detect the invisible heterogeneity of tumors and were available to predict HER2 status in patients with BC.

Generally, one feature selection method is adopted in conventional radiomics analysis. In the study by Xu et al. [37], six features based on ultrasound radiomics were selected by the recursive feature elimination, and a random forest model including 90 trees was built for prediction of HER2 status, with the AUC of 0.780 and 0.740 in the training and validation sets. In order to reduce overfitting effectively, we used the ICC and Mann–Whitney U test for feature selection in the first step and LASSO regression in the second step, and we achieved better predictive performance with the LR classifier than the study by Xu et al., with AUC values of 0.804 and 0.786 in the training and validation sets, respectively. In addition, the statistical power of our study might be more robust because the sample size in our study was significantly larger than theirs (309 vs. 114).

A prior study by Wu et al. based on ultrasound radiomics developed models to predict the expression of molecular biomarkers of the mass type of breast ductal carcinoma in situ (DCIS) [29]. Based on 41 ultrasound radiomics features, they generated a model predictive of HER2+ type in BC patients with AUC values of 0.940 in the training set and 0.740 in the validation set. As the significantly reduced AUC value in the validation set and 41 ultrasound radiomics features (much more than 10% of the sample size of the training set) were selected to establish the model, we speculated that the overfitting problem should be taken into account. Moreover, in their study, only patients with a mass type of DCIS were enrolled, whereas in this study, tumors such as invasive ductal carcinoma, invasive lobular carcinoma, and mucinous breast carcinoma were included, which expanded the range of tumor types. Furthermore, the sample size of their retrospective study was much smaller than ours (116 vs. 309). Hence, compared with the study by Wu et al., a major highlight in our study was the larger sample size and diversity of tumor types, which might increase the universality of the nomogram model. We obtained a higher AUC value compared to the aforementioned studies with regards to prediction of HER2 status by using radiomics and a machine-learning algorithm [29,37]. The most probable explanation for this is that we adopted seven machine learning classifiers to develop seven prediction models and selected the one with the highest AUC value. Furthermore, the nomogram model combining the Rad-score with the clinical risk factor of tumor size was constructed and achieved better predictive performance than the LR classifier.

Despite the significance of the present research, there are several shortcomings in our study. Firstly, the prediction model based on ultrasound radiomics features was established and tested for identifying between HER2+ and HER2− BC in a single hospital with only 216 patients in the training set and 93 patients in the validation set. In addition, as all data was collected retrospectively and limited to Chinese patients, bias was inevitable. Therefore, further prospective studies need to involve a larger patient population and perform multicenter external validation. Secondly, in our study, the extraction of radiomics features required time-consuming tumor boundary segmentation and human-defined features, and we believe that a deep learning algorithm might accurately and automatically detect, segment, and achieve more objective results [38,39]. Thirdly, only gray-scale

ultrasound images were adopted to develop the radiomics model, and other types of images like elastosonography or color Doppler ultrasound might be taken into account for multi-modal imaging to improve the predictive performance. Finally, radiomics studies based on gray-scale ultrasound images still lack reproducibility, as researchers always select different ultrasound images of the same lesion for radiomics analysis. Three-dimensional ultrasound images for feature extraction might be more objective than the conventional two-dimensional images, which could be considered in future studies.

5. Conclusions

In summary, the Rad-score model performs best among the seven classifiers. The nomogram model based on Rad-score and tumor size has slightly better predictive performance than the Rad-score model, and it has the potential to be utilized as a routine modality for preoperatively determining HER2 status in BC patients non-invasively. However, further studies with a prospective design and a larger population are required to validate the conclusions.

Supplementary Materials: The following supporting information can be downloaded at: https://www.mdpi.com/article/10.3390/diagnostics12123130/s1. Data S1: The ultrasound radiomics extraction settings; Data S2: The morphological characteristics of the randomly selected 50 lesions for ICC assessment; Data S3: The corresponding fitting formula for calculating the Rad-score.

Author Contributions: Conceptualization, J.W.; Methodology, Y.W.; Software, Y.G. and Y.W.; Data curation, Y.G., Y.W. and Y.J.; Writing—original draft, Y.G., J.W. and Y.J.; Writing—review & editing, J.W. All authors have read and agreed to the published version of the manuscript.

Funding: This research was funded by the Jinhua Science and Technology Bureau Scientific Research Project (2022-3-019).

Institutional Review Board Statement: The study was approved by the Ethics Committee of Dongyang People's Hospital (2022-YX-024).

Informed Consent Statement: This retrospective study was approved by our institutional review board, which waived the requirement for patients' informed consent.

Data Availability Statement: The data presented in this study are available on request from the corresponding author.

Conflicts of Interest: The authors declare no conflict of interest.

References

1. Sung, H.; Ferlay, J.; Siegel, R.L.; Laversanne, M.; Soerjomataram, I.; Jemal, A.; Bray, F. Global Cancer Statistics 2020: GLOBOCAN Estimates of Incidence and Mortality Worldwide for 36 Cancers in 185 Countries. *CA Cancer J. Clin.* **2021**, *71*, 209–249. [CrossRef] [PubMed]
2. Harbeck, N.; Gnant, M. Breast cancer. *Lancet* **2017**, *389*, 1134–1150. [CrossRef] [PubMed]
3. Liang, Y.; Zhang, H.; Song, X.; Yang, Q. Metastatic heterogeneity of breast cancer: Molecular mechanism and potential therapeutic targets. *Semin. Cancer Biol.* **2020**, *60*, 14–27. [CrossRef] [PubMed]
4. Hammerl, D.; Smid, M.; Timmermans, A.M.; Sleijfer, S.; Martens, J.W.M.; Debets, R. Breast cancer genomics and immuno-oncological markers to guide immune therapies. *Semin. Cancer Biol.* **2018**, *52 Pt 2*, 178–188. [CrossRef] [PubMed]
5. Carey, L.A.; Perou, C.M.; Livasy, C.A.; Dressler, L.G.; Cowan, D.; Conway, K.; Karaca, G.; Troester, M.A.; Tse, C.K.; Edmiston, S.; et al. Race, breast cancer subtypes, and survival in the Carolina Breast Cancer Study. *JAMA* **2006**, *295*, 2492–2502. [CrossRef] [PubMed]
6. Goldhirsch, A.; Wood, W.C.; Coates, A.S.; Gelber, R.D.; Thürlimann, B.; Senn, H.-J. Strategies for subtypes—Dealing with the diversity of breast cancer: Highlights of the St. Gallen International Expert Consensus on the Primary Therapy of Early Breast Cancer 2011. *Ann. Oncol.* **2011**, *22*, 1736–1747. [CrossRef]
7. Mustacchi, G.; Biganzoli, L.; Pronzato, P.; Montemurro, F.; Dambrosio, M.; Minelli, M.; Molteni, L.; Scaltriti, L. HER2-positive metastatic breast cancer: A changing scenario. *Crit. Rev. Oncol. Hematol.* **2015**, *95*, 78–87. [CrossRef]
8. Singla, H.; Ludhiadch, A.; Kaur, R.P.; Chander, H.; Kumar, V.; Munshi, A. Recent advances in HER2 positive breast cancer epigenetics: Susceptibility and therapeutic strategies. *Eur. J. Med. Chem.* **2017**, *142*, 316–327. [CrossRef]
9. Guarneri, V.; Dieci, M.; Barbieri, E.; Piacentini, F.; Omarini, C.; Ficarra, G.; Bettelli, S.; Conte, P. Loss of HER2 positivity and prognosis after neoadjuvant therapy in HER2-positive breast cancer patients. *Ann. Oncol.* **2013**, *24*, 2990–2994. [CrossRef]

10. Von Minckwitz, G.; Procter, M.; de Azambuja, E.; Zardavas, D.; Benyunes, M.; Viale, G.; Suter, T.; Arahmani, A.; Rouchet, N.; Clark, E.; et al. Adjuvant Pertuzumab and Trastuzumab in Early HER2-Positive Breast Cancer. *N. Engl. J. Med.* **2017**, *377*, 122–131. [CrossRef]
11. Schneeweiss, A.; Chia, S.; Hickish, T.; Harvey, V.; Eniu, A.; Hegg, R.; Tausch, C.; Seo, J.; Tsai, Y.-F.; Ratnayake, J.; et al. Pertuzumab plus trastuzumab in combination with standard neoadjuvant anthracycline-containing and anthracycline-free chemotherapy regimens in patients with HER2-positive early breast cancer: A randomized phase II cardiac safety study (TRYPHAENA). *Ann. Oncol.* **2013**, *24*, 2278–2284. [CrossRef] [PubMed]
12. Gianni, L.; Pienkowski, T.; Im, Y.-H.; Tseng, L.-M.; Liu, M.-C.; Lluch, A.; Starosławska, E.; De La Haba-Rodríguez, J.R.; Im, S.-A.; Pedrini, J.L.; et al. 5-year analysis of neoadjuvant pertuzumab and trastuzumab in patients with locally advanced, inflammatory, or early-stage HER2-positive breast cancer (NeoSphere): A multicentre, open-label, phase 2 randomised trial. *Lancet Oncol.* **2016**, *17*, 791–800. [CrossRef] [PubMed]
13. Bonacho, T.; Rodrigues, F.; Liberal, J. Immunohistochemistry for diagnosis and prognosis of breast cancer: A review. *Biotech. Histochem.* **2020**, *95*, 71–91. [CrossRef] [PubMed]
14. Bruening, W.; Fontanarosa, J.; Tipton, K.; Treadwell, J.R.; Launders, J.; Schoelles, K. Systematic review: Comparative effectiveness of core-needle and open surgical biopsy to diagnose breast lesions. *Ann. Intern. Med.* **2010**, *152*, 238–246. [CrossRef] [PubMed]
15. Ebner, F.; Friedl, T.W.P.; de Gregorio, A.; Lato, K.; Bekes, I.; Janni, W.; de Gregorio, N. Seroma in breast surgery: All the surgeons fault? *Arch. Gynecol. Obstet.* **2018**, *298*, 951–959. [CrossRef]
16. Juan, M.; Yu, J.; Peng, G.; Jun, L.; Feng, S.; Fang, L. Correlation between DCE-MRI radiomics features and Ki-67 expression in invasive breast cancer. *Oncol. Lett.* **2018**, *16*, 5084–5090. [CrossRef]
17. Gillies, R.J.; Kinahan, P.E.; Hricak, H. Radiomics: Images Are More than Pictures, They Are Data. *Radiology* **2016**, *278*, 563–577. [CrossRef]
18. Lambin, P.; Rios-Velazquez, E.; Leijenaar, R.; Carvalho, S.; van Stiphout, R.G.P.M.; Granton, P.; Zegers, C.M.L.; Gillies, R.; Boellard, R.; Dekker, A.; et al. Radiomics: Extracting more information from medical images using advanced feature analysis. *Eur. J. Cancer* **2012**, *48*, 441–446. [CrossRef]
19. Leithner, D.; Mayerhoefer, M.E.; Martinez, D.F.; Jochelson, M.S.; Morris, E.A.; Thakur, S.B.; Pinker, K. Non-Invasive Assessment of Breast Cancer Molecular Subtypes with Multiparametric Magnetic Resonance Imaging Radiomics. *J. Clin. Med.* **2020**, *9*, 1853. [CrossRef]
20. Shin, J.; Seo, N.; Baek, S.-E.; Son, N.-H.; Lim, J.S.; Kim, N.K.; Koom, W.S.; Kim, S. MRI Radiomics Model Predicts Pathologic Complete Response of Rectal Cancer Following Chemoradiotherapy. *Radiology* **2022**, *303*, 351–358. [CrossRef]
21. Yu, Y.; He, Z.; Ouyang, J.; Tan, Y.; Chen, Y.; Gu, Y.; Mao, L.; Ren, W.; Wang, J.; Lin, L.; et al. Magnetic resonance imaging radiomics predicts preoperative axillary lymph node metastasis to support surgical decisions and is associated with tumor microenvironment in invasive breast cancer: A machine learning, multicenter study. *EBioMedicine* **2021**, *69*, 103460. [CrossRef] [PubMed]
22. Liu, B.; Liu, H.; Zhang, L.; Song, Y.; Yang, S.; Zheng, Z.; Zhao, J.; Hou, F.; Zhang, J. Value of contrast-enhanced CT based radiomic machine learning algorithm in differentiating gastrointestinal stromal tumors with KIT exon 11 mutation: A two-center study. *Diagn. Interv. Radiol.* **2022**, *28*, 29–38. [CrossRef] [PubMed]
23. Berg, W.A.; Bandos, A.I.; Mendelson, E.B.; Lehrer, D.; Jong, R.A.; Pisano, E.D. Ultrasound as the Primary Screening Test for Breast Cancer: Analysis From ACRIN 6666. *J. Natl. Cancer Inst.* **2015**, *108*, djv367. [CrossRef]
24. Zhou, S.-C.; Liu, T.-T.; Zhou, J.; Huang, Y.-X.; Guo, Y.; Yu, J.-H.; Wang, Y.-Y.; Chang, C. An Ultrasound Radiomics Nomogram for Preoperative Prediction of Central Neck Lymph Node Metastasis in Papillary Thyroid Carcinoma. *Front. Oncol.* **2020**, *10*, 1591. [CrossRef] [PubMed]
25. Li, C.; Song, L.; Yin, J. Intratumoral and Peritumoral Radiomics Based on Functional Parametric Maps from Breast DCE-MRI for Prediction of HER-2 and Ki-67 Status. *J. Magn. Reson. Imaging* **2021**, *54*, 703–714. [CrossRef]
26. Zhou, W.J.; Zhang, Y.D.; Kong, W.T.; Zhang, C.X.; Zhang, B. Preoperative prediction of axillary lymph node metastasis in patients with breast cancer based on radiomics of gray-scale ultrasonography. *Gland. Surg.* **2021**, *10*, 1989–2001. [CrossRef]
27. Lee, S.E.; Han, K.; Kwak, J.Y.; Lee, E.; Kim, E.-K. Radiomics of US texture features in differential diagnosis between triple-negative breast cancer and fibroadenoma. *Sci. Rep.* **2018**, *8*, 13546. [CrossRef]
28. Hammond, M.E.H.; Hayes, D.F.; Dowsett, M.; Allred, D.C.; Hagerty, K.L.; Badve, S.; Fitzgibbons, P.L.; Francis, G.; Goldstein, N.S.; Hayes, M.; et al. American Society of Clinical Oncology/College Of American Pathologists guideline recommendations for immunohistochemical testing of estrogen and progesterone receptors in breast cancer. *J. Clin. Oncol.* **2010**, *28*, 2784–2795. [CrossRef]
29. Wu, L.; Zhao, Y.; Lin, P.; Qin, H.; Liu, Y.; Wan, D.; Li, X.; He, Y.; Yang, H. Preoperative ultrasound radiomics analysis for expression of multiple molecular biomarkers in mass type of breast ductal carcinoma in situ. *BMC Med. Imaging* **2021**, *21*, 84. [CrossRef]
30. Luo, W.Q.; Huang, Q.X.; Huang, X.W.; Hu, H.T.; Zeng, F.Q.; Wang, W. Predicting Breast Cancer in Breast Imaging Reporting and Data System (BI-RADS) Ultrasound Category 4 or 5 Lesions: A Nomogram Combining Radiomics and BI-RADS. *Sci. Rep.* **2019**, *9*, 11921. [CrossRef]
31. Romeo, V.; Cuocolo, R.; Apolito, R.; Stanzione, A.; Ventimiglia, A.; Vitale, A.; Verde, F.; Accurso, A.; Amitrano, M.; Insabato, L.; et al. Clinical value of radiomics and machine learning in breast ultrasound: A multicenter study for differential diagnosis of benign and malignant lesions. *Eur. Radiol.* **2021**, *31*, 9511–9519. [CrossRef] [PubMed]

32. Qiu, X.; Jiang, Y.; Zhao, Q.; Yan, C.; Huang, M.; Jiang, T. Could Ultrasound-Based Radiomics Noninvasively Predict Axillary Lymph Node Metastasis in Breast Cancer? *J. Ultrasound Med.* **2020**, *39*, 1897–1905. [CrossRef] [PubMed]
33. Gao, Y.; Luo, Y.; Zhao, C.; Xiao, M.; Ma, L.; Li, W.; Qin, J.; Zhu, Q.; Jiang, Y. Nomogram based on radiomics analysis of primary breast cancer ultrasound images: Prediction of axillary lymph node tumor burden in patients. *Eur. Radiol.* **2021**, *31*, 928–937. [CrossRef]
34. Guo, Y.; Hu, Y.; Qiao, M.; Wang, Y.; Yu, J.; Li, J.; Chang, C. Radiomics Analysis on Ultrasound for Prediction of Biologic Behavior in Breast Invasive Ductal Carcinoma. *Clin. Breast Cancer* **2018**, *18*, e335–e344. [CrossRef] [PubMed]
35. Liu, Z.; Zhu, G.; Jiang, X.; Zhao, Y.; Zeng, H.; Jing, J.; Ma, X. Survival Prediction in Gallbladder Cancer Using CT Based Machine Learning. *Front. Oncol.* **2020**, *10*, 604288. [CrossRef] [PubMed]
36. Velichko, Y.S.; Mozafarykhamseh, A.; Trabzonlu, T.A.; Zhang, Z.; Rademaker, A.W.; Yaghmai, V. Association Between the Size and 3D CT-Based Radiomic Features of Breast Cancer Hepatic Metastasis. *Acad. Radiol.* **2021**, *28*, e93–e100. [CrossRef]
37. Xu, Z.; Yang, Q.; Li, M.; Gu, J.; Du, C.; Chen, Y.; Li, B. Predicting HER2 Status in Breast Cancer on Ultrasound Images Using Deep Learning Method. *Front. Oncol.* **2022**, *12*, 829041. [CrossRef]
38. Sun, Q.; Lin, X.; Zhao, Y.; Li, L.; Yan, K.; Liang, D.; Sun, D.; Li, Z.-C. Deep Learning vs. Radiomics for Predicting Axillary Lymph Node Metastasis of Breast Cancer Using Ultrasound Images: Don't Forget the Peritumoral Region. *Front. Oncol.* **2020**, *10*, 53. [CrossRef]
39. Wei, P. Radiomics, deep learning and early diagnosis in oncology. *Emerg. Top. Life Sci.* **2021**, *5*, 829–835. [CrossRef]

cancers

MDPI

Article

The Development of an Intelligent Agent to Detect and Non-Invasively Characterize Lung Lesions on CT Scans: Ready for the "Real World"?

Martina Sollini [1,2], Margarita Kirienko [3], Noemi Gozzi [4], Alessandro Bruno [1], Chiara Torrisi [2], Luca Balzarini [2], Emanuele Voulaz [1,2], Marco Alloisio [1,2] and Arturo Chiti [1,2,*]

1 Department of Biomedical Sciences, Humanitas University, Via Rita Levi Montalcini 4, Pieve Emanuele, 20090 Milan, Italy
2 IRCCS Humanitas Research Hospital, Via Manzoni 56, Rozzano, 20089 Milan, Italy
3 Fondazione IRCCS Istituto Nazionale Tumori, Via G. Venezian 1, 20133 Milan, Italy
4 Laboratory for Neuroengineering, Department of Health Sciences and Technology, Institute for Robotics and Intelligent Systems, ETH Zurich, 8092 Zurich, Switzerland
* Correspondence: arturo.chiti@hunimed.eu

Simple Summary: An "intelligent agent" based on deep learning solutions is proposed to detect and non-invasively characterize lung lesions on computed tomography (CT) scans. Our retrospective study aimed to assess the effectiveness of Retina U-Net and the convolutional neural network for computer-aided detection (CADe) and computer-aided diagnosis (CADx) purposes. CADe and CADx were trained, validated, and tested on the publicly available LUNA challenge dataset and two local low-dose CT datasets from the IRCCS Humanitas Research Hospital.

Abstract: (1) Background: Once lung lesions are identified on CT scans, they must be characterized by assessing the risk of malignancy. Despite the promising performance of computer-aided systems, some limitations related to the study design and technical issues undermine these tools' efficiency; an "intelligent agent" to detect and non-invasively characterize lung lesions on CT scans is proposed. (2) Methods: Two main modules tackled the detection of lung nodules on CT scans and the diagnosis of each nodule into benign and malignant categories. Computer-aided detection (CADe) and computer aided-diagnosis (CADx) modules relied on deep learning techniques such as Retina U-Net and the convolutional neural network; (3) Results: Tests were conducted on one publicly available dataset and two local datasets featuring CT scans acquired with different devices to reveal deep learning performances in "real-world" clinical scenarios. The CADe module reached an accuracy rate of 78%, while the CADx's accuracy, specificity, and sensitivity stand at 80%, 73%, and 85.7%, respectively; (4) Conclusions: Two different deep learning techniques have been adapted for CADe and CADx purposes in both publicly available and private CT scan datasets. Experiments have shown adequate performance in both detection and diagnosis tasks. Nevertheless, some drawbacks still characterize the supervised learning paradigm employed in networks such as CNN and Retina U-Net in real-world clinical scenarios, with CT scans from different devices with different sensors' fingerprints and spatial resolution. Continuous reassessment of CADe and CADx's performance is needed during their implementation in clinical practice.

Keywords: CT scans; lung nodules; artificial intelligence; deep learning

Citation: Sollini, M.; Kirienko, M.; Gozzi, N.; Bruno, A.; Torrisi, C.; Balzarini, L.; Voulaz, E.; Alloisio, M.; Chiti, A. The Development of an Intelligent Agent to Detect and Non-Invasively Characterize Lung Lesions on CT Scans: Ready for the "Real World"? *Cancers* **2023**, *15*, 357. https://doi.org/10.3390/cancers15020357

Academic Editor: Yasushi Shintani

Received: 1 November 2022
Revised: 4 December 2022
Accepted: 2 January 2023
Published: 5 January 2023

1. Introduction

Lung lesions are common. The overall incidence of lung nodules has increased 10-fold from 1959 to 2015 [1], but–fortunately—the diagnosis of lung cancer has not risen accordingly [2]. The increasing use of "modern" imaging techniques, the higher adherence to screening programs, and the regular follow-up of patients suffering from other cancers

result in a more significant number of lung lesions being incidentally detected in asymptomatic people [2]. Several factors should be considered dealing with the first diagnosis of lung nodules, including the patient's pre-test probability of malignancy (e.g., smoking habits and familiar or previous history of lung cancer), and the lesion's characteristics (e.g., size, spiculation, and pleura indentation) [2]. Based on these risk assessments, patients are assigned to a class of risk and are managed accordingly [2]. The workup of patients with incidentally detected pulmonary lesions comprises actions from no further steps to computed tomography (CT) surveillance, to [^{13}F]FDG positron emission tomography (PET)/CT, to invasive procedures (biopsy, surgery, radiation therapy, or interventional radiology treatment). From a practical point of view, once identified, lung lesions must be characterized by assessing the risk of malignancy. Several qualitative CT features have been reported to be associated with malignancy (e.g., size and attenuation characteristics) [2,3], and standardized criteria to describe pulmonary nodules have been proposed (number, size, and pattern) [3]. Nonetheless, there are still several hurdles to be overcome concerning the applicability and reproducibility of these criteria (i.e., inter-operator and intra-operator variability due to misinterpretation and different experiences and expertise), ultimately affecting the management of patients diagnosed with lung nodule(s).

In recent years, artificial intelligence, acting as "another pair of eyes", has gained popularity. Computer-aided detection (CADe) and computer-aided diagnosis (CADx) systems have been recently developed [4–6] to support imagers in both lung lesion detection and diagnosis tasks. A number of models have been developed for the purpose of lung nodule detection and segmentation [7,8]. Many lung nodule segmentation algorithms based on either general or multiview neural network architecture have been proposed. Most studies adopting multiview neural networks have introduced new architectures by taking multiple lung nodule views. Subsequently, they use those views as inputs to the neural networks. On the contrary, the general neural-network-based methods rely primarily on U-Net architecture. Moreover, different lung nodule segmentation methods can be used for different types of lung nodules. Additionally, many techniques have been proposed for the classification of lung nodules (e.g., whether they are benign or malignant) focused on supervised, as opposed to semi-supervised, learning [7,8]. Despite the promising performance of these computer-aided systems, there are still limitations related to the study design (e.g., retrospective trial), technical issues (e.g., the manual labeling of images and high cost) and the efficiency (e.g., low calculation efficiency) of these tools.

The study presented in this paper aimed to develop an "intelligent agent" to detect and non-invasively characterize lung lesions on CT scans. Our goal was to apply CNN for lung cancer identification on the CT scans inspired by the available literature, but more importantly we aimed to test the tool in a "real-world" setting. In greater detail, the project involved two main modules: the first one addressed the detection of lung nodules on CT scans; the second dealt with the diagnosis (CADx) of each nodule into benign and malignant categories. The "intelligent agent" relied on deep learning techniques, which are described in the following sections.

2. Materials and Methods

2.1. Study Design

The study was a retrospective, single-institution trial.

We used public and local datasets to develop the CADe-CADx. CADe and CADx were independently developed. The study was approved by the institutional Ethics Committee.

2.2. Datasets and Image Analysis

This subsection provides details for both publicly available and local datasets for our CADe-CADx. Tables 1–3 set out lung abnormalities within the LUNA challenge dataset, CT scans used for CADs' development, and the number of nodules used for CADx.

Table 1. Lung abnormalities annotated within the LUNA challenge dataset.

Nodule $\geq$ 3 mm	Complete region of interest (ROI) boundary (>1 point) Nodule characteristics (e.g., roundness, sharpness of the margin, internal structure, etc.)
Nodule < 3 mm	The approximate centroid of the nodule No characteristics
Non nodule > 3 mm	The approximate centroid of the nodule No characteristics

Table 2. CT series datasets used for the CADs' development.

Dataset	Training	Validation	Test	Total
LUNA	603	202	-	805
ICH_s1	764	191	234	1189
ICH_s2	54	19	19	92
Total	1421	412	253	2086

Table 3. The number of lung nodules included in each dataset used for the CADx development.

Final Diagnosis	Training	Validation	Test	Total
Benign nodule	381	192	59	632
Malignant nodule	439	198	77	714
Total	820	390	136	1346

2.2.1. LUNA Challenge Dataset

The open-source LUNA challenge dataset [9] and the local ICH_s1 and ICH_s2 datasets were used for the detection task.

The LUNA dataset consists of 805 series of diagnostic and lung cancer screening chest CT scans along with XML annotation files. Lung abnormalities have been annotated by four thoracic radiologists. Each abnormality is classified as a nodule or not, and annotated according to size, as detailed in Table 1.

The mask of the region of interest (ROI) for nodules of at least 3 mm was based on a 50% consensus criterion on four radiologists' segmentations.

2.2.2. Local Datasets—ICH_s1 and ICH_s2

ICH_s1 is a local dataset consisting of 1189 low-dose CT series. The images were independently analyzed by two expert chest radiologists, and all of the nodules were segmented on non-contrast-enhanced images regardless of size. ICH_s2 consisted of 92 annotated lesions close to the mediastinum. The "ground truth" for the CADe was the segmentation performed by imagers (full concordance between radiologists). Collectively, local datasets included 1281 CT scans (441 with at least one nodule). The above-mentioned datasets were split into three subsets (training, validation, and test), as detailed in Table 2. Therefore, test set images for both ICH_s1 and ICH_s2 were used neither for training nor validation purposes.

The 234-test series from the ICH_s1 dataset comprises 104 nodules. One nodule per series is present in the 19-test series from the ICH_s2 dataset. Image segmentation and labelling were performed using a dedicated plug-in implemented for the 3D-slicer software tool (version 4.10.2, Slicer.org, Boston, MA, USA) [8].

2.2.3. CADx—Datasets and Image Analysis

The local datasets, ICH_x1 and ICH_x2, were used for classification tasks. The ICH_x1 subset comprised 349 low-dose CT images with nodules, with 29 confirmed to be malig-

nant. The images were analyzed by an expert chest radiologist (CT), and all of the nodules were segmented on non-contrast-enhanced images regardless of size. There was a partial overlap between the series included in ICH_s1 and ICH_x1. The ICH_x2 subset consists of 957 CT scans (all with at least one nodule) annotated by marking the lesion centroid. ICH_x2 samples were annotated on non-contrast-enhanced images by experienced imagers (CT and MS). ICH_x2 comprises any type of CT scan acquired at our institution, including co-registered images of PET/CT (n = 301), biopsy-guiding CT scans (n = 305), and diagnostic CT scans (n = 351, respectively). Collectively, 1346 nodules in 1306 CT scans were segmented and labelled. Radiological follow-up and pathology were used as reference standards in 350/1346 and 996/1346 cases, respectively (Table 3). Specifically, complete resolution of lung lesions was used as a radiological reference standard to define a nodule as benign. The final radiological diagnosis was used to classify 567/632 benign nodules. In the other 65/632 cases, benign nodules were pathologically confirmed. All malignant nodules were pathologically confirmed. Malignancy included primary lung cancer (adenocarcinoma = 392/714, squamous cell carcinoma = 113/714, carcinoid tumor = 31/714, and other = 35/714) and lung metastases (n = 133/714). In ten patients, the primary lung tumor subtype was not specified. The final diagnosis was collected from electronic medical records. Image segmentation and labelling were performed using a dedicated plug-in implemented with the 3D slicer tool.

2.3. CADe and CADx Architectures

As briefly mentioned in the previous sections, deep learning paradigms are behind the proposed CADe and CADx systems. One of the main challenges in our work was to test the effectiveness of deep learning architectures in real scenarios accounting for several variables, such as different CT devices, images with different spatial resolutions, and device fingerprints.

Due to the different nature of detection and diagnosis tasks, we opted for two different deep neural network architectures. CADe relies on pixel-wise segmentation to reveal whether a pixel is part of a lung lesion. To this end, it is necessary to obtain a full-resolution output binary mask to retrieve both the coordinates and the region of the lung lesion.

Conversely, CADx focuses on the final diagnosis of a given lung lesion. The system is meant to return a label indicating 'benign nodule' or 'malignant nodule'. Then, it is not necessary to make the system to return a full-resolution output mask while only an output label is needed. The following two subsections provide further technicalities regarding the two different architectures for CADe and CADx.

Furthermore, it is necessary to point out that deep learning networks must ingest many images to deliver a model with knowledge inference and generalization that can accomplish a specific domain task. The biomedical image analysis scenario is afflicted by a dimensionality problem due to the lack of manually annotated data. To be more accurate the dimensionality issue refers to the size of hand-labelled data, which is not reasonably big enough to have a deep neural network trained from scratch.

That is where data augmentation comes into play; applying image transformations without altering the meaningful content of the image itself makes a given dataset bigger in size by generating new samples. Examples of primary data augmentation are the following: flipping, mirroring, rotation, translation, and scaling.

In the following two subsections, a further description of the deep learning techniques for CADe and CADx tasks is given.

2.3.1. CADe Architecture and Development

The main goal of a CADe system is to return a full-resolution mask highlighting the suggested regions of interest for a given input image. That is why we opted for the fully convolutional neural network (FCNN) architecture. CADe tasks are, therefore, accomplished in a pixel-wise manner to extract information related to both the ROI (region of interest) and the corresponding targets. FCNN allows for return of a full-resolution

mask for a given input image. In simpler terms, an FCNN ingests an input image with size M × N and returns an output mask with the exact dimensions. The latter makes it suitable for critical biomedical image analysis tasks, such as segmentation and detection.

One of the most popular and cited FCNNs for biomedical image segmentation is the so-called U-Net [10] which owes its name to the U-shape of the network architecture. In this section, we provide readers with the overall description of U-Net, including the main layers and operations throughout the network. For the sake of clarity, we do not address the most complex mathematical concepts, and instead point the readers toward to the reference articles for further details [10].

The overall U-Net architecture is depicted in Figure 1. The encoder is responsible for extracting hidden information within the pixel domain. The latter is achieved with a stack of filters that down-sample input images in the first place. In simpler terms, the network architecture is organized in levels, with each level consisting of two Conv (convolutional layers) followed by a ReLU (rectified linear unit), a max pooling layer characterized by a parameter, namely 'stride', tuning the down-sampling factor for the input image.

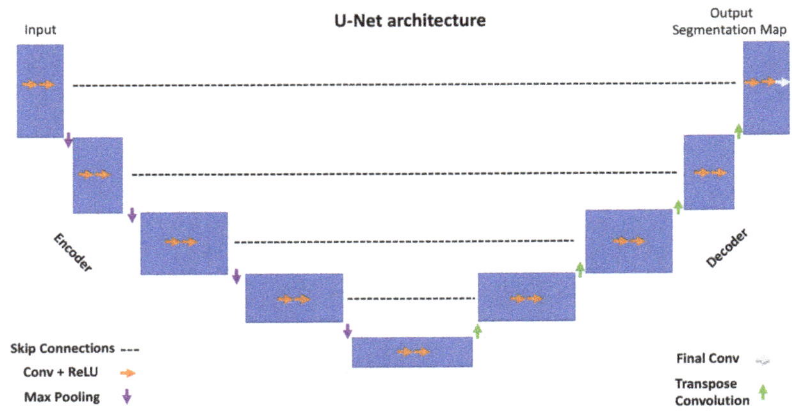

Figure 1. U-Net architecture.

All of the encoder levels are meant to extract the most meaningful features from the input images all the way to the network bottom level. Each level returns outputs through feature maps (or channels). They represent intermediate stages of the network layers that feed the following level in the stack. From a graphical viewpoint, blue rectangles indicate the input, feature maps, and output of the network. Going through consecutive layers through the encoder, it is noticeable how rectangles change in size, turning into shorter but wider blue rectangles. This is a descriptive representation showing what happens inside the network: convolutional layers work as image feature extractors; ReLU is an activation function whose primary role is to give neural networks non-linearity representation capabilities to represent results with more accuracy. Max pooling is a "pooling" operator extracting the max value from image patches and bringing down down-sampled patches.

Purple downward arrows in Figure 1 show max pooling coming into play, while orange arrows represent the sequence Conv + ReLU. The encoder is responsible for extracting "what" is in the images, while the decoder deals with the "where".

The features extracted by the contracting path are then progressively reconstructed by the expanding path (decoder) with layers consisting of transpose convolution (deconvolution), Conv + ReLU and Final Conv. Transpose convolution allows upsampling of the feature maps out of the previous layers; Conv + ReLU are then applied in combination with skip connections to refine the results in each level. Skip connections help to retrieve missing information from the encoder feature maps standing on the same level. The top left corner

of the network returns a segmentation map by adopting a one-dimensional convolutional layer. The latter can return labels in a pixel-wise fashion.

The network employed for our CADe, namely, Retina U-Net [11], is a variant of two pre-existing networks, Retina Net [12] and U-Net [10].

2.3.2. Retina U-Net

Retina U-Net [11] integrates elements from Retina Net and U-Net to combine object detection and semantic segmentation. Taking after most of the state-of-the-art object detectors, Retina U-Net complements U-Net architecture by introducing object-level predictions through feature pyramid networks (FPNs) [13]. FPNs are feature extractors with bottom-up and top-down paths. The overall Retina U-Net architecture is graphically represented in Figure 2. The overall pipeline is mainly characterized by FPNs, coarse features detectors, skip connections, Conv + Softmax, Conv + ReLu + MaxPool.

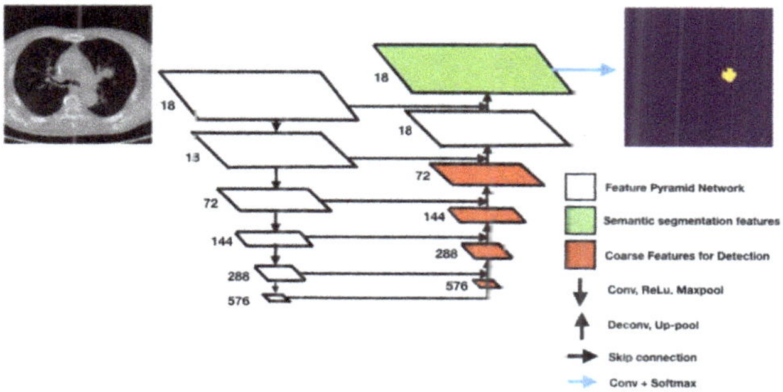

Figure 2. Architecture of the U-Net neural network used to segment lung nodules in CT scans. The number left on each layer represents the number of output channels.

Coarse feature detectors, indicated by red rectangles in Figure 2, are responsible for detecting small-sized objects using sub-network operations such as the so-called bounding box regressor (a well-known object detection technique) [14]. Skip connections support the network in retrieving missing information from the encoder feature maps standing on the same level. The Conv + ReLU + MaxPool stack consists of convolutional filter, a rectified linear unit function, and a max pooling filter. They are key to the contracting path of the FCNN as Conv filters and MaxPool filters down-sample the input feature map while ReLU allows for generalization and inference of knowledge from a non-linear input (as it is a piecewise linear function).

Conv + SoftMax consists of a sequence of a convolutional filter and a SoftMax function returning a probability map for every possible class to be detected in the images. The Up-pool and Deconv layers are responsible for the image reconstruction starting from the network bottleneck (the bottom layer in the U-shaped architecture).

In this work, the Retina U-Net was implemented to segment lung nodules. It sums up 6 layers in the contracting path (see Figure 2), 18 feature maps in the first layer and 576 in the deeper one. In the expansive path, on the other hand, the number of channels is half the ones in the first 4 layers, starting from 576, but then it is kept to 18 in the last 2 upper layers, consistently with the contracting path.

2.3.3. CADx Architecture and Development

The neural network architecture adopted to classify lung nodules is a convolutional neural network (CNN) adapted from [15] (Figure 3).

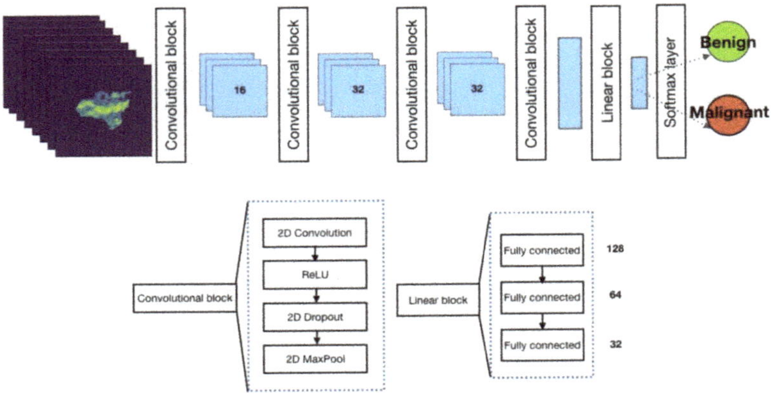

Figure 3. Architecture of the convolutional neural network used to classify lung nodules as benign or malignant. The number in each layer represents the number of output channels in that layer.

CNN consists of several layers responsible for feature extraction steps (four convolutional blocks) and classification (three fully connected layers and a SoftMax layer). The SoftMax function returns probability values for a given lung lesion, which is then classified as benign or malignant.

In Figure 3, the CNN layers are grouped into three blocks: the convolutional block, linear block, and SoftMax layer.

The convolution block consists of a convolutional layer, ReLU (rectified linear unit), and 2D dropout. Unlike FCNN, CNN does not account for an expanding path because it is not designed to return full-resolution images; its output labels are related to the classification task. As noticeable in Figure 3, a stack of convolution blocks allows for down-sampling of the input image (CT scan) into feature maps that are subsequently ingested by a linear block. The latter consists of fully connected layers paramount to the classification task and ingests high-level features out of down-sampled feature maps from the previous layers. The last layer is characterized by the SoftMax function returning probability values for the input belonging to the category of interest.

Training was performed using an equally balanced cross-entropy loss and Adam optimizer. Each series was preprocessed to extract the pixels belonging to lung nodules; indeed, the series was multiplied by the binary segmentation of each nodule.

As a result, any pixel not belonging to lung nodules is considered a background pixel. In the inference phase, the binary mask of each nodule is the result of the segmentation network described in the previous section, followed by the CNN.

The input volumes are centrally cropped around the lesion to a target size of eight slices, with a 100 $\times$ 100-pixel mask. During training, image augmentation is performed by applying random rotations, flipping, and brightness variation. The latter step is to increase the size of the training set to prevent the output model from being prone to overfitting.

As can be noticed in Figure 3 the latest layer from the network stack is a SoftMax function, which is responsible for returning probability values. The likelihood value is then adopted to extract the classification target, which is the network output.

The following section focuses on the system infrastructure and depicts the healthcare scenario we adopted in this study.

3. System Infrastructure

DICOM series identified from the institutional PACS as chest CT scan acquired and stored according to good clinical practice were downloaded and retrieved from the PACS AI Invariant. Data were anonymously stored in this layer to address privacy requirements compliance. Each series retrieved from the PACS AI Invariant was added daily to a DICOM

series queue preprocessed in a cascade by the neural networks previously described. The Invariant AI Runtime module (Figure 4) was used to run the models. The results were then re-transferred to PACS AI Invariant to be processed, consulted, and envisaged on radiological workstations. The model results and manual annotations performed using the 3D-slicer plugin were stored in PACS AI Invariant and a data warehouse.

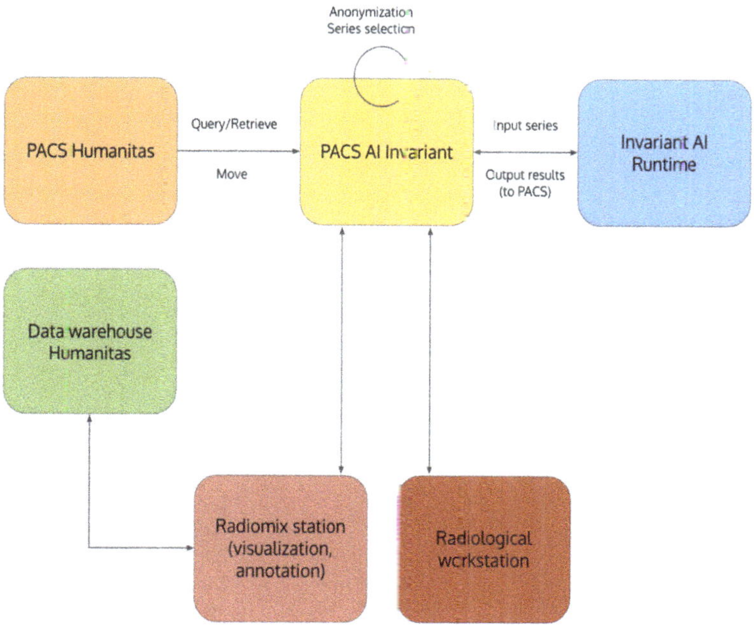

Figure 4. System infrastructure components: PACS Humanitas; PACS AI Invariant, Invariant AI Runtime; Data Warehouse; Radiomix Station, Radiological Workstation.

4. Metrics

The detection rate, accuracy, specificity, and sensitivity were computed to evaluate the performance of the CADs and the CADx, respectively. Specifically, the "ground truth" for the CADs was the segmentation performed by the imagers (complete concordance between the imagers). The detection rate was calculated as the number of nodules correctly identified by the CADe and the total number of nodules segmented by the imagers. The Dice score was calculated to compare CADe's and imager's segmentation. The final diagnosis (radiological follow-up or pathology) represented the reference standard to evaluate CADx's performance. Accordingly, each CADx prediction was classified as true positive, true negative, false positive, or false negative. The confidence analysis was used to evaluate the distribution of the probability values of each predicted nodule to belong to its class. The abovementioned metrics were calculated for training, validation, and test sets.

5. Results

As mentioned above, CADe and CADx were independently developed, trained, and tested. The results of CADx (i.e., classification) were not related to the CADe's prediction (i.e., segmentation). We reported the results of the performance obtained in the test set.

5.1. CADe

CADe correctly identified 96/123 nodules (78%) and missed 27/123 nodules. Specifically, 90/104 and 6/19 nodules of the ICH_s1 and ICH_s2 datasets, respectively, were

detected correctly. Failures were relayed mainly on ground glass opacity ($n = 6$) and very small or very large nodules close to vessels, pleura (Figure 5), and/or mediastinum ($n = 6$, $n = 9$, and $n = 4$, respectively).

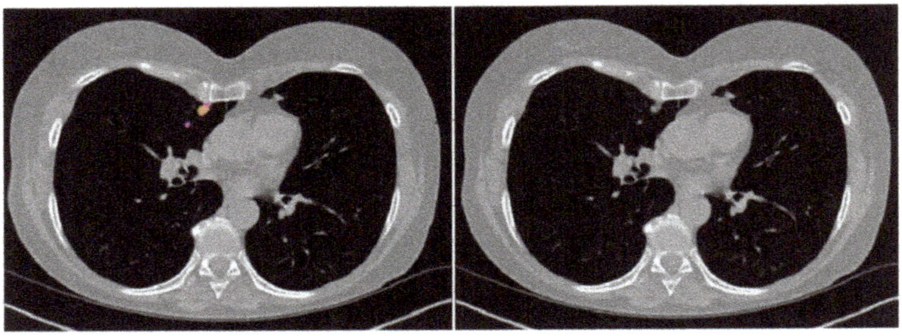

Figure 5. Example of a nodule close to pleura in the right lung correctly predicted by the CADe and of a small nodule, near to the previous one, missed by the CADe. Left panel: axial CT slice with prediction (yellow) and/or mask (pink); right panel: original CT image.

An average of 10.84 nodules per series were falsely identified. The number of false positives was reduced to 6.5 nodules per series when excluding nodules smaller than 3 mm.

5.2. CADx

CADx correctly classified 109/136 nodules (43 true negatives and 66 true positives). The CADx failed in classifying 27 nodules (11 false negatives and 16 false positives, Figures 6 and 7). The size of nodules wrongly classified was between 3 and 6 mm in 7/27 cases (6/7 solid and all falsely classified as benign), greater than 6 mm but smaller than 8 mm in 5/27 cases (3/5 solid and 2/5 falsely classified as malignant), between 8 and 10 mm in 2/27 cases (both ground glass opacity resulted false positive), bigger than 10 mm but less than 15 mm in 2/27 cases (both solid, one resulted in a false negative and one resulted in a false positive), between 15 and 25 mm in 9/27 cases, and greater than 25 mm in the remaining 2/27 cases. Specifically, false negative nodules were small nodules with a median size of 4.85 mm (range 3–11.3 mm) and solid in the majority of the cases (8/11). Considering only solid nodules, the median size of lesions falsely classified as negative was 4.7 mm (range 3–11.2 mm). Three round glass opacities (median size of 7 mm, range 3.3–7) were wrongly classified as benign. False positive nodules were quite big nodules with a median size of 20 mm (range 7.2–55 mm). Nodules wrongly classified as malignant were mainly solid (10/16) with a median size of 22 mm (range 7.2–55 mm). Considering only this class (i.e., solid nodules resulted in false positives), a consistent number of nodules (7/10) were bigger than 15 mm. Other false positive results accounted for ground glass opacity ($n = 3/16$) and part-solid nodules (3/16) with a median size of 10 mm (range 9–15 mm) and 23 mm (range 20–23 mm), respectively.

Our CADx system achieved an 80% accuracy rate. The sensitivity and specificity rates were equal to 85.7% and 73%, respectively.

The graphs in Figure 8 show the probability of each predicted nodule belonging to its class being similar for correctly classified lesions and nodules misclassified as benign (mean = 0.84 and standard deviation = 0.09 and mean = 0.84 and standard deviation = 0.10, respectively). In contrast, the confidence mean of the CAD in incorrectly predicted malignant lung lesions was lower (mean = 0.72 and standard deviation = 0.08, Figure 7).

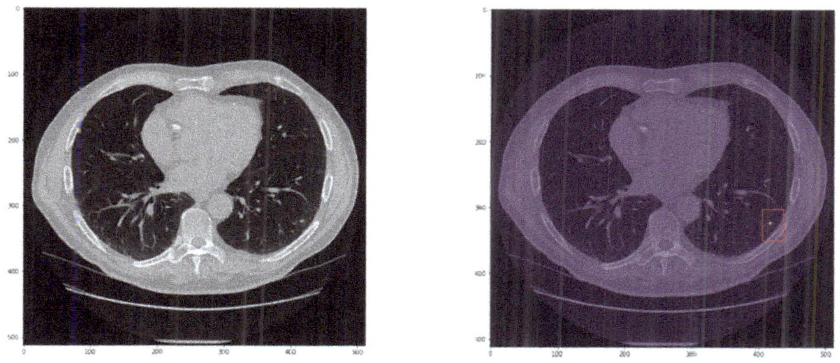

Figure 6. Example of a solid nodule of 6.3 mm (inside the red box) wrongly predicted as benign by the CADx.

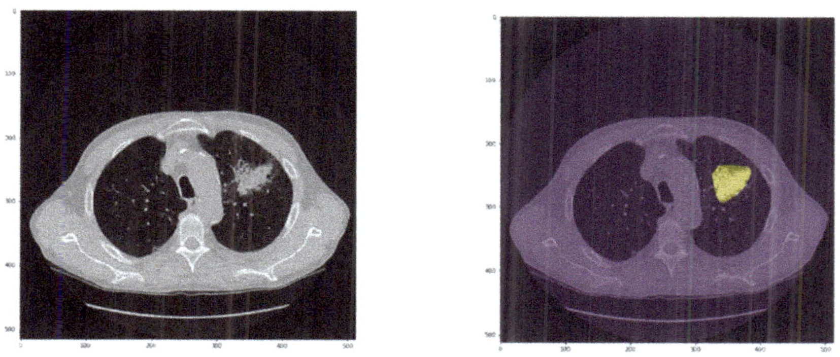

Figure 7. Example of a lesion of 55 mm (in yellow) wrongly predicted as malignant by the CADx.

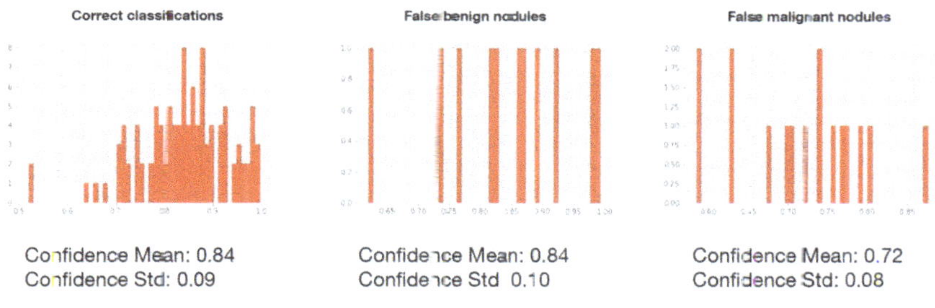

Figure 8. Confidence mean and standard deviation for correct classifications, false benign nodules and false malignant nodules.

6. Discussion

We developed an "intelligent agent" to detect and non-invasively characterize lung lesions using any type of CT scan. Big nodules detected incidentally are typically not a challenge for clinicians since the size and radiological characteristics rarely leave room for doubt. In contrast, nodules of less than 1 cm may be uncertain and difficult to characterize. In this setting, based on patient risk assessment (low versus high), number (solitary versus multiple), pattern (solid, part-solid, and ground glass), and the size of the nodule, radiological follow-up, [18F]FDG PET/CT and biopsy are recommended [3]. However, these actions might be not feasible and/or can result in inconclusive results. Therefore, a tool able

to correctly classify at least small nodule (3–8 mm) as benign or malignant is actually an unmet clinical need. As mentioned, our CADe missed some nodules (22%), mainly ground glass opacity or nodules close to vessels, pleura, or mediastinum. Notably, all nodules of 15 mm or greater were wrongly classified as false positives, while the majority of nodules smaller than 10 mm (77%) resulted in false negatives. Collectively, our CADx was more sensitive than specific and wrongly classified 20% of nodules (8% as false negatives and 12% as false positives).

Performant algorithms capable of detecting lung lesions and discriminating benign from malignant nodules with great accuracy have been described [4,6]. Our CADe and CADx exhibited lower accuracy for both detection and classification (78% and 80%, respectively) tasks than those achieved by the algorithms reported in the literature (up to 95% [6] and 96% [4], respectively). Our CADe missed some ground glass opacities and close-to-vessel nodules, pleura, and/or mediastinum. Similar failures have been reported for deep learning-based algorithms in the literature [16]. Nonetheless, our CADs-CADx benefitted in some respects. Firstly, they were developed and tested using a local dataset from real-scenario data including different types of CT images (co-registered CT from PET/CT = 23%, biopsy-guiding CT scans = 23%, low-dose CT = 27%, and fully diagnostic CT = 27%). The performance achieved in highly selected and homogeneous datasets may lead to overestimated model reliability. Therefore, continuous "real-world" re-validation is necessary for clinical implementation of DL-based tools.

Secondly, our dataset consists of well-balanced classes of benign and malignant nodules (47% and 53%, respectively). Thirdly, the final diagnosis does not rely on subjective interpretative criteria to assess malignancy risks.

Conversely, we used pathology or a rigorous radiological criterion to determine whether a nodule was benign or malignant (approximately 60% and 40% of cases, respectively). Several deep-learning-based algorithms developed to detect and classify lung nodules relied on public datasets consisting of low-dose CT images collected within lung screening programs [4,6], which dealt with a low prevalence of relatively small nodules. Many publicly available databases see the risk of malignancy assessment by expert imagers as the "ground truth" [17–19]. Nonetheless, the latter has been recently shown to affect CADx's reliability and performance [16]. Moreover, in many experiments, malignant nodules accounted for approximately one-third of the total number of nodules [20–22], potentially causing overfitting and ultimately affecting the model's reliability. Lastly, malignancy in our datasets comprised primary lung tumors and lung metastases (81% and 19%, respectively). The pattern recognition out of CNN has shown similarities to typical image-feature-based learning [23]. Still, different imaging-based features in primary lung tumors and metastases have been reported [24], suggesting specific histology-based descriptors.

On one hand, all these factors, although theoretically positive, generated a widely heterogenous dataset which was analyzed using the gold standard as a reference, which possibly explains why our tool was less performant than those reported in the literature. On the other hand, with the dataset being more heterogeneous, it positively impacted the overfitting and the generalizability of the CADs-CADx in the "real world". Therefore, we can realistically consider our CADx as a tool—albeit to be further improved—for a "virtual biopsy". It could result in several worthwhile circumstances, including, among others, lung nodules of undetermined significance. Giles et al. [25] reported that lung nodules of unknown significance were malignant in 86% of cases. Notably, in this series of 500 surgically treated patients, the percentage of lung metastases was not negligible concerning the total number of malignant lesions (22% metastases versus 78% primary lung tumors) [25], thus underlying the potential additional value of our CADx. Moreover, synchronous and metachronous tumors incidentally detected during staging or follow-up examinations have increased [26], making it imperative to exclude malignancy in a patient with a newly diagnosed lung nodule and a history of cancer.

Despite the abovementioned positive aspects, this study also presented some limitations. Firstly, the CADs-CADx were independently developed, and the presented results

refer to the detection and classification tasks separately. The next step will be to test the end-to-end tool on independent data. Furthermore, the algorithms' architectures used for the CADs-CADx were modified from pre-existing neural networks. That is common for real scenario-oriented deep learning, with fewer methodological and theoretical contributions than new, application-oriented results; the novelty is often represented by the employment of pre-existing deep learning techniques applied in new scenarios and research fields through context-based modifications.

The consideration above paves the way to a crucial point in the reliability of so-called supervised deep learning for some specific tasks. Two main questions arise from our experimental results: Can CNNs and FCNNs be considered as reliable tools for CADe and CADx? Is the supervised learning paradigm gradually going to be left behind in favor of semi-self-supervised deep learning architectures?

The paradigm adopted might not be the most suitable for a scenario with several constraints: images with different spatial resolutions and various sensors' fingerprints. The latest progress in AI sees new architectures reliant on self-supervised learning, which move toward AGI (artificial general intelligence) capable of inferring hidden properties from input data to be fine-tuned over a specific target with only a limited number of annotated samples. The results bring up some other aspects that deserve further investigation. For example, our experimental campaign ran essential data augmentation to prevent lung lesion shape distortion. Nonetheless, more advanced augmentation techniques based on generative deep learning, such as GANs (generative adversarial networks), appear to be promising to provide datasets with many more samples to be re-utilized for training purposes.

All that said, as for other domains of image patter recognition (e.g., animal photos) [27], we are convinced that sophisticated algorithms are insufficient in the setting of "real-world" data, and a huge number of observations (A million? A billion?) are needed to reach satisfactory results in terms of sensitivity and specificity. Moreover, we should keep in mind that our final goal is to develop a tool able to reach 100% accuracy, since even only one misclassified case is a misdiagnosed patient.

7. Conclusions

We have presented a specific case study on the detection and classification of lung lesions on CT scans to test the effectiveness of two of the most popular deep learning architectures, FCNN and CNN. To this end, we employed data from datasets with different features and specs. The first one was the LUNA 16 Challenge dataset; the second one consisted of images locally acquired and labelled. Furthermore, CT scans were acquired with different scanners making the case study close to real scenarios with the probability of unknown information about the sensors generating the images undergoing CADe and CADx checks. The experimental campaign confirmed the promise of these approaches in automated lung nodule assessment on CT, alongside with some drawbacks of the supervised learning paradigm employed in networks such as CNN and Retina U-Net in real-world clinical scenarios, with CT scans from different devices with different sensors' fingerprints. Collectively, we proved that these tools, although promising, are not "mature" enough to successfully analyze "real-world" data and to be finally implemented in clinical practice.

Author Contributions: Conceptualization, M.S., A.C. and M.K.; study design, M.S., M.K. and A.C.; patient screening, identification, and management, M.A., L.B. and E.V.; data analysis M.S., M.K., N.G. and C.T.; writing—original draft preparation, M.S., M.K. and N.G.; writing—review and editing, A.B. and A.C.; supervision, A.C.; project administration, A.C.; funding acquisition M.A. and M.K. All authors have read and agreed to the published version of the manuscript.

Funding: This research was funded by Lega Italiana per la Lotta contro i Tumori (LILT) "Bando di ricerca sanitaria LILT 2017 Prof. Umberto Veronesi Bando 5xmille anno 2015" and by the "Bando 5xmille Ministero della Salute 2019" assigned to Arturo Chiti from the IRCCS Humanitas.

Institutional Review Board Statement: The study was approved by the institutional Ethics Committee of IRCCS Humanitas Research Hospital (approval number 3/18, on 17 April 2018, protocol ADVIMAG-Thorax). All of the procedures performed in the studies involving human participants were in accordance with the ethical standards of the institutional and/or national research committee and in accordance with the principles of the 1964 Declaration of Helsinki and its later amendments or comparable ethical standards.

Informed Consent Statement: In view of the retrospective study design, informed consent was waived.

Data Availability Statement: Four local and private datasets, namely, ICH_s1, ICH_s2, ICH_x1, and ICH_x2, were collected and labelled in IRCCS Humanitas. They are stored in the institutional archive; they are available from the corresponding author on reasonable request. The LUNA Challenge dataset is publicly available [9].

Acknowledgments: Luca Antiga, Federica Moretti, Pietro Rota, and Gerlando Scibetta from Orobix SpA are acknowledged for the development of algorithms. The authors acknowledge the National Cancer Institute and the Foundation for the National Institutes of Health and their critical role in the creation of the free, publicly available LIDC/IDRI Database (LUNA Challenge dataset) used in this study.

Conflicts of Interest: Chiti reports unrestricted grants from Novartis, personal fees from AAA, Blue Earth Diagnostics, and General Electric Healthcare, outside of the submitted work. The other authors do not report any conflict of interest.

References

1. Tanner, N.T.; Silvestri, G.A. What's in a number? When it comes to pulmonary nodules, it's all about the number. *Am. J. Respir. Crit. Care Med.* **2015**, *192*. [CrossRef]
2. Loverdos, K.; Fotiadis, A.; Kontogianni, C.; Iliopoulou, M.; Gaga, M. Lung nodules: A comprehensive review on current approach and management. *Ann. Thorac. Med.* **2019**, *14*, 226–238.
3. MacMahon, H.; Naidich, D.P.; Goo, J.M.; Lee, K.S.; Leung, A.N.C.; Mayo, J.R.; Mehta, A.C.; Ohno, Y.; Powell, C.A.; Prokop, M.; et al. Guidelines for Management of Incidental Pulmonary Nodules Detected on CT Images: From the Fleischner Society 2017. *Radiology* **2017**, *284*, 228–243. [CrossRef]
4. Zhang, G.; Yang, Z.; Gong, L.; Jiang, S.; Wang, L.; Cao, X.; Wei, L.; Zhang, H.; Liu, Z. An Appraisal of Nodule Diagnosis for Lung Cancer in CT Images. *J. Med. Syst.* **2019**, *43*, 181. [CrossRef]
5. Kirienko, M.; Biroli, M.; Gelardi, F.; Seregni, E.; Chiti, A.; Sollini, M. Deep learning in Nuclear Medicine—Focus on CNN-based approaches for PET/CT and PET/MR: Where do we stand? *Clin. Transl. Imaging* **2021**, *9*, 37–55. [CrossRef]
6. Tandon, Y.K.; Bartholmai, B.J.; Koo, C.W. Putting artificial intelligence (AI) on the spot: Machine learning evaluation of pulmonary nodules. *J. Thorac. Dis.* **2020**, *12*, 6954–6965. [CrossRef]
7. Li, R.; Xiao, C.; Huang, Y.; Hassan, H.; Huang, B. Deep Learning Applications in Computed Tomography Images for Pulmonary Nodule Detection and Diagnosis: A Review. *Diagnostics* **2022**, *12*, 298. [CrossRef]
8. Lee, J.H.; Hwang, E.J.; Kim, H.; Park, C.M. A narrative review of deep learning applications in lung cancer research: From screening to prognostication. *Transl. Lung Cancer Res.* **2022**, *11*, 1217–1229. [CrossRef]
9. Armato, S.G.I.; McLennan, G.; Bidaut, L.; McNitt-Gray, M.F.; Meyer, C.R.; Reeves, A.P.; Clarke, L.P. The Cancer Imaging Archive. *Cancer Imaging Arch.* **2015**, *10*, K9.
10. Ronneberger, O.; Fischer, P.; Brox, T. U-Net: Convolutional Networks for Biomedical Image Segmentation. In Proceedings of the Medical Image Computing and Computer-Assisted Intervention—MICCAI 2015, Munich, Germany, 5–9 October 2015; Navab, N., Hornegger, J., Wells, W.M., Frangi, A.F., Eds.; Springer International Publishing: Cham, Switzerland, 2015; pp. 234–241.
11. Jaeger, P.F.; Kohl, S.A.A.; Bickelhaupt, S.; Isensee, F.; Kuder, T.A.; Schlemmer, H.-P.; Maier-Hein, K.H. *Retina U-Net: Embarrassingly Simple Exploitation of Segmentation Supervision for Medical Object Detection*; PMLR: Westminster, UK; London, UK, 2020.
12. Lin, T.Y.; Goyal, P.; Girshick, R.; He, K.; Dollar, P. Focal Loss for Dense Object Detection. *IEEE Trans. Pattern Anal. Mach. Intell.* **2020**, *42*, 318–327. [CrossRef]
13. Lin, T.-Y.; Dollar, P.; Girshick, R.; He, K.; Hariharan, B.; Belongie, S. Feature Pyramid Networks for Object Detection. In Proceedings of the 2017 IEEE Conference on Computer Vision and Pattern Recognition (CVPR), Honolulu, HI, USA, 21–26 July 2017; IEEE: New York, NY, USA, 2017; pp. 936–944.
14. He, Y.; Zhu, C.; Wang, J.; Savvides, M.; Zhang, X. Bounding box regression with uncertainty for accurate object detection. In Proceedings of the IEEE Computer Society Conference on Computer Vision and Pattern Recognition 2019, Long Beach, CA, USA, 15–20 June 2019; pp. 2888–2897.
15. Song, Q.Z.; Zhao, L.; Luo, X.K.; Dou, X.C. Using Deep Learning for Classification of Lung Nodules on Computed Tomography Images. *J. Healthc. Eng.* **2017**, *2017*, 8314740. [CrossRef]

16. Liu, Z.; Li, L.; Li, T.; Luo, D.; Wang, X.; Luo, D. Does a Deep Learning–Based Computer-Assisted Diagnosis System Outperform Conventional Double Reading by Radiologists in Distinguishing Benign and Malignant Lung Nodules? *Front. Oncol.* **2020**, *10*, 545862. [CrossRef]
17. Ardila, D.; Kiraly, A.P.; Bharadwaj, S.; Choi, B.; Reicher, J.J.; Peng, L.; Tse, D.; Etemadi, M.; Ye, W.; Corrado, G.; et al. End-to-end lung cancer screening with three-dimensional deep learning on low-dose chest computed tomography. *Nat. Med.* **2019**, *25*, 954–961. [CrossRef]
18. Shen, W.; Zhou, M.; Yang, F.; Yu, D.; Dong, D.; Yang, C.; Zang, Y.; Tian, J. Multi-crop Convolutional Neural Networks for lung nodule malignancy suspiciousness classification. *Pattern Recognit.* **2017**, *61*, 663–673. [CrossRef]
19. Xie, Y.; Zhang, J.; Xia, Y.; Fulham, M.; Zhang, Y. Fusing texture, shape and deep model-learned information at decision level for automated classification of lung nodules on chest CT. *Inf. Fusion* **2018**, *42*, 102–110. [CrossRef]
20. de Carvalho Filho, A.O.; Silva, A.C.; de Paiva, A.C.; Nunes, R.A.; Gattass, M. Classification of patterns of benignity and malignancy based on CT using topology-based phylogenetic diversity index and convolutional neural network. *Pattern Recognit.* **2018**, *81*, 200–212. [CrossRef]
21. Liu, Y.; Hao, P.; Zhang, P.; Xu, X.; Wu, J.; Chen, W. Dense Convolutional Binary-Tree Networks for Lung Nodule Classification. *IEEE Access* **2018**, *6*, 49080–49088. [CrossRef]
22. Tajbakhsh, N.; Suzuki, K. Comparing two classes of end-to-end machine-learning models in lung nodule detection and classification: MTANNs vs. CNNs. *Pattern Recognit.* **2017**, *63*, 476–486. [CrossRef]
23. Chen, M.; Shi, X.; Zhang, Y.; Wu, D.; Guizani, M. Deep Features Learning for Medical Image Analysis with Convolutional Autoencoder Neural Network. *IEEE Trans. Big Data* **2017**, *7*, 750–758. [CrossRef]
24. Kirienko, M.; Cozzi, L.; Rossi, A.; Voulaz, E.; Antunovic, L.; Fogliata, A.; Chiti, A.; Sollini, M. Ability of FDG PET and CT radiomics features to differentiate between primary and metastatic lung lesions. *Eur. J. Nucl. Med. Mol. Imaging* **2018**, *45*, 1649–1660. [CrossRef]
25. Giles, A.E.; Teferi, Y.; Kidane, B.; Bayaraa, B.; Tan, L.; Buduhan, G.; Srinathan, S. Lung resection without tissue diagnosis: A pragmatic perspective on the indeterminate pulmonary nodule. *Clin. Lung Cancer* **2021**, *22*, E774–E781. [CrossRef]
26. Loukeri, A.A.; Kampolis, C.F.; Ntokou, A.; Tsoukalas, G.; Syrigos, K. Metachronous and synchronous primary lung cancers: Diagnostic aspects, surgical treatment, and prognosis. *Clin. Lung Cancer* **2015**, *16*, 15–23. [CrossRef]
27. Golle, P. Machine learning attacks against the asirra CAPTCHA. In Proceedings of the ACM Conference on Computer and Communications Security 2008, Alexandria, VA, USA, 27–31 October 2008.

cancers

MDPI

Article

Predictive Value of ^{18}F-FDG PET/CT Using Machine Learning for Pathological Response to Neoadjuvant Concurrent Chemoradiotherapy in Patients with Stage III Non-Small Cell Lung Cancer

Jang Yoo [1], Jaeho Lee [2], Miju Cheon [1], Sang-Keun Woo [3], Myung-Ju Ahn [4], Hong Ryull Pyo [5], Yong Soo Choi [6], Joung Ho Han [7] and Joon Young Choi [8,*]

1 Department of Nuclear Medicine, Veterans Health Service Medical Center, Seoul 05368, Korea; jang8214.yoo@gmail.com (J.Y.); diva1813@naver.com (M.C.)
2 Department of Preventive Medicine, Seoul National University College of Medicine, Seoul 03080, Korea; hoyajh21@gmail.com
3 Department of Nuclear Medicine, Korea Cancer Center Hospital, Korea Institute of Radiological and Medical Sciences (KIRAMS), Seoul 01812, Korea; skwoo@kirams.re.kr
4 Division of Hematology-Oncology, Department of Medicine, Samsung Medical Center, Sungkyunkwan University School of Medicine, Seoul 06351, Korea; silk.ahn@samsung.com
5 Department of Radiation Oncology, Samsung Medical Center, Sungkyunkwan University School of Medicine, Seoul 06351, Korea; hr.pyo@samsung.com
6 Department of Thoracic and Cardiovascular Surgery, Samsung Medical Center, Sungkyunkwan University School of Medicine, Seoul 06351, Korea; ysooyah.choi@samsung.com
7 Department of Pathology, Samsung Medical Center, Sungkyunkwan University School of Medicine, Seoul 06351, Korea; joungho.han@samsung.com
8 Department of Nuclear Medicine, Samsung Medical Center, Sungkyunkwan University School of Medicine, Seoul 06351, Korea
* Correspondence: jynm.choi@samsung.com; Tel.: +82-2-3410-2648; Fax: +82-2-3410-2639

Citation: Yoo, J.; Lee, J.; Cheon, M.; Woo, S.-K.; Ahn, M.-J.; Pyo, H.R.; Choi, Y.S.; Han, J.H.; Choi, J.Y. Predictive Value of ^{18}F-FDG PET/CT Using Machine Learning for Pathological Response to Neoadjuvant Concurrent Chemoradiotherapy in Patients with Stage III Non-Small Cell Lung Cancer. *Cancers* **2022**, *14*, 1987. https://doi.org/10.3390/cancers14081987

Academic Editors: Hamid Khayyam, Ali Madani, Rahele Kafieh and Ali Hekmatnia

Received: 17 February 2022
Accepted: 13 April 2022
Published: 14 April 2022

Simple Summary: The pathological complete response (pCR) after neoadjuvant chemoradiotherapy (CCRT) is an independent prognostic factor for progression-free and overall survival in non-small cell lung cancer (NSCLC). ^{18}F-FDG PET/CT has been performed for initial staging work-up, treatment response, and follow-up in patients with NSCLC. Machine learning (ML) as an empirical data science has become relevant to nuclear medicine. We investigated the predictive performance of ^{18}F-FDG PET/CT using an ML model to assess the treatment response to neoadjuvant CCRT in patients with stage III NSCLC, and compared the performance of the ML model predictions to predictions from conventional PET parameters and from physicians. The predictions from the ML model using radiomic features of ^{18}F-FDG PET/CT provided better accuracy than predictions from conventional PET parameters and from physicians for the neoadjuvant CCRT response of stage III non-small cell lung cancer.

Abstract: We investigated predictions from ^{18}F-FDG PET/CT using machine learning (ML) to assess the neoadjuvant CCRT response of patients with stage III non-small cell lung cancer (NSCLC) and compared them with predictions from conventional PET parameters and from physicians. A retrospective study was conducted of 430 patients. They underwent ^{18}F-FDG PET/CT before initial treatment and after neoadjuvant CCRT followed by curative surgery. We analyzed texture features from segmented tumors and reviewed the pathologic response. The ML model employed a random forest and was used to classify the binary outcome of the pathological complete response (pCR). The predictive accuracy of the ML model for the pCR was 93.4%. The accuracy of predicting pCR using the conventional PET parameters was up to 70.9%, and the accuracy of the physicians' assessment was 80.5%. The accuracy of the prediction from the ML model was significantly higher than those derived from conventional PET parameters and provided by physicians ($p < 0.05$). The ML model is useful for predicting pCR after neoadjuvant CCRT, which showed a higher predictive accuracy than those achieved from conventional PET parameters and from physicians.

Keywords: non-small cell lung cancer; neoadjuvant concurrent chemoradiotherapy; ^{18}F-FDG PET/CT; machine learning; random forest; pathologic complete response

1. Introduction

Lung cancer is the most common malignant tumor and remains the leading cause of cancer-related death worldwide in spite of major advances in prevention and multimodal treatment [1]. Non-small cell lung cancer (NSCLC) accounts for more than 85% of all lung cancers and about 30% of NSCLC present with locally advanced disease in stage III [2]. Patients with stage III NSCLC are usually considered as inoperable. Neoadjuvant concurrent chemoradiotherapy (CCRT) followed by surgery has been established as being able to improve the overall outcome by reducing the rate of local failures and distant metastasis [3,4].

In patients receiving neoadjuvant CCRT for stage III NSCLC, surgical resection allows for the identification of the histopathologic tumor response to determine the prognosis and to evaluate postoperative therapeutic options. According to previous studies, the pathologic complete response (pCR) after neoadjuvant CCRT is an independent prognostic factor for progression-free and overall survival in NSCLC [5,6]. Although several papers have reported a wide range of pCR values of 16–27%, it is clear that the pCR is highly correlated with patient survival [7–10].

^{18}F-fluorodeoxyglucose positron emission tomography/computed tomography (^{18}F-FDG PET/CT) has been performed for initial staging work-up, treatment response, and follow-up in patients with NSCLC. It has also been viewed as appropriate for the precise investigation of treatment response after CCRT [11,12]. Previous studies have focused on the comparison of quantitative PET parameters such as the standard uptake value (SUV) after neoadjuvant treatment and histopathologic findings after surgery [13,14]. Moreover, the application of the PET response criteria in solid tumors (PERCIST 1.0) as an evaluation for ^{18}F-FDG PET/CT has been performed to enhance the limitation of anatomic tumor response metrics [15,16]. The role of ^{18}F-FDG PET/CT still needs to be explored because possible misinterpretations due to radiation-induced inflammation such as pneumonitis can cause problems in ^{18}F-FDG PET/CT images [17,18].

Machine learning (ML) as an empirical data science, which can learn patterns or characteristics from one set of given data and use them to evaluate new data, has become relevant to nuclear medicine. Our previous study demonstrated that ML is well suited to performing analyses of high dimensionality radiomic feature extraction from ^{18}F-FDG PET/CT, and ML analysis provided better diagnostic performance than physicians for evaluating metastatic mediastinal lymph nodes in NSCLC [19]. Although assessing the radiomic features of a tumor in clinical practice has some challenges because of the time, effort, and skill involved, we have shown that ML can improve the diagnostic accuracy and its availability in NSCLC. However, there is still no study that has evaluated the predictive performance of ML for the neoadjuvant CCRT response using the radiomic features of ^{18}F-FDG PET/CT.

Therefore, we investigated the predictive performance of ^{18}F-FDG PET/CT using an ML model to assess the treatment response to neoadjuvant CCRT in patients with stage III NSCLC, and compared the performance of the ML model predictions to predictions from conventional PET parameters and from physicians.

2. Materials and Methods

2.1. Subjects

We retrospectively reviewed the medical records of all patients newly diagnosed with stage III NSCLC through imaging studies such as chest X-ray, enhanced chest CT, and ^{18}F-FDG PET/CT, as well as pathologic studies including endobronchial ultrasound-guided transbronchial needle aspiration, mediastinoscopic biopsy, or thoracotomy, between

November 2008 and October 2020. To be included in the study population, patients needed to complete a planned neoadjuvant CCRT and undergo curative-intent surgical treatment for stage III NSCLC according to the 7th edition of the TNM classification [20], and undergo a second ^{18}F-FDG PET/CT within approximately 3 weeks following the completion of neoadjuvant CCRT for restaging work-up. Patients in poor cardiopulmonary condition that precluded surgery or who had previously been treated because of another malignant disease were excluded from the study population. Patients who received neoadjuvant chemotherapy or radiotherapy alone were also excluded.

This study was approved by the institutional review board of our institution (IRB No. 2020-09-185), and the requirement for informed patient consent was waived due to its retrospective design.

2.2. Neoadjuvant CCRT and Histopathologic Evaluation

The neoadjuvant CCRT consisted of chemotherapy and concurrent thoracic radiotherapy. Thoracic radiotherapy was delivered to patients with a total dose of 45 Gy with 1.8 Gy/fraction over 5 weeks from November 2008 to October 2009 or 44 Gy with 2.0 Gy/fraction over 4.5 weeks using 10-MV X-rays from October 2009 and thereafter. The radiotherapy target volume included the known gross and clinical disease plus adequate peripheral margins. The chemotherapy regimens mostly consisted of intravenous administration of paclitaxel (50 mg/m^2 per week) or docetaxel (20 mg/m^2 per week) plus either cisplatin (25 mg/m^2 per week) or carboplatin (AUC, 1.5/week) for 5 weeks. The first dose of chemotherapy was delivered on the first day of thoracic radiotherapy [3,4,21].

Surgical procedures were planned for 4~6 weeks following the completion of neoadjuvant CCRT and comprised resection of the affected lung plus mediastinal lymph nodes dissection, depending on the clinical stage. Pulmonary resection included lobectomy, bilobectomy, pneumonectomy, or lobectomy with en bloc wedge resection according to the extent of the primary tumor. After surgical resection, the specimens were examined by pathologists for residual tumors based on hematoxylin and eosin-stained slides. They reported the percentage of residual tumor, which was determined by comparing the estimated cross-sectional area of the viable tumor foci with the estimated cross-sectional areas of necrosis, fibrosis, and inflammation on each slide. The absolute viable tumor extent was also assessed based on their calculation, and pathologic complete response (pCR) was defined as no residual viable tumor remaining in the post-therapy pathology specimen [22,23].

2.3. ^{18}F-FDG PET/CT Analysis

All patients fasted for at least 6 h before ^{18}F-FDG PET/CT was performed to keep their blood glucose level below 200 mg/dL. Torso PET and unenhanced CT images were acquired using a dedicated PET/CT scanner (Discovery STe, GE Healthcare, Waukesha, WI, USA) approximately 60 min after intravenous injection of 5.5 MBq/kg of ^{18}F-FDG. CT images were obtained using a 16-slice helical CT with the following settings: 140 keV, 30–170 mAs with Auto A mode, and a slice section of 3.75 mm. PET images were acquired from head to thigh and attenuation-corrected PET images (voxel size, $3.9 \times 3.9 \times 3.3$ mm^3) were reconstructed using a 3D ordered-subset expectation-maximization algorithm (20 subsets, 2 iterations).

For quantitative analysis, the volume of interest (VOI) from the primary tumor was delineated using the gradient-based segmentation method (PET Edge) in MIM version 6.4 (MIM Software Inc., Cleveland, OH, USA) [19]. These VOIs were saved as a DICOM-RT structure that was imported into the Chang-Gung Image Texture Analysis toolbox (CGITA, http://code.google.com/p/cgita, accessed on 1 March 2020) facilitated by MATLAB software (version 2014b; MathWorks, Inc., Natick, MA, USA) to extract the radiomic features from the PET images (Supplemental Table S1) as well as conventional PET parameters, including the maximum SUV (SUVmax), mean SUV (SUVmean), metabolic tumor volume (MTV), and total lesion glycolysis (TLG). We also calculated the differences of these con-

ventional parameters between PET1 and PET2 by subtracting PET2 parameters from those of PET1 and dividing by those of PET1.

Two nuclear medicine physicians (J.Y.C. and B.T.K) with more than 15 years of experience in PET/CT interpretation assessed the neoadjuvant treatment response according to PERCIST 1.0 [16] by means of a baseline ^{18}F-FDG PET/CT (PET1) and second PET/CT (PET2) undertaken before surgery. They categorized all patients into four response criteria: complete metabolic response (CMR), partial metabolic response (PMR), stable metabolic disease (SMD), and progressive metabolic disease (PMD). After that, the accuracy of the predicted CMR results were compared to histopathologic pCR.

2.4. Machine Learning (ML) Model

The ML model was developed as a binary classification. First, data were partitioned into a training dataset (70%) for model building and an independent testing dataset (30%) for internal validation. We developed an ML tree-based boosting model for pCR prediction using a random forest (RF) algorithm, which consisted of a multitude of decision trees and used an ensemble method to decide the outcome. Our model was trained with the bagging method to predict the pCR. Different numbers of trees were used to classify the binary decision of the result to achieve the best performance score. The Gini impurity was measured to the quality of a split. The maximum depth of the tree was 5, and the square root of the number of the features was considered for the max. number of features to look for the best split of the model. We applied a random grid search method to determine the optimal hyperparameter of the RF model [24–27]. A 10-fold cross-validation in the training dataset, a technique for reducing the bias that can occur as a result of using a single training set, was applied for method validation. All ML statistical analyses were performed using Python (version 3.8.3).

In classic oversampling techniques, the minority data are simply replicated from the minority data population. The ML model does not reflect on variation from the oversampling data. Therefore, we tried to use SMOTE (Synthetic Minority Oversampling Technique) to deal with this class problem. This technique helped with unbalanced data by creating new synthetic data to provide balance in the distribution. SMOTE starts by choosing random data from the minority class. Then it uses a K-Nearest Neighbor (KNN) algorithm to set new points of the data. Next, new synthetic data are created between the random data and new point, which is derived from KNN algorithm. This process is repeated until the minority class reaches the same size as the majority class. Therefore, we added 322 more participants from the existing raw data. A total of 752 participants were analyzed using this oversampling technique.

Several useful scaling techniques (Min–Max scaler, Normalization, Standardization) prevent overflow and underflow of the data. They help to compare dimensional data more efficiently through a scaling process. The process reduces the conditional number of covariance matrices from the independent variables. This reduction enhances the speed of conversion and stability of the model during the optimization process. We used a standard scaler, which removes the mean and helps to scale the value's unit variance. To adjust for the different scales of the features, standardization of the variables is necessary for the preprocessing steps.

For feature selection, top 10, 20, and 30 variables among 144 variables were selected according to the importance of the variables based on the mean decrease impurity (MDI). MDI or Gini importance was calculated as the decrease in node impurity weighted by the probability of reaching the node. The sum over the number of splits decided the variable importance of the model. The higher value of MDI meant the critical feature in the model.

2.5. Statistical Analysis

The association between conventional PET parameters and pCR was determined by an independent *t*-test or the Mann–Whitney test according to the Kolmogorov–Smirnov test. Receiver operating characteristic (ROC) curve analysis was performed to assess

optimal cutoff values of continuous variables using the MedCalc software package (Ver. 9.5, MedCalc Software, Mariakerke, Belgium). The predictive performance of conventional PET parameters and physicians' diagnostic results were reported using sensitivity (Sen), specificity (Spe), positive predictive value (PPV), negative predictive value (NPV), and accuracy (ACC).

For predictive performance of the ML model, we measured the areas under curve (AUCs), ACC, F1 score, precision (also called PPV), and recall (also known as Sen). We compared the measured values with those of predictions from conventional PET parameters and from physicians by using a McNemar test or Fisher's exact test. A *p*-value of less than 0.05 was considered statistically significant.

3. Results

3.1. Subject Characteristics

Among 484 consecutive patients, 430 patients were enrolled in this study. Fifty-four patients were excluded from the analysis due to a lack of surgical treatment after completion of neoadjuvant CCRT (Figure 1). The clinical characteristics of the 430 patients are summarized in Table 1. The patients were predominantly male (71.9%), and there was a high prevalence (67.2%) of adenocarcinoma among the patients. After neoadjuvant CCRT followed by surgery, the mean percentage of viable tumor in the pathologic specimen was 28.8% (range 0–95%). The pCR was observed in 54 patients (12.6%). According to PERCIST criteria, 16.7% of patients had CMR ($n = 72$).

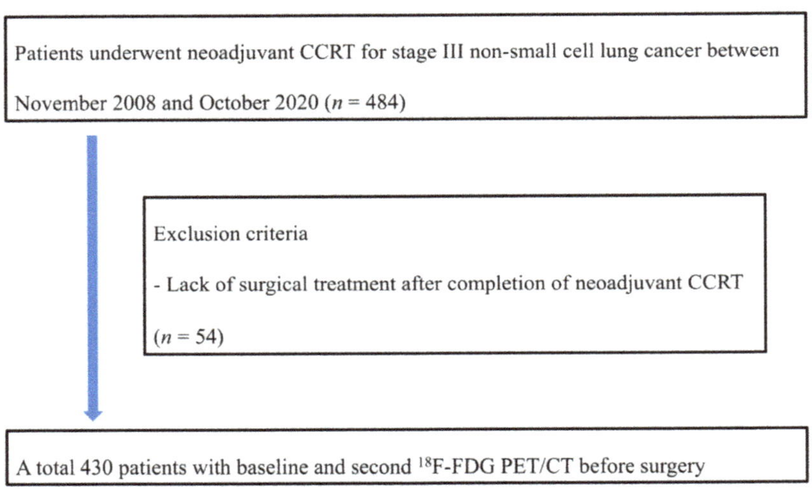

Figure 1. Flowchart of the inclusion and exclusion criteria for the patients.

3.2. Predictive Performance of ML Model for pCR

The radiomic feature importance was obtained using a Gini index representing the coefficient of the attributes on the prediction model, as listed in Figure 2. The overall prediction performance of the ML model was compared by calculating each of the PET1 and PET2 features separately, and all variables from both PET1 and PET2 (PET3) were analyzed (Table 2). The AUCs determined by the ML model were 0.934 in PET1, 0.975 in PET2, and 0.977 in PET3. For comparison ROC curve analysis (Figure 3), the AUCs of PET2 and PET3 were significantly higher than that of PET1 ($p = 0.009$, $p = 0.006$, respectively). However, there was no significant difference between the AUCs of PET2 and PET3 ($p = 0.805$). According to other indices, PET3 revealed a better predictive performance than those results with either PET1 or PET2 variables.

Table 1. Subjects' characteristics.

Characteristics		No.
Sex	Male	309 (71.9%)
	Female	121 (28.1%)
Age (years)	Mean (range)	61.8 (31.1–79.5)
Tumor pathology	Adenocarcinoma	289 (67.2%)
	Squamous cell carcinoma	125 (29.1%)
	Others	16 (3.7%)
Stage (AJCC 7th)	IIIa	415 (96.5%)
	IIIb	15 (3.5%)
Type of surgery	Lobectomy	339 (78.8%)
	Bilobectomy	32 (7.4%)
	Pneumonectomy	23 (5.4%)
	Lobectomy with en bloc wedge resection	36 (8.4%)
Viable tumor on pathologic specimen	Mean % (range)	28.8 (0–95.0)
Pathologic response	pCR	54 (12.6%)
	Non-pCR	376 (87.4%)
Response by PERCIST	CMR	72 (16.7%)
	PMR	281 (65.4%)
	SMD	72 (16.7%)
	PMD	5 (1.2%)

pCR, pathologic complete response; PERCIST, PET response criteria in solid tumors; CMR, complete metabolic response; PMR, partial metabolic response; SMD, stable metabolic disease; PMD, progressive metabolic disease.

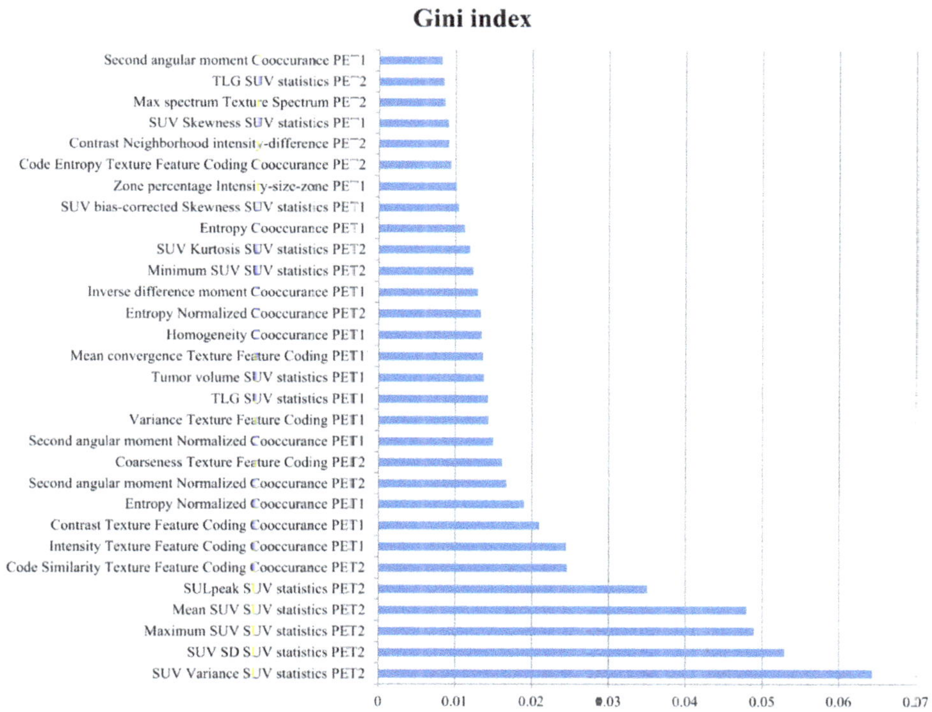

Figure 2. The top 30 important radiomic features from ^{18}F-FDG PET/CT for pCR prediction after neoadjuvant CCRT

Table 2. Comparisons in predictive performance of the ML models using a random forest algorithm for pCR prediction with the included PET data.

ML Model	AUC	ACC	F1	Precision	Recall
PET1	0.934 *,†	0.827 *,†	0.853 *,†	0.802 *,†	0.912 †
PET2	0.975 *	0.902 *,‡	0.912 *,‡	0.905 *,‡	0.920
PET3	0.977 †	0.934 †,‡	0.940 †,‡	0.937 †,‡	0.944 †

AUC, area under curve; ACC, accuracy; PET3, combining PET1 and PET2; *, †, ‡, $p < 0.05$.

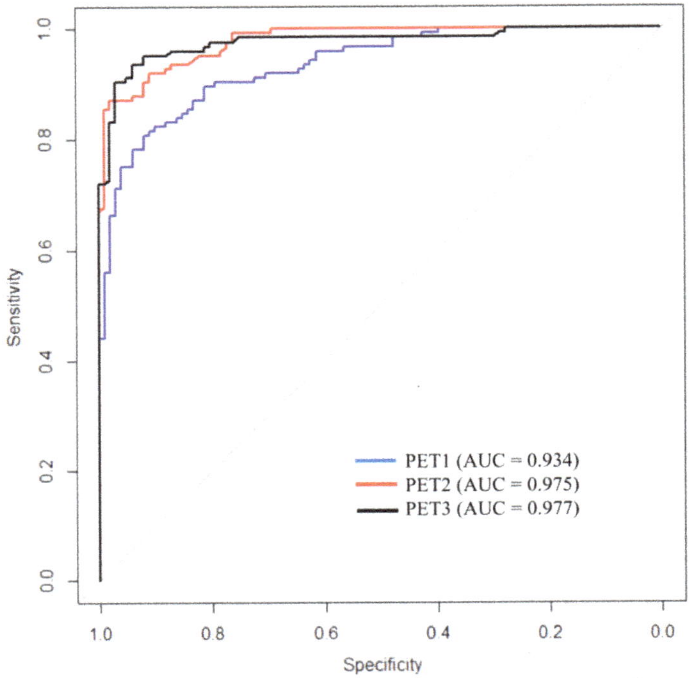

Figure 3. Comparisons of the ROC curves of the ML models according to the included PET data. It showed that the AUC of ML using PET/CT data obtained after neoadjuvant CCRT was significantly higher than that of using only baseline PET/CT data ($p < 0.05$).

Additionally, we investigated the predictive results from the ML model using four feature subsets with the top 10, 20, 30, and all features from PET3 (Supplemental Table S2 and Supplemental Figure S1). The ML model outperformed other methods when all features were selected (AUC = 0.977, ACC = 0.934, F1 = 0.940, Precision = 0.937, Recall = 0.944).

3.3. Predictive Performances of Conventional PET Parameters and Physicians for pCR Prediction

In conventional PET parameters, the SUVmax, SUVmean, MTV, and TLG of PET1 and the SUVmax and SUVmean of PET2 were significantly associated with the pCR ($p < 0.05$). The difference between PET1 and PET2 of the SUVmax ($p < 0.001$), SUVmean ($p < 0.001$), MTV ($p = 0.003$), and TLG ($p < 0.001$) were also significantly associated with the pCR. In contrast, the MTV and TLG of PET2 were not statistically associated with the pCR (Table 3).

Table 3. Comparisons in conventional PET parameters according to the presence of pCR.

			Pathologic Response		*p*-Value
			pCR	Non-pCR	
PET1	SUVmax	Median	13.59	11.58	0.029 *
		IQR	10.01–17.47	8.35–15.53	
	SUVmean	Median	5.91	5.28	0.037 *
		IQR	4.86–7.48	3.97–6.69	
	MTV (cm^3)	Median	42.96	21.13	0.003 *
		IQR	16.02–74.89	7.38–47.48	
	TLG	Median	223.26	113.63	0.001 *
		IQR	96.29–436.26	30.77–279.36	
PET2	SUVmax	Median	3.17	4.57	<0.001 *
		IQR	2.22–4.13	2.92–6.98	
	SUVmean	Median	1.69	2.35	<0.001 *
		IQR	1.43–2.15	1.74–3.33	
	MTV (cm^3)	Median	10.40	8.71	0.327
		IQR	3.64–27.11	3.64–19.46	
	TLG	Median	19.42	22.00	0.475
		IQR	6.32–47.35	8.61–56.52	
Delta PET parameters (%)	dSUVmax	Median	74.68	58.14	<0.001 *
		IQR	64.25–84.25	36.07–74.20	
	dSUVmean	Median	70.17	50.79	<0.001 *
		IQR	54.34–78.57	31.58–66.28	
	dMTV (cm^3)	Median	68.63	48.18	0.003 *
		IQR	42.81–82.49	14.76–71.75	
	dTLG	Median	89.52	73.68	<0.001 *
		IQR	79.40–95.47	50.80–88.83	

pCR, pathologic complete response; PET, positron emission tomography; SUV, standard uptake value; MTV, metabolic tumor volume; TLG, total lesion glycolysis; IQR, interquartile range; *, $p < 0.05$.

The optimal cutoff values that allowed significant association with the pCR were PET1-SUVmax = 13.15, PET1-SUVmean = 4.70, PET1-MTV = 41.11, PET1-TLG = 142.97, PET2-SUVmax = 3.97, PET2-SUVmean = 1.83, dSUVmax = 56.5%, dSUVmean = 43.9%, dMTV = 55.4%, and dTLG = 86.2%. Using these cutoff values, the predictive performance of the PET parameters are listed in Table 4. The predictive performance of the physicians based on their diagnostic result are also presented in Table 4.

Table 4. Comparisons of predictive performance from conventional PET parameters, from physicians and from the ML model.

	Cutoff	AUC	Sen (%)	Spe (%)	PPV (%)	NPV (%)	ACC (%)
PET1-SUVmax	>13.15	0.592	57.4	61.7	17.7	90.9	61.2
PET1-SUVmean	>4.70	0.588	79.6	39.1	15.8	93.0	44.2
PET1-MTV (cm^3)	>41.11	0.627	53.7	70.2	20.6	91.3	68.1
PET1-TLG	>142.97	0.635	68.5	57.1	18.9	92.7	59.1
PET2-SUVmax	≤3.97	0.687	74.1	58.8	20.5	94.0	60.7
PET2-SUVmean	≤1.83	0.726	66.7	71.5	25.2	93.7	70.9
dSUVmax	>56.5%	0.737	88.9	48.7	19.9	96.8	53.7
dSUVmean	>43.9%	0.745	94.4	42.8	19.2	98.2	49.3
dMTV (cm^3)	>55.4%	0.625	68.5	56.6	18.5	92.6	58.1
dTLG	>86.2%	0.703	68.5	69.1	24.2	93.9	69.1
Physicians			33.9	86.4	29.2	90.8	80.5
ML model		0.977	94.4	92.2	93.7	93.1	93.4

AUC, area under curve; Sen, sensitivity; Spe, specificity; PPV, positive predictive value; NPV, negative predictive value; ACC, accuracy.

3.4. Comparisons of the ML Model with Conventional PET Parameters and Physicians

A comparison of the predictive performances between conventional PET parameters, physicians, and the ML model are shown in Table 4. First, the performance of the ML

model for pCR prediction was compared with those of conventional PET parameters by analyzing the AUCs. The ML model revealed higher AUC values than all of the single PET parameters ($p < 0.001$). When the pCR was predicted with the conventional single PET parameter, the AUC was only 0.588 to 0.745. By applying the ML model using variable radiomic features, however, the AUC improved to 0.977. In terms of predictive performance, the ML model showed significantly higher performance in Spe, PPV, and ACC than was achieved with any of the conventional PET parameters ($p < 0.001$). When comparing the predictive performances of physicians and of the ML model, the ACC of the ML model was significantly higher than that of physicians (93.4 vs. 80.5%, $p < 0.001$). Not only ACC, but also Sen, Spe, and PPV showed that the ML model significantly increased the results of physicians (94.4 vs. 33.9%, $p < 0.001$; 92.2 vs. 86.4%, $p = 0.001$; 93.7 vs. 29.2%, $p < 0.001$; respectively). NPV was the only case where there was no significant difference between the ML model and prediction by physicians (93.1 vs. 90.8%, $p = 0.155$).

4. Discussion

We have demonstrated that the ML model using an RF algorithm could be robust and useful in determining the pCR following neoadjuvant CCRT by radiomic features of ^{18}F-FDG PET/CT. Although several studies evaluating ML for treatment response have been published recently [28–31], they mainly conducted research with multiparametric MRI features and not with ^{18}F-FDG PET/CT. Only a few studies have used ^{18}F-FDG PET/CT features to assess neoadjuvant treatment response in breast and rectal cancer using ML models [26,27]. To the best of our knowledge, this is the first study to predict the response to neoadjuvant CCRT in patients with NSCLC using an ML model.

The response to neoadjuvant CCRT is critical because it affects postoperative treatment and individual prognosis. Furthermore, the correct prediction of the pCR can determine which patients will require more or less aggressive adjuvant treatment to reduce the risk of complications. Despite improvements in therapeutic modalities of neoadjuvant CCRT, the pCR rate still remains with a variety of outcomes. The gold standard for assessing the pCR is based on postoperative histopathologic findings, which could be inefficient to implement in all patients with advanced NSCLC. Therefore, it is necessary to develop a method of improving the predictive significance of non-invasive imaging modalities for establishing a personalized therapeutic strategy.

Radiomics is an emerging field where various imaging modalities are performed to extract features that may reflect changes in human tissues at the cellular levels and estimate detailed information on tumor biology and microenvironment in nuclear medicine [32,33]. The radiomic features delineated on PET/CT images can represent tumor heterogeneity including fractal dimension, tumor shape, and proliferation [34]. In our experiments, voxel statistics of radiomic features were highly ranked in the prediction for the pCR, followed by texture spectrum and co-occurrence matrix. Although there are differences in the feature importance of many radiomic variables, the ML model using them demonstrated better predictive performance for the pCR than the single conventional PET/CT parameters. Conventional PET parameters and their changes in FDG uptake before and after CCRT have been previously evaluated in determining the treatment response in patients with NSCLC [11]. We also performed these analyses; however, the ACC of the predictive performance using them was only shown to be 44.2–70.9%. Therefore, it seemed unfavorable to evaluate the predictive performance using single PET parameters even though they were statistically significantly correlated with the pCR.

The ML model significantly outperformed the physicians in terms of Sen, Spe, PPV, and ACC. The outcomes of conducting the ML model with PET2 data revealed higher predictive performance than those of the ML model with PET1 data. It appears that radiomic features obtained from PET/CT after neoadjuvant CCRT have more relevant clinical value in the prediction of the pCR. Compared to the results of the ML model with only the variables from each time of PET/CT images, the predictive performance also increased by inputting all variables from both PET1 and PET2. We assumed that the improvement in

performance is probably because of the feature importance for predicting the pCR, which is somewhat different between radiomics of PET1 and PET2. If more significant variables were input into the ML model, the predictive performance may be further improved. The PET-based radiomics can provide the potential to characterize intratumoral heterogeneity indicating resistance to neoadjuvant CCRT. Therefore, it is clinically important to evaluate treatment response not only to obtain baseline PET/CT images but also to examine PET/CT after neoadjuvant CCRT. As the current study demonstrated, the use of ML with radiomics features could be predictive of treatment response and thus help to select a more aggressive treatment for those with high-risk factors after curative surgery in patients with stage III NSCLC.

This study had several limitations. First, this study was conducted in a retrospective manner with a limited sample size from a single center. Because radiomic features can be highly dependent on reconstruction methods and imaging parameters [35], it is planned to obtain a prospective multicenter trial to be more generalizable in the future. Second, the study population was composed of patients with different therapeutic schemes. Although we addressed a homogeneous population of patients with stage III NSCLC, it is also needed to select patients with a more uniform therapeutic modality based on the consistent guideline. Third, various pulmonary side effects can arise after radiotherapy, such as pneumonitis or fibrosis, which may challenge the response assessment, although we tried our best to exclude the possibility of treatment-induced inflammatory changes based on the relative intensity and distribution of FDG uptake in the lung parenchyma and automatically generated tumor VOI [36]. Finally, although the proposed ML model was analyzed using a 10-fold cross-validation for minimizing overfitting instead of splitting the dataset into training and test sets, external validation using an independent dataset is necessary to verify the clinical significance using a larger cohort.

5. Conclusions

In conclusion, the developed ML model using an RF algorithm and ^{18}F-FDG PET/CT radiomics features was useful for predicting the pCR after neoadjuvant CCRT in NSCLC. The predictions of the ML model had higher accuracy than predictions from conventional PET parameters and from physicians. The ML model using radiomics features can be used to facilitate the preoperative individualized prediction for the pCR. Our findings further highlight the potential, non-invasive, and effective clinical significance of an ML model to predict the pCR in patients with stage III NSCLC who had received neoadjuvant CCRT followed by surgery.

Supplementary Materials: The following supporting information can be downloaded at: https://www.mdpi.com/article/10.3390/cancers14081987/s1, Figure S1: Comparison of ROC curves from random forest according to ranking-based feature selection; Table S1: List of quantitative PET-based radiomic features from CGITA; Table S2: Predictive performance of random forest according to ranking-based feature selection

Author Contributions: All authors contributed equally to this work. Conceptualization, J.Y. and J.Y.C.; data curation, J.Y.; formal analysis, J.Y., J.L. and J.Y.C.; funding acquisition, J.Y. and J.Y.C.; investigation, J.Y. and J.L.; methodology, J.Y., J.L. and J.Y.C.; project administration, J.Y.C.; resources, M.C., S.-K.W., M.-J.A., H.R.P., Y.S.C. and J.H.H.; software, J.Y. and J.L.; supervision, J.Y.C.; validation, J.Y. and J.Y.C.; visualization, J.Y. and J.L.; writing—original draft preparation, J.Y.; writing—review and editing, M.C., S.-K.W., M.-J.A., H.R.P., Y.S.C., J.H.H. and J.Y.C. All authors have read and agreed to the published version of the manuscript.

Funding: This work was supported by the National Research Foundation of Korea (NRF) grant funded by the Korea government (Ministry of Science and ICT) (No. NRF-2020M2D9A1094072), Future Medicine 20*30 Project of the Samsung Medical Center (#SMO1220071), and VHS Medical Center Research Grant (No. VHSMC 22001).

Institutional Review Board Statement: This study was conducted according to the guidelines of the Declaration of Helsinki and approved by the Institutional Review Board of the Samsung Medical Center (IRB No. 2020-09-185).

Informed Consent Statement: Patient consent was waived due to the retrospective design of this study.

Data Availability Statement: Restrictions apply to the availability of these data. Data were obtained from the Samsung Medical Center and are available from the corresponding author with the permission of the Samsung Medical Center.

Conflicts of Interest: The authors declare no conflict of interest.

References

1. DeSantis, C.E.; Lin, C.C.; Mariotto, A.B.; Siegel, R.L.; Stein, K.D.; Kramer, J.L.; Alteri, R.; Robbins, A.S.; Jemal, A. Cancer treatment and survivorship statistics, 2014. *CA Cancer J. Clin.* **2014**, *64*, 252–271. [CrossRef]
2. Arbour, K.C.; Riely, G.J. Systemic therapy for locally advanced and metastatic non-small cell lung cancer: A review. *JAMA* **2019**, *322*, 764–774. [CrossRef]
3. Kim, H.K.; Cho, J.H.; Choi, Y.S.; Zo, J.I.; Shim, Y.M.; Park, K.; Ahn, M.-J.; Ahn, Y.C.; Kim, K.; Kim, J. Outcomes of neoadjuvant concurrent chemoradiotherapy followed by surgery for non-small-cell lung cancer with N2 disease. *Lung Cancer* **2016**, *96*, 56–62. [CrossRef]
4. Hyun, S.H.; Ahn, H.K.; Ahn, M.J.; Ahn, Y.C.; Kim, J.; Shim, Y.M.; Choi, J.Y. Volume-based assessment with 18F-FDG PET/CT improves outcome prediction for patients with stage IIIA-N2 non-small cell lung cancer. *AJR Am. J. Roentgenol.* **2015**, *205*, 623–628. [CrossRef]
5. Schreiner, W.; Gavrychenkova, S.; Dudek, W.; Rieker, R.J.; Lettmaier, S.; Fietkau, R.; Sirbu, H. Pathologic complete response after induction therapy-the role of surgery in stage IIIA/B locally advanced non-small cell lung cancer. *J. Thorac. Dis.* **2018**, *10*, 2795–2803. [CrossRef]
6. Tenahashi, M.; Niwa, H.; Yukiue, H.; Suzuki, E.; Yoshii, N.; Watanabe, T.; Kaminuma, Y.; Chiba, K.; Tsuchida, H.; Yobita, S. Feasibility and prognostic benefit of induction chemoradiotherapy followed by surgery in patients with locally advanced non-small cell lung cancer. *J. Thorc. Dis.* **2020**, *12*, 2644–2653. [CrossRef]
7. Kim, A.W.; Liptay, M.J.; Bonomi, P.; Kim, A.W.; Liptay, M.J.; Bonomi, P.; Warren, W.H.; Basu, S.; Farlow, E.C.; Faber, L.P. Neoadjuvamt chemoradiation for clinically advanced non-small-cell lung cancer: An analysis of 233 patients. *Ann. Thorac. Surg.* **2011**, *92*, 233–241, discussion 241–243. [CrossRef]
8. Pottgen, C.; Eberhardt, W.; Graupner, B.; Theegarten, D.; Gauler, T.; Freitag, L.; Jawad, J.A.; Wohlschlaeger, J.; Welter, S.; Stamatis, G.; et al. Accelerated hyperfractionated radiotherapy within trimodality therapy concepts for stage IIIA/B non-small-cell lung cancer: Markedly higher rates of pathologic complete remissions than with conventional fractionation. *Eur. J. Cancer* **2013**, *49*, 2107–2115. [CrossRef]
9. D'Angelillo, R.M.; Trodella, L.; Ciresa, M.; Cellini, F.; Fiore, M.; Greco, C.; Pompeo, E.; Mineo, T.C.; Paleari, L.; Granone, P.; et al. Multimodality treatment of stage III non-small cell lung cancer: Analysis of a phase III trial using preoperative cisplatin and gemcitabine with concurrent radiotherapy. *J. Thorac. Oncol.* **2009**, *4*, 1517–1523. [CrossRef]
10. Stupp, R.; Mayer, M.; Kann, R.; Weder, W.; Zouhair, A.; Betticher, D.C.; Pless, M. Neoadjuvant chemotherapy and radiotherapy followed by surgery in selected patients with stage IIIB non-small-cell lung cancer: A multicetre phase III trial. *Lancet Oncol.* **2009**, *10*, 785–793. [CrossRef]
11. Cremonesi, M.; Gilardi, L.; Ferrari, M.E.; Piperno, G.; Travaini, L.L.; Timmerman, R.; Botta, F.; Baroni, G.; Grana, C.M.; Ronchi, S.; et al. Role of interim 18F-FDG PET/CT for the early prediction of clinical outcomes of non-small cell lung cancer (NSCLC) during radiotherapy or chemo-radiotherapy. A systematic review. *Eur. J. Nucl. Med. Mol. Imaging* **2017**, *44*, 1915–1927. [CrossRef]
12. Roengvoraphoj, O.; Wijaya, C.; Eze, C.; Li, M.; Dantes, M.; Taugner, J.; Tufman, A.; Huber, R.M.; Belka, C.; Manapov, F. Analysis of primary tumor metabolic volume during chemoradiotherapy in locally advanced non-small cell lung cancer. *Strahlenther. Onkol.* **2018**, *194*, 107–115. [CrossRef]
13. Pöttgen, C.; Levegrün, S.; Theegarten, D.; Marnitz, S.; Grehl, S.; Pink, R.; Eberhardt, W.; Stamatis, G.; Gauler, T.; Antoch, G.; et al. Value of ^{18}F-fluoro-2-deoxy-D-glucose-positron emission tomography/computed tomography in non-small-cell lung cancer for prediction of pathologic response and times to relapse after neoadjuvant chemoradiotherapy. *Clin. Cancer Res.* **2006**, *12*, 97–106. [CrossRef]
14. Cerfolio, R.J.; Bryant, A.S.; Winokur, T.S.; Ohja, B.; Bartolucci, A.A. Repeat FDG-PET after neoadjuvant therapy is a predictor of pathologic response in patients with non-small cell lung cancer. *Ann. Thorac. Surg.* **2004**, *78*, 1903–1909. [CrossRef]
15. Wahl, R.L.; Jacene, H.; Kasamon, Y.; Lodge, M.A. From RECIST to PERCIST: Evolving considerations for PET response criteria in solid tumors. *J. Nucl. Med.* **2009**, *50*, 122S–150S. [CrossRef]
16. Joo Hyun, O.; Lodge, M.A.; Wahl, R.L. Practical PERCIST: A simplified guide to PET response criteria in solid tumors 1.0. *Radiology* **2016**, *280*, 576–584.

17. De Ruysscher, D.; Houben, A.; Aerts, H.J.; Dehing, C.; Wanders, R.; Öllers, M.; Lambin, P. Increased (18)F-deoxyglucose uptake in the lung during the first weeks of radiotherapy is correlated with subsequent radiation-induced lung toxicity (RILT): A prospective pilot study. *Radiother. Oncol.* **2009**, *91*, 415–420. [CrossRef]
18. Szyszko, T.A.; Yip, C.; Szlosarek, P.; Goh, V.; Cook, G.J. The role of new PET tracers for lung cancer. *Lung Cancer* **2019**, *94*, 7–14. [CrossRef] [PubMed]
19. Yoo, J.; Cheon, M.; Park, Y.J.; Hyun, S.H.; Zo, J.I.; Um, S.W.; Choi, J.Y. Machine learning-based diagnostic method of pre-therapeutic ^{18}F-FDG PET/CT for evaluating mediastinal lymph nodes in non-small cell lung cancer. *Eur. Radiol.* **2021**, *31*, 4184–4194. [CrossRef]
20. Rami-Porta, R.; Crowley, J.J.; Goldstraw, P. The revised TNM staging system for lung cancer. *Ann. Thorac. Cardiovasc. Surg.* **2009**, *15*, 4–9.
21. Shin, S.; Kim, H.K.; Cho, J.H.; Choi, Y.S.; Kim, K.; Kim, J.; Zo, J.I.; Sun, J.; Ahn, M.; Park, K.; et al. Adjuvant therapy in stage IIIA-N2 non-small cell lung cancer after neoadjuvant concurrent chemotherapy followed surgery. *J. Thorac. Dis.* **2020**, *12*, 2602–2613. [CrossRef] [PubMed]
22. Cottrell, T.R.; Thompson, E.D.; Forde, P.M.; Stein, J.E.; Duffield, A.S.; Anagnostou, V.; Rekhtman, N.; Anders, R.A.; Cuda, J.D.; Illei, P.B.; et al. Pathologic features of response to neoadjuvant anti-PD-1 in resected non-small-cell lung carcinoma: A proposal for quantitative immune-related pathologic response criteria (irPRC). *Ann. Oncol.* **2018**, *29*, 1853–1860. [CrossRef] [PubMed]
23. Mouillet, G.; Monnet, E.; Milleron, B.; Puyraveau, M.; Quoix, E.; David, P.; Ducoloné, A.; Molinier, O.; Zalcman, G.; Depierre, A.; et al. Pathologic complete response to preoperative chemotherapy predicts cure in early-stage non-small-cell lung cancer: Combined analysis of two IFCT radomized trials. *J. Thorac. Oncol.* **2012**, *7*, 841–849. [CrossRef] [PubMed]
24. Li, P.; Wang, X.; Xu, C.; Liu, C.; Zheng, C.; Fulham, M.J.; Feng, D.; Wang, L.; Song, S.; Huang, G. (18)F-FDG PET/CT radiomic predictors of pathologic complete response (pCR) to neoadjuvant chemotherapy in breast cancer patients. *Eur. J. Nucl. Med. Mol. Imaging* **2020**, *47*, 1116–1126. [CrossRef] [PubMed]
25. Shen, W.C.; Chen, S.W.; Wu, K.C.; Lee, P.Y.; Feng, C.L.; Hsieh, T.C.; Yen, K.; Kao, C. Predicting pathological complete response in rectal cancer after chemoradiotherapy with a random forest using 18F-fluorodeoxyglucose positron emission tomography and computed tomography radiomics. *Ann. Transl. Med.* **2020**, *8*, 207. [CrossRef] [PubMed]
26. Tahmassebi, A.; Wengert, G.J.; Helbich, T.H.; Bago-Horvath, Z.; Alaei, S.; Bartsch, R.; Dubsky, P.; Baltzer, P.; Clauser, P.; Kapetas, P.; et al. Impact of machine learning with parametric magnetic resonance imaging of the breast for early prediction of response to neoadjuvant chemotherapy and survival outcomes in breast cancer patients. *Investig. Radiol.* **2019**, *54*, 110–117. [CrossRef]
27. Yakar, M.; Etiz, D.; Metintas, M.; Ak, G.; Celik, O. Prediction of radiation pneumonitis with machine learning in stage III lung cancer: A pilot study. *Technol. Cancer Res. Treat.* **2021**, *20*, 15330338211016373. [CrossRef]
28. Meti, N.; Saednia, K.; Lagree, A.; Tabbarah, S.; Mohebpour, M.; Kiss, A.; Lu, F.; Slodkowska, E.; Gandhi, S.; Jerzak, K.J.; et al. Machine learning frameworks to predict neoadjuvant chemotherapy response in breast cancer using clinical and pathological features. *JCO Clin. Cancer Inform.* **2021**, *5*, 66–80. [CrossRef]
29. Lo Gullo, R.; Eskreis-Winkler, S.; Morris, E.A.; Pinker, K. Machine learning with multiparametric magnetic resonance imaging of the breast for early prediction of response to neoadjuvant chemotherapy. *Breast* **2020**, *49*, 115–122. [CrossRef]
30. Huang, C.; Huang, M.; Huang, C.; Tsai, H.; Su, W.; Chang, W.; Wang, J.; Shi, H. Machine learning for predicting pathological complete response in patients with locally advanced rectal cancer after neoadjuvant chemoradiotherapy. *Sci. Rep.* **2020**, *10*, 12555. [CrossRef]
31. Eun, N.L.; Kang, D.; Son, E.J.; Park, J.S.; Youk, J.H.; Kim, J.A.; Gweon, H.M. Texture analysis with 3.0-T MRI for association of response to neoadjuvant chemotherapy in breast cancer. *Radiology* **2020**, *294*, 31–41. [CrossRef] [PubMed]
32. Antunovic, L.; de Sanctis, R.; Cozzi, L.; Kirienko, M.; Sagona, A.; Torrisi, R.; Tinterri, C.; Santoro, A.; Chiti, A.; Zelic, R.; et al. PET/CT radiomics in breast cancer: Promising tool for prediction of pathological response to neoadjuvant chemotherapy. *Eur. J. Nucl. Med. Mol. Imaging* **2019**, *46*, 1468–1477. [CrossRef] [PubMed]
33. Ha, S.; Park, S.; Bang, J.I.; Kim, E.K.; Lee, H.Y. Metabolic radiomics for pretreatment (18)F-FDG PET/CT to characterize locally advanced breast cancer: Histopathologic characteristics, response to neoadjuvant chemotherapy, and prognosis. *Sci. Rep.* **2017**, *7*, 1556. [CrossRef] [PubMed]
34. Hoffmann, B.; Frenzel, T.; Schmitz, R.; Schumacher, U.; Wedemann, G. Modeling growth of tumors and their spreading behavior using mathematical functions. *Methods Mol. Biol.* **2019**, *1878*, 263–277. [PubMed]
35. Sollini, M.; Cozzi, L.; Antunovic, L.; Chiti, A.; Kirienko, M. PET radiomics in NSCLC: State of the art and a proposal for harmonization of methodology. *Sci. Rep.* **2017**, *7*, 358. [CrossRef] [PubMed]
36. Iravani, A.; Turgeon, G.A.; Akhurst, T.; Callahan, J.W.; Bressel, M.; Everitt, S.J.; Siva, S.; Hofman, M.S.; Hicks, R.J.; Ball, D.L.; et al. PET-detected pneumonitis following curative-intent chemoradiation in non-small cell lung cancer (NSCLC): Recognizing patterns and assessing the impact on the predictive ability of FDG-PET/CT response assessment. *Eur. J. Nucl. Med. Mol. Imaging* **2019**, *46*, 1869–1877. [CrossRef]

cancers

MDPI

Article

Synaptophysin, CD117, and GATA3 as a Diagnostic Immunohistochemical Panel for Small Cell Neuroendocrine Carcinoma of the Urinary Tract

Gi Hwan Kim [1], Yong Mee Cho [1], So-Woon Kim [2], Ja-Min Park [3], Sun Young Yoon [3], Gowun Jeong [4], Dong-Myung Shin [5], Hyein Ju [5] and Se Un Jeong [1,*]

1 Asan Medical Center, Department of Pathology, University of Ulsan College of Medicine, 88, Olympic-ro 43 Gil, Songpa-gu, Seoul 05505, Korea; standupbau@hanmail.net (G.H.K.); yongcho@amc.seoul.kr (Y.M.C.)
2 Department of Pathology, Kyung Hee University Medical Center, Kyung Hee University College of Medicine, Seoul 02447, Korea; sowoonkim86@gmail.com
3 Asan Medical Center, Asan Institute of Life Science, Seoul 05505, Korea; parkja09@naver.com (J.-M.P.); mysunyoung14@naver.com (S.Y.Y.)
4 AI Recommendation, T3K, SK Telecom, 65, Eulji-ro, Jung-gu, Seoul 04539, Korea; gowun.jeong@googlemail.com
5 Asan Medical Center, Departments of Biomedical Sciences and Physiology, University of Ulsan College of Medicine, Seoul 05505, Korea; d0shin03@amc.seoul.kr (D.-M.S.); alal0903@naver.com (H.J.)
* Correspondence: redmelon44@hanmail.net; Tel.: +82-2-3010-4560; Fax: +82-2-3010-7898

Simple Summary: While diagnosing a case of small cell neuroendocrine carcinoma (SCNEC) in the urinary tract, we found that the previous biopsy had been misdiagnosed as urothelial carcinoma (UC) because only chromogranin and synaptophysin were tested to define neuroendocrine differentiation and both tests were negative. This case led us to conduct this present study to define a panel of neuroendocrine markers to ensure the diagnosis of traditional neuroendocrine marker-negative SCNEC. We employed a decision tree classifier algorithm to analyze the expression of 17 immunohistochemical markers and found that the extent of synaptophysin (>5%) and CD117 (>20%) and the intensity of GATA3 (negative or weak) are major parameters. Since SCNEC is an aggressive tumor type and requires therapeutic approaches that differ from those used for UC, an accurate diagnosis of SCNEC is critical and this model may help pathologists accurately diagnose SCNEC in daily practice.

Abstract: Although SCNEC is based on its characteristic histology, immunohistochemistry (IHC) is commonly employed to confirm neuroendocrine differentiation (NED). The challenge here is that SCNEC may yield negative results for traditional neuroendocrine markers. To establish an IHC panel for NED, 17 neuronal, basal, and luminal markers were examined on a tissue microarray construct generated from 47 cases of 34 patients with SCNEC as a discovery cohort. A decision tree algorithm was employed to analyze the extent and intensity of immunoreactivity and to develop a diagnostic model. An external cohort of eight cases and transmission electron microscopy (TEM) were used to validate the model. Among the 17 markers, the decision tree diagnostic model selected 3 markers to classify NED with 98.4% accuracy in classification. The extent of synaptophysin (>5%) was selected as the initial parameter, the extent of CD117 (>20%) as the second, and then the intensity of GATA3 ($\leq$1.5, negative or weak immunoreactivity) as the third for NED. The importance of each variable was 0.758, 0.213, and 0.029, respectively. The model was validated by the TEM and using the external cohort. The decision tree model using synaptophysin, CD117, and GATA3 may help confirm NED of traditional marker-negative SCNEC.

Keywords: carcinoma; neuroendocrine; urinary bladder; decision trees; immunohistochemistry; synaptophysin; negative results

Citation: Kim, G.H.; Cho, Y.M.; Kim, S.-W.; Park, J.-M.; Yoon, S.Y.; Jeong, G.; Shin, D.-M.; Ju, H.; Jeong, S.U. Synaptophysin, CD117, and GATA3 as a Diagnostic Immunohistochemical Panel for Small Cell Neuroendocrine Carcinoma of the Urinary Tract. *Cancers* **2022**, *14*, 2495. https://doi.org/10.3390/cancers14102495

Academic Editors: Hamid Khayyam, Ali Madani, Rahele Kafieh and Ali Hekmatnia

Received: 21 April 2022
Accepted: 16 May 2022
Published: 19 May 2022

1. Introduction

Small cell neuroendocrine carcinoma (SCNEC) is a rare entity in the urinary tract, representing 0.5–1% of urinary bladder cancers [1,2]. It usually presents as a high stage tumor with frequent muscularis propria invasion and metastasis compared to conventional urothelial carcinoma (UC) [3]. SCNEC requires an aggressive clinical course, and its 5-year survival rate is as low as 8% [4]. A recently reported combined therapeutic approach included neoadjuvant chemotherapy with cisplatin and etoposide, followed by either radiation therapy or cystectomy if no systemic disease is present; the overall survival was higher in patients who received the neoadjuvant chemotherapy than in those who did not receive it [5,6]. Therefore, accurate diagnosis of SCNEC is critical because of its poor prognosis and therapeutic approaches differing from those used for UC.

SCNEC is defined by its characteristic histology: sheets and large nests of relatively small cells with scant cytoplasm, speckled nuclei, and indistinct nucleoli. In the urinary bladder, SCNEC presents as a pure form or more frequently as a component of combined SCNEC and non-SCNEC [4,7]. The non-SCNEC component includes UC, invasive or in situ, and other divergent differentiation and histologic variants such as squamous, glandular, nested, plasmacytoid, sarcomatoid, and trophoblastic.

The diagnosis of SCNEC is classically based on the histologic features, but immunohistochemical (IHC) staining is commonly employed to confirm the diagnosis or to exclude an alternative diagnosis in cases with ambiguous histology. Similar to its more common counterpart in the lungs, synaptophysin, chromogranin, and CD56 are widely used neuroendocrine (NE) markers in a panel to compensate the suboptimal sensitivity and specificity of each marker [8]. Synaptophysin has a relatively reliable diagnostic potential; chromogranin is less sensitive with weak and focal positivity; and CD56 is most sensitive but less specific [3,9]. However, SCNEC may yield negative results for all three of these markers [10]. In fact, up to two-thirds of small cell lung cancer could provide negative results for the relatively specific NE markers synaptophysin and chromogranin A [10,11]. The challenge is that SCNEC may have ambiguous or overlapping features with UC, especially in cases of combined SCNEC and UC [5]. In such cases, it might be difficult to accurately diagnose SCNEC, and when the traditional NE markers are negative, it could result in misdiagnosis as UC.

Follow-up biopsies are scheduled for bladder cancer patients to estimate treatment response and detect tumor recurrence. While diagnosing a case of SCNEC in the urinary bladder, we found that the previous bladder biopsy had been misdiagnosed as UC because only chromogranin and synaptophysin were tested to define NE differentiation and both tests were negative. This case led us to conduct this present study to define a panel of NE markers to ensure the diagnosis of traditional NE marker-negative SCNEC. We employed a decision tree classifier algorithm to analyze the expression of 17 IHC markers and finally propose a decision tree model using three markers synaptophysin, CD117, and GATA3.

2. Materials and Methods

2.1. Study Samples

This retrospective study was approved by the Asan Medical Center Institutional Review Board (2013–0107). Initially, the cohort consisted of 47 patients who were diagnosed with SCNEC of the urinary tract (urinary bladder and ureter) as a pure form or combined with UC between May 2002 and October 2020 at Asan Medical Center, Seoul, Republic of Korea. The diagnosis of SCNEC was based on histologic features only or IHC expression analysis of NSE, CD56, chromogranin, and synaptophysin (alone or in combination). After exclusion of 13 patients for which glass slides or paraffin blocks were not available, 34 patients of SCNEC were included in the discovery cohort. Among the 34 patients, 23 patients were biopsied once and accounted for one case each. Nine patients were biopsied twice (accounting for two cases each), and two patients were biopsied thrice (accounting for three cases each). Among the 11 patients who had been biopsied more than once, six patients had specimens diagnosed with UC during the period. The UC cases of

these patients were also included in the analysis to compare their immunoprofile with that of SCNEC. Therefore, 34 patients and their 47 cases (40 cases of pure and combined SCNEC and 7 cases of UC) were finally included in the discovery cohort.

For an external validation of the diagnostic model, data for eight patients were retrieved at the Kyung Hee University Medical Center (KHMC), Seoul, Republic of Korea from 2000 to 2020. They had a confirmed or suspected diagnosis of SCNEC of the urinary bladder based on the IHC staining of NE markers.

Patients' clinicopathological information was obtained from electronic medical records and surgical pathology reports. Pathologic materials of both discovery and external validation cohorts were reassessed according to the 2016 World Health Organization Tumor Classification criteria and staged according to the American Joint Committee on Cancer Staging System, 8th edition.

2.2. Tissue Microarray Construction

Tissue microarray blocks with 2-mm-diameter cores were constructed from 10% neutrally buffered formalin-fixed, paraffin-embedded urinary bladder tumor blocks using a tissue microarrayer (Quick-Ray, Unitma Co. Ltd., Seoul, Republic of Korea). In general, three representative cores from each case were generated while trying to exclude necrotic and degenerative areas and to maximize tumor cell content. In cases showing histologically divergent or variant features of UC, each representative area was included, resulting in up to 11 cores generated for one case. As a result, a total of 211 cores were generated.

2.3. IHC

IHC analysis was performed using NE, basal, and luminal markers of bladder cancer [11]. The NE markers included in the present study were CD56, CD117, chromogranin, insulinoma-associated protein 1 (INSM1), neuron specific enolase (NSE), SRY (sex determining region Y)-box 2 (SOX2), synaptophysin, somatostatin receptor 2 (SSTR2), and tubulin beta 2B class IIB (TUBB2B). The loss of retinoblastoma-associated protein (Rb) and p53 was reported in bladder cancers with NE differentiation [11–14]. The basal markers were cytokeratin 5/6 (CK5/6) and cytokeratin 14 (CK14). High expression of epidermal growth factor receptor (EGFR) was reported in the basal subtype of bladder cancer [15]. Luminal markers were cytokeratin 20 (CK20), GATA binding protein 3 (GATA3), and forkhead box A1 (FOXA1) [11,16]. The primary antibodies used in this study, their dilutions, and the subcellular location of each antigen are summarized in Supplementary Table S1. IHC staining was performed using an automated staining system (BenchMark XT, Ventana Medical Systems, Tucson, AZ, USA). The nuclei were counterstained with hematoxylin.

The IHC staining results were assessed in a semiquantitative manner by two pathologists (G.H.K. and S.U.J). The immunoreactivity of the markers was evaluated according to the intensity (negative (0), weak (1), moderate (2), or strong (3)) and the extent of positive tumor cells (percentage). A diffuse expression in a core was defined as immunoreactivity in more than half of tumor cells. The intensity and extent of marker expression were independently assessed in the decision tree analysis.

2.4. Establishment of the Decision Tree Model

All 17 IHC markers were included as variables and analyzed for their intensity and extent to classify the cases as neuroendocrine differentiation (NED) and non-neuroendocrine differentiation (non-NED). NED was defined as immunoreactivity to one or more NE markers in cores with SCNEC histology [11]. Based on histologic features and IHC results, the 211 cores were classified into 146 NED cores and 65 non-NED cores. In an attempt to overcome the small number of cases, each core type was analyzed separately to represent NED and non-NED. In cores with simultaneous expression of NE markers with luminal or basal markers, the core was classified as NED when it showed histologic features of SCNEC.

A decision tree model was constructed using a decision tree classifier algorithm on python-3.8, sklearn-1.0.2, and dtreeviz-1.3.2. The algorithm randomly selected 147 cores for the training set and 64 cores for the validation set at odds of 7 to 3. To select a diagnostic IHC panel for NED using the intensity and extent of immunoreactivity of 17 markers, the algorithm repeatedly classified all cores into NED and non-NED to minimize incorrect classifications [17]. A decision tree-derived diagnostic model was visualized after the training procedure was finished. The finally classified cores are colored yellow for NED and green for non-NED in all plots.

2.5. Transmission Electron Microscopy (TEM) Analysis

TEM analysis was performed using standard techniques. The submitted tissues were retrieved from paraffin blocks, deparaffinized, post-fixed in 1% buffered osmium tetroxide, dehydrated, and embedded in Epon. Ultrathin sections (1 μm) were stained with uranyl acetate-lead citrate and examined using a JEOL 1200 EX-II TEM (Jeol, Tokyo, Japan) [18].

3. Results

3.1. Patients' Characteristics

The clinicopathological features of the 47 cases from the 34 patients are summarized in Table 1. The median age at the initial diagnosis of bladder cancer of the 34 patients was 66 years (range, 31–86 years) with a 6:1 male to female ratio. Most cases were diagnosed by transurethral resection (34 cases, 72.3%) and followed by partial or radical cystectomy (10 cases, 21.3%), ureterectomy (2 cases, 4.3%), and cystoscopic biopsy (1 case, 2.1%). The mean tumor size was 4.36 cm in its greatest dimension (range, 1.0–11.4 cm).

Table 1. Clinicopathological features of the discovery cohort.

Features		Value
Patients (n = 34)		
Age at initial diagnosis (years)		66.1 (31–86)
Sex	Male	29 (85.3)
	Female	5 (14.7)
All cases (n = 47)		
Tumor size (cm)		4.36 (1.0–11.4)
Location	Urinary bladder	45 (95.7)
	Ureter	2 (4.3)
Procedure	Cystoscopic biopsy	1 (2.1)
	Transurethral resection	34 (72.3)
	Partial cystectomy	2 (4.3)
	Radical cystectomy/ureterectomy	10 (21.3)
Histology	Pure NEC	29 (61.7)
	Mixed NEC and non-NEC	15 (31.9)
	Non-NEC	3 (6.4)
Invasion depth	Non-invasive	0 (0.0)
	Subepithelial connective tissue	9 (19.1)
	Muscularis propria	28 (59.6)
	Perivesical tissue	9 (19.1)
	Other organs *	1 (2.1)
Lymphovascular invasion	Present	25 (53.2)
	Absent	22 (46.8)
Cystectomy cases (n = 10)		
	pT2	1 (10.0)
	pT3	8 (80.0)
	pT4	1 (10.0)
N stage	NX	1 (10.0)

Table 1. *Cont.*

Features		Value
Tumor stage	pT1	0 (0.0)
	N0	4 (40.0)
	N1-3	5 (50.0)

* Other organs: prostate, both seminal vesicles, and right vas deferens.

During the reassessment of the cases, we noted that four SCNEC cases from four patients had been misdiagnosed as UC. In three cases, the SCNEC histology was not recognized and IHC for NE markers was not performed. In the remaining case, the SCNEC with ambiguous histology was recognized but chromogranin and synaptophysin staining were negative (Figure 1).

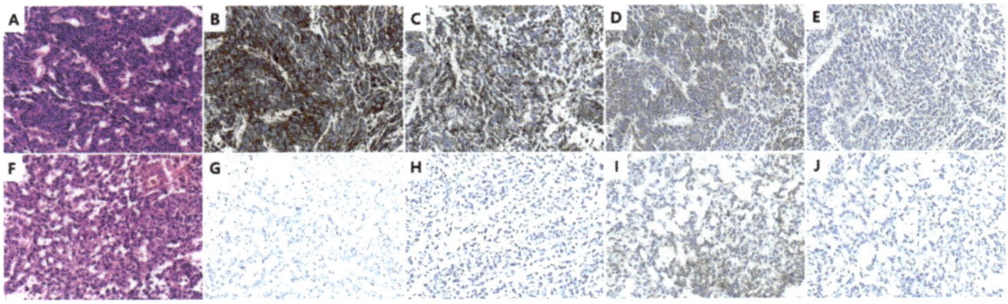

Figure 1. Representative H&E and immunohistochemical images of small cell neuroendocrine carcinoma (SCNEC) of classic histology (**A–E**) and with ambiguous histology (**F–J**). SCNEC shows sheets of relatively small cells with scant cytoplasm, speckled nuclei, and indistinct nucleoli (**A**). It is typically immunoreactive for synaptophysin (**B**), chromogranin (**C**), and CD117 (**D**) and negative for GATA3 (**E**). SCNEC with ambiguous histology shows sheets of cells with small to medium nuclei, relatively abundant cytoplasm, mild pleomorphism and occasional nucleoli (**F**). Although this case is immunonegative for synaptophysin (**G**) and chromogranin (**H**), the tumor is diffusely immunoreactive for CD117 (**I**) and negative for GATA3 (**J**). (Original magnification: A–I, ×400).

After the reassessment of H&E slides and immune-stained slides, the cases were classified as pure SCNEC (29 cases, 61.7%), combined SCNEC and UC (15 cases, 31.9%), and UC (3 cases, 6.4%). Divergent differentiation and variant histology were frequently noted and included glandular (6 cases, 12.7%) and squamous (3 cases, 6.4%) differentiation and micropapillary (4 cases, 8.5%), rhabdoid (1 case, 2.1%), and giant cell (1 case, 2.1%) variants. Tumor invasion into the muscularis propria was noted in 38 cases (80.9%). Twenty-five patients were treated with chemotherapy. Among the 10 cases involving partial or radical cystectomy, most were of high pathologic stages with pT3 (8 cases, 80%) and pT4 (1 case, 10%), and half of the patients had lymph node metastasis (5 patients, 50.0%).

3.2. Expression of NE, Luminal, and Basal Markers in the Discovery Cohort

The expression profile of 17 IHC markers in the 146 NED cores and 65 non-NED cores is summarized in Table 2. Detailed information on the IHC markers is presented in Supplementary Table S1. Representative IHC images are presented in Supplementary Figure S1.

Table 2. Immunoprofile of neuroendocrine cores and non-neuroendocrine cores from small cell neuroendocrine carcinomas of the urinary tract.

	Neuroendocrine Cores (n = 146)										Non-Neuroendocrine Cores (n = 65)									
	Intensity				Extent						Intensity				Extent					
	0 and 1		2 and 3		≤5%		>5–≤50%		>50%		0 and 1		2 and 3		≤5%		>5–≤50%		>50%	
SYP	29	(19.9)	117	(80.1)	12	(8.2)	18	(12.3)	116	(79.5)	65	(100)	0	(0.0)	65	(100)	0	(0.0)	0	(0.0)
CGA	84	(57.5)	62	(42.5)	89	(61.0)	27	(18.5)	30	(20.5)	65	(100)	0	(0.0)	65	(100)	0	(0.0)	0	(0.0)
CD56	47	(32.2)	99	(67.8)	32	(21.9)	25	(17.1)	89	(61.0)	63	(96.9)	2	(3.1)	64	(98.5)	0	(0.0)	1	(1.5)
CD117	74	(50.7)	72	(49.3)	38	(26.0)	23	(15.8)	85	(58.2)	61	(93.8)	4	(6.2)	62	(95.4)	3	(4.6)	0	(0.0)
INSM1	43	(29.5)	103	(70.5)	33	(22.6)	49	(33.6)	64	(43.8)	65	(100)	0	(0.0)	64	(98.5)	1	(1.5)	0	(0.0)
NSE	35	(24.0)	111	(76.0)	20	(13.7)	15	(10.3)	111	(76.0)	45	(69.2)	20	(30.8)	37	(56.9)	19	(29.2)	9	(13.8)
SOX2	25	(17.1)	121	(82.9)	30	(20.5)	16	(11.0)	100	(68.5)	27	(41.5)	38	(58.5)	36	(55.4)	21	(32.3)	8	(12.3)
TUBB2B	78	(53.4)	68	(46.6)	68	(46.6)	27	(18.5)	51	(34.9)	54	(83.1)	11	(16.9)	56	(86.2)	8	(12.3)	1	(1.5)
SSTR2	78	(53.4)	68	(46.6)	81	(55.5)	23	(15.8)	42	(28.8)	62	(95.4)	3	(4.6)	63	(96.9)	2	(3.1)	0	(0.0)
p53	17	(11.6)	129	(88.4)	26	(17.8)	9	(6.2)	111	(76.0)	15	(23.1)	50	(76.9)	9	(13.8)	0	(0.0)	56	(86.2)
Rb	131	(89.7)	15	(10.3)	130	(89.0)	8	(5.5)	8	(5.5)	65	(100)	0	(0.0)	65	(100)	0	(0.0)	0	(0.0)
EGFR	95	(65.1)	51	(34.9)	81	(55.5)	19	(13.0)	46	(31.5)	10	(15.4)	55	(84.6)	6	(9.2)	11	(16.9)	48	(73.8)
CK5/6	138	(94.5)	8	(5.5)	142	(97.3)	4	(2.7)	0	(0.0)	41	(63.1)	24	(36.9)	46	(70.8)	11	(16.9)	8	(12.3)
CK14	137	(93.8)	9	(6.2)	143	(97.9)	3	(2.1)	0	(0.0)	43	(66.2)	22	(33.8)	50	(76.9)	9	(13.8)	6	(9.2)
CK20	119	(81.5)	27	(18.5)	135	(92.5)	4	(2.7)	7	(4.8)	17	(26.2)	43	(73.8)	21	(32.3)	22	(33.8)	22	(33.8)
FOXA1	39	(26.7)	107	(73.3)	18	(12.3)	23	(15.8)	105	(71.9)	23	(35.4)	42	(64.6)	14	(21.5)	15	(23.1)	36	(55.4)
GATA3	131	(89.7)	15	(10.3)	134	(91.8)	8	(5.5)	4	(2.7)	8	(12.3)	57	(87.7)	9	(13.8)	4	(6.2)	52	(80.0)

Data are expressed as number (%). Abbreviations: SYP, synaptophysin; CGA, chromogranin; INSM1, insulinoma-associated protein 1; NSE, neuron specific enolase; SOX2, SRY (sex determining region Y)-box 2; TUBB2B, tubulin beta 2B class IIb, SSTR2, somatostatin receptor 2; p53, tumor protein p53; Rb, retinoblastoma-associated protein; EGFR, epidermal growth factor receptor; CK5/6, cytokeratin 5/6; CK14, cytokeratin 14; CK20, cytokeratin 20; FOXA1, forkhead box A1; GATA3, GATA binding protein 3.

In the NED cores, synaptophysin was the most strongly and widely expressed NE marker, and approximately 80% of NED cores showed diffuse expression. CD56 and CD117 were also diffusely expressed in 61.0% and 58.2% of NED cores, respectively. However, a subset of NED cores was negative for the NE markers synaptophysin (12 cores, 8.2%), CD56 (30 cores, 20.5%), and CD117 (38 cores, 26.0%). Chromogranin and INSM1were expressed less widely, and their diffuse expression was noted in 20.5% and 43.8% of NED cores, respectively. As expected, the expression of luminal (CK20 and GATA3) and basal (CK5/6 and CK14) markers was negative or weak in ≤5% NED cores. However, EGFR and FOXA1 were expressed in a significant number of NED cores and immunoreactive in 31.5% and 71.9% of NED cores, respectively, with varying intensities.

In the non-NED cores, most of the NE markers such as synaptophysin, chromogranin, CD56, INSM1, SSTR2, and CD117 were negative or weakly expressed (≤5%) in more than 95% of such cores. NSE, SOX2, and TUBB2 were immunoreactive in a significant extent (>5%) of non-NED cores (43.0%, 44.6%, and 13.8%, respectively) with varying intensities, although they were expressed as such in most NED cores (86.3%, 79.5%, 53.4%, respectively). GATA3 and EGFR showed diffuse expression in 80.0% and 73.9% of non-NED cores, respectively.

3.3. Decision Tree-Based Diagnostic NE IHC Model

Given the lack of expression of NE markers in a significant number of NED cores, the decision tree classifier algorithm was employed to define a diagnostic IHC panel for NED. Among multiple models suggested by the algorithm, this model was selected because it was relatively simple, highly reproducible, and easy to apply in routine clinical practice. It consisted of three markers synaptophysin (cutoff >5% immunoreactive area), CD117 (cutoff >20% immunoreactive area), and GATA3 (cutoff of negative/weak intensity to be classified as NED) and applied in that order. The relative importance of the markers was 0.758 for synaptophysin, 0.213 for CD117, and 0.029 for GATA3 in the model.

An overview of the decision tree model using 147 cores of the training set is shown in Figure 2. The synaptophysin immunoreactivity was noted in >5% tumor area in 94 cores and was classified as NED (64.0%). Among 53 cores with ≤5% synaptophysin-immunoreactive area, 43 cores were of CD117-immunoreactive area ≤20% and classified as non-NED (81.1%). In cores with the CD117-immunoreactive area >20%, the intensity

of GATA3 immunoreactivity was considered, being classified as NED in 9 cores with negative/weak intensity (90.0%) and non-NED in 1 core with moderate to strong intensity (10.0%) (Supplementary Figure S2). The overall accuracy and area under the receiver operating characteristic curve were 98.4% and 98.8% according to the internal validation.

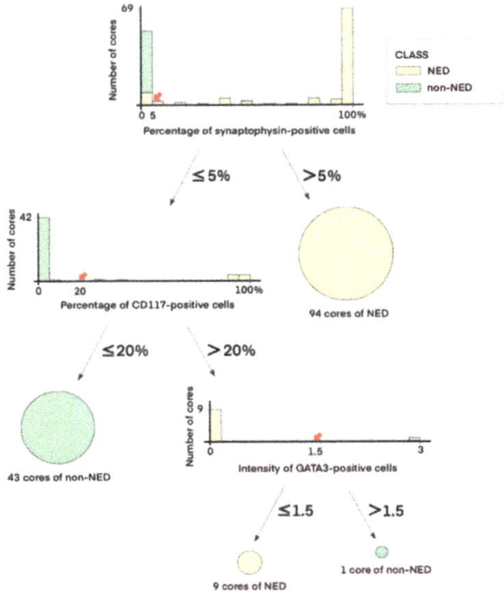

Figure 2. Decision tree model of the discovery cohort. Diagnostic flow of the training set is demonstrated with cutoff values (bold red arrow) and distribution plots of NED and non-NED cores. Each distribution plot stands for a split-by-condition node. The *x*-axis and *y*-axis represent the extent or intensity of the corresponding IHC marker and the number of NED or non-NED cores, respectively. The finally classified cores are colored yellow for NED and green for non-NED. The degrees of intensity of GATA3 are represented as follows: 0, negative; 1, weak; 2, moderate; 3, strong.

The distribution of expression and association of each marker in all cores of the discovery cohort are presented in Figure 3. When the decision tree model was applied to all 211 cores, 11 cores with ≤5% of synaptophysin-immunoreactive area were classified as NED. They expressed one or more NE markers such as CD117 (11/11 cores, 100%), CD56 (9/11 cores; 81.8%), TUBB2B (6/11 cores, 54.6%), SOX2 (9/11 cores, 81.8%), NSE (7/11 cores, 63.6%), SSTR2 (5/11 cores, 45.5%), and INSM1 (3/11 cores, 27.3%). According to the model, CD117 expression was identified in all NED cores with ≤5% of synaptophysin-immunoreactive area and showed a weak relationship with synaptophysin compared to other NE markers.

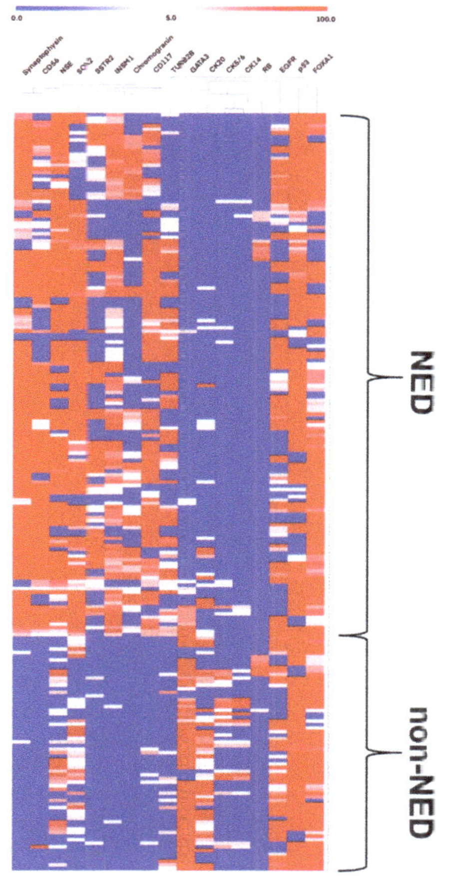

Figure 3. Distribution of the expression of 17 markers in NED and non-NED cores. Heatmap of 17 markers is presented. The white to red shades show increasing immunoreactivity from 5% to 100%, and the blue color represents less than 5% immunoreactivity of IHC markers including no expression. See color scale.

3.4. Application of the Diagnostic NE IHC Model on an External Cohort

Six SCNEC cases and two UC cases from the external cohort were immunostained for synaptophysin, CD117, and GATA3 using whole tumor sections in our institution. According to the model, five SCNEC cases were immunoreactive for synaptophysin in more than 20% of tumor cells and classified as NED. The remaining SCNEC case was negative for synaptophysin but immunoreactive for CD117 in more than 90% of tumor cells, being classified as NED. The two UC cases were immunonegative for all three markers and classified as non-NED. These results were consistent with the original diagnosis.

3.5. Ultrastructural Validation of NE Differentiation

TEM was performed on samples from five SCNEC cases (four cases in the discovery cohort from which the 11 cores with $\leq$ 5% of synaptophysin-immunoreactive area were derived and one such case from the external cohort). Two SCNEC cases with diffuse synaptophysin expression and two UC cases were also included as positive and negative controls, respectively.

All five cases showed varied numbers of electron dense neurosecretory granules in the cytoplasm of the tumor cells, similar to those of SCNEC (Figure 4). They ranged from 144.5 to 582.2 nm. The granules were round with a dense core, although the delimiting outer membrane and peripheral halos were not clearly observed probably due to the deparaffinization process. There were no neurosecretory granules in the two UC cases (data not shown).

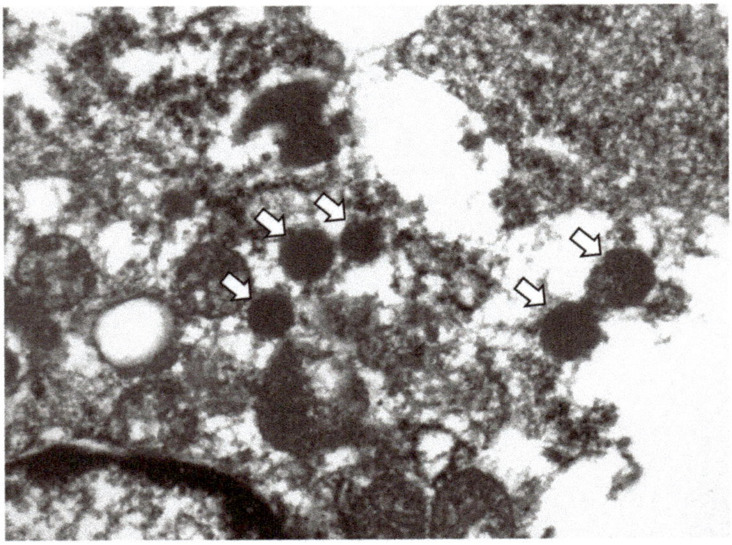

Figure 4. Transmission electron microscopy image of synaptophysin-negative SCNEC. Arrows indicate neurosecretory granules (218.31–275.16 nm). (Original magnification, ×20,000).

4. Discussion

Herein, we propose a decision tree-based IHC model consisting of two inclusion markers synaptophysin and CD117 and one exclusion marker GATA3 for the diagnosis of SCNEC of the urinary bladder. It could detect NED of not only NE marker-positive SCNEC but also traditional marker-negative SCNEC. The model was validated using an external cohort and by TEM analysis.

Through this study, we emphasize the following points for the diagnosis of SCNEC. First, it is crucial to be familiar with the histological features of SCNEC. In cases with ambiguous histological features that are difficult to differentiate from UC, IHC for NE markers should be performed with a low threshold. Second, even focal (>5%) and weak synaptophysin immunoreactivity would be sufficient for the diagnosis of SCNEC. Third, in synaptophysin-negative cases, CD117 and GATA3 may be helpful to distinguish between SCNEC and non-SCNEC.

SCNEC is mainly diagnosed based on histology and may not require IHC confirmation. As reported previously, most of our cases including traditional NE marker-negative cases showed classic histological features of SCNEC. The tumor presented as solid sheets, nests, or trabeculae of small cells. Tumor cells have sparse cytoplasm, nuclear molding, finely granular stippled chromatin, inconspicuous nucleoli, high mitotic count, and frequent individual and geographic necrosis [4]. However, ambiguous histological features such as relatively abundant cytoplasm and the presence of nucleoli albeit inconspicuous were noted as shown in Figure 1. In such cases, IHC for NE markers might be useful to confirm NED.

Synaptophysin, chromogranin, and CD56 are widely used clinically in a diagnostic panel because of their suboptimal sensitivity and specificity as individual markers [9]. In the more common counterpart lung cancer, synaptophysin is expressed in 41–75% of small

cell lung carcinoma (SCLC) and 58–85% of large cell neuroendocrine carcinomas (LCNEC). Chromogranin may show weak and focal positivity and less sensitivity, being expressed in only 23–58% of SCLC and 42–69% of LCNEC. CD56 is expressed in most SCLC (72–99%) and LCNEC (72–94%) cases but at the cost of relatively low specificity (72%). As expected synaptophysin was chosen as the most important NE marker in our model.

CD117 was chosen as the second most important marker for the diagnosis of SCNEC in preference to other traditional or emerging NE markers. This could be explained, at least in part, by the fact that other NE markers were often expressed simultaneously whereas CD117 was expressed in those NE marker-negative SCNEC cases. CD117 expression has been reported in SCNEC of various organs such as the lung, uterine cervix, and esophagus [19–21]. CD117 expression was also noted in 27% cases of SCNEC in the urinary bladder [22]. The mechanisms of CD117 expression in NE carcinoma are largely unknown, but an autocrine growth loop has been suggested in SCLC cell lines [23]. As a member of the type III receptor tyrosine kinase family, CD117 activates several signaling pathways, such as the JAK/STAT, RAS/MAP kinase pathway, PI3 kinase, PLCγ pathway, and SRC pathway [24]. Consequently, it plays an important role in the proliferation, survival, differentiation, apoptosis, and migration of tumor cells [24]. Another hypothesis is that CD117 may increase cancer stem cell phenotype in SCNEC since it plays a key role in maintaining the stemness of cancer stem cells [24]. Because both UC and SCNEC arise from common multipotential cancer stem cells, SCNEC frequently coexists with conventional UC [25]. Therefore, CD117 expression may represent a marker of aggressive biologic behavior of SCNEC instead of NED in the model.

According to previous reports, a novel pan-NE marker INSM1 was superior to traditional NE markers with high sensitivity (93.9%) and specificity (97.4%) in the SCNEC of the genitourinary tract [26,27]. In our cases, INSM1 showed relatively lower sensitivity (78.1%) but similar high specificity (96.9%) compared to the previous report. Nevertheless, this novel marker was not selected in our model. The decision tree model suggests variables based on the causal relationship and selects the best one if multiple variables are correlated. As shown in Figure 3, when there is a strong relationship between INSM1 and synaptophysin immunoreactivity, synaptophysin might be selected in the model.

Among non-NE markers employed in the present study, GATA3 immunoreactivity was selected as an exclusion marker for NE differentiation probably because of its relatively higher specificity than that of the other non-NE markers. The basal markers CK5/6 and CK14 were not only negative in most NE cores (94.5% and 93.8%, respectively) but also not expressed in more than half of non-NE cores (63.1% and 66.2%, respectively). The luminal marker FOXA1 was expressed similarly in NE cores and non-NE cores (88.4% and 83.1%, respectively). In the remaining luminal markers, GATA3 was negative in more NE cores than CK20 (89.7% and 81.5%, respectively) and had stronger immunoreactivity in the non-NE cores (moderate to strong immunoreactivity in 89.3% and 75.3%, respectively). Therefore, basal markers CK5/6 and CK14 and luminal marker FOXA1 might offer suboptimal distinguishing power between NE cores and non-NE cores, and GATA3 might be a better exclusion marker than CK20.

Although the demand for TEM has decreased due to the development of IHC staining and molecular pathology, this technique is still used for accurate diagnosis. TEM is particularly useful for the differential diagnosis between malignant mesothelioma and serous carcinoma, whereas immunostaining results alone cannot achieve an accurate diagnosis [28]. In the present study, neurosecretory granules were found in all synaptophysin-negative and inconspicuous ($\leq$5%) cases and were useful for confirming NED in those cases, although the number of granules was fewer than that in classic SCNEC cases.

Genomic analyses of bladder cancer have been used for the molecular characterization of variant histologic subtypes. The Cancer Genome Atlas (TCGA) and a report by Lund et al. have identified neuronal subtype or small cell/neuroendocrine (SC/NE) consensus cluster, accounting for 3–15% of bladder cancer by RNA-sequencing analysis [16,29,30]. A TCGA report has shown that tumors representing NED at the molecular level were not

similar in histology to SCNEC in 85% of cases (17/20) [16]. A report by Lund et al. showed that only half of the SC/NE consensus cluster represented the enriched expression of neuronal markers such as synaptophysin, chromogranin, and CD56 [29]. Phenotypical UC with the absence of NE histology may also reveal transcriptomic patterns of NE carcinoma and be defined as neuroendocrine-like (NE-like) tumors [11]. These reports suggest that histological, molecular, and IHC results of SCNEC may not agree completely with each other. Combining our findings with previous results, continuous efforts should be made to define the diagnostic criteria for aggressive NE carcinoma that requires therapeutic approaches different from those used for UC.

The present study has limitations. Although the performance of the decision tree diagnostic model was excellent, the possibility of overfitting cannot be excluded. Since we performed core-based analysis to compensate for the small number of SCNEC cases, this model needs to be validated with larger numbers of SCNEC cases, preferably in a multicenter study.

5. Conclusions

Our study demonstrated that the decision tree model using synaptophysin, CD117, and GATA3 may help confirm NED of not only NE marker-positive SCNEC but also traditional marker-negative SCNEC.

Supplementary Materials: The following supporting information can be downloaded at: https://www.mdpi.com/article/10.3390/cancers14102495/s1, Figure S1: Representative immunohistochemical analysis of 17 markers used in the present study.; Figure S2: Representative immunohistochemistry of GATA3. Table S1: Antibodies used in the study.; Table S2: Immunoprofile of neuroendocrine cores and non-neuroendocrine cores from small cell neuroendocrine carcinomas of the urinary tract.

Author Contributions: Conceptualization, Y.M.C.; methodology, S.U.J. and Y.M.C.; software, G.J. and S.U.J.; validation, G.H.K. and S.-W.K.; formal analysis, S.U.J.; investigation, G.H.K.; resources, J.-M.P. and S.Y.Y.; data curation, G.J.; writing—original draft preparation, G.H.K.; writing—review and editing, S.U.J. and Y.M.C.; visualization, G.J., D.-M.S. and H.J.; supervision, Y.M.C.; project administration, Y.M.C.; funding acquisition, Y.M.C. All authors have read and agreed to the published version of the manuscript.

Funding: This work is funded by the Ministry of Science, ICT and Future Planning (2019R1A2C1088246) and a grant (2019IP0870-2) from the Asan Institute for Life Sciences, Asan Medical Centre, Seoul, Korea.

Institutional Review Board Statement: The study was approved by the Institutional Review Board of Asan Medical Center (2013-0107).

Informed Consent Statement: The patient consent was waived due to retrospective nature of the study.

Data Availability Statement: The data are available on request from the corresponding author.

Acknowledgments: This work is supported by the Basic Science Research Program through the National Research Foundation of Korea (NRF).

Conflicts of Interest: The authors declare no conflict of interest.

References

1. Chau, C.; Rimmer, Y.; Choudhury, A.; Leaning, D.; Law, A.; Enting, D.; Lim, J.H.; Hafeez, S.; Khoo, V.; Huddart, R.; et al. Treatment outcomes for small cell carcinoma of the bladder: Results from a UK patient retrospective cohort study. *Int. J. Radiat. Oncol. Biol. Phys.* **2021**, *110*, 1143–1150. [CrossRef] [PubMed]
2. Kouba, E.; Cheng, L. Neuroendocrine tumors of the urinary bladder according to the 2016 World Health Organization Classification: Molecular and clinical characteristics. *Endocr. Pathol.* **2016**, *27*, 188–199. [CrossRef]
3. Royce, T.J.; Lin, C.C.; Gray, P.J.; Shipley, W.U.; Jemal, A.; Efstathiou, J.A. Clinical characteristics and outcomes of nonurothelial cell carcinoma of the bladder: Results from the National Cancer Data Base. *Urol. Oncol.* **2018**, *36*, e71–e78. [CrossRef] [PubMed]
4. Zhao, X.; Flynn, E.A. Small cell carcinoma of the urinary bladder: A rare, aggressive neuroendocrine malignancy. *Arch. Pathol. Lab. Med.* **2012**, *136*, 1451–1459. [CrossRef] [PubMed]

5. Gupta, S.; Thompson, R.H.; Boorjian, S.A.; Thapa, P.; Hernandez, L.P.; Jimenez, R.E.; Costello, B.A.; Frank, I.; Cheville, J.C. High grade neuroendocrine carcinoma of the urinary bladder treated by radical cystectomy: A series of small cell, mixed neuroendocrine and large cell neuroendocrine carcinoma. *Pathology* **2015**, *47*, 533–542. [CrossRef] [PubMed]
6. Lynch, S.P.; Shen, Y.; Kamat, A.; Grossman, H.B.; Shah, J.B.; Millikan, R.E.; Dinney, C.P.; Siefker-Radtke, A. Neoadjuvant chemotherapy in small cell urothelial cancer improves pathologic downstaging and long-term outcomes: Results from a retrospective study at the MD Anderson Cancer Center. *Eur. Urol.* **2013**, *64*, 307–313. [CrossRef]
7. Al-Ahmadie, H.; Netto, G.J. Updates on the genomics of bladder cancer and novel molecular taxonomy. *Adv. Anat. Pathol.* **2020**, *27*, 36–43. [CrossRef]
8. WHO Classification of Tumours Editorial Board. *Thoracic Tumours (WHO Classification of Tumours Series)*, 5th ed.; International Agency for Research on Cancer (IARC): Lyon, France, 2021; Volume 5.
9. Rooper, L.M.; Sharma, R.; Li, Q.K.; Illei, P.B.; Westra, W.H. INSM1 demonstrates superior performance to the individual and combined use of synaptophysin, chromogranin and CD56 for diagnosing neuroendocrine tumors of the thoracic cavity. *Am. J. Surg. Pathol.* **2017**, *41*, 1561–1569. [CrossRef]
10. Travis, W.D. Update on small cell carcinoma and its differentiation from squamous cell carcinoma and other non-small cell carcinomas. *Mod. Pathol.* **2012**, *25* (Suppl. 1), S18–S30. [CrossRef]
11. Batista da Costa, J.; Gibb, E.A.; Bivalacqua, T.J.; Liu, Y.; Oo, H.Z.; Miyamoto, D.T.; Alshalalfa, M.; Davicioni, E.; Wright, J.; Dall'Era, M.A.; et al. Molecular characterization of neuroendocrine-like bladder cancer. *Clin. Cancer Res.* **2019**, *25*, 3908–3920. [CrossRef]
12. Erler, B.S.; Presby, M.M.; Finch, M.; Hodges, A.; Horowitz, K.; Topilow, A.A.; Matulewicz, T. CD117, Ki-67, and p53 predict survival in neuroendocrine carcinomas, but not within the subgroup of small cell lung carcinoma. *Tumour Biol.* **2011**, *32*, 107–111. [CrossRef] [PubMed]
13. Papouchado, B.; Erickson, L.A.; Rohlinger, A.L.; Hobday, T.J.; Erlichman, C.; Ames, M.M.; Lloyd, R.V. Epidermal growth factor receptor and activated epidermal growth factor receptor expression in gastrointestinal carcinoids and pancreatic endocrine carcinomas. *Mod. Pathol.* **2005**, *18*, 1329–1335. [CrossRef] [PubMed]
14. Qian, Z.R.; Li, T.; Ter-Minassian, M.; Yang, J.; Chan, J.A.; Brais, L.K.; Masugi, Y.; Thiaglingam, A.; Brooks, N.; Nishihara, R.; et al. Association between somatostatin receptor expression and clinical outcomes in neuroendocrine tumors. *Pancreas* **2016**, *45*, 1386–1393. [CrossRef] [PubMed]
15. Choi, W.; Porten, S.; Kim, S.; Willis, D.; Plimack, E.R.; Hoffman-Censits, J.; Roth, B.; Cheng, T.; Tran, M.; Lee, I.L.; et al. Identification of distinct basal and luminal subtypes of muscle-invasive bladder cancer with different sensitivities to frontline chemotherapy. *Cancer Cell* **2014**, *25*, 152–165. [CrossRef] [PubMed]
16. Robertson, A.G.; Kim, J.; Al-Ahmadie, H.; Bellmunt, J.; Guo, G.; Cherniack, A.D.; Hinoue, T.; Laird, P.W.; Hoadley, K.A.; Akbani, R.; et al. Comprehensive molecular characterization of muscle-invasive bladder cancer. *Cell* **2017**, *171*, 540–556.e525. [CrossRef]
17. De Felice, F.; Crocetti, D.; Parisi, M.; Maiuri, V.; Moscarelli, E.; Caiazzo, R.; Bulzonetti, N.; Musio, D.; Tombolini, V. Decision tree algorithm in locally advanced rectal cancer: An example of over-interpretation and misuse of a machine learning approach. *J. Cancer Res. Clin. Oncol.* **2020**, *146*, 761–765. [CrossRef]
18. Kim, M.; Ro, J.Y.; Amin, M.B.; de Peralta-Venturina, M.; Kwon, G.Y.; Park, Y.W.; Cho, Y.M. Urothelial eddies in papillary urothelial neoplasms: A distinct morphologic pattern with low risk for progression. *Int. J. Clin. Exp. Pathol.* **2013**, *6*, 1458–1466.
19. Ohwada, M.; Wada, T.; Saga, Y.; Tsunoda, S.; Jobo, T.; Kuramoto, H.; Konno, R.; Suzuki, M. C-kit overexpression in neuroendocrine small cell carcinoma of the uterine cervix. *Eur. J. Gynaecol. Oncol.* **2006**, *27*, 53–55.
20. Terada, T. Small cell neuroendocrine carcinoma of the esophagus: Report of 6 cases with immunohistochemical and molecular genetic analysis of *KIT* and *PDGFRA*. *Int. J. Clin. Exp. Pathol.* **2013**, *6*, 485–491.
21. Potti, A.; Moazzam, N.; Ramar, K.; Hanekom, D.S.; Kargas, S.; Koch, M. CD117 (c-KIT) overexpression in patients with extensive-stage small-cell lung carcinoma. *Ann. Oncol.* **2003**, *14*, 894–897. [CrossRef]
22. Pan, C.X.; Yang, X.J.; Lopez-Beltran, A.; MacLennan, G.T.; Eble, J.N.; Koch, M.O.; Jones, T.D.; Lin, H.; Nigro, K.; Papavero, V.; et al. *c-kit* Expression in small cell carcinoma of the urinary bladder: Prognostic and therapeutic implications. *Mod. Pathol.* **2005**, *18*, 320–323. [CrossRef] [PubMed]
23. Heinrich, M.C. Is KIT an important therapeutic target in small cell lung cancer? *Clin. Cancer Res.* **2003**, *9*, 5825–5828. [PubMed]
24. Foster, B.M.; Zaidi, D.; Young, T.R.; Mobley, M.E.; Kerr, B.A. CD117/c-kit in cancer stem cell-mediated progression and therapeutic resistance. *Biomedicines* **2018**, *6*, 31. [CrossRef] [PubMed]
25. Wang, G.; Xiao, L.; Zhang, M.; Kamat, A.M.; Siefker-Radtke, A.; Dinney, C.P.; Czerniak, B.; Guo, C.C. Small cell carcinoma of the urinary bladder: A clinicopathological and immunohistochemical analysis of 81 cases. *Hum. Pathol.* **2018**, *79*, 57–65. [CrossRef]
26. Chen, J.F.; Yang, C.; Sun, Y.; Cao, D. Expression of novel neuroendocrine marker insulinoma-associated protein 1 (INSM1) in genitourinary high-grade neuroendocrine carcinomas: An immunohistochemical study with specificity analysis and comparison to chromogranin, synaptophysin, and CD56. *Pathol. Res. Pract.* **2020**, *216*, 152993. [CrossRef]
27. Kim, I.E., Jr.; Amin, A.; Wang, L.J.; Cheng, L.; Perrino, C.M. Insulinoma-associated protein 1 (INSM1) expression in small cell neuroendocrine carcinoma of the urinary tract. *Appl. Immunohistochem. Mol. Morphol.* **2020**, *28*, 687–693. [CrossRef]
28. Fortarezza, F.; Della Barbera, M.; Pezzuto, F.; Lunardi, F.; Faccioli, E.; Pasello, G.; Rea, F.; Rizzo, S.; Calabrese, F. Diagnostic challenges in epithelioid pleural mesothelioma: Case series with support from electron microscopy. *Diagnostics* **2021**, *11*, 841. [CrossRef]

29. Sjödahl, G.; Eriksson, P.; Liedberg, F.; Höglund, M. Molecular classification of urothelial carcinoma: Global mRNA classification versus tumour-cell phenotype classification. *J. Pathol.* **2017**, *242*, 113–125. [CrossRef]
30. Kamoun, A.; de Reyniès, A.; Allory, Y.; Sjödahl, G.; Robertson, A.G.; Seiler, R.; Hoadley, K.A.; Groeneveld, C.S.; Al-Ahmadie, H.; Choi, W.; et al. A consensus molecular classification of muscle-invasive bladder cancer. *Eur. Urol.* **2020**, *77*, 420–433. [CrossRef]

diagnostics

MDPI

Article

Integrated Analysis of Tumor Mutation Burden and Immune Infiltrates in Hepatocellular Carcinoma

Yulan Zhao, Ting Huang and Pintong Huang *

Department of Ultrasound in Medicine, Second Affiliated Hospital of Zhejiang University School of Medicine, Hangzhou 310000, China

* Correspondence: huangpintong@zju.edu.cn; Tel.: +86-18857168333; Fax: +86-0571-87783934

Abstract: Tumor mutation burdens (TMBs) act as an indicator of immunotherapeutic responsiveness in various tumors. However, the relationship between TMBs and immune cell infiltrates in hepatocellular carcinoma (HCC) is still obscure. The present study aimed to explore the potential diagnostic markers of TMBs for HCC and analyze the role of immune cell infiltration in this pathology. We used OA datasets from The Cancer Genome Atlas database. First, the "maftools" package was used to screen the highest mutation frequency in all samples. R software was used to identify differentially expressed genes (DEGs) according to mutation frequency and perform functional correlation analysis. Then, the gene ontology (GO) enrichment analysis was performed with "clusterProfiler", "enrichplot", and "ggplot2" packages. Finally, the correlations between diagnostic markers and infiltrating immune cells were analyzed, and CIBERSORT was used to evaluate the infiltration of immune cells in HCC tissues. As a result, we identified a total of 359 DEGs in this study. These DEGs may affect HCC prognosis by regulating fatty acid metabolism, hypoxia, and the P53 pathway. The top 15 genes were selected as the hub genes through PPI network analysis. *SRSF1*, *SNRPA1*, and *SRSF3* showed strong similarities in biological effects, NCBP2 was demonstrated as a diagnostic marker of HCC, and high NCBP2 expression was significantly correlated with poor over survival (OS) in HCC. In addition, NCBP2 expression was correlated with the infiltration of B cells (r = 0.364, $p = 3.30 \times 10^{-12}$), CD8$^+$ T cells (r = 0.295, $p = 2.71 \times 10^{-8}$), CD4$^+$ T cells, (r = 0.484, $p = 1.37 \times 10^{-21}$), macrophages (r = 0.551, $p = 1.97 \times 10^{-28}$), neutrophils (r = 0.457, $p = 3.26 \times 10^{-19}$), and dendritic cells (r = 0.453, $p = 1.97 \times 10^{-18}$). Immune cell infiltration analysis revealed that the degree of central memory T-cell (Tcm) infiltration may be correlated with the HCC process. In conclusion, NCBP2 can be used as diagnostic markers of HCC, and immune cell infiltration plays an important role in the occurrence and progression of HCC.

Keywords: hepatocellular carcinoma; tumor mutation burden; immune cells; The Cancer Genome Atlas; CIBERSORT

Citation: Zhao, Y.; Huang, T.; Huang, P. Integrated Analysis of Tumor Mutation Burden and Immune Infiltrates in Hepatocellular Carcinoma. *Diagnostics* **2022**, *12*, 1918. https://doi.org/10.3390/diagnostics12081918

Academic Editor: Gian Paolo Caviglia

Received: 26 June 2022
Accepted: 1 August 2022
Published: 8 August 2022

Publisher's Note: MDPI stays neutral with regard to jurisdictional claims in published maps and institutional affiliations.

1. Introduction

Hepatocellular carcinoma (HCC) is one of the most common and aggressive malignancies in the digestive system and contributes to a severe global disease burden worldwide [1]. It ranked sixth in global incidence (4.7%) and was the third leading cause of cancer-related deaths (8.3%) in 2020, according to a recent study [2]. The prognosis of patients is usually driven by the tumor stage. The 5-year survival rates for local disease exceed 70%; however, the median survival time of advanced-stage HCC patients is only 1 year [3]. Although the survival situation has improved, benefiting from advancements in medical treatments [4], approximately 2/3 of HCC patients are diagnosed at advanced stages, and the median overall survival rate remains at a low level [5]. Therefore, there is an urgent need to explore the potential molecular mechanisms of tumor progression to develop better therapeutic strategies and investigate the potential benefits of adjuvant systemic therapies.

The molecular mechanisms contributing to the development of HCC are extremely complex and involve various genetic abnormalities, such as the dysregulation of signaling

pathways, genomic instability, single-nucleotide polymorphisms (SNPs), and somatic mutations [6,7]. The somatic mutations were reported frequently among HCC patients, and the landscape was complicated, including somatic mutations that occur in multitudes of genes accompanied by the changes of multiple signaling pathways [8], which contribute to various molecular heterogeneities that remain poorly understood. With the rise of high-throughput sequencing technology, a large number of databases based on TCGA (The Cancer Genome Atlas) and GEO (Gene Expression Omnibus) datasets have emerged, making it convenient for us to investigate the complex relationships between HCC and the underlying oncogenic somatic mutation molecular mechanisms. Our results may provide new insight into novel diagnostic and prognostic values for HCC.

In addition, recent studies have demonstrated that TMB(Tumor mutation burden) was correlated with immune cell infiltration and subtypes [9,10]. TMB is defined as the frequency of gene mutations (total count of variants/the whole length of exons), including translocation, deletion, and insertion mutations, in addition to other mutations that appear in the somatic-gene-coding region, with an average 1 Mb-base range for the tumor genome, and it is used as a biomarker to predict the sensitivity, efficacy, and treatment outcomes of immune checkpoint inhibitors (ICPIs) [11,12]. The tumor cell carries new antigens generated by somatic mutations on the cell surface that may be recognized by the immune system, further making the tumor cell a target for activated immune cells [13]. To date, there have been numerous studies focusing on the relationship between TMB and immunotherapy in diverse cancers [14–16], and accumulating evidence indicates that a high tumor mutation burden confers an increased immune reaction to tumors and a better response to ICPI treatment [17]. However, the prognostic value of TMB in HCC has not yet been clearly determined.

In the present study, we downloaded The Cancer Genome Atlas HCC data sets using R software package and other online databases to investigate the association of genes bearing important mutations contributing to TMBs with clinical and genomic features in HCC patients. We performed gene ontology (GO) term enrichment and protein–protein interaction (PPI) analysis and constructed functional networks related to NCBP2 in HCC. Finally, the relationship between NCBP2 and immune cell infiltration in the HCC was also analyzed. The findings from the present study suggest that NCBP2 influences the prognosis of HCC patients via its interaction with infiltrating immune cells.

2. Materials and Methods

2.1. Data Download

The Cancer Genome Atlas (TCGA, https://cancergenome.nih.gov/) (accessed on 1 March 2021) database provides publicly available cancer genome datasets. TCGA database contains 369 cases of LICH tissue samples. We used R language RTCGToolbox package from TCGA database (https://portal.gdc.cancer.gov/) (accessed on 1 March 2021) to download Liver Cancer (LIHC) gene expression spectrum and clinical data as the training sets. We included a total of 364 cases of LIHC samples in the present study. We used the maftools package to screen the 20 genes with the highest mutation frequencies in all samples, and we visualized the mutation situations and frequencies of all samples. We grouped all samples according to the genes with the highest mutation frequencies.

2.2. Data Preprocessing and Differentially Expressed Gene (DEG) Screening

We used affy package (R version 3.6.3; TUNA Team, Tsinghua University, Beijing, China) to perform background correction and data normalization, and we screened differentially expressed genes (DEGs) by using Limma software package. The screening criteria were: $|\log2$ fold change $(\log2FC)| > 1$, adjust $p < 0.05$. We used univariate Cox regression to screen out prognostic Genes. We used the intersection Search Tool (http://string-db.org; Version: 11.0) (accessed on 1 March 2021) for the Retrieval of Separated Genes (STRING) to predict the protein–protein interaction (PPI) network. We used Cytoscape to visualize complex networks and integrate them with data of any attribute type. Gene ontology (GO)

is a common method used to annotate genes and their products. This method is often used to annotate large-scale genes, determining molecular function (MF) and biological process (BP). We used cellular components (CCs) for a GO analysis of intersecting genes.

2.3. GSEA and GSVA Analysis

We performed GSEA and GSVA analysis to explore the important pathway of enrichment between the two groups. The reference gene set was H.all.v.7.1.symbols.gmt. We replaced 1000 genomes to achieve standardized enrichment scores for each analysis. We considered a nominal $p < 0.05$ and a false discovery rate < 0.05 as significant results. We used clusterProfiler and GSVA packages for GSVA analysis, and we considered adj.p value < 0.05 as a meaningful pathway.

2.4. Verification of Differential Expression of NCBP2

We used GEPIA2 (http://gepia2.cancer-pku.cn/) (accessed on 1 March 2021) to verify the differential expression between liver cancer and other cancer and paracancer samples in the database. We applied the box plot module of the GEPIA2 database to explore the expression level of NCBP2 in various cancer datasets, including the GTEx and TCGA databases, and we also analyzed the expression levels of NCBP2 in different stages of liver cancer through a Stage Plot module. Then, we used the Survival Map module to investigate the overall survival (OS) rates in liver and other cancers. Significance level is 0.05.

2.5. Prognostic Analysis

The Kaplan–Meier mapping platform is able to assess the effects of more than 50,000 genes on survival in 21 cancer types. The primary purpose of this tool is the discovery and validation of survival biomarkers based on meta-analysis. We explored the correlation between NCBP2 and prognosis of liver cancer in Kaplan–Meier mapping platform to verify the relationship between NCBP2 and liver cancer prognosis.

2.6. Expression Verification of NCBP2 in Cells and Tissues

The Human Protein Atlas is an open-access database used to map all human proteins in organ tissues and cells, and integrates various omics techniques. We detected the mRNA expression of NCBP2 in organ tissues and large tumors using the Human Protein Atlas and TIMER database. We used this database to preliminarily verify the expression levels of NCBP2 in cells and tissues.

2.7. Correlation Analysis between NCBP2 and Immunity

We applied "corrplot package" to further investigate the infiltration conditions of immune cells and the relationship between NCBP2 and immune cells in liver cancer. We constructed a correlation heatmap to visualize the correlation of 22 types of infiltrating immune cells in liver cancer. Then, we performed Spearman correlation analyses using "ggstatsplot" package (https://github.com/IndrajeetPatil/ggstatsplot) (accessed on 1 March 2021) to investigate the relationship between the levels of NCBP2 and immune cells.

3. Results

3.1. Landscape of Gene Mutation Files in LIHC

To investigate the mutation profile among the TCGA-LIHC cohort, we used the RTCGToolbox package of R language to acquire the LIHC gene expression spectrum and clinical data as the training set from TCGA database (https://portal.gdc.cancer.gov/) (accessed on 1 March 2021). The maftools package was used to screen the top 20 genes with high mutation frequencies in all samples, and waterfall plots were utilized to visualize the mutation landscapes of the genes. The results of the somatic mutation profiles in 364 cases of LIHC samples included in the present study showed that around 312 (85.71%) samples possessed somatic mutations. As for the top 20 mutated genes shown in Figure 1, we

discovered that gene *TP53* mutated most frequently, approximately accounting for 28% of mutations, followed by *TTN* (25%), *CTNNB1* (24%), *MUC16* (16%), *ALB* (11%), *PCLO* (11%), *MUC4* (10%), *RYR2* (10%), *ABCA13* (9%) and *APOB* (9%), *CSMD3* (8%), *FLG* (8%), *LRP1B* (8%), *OBSCN* (8%), *AXIN1* (8%), *XIRP2* (8%), *ARID1A* (7%), *HMCN1* (7%), *CACNA1E* (7%), and *SPTA1* (7%). Missense mutations were the most frequent among these alterations.

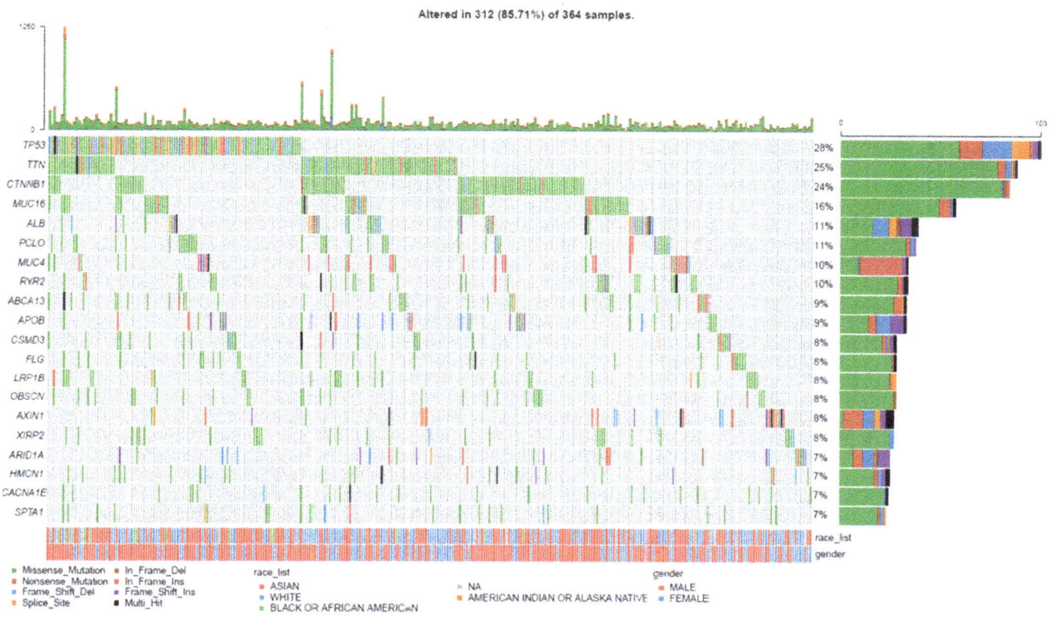

Figure 1. Landscape profile of top 20 mutated genes in 364 LIHC from TCGA database. Mutations of each gene in each sample are shown in waterfall plot. Each column presents specific sample, each line presents mutated gene, and name is listed on left. Different forms of somatic mutations and percentages of gene mutation types are shown on right (color version of figure is available online). LIHC: Liver Cancer; TCGA: The Cancer Genome Atlas.

3.2. Data Preprocessing and Screening of DEGs

All samples obtained from above were divided into high- and low-TMB groups according to the median TMB threshold, and we further evaluated the missing data and normalization for data preprocessing. The box chart results showed that similar levels of data points were achieved after correcting the mean value of the gene expression, and the data homogenization was credible (Figure 2A,B). The gene expression matrix was then merged for further normalization. The PCA results indicated that the clustering of samples was more obvious between the two groups after homogenization (Figure 2C,D), and the results suggested that the sample data source included in the present study was reliable and could be used for further analysis. After data preprocessing, we identified 2171 DEGs between high- and low-TMB groups with |Log FC| > 1 and p value < 0.05 through the limma package of R software. The result was presented via a volcano map (Figure 2E), in which green dots represent downregulated genes, red dots represent upregulated genes, and black dots represent unchanged genes.

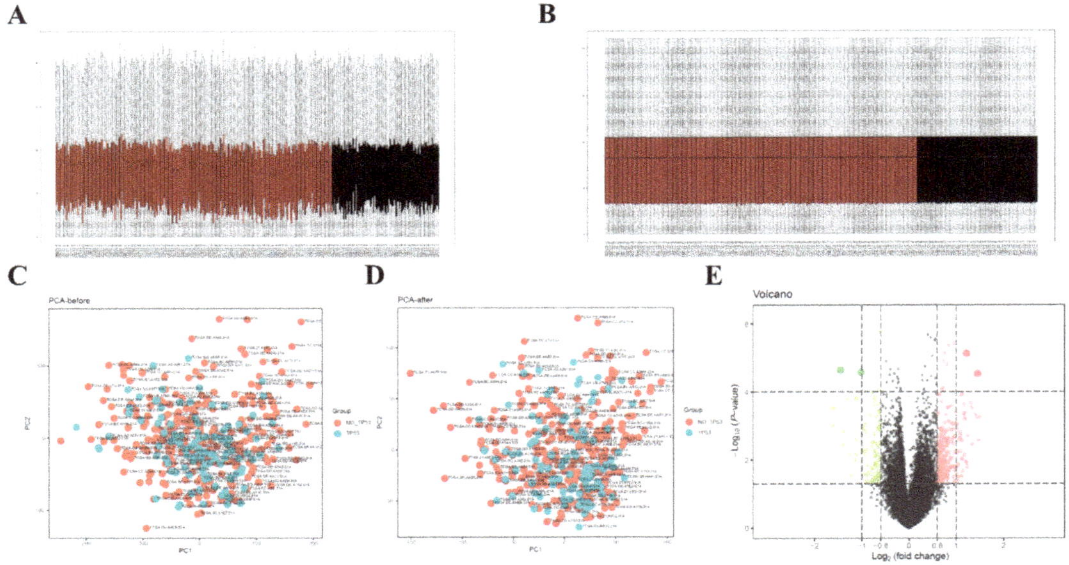

Figure 2. Data preprocessing and differential expression analysis. (**A**,**B**) Box chart of gene expression among high- and low-TMB groups. Black dots represent mean values of gene expression after sample normalization before (**A**) and after (**B**) sample normalization. (**C**,**D**) before (**C**) and after (**D**) principal component analyses (PCA) of gene expression between high- and low-TMB groups. (**E**) Volcano map of DEGs; red represents upregulated differential genes, green represents downregulated differential genes, and grey represents no-significant-difference genes. TMB: Tumor Mutation Burden.

3.3. Joint Screening of Genes, PPI Network Construction, Hub Genes Screening, and Similarities

In order to explore more accurate genes related to the prognosis of patients with HCC, intersection analysis was conducted on the identified differentially expressed genes between the high- and low-TMB groups, and the prognosis-related genes with p values < 0.05 in univariate Cox analysis were obtained from TCGA database. The combined results revealed that a total of 359 differentially expressed genes were identified following the intersection of 2171 DEGs between high- and low-TMB groups, with 2250 genes related to prognosis and survival (Figure 3A). Search Tool for the Retrieval of Interacting Genes (STRING) (http://string db.org; Version: 11.0) (accessed on 1 March 2021) is an online tool for predicting protein–protein interaction (PPI) networks. An analysis of functional interactions between proteins can provide more information into the mechanisms of disease occurrence or development. Through Cytoscape and its plug-in cytoHubba, we constructed the PPI network of DEGs related to prognosis obtained above (Figure 3B). The top 15 genes were selected as the hub genes through the MCC cytoHubba plugin with the highest correlation scores in this PPI network: *USP39*, *RBM22*, *SNRPD1*, *CPSF3*, *SRSF1*, *SRSF3*, *HSPA8*, *HNRNPU*, *SRSF4*, *CWC27*, *EFTUD2*, *ALYREF*, *NCBP2*, *SNRPA1*, and *POLR2D* (Figure 3C). To further explore the closeness of the correlation between hub DEGs, which were ranked on the basis of average functional similarity, the results suggested that *SRSF1*, *SNRPA1*, *SRSF3*, *SRSF4*, *ALYREF*, *NCBP2*, *SNRPD1*, and *EFTUD2* were found to be hub genes with cut-off values greater than 0.7, and *SRSF1*, *SNRPA1*, and *SRSF3* showed a strong similarity in biological effects (Figure 3D).

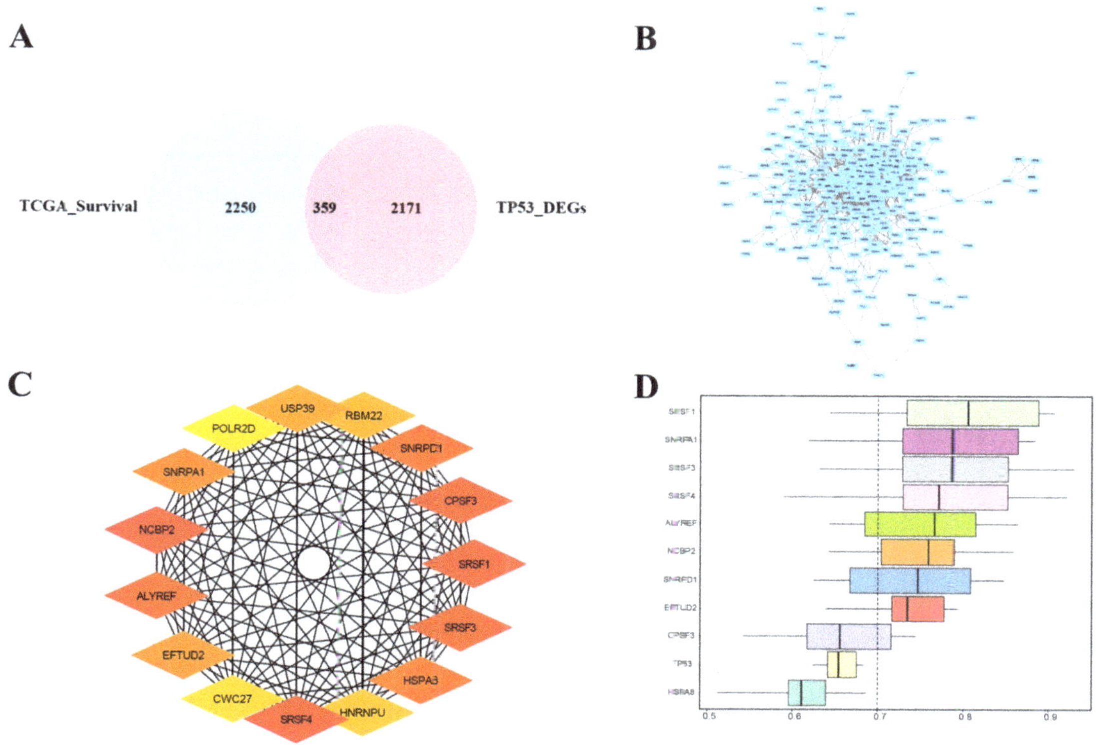

Figure 3. Joint screening of DEGs, Protein–protein interaction (PPI), hub DEGs, and functional similarity analysis of DEGs. (**A**) Venn diagram of DEGs between high- and low-TMB groups and the prognosis-related genes with *p* value less than 0.05 in Cox univariate analysis obtained from TCGA. Middle part represents overlap of two groups of data. (**B**) Gene interaction network of 359 prognosis-related DEGs visualized with PPI network. (**C**) Interaction network of 15 DEGs scored by maximum correlation coefficient; the darker the color, the higher the MCC algorithm score. (**D**) Functional similarities of 11 hub genes—dashed line represents cut-off value of similarity. DEGs: Differentially Expressed Genes; TCGA: The Cancer Genome Atlas; MCC: Matthews correlation coefficient.

3.4. Functional Correlation Analysis

A total of 359 differentially expressed genes related to prognosis in HCC samples were further subjected to GO analysis. The results suggested that in the biological process (BP) category, these prognosis-related differentially expressed genes were mainly correlated with RNA localization and the transport and export of components in the nucleus (Figure 4A). In order to explore the important pathway of enrichment between the two groups, the gene set enrichment analysis (GSEA) of gene expression profiles was used to identify differentially enriched signaling pathways between patients in high- and low-TMB groups. The results suggested that the enriched functions and pathways in the high-TMB group mainly involved fatty acid metabolism, hypoxia, and the P53 pathway (Figure 4B). The results of gene set variation analysis (GSVA) revealed that androgen response, coagulation, bile acid metabolism, angiogenesis, pancreas beta cells, fatty acid metabolism, TNFA signaling via NFKB and adipogenesis were enriched in the high-TMB group (Figure 4C).

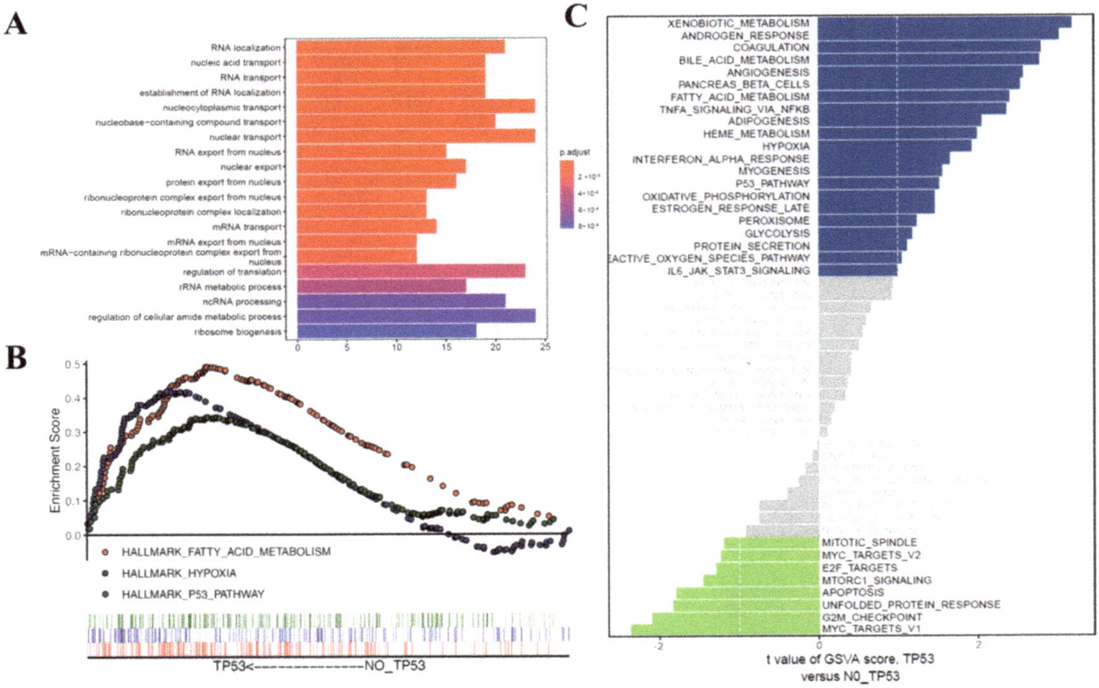

Figure 4. GO, GSEA, and GSVA analyses. (**A**) Significantly enriched gene ontology terms in categories BP. (**B**) GSEA analysis based on h.all.v7.1.symbols.gmt. (**C**) GSVA analysis based on h.all.v7.1.symbols.gmt. GO: Gene ontology; GSEA: gene set enrichment analysis; GSVA: gene set variation analysis.

3.5. The mRNA Expression Level of NCBP2 in Hepatocellular Carcinoma

To further explore the mRNA expression level of NCBP2 in hepatocellular carcinoma, we performed a verification to investigate the differential mRNA expression between HCC tumor samples and adjacent normal samples in the GEPIA2 (http://gepia2.cancer-pku.cn/) (accessed on 1 March 2021) and TIMER databases (https://cistrome.shinyapps.io/timer/) (accessed on 1 March 2021). As a result, the GEPIA-based analysis indicated that NCBP2 was upregulated in 17 of 33 cancer types, including hepatocellular carcinoma, which was computed in the form of transcripts per million compared with adjacent tissues (Figure 5A). In addition, the mRNA expression of NCBP2 was significantly different among different stages of HCC (F value = 0.53, Pr(>F) = 0.0014) (Figure 5B). Finally, we evaluated the NCBP2 mRNA expression using the RNA-seq data in TIMER database. The result also indicated that the mRNA expression of NCBP2 was overexpressed in hepatocellular carcinoma tissues compared with adjacent tissues, and NCBP2 mRNA expression was also overexpressed in other cancer types, such as BLCA (bladder urothelial carcinoma), BRCA (breast invasive carcinoma), CHOL (cholangiocarcinoma), COAD (colon adenocarcinoma), ESCA (esophageal carcinoma), GBM (glioblastoma multiforme), HNSC (head and neck squamous cell carcinoma), KIRP (kidney renal papillary cell carcinoma), LUAD (lung adenocarcinoma), LUSC (lung squamous cell carcinoma), PRAD (prostate adenocarcinoma), READ (rectum adenocarcinoma), STAD (stomach adenocarcinoma), and UCEC (uterine corpus endometrial carcinoma), but downregulated in KICH (kidney chromophobe) and KIRC (kidney renal clear cell carcinoma) (Figure 5C). In summary,

all these results indicate that the mRNA expression level of NCBP2 is significantly overexpressed in HCC.

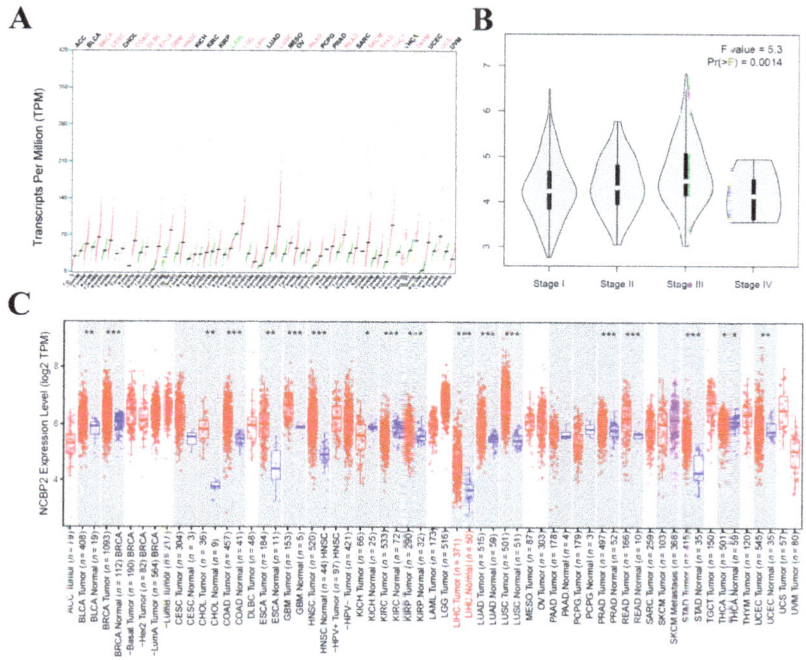

Figure 5. NCBP2 expression levels in HCC. (**A**) Expression patterns of NCBP2 in 33 cancer types and paired non-tumor samples. (**B**) Violin plots reveal relationship between NCBP2 expression and LIHC staging. (**C**) Human NCBP2 expression levels in different tumor types determined by TIMER (* $p < 0.05$, ** $p < 0.01$, *** $p < 0.001$). TIMER: Tumor Immune Estimation Resource.

3.6. Correlations between the mRNA Expression Level of NCBP2 and Survival in HCC Patients

To further investigate the relationship of the mRNA expression level of NCBP2 with the survival situation in HCC patients, the Kaplan–Meier Plotter, which is based on the transcriptome data mainly extracted from GEO, EGA, and TCGA, was used to assess the NCBP2-related survival rate. As a result, we firstly identified NCBP2 as a detrimental prognostic factor in LIHC (Overall Survival (OS): HR = 1.86, 95% CI from 1.31 to 2.63, log-rank $p = 4 \times 10^{4}$) (Figure 6A). Then, we further investigated the prognostic value of NCBP2 expression for pan-cancer in another database. The correlation between NCBP2 expression and the prognosis of each cancer were investigated, and the result suggested that NCBP2 expression was significantly related to a total of six cancer types, including KICH, KIRP, LICH, LUAD, PAAD, and PRAD (Figure 6B), and the expression level of NCBP2 was negatively correlated with over survival. Among those cancers, NCBP2 played a detrimental role in LIHC according to the GEPIA2 database (OS: total number = 364, HR = 1.9, log-rank $p = 0.00026$) (Figure 6C). In summary, we identified NCBP2 as a detrimental biomarker for the survival prognosis of HCC.

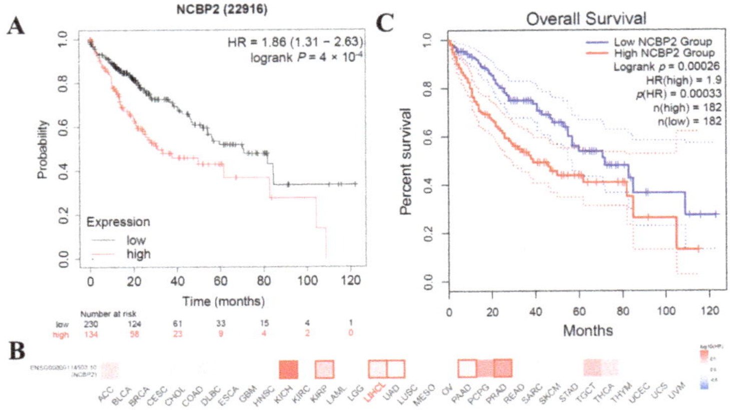

Figure 6. Kaplan–Meier survival curves comparing high and low expressions of NCBP2 in different databases. (**A**) Kaplan–Meier survival curves of LIHC in PrognoScan. (**B**) Relationship between NCBP2 expression and survival prognosis of each cancer in TCGA. (**C**) Kaplan–Meier survival curves of LIHC in Kaplan–Meier Plotter. Number at risk represent number of people exposed to outcome risk at each time point.

3.7. Protein Expression Level of NCBP2 in Human Tissue and Cell Lines

After investigating the mRNA expression pattern of NCBP2 in various databases, we further explored the protein expression pattern of NCBU2 in cell lines and human tissue in The Human Protein Atlas database (THPA), including tumor samples and normal adjacent specimens. The results confirmed that the protein level of NCBP2 was expressed moderately less in normal liver tissues compared with other normal tissues (Figure 7A), and the immunohistochemical analysis demonstrated that NCBP2 was overexpressed in HCC tissue relative to the normal adjacent sample (Figure 7B). The expression level of NCBP2 in liver cancer cell lines was analyzed using the CCLE online platform, and the result showed that the liver cancer cell lines with the highest expression of NCBP2 was from the HEP3B cell, and the lowest was from the JHH6 cell (Figure 7C).

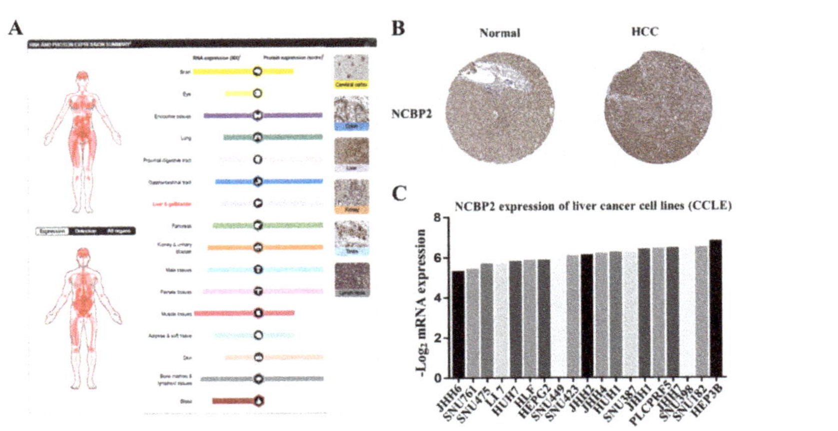

Figure 7. NCBP2 protein expression in human tissues and cell lines. (**A**) NCBP2 protein expression in normal human tissues based on The Human Protein Atlas (THPA). (**B**) NCBP2 expression assessed using immunohistochemistry in normal and liver cancer tissues. (**C**) NCBP2 gene expression profiles of 19 liver cancer cell lines based on Cancer Cell Line Encyclopedia (CCLE) database.

3.8. Relationship between the NCBP2 Expression and TP53 Mutation with Immune Makers

Immune infiltration was involved with hepatocellular carcinoma progression. Since NCBP2 expression was related to the prognostic of hepatocellular carcinoma, the relationship between 22 infiltrating immune cells and the NCBP2 expression was investigated by the TIMER database. The results suggested that, after adjustments for tumor purity, the NCBP2 expression was positively associated with all immune cells, including B cells ($r = 0.364$, $p = 3.30 \times 10^{-12}$), $CD8^{+}$ T cells ($r = 0.295$, $p = 2.71 \times 10^{-8}$), $CD4^{+}$ T cells, ($r = 0.484$, $p = 1.37 \times 10^{-21}$), macrophages ($r = 0.551$, $p = 1.97 \times 10^{-28}$), neutrophils ($r = 0.457$, $p = 3.26 \times 10^{-19}$), and dendritic cells ($r = 0.453$, $p = 1.97 \times 10^{-18}$) (Figure 8A). Intriguingly, we also found that the expression of NCBP2 was positively associated with TP53 (Figure 8B). After the prognosis of hepatocellular carcinoma related to the genetic mutations, among which TP53 represented a primary concern, we further investigated the relationship between the TP53 mutation and immune infiltration. The results showed that B cells and macrophages were significantly higher in the TP53 mutant than the wildtype; however, the rest of the immune cells, including $CD8^{+}$ T cells, $CD4^{+}$ T cells, neutrophils, and dendritic cells, were not statistically significant with TP53 (Figure 8C). We further analyzed the relationship between NCBP2 expression with macrophages and $CD4^{+}$ T cell infiltration levels in diverse cancer types using the TIMER 2.0 database. The results indicated that NCBP2 expression was positively correlated with the immune infiltration levels of macrophages (Figure 9A) and $CD4^{+}$ T cells (Figure 9B) across most tumor types, with the highest correlation shown in LIHC. Univariate and multivariate COX regression also showed that the stage of HCC, $CD8^{+}$ T cells, and the expression of NCBP2 were the independent indicators for predicting the prognosis of OS patients (Table 1).

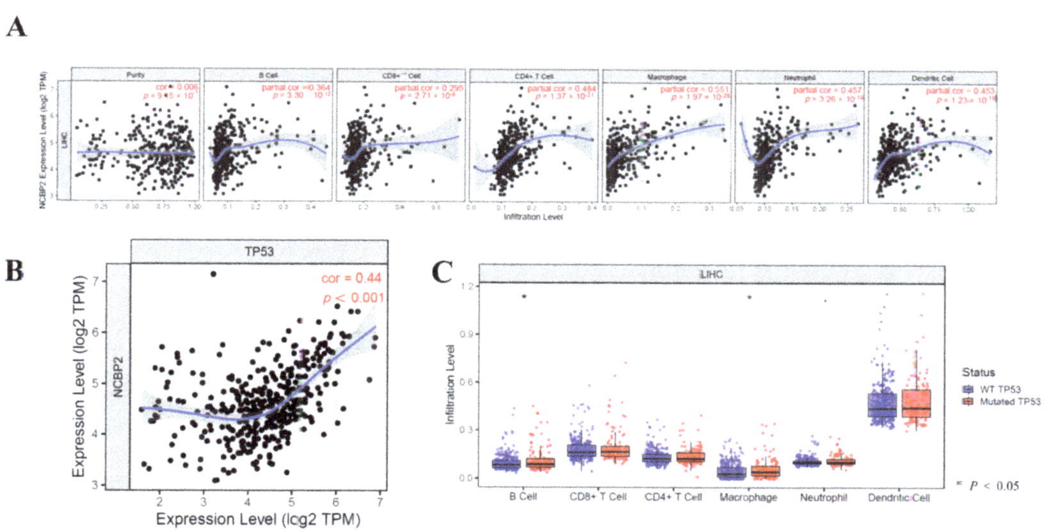

Figure 8. Correlation of NCBP2 expression and TP53 mutation with immune infiltration levels in LIHC. (**A**) Relationship of NCBP2 expression with immune infiltration. (**B**) Relationship of NCBP2 expression with TP53. (**C**) Correlation between TP53 mutation and immune infiltration.

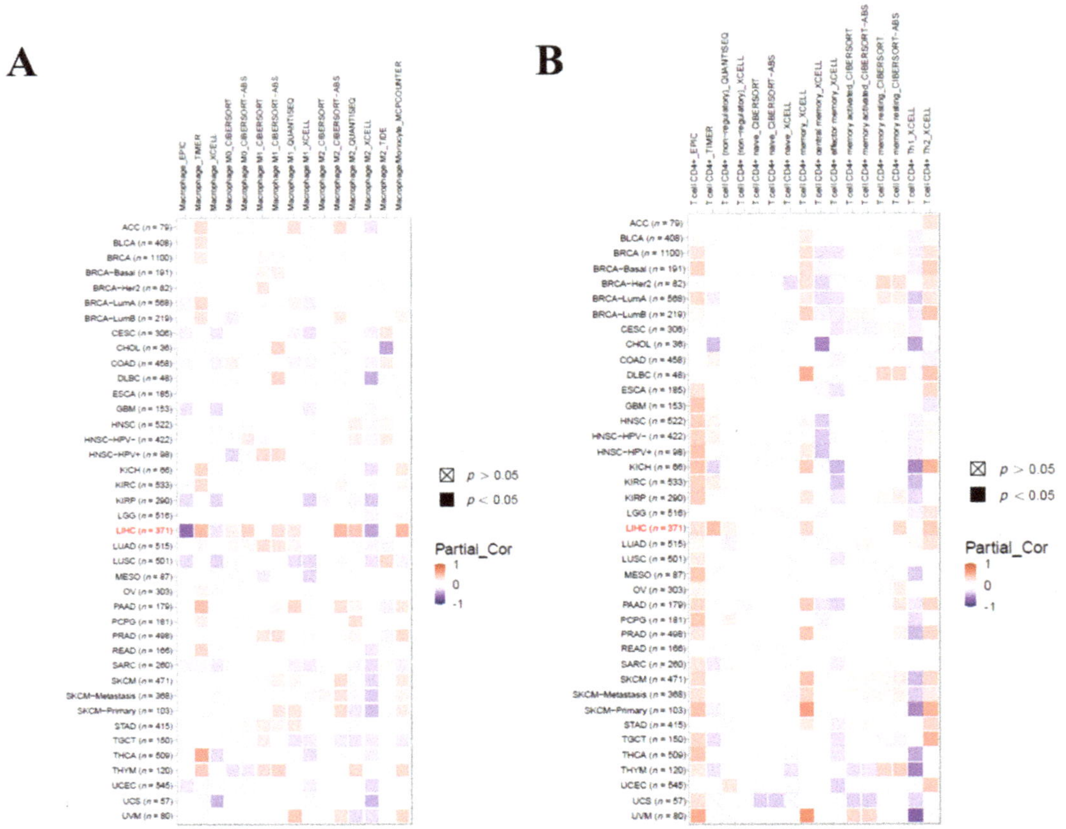

Figure 9. Relationship of NCBP2 expression with immune infiltration level in diverse cancer types (TIMER 2.0). (**A**) Macrophage immune infiltration level. (**B**) $CD4^+$ T-cell immune infiltration level.

Table 1. Univariate and multivariate Cox regressions on clinicopathological characteristics and NCBP2 expression signature.

Variables	Univariate Cox		Multivariate Cox	
	HR (95% CI)	*p* Value	HR (95% CI)	*p* Value
stage2	1.576	0.345	1.367	0.215
stage3	2.205	0.001 **	2.205	0.001 **
stage4	4.575	0.005 *	4.575	0.015 *
Gender male	0.907	0.789	0.907	0.632
B_cell	0.008	0.235	0.008	0.182
$CD8^+$ _ T cell	0.005	0.045 *	0.005	0.037 *
$CD4^+$ _ T cell	0.014	0.34	0.014	0.22
NCBP2	1.467	0.006 **	1.437	0.046 *

* $p < 0.05$; ** $p < 0.01$; CI: Confidence Interval; HR Hazard Ratio.

3.9. Immune Cell Infiltration Analysis in LIHC

Finally, we evaluated the infiltration of immune cells in LIHC. The results of the correlation heatmap between the 22 types of immune cells revealed that T cells had a significant positive correlation with cytotoxic cells and type 1 T-helper cells (Th1), and the macrophages and immature dendritic cells (iDC) also had a positive correlation. Type 2

T-helper cells (Th2) had a significant negative correlation with dendritic cells (DCs) and neutrophils, and the T-helper cells also had a negative correlation with DCs (Figure 10A). The immune cell interaction network results suggested that neutrophils, T cells, and follicular helper T cells (TFH) have strong relationships with other immune cells, but that regulatory cells (TReg) and plasmacytoid dendritic cells (pDC) have a weak relationship with other immune cells (Figure 10B). The violin plot of the immune cell infiltration results revealed that the degree of central memory T-cell (Tcm) infiltration was higher than in the low mutation frequencies of the TP53 samples ($p < 0.05$) (Figure 10C).

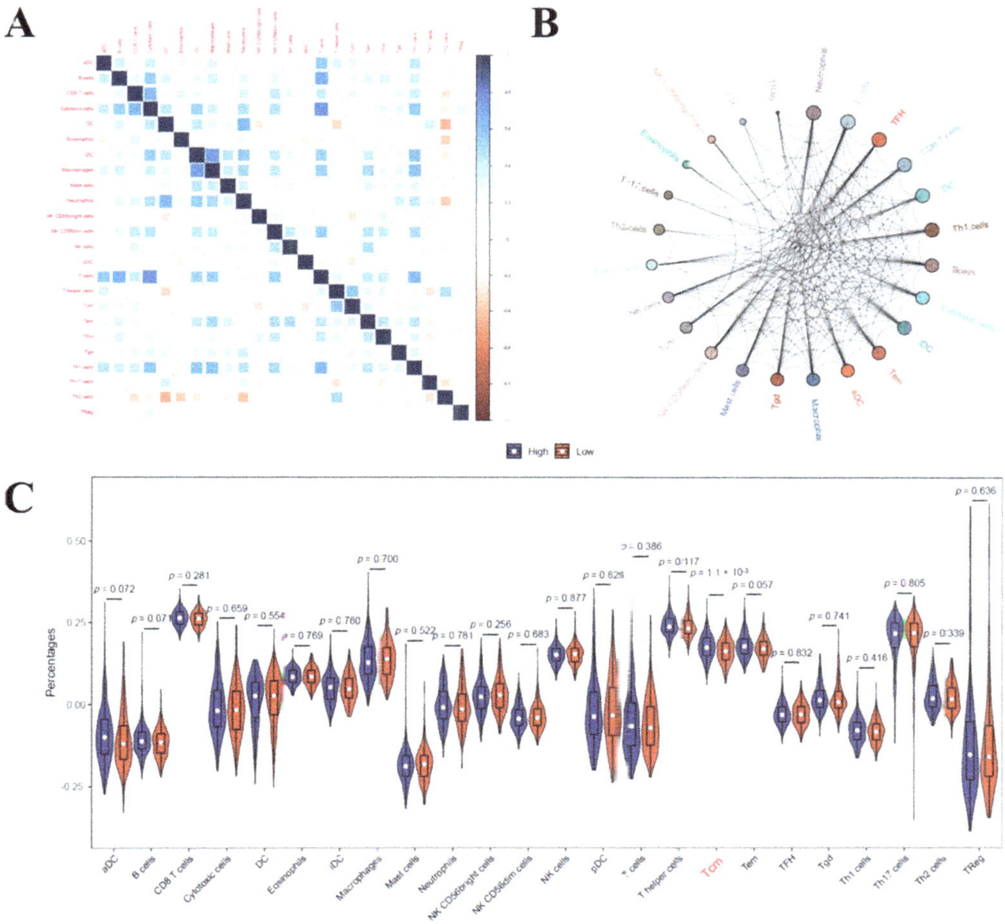

Figure 10. Correlation plots of immune cell infiltration analysis in LIHC. (**A**) Correlation heat map of 22 immune cells. Blue indicates positive correlation, red indicates negative correlation. Size of colored squares indicates strength of correlation. (**B**) Network diagram of 24 immune cell types. The circle size indicates the strength of interaction. (**C**) Violin diagram shows the difference of 24 types of immune cell infiltration in high mutation frequency of TP53 versus low mutation frequency of TP53.

4. Discussion

HCC is one of the most common malignant tumors. According to Global Cancer Statistics 2020, there were 906,000 new cases of HCC worldwide each year, causing about 830,000 deaths [2]. The main risk factors for HCC are chronic infection with the hepatitis B

(HBV) or C virus (HCV), alcoholic cirrhosis, aflatoxin-contaminated foods, and excess body weight [18,19]. Due to early detection and a systemic therapy of surgery combined with adjuvant chemotherapy, targeted treatment, or immunotherapy, the mortality rate of HCC has declined in the last three decades [4]. However, the 5-year survival rate of patients with advanced HCC is still low, which is mainly due to tumor advances [20]. Therefore, it is important to understand the molecular mechanisms underlying HCC to identify an effective target for prevention and treatment. Recent studies have focused on the relationships between HCC, TMBs, and immunity and have confirmed that HCC with a high tumor mutation burden (TMB-H) may generate immunogenic neoantigens. The increased production of neoantigens is positively related to the infiltration of immune cells, especially for the count of macrophages and $CD4^+$ and central memory T cells [21,22]. The infiltration changes of immune cells are the basis for a good response to immunotherapy [23]. However, none of these have been applied clinically; therefore, we used bioinformatics tools to analyze HCC-associated TMBs and to identify potential immune biomarkers for the diagnosis and prognosis of HCC.

In the present study, we performed a comprehensive biological analysis on the relationship between tumor somatic mutational profiles and immunity for HCC. To understand the functions and associations of these TMB-associated DEGs, GO analyses were performed. The result showed that DEGs are mainly enriched in nucleocytoplasmic and nuclear transport, and previous studies have confirmed that nucleocytoplasmic and nuclear transport are closely associated with the development of tumorigenesis [24,25]. Further studies have confirmed that nucleocytoplasmic and nuclear transport are closely related to HCC metastasis [26]. These studies suggested that the DEGs of TMBs may be closely correlated with the metastasis of HCC. By constructing a PPI network, we found that *USP39*, *RBM22*, *SNRPD1*, *CPSF3*, *SRSF1*, *SRSF3*, *HSPA8*, *HNRNPU*, *SRSF4*, *CWC27*, *EFTUD2*, *ALYREF*, *NCBP2*, *SNRPA1*, and *POLR2D* may play pivotal roles in the development of HCC. There was no research to investigate the relationship between HCC and the genes of *RBM22*, *SRSF4*, *CWC27*, and *POLR2D*, which would provide us a new research direction. In addition, we further used GO annotation semantics to investigate the functional similarity of key DEGs, and a strong biological functional similarity was found between *SRSF1*, *SNRPA1*, *SRSF3*, *SRSF4*, and *ALYREF*. *SNRPA1* was reported to promote HCC proliferation through activating the mTOR-signaling pathway [27], and the phosphorylation of *SRSF3* by *PPM1G* could result in the proliferation, invasion, and metastasis of HCC [28]; furthermore, *ALYREF* was significantly correlated to both advanced tumor-node-metastasis stages and poor HCC prognosis [29], which is similar to our results. However, we have not found any reports focused on the effects of *NCBP2* in HCC, which may have helped us to find new immunotherapy targets in HCC; however, it is worth considering for further investigation in future studies. In addition, the pathway enriched by GSEA mainly involved fatty acid metabolism, hypoxia, and the P53 pathway. Fatty acid metabolism was reported to be correlated with the advance of HCC and simultaneously influenced the infiltration of immune cells [30]. Both hypoxia and the mutation of P53 were also reported to lead to the metastasis of HCC [31,32]. The above studies are similar to our results, suggesting that the conclusions of the present study are accurate.

NCBP2, also known as *CBP20* or *NIP1*, can bind to the monomethylated 5′ cap of nascent pre-mRNA. *NCBP2* has an RNP domain usually found in RNA-binding proteins and contains the cap-binding activity [33,34]. It has been reported that *NCBP2* regulates proliferation, metastasis, and apoptosis in multiple cancers [35,36], and accumulating evidence suggests that *NCBP2* may serve as a biomarker for carcinogenesis and cancer progression. For example, NCBP2 was upregulated in an acute lymphoblastic leukemia rearrangement child patient (r ALL) compared with non-r ALL patients. Childhood ALL patients with high expressions of NCBP2 had significantly poorer overall survival rates [37]. The latest study revealed that NCBP2 was overexpressed in the high-risk group of acute myeloid leukemia (AML) and was negatively correlated with survival [38]. In the present study, the results showed that NCBP2 was upregulated in multiple cancers and played a detrimental role at

the LIHC stage, and NCBP2 expression was significantly related to another five cancers, including KICH, KIRP, LUAD, PAAD, and FRAD, and was negatively correlated with the over survival of those cancers. Moreover, the present study revealed that the expression of NCBP2 was significantly upregulated in HCC compared with adjacent liver tissues according to the Human Protein Atlas database, and NCBP2 played a detrimental role in the OS of HCC patients. The antisense gene protein NCBP2-AS2 (transcribed from the antisense DNA strand of the gene NCBP2) also plays an important role in multiple tumors. A study has revealed that NCBP2-AS2 was overexpressed in hypoxic-cancer-associated fibroblasts, and it can promote the secretion of pro-angiogenic factor VEGFA, consequently reducing VEGF/VEGFR downstream signaling, which leads to tumor metastasis and reduces the efficacy of therapy [39]. Furthermore, LncRNA NCBP2-AS2 was upregulated in lung squamous cell carcinoma samples compared with lung adenocarcinoma samples and adjacent tissues and promoted cell proliferation and metastasis, as well as the invasive and inhibited apoptosis of SCC cells via the TAp63/ZEB1-regulating pathway [40]. LncRNA NCBP2-AS2 also could promote HCC cell growth and proliferation through regulating KRASIM [41]. In conclusion, NCBP2 is overexpressed in multiple cancers compared with adjacent normal tissues, and high expressions of NCBP2 were significantly correlated with poor OS in HCC. However, further research is needed to establish diagnostic accuracy and treatment with NCBP2 in liver cancer.

To further investigate the role of immune cell infiltration in HCC, TIMER database analysis revealed that the NCBP2 expression was most positively correlated with macrophages (r = 0.551, p = 1.97 × 10^{-28}) and CD4$^+$ T cells (r = 0.484, p = 1.37 × 10^{-21}). Studies have demonstrated that by infiltrating tumor-associated macrophages (TAMs) at a high level in HCC, target TAM infiltration results in tumor growth inhibition in a mouse HCC model [42,43]. Higher infiltrating fractions of activated memory CD4$^+$ T cells were also found in high-risk groups of HCC patients [44,45]. These results showed that the expression level of NCBP2 may be associated with the immune response to the tumor microenvironment of HCC, especially with CD4$^+$ T cells and macrophages. In addition, our study investigates the details of 22 types of immune cell infiltrations in HCC, and the results showed that T cells were closely related to follicular helper T cells (TFH), whereas regulatory cells (TReg) showed the weakest interactions with plasmacytoid dendritic cells (pDC), which provided ideas for further investigations regarding the regulation mechanisms of HCC in immune cells, for which no research currently exists. The degree of central memory T cell (Tcm) infiltration was higher in the high-mutation-frequency TP53 samples. Accumulating research has demonstrated that the infiltration of Tcm may help to discover novel treatments for more effective cancer immunotherapies [46,47]. Tcm are functionally and phenotypically distinct monitoring points in the liver, capable of long-lived retention, and well positioned for rapid and potent front-line immunosurveillance [48]. The above studies, combined with our research, have shown that immune cells, especially CD4$^+$ T cells, macrophages, and central memory T cells, play important roles in HCC and should be the focus of further studies.

In summary, comprehensive bioinformatic analyses were performed to analyze the predictive value of TMB in HCC prognosis and identified that the expression of NCBP2 was strongly correlated to HCC prognosis. Moreover, immune cell infiltration investigations also suggested that immune cells, especially CD4$^+$ T cells, macrophages, and central memory T cells, play important roles in HCC. It is noteworthy that the systematic analysis of TMB-status hub genes in the present study will facilitate an understanding of the role played by TMEs in HCC and contribute to accurate immunotherapeutic treatment. Our findings may serve as a potential guide for targeted immunotherapy and provide ideas for the further development of new immunotherapies. Notwithstanding, more clinical studies and experimental research are needed to verify our findings and explore the molecular mechanisms of TMBs in HCC.

Author Contributions: Conceptualization, P.H.; data curation, T.H.; formal analysis, Y.Z.; funding acquisition, P.H.; investigation, Y.Z.; methodology, T.H.; project administration, P.H.; software, Y.Z.; supervision, P.H.; validation, T.H.; writing—original draft, Y.Z.; writing—review and editing, T.H. All authors have read and agreed to the published version of the manuscript.

Funding: This research received no external funding.

Institutional Review Board Statement: Not applicable.

Informed Consent Statement: Not applicable.

Data Availability Statement: TCGA-LIHC dataset is available at The Cancer Genome Atlas (https://cancergenome.nih.gov/).

Conflicts of Interest: The authors declare no conflict of interest.

References

1. McGlynn, K.A.; Petrick, J.L.; El-Serag, H.B. Epidemiology of Hepatocellular Carcinoma. *Hepatology* **2021**, *73* (Suppl. 1), 4–13. [CrossRef] [PubMed]
2. Sung, H.; Ferlay, J.; Siegel, R.L.; Laversanne, M.; Soerjomataram, I.; Jemal, A.; Bray, F. Global Cancer Statistics 2020: GLOBOCAN Estimates of Incidence and Mortality Worldwide for 36 Cancers in 185 Countries. *CA Cancer J. Clin.* **2021**, *71*, 209–249. [CrossRef] [PubMed]
3. Kanwal, F.; Singal, A.G. Surveillance for Hepatocellular Carcinoma: Current Best Practice and Future Direction. *Gastroenterology* **2019**, *157*, 54–64. [CrossRef]
4. Forner, A.; Reig, M.; Bruix, J. Hepatocellular carcinoma. *Lancet* **2018**, *391*, 1301–1314. [CrossRef]
5. Llovet, J.M.; Kelley, R.K.; Villanueva, A.; Singal, A.G.; Pikarsky, E.; Roayaie, S.; Lencioni, R.; Koike, K.; Zucman-Rossi, J.; Finn, R.S. Hepatocellular carcinoma. *Nat. Rev. Dis. Primers* **2021**, *7*, 6. [CrossRef]
6. Balague-Dobon, L.; Caceres, A.; Gonzalez, J.R. Fully exploiting SNP arrays: A systematic review on the tools to extract underlying genomic structure. *Brief. Bioinform.* **2022**, *23*, bbac043. [CrossRef]
7. Vokes, N.I.; Chambers, E.; Nguyen, T.; Coolidge, A.; Lydon, C.A.; Le, X.; Sholl, L.; Heymach, J.V.; Nishino, M.; Van Allen, E.M.; et al. Concurrent TP53 mutations facilitate resistance evolution in EGFR mutant lung adenocarcinoma. *J. Thorac. Oncol.* **2022**, *17*, 779–792. [CrossRef]
8. Jang, J.W.; Kim, J.S.; Kim, H.S.; Tak, K.Y.; Lee, S.K.; Nam, H.C.; Sung, P.S.; Kim, C.M.; Park, J.Y.; Bae, S.H.; et al. Significance of TERT Genetic Alterations and Telomere Length in Hepatocellular Carcinoma. *Cancers* **2021**, *13*, 2160. [CrossRef]
9. Shao, W.; Ding, Q.; Guo, Y.; Xing, J.; Huo, Z.; Wang, Z.; Xu, Q.; Guo, Y. A Pan-Cancer Landscape of HOX-Related lncRNAs and Their Association with Prognosis and Tumor Microenvironment. *Front. Mol. Biosci.* **2021**, *8*, 767856. [CrossRef]
10. McGrail, D.J.; Pilie, P.G.; Rashid, N.U.; Voorwerk, L.; Slagter, M.; Kok, M.; Jonasch, E.; Khasraw, M.; Heimberger, A.B.; Lim, B.; et al. High tumor mutation burden fails to predict immune checkpoint blockade response across all cancer types. *Ann. Oncol.* **2021**, *32*, 661–672. [CrossRef]
11. Yarchoan, M.; Hopkins, A.; Jaffee, E.M. Tumor Mutational Burden and Response Rate to PD-1 Inhibition. *N. Engl. J. Med.* **2017**, *377*, 2500–2501. [CrossRef] [PubMed]
12. Klempner, S.J.; Fabrizio, D.; Bane, S.; Reinhart, M.; Peoples, T.; Ali, S.M.; Sokol, E.S.; Frampton, G.; Schrock, A.B.; Anhorn, R.; et al. Tumor Mutational Burden as a Predictive Biomarker for Response to Immune Checkpoint Inhibitors: A Review of Current Evidence. *Oncologist* **2020**, *25*, e147–e159. [CrossRef] [PubMed]
13. Brown, Z.J.; Heinrich, B.; Greten, T.F. Mouse models of hepatocellular carcinoma: An overview and highlights for immunotherapy research. *Nat. Rev. Gastroenterol. Hepatol.* **2018**, *15*, 536–554. [CrossRef] [PubMed]
14. Wang, L.; Liu, Z.; Liu, L.; Guo, C.; Jiao, D.; Li, L.; Zhao, J.; Han, X.; Sun, Y. CELF2 is a candidate prognostic and immunotherapy biomarker in triple-negative breast cancer and lung squamous cell carcinoma: A pan-cancer analysis. *J. Cell Mol. Med.* **2021**, *25*, 7559–7574. [CrossRef]
15. Teo, M.Y.; Seier, K.; Ostrovnaya, I.; Regazzi, A.M.; Kania, B.E.; Moran, M.M.; Cipolla, C.K.; Bluth, M.J.; Chaim, J.; Al-Ahmadie, H.; et al. Alterations in DNA Damage Response and Repair Genes as Potential Marker of Clinical Benefit From PD-1/PD-L1 Blockade in Advanced Urothelial Cancers. *J. Clin. Oncol.* **2018**, *36*, 1685–1694. [CrossRef]
16. Hellmann, M.D.; Ciuleanu, T.E.; Pluzanski, A.; Lee, J.S.; Otterson, G.A.; Audigier-Valette, C.; Minenza, E.; Linardou, H.; Burgers, S.; Salman, P.; et al. Nivolumab plus Ipilimumab in Lung Cancer with a High Tumor Mutational Burden. *N. Engl. J. Med.* **2018**, *378*, 2093–2104. [CrossRef]
17. Chan, T.A.; Yarchoan, M.; Jaffee, E.; Swanton, C.; Quezada, S.A.; Stenzinger, A.; Peters, S. Development of tumor mutation burden as an immunotherapy biomarker: Utility for the oncology clinic. *Ann. Oncol.* **2019**, *30*, 44–56. [CrossRef]
18. Kim, H.N.; Newcomb, C.W.; Carbonari, D.M.; Roy, J.A.; Torgersen, J.; Althoff, K.N.; Kitahata, M.M.; Reddy, K.R.; Lim, J.K.; Silverberg, M.J.; et al. Risk of HCC with Hepatitis B Viremia Among HIV/HBV-Coinfected Persons in North America. *Hepatology* **2021**, *74*, 1190–1202. [CrossRef]

19. Kanwal, F.; Khaderi, S.; Singal, A.G.; Marrero, J.A.; Loo, N.; Asrani, S.K.; Amos, C.I.; Thrift, A.P.; Gu, X.; Luster, M.; et al. Risk Factors for Hepatocellular Cancer in Contemporary Cohorts of Patients with Cirrhosis. *Hepatology* **2022**. [CrossRef]
20. Ahn, J.C.; Teng, P.C.; Chen, P.J.; Posadas, E.; Tseng, H.R.; Lu, S.C.; Yang, J.D. Detection of Circulating Tumor Cells and Their Implications as a Biomarker for Diagnosis, Prognostication, and Therapeutic Monitoring in Hepatocellular Carcinoma. *Hepatology* **2021**, *73*, 422–436. [CrossRef]
21. Rizzo, A.; Ricci, A.D. PD-L1, TMB, and other potential predictors of response to immunotherapy for hepatocellular carcinoma: How can they assist drug clinical trials? *Expert Opin. Investig. Drugs* **2021**, *31* 415–423. [CrossRef] [PubMed]
22. Liu, T.; Tan, J.; Wu, M.; Fan, W.; Wei, J.; Zhu, B.; Guo, J.; Wang, S.; Zhou, P.; Zhang, H.; et al. High-affinity neoantigens correlate with better prognosis and trigger potent antihepatocellular carcinoma (HCC) activity by activating CD39(+)CD8(+) T cells. *Gut* **2021**, *70*, 1965–1977. [CrossRef] [PubMed]
23. Bersanelli, M. Tumour mutational burden as a driver for treatment choice in resistant tumours (and beyond). *Lancet Oncol.* **2020**, *21*, 1255–1257. [CrossRef]
24. Nataraj, N.B.; Noronha, A.; Lee, J.S.; Ghosh, S.; Mohan, R.H.; Sekar, A.; Zuckerman, B.; Lindzen, M.; Tarcitano, E.; Srivastava, S.; et al. Nucleoporin-93 reveals a common feature of aggressive breast cancers: Robust nucleocytoplasmic transport of transcription factors. *Cell Rep.* **2022**, *38*, 110418. [CrossRef] [PubMed]
25. Li, Y.; Huang, Y.; Ren, S.; Xiao, X.; Cao, H.; He, J. A Pan-Cancer Analysis of the Oncogenic Role of Nuclear Transport Factor 2 in Human Cancers. *Front. Oncol.* **2022**, *12*, 829389. [CrossRef]
26. Zhong, F.J.; Sun, B.; Cao, M.M.; Xu, C.; Li, Y.M.; Yang, L.Y. STMN2 mediates nuclear translocation of Smad2/3 and enhances TGFbeta signaling by destabilizing microtubules to promote epithelial-mesenchymal transition in hepatocellular carcinoma. *Cancer Lett.* **2021**, *506*, 128–141. [CrossRef]
27. Feng, J.; Guo, J.; Zhao, P.; Shen, J.; Chai, B.; Wang, J. mTOR up-regulation of SNRPA1 contributes to hepatocellular carcinoma development. *Biosci. Rep.* **2020**, *40*, BSR20193815. [CrossRef]
28. Chen, D.; Zhao, Z.; Chen, L.; Li, Q.; Zou, J.; Liu, S. PPM1G promotes the progression of hepatocellular carcinoma via phosphorylation regulation of alternative splicing protein SRSF3. *Cell Death Dis.* **2021**, *12*, 722. [CrossRef]
29. Xue, C.; Zhao, Y.; Li, G.; Li, L. Multi-Omic Analyses of the m(5)C Regulator ALYREF Reveal Its Essential Roles in Hepatocellular Carcinoma. *Front. Oncol.* **2021**, *11*, 633415. [CrossRef]
30. Behary, J.; Amorim, N.; Jiang, X.T.; Raposo, A.; Gong, L.; McGovern, E.; Ibrahim, R.; Chu, F.; Stephens, C.; Jebeili, H.; et al. Gut microbiota impact on the peripheral immune response in non-alcoholic fatty liver disease related hepatocellular carcinoma. *Nat. Commun.* **2021**, *12*, 187. [CrossRef]
31. Liu, X.; Zhang, X.; Peng, Z.; Li, C.; Wang, Z.; Wang, C.; Deng, Z.; Wu, B. Cui, Y.; Wang, Z.; et al. Deubiquitylase OTUD6B Governs pVHL Stability in an Enzyme-Independent Manner and Suppresses Hepatocellular Carcinoma Metastasis. *Adv. Sci.* **2020**, *7*, 1902040. [CrossRef] [PubMed]
32. Luo, Y.D.; Fang, L.; Yu, H.Q.; Zhang, J.; Lin, X.T.; Liu, X.Y.; Wu, D.; Li, G.X.; Huang, D.; Zhang, Y.J.; et al. p53 haploinsufficiency and increased mTOR signalling define a subset of aggressive hepatocellular carcinoma. *J. Hepatol.* **2021**, *74*, 96–108. [CrossRef] [PubMed]
33. Singh, M.D.; Jensen, M.; Lasser, M.; Huber, E.; Yusuff, T.; Pizzo, L.; Lifschutz, B.; Desai, I.; Kubina, A.; Yennawar, S.; et al NCBP2 modulates neurodevelopmental defects of the 3q29 deletion in Drosophila and Xenopus laevis models. *PLoS Genet.* **2020**, *16*, e1008590. [CrossRef] [PubMed]
34. Gebhardt, A.; Bergant, V.; Schnepf, D.; Moser, M.; Meiler, A.; Togbe, D.; Mackowiak, C.; Reinert, L.S.; Paludan, S.R. Ryffel, B.; et al. The alternative cap-binding complex is required for antiviral defense in vivo. *PLoS Pathog.* **2019**, *15*, e1008155. [CrossRef]
35. Hu, G.; Jiang, Q.; Liu, L.; Peng, H.; Wang, Y.; Li, S.; Tang, Y.; Yu, J.; Yang, J.; Liu, Z. Integrated Analysis of RNA-Binding Proteins Associated With the Prognosis and Immunosuppression in Squamous Cell Carcinoma of Head and Neck. *Front. Genet.* **2020**, *11*, 571403. [CrossRef]
36. Nastase, A.; Lupo, A.; Laszlo, V.; Damotte, D.; Dima, S.; Canny, E.; Alifano, M.; Popescu, I.; Klepetko, W.; Grigoroiu, M. Platinum Drug Sensitivity Polymorphisms in Stage III Non-small Cell Lung Cancer with Invasion of Mediastinal Lymph Nodes. *Cancer Genom. Proteom.* **2020**, *17*, 587–595. [CrossRef]
37. Wang, L.L.; Yan, D.; Tang, X.; Zhang, M.; Liu, S.; Wang, Y.; Zhang, M.; Zhou, G.; Li, T.; Jiang, F.; et al. High Expression of BCL11A Predicts Poor Prognosis for Childhood MLL-r ALL. *Front. Oncol.* **2021**, *11*, 755188. [CrossRef]
38. Zhang, H.; Cheng, L.; Liu, C. Regulatory Networks of Prognostic mRNAs in Pediatric Acute Myeloid Leukemia. *J. Healthc. Eng.* **2022**, *2022*, 2691997. [CrossRef]
39. Kugeratski, F.G.; Atkinson, S.J.; Neilson, L.J.; Lilla, S.; Knight, J.; Serneels, J.; Juin, A.; Ismail, S.; Bryant, D.M.; Markert, E.K.; et al. Hypoxic cancer-associated fibroblasts increase NCBP2-AS2/HIAR to promote endothelial sprouting through enhanced VEGF signaling. *Sci. Signal.* **2019**, *12*, eaan8247. [CrossRef]
40. Zhang, H.Y.; Yang, W.; Zheng, F.S.; Wang, Y.B.; Lu, J.B. Long non-coding RNA SNHG1 regulates zinc finger E-box binding homeobox 1 expression by interacting with TAp63 and promotes cell metastasis and invasion in Lung squamous cell carcinoma. *Biomed. Pharmacother.* **2017**, *90*, 650–658. [CrossRef]

41. Xu, W.; Deng, B.; Lin, P.; Liu, C.; Li, B.; Huang, Q.; Zhou, H.; Yang, J.; Qu, L. Ribosome profiling analysis identified a KRAS-interacting microprotein that represses oncogenic signaling in hepatocellular carcinoma cells. *Sci. China Life Sci.* **2020**, *63*, 529–542. [CrossRef] [PubMed]
42. Chen, J.; Lin, Z.; Liu, L.; Zhang, R.; Geng, Y.; Fan, M.; Zhu, W.; Lu, M.; Lu, L.; Jia, H.; et al. GOLM1 exacerbates CD8(+) T cell suppression in hepatocellular carcinoma by promoting exosomal PD-L1 transport into tumor-associated macrophages. *Signal Transduct. Target. Ther.* **2021**, *6*, 397. [CrossRef] [PubMed]
43. Jiang, J.; Mei, J.; Yi, S.; Feng, C.; Ma, Y.; Liu, Y.; Liu, Y.; Chen, C. Tumor associated macrophage and microbe: The potential targets of tumor vaccine delivery. *Adv. Drug Deliv. Rev.* **2022**, *180*, 114046. [CrossRef] [PubMed]
44. Tang, B.; Zhu, J.; Zhao, Z.; Lu, C.; Liu, S.; Fang, S.; Zheng, L.; Zhang, N.; Chen, M.; Xu, M.; et al. Diagnosis and prognosis models for hepatocellular carcinoma patient's management based on tumor mutation burden. *J. Adv. Res.* **2021**, *33*, 153–165. [CrossRef] [PubMed]
45. Zheng, B.; Wang, D.; Qiu, X.; Luo, G.; Wu, T.; Yang, S.; Li, Z.; Zhu, Y.; Wang, S.; Wu, R.; et al. Trajectory and Functional Analysis of PD-1(high) CD4(+)CD8(+) T Cells in Hepatocellular Carcinoma by Single-Cell Cytometry and Transcriptome Sequencing. *Adv. Sci.* **2020**, *7*, 2000224. [CrossRef]
46. Park, S.L.; Gebhardt, T.; Mackay, L.K. Tissue-Resident Memory T Cells in Cancer Immunosurveillance. *Trends Immunol.* **2019**, *40*, 735–747. [CrossRef]
47. Okla, K.; Farber, D.L.; Zou, W. Tissue-resident memory T cells in tumor immunity and immunotherapy. *J. Exp. Med.* **2021**, *218*, e20201605. [CrossRef]
48. Pinato, D.J.; Guerra, N.; Fessas, P.; Murphy, R.; Mineo, T.; Mauri, F.A.; Mukherjee, S.K.; Thursz, M.; Wong, C.N.; Sharma, R.; et al. Immune-based therapies for hepatocellular carcinoma. *Oncogene* **2020**, *39*, 3620–3637. [CrossRef]

cancers

MDPI

Article

Clinically Applicable Pathological Diagnosis System for Cell Clumps in Endometrial Cancer Screening via Deep Convolutional Neural Networks

Qing Li [1,2,†], Ruijie Wang [3,†], Zhonglin Xie [3], Lanbo Zhao [1], Yiran Wang [1], Chao Sun [1], Lu Han [1], Yu Liu [4], Huilian Hou [4], Chen Liu [2], Guanjun Zhang [4], Guizhi Shi [5], Dexing Zhong [3,6,7,*] and Qiling Li [1,2,*]

1 Department of Obstetrics and Gynecology, The First Affiliated Hospital of Xi'an Jiaotong University, Xi'an 710061, China
2 Department of Obstetrics and Gynecology, Northwest Women's and Children's Hospital, Xi'an 710061, China
3 School of Automation Science and Engineering, Xi'an Jiaotong University, Xi'an 710049, China
4 Department of Pathology, The First Affiliated Hospital of Xi'an Jiaotong University, Xi'an 710061, China
5 Laboratory Animal Center, Institute of Biophysics, Chinese Academy of Sciences, University of Chinese Academy of Sciences, Beijing 100101, China
6 State Key Laboratory for Novel Software Technology, Nanjing University, Nanjing 210093, China
7 Pazhou Lab, Guangzhou 510335, China
* Correspondence: bell@xjtu.edu.cn (D.Z.); liqiling@mail.xjtu.edu.cn (Q.L.); Tel.: +86-029-8532-3849 (D.Z.); +86-029-8266-8665 (Q.L.); Fax: +86-029-8821-4667 (D.Z.); +86-029-8266-8665 (Q.L.)
† These authors contributed equally to this work.

Simple Summary: The soaring demand for endometrial cancer screening has exposed a huge shortage of cytopathologists worldwide. Deep learning algorithms, based on convolutional neural networks, have been successfully applied to the classification and segmentation of medical images. The aim was to establish an artificial intelligence system that automatically recognizes and diagnoses pathological images of endometrial cell clumps (ECCs). Total 39,000 ECCs (26,880 for training, 11,520 for testing and 600 malignant for verification) patches were obtained by the segmentation network. The training set reached 100% accuracy, the testing set gained 93.5% accuracy, 92.2% specificity, and 92.0% sensitivity. Therefore, an artificial intelligence system was successfully built to classify malignant and benign ECCs for reducing pathologists' workload, providing decision-making assistance and promoting the development of endometrial cancer screening.

Abstract: Objectives: The soaring demand for endometrial cancer screening has exposed a huge shortage of cytopathologists worldwide. To address this problem, our study set out to establish an artificial intelligence system that automatically recognizes and diagnoses pathological images of endometrial cell clumps (ECCs). Methods: We used Li Brush to acquire endometrial cells from patients. Liquid-based cytology technology was used to provide slides. The slides were scanned and divided into malignant and benign groups. We proposed two (a U-net segmentation and a DenseNet classification) networks to identify images. Another four classification networks were used for comparison tests. Results: A total of 113 (42 malignant and 71 benign) endometrial samples were collected, and a dataset containing 15,913 images was constructed. A total of 39,000 ECCs patches were obtained by the segmentation network. Then, 26,880 and 11,520 patches were used for training and testing, respectively. On the premise that the training set reached 100%, the testing set gained 93.5% accuracy, 92.2% specificity, and 92.0% sensitivity. The remaining 600 malignant patches were used for verification. Conclusions: An artificial intelligence system was successfully built to classify malignant and benign ECCs.

Keywords: endometrial cancer; deep learning; screening; pathological diagnosis system; cell clumps

Citation: Li, Q.; Wang, R.; Xie, Z.; Zhao, L.; Wang, Y.; Sun, C.; Han, L.; Liu, Y.; Hou, H.; Liu, C.; et al. Clinically Applicable Pathological Diagnosis System for Cell Clumps in Endometrial Cancer Screening via Deep Convolutional Neural Networks. *Cancers* **2022**, *14*, 4109. https://doi.org/10.3390/cancers14174109

Academic Editor: David Wong

Received: 26 July 2022
Accepted: 22 August 2022
Published: 25 August 2022

1. Introduction

Endometrial cancer (EC) has become the second most common malignant tumor in the female reproductive system, with about 378,400 new cases in 2018 worldwide [1]. With increasing life expectancy and altered living habits, the incidence of EC is on the rise, and patients tend to be younger [2,3]. The 5-year survival rate with appropriate treatment is more than 85% for localized, 49% to 71% for regional, and less than 17% for distant stages of EC [4]. Women exposed to high risks have been recommended to be screened. Screening for EC and precancerous changes has been strongly suggested for early diagnosis and to reduce morbidity and mortality [5].

Researchers on the early detection of EC focus on minimally invasive histopathologic and cytopathologic procedures [6]. An endometrial cytologic test (ECT) has been carried out in many countries, including Italy, the United States, and Japan. ECT was added into Japanese Law on health care for the elderly in 1987. The mortality from EC among Japanese high-risk women fell from 20% in 1950 to 8% in 1999 [7]. In the past 20 years, academics from different regions have put forward the invention and improvement of endometrial samplers and have recommended diagnosis systems for endometrial cytopathology [8–10]. Confirmed by diagnostic curettage, the sensitivity, specificity, and coincidence rate of a well-designed endometrial sampling device, Li Brush, were 92.73%, 98.15%, and 92.73%, respectively [11]. On the other hand, a large number of endometrial cytopathological slides need to be identified, which exposes the lack of pathologists.

With the development of artificial intelligence (AI) technology and the improvement of hardware computing power in recent years, deep learning (DL) in medical analysis is considered as a third eye for doctors [12]. DL algorithms, based on deep convolutional neural networks (CNNs), have been proven to strongly boost the development of biomedical image analysis [13,14]. CNNs are becoming a reference tool for pathologists and have been successfully applied to the classification and segmentation of medical images, reducing the workload of pathologists and providing decision-making assistance [15–17].

AI has been successfully used in recognizing pathologic images and identifying malignant and benign tumors. However, there are relatively few studies on EC recognition. In one study, a computer-aided morphology program was established to distinguish benign and malignant cells. Geometric and densitometric nuclear features were measured for analysis. However, the typical three-dimensional shape (crowded and overlapping nuclei) of the endometrium increased miscalculation [18]. In another experiment, an endometrial histopathological AI recognition system was built, though it had a relatively high false-negative rate because a few subtle features were undetectable at the cellular level [19]. Inspired by these studies, we developed a recognition system based on CNNs to automatically identify benign and malignant endometrial cell clumps (ECCs). The shortcomings of the two above studies will be overcome by analyzing the cellular clump's structure and cytological characteristics.

2. Materials and Methods

2.1. Ethics Statement and Patients

The patients, who underwent curettage or hysterectomy, were recruited in the First Affiliated Hospital of Xi'an Jiaotong University from July 2015 to July 2020. This study was approved by the Ethics Committee of the First Affiliated Hospital of Xi'an Jiaotong University (XJTU1AHCR2014-007), and all patients signed written informed consent. The protocols were in compliance with the ethical principles for research that involves human subjects of the Helsinki Declaration for medical research [20].

Patients were excluded who had been diagnosed with suspected pregnancy or pregnancy, acute inflammation of the reproductive system, cervical cancer, or dysfunctional clotting diseases. Women with body temperature at or more than 37.5 °C were also excluded after being measured twice a day.

2.2. Preparation of Pathological Slides

We chose Li Brush (20152660054, Xi'an Meijiajia Medical Technology Co., Ltd., China) for endometrial cytological sampling (Figure 1a). Liquid-based cytology combined with Hematoxylin and Eosin staining was used for pathological slides of endometrial cells. The sampling, pathological slide, and staining procedures were described by Lu Han et al. [11]. Based on the endometrial cytological diagnostic criteria proposed by Chinese Expert Consensus [21], two experienced pathological professors (H.H. and G.S., with over 20 years of endometrial cytopathology experience) labeled all cytopathological slides and divided them into two classes: malignant (atypical cells of undetermined significance, suspected malignant tumor cells, and malignant tumor cells), and benign (non-malignant tumor cells). Slides with fewer than 10 or 5 ECCs were judged to be "unsatisfactory for evaluation" for premenopausal or postmenopausal women, respectively. Only a few isolated atypical or cancerous cells present were considered as satisfactory [22]. Histopathological diagnosis, acquired from the endometrium by curettage or hysterectomy, was regarded as the gold standard. Normal endometrium and endometrial hyperplasia without atypia were considered as benign; endometrial atypical hyperplasia and endometrial cancer were malignant. Only when consistent classification was reached between histology and the two pathologists' cytology on a sample was the sample considered for the study. Otherwise, it was suspended [22].

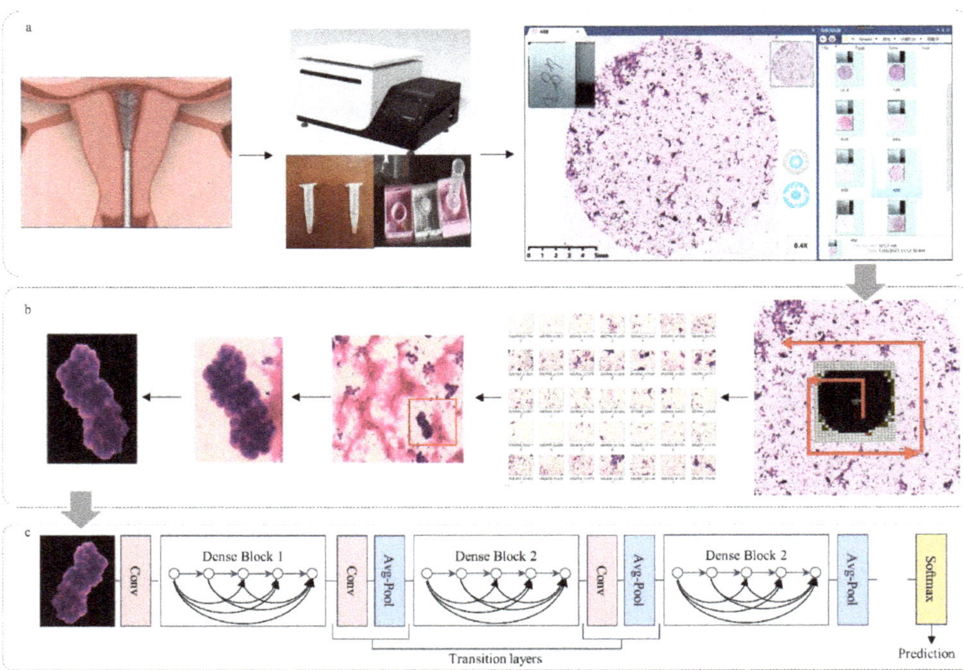

Figure 1. The process of obtaining images and recognition. (**a**) Sampling procedure; (**b**) cytological slides diagnosis; (**c**) classification using endometrial cytological images feature.

2.3. Cytopathological Image Acquisition

We used a MOTIC digital biopsy scanner (EasyScan 60, 20192220065, Motic, Xiamen, China) to scan cytopathological slides (Figure 1b), using a lens with 200 times magnification (20×) to obtain whole slide images. A counterclockwise spiral scan was performed with a camera exposure time of 0.65 s per slide and automatic focal adjustment. Each scanned slide image was segmented into 1360 small images (1816 × 1519 pixels) (Figure 1b).

2.4. ECCs Image Annotation

Adobe Photoshop CC (2019 v20.0.2.30, Adobe Inc., San Jose, CA, USA) was engaged to sketch the edge of the ECCs. There is no doubt that ECCs from negative slides were all negative, but some ECCs were negative in positive slides. Thus, the two pathologists voted on the labels of each ECC again; when discordant voting results happened, they would have a discussion. If the discussion failed to conclude with an accurate diagnosis, the ECC was discarded. A benign diagnosis was defined as cell clumps with neat edges, nuclei with oval or spindle shape, and evenly distributed, finely granular chromatin [23,24]. Malignant diagnosis referred to a three-dimensional appearance, irregular (including dilated, branched, protruding, and papillotubular) edge, with the nucleus poloidal disordering or disappearing (including megakaryocyte appearance, nuclear membrane thickness, and coarse granular or coarse block chromatin) [23,25].

2.5. Segmentation Networks

The U-Net with jumping connection structure was selected to eliminate the interference of neutrophils and single cells, facilitating ECC extraction from each image. Figure 2 shows the U-Net architecture based on full convolutional networks. The U-Net architecture combined a down-sampling path to capture context and an up-sampling path to achieve precise localization. We calculated the probability that each pixel belonged to the cell clumps and normalized it. The collection of a detected cell clumps image was automatically marked as a region of interest (ROI) area. A total of 1000 images and their corresponding masks marked by pathologists were randomly selected for training. In order to describe the effect of the U-Net, we selected the Dice coefficient (a verification index of image segmentation accuracy) for evaluation.

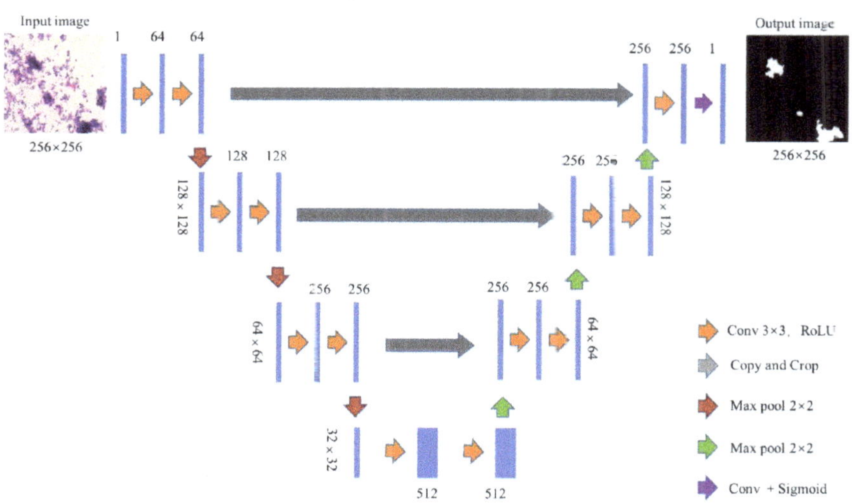

Figure 2. Segmentation network. The blue box represents the feature map. The yellow arrow represents 3 × 3 convolution and striding of 1 used for feature extraction; we set the padding as 1 to ensure that the size of the convolutional image at the same steps was stable. The gray arrow indicates skip-connection, which is used for feature fusion, and pure up-sampling will cause the loss of information. The red arrow indicates the 2 × 2 maximum pooling, which is used to reduce the dimensionality. The green arrow indicates up-sampling, which is used to restore the dimension. The cyan arrow indicates the convolution plus activation function, which is used to output the result.

$$\text{Dice} = \frac{2|A \cap B|}{A + B}$$

The Dice coefficient is at the pixel level; A represents the area where the real target appears, and B signifies the target area that showed the predicted result (Figure 3a).

The segmented mask often has small holes and residues (Figure 3b). We used morphological operations (first corrosion and then expansion) to eliminate small holes. The ROI set was input into a subsequent neural network for endometrial cytopathological screening.

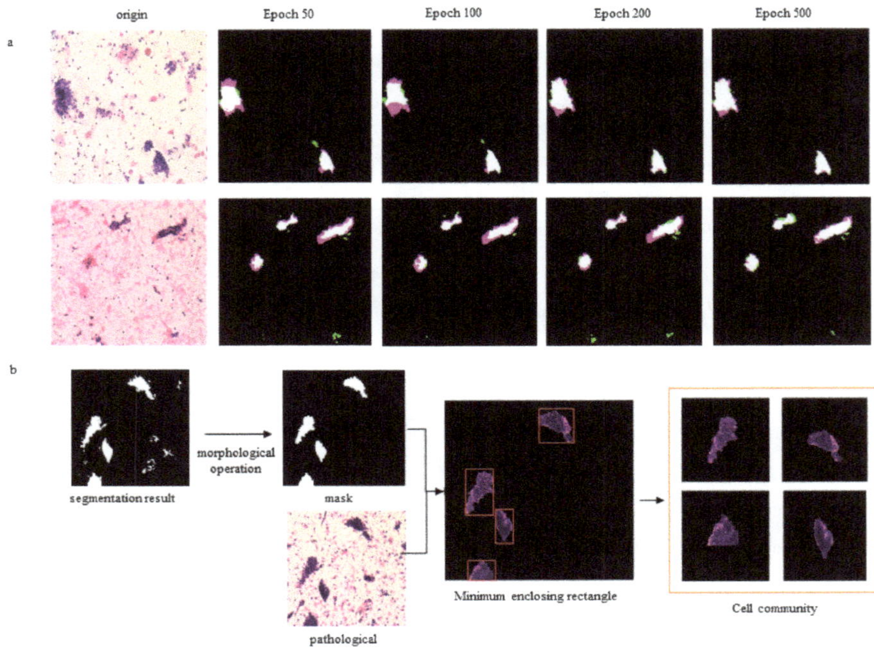

Figure 3. The effect of segmentation. (**a**) Variation of segmentation accuracy with training epochs. Compared with the ground truth (mask was manually marked by the physician), the red areas were not predicted in the mask of the model training; compared with the ground truth, the green areas represent other predicted areas in the mask of model training. (**b**) The process of ECC acquisition.

2.6. Data Preprocessing

We input the cytopathologic images into a trained U-Net to obtain the patch set of cell clumps. The segmentation results were first obtained by the U-Net, and background images (free single cells and white cells) were removed. Then, we extracted all the cell clumps using the minimum outer rectangle. The size of all cell clusters was uniformly resized to 256×256 by filling the surrounding area with pixels of value 0.

2.7. Classification Network

The CNNs were used to capture the characteristics of ECCs: nuclear heterogeneity, nuclear size, ratio between nucleus and plasma, chromatin homogeneity, cell polarity, isolation and aggregation of cell clumps, regularity of cell clump's edge, etc. We constructed a DL model with DenseNet201 being the backbone to classify malignant and benign cell communities. The training set was annotated by two cytopathologists. The final fully connected layer of DenseNet201 was replaced by a global average pooling layer, then a single fully connected layer. The specific architecture is shown in Figure 4, and the output results were classified into two categories (Figure 1c). Then, the classification network was pre-trained on ImageNet. Several groups were carried out for comparative experiments to find the best patch input size and iteration time. The iteration was set to be 50, 100, 150, and 500 epochs in the training process. The results showed that the network converged at

100 epochs, and a longer training time was not necessary (Figure 5a). We changed the size of the input patch to 32 × 32, 64 × 64, 128 × 128, and 256 × 256, respectively (Figure 5b). When the input patch size was 256 × 256, the best result was achieved.

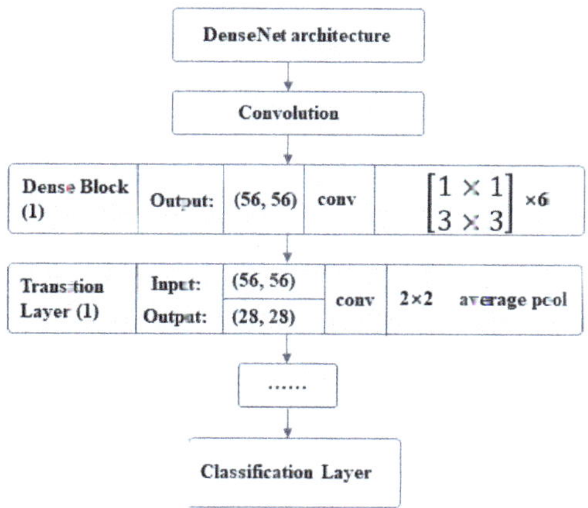

Figure 4. The recognition network architecture for classifying endometrial cell clusters. The size of the input image is 256 × 256, and each 3 × 3 convolution is preceded by a 1 × 1 convolution operation.

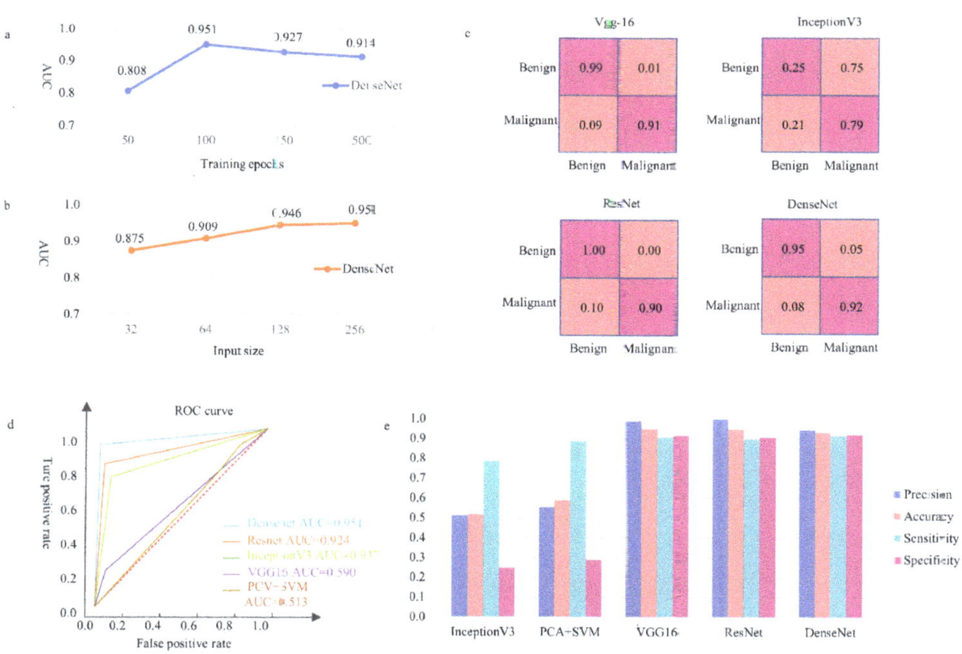

Figure 5. The performance of our model and four other common DL models on the same validation set. (**a**) Description of the AUC corresponding to the network with different numbers of iterations. (**b**) Description of the AUC corresponding to the network with different image input sizes. (**c**) The confusion matrix of different networks under the same hyperparameter conditions. The horizontal axis was a true label, the vertical axis was the predicted label, and the lower false-negative rate was preferred. (**d**) The ROC curves of different models. (**e**) The precision, accuracy, sensitivity, and specificity of different models.

2.8. Network Evaluation

We conducted comparative experiments on four CNNs (VGG16, InceptionV3, ResNet, and DenseNet) and one Support Vector Machine (SVM). The hyperparameters, all kept consistent, were as follows: Loss function (Binary Cross-Entropy), Initial learning rate (0.0001), Learning rate delay (0.5), Batch-size (8), and Adam optimizer. In addition, the SVM classifier used a radial basic function kernel with parameters of 0.0078 and 2. DenseNet gained the best result due to its advantage of featured graph jump connection (Figure 5c–e).

All experiments were performed on a personal computer equipped with a GeForce GTX2080 super (NVIDIA) graphics processing unit. Python programming language 3.6.12 (Python Software Foundation, Wilmington, DE, USA) with keras 2.4.3 (Google Brain, Mountain View, CA, USA) and Tensor Flow 2 2.0 (Google Brain, Mountain View, CA, USA) for neural networks was used for the training.

2.9. Statistical Analysis

The following indexes were calculated by the four-lattice paired hypothesis test for statistical analysis: accuracy (Acc), sensitivity (Se), and specificity (Sp). The confusion matrix and receiver operating characteristic (ROC) curve were used to visualize the classification effect. The definition criteria were as follows:

$$\text{Acc} = \frac{(\text{TP} + \text{TN})}{(\text{TP} + \text{FP} + \text{TN} + \text{FN})}$$

$$\text{Se} = \frac{\text{TP}}{(\text{TP} + \text{FN})}$$

$$\text{Sp} = \frac{\text{TN}}{(\text{TN} + \text{FP})}$$

2.10. Plots and Charts

All the drawings were performed using the matplotlib package in Python and Matlab. The ROC curve of model performance was shown with specificity being the X axis and sensitivity being the Y axis. We used a bar chart to show the predictions from different CNNs and SVM. Line graphs were drawn to illustrate the results and compare performance between different groups.

3. Results

3.1. Baseline Characteristics

A total of 113 patients who met the criteria were enrolled for final analysis, among which 42 were malignant and 71 were benign. Table 1 lists the demographic data of these patients.

3.2. Dataset

A total of 15,913 annotated cell clump images were segmented on ×20 magnification digital slides. The average image size was 1816 × 1519 pixels by width and height. We used a trained U-Net to extract ECC patches from the 15,913 images and obtained 39,000 ECC patches. Divided in 7:3, 26,880 and 11,520 patches were used for training and testing. The remaining 300 benign patches and 300 malignant patches were included in a verification set.

3.3. Verification Set and Test Set

The prediction results of ECC patches were completely in accordance with the labels given by the pathologists. We randomly exhibit the results of eight (three malignant and five benign) validation patches (Figure 6A)

Table 1. Patients characteristics.

Characteristics	n
Source	
Inpatient Department	66
Outpatient Department	47
Age	
<40 years old	13
≥40 years old	100
Menstrual Status	
Premenopausal	51
Postmenopausal	66
Abnormal uterine bleeding	35
Other Disease	
Ovarian cancer	0
Hypertension	10
Diabetes	4
Hormone replacement therapy	1

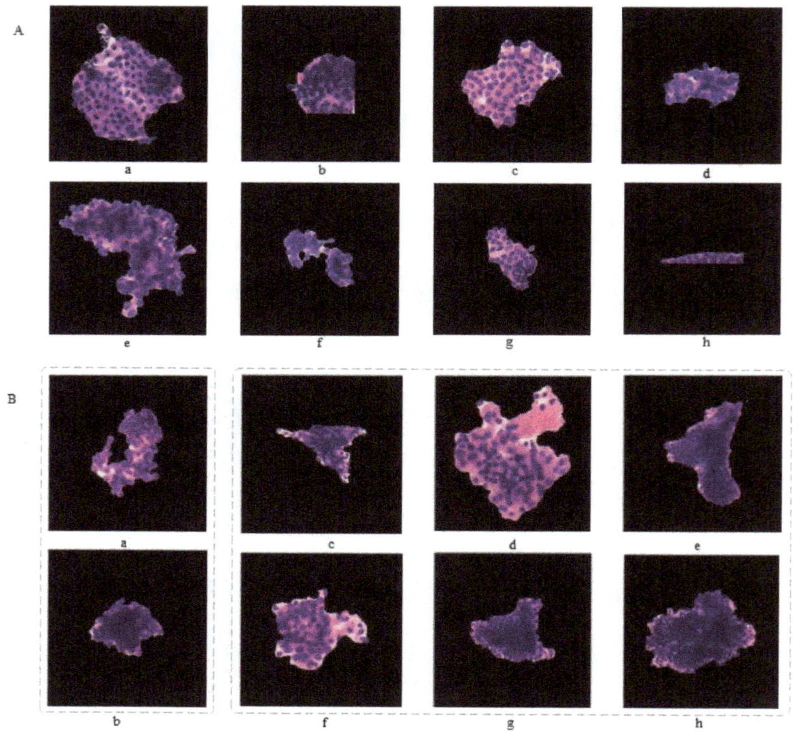

Figure 6. Presentation of true and false results. (**A**) A 100% consistency of results was achieved in the training set. Patches (**a–c**) showed the true positive, and patches (**d–h**) showed the true negative. (**B**) Analysis of false results in test set. The two false-positive (over diagnosis) patches (**a**,**b**) are exhibited. The six false-negative patches included one well-differentiated endometrial adenocarcinoma (**c**), three atypical hyperplasia (**d–f**), and two poorly differentiated adenocarcinomas (**g**,**h**).

In the test set, the accuracy and specificity of the classifier were 93.5% and 92.2%, respectively. The DenseNet achieved a 95.1% area under the curve score (AUC). In addition, we compared the results with four other common classification models (Figure 5c–e).

3.4. False Results

DenseNet obtained a 5% false-positive rate and an 8% false-negative rate in the test set (Figure 5c). We randomly listed eight common failure patches in the test set. The six false-negative (missed diagnosis) patches included one well-differentiated endometrial adenocarcinoma, three endometrial atypical hyperplasia, and two poorly differentiated endometrial adenocarcinomas. In addition, two over-diagnoses occurred (Figure 6B).

3.5. Data Supporting

The results of this study are available from the corresponding authors (Qiling Li and Dexing Zhong). Because of hospital policy, the data cannot be made public.

4. Discussion

Principal Findings

For the first time, we introduced two neural networks based on deep convolution, namely U-Net and DenseNet, to segment ECC images and recognize patches, respectively. The DenseNet achieved 93.5% accuracy and 92.2% specificity. At the same time, this system was developed for screening, and the sensitivity of our algorithm was better than that of all the comparison ones, reaching 92.0%. The results indicated that the neural network has great feasibility and potentiality in endometrial pathological image recognition.

5. Results

It is well-known that a large amount of labeled data is often required to train a high-quality machine learning classifier through DL to complete a specific cancer classification task [26]. Due to the high amount of time and effort required for image annotation work, as well as the protection of patients' privacy, there are currently few endometrial image datasets available to the public. Despite the limited dataset, our classifier performed well in the 10-fold cross-validation and in the external validation of 15,913 images.

5.1. Clinical Implications

At the beginning of the experiment, we considered that DL was able to automatically learn cancer's information from pathological images [27]. We put the unlabeled benign and malignant images into the network for recognition and obtained 40–70% specificity (data not shown) from multiple networks, proving the method to be a failure. ECCs are quite different from non-cellular clumps in ecological appearance, cell morphological structure and other pathological characteristics. U-net combines low-resolution information (to provide the basis for object category recognition) and high-resolution information (to provide the basis for precise segmentation and positioning), which is perfectly suitable for medical image segmentation. Combined with the pathological features in patients with ECCs, we chose the U-Net as the segmentation network to analyze and calculate the probability that each pixel belonged to the cell clumps. The detected cell clump images were automatically marked as ROI areas. The obtained ROI set was processed by a traditional image-processing algorithm to eliminate small holes. The ROI set was input into a subsequent neural network for cytopathological screening of the endometrium. We built a DL model with DenseNet201 as the backbone. The DL model was trained by the dataset annotated by cytopathologists, and the model was built to classify malignant and benign cell clumps. It turned out that our model alleviated the vanishing gradient problem, strengthened feature propagation, encouraged feature reuse, and outperformed ResNet50 with the same number of parameters. In order to compare the prediction performance of various DL algorithms on an experimental dataset, four commonly used CNNs were used to train different classifiers, namely VGG16, Inception-v3, ResNet, and DenseNet. The SVM classifier, which used features extracted by the CNN as input, had a better performance than the end-to-end CNN classifier [28,29]. Therefore, on the basis of previous experiments, the DenseNet had the best performance in extracting sample features to train the SVM classifier. In addition, a group of comparative experiments were also performed with traditional PCA + SVM machine learning method.

The results of the test set showed that the false-negative rate was twice as high as the false-positive rate. We analyzed all the missed and over-diagnosis images and randomly selected eight patches to illustrate the common error that occurred. There were two false positives: one patch of secretory phase endometrium and one patch of complex hyperplasia. One reason for this was that the endometrial cells were clustered and seriously overlapped. It was difficult to distinguish well-differentiated EC from the proliferative endometrium, and it was difficult to distinguish complex hyperplasia from atypical hyperplasia. Another reason was that the dysplasia coincidence rate between the cytological and histological pathological diagnosis was relatively low, which was 56% in some studies [30]. This was the main reason for their miscalculation.

5.2. Research Implications

Due to the development of liquid-based cytology and endometrial cell sampling in recent years, ECT has been gradually accepted as a simple, rapid, and economical endometrial screening method [31]. Moreover, AI can be applied to the pathological recognition of endometrial cells to promote screening. AI works steadily and indefatigably, and can quickly screen out suspicious malignant results, allowing pathologists to focus on the malignant results and improve the accuracy and efficiency of diagnosis [32].

5.3. Strengths and Limitations

This study had some limitations. First, although our images were labeled in a randomized and blind way, and histological diagnosis was used as control, and the two pathologists' diagnoses were still somehow subjective. We hope that more recommendations from pathologists in different treatment centers will be included in follow-up studies regarding the proposed diagnostic system. Second, liquid endometrial cytological smear was used in our diagnostic system. At present, cell block technology can prepare slides with cell clumps and micro tissues, which is expected to further refine the diagnostic results and provide better diagnosis and treatment suggestions for clinical work [33]. We will focus on improving the performance of the classifier by training it with more samples, aiming at subdividing endometrial pathological types in future research.

6. Conclusions

This study confirmed that the recognition of DL has similar specificity and sensitivity to manual diagnosis. At the same time, the DL saves time and manpower. Therefore, the use of endometrial liquid-based cytology in combination with AI to identify ECC is reliable for EC screening and is able to reduce pathologists' workload. By carrying out this form of screening work, cross-population, big data will be rapidly established, and the participation of scholars from different regions will greatly promote the development of precision medicine.

Author Contributions: Conceptualization, D.Z. and Q.L. (Qiling Li); Data curation, R.W. and Z.X.; Funding acquisition, Q.L. (Qiling Li); Methodology, Y.L., H.H., C.L., G.Z. and G.S.; Resources, Y.W., C.S. and L.H.; Writing—original draft, Q.L. (Qing Li) and R.W.; Writing—review & editing, L.Z. All authors have read and agreed to the published version of the manuscript.

Funding: This study was supported by the Clinical Research Award of the First Affiliated Hospital of Xi'an Jiaotong University, China (No. XJTU1AF-CRF-2019-002); the Natural Science Basic Research Program of Shaanxi (2017ZDJC-11, 2018JM7073); the Clinical Research Award of the First Affiliated Hospital of Xi'an Jiaotong University, China (XJTU1AF-2018-017); the Key Research and Development Program of Shaanxi (2017ZDXM-SF-068, 2019QYPY-138); the Innovation Capability Support Program of Shaanxi (2017XT-026, 2018XT-002); and the Medical Research Project of Xi'an Social Development Guidance Plan (2017117SF/YX011-3). The sponsors had no involvement in the study's design, data collection, analysis, and interpretation, or in writing of the manuscript.

Institutional Review Board Statement: This study was approved by the Ethics Committee of the First Affiliated Hospital of Xi'an Jiaotong University (XJTU1AHCR2014-007), and informed written consent of all subjects was obtained prior to this study.

Informed Consent Statement: The participants provided their written informed consent to participate in this study.

Data Availability Statement: The raw data supporting the conclusions of this manuscript will be made available by the authors without undue reservation to any qualified researcher. All data generated or analyzed during this study are included either in this article or available from the correspondence authors.

Acknowledgments: We thank the staff in the Department of Obstetrics and Gynecology, the First Affiliated Hospital of Xi'an Jiaotong University for supporting with data collection.

Conflicts of Interest: The authors declare no conflict of interest.

References

1. Ferlay, J.; Colombet, M.; Soerjomataram, I. Estimating the global cancer incidence and mortality in 2018: GLOBOCAN sources and methods. *Int. J. Cancer* **2019**, *144*, 1941–1953. [CrossRef]
2. Feng, R.M.; Zong, Y.N.; Cao, S.M.; Xu, R.H. Current cancer situation in China: Good or bad news from the 2018 Global Cancer Statistics? *Cancer Commun.* **2019**, *39*, 22. [CrossRef]
3. Ganz, P.A. Current US Cancer Statistics: Alarming Trends in Young Adults? *J. Natl. Cancer Inst.* **2019**, *111*, 1241–1242. [CrossRef]
4. Siegel, R.L.; Miller, K.D.; Fuchs, H.E.; Jemal, A. Cancer Statistics, 2021. *CA A Cancer J. Clin.* **2021**, *71*, 7–33. [CrossRef]
5. Amant, F.; Mirza, M.R.; Koskas, M.; Creutzberg, C.L. Cancer of the corpus uteri. *Int. J. Gynecol. Obstet.* **2018**, *143*, 37–50. [CrossRef]
6. Nishida, N.; Murakami, F.; Kuroda, A. Clinical Utility of Endometrial Cell Block Cytology in Postmenopausal Women. *Acta Cytol.* **2017**, *61*, 441–446. [CrossRef]
7. Takeda, T.; Wong, T.F.; Adachi, T. Guidelines for office gynecology in Japan: Japan Society of Obstetrics and Gynecology and Japan Association of Obstetricians and Gynecologists 2011 edition. *J. Obstet. Gynaecol. Res.* **2012**, *38*, 615–631. [CrossRef] [PubMed]
8. Nakagawa-Okamura, C.; Sato, S.; Tsuji, I. Effectiveness of mass screening for endometrial cancer. *Acta Cytol.* **2002**, *46*, 277–283. [CrossRef]
9. Kipp, B.R.; Medeiros, F.; Campion, M.B. Direct uterine sampling with the Tao brush sampler using a liquid-based preparation method for the detection of endometrial cancer and atypical hyperplasia: A feasibility study. *Cancer* **2008**, *114*, 228–235. [CrossRef]
10. Remondi, C.; Sesti, F.; Bonanno, E.; Pietropolli, A.; Piccione, E. Diagnostic accuracy of liquid-based endometrial cytology in the evaluation of endometrial pathology in postmenopausal women. *Cytopathology* **2013**, *24*, 365–371. [CrossRef]
11. Han, L.; Du, J.; Zhao, L. An Efficacious Endometrial Sampler for Screening Endometrial Cancer. *Front. Oncol.* **2019**, *9*, 67. [CrossRef] [PubMed]
12. Fourcade, A.; Khonsari, R.H. Deep learning in medical image analysis: A third eye for doctors. *J. Stomatol. Oral Maxillofac. Surg.* **2019**, *120*, 279–288. [CrossRef] [PubMed]
13. Soenksen, L.R.; Kassis, T.; Conover, S.T. Using deep learning for dermatologist-level detection of suspicious pigmented skin lesions from wide-field images. *Sci. Transl. Med.* **2021**, *13*, eabb3652. [CrossRef] [PubMed]
14. Gulshan, V.; Peng, L.; Coram, M. Development and Validation of a Deep Learning Algorithm for Detection of Diabetic Retinopathy in Retinal Fundus Photographs. *JAMA* **2016**, *316*, 2402–2410. [CrossRef]
15. Colling, R.; Pitman, H.; Oien, K. Artificial intelligence in digital pathology: A roadmap to routine use in clinical practice. *J. Pathol.* **2019**, *249*, 143–150. [CrossRef] [PubMed]
16. Jha, S.; Topol, E.J. Adapting to Artificial Intelligence: Radiologists and Pathologists as Information Specialists. *JAMA* **2016**, *316*, 2353–2354. [CrossRef]
17. Bera, K.; Schalper, K.A.; Rimm, D.L.; Velcheti, V.; Madabhushi, A. Artificial intelligence in digital pathology—new tools for diagnosis and precision oncology. *Nat. Rev. Clin. Oncol.* **2019**, *16*, 703–715. [CrossRef]
18. Makris, G.M.; Pouliakis, A.; Siristatidis, C. Image analysis and multi-layer perceptron artificial neural networks for the discrimination between benign and malignant endometrial lesions. *Diagn. Cytopathol.* **2017**, *45*, 202–211. [CrossRef]
19. Sun, H.; Zeng, X.; Xu, T.; Peng, G.; Ma, Y. Computer-Aided Diagnosis in Histopathological Images of the Endometrium Using a Convolutional Neural Network and Attention Mechanisms. *IEEE J. Biomed. Health Inform.* **2020**, *24*, 1664–1676. [CrossRef]
20. The American Society for Bone and Mineral Research. Issue Information-Declaration of Helsinki. *J. Bone Miner. Res.* **2018**, *33*, BM i–BM ii. [CrossRef]
21. Yu, M.; Xiang, Y.; Ma, X.X. Advices on standards of endometrial cancer screening. *Zhonghua Fu Chan Ke Za Zhi.* **2020**, *55*, 307–311. [PubMed]
22. Margari, N.; Pouliakis, A.; Anoinos, D. A reporting system for endometrial cytology: Cytomorphologic criteria-Implied risk of malignancy. *Diagn. Cytopathol.* **2016**, *44*, 888–901. [CrossRef]

23. Norimatsu, Y.; Shimizu, K.; Kobayashi, T.K. Cellular features of endometrial hyperplasia and well differentiated adenocarcinoma using the Endocyte sampler: Diagnostic criteria based on the cytoarchitecture of tissue fragments. *Cancer* **2006**, *108*, 77–85. [CrossRef] [PubMed]
24. Yanoh, K.; Norimatsu, Y.; Hirai, Y. New diagnostic reporting format for endometrial cytology based on cytoarchitectural criteria. *Cytopathology* **2009**, *20*, 388–394. [CrossRef] [PubMed]
25. Cunningham, M.L.; Seto, M.L.; Hing, A.V.; Bull, M.J.; Hopkin, R.J.; Leppig, K.A. Cleidocranial dysplasia with severe parietal bone dysplasia: C-terminal RUNX2 mutations. *Birth Defects Res. Part A Clin. Mol. Teratol.* **2006**, *76*, 78–85. [CrossRef] [PubMed]
26. Dance, A. AI spots cell structures that humans can't. *Nature* **2021**, *592*, 154–155. [CrossRef]
27. Yamamoto, Y.; Tsuzuki, T.; Akatsuka, J. Automated acquisition of explainable knowledge from unannotated histopathology images. *Nat. Commun.* **2019**, *10*, 5642. [CrossRef]
28. Araujo, T.; Aresta, G.; Castro, E. Classification of breast cancer histology images using Convolutional Neural Networks. *PLoS ONE* **2017**, *12*, e0177544. [CrossRef]
29. Xu, Y.; Jia, Z.; Wang, L.B. Large scale tissue histopathology image classification, segmentation, and visualization via deep convolutional activation features. *BMC Bioinform.* **2017**, *18*, 281. [CrossRef]
30. Papaefthimiou, M.; Symiakaki, H.; Mentzelopoulou, P. Study on the morphology and reproducibility of the diagnosis of endometrial lesions utilizing liquid-based cytology. *Cancer* **2005**, *105*, 56–64. [CrossRef]
31. Wang, Q.; Wang, Q.; Zhao, L. Endometrial Cytology as a Method to Improve the Accuracy of Diagnosis of Endometrial Cancer: Case Report and Meta-Analysis. *Front. Oncol.* **2019**, *9*, 256. [CrossRef] [PubMed]
32. Song, Z.; Zou, S.; Zhou, W. Clinically applicable histopathological diagnosis system for gastric cancer detection using deep learning. *Nat. Commun.* **2020**, *11*, 4294. [CrossRef] [PubMed]
33. Denny, J.C.; Collins, F.S. Precision medicine in 2030-seven ways to transform healthcare. *Cell* **2021**, *184*, 1415–1419. [CrossRef] [PubMed]

cancers

MDPI

Article

Learn to Estimate Genetic Mutation and Microsatellite Instability with Histopathology H&E Slides in Colon Carcinoma

Yimin Guo [1,†], Ting Lyu [1,†], Shuguang Liu [1], Wei Zhang [1], Youjian Zhou [1], Chao Zeng [1,*] and Guangming Wu [2,*]

1 The Eighth Affiliated Hospital, Sun Yat-sen University, Shenzhen 518000, China
2 Center for Spatial Information Science, The University of Tokyo, Kashiwa 277-8568, Japan
* Correspondence: zengch35@mail.sysu.edu.cn (C.Z.); huster-wgm@csis.u-tokyo.ac.jp (G.W.)
† These authors contributed equally to this work.

Simple Summary: Colorectal cancer is one of the most common malignancies and the third leading cause of cancer-related mortality worldwide. Identifying KRAS, NRAS, and BRAF mutations and MSI status are closely related to the individualized therapeutic judgment and oncologic prognosis of CRC patients. In this study, we introduced a cascaded network framework with an average voting ensemble strategy to sequentially identify the tumor regions and predict gene mutations & MSI status from whole-slide H&E images. Experiments on a colorectal cancer dataset indicated that the proposed method can achieve high fidelity in both gene mutation prediction and MSI status estimation. In our testing set, the AUCs for KRAS, NRAS, BRAF, and MSI were ranged from 0.794 to 0.897. The results suggested that the deep convolutional networks have the potential to assist pathologists in prediction of gene mutation & MSI status in colorectal cancer.

Abstract: Colorectal cancer is one of the most common malignancies and the third leading cause of cancer-related mortality worldwide. Identifying KRAS, NRAS, and BRAF mutations and estimating MSI status is closely related to the individualized therapeutic judgment and oncologic prognosis of CRC patients. In this study, we introduce a cascaded network framework with an average voting ensemble strategy to sequentially identify the tumor regions and predict gene mutations & MSI status from whole-slide H&E images. Experiments on a colorectal cancer dataset indicate that the proposed method can achieve higher fidelity in both gene mutation prediction and MSI status estimation. In the testing set, our method achieves 0.792, 0.886, 0.897, and 0.764 AUCs for KRAS, NRAS, BRAF, and MSI, respectively. The results suggest that the deep convolutional networks have the potential to provide diagnostic insight and clinical guidance directly from pathological H&E slides

Keywords: deep convolutional network; H&E slide; gene mutation prediction; microsatellite instability; colon carcinoma

Citation: Guo, Y.; Lyu, T.; Liu, S.; Zhang, W.; Zhou, Y.; Zeng, C.; Wu, G. Learn to Estimate Genetic Mutation and Microsatellite Instability with Histopathology H&E Slides in Colon Carcinoma. *Cancers* **2022**, *14*, 4144. https://doi.org/10.3390/cancers14174144

Academic Editor: Luca Roncucci

Received: 30 July 2022
Accepted: 22 August 2022
Published: 27 August 2022

1. Introduction

Colorectal cancer (CRC) is one of the most common lower gastrointestinal malignancies and is currently the third leading cause of cancer-related mortality worldwide [1,2]. Despite the over survival rate of colorectal cancer has increased in recent years due to the improved treatment strategies [3], distant metastasis is still a significant cause of high morbidity and mortality for CRC patients [4]. So far, various predominant environmental risk factors for the development of CRC have been identified, including diet, obesity, lack of physical activity, and inflammatory bowel disease [5]. However, a module formed by the interaction of multiple genetic alterations determines individual differences and tumor progression in CRC patients.

In the past decades, a deep understanding of molecular profiles has been more significant for selecting appropriate therapies for metastatic CRC patients [6]. Numerous frequent genetic mutations have been identified as critical drivers responsible for comprehensive therapeutic judgment and oncologic prognosis [7]. Mutations of RAS (i.e., exon 2, 3, and 4 of KRAS, exon 2 and 3 of NRAS) are considered negative predictors for targeted therapy with anti-EGFR monoclonal antibodies (e.g., cetuximab and panitumumab) [8,9]. Mutation of BRAF V600E is a worse prognostic biomarker. Patients with BRAF V600E mutation will be less likely to respond to treatment with cetuximab and panitumumab unless combined with a BRAF inhibitor [10,11]. Moreover, the microsatellite instability (MSI) status of CRC patients is also an important marker closely related to the assessment of prognosis, the efficacy of chemotherapeutic and immunity therapy [12,13]. Therefore, all metastatic CRC patients are suggested to detect the KRAS, NRAS, and BRAF mutations and MSI status according to the National Comprehensive Cancer Network (NCCN) clinical practice guidelines in oncology (Colon Cancer, Version 2.2021) [14].

The general diagnosis procedure of molecular pathology includes Sanger sequencing, Next-Generation Sequencing (NGS), ARMS-PCR, and digital PCR , etc. [15]. In recent years, the accuracy and sensitivity of those methods have been significantly improved. However, molecular detection remains limited by various factors such as sample quality, mutated gene abundance, and laboratory conditions. Moreover, in a short period of time, high testing prices are also a heavy burden for most families.

With the development of big data and deep convolutional network, artificial intelligence (AI)-assisted pathological diagnosis has attracted more and more attention. In 2018, Coudray et al. trained a deep convolutional neural network on Whole-Side Images (WSIs) to predict the cancer subtype and gene mutations in lung cancer [16]. Later, MSI status estimation of CRC from H&E histology was reported [17,18]. Furthermore, Skrede et al. exhibited a promising result in the survival risk interpretation of tumor patients based on artificial intelligence [19]. These methods have significantly extended the application capability of deep convolutional networks. However, genetic mutation prediction from H&E slices in CRC, which has more clinical significance in precision diagnosis, is still very challenging. To fulfill this demand and further explore the potential of H&E slides, we propose a cascaded deep convolutional framework to simultaneously generate gene mutation predicting and MSI status estimation using WSIs in colorectal cancer. The proposed method consists of two tumor region classification models, gene mutation& MSI status estimation models, and an average voting ensemble strategy. The effectiveness of the proposed method is demonstrated by a CRC dataset collected from GDC Data Portal and Eighth Affiliated Hospital, Sun Yat-sen University (see Section 2.1). In qualitative and quantitative evaluation, the proposed method reveals promising accuracy in tumor classification (0.939–0.976 AUC), gene mutation prediction (0.792–0.897 AUC), and MSI status estimation (0.764 AUC).

The main contributions of this study can be summarized as follows:

- We proposed a cascaded deep convolutional framework to simultaneously generate gene mutation prediction and MSI status estimation in colorectal cancer.
- We introduced a simple yet efficient average voting ensemble strategy to produce high fidelity gene mutation prediction and MSI status estimation of the WSI.
- We further analyzed the effectiveness of the number of features selected for model ensembling to understand its effects on the performances of deep CNN models.

The rest of the paper is organized as follows: Firstly, we present the datasets and methods used for this research in Section 2. Then, we illustrate the quantitative and qualitative results in Section 3. Finally, discussion and conclusion are presented in the Sections 4 and 5, respectively.

2. Materials and Methods

2.1. Data

To explore the possibility of estimating somatic mutations and microsatellite instability (MSI) using Hematoxylin-Eosin(H&E) stained whole-slide image (WSI), we downloaded diagnostic slides and corresponding clinical data of the TCGA-COAD cohort from GDC Data Portal (https://portal.gdc.cancer.gov/projects/TCGA-COAD, accessed at 20 February 2022). The pre-compiled somatic mutation data and MSI status data were acquired from UCSC Xena (https://xenabrowser.net/datapages/, accessed at 10 March 2022) and MSIsensor-pro [20], respectively. The original WSIs were formated in a magnification ratio of either 20× or 40×. Prior to performing our experiments, we manually resize the 40× images to 20× using libvips (https://github.com/libvips/libvips) (see Figure 1A–D). There were 292 WSIs with corresponding somatic mutations and MSI statuses in the TCGA-COAD dataset. To achieve better generalization, we also collected the SYSU8H dataset with the cooperation of The Eighth Affiliated Hospital, Sun Yat-sen University. The selected pathological specimens were fixed in formalin, embedded in paraffin wax block, and cut by several consecutive slices in 3–5 um by a Leica HistoCore Autocut. Later the slices were used for Hematoxylin-Eosin (H&E) staining, IHC staining, or gene sequencing, separately. Compared with the scanned H&E slices, the tumor areas for the sequencing slices are in micron-level drifts that tumor genomic heterogeneity among these slices is negligible. There were total 104 WSIs captured with 20× magnification ratio by PANNORAMIC 1000, 3DHISTECH Ltd.(see Figure 1E–H). Unlike next-generation sequencing (NGS) of TCGA-COAD, in the SYSU8H dataset, the genetic information was obtained by sanger sequencing. The binary masks of tumor areas of the WSIs were carefully annotated by experienced pathologists using QGIS (v3.22.7 LTR, https://qgis.org/).

As shown in Table 1, the 396 WSIs samples were randomly divided into training, validation, and testing groups with the ratios of 70%(278), 15%(59), and 15%(59), respectively. At 5× magnification WSIs, there were 283,126, 49,988, and 55,787 tiles within the corresponding training, validating, and testing set. At 10× magnification WSIs, 1,152,481, 203,183, and 2,275,595 tiles were within the corresponding training, validating, and testing set. In our experiment, the size of each tile was set to 512 × 512 pixels.

Table 1. Distribution of patients and whole-side images samples.

		WSIs			
		Train (n = 278)	**Val (n = 59)**	**Test (n = 59)**	**Overall (n = 396)**
Age(year)	Min.	22	29	36	22
	Max.	90	90	90	90
	Median	65.5	67	68	66
Gender	Male	138	30	33	201
	Female	140	29	26	195
KRAS	W.T.	162	24	35	221
	M.T	116	35	24	175
NRAS	W.T.	267	57	57	381
	M.T	11	2	2	15
BRAF	W.T.	251	51	51	353
	M.T.	27	8	8	43
MSI	MSI-H	236	49	52	337
	MSS/MSI-L	42	10	7	59
Tiles	5× Mag.	283,126	49,988	55,787	388,901
	10× Mag.	1,152,481	203,183	227,595	1,533,259

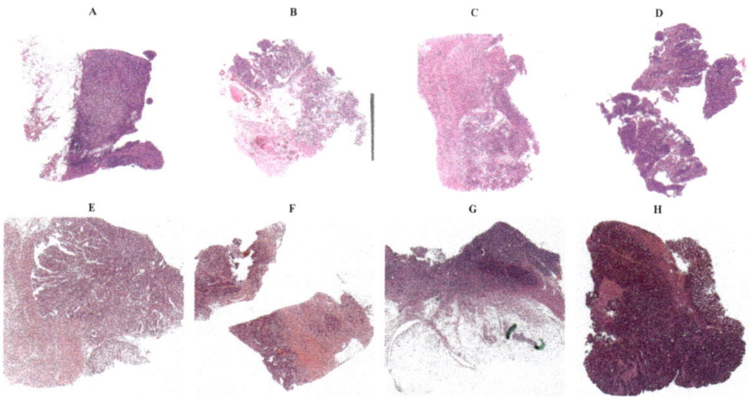

Figure 1. Representative H&E stained whole-side images (WSIs) from SYSU8H and TCGA-COAD dataset. The (**A–D**) and (**E–H**) samples are randomly selected from SYSU8H and TCGA-COAD datasets, respectively.

2.2. Methodology

In this study, we proposed a cascaded network framework to directly estimate somatic gene mutation and microsatellite instability status from the H&E stained whole-side image.

As shown in Figure 2, at the training stage, WSIs and corresponding binary masks of the training and validation set were partitioned into 5× or 10× tiles for training and validating the tumor classifier. The annotated tumor tiles and their somatic gene mutations or microsatellite instability (MSI) were used for training a binary classifier to discriminate wild type (i.e., W.T.) vs. mutant type (i.e., M.T.) of the gene or MSI-H vs. MSS/MSI-L, respectively. The top N highest probabilities of all tiles within a WSI were used to generate the final prediction for the patient.

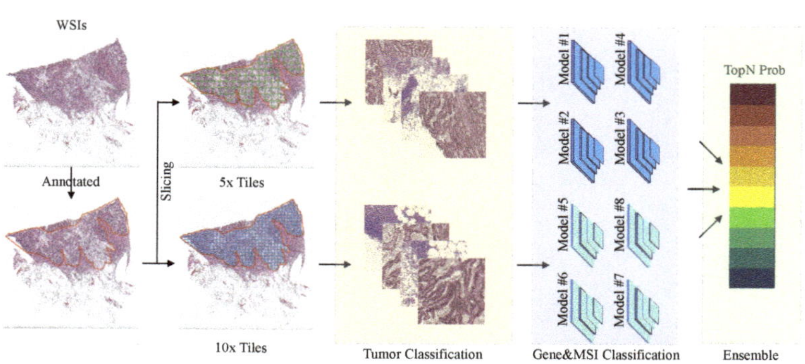

Figure 2. Experimental workflow for estimating somatic gene mutation and microsatellite instability with H&E stained whole-side images. The 5× or 10× tiles from WSIs will be accessed by a tumor classifier, a gene&MSI classifier, and a TopN ensemble classifier.

Through several cycles of training and validation, the hyperparameters, including batch size, the number of iterations, and learning rate, were optimized with the Adam stochastic optimizer [21]. Subsequently, the predictions generated by the optimized models were evaluated using the WSIs of the test set (see details in Table 1). For performance evaluations, we carefully measured the area under the receiver operator characteristic (ROC) curve [22] and its confidence interval (CI) [23].

2.2.1. Data Preprocessing

At first, the 396 pairs of whole-side images (WSIs) and their corresponding clinical records were shuffled and partitioned into three groups: training (70%), validating (15%), and testing (15%). Within each pair, a binary tumor mask of WSI was generated through polygon rasterization of its manually created tumor annotation. Later, a square window of 512 × 512 pixels was applied to the whole-side image and the corresponding tumor mask to extract paired tiles of WSI and mask. Then, each tile of WSI was labeled according to the positive ratio of pixels of the tumor mask. To focus on the tumor regions, tiles with positive ratios less than 80% were marked as 0. Otherwise, tiles were marked as 1. There were 388,901 and 1,583,259 tiles extracted from 5× and 10× magnification. As shown in Table 1, at 5× magnification, there were 283,126, 49,988, and 55,787 tiles within the training, validating, and testing set. While at 10× magnification, the number of tiles used for training, validation, and testing was 1,152,481, 203,183, and 2,275,595, respectively.

2.2.2. Network Architectures

For simplicity and efficiency, we adopted an advanced convolutional neural network (CNN) architecture, i.e., EfficientNet [24], as a backbone for tumor classification and gene&MSI classification.

In 1998, Lecun et al. introduced the classic CNN architecture, LetNet-5 [25], which consists of two sets of convolutional & pooling layers, a flattening convolutional layer, and two fully-connected layers. The CNN reveals two important concepts, sparse connectivity and shared weights, significantly reducing memory occupation and promoting computational efficiency. With the growing complexity of the dataset and rapid development of computational capacity, computer scientists have proposed more advanced CNN architectures for better generalization capacity and computational efficiency [26]. These architectures significantly promote CNN performance by introducing well-designed novel strategies, such as network in network (i.e. NIN) [27], residual learning (i.e., ResNet) [28], inception architecture [29], and dense connection (i.e., DenseNet) [30]. Differ from the above-mentioned models, which mainly focus on model accuracy, the EfficientNet architecture is designed to get a present accuracy level with limited computational operations. The EfficientNet introduces a uniformed scaling method that scales all dimensions of depth, width, and resolution with a set of fixed scaling coefficients [24].

In our experiments, we chose an ImageNet-1K [31] pretrained EfficientNet B0 (https://pytorch.org/vision/master/models/generated/torchvision.models.efficientnet_b0.html, accessed at 4 March 2022) as the backbone for both tumor classification and Gene&MSI classification. As shown in Table 2, we introduced a dropout layer ($p = 0.5$) [32] to prevent overfitting. Then, we replaced the dimensions of fully-connected (FC) layer from 1280 × 1000 to 1280 × 1.

Subsequently, the activation function was changed from softmax to sigmoid.

$$\begin{aligned} z_i &= b + \sum_{j=1}^{c} w_j \times x_{i,j} \\ p_i &= \frac{1}{1 - e^{-z_i}} \end{aligned} \tag{1}$$

The $w \in \mathrm{R}^c$ and $b \in \mathrm{R}^1$ denote the weights and bias, respectively. The range of prediction p_i is limited to [0, 1].

Instead of binary cross entropy [33], we adopted focal loss [34] as our object function to focus learning on hard misclassified examples and address class imbalance. The equation can be formulated as:

$$\begin{aligned} p_t &= \begin{cases} p_i, & \text{if } y_i = 1 \\ 1 - p_i, & \text{if } y_i = 0 \end{cases} \\ Loss_{focal} &= -(1 - p_t)^{\gamma} log(p_t) \end{aligned} \tag{2}$$

where p_i and y_i is the ith prediction and corresponding ground truth. The value of p_t is p_i if the observation is in class 1; otherwise, the value is $1 - p_i$. The γ (≥ 0) is a tunable focusing parameter which reduces the relative loss for well-classified examples (i.e., $p_t > 0.5$) and puts more focus on hard, misclassified examples.

Table 2. The backbone network for both tumor classification and Gene&MSI classification. Each row describes the stage, operation, input resolution, output channel, and the number of layers.

Stage	Operator	Resolution	Channels	Layers
0		512 × 512	3	0
1	Conv3 × 3	512 × 512	32	1
2	MBConv1, k3 × 3	256 × 256	16	1
3	MBConv6, k3 × 3	256 × 256	24	2
4	MBConv6, k5 × 5	128 × 128	40	2
5	MBConv6, k3 × 3	64 × 64	80	3
6	MBConv6, k5 × 5	32 × 32	112	3
7	MBConv6, k5 × 5	32 × 32	192	4
8	MBConv6, k3 × 3	16 × 16	320	1
9	Conv1 × 1&Pooling	16 × 16	1280	1
10	Dropout&FC	1280 × 1	1	1

With all of the above layers being trained by mini-batch stochastic gradient descent (SGD) [35] to minimize the focal loss, the model learns how to map from the input 512 × 512 RGB image to a binary prediction.

2.2.3. Model Ensemble

To make a decisive conclusion on the whole-slide-image (WSIs) using the separated predictions of 5× and 10× tiles, we introduced a simple yet efficient average voting strategy using the top N number of features to ensemble models. To ensure the high fidelity of selected features, a high threshold (i.e., 0.8) was used to filter out tiles with a low probability of being a tumor region. Later, tiles with a high probability of being tumor regions were passed to corresponding gene&MSI classification models to generate predictions of 5× tiles (P_{x5}) and 10× tiles (P_{x10}). Then, the top N highest probabilities of predictions from both 5× and 10× tiles were selected for the final estimation of the WSI (P_{wsi}). Finally, the P_{wsi} and corresponding ground truth (Y_{wsi}) were used to calculate the area under the curve (AUC) for performance estimation.

$$
\begin{aligned}
P_{topN} &= max([P_{x5}, P_{x10}], N) \\
P_{wsi} &= \frac{1}{N}\sum_{i=1}^{N} P_{topN}
\end{aligned}
\tag{3}
$$

3. Results

A total of 396 colorectal cancer (CRC) patients with various gene mutations and MSI status from the SYSU8H and TCGA-COAD datasets were recruited in this study. The collected WSIs were randomly split into three sets: training, validation, and testing with the ratio of 70%, 15%, and 15%, respectively. The tiles extracted from the training and validation set wereused for training and optimizing hyperparameters of the proposed classification models. In order to estimate the performance of the proposed classification models, we have conducted heavy quantitative and qualitative comparisons on the testing set. All experiments were performed on the same dataset and processing platform.

3.1. Tumor Classification

The tumor regions annotated by the pathologist and probability maps generated by the tumor classification models using 5× and 10× tiles of WSIs are presented in Figure 3. Both 5× and 10× models display high fidelity in tumor recognition compared to manual annotations. Compared with the 5× model, the model trained with 10× tiles shows fewer false positives (e.g., orange and red patches outside the blue dashed curve of A, B, and C), fewer false negatives (e.g., blue and green patches inside the blue dashed curve of E and F), and better boundaries (e.g., around the blue dashed curve of A, B, and D). We selected 5 tiles from each of the four randomly selected whole slide images in the testing set, which present the highest probabilities to be the tumor regions according to our trained 5× or 10× tumor classification models (Figures 4 and 5). The selected tiles show high consensus with the annotations by the pathologist. The receiver operator characteristic (ROC) curve and area under the curve (AUC), are used to evaluate the performance of tumor classification models using 5× and 10× tiles of the WSIs (Figure 6). The AUCs of 5× classification model have achieved 0.939 (95% CI of 0.937–0.940), 0.910 (95% CI of 0.905–0.914), and 0.959 (95% CI of 0.957–0.961) for training, validating, and testing set, respectively. Slightly better than the 5× model, the AUCs of 10× classification model are up to 0.971 (95% CI of 0.971–0.972), 0.973 (95% CI of 0.972–0.973), and 0.976 (95% CI of 0.975–0.977) for training, validating, and testing set, respectively. These values are consistent with our observation in Figures 3–5.

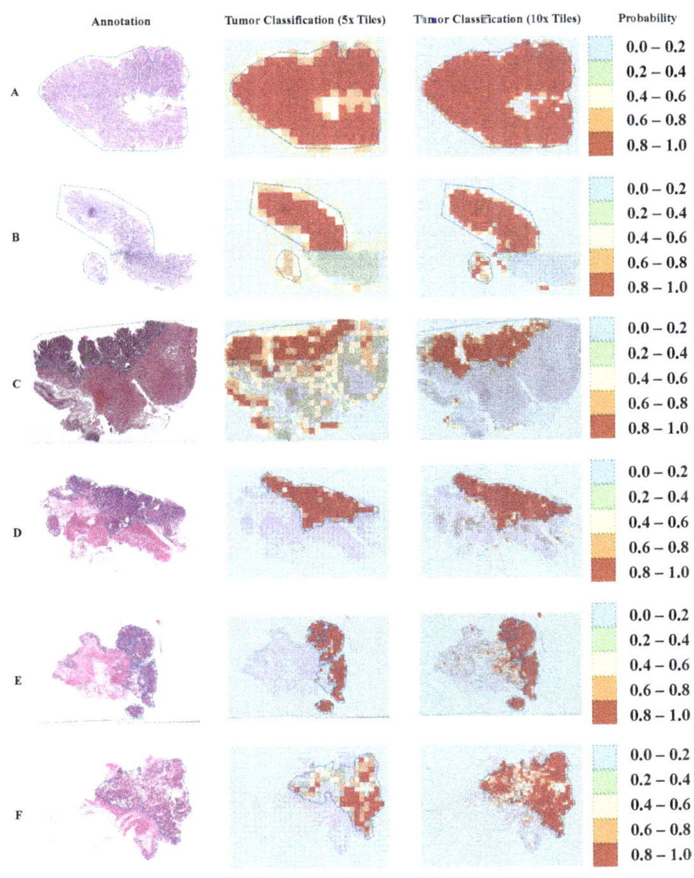

Figure 3. Probability maps of tumor classification using 5× and 10× tiles of the whole slide images (WSIs). The annotations created by the pathologist are marked with the blue dashed curve. The probability values are categorized into five groups with different color representations. The (**A–C**) and (**D–F**) samples are randomly selected from the testing set of TCGA-COAD and SYSU8H, respectively.

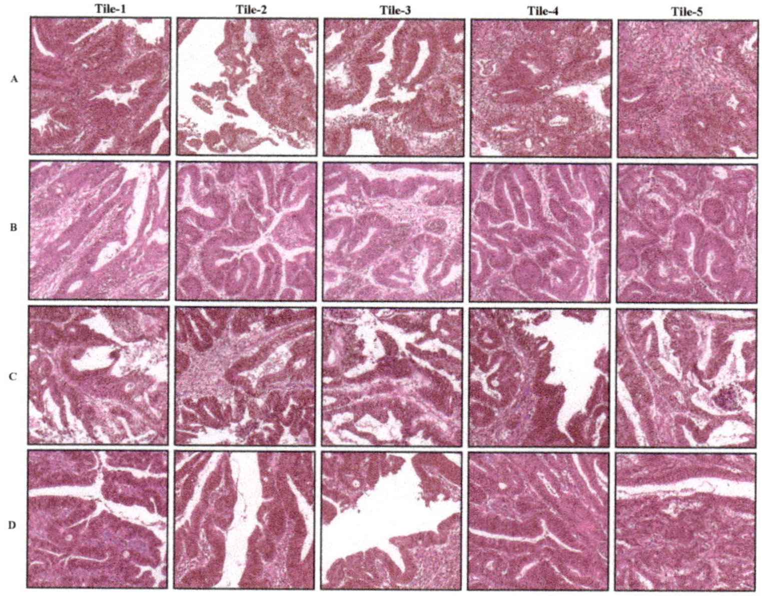

Figure 4. Representative tiles of tumor classification using 5× tiles of the whole slide images (WSIs). The (**A–D**) samples are randomly selected from the testing set. In each row, tiles 1–5 are patches from the same WSI.

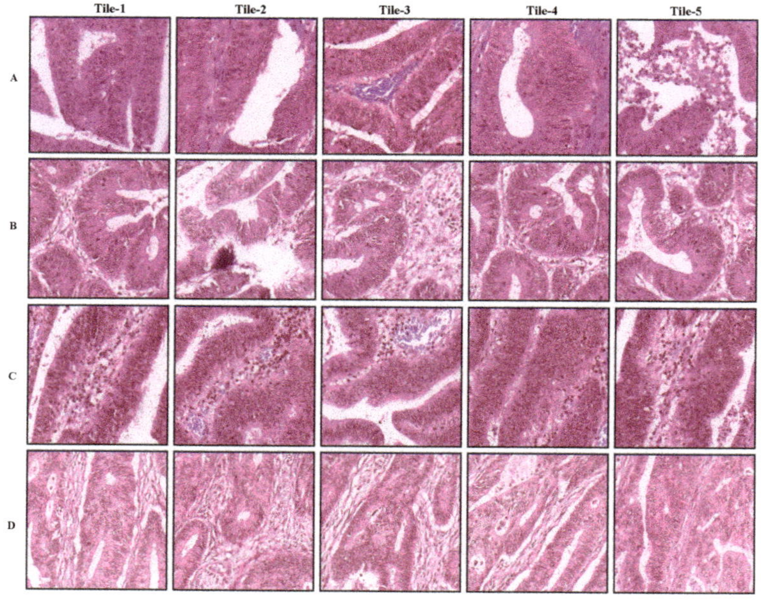

Figure 5. Representative tiles of tumor classification using 10× tiles of the whole slide images (WSIs). The (**A–D**) samples are randomly selected from the testing set. In each row, tiles 1–5 are patches from the same WSI.

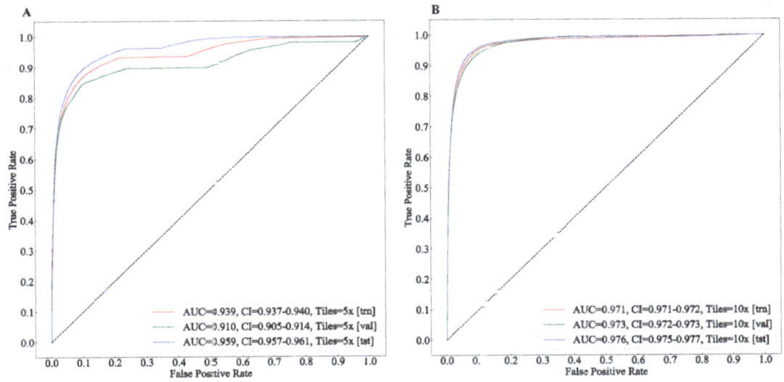

Figure 6. The receiver operator characteristic (ROC) curve and area under the curve(AUC) of tumor classification using 5× and 10× tiles of the whole slide images (WSIs). (**A**) The curves of training, validating, and testing set using 5× tiles. (**B**) The curves of training, validating, and testing set using 10× tiles.

3.2. Gene&MSI Classification

After model ensembling, the proposed method generates probabilities of gene mutations (i.e., KRAS, NRAS, and BRAF) and MSI status of every WSI.

As shown in Figure 7a–c, in the testing set, the proposed method reaches 0.792 (95% CI of 0.669–0.914), 0.886 (95% CI of 0.688–1.00), and 0.897 (95% CI of 0.800–0.994) AUCs for gene mutation predictions of KRAS, NRAS, and BRAF, respectively. In Figure 7d, our method shows high accuracy (i.e., 0.764 AUC, 95% CI 0.563–0.965) on the MSI status estimating in colorectal cancer.

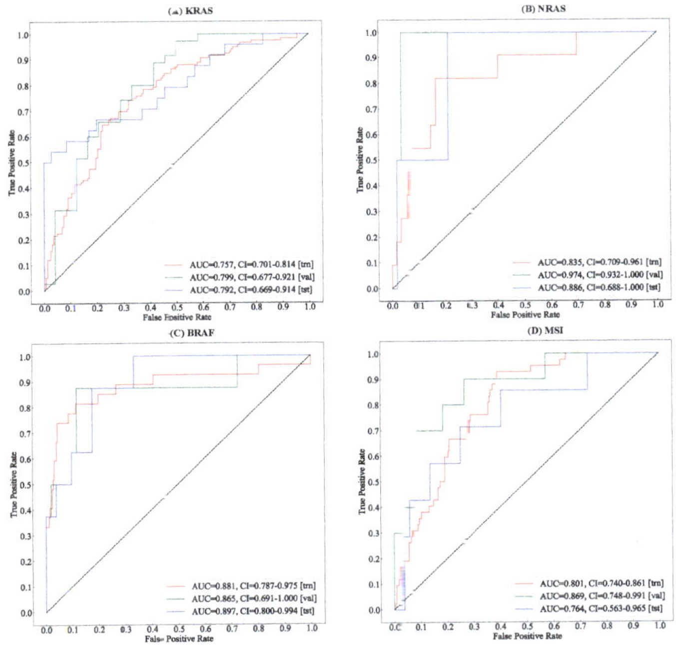

Figure 7. The receiver operator characteristic (ROC) curve and area under the curve(AUC) of Gene&MSI classification using 5&10× tiles of the whole slide images (WSIs). (**A**) The curves of KRAS gene mutation classification. (**B**) The curves of NRAS gene mutation classification. (**C**) The curves of BRAF gene mutation classification. (**D**) The curves of MSI status classification.

To investigate the effect of the selected number of features (i.e., topN) used for model ensembling, we conducted a comparison experiment on the testing set using sequential values (i.e., [1, 3, 5, 7, 9]) of topN. Figure 8 shows the trend of the AUC values under sequential values of topN in the testing set. Among all values, the proposed method achieves the highest KRAS, NRAS, and BRAF gene mutation prediction accuracy while topN equals 7. In gene mutation predictions, as the value of topN increases, the AUC value will firstly increase and then decrease. In MSI status estimation, the AUC increases gradually as the value of topN increases. As the value of topN passes 7, the increment of AUC narrows down.

Figure 9 shows the top weighted tiles of whole slide images (WSIs) in gene mutation prediction and MSI status estimation by the proposed models.

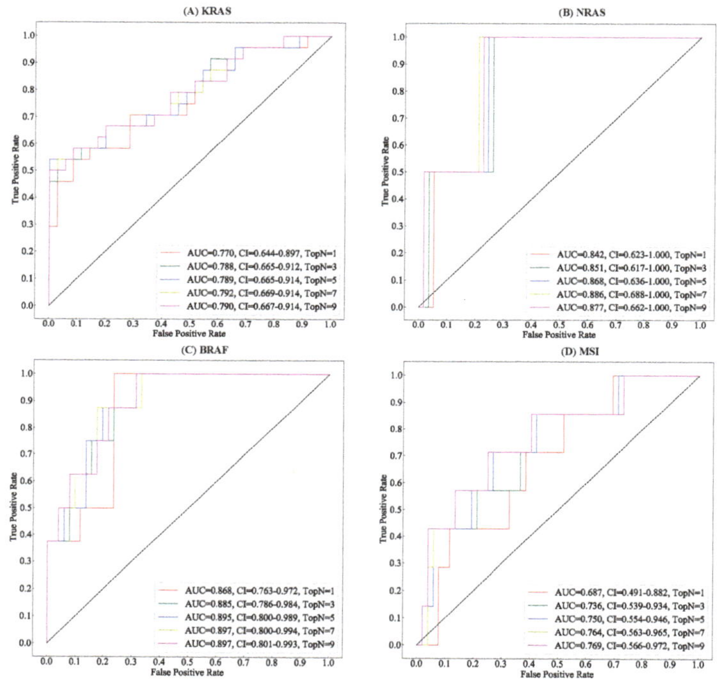

Figure 8. The receiver operator characteristic (ROC) curve and area under the curve(AUC) of Gene&MSI classification using sequential values of topN. (**A**) The trend of KRAS gene mutation classification. (**B**) The trend of NRAS gene mutation classification. (**C**) The trend of BRAF gene mutation classification. (**D**) The trend of MSI status classification.

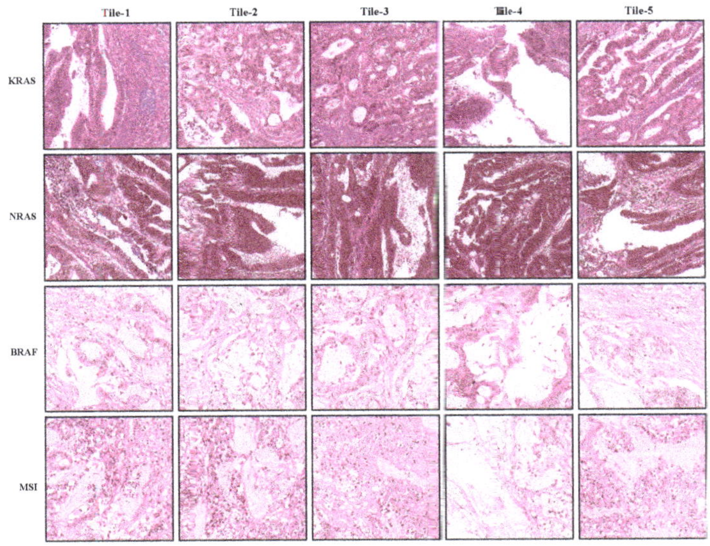

Figure 9. Top weighted tiles of whole slide images (WSIs) in gene mutation and MSI status estimation. In each row, tiles 1–5 are either 5× or 10× tiles extracted from the same WSI.

4. Discussion

4.1. Regarding the Cascaded Framework

In recent years, deep convolutional networks have demonstrated their potential in computer-aided cancer identification using clinical images such as CT scan [36], ultrasonic [37], and MRI images [38]. Other than tumor recognization, a growing number of researches are trying to look deeper into microsatellite instability estimation [39,40], gene mutation prediction [41] or survival risk evaluation [19], which are vital for precision pathological diagnosis and treatment.

To the best of our knowledge, the proposed cascaded framework is the first end-to-end method that simultaneously generates gene mutation prediction and MSI status estimation using the whole slide image (WSI) in colorectal cancer. Our method can produce high-fidelity gene mutation prediction and MSI status estimation for each WSI through a simple yet efficient average voting strategy to ensemble models. Predicting the gene mutations (KRAS, NRAS, and BRAF) and MSI status from deep convolutional networks provides pathologists with a more convenient way to evaluate prognosis and guide medication. For example, advanced metastatic CRC patients with KRAS and NRAS mutations are not recommended to choose anti-EGFR monoclonal drugs (cetuximab and panimab) for treatment. The evaluation of BRAF mutation can stratify the prognosis and guide clinical treatment. Patients with BRAF genetic mutation are unlikely to respond to the treatment of cetuximab or panimab. MSI is a predictor of the efficacy of immune checkpoint inhibitors, CRC patients with MSI-H are more likely to benefit from the treatment of immune checkpoint inhibitors (e.g., pabolizumab). Qualitative and quantitative results of the experiment data demonstrated the effectiveness of our proposed framework. These results suggest that the deep learning models have the potential to provide diagnostic insight and clinical guidance directly from pathological H&E slides. Additionally, as the gene mutation prediction and MSI status estimation are directly computed from histopathology H&E slides, in principle, the proposed method should apply not only to colorectal cancer but also to other malignant cancers (e.g., lung, breast, and liver cancer).

4.2. Accuracies, Uncertainties, and Limitations

The proposed framework revealed high values of area under the curve (AUC) in both tumor classification and gene&MSI classification tasks. In tumor classification,the 5× and 10× classification models achieved 0.959 (95% CI of 0.957–0.961) and 0.976 (95% CI of 0.975–0.977) AUCs in the testing set, respectively. The values show a very close judgment between the pathologist and the proposed method, which suggest that the AI-algorithm can potentially serve as a pre-screening tool. The performance will be further evaluated using a larger dataset with multiple tissue samples collected from varied pathology departments.

In gene mutation prediction, the proposed method achieved 0.792 (95% CI of 0.669–0.914), 0.886 (95% CI of 0.688–1.00), and 0.897 (95% CI of 0.800–0.994) AUCs for gene mutation predictions of KRAS, NRAS, and BRAF, respectively. Because of the extremely biased ratio of mutant type / wild type distribution (i.e., 15 vs. 381 of NRAS, 43 vs. 353 of BRAF), the value of AUCs fluctuates in a large range within 95% confidence interval (see details in Figure 7). In terms of MSI status estimation, recent researches [39,40] had reported higher performance than ours(i.e., 0.764 AUC, 95% CI 0.563–0.965). Compared with these methods, our method is able to simultaneously generate gene mutation prediction (KRAS, NRAS, and BRAF) and MSI status estimation, which are all mandatory for metastatic CRC patients. As for future clinical application, improving the accuracy level of our algorithm remains one of the main future goals.

With the current cascaded classification-based scheme, the models are trained to generate tile-to-label predictions using features extracted from sequential convolutional layers. The lack of internal connectivity with adjacent tiles within the same WSI might lead to partial misclassification (e.g., red patches outside the blue dashed curve and green patches within the blue dashed curve in Figure 1B,D). Since the models are trained and optimized separately, the proposed framework requires extra computational time and storage for training and saving checkpoints of multiple models. Considering the computational efficiency, a unified model with shared parameters and object functions should be explored in further work.

Considering the type of H&E used for staining, varied types of hematoxylin have certain differences in stability, durability, and dyeing time, which may lead to distinct visual patterns. In the SYSU8H dataset, the H&E slices were stained using an identical form of hematoxylin (i.e., Harris hematoxylin) to make sure both the nucleus and cytoplasm can be clearly visible and discriminated. Due to the fact that the TCGA-COAD dataset was collected from multiple centers, the forms of hematoxylin used for staining were very likely to be different. However, as shown in Figure 3, in tumor classification, prediction accuracies among slices were not so significant. The result indicates that our method can be adapted to different forms of H&E staining approaches.

Another issue that should not be ignored is the tumor heterogenity of the primary and metastatic lesions. Clinically, whether it is pathological diagnosis or target gene detection, the tumor specimen of the primary lesion is the first choice. However, there may be discrepancies in the gene mutation between the primary and metastatic tumor. For advanced metastatic tumors,when the target gene mutation of the primary tumor is negative, the target gene detection of the metastatic tumor can be carried out if conditions permitted, which can increase the opportunity for patients to receive one more targeted drug treatment. In this study, limited by the publicly available clinical samples attached with the gene mutation information of the primary tumor and the corresponding metastases, our method focused exclusively on primary tumors. Further evaluation is still necessary to clarify the reliability and generalization of our model performance.

5. Conclusions

For colon carcinoma, we design a cascaded deep convolutional framework to simultaneously generate gene mutation predicting and MSI status estimation based on the whole-slide images. The proposed method introduces a simple yet efficient average voting ensemble strategy to produce a high-fidelity prediction of the WSI. In gene mutation&MSI

status classification task, the proposed method achieves 0.792 (95% CI of 0.669–0.914), 0.886 (95% CI of 0.688–1.00), 0.897 (95% CI of 0.800–0.994), and 0.764 (95% CI 0.563–0.965) AUCs for KRAS, NRAS, BRAF, and MSI, respectively. These results suggest that the deep learning models have the potential to provide diagnostic insight and clinical guidance directly from pathological H&E slides. We plan to improve the architecture of the framework and apply it to other data sources to achieve better generalization capacity and diagnostic reliability.

Author Contributions: Conceptualization, C.Z. and G.W.; Formal analysis, S.L., W.Z. and Y.Z.; Funding acquisition, C.Z.; Investigation, Y.G. and T.L.; Methodology, Y.G. and T.L.; Writing—original draft, Y.G. and T.L.; Writing—review & editing, G.W. All authors have read and agreed to the published version of the manuscript.

Funding: This research received no external funding.

Institutional Review Board Statement: The study was conducted in accordance with the Declaration of Helsinki, and approved by the Institutional Review Board (or Ethics Committee) of The Eighth Affiliated Hospital, Sun Yat-sen University (protocol code 2022d013, 20220222).

Informed Consent Statement: Informed consent was obtained from all subjects involved in the study.

Data Availability Statement: Not applicable.

Acknowledgments: The authors would like to thank all investigators and contributing pathologists from the TCGA (http://portal.gdc.cancer.gov), UCSC Xena (https://xenabrowser.net), and SYSU8H (http://www.sysu8h.com.cn/) for providing us samples and tools in this study.

Conflicts of Interest: The authors declare no conflict of interest.

Sample Availability: Samples of the SYSU8H whole-side images and corresponding clinical information are available from the authors.

Abbreviations

The following abbreviations are used in this manuscript:

CRC	Colorectal Cancer
MSI	Microsatellite Instability
WSI	Whole-Side Image
CNN	Convolutional Neural Network

References

1. Wang, Z.; Zhou, C.; Feng, X.; Mo, M.; Shen, J.; Zheng, Y. Comparison of cancer incidence and mortality between China and the United States. *Precis. Cancer Med.* **2021**, *4*. [CrossRef]
2. Xia, C.; Dong, X.; Li, H.; Cao, M.; Sun, D.; He, S.; Yang, F.; Yan, X.; Zhang, S.; Li, N.; et al. Cancer statistics in China and United States, 2022: Profiles, trends, and determinants. *Chin. Med. J.* **2022**, *135*, 584–590. [CrossRef] [PubMed]
3. Lee, R.M.; Cardona, K.; Russell, M.C. Historical perspective: Two decades of progress in treating metastatic colorectal cancer. *J. Surg. Oncol.* **2019**, *119*, 549–563. [CrossRef] [PubMed]
4. Zhou, Z.; Mo, S.; Dai, W.; Xiang, W.; Han, L.; Li, Q.; Wang, R.; Liu, L.; Zhang, L.; Cai, S.; et al. Prognostic nomograms for predicting cause-specific survival and overall survival of stage I–III colon cancer patients: A large population-based study. *Cancer Cell Int.* **2019**, *19*, 1–15. [CrossRef]
5. Carr, P.R.; Weigl, K.; Edelmann, D.; Jansen, L.; Chang-Claude, J.; Brenner, H.; Hoffmeister, M. Estimation of absolute risk of colorectal cancer based on healthy lifestyle, genetic risk, and colonoscopy status in a population-based study. *Gastroenterology* **2020**, *159*, 129–138. [CrossRef]
6. Xie, Y.H.; Chen, Y.X.; Fang, J.Y. Comprehensive review of targeted therapy for colorectal cancer. *Signal Transduct. Target. Ther.* **2020**, *5*, 1–30. [CrossRef]
7. Mizukami, T.; Izawa, N.; Nakajima, T.E.; Sunakawa, Y. Targeting EGFR and RAS/RAF signaling in the treatment of metastatic colorectal cancer: From current treatment strategies to future perspectives. *Drugs* **2019**, *79*, 633–645. [CrossRef]
8. Van Cutsem, E.; Köhne, C.H.; Hitre, E.; Zaluski, J.; Chang Chien, C.R.; Makhson, A.; D'Haens, G.; Pintér, T.; Lim, R.; Bodoky, G.; et al. Cetuximab and chemotherapy as initial treatment for metastatic colorectal cancer. *N Engl. J. Med.* **2009**, *360*, 1408–1417. [CrossRef]

9. De Roock, W.; Claes, B.; Bernasconi, D.; De Schutter, J.; Biesmans, B.; Fountzilas, G.; Kalogeras, K.T.; Kotoula, V.; Papamichael, D.; Laurent-Puig, P.; et al. Effects of KRAS, BRAF, NRAS, and PIK3CA mutations on the efficacy of cetuximab plus chemotherapy in chemotherapy-refractory metastatic colorectal cancer: A retrospective consortium analysis. *Lancet Oncol.* **2010**, *11*, 753–762. [CrossRef]
10. Tol, J.; Nagtegaal, I.D.; Punt, C.J. BRAF mutation in metastatic colorectal cancer. *N. Engl. J. Med.* **2009**, *361*, 98–99. [CrossRef]
11. Taieb, J.; Lapeyre-Prost, A.; Laurent Puig, P.; Zaanan, A. Exploring the best treatment options for BRAF-mutant metastatic colon cancer. *Br. J. Cancer* **2019**, *121*, 434–442. [CrossRef]
12. Copija, A.; Waniczek, D.; Witkoś, A.; Walkiewicz, K.; Nowakowska-Zajdel, E. Clinical significance and prognostic relevance of microsatellite instability in sporadic colorectal cancer patients. *Int. J. Mol. Sci.* **2017**, *18*, 107. [CrossRef]
13. Battaglin, F.; Naseem, M.; Lenz, H.J.; Salem, M.E. Microsatellite instability in colorectal cancer: Overview of its clinical significance and novel perspectives. *Clin. Adv. Hematol. Oncol. H&O* **2018**, *16*, 735.
14. Benson, A.B.; Venook, A.P.; Al-Hawary, M.M.; Arain, M.A.; Chen, Y.J.; Ciombor, K.K.; Cohen, S.; Cooper, H.S.; Deming, D.; Farkas, L.; et al. Colon cancer, version 2.2021, NCCN clinical practice guidelines in oncology. *J. Natl. Compr. Cancer Netw.* **2021**, *19*, 329–359. [CrossRef]
15. McDonough, S.J.; Bhagwate, A.; Sun, Z.; Wang, C.; Zschunke, M.; Gorman, J.A.; Kopp, K.J.; Cunningham, J.M. Use of FFPE-derived DNA in next generation sequencing: DNA extraction methods. *PLoS ONE* **2019**, *14*, e0211400. [CrossRef]
16. Coudray, N.; Ocampo, P.S.; Sakellaropoulos, T.; Narula, N.; Snuderl, M.; Fenyö, D.; Moreira, A.L.; Razavian, N.; Tsirigos, A. Classification and mutation prediction from non–small cell lung cancer histopathology images using deep learning. *Nat. Med.* **2018**, *24*, 1559–1567. [CrossRef]
17. Kather, J.N.; Pearson, A.T.; Halama, N.; Jäger, D.; Krause, J.; Loosen, S.H.; Marx, A.; Boor, P.; Tacke, F.; Neumann, U.P.; et al. Deep learning can predict microsatellite instability directly from histology in gastrointestinal cancer. *Nat. Med.* **2019**, *25*, 1054–1056. [CrossRef]
18. Park, J.H.; Kim, E.Y.; Luchini, C.; Eccher, A.; Tizaoui, K.; Shin, J.I.; Lim, B.J. Artificial Intelligence for Predicting Microsatellite Instability Based on Tumor Histomorphology: A Systematic Review. *Int. J. Mol. Sci.* **2022**, *23*, 2462. [CrossRef]
19. Skrede, O.J.; De Raedt, S.; Kleppe, A.; Hveem, T.S.; Liestøl, K.; Maddison, J.; Askautrud, H.A.; Pradhan, M.; Nesheim, J.A.; Albregtsen, F.; et al. Deep learning for prediction of colorectal cancer outcome: A discovery and validation study. *Lancet* **2020**, *395*, 350–360. [CrossRef]
20. Jia, P.; Yang, X.; Guo, L.; Liu, B.; Lin, J.; Liang, H.; Sun, J.; Zhang, C.; Ye, K. MSIsensor-pro: Fast, accurate, and Matched-normal-sample-free detection of microsatellite instability. *Genom. Proteom. Bioinform.* **2020**, *18*, 65–71. [CrossRef]
21. Kingma, D.P.; Ba, J. Adam: A method for stochastic optimization. *arXiv* **2014**, arXiv:1412.6980.
22. Bradley, A.P. The use of the area under the ROC curve in the evaluation of machine learning algorithms. *Pattern Recognit.* **1997**, *30*, 1145–1159. [CrossRef]
23. Sun, X.; Xu, W. Fast implementation of DeLong's algorithm for comparing the areas under correlated receiver operating characteristic curves. *IEEE Signal Process. Lett.* **2014**, *21*, 1389–1393. [CrossRef]
24. Tan, M.; Le, Q. Efficientnet: Rethinking model scaling for convolutional neural networks. In Proceedings of the International Conference on Machine Learning, PMLR, Long Beach, CA, USA, 9–15 June 2019; pp.6105–6114.
25. LeCun, Y.; Bottou, L.; Bengio, Y.; Haffner, P. Gradient-based learning applied to document recognition. *Proc. IEEE* **1998**, *86*, 2278–2324. [CrossRef]
26. Simonyan, K.; Zisserman, A. Very deep convolutional networks for large-scale image recognition. *arXiv* **2014**, arXiv:1409.1556.
27. Lin, M.; Chen, Q.; Yan, S. Network in network. *arXiv* **2013**, arXiv:1312.4400.
28. He, K.; Zhang, X.; Ren, S.; Sun, J. Deep residual learning for image recognition. In Proceedings of the IEEE Conference on Computer Vision and Pattern Recognition, Las Vegas, NV, USA, 27–30 June 2016; pp. 770–778.
29. Szegedy, C.; Vanhoucke, V.; Ioffe, S.; Shlens, J.; Wojna, Z. Rethinking the inception architecture for computer vision. In Proceedings of the IEEE Conference on Computer Vision and Pattern Recognition, Las Vegas, NV, USA, 27–30 June 2016; pp. 2818–2826.
30. Huang, G.; Liu, Z.; Van Der Maaten, L.; Weinberger, K.Q. Densely connected convolutional networks. In Proceedings of the IEEE Conference on Computer Vision and Pattern Recognition, Honolulu, HI, USA, 21–26 July 2017; pp. 4700–4708.
31. Krizhevsky, A.; Sutskever, I.; Hinton, G.E. Imagenet classification with deep convolutional neural networks. *Adv. Neural Inf. Process. Syst.* **2012**, *25* . [CrossRef]
32. Baldi, P.; Sadowski, P.J. Understanding dropout. *Adv. Neural Inf. Process. Syst.* **2013**, *26* .
33. Shore, J.; Johnson, R. Properties of cross-entropy minimization. *IEEE Trans. Inf. Theory* **1981**, *27*, 472–482. [CrossRef]
34. Lin, T.Y.; Goyal, P.; Girshick, R.; He, K.; Dollár, P. Focal loss for dense object detection. In Proceedings of the IEEE International Conference on Computer Vision, Venice, Italy, 22–29 October 2017; pp. 2980–2988.
35. Hinton, G.; Srivastava, N.; Swersky, K. Overview of mini-batch gradient descent. *Neural Netw. Mach. Learn.* **2012**, *575* .
36. Jin, Q.; Meng, Z.; Sun, C.; Cui, H.; Su, R. RA-UNet: A hybrid deep attention-aware network to extract liver and tumor in CT scans. *Front. Bioeng. Biotechnol.* **2020**, 1471 . [CrossRef]
37. Liu, Z.; Yang, C.; Huang, J.; Liu, S.; Zhuo, Y.; Lu, X. Deep learning framework based on integration of S-Mask R-CNN and Inception-v3 for ultrasound image-aided diagnosis of prostate cancer. *Future Gener. Comput. Syst.* **2021**, *114*, 358–367. [CrossRef]
38. Ismael, S.A.A.; Mohammed, A.; Hefny, H. An enhanced deep learning approach for brain cancer MRI images classification using residual networks. *Artif. Intell. Med.* **2020**, *102*, 101779. [CrossRef]

39. Hildebrand, L.A.; Pierce, C.J.; Dennis, M.; Paracha, M.; Maoz, A. Artificial intelligence for histology-based detection of microsatellite instability and prediction of response to immunotherapy in colorectal cancer. *Cancers* **2021**, *13*, 391. [CrossRef]
40. Echle, A.; Laleh, N.G.; Quirke, P.; Grabsch, H.; Muti, H.; Saldanha, O.; Brockmoeller, S.; van den Brandt, P.; Hutchins, G.; Richman, S.; et al. Artificial intelligence for detection of microsatellite instability in colorectal cancer—A multicentric analysis of a pre-screening tool for clinical application. *ESMO Open* **2022**, *7*, 100400. [CrossRef]
41. Chen, M.; Zhang, B.; Topatana, W.; Cao, J.; Zhu, H.; Juengpanich, S.; Mao, Q.; Yu, H.; Cai, X. Classification and mutation prediction based on histopathology H&E images in liver cancer using deep learning. *NPJ Precis. Oncol.* **2020**, *4*, 1–7.

Current Oncology

MDPI

Article

Prediction of Postoperative Pathologic Risk Factors in Cervical Cancer Patients Treated with Radical Hysterectomy by Machine Learning

Zhengjie Ou [1,†], Wei Mao [1,†], Lihua Tan [1], Yanli Yang [2], Shuanghuan Liu [1], Yanan Zhang [1], Bin Li [1] and Dan Zhao [1,*]

[1] Department of Gynecology Oncology, National Cancer Center, National Clinical Research Center for Cancer, Cancer Hospital, Chinese Academy of Medical Sciences, Peking Union Medical College, Beijing 100021, China
[2] Department of Gynecology Oncology, The Fifth People's Hospital of Qinghai Province, Xining 310007, China
* Correspondence: zhaodan@cicams.ac.cn; Tel.: +86-010-8778-7384; Fax: +86-097-1636-0700
† These authors contributed equally to this work.

Abstract: Pretherapeutic serological parameters play a predictive role in pathologic risk factors (PRF), which correlate with treatment and prognosis in cervical cancer (CC). However, the method of pre-operative prediction to PRF is limited and the clinical availability of machine learning methods remains unknown in CC. Overall, 1260 early-stage CC patients treated with radical hysterectomy (RH) were randomly split into training and test cohorts. Six machine learning classifiers, including Gradient Boosting Machine, Support Vector Machine with Gaussian kernel, Random Forest, Conditional Random Forest, Naive Bayes, and Elastic Net, were used to derive diagnostic information from nine clinical factors and 75 parameters readily available from pretreatment peripheral blood tests. The best results were obtained by RF in deep stromal infiltration prediction with an accuracy of 70.8% and AUC of 0.767. The highest accuracy and AUC for predicting lymphatic metastasis with Cforest were 64.3% and 0.620, respectively. The highest accuracy of prediction for lymphavascular space invasion with EN was 59.7% and the AUC was 0.628. Blood markers, including D-dimer and uric acid, were associated with PRF. Machine learning methods can provide critical diagnostic prediction on PRF in CC before surgical intervention. The use of predictive algorithms may facilitate individualized treatment options through diagnostic stratification.

Keywords: blood biomarker; cervical cancer; deep stromal infiltration; lymph node metastasis; lymph-vascular space invasion; machine learning methods

Citation: Ou, Z.; Mao, W.; Tan, L.; Yang, Y.; Liu, S.; Zhang, Y.; Li, B.; Zhao, D. Prediction of Postoperative Pathologic Risk Factors in Cervical Cancer Patients Treated with Radical Hysterectomy by Machine Learning. *Curr. Oncol.* **2022**, *29*, 9613–9629. https://doi.org/10.3390/curroncol29120755

Received: 7 October 2022
Accepted: 29 November 2022
Published: 6 December 2022

1. Introduction

Cervical cancer remains one of the most frequent malignant tumors in women [1]. With the widespread application of human papillomavirus (HPV) vaccination and the popularity of screening, patients diagnosed at early stages have accounted for the majority. Radical hysterectomy (RH) is the standard-of-care treatment for these patients [2]. The unavoidable problem after surgery is whether adjuvant treatment is required, which is judged in accordance with postoperative pathological risk factors. The likelihood of risk factors that increase the risk of recurrence is high, especially in stage IB3-IIA2 (the 2018 International Federation of Gynecology and Obstetrics, FIGO) due to large tumor bulk [2]. Previous studies have illustrated that neoadjuvant chemotherapy (NACT) plus surgery inhibited micro-metastasis and distant metastasis of tumors, and was associated with a declined incidence of pathologic risk factors [3]. However, despite the fact that NACT reduces the rate of adjuvant therapy after surgery, patients treated with NACT cannot be thoroughly free from radiotherapy and the adverse effects that radiotherapy brings.

In addition, concurrent chemoradiotherapy (CCRT) is also an alternative initial treatment for early-stage cervical cancer, particularly for locally advanced cervical cancer. As for a patient with several pathologic risk factors, conformed to the adjuvant therapy standard.

CCRT should be considered as the initial therapy but not RH, which shortens the treatment process for the same effect and reduces treatment costs [4]. With regard to patients staged IB-IIA, according to the National Comprehensive Cancer Network (NCCN) guidelines, concurrent chemoradiation and RH both serve as alternative primary treatment options, sharing nearly therapeutic equivalence. However, increased morbidity and complications have been specifically illustrated when surgery and radiotherapy are combined [5,6]. This multimodal treatment modality has caused them to bear a double treatment burden and increased medical cost. In addition, the successive therapeutic process also prolongs the treatment period, aggregates their side effects and affects quality of life in the long run. Accordingly, it is necessary to construct a model to predict pathologic risk factors before primary treatment, which will help select those for whom it is more appropriate to receive direct chemoradiation therapy rather than RH. Additionally, the development of model to predict postoperative pathologic risk factors is an important element for individual prognosis stratification and personalized medicine.

Pathologic risk factors in cervical cancer include lymph node metastasis (LNM), parametria infiltration, positive surgical margins, lymph-vascular space invasion (LVSI), tumor size >4 cm and deep stromal infiltration (DSI) [2]. Previous studies illustrated that many clinicopathologic factors were related to pathologic risk factors by common statistical methods, but these methods were not suited to handle more complex data [7–9]. Machine learning is a branch of artificial intelligence (AI) technology that allows the computer to conclude potential rules from complicated data of retrospective examples. AI technology has been widely used to analyze clinical material to construct a model to predict clinicopathological factors and treatment outcome, acquiring a properly higher accuracy compared with traditional statistical methods [10–12]. Therefore, it is feasible and reasonable to apply machine learning to the prediction of postoperative pathologic risk factors.

Based on the successful application of AI technology and the discovery of related factors with pathologic risk factors, we hypothesized that pretreatment of clinicopathological factors would be effective in the prediction of postoperative pathologic risk factors by machine learning analysis in FIGO stage IB-IIA cervical cancer. In addition, because of the low incidence rate of positive margins and parametria infiltration in primary cohorts and preoperative confirmation of tumor size via clinical palpation, this study's outcome contained a prediction of other pathologic risk factors. Therefore, in the present study, we aimed to explore the construction of a model for predicting LNM, LVSI and DSI through machine learning combing of clinicopathological biomarkers and explore unreported significant parameters associated with these factors.

2. Materials and Methods

2.1. Patients and Considered Features

This was a retrospective cohort study of 1260 patients with FIGO stage (2003) IB and IIA cervical cancer who were treated with RH with retroperitoneal lymphadenectomy between 2003 and 2017 in our institution (National Cancer Center/Cancer Hospital, Chinese Academy of Medical Sciences; CICAMS). We retrospectively collected clinicopathological parameters, including age at diagnosis, body mass index (BMI), menopausal status, clinical FIGO stage, gross type, histologic grade, clinical tumor diameter, 75 preoperative peripheral blood biomarkers, etc. (Table 1 and Table S1). Tumor diameter was obtained via clinical palpation before surgical intervention.

Table 1. Clinical and pathologic characteristics of 1260 patients with cervical cancer.

Variables	All Patients (n = 1260)	Training Cohort (n = 630)	Test Cohort (n = 630)	p Value
Age (years)	45 (18–74)	45 (18–74)	45 (21–73)	0.777
BMI (kg/m^2)	23.6 (16.0–42.7)	23.6 (16.0–47.5)	23.7 (16.5–42.7)	0.453
Menopausal status				
Yes	353 (28.0%)	446 (70.8%)	461 (73.2%)	0.347
No	907 (72.0%)	184 (29.2%)	169 (26.8%)	
Clinical tumor diameter (cm)	3.5 (0.5–8.0)	3.5 (0.5–10.0)	3.5 (0.5–8.0)	0.211
Histology				
Squamous carcinoma	1053 (83.6%)	525 (83.3%)	528 (83.8%)	0.82
Adenocarcinoma	133 (10.6%)	69 (11.0%)	64 (10.2%)	0.647
Others	74 (5.8%)	36 (5.7%)	38 (6.0%)	0.811
FIGO stage (2003)				
IB1	707 (56.1%)	361 (57.3%)	346 (54.9%)	0.394
IB2	289 (22.9%)	142 (22.5%)	147 (23.3%)	0.738
IIA1	135 (10.7%)	60 (9.5%)	75 (11.9%)	0.172
IIA2	129 (10.3%)	67 (10.6%)	62 (9.8%)	0.642
Gross type				
Exophytic	1163 (92.3%)	587 (93.2%)	576 (91.4%)	0.245
Endophytic	97 (7.7%)	43 (6.8%)	54 (8.6%)	
Previous abdominal surgery				
Yes	255 (20.2%)	133 (21.1%)	122 (19.4%)	0.441
No	1005 (79.8%)	497 (78.9%)	508 (80.6%)	
Histologic grade				
Good	87 (6.9%)	43 (6.8%)	44 (7.0%)	0.912
Moderate	506 (40.2%)	256 (40.6%)	250 (39.7%)	0.73
Poor	667 (52.9%)	331 (52.5%)	336 (53.3%)	0.778
Deep stromal infiltration				
Negative	653 (51.8%)	335 (53.2%)	318 (50.5%)	0.338
Positive	607 (48.2%)	295 (46.8%)	312 (49.5%)	
Lymph-vascular space invasion				
Negative	829 (65.8%)	415 (65.9%)	414 (65.7%)	0.953
Positive	431 (34.2%)	215 (34.1%)	216 (34.3%)	
Lymph node metastasis				
Negative	1017 (80.7%)	496 (78.7%)	521 (82.7%)	0.074
Positive	243 (19.3%)	134 (21.3%)	109 (17.3%)	

2.2. Data Splitting

We obtained 1260 samples after preliminary preprocessing: removing medically impossible data (containing obvious record error), removing the features with 10% missing values and the samples with missing values. Variables of age, BMI, menopausal status, clinical tumor diameter, histology, FIGO stage, gross type, previous abdominal surgery, histologic grade (obtained via cervical biopsy preoperatively) and 75 pretreatment peripheral blood markers were all incorporated into the model construction. We started to handle the features: the continuous features were normalized and categorical features were one-hot coded, and LinearSVC method with L1 penalty was used to choose features.

The dataset was split into training and test cohorts according to a ratio of 1:1 by repeated random sampling until there was no significant difference (p value > 0.05) between the two cohorts with respect to the three tasks (Table 1). The p values were calculated using Chi-square or Fisher exact test for categorical variables, and the student's t-test or the Mann–Whitney U test were conducted for analyzing normally distributed or non-normally distributed continuous variables. This resulted in the training cohort and the test cohort both having 630 patients.

2.3. Supervised Machine Learning Classifiers

In this study, we evaluated six types of supervised machine learning classifiers, including GBM (Gradient Boosting Machine) [13,14], SVMRadial (Support Vector Machine with Gaussian kernel) [15], RF (Random Forest) [16], Cforest (Conditional Random Forest) [17], NB (Naive Bayes) [18] and EN (Elastic Net) [19]. In addition, a logistic regression classifier was used as a baseline. R software version 4.2.1 with R package caret was used to implement all classifiers. One hundred independent training sets were conducted using different random seeds in order to calculate variable importance for prediction. We used the median of variable importance acquired from each training as a representative value. The importance of each variable was calculated using the varImp function of the caret package. A RF classifier combines two machine learning techniques: bagging and random feature selection consisting of a group of decision trees. Cforest is an algorithm using conditional inference trees as base learners, implementing both the random forest and the bagging ensemble algorithm. EN is a logistic regression classifier trained by using a regularized method that linearly combines the L1 and L2 penalties of the lasso and ridge methods.

2.4. Model Assessment

To assess the performance of different models, we computed the accuracy (ACC) and the area under the ROC curve (AUC) on the test cohort as our evaluation metrics. Here, ACC was obtained by setting the threshold corresponding to the top left point of the ROC curve. As the AUC is independent of the chosen threshold, we used it as the main evaluation metric.

2.5. Confidence of Prediction and Shannon's Information Gain

Shannon's information gain was used to assess the prediction confidence [20]. If a patient, i, is lacking the information concerning the class that the patient is included in (k-class), the Shannon's information entropy representing uncertainty is expressed with:

$$H(i) = \log_2 k$$

If a classifier provides prediction probabilities for each class, the entropy will be:

$$H_c(i) = \sum_{j=1}^{k} p_j(i) \log_2(p_j(i))$$

Here, $p_j(i)$ is the predicted probability that the patient i is included in class j. Thus, we obtain the information gain, i.e., information gained by the prediction:

$$IG(i) = H(i) - H_c(i)$$

The individual information gain for each class is given by:

$$IG_j(i) = p_j(i) \times IG(i)$$

3. Results

3.1. Prediction of Deep Stromal Infiltration of Cervical Cancer Based on Multiple Preoperative Blood Markers Using Machine Learning Methods

Depth of stromal invasion was evaluated by an experienced pathologist and was recognized as significant, with more than one millimeter of invasion in the depth of the stroma in a microscopic examination. The status of the depth of stromal infiltration was classified into two groups: "non-deep" and "deep". The "deep" group referred to patients who had an invasive carcinoma with greater than one-third stromal invasion according to the pathologic findings. "Non-deep" indicated a carcinoma infiltrating no more than one third of the cervical stroma. The values for the highest ACC of the prediction and the AUC were 70.8% and 0.767 with RF classifier, which achieved a 5.4% higher score than the traditional method of multiple logistic regression analysis in AUC (Figure 1A; Supplemental Table S2). It is notable that the best two classifiers, RF and GBM, both used ensemble methods that combine weak decision trees.

Next we focused on the best model, RF and understood the variables. The relative importance of each variable for segregating deep stromal infiltration patients from non-deep infiltration ones was calculated for RF (Figure 1B). We identified the top eight factors, including SCC, D-D, tumor diameter, URIC, age, neut%, ALP and TP, as important RF predictors for distinguishing deep infiltration from non-deep infiltration. Standard box plots that presented the distribution of each variable between deep and non-deep samples are shown in Figure 1C.

Interestingly, we found that D-D was a critical variable, in addition to SCC. From the confusion matrix (Figure 1D), RF predicted 81 patients with deep infiltration as ones with non-deep infiltration and predicted 103 patients with non-deep infiltration as ones with deep infiltration. When we considered the Shannon gain to represent the confidence of predictions and chose those patients with certain higher confidence of predictions, the predictions designated as higher confidence (>0.2 bits from Shannon information gain computation) contained only 21 mispredictions out of 148 instances (Figure 1E). In particular, for the predictions with higher confidence, if a patient was predicted as non-deep, this was right at a rate of $1 - 7/52 = 86.5\%$.

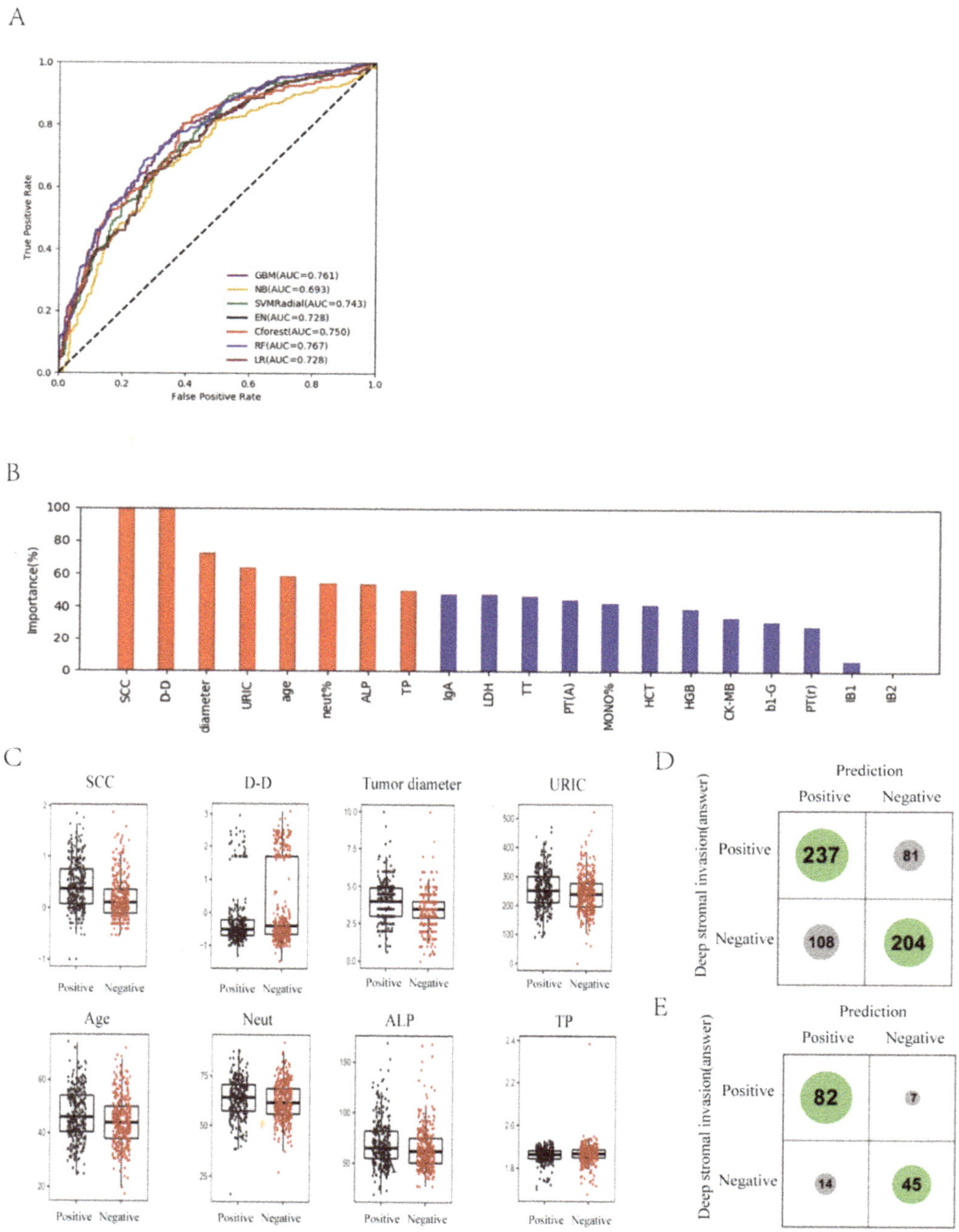

Figure 1. Prediction of deep stromal infiltration of cervical cancer based on multiple preoperative blood markers using machine learning methods. (**A**) ROC curves derived from logistic regression for predicting deep stromal infiltration of cervical cancer based on all 75 peripheral blood markers using machine learning methods compared with logistic regression. (**B**) Relative importance of variables for prediction of deep stromal infiltration calculated in the RF. Variable importance is represented as

a percentage of the highest value. (**C**) Box and jitter plots representing the distribution of top eight important parameters for distinguishing infiltration from non-infiltration. (**D**,**E**), Confusion matrix indicating the prediction quality of the RF classification for all predictions (**D**) and for those predictions with high (>0.2 bits) confidence (**E**). Notes: SCC, squamous cell carcinoma antigen; D-D, D-dimer; URIC, uric acid; ALP, alkaline phosphatase; TP, total protein; IgA, immunoglobulin A; LDH, lactate dehydrogenase; TT, thrombin time; PT(A), plasma prothrombin time ratio (A); MONO%, percentage of monocytes; HCT, hematocrit; HGB, hemoglobin; CK-MB, creatine kinase-MB isoenzyme; b1-G, beta 1 globulin; PT(r), plasma prothrombin time ratio (r).

3.2. Differentiation of Lymph Node Metastasis of Cervical Cancer with Machine Learning Methods

The status of lymph node metastasis was classified into two groups: "metastasis" and "non-metastasis". We found that Cforest showed the best prediction performance with an ACC of 64.3% and an AUC of 0.620 (Figure 2A; Supplemental Table S2), which achieved a 5.8% higher score than LR in AUC.

Next, the relative importance of a variable for segregating metastatic patients from non-metastatic ones was calculated for Cforest (Figure 2B). We identified the top eight factors, including SCC, IB2, IB1, MONO%, diameter, PT(A), HCT and TT, as important Cforest predictors for distinguishing metastatic patients from non-metastatic ones. It should be noted that as the clinical stage progresses, SCC and tumor diameter can increase. Standard box plots that presented the distribution of each variable between metastatic and non-metastatic samples are shown in Figure 2C.

Interestingly, we found that SCC was a critical variable. From the confusion matrix (Figure 2D), RF predictions had 105 false negative samples and 13 false positive samples. However, predictions designated as higher confidence (>0.2 bits from Shannon information gain computation) contained only 29 misprediction out of 230 instances (Figure 3E). In particular, for the predictions with higher confidence, if a patient was predicted as non-metastasis, this was right at a rate of 1 − 29/230 = 87.4%.

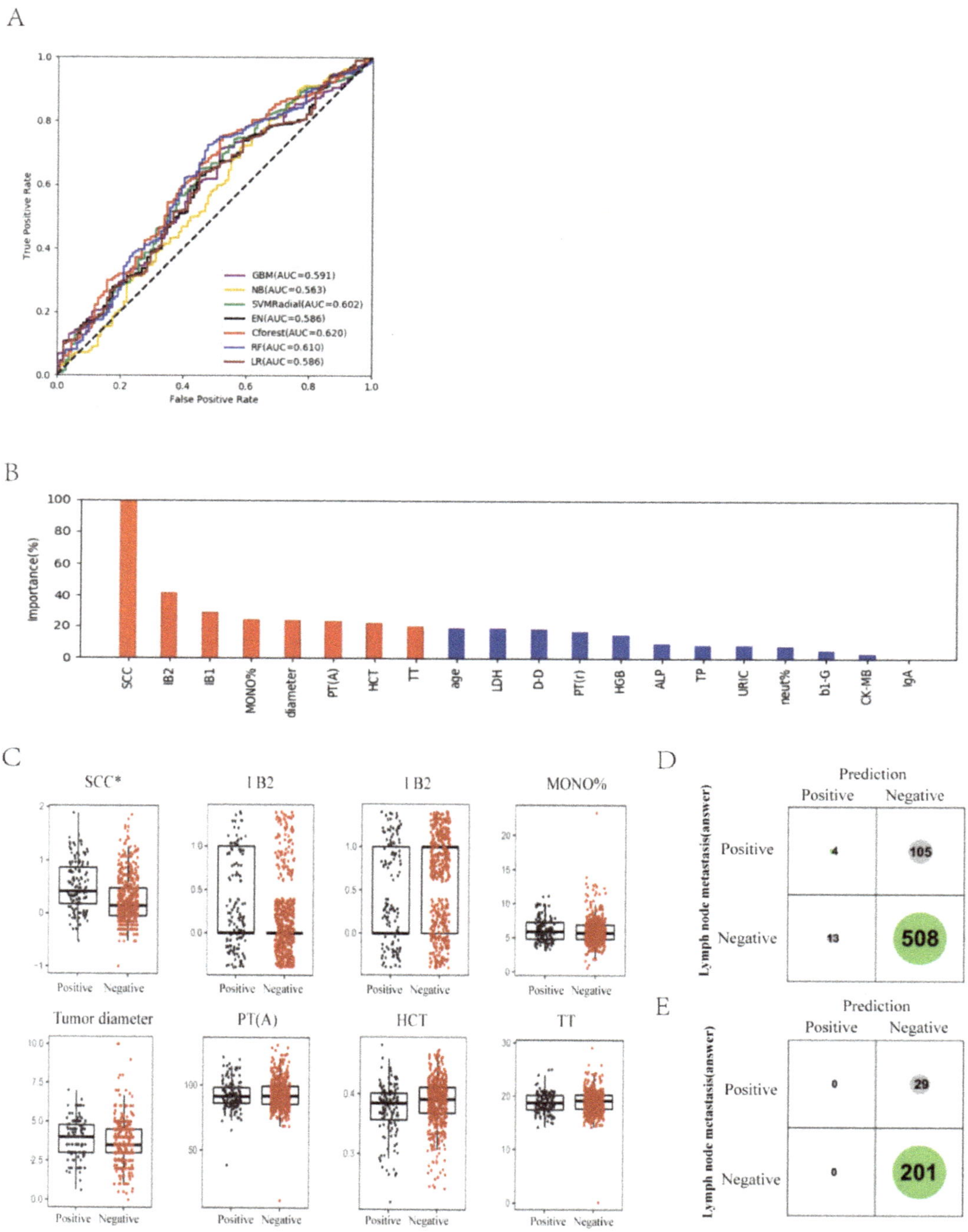

Figure 2. Differentiation of lymph node metastasis of cervical cancer with machine learning methods. **(A)** ROC curves derived from logistic regression for predicting lymph node metastasis of cervical cancer based on all 75 peripheral blood markers using machine learning methods compared with

logistic regression. (**B**) Relative importance of variables for prediction of lymph node metastasis calculated in the Cforest. Variable importance is represented as a percentage of the highest value. (**C**) Box and jitter plots representing the distribution of top eight important parameters for distinguishing metastasis from non-metastasis. (**D**,**E**), Confusion matrix indicating the prediction quality of the Cforest classification for all predictions (**D**) and for those predictions with high (>0.2 bits) confidence (**E**). Notes: SCC, squamous cell carcinoma antigen; MONO%, percentage of monocytes; PT(A), plasma prothrombin time ratio (A); HCT, hematocrit; TT, thrombin time; LDH, lactate dehydrogenase; D-D, D-dimer; PT(r), plasma prothrombin time ratio (r); HGB, hemoglobin; ALP, alkaline phosphatase; TP, total protein; URIC, uric acid; neut%, percentage of neutrophils; b1-G, beta 1 globulin; CK-MB, creatine kinase-MB isoenzyme; IgA, immunoglobulin A.

3.3. *Prediction of Lymph-Vascular Space Invasion of Cervical Cancer Based on Preoperative Blood Markers Using Machine Learning Methods*

In the task of lymph-vascular space invasion, patients were labeled as "invasion" or "non-invasion". LVSI refers to the presence of epithelial tumor cells in the lumen of vessels. "Invasion" indicated positive pathologic findings of LVSI and "non-invasion" indicated no pathologic proof of LVSI. We found that EN showed the best prediction performance, with ACC of 59.7% and AUC of 0.628, and the traditional method of multiple logistic regression analysis was comparative with ACC of 59.5% and AUC of 0.627 (Figure 3A; Supplemental Table S2).

Next, the relative importance of each variable for segregating invasion from non-invasion was calculated for EN (Figure 3B). We identified the top eight factors, including RDW-SD, CK-MB, PCT, A/G, PT(A), IB1, TT and TBIL, as important EN predictors for distinguishing invasion patients from non-invasion ones. Standard box plots that present the distribution of each variable between invasion and non-invasion are shown in Figure 3C.

Interestingly, we found that RDW-SD was a critical variable. From the confusion matrix (Figure 3D), EN predictions had 180 false negative samples and 36 false positive samples. However, predictions designated as higher confidence (>0.2 bits from Shannon information gain computation) contained only 15 misprediction out of 98 instances (Figure 3D,E). In particular, for the predictions with higher confidence, if a patient was predicted as non-invasion, it was right at a rate of $1 - 15/98 = 84.7\%$.

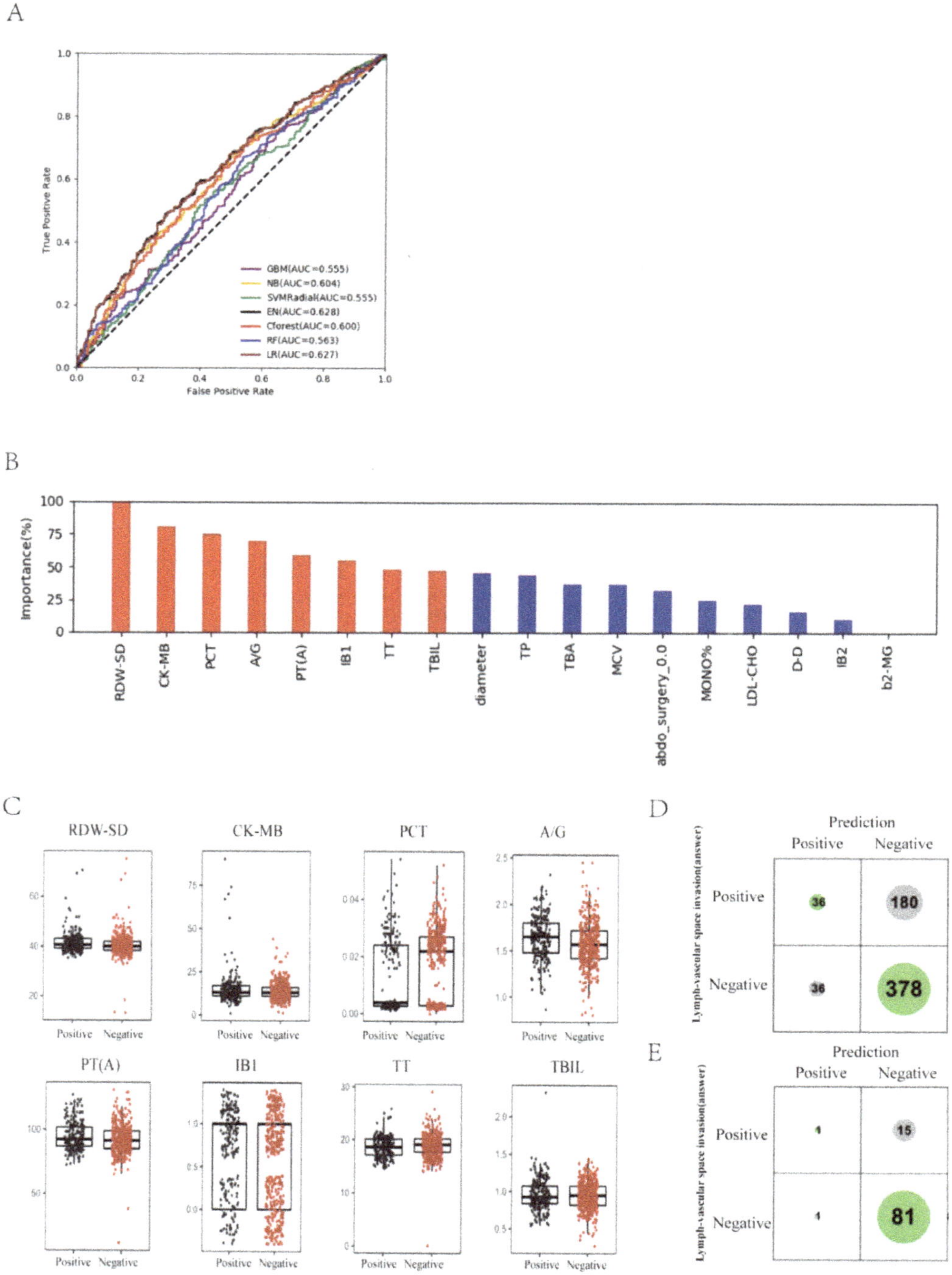

Figure 3. Prediction of lymph-vascular space invasion of cervical cancer based on preoperative blood markers using machine learning methods. (**A**) ROC curves derived from logistic regression for predicting lymph-vascular space invasion of cervical cancer based on all 75 peripheral blood markers

using machine learning methods compared with logistic regression. (**B**) Relative importance of variables for prediction of lymph-vascular space invasion calculated in the EN. Variable importance is represented as a percentage of the highest value. (**C**) Box and jitter plots representing the distribution of top eight important blood markers for distinguishing invasion from non-invasion. (**D**,**E**) Confusion matrix indicating the prediction quality of the EN classification for all predictions (**D**) and for those predictions with high (>0.2 bits) confidence (**E**). Notes: RDW-SD, standard deviation of red blood cell distribution width; CK-MB, creatine kinase-MB isoenzyme; PCT, plateletcrit; A/G, albumin to globulin ratio; PT(A), plasma prothrombin time ratio (A); TT, thrombin time; TBIL, total bilirubin; TP, total protein; TBA, total bile acid; MCV, mean corpuscular volume; abdo_surgery_0.0, previous abdominal surgery; MONO%, percentage of monocytes; LDL-CHO, low density lipoprotein cholesterol; D-D, D-dimer; b2-MG, beta 2 microglobulin.

4. Discussion

In recent years, machine learning algorithms based on AI technology have been widely accepted and extensively utilized for diagnostic and prognostic assessment of various types of cancers in the context of precision medicine [11,21,22]. This innovative approach, serving as an important tool with high accuracy and efficient ability to process complex data, can explore the key related factors to effectively assist in the clinical decision making of cervical cancer treatment. More importantly, hidden and embedded patterns within familiar clinical data can be revealed with the aid of AI models. However, so far, no studies have been conducted on integrating readily accessible clinical blood markers into the model construction of predicting pathologic risk factors in cervical cancer based on AI technology. Our study allowed for the comparison of various machine learning algorithms with the traditional logistic regression analysis to identify the approach with the most favorable performance and explore the serologic biomarkers with potential diagnostic potency. In cervical cancer with FIGO stage IB-IIA, radical hysterectomy followed by tailored adjuvant radiotherapy and concurrent chemoradiotherapy are both recommended for suitable treatment modalities [21]. Postoperative adjuvant radiotherapy is warranted for women with histopathologically verified risk factors, such as LVSI, LNM, DSI, etc., to improve prognosis [22–24], which led to an increase in the risk of higher morbidity [25–27]. It is beneficial and meaningful to predict pathologic risk factors so as to identify those more likely to receive postoperative adjuvant radiotherapy to avoid compounding treatment-related morbidity. Currently, the lack of ability to accurately identify those with a higher chance to receive postoperative radiotherapy and achieve individualized medical management instead of a "one-size fits all" approach has been a primary clinical limitation. Therefore, predicting pathologic risk factors by comprehensive utility of laboratory blood tests and other pretreatment information is a fundamental way toward individualized optimal medical care. In this study, we explored the ability of multiple machine learning methods to predict pathologic risk factors of patients with cervical cancer by incorporating readily available blood biomarkers. We found that three ensemble classifiers, RF, Cforest and EN, were able to predict pathologic risk factors of early-stage cervical cancer, in which RF showed the best predictive performance with an appreciable accuracy of 70.8% and AUC of 0.767 for DSI. Cforest showed the most accurate predictive value for LNM (64.3% accuracy and 0.620 AUC), and EN for LVSI (59.7% accuracy and 0.628 AUC). Compared to the traditional approach of logistic regression analysis, the RF classifier achieved a 5.4% higher score of AUC in DSI prediction, Cforest achieved a 3.4% higher score of AUC in LNM prediction and EN showed almost the same performance in LVSI prediction. The underperformance of these classifiers with regard to LNM and LVSI may be attributable to the lack of particularly strong distinctions of cervical cancer at the level of an early stage based on serum biomarkers. Nevertheless, the results indicate that AI technology can provide valuable predictive information before primary treatment to facilitate individualized medical strategy. In addition, based on the optimal results of machine learning algorithms, this study may offer useful clinical information concerning variables that are of most importance for identification of pathologic risk factors, like DSI, in early-stage patients.

Previous evidence has suggested that cancer is a metabolic disease associated with inflammation [28]. Cervical cancer harbors a unique collection of inflammatory and metabolic molecules in the serum [29]. In early-stage cervical cancer, local inflammatory processes may be at an initial state in which the peritumoral microenvironment perhaps alters the most, while distant and systemic metabolic features and cancer-target responses are immunosuppressed [30], leading to the slight distinction of cancer invasiveness, which was obscured in serum markers. Understandably, as tumor debulk progresses, tumor burden aggravates, leading to cancer invasiveness. In this study, we found that squamous cell carcinoma antigen (SCC), D-dimer and uric acid (UA) levels were the top five significant plasma biomarkers for predicting DSI. SCC has been considered as the most important diagnostic and prognostic tumor marker in cervical cancer. Many studies demonstrated that an elevated level of pretreatment serum SCC was closely associated with disease progression and recurrence [31,32]. UA is a powerful antioxidant and considered as a protective factor against cancer [33]. It has been reported that an elevated level of UA was associated with cancer risk, aggressiveness and poor oncologic outcomes in various cancer types [34–36], but few studies have focused on gynecologic cancer. Interestingly, previous studies have also shown a prooxidant role of UA [37] and lower levels of UA were associated with elevated risk of cancer-related mortality compared with high levels [38]. The precise relation of UA with cancer, especially cervical cancer, needs further study. D-dimer serves as a valuable marker of activation of coagulation and fibrinolysis, and is also known as a biomarker of cancer prognosis, especially in metastasized patients [39–41]. The pretreatment prediction model of DSI in cervical cancer performed well and revealed potential meaningful serum biomarkers that were readily available in clinical settings, which is also consistent with previous studies. This study's findings suggest that the supervised machine learning analysis serves as a feasible and effective approach that can aid in discovering more meaningful biomarkers that are correlated with PRF in cervical cancer and are not identified by conventional multiple regression analysis.

Identification of reliable pretreatment blood markers associated with pathologic risk factors helps clinicians in clinical decision making [42]. In this study, we found some serologic indicators, such as RDW-SD and other indicators, that had scarcely been found to be related to the diagnosis and prognosis of cervical cancer in previous studies. We found that RDW was the top predictive indicator for LVSI. RDW is a routinely measured hematological index, primarily reflecting the degree of anisocytosis. It has been reported that this simple and inexpensive parameter is a strong and independent risk factor for death in the general population [43]. Research has demonstrated that an aberrant elevation level of RDW leads to poor survival outcomes in most tumor types and stages, independent of age, gender or region [44]. However, little is known about RDW in cervical cancer. One recent study indicated that RDW was associated with worse prognosis in cervical cancer [45]. Excessive oxidative stress, inflammation, and cell senescence were proposed as the conditions that RDW associates closely with mortality [46,47]. More dataset analysis is still needed to confirm the predictive ability of these factors. Based on the high efficiency of pretreatment blood markers, the dynamic detection of serological indicators in multiple time periods may be more powerful in prediction. As the dynamic analysis of serological indicators is more complex, future studies should develop the use of artificial intelligence-based machine learning algorithms to identify the predictive features of preoperative blood variable time series, which might significantly facilitate the accuracy of clinical characteristics prediction and deserve further study.

As tumors progress over time, the signal transduction and correlation between the tumor and its microenvironment, including fibroblasts, tumor-related immune cells and endothelial cells, will become increasingly closer [48]. The changes of peripheral blood parameters before surgery were inherently a combination of tumor-specific and microenvironment-specific factors and the result of the interaction between tumor and microenvironment. Given the importance of tumor microenvironment in the process of tumor development, clinicians should make full use of preoperative peripheral blood indicators for treatment decision making, cancer progression evaluation and prognosis assessment. In previous studies, clinicians often ignored the reflection of regular blood biomarkers on the biological characteristics of tumors and relied almost exclusively on tumor-specific factors as included indicators for assessment, which was also a common problem in previous retrospective analysis of tumors. In this study, we identified a series of blood indicators that were readily available and necessary for preoperative evaluation related to pathologic risk factors by machine learning methods, such as UA, D-dimer, thrombin time, AST, MONO%, RDW-SD, etc. These parameters have the potential to be related to the microenvironment in cancer progression or metastasis, and their changes will also influence treatment timing and selection.

There have been a few previous studies exploring the use of serologic biomarkers to predict PRF. One study [49] in 2016 incorporated clinical factors and three blood markers derived from pretreatment blood routine examination to predict LNM, patients' overall survival and recurrence-free survival. They found platelet/lymphocyte ratio were significantly associated with LNM. Another study [50] in 2020 found that pretreatment albumin to fibrinogen ratio was significantly related to lymph node metastasis, depth of stromal infiltration, etc. Many studies focused on prediction for survival outcomes or a single PRF of cervical cancer based on clinical factors [51–53] and/or radiomic parameters [54,55]. However, no studies have made an attempt to predict three PRFs based on a series of clinically readily available blood markers. In addition to critical data analysis methods based on clinical factors, there are still many studies exploring new approaches of post-operative pathologic risk factors prediction. It is clear that the diagnosis of pathologic risk factors could only be accurately judged from the postoperative report of cervical cancer. Identification of reliable approaches that are able to predict pathologic risk factors in advance would facilitate the identification of more accurate diagnostic stratification and a more appropriate treatment strategy. A previous study indicated that DSI can be determined by combining the 2D or 3D ultrasound with clinical variables before treatment, with over 70% accuracy and AUC [56]. However, this diagnostic approach depended more on subjective judgment rather than objective parameters based on relatively few cases. It was reported that the assessment of cervical cancer with full-thickness stromal invasion by MRI examination was limited [57]. In Bidus's study, the conical method combined with clinical factors to determine DSI and LVSI before treatment also achieved good accuracy but this method is a destructive examination and may easily interfere with the complete resection of radical surgery [58]. In the study of LNM diagnosis, sentinel node staining is currently the most commonly developed method, but it is only used to determine whether complete lymph node resection is performed before surgery [59,60]. In this study, LNM was associated closely with primary tumor size as staging and tumor diameter were among the top five predictors for LNM. Results indicated that imaging materials, such as MRI, reflecting the visual size of the tumor itself and enlarged lymph nodes would potentially provide more accurate predictive information preoperatively. However, previous studies also used magnetic resonance imaging (MRI) and ultrasound to determine lymph node metastasis, but imaging data could only determine lymphadenectasis rather than tumor cell metastases in most cases, which leads to the unsatisfactory accuracy of the prediction model [56,61]. This is a reminder that traditional data analysis on simple integration of imaging information is not adequate enough to achieve LNM prediction. It is promising to achieve more comprehensive and precise prediction by virtue of effective integration of high-throughput extraction of a large amount of information from images based on AI technology, which will be the focus of our subsequent research. As the approach used in this study did not consider any information from pretreatment biopsies or imaging studies, there may be a limitation of the ability to predict pathologic risk factors before initial treatment; indeed, more independent datasets from other institutions are required to investigate how pretreatment blood signatures can be utilized for more accurate assessment of pathologic risk factors. Manipulation of high-throughput sequencing analysis, such as RNA sequencing, of pretreatment peripheral blood may improve predictive performance, however, from another perspective, it may become more complicated and expensive to incorporate RNA analysis information into the process of preoperative assessment in the current context of clinical settings. Further comprehensive investigation is needed in the hope of achieving the best clinical and socioeconomic benefits.

Our study has some limitations. Firstly, this study was a single-center retrospective study. The retrospective nature may result in inherent bias. Secondly, results from our database should be supplemented with external and prospective validation for prevention of overfitting as well as further spread of application in clinical practice. Thirdly, other machine learning approaches should be undertaken to manage the missing data in future work. Fourthly, our assessment of diagnostic ability to predict pathological risk factors was preliminary, and further study is warranted to better validate the accuracy of blood biomarkers. At present, our model is not sufficiently powerful and accurate to predict LVSI and LNM, but some blood biomarkers have been revealed for the first time that may be potentially useful predictors from a large number of variables. However, a positive prediction is not trivial; compared with traditional methods, the machine learning algorithms could serve as a feasible tool for clinicians to predict oncologic outcomes based solely on pretherapeutic information.

5. Conclusions

This study indicates that AI-based algorithms are useful tools that may aid in providing critical information for diagnostic evaluation of pathologic risk factors in patients with cervical cancer before initial treatment. The use of predictive algorithms may facilitate personalized treatment selection through pretherapeutic assessment.

Supplementary Materials: The following supporting information can be downloaded at: https://www.mdpi.com/article/10.3390/curroncol29120755/s1, Table S1: Pretreatment peripheral blood tests of 1260 cervical cancer patients included in the primary cohort; Table S2: Diagnostic accuracy of clinicopathological factors using machine learning algorithms.

Author Contributions: Conceptualization, D.Z., Y.Y. and B.L.; methodology, D.Z.; formal analysis, Z.O.; investigation, Z.O., W.M., L.T., S.L. and Y.Z.; resources, D.Z. and B.L.; data curation, Z.O., W.M., S.L. and Y.Z.; writing—original draft preparation, Z.O. and W.M.; writing—review and editing, D.Z. and W.M.; visualization, W.M.; supervision, D.Z.; project administration, D.Z. and B.L.; funding acquisition, D.Z. All authors have read and agreed to the published version of the manuscript.

Funding: This research was funded by the National Natural Science Foundation of China (D.Z., grant number 62176267), the Natural Science Foundation of Qinghai Province (D.Z., grant number 2021-ZJ-922); the CAMS Innovation Fund for Medical Sciences (D.Z., grant number 2021-I2M-C&T-B-048), the Beijing Hope Run Special Fund of Cancer Foundation of China (D.Z., grant number LC2021A10) and Capital's Funds for Health Improvement and Research (D.Z., grant number 2022-2-4026).

Institutional Review Board Statement: Ethical review and approval were waived for this study due to the retrospective nature of the data.

Informed Consent Statement: Patient consent was waived due to the retrospective nature of the study.

Data Availability Statement: The datasets used and/or analyzed during the current study are available from the corresponding author on reasonable request.

Conflicts of Interest: The authors declare no conflict of interest. The funders had no role in the design of the study; in the collection, analyses or interpretation of data; in the writing of the manuscript; or in the decision to publish the results.

References

1. Bray, F.; Ferlay, J.; Soerjomataram, I.; Siegel, R.L.; Torre, L.A.; Jemal, A. Global cancer statistics 2018: GLOBOCAN estimates of incidence and mortality worldwide for 36 cancers in 185 countries. *CA Cancer J. Clin.* **2018**, *68*, 394–424. [CrossRef] [PubMed]
2. Bhatla, N.; Aoki, D.; Sharma, D.N.; Sankaranarayanan, R. Cancer of the cervix uteri. *Int. J. Gynaecol. Obstet.* **2018**, *143* (Suppl. 2), 22–36. [CrossRef] [PubMed]
3. Peng, Y.H.; Wang, X.X.; Zhu, J.S.; Gao, L. Neo-adjuvant chemotherapy plus surgery versus surgery alone for cervical cancer: Meta-analysis of randomized controlled trials. *J. Obstet. Gynaecol. Res.* **2016**, *42*, 128–135. [CrossRef] [PubMed]
4. Landoni, F.; Colombo, A.; Milani, R.; Placa, F.; Zanagnolo, V.; Mangioni, C. Randomized study between radical surgery and radiotherapy for the treatment of stage IB-IIA cervical cancer: 20-year update. *J. Gynecol. Oncol.* **2017**, *28*, e34. [CrossRef] [PubMed]
5. Barter, J.F.; Soong, S.J.; Shingleton, H.M.; Hatch, K.D.; Orr, J.W., Jr. Complications of combined radical hysterectomy-postoperative radiation therapy in women with early stage cervical cancer. *Gynecol. Oncol.* **1989**, *32*, 292–296. [CrossRef] [PubMed]
6. Ayhan, A.; Al, R.A.; Baykal, C.; Demirtas, E.; Ayhan, A.; Yüce, K. Prognostic factors in FIGO stage IB cervical cancer without lymph node metastasis and the role of adjuvant radiotherapy after radical hysterectomy. *Int. J. Gynecol. Cancer* **2004**, *14*, 286–292. [CrossRef]
7. Kim, D.Y.; Shim, S.H.; Kim, S.O.; Lee, S.W.; Park, J.Y.; Suh, D.S.; Kim, J.H.; Kim, Y.M.; Kim, Y.T.; Nam, J.H. Preoperative nomogram for the identification of lymph node metastasis in early cervical cancer. *Br. J. Cancer* **2014**, *110*, 34–41. [CrossRef]
8. Hutchcraft, M.L.; Smith, B.; McLaughlin, E.M.; Hade, E.M.; Backes, F.J.; O'Malley, D.M.; Cohn, D.E.; Fowler, J.M.; Copeland, L.J.; Salani, R. Conization pathologic features as a predictor of intermediate and high risk features on radical hysterectomy specimens in early stage cervical cancer. *Gynecol. Oncol.* **2019**, *153*, 255–258. [CrossRef]
9. Li, X.; Zhou, J.; Huang, K.; Tang, F.; Zhou, H.; Wang, S.; Jia, Y.; Sun, H.; Ma, D.; Li, S. The predictive value of serum squamous cell carcinoma antigen in patients with cervical cancer who receive neoadjuvant chemotherapy followed by radical surgery: A single-institute study. *PLoS ONE* **2015**, *10*, e0122351. [CrossRef]
10. Obrzut, B.; Kusy, M.; Semczuk, A.; Obrzut, M.; Kluska, J. Prediction of 5-year overall survival in cervical cancer patients treated with radical hysterectomy using computational intelligence methods. *BMC Cancer* **2017**, *17*, 840. [CrossRef]
11. Matsuo, K.; Purushotham, S.; Jiang, B.; Mandelbaum, R.S.; Takiuchi, T.; Liu, Y.; Roman, L.D. Survival outcome prediction in cervical cancer: Cox models vs deep-learning model. *Am. J. Obstet. Gynecol.* **2019**, *220*, 381.e1–381.e14. [CrossRef]
12. Papadia, A.; Bellati, F.; Bogani, G.; Ditto, A.; Martinelli, F.; Lorusso, D.; Donfrancesco, C.; Gasparri, M.L.; Raspagliesi, F. When Does Neoadjuvant Chemotherapy Really Avoid Radiotherapy? Clinical Predictors of Adjuvant Radiotherapy in Cervical Cancer. *Ann. Surg. Oncol.* **2015**, *22* (Suppl. 3), S944–S951. [CrossRef]
13. Friedman, J.H. Greedy function approximation: A gradient boosting machine. *Ann. Stat.* **2001**, *29*, 1189–1232. [CrossRef]
14. Natekin, A.; Knoll, A. Gradient boosting machines, a tutorial. *Front. Neurorobot.* **2013**, *7*, 21. [CrossRef]
15. Cortes, C.; Vapnik, V. Support-Vector Networks. *Mach. Learn.* **1995**, *20*, 273–297. [CrossRef]
16. Breiman, L. Random forests. *Mach. Learn.* **2001**, *45*, 5–32. [CrossRef]
17. Liu, L.; Chen, L.; Zhang, K.; Liusan, N.; Yang, Z. Conditional Random Forest Based Smiling Face Detector, Has Random Forest Smile Classification Module for Detecting Dynamic Smiling Face Classifying Random Forest Non-Classification Face Area of Smiling Face. China Patent CN106650637-A, 10 May 2017.
18. Dv, L. Fiducial distributions and Bayes' theorem. *J. R. Stat. Soc.* **1958**, *1*, 102–107.
19. Zou, H.; Hastie, T. Regularization and variable selection via the elastic net. *J. R. Stat. Soc.* **2005**, *67*, 301–320. [CrossRef]
20. Feutrill, A.; Roughan, M. A Review of Shannon and Differential Entropy Rate Estimation. *Entropy* **2021**, *23*, 1046. [CrossRef]
21. Bhatla, N.; Aoki, D.; Sharma, D.N.; Sankaranarayanan, R. Cancer of the cervix uteri: 2021 update. *Int. J. Gynecol. Obstet.* **2021**, *155*, 28–44. [CrossRef]
22. Sedlis, A.; Bundy, B.N.; Rotman, M.Z.; Lentz, S.S.; Muderspach, L.I.; Zaino, R.J. A randomized trial of pelvic radiation therapy versus no further therapy in selected patients with stage IB carcinoma of the cervix after radical hysterectomy and pelvic lymphadenectomy: A Gynecologic Oncology Group Study. *Gynecol. Oncol.* **1999**, *73*, 177–183. [CrossRef] [PubMed]
23. Pieterse, Q.D.; Trimbos, J.B.M.Z.; Dijkman, A.; Creutzberg, C.L.; Gaarenstroom, K.N.; Peters, A.A.W.; Kenter, G.G. Postoperative radiation therapy improves prognosis in patients with adverse risk factors in localized, early-stage cervical cancer: A retrospective comparative study. *Int. J. Gynecol. Cancer* **2006**, *16*, 1112–1118. [CrossRef] [PubMed]
24. Ryu, S.-Y.; Park, S.-I.; Nam, B.-H.; Cho, C.-K.; Kim, K.; Kim, B.-J.; Kim, M.-H.; Choi, S.-C.; Lee, E.-D.; Lee, K.-H. Is adjuvant chemoradiotherapy overtreatment in cervical cancer patients with intermediate risk factors? *Int. J. Radiat. Oncol. Biol. Phys.* **2011**, *79*, 794–799. [CrossRef] [PubMed]
25. Peters, W.A.; Liu, P.Y.; Barrett, R.J.; Stock, R.J.; Monk, B.J.; Berek, J.S.; Souhami, L.; Grigsby, P.; Gordon, W.; Alberts, D.S. Concurrent Chemotherapy and Pelvic Radiation Therapy Compared With Pelvic Radiation Therapy Alone as Adjuvant Therapy After Radical Surgery in High-Risk Early-Stage Cancer of the Cervix. *J. Clin. Oncol.* **2000**, *18*, 1606–1613. [CrossRef] [PubMed]
26. Landoni, F.; Maneo, A.; Colombo, A.; Placa, F.; Milani, R.; Perego, P.; Favini, G.; Ferri, L.; Mangioni, C. Randomised study of radical surgery versus radiotherapy for stage Ib-IIa cervical cancer. *Lancet* **1997**, *350*, 535–540. [CrossRef]
27. Kong, T.-W.; Lee, J.-D.; Son, J.-H.; Paek, J.; Chun, M.; Chang, S.-J.; Ryu, H.-S. Treatment outcomes in patients with FIGO stage IB–IIA cervical cancer and a focally disrupted cervical stromal ring on magnetic resonance imaging: A propensity score matching study. *Gynecol. Oncol.* **2016**, *143*, 77–82. [CrossRef]

28. Wishart, D.S.; Mandal, R.; Stanislaus, A.; Ramirez-Gaona, M. Cancer Metabolomics and the Human Metabolome Database. *Metabolites* **2016**, *6*, 10. [CrossRef]
29. Yang, K.; Xia, B.; Wang, W.; Cheng, J.; Yin, M.; Xie, H.; Li, J.; Ma, L.; Yang, C.; Li, A.; et al. A Comprehensive Analysis of Metabolomics and Transcriptomics in Cervical Cancer. *Sci. Rep.* **2017**, *7*, 43353. [CrossRef]
30. Yuan, Y.; Cai, X.; Shen, F.; Ma, F. HPV post-infection microenvironment and cervical cancer. *Cancer Lett.* **2021**, *497*, 243–254. [CrossRef]
31. Charakorn, C.; Thadanipon, K.; Chaijindaratana, S.; Rattanasiri, S.; Numthavaj, P.; Thakkinstian, A. The association between serum squamous cell carcinoma antigen and recurrence and survival of patients with cervical squamous cell carcinoma: A systematic review and meta-analysis. *Gynecol. Oncol.* **2018**, *150*, 190–200. [CrossRef]
32. Choi, K.H.; Lee, S.W.; Yu, M.; Jeong, S.; Lee, J.W.; Lee, J.H. Significance of elevated SCC-Ag level on tumor recurrence and patient survival in patients with squamous-cell carcinoma of uterine cervix following definitive chemoradiotherapy: A multi-institutional analysis. *J. Gynecol. Oncol.* **2019**, *30*, e1. [CrossRef]
33. Ames, B.N.; Cathcart, R.; Schwiers, E.; Hochstein, P. Uric acid provides an antioxidant defense in humans against oxidant- and radical-caused aging and cancer: A hypothesis. *Proc. Natl. Acad. Sci. USA* **1981**, *78*, 6858–6862. [CrossRef]
34. Xu, Y.; Wu, Z.; Ye, W.; Xiao, Y.; Zheng, W.; Chen, Q.; Bai, P.; Lin, Z.; Chen, C. Prognostic value of serum uric acid and tumor response to induction chemotherapy in locally advanced nasopharyngeal carcinoma. *BMC Cancer* **2021**, *21*, 519. [CrossRef]
35. Hayashi, M.; Yamada, S.; Tanabe, H.; Takami, H.; Inokawa, Y.; Sonohara, F.; Shimizu, D.; Hattori, N.; Kanda, M.; Tanaka, C.; et al. High Serum Uric Acid Levels Could Be a Risk Factor of Hepatocellular Carcinoma Recurrences. *Nutr. Cancer* **2021**, *73*, 996–1003. [CrossRef]
36. Yan, S.; Zhang, P.; Xu, W.; Liu, Y.; Wang, B.; Jiang, T.; Hua, C.; Wang, X.; Xu, D.; Sun, B. Serum Uric Acid Increases Risk of Cancer Incidence and Mortality: A Systematic Review and Meta-Analysis. *Mediat. Inflamm.* **2015**, *2015*, 764250. [CrossRef]
37. Kang, D.H.; Ha, S.K. Uric Acid Puzzle: Dual Role as Anti-oxidantand Pro-oxidant. *Electrolyte Blood Press.* **2014**, *12*, 1–6. [CrossRef]
38. Kuo, C.F.; See, L.C.; Yu, K.H.; Chou, I.J.; Chiou, M.J.; Luo, S.F. Significance of serum uric acid levels on the risk of all-cause and cardiovascular mortality. *Rheumatology* **2013**, *52*, 127–134. [CrossRef]
39. Watanabe, A.; Araki, K.; Harimoto, N.; Kubo, N.; Igarashi, T.; Ishii, N.; Yamanaka, T.; Hagiwara, K.; Kuwano, H.; Shirabe, K. D-dimer predicts postoperative recurrence and prognosis in patients with liver metastasis of colorectal cancer. *Int. J. Clin. Oncol.* **2018**, *23*, 689–697. [CrossRef]
40. Kim, E.Y.; Song, K.Y. Prognostic value of D-dimer levels in patients with gastric cancer undergoing gastrectomy. *Surg. Oncol.* **2021**, *37*, 101570. [CrossRef]
41. Lin, Y.; Liu, Z.; Qiu, Y.; Zhang, J.; Wu, H.; Liang, R.; Chen, G.; Qin, G.; Li, Y.; Zou, D. Clinical significance of plasma D-dimer and fibrinogen in digestive cancer: A systematic review and meta-analysis. *Eur. J. Surg. Oncol.* **2018**, *44*, 1494–1503. [CrossRef]
42. Ma, J.Y.; Ke, L.C.; Liu, Q. The pretreatment platelet-to-lymphocyte ratio predicts clinical outcomes in patients with cervical cancer: A meta-analysis. *Medicine* **2018**, *97*, e12897. [CrossRef] [PubMed]
43. Montagnana, M.; Danese, E. Red cell distribution width and cancer. *Ann. Transl. Med.* **2016**, *4*, 399. [CrossRef] [PubMed]
44. Wang, P.F.; Song, S.Y.; Guo, H.; Wang, T.J.; Liu, N.; Yan, C.X. Prognostic role of pretreatment red blood cell distribution width in patients with cancer: A meta-analysis of 49 studies. *J. Cancer* **2019**, *10*, 4305–4317. [CrossRef] [PubMed]
45. Lima, P.S.V.d.; Mantoani, P.T.S.; Murta, E.F.C.; Nomelini, R.S. Laboratory parameters as predictors of prognosis in uterine cervical neoplasia. *Eur. J. Obstet. Gynecol. Reprod. Biol.* **2021**, *256*, 391–396. [CrossRef]
46. Salvagno, G.L.; Sanchis-Gomar, F.; Picanza, A.; Lippi, G. Red blood cell distribution width: A simple parameter with multiple clinical applications. *Crit. Rev. Clin. Lab. Sci.* **2015**, *52*, 86–105. [CrossRef] [PubMed]
47. Pan, J.; Borné, Y.; Engström, G. The relationship between red cell distribution width and all-cause and cause-specific mortality in a general population. *Sci. Rep.* **2019**, *9*, 16208. [CrossRef]
48. Whiteside, T.L. The tumor microenvironment and its role in promoting tumor growth. *Oncogene* **2008**, *27*, 5904–5912. [CrossRef]
49. Chen, L.; Zhang, F.; Sheng, X.G.; Zhang, S.Q.; Chen, Y.T.; Liu, B.W. Peripheral platelet/lymphocyte ratio predicts lymph node metastasis and acts as a superior prognostic factor for cervical cancer when combined with neutrophil: Lymphocyte. *Medicine* **2016**, *95*, e4381. [CrossRef]
50. Huang, L.; Mo, Z.; Zhang, L.; Qin, S.; Qin, S.; Li, S. Diagnostic Value of Albumin to Fibrinogen Ratio in Cervical Cancer. *Int. J. Biol. Markers* **2020**, *35*, 66–73. [CrossRef]
51. Chen, X.; Duan, H.; Liu, P.; Lin, L.; Ni, Y.; Li, D.; Dai, E.; Zhan, X.; Li, P.; Huo, Z.; et al. Development and validation of a prognostic nomogram for 2018 FIGO stages IB1, IB2, and IIA1 cervical cancer: A large multicenter study. *Ann. Transl. Med.* **2022**, *10*, 121. [CrossRef]
52. Chu, R.; Zhang, Y.; Qiao, X.; Xie, L.; Chen, W.; Zhao, Y.; Xu, Y.; Yuan, Z.; Liu, X.; Yin, A.; et al. Risk Stratification of Early-Stage Cervical Cancer with Intermediate-Risk Factors: Model Development and Validation Based on Machine Learning Algorithm. *Oncologist* **2021**, *26*, e2217–e2226. [CrossRef] [PubMed]
53. Yang, H.S.; Li, B.; Liu, S.H.; Ao, M. Nomogram model for predicting postoperative survival of patients with stage IB-IIA cervical cancer. *Am. J. Cancer Res.* **2021**, *11*, 5559–5570. [PubMed]
54. Du, W.; Wang, Y.; Li, D.; Xia, X.; Tan, Q.; Xiong, X.; Li, Z. Preoperative Prediction of Lymphovascular Space Invasion in Cervical Cancer With Radiomics–Based Nomogram. *Front. Oncol.* **2021**, *11*, 637794. [CrossRef] [PubMed]

55. Huang, G.; Cui, Y.; Wang, P.; Ren, J.; Wang, L.; Ma, Y.; Jia, Y.; Ma, X.; Zhao, L. Multi-Parametric Magnetic Resonance Imaging-Based Radiomics Analysis of Cervical Cancer for Preoperative Prediction of Lymphovascular Space Invasion. *Front. Oncol.* **2021**, *11*, 663370. [CrossRef] [PubMed]
56. Palsdottir, K.; Fischerova, D.; Franchi, D.; Testa, A.; Di Legge, A.; Epstein, E. Preoperative prediction of lymph node metastasis and deep stromal invasion in women with invasive cervical cancer: Prospective multicenter study using 2D and 3D ultrasound. *Ultrasound Obstet. Gynecol.* **2015**, *45*, 470–475. [CrossRef]
57. Okuno, K.; Joja, I.; Miyagi, Y.; Sakaguchi, Y.; Notohara, K.; Kudo, T.; Hiraki, Y. Cervical carcinoma with full-thickness stromal invasion: Relationship between tumor size on T2-weighted images and parametrial involvement. *J. Comput. Assist. Tomogr.* **2002**, *26*, 119–125. [CrossRef]
58. Bidus, M.A.; Caffrey, A.S.; You, W.B.; Amezcua, C.A.; Chernofsky, M.R.; Barner, R.; Seidman, J.; Rose, G.S. Cervical biopsy and excision procedure specimens lack sufficient predictive value for lymph-vascular space invasion seen at hysterectomy for cervical cancer. *Am. J. Obstet. Gynecol.* **2008**, *199*, 151.e1–151.e4. [CrossRef]
59. Salvo, G.; Ramirez, P.T.; Levenback, C.F.; Munsell, M.F.; Euscher, E.D.; Soliman, P.T.; Frumovitz, M. Sensitivity and negative predictive value for sentinel lymph node biopsy in women with early-stage cervical cancer. *Gynecol. Oncol.* **2017**, *145*, 96–101. [CrossRef]
60. Gortzak-Uzan, L.; Jimenez, W.; Nofech-Mozes, S.; Ismiil, N.; Khalifa, M.A.; Dube, V.; Rosen, B.; Murphy, J.; Laframboise, S.; Covens, A. Sentinel lymph node biopsy vs. pelvic lymphadenectomy in early stage cervical cancer: Is it time to change the gold standard? *Gynecol. Oncol.* **2010**, *116*, 28–32. [CrossRef]
61. Chen, X.L.; Chen, G.W.; Xu, G.H.; Ren, J.; Li, Z.L.; Pu, H.; Li, H. Tumor Size at Magnetic Resonance Imaging Association With Lymph Node Metastasis and Lymphovascular Space Invasion in Resectable Cervical Cancer: A Multicenter Evaluation of Surgical Specimens. *Int. J. Gynecol. Cancer* **2018**, *28*, 1545–1552. [CrossRef]

cancers

MDPI

Article

Using Whole Slide Gray Value Map to Predict HER2 Expression and FISH Status in Breast Cancer

Qian Yao [1,†], Wei Hou [1,†], Kaiyuan Wu [2], Yanhua Bai [1], Mengping Long [1], Xinting Diao [1], Ling Jia [1], Dongfeng Niu [1,*] and Xiang Li [2,*]

1 Department of Pathology, Key Laboratory of Carcinogenesis and Translational Research (Ministry of Education), Peking University Cancer Hospital and Institute, Beijing 100142, China
2 PingAn Technology, Beijing 100016, China
* Correspondence: dongfengniu@foxmail.com (D.N.); lixiang453@pingan.com.cn (X.L.); Tel.: +86-15801674868 (D.N.); +86-13810256327 (X.L.)
† These authors contributed equally to this work.

Simple Summary: HER2 expression is important for target therapy in breast cancer patients, however, accurate evaluation of HER2 expression is challenging for pathologists owing to the ambiguities and subjectivities of manual scoring. We proposed a deep learning framework using a Whole Slide gray value map and convolutional neural network model to predict HER2 expression level on immunohistochemistry (IHC) assay and predict HER2 gene status on fluorescence in situ hybridization (FISH) assay. Our results indicated that the proposed model is feasible for predicting HER2 expression and gene amplification and achieved high consistency with the experienced pathologists' assessment. This unique HER2 scoring model did not rely on challenging manual intervention and proved to be a simple and robust tool for pathologists to improve the accuracy of HER2 interpretation and provided a clinical aid to target therapy in breast cancer patients.

Abstract: Accurate detection of HER2 expression through immunohistochemistry (IHC) is of great clinical significance in the treatment of breast cancer. However, manual interpretation of HER2 is challenging, due to the interobserver variability among pathologists. We sought to explore a deep learning method to predict HER2 expression level and gene status based on a Whole Slide Image (WSI) of the HER2 IHC section. When applied to 228 invasive breast carcinoma of no special type (IBC-NST) DAB-stained slides, our GrayMap+ convolutional neural network (CNN) model accurately classified HER2 IHC level with mean accuracy 0.952 ± 0.029 and predicted HER2 FISH status with mean accuracy 0.921 ± 0.029. Our result also demonstrated strong consistency in HER2 expression score between our system and experienced pathologists (intraclass correlation coefficient (ICC) = 0.903, Cohen's κ = 0.875). The discordant cases were found to be largely caused by high intra-tumor staining heterogeneity in the HER2 IHC group and low copy number in the HER2 FISH group.

Keywords: breast cancer; HER2; artificial intelligence; deep learning; immunohistochemical (IHC) scoring

Citation: Yao, Q.; Hou, W.; Wu, K.; Bai, Y.; Long, M.; Diao, X.; Jia, L.; Niu, D.; Li, X. Using Whole Slide Gray Value Map to Predict HER2 Expression and FISH Status in Breast Cancer. *Cancers* **2022**, *14*, 6233. https://doi.org/10.3390/cancers14246233

Academic Editors: Hamid Khayyam, Ali Madani, Rahele Kafieh and Ali Hekmatnia

Received: 10 November 2022
Accepted: 14 December 2022
Published: 17 December 2022

1. Introduction

Breast cancer is the most diagnosed cancer that seriously threatens the life and health of women all over the world, with high morbidity and mortality rates of 24.5% and 15.5%, respectively [1]. The HER2 (human epidermal growth factor receptor-2) gene, located at chromosome 17q12–21[2], plays an important role in the development of breast cancer. Fifteen to twenty percent of breast cancer patients are HER2 positive, including HER2 gene amplification and/or overexpression. HER2-positive breast cancer has poor clinical outcomes [2,3], but fortunately, there is a targeted drug-Trastuzumab (Herceptin), which can effectively improve the prognosis [4,5]. HER2 gene amplification assessed by in situ hybridization (ISH) or protein overexpression assessed by IHC remains the primary predictor of responsiveness to HER2- targeted therapies and a key prognostic biomarker in breast cancer [6]. According to the latest American Society of Clinical Oncology (ASCO)/College of American Pathologists (CAP) guideline [6], all newly diagnosed patients with breast cancer must have a HER2 test performed. In routine clinical practice, the IHC test is first performed. The IHC test gives a score of 0, 1+, 2+, or 3+ that measures the amount of HER2 receptor protein on the surface of cells in a breast cancer tissue sample. The 3+ is the strongest staining, with which the patient must be diagnosed as HER2 positive. 2+ is also known as the equivocal level. Fluorescence in situ hybridization (FISH) must be performed to further decide the HER2 status for patients with IHC 2+ score. Therefore, accurate and efficient HER2 IHC evaluation is important for the diagnosis and treatment of breast cancer patients. In the HER2 IHC test, the HER2-receptor protein is commonly stained with 3,3′-diaminobenzidine (DAB), which has a brown color, meanwhile, hematoxylin staining which has blue color is also applied to visualize the cell nuclei. The stained slide is manually accessed by pathologists under the microscope. Although many countries have implemented national testing guidelines to standardize testing procedures and make results more accurate, the procedure is subjective and semi-quantitative and quite often leads to high inter- and intra-observer variation [7–9]. Therefore, there is an urgent need for an objective and consistent HER2 evaluation system.

Many researchers are devoted to developing computer-aided solutions, semi-automatically or fully automatically, to address the ambiguities and subjectivities of manual scoring. Compared to manual scoring, the computer-aided solution can decrease human error, increase the accuracy of diagnosis, reduce the workload of pathologists, and standardize the scoring systems [10,11]. The pathology whole slide images (WSI) have trillions of pixels, which are too large to process in a single-shot end-to-end way, i.e., processing WSI as a traditional image, even on modern computers. Usually, the fully automatic methods have the following three steps: WSI is first split into small size, i.e., 512 $\times$ 512, image patches; then information of single patch image are extracted; and at last single patch information are summarized to conclude the WSI level result. While the semiautomatic methods need pathologists to manually select regions of interest in the WSI. Masmoudi, et al. [12] presented a method for automated assessment of HER2 IHC staining. They first used a linear classification model on the color information of pixels to discriminate the membrane pixels and nuclei pixels, then watershed algorithm and adaptive ellipse fitting were applied to segment the nuclei and cell membrane. At last, slides were classified into one of the three scoring groups based on features describing the membrane staining intensity and completeness. In contrast to Masmoudi et al. work, HER2CONNECT found the distribution of the area of the connected brown color components (the stained membranes) in the core invasive cancer region had a good correlation with the HER2 expression level, therefore can be used to predict HER2 score. Their method reached 92.3% between the software and the score by the pathologist [13]. Ruifrok et al. [14] proposed a color deconvolution method to deconvolute and quantify the contributions of each staining in the histochemical slide. Motivated by the color convolution method, many researchers were devoted to quantifying the gray level of the HER2 IHC slide. ImmunoMembrane, a web-based application, utilized color deconvolution to separate stained membranes and then designed the IM-score, which is the sum of membrane completeness score and membrane intensity score to classify HER2 scores [15]. Kabakci et al. [16] characterized the cell membrane staining intensity in a comprehensive way using the so call Membrane Intensity Histogram (MIH) method which described the distribution of the staining intensity in different directions.

Deep Learning (DL) models are increasingly being used in various application areas such as computer vision, natural language processing, text or image classification, sentiment analysis, recommender systems, user profiling, etc. [17,18]. Compared to handcraft feature engineering, one of the major advantages of the DL model is the automatic learning feature representation and high representability, which bring the DL model much more versatility when dealing with large datasets and complex problems. Saha et al. [11] developed a cell segmentation model using Trapezoidal LSTM units and HER2 scoring based on the segmented membranes. However, Saha uses 2048 × 2048 patches, rather than the entire WSI. Qaiser et al. [19] also achieved patch-level HER2 scoring with the help of reinforcement learning. Zhen Chen, et al. [20] proposed a Focal-Aware Module to estimate diagnosis-related regions and a Relevance-enhanced Graph Convolutional Network to summarize information extracted from different levels of the original WSI.

Recently DL models are attracting increasing attention to predicting gene expression status using the WSI image [21–24]. The diagnosis label is usually provided at the WSI level, which cannot be treated as a cluster label of the inputs of the underline model. Therefore, multiple instance learning (MIL) is often implemented to overcome the issue. In this paper, we propose a new artificial intelligence (AI) method to predict HER2 protein expression level and gene status using the WSIs. Instead of using a manual strong label of patch level image or using MIL on the slide-level labeled dataset, we first calculate the unsupervised feature for each patch image, i.e., the gray level, the gray level area fraction, and generate a slide-level feature map using the patch-level feature to represent each patch. In this way, we can reduce the input size of the original slide. Then we build a multi-task deep learning model to predict HER2 protein expression level and gene amplification status simultaneously.

2. Material and Methods

Figure 1 shows the workflow of our study.

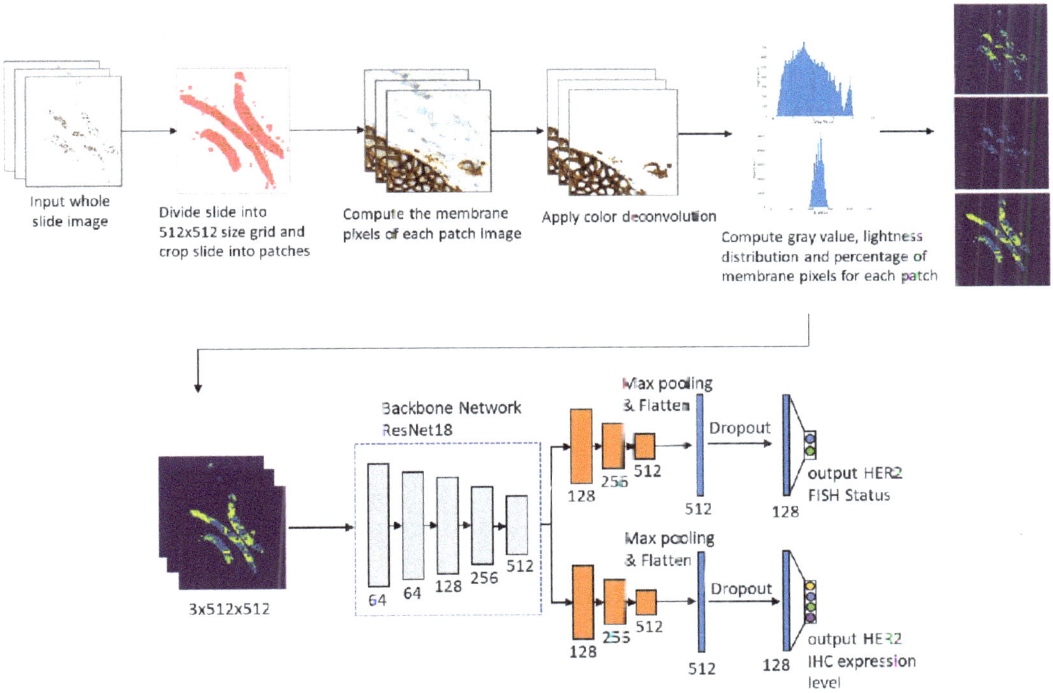

Figure 1. The workflow of our study includes the main steps for preprocessing slides and training the deep learning model. The numbers below the model block give the channel number respectively.

2.1. Human Subjects

We selected 228 biopsy cases of IBC-NST with both IHC and FISH information which were collected between 2010 and 2021 from the department of pathology, Peking University Cancer Hospital & Institute. All subjects were female. Our study obtained permission from the Peking University Cancer Hospital Institutional Review Board and Ethics Committee (Grant: 2022KT15).

2.2. ImmunohistoChemical Staining

Commercially available primary antibody HER2 (4B5, Roche Ventana) was applied. Immunohistochemical stains were performed on Ventana Benchmark automated immune-Stainer (Tucson, Arizona), following the vendor's protocol. The appropriate positive and negative controls were included for each run. HER2 immunoexpressing was evaluated as 0, 1+, 2+, and 3+ based on the 2018 ASCO/CAP guideline [6] by three experienced pathologists (Q.Y., D.N., and Y.B.). To prevent intra-rater variability, three pathologists were blind to the initial manual evaluation and AI-based scores, and all the cases were reviewed a second time after a 4-week washout period. The discrepant cases were reviewed again to get the final score.

2.3. Fluorescence In Situ Hybridization

HER2 FISH was carried out using the Path Vysion HER2 DNA Probe Kit (Abbott Molecular, Abbott Park, Illinois) and followed the manufacturer's instructions. Two experienced pathologists (DFN and Y.B.) evaluated the HER2 copy number, CEP17 copy number, and their ratios of 20 tumor cells independently and blinded to IHC results. FISH results were recorded as negative and positive according to the 2018 ASCO/CAP guideline. In detail, HER2 FISH results were designated into five groups: group one (G1, HER2/CEP17 ratio $\geq$ 2.0; average HER2 copy number $\geq$ 4.0/cell); group two (G2, HER2/CEP17 ratio $\geq$ 2.0; average HER2 copy number < 4.0/cell); group three (G3, HER2/CEP17 ratio < 2.0; average HER2 copy number $\geq$ 6.0/cell); group four (G4, HER2/CEP17 ratio < 2.0; 4.0 $\leq$ average HER2 copy number < 6.0/cell); and group five (G5, HER2/CEP17 ratio < 2.0; average HER2 copy number < 4.0/cell) [6]. G1 was considered FISH positive and G5 was FISH negative. However, G2 and G4 should evaluate the HER2 IHC results in addition, if not 3+, then those cases should be considered HER2 negative. In G3 cases, when concurrent IHC results are negative (0 or 1+), it is recommended that the specimen be considered HER2 negative.

2.4. Image Processing

The digitized whole-slide images (WSIs) were acquired using a Leica Aperio Versa pathologic scanner (Aperio, Leica Biosystems Imaging, Inc.) viewed at 400$\times$ magnification using Leica ImageScope software. The order of magnitude of pixels was $10^9 \sim 10^{10}$.

Figure 1 shows the flowchart of the method. The whole slide image was first partitioned into 512 $\times$ 512 patches. Then for each small patch image, we segment the membrane pixels using color deconvolution and the k-means method (k-means parameters: number of clusters is 3, the maximum number of iterations is 50, number of redos is 10). After the membrane segmentation, we evaluate the gray value and membrane pixels fraction of each patch. The original WSI is profiled into three maps. In the following, we describe the procedure in detail.

2.5. Membrane Segmentation

The DAB signal is mainly located at the membrane. In the following, we introduce the membrane segmentation method which is based on the color deconvolution and k-means method. Ruifrok etc. applied the Beer-Lambert law to model the stained slide image and proposed the color deconvolution method to separate and quantify immunohistochemical staining [14]. According to the Beer-Lambert law,

$$I_c = I_{0,c} 10^{-AC_c} \tag{1}$$

where I_c is the intensity of light detected after passing the specimen, $I_{0,c}$ is the intensity of light entering the specimen and A is the amount of the stain with absorption factor C. The subscript c indicates the detection channel. By assuming a linear relation between stain concentration and absorbance, Ruifrok proposed the following color deconvolution method,

$$A = -\log 10\left(\frac{I}{I_0}\right) \times OD^{-1} \quad (2)$$

where A is a vector representing the amount of different stains, I is the transmitted light intensity, i.e., the detected slide image, OD is the normalized optical density matrix, which can be measured experimentally. In the analysis of the HER2 IHC slide, because there are only two kinds of stains, we use the following normalized OD matrix

$$OD = \begin{pmatrix} 0.650 & 0.704 & 0.286 \\ 0.268 & 0.570 & 0.776 \\ 0.636 & -0.710 & 0.302 \end{pmatrix} \quad (3)$$

where the first two row vectors correspond to the OD vectors of hematoxylin and DAB[14] and the last row vector is the normalized cross product of hematoxylin and DAB OD vectors. Following the convention of color deconvolution code given in the Color Deconvolution 2 ImageJ plugin, we use $A = -\log 10\left(\frac{I}{255}\right) \times OD^{-1}$ to deconvolute the original slide image.

After color deconvolution, the value of the 2nd channel corresponds to the intensity of the DAB stain. We then apply the k-means method to the original image. The image is first converted from RGB to Luv color to get better perceptual uniformity which is more suitable for clustering analysis. Define the distance between pixels p, q:

$$D(p,q) = \sqrt{(L_p - L_q)^2 + (u_p - u_q)^2 + (v_p - v_q)^2} \quad (4)$$

where (L_p, u_p, v_p) and (L_q, u_q, v_q) are Luv values of pixel p and q, respectively. Based on the distance $D(p,q)$, we use the k-means algorithm to cluster the pixels in the slice into three clusters, which correspond to the stained cell membrane region, the nuclei region, and the complementary region respectively. At last, we calculate the mean gray values of each pixel group according to the DAB channel calculated previously. We select the group with the highest mean gray value as the cell membrane. Figure 2A–D gives an illustration of the cell membrane segmentation.

2.6. Gray Value Map

In this section, we describe the gray value map which integrates patch-level gray value information to get slide-level gray value information. After segmentation of the cell membrane of each patch image, we calculate the mean gray value and membrane pixel fraction of each patch image. We find that the value of the DAB channel cannot reflect well when the visual gray value is greater than 8, as shown in Figure 2E. By checking the RGB channel value of the membrane pixels, we find that this effect is partially caused by the saturation of the blue channel. It is unclear whether this is truly caused by the stain absorbing all blue light or whether there are some other effects of the hardware device. We notice that the Lightness channel of Luv color space generally reflects the visual gray level except the low gray value range. Therefore, we add the Lightness channel value to the gray value map and build the model to automatically fuse the information. In summary, the gray value A, membrane pixel fraction F, and Lightness value L at patch level are defined as:

$A = \text{mean}_i A_i$ where mean is over all pixels in the membrane cluster,

$F = \frac{\text{number of pixels in membrane cluster}}{\text{total number of pixels}}$,

$L = \text{mean}_i L_i$ where mean is over all pixels in the membrane cluster.

Figure 3 shows the gray value map of IHC HER2 expression 0/1+, 2+, and 3+ cases

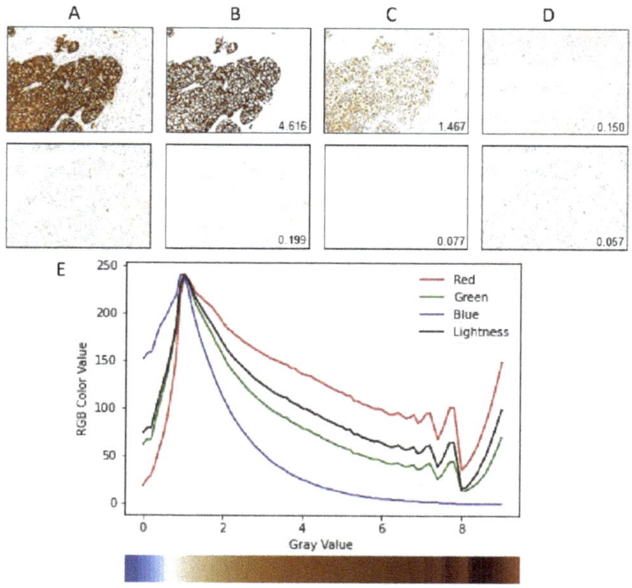

Figure 2. Cell membrane segmentation and the schematic of Graymap. (**A**) raw section of HER2 3+ and HER2 0/1+. (**B–D**) are three groups of K-Means output. The gray values are labeled on the images respectively. (**E**) The mean RGB value of different gray value membrane pixels. The bottom color bar is an RGB color map of different gray values.

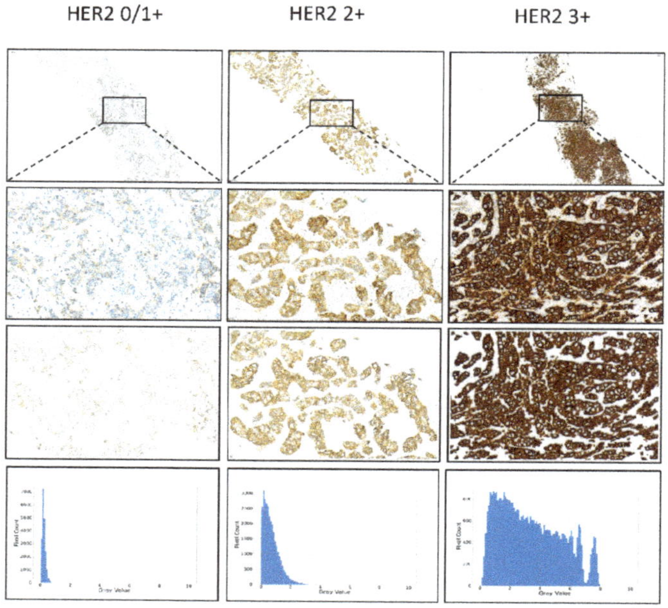

Figure 3. Examples of GrayMap of HER2 IHC expression. Typical examples of HER2 0/1+, 2+, 3+ cases in IBC-NST. From top to bottom: HER2 IHC raw images, magnified images, cell membrane segmentation, and pixels' gray value's distribution of the images.

2.7. Multitask Convolutional Neural Network (CNN)

After getting the gray value map of the whole slide, we further utilize a multi-task CNN model to classify the IHC HER2 expression level and the FISH status simultaneously. We use Resnet18 with base channel number 64 as our backbone network. After the backbone network, we concatenate two task branches corresponding to the IHC HER2 expression classification and the FISH status classification respectively. For each task branch, we use the sigmoid cross-entropy loss as the classification loss and add the dropout layer before the last fully connected layer. All Relu activations are replaced with PRelu to avoid the Relu blow-up issue due to a lack of pretrained weight initialization.

Data augment techniques and manually synthesized images are used to overcome the overfit issue due to the lack of training data samples. We add random rotation (−180, +180), random crop (512, 512) (raw training input size is (680, 680)), random horizontal flip, and random vertical flip data augmentations. We also manually synthesize the image for each original data sample by first manually drawing a mask of a random sample that has the same FISH status, and the same fold-id, but a lower HER2 expression level of the target sample and then paste the masked part of the selected sample into the target sample's blank space. In this way, we partially increase our training dataset.

The model is implemented in Pytorch using the MMDetection framework and trained with the Adam optimizer with Cosine learning rate policy (learning rate parameters base learning rate is 0.001, the minimum learning rate is 1.0×10^{-8}). We utilized the 5-fold cross-validation method to evaluate the model. The mean and standard deviation were calculated using prediction on each fold to demonstrate the model performance and stability. Evaluation metrics including precision, recall, F1-score, Jaccard Index, specificity, accuracy, and Area Under Curve of receiver operating characteristic curve (ROC) (AUC) were calculated for binary FISH status prediction. Evaluation metrics including accuracy, F1-score, Cohen's kappa coefficient (κ), and Mathews correlation coefficient (MCC) were calculated for multiclass IHC prediction using macro average mode.

3. Results

3.1. HER2 IHC Status Classification Using GrayMax Model

In the first step, we obtained the manual results of HER2 IHC and HER2 FISH. HER2 IHC was evaluated by three experienced pathologists. We used the median score of three pathologists to further reduce the inter-observer variability, which meant if there was a difference between the three scores, we used the median value of three scores. The details of the HER2 status including IHC and FISH results are shown in Table 1. According to the 2018 ASCO/CAP clinical practice guideline, the cutoff of HER2 IHC staining is 10%, which means the 10% strongest staining of HER2 IHC can be chosen as the represent score of the whole slice. So, we first use the maximum gray value of all patches to represent the gray value of WSI Then we compared the GrayMax model with the median HER2 scores of pathologists. However, after utilization of the 5-fold cross-validation method, the GrayMax model showed relatively inferior performance with an average accuracy of 0.842 ± 0.023, F1-score of 0.665 ± 0.078, *Cohen's* κ of 0.640 ± 0.063 and MCC of 0.663 ± 0.058 (Table 2) We analysed the details of our model and found the errors in the cases with a heterogeneity of staining, nonspecific cytoplasmic staining, and in cases with invasive micropapillary carcinoma component, mucinous carcinoma component and ductal carcinoma in situ (DCIS) component and interference by necrosis region.

Table 1. Summary of the cohort of the different HER2 statuses.

HER2 Expression Score					
Fish Status	n	0	1+	2+	3+
Negative	128	5	19	104	0
Positive	100	0	2	53	45

Table 2. Performance comparison of GrayMax and GrayMap + CNN methods by cross-validation classification.

Method	Fold	Accuracy	F1	Kappa	MCC
GrayMax	0	84.78%	69.71%	67.83%	68.51%
	1	84.78%	70.79%	60.92%	67.47%
	2	86.96%	76.87%	72.23%	73.87%
	3	80.00%	55.38%	53.71%	56.18%
	4	84.44%	59.93%	65.27%	65.49%
	Avg.	84.19%	66.54%	63.99%	66.30%
	Std.	2.28%	7.78%	6.31%	5.77%
GrayMap + CNN	0	93.48%	63.63%	83.13%	84.38%
	1	91.30%	84.65%	80.55%	82.54%
	2	95.65%	94.13%	91.54%	91.96%
	3	100.00%	100.00%	100.00%	100.00%
	4	95.56%	87.81%	90.36%	90.36%
	Avg.	95.20%	86.04%	89.12%	89.85%
	Std.	2.88%	12.39%	6.86%	6.18%

Abbreviation: Avg, Average value; Std, Standard deviation.

3.2. HER2 IHC Status Classification Using GrayMap + CNN Model

To solve the issues of the GrayMax model, we developed a new method to classify the HER2 IHC status. The main issue of the GrayMax model is that a single maximum gray value cannot represent the information of the whole slide. Therefore, we first used the GrayMap of the original whole slide, which contained the gray value information of all the patches, as described in the materials and methods section. Figure 2 showed the segmentation of the cell membrane and the schematic of GrayMap. Figure 3 showed typical examples of GrayMap in a subgroup of 0/1+, 2+, and 3+. Next, we utilized a multi-task CNN model to classify the IHC HER2 expression level as described in the material and methods section (Figure 1). We evaluated the model through a 5-fold cross-validation method and compared the results with three experienced pathologists. The experiment results show that the GrayMap model has much better performance than the GrayMax model with an average accuracy of 0.952 ± 0.029, F1-score of 0.860 ± 0.12, *Cohen's* κ of 0.891 ± 0.069 and MCC of 0.899 ± 0.062 (Table 2). Parameters of evaluation metrics on a subgroup of 0/1+, 2+, and 3+ showed in Figure 4A and Table S1. We further analyzed the intraclass correlation coefficient (ICC) among pathologists and found the ICC value was 0.791 (95% confidence interval [CI], 0.749–0.829) (Figure 4B). It indicated the presence of inter-observer variability and suggested that manual interpretation by the single pathologist may face a high risk of misdiagnosis. Then HER2-AI and HER2-pathologists were compared to show consistency between the AI system and pathologists. The median variables of HER2 pathologists were used in the comparison. The results showed a high consistency between the HER2-AI and HER2-pathologists (ICC = 0.903) (Figure 4C).

3.3. HER2 Gene Status Prediction Using GrayMap+ CNN Model

Since HER2 IHC expression largely represents the HER2 gene amplification status [25]. We also utilized the GrayMap model to predict HER2 gene status and compared the data with the FISH results. Our system demonstrated high performance in predicting HER2 gene status with an accuracy of 0.921, specificity of 0.945, precision of 0.927, recall of 0.89, F1-score of 0.908, and Jaccard Index of 0.832 (Figure 5A and Table S2) and AUC value of 0.936 in the ROC curve which presented the high quality in FISH classification via 5-fold cross-validation method (Figure 5B). This data further confirmed our model as a robust high-performance system not only in HER2 IHC classification but also in HER2 gene status prediction.

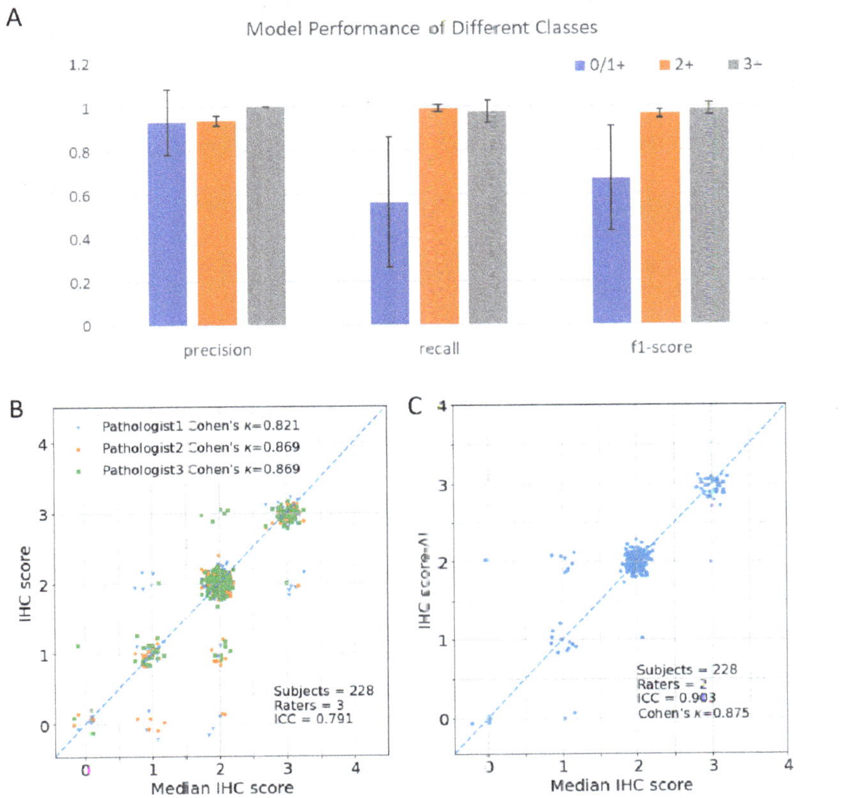

Figure 4. Consistency of the pathologists and the AI system on HER2 IHC classification. (**A**) Histograms of GrayMap model performance in a subgroup of 0/1+, 2+, and 3+. (**B**) The intraclass consistency of HER2 IHC scores in pathologists. (**C**) Consistency of HER2 between AI system (IHC score-AI) and median IHC score in pathologists (median IHC score).

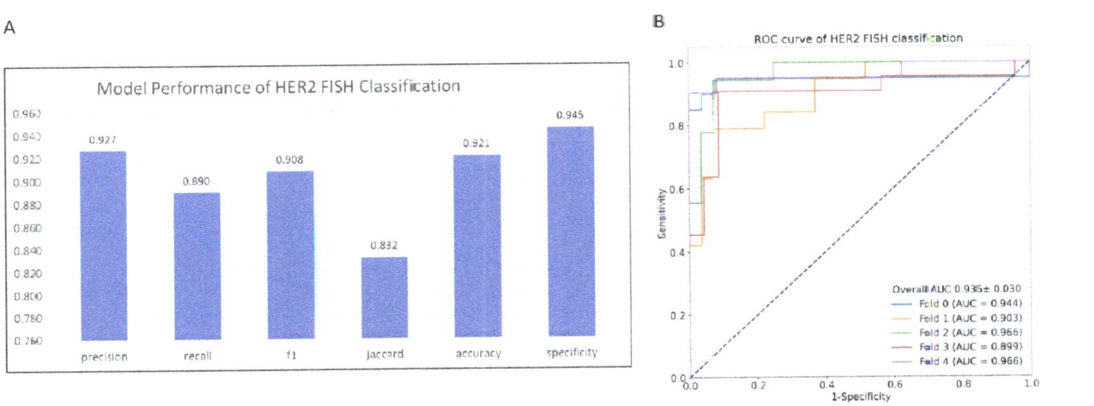

Figure 5. Performance of AI system on HER2 FISH classification. (**A**) Histograms of GrayMap model performance. (**B**) ROC curve of HER2 FISH status classification by cross-validation classification.

3.4. The Analysis of Discordant Cases

The proposed system correctly classified most of the WSIs. However, there were several discordant cases with false positive and negative samples (Figure 6A). We further analyzed the difference between AI systems and pathologists. As for the HER2 IHC results, 13 (13/228, 5.70%) cases were discordant between AI and pathologists. We investigated each case to identify the causes of the variability. Intra-tumor cell heterogeneity of HER2 staining was detected in six cases (6/13, 46.15%) (Figure 6B). Nonspecific cytoplasmic staining was found in four cases (Figure 6C). Another one was due to the nonspecific staining in DCIS (Figure 6D). Our result provided that HER2 staining heterogeneity was identified as the main driver of disagreement between AI and pathologists. Furthermore, the cytoplastic staining can interfere with the machine's extraction of cell membrane staining, resulting in misinterpretation. The nonspecific HER2 expression on DCIS will also lead to error, especially on biopsy tissue with a substantial amount of DCIS. HER2 validation is supposed to be performed only in the IBC-NST component. Since we did not annotate the IBC-NST region on WSIs, we calculated the DCIS component and found 75 cases (75/228, 32.89%) of samples had a DCIS component with a ratio of 5–35%. Only one case (1/75, 1.33%) was included in discordant cases, thus, our model had the ability to resolve the hidden trouble of DCIS. Only two cases could not find a clear explanation for discordance. According to HER2 FISH status, there were 18 (18/228, 7.89%) discordant cases. Five cases were identified intra-tumor cell heterogeneity of dual-color probes. For example, one case with only 2% tumor cells HER2 amplification and one case with 5%. Seven cases have low HER2 copy numbers (average copy number range 4–6 signals/cell). Three cases that were manually evaluated as negative belonged to the G2 and G4 groups, which were the new FISH group according to the 2018 ASCO/CAP guideline. Though the seven low-copy number cases were evaluated as positive and the new FISH group was regarded as negative, the efficacy of HER2-targeted therapy on these groups still needs to be investigated because of the limited evidence with a small subset of cases [6]. Only five cases were left without any explanation for discordance. Our results indicated that AI-based classification guaranteed high diagnostic accuracy and enabled us to reduce misinterpretation.

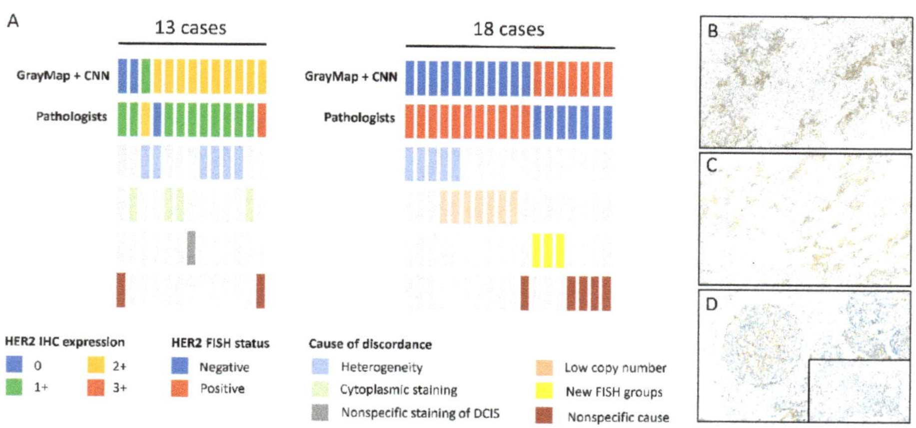

Figure 6. HER2 scoring discordance between pathologists and AI system and the possible causes of the variability. (**A**) Top 2 lines: Comparison between GrayMap model and the pathologist assessment; Bottom 4 lines: The possible causes of the variability; Left: The discordant cases on HER2 IHC classification; Right: The discordant cases on HER2 FISH classification. Vertical bars represent single cases and the representation of different colors are listed at the bottom. The typical image of (**B**) HER2 staining heterogeneity, (**C**) nonspecific cytoplasmic staining, (**D**) nonspecific staining in ductal carcinoma in situ (DCIS) with negative staining of the invasive component.

4. Discussion

In this paper, we proposed a new AI method to tackle the subjectivity and inter-observer disagreement issues of manual interpretation of HER2 IHC slides. The experiments' results showed that the new method could accurately predict HER2 protein expression level (Accuracy 0.95 $\pm$ 0.029, Cohen's κ 0.891 $\pm$ 0.069) and FISH status (AUC 0.936 $\pm$ 0.030). The test of concordance with the three pathologists' interpretation showed that the new method has the highest ICC (ICC 0.903, 95%-Confidence Interval 0.875 $\sim$ 0.924). Breast cancer (BC) has become the most common cancer diagnosed in women. Personalized medicine, especially drugs focused on target genes in BC, such as trastuzumab, has greatly improved survival. HER2 protein expression level and gene amplification status are the most important indicators for the targeted therapy of BC. However, traditional manual interpretation of HER2 slide has been criticized for subjectivity and inter-observer disagreement among pathologists. This is not only caused by the subjective decision that needs clinic pathologists to take, such as completeness of the membrane staining, intensity of staining, and percentage of positive cells, according to the ASCO/CAP guideline, but also caused by the heterogeneity of BC. AI-based methods, because of the nature of the parametrized model and deterministic behavior, are a prospective approach to solving the pool reproducibility issue of manual interpretation. However, on one hand, the whole slide image is too large to be processed by a single model directly, on the other hand, a single patch-level image of WSI is not able to capture the heterogeneity property of BC. Currently, there are several approaches to solving this issue. The first approach predicts the HER2 expression of each patch and uses the statistical average method to summarize the patch-level results. Compared to this approach, the method proposed in this work adopts a deep learning model to do slide-level predictions, which are more flexible and powerful than the simple statistical average method. Another approach generally follows the ASCO/CAP guideline, making predicting at the cell level. This approach needs considerable human labeling which is not only tedious but also prone to label error, especially for weak staining samples. The weakly Supervised Learning (WSL) method is an attractive method to alleviate patch-level labeling [26]. However, WSL needs a considerable amount of slide-level data. Currently, the performance of WSL on a large HER2 IHC dataset is unclear yet. The method proposed in this work could be another prospective approach to do slide-level predictions.

The proposed AI system can be applied in our actual work in the pathology department. After uploading the WSIs into the system, our model can automatically process patches splitting, cell segmentation, gray value map information extraction, and HER2 IHC and FISH results prediction. The system assists pathologists by pre-reading HER2 IHC slides and presenting calculated results as second opinions to pathologists, especially those with equivocal results as 2+. Our system will significantly mitigate the interobserver discrepancy and contribute to the efficacy and safety of HER2-targeted therapies on BC. At present, a new HER2-low subtype was defined by a score of IHC 1 +or IHC 2+/FISH −, who may benefit from the new HER2-ADC drugs, such as trastuzumab deruxtecan (T-DXd) [27]. The current system has the potential to recognize HER2-low cases with an accurate prediction of both IHC and FISH status.

In our study, compared to the former GrayMax algorithm, the upgraded GrayMap + CNN model can get rid of the most nonspecific and heterogeneous staining problem as well as the special staining pattern of specific breast cancer subtypes in HER2 IHC classification. However, inconsistency between AI systems and pathologists still exists. Consistent with the previous study, HER2 staining heterogeneity was identified as the main driver of disagreement [28]. Intratumoral heterogeneity of HER2 may be due to intrinsic the characteristics of BC, defined as regional heterogeneity and genetic heterogeneity [29]. It may also be caused by IHC procedures, tissue collection, and processing, or slide scanning procedure. In our dataset, most heterogeneity staining cases of the discordant cohort were weak staining thus our model need to improve its capability in dealing with weak HER2 staining. As for HER2 FISH classification, in addition to heterogeneity, a low copy number (average copy number range 4–6 signals/cell) was the most common cause of inconsistency. According to the 2018 guideline, an average HER2 copy number ≥4 signal/cell is regarded as FISH positive. However, the study showed a clear difference on HER2 copy levels using droplet digital PCR (ddPCR) and targeted next-generation sequencing (NGS) method between the 4–6 copy number groups and ≥6 groups. However, it remains unclear if patients of the 4–6 copy number group derive the same level of benefit as the≥6 groups in HER2-targeted therapy [30]. Futhermore, there were three cases belonging to G2 and G4 groups according to the 2018 guideline, which was the new FISH and should be recognized as FISH negative. However, the researcher showed the G2 group represents a biologically heterogeneous subset, which is different from those in G1 (FISH positive) and G5 (FISH negative) [31]. The G4 group was also proved to be a distinct group with intermediate levels of RNA/protein expression, close to positive/negative cut points [32]. Additional outcome information after HER2-targeted treatment is needed for the new FISH groups.

To improve the accurate, precise, and reproducible interpretation of HER2 IHC results for BC, where quantitative image analysis (QIA) is applied, The College of American Pathologists (CAP) developed the guideline with eleven recommendations [33]. The recommendations suggested that QIA and procedures must be validated before implementation, followed by regular maintenance and ongoing evaluation of quality control and quality assurance. In addition, HER2 QIA performance, interpretation, and reporting should be supervised by pathologists with expertise in QIA. We studied the detailed description of the guideline and found our AI model and procedures met most of the criteria, which suggested the present model is a promising tool for HER2 interpretation. However, this study still had some limitations. First, this work uses the k-means method to segment the cell membrane. It may wrongly classify the cytoplasmic pixels into membrane when the cell is weakly stained or cytoplastic immunohistochemical staining. For most of the weakly stained cases, the method is still able to do correct predictions, because the intensity and percentage of positive cells are major discrimination factors. However, for cytoplastic staining cases, as also demonstrated in the analysis of discordant cases section (four out of 13 total error cases), more local features are needed to discriminate the wrong cases. Secondly, we did not segment the invasive carcinoma region first. The current method relies on the deep learning model to automatically learn features from the data. In future works, we will collect more data and investigate the performance difference between the current method and model which makes predictions only rely on carcinoma region. Third, the completeness of the cell membrane is not represented in the current method. 2018 ASCO/CAP guidelines lay more emphasis on the completeness of cell membrane staining on HER2 2+ and 3+ cases in order to reduce the confusion of pathologists and allow greater discrimination between positive and negative results [6]. Our AI system promised high performance without calculating membrane completeness, however, a feature still needed to be found to represent the completeness of cell membrane staining according to the ASCO/CAP guideline to get a better result.

In conclusion, experimental results indicated that the proposed AI model is feasible for predicting HER2 expression score and HER2 gene amplification using IHC WSI and achieved high consistency with the experienced pathologists' assessments. This unique HER2 scoring model does not rely on challenging manual intervention and is proven to be a simple and robust tool for pathologists to improve the accuracy of HER2 interpretation and provides a clinical aid to target therapy in BC patients.

Supplementary Materials: The following supporting information can be downloaded at: https://www.mdpi.com/article/10.3390/cancers14246233/s1, Table S1: HER2 IHC classification performance of GrayMap methods by cross-validation classification in the subgroup of 0/1+, 2+, and 3+. Table S2: HER2 FISH prediction performance of GrayMap methods on the subgroup of 0/1+, 2+, and 3+.

Author Contributions: Conceptualization, D.N., K.W., Q.Y., and W.H.; Methodology, X.D., and L.J.; Investigation, W.H., K.W., Y.B., and M.L.; Writing—Original Draft, K.W., Q.Y., and D.N.; Writing—Review & Editing, K.W., and Q.Y.; Funding Acquisition, W.H.; Resources, W.H., and Q.Y.; Supervision, D.N., and X.L. All authors have read and agreed to the published version of the manuscript.

Funding: This work was supported by the National Natural Science Foundation of China (No. 81301879 and No. 81702839), Hygiene and Health Development Scientific Research Fostering Plan of Haidian District Beijing (No. HP2022-31-503001), Science Foundation of Peking University Cancer Hospital (No. 2021-11).

Institutional Review Board Statement: This study obtained permission from the Peking University Cancer Hospital Institutional Review Board and Ethics Committee (Grant: 2022KT15).

Informed Consent Statement: The requirements for informed consent were waived due to the noninvasive nature of the study.

Data Availability Statement: All image data associated with this study can be downloaded at https://data.mendeley.com/datasets/3njjk252vc/draft?a=29e5963c-e2d6-4bdb-9c7b-b51d3741b6f0. The source code and the guideline are publicly available at https://github.com/KaiyuanWu/Her2GrayMap. Any further information and requests for resources and materials should be directed to and will be fulfilled by the lead contact, Dongfeng Niu (dongfengniu@foxmail.com).

Conflicts of Interest: The authors declare no conflict of interests.

References

1. Sung, H.; Ferlay, J.; Siegel, R.L.; Laversanne, M.; Soerjomataram, I.; Jemal, A. Bray, F. Global Cancer Statistics 2020: GLOBOCAN Estimates of Incidence and Mortality Worldwide for 36 Cancers in 185 Countries. *CA Cancer J. Clin.* **2021**, *71*, 209–249. [CrossRef] [PubMed]
2. Slamon, D.J.; Clark, G.M.; Wong, S.G.; Levin, W.J.; Ullrich, A.; McGuire, W.L. Human breast cancer: Correlation of relapse and survival with amplification of the HER-2/neu oncogene. *Science* **1987**, *235*, 177–182. [CrossRef] [PubMed]
3. Tandon, A.K.; Clark, G.M.; Chamness, G.C.; Ullrich, A.; McGuire, W.L. HER-2/neu oncogene protein and prognosis in breast cancer. *J. Clin. Oncol.* **1989**, *7*, 1120–1128. [CrossRef] [PubMed]
4. Cameron, D.; Piccart-Gebhart, M.J.; Gelber, R.D.; Procter, M.; Goldhirsch, A.; de Azambuja, E.; Castro, G., Jr.; Untch, M.; Smith, I.; Gianni, L.; et al. 11 years' follow-up of trastuzumab after adjuvant chemotherapy in HER2-positive early breast cancer: Final analysis of the HERceptin Adjuvant (HERA) trial. *Lancet* **2017**, *389*, 1195–1205. [CrossRef]
5. Woo, J.W.; Lee, K.; Chung, Y.R.; Jang, M.H.; Ahn, S.; Park, S.Y. The updated 2018 American Society of Clinical Oncology/College of American Pathologists guideline on human epidermal growth factor receptor 2 interpretation in breast cancer: Comparison with previous guidelines and clinical significance of the proposed in situ hybridization groups. *Hum. Pathol.* **2020**, *98*, 10–21.
6. Wolff, A.C.; Hammond, M.E.H.; Allison, K.H.; Harvey, B.E.; Mangu, P.B.; Bartlett, J.M.S.; Bilous, M.; Ellis, I.O.; Fitzgibbons, P.; Hanna, W.; et al. Human Epidermal Growth Factor Receptor 2 Testing in Breast Cancer: American Society of Clinical Oncology/College of American Pathologists Clinical Practice Guideline Focused Update. *J. Clin. Oncol.* **2018**, *36*, 2105–2122. [CrossRef]
7. Lacroix-Triki, M.; Mathoulin-Pelissier, S.; Ghnassia, J.P.; Macgrogan, G.; Vincent-Salomon, A.; Brouste, V.; Mathieu, M.C.; Roger, P.; Bibeau, F.; Jacquemier, J.; et al. High inter-observer agreement in immunohistochemical evaluation of HER-2/neu expression in breast cancer: A multicentre GEFPICS study. *Eur. J. Cancer* **2006**, *42*, 2946–2953. [CrossRef]
8. Thomson, T.A.; Hayes, M.M.; Spinelli, J.J.; Hilland, E.; Sawrenko, C.; Phillips, D.; Dupuis, B.; Parker, R.L. HER-2/neu in breast cancer: Interobserver variability and performance of immunohistochemistry with 4 antibodies compared with fluorescent in situ hybridization. *Mod. Pathol.* **2001**, *14*, 1079–1086. [CrossRef]
9. Press, M.F.; Sauter, G.; Bernstein, L.; Villalobos, I.E.; Mirlacher, M.; Zhou, J.-Y.; Wardeh, R.; Li, Y.-T.; Guzman, R.; Ma, Y.; et al. Diagnostic evaluation of HER-2 as a molecular target: An assessment of accuracy and reproducibility of laboratory testing in large, prospective, randomized clinical trials. *Clin. Cancer Res.* **2005**, *11*, 6598–6607. [CrossRef]
10. Khameneh, F.D.; Razavi, S.; Kamasak, M. Automated segmentation of cell membranes to evaluate HER2 status in whole slide images using a modified deep learning network. *Comput. Biol. Med.* **2019**, *110*, 164–174. [CrossRef]
11. Saha, M.; Chakraborty, C. Her2Net: A Deep Framework for Semantic Segmentation and Classification of Cell Membranes and Nuclei in Breast Cancer Evaluation. *IEEE Trans. Image Process.* **2018**, *27*, 2189–2200. [CrossRef] [PubMed]
12. Masmoudi, H.; Hewitt, S.M.; Petrick, N.; Myers, K.J.; Gavrielides, M.A. Automated quantitative assessment of HER-2/neu immunohistochemical expression in breast cancer. *IEEE Trans. Med. Imaging* **2009**, *28*, 916–925. [CrossRef] [PubMed]
13. Brügmann, A.; Eld, M.; Lelkaitis, G.; Nielsen, S.; Grunkin, M.; Hansen, J.D.; Foged, N.T.; Vyberg, M. Digital image analysis of membrane connectivity is a robust measure of HER2 immunostains. *Breast Cancer Res. Treat.* **2012**, *132*, 41–49. [CrossRef]

14. Ruifrok, A.C.; Johnston, D.A. Quantification of histochemical staining by color deconvolution. *Anal. Quant. Cytol. Histol.* **2001**, *23*, 291–299.
15. Lodato, R.F.; Maguire, H.C., Jr.; Greene, M.I.; Weiner, D.B.; LiVolsi, V.A. Immunohistochemical evaluation of c-erbB-2 oncogene expression in ductal carcinoma in situ and atypical ductal hyperplasia of the breast. *Mod. Pathol.* **1990**, *3*, 449–454. [PubMed]
16. Kabakci, K.A.; Cakir, A.; Turkmen, I.; Toreyin, B.U.; Capar, A. Automated scoring of CerbB2/HER2 receptors using histogram based analysis of immunohistochemistry breast cancer tissue images. *Biomed Signal Proces. Control* **2021**, *69*, 102924. [CrossRef]
17. LeCun, Y.; Bengio, Y.; Hinton, G. Deep learning. *Nature* **2015**, *521*, 436–444. [CrossRef]
18. Echle, A.; Rindtorff, N.T.; Brinker, T.J.; Luedde, T.; Pearson, A.T.; Kather, J.N. Deep learning in cancer pathology: A new generation of clinical biomarkers. *Br. J. Cancer* **2021**, *124*, 686–696. [CrossRef]
19. Qaiser, T.; Rajpoot, N.M. Learning Where to See: A Novel Attention Model for Automated Immunohistochemical Scoring. *IEEE Trans. Med. Imaging* **2019**, *38*, 2620–2631. [CrossRef]
20. Chen, Z.; Zhang, J.; Che, S.L.; Huang, J.Z.; Han, X.; Yuan, Y.X. Diagnose Like A Pathologist: Weakly-Supervised Pathologist-Tree Network for Slide-Level Immunohistochemical Scoring. *Proc. AAAI Conf. Artif. Intell.* **2021**, *35*, 47–54. [CrossRef]
21. Chen, M.; Zhang, B.; Topatana, W.; Cao, J.; Zhu, H.; Juengpanich, S.; Mao, Q.; Yu, H.; Cai, X. Classification and mutation prediction based on histopathology H&E images in liver cancer using deep learning. *NPJ Precis. Oncol.* **2020**, *4*, 14. [PubMed]
22. Kather, J.N.; Heij, L.R.; Grabsch, H.I.; Loeffler, C.; Echle, A.; Muti, H.S.; Krause, J.; Niehues, J.M.; Sommer, K.A.J.; Bankhead, P.; et al. Pan-cancer image-based detection of clinically actionable genetic alterations. *Nat. Cancer* **2020**, *1*, 789–799. [CrossRef] [PubMed]
23. Wang, X.; Zou, C.; Zhang, Y.; Li, X.; Wang, C.; Ke, F.; Chen, J.; Wang, W.; Wang, D.; Xu, X.; et al. Prediction of BRCA Gene Mutation in Breast Cancer Based on Deep Learning and Histopathology Images. *Front. Genet.* **2021**, *12*, 661109. [CrossRef] [PubMed]
24. Yamashita, R.; Long, J.; Longacre, T.; Peng, L.; Berry, G.; Martin, B.; Higgins, J.; Rubin, D.L.; Shen, J. Deep learning model for the prediction of microsatellite instability in colorectal cancer: A diagnostic study. *Lancet Oncol.* **2021**, *22*, 132–141. [CrossRef]
25. Owens, M.A.; Horten, B.C.; Da Silva, M.M. HER2 amplification ratios by fluorescence in situ hybridization and correlation with immunohistochemistry in a cohort of 6556 breast cancer tissues. *Clin. Breast Cancer* **2004**, *5*, 63–69. [CrossRef]
26. Li, Y.F.; Guo, L.Z.; Zhou, Z.H. Towards Safe Weakly Supervised Learning. *IEEE Trans. Pattern Anal. Mach. Intell.* **2021**, *43*, 334–346. [CrossRef]
27. Modi, S.; Jacot, W.; Yamashita, T.; Sohn, J.; Vidal, M.; Tokunaga, E.; Tsurutani, J.; Ueno, N.T.; Prat, A.; Chae, Y.S.; et al. Trastuzumab Deruxtecan in Previously Treated HER2-Low Advanced Breast Cancer. *N. Engl. J. Med.* **2022**, *387*, 9–20. [CrossRef]
28. Vandenberghe, M.E.; Scott, M.L.; Scorer, P.W.; Soderberg, M.; Balcerzak, D.; Barker, C. Relevance of deep learning to facilitate the diagnosis of HER2 status in breast cancer. *Sci. Rep.* **2017**, *7*, 45938. [CrossRef]
29. Seol, H.; Lee, H.J.; Choi, Y.; Lee, H.E.; Kim, Y.J.; Kim, J.H.; Kang, E.; Kim, S.-W.; Park, S.Y. Intratumoral heterogeneity of HER2 gene amplification in breast cancer: Its clinicopathological significance. *Mod. Pathol.* **2012**, *25*, 938–948. [CrossRef]
30. Yang, S.R.; Bouhlal, Y.; De La Vega, F.M.; Ballard, M.; Kuo, C.J.; Vilborg, A.; Jensen, G.; Allison, K. Integrated genomic characterization of ERBB2/HER2 alterations in invasive breast carcinoma: A focus on unusual FISH groups. *Mod. Pathol.* **2020**, *33*, 1546–1556. [CrossRef]
31. Wang, X.; Teng, X.; Ding, W.; Sun, K.; Wang, B. A clinicopathological study of 30 breast cancer cases with a HER2/CEP17 ratio of >/=2.0 but an average HER2 copy number of <4.0 signals per cell. *Mod. Pathol.* **2020**, *33*, 1557–1562. [PubMed]
32. Gupta, S.; Neumeister, V.; McGuire, J.; Song, Y.S.; Acs, B.; Ho, K.; Weidler, J.; Wong, W.; Rhees, B.; Bates, M.; et al. Quantitative assessments and clinical outcomes in HER2 equivocal 2018 ASCO/CAP ISH group 4 breast cancer. *NPJ Breast Cancer* **2019**, *5*, 28. [CrossRef] [PubMed]
33. Bui, M.M.; Riben, M.W.; Allison, K.H.; Chlipala, E.; Colasacco, C.; Kahn, A.G.; Lacchetti, C.; Madabhushi, A.; Pantanowitz, L.; Salama, M.E.; et al. Quantitative Image Analysis of Human Epidermal Growth Factor Receptor 2 Immunohistochemistry for Breast Cancer: Guideline From the College of American Pathologists. *Arch. Pathol. Lab. Med.* **2019**, *143*, 1180–1195. [CrossRef] [PubMed]

cancers

MDPI

Article

Predicting Tumor Perineural Invasion Status in High-Grade Prostate Cancer Based on a Clinical–Radiomics Model Incorporating T2-Weighted and Diffusion-Weighted Magnetic Resonance Images

Wei Zhang [1,2,†], Weiting Zhang [2,3,†], Xiang Li [2,3], Xiaoming Cao [1], Guoqiang Yang [2,3,4,*] and Hui Zhang [2,3,4,*]

1 Department of Urology, First Hospital of Shanxi Medical University, Taiyuan 030001, China
2 College of Medical Imaging, Shanxi Medical University, Taiyuan 030001, China
3 Department of Radiology, First Hospital of Shanxi Medical University, Taiyuan 030001, China
4 Intelligent Imaging Big Data and Functional Nano-Imaging Engineering Research Center of Shanxi Province, First Hospital of Shanxi Medical University, Taiyuan 030001, China
* Correspondence: yangguoqiang@sxmu.edu.cn (G.Y.); zhang_hui@sxmu.edu.cn (H.Z.); Tel.: +86-18734198876 (G.Y.); +86-18635580000 (H.Z.)
† These authors contributed equally to this work.

Simple Summary: Perineural invasion (PNI) is present in 17–75% of prostate cancer patients and is an important mechanism for cancer progression, leading to poor prognoses. An optimized preoperative technique is needed to detect PNI in prostate cancer patients and administer the best treatment. The aim of our retrospective study was to develop a model based on high-throughput radiomic features of bi-parametric MRI combined with clinical factors that can predict PNI status in high-grade prostate cancers. In total, 183 high-grade PCa patients were included in this retrospective study, and the radiomics model based on 13 selected features of bi-parametric MRI showed better discrimination than did the conventional model in the test cohort (area under the curve (AUC): 0.908). Discrimination efficiency improved when the radiomics and clinical models were combined (AUC: 0.947). This improved model may help predict PNI in prostate cancer patients and allow more personalized clinical decision-making.

Abstract: Purpose: To explore the role of bi-parametric MRI radiomics features in identifying PNI in high-grade PCa and to further develop a combined nomogram with clinical information. Methods: 183 high-grade PCa patients were included in this retrospective study. Tumor regions of interest (ROIs) were manually delineated on T2WI and DWI images. Radiomics features were extracted from lesion area segmented images obtained. Univariate logistic regression analysis and the least absolute shrinkage and selection operator (LASSO) method were used for feature selection. A clinical model, a radiomics model, and a combined model were developed to predict PNI positive. Predictive performance was estimated using receiver operating characteristic (ROC) curves, calibration curves, and decision curves. Results: The differential diagnostic efficiency of the clinical model had no statistical difference compared with the radiomics model (area under the curve (AUC) values were 0.766 and 0.823 in the train and test group, respectively). The radiomics model showed better discrimination in both the train cohort and test cohort (train AUC: 0.879 and test AUC: 0.908) than each subcategory image (T2WI train AUC: 0.813 and test AUC: 0.827; DWI train AUC: 0.749 and test AUC: 0.734). The discrimination efficiency improved when combining the radiomics and clinical models (train AUC: 0.906 and test AUC: 0.947). Conclusion: The model including radiomics signatures and clinical factors can accurately predict PNI positive in high-grade PCa patients.

Keywords: prostate cancer; PNI; bi-parametric MRI; radiomics; nomogram

Citation: Zhang, W.; Zhang, W.; Li, X.; Cao, X.; Yang, G.; Zhang, H. Predicting Tumor Perineural Invasion Status in High-Grade Prostate Cancer Based on a Clinical–Radiomics Model Incorporating T2-Weighted and Diffusion-Weighted Magnetic Resonance Images. *Cancers* **2023**, *15*, 86. https://doi.org/10.3390/cancers15010086

Academic Editor: Hamid Khayyam

Received: 12 November 2022
Revised: 3 December 2022
Accepted: 17 December 2022
Published: 23 December 2022

1. Introduction

Prostate cancer (PCa) is the most frequent malignant tumor in 105 countries worldwide and the first leading cause of cancer-related death in 46 countries among males [1]. Often, there are significant differences in the prognosis of patients with the same stratification who adopt the same treatment plan [2]. In addition, many localized PCa cases, especially high-grade cases, are not truly localized tumors when they are diagnosed. The reasons for this situation are that cancer cells have already spread beyond the scope of surgery or radiotherapy, and these patients are prone to developing biochemical recurrence [3]. It is widely accepted that prostate-specific antigen (PSA), Gleason score (GS), and T stage are the main variables for evaluating the prognosis of localized PCa. Among the factors causing tumor spread, perineural invasion (PNI), which is invasion along or around nerves within the perineural space, also plays an important role in cancer [4]. PNI can be evaluated in a biopsy specimen or radical prostatectomy specimen, and it is present in 17–75% of prostate cancer patients [5]. The College of American Pathologists published a consensus statement on prognostic factors for PCa in which PNI was identified as a potential prognostic factor (category III) that needed additional study [6]. Therefore, identifying the PNI status of high-grade PCa is an urgent problem to be solved.

At present, magnetic resonance imaging (MRI) is widely used for diagnosing PCa and can help detect several prognostic factors; it has been used to increase T staging accuracy and predict positive surgical margins (PSMs) by detecting and localizing extracapsular extension (ECE) [7,8]. Radiomics, as an extension concept of texture analysis, can convert medical images into high-dimensional mineable and quantitative features by using high-throughput extraction algorithms of these characterizations. In recent years, qualitative analysis of prostate MRI images by means of radiomics plays a crucial role at the pretreatment staging step and is increasingly applied to determine invasion and prognosis for prostate cancer [9,10]. PNI is a pathological feature that can only be detected after an invasive biopsy or prostatectomy. This form of metastasis can affect peri-prostatic neurovascular fibers, the lumbosacral plexus, and the sciatic nerve, and MRI can visualize involvement of these nerve fibers as direct evidence of cancer cell spreading [11,12]. In the age of high-resolution imaging, developing a method based on radiomics to accurately assess the PNI status of PCa is urgently needed.

In this study, we evaluated the relationship between MRI radiomics signature, as well as other clinical and pathological factors, and PNI in high-grade PCa. We hypothesized that the MRI radiomics signature may provide effective information and established a model for preoperatively predicting the probability of PNI in high-grade PCa patients.

2. Materials and Methods

2.1. Patients

This retrospective study received Institutional Review Board approval of the First Hospital of Shanxi Medical University, ethic code: (K131). We retrospectively selected PCa patients with clinical and imaging data from January 2016 to May 2021 who underwent prostate MR examination before systematic prostate biopsy or radical prostatectomy (RP). Clinical data, including age, PSA level, prostate volume, prostate-specific antigen density (PSAD), GS, grading groups (GGs), and tumor location in the prostate, were collected from patient medical records. The study inclusion criteria were as follows: (a) high-grade PCa patients who underwent prostate MRI examination; and (b) tumor perineural invasion status obtained on histopathology by biopsy or RP. The following exclusion criteria were applied: (a) PCa patients who received other treatments before MRI examination, such as androgen suppression therapy or any previous transurethral surgery; (b) poor image quality due to artifacts; (c) incomplete MR sequence; and (d) incomplete clinical data collection; (e) the lesions were too small for segmentation and analysis (maximum diameter <3 mm). A total of 208 high-grade prostate cancer patients' data were collected. According to the exclusion criteria, 25 patients were excluded. Ultimately, 183 high-grade PCa patients were enrolled in the study. The patients were randomly divided into training and test groups at a ratio of 7 to 3 (training group: 128 patients, test group: 55 patients).

2.2. MR Image Data

The prostate MRI examination was performed according to PI-RADS v2.1 protocol and the process was as follows. We utilized a 3.0-T scanner (GE Signa HDxt) with an 8-channel array coil to acquire the images of multiplanar T2-weighted imaging (T2WI) and diffusion-weighted imaging (DWI), which were obtained with a turbo spin-echo sequence and the following parameters: repetition time/echo time (TR/TE): 3360/68.16 ms; field of view (FOV): 220 × 220 mm; matrix: 320 × 256; slice thickness: 5 mm; and spacing between slices: 5.5 mm. A single-shot echo-planar sequence with four b-values was also acquired: 0 and 1500 s/mm (TR/TE: 5250/78.6 ms; FOV: 100 × 100 mm; matrix: 128 × 160; and slice thickness: 5 mm).

2.3. Histopathologic Analysis

All patients underwent transrectal ultrasound-guided 12-core systematic prostate biopsy or RP after prostate MRI examination. The specimen pathological diagnosis was made by two pathologists with more than three years of experience in diagnosis of prostate diseases. The GS was updated according to the 2014 International Society of Urological Pathology criteria. PNI was diagnosed when PCa infiltration was identified in any layer of the nerve sheath or tumor invasion involved at least one-third of the nerve circumference. Pathologic information was collected, and, according to the outcomes, all patients were divided into two groups: one group had positive prostate cancer cell PNI and the other group had negative prostate cancer cell PNI (Figure 1).

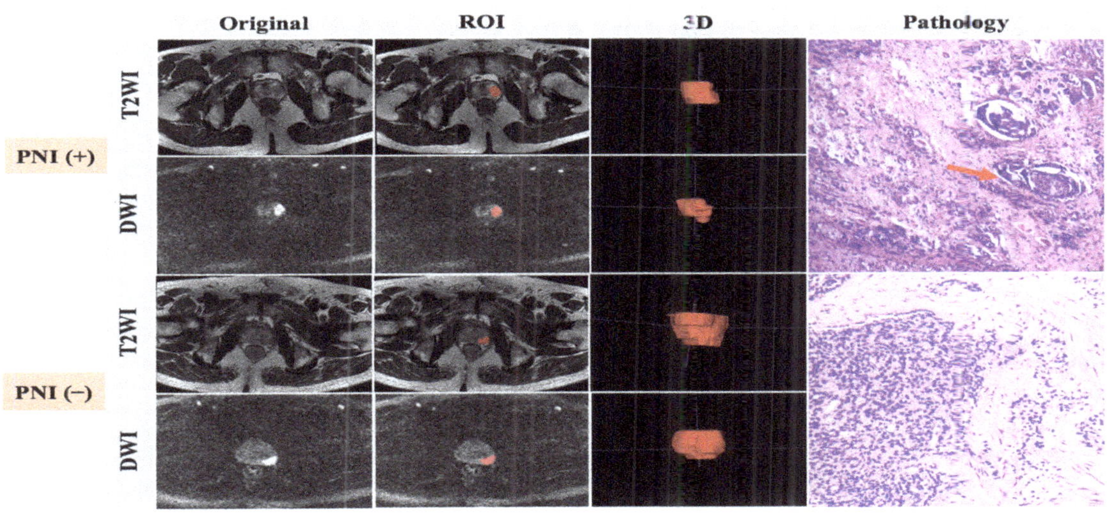

Figure 1. Preoperative MRI images, ROI delineation, and pathological comparison of prostate cancer with and without PNI, as indicated by the arrow.

2.4. Tumor Segmentation

All MR images were manually delineated by two independent readers with more than 5 years' experience in reading prostate MR images. ITK-SNAP software was used to process T2WI and high-b-value (b = 1500) DWI images. Tumors were targeted as the regions of interest (ROIs), defined as hypointense signal areas compared with the normal prostate area on T2WI and a higher signal intensity than that of the normal prostate area on DWI. For consistency between ROIs in both T2WI and DWI images, all depicted ROIs were strictly delineated with the same criteria and visually validated by the same expert. The ROIs were manually delineated layer-by-layer along the lesion boundary, obtaining three-dimensional data (Figure 1).

2.5. Extraction of Radiomic Features

Software of FAE (FAE version is 0.5.2 and PyRadiomics version is 3.0.1. The software was soured from East China Normal University, Shanghai, China. https://github.com/salan668/FAE accessed on 16 December 2022), which was developed based on the PyRadiomics package (https://github.com/Radiomics/pyradiomics, accessed on 2 June 2022), was used to extract features from the T2WI ROIs and DWI ROIs. The parameters of feature extraction were: first order statistics, shape-based, GLCM, GLRLM, GLSZM, GLDM, NGTDM. A total of 1702 features were extracted from the MRI data and 851 features each from T2WI and DWI, including 14 shape features, 18 first-order features, 24 gray level co-occurrence matrix (GLCM) features, 16 gray level run length matrix (GLRLM) features, 16 gray level size zone matrix (GLSZM) features, 5 neighboring gray tone difference matrix (NGTDM) features, and 14 gray level dependence matrix (GLDM) features and 744 wavelet features [13].

2.6. Feature Selection and Model Building

The process of feature selection was based on training set. Thirty patients were randomly selected for a double-blinded comparison of manual segmentations by two radiologists. Inter- and intraclass correlation coefficients (ICCs) between groups and within groups were calculated to select features with high stability and reproducibility, and ICCs greater than or equal to 0.75 were considered to have good agreement. To remove the imbalance of the training dataset, we used the synthetic minority oversampling technique (SMOTE) to balance the positive/negative samples. Before feature selection, we subtracted by the mean value and divided by the standard deviation to normalize the feature matrix for each feature vector. Next, the feature selection process was divided into two steps. In the first step, the features with statistical significance for identifying PNI positivity were selected by univariate logistic regression analysis. In addition, the first stage of dimensionality reduction of the data was achieved to ensure that each feature had a significant effect on the outcome. In the second step, least absolute shrinkage and selection operator (LASSO) regression analysis was used for further data dimensionality reduction, and the best features were determined for establishment of the radiomics model. The hyperparameter lambda value and the number of selected features were determined by tenfold cross-validation. After the radiomics model was established, each feature was multiplied by its corresponding coefficient, and an intercept value was added to calculate the radiomics score (Rad-score) for each patient, which was establishment of the radiomics signature (Appendix A).

For clinical features, we used the univariate analysis method, and the features with statistical significance for the results were selected to construct a clinical model. Finally, the combined model of clinical and radiomics features was established by multiple logistic regression analysis method.

2.7. Model Evaluation

After the models were built, their performance was evaluated using receiver operating characteristic (ROC) curve analysis. The area under the ROC curve (AUC) was calculated for quantification of the performance. The accuracy, sensitivity, and specificity were also calculated at a cutoff value that maximized the value of the Youden index. A radiomic nomogram combining the Rad-score derived from T2WI and DWI scans and clinical factors was developed for predicting PNI. The calibration curves measured the consistency between the predicted probability of PNI and the actual probability of PNI. Decision curve analysis was applied to measure the clinical utility of the nomogram.

2.8. Statistical Analysis

Demographic data were compared by chi-squared test, Mann-Whitney test, or *t*-test. Continuous variables are expressed as mean ± standard deviation, and categorical variables are expressed as median (25 quantile, 75 quantile). A value of $p < 0.05$ was considered statistically significant. Statistical analyses were performed using SPSS v22.0 (IBM SPSS Statistics, IBM Corp., Armonk, NY, USA) and R software (R is a language and environment for statistical computing and graphics. It is a GNU project which is similar to the S language and environment which was developed at Bell Laboratories (formerly AT&T, now Lucent Technologies) by John Chambers and colleagues, version 4.1.2; http://www.Rproject.org, accessed on 17 December 2022).

3. Results

3.1. Patient Characteristics

PNI was diagnosed histologically based on RP or biopsy specimen tissues. In total, 183 patients were then divided into the PNI positive [PNI (+)] group and the PNI negative [PNI (−)] group. The PNI (+) group contained 54 patients (29.51%), while the PNI (−) group contained 129 patients (70.49%). In the PNI positive group, 42 were detected on RP and 12 on biopsy. Twenty-seven of the forty-two cases were confirmed PNI positive both on preoperative biopsy and RP; eight of the forty-two cases had no PNI positive results on biopsy, but the RP outcomes were determinative; seven of the forty-two cases obtained a biopsy at another center, and we only had PNI positive results after RP in our center. Twelve PNI positive cases confirmed by biopsy did not undergo RP after biopsy in our center. The concordance rate of PNI positive results between biopsy and RP was 64.29%. In the PNI negative group, 98 cases were diagnosed as PNI negative both on preoperative biopsy and RP; 31 cases obtained a biopsy at another center; we only had their PNI negative outcomes of RP in our center. The concordance rate was 75.97%. The average ages were 69.7 ± 8.2 years and 72.0 ± 9.0 years in the two respective groups. The PSA levels were 15.9 ng/mL and 17.4 ng/mL in the two respective groups. In the PNI (+) group, the GS proportions were distributed as follows: 22.2% of patients (12/54) had a score of 8, 42.6% (23/54) had a score of 9, and 11.1% (6/54) had a score of 10. In the PNI (−) group, the GS proportions were distributed as follows: 41.1% of patients (53/129) had a score of 8, 39.5% (51/129) had a score of 9, and 19.4% (25/129) had a score of 10. The radiological and other clinical characteristics of the two groups are summarized in Table 3. There were no significant differences between these two groups in terms of age, PSA level, PSAD, or tumor location. However, there were significant differences in prostate volume, GS, and GG ($p < 0.05$). There were no significant differences between the training and test cohorts in terms of all clinical characteristics, which are summarized in Table 2 ($p > 0.05$).

Table 1. Patient clinic radiological characteristics between groups of PNI (+) and PNI (−).

Characteristics	PNI (+) (N = 54)	PNI (−) (N = 129)	*p* Value
Age (years)	69.7 ± 8.2	72.0 ± 9.0	0.121
PSA level (ng/mL)	15.9 (10–23)	17.4 (11.4–25.7)	0.406
Prostate volume (mL)	43.7 (31.3–59.7)	53.7 (38.1–87.7)	0.006
Foot–head (FH) (cm)	4.4 (3.5–5.1)	4.7 (3.9–5.8)	0.02
Right–left (RL) (cm)	4.7 (4–5)	5.1 (4.5–5.9)	<0.001
Anterior–posterior (AP) (cm)	4.1 (3.6–4.9)	4.3 (3.7–5.2)	0.247
PSAD (ng/mL/cm^3)	0.4 (0.2–0.5)	0.3 (0.2–0.5)	0.176
Gleason Score (GS)	9.13 (9–10)	8.78 (8–9)	0.005
Grading Groups (GG)			<0.001
Grade 1	0.0% (0/54)	0.0% (0/129)	
Grade 2	0.0% (0/54)	0.0% (0/129)	
Grade 3	0.0% (0/54)	0.0% (0/129)	

Table 1. *Cont.*

Characteristics	PNI (+) (N = 54)	PNI (−) (N = 129)	*p* Value
Grade 4	22.2% (12/54)	41.1% (53/129)	
Grade 5	77.8% (42/54)	58.9% (76/129)	
Location			0.196
Central zone	1.9% (1/54)	2.3% (3/129)	
Transition zone	13.0% (7/54)	7.0% (9/129)	
Peripheral zone	25.9% (14/54)	17.1% (22/129)	
Multiple zone	59.3% (32/54)	73.6% (95/129)	
Rad-score	1.52 ± 2.649	−1.815 ± 2.065	<0.001

Table 2. Patient clinic radiological characteristics between training and test cohort.

Characteristics	Training (N = 128)	Test (N = 55)	*p* Value
Age (years)	72.0 ± 8.6	69.8 ± 9.1	0.117
PSA level (ng/mL)	42.4 (14.3–138.6)	49.8 (13.9–169)	0.716
Prostate volume (mL)	48.6 (35.2–77.4)	52.9 (36.6–71.0)	0.797
Foot–head (FH) (cm)	4.7 (3.8–5.7)	4.6 (3.8–5.3)	0.484
Right–left (RL) (cm)	4.9 (4.4–5.5)	4.9 (4.2–5.5)	0.796
Anterior–posterior (AP) (cm)	4.3 (3.7–5.2)	4.1 (3.4–4.9)	0.157
PSAD (ng/mL/cm^3)	0.9 (0.3–2.9)	0.9 (0.3–2.8)	0.861
Gleason Score (GS)	9.0 (8–9)	9.0 (8–9)	0.092
Location			0.193
Central zone	1.6% (2/128)	3.6% (2/55)	
Transition zone	10.9% (14/128)	3.6% (2/55)	
Peripheral zone	21.1% (27/128)	14.5% (8/55)	
Multiple zone	66.4% (85/128)	78.2% (43/55)	
Rad-score	−0.542 ± 2.518	−1.503 ± 3.046	0.052

PSA: prostate-specific antigen. Prostate volume: foot–head (FH) length × right–left (RL) length × anterior–posterior (AP) length × $\pi/6$. PSAD: prostate-specific antigen density, PSA value divided by MRI-estimated prostate volume. Grading groups (GG): GG1: Gleason scores ≤ 6; GG2: Gleason scores 3 + 4; GG3: Gleason scores 4 + 3; GG4: Gleason scores 4 + 4, 3 + 5, 5 + 3; GG5: Gleason scores 4 + 5, 5 + 4, 5 + 5. $p < 0.05$ indicates a statistically significant difference.

3.2. Feature Selection and Comparison of Models

Further, 1193 stable features with ICCs ≥ 0.75 were retained (611 features from T2WI, and 582 features from DWI). The T2WI sequence selected 10 features when the λ_{1se} was equal to 0.06478 and obtained the highest AUC on the testing dataset. The AUC and accuracy of the model were 0.827 (95% CI 0.707–0.947) and 0.818, respectively. The DWI sequence selected four features when the λ_{1se} was equal to 0.11225 and obtained the highest AUC on the testing dataset. The AUC and accuracy of the model were 0.734 (95% CI 0.593–0.975) and 0.746, respectively. The T2WI + DWI sequence selected 13 features when the λ_{1se} was equal to 0.06787 and obtained the highest AUC on the validation dataset. The AUC and accuracy of the model were 0.908 (95% CI 0.821–0.996) and 0.855, respectively. Thirteen features were found to have high stability for prediction of PNI and were chosen to construct the final model. The details of feature selection and comparison of models were shown in Figures 2 and 3 and Tables 3 and 4.

The clinical model based on features including FH, RL, prostate volume, and GS obtained the highest AUC on the test dataset. The AUC and accuracy of the model were 0.823 (95% CI 0.712–0.933) and 0.673, respectively, on the testing dataset (Figures 2 and 3 and Table 4).

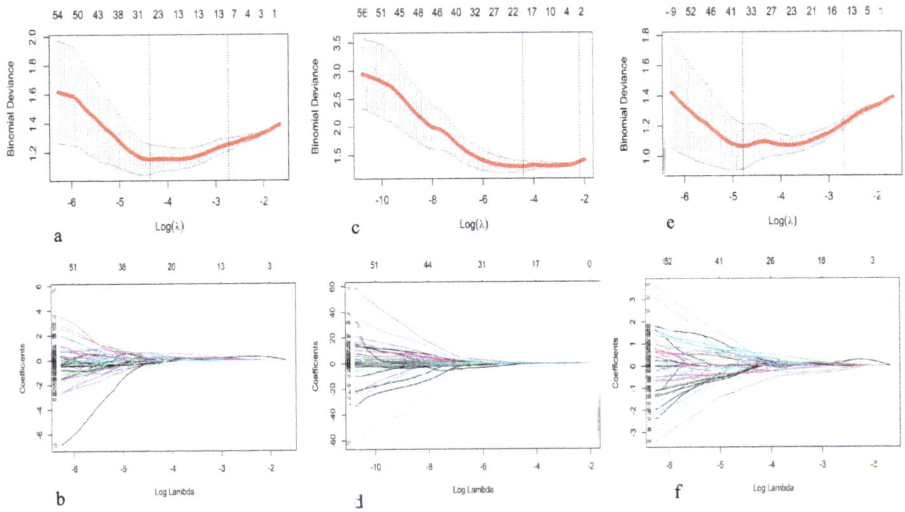

Figure 2. The lasso plots for radiomics feature selection: (**a**,**b**) for T2WI, 10 features were selected when the λ_{1se} = 0.06478, (**c**,**d**) for DWI, 4 features were selected when the λ_{1se} = 0.11225, and (**e**,**f**) for T2WI + DWI sequences, 13 features were selected when the λ_{1se} = 0.06787.

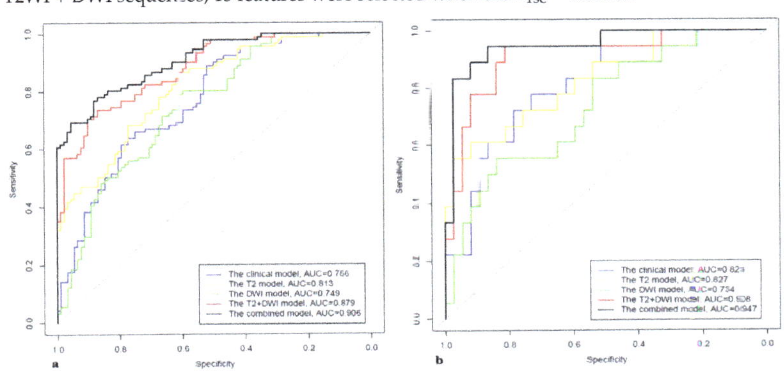

Figure 3. The AUCs of different models in the training (**a**) and test (**b**), respectively.

Table 3. The selected radiomics features of T2WI, DWI, and T2WI + DWI models.

	Radiomics Features	Coefficient	Odds Ratio (95% CI)	*p*-Value
T2WI	T2_wavelet.HHH_glrlm_RunPercentage	−0.220	0.802 (0.533–1.236)	0.298
	T2_wavelet.HHH_ngtdm_Coarseness	1.471	4.355 (0.800–29.392)	0.106
	T2_wavelet.HLH_gldm_ SmallDependenceHighGrayLevelEmphasis	−5.081	0.006 (5.54×10^{-6}–0.687)	0.080
	T2_wavelet.HLH_glrlm_RunPercentage	1.443	4.235 (1.481–26.510)	0.045
	T2_wavelet.HLL_ngtdm_Coarseness	−1.294	0.274 (0.043–1.324)	0.134
	T2_wavelet.LHH_gldm_ DependenceNonUniformityNormalized	5.107	1.652 (1.358–4.033)	0.104
	T2_wavelet.LHH_glszm_ SizeZoneNonUniformityNormalized	0.860	2.362 (1.187–5.205)	0.022
	T2_wavelet.LHH_ngtdm_Contrast	0.722	2.058 (1.291–3.564)	0.005
	T2_wavelet.LHL_firstorder_RootMeanSquared	0.270	1.310 (0.808–2.146)	0.268
	T2_wavelet.LLL_gldm_ SmallDependenceLowGrayLevelEmphasis	0.025	1.025 (0.637–1.626)	0.916

Table 3. *Cont.*

	Radiomics Features	Coefficient	Odds Ratio (95% CI)	*p*-Value
DWI	DWI_original_glszm_SizeZoneNonUniformityNormalized	0.378	1.460 (1.0109–2.229)	0.061
	DWI_original_shape_SurfaceArea	−0.443	0.642 (0.257–1.511)	0.324
	DWI_wavelet.HLH_glcm_MaximumProbability	−0.731	0.481 (0.272–0.763)	0.005
	DWI_wavelet.LLL_glrlm_RunLengthNonUniformity	−0.700	0.496 (0.200–1.136)	0.109
T2WI + DWI	T2_wavelet.HLH_gldm_ SmallDependenceHighGrayLevelEmphasis	0.947	2.579 (1.255–7.864)	0.030
	T2_wavelet.HLH_glrlm_RunPercentage	−0.509	0.601 (0.278–1.236)	0.176
	T2_wavelet.HLL_ngtdm_Coarseness	0.703	2.020 (0.844–6.290)	0.181
	T2_wavelet.LHH_gldm_ DependenceNonUniformityNormalized	0.834	2.303 (1.171–5.080)	0.023
	T2_wavelet.LHH_glszm_ SizeZoneNonUniformityNormalized	0.537	1.710 (1.059–2.955)	0.039
	T2_wavelet.LHH_ngtdm_Contrast	0.304	1.355 (0.808–2.315)	0.249
	T2_wavelet.LHL_firstorder_RootMeanSquared	0.343	1.409 (0.859–2.375)	0.180
	DWI_original_glszm_SizeZoneNonUniformityNormalized	0.271	1.311 (0.829–2.266)	0.289
	DWI_original_shape_SurfaceArea	−0.896	0.408 (0.162–0.896)	0.039
	DWI_wavelet.HHH_glcm_DifferenceEntropy	0.687	1.988 (1.010–4.306)	0.064
	DWI_wavelet.HLH_glcm_MaximumProbability	−0.494	0.610 (0.299–1.178)	0.151
	DWI_wavelet.HLL_gldm_ LargeDependenceLowGrayLevelEmphasis	0.377	1.457 (0.873–2.460)	0.152
	DWI_wavelet.LHH_glszm_ZoneEntropy	−0.127	0.881 (0.463–1.668)	0.697

Table 4. The diagnostic performance of models.

Model	Train				Test			
	AUC	Sensitivity	Specificity	P	AUC	Sensitivity	Specificity	P
Clinical	0.766 (0.698–0.834)	0.890	0.522		0.823 (0.712–0.933)	1	0.514	
T2WI	0.813 (0.753–0.873)	0.868	0.609	0.276	0.827 (0.707–0.947)	0.611	0.919	0.959
DWI	0.749 (0.678–0.819)	0.802	0.598	0.709	0.734 (0.593–0.975)	0.556	0.838	0.269
T2WI + DWI	0.879 (0.832–0.926)	0.736	0.870	0.003	0.908 (0.821–0.996)	0.944	0.811	0.197
Combined	0.906 (0.866–0.947)	0.780	0.870	<0.01	0.947 (0.884–1)	0.944	0.865	0.01

P: AUC value of T2WI model, DWI model, T2WI + DWI model, and radiomic combined clinical model, respectively, compared to AUC value of clinical model.

3.3. Development of the Clinical–Radiomics Predictive Model

After the independently associated risk factors of FH, RL, volume, and GS were selected, we combined them with the Rad-score of the 13 features to form a PNI predictive nomogram. This nomogram had better performance in predicting PNI: the AUCs were 0.906 (95% CI 0.866–0.947) in the training group and 0.947 (95% CI 0.884–1) in the test group (Figure 4 and Table 4).

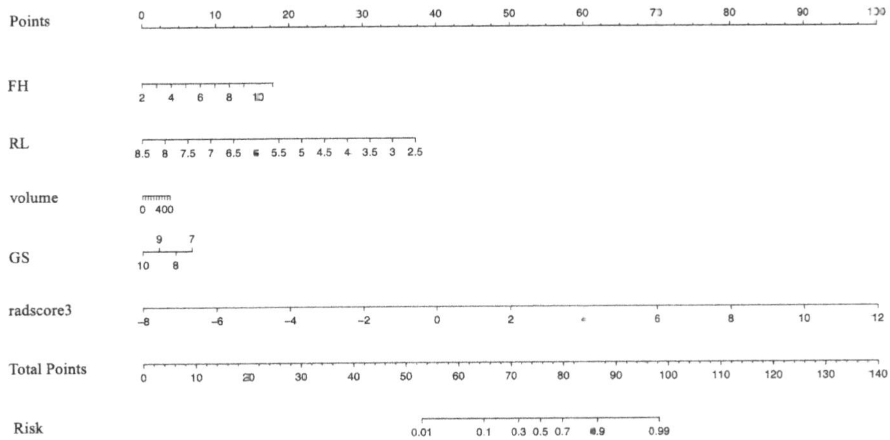

Figure 4. Nomogram developed for prediction of PNI. Radiomic nomogram combining the Rad-score derived from T2WI and DWI scans and clinical–radiological factors for predicting PNI. PNI: perineural invasion.

3.4. *Validation of the Clinical–Radiomics Predictive Nomogram*

The calibration charts showed that the actual probability of PNI occurrence was consistent with the predicted probability, and the Hosmer-Leme show test yielded P values of 0.907 and 0.689 in the training and test cohorts, respectively. As shown in Figure 5, decision curve analysis indicated that the PNI predictive nomogram model was the best method across the full range of reasonable threshold probabilities. In the training group, the net reclassification index (NRI) was 1.1252 (0.8659–1.3644, $p < 0.01$) comparing the clinical model and combined model, while the NRI was 0.386 (0.6271–1.449, $p < 0.01$) comparing the radiomic model and combined model. In the test group, the NRI was 1.2312 (0.7796–1.6829, $p < 0.01$) comparing the clinical model and combined model, while the NRI was 1.0691 (0.5958–1.5424, $p < 0.01$) comparing the radiomic model and combined model (Figure 6).

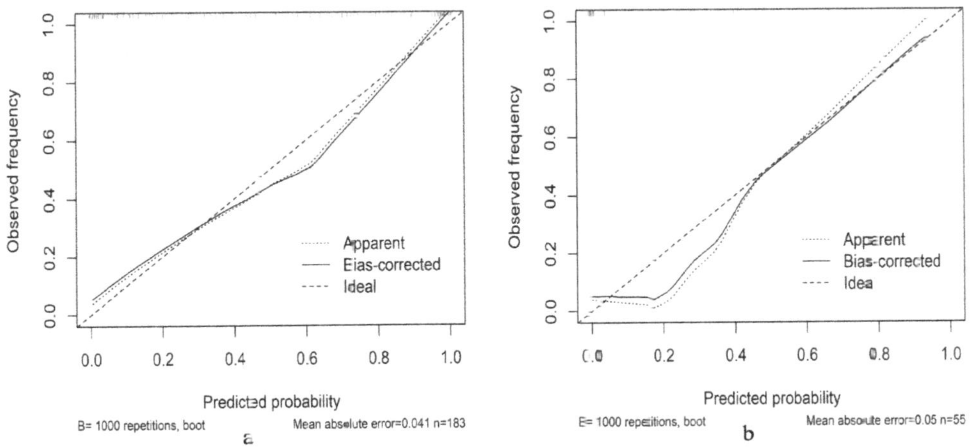

Figure 5. Calibration curve of the nomogram in the training (**a**) and test (**b**) groups.

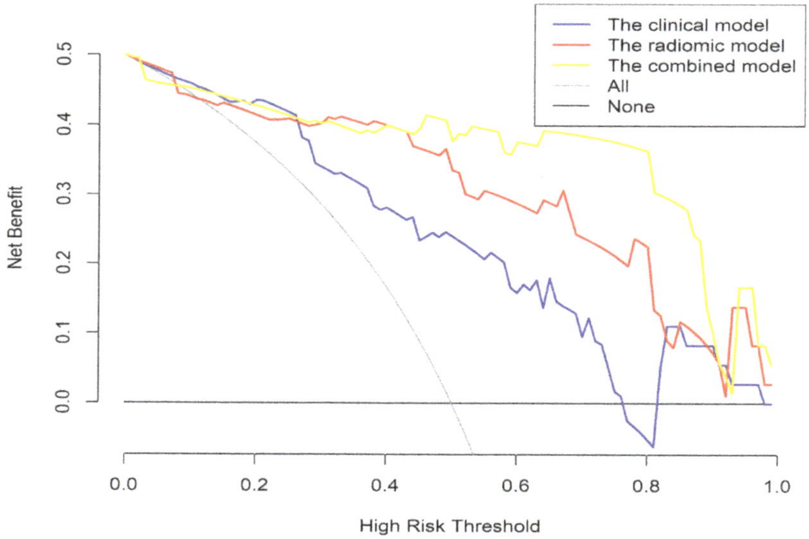

Figure 6. Decision curve analysis.

4. Discussion

PNI is a histological phenomenon in which cancer cells surround and invade nerves in the tumor microenvironment and play a role in development and regeneration of cancer cells. Nerves and cancer cells communicate bidirectionally to each other, providing a mechanism that could induce cancer invasion and spread. Studies have shown that the sympathetic nervous system in cancer can regulate pathological gene expression, leading to DNA damage repair inhibition and oncogene activation to increase cancer cell metastasis and tumorigenesis [14,15]. On the other hand, cancer cells can secrete neurotrophic growth factors or chemokines, such as CCL2 and CXCL12, to promote development of neural progenitors, causing nerve growth [16,17]. PNI in cancer is associated with poor prognosis, likely because neoplastic cells hidden in the perineural space cannot be removed during tumor resection and cause recurrence.

In 1999, the College of American Pathologists published a consensus statement on prognostic factors for PCa in which PNI was classified as category III for risk of recurrence and needed additional study [6]. In multivariate analysis, PNI on biopsy showed significance for recurrence. The presence of PNI on target-biopsy associated with worse histopathologic features on RP and poorer outcomes might thus be useful for risk stratification [18]. As primary treatment decisions are often based on biopsy results, the additional PNI information may be relevant for optimal patient care [19]. PNI found on prostate biopsies has been shown to be an independent predictor of high-grade disease associated with a higher mean PSA, adverse pathologic features of higher GS, and extra-prostatic extension [20,21]. In our study, 54 PNI (+) patients among 183 high-grade PCa patients had higher GG and GS than PNI (−) patients, and the outcome was consistent with these studies. PCa patients with PNI positivity showed an increased risk of biochemical recurrence after prostatectomy or radiotherapy and worse survival outcomes, which have important implications for treatment decision-making and management of PCa [22–24].

The slowly progressive nature of nerve involvement can often make PNI difficult to diagnose, and PNI is always detected based on the pathological results of the biopsy and prostatectomy specimens of PCa patients. As not all PCa cases are diagnosed at the initial biopsy, PNI as an independent prognostic factor remains difficult to quantitatively measure in pathological samples because of its heterogenous presentations and the multifocal nature of RP specimens [25]. Recent research has shown that the distribution of nerves within the tumor-infiltrating microenvironment is not homogeneous. The neural density was significantly higher in the cancer periphery close to cancer infiltration than in the cancer core area, which suggests that nerves may drive tumor progression and invasion [26]. Many factors may influence the true pathological positive rate of PNI, such as the needle core number of biopsy and the processing method of RP specimen tissues [27]. Thus, the prognostic value of PNI evaluation in pathological analysis should be further assessed and a better method should be developed to provide a detailed spatial representation of heterogeneity.

MRI is a noninvasive diagnostic tool that can acquire entire anatomical images of the prostate for cancer staging, such as extra-prostatic extension. This is important for urologists to determine a treatment plan before surgery, such as preservation of the neurovascular bundle (NVB) [28]. In the era of high-resolution imaging, extra-prostatic extension on MR images already has a better ability to predict locally advanced-stage PCa than PNI positivity on biopsy [29]. Whether PNI, as a predominant mechanism and a predictor of PCa progression to an advanced stage, can be directly assessed on imaging measures needs further study to develop a visualization method. Jonathan J. Stone retrospectively reviewed the data of 3733 PCa patients from a medical database who had undergone both MRI and PET before surgery to identify direct radiological evidence of PNI. Fifteen patients who had perineural spread found on MRI presented enlargement of the spinal nerves, lumbosacral plexus, sciatic nerve on T1-weighted sequences, hyperintensity on T2-weighted sequences, and/or abnormal nerve enhancement after gadolinium administration [30]. Salvatore Siracusano evaluated a new MRI modality called diffusion tensor imaging (DTI), which can provide sharp depiction of peripheral nervous fibers to detect changes in peri-prostatic neuro-vasculature (PNF) before and after RP. DTI was able to detect quantitative changes in the number, length, and fractional anisotropy values of the PNF, and they observed that the fiber number in MRI images can serve as a recovery indicator of erectile dysfunction in nerve-sparing prostatectomy [31]. However, PNI is a microscopic-level finding in PCa. Huijuan You combined MRI and magnetic particle imaging involving superparamagnetic iron oxide nanoparticles to precisely distinguish high and low nerve densities of the PCa tissue microenvironment in a mouse model. Their method could visualize the nerve density, and they observed a positive correlation with the aggressiveness of PCa cancer cells, which can be a novel strategy for discovering biomarkers for neural tissue and tumor aggressiveness in PCa [32].

Although MR plays an important role in detecting and accurately evaluating PCa, image outcome reporting depends on the subjective judgment of radiologists, which causes high inter-reader variability. Recently, the quantitative analysis method based on machine learning techniques called radiomics was shown to automatically obtain high-throughput imaging features to overcome the above limitations and assess tumor biology characteristics. Several studies have reported use of MR-based radiomics to detect clinically significant PCa and assess aggressiveness and tumor staging [33]. Shuai Ma developed and validated a radiomics model that contains 17 stable radiomics features extracted from 1619 features based on T2WI to predict ECE in PCa. The AUC was 0.883 in the validation cohort, and the model was more sensitive than the radiologists' interpretations, especially for apical tumors, which would influence a nerve-sparing surgical plan [34]. PNI is a predominant mechanism of ECE in PCa; to the best of our knowledge, there is no radiomics model based on MRI for preoperatively predicting this histopathological phenomenon.

In our study, we constructed a model derived from clinical and imaging data, including radiomic features from T2WI and DWI, based on computer-aided analysis to evaluate the PNI status in high-grade PCa. Our best radiomics model contained three GLDM features, one GLRLM feature, two NGTDM features, three GLSZM features, two GLCM features, one first-order feature, and one shape feature from T2WI and DWI images, which have the best predictive ability for PNI status in high-grade PCa. Our results demonstrated that the NGTDM feature had the greatest weight of the features in the T2WI model, while, in the DWI model, it was the GLCM feature, which is associated with tumor invasion and is a predictor of PCa aggressiveness, consistent with recently published findings concerning risk stratification for Pca. This finding suggests that invading nerves in the tumor microenvironment may affect the homogeneous texture features and that these radiomics features associated with PNI positivity may provide some additional information related to Pca aggressiveness, as previous studies reported [35,36]. The feature with the greatest weight in the T2WI + DWI model was the higher-order feature GLDM; this feature describes the gray level intensity within the ROI between the PNI positive and PNI negative groups and is used to highlight local heterogeneity information. This texture feature was rarely mentioned in previous radiomics studies for Pca, but, for other tumors, such as rectal cancer and cervical cancer, GLDM was thought to be associated with locally advanced tumors and poor prognosis in recent studies [37,38]. Similar to those in nontumor tissues, the GLDM metrics were found to be significantly different among peritumoral fat between high-grade and low-grade clear cell renal carcinoma and urothelial carcinoma [39,40]. Therefore, whether radiomics feature GLDM could be a biomarker for predicting the heterogeneity of interstitial composition in urologic cancers requires more research. Similar to the study of B. De Santi, which showed that a difference in voxel intensity distribution could distinguish cancerous and normal prostatic tissues [41], our model led to the conclusion that differences in heterogeneity between PNI positive and PNI negative samples can be detected and, therefore, can help depict the tissue microstructure as PNI positive or PNI negative before surgery.

Our clinical–radiomics prediction model, which integrates clinical characteristics and the Rad-score derived from MRI, had good sensitivity (0.944) and good specificity (0.865) in the test cohort, indicating that it is superior to all the above-mentioned models for predicting PNI status. Comparing the AUC values in the independent test cohort, our clinical–radiomics prediction model (AUC 0.947; 95% CI 0.884–1) performed better than the radiomics model alone (AUC 0.908; 95% CI 0.821–0.996) and the clinical model alone (AUC 0.823; 95% CI 0.712–0.933). Decision curve analysis showed that the clinical–radiomics model had a better ability to predict PNI than the other two models at any given threshold probability. This finding confirms that assessment of PNI with clinical or radiomic information alone will not be comprehensive.

Several limitations should be noted when considering this study. First, we included GGs of high-grade patients only; those with GS ≤ 7 patterns were excluded, especially patients with GS 4 + 3 who have a much worse prognosis, and their PNI status was not assessed. Second, some GS values were based on biopsy rather than on RP in our study, possibly causing sampling error. Third, there was a lack of spatial co-registration of the histopathology slides and MR images, which may cause a mismatch in delineating the ROIs directly on the T2WI and DWI images. Fourth, FAE software can be used conveniently for binary classification, but it has not yet provided an integrated UI for multilabel classification and regression problems. Fifth, this study was a single-institutional retrospective study design without external validation.

5. Conclusions

In our study, the results showed that MRI-derived radiomic features can be independent predictors of PNI in high-grade PCa. The combination of radiomic features extracted from T2WI and DWI maps produced higher diagnostic power to predict PNI than a single pattern. Additionally, our clinical–radiomics model was superior to a single radiomics model and a clinical model, suggesting that, combined, the radiomic features and clinical pathology information may have considerable value in predicting PNI in high-grade PCa, which can aid clinicians in choosing appropriate treatment options and estimating prognoses for such patients.

Author Contributions: Conceptualization: W.Z. (Wei Zhang) and H.Z.; methodology: W.Z. (Wei Zhang) and G.Y.; software: W.Z. (Wei Zhang), W.Z. (Weiting Zhang), and X.L.; validation: W.Z. (Wei Zhang) and W.Z. (Weiting Zhang); formal analysis: W.Z. (Weiting Zhang) and G.Y.; investigation: W.Z. (Wei Zhang) and W.Z. (Weiting Zhang); resources: H.Z. and X.C.; data curation: W.Z. (Weiting

Zhang); writing—original draft preparation: W.Z. (Wei Zhang); writing—review and editing: W.Z. (Wei Zhang) and G.Y.; project administration: H.Z. All authors have read and agreed to the published version of the manuscript.

Funding: This research received no external funding.

Institutional Review Board Statement: This retrospective study received Institutional Review Board approval of the First Hospital of Shanxi Medical University, ethic code: K131.

Informed Consent Statement: Informed consent was obtained from all subjects involved in the study.

Data Availability Statement: Not applicable.

Conflicts of Interest: The authors declare no conflict of interest.

Appendix A

$$\begin{gathered}\text{Rad-score} = -0.6252 + \text{T2\_wavelet.HLH\_gldm\_SmallDependenceHighGrayLevelEmphasis} \times 0.9473 + \\ \text{T2\_wavelet.HLH\_glrlm\_RunPercentage} \times (-0.5091) + \text{T2\_wavelet.HLL\_ngtdm\_Coarseness} \times 0.7033 + \\ \text{T2\_wavelet.LHH\_gldm\_DependenceNonUniformityNormalized} \times 0.8344 + \\ \text{T2\_wavelet.LHH\_glszm\_SizeZoneNonUniformityNormalized} \times 0.5365 + \\ \text{T2\_wavelet.LHH\_ngtdm\_Contrast} \times 0.3040 + \text{T2\_wavelet.LHL\_firstorder\_RootMeanSquared} \times 0.3430 + \\ \text{DWI\_original\_glszm\_SizeZoneNonUniformityNormalized} \times 0.2708 + \\ \text{DWI\_original\_shape\_SurfaceArea} \times (-0.8964) + \text{DWI\_wavelet.HHH\_glcm\_DifferenceEntropy} \times 0.6870 + \\ \text{DWI\_wavelet.HLH\_glcm\_MaximumProbability} \times (-0.4943) + \\ \text{DWI\_wavelet.HLL\_gldm\_LargeDependenceLowGrayLevelEmphasis} \times 0.3766 + \\ \text{DWI\_wavelet.LHH\_glszm\_ZoneEntropy} \times (-0.1266)\end{gathered}$$

References

1. Bray, F.; Ferlay, J.; Soerjomataram, I.; Siegel, R.L.; Torre, L.A.; Jemal, A. Global cancer statistics 2018: GLOBOCAN estimates of incidence and mortality worldwide for 36 cancers in 185 countries. *CA Cancer J. Clin.* **2018**, *68*, 394–424. [CrossRef] [PubMed]
2. D'Amico, A.V.; Whittington, R.; Malkowicz, S.B.; Schultz, D.; Blank, K.; Broderick, G.A.; Tomaszewski, J.E.; Renshaw, A.A.; Kaplan, I.; Beard, C.J.; et al. Biochemical Outcome After Radical Prostatectomy, External Beam Radiation Therapy, or Interstitial Radiation Therapy for Clinically Localized Prostate Cancer. *JAMA* **1998**, *280*, 969–974. [CrossRef] [PubMed]
3. Ward, J.F.; Blute, M.L.; Slezak, J.; Bergstralh, E.J.; Zincke, H. The Long-Term Clinical Impact of Biochemical Recurrence of Prostate Cancer 5 or More Years After Radical Prostatectomy. *J. Urol.* **2003**, *170*, 1872–1876. [CrossRef] [PubMed]
4. Ahmad, A.S.; Parameshwaran, V.; Beltran, L.; Fisher, G.; North, B.V.; Greenberg, D.; Soosay, G.; Møller, H.; Scardino, P.; Cuzick, J.; et al. Should reporting of peri-neural invasion and extra prostatic extension be mandatory in prostate cancer biopsies? correlation with outcome in biopsy cases treated conservatively. *Oncotarget* **2018**, *9*, 20555–20562. [CrossRef] [PubMed]
5. Meng, Y.; Liao, Y.-B.; Xu, P.; Wei, W.-R.; Wang, J. Perineural invasion is an independent predictor of biochemical recurrence of prostate cancer after local treatment: A meta-analysis. *Int. J. Clin. Exp. Med.* **2015**, *8*, 13267–13274.
6. Bostwick, D.G.; Grignon, D.J.; Hammond, M.E.H.; Amin, M.B.; Cohen, M.; Crawford, D.; Gospadarowicz, M.; Kaplan, R.S.; Miller, D.S.; Montironi, R.; et al. Prognostic Factors in Prostate Cancer: College of American Pathologists Consensus Statement 1999. *Arch. Pathol. Lab. Med.* **2000**, *124*, 995–1000. [CrossRef]
7. Pasoglou, V.; Larbi, A.; Collette, L.; Annet, L.; Jamar, F.; Machiels, J.; Michoux, N.; Berg, B.C.V.; Tombal, B.; Lecouvet, F.E. One-step TNM staging of high-risk prostate cancer using magnetic resonance imaging (MRI): Toward an upfront simplified "all-in-one" imaging approach? *Prostate* **2014**, *74*, 469–477. [CrossRef]
8. Tamada, T.; Sone, T.; Kanomata, N.; Miyaji, Y.; Kido, A.; Jo, Y.; Yamamoto, A.; Ito, K. Value of preoperative 3T multiparametric MRI for surgical margin status in patients with prostate cancer. *J. Magn. Reson. Imaging* **2016**, *44*, 584–593. [CrossRef]
9. Liu, B.; Cheng, J.; Guo, D.J.; He, X.J.; Luo, Y.D.; Zeng, Y.; Li, C.M. Prediction of prostate cancer aggressiveness with a combination of radiomics and machine learning-based analysis of dynamic contrast-enhanced MRI. *Clin. Radiol.* **2019**, *74*, 896.e1–896.e8. [CrossRef]
10. Bourbonne, V.; Vallières, M.; Lucia, F.; Doucet, L.; Visvikis, D.; Tissot, V.; Pradier, O.; Hatt, M.; Schick, U. MRI-Derived Radiomics to Guide Post-operative Management for High-Risk Prostate Cancer. *Front. Oncol.* **2019**, *9*, 807. [CrossRef]
11. Hébert-Blouin, M.N.; Amrami, K.K.; Myers, R.P.; Hanna, A.S.; Spinner, R.J. Adenocarcinoma of the prostate involving the lumbosacral plexus: MRI evidence to support direct perineural spread. *Acta Neurochir.* **2010**, *152*, 1567–1576. [CrossRef]
12. Capek, S.; Howe, B.M.; Amrami, K.K.; Spinner, R.J. Perineural spread of pelvic malignancies to the lumbosacral plexus and beyond: Clinical and imaging patterns. *Neurosurg. Focus* **2015**, *39*, E14. [CrossRef]
13. Song, Y.; Zhang, J.; Zhang, Y.-D.; Hou, Y.; Yan, X.; Wang, Y.; Zhou, M.; Yao, Y.-F.; Yang, G. FeAture Explorer (FAE): A tool for developing and comparing radiomics models. *PLoS ONE* **2020**, *15*, e0237587. [CrossRef]

14. Kuol, N.; Stojanovska, L.; Apostolopoulos, V.; Nurgali, K. Role of the nervous system in cancer metastasis. *J. Exp. Clin. Cancer Res.* **2018**, *37*, 5. [CrossRef]
15. Cole, S.W.; Nagaraja, A.; Lutgendorf, S.K.; Green, P.; Sood, A.K. Sympathetic nervous system regulation of the tumour microenvironment. *Nat. Rev. Cancer* **2015**, *15*, 563–572. [CrossRef]
16. He, S.; He, S.; Chen, C.-H.; Deborde, S.; Bakst, R.L.; Chernichenko, N.; McNamara, W.F.; Lee, S.Y.; Barajas, F.; Yu, Z.; et al. The Chemokine (CCL2–CCR2) Signaling Axis Mediates Perineural Invasion. *Mol. Cancer Res.* **2015**, *13*, 380–390. [CrossRef]
17. Zhang, S.; Qi, L.; Li, M.; Zhang, D.; Xu, S.; Wang, N.; Sun, B. Chemokine CXCL12 and its receptor CXCR4 expression are associated with perineural invasion of prostate cancer. *J. Exp. Clin. Cancer Res.* **2008**, *27*, 62. [CrossRef]
18. Suresh, N.; Teramoto, Y.; Goto, T.; Wang, Y.; Miyamoto, H. Clinical significance of perineural invasion by prostate cancer on magnetic resonance imaging–targeted biopsy. *Hum. Pathol.* **2022**, *121*, 65–72. [CrossRef]
19. Niu, Y.; Förster, S.; Muders, M. The Role of Perineural Invasion in Prostate Cancer and Its Prognostic Significance. *Cancers* **2022**, *14*, 4065. [CrossRef]
20. Truong, M.; Rais-Bahrami, S.; Nix, J.W.; Messing, E.M.; Miyamoto, H.; Gordetsky, J.B. Perineural invasion by prostate cancer on MR/US fusion targeted biopsy is associated with extraprostatic extension and early biochemical recurrence after radical prostatectomy. *Hum. Pathol.* **2017**, *66*, 206–211. [CrossRef]
21. Lee, I.H.; Roberts, R.; Shah, R.B.; Wojno, K.J.; Wei, J.T.; Sandler, H.M. Perineural Invasion is a Marker for Pathologically Advanced Disease in Localized Prostate Cancer. *Int. J. Radiat. Oncol. Biol. Phys.* **2007**, *68*, 1059–1064. [CrossRef] [PubMed]
22. Zhang, L.-J.; Wu, B.; Zha, Z.-L.; Qu, W.; Zhao, H.; Yuan, J.; Feng, Y.-J. Perineural invasion as an independent predictor of biochemical recurrence in prostate cancer following radical prostatectomy or radiotherapy: A systematic review and meta-analysis. *BMC Urol.* **2018**, *18*, 5. [CrossRef]
23. Dell'Atti, L. Prognostic significance of perineural invasion in patients who underwent radical prostatectomy for localized prostate cancer. *J. B.U.ON. Off. J. Balk. Union Oncol.* **2016**, *21*, 1219–1223.
24. DeLancey, J.O.; Wood, D.P.; He, C.; Montgomery, J.S.; Weizer, A.Z.; Miller, D.C.; Jacobs, B.L.; Montie, J.E.; Hollenbeck, B.K.; Skolarus, T.A. Evidence of Perineural Invasion on Prostate Biopsy Specimen and Survival After Radical Prostatectomy. *Urology* **2013**, *81*, 354–357. [CrossRef] [PubMed]
25. Wu, S.; Xie, L.; Lin, S.X.; Wirth, G.J.; Lu, M.; Zhang, Y.; Blute, M.L.; Dahl, D.M.; Wu, C.-L. Quantification of perineural invasion focus after radical prostatectomy could improve predictive power of recurrence. *Hum. Pathol.* **2020**, *104*, 96–104. [CrossRef]
26. Sigorski, D.; Gulczyński, J.; Sejda, A.; Rogowski, W.; Iżycka-Świeszewska, E. Investigation of Neural Microenvironment in Prostate Cancer in Context of Neural Density, Perineural Invasion, and Neuroendocrine Profile of Tumors. *Front. Oncol.* **2021**, *11*, 710899. [CrossRef]
27. Billis, A.; De Quintal, M.M.; Meirelles, L.; Freitas, L.L.L.; Magna, L.A.; Ferreira, U. Does tumor extent on needle prostatic biopsies influence the value of perineural invasion to predict pathologic stage > T2 in radical prostatectomies? *Int. braz j urol* **2010**, *36*, 439–447. [CrossRef]
28. Lee, H.; Kim, C.K.; Park, B.K.; Sung, H.H.; Han, D.H.; Jeon, H.G.; Jeong, B.C.; Seo, S.I.; Jeon, S.S.; Choi, H.Y.; et al. Accuracy of preoperative multiparametric magnetic resonance imaging for prediction of unfavorable pathology in patients with localized prostate cancer undergoing radical prostatectomy. *World J. Urol.* **2017**, *35*, 929–934. [CrossRef]
29. Griffiths, L.; Kotamarti, S.; Mikhail, D.; Sarcona, J.; Rastinehad, A.R.; Villani, R.; Kreshover, J.; Hall, S.J.; Vira, M.A.; Schwartz, M.J.; et al. Extracapsular extension on multiparametric magnetic resonance imaging better predicts pT3 disease at radical prostatectomy compared to perineural invasion on biopsy. *Can. Urol. Assoc. J.* **2021**, *15*, 261–266. [CrossRef]
30. Stone, J.J.; Adamo, D.A.; Khan, D.Z.; Packard, A.T.; Broski, S.M.; Nathan, M.A.; Howe, B.M.; Spinner, R.J. Multimodal Imaging Aids in the Diagnosis of Perineural Spread of Prostate Cancer. *World Neurosurg.* **2019**, *122*, e235–e240. [CrossRef]
31. Siracusano, S.; Porcaro, A.B.; Tafuri, A.; Pirozzi, M.; Cybulski, A.; Shakir, A.; Tiso, L.; Talamini, R.; Mucelli, R.P. Visualization of peri-prostatic neurovascular fibers before and after radical prostatectomy by means of diffusion tensor imaging (DTI) with clinical correlations: Preliminary report. *J. Robot. Surg.* **2020**, *14*, 357–363. [CrossRef]
32. You, H.; Shang, W.; Min, X.; Weinreb, J.; Li, Q.; Leapman, M.; Wang, L.; Tian, J. Sight and switch off: Nerve density visualization for interventions targeting nerves in prostate cancer. *Sci. Adv.* **2020**, *6*, eaax6040. [CrossRef]
33. Sun, Y.; Reynolds, H.M.; Parameswaran, B.; Wraith, D.; Finnegan, M.E.; Williams, S.; Haworth, A. Multiparametric MRI and radiomics in prostate cancer: A review. *Australas. Phys. Eng. Sci. Med.* **2019**, *42*, 3–25. [CrossRef]
34. Ma, S.; Xie, H.; Wang, H.; Han, C.; Yang, J.; Lin, Z.; Li, Y.; He, Q.; Wang, R.; Cui, Y.; et al. MRI-Based Radiomics Signature for the Preoperative Prediction of Extracapsular Extension of Prostate Cancer. *J. Magn. Reson. Imaging* **2019**, *50*, 1914–1925. [CrossRef]
35. Peng, Y.; Jiang, Y.; Antic, T.; Giger, M.L.; Eggener, S.E.; Oto, A. Validation of Quantitative Analysis of Multiparametric Prostate MR Images for Prostate Cancer Detection and Aggressiveness Assessment: A Cross-Imager Study. *Radiology* **2014**, *271*, 461–471. [CrossRef]
36. Vignati, A.; Mazzetti, S.; Giannini, V.; Russo, F.; Bollito, E.; Porpiglia, F.; Stasi, M.; Regge, D. Texture features on T2-weighted magnetic resonance imaging: New potential biomarkers for prostate cancer aggressiveness. *Phys. Med. Biol.* **2015**, *60*, 2685–2701. [CrossRef]
37. Jajodia, A.; Gupta, A.; Prosch, H.; Mayerhoefer, M.; Mitra, S.; Pasricha, S.; Mehta, A.; Puri, S.; Chaturvedi, A. Combination of Radiomics and Machine Learning with Diffusion-Weighted MR Imaging for Clinical Outcome Prognostication in Cervical Cancer. *Tomography* **2021**, *7*, 344–357. [CrossRef]

38. Linsalata, S.; Borgheresi, R.; Marfisi, D.; Barca, P.; Sainato, A.; Paiar, F.; Neri, E.; Traino, A.C.; Giannelli, M. Radiomics of Patients with Locally Advanced Rectal Cancer: Effect of Preprocessing on Features Estimation from Computed Tomography Imaging. *BioMed Res. Int.* **2022**, *2022*, 1–21. [CrossRef]
39. Gill, T.S.; Varghese, B.A.; Hwang, D.H.; Cen, S.Y.; Aron, M.; Aron, M.; Duddalwar, V.A. Juxtatumoral perinephric fat analysis in clear cell renal cell carcinoma. *Abdom. Radiol.* **2019**, *44*, 1470–1480. [CrossRef]
40. Fan, T.-W.; Malhi, H.; Varghese, B.; Cen, S.; Hwang, D.; Aron, M.; Rajarubendra, N.; Desai, M.; Duddalwar, V. Computed tomography-based texture analysis of bladder cancer: Differentiating urothelial carcinoma from micropapillary carcinoma. *Abdom. Radiol.* **2018**, *44*, 201–208. [CrossRef]
41. De Santi, B.; Salvi, M.; Giannini, V.; Meiburger, K.M.; Marzola, F.; Russo, F.; Bosco, M.; Molinari, F. Comparison of Histogram-based Textural Features between Cancerous and Normal Prostatic Tissue in Multiparametric Magnetic Resonance Images. In Proceedings of the 2020 42nd Annual International Conference of the IEEE Engineering in Medicine & Biology Society (EMBC), Montreal, QC, Canada, 20–24 July 2020; Volume 2020, pp. 1671–1674. [CrossRef]

diagnostics

MDPI

Article

Gut Microbial Shifts Indicate Melanoma Presence and Bacterial Interactions in a Murine Model

Marco Rossi [1,†], Salvatore M. Aspromonte [2,3,†], Frederick J. Kohlhapp [3], Jenna H. Newman [3], Alex Lemenze [4], Russell J. Pepe [2], Samuel M. DeFina [3], Nora L. Herzog [3], Robert Donnelly [4], Timothy M. Kuzel [1], Jochen Reiser [1], Jose A. Guevara-Patino [5,*] and Andrew Zloza [1,*]

1. Rush University Medical Center, Chicago, IL 60612, USA; marco_rossi@rush.edu (M.R.); timothy_kuzel@rush.edu (T.M.K.); jochen_reiser@rush.edu (J.R.)
2. Rutgers Robert Wood Johnson Medical School, Rutgers, The State University of New Jersey, New Brunswick, NJ 08901, USA; saa9216@nyp.org (S.M.A.); rjpepe19@gmail.com (R.J.P.)
3. Rutgers Cancer Institute of New Jersey, Rutgers, The State University of New Jersey, New Brunswick, NJ 08901, USA; kohlhapp@gmail.com (F.J.K.); jenna.newman@mssm.edu (J.H.N.); sam.defina@yale.edu (S.M.D.); herzog.nora@gmail.com (N.L.H.)
4. Rutgers New Jersey Medical School, Rutgers, The State University of New Jersey, Newark, NJ 07103, USA; alemenze@gmail.com (A.L.); donnelly@rutgers.edu (R.D.)
5. Moffitt Cancer Center, Tampa, FL 33612, USA

* Correspondence: josealejandro.guevara@moffitt.org (J.A.G.-P.); andrew_zloza@rush.edu (A.Z.)

† These authors contributed equally to this work.

Abstract: Through a multitude of studies, the gut microbiota has been recognized as a significant influencer of both homeostasis and pathophysiology. Certain microbial taxa can even affect treatments such as cancer immunotherapies, including the immune checkpoint blockade. These taxa can impact such processes both individually as well as collectively through mechanisms from quorum sensing to metabolite production. Due to this overarching presence of the gut microbiota in many physiological processes distal to the GI tract, we hypothesized that mice bearing tumors at extraintestinal sites would display a distinct intestinal microbial signature from non-tumor-bearing mice, and that such a signature would involve taxa that collectively shift with tumor presence. Microbial OTUs were determined from 16S rRNA genes isolated from the fecal samples of C57BL/6 mice challenged with either B16-F10 melanoma cells or PBS control and analyzed using QIIME. Relative proportions of bacteria were determined for each mouse and, using machine-learning approaches, significantly altered taxa and co-occurrence patterns between tumor- and non-tumor-bearing mice were found. Mice with a tumor had elevated proportions of *Ruminococcaceae*, *Peptococcaceae*.g_rc4.4, and *Christensenellaceae*, as well as significant information gains and ReliefF weights for *Bacteroidales.f__S24.7*, *Ruminococcaceae*, *Clostridiales*, and *Erysipelotrichaceae*. *Bacteroidales.f__S24.7*, *Ruminococcaceae*, and *Clostridiales* were also implicated through shifting co-occurrences and PCA values. Using these seven taxa as a melanoma signature, a neural network reached an 80% tumor detection accuracy in a 10-fold stratified random sampling validation. These results indicated gut microbial proportions as a biosensor for tumor detection, and that shifting co-occurrences could be used to reveal relevant taxa.

Keywords: gut microbiota; machine learning; statistical algorithms; co-occurrence patterns; melanoma

Citation: Rossi, M.; Aspromonte, S.M.; Kohlhapp, F.J.; Newman, J.H.; Lemenze, A.; Pepe, R.J.; DeFina, S.M.; Herzog, N.L.; Donnelly, R.; Kuzel, T.M.; et al. Gut Microbial Shifts Indicate Melanoma Presence and Bacterial Interactions in a Murine Model. *Diagnostics* **2022**, *12*, 958. https://doi.org/10.3390/diagnostics12040958

Academic Editor: Sung Chul Lim

Received: 4 February 2022
Accepted: 4 March 2022
Published: 12 April 2022

Publisher's Note: MDPI stays neutral with regard to jurisdictional claims in published maps and institutional affiliations.

1. Introduction

The gastrointestinal microbiota contains a diverse and dense collection of symbiotic organisms that contribute to intestinal homeostasis. Nutrient digestion, synthesis of vitamins, protection against pathologic organisms, and production of neurotransmitters are just a few of the biological functions that these organisms provide [1–3]. The host's immune system plays an essential role in controlling microbial growth and development in the microbiome to ensure that a mutual relationship is maintained between the host and organism.

At the same time, the microbiota plays a role in adapting the host's immune system to various stressors [4]. In fact, evidence is accumulating that the intestinal microflora can respond to changes in host health status by sensing soluble host elements and local micro-environmental cues [5]. For this reason, the gastrointestinal microbiota is affected by the pathological immune responses derived from diseases such as diabetes mellitus, cancer, obesity, and inflammatory diseases, which impacts the body's immune response against disease [2,6,7].

It is increasingly being recognized that the gut microbiome composition differs significantly between healthy individuals and those with various pathological conditions. Dongmei et al. found that healthy individuals have a more diverse gut flora than those with colorectal cancer. In addition, certain bacterial populations were more likely to co-occur in patients with colorectal cancer than in healthy individuals [3]. While alterations in microbiome composition can be seen in pathologic conditions such as cancer, it is unclear whether these changes are a cause or a consequence of the disease [6]. Multiple studies that analyzed the composition of the gut microbiota in colorectal cancer patients suggested the presence of both "driver bacteria", or those that promote cancer growth, and "passenger bacteria", or those that solely flourish in the proinflammatory environment, but do not impact tumor progression. Geng et al. found that in their colorectal cancer patients, members of the *Enterobacteriaceae* family promoted cancer growth, whereas members of the *Streptococcaceae* family merely flourished in a proinflammatory environment [7].

The presence of these microbial mechanisms in which bacterial taxa have a certain level of dependency have wide implications for their use in modeling respective pathological conditions. Typically, connectivity and dependency between variables such as bacterial taxa in the context of predictive modeling has typically been a hindrance to model performance [8–10]. It is widely understood with many kinds of algorithms that, in various circumstances, variables with some manner of co-occurrence provide a certain level of redundant information, and therefore reduce the variability explained in models [8]. This presence of redundant information decreases the model's fit to the training dataset, as well as its prediction accuracy in the testing dataset [10–12].

Despite these limitations, co-occurrences in the context of pathological prediction with microbial taxa may still hold significance in the application of diagnostic signatures [8,13]. When co-occurrences shift between conditions, so does the direction of variability represented by relevant taxa in planes of higher dimensionality [9,10,14]. These shifts are reflected in principal component analysis, in which each principal component represents a different proportion of the total variability present [8,13]. They are also represented in ReliefF and information gain values, in which microbial taxa with these differences in variability have increased reliability as predictors [11,15]. Therefore, the identification of these shifts in co-occurrences in pathological conditions such as cancer is optimal for the implementation of gut microbial diagnostic signatures.

The implementation of machine-learning algorithms for the prediction of the presence of various cancers using the gut microbiome has been widely studied [16–18]. However, to date, relatively little work has been done regarding the use of the gut microbiome to predict the presence of melanoma. In addition, one of the challenges of predicting the presence of a specific disease with the gut microbiota is the variability in relative proportions of specific gut bacteria that can exist between patients and populations [12]. Through our analyses, we have indicated shifts in microbial co-occurrences as a potential method in accounting for such variability. Therefore, we hypothesized that models based on gut microbial proportion profiles of taxa involved in co-occurrence shifts could form a distinct diagnostic signature that effectively differentiated mice bearing mouse melanoma tumors from non-tumor-bearing mice. This implies that the intestinal microflora may function as a biosensor for the presence of cancer, and that its manipulation may alter cancer prognoses.

2. Results

2.1. Shifts in Microbial Taxon Proportions of Melanoma-Bearing Mice

Mice bearing melanoma tumors displayed significant shifts in gut microbial proportions compared to non-tumor-bearing mice, which: (1) implicated consistency in changes in gut microbiota data with tumors in the skin distal to the gut; and (2) implied that such changes could be used by an algorithm to detect the presence of cancer. We compared the microbial composition of fecal samples of melanoma-bearing and tumor-free mice by terminal restriction fragment length polymorphism (T-RFLP) analysis [14,16]. This technique is commonly used to study complex microbial communities based on 16S rRNA gene variation, and has been applied in the study of microbial communities in soil and sludge systems [19]. T-RFLP analysis was carried out in a blinded fashion as previously described [4]. It was readily seen for the two mouse experiments (Figure 1) that the co-occurrences of relative taxon proportions shifted in the presence of B16 melanoma. In addition, *Peptococcaceae*.g_rc4.4 was significantly increased (Wilcoxon $p < 0.05$) in both groups of mice (Figure 1). These data demonstrated that the intestinal flora developed detectable changes that discriminated a tumor-bearing from a tumor-free host. In order to more fully determine the extent to which these results distinguished between hosts that had a tumor and those that did not, the two mouse groups were combined and further analyzed as a single dataset ($n = 56$).

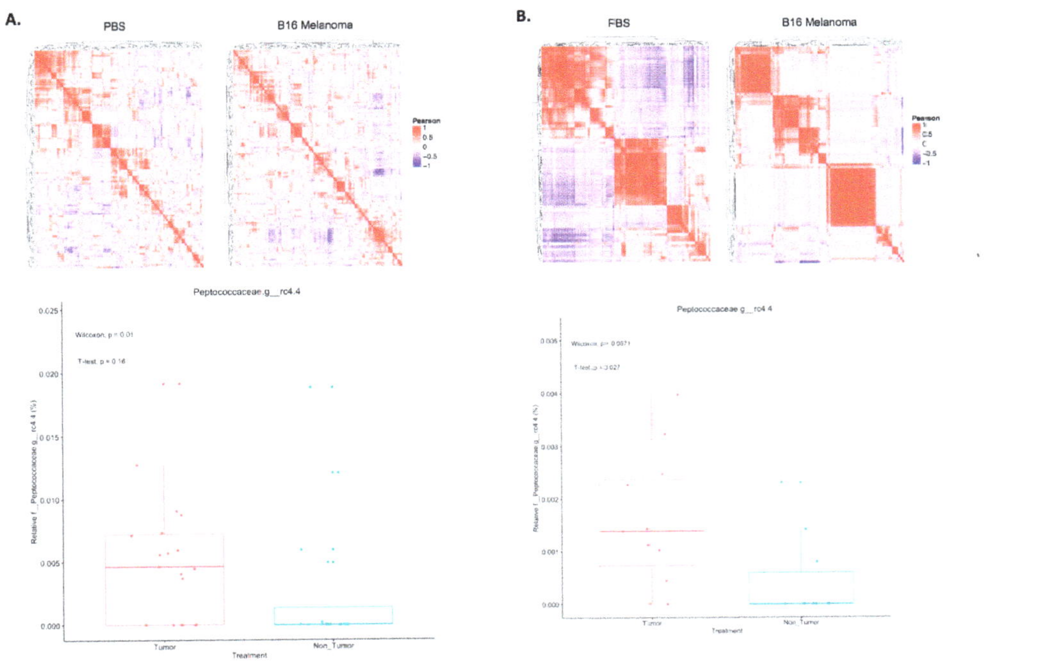

Figure 1. Shifted co-occurrences of microbial taxa and increased *Peptococcaceae*.g_rc4.4 characterize tumor presence. (**A**) C57BL/6 (B6) male mice were injected with either 10^5 B16 melanoma cells ($n = 19$) or PBS ($n = 16$). After 10 days, fecal samples were collected and 16S rRNA genes were analyzed using terminal restriction fragment length polymorphism (T-RFLP) analysis. From individual taxon proportion and co-occurrence patterns, it could be seen that such patterns shifted with melanoma presence, and *Peptococcaceae*.g_rc4.4 levels increased. (**B**) B6 male mice were injected with either 10^5 B16 melanoma cells ($n = 11$) or PBS ($n = 10$). After 16 days, fecal samples were collected and 16S rRNA genes were analyzed using terminal restriction fragment length polymorphism (T-RFLP) analysis. The results of these data directly corresponded with the mice in (**A**).

2.2. Co-Occurrence between Bacteroidales.f__S24.7, Clostridiales, and Ruminococcaceae Proportions in Mouse Melanoma

Seeking to identify the specific bacterial co-occurrences that were altered in the presence of a tumor, we first used Cytoscape to map them in the B16-melanoma- and PBS-treated mice. From these diagrams (Figure 2A,B), it was found that the co-occurrences of *Bacteroidales.f__S24.7* greatly differed between the two treatments. When looking further into this taxon, it was found that its co-occurrences with *Clostridiales* and *Ruminococcaceae* had changed the most between tumor and nontumor/PBS (Figure 2C,D), with Pearson correlation values of approximately −0.9 and −0.8 for tumor, as well as −0.15 and −0.13 for nontumor, respectively. Interestingly, however, when looking at the individual relative amounts of these taxa, the only one that was significantly different between tumor and nontumor was *Ruminococcaceae* (Wilcoxon $p < 0.05$, *T*-test $p < 0.05$; Figure 2E). Thus, we concluded that the potential for these taxa to predict tumor presence relied heavily on the extent to which their co-occurrences shifted in that condition, rather than changes in their individual relative amounts.

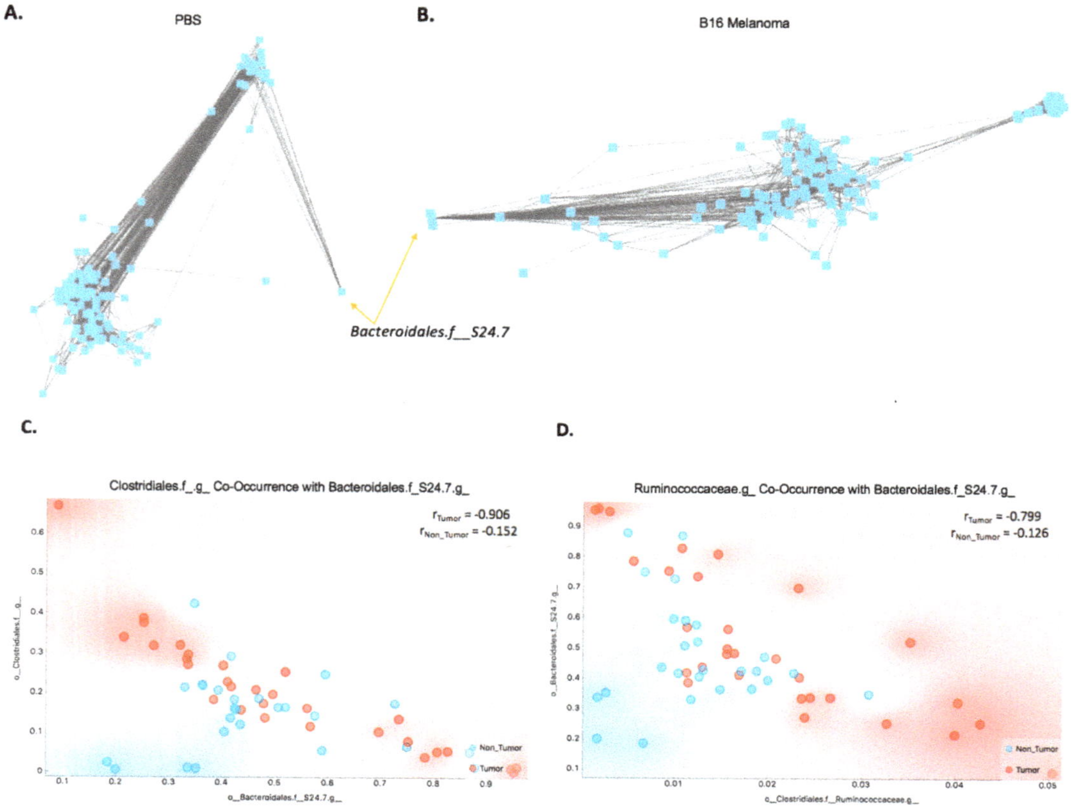

Figure 2. *Cont.*

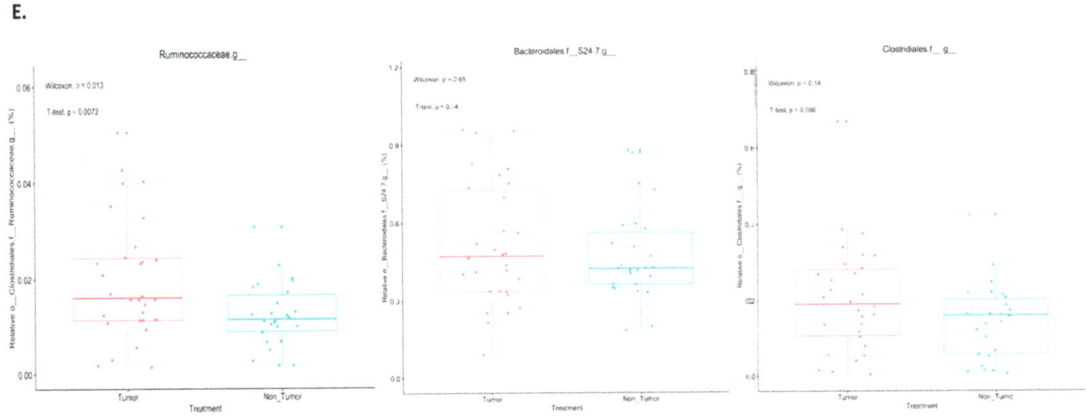

Figure 2. Co-occurrence changes between *Bacteroidales.f__S24.7*, *Clostridiales*, and *Ruminococcaceae* occur with tumor presence. (**A**,**B**) Pearson correlation matrices were determined for microbiotas from tumor and nontumor mice and displayed using Cytoscape. From these visualizations, *Bacteroidales.f__S24.7* co-occurrences greatly changed with tumor presence. (**C**,**D**) Using the R programming language, it was found that the most dramatic shifts of *Bacteroidales.f__S24.7* were in conjunction with *Clostridiales* and *Ruminococcaceae*. (**E**) When comparing each taxon individually between tumor and nontumor, only *Ruminococcaceae* was significantly different.

2.3. Differences in Principal Components between Tumor and Nontumor

Considering our results for both individual microbial taxa and co-occurrence shifts, we wanted to assess the relevance of each taxon in the context of predictive modeling. Thus, we calculated the information gains and ReliefF weights for each taxon (Figure 3A,B). In the scoring for information gains, *Ruminococcaceae*, *Peptococcaceae.g_rc4.4*, and *Christensenellaceae* consistently scored higher than the majority of taxa (Figure 3A). For the ReliefF algorithm, *Bacteroidales.f__S24.7* had a fairly high weight, along with *Peptococcaceae.g_rc4.4* and *Christensenellaceae* (Figure 3A). Further, *Christensenellaceae* was found to be significantly different between tumor and nontumor (Wilcoxon $p < 0.05$, Figure 3A,B). Considering that *Bacteroidales.f__S24.7* shifted its co-occurrences and its ReliefF weight indicated variable importance, we performed a principal component analysis (PCA) using this taxon (Figure 3C,D). Two PCAs were performed, one with *Clostridiales* and the other with *Ruminococcaceae* (Figure 3C,D). After performing the PCAs, we compared the resulting principal component coordinates between tumor and nontumor mice. From this comparison, we found that, although the first principal components did not differ between the two groups (Figure 3C), the second ones did (Wilcoxon $p < 0.05$, *T*-test $p < 0.05$; Figure 3D). These results indicated that the coordinates of these second principal components could be implemented in predictive modeling.

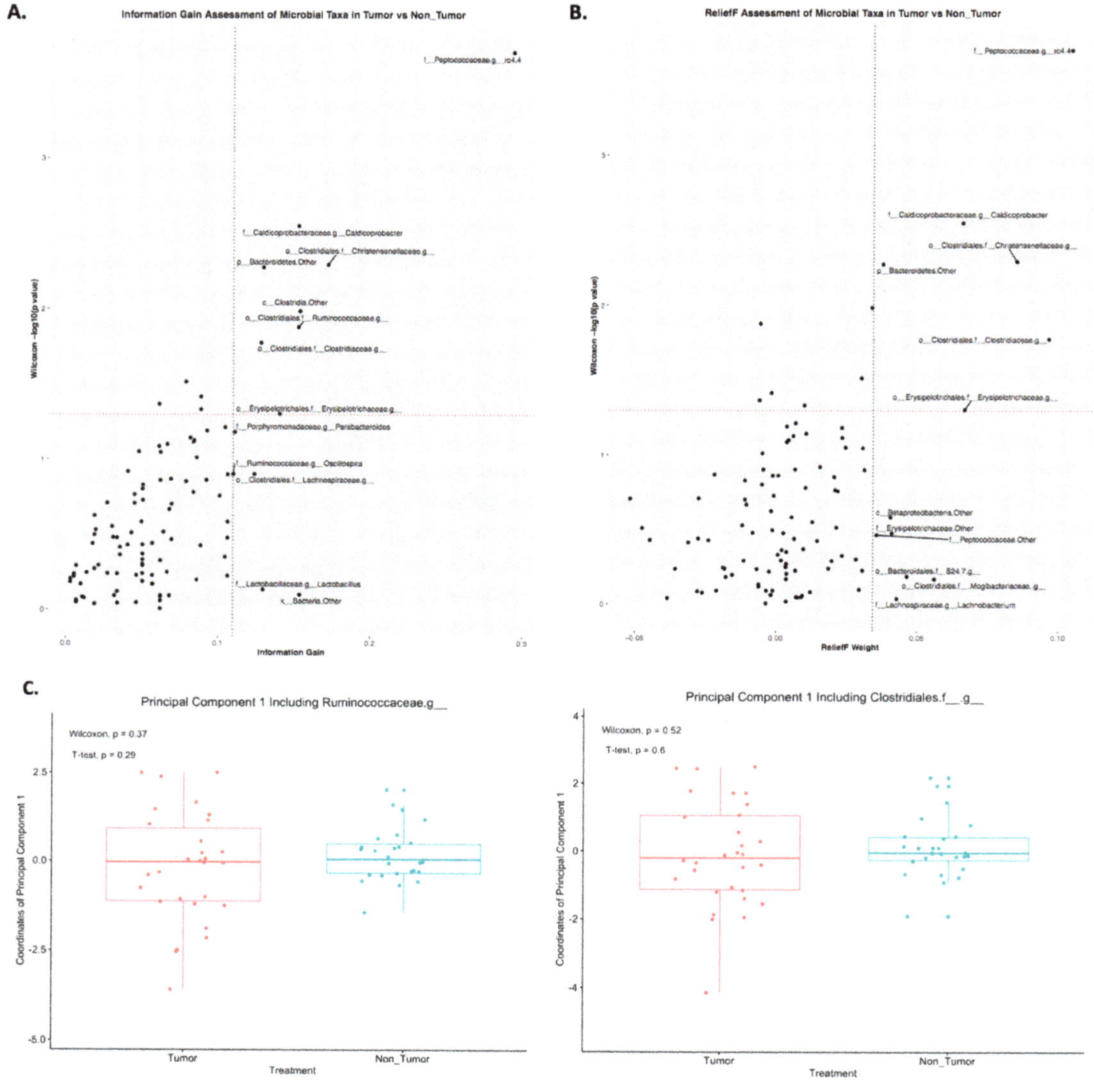

Figure 3. *Cont.*

D.

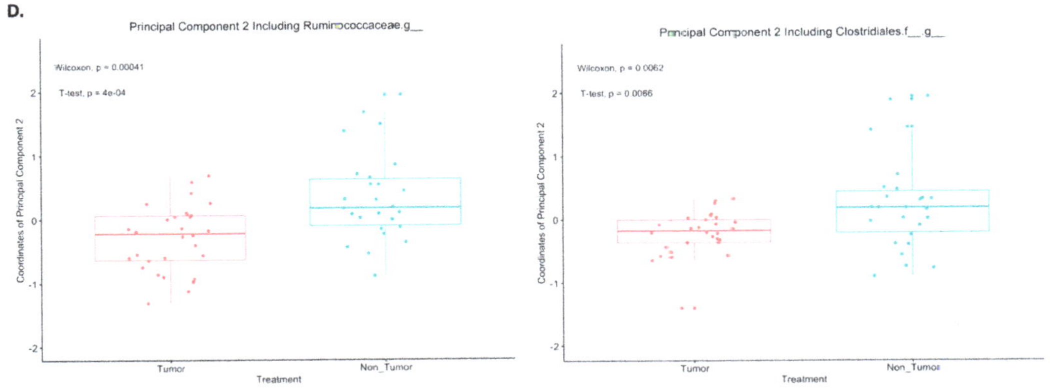

Figure 3. Significant predictors of tumor presence include the second principal components involving *Bacteroidales.f__S24.7*, *Clostridiales*, and *Ruminococcaceae*. (**A**,**B**) Using the CORElearn package in the R programming language, the information gains and ReliefF weights were calculated for each taxon. (**A**) *Ruminococcaceae*, *Peptococcaceae.g_rc4.4*, and *Christensenellaceae* were found significantly altered with tumor presence and having high information gains. (**B**) Along with *Peptococcaceae.g_rc4.4* and Christensenellaceae, *Bacteroidales.f__S24.7* and *Erysipelotrichaceae* had high ReliefF weights. (**C**,**D**) Two PCAs using *Bacteroidales.f__S24.7*, one with *Ruminococcaceae* and the other with *Clostridiales*, were conducted using R. While their first principal components did not change with tumor, their second ones did (Wilcoxon $p < 0.05$, *T*-test $p < 0.05$ (**D**)).

2.4. Prediction of Tumor Presence Using Microbial Taxa Involved in Altered Co-Occurrences

Since the second principal components involving *Bacteroidales.f__S24.7*, *Ruminococcaceae*, and *Clostridiales* were found to significantly differ with tumor presence, the proportions of those taxa, along with those of *Peptococcaceae.g_rc4.4*, *Christensenellaceae*, and *Erysipelotrichaceae*, were implemented as a mouse melanoma signature (Figure 4A,B). The 10-fold stratified random sampling used to obtain melanoma prediction results with machine-learning algorithms was performed by randomly selecting 90% of the mouse samples to train the algorithms and then testing them with the remaining 10% of samples (Figure 4A). This process was repeated 10 times, and the prediction results were averaged over those repeats (Figure 4A). Using this protocol, the highest percent accuracy in melanoma prediction was achieved by the neural network, with 80% (Figure 4A,B). Thus, the implementation of microbial taxa indicated by the second principal components in the prediction signature allowed for the identification of melanoma presence.

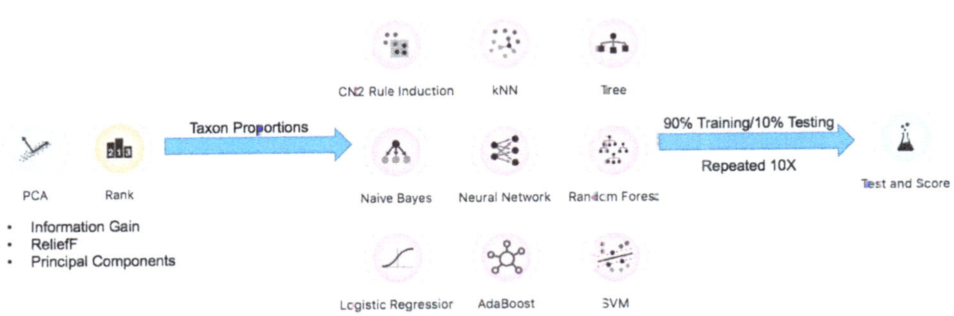

(**A**)

Figure 4. *Cont.*

10-fold Stratified Random Sampling with Taxon Proportions

Model	*AUC*	*CA*	*F1*	*Precision*	*Recall*
Neural Network	0.792	0.800	0.794	0.838	0.800
SVM	0.771	0.783	0.773	0.849	0.783
AdaBoost	0.750	0.750	0.747	0.764	0.750
CN2 rule inducer	0.775	0.733	0.733	0.734	0.733
Random Forest	0.861	0.733	0.729	0.751	0.733
kNN	0.660	0.667	0.665	0.670	0.667
Naive Bayes	0.852	0.667	0.661	0.679	0.667
Tree	0.686	0.650	0.645	0.659	0.650
Logistic Regression	0.590	0.550	0.436	0.763	0.550

(B)

Figure 4. Implementation of microbial taxa implicated in second principal components accurately predict tumor presence. (**A**) Using Orange3, 10-fold stratified shuffle splits were performed. (**B**) Using a prediction signature which included *Bacteroidales.f__S24.7*, *Ruminococcaceae*, and *Clostridiales*, implicated in the second principal components, resulted in an average accuracy of 80% achieved with a Neural Network classifier. AUC, area under the curve; CA, classification accuracy; F1, F1 score).

3. Discussion

Our findings demonstrated that the presence of a mouse melanoma tumor can be detected through the altered gut microbial proportions using classification algorithms. By using the gut microbial taxa to model tumor presence, it became apparent that such a condition manifested in more ways than just changes in individual amounts of certain taxa. Indeed, one of the main implications of this study is that considering gut microbial taxa co-occurrences and dependencies in predictive modeling can significantly increase predictive power in melanoma, more so than analyzing only statistical significance between groups. This concept of intertaxa correlations in modeling microbial-based conditions has wide applications in the interpretation of the gut microbiota, as it suggests that the role of an individual taxon in manifesting a biological phenotype is not solely attributed to its unique characteristics [17,18]. Rather, this role also depends on the extent to which a single taxon can communicate and affect other taxa through various mechanisms, from quorum sensing to metabolite production [20–23].

Despite this apparent, predictive relationship between murine melanoma and the gut microbiota, certain experimental limitations still existed. The primary limitation for consideration was the external validity of these results. It is often the case that gut microbiota data do not directly correspond between murine and human subjects, with various mechanisms implicated, from general differences in GI physiology to lifestyle, epigenetics, and immune responses [24–26]. Thus, in order for gut microbial associations to be implemented in clinical cancer diagnoses, further work needs to be done to elucidate pertinent taxa in a variety of human populations and pathophysiological states, including cancer, as well as the interaction between shifts in gut microbial content and certain factors such as diet and lifestyle. Most pertinent to patient treatment is the level of interaction between host immune responses and the gut microbiota, as antitumor immunity and immunotherapies may affect prediction outcomes [27,28]. These studies would also need to consider the correlation between patient stool sampling and gut microbial content with cancer presence, as sampling variation may be a confound [24]. Finally, since our gut microbiota data had a certain level of variation, other parameters should be considered in the future predictive modeling of human melanoma, such as biochemical and clinical observations [29].

In the statistical analysis of gut microbial taxa, algorithms have been developed to accurately detect the presence of these intertaxa co-occurrences [30–32]. Such algorithms for the detection of microbial "co-occurrence networks" include Sparse Inverse Covariance Estimation for Ecological Association Inference (SPIEC-EASI) and Sparse Correlations for Compositional Data (SparCC) [31–33]. However, despite these advances in the statistical detection of these interactions, there has not been as much work to determine their efficacy in different types of classification algorithms in conditions such as melanoma. In fact, their presence in predictive models has generally been discouraged, as the collinearity they create have been shown to compromise the performance of many model types [34–36]. Further, even for models that can more readily account for collinearity, the use of such interactions in these models does not consistently increase the performance of those models [34–36]. Thus, there is a necessity for a new statistical interpretation of intertaxa co-occurrences in order for them to be optimally utilized in a predictive model. Perhaps new insights into such interpretations can be eventually made when taxa indicated by shifts in co-occurrence networks are further tested in more architecturally complex algorithms such as deep-learning neural networks.

Traditionally, one of the most common procedures in dealing with collinearity between variables such as microbial taxa is the use of principal components in principal component analysis (PCA) [34–37]. By definition, the resulting principal components do not significantly correlate with each other, and are thus used in various model types [34–37]. These components are not usually interpretable from the perspective of the original data because they are linear transformations of that data [34–37]. However, if a small number of variables (e.g., two or three) is used, the principal components can be more easily interpreted [34–37] In this study, PCA analysis was able to differentiate the two groups of mice successfully, however, much work still needs to be done to characterize the significance of individual PCs in different situations, such as in other clinically relevant tumor types.

4. Methods

4.1. Cell Culture

B16-F10 cells (ATCC) were cultured in RPMI 1640 plus 10% heat-inactivated fetal bovine serum (Atlanta Biologicals, Flowery Branch, GA, USA), 2 mM L-glutamine (Mediatech, Manassas, VA, USA), and 1% penicillin/streptomycin (Mediatech).

4.2. Mouse Experiments

C57BL/6 mice (B6; no. 00664; Jackson Laboratory) were housed in a specific pathogen-free facility at the Rutgers Cancer Institute of New Jersey. Experiments involving animals were carried out in accordance with respective Institutional Animal Care and Use Committee (IACUC) and Institutional Biosafety Committee (IBC) guidelines.

In the first experiment, 35 B6 male mice, aged 6 to 8 weeks old from the Jackson Laboratory were intradermally challenged in the right flank with 10^5 cells of the highly aggressive and poorly immunogenic melanoma B16 cell line (n = 19) [17] or phosphate buffered saline (PBS) (n = 16) under isoflurane anesthesia. Mice were fed regular chow according to animal care institutional guidelines. Fecal sample collection to compare tumor-bearing to non-tumor-bearing mice was carried out on day 10, when tumors were approximately 25–50 mm^2. Samples were stored immediately at −80 °C until DNA extraction [38] and sequencing.

The second experiment at this facility followed the identical protocol, using 21 B6 male mice aged 6 to 8 weeks old that were intradermally challenged in the right flank with 10^5 cells of the highly aggressive and poorly immunogenic melanoma B16 cell line (n = 11) [17] or phosphate buffered saline (PBS) (n = 10) under isoflurane anesthesia. Fecal sample collection to compare tumor-bearing to non-tumor-bearing mice was carried out on day 16, when tumors were approximately 25–50 mm^2 in diameter. Samples were stored immediately at −80 °C until DNA extraction [38] and sequencing.

4.3. DNA Extraction

Fecal pellets were homogenized and extracted using the QIAamp PowerFecal DNA Extraction kit following the manufacturer's protocols [39].

4.4. 16S rRNA Gene Sequencing and Data Analysis

The 16S rRNA genes were amplified from purified DNA using PCR primers specific to the V3–V4 region of the 16S rRNA gene and sequenced by Illumina MiSeq in a 2 × 150 bp configuration at the Rutgers New Jersey Medical School Genomics Core. Quantitative Insights Into Microbial Ecology (QIIME) software was used for open-reference operational taxonomic unit (OTU) classification with OTU clustering at 0.97, followed by rarefaction and taxonomic classification of de novo OTUs [40].

4.5. qPCR for Bacterial Load and Taxa Assays

Bacterial loads of extracted fecal DNA were determined by qPCR. DNA were quantified against a standard curve, and the results were normalized to the weight of fecal samples [40].

4.6. Taxon Comparisons, Analyses, and Statistical Modeling

Using the R programming language, microbial taxa between tumor-bearing and PBS control mice were compared using Welch's t-test as well as the Mann–Whitney U test (a p-value of <0.05 was considered to denote statistically significant differences). Between these two groups of mice, general taxa and comparison attributes were determined using the Orange3 v3.27.1 data-mining program and the CORElearn package in CRAN. PCA analysis and principal components were determined using the prcomp function in R. General machine-learning model analyses and cross-validation procedures were performed using the Orange3 program with these settings:

The neural network was a 100-neuron single hidden layer that used the ReLu activation function and the Adam solver.

The support vector machine (SVM) used a radial basis function (RBF) kernel with a cost of 1.0 and a regression loss epsilon of 0.1.

The AdaBoost used a SAMME.R classification algorithm with a linear regression loss function, 50 estimators, and learning rate of 1.0.

The CN2 rule inducer used entropy as the evaluation measure, a beam width of 5, and a maximum rule length of 5.

The random forest used a 12-tree ensemble with subsets split no smaller than 5.

The k-nearest neighbor (kNN) used 5 neighbors and considered the Euclidean distance and uniform weights.

For the naïve Bayes, the attributes were not weighted.

Tree used a maximal tree depth of 100 and subsets not split smaller than 5.

In the logistic regression, a ridge regularization was implemented.

Quality parameters for this model were determined using an internal 10-fold stratified shuffle split, with 90% of the samples selected for training and the remaining 10% for testing in Orange3. Results were graphed using the ggplot2, ggrepel, and ggpubr packages in CRAN, as well as Orange3 and Cytoscape v3.7.2. Heatmaps were generated using the ComplexHeatmap package in CRAN. Tables were formatted using the sjPlot package in CRAN.

Author Contributions: Conceptualization, J.A.G.-P. and A.Z.; formal analysis, M.R.; investigation, S.M.A., F.J.K., J.H.N., A.L., R.J.P., S.M.D., N.L.H. and R.D.; resources, T.M.K. and J.R.; data curation, M.R.; writing—original draft preparation, M.R.; writing—review and editing, S.M.A., F.J.K., J.H.N., A.L., R.J.P., S.M.D., N.L.H., R.D., T.M.K., J.R., J.A.G.-P. and A.Z.; supervision, J.A.G.-P. and A.Z. All authors have read and agreed to the published version of the manuscript.

Funding: This research received no external funding.

Institutional Review Board Statement: Not applicable.

Informed Consent Statement: Not applicable.

Data Availability Statement: The data supporting the findings of this study are available from the corresponding authors upon reasonable request.

Acknowledgments: We would like to thank those at the Rutgers Robert Wood Johnson Medical School, Rutgers New Jersey Medical School, and Rutgers Cancer Institute of New Jersey who offered technical assistance. We would also like to thank those at Rush University Medical Center who provided input and support in the writing of this manuscript.

Conflicts of Interest: The authors declare no conflict of interest.

References

1. Lee, Y.K.; Mazmanian, S.K. Has the microbiota played a critical role in the evolution of the adaptive immune system? *Science* **2010**, *330*, 1768–1773. [CrossRef] [PubMed]
2. Faust, K.; Raes, J. Microbial interactions: From networks to models. *Nat. Rev. Genet.* **2012**, *10*, 538–550. [CrossRef] [PubMed]
3. Ai, D.; Pan, H.; Li, X.; Gao, Y.; Liu, G.; Xia, L.C. Identifying gut microbiota associated with colorectal cancer using a zero-inflated lognormal model. *Front. Microbiol.* **2019**, *10*, 826. [CrossRef] [PubMed]
4. Hooper, L.V.; Littman, D.R.; MacPherson, A.J. Interactions between the microbiota and the immune system. *Science* **2012**, *336*, 1268–1273. [CrossRef]
5. Wu, L.; Estrada, O.; Zaborina, O.; Bains, M.; Shen, L.; Kohler, J.E.; Patel, N.; Musch, M.W.; Chang, E.B.; Fu, Y.-X.; et al. Recognition of host immune activation by *Pseudomonas aeruginosa*. *Science* **2005**, *309*, 774–777. [CrossRef]
6. Schwabe, R.F.; Jobin, C. The microbiome and cancer. *Nat. Rev. Cancer* **2013**, *13*, 800–812. [CrossRef]
7. Geng, J.; Song, Q.; Tang, X.; Liang, X.; Fan, H.; Peng, H.; Guo, Q.; Zhang, Z. Co-occurrence of driver and passenger bacteria in human colorectal cancer. *Gut Pathog.* **2014**, *6*, 26. [CrossRef]
8. Ruiz-Perez, D.; Guan, H.; Madhivanan, P.; Mathee, K.; Narasimhan, G. So you think you can PLS-DA? *BMC Bioinform.* **2020**, *21*, 2. [CrossRef]
9. Miller, D.D. The medical AI insurgency: What physicians must know about data to practice with intelligent machines. *NPJ Digit. Med.* **2019**, *2*, 62. [CrossRef]
10. Chou, I.-C.; Voit, E.O. Recent developments in parameter estimation and structure identification of biochemical and genomic systems. *Math. Biosci.* **2009**, *219*, 57–83. [CrossRef]
11. Saeys, Y.; Inza, I.; Larrañaga, P. A review of feature selection techniques in bioinformatics. *Bioinformatics* **2007**, *23*, 2507–2517. [CrossRef] [PubMed]
12. Half, E.; Keren, N.; Reshef, L.; Dorfman, T.; Lachter, I.; Kluger, Y.; Reshef, N.; Knobler, H.; Maor, Y.; Stein, A.; et al. Fecal microbiome signatures of pancreatic cancer patients. *Sci. Rep.* **2019**, *9*, 16081. [CrossRef] [PubMed]
13. Ringnér, M. What is principal component analysis? *Nat. Biotechnol.* **2008**, *26*, 303–304. [CrossRef] [PubMed]
14. Liu, W.-T.; Marsh, T.; Cheng, H.; Forney, L.J. Characterization of microbial diversity by determining terminal restriction fragment length polymorphisms of genes encoding 16S rRNA. *Appl. Environ. Microbiol.* **1997**, *63*, 4516–4522. [CrossRef] [PubMed]
15. Lu, M.; Fan, Z.; Xu, B.; Chen, L.; Zheng, X.; Li, J.; Znati, T.; Mi, Q.; Jiang, J. Using machine learning to predict ovarian cancer. *Int. J. Med. Inform.* **2020**, *141*, 104195. [CrossRef] [PubMed]
16. Osborn, A.M.; Moore, E.R.B.; Timmis, K.N. An evaluation of terminal-restriction fragment length polymorphism (T-RFLP) analysis for the study of microbial community structure and dynamics. *Environ. Microbiol.* **2000**, *2*, 39–50. [CrossRef]
17. Sivan, A.; Corrales, L.; Hubert, N.; Williams, J.B.; Aquino-Michaels, K.; Earley, Z.M.; Benyamin, F.W.; Lei, Y.M.; Jabri, B.; Alegre, M.-L.; et al. Commensal bifidobacterium promotes antitumor immunity and facilitates anti-PD-L1 efficacy. *Science* **2015**, *350*, 1084–1089. [CrossRef]
18. Derrien, M.; Belzer, C.; de Vos, W.M. *Akkermansia muciniphila* and its role in regulating host functions. *Microb. Pathog.* **2017**, *106*, 171–181. [CrossRef]
19. Zeng, H.; Ishaq, S.; Liu, Z.; Bukowski, M. Colonic aberrant crypt formation accompanies an increase of opportunistic pathogenic bacteria in C57BL/6 mice fed a high-fat diet. *J. Nutr. Biochem.* **2018**, *54*, 18–27. [CrossRef]
20. Miller, M.B.; Bassler, B.L. Quorum sensing in bacteria. *Annu. Rev. Microbiol.* **2001**, *55*, 165–199. [CrossRef]
21. Mulcahy, L.R.; Isabella, V.M.; Lewis, K. *Pseudomonas aeruginosa* biofilms in disease. *Microb. Ecol.* **2014**, *68*, 1–12. [CrossRef] [PubMed]
22. Li, F.; Cimdins, A.; Rohde, M.; Jänsch, L.; Kaever, V.; Nimtz, M.; Römling, U. DncV synthesizes cyclic GMP-AMP and regulates biofilm formation and motility in *Escherichia coli* ECOR31. *mBio* **2019**, *10*, e02492-13. [CrossRef] [PubMed]
23. Chen, L.; Li, X.; Zhou, X.; Zeng, J.; Ren, Z.; Lei, L.; Kang, D.; Zhang, K.; Zou, J.; Li, Y. Inhibition of *Enterococcus faecalis* growth and biofilm formation by molecule targeting cyclic di-AMP synthetase activity. *J. Endod.* **2018**, *44*, 1381–1388.e2. [CrossRef] [PubMed]

24. Zmora, N.; Zilberman-Schapira, G.; Suez, J.; Mor, U.; Dori-Bachash, M.; Bashiardes, S.; Kotler, E.; Zur, M.; Regev-Lehavi, D.; Brik, R.B.-Z.; et al. Personalized gut mucosal colonization resistance to empiric probiotics is associated with unique host and microbiome features. *Cell* **2018**, *174*, 1388–1405.e21. [CrossRef] [PubMed]
25. Thevaranjan, N.; Puchta, A.; Schulz, C.; Naidoo, A.; Szamosi, J.; Verschoor, C.P.; Loukov, D.; Schenck, L.P.; Jury, J.; Foley, K.P.; et al. Age-associated microbial dysbiosis promotes intestinal permeability, systemic inflammation, and macrophage dysfunction. *Cell Host Microbe* **2017**, *21*, 455–466.e4. [CrossRef]
26. Gibbons, S.M.; Kearney, S.M.; Smillie, C.S.; Alm, E.J. Two dynamic regimes in the human gut microbiome. *PLoS Comput. Biol.* **2017**, *13*, e1005364. [CrossRef]
27. Wong, S.H.; Zhao, L.; Zhang, X.; Nakatsu, G.; Han, J.; Xu, W.; Xiao, X.; Kwong, T.N.Y.; Tsoi, H.; Wu, W.K.K.; et al. Gavage of fecal samples from patients with colorectal cancer promotes intestinal carcinogenesis in germ-free and conventional mice. *Gastroenterology* **2017**, *153*, 1621–1633.e6. [CrossRef]
28. Elkrief, A.; Derosa, L.; Zitvogel, L.; Kroemer, G.; Routy, B. The intimate relationship between gut microbiota and cancer immunotherapy. *Gut Microbes* **2019**, *10*, 424–428. [CrossRef]
29. McDonald, D.; Hyde, E.; Debelius, J.W.; Morton, J.T.; Gonzalez, A.; Ackermann, G.; Aksenov, A.A.; Behsaz, B.; Brennan, C.; Chen, Y.; et al. American gut: An open platform for citizen science microbiome research. *mSystems* **2018**, *3*, e00031-18. [CrossRef]
30. San-Juan-Vergara, H.; Zurek, E.; Ajami, N.J.; Mogollon, C.; Peña, M.; Portnoy, I.; Velez, J.; Cadena-Cruz, C.; Diaz-Olmos, Y.; Hurtado-Gómez, L.; et al. A Lachnospiraceae-dominated bacterial signature in the fecal microbiota of HIV-infected individuals from Colombia, South America. *Sci. Rep.* **2018**, *8*, 4479. [CrossRef]
31. Kurtz, Z.D.; Mueller, C.L.; Miraldi, E.R.; Littman, D.R.; Blaser, M.J.; Bonneau, R.A. Sparse and compositionally robust inference of microbial ecological networks. *PLoS Comput. Biol.* **2015**, *11*, e1004226. [CrossRef] [PubMed]
32. Kim, S.; Thapa, I.; Zhang, L.; Ali, H. A novel graph theoretical approach for modeling microbiomes and inferring microbial ecological relationships. *BMC Genom.* **2019**, *20*, 945. [CrossRef] [PubMed]
33. Hirano, H.; Takemoto, K. Difficulty in inferring microbial community structure based on co-occurrence network approaches. *BMC Bioinform.* **2019**, *20*, 329. [CrossRef]
34. De Marco, P.J.; Nobrega, C.C. Evaluating collinearity effects on species distribution models: An approach based on virtual species simulation. *PLoS ONE* **2018**, *13*, e0202403. [CrossRef] [PubMed]
35. Pasanen, L.; Holmström, L.; Sillanpää, M.J. Bayesian LASSO, scale space and decision making in association genetics. *PLoS ONE* **2015**, *10*, e0120017. [CrossRef]
36. Schisterman, E.F.; Perkins, N.J.; Mumford, S.L.; Ahrens, K.A.; Mitchell, E.M. Collinearity and causal diagrams: A lesson on the importance of model specification. *Epidemiology* **2017**, *28*, 47–53. [CrossRef]
37. Jiang, M.; Zhu, L.; Wang, Y.; Xia, L.; Shou, G.; Liu, F.; Crozier, S. Application of kernel principal component analysis and support vector regression for reconstruction of cardiac transmembrane potentials. *Phys. Med. Biol.* **2011**, *56*, 1727–1742. [CrossRef]
38. Matson, V.; Fessler, J.; Bao, R.; Chongsuwat, T.; Zha, Y.; Alegre, M.-L.; Luke, J.J.; Gajewski, T.F. The commensal microbiome is associated with anti–PD-1 efficacy in metastatic melanoma patients. *Science* **2018**, *359*, 104–108. [CrossRef]
39. McGaughey, K.D.; Yilmaz-Swenson, T.; Elsayed, N.; Cruz, D.A.; Rodriguez, R.R.; Kritzer, M.D.; Peterchev, A.V.; Gray, M.; Lewis, S.; Roach, J.; et al. Comparative evaluation of a new magnetic bead-based DNA extraction method from fecal samples for downstream next-generation 16S rRNA gene sequencing. *PLoS ONE* **2018**, *13*, e0202858. [CrossRef]
40. Lei, Y.M.; Chen, L.; Wang, Y.; Stefka, A.; Molinero, L.L.; Theriault, B.; Aquino-Michaels, K.; Sivan, A.S.; Nagler, C.R.; Gajewski, T.F.; et al. The composition of the microbiota modulates allograft rejection. *J. Clin. Investig.* **2016**, *126*, 2736–2744. [CrossRef]

Current Oncology

Communication

Method for the Intraoperative Detection of IDH Mutation in Gliomas with Differential Mobility Spectrometry

Ilkka Haapala [1,*], Anton Kondratev [2], Antti Roine [2,3], Meri Mäkelä [2,3], Anton Kontunen [2,3], Markus Karjalainen [2,3], Aki Laakso [4], Päivi Koroknay-Pál [4], Kristiina Nordfors [5], Hannu Haapasalo [6], Niku Oksala [2,3], Antti Vehkaoja [2] and Joonas Haapasalo [1]

1. Department of Neurosurgery, Tampere University Hospital, 33520 Tampere, Finland; joonas.haapasalo@gmail.com
2. Faculty of Medicine and Health Technology, Tampere University, 33014 Tampere, Finland; anton.kondratev@tuni.fi (A.K.); antti.roine@olfactomics.fi (A.R.); meri.makela@olfactomics.fi (M.M.); anton.kontunen@tuni.fi (A.K.); markus.karjalainen@tuni.fi (M.K.); niku.oksala@olfactomics.fi (N.O.); antti.vehkaoja@tuni.fi (A.V.)
3. Olfactomics Ltd., 33720 Tampere, Finland
4. Department of Neurosurgery, Helsinki University Hospital, 00260 Helsinki, Finland; aki.laakso@hus.fi (A.L.); paivi.koroknay-pal@hus.fi (P.K.-P.)
5. Faculty of Pediatrics, Tampere University Hospital, 33520 Tampere, Finland; kristiina.nordfors@gmail.com
6. Fimlab Laboratories Ltd., 33520 Tampere, Finland; hannu.haapasalo@fimlab.fi

* Correspondence: ilkka.haapala@fimnet.fi

Abstract: Isocitrate dehydrogenase (IDH) mutation status is an important factor for surgical decision-making: patients with IDH-mutated tumors are more likely to have a good long-term prognosis, and thus favor aggressive resection with more survival benefit to gain. Patients with IDH wild-type tumors have generally poorer prognosis and, therefore, conservative resection to avoid neurological deficit is favored. Current histopathological analysis with frozen sections is unable to identify IDH mutation status intraoperatively, and more advanced methods are therefore needed. We examined a novel method suitable for intraoperative IDH mutation identification that is based on the differential mobility spectrometry (DMS) analysis of the tumor. We prospectively obtained tumor samples from 22 patients, including 11 IDH-mutated and 11 IDH wild-type tumors. The tumors were cut in 88 smaller specimens that were analyzed with DMS. With a linear discriminant analysis (LDA) algorithm, the DMS was able to classify tumor samples with 86% classification accuracy, 86% sensitivity, and 85% specificity. Our results show that DMS is able to differentiate IDH-mutated and IDH wild-type tumors with good accuracy in a setting suitable for intraoperative use, which makes it a promising novel solution for neurosurgical practice.

Keywords: differential mobility spectrometry; neuro-oncology; neurosurgery; glioma; classification; isocitrate dehydrogenase (IDH)

Citation: Haapala, I.; Kondratev, A.; Roine, A.; Mäkelä, M.; Kontunen, A.; Karjalainen, M.; Laakso, A.; Koroknay-Pál, P.; Nordfors, K.; Haapasalo, H.; et al. Method for the Intraoperative Detection of IDH Mutation in Gliomas with Differential Mobility Spectrometry. *Curr. Oncol.* **2022**, *29*, 3252–3258. https://doi.org/10.3390/curroncol29050265

Received: 21 February 2022
Accepted: 29 April 2022
Published: 4 May 2022

1. Introduction

Gliomas represent the most clinically important group of primary brain tumors. Traditionally, they have been classified into WHO groups I–IV to evaluate their malignant potential by analysis of their morphological features. However, the past decades of research have led to the discovery of many molecular alterations in gliomas that have a great impact on the tumor's malignancy and, accordingly, to the patient's prognosis [1]. Among such alterations, the mutation of isocitrate dehydrogenase (IDH) enzymes 1 or 2 is highly correlated with the patient's overall survival, and the effect is present regardless of the tumor's histopathological WHO grade [2–5]. IDH mutation also seems to play a pivotal role in the carcinogenesis of other solid tumors, such as cholangiocarcinoma, where it is also a major target for medical therapy [6–8].

Normally, IDH enzymes catalyze the oxidative decarboxylation of isocitrate to form a-ketoglutarate (aKG) in the Krebs cycle. IDH1 and IDH2 localize differently in the cell but share the same function; hence, they are hereafter referred to collectively as IDH. The mutation of IDH confers a neomorphic enzyme activity that catalyzes the reduction of aKG into the putative oncometabolite D-2-hydroxyglutarate (D2HG) [9]. The accumulation of D2HG is further associated with the hypermethylation of DNA and chromatin, which is thought to dysregulate cell epigenetics [10,11].

IDH mutation status is an important factor for surgical decision-making: patients with IDH-mutated tumors are more likely to have a good long-term prognosis, and thus favor aggressive gross total resection with more survival benefit to gain. Patients with IDH wild-type tumors have a generally poorer prognosis and, therefore, conservative resection to avoid neurological deficit is favored [12–14]. The effect of gross total resection on survival remains also in recurrent diseases [15,16]. Current histopathological analysis based on frozen sections is unable to identify molecular characteristics, including IDH mutation, within the time frame of surgery [17], thus creating an imminent need for new solutions.

We have previously shown that differential mobility spectrometry (DMS) is able to identify different brain tumors ex vivo [18]. DMS characterizes substances based on the mobility differences of ionized particles in high-frequency electrical fields, resulting in a substance-specific dispersion spectrum, or "smell fingerprint" [19]. The simplicity, quickness and cost-effectiveness of DMS makes it a compelling emerging technology for clinical applications [18]. In this study, we demonstrate the rapid, preparation-free analysis of a tumor's IDH mutation status with DMS.

2. Materials and Methods

We prospectively obtained tumor samples from 22 patients who had neurosurgical operations at Tampere University Hospital between the years 2018 and 2021, and at Helsinki University Hospital in 2020. Patient recruitment was continued until we had a sufficient number of IDH-mutated tumors, which are rarer. To make balanced classes, an equal number of IDH wild-type tumors were randomly selected for the experiment. Eventually, we had 11 IDH-mutated tumors and 11 IDH wild-type tumors. IDH-mutated tumors included 5 WHO gr. II–III astrocytomas, 3 gr. II–III oligodendrogliomas, and 3 gr. IV glioblastomas (GBM). IDH wild-type tumors included 1 gr. III astrocytoma and 10 GBMs. Diagnoses were made by an experienced neuropathologist and IDH mutation was identified with immunohistochemistry. The study was approved by the ethics review board of Pirkanmaa Hospital District, Finland. The patients gave their written consent for the study.

All samples were stored in a freezer at −70 °C. The samples were carefully cut into 88 (44 IDH-mutated and 44 IDH wild-type) smaller specimens of macroscopically equal sizes. Blood, if macroscopically visible, was carefully rinsed from the samples before the analysis. The samples were randomly placed in a plastic well plate with each well containing 0.18 mL of agar in the bottom. Each sample was incised with a custom-built, computer-controlled, 40 W, 10.6 μm CO_2 laser evaporator four times in a quadratic manner, with 1 mm gaps between the incisions. The total number of incisions was 352. The laser sampling was controlled by a graphical user interface. To provide a clean and controlled supply of carrier gas for the analyte gas, purified and humidified pressurized air was introduced to the sampling stage via a sampling nozzle. The sampling nozzle provided a protective stream of carrier gas around the sampling area and, after sample vaporization, transported the sample gas to the DMS inlet. The DMS used in the study was a commercial IonVision instrument (Olfactomics Oy, Finland). The measurement parameters for the DMS spectrum were: separation voltage (Usv), 200–1000 V with 20 increments; compensation voltage (Ucv), −2–10 V with 60 increments; separation field frequency, 1 MHz; and duty cycle of the field, 22%. With these parameters, the DMS measurement produced a total of 1200 data points and the duration of the measurement was approximately 13 s, during which 250 2 ms laser pulses were used to provide a sample stream of vaporized tissue to the DMS.

A gross appearance of the setup (A–D) and examples of the dispersion spectra (G) are presented in Figure 1.

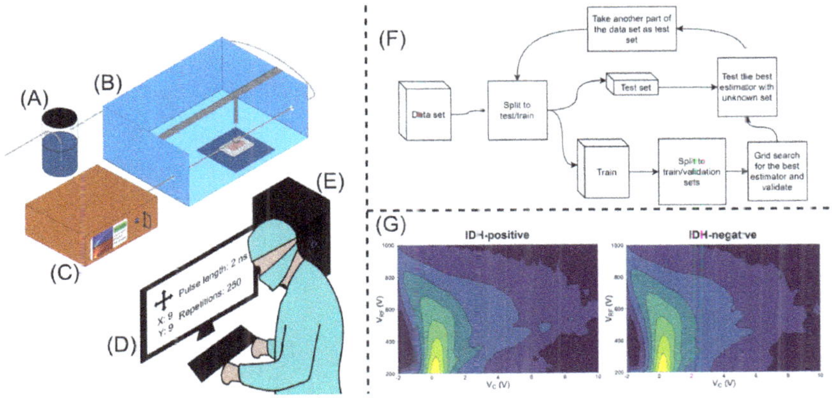

Figure 1. The setup: (**A**) humidifier; (**B**) sampling unit; (**C**) DMS analyzer (**D**); graphical user interface; (**E**) computing unit for data analytics; (**F**) workflow of the algorithm; (**G**) examples of IDH−positive and −negative dispersion spectra. Vc = compensation voltage; Vrf = peak-to-peak amplitude of the radiofrequency waveform voltage.

We evaluated the accuracy of several machine learning algorithms for the detection of differences in dispersion spectra and the classification of the analyzed samples. Linear discriminant analysis (LDA) was found to be the best performing algorithm. The main idea of training an LDA algorithm is the projection of data points to a lower dimensional space so that the between-class distance of class centers is maximized, and the within-class distance of data points is minimized, defining a decision boundary between the classes that is used to classify new samples. The other algorithms tested were K-nearest neighbors (KNN), random forest (RF), decision tree (DT), support vector machines (SVM) and XGBoost (XGB).

3. Results

The data set revealed a temperature rise, which caused baseline drift during the measurement of one well plate, making the data biased. Thus, a necessary preprocessing method was to remove the dimension-wise linear trend which belonged the well plate from each part of the data set. This preprocessing step improved the classification results compared to the classification of the raw data. The data set contained 352 samples taken from 22 patients. Group cross-validation was utilised to estimate the classification performance. Group cross-validation is implemented so that, at every iteration, it leaves one group of samples only for testing. The other groups are used for training. In this case, the nested group cross-validation technique was used. This algorithm leaves one group for testing and the other groups are used for training and validating. For the next iteration, the second group is used for testing and the others for training and validating, and so on. This approach ensures that there are no data leakages into the training phase. With the nested group cross-validation training, the LDA algorithm reached a classification accuracy of 86%, with 86% sensitivity and 85% specificity (Table 1). The workflow of the LDA algorithm is presented in Figure 1F. Further details of the cross-validation and classification results reached with other algorithms are presented in the Supplementary File.

In terms of the samples, out of the original 22 tumor samples (352 incisions), 8 samples had all their incisions correctly classified. In five samples, less than 10% of incisions were erroneous. In four samples, 10–20% were wrong. In five samples, 20–50% of the incisions were incorrectly classified. The tumors that had incorrectly clustered incisions included eight IDH wild-type tumors and six IDH-mutated tumors. The most difficult tumor type for the classifier was gr. IV GBM.

Table 1. Cross tabulation of the classification results (LDA).

IDH Mutation	−	**150**	**26**
	+	**25**	**151**
		−	+
		Classification result	
Sens. 0.85		Spec. 0.85	

4. Discussion

Our results show that the smoke generated from the IDH-mutated and IDH wild-type gliomas had distinct DMS profiles, and the DMS could differentiate them with good sensitivity and specificity. The laser evaporator platform is compact enough to be placed in the operating room and used for intermittent analysis of the tumor samples during surgery. The duration of measurement was approximately 13 s, so the DMS operates in almost real time. The DMS is also simpler and more economical than conventional mass spectrometer-based solutions. Conventional frozen section analysis is unable to identify molecular alterations in tumors, such as IDH mutation. In the latest WHO tumor classification, these alterations have become ever more prominent. This creates an increasing need for novel tumor identification methods in neurosurgical departments worldwide.

Recently, Raman spectroscopy has also been used for genotyping unprocessed glioma samples [20]. Raman spectroscopy is a modality that gives spectral tissue characteristics based on molecular signatures resulting from the inelastic scattering of incident light. Our results equal those achieved with Raman spectroscopy, and the workflow in DMS is at least as fast and straightforward.

Our tumor sample set included both IDH-mutated and IDH wild-type gr. IV GBMs and gr. III malignant astrocytomas. Out of the tumors with an unusual IDH mutation status given their histology, one GBM had 25% (9 out of 36) of the incisions erroneously classified, but all the other tumors (two IDH-mutated gr. IV GBMs and one IDH wild-type gr. III astrocytoma) had all their incisions correct classified, even though the opposite cluster had multiple histologically similar tumors. This indirectly indicates that the divisive features in the classification process were actually due to the cellular metabolic changes driven by an IDH mutation. The phospholipid content of tissue has previously been identified as a key distinguishing factor in DMS analysis [18]. The metabolic changes associated with an IDH mutation include aberrations in phospholipid composition [10], which constitutes a plausible theoretical basis for the detection of IDH mutation by DMS.

A potential source of error in DMS analysis is intratumoral heterogeneity. This is especially true in GBMs, which vary in terms of cellular density, nuclear pleomorphism, necrosis, histologic architecture, vasculature, mitoses, and multifaceted microenvironments [21,22]. This can cause variance in tissue impedance and disturb the classifier [23]. An additional confounding factor in our study was 5-ALA, which was used only in the resections of tumors that radiologically appeared as malignant. However, all three IDH-mutated GBMs were resected with 5-ALA guidance, and still the classifier was able to classify them correctly.

Our study was limited by a relatively small number of samples that we multiplied into smaller specimens. In order to achieve a setup resembling actual intraoperative use, we only minimally prepared the tumor samples for the analysis. This inevitably caused spatial variance in the specimens that affected the DMS signal strength, thus creating an additional confounding factor to the classifier. This issue could be addressed in future studies by processing the samples into a more homogeneous cell suspension by a centrifuge before the analysis. The suspension could then be pipetted into the well plate to obtain precisely equal sample sizes. We also used frozen samples instead of fresh tumors. In our earlier unpublished experiments, freezing of the samples was not found to affect the classification results. However, this should be verified in peer-reviewed studies in the future.

5. Conclusions

Our results show that the DMS is able to differentiate IDH-mutated and IDH wild-type tumors with good accuracy in a setting suitable for intraoperative use. The role of molecular alterations in classifying brain tumors and evaluating their prognosis is increasing. Additionally, the degree of survival benefit achieved with a gross-total resection varies even in histologically similar tumors based on their IDH mutation status, which is impossible to identify with conventional frozen section analysis. This makes the DMS a promising novel tool for neurosurgical practice.

Supplementary Materials: The following supporting information can be downloaded at: https://www.mdpi.com/article/10.3390/curroncol29050265/s1. The work includes a supplementary file; detailed description of data analysis and classification results achieved with other algorithms. Figure S1: Nested cross-validation.

Author Contributions: Conceptualization, I.H., J.H., A.R., A.V. and N.O.; data curation, A.K. (Anton Kondratev), A.K. (Anton Kontunen), M.K. and M.M.; formal analysis, A.K. (Anton Kondratev), A.K. (Anton Kontunen), M.K. and M.M.; funding acquisition, J.H., K.N., A.R. and N.O.; investigation, I.H., A.K. (Anton Kontunen) and M.M.; methodology I.H., J.H., A.K. (Anton Kontunen), M.K., A.R., N.O. and A.V.; project administration, J.H., K.N., A.R., A.V. and N.O.; resources, H.H., A.R., A.V., N.O., A.L. and P.K.-P.; software, A.K. (Anton Kontunen), M.K., A.R. and N.O.; supervision, J.H., K.N., A.R., A.V., N.O., A.L. and P.K.-P.; validation, A.V., A.R. and N.O.; visualization, I.H., A.K. (Anton Kondratev), A.K. (Anton Kontunen) and M.M.; writing—original draft, I.H. and A.K.; writing—review and editing, all authors. All authors have read and agreed to the published version of the manuscript.

Funding: Ilkka Haapala declares funding from The Finnish Medical Foundation, Eka -grant, decision number 3535. Anton Kondratev declares funding from Academy of Finland, Programmable Scent Environments -project, decision number 323530. Anton Kontunen declares funding from the Doctoral School of Tampere University and Emil Aaltonen Foundation (Grant number 210073). Markus Karjalainen declares funding from The Finnish Cultural Foundation, Pirkanmaa Regional Fund. Kristiina Nordfors declares funding from Finnish Pediatric Research Foundation, Pediatric Cancer Research Foundation Väre and The Finnish Ministry of Social Affairs and Health. Hannu Haapasalo declares funding from Competitive State Research Financing of the Expert Responsibility area of Tampere University Hospital. Niku Oksala declares funding from Competitive State Research Financing of the Expert Responsibility area of Tampere University Hospital (Grant numbers 9s045, 9T044, 9U042, 150618, 9V044, 9X040, 9AA057, 9AB052, and MK301); from Competitive funding to strengthen university research profiles funded by Academy of Finland, decision number 292477; and from Tampere Tuberculosis Foundation. Joonas Haapasalo declares funding from Emil Aaltonen Foundation, Finnish Pediatric Research Foundation, Competitive State Research Financing of the Expert Responsibility area of Tampere University Hospital, Pediatric Cancer Research Foundation Väre and The Finnish Ministry of Social Affairs and Health. The study sponsors did not have any involvement in the study design; collection, analysis, and interpretation of data; the writing of the manuscript; or the decision to submit the manuscript for publication.

Institutional Review Board Statement: The study was approved by the Ethics Committee of Tampere University Hospital (ETL R10066) but was not listed in ClinicalTrials Database, because it was a non-randomized, non-interventional study.

Informed Consent Statement: Written informed consent has been obtained from the patients.

Data Availability Statement: The data presented in this study are available on request from the corresponding author.

Conflicts of Interest: A.R., N.O., A.Kontunen and M.K. are shareholders and employees of Olfactomics Ltd. M.M. is an employee of Olfactomics. Other authors do not declare any conflict of interest with respect to this work.

References

1. Eckel-Passow, J.E.; Lachance, D.H.; Molinaro, A.M.; Walsh, K.M.; Decker, P.A.; Sicotte, H.; Pekmezci, M.; Rice, T.W.; Kosel, M.L.; Smirnov, I.V.; et al. Glioma Groups Based on 1p/19q, IDH, and TERT Promoter Mutations in Tumors. *N. Engl. J. Med.* **2015**, *372*, 2499–2508. [CrossRef] [PubMed]
2. Houillier, C.; Wang, X.; Kaloshi, G.; Mokhtari, K.; Guillevin, R.; Laffaire, J.; Paris, S.; Boisselier, B.; Idbaih, A.; Laigle-Donadey, F.; et al. IDH1 or IDH2 mutations predict longer survival and response to temozolomide in low-grade gliomas. *Neurology* **2010**, *75*, 1560–1566. Available online: https://n.neurology.org/content/75/17/1560.short (accessed on 7 April 2021). [CrossRef] [PubMed]
3. Metellus, P.; Coulibaly, B.; Colin, C.; de Paula, A.M.; Vasiljevic, A.; Taieb, D.; Barlier, A.; Boisselier, B.; Mokhtari, K.; Wang, X.W.; et al. Absence of IDH mutation identifies a novel radiologic and molecular subtype of WHO grade II gliomas with dismal prognosis. *Acta Neuropathol.* **2010**, *120*, 719–729. [CrossRef] [PubMed]
4. Songtao, Q.; Lei, Y.; Si, G.; Yanqing, D.; Huixia, H.; Xuelin, Z.; Lanxiao, W.; Fei, Y. IDH mutations predict longer survival and response to temozolomide in secondary glioblastoma. *Wiley Online Libr.* **2012**, *103*, 269–273. [CrossRef] [PubMed]
5. Juratli, T.A.; Kirsch, M.; Geiger, K.; Klink, B.; Leipnitz, E.; Pinzer, T.; Soucek, S.; Schrok, E.; Schackert, G.; Krex, D. The prognostic value of IDH mutations and MGMT promoter status in secondary high-grade gliomas. *J. Neurooncol.* **2012**, *110*, 325–333. [CrossRef] [PubMed]
6. Rizzo, A.; Ricci, A.D.; Tober, N.; Nigro, M.C.; Mosca, M.; Palloni, A.; Abbati, F.; Frega, G.; De Lorenzo, S.; Tavolari, S.; et al. Second-line Treatment in Advanced Biliary Tract Cancer: Today and Tomorrow. *Anticancer Res.* **2020**, *40*, 3013–3030. [CrossRef] [PubMed]
7. Rizzo, A.; Ricci, A.D.; Brandi, G. IDH inhibitors in advanced cholangiocarcinoma: Another arrow in the quiver? *Cancer Treat. Res. Commun.* **2021**, *27*, 100356. [CrossRef]
8. Rizzo, A.; Brandi, G. First-line Chemotherapy in Advanced Biliary Tract Cancer Ten Years after the ABC-02 Trial: "And Yet It Moves!". *Cancer Treat. Res. Commun.* **2021**, *27*, 100335. [CrossRef]
9. Waitkus, M.S.; Diplas, B.H.; Yan, H. Isocitrate dehydrogenase mutations in gliomas. *Neuro-Oncology* **2015**, *18*, 16–26. Available online: https://academic.oup.com/neuro-oncology/article-abstract/18/1/16/2509155 (accessed on 7 April 2021). [CrossRef]
10. Fack, F.; Tardito, S.; Hochart, G.; Oudin, A.; Zheng, L.; Fritah, S.; Golebiewska, A.; Nazarov, P.; Bernard, A.; Hau, A.; et al. Altered metabolic landscape in IDH-mutant gliomas affects phospholipid, energy, and oxidative stress pathways. *EMBO Mol. Med.* **2017**, *9*, 1681–1695. [CrossRef]
11. Garrett, M.; Sperry, J.; Braas, D.; Yan, W.; Le, T.M.; Mottahedeh, J.; Ludwig, K.; Eskin, A.; Qin, Y.; Levy, R.; et al. Metabolic characterization of isocitrate dehydrogenase (IDH) mutant and IDH wildtype gliomaspheres uncovers cell type-specific vulnerabilities. *Cancer Metab.* **2018**, *6*, 4. [CrossRef] [PubMed]
12. Beiko, J.; Suki, D.; Hess, K.R.; Fox, B.D.; Cheung, V.; Cabral, M.; Shonka, N.; Gilbert, M.R.; Sawaya, R.; Prabhu, S.S.; et al. IDH1 mutant malignant astrocytomas are more amenable to surgical resection and have a survival benefit associated with maximal surgical resection. *Neuro-Oncology* **2014**, *16*, 81–91. [CrossRef] [PubMed]
13. Kawaguchi, T.; Sonoda, Y.; Shibahara, I.; Saito, R.; Kanamori, M.; Kumabe, T.; Tominaga, T. Impact of gross total resection in patients with WHO grade III glioma harboring the IDH 1/2 mutation without the 1p/19q co-deletion. *J. Neurooncol.* **2016**, *129*, 505–514. [CrossRef] [PubMed]
14. Patel, T.; Bander, E.D.; Venn, R.A.; Powell, T.; Cederquist, G.Y.-M.; Schaefer, P.M.; Puchi, L.A.; Akhmerov, A.; Ogilvie, S.; Reiner, A.S.; et al. The role of extent of resection in IDH1 wild-type or mutant low-grade gliomas. *Neurosurgery* **2018**, *82*, 808–814. [CrossRef]
15. Montemurro, N.; Fanelli, G.N.; Scatena, C.; Ortenzi, V.; Pasqualetti, F.; Mazzanti, C.M.; Morganti, R.; Paiar, F.; Naccarato, A.G.; Perrini, P. Surgical outcome and molecular pattern characterization of recurrent glioblastoma multiforme: A single-center retrospective series. *Clin. Neurol Neurosurg.* **2021**, *207*, 106735. [CrossRef]
16. Van Tellingen, O.; Yetkin-Arik, B.; De Gooijer, M.C.; Wesseling, P.; Wurdinger, T.; De Vries, H.E. Overcoming the blood–brain tumor barrier for effective glioblastoma treatment. *Drug Resist. Updat.* **2015**, *19*, 1–12. [CrossRef]
17. Longuespée, R.; Wefers, A.K.; De Vita, E.; Miller, A.K.; Reuss, D.E.; Wick, W.; Herold-Mende, C.; Kriegsmann, M.; Schirmacher, P.; Von Deimling, A.; et al. Rapid detection of 2-hydroxyglutarate in frozen sections of IDH mutant tumors by MALDI-TOF mass spectrometry. *Acta Neuropathol. Commun.* **2018**, *6*, 21. [CrossRef]
18. Haapala, I.; Karjalainen, M.; Kontunen, A.S.; Vehkaoja, A.; Nordfors, K.; Haapasalo, H.; Haapasalo, J.; Oksala, N.; Roine, A. Identifying brain tumors by differential mobility spectrometry analysis of diathermy smoke. *J. Neurosurg.* **2020**, *133*, 100–106. [CrossRef]
19. Schneider, B.B.; Nazarov, E.G.; Londry, F.; Vouros, P.; Covey, T.R. Differential mobility spectrometry/mass spectrometry history, theory, design optimization, simulations, and applications: Differential Mobility Spectrometry/Mass Spectrometry Non-radioactive Ion Source for Operation in Ambient Pressure View project Properties of Gas Phase Molecular Clusters View project Differential Mobility Spectrometry/Mass Spectrometry History, Theory, Design Optimization, Simulations, and Applications. *Wiley Online Libr.* **2016**, *35*, 687–737. [CrossRef]
20. Sciortino, T.; Secoli, R.; D'Amico, E.; Moccia, S.; Nibali, M.C.; Gay, L.; Rossi, M.; Pecco, N.; Castellano, A.; De Momi, E.; et al. Raman Spectroscopy and Machine Learning for IDH Genotyping of Unprocessed Glioma Biopsies. *Cancers* **2021**, *13*, 4196. [CrossRef]

21. Decordova, S.; Shastri, A.; Tsolaki, A.G.; Yasmin, H.; Klein, L.; Singh, S.K.; Kishore, U. Molecular Heterogeneity and Immunosuppressive Microenvironment in Glioblastoma. *Front. Immunol.* **2020**, *11*, 1402. [CrossRef] [PubMed]
22. Campos, B.; Olsen, L.R.; Urup, T.; Poulsen, H.S. A comprehensive profile of recurrent glioblastoma. *Oncogene* **2016**, *35*, 5819–5825. [CrossRef] [PubMed]
23. Zhang, Y.; Xu, S.; Min, W.; Shen, L.; Zhang, Y.; Yue, Z. SURG-25. A novel bio-impedance spectroscopy system real-time intraoperatively discriminates glioblastoma from brain tissue in mice. *Neuro-Oncology* **2017**, *19* (Suppl. 6), vi240. [CrossRef]

cancers

MDPI

Article

Artificial Intelligence Predicted Overall Survival and Classified Mature B-Cell Neoplasms Based on Immuno-Oncology and Immune Checkpoint Panels

Joaquim Carreras [1,*], Giovanna Roncador [2] and Rifat Hamoudi [3,4]

1 Department of Pathology, School of Medicine, Tokai University, 143 Shimokasuya, Isehara 259-1193, Kanagawa, Japan
2 Monoclonal Antibodies Unit, Spanish National Cancer Research Center (Centro Nacional de Investigaciones Oncologicas, CNIO), Melchor Fernandez Almagro 3, 28029 Madrid, Spain
3 Department of Clinical Sciences, College of Medicine, University of Sharjah, Sharjah P.O. Box 27272, United Arab Emirates
4 Division of Surgery and Interventional Science, University College London, Gower Street, London WC1E 6BT, UK
* Correspondence: joaquim.carreras@tokai-u.jp; Tel.: +81-463-93-1121 (ext. 3170); Fax: +81-463-91-1370

Simple Summary Artificial intelligence (AI) is a field that combines computer science with robust datasets to solve problems. AI in medicine uses machine learning and deep learning to analyze medical data and gain insight into the pathogenesis of diseases. This study summarizes and integrates our previous research and advances the analyses of macrophages. We used artificial neural networks and several types of machine learning to analyze the gene expression and protein levels by immunohistochemistry of several hematological neoplasia and pan-cancer series. As a result, the patients' survival and disease subtype classification were achieved with high accuracy. Additionally, a review of the literature on the latest progress made by AI in the hematopathology field and future perspectives are given.

Abstract: Artificial intelligence (AI) can identify actionable oncology biomarkers. This research integrates our previous analyses of non-Hodgkin lymphoma. We used gene expression and immunohistochemical data, focusing on the immune checkpoint, and added a new analysis of macrophages, including 3D rendering. The AI comprised machine learning (C5, Bayesian network, C&R, CHAID, discriminant analysis, KNN, logistic regression, LSVM, Quest, random forest, random trees, SVM, tree-AS, and XGBoost linear and tree) and artificial neural networks (multilayer perceptron and radial basis function). The series included chronic lymphocytic leukemia, mantle cell lymphoma, follicular lymphoma, Burkitt, diffuse large B-cell lymphoma, marginal zone lymphoma, and multiple myeloma, as well as acute myeloid leukemia and pan-cancer series. AI classified lymphoma subtypes and predicted overall survival accurately. Oncogenes and tumor suppressor genes were highlighted (MYC, BCL2, and TP53), along with immune microenvironment markers of tumor-associated macrophages (M2-like TAMs), T-cells and regulatory T lymphocytes (Tregs) (CD68, CD163, MARCO, CSF1R, CSF1, PD-L1/CD274, SIRPA, CD85A/LILRB3, CD47, IL10, TNFRSF14/HVEM, TNFAIP8, IKAROS, STAT3, NFKB, MAPK, PD-1/PDCD1, BTLA, and FOXP3), apoptosis (BCL2, CASP3, CASP8, PARP, and pathway-related MDM2, E2F1, CDK6, MYB, and LMO2), and metabolism (ENO3, GGA3). In conclusion, AI with immuno-oncology markers is a powerful predictive tool. Additionally, a review of recent literature was made.

Keywords: non-Hodgkin lymphoma; mature B-cell neoplasms; immune checkpoint; immuno-oncology; immune microenvironment; 3D macrophages; artificial intelligence; machine learning; artificial neural networks; deep learning

Citation: Carreras, J.; Roncador, G.; Hamoudi, R. Artificial Intelligence Predicted Overall Survival and Classified Mature B-Cell Neoplasms Based on Immuno-Oncology and Immune Checkpoint Panels. *Cancers* **2022**, *14*, 5318. https://doi.org/10.3390/cancers14215318

Academic Editors: Kentaro Inamura and Sam Payabvash

Received: 23 September 2022
Accepted: 24 October 2022
Published: 28 October 2022

1. Introduction

Lymphoid neoplasms are tumors of the hematopoietic system derived from immature and mature B lymphocytes, T lymphocytes, and natural killer (NK) cells that evoke the normal stages of cell differentiation. Nevertheless, some neoplasms (such as hairy cell leukemia) show lineage heterogeneity and plasticity, and their normal counterparts cannot be found [1–7]. The 2016 revision of the World Health Organization (WHO) classification of lymphoid neoplasms [3] and the International Consensus Classification (ICC) [6] describe around 45 different subtypes of mature lymphoid neoplasms [3,6,7]. In this research, we analyzed the gene expression of some of the most relevant and frequent ones.

Chronic lymphocytic leukemia/small lymphocytic lymphoma (CLL/SLL) develops from small mature CD5+ and CD23+ B-cells with mutated or unmutated *IGHV* genes [3,8].

Follicular lymphoma (FL) is a neoplasia of the germinal centers of follicles (centrocytes and centroblasts), with a follicular (nodular) pattern, and is frequently associated with the *IGH/BCL2* translocation (t14;18)(q32;q21) that occurs in the bone marrow [3,9,10].

Extranodal marginal zone lymphoma of mucosa-associated lymphoid tissue is an extranodal lymphoma (MALT lymphoma) composed of a heterogeneous population of small B-cells [3]. It originates in the marginal zones, but it extends into the interfollicular and follicular regions and infiltrates the epithelium, forming the lymphoepithelial lesions [3,11].

Mantle cell lymphoma (MCL) is characterized by monomorphic small to medium-sized lymphoid cells with irregular nuclei and the *CCND1* translocation, originating from peripheral B lymphocytes of the inner mantle zone, CD5+, and SOX11+ in the classical form [3,12,13].

Diffuse large B-cell lymphoma (DLBCL) is a neoplasm of medium or large B lymphoid cells that originate from the germinal center in the germinal center B-cell-like type, or from the post-germinal center in the activated B-cell-like type [3,14,15]. According to the clinical, morphological, and biological features, DLBCL can be subdivided into different subtypes; the remaining ones are not otherwise specified (NOS).

Burkitt lymphoma is a highly aggressive but curable lymphoma that often appears at extranodal sites or as acute leukemia. It is characterized by a monomorphic proliferation of medium-size B-cells, mitotic figures, and the *MYC* translocation to the immunoglobulin (IG) locus. It originates from the germinal centers. There are three epidemiological variants, with variable association with the Epstein-Barr virus (EBV): endemic, sporadic, and immunodeficiency-associated [3,16–18].

Figure 1 shows the stages of the B-lymphocyte differentiation, and the relationship with the different lymphoma subtypes [19].

Nowadays, there has been rapid advance in the field of artificial intelligence (AI), and its role in medicine is gaining relevance. AI integrates computer science and datasets to make predictions or classifications based on input data.

There are two types of artificial intelligence, weak and strong AI. Weak AI, also known as narrow AI (NAI), is trained to perform specific tasks. Conversely, strong AI includes artificial general intelligence (AGI) or artificial super intelligence (ASI), and it is expected to surpass human abilities in the future [20–26].

In this research, we used weak artificial intelligence to predict the prognosis of the patients and to classify several subtypes of mature B-cell neoplasms (output). Gene expression (transcriptomics) and protein immunohistochemical data were used as predictors (input data). The research focused on artificial neural networks (mainly multilayer perceptron), but also used other neural networks such as the radial basis function and other machine learning techniques. Regarding the neural networks, "basic" but robust and reliable architectures were chosen as an elemental part of the analysis. Then, the "basic" networks were combined in more complex, multivariate analysis algorithms. Figure 2 describes the basic structure of the neural network.

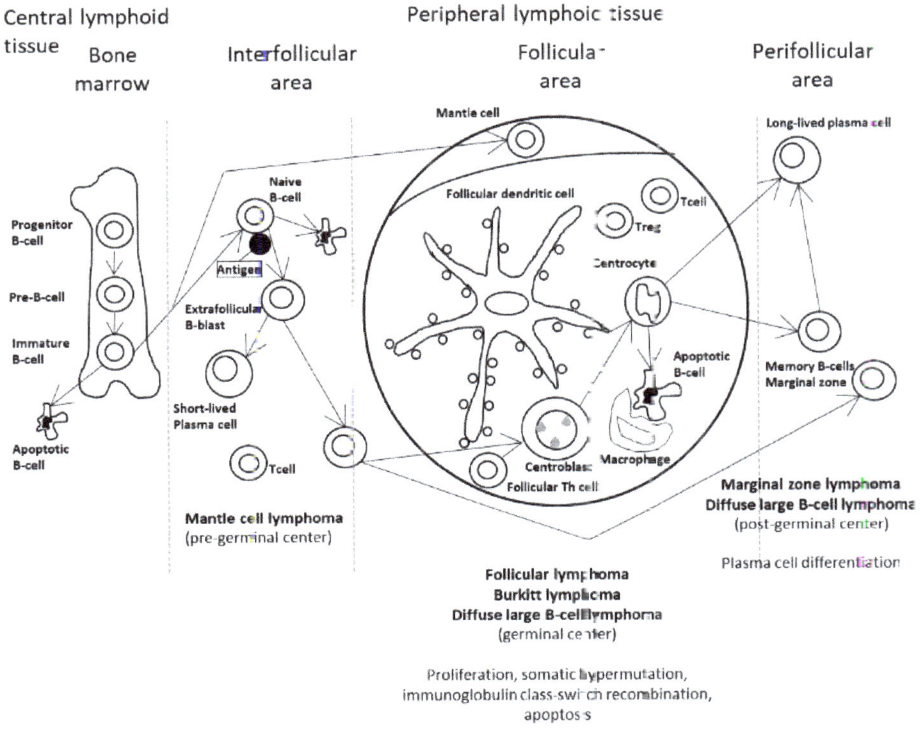

Figure 1. Postulated cell of origin of the non-Hodgkin lymphoma subtypes. In the current theory of the pathogenesis of hematopoietic and lymphoid tissues, B-cell neoplasms correspond to various stages of B-cell differentiation. For example, follicular lymphoma, Burkitt lymphoma, and diffuse large B-cell lymphoma develop (or have a stage of differentiation) from mature B lymphocytes from the germinal centers of follicles of peripheral lymphoid tissues. Of note, follicular lymphoma is characterized by the IGH/BCL2 translocation (t14;18)(q32;q21) that occurs in the bone marrow. Nevertheless, this genetic alteration is not sufficient to generate lymphoma, and additional cumulative changes are necessary.

The immune checkpoints are regulators of the immune system that belong to the self-tolerance pathways. Without them, the immune system would attach to cells indiscriminately. Cancer uses several mechanisms to proliferate, including evading the host immune response using immune checkpoint molecules. There are two types of immune checkpoint molecules: stimulatory and inhibitory. Inhibitory checkpoint molecules inhibit the immune response and include several markers such as B7-H3 (CD276), BTLA, CTLA-4, LAG3, PD-1, TIM-3, and VISTA. Nowadays, immune checkpoints are important because they are the basis of cancer immunotherapy. Currently approved checkpoint inhibitors are anti CTLA-4, PD-1, and PD-L1 [19,27–35]. In this research, artificial intelligence was used to classify and to predict the overall survival of different lymphoma subtypes using gene expression data, all the genes of the arrays, and specific panels of the immune checkpoint.

This manuscript integrates our previous publications to provide a general view of the results and adds new analysis on tumor-associated macrophages (TAMs).

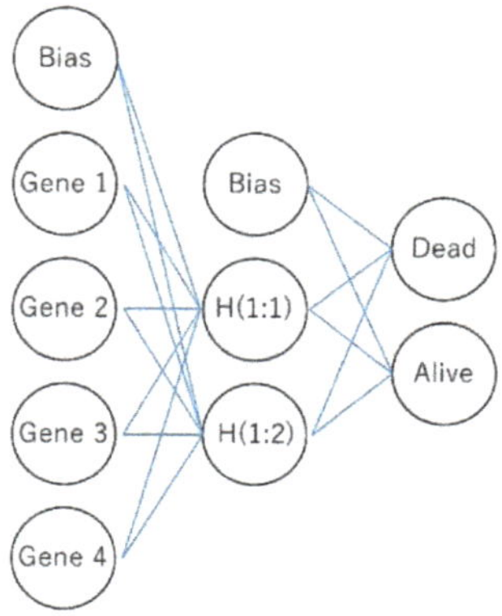

Figure 2. The basic structure of a neural network. The network is a function of predictors (also called inputs or independent variables) that minimize the prediction error of target variables (outputs). In the case of a multilayer perceptron, it is a feed-forward architecture because the connections flow from the input to the output layer without loops. Here, four genes predict the overall survival of patients. The input layer contains these genes. The hidden layer contains the unobservable nodes (units). The output layer contains the responses; the overall survival is a categorical variable (dead vs alive).

2. Materials and Methods

2.1. Machine Learning and Neural Networks

This research integrates all the previous analyses that were obtained using conventional biostatistics, machine learning, and artificial neural networks. Machine learning included Bayesian network, C&R tree, C5 tree, CHAID tree, discriminant analysis, KNN algorithm, logistic regression, LSVM, Quest tree, random forest, random trees, SVM, tree-AS, XGBoost linear, and XGBoost tree. Two types of artificial neural networks were used: the multilayer perceptron and radial basis function. The digital image quantification of markers was performed using the Waikato Environment for Knowledge Analysis (Weka), and the training of the classifier included fast random forest. All the materials and methods were thoroughly described in the previous publications [19,27–35].

2.2. Multilayer Perceptron Artificial Neural Network

The multilayer perceptron architecture was chosen in most cases. Several parameters were chosen to optimize the neural network. The predictors were included in the input layer, the unobservable nodes or units in the hidden layer, and the responses in the output layer. Scale-dependent variables and covariates were rescaled to improve network training. The method for rescaling of covariates was standardized: subtract the mean and divide by the standard deviation, $(x-\text{mean})/s$.

The series of cases were randomly partitioned into training (70%) and testing (30%) datasets. The best performance was found using one hidden layer. The activation function linked the weighted sums of units in a layer to the values of units in the succeeding layer. The hyperbolic tangent was usually used. This function has the form $\gamma(c) = \tanh(c) = (e^c - e^{-c})/(e^c + e^{-c})$. It takes real-valued arguments and transforms them into the range (−1, 1). When automatic architecture selection is used, this is the activation function for all units in the hidden layers. The number of units in each hidden layer was determined automatically by an estimation algorithm.

The output layer contained the target (dependent) variables and the activation function was softmax. This function has the form: $\gamma(c_k) = \exp(c_k)/\Sigma_j \exp(c_j)$. It takes a vector of real-valued arguments and transforms it into a vector whose elements fall in the range (0,1) and sum to 1. Softmax is available only if all dependent variables are categorical.

The training type determines how the network processes the records; the training type was batch. The training options were initial lambda (0.0000005), initial sigma (0.00005), interval center (0), and interval offset (+/−0.5). The network performance was assessed by the classification results, receiver operating characteristic (ROC) curve, cumulative gains chart, lift chart, predicted by observed chart, and residual by predicted chart. Using a sensitivity analysis, the independent variables were ranked according to their importance for predicting the dependent variable and in determining the neural network (Figure 3).

Sensitivity Analysis

For each predictor p and each input pattern m, compute:

$$d_{pm} = \max_{x_{p_1}, x_{p_2} \in S_p} \| \hat{Y}_{p_1}^{(m)} - \hat{Y}_{p_2}^{(m)} \|$$

where $\hat{Y}_p^{(m)}$ is the predicted output vector (standardized if standardization of output variable is used in training) using $\left(x_1^{(m)}, \ldots, x_{p-1}^{(m)}, x_p, x_{p+1}^{(m)}, \ldots, x_P^{(m)}\right)$ as its input, and $S_p = \left\{x_p^{\min}, x_p^{(2)}, x_p^{(3)}, x_p^{(4)}, x_p^{\max}\right\}$ for scale predictors and $\{(1, 0, \ldots, 0), (0, 1, 0, \ldots, 0), \ldots, (0, 0, \ldots, 1)\}$ for categorical predictors.

Then compute:

$$d_p = \frac{1}{M} \sum_{m=1}^{M} d_{pm}$$

and normalize the d_ps to sum to 1, and report these normalized values as the sensitivity values for the predictors. This is the average maximum amount we can expect the output to change based on changes in the pth predictor. The greater the sensitivity, the more we expect the output to change when the predictor changes.

Figure 3. Sensitivity analysis. Using a sensitivity analysis, the independent variables were ranked according to their importance for predicting the dependent variable and in determining the neural network.

The general architecture for a multilayer perceptron is as follows [34]:

Input layer: $J_0 = P$ units, $a_{0:1}, \ldots, a_{0:J0}$ with $a_{0:j} = x_j$.

Hidden layer: J_i units, $a_{i:1}, \ldots, a_{i:Ji}$; with $a_{i:k} = \gamma_i(c_{i:k})$ and $c_{i:k} = \sum_{j=0}^{Ji-1} w_{i:j,k} a_{i\_1:j}$ where $a_{i-1:0} = 1$

Output layer: $J_I = R$ units, $a_{I:1}, \ldots, a_{I:Ji}$; with $a_{I:k} = \gamma_I(c_{I:k})$ and $c_{I:k} = \sum_{j=0}^{J1} w_{I:j,k} a_{i\_1:j}$ where $a_{i-1:0} = 1$

Notation [34]:

I	Number of layers, discounting the input layer.
J_i	Number of units in layer i. $J_0 = P, J_i = R$, discounting the bias unit.
$w_{i:j,k}$	Weight leading from layer i–1, unit j to layer i, unit k. No weights connect $a_{i-1:j}^m$ and the bias $a_{i-j:0}^m$; that is, there is no $w_{i:j,0}$ for any j.
$\gamma_i(c)$	Activation function for layer i.
w	Weight vector containing all weights $(w_{1:0,1}, w_{1:0,2}, \ldots, w_{I:JI-1,JI})$.

2.3. Differential Gene Expression Using the GEO2R Software

The GEO2R 1.0 software was used to compare the differential gene expression between subtypes simply. The Benjamini–Hochberg false discovery rate was applied to adjust the p values. Log transformation was applied if necessary. Limma precision weights and force normalization were not applied. The data were visualized using volcano and mean difference (MA) plots, contrasted with a level of cut-off significance set a priori at 0.05. This software runs in R 3.2.3, Biobase 2.30.0, GEOquery 2.40.0, limma 3.26.8. Webpage: https://www.ncbi.nlm.nih.gov/geo/info/geo2r.html (accessed on 23 July 2022).

2.4. Gene Set Enrichment Analysis

The Gene Set Enrichment Analysis (GSEA) was used to determine if a pathway of interest was associated with a particular biological state (for example, dead vs alive) [36,37]. The pathways were obtained from the Molecular Signatures Database (MSigDB 7.0 and greater) or designed in-house. The software GSEA v4.2.3 was downloaded from the webpage of UC San Diego, Broad Institute: http://www.gsea-msigdb.org/gsea/index.jsp (accessed on 23 July 2022).

2.5. Conventional Statistical Analyses

Comparisons between groups were performed using crosstabulation with Pearson Chi-Square and Fisher's exact tests, and with nonparametric Mann–Whitney U (2 groups) and Kruskal-Wallis H ($\geq$3 groups) tests. Survival analyses used the Kaplan–Meier and Log-rank tests, and the univariate and multivariate Cox Regression. The criteria of survival and response were the standard [38]. Overall survival was calculated from the time of diagnosis to the last contact with the patient (event recorded as alive vs dead).

2.6. Risk Groups

Risk groups were created using the risk score (prognostic index), which was calculated by multiplying the beta coefficients of the Cox model by the gene expression values (Risk score = $B_1X_1 + B_2X_2 + \ldots + B_pX_p$, where x_i is the expression value and B_I is the beta value of the Cox table). In the Cox, all the genes are included in a unique model [39].

2.7. Hardware

The analyses were performed on a desktop equipped with an AMD Ryzen 5 1600 and NVIDIA GeForce GTX 1050 Ti [27], Ryzen 7 3700X and GeForce GTX 1650 [30,33,34], and a Ryzen 9 5900X and GeForce GTX 3060 Ti [35], all with 16.0 GB of RAM.

Appendix A describes all the software that was used to perform the biostatistical analyses, including machine learning and artificial neural networks [19,27–35].

2.8. Datasets and Immunohistochemical Procedures

We used publicly available datasets downloaded from the Gene Expression Omnibus (GEO) repository, webpage: https://www.ncbi.nlm.nih.gov/geo/ (accessed on 23 July 2022) (Appendix B Table A1) [40–57], and own Tokai University Hospital gene expression (transcriptomic) and immunohistochemical (proteomic) datasets for this research.

Several of the markers that were highlighted in the AI analyses (both machine learning and artificial neural network) were validated by immunohistochemistry at the protein level. The cases were selected from the lymphoma series of Tokai University Hospital. The series of cases ranged from 100 to 293 cases, depending on the project. Immunohistochemistry was performed using a Leica Bond Max autostainer following the manufacturer's instructions (Leica K.K., Tokyo, Japan). Table 1 details the primary antibodies that were used. The review section was made on the basis of PRISMA guidelines: https://prisma-statement.org/ (accessed on 29 September 2022), Carreras, J. (20 October 2022). Systematic review. https://doi.org/10.17605/OSF.IO/436JQ. The manuscripts were selected in PubMed using the keywords "lymphoma" and "artificial intelligence", and were organized according to the type of input data as PET/CT scan, histological images, immunophenotype, clinicopathological variables, and gene expression, mutational, and integrative analysis-based artificial intelligence.

Table 1. Immunohistochemical markers used in lymphoma cases of Tokai University, School of Medicine.

Marker	Target/Pathway	Primary Antibody	Company
BCL2	Apoptosis	bcl2/100/D5	Novocastra
BCL6	Germinal center	LN22	Novocastra
cCASP3	Apoptosis	Asp175, #9661	Cell Signaling
CASP8	Apoptosis	active subunit p18, 11B6	Novocastra
CD3	T lymphocytes	CD3 epsilon, LN10	Novocastra
CD5	T lymphocytes	4C7	Novocastra
CD10	Germinal center	56C6	Novocastra
CD16	M1-like macrophages	2H7	Novocastra
CD20	B lymphocytes	L26	Novocastra
CD47	B lymphocytes	D3O7P	Cell Signaling
CD68	Pan-macrophages	514H12	Novocastra
CD85A/LILRB3	M2-like macrophages	FRAS92B	CNIO
CD163	M2-like macrophages	10D6	Novocastra
CDK6	Cell cycle	98D	CNIO
CSF1	CSF1R pathway	2D10	LSBio
CSF1R	M2-like macrophages	2D10	LSBio
Cyclin D1	Cell cycle	P2D11F11	Novocastra
E2F1	Cell cycle	Agro368V	CNIO
EBER	Epstein-Barr virus	#PB0589, #AR0833	Novocastra
IKAROS	Cytokine signaling	D6N9Y	Cell Signaling
IL10	M2c-like macrophages	LS-B7432	Lifespan Bioscience
Ki67	Cell cycle	MM1	Novocastra
LMO2	Proto-oncogene	299B	CNIO
MARCO	Macrophages	HPA063793	Atlas antibodies
MDM2	p53 signaling	1F2	Invitrogen
MITF	M2-like macrophages	C5/D5, MAB10775	Abnova
MUM1	Plasma cells	IRF4, EAU32	Novocastra
MYC	Proto-oncogene	Y69	Abcam
NFKB p105/p50	NFKB pathway	#3035	Cell Signaling
cPARP	Apoptosis	Asp214, D64E10	Cell Signaling
PD-L1	Immune checkpoint	E1J2J	Cell Signaling
p-p44/42 MAPK	MAPK pathway	Thr202/Tyr204, #4370	Cell Signaling
pSTAT3	STAT3 pathway	Tyr705, D3A7	Cell Signaling
PTX3	M2c-like macrophages	PPZ1228	Perseus Proteomics
RGS1	Signal transduction	Rabbit polyclonal	Thermo Fisher
SIRPA	M2-like macrophages	D6I3M	Cell Signaling
TNFAIP8	Apoptosis	#14559-MM01	Sino Biological
TP53	Cell cycle, apoptosis	DO-7	Novocastra
RGS1	Signal transduction	Rabbit polyclonal	Thermo Fisher

CNIO, Centro Nacional de Investigaciones Oncológicas (Spanish National Cancer Research Center).

3. Results

The different subtypes of hematological neoplasia (mainly non-Hodgkin lymphomas) were predicted using artificial neural networks, machine learning, and conventional bio-statistics. The analysis used transcriptomic data and protein levels assessed by immunohistochemistry. The results are summarized as a bulleted list.

3.1. Predictive Classification of Non-Hodgkin Lymphomas

- Using the whole array of 20,863 and a cancer transcriptome panel, the lymphoma subtypes were predicted by a neural network with high accuracy [19].
- A set of 30 genes derived from the neural network also predicted the overall survival of an independent series of diffuse large B-cell lymphoma, and a pan-cancer series of 7441 cases of The Cancer Genome Atlas (TCGA) [19] (Figure 4).

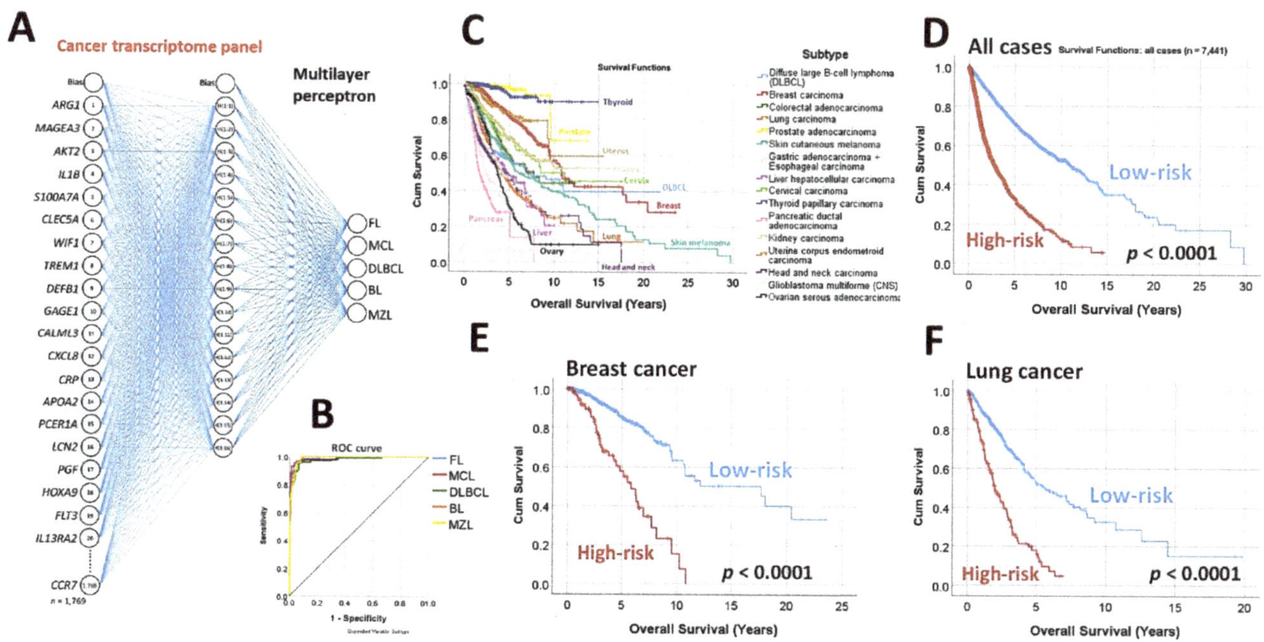

Figure 4. Prediction of lymphoma subtype by a neural network with high accuracy. (**A**) A multilayer perceptron predicted the different non-Hodgkin lymphoma subtypes, including follicular lymphoma, mantle cell lymphoma, diffuse large B-cell lymphoma, Burkitt's lymphoma, and marginal zone lymphoma. The predictors (inputs) were the gene expression values of a pan-cancer transcriptome panel. The architecture of the network had 1769 nodes in the input layer, a hidden layer of 16 nodes, and an output layer with 5 nodes (5 lymphoma subtypes). In this figure, the top 20 most relevant genes for predicting the lymphoma subtype are shown, based on their average normalized importance for prediction. The most relevant gene was *ARG1*, followed by *MAGEA3*, *AKT2*, and *IL1B*. (**B**) This multilayer perceptron had a high performance, as shown in the ROC curve that had an area under the curve near 1. (**C–F**) Interestingly, the top 30 genes of the neural network not only predicted the lymphoma subtype but also managed to predict the overall survival of a large pan-cancer series from the TCGA of 7441 cases. Using a risk score formula, the cases of each series were stratified into high- and low-risk groups. The risk scores were calculated by multiplying the beta values of the Cox regression per gene expression values for each gene. The overall survival was calculated using the Kaplan–Meier and log-rank test and Cox regression analyses. These top 30 genes belonged to a pan-cancer transcriptome panel. Therefore, this may explain why they have predictive value in a pan-cancer series, and points out that there may be common cancer mechanisms in all human neoplasia.

3.2. Follicular Lymphoma, Immune Response, and Microenvironment

- An algorithm combined two types of neural networks (multilayer perceptron and radial basis function) to predict the overall survival, in combination with other clinically relevant variables [29].
- These variables were more than 60 years, the number of extranodal sites > 1, LDH-level ratio > 1, stage > 2, IPI score 2–3, with translocation (14;18) positive, immune response ratio 2:1 high (≥0.97), and overall survival up to 5 years vs alive from 10 years [29].
- As a result, new poor and favorable prognostic genes were identified, and were correlated with the immune microenvironment (M2-like tumor-associated macrophages) [29] (Figures 5 and 6).

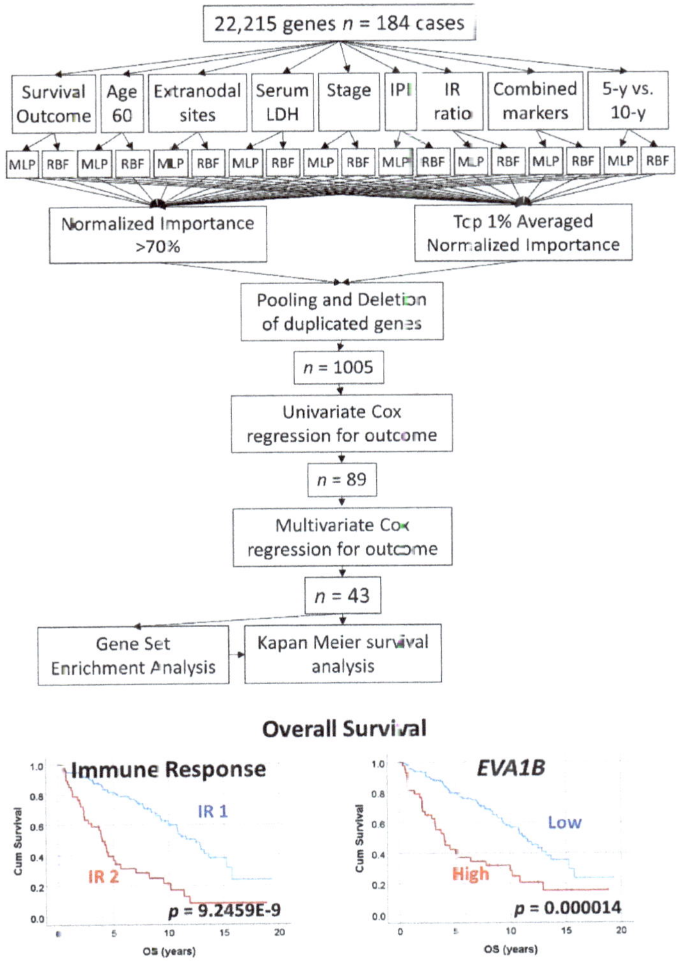

Figure 5. Prediction of the overall survival of follicular lymphoma using an algorithm based on neural networks. The algorithm combined multilayer perceptron (MLP), radial basis function (RBF), and COX regression to highlight 43 genes with prognostic relevance; finally, a correlation with immuno-oncology genes was also performed. This figure shows the algorithm (method) that was used to analyze the gene expression data of follicular lymphoma using artificial neural networks. From an initial set of 22,215 genes, a strategy of dimensionality reduction highlighted 43 genes, of which 18 were associated with poor and 25 with good overall survival of the patients. The first step

consisted of several independent artificial neural networks. The network architecture included the 22,215 genes as predictors (inputs), a hidden layer, and an output layer with the predicted variable. The predicted variables were the overall survival of the patients (outcome dead vs alive), and other relevant clinicopathological variables of follicular lymphoma. The result of the neural network ranked all the genes according to their normalized importance for predicting the target variable. The results of the independent multiple neural networks were pooled resulting in 1005 genes, and the most relevant ones were highlighted using univariate and multivariate Cox regression analyses. The relevance of these genes was confirmed using gene set enrichment analysis (GSEA). Finally, these genes were also correlated with several immuno-oncology genes. The 43 genes were the following: 18 were associated with a poor prognosis (*FRYL, KIAA0100, CDC40, MED8, PTP4A2, BNIP2, TMEM70, MED6, SLC24A2, KLK10, RANBP9, PRB1, EVA1B, CBFA2T2, ALDH1L1, KRT19, BTN2A3P,* and *TRPM4*) and 25 were associated with a good prognosis of the patients (*HSF2, ATPAF2, SLC7A11, PTAFR, TTLL3, TCP10L, DNAAF1, PRH1, NSDHL, TAF12, TSPAN3, AKIRIN1, ITK, TDRD12, LPP, BTD, SIRT5, ZNF230, ABHD6, TOP2B, ARPC2, ASAP2, IDH3A, PSMF1,* and *ARFGEF1*) (Supplementary Tables S1–S5). LDH, lactate dehydrogenase; IPI, international prognostic index; IR ratio, immune response ratio; 5-y, five years; MLP, multilayer perceptron; RBF, radial basis function.

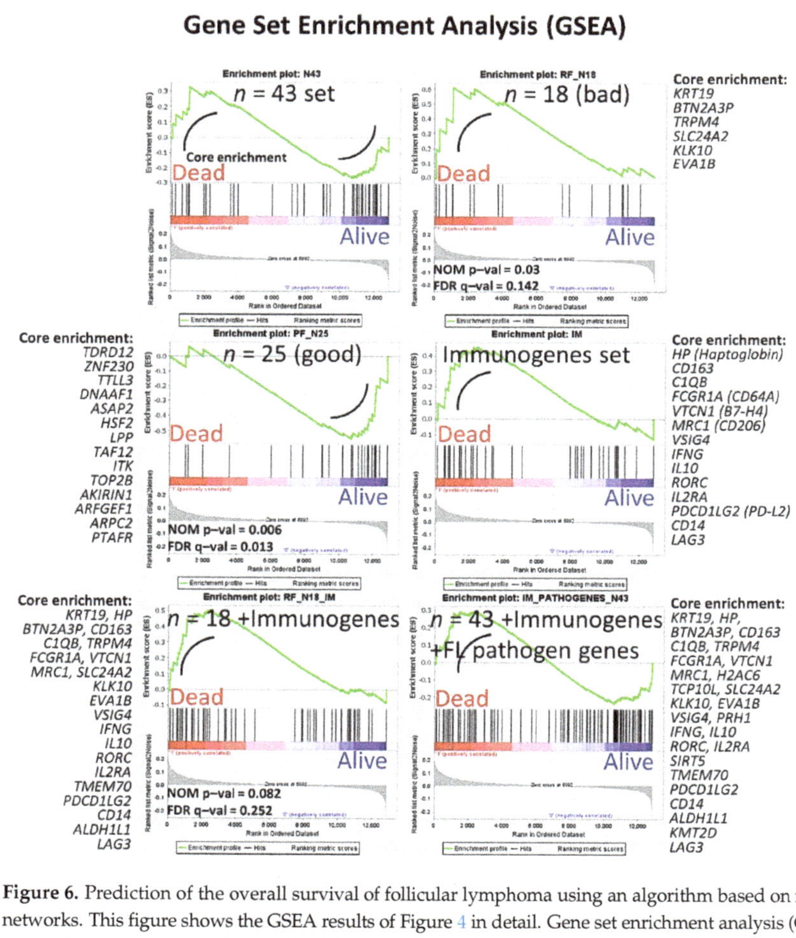

Figure 6. Prediction of the overall survival of follicular lymphoma using an algorithm based on neural networks. This figure shows the GSEA results of Figure 4 in detail. Gene set enrichment analysis (GSEA) was performed to confirm the results of the multivariate Cox regression for the overall survival analysis.

The set of 43 was used in addition to genes of the immune response as well as oncogenes and tumor suppressor genes related to the pathogenesis of follicular lymphoma. Of note, genes related to macrophages were highlighted, such as *CD163*. NOM p–val, nominal p value (the nominal *p* value estimates the statistical significance of the enrichment score for a single gene set). FDR q–val, false discovery rate.

- Tridimensional (3D) analysis of tumor-associated macrophages (TAMs) of follicular lymphoma and transformation to diffuse large B-cell lymphoma was associated with increased numbers of TAMs, which created a network-like structure (Figure 7).

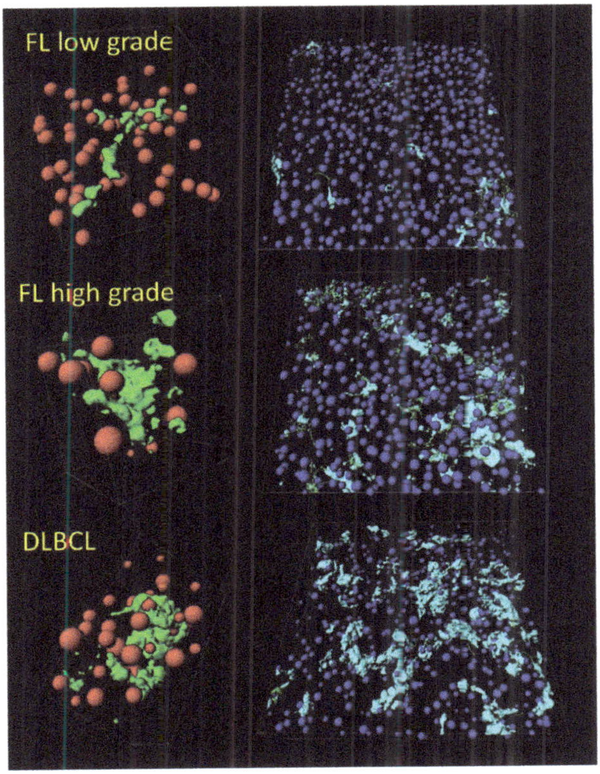

Figure 7. Tridimensional analysis of tumor-associated macrophages (TAMs) in follicular lymphoma The analysis of M2-like TAMs in follicular lymphoma showed that the progression from low grade to high grade, and the transformation to diffuse large B-cell lymphoma, were associated with increased numbers of TAMs, which created a physical network-like structure. This result points out that TAMs may contribute to the disease pathogenesis. In this figure, the macrophages are highlighted in pale blue (right) and green (left). B and T lymphocytes are in dark blue and red. The images were obtained using a LSM 700 laser scanning confocal microscope from Carl Zeiss (Carl-Zeiss-Strasse 22, 73447 Oberkochen, Germany), and Imaris software (version 8.4, Oxford Instruments, Belfast, United Kingdom). FL, follicular lymphoma; DLBCL, diffuse large B-cell lymphoma.

3.3. *Follicular Lymphoma, Random Number Generator-Based Strategy*

- The random number generation created 120 independent multilayer perceptron solutions and 22,215 gene probes were ranked according to their averaged normalized importance for predicting the overall survival [35].

- The analysis identified new predictor genes, which were related to cell adhesion and migration, cell signaling, and metabolism. These genes were also correlated to the immuno-oncology markers of *CD163, CSF1R, FOXP3, PDCD1 (PD-1), TNFRSF14 (HVEM)*, and *IL10* [35].
- A comparison with other machine learning techniques was also performed. Machine learning included the following techniques: Bayesian network, C&R tree, C5 tree, CHAID tree, discriminant analysis, KNN algorithms, logistic regression, LSVM, Quest tree, random forest, random trees, SVM, tree-AS, XGBoost linear, and XGBoost tree. A neural network analysis was also made [35] (Figure 8).

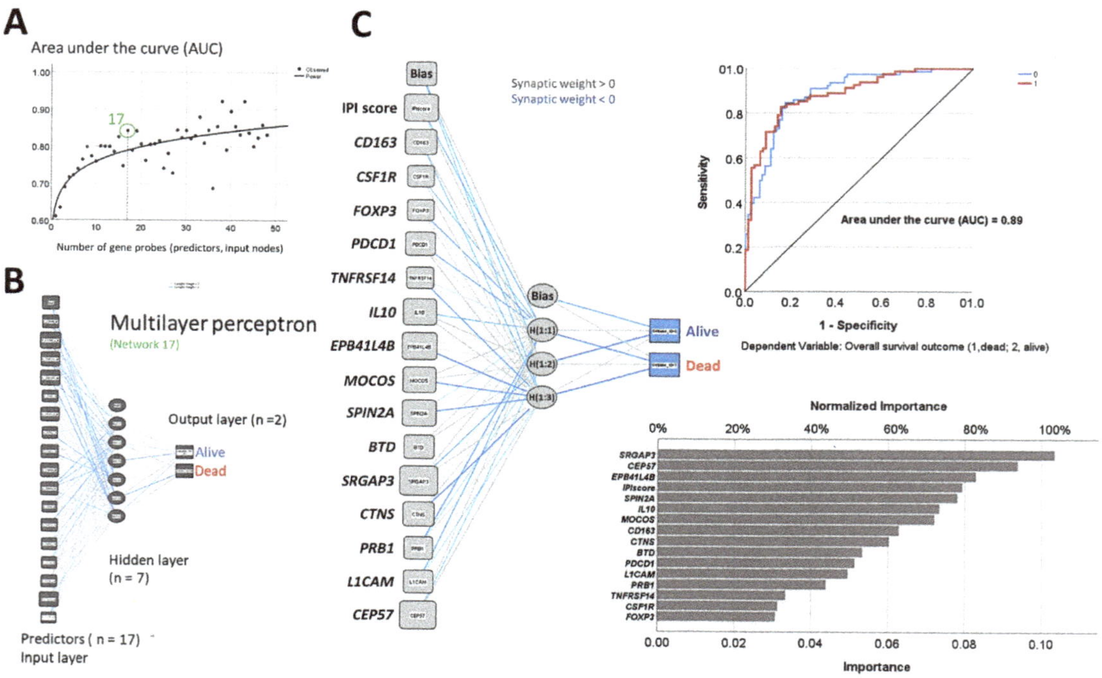

Figure 8. Prediction of the overall survival of follicular lymphoma taking advantage of the random number generator. (**A**) By using the random generator, 120 independent and different neural network solutions were calculated, and the averaged normalized importance of each gene for predicting the overall survival was recorded. Then, the minimal number of genes of a neural network with sufficient performance was selected, and a final neural network with 17 genes was defined. (**B**) This neural network (multilayer perceptron type) included 17 genes in the input layer, a hidden layer of 7 nodes, and an output layer of 2 nodes (overall survival, death vs alive). (**C**) A new neural network was created with the highlighted 17 genes and known immuno-oncology genes. The resulting model had an acceptable accuracy, with an area under the curve (AUC) of 0.89. The predictors (inputs) were ranked according to their normalized importance in predicting the overall survival.

3.4. Mantle Cell Lymphoma, Use of Immuno-Oncology Panels to Predict Survival

- An analysis algorithm included several analysis techniques such as neural networks (both the multilayer perceptron artificial and radial basis function), GSEA, and conventional statistics. In this analysis, 20,862 genes were correlated with 28 prognostic genes of mantle cell lymphoma. After dimensionality reduction, the patients' overall survival was predicted, and new markers were highlighted (Figure 9) [34].

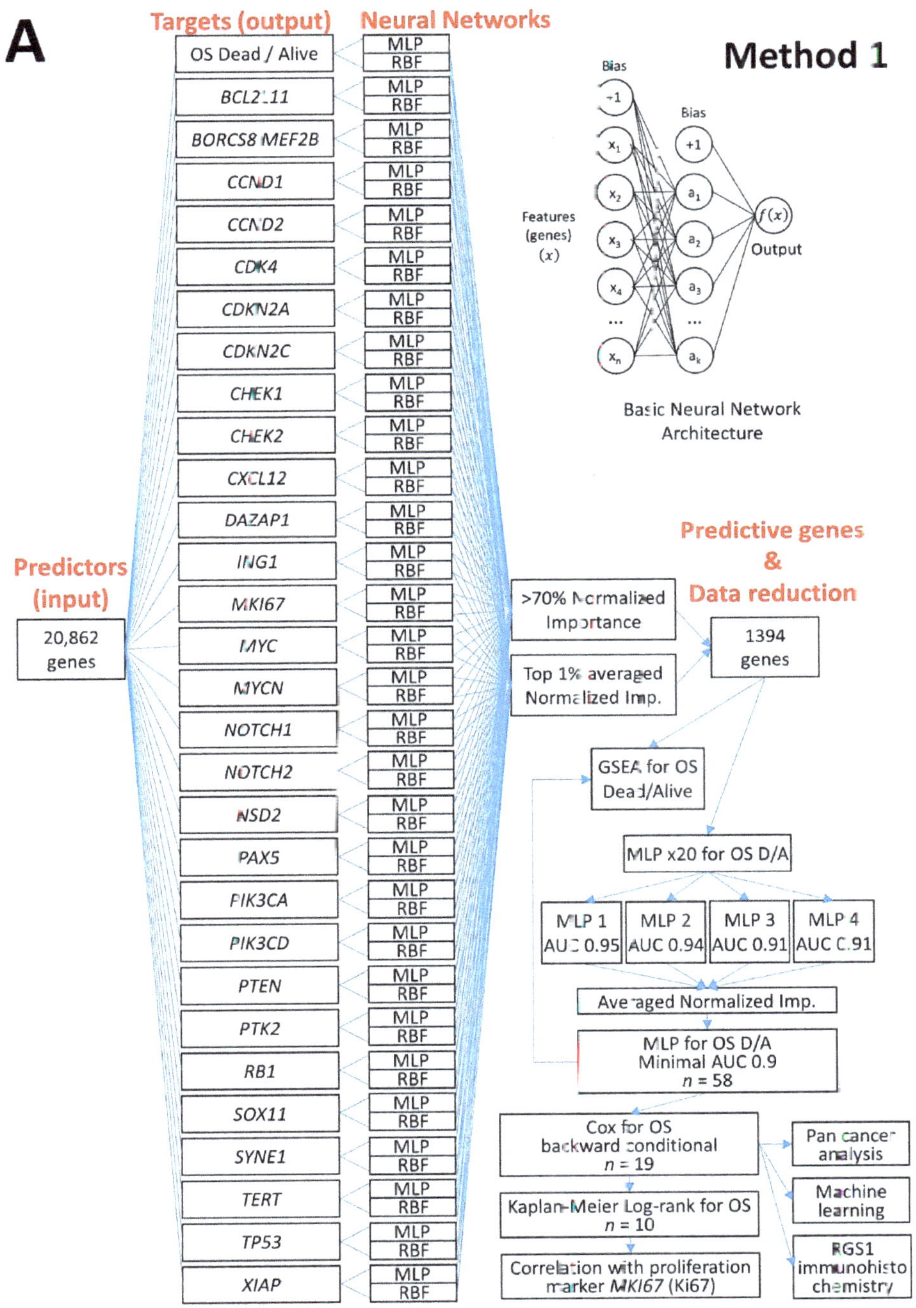

Figure 9. *Cont.*

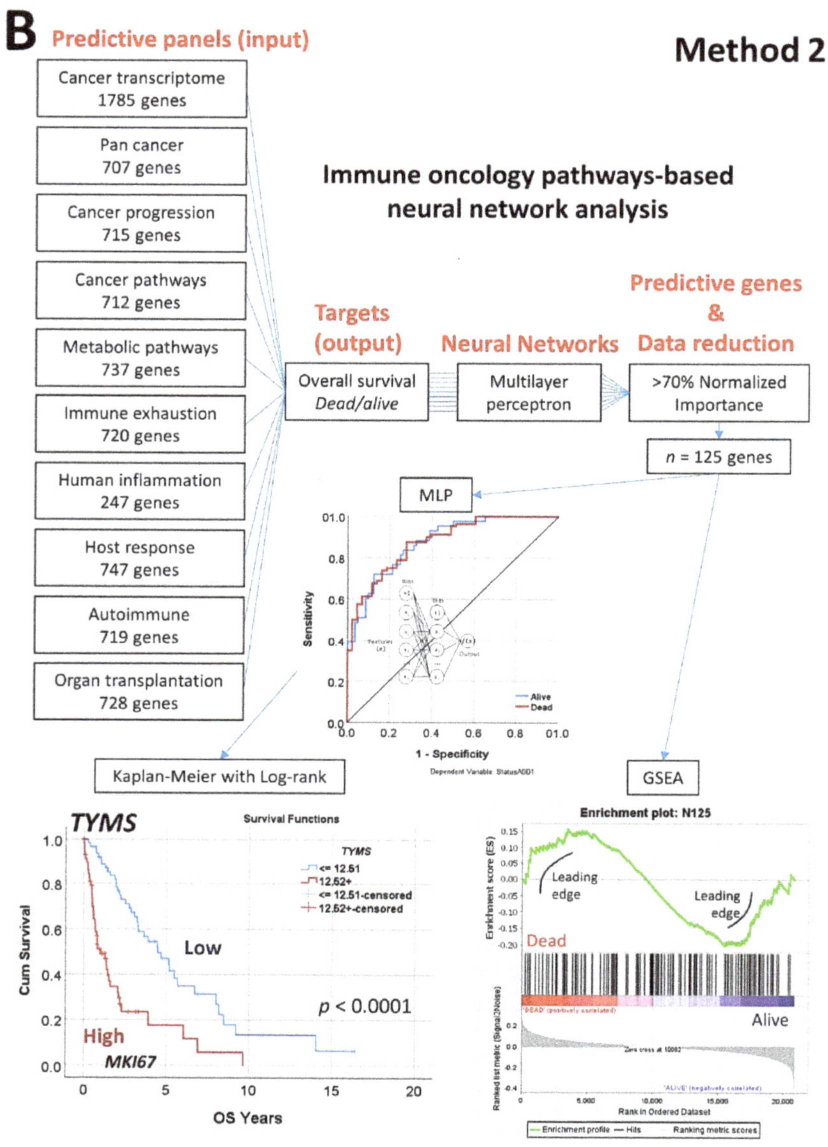

Figure 9. Prediction of the overall survival of mantle cell lymphoma using an algorithm based on neural networks. Two methods (**A** and **B** algorithms) were designed. Method 1 used as input 20,862 genes to predict the overall survival outcome (dead vs. alive) and other prognostic markers; because of dimensionality reduction, a final set of 19 genes were highlighted. The analysis also included testing the final 19 genes with other machine learning analysis, and conventional overall survival with log-rank test. Method 2 used as input several gene panels to predict the overall survival. As a result, 125 pan-cancer and immuno-oncology genes were highlighted. The association with the patients overall survival was confirmed by GSEA and conventional overall survival with log-rank test. OS, overall survival; MLP, multilayer perceptron; RBF, radial basis function; GSEA, gene set enrichment analysis; D/A, dead/Alive; AUC, area under the curve; NI, normalized importance.

- The highlighted genes were related to the cell cycle, apoptosis, and metabolism. The genes not only predicted the survival of mantle cell lymphoma, but also of diffuse large B-cell lymphoma and a large pan-cancer series of the TCGA [34].
- A neural network algorithm that combined 10 oncology and immuno-oncology panels predicted overall survival (Figure 9) [34].
- Other machine learning techniques were used. Additionally, a correlation with the MCL35 proliferation assay, which was created by the Lymphoma/Leukemia Molecular Profiling Project, was made [34] (Figure 9).

3.5. Diffuse Large B-Cell Lymphoma, Identification of the 25 Genes Set

- A multilayer perceptron analysis predicted the overall survival of 100 cases using as input 54,614 gene probes, and highlighted 25 genes with prognostic value [27].
- Correlation with known diffuse large B-cell lymphoma markers showed that high expression of MYC, BCL2, and ENO3 was associated with worse outcome [27] (Figures 10 and 11).

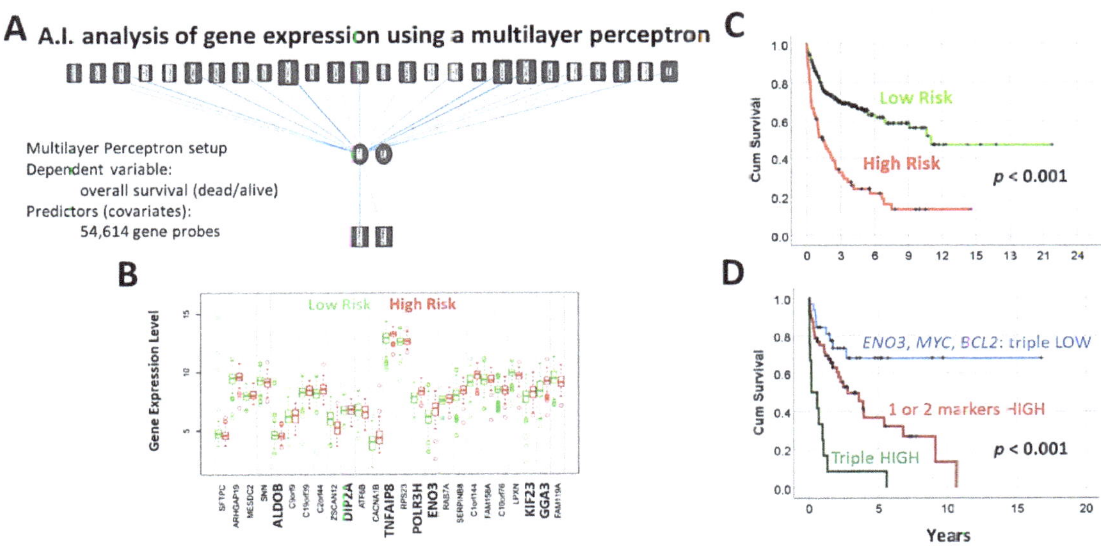

Figure 10. A neural network predicted the overall survival of diffuse large B-cell lymphoma using gene expression data. (**A**) A multilayer perceptron predicted the overall survival and highlighted the most important 25 genes. (**B**) Using a risk score formula and the gene expression of the 25 genes, two groups of patients with different overall survival were found; this figure shows the different gene expression of the 25 genes between the two risk groups. (**C**) The two risk groups had different overall survival. (**D**) Among the 25 genes, *ENO3*, *MYC*, and *BCL2* were the most important, and only with these 3 genes the survival of the patients could be determined.

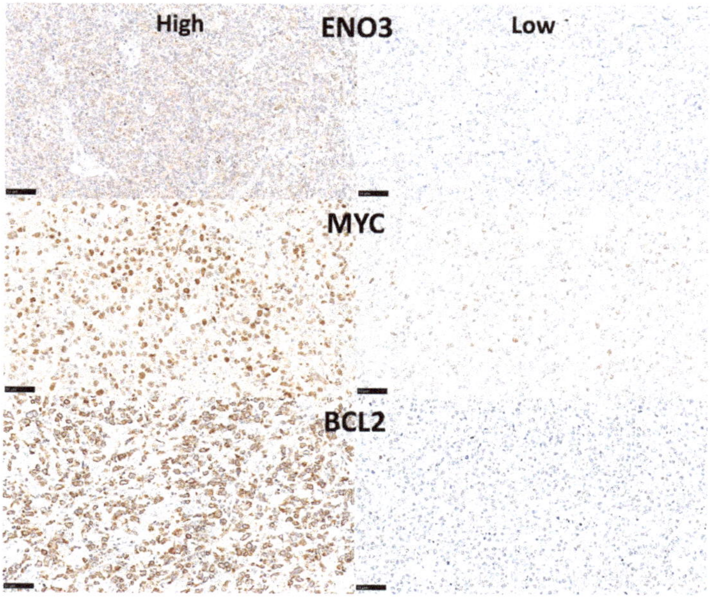

Figure 11. Immunohistochemical staining of ENO3, MYC, and BCL2 in diffuse large B-cell lymphoma. This figure shows six different lymphoma cases, with high or low expression of the 3 markers. Original magnification: 400× (scale bar = 50 um).

3.6. Diffuse Large B-Cell Lymphoma, Prognostic Value of the 25 Genes in Hematological Neoplasia, and TNFAIP8 Validation

- The previously identified set of 25 genes not only predicted the prognosis of 741 cases of diffuse large B-cell lymphoma, but also predicted other hematological neoplasia, including chronic lymphocytic leukemia (n = 308), mantle cell lymphoma (n = 92), follicular lymphoma (n = 180), multiple myeloma (n = 559), and acute myeloid leukemia (n = 149) [28].
- The TNFAIP8 marker was highlighted in this analysis. Because of TNFAIP8's importance in the apoptotic pathway, it was validated by immunohistochemistry (i.e., at protein level) in an independent series of 97 cases from Tokai University. Digital image quantification of TNFAIP8 was performed using an AI-based method. Correlations with the prognosis of the patients showed that high TNFAIP8 is associated with poor survival [28].
- TNFAIP8 correlated positively with high M2-like CD163-positive tumor-associated macrophages (TAMs) and non-GCB cell of origin phenotype [28] (Figure 12).

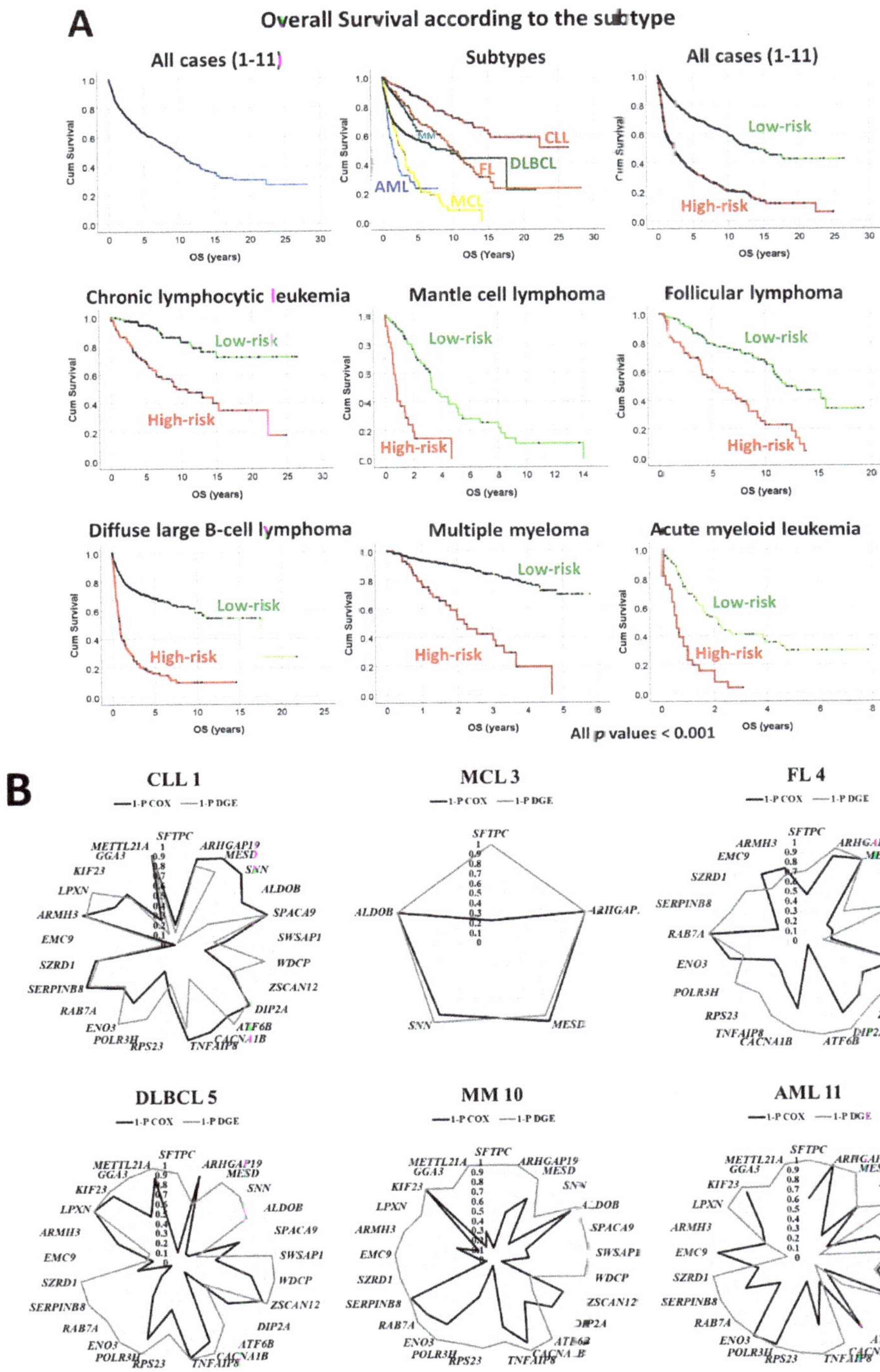

Figure 12. *Cont.*

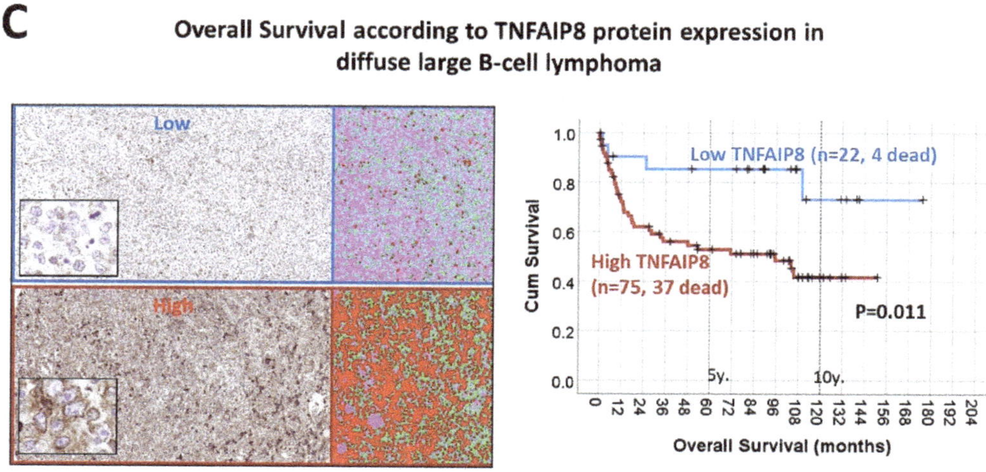

Figure 12. A set of 25 genes derived from a neural network predicted the overall survival of several lymphoma subtypes and acute myeloid leukemia, and high protein expression of TNFAIP8 correlated with poor survival of diffuse large B-cell lymphoma patients. (**A**) Using the gene expression values of 25 genes, previously identified using artificial neural networks, and a risk score formula, it was possible to predict the overall survival of several hematological neoplasia (lymphomas and acute myeloid leukemia). All Kaplan–Meier analyses with log-rank tests were statistically significant and had a $p < 0.001$. (**B**) Although all 25 genes were relevant, the strength and direction of the association was different in each subtype of hematological neoplasia. For example, *TNFAIP8* was more relevant for the overall survival of diffuse large B-cell lymphoma and chronic lymphocytic leukemia, but less relevant for acute myeloid leukemia and multiple myeloma. Nevertheless, *TNFAIP8* contributed to the survival of all these hematological neoplasia. (**C**) High TNFAIP8 protein expression, evaluated by immunohistochemistry using both conventional digital image analysis and AI-based methods, correlated with poor overall survival of diffuse large B-cell lymphoma patients. This figure shows two cases of diffuse large B-cell lymphoma. The figure at the top express low TNFAIP8. On the left, the hematoxylin (dark blue) and DAB-based (brown) immunohistochemical image is shown. As shown in the inset, the TNFAIP8 staining was cytoplasmic. On the right, the AI-based digital image analysis is shown for the same case and area. TNFAIP8 is highlighted in red, cellular structures (B lymphocytes of the lymphoma, T lymphocytes, and macrophages) in pink, and intercellular tissue in green. The figure at the bottom is characterized by high TNFAIP8 expression. After staining procedures, the immunohistochemical slides were digitalized and visualized (NanoZoomer S360 scanner and NDP.view2 viewing software, Hamamatsu KK.). Original magnification: 200×. High TNFAIP8 correlated with age > 60 years, high serum IL2RA, non-GCB phenotype, and high infiltration of CD163+ M2-like tumor-associated macrophages (CD163+TAMs). TNFAIP8 also moderately correlated with MYC (Spearman's correlation coefficient 0.389, $p = 0.009$) and Ki67 (proliferation index; Spearman's correlation coefficient 0.48, $p = 0.001$). High TNFAIP8 was also associated (trend) with worse progression-free survival ($p = 0.052$). Finally, a multivariate COX analysis between TNFAIP8 (high vs low) and the international prognostic index (IPI) (low+low/intermediate vs high/intermediate + high) showed that only TNFAIP8 retained the prognostic value (HR = 3.5, $p = 0.040$). CLL, chronic lymphocytic leukemia; DLBCL, diffuse large B-cell lymphoma; FL, follicular lymphoma; MM, multiple myeloma; MCL, mantle cell lymphoma; AML, acute myeloid leukemia.

3.7. Diffuse Large B-Cell Lymphoma, Prediction of Survival by Caspase-8

- The protein expression of caspase-8 (which is inhibited by TNFAIP8) was analyzed by immunohistochemistry in a series of 97 cases of diffuse large B-cell lymphoma, and high expression correlated with a favorable overall and progression-free survival [31].
- Based on an immunohistochemical analysis, caspase-8 was correlated with other markers of its pathway, including BCL2, caspase-3, CDK6, cleaved PARP, E2F1, Ki67, LMO2, MDM2, MYB, MYC, TNFAIP8, and TP53 [31].
- The caspase-8 protein expression was also modeled using several machine learning and artificial neural networks [31] (Figures 13 and 14).

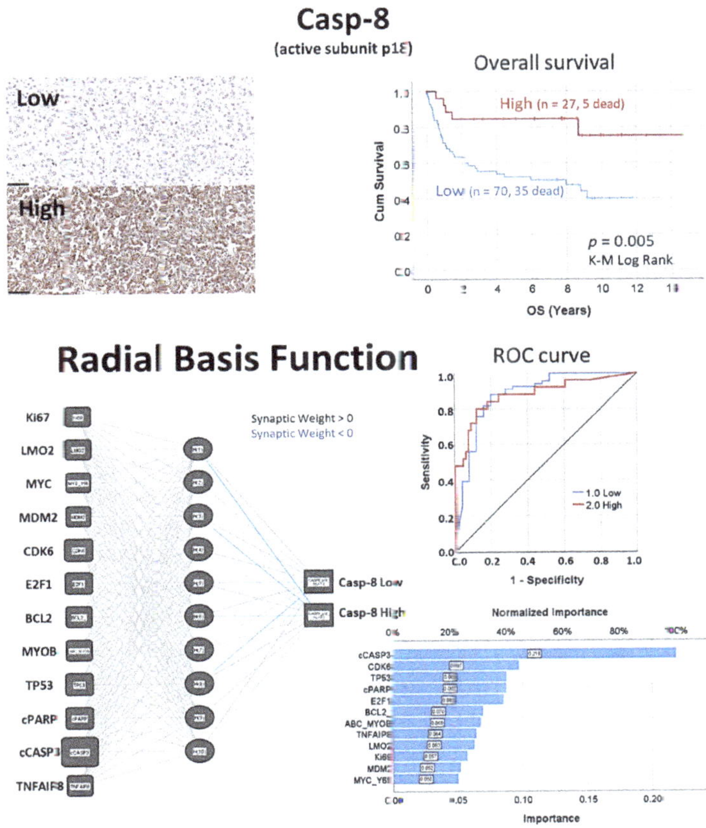

Figure 13. High caspase-8 correlated with favorable survival of diffuse large B-cell lymphoma patients. The protein levels of caspase-8 (*CASP8*) were evaluated by immunohistochemistry, and later correlated with the survival of the patients. Two types of immunohistochemical staining were observed, low and high. In diffuse large B-cell lymphoma, high caspase-8 expression is associated with a favorable overall survival ($p = 0.005$). Additionally, other markers of the capsase-8 pathway, including caspase-3, cleaved PARP, BCL2, TP53, MDM2, MYC, Ki67, E2F1, CDK6, MYB, LMO2, and TNFAIP8, were evaluated by immunohistochemistry and quantified using digital image analysis. Caspase-8 was successfully predicted by the pathway markers, both using conventional statistics and several machine learning techniques and artificial neural networks. Of note, after staining procedures, the immunohistochemical slides were digitalized and visualized (NanoZoomer S360 scanner and NDP.view2 viewing software, Hamamatsu K.K.). Original magnification: 400× (scale bar = 50 um). OS, overall survival; ROC curve, the receiver operating characteristic curve.

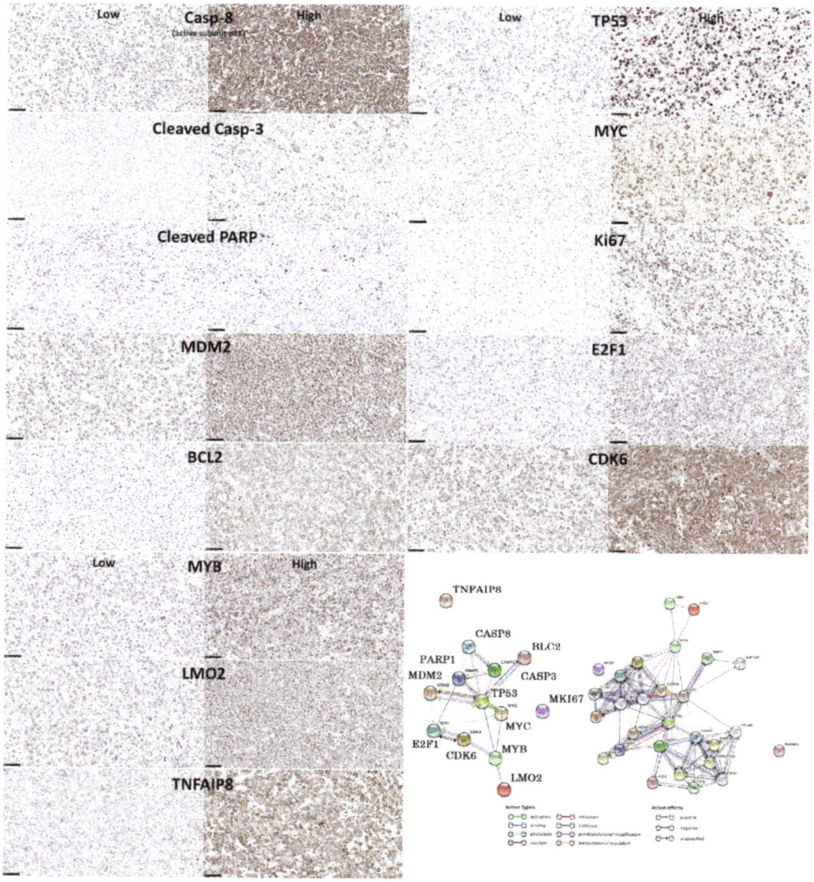

Figure 14. High caspase-8 correlated with favorable survival of diffuse large B-cell lymphoma patients. This figure shows the immunohistochemical expression of active subunit p18 casp-8 (CASP8), which correlated with good prognosis of the patients when high. Other related markers, as shown in the protein–protein interaction analysis, were also analyzed by immunohistochemistry. After staining procedures, the immunohistochemical slides were digitalized and visualized (NanoZoomer S360 scanner and NDP.view2 viewing software, Hamamatsu KK.). All the markers were quantified using digital image analysis. This figure shows examples of low and high expressions for each marker. Original magnification: 400× (scale bar = 50 um).

3.8. Diffuse Large B-Cell Lymphoma, CD274 (PD-L1) and IKAROS

- An algorithm included multilayer perceptron, radial basis function, GSEA, COX regression, and several machine learning techniques to predict the overall survival of 414 cases of diffuse large B-cell lymphoma [30].
- The machine learning techniques were Bayesian network, C5.0 algorithm, chi-squared automatic interaction detection CHAID tree, classification and regression (C&R) tree, discriminant analysis, logistic regression, Quest tree, random trees, and tree-AS. The neural network was the multilayer perceptron [30].
- The association of PD-L1 (CD274) and IKAROS with the overall survival was validated in an independent series of 113 cases by immunohistochemistry. The quantification included an AI-based method [30] (Figure 15).

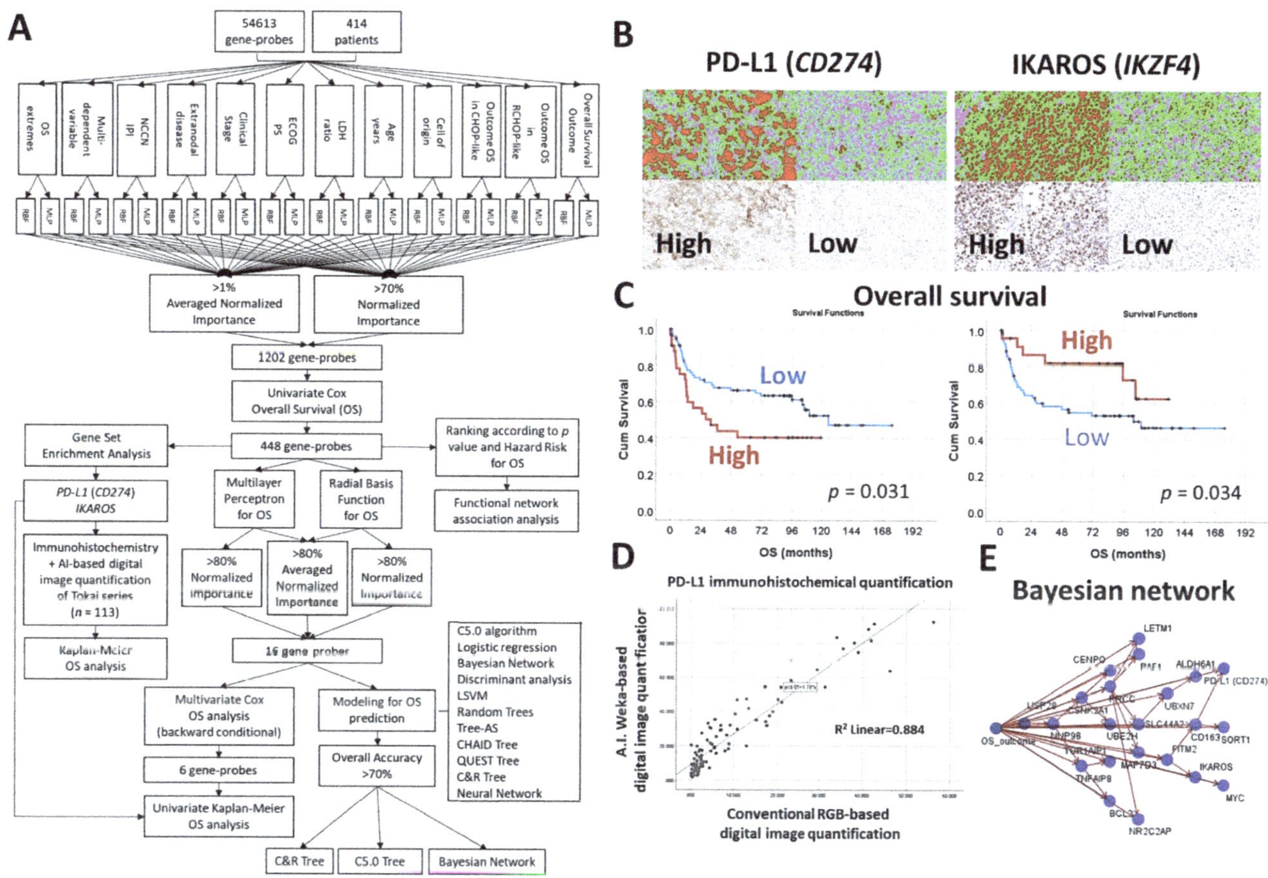

Figure 15. An algorithm that included artificial neural networks and machine learning predicted the survival of diffuse large B-cell lymphoma, and highlighted *PD-L1* and *IKAROS* as prognostic markers. (**A**) Algorithm: This algorithm is similar to that one of follicular lymphoma and mantle cell lymphoma. The basic structure

analysis is an artificial neural network (multilayer perceptron). In this analysis, 54,613 gene probes were used as predictors for the overall survival, but also for other relevant clinicopathological variables. The basic neural network was composed of the input layer (predictors, 54,613 gene probes), a hidden layer (automatically computed), and an output layer (predicted variable; for example, the overall survival outcome as a dichotomic variable dead vs alive, or the cell of origin classification (GCB vs ABC), etc.). The dimensionality reduction included additional steps of machine learning, Cox regression, and GSEA. (**B**) Digital image quantification using AI-based strategy for PD-L1 (CD274) and IKAROS. (**C**) High protein expression of PD-L1 correlated with poor survival of the patients. Conversely, high IKAROS was associated with favorable survival. (**D**) AI-based quantification correlated well with conventional digital image quantification. Therefore, both techniques provide comparable results. (**E**) Modeling of the overall survival using a Bayesian network. The Bayesian network builds a probability model, a graphical model that shows variables (nodes) of the dataset, and the probabilistic (conditional) independences between them. The links of the network are called arcs and represent the relationship between the variables, but do not necessarily mean cause and effect. Original magnification: 200×. OS, overall survival; NCCN IPI, National Comprehensive Cancer Network International Prognostic Index; ECOG PS, Eastern Cooperative Oncology Group Performance Status; LDH, lactate dehydrogenase; R-CHOP, rituximab, cyclophosphamide, doxorubicin hydrochloride, vincristine, and prednisolone; AI, artificial intelligence.

3.9. Diffuse Large B-Cell Lymphoma, CSF1R

- The protein expression of CSF1R was analyzed by immunohistochemistry in 198 cases of diffuse large B-cell lymphoma, and it was found that high CSF1R-positive TAMs were associated with poor progression-free survival (Figure 16) [32].

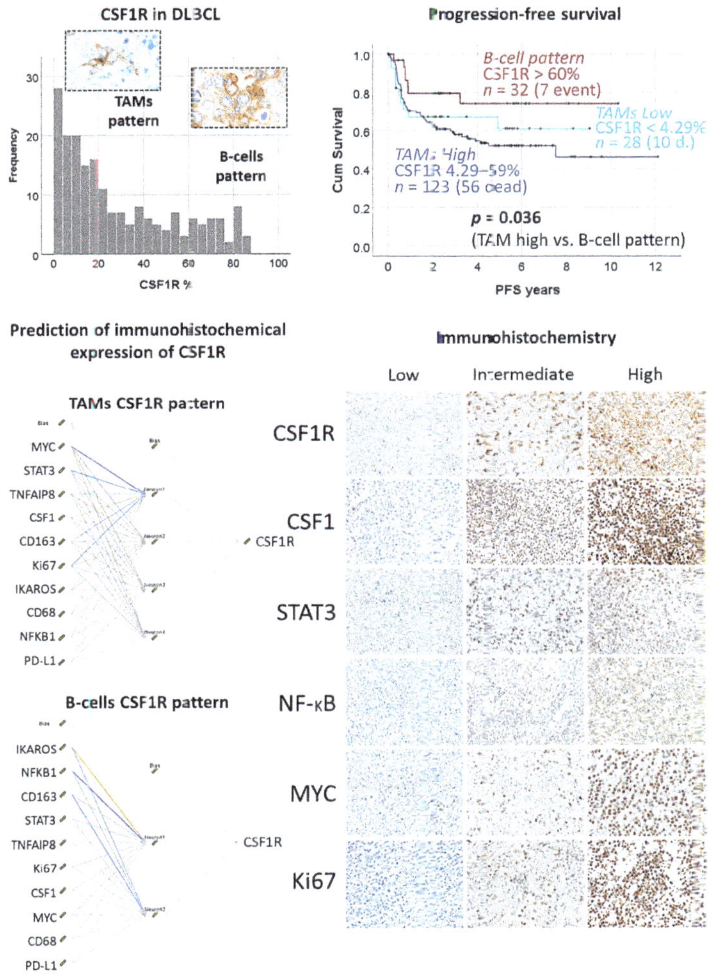

Figure 16. Role of CSF1R in the prognosis of diffuse large B-cell lymphoma. CSF1R was analyzed by immunohistochemistry in a series of 198 cases, and two histological patterns were found. A CSF1R-positive B-cell pattern was characterized by favorable progression-free survival; this pattern was less frequent (around 30% of the cases). Conversely, the most frequent pattern was of CSF1R-positive tumor-associated macrophages (TAMs) and was associated with an unfavorable outcome. Additionally, the prediction of the immunohistochemical expression of CSF1R by other CSF1R-related markers was performed using neural networks. The CSF1R-related markers were CSF1, STAT3, NFKB, MYC, and Ki67. All markers were quantified using digital image analysis. Of note, the multilayer perceptron network analyses were performed to predict both the TAM and the B-cell patterns. Our data suggested that the use of a CSF1R inhibitor such as Pexidartinib could be used in the CSF1R + TAM pattern. CSF1R, macrophage colony-stimulating factor 1 receptor; DLBCL, diffuse large B-cell lymphoma; TAM, tumor-associated macrophage; PFS, progression-free survival.

- Using a neural network, CSF1R protein expression was predicted by 10 CSF1R-related markers (CSF1, STAT3, NFKB1, Ki67, MYC, PD-L1, TNFAIP8, IKAROS, CD163, and CD68) (Figure 16) [32].
- The gene expression of *CSF1R* was predicted by all the genes, and by an immuno-oncology pattern, and correlated with *SIRPA* and *CD47* [32] (Figures 17 and 18).

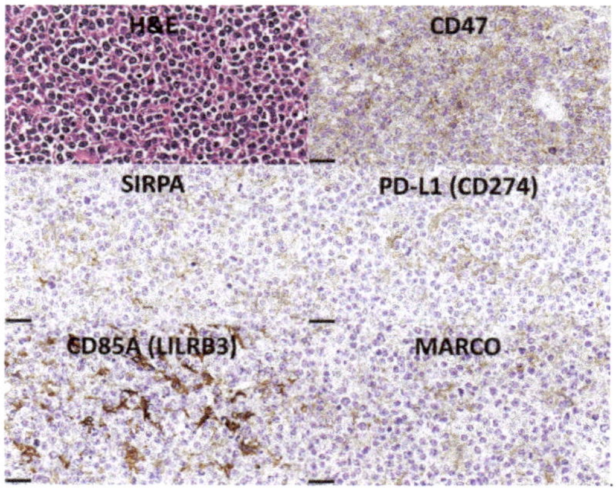

Figure 17. Correlation between expression levels of CSF1R and SIRPA/CD47 in diffuse large B-cell lymphoma. The immunohistochemical pattern of CSF1R-positive tumor-associated macrophages (TAMs) suggested a relationship with other makers such as SIRPA. SIRPA is a relevant immune checkpoint marker that mediates negative regulation of phagocytosis. The histological pattern of SIRPA was of TAMs, similar to PD-L1, CD85A, and MARCO. A ligand for SIRPA is CD47. In our series, the histological pattern of CD47 was of B lymphocytes of the diffuse large B-cell lymphoma.

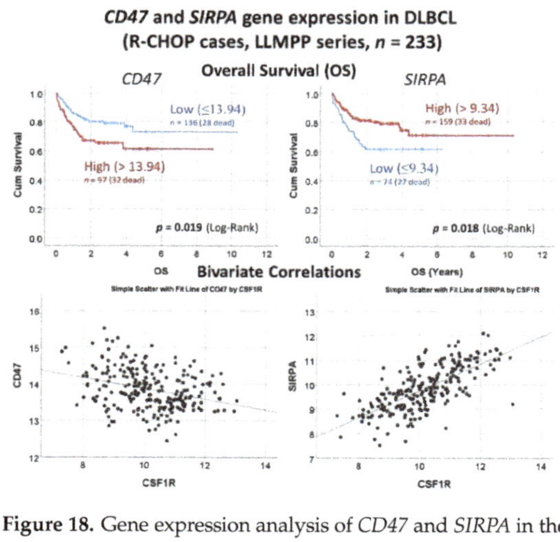

Figure 18. Gene expression analysis of *CD47* and *SIRPA* in the diffuse large B-cell lymphoma. In the series of the Lymphoma/Leukemia Molecular Profiling Project (LLMPP), when analyzing only the cases with R-CHOP-like treatment, high *CD47* but low *SIRPA* correlated with poor overall survival of the patients, and *SIRPA* positively correlated with *CSF1R*. CD47 is a ligand for SIRPA (SIRPα), a protein expressed by macrophages and dendritic cells. These two markers belong to the immune checkpoint pathway, and mediate a negative regulation of phagocytosis. R-CHOP, rituximab, cyclophosphamide, doxorubicin hydrochloride, vincristine, and prednisolone; LLMPP, Lymphoma/Leukemia Molecular Profiling Project; OS, overall survival.

3.10. Diffuse Large B-Cell Lymphoma, Pan-Cancer Immuno-Oncology Panel

- An immuno-oncology panel of 730 genes predicted the overall survival and cell-of-origin phenotype (Lymph2Cx assay) of a series of 106 diffuse large B-cell lymphoma cases, using artificial neural networks and machine learning [33].
- The association of MAPK3 with the GCB phenotype was confirmed by immunohistochemistry [33] (Figure 19).

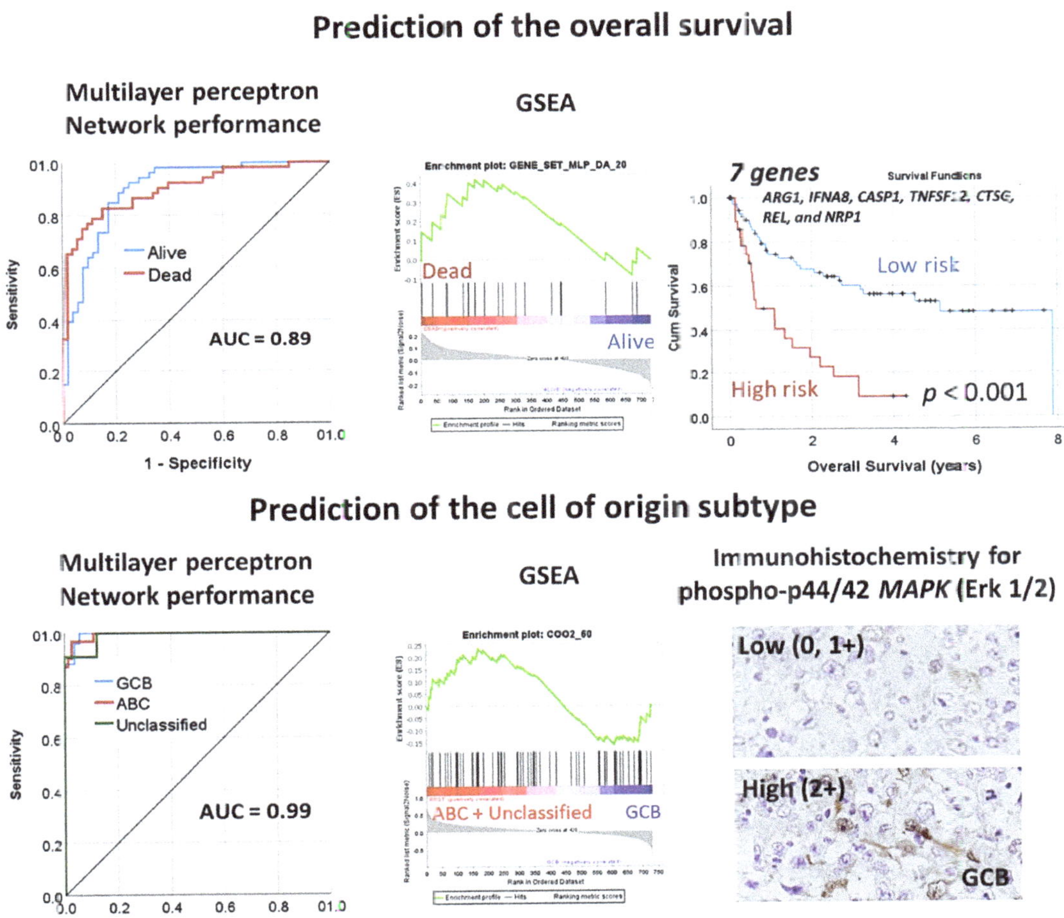

Figure 19. An artificial neural network predicted the overall survival of the diffuse large B-cell lymphoma patients, and the cell of origin subtype using a pan-cancer immuno-oncology gene expression panel. The analysis consisted of the multilayer perceptron. The cell of origin characterization was assessed with the NanoString Lymph2Cx assay. The performance of the network was high, 0.89 for overall survival and 0.99 for the cell of origin phenotype. GSEA analysis confirmed enrichment toward the survival outcome of the dead and the cell of origin subtype of activated (ABC) + unspecified. Using a risk score formula, with 7 genes it was possible to predict the survival of diffuse large B-cell lymphoma. The association of phospho-MAPK with the germinal center B-cell (GCB) phenotype was also noted and confirmed by immunohistochemistry. GSEA, gene set enrichment analysis. ABC, activated B-cell type; GCB, germinal center B-cell type.

3.11. Diffuse Large B-Cell Lymphoma, Integrative Analysis of Macrophage Markers

Gene expression profiling of 233 DLBCL patients treated with chemotherapy plus Rituximab was obtained from the series GSE10846, present in the NCBI Gene Expression Omnibus database. The prognostic value for overall survival of the gene expression of *CD163* was first tested and 100 representative cases were selected, which contained high-risk (i.e., high *CD163*) and low-risk cases (i.e., low *CD163*) (Figure 20).

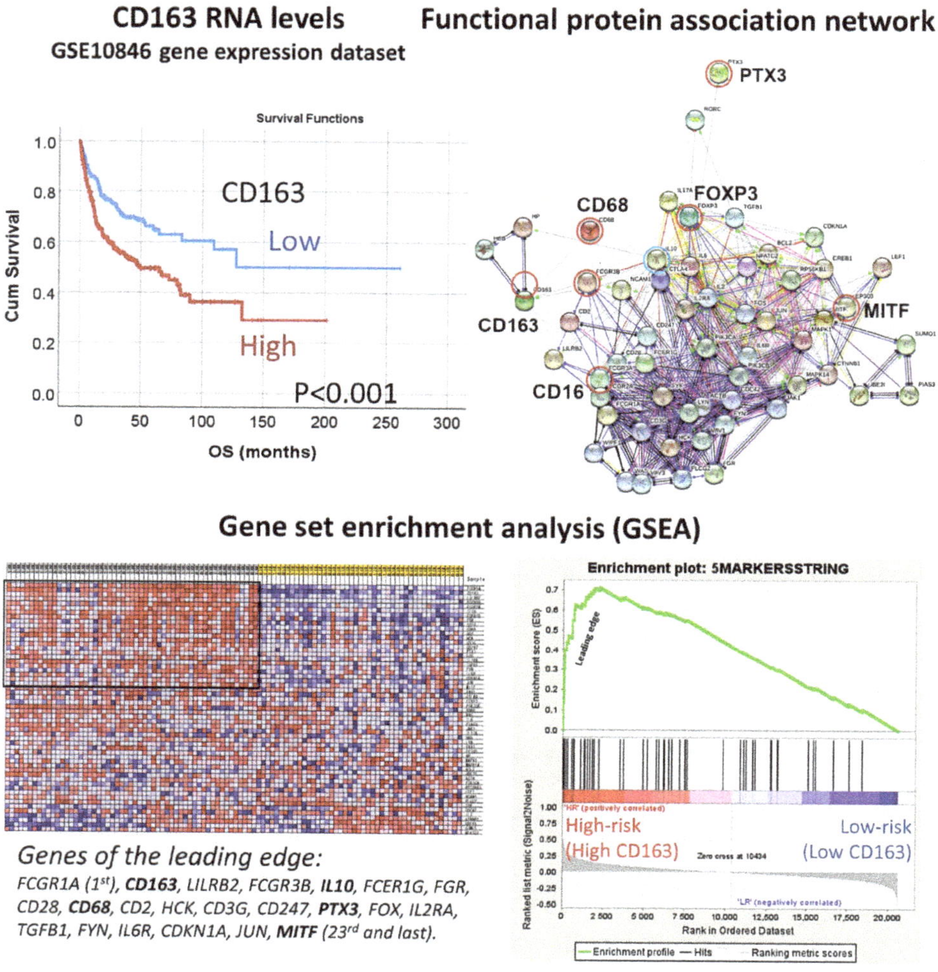

Figure 20. Analysis of macrophages in diffuse large B-cell lymphoma. The overall survival of diffuse large B-cell lymphoma was assessed based on the expression of *CD163*, which is an M2-like macrophage marker. High expression was associated with a poor prognosis of the patients. Then, a protein–protein functional network association analysis was performed using the macrophage markers of CD68 (pan-macrophages), CD16 (M1-like macrophages), CD163 (M2-like), PTX3 (M2c-like), and MITF (M2-like), and the regulatory T lymphocytes (Tregs) marker of FOXP3. The network created a macrophage pathway that was subsequently applied to a gene set enrichment analysis (GSEA). The GSEA confirmed the association of the macrophage pathway with the high-risk group, which was characterized by poor overall survival and high CD163-positive macrophages.

A functional protein association network was created using the five macrophage and one regulatory T lymphocyte (Treg) markers: CD68, CD16, CD163, PTX3, MITF, and FOXP3 as the initial nodes (identifies). Then, the resulting network (i.e., pathway) that contained 57 markers was tested for GSEA analysis in the GSE10846 series of gene expression of diffuse large B-cell lymphoma. We identified the most relevant pathological markers (i.e., genes) that are associated with the prognosis of the patients as follows: high-risk (bad prognosis, and with high *CD163* expression) vs low-risk (good prognosis, low *CD163*). We found that this pathway was enriched in the high-risk phenotype with a NOM p-val < 0.001 and FDR q-val < 0.001. In the enrichment score, we could identify the markers: *CD163* (2nd in the list with a rank metric score of 0.515), *CD16* (FCGR3B, 4th), *CD68* (10th), *PTX3* (15th), and *MITF* (23rd). Of note, *FOXP3* was outside the enrichment set of genes so it was not associated with the high-risk group. Importantly, at fifth position, IL10, was identified. GSEA with markers belonging to the immune regulatory M2c-like TAM pathway was also tested with similar results (Figure 20).

The macrophage markers were analyzed at protein level by immunohistochemistry in the series of Tokai University (n = 132) (Figure 21). The distribution of the markers in the normal reactive tonsil was also evaluated.

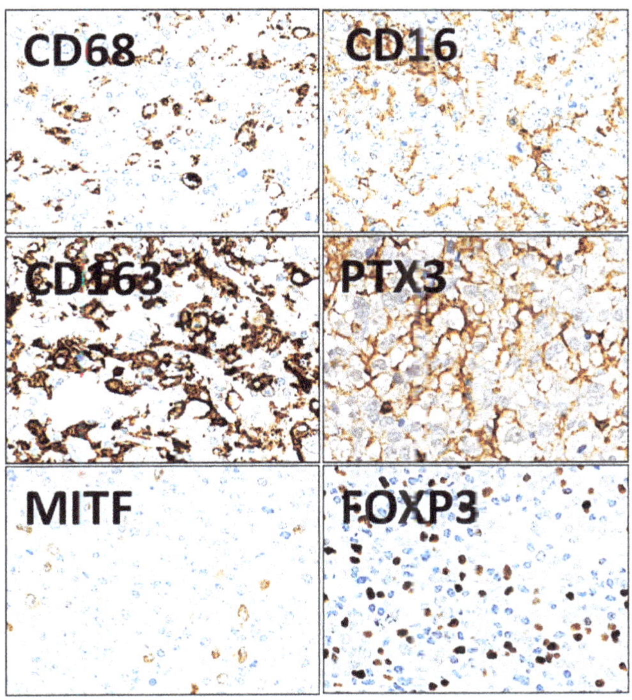

Figure 21. Immunohistochemical staining of macrophage markers and regulatory T lymphocytes (Tregs) in diffuse large B-cell lymphoma. The expression of macrophage markers and Tregs was evaluated using immunohistochemical procedures. The staining confirmed that when macrophages are present at a high concentration in the tissues, their shape is more elongated and dendriform-like. CD68 is a pan-macrophage marker, CD16 is macrophage polarization M1-like, and CD163, PTX3, and MITF are M2-like. FOXP3 is a specific marker of Tregs. Original magnification: 400×.

The histological analysis in reactive tonsil, a secondary lymphoid organ, showed a different distribution of the different markers. CD68-positive and MITF-positive macrophages were widely distributed in all areas. CD16-positive cells were scarce and only identified in the lympho-epithelium, the epithelial barrier. CD163-positive macrophages were mainly present in the interfollicular regions and infrequently in the germinal centers of the follicles. PTX3-positive cells were of macrophage morphology in all areas and in the germinal centers PTX3-positive cells also had a morphology of B lymphocytes (mainly centroblasts). IL10-positive macrophages were scarce but present in all areas. Double IHC showed mutually exclusive distribution between CD163 and CD16 and partially exclusive with MITF.

The multilayer perceptron (MLP) procedure was performed to produce a predictive model for one target variable, using the values of several predictors. The target was the dead or alive variable for overall survival. The predictors were the same categorical variables used in the COX multivariate analysis: CD163, PTX3 Total, MITF, FOXP3, and IL10. The independent variables normalized importance were as follows: PTX3 Total (100%), IL10 (95.9%), FOXP3 (48.9%), MITF (35.8%), and CD163 (6.3%) (Figure 22). This result is compatible with COX. The same procedure was performed to predict the Hans classifier and the importance was IL10 (100%), PTX3 Total (67.1%), FOXP3 (44.8%), CD163 (39.8%), and MITF (32.8%) (Figure 22).

Additional analysis consisted of validation the macrophage markers in an independent series of cases of diffuse large B-cell lymphoma, from the Lymphoma/Leukemia Molecular Profiling Project (LLMPP), the GSE10846 (webpage: https://www.ncbi.nlm.nih.gov/geo/query/acc.cgi?acc=GSE10846, accessed on 21 September 2022). Only the cases treated with R-CHOP-like therapy were selected (n = 233). Several machine learning and artificial neural networks (multilayer perceptron) were used. The dependent (target) variable was the overall survival (outcome dead vs alive). As predictors, the macrophage genes of *CD163, CSF1R, PTX3, CD274 (PD-L1)*, and *IL10* were used. Additional immuno-oncology predictors were markers previously highlighted in the analyses, including *MYC, BCL2, TP53, FOXP3, CSF1, IL34, PDCD1 (PD-1), TNFRSF14, TNFAIP8, IKZF1, STAT3, NFKB1, MYD88, RELA, CASP8, CASP3, PARP1, BCL2, MKI67, ENO3*, and *GGA3*. In total, 25 genes were analyzed and the overall survival was successfully predicted. Table 2 shows the machine learning and neural network models, the number of predictors used in the models, and the overall accuracy. Figure 16 shows the most relevant models and the most relevant genes. The models confirmed the importance of the immuno-oncology markers (Figure 23).

Table 2. Machine learning and artificial neural network analysis using gene expression data.

Model	No. of Predictors	Overall Accuracy (%)
XGBoost Tree	25	100
Random Forest	25	98.3
Random Trees	25	97.1
Bayesian Network	25	89.3
SVM	25	84.5
KNN Algorithm	25	81.9
CHAID	6	79.8
LSVM	25	78.5
Logistic Regression	25	78.1
C5 Tree	3	75.9
Tree-AS	2	74.3
XGBoost Linear	25	74.3
Quest	25	74.3
C&R Tree	25	74.3
Neural Net	25	74.3
Discriminant	25	72.9

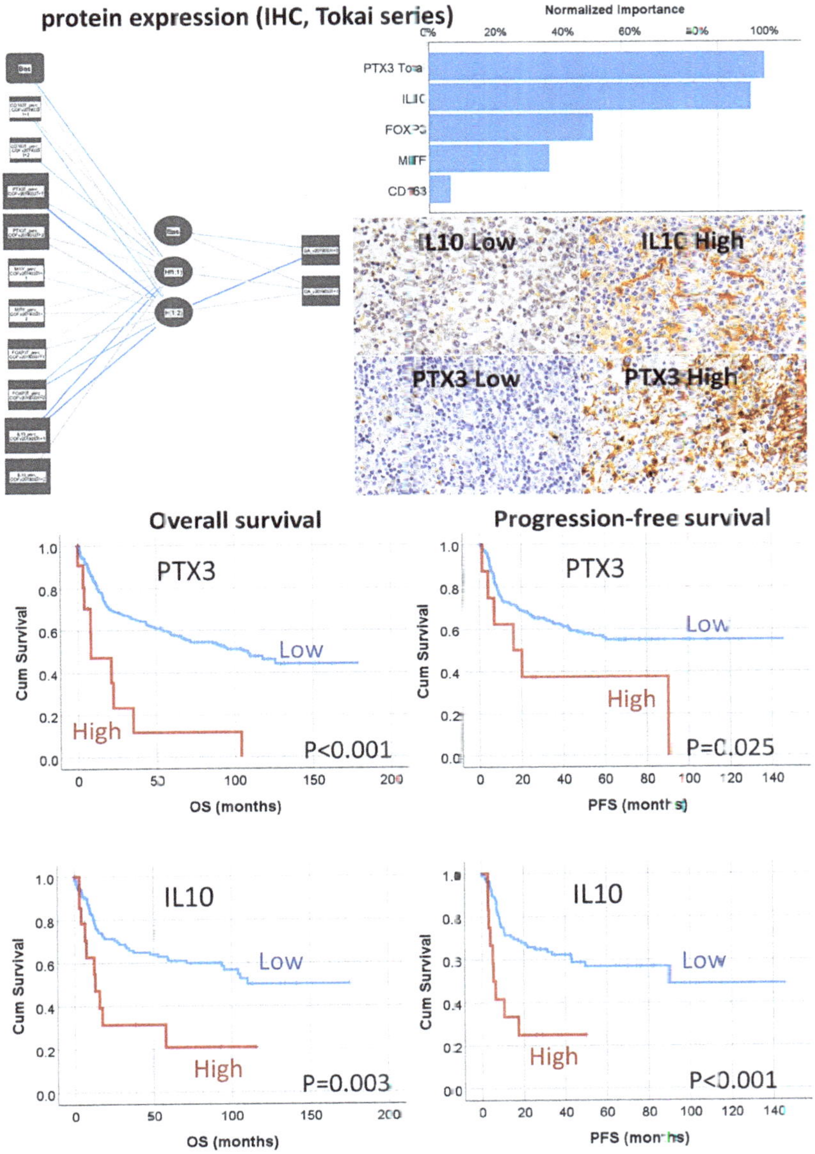

Figure 22. Prediction of the overall survival of diffuse large B-cell lymphoma by M2c-like macrophages using an artificial neural network. The overall survival of the patients was predicted using an artificial neural network using the histochemical data of the tissue samples. The network confirmed that the most relevant markers were PTX3 and IL10, which characterized the immune regulatory M2c-like macrophages. A conventional survival analysis using the Kaplan–Meier with log-rank test confirmed the association of high M2c-like macrophages with poor overall and progression-free survival of the patients. Original magnification: 400×.

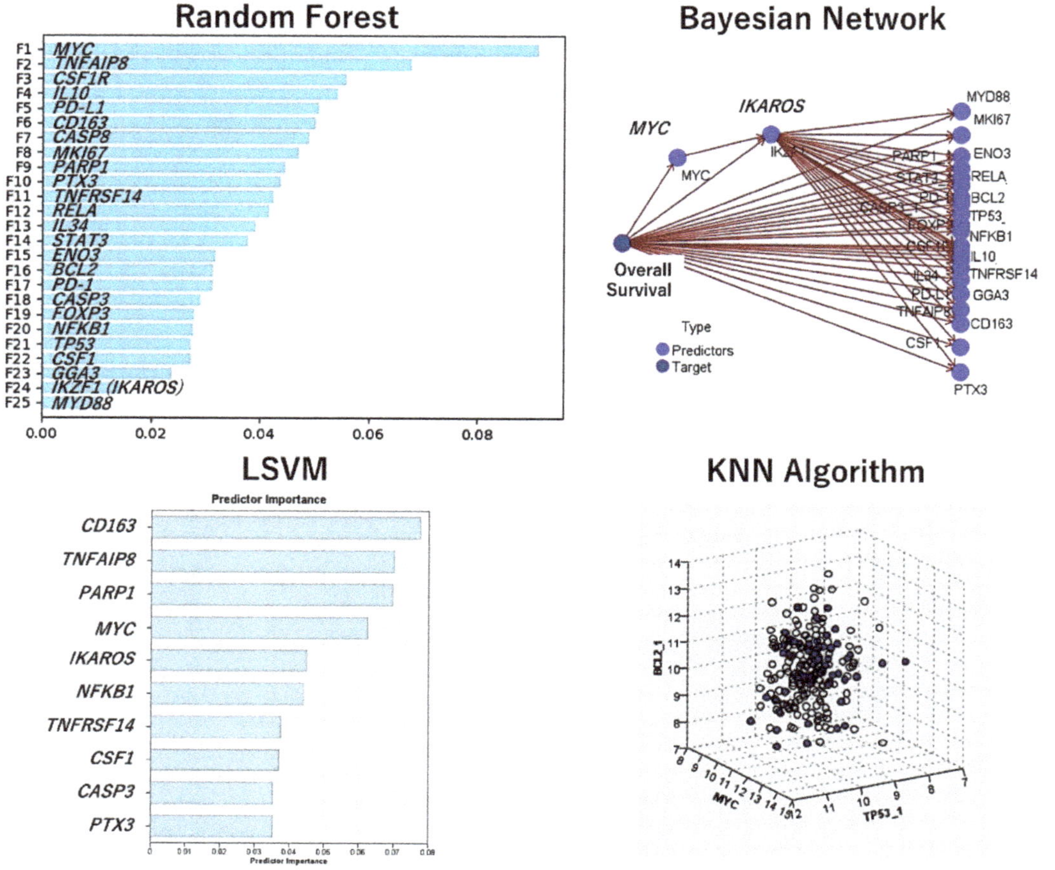

Figure 23. Prediction of the overall survival of diffuse large B-cell lymphoma using immune checkpoint and immuno-oncology markers. Using gene expression data of the GSE10846 dataset, the association of markers of immune regulatory M2c-like tumor-associated macrophages and other immune checkpoint markers was assessed. The methodology included several machine learning and artificial neural networks. The overall accuracy of each method is shown in Table 2.

Using the random forest, the markers were ranked according to their significance for predicting the patients' overall survival. The random forest uses a tree model and a bagging method.

The Bayesian network is a graphical model that shows variables (nodes) in a dataset and the probabilistic, or conditional, independences between them. It constructs a probability model by combining observed and recorded evidence. The network's links (arcs) do not always depict cause and effect.

The LSVM method permits the classification of data using a linear support vector machine. With large datasets, or ones with numerous predictor fields, LSVM is an especially adequate method. In this LSVM analysis, the predictors were ranked in order of relevance.

Nearest Neighbor Analysis classifies the cases based on the resemblance to others and patterns; this chart is a lower-dimensional projection of the predictor space, which contains 25 predictors (genes).

4. Discussion

Artificial intelligence (AI) is a recently developed field that integrates computer science with datasets to perform out calculations. In medicine, both machine learning and deep learning analyze medical data and gain insights on diseases. Artificial intelligence has many applications, including diagnosis, disease classification, image analysis, etc. [20–24].

Machine learning is a specialty in artificial intelligence. By using statistics, algorithms are trained to make classifications or predictions [20–23]. An algorithm of machine learning is composed of three parts:

(1) Decision process. Based on the labeled or unlabeled input data, an estimated pattern is produced by the algorithm.
(2) Error function, which evaluates the prediction of the model.
(3) Model optimization process. During the fitting, the weights are adjusted to reduce discrepancy between the known and the estimates, and weights are updated autonomously until a threshold of accuracy is met.

There are three categories of machine learning models:

(1) Supervised, which use labeled datasets, such as linear regression, logistic regression, random forest, and support vector machine (SVM).
(2) Unsupervised, which use unlabeled datasets and discover hidden patterns or data groupings without the need of human intervention, such as principal component analysis (PCA), singular value decomposition (SVD), and k-means clustering.

A linear regression algorithm is used to predict numerical values based on a linear relationship between predictors. Logistic regression is a type of supervised learning that predicts a categorical variable (binary). The clustering analysis uses unsupervised learning and identifies patterns to group the cases. Decision trees can be used to predict numerical values or to classify the data into categories; they use a branching sequence of link decisions that are represented in a tree diagram. Random forests predict a value or category by combining the results of decision trees [20].Artificial neural networks (ANNs) are algorithms that, in essence, mimic the human brain. Many data mining applications use neural networks because they are flexible and powerful for complex processes [25].

A neural network is composed of an input layer, multiple hidden layers (deep neural network), and an output layer. Most neural networks are feed-forward, which means that the flow moves in one direction from the input to the output [20–24]. The "deep" term refers to the number of layers (inclusive of input, hidden, and output layer); more than three layers can be considered in a deep learning algorithm [21]. The multilayer perceptron (MLP) and radial basis function (RBF) are used in predictive applications, and are supervised because the results can be compared with the known values of the target variables [20–26]. The input layer contains the predictors (for example, the genes). The hidden layer contains unobservable nodes (units). The value of each hidden unit is some function of the predictors. The output layer contains the responses (Figure 2).

This research predicted the prognosis (mainly the overall survival) and classified the different subtypes of mature B-cell neoplasms (non-Hodgkin lymphomas) with high accuracy. Therefore, machine learning and artificial neural networks are useful biostatistical tools in biomedical research, and it is expected that the importance of artificial intelligence in medicine will increase in the future.

This research used basic types of neural networks to obtain reliable results. The single neural networks created the basis for more complex algorithms, making the analysis similar to a classical multivariate analysis. The neural networks were also complemented with other conventional biostatistical analyses, such as gene set enrichment analysis (GSEA) and Cox regression. Additionally, other machine learning techniques were used to complement the results. Each type of machine learning has special uses, and in the results, the information that is provided was complementary.

In the different algorithms, the input data comprised all the genes of the array or specific panels. The panels that were used were carefully selected, and included cancer transcriptome, pan-cancer, cancer progression, and metabolic pathways that incorporate many oncogenes and tumor suppressor genes, but also immune-related panels such as immune exhaustion, human inflammation, host response, autoimmune, and immuno-oncology. Nowadays, immuno-oncology panels are particularly relevant. This research highlighted many important immuno-oncology markers such as CD163, CSF1R, CSF1, PD-L1, IL10, TNFRSF14, TNFAIP8, PD-1, and FOXP3 which are markers of tumor-associated macrophages (TAMs), T lymphocytes, and regulatory T lymphocytes (Tregs). A complete discussion can be found in the previous publications [19,27–35]. Most of these markers can be targeted using inhibitors. In diffuse large B-cell lymphoma, the use of immunomodulatory drugs and immune checkpoint inhibitors is a new and promising field for treating the patients beyond the classical R-CHOP [58] (Table 3).

Table 3. Immuno-oncology and pathway-related markers that were highlighted in this research.

Marker	Target Cell/Pathway	Function/Prognostic Association
FOXP3	Tregs	Immune tolerance and homeostasis of the immune system. High frequency associated with a favorable prognosis of DLBCL.
PD-1	T lymphocytes	Co-inhibition
BTLA	B and T lymphocytes	Co-inhibition
CD163	M2-like TAMs	Pro-tumoral. High frequency is associated with poor prognosis of DLBCL and FL.
CSF1R	M2-like TAMs	Pro-tumoral. High CSF1R + TAMs associated with poor prognosis, but high CSF1R + B-cells of DLBCL with favorable prognosis.
CSF1	B lymphocytes	Ligand of CSF1R
PD-L1	M2c-like TAMs	Pro-tumoral, immune regulatory macrophages (M2c-like). High expression associated with poor prognosis of DLBCL.
SIRPA	M2-like TAMs	Limit phagocytosis
CD47	B lymphocytes	Limit phagocytosis
IL10	M2c-like TAMs	Pro-tumoral, immune regulatory macrophages (M2c-like). High expression associated with poor prognosis of DLBCL and FL.
TNFRSF14	Antigen-presenting cells	Ligand of BTLA, co-inhibitory pathway
IKAROS	Pathway-related	Transcription factor, chromatin remodeling, hemolymphopoietic system. High expression associated with a favorable prognosis of DLBCL.
STAT3	Pathway-related	Cell growth and apoptosis
NFKB1	Pathway-related	Activated by cytokines, oxidant-free radicals, ultraviolet irradiation, and bacterial or viral products. Activated NFKB translocates into the nucleus and stimulates expression multiple genes of wide variety of biological functions.
MAPK	Pathway-related	p44/42 MAPK (Erk1/2) signaling pathway. High expression associated with GCB phenotype of DLBCL (and a favorable prognosis).
TNFAIP8	Pathway-related	Anti-apoptosis. High expression associated with poor prognosis of DLBCL.
BCL2	Pathway-related	Anti-apoptosis
CASP8	Pathway-related	Pro-apoptosis. High expression associated with a favorable prognosis of DLBCL.
CASP3	Pathway-related	Pro-apoptosis
PARP	Pathway-related	Pro-apoptosis
MDM2	Pathway-related	TP53 in inhibitor
E2F1	Pathway-related	Transcription factor, cell cycle, tumor suppressor
CDK6	Pathway-related	Cell cycle
MYB	Germinal center B-cells	Transcriptional transactivator
LMO2	Germinal center B-cells	Hematopoietic development
ENO3	Pathway-related	Glycolysis and glycosaminoglycan metabolism. High expression associated with a poor prognosis of DLBCL.
GGA3	Pathway-related	Positive regulation of protein catabolic processes

Tregs, regulatory T lymphocytes; TAMs, tumor-associated macrophages; DLBCL, diffuse large B-cell lymphoma; FL, follicular lymphoma. Information based on UniProt and GeneCards, and our results.

Interestingly, some of the identified markers were also relevant for the prognosis of nonhematological neoplasia, which suggests that there are common pathogenic mechanisms in all types of neoplasia.

AI analysis combined neural networks such as multilayer perceptron and radial basis function, and several machine learning techniques such as Bayesian network, C&R tree, C5 tree, CHAID tree, discriminant analysis, KNN algorithm, logistic regression, LSVM, Quest tree, random forest, random trees, SVM, tree-AS, XGBoost linear, XGBoost tree. It is impossible to decide which the best technique is because each method has some strengths and weaknesses, and its applicability depends on the type of data, number of cases, and number of variables (inputs).

The term neural network refers to a family of loosely related models that are characterized by large parameter spaces and flexible structures, derived from the study of brain function. Neural networks are the tools of choice in many data mining applications because of their power and flexibility, especially if the underlying process is complex [28].

Artificial neural networks used in prediction applications, such as multilayer perceptron (MLP) and radial basis function (RBF) networks, are supervised in the sense that the results predicted by the model are compared to known values of target variables. The choice between the MLP and RBF methods depends on the type of data and the level of complexity of the problem. The MLP method can find more complex relationships, while RBF is generally faster [30]. Deep neural networks have been criticized for being opaque because their predictions are incomprehensible to humans; their multi-layered nonlinear structure is a "black box model" [31].

We recently modeled celiac disease and ulcerative colitis using AI [59,60]. In the case of ulcerative colitis, we analyzed a series of 43 cases, including 13 healthy controls, 8 inactive ulcerative colitis, 7 non-involved active ulcerative colitis, and 15 involved active ulcerative colitis. As input, 734 genes were included. A total of 16 models were used to predict ulcerative colitis. The overall accuracy was as follows: C5 decision tree (100%, 2 fields used); logistic regression, discriminant analysis, LSVM, SVM, XGBoost linear, XGBoost tree, and neural network (100%, 734 fields); CHAID (97.7%, 2 fields); random forest (97.7%, 734); KNN algorithm (95.4%, 734); C&R tree (95.4%, 12); Quest (83.7%, 6); Bayesian network (65.1%, 734); random trees (0%, 734). In this research, most of the machine learning methods and neural networks had accuracy above 85%. Nevertheless, the number of fields that were used was variable. As also observed in the data of mature B-cell neoplasms, decision trees have difficulties in handling a large set of variables. Bayesian networks provide acceptable results, but are not superior to neural networks. Logistic regression accuracy is usually high and uses many variables. In the end, the most practical strategy is to test all methods and select the ones that predict better. In Table 2, the same 16 models are applied to our data of diffuse large B-cell lymphoma. Generally, the machine learning methods successfully predicted the overall survival of patients with diffuse large B-cell lymphoma using immuno-oncology and immune checkpoint markers. In this particular experiment, neural networks did not have high accuracy.

In conclusion, artificial intelligence analysis is a useful tool for analyzing the prognosis and classification of non-Hodgkin lymphomas.

5. Review of the Literature and Future Perspective in Hematological Neoplasia Using AI

Other groups have also used artificial intelligence in the field of hematopathology research. Table 4 provides precise updates on the latest progress made in hematological malignancies using machine learning and neural networks. The manuscripts were selected in PubMed using the keywords "lymphoma" and "artificial intelligence". Among all articles that were found within the past 3–4 years, a selection of the most recent research was made. Because of limited space, not all relevant manuscripts are included in Table 4.

Table 4. Update on the latest progress made in hematological malignancies using artificial intelligence.

Authors (Year)	Journal	Research Title	Summary	Technique Used	Reference
(1) PET/CT scan-based AI					
Lisson CS et al. (2022)	*Cancers (Basel)*	Deep neural networks and machine learning radiomics modeling for the prediction of relapse in mantle cell lymphoma	This research predicted the relapse of mantle cell lymphoma (MCL) using baseline CT scans. The accuracies of predictions ranged from 64% to 70%.	3D SEResNet50, 3D DenseNet, optimized 3D CNN, K-nearest Neighbor (KNN), and Random Forest (RF)	[61]
Sadik M et al. (2021)	*Sci Rep.*	Artificial intelligence could alert for focal skeleton/bone marrow uptake in patients with Hodgkin's lymphoma staged with FDG-PET/CT	Detection of focal skeleton/bone marrow uptake (BMU) in patients with Hodgkin's lymphoma (HL) undergoing staging with FDG-PET/CT. Training set, $n = 153$; validation set, $n = 48$.	Convolutional neural network (CNN)	[62]
Wang YJ et al. (2021)	*Eur J Nucl Med Mol Imaging*	Artificial intelligence enables whole-body positron emission tomography scans with minimal radiation exposure	Thirty-three diagnostic ^{18}F-FDG PET images of patients with pediatric cancer were generated from ultra-low dose ^{18}F-FDG PET input images using an AI algorithm. Then, the AI-generated PET scans were compared with clinical standard PET scans.	Convolutional neural network (CNN)	[63]
Pinochet P et al. (2021)	*Front Med (Lausanne)*	Evaluation of an automatic classification algorithm using convolutional neural networks in oncological positron emission tomography	This research measured the efficiency and performance in both clinical and research environments of a system called positron emission tomography (PET)-assisted reporting system (PARS) (Siemens Healthineers). The method was based on a convolutional neural network (CNN) that identified suspected cancer sites in fluorine-18 fluorodeoxyglucose (18F-FDG) PET/computed tomography. These data were correlated with the survival of the patients. Two cohorts were evaluated: 119 cases of DLBCL, and 430 cases of DLBCL and other tumors.	Dice score	[64]
(2) Histological images-based AI					
El Hussein S et al. (2022)	*J Pathol.*	Artificial intelligence strategy integrating morphologic and architectural biomarkers provide robust diagnostic accuracy for disease progression in chronic lymphocytic leukemia	Cytologic and architectural features obtained from whole-slides images were used to classify 125 samples into three subtypes: chronic lymphocytic leukemia (CLL, $n = 69$), progression to accelerated CLL (aCLL, $n = 44$), and transformation to diffuse large B-cell lymphoma (Richter transformation; RT, $n = 80$).	Hover-Net	[65]
Swiderska-Chadaj Z et al. (2021)	*Virchows Arch.*	Artificial intelligence to detect *MYC* translocation in slides of diffuse large B-cell lymphoma	The H&E slides of 287 cases were evaluated using a deep learning algorithm to identify *MYC* rearrangement by DNA in situ hybridization (FISH).	Deep learning neural network (U-Net) and classical machine learning (random forest classification)	[66]

Table 4. *Cont.*

Authors (Year)	Journal	Research Title	Summary	Technique Used	Reference
Steinbuss G et al. (2021)	*Cancers (Basel)*	Deep learning for the classification of non-Hodgkin lymphoma on histopathological images	In this research, the training set included 84,139 image patches from 629 patients that were classified as reactive lymph nodes, nodal small lymphocytic lymphoma/chronic lymphocytic leukemia, and nodal diffuse large B-cell lymphoma. The validation set included 16,960 image patches from 125 patients. The final model had an accuracy of 96%.	EfficientNet convolutional neuronal network (CNN)	[67]
Zhang X et al. (2021)	*Technol Health Care*	Research on the classification of lymphoma pathological images-based on deep residual neural networks	The analysis used 374 pathological images, including chronic lymphocytic leukemia, follicular lymphoma, and mantle cell lymphoma.	BP neural network and BP neural network optimized by genetic algorithm (GA-BP), deep residual neural network model (ResNet50), softmax layer	[68]
Tang G et al. (2021)	*Acta Cytol.*	A machine learning tool using digital microscopy (Morphogo) for the identification of abnormal lymphocytes in the bone marrow	Morphological differentiation of abnormal lymphocytes in bone marrow was evaluated in 53 cases of different subtypes of B-cell lymphomas, using automated digital images.	"Morphogo" system	[69]
Yu WH et al. (2021)	*Cancers (Basel)*	Machine learning based on morphological features enables the classification of primary intestinal T-cell lymphomas.	A total of 40 primary intestinal T-cell lymphomas (PITL), including 26 monomorphic epitheliotropic intestinal T-cell lymphoma (MEITL), 10 intestinal T-cell lymphoma, not otherwise specified (ITCL-NOS), and 4 borderline cases were analyzed. The inputs were the morphological features and the immunophenotypes (CD8 and CD56).	XGBoost and CNN (HTC-RCNN with ResNet50)	[70]
Zhou M et al. (2021)	*Front Pediatr.*	Development and evaluation of a leukemia diagnosis system using deep learning in real clinical scenarios	A total of 1732 bone marrow, raw images of 89 children with leukemia were analyzed with convolutional neural networks, with a performance accuracy of 89%. Apart from detecting leukocytes, the system also detected bone marrow metastasis of lymphoma and neuroblastomas.	RetinaNet, VGG, Feature Pyramid Network, ResNet, convolutional neural network (CNN)	[71]
Zhang J et al. (2020)	*Med Phys.*	Classification of digital pathological images of non-Hodgkin's lymphoma subtypes based on the fusion of transfer learning and principal component analysis	Digital pathology images of non-Hodgkin lymphoma, including chronic lymphocytic leukemia (CLL), follicular lymphoma (FL), and mantle cell lymphoma (MCL) tumor were analyzed and classified. The model had an overall accuracy of 98.9%.	Transfer learning (TL) and principal component analysis (PCA)	[72]
Mohlman JS et al. (2020)	*Am J Clin Pathol*	Improving augmented human intelligence to distinguish Burkitt lymphoma from diffuse large B-cell lymphoma cases	A total of 10,818 H&E images from 34 cases of Burkitt lymphoma and 36 cases of diffuse large B-cell lymphoma were used to train and differentiate the two lymphoma subtypes.	Convolutional neural network (CNN)	[73]

Table 4. *Cont.*

Authors (Year)	Journal	Research Title	Summary	Technique Used	Reference
Li D et al. (2020)	*Nat Commun.*	A deep learning diagnostic platform for diffuse large B-cell lymphoma with high accuracy across multiple hospitals	This research used histological images of H&E to classify diffuse large B-cell lymphoma (DLBCL) vs non-DLBCL. Non-DLBCL included metastatic carcinoma, melanoma, and other lymphomas including small lymphocytic lymphoma/chronic lymphocytic leukemia, mantle cell lymphoma, follicular lymphoma, and classical Hodgkin lymphoma. The GOTDP-MP-CNNs (with combined 17 CNNs) model had an accuracy of 99.7% to 100%.	17 types of CNN: AlexNet, GoogLeNet (ImageNet), GoogLeNet (Places365), ResNet18, ResNet50, ResNet101, Vgg16, Vgg19, Inceptionv3, InceptionResNetv2, SqueezeNet, DenseNet201, MobileNetv2, ShuffleNet, Xception, NasNetmobile, Nasnetlarge	[74]
Miyoshi H et al. (2020)	*Lab Invest.*	Deep learning shows the capability of high-level computer-aided diagnosis of malignant lymphoma.	The H&E images of 388 cases, including 259 with diffuse large B-cell lymphoma, 89 with follicular lymphoma, and 40 with reactive lymphoid hyperplasia, were analyzed using deep learning. The accuracy of the model was 97%.	Convolutional neural network (CNN)	[75]
Zorman M et al. (2011)	*Wien Klin Wochenschr.*	Classification of follicular lymphoma images: a holistic approach with symbol-based machine learning methods.	Analysis of follicular lymphoma images, focusing on the identification of follicles.	Decision trees (MtDeciT 3.1, RSES 2.2, and Weka 3) and artificial neural networks (multilayer perceptron)	[76]
(3) Immunophenotype-based AI					
Zhao M et al. (2020)	*Cytometry A.*	Hematologist-level classification of mature B-cell neoplasms using deep learning on multiparameter flow cytometry data	Information captured by multiparameter flow cytometry (MFC) of 18,274 cases, including chronic lymphocytic leukemia and its precursor monoclonal B-cell lymphocytosis, marginal zone lymphoma, mantle cell lymphoma, prolymphocytic leukemia, follicular lymphoma, hairy cell leukemia, lymphoplasmacytic lymphoma were analyzed; the model was tested on a set of 2346 cases. The model performance had an F1 score of 0.94.	Self-organizing maps and convolutional neural networks	[77]
Gaidano V et al. (2020)	*Cancers (Basel)*	A clinically applicable approach to the classification of B-cell non-Hodgkin lymphomas with flow cytometry and machine learning	The immunophenotype data from flow cytometry of 1465 B-cell non-Hodgkin lymphoma (NHL) cases were analyzed. The cases included chronic lymphocytic leukemia (CLL), diffuse large B-cell lymphoma (DLBCL), Burkitt lymphoma (BL), follicular cell lymphoma (FCL), hairy cell leukemia (HCL), splenic lymphoma (SL), mantle cell lymphoma (MCL), marginal zone lymphoma (MZL), and lymphoplasmacytic lymphoma (LPL). The accuracy of the classification ranged from 92% to 100%.	Classification trees	[78]

Table 4. *Cont.*

Authors (Year)	Journal	Research Title	Summary	Technique Used	Reference
(4) Clinicopathological variables-based AI					
Zhan M et al. (2021)	*Leuk Lymphoma*	Machine learning to predict high-dose methotrexate-related neutropenia and fever in children with B-cell acute lymphoblastic leukemia	A model included 57 SNPs of 16 genes and clinical variables to predict neutropenia and fever in 139 pediatric cases of acute lymphoblastic leukemia treated with high-dose methotrexate (MTX).	Random forest	[79]
Buciński A et al. (2010)	*Eur J Cancer Prev.*	Contribution of artificial intelligence to the knowledge of prognostic factors in Hodgkin's lymphoma	A total of 31 variables from 114 patients with Hodgkin's lymphoma were used to predict the prognosis of the patients.	Artificial neural network (ANN)	[80]
(5) Gene expression, mutational, and integrative analysis-based AI					
Carreras J et al. (2022)	*Healthcare (Basel)*	Artificial intelligence analysis of gene expression predicted the overall survival of mantle cell lymphoma and a large pan-cancer Series	The gene expression data of 123 cases of mantle cell lymphoma (MCL) were analyzed with artificial neural networks to predict the overall survival of the patients with high accuracy. The survival of diffuse large B-cell lymphoma (DLBCL), and a pan-cancer series was also predicted.	Several machine learning techniques, and artificial neural networks	[34]
Carreras J et al. (2021)	*Cancers (Basel)*	Artificial neural networks predicted the overall survival and molecular subtypes of diffuse large B-cell lymphoma using a pan-cancer immune-oncology panel	The gene expression of an immuno-oncology panel of a series of 106 cases of diffuse large B-cell lymphoma was analyzed using artificial intelligence to predict the overall survival and the cell of origin molecular subtypes. The model had a high accuracy of classification.	Several machine learning techniques, and artificial neural networks	[33]
Carreras J et al. (2021)	*Tokai J Exp Clin Med.*	Artificial intelligence analysis of gene expression data predicted the prognosis of patients with diffuse large B-cell lymphoma	The gene expression of a series of 414 cases of diffuse large B-cell lymphoma (DLBCL) was analyzed to predict the overall survival, and was correlated with other known pathogenic genes such as *BCL2* and *MYC*.	Artificial neural networks (ANN)	[27]
Xu-Monette ZY et al. (2020)	*Blood Adv.*	A refined cell of origin classifier with targeted NGS and artificial intelligence showed robust predictive value in DLBCL	The series of diffuse large B cell lymphoma of 418 cases included immunohistochemical, gene expression, DNA in situ hybridization, array CGH, and NGS sequencing. Using an autoencoder, the cases were classified according to the cell of origin and the survival (overall survival and progression-free survival).	Autoencoder, logistic regression, and CPH models	[81]
Zhang W et al. (2020)	*BMC Cancer*	Novel bioinformatic classification system for genetic signature identification in diffuse large B-cell lymphoma	A total of 342 cases of diffuse large B-cell lymphoma were analyzed using mutational data from a panel of 46 genes by NGS.	Random forest	[82]
Parodi S et al. (2018)	*Health Informatics J.*	Logic learning machine and standard supervised methods for Hodgkin's lymphoma prognosis using gene expression data and clinical variables	The data of 130 patients diagnosed with Hodgkin's lymphoma, including a small set of clinical variables and more than 54,000 gene features, were used to predict the prognosis.	K-nearest neighbor (KNN), artificial neural network (ANN), support vector machine (SVM), decision tree, and the innovative logic learning machine method	[83]

Table 4. *Cont.*

Authors (Year)	Journal	Research Title	Summary	Technique Used	Reference
Schmitz R et al. (2018)	*N Engl J Med.*	Genetics and pathogenesis of diffuse large B-cell lymphoma	The data of 574 diffuse large B-cell lymphoma cases, which included exome and transcriptome sequencing, array-based DNA copy-number analysis, and targeted amplicon resequencing of 372 genes, were used to identify genetic subtypes.	Random forest	[84]

H&E, hematoxylin and eosin. The publications were selected from PubMed using the keywords "artificial intelligence" and "lymphoma".

The manuscripts were organized according to the type of input data, i.e., PET/CT scan, histological images, immunophenotype, clinicopathological variables, and gene expression, mutational, and integrative analysis-based artificial intelligence [61–84].

Worth mentioning is the work of Schmitz R et al. published in the *New England Journal of Medicine* in 2018. The genetics and pathogenesis of diffuse large B-cell lymphoma were analyzed using random forest. The input data from 574 diffuse large B-cell lymphoma cases included exome and transcriptome sequencing, whole-genome copy-number array-based DNA analysis, and targeted amplicon resequencing of 372 genes to identify genetic subtypes [84].

A similar work was published by Xu-Monette ZY et al. in 2020 in *Blood Advances*. Based on targeted next-generation sequencing (NGS), a correlation with the cell of origin subtypes was made using AI in diffuse large B-cell lymphoma. The series of 418 cases included immunohistochemical, gene expression, DNA in situ hybridization, array CGH, and NGS sequencing. Using autoencoders and CPH models, the cases were classified according to the cell of origin and the patients' survival (overall survival and progression-free survival) [81].

Li D et al. reported in 2020 in *Nature Communications* a deep learning diagnostic platform for diffuse large B-cell lymphoma. The method included data from multiple hospitals. This research used histological images of H&E to classify diffuse large B-cell lymphoma (DLBCL) vs non-DLBCL. Non-DLBCL included cases of metastatic carcinoma, melanoma, and other lymphomas. The lymphoma subtypes were chronic lymphocytic leukemia, mantle cell lymphoma, follicular lymphoma, and classical Hodgkin lymphoma. Seventeen types of convolutional neural networks were used, and the model had an accuracy of 99.7–100% [74].

In the past five years, there has been a significant increase in the use of artificial intelligence in cancer research, and many applications in hematological neoplasia have been published [85]. Many studies have used convolutional neural networks to classify digitalized histological images. Machine learning and artificial neural networks have also been used to analyze gene expression and mutational data. It is expected that in the future, artificial intelligence techniques will become a standard part of the biostatistical analysis, and complementary to "conventional" bioinformatics.

Supplementary Materials: The following supporting information can be downloaded at: https://www.mdpi.com/article/10.3390/cancers14215318/s1, Table S1: Multilayer perceptron analysis (MLP). Table S2: Radial basis function analysis (RBF). Table S3: Genes associated to poor prognosis in the multivariate Cox survival analysis. Table S4: Genes associated to good prognosis in the multivariate Cox survival analysis. Table S5: Clinicopathological correlations with the final set of seven prognostic genes.

Author Contributions: Conceptualization, methodology, formal analysis, investigation, and writing, J.C. Resources, G.R. Resources, software, validation, R.H. All authors have read and agreed to the published version of the manuscript.

Funding: Joaquim Carreras was funded by the Ministry of Education, Culture, Sports, Science and Technology (MEXT) and the Japan Society for the Promotion of Science (JSPS) (grant numbers KAKEN 15K19061, 18K15100, and 24590430) and the Tokai University School of Medicine research incentive assistant plan (grant number 2021-B04). Rifat Hamoudi was funded by Al-Jalila Foundation (grant number AJF201090) and University of Sharjah (grant number 22010902103).

Institutional Review Board Statement: The study was conducted in accordance with the Declaration of Helsinki, and approved by the Institutional Review Board of Tokai University, School of Medicine (protocol code IRB14R-080, and IRB20-156).

Informed Consent Statement: Informed consent was obtained from all subjects involved in the study.

Data Availability Statement: All the data are available upon request to Joaquim Carreras.

Acknowledgments: The authors thank all the colleagues who had previously contributed to the research.

Conflicts of Interest: The authors declare no conflict of interest.

Appendix A

The analyses used several software applications, including EditPad Lite (version 8.4.0 x64, Just Great Software Co. Ltd.); Fiji (version ImageJ 1.53u, NIH); GSEA (version 4.3.2, Broad Institute); GIMP (version 2.10.8, GNU); IBM SPSS 25 to 27; IBM modeler 18 (IBM); JMP Pro 14 (JMP Statistical Discovery LLC, SAS); Microsoft excel 2016 (version 16.0.5317.1000, Microsoft Corporation); Minitab (version 21.1.0, Minitab, LLC); Morpheus matrix visualization and analysis software (version 1, https://github.com/cmap/morpheus.js, Broad Institute) (accessed date 25 October 2022); NSolver (version 4.0, NanoString); RapidMiner Studio (version 9.10.011, RapidMiner); R (version 4.2.1) (http://cran.r-project.org) (accessed date 25 October 2022); RStudio (version 2022.07.2, Build 576, RStudio, PBC); STRING protein–protein interaction networks (version 11.5, STRING Consortium 2022); and Xlstat (Premium 2018.1, Build 49320 x64, multilingual, Addinsoft).

Appendix B

Table A1. Publicly available datasets used in addition to the Tokai University series.

Diagnosis	Dataset	No. of Cases	Reference
Non-Hodgkin lymphomas	GSE132929	290	[40]
Follicular lymphoma		65	
Mantle cell lymphoma		43	
Diffuse large B-cell lymphoma		100	
Burkitt lymphoma		59	
Marginal zone lymphoma		23	
Chronic lymphocytic leukemia	GSE22762	107	[41,42]
	ICGC CLLE-ES	201	
Diffuse large B-cell lymphoma	GSE10846	414	[43,44]
	GSE23501	69	[45]
	GSE4475	159	[46,47]
	TCGA-DLBCL v.2016	47	
Follicular lymphoma	E-TABM-346	52	[48]
	GSE16131	180	[49]
Mantle cell lymphoma	LLMPP Rosenwald 2003	92	[50]
	GSE93291	123	[51]
Multiple myeloma	GSE2658	559	[52–57]
Acute Myeloid Leukemia	TCGA-AML v.2016	149	

Appendix C. Comments and Analysis Of breast Cancer Detection Using Deep Neural Networks

Breast cancer is the second most frequent type of cancer in women, just before skin cancer. Worldwide, breast cancer represents the 30% of all female cancers, and it has a mortality of about 15%, but in emergent countries can reach up to 70% [86,87]. The worldwide incidence ranges from 27 to 97 cases for 100,000 [87], and in about 10% of the cases, there is a genetic predisposition or family history [87]. The most frequently associated germline mutations affect the *BRCA1* and *BRCA2* genes [88,89].

The development of strategies for the early detection of breast cancer is necessary to improve access to treatment and reduce the mortality rate. As described by Basurto-Hurtado JA et al. [90], breast cancer detection includes four steps: image acquisition, segmentation and pre-processing, feature extraction, and classification [90].

Image acquisition can be obtained through several methods, such as mammography, ultrasound, magnetic resonance imaging (MRI), and other approaches, including microwave, computed tomography (TC), and positron emission tomography (PET) [90].

The image processing and classification strategies include several steps: region of interest (ROI) estimation, and feature extraction. The classifiers can be both unsupervised and supervised. Examples of unsupervised classifiers include K-means and hierarchical clustering. Examples of supervised classifiers are decision trees, random forests, AdaBoost, support vector machines, artificial neural networks, and convolutional neural networks [90]. Recently, new image generation techniques have developed, such as infrared thermography (IRT). This technique has been successfully applied to breast cancer; the classification methods included several machine learning and artificial neural networks, and the accuracy ranged from 90% to 100% [90–101].

Recently, new classification algorithms have been developed, including autoencoders, deep belief networks, ladder networks, and deep neural network (DNN)-based algorithms such as the deep Kronecker neural network [90,102].

Gene expression profiling is a useful tool in medical research, both for diagnosis and for the elucidation of the disease pathogenesis. Artificial neural networks can handle gene expression profiling data successfully, and we recently described their usability in hematological neoplasia [27–35]. In our research, we used conventional machine learning techniques and artificial neural networks because the aim was to identify prognostic factors in a reliable and systematic manner instead of developing new advanced mathematical algorithms. Nevertheless, the performance of the artificial neural networks can be improved with the use of adaptive activation functions (AAFs). Kronecker neural networks (KNNs) are a new type of neural network with adaptive activation functions described by Jagtap AD et al. [103]. Unlike the traditional neural network architecture, in a KNN, the output of the neuron passes to more than one activation function [103]. The use of the Kronecker product in the KNN made the network wide, while at the same time, the number of trainable parameters remained low [103]. Recently, a multi-level KNN approach was used in the analysis of MRI images of brain tumors (glioma) to develop an automated glioma segmentation system [104].

The research in this manuscript focuses on immuno-oncology markers, as we have recently described [85]. In relation to breast cancer, we tested the prognostic value of a set of 718 genes from a pan-cancer immune profiling panel on the overall survival of the patients. A series of 1215 breast cancer patients from The Cancer Genome Atlas (TCGA) was selected. Unfortunately, in this model, a multilayer perceptron analysis failed to properly predict the overall survival of the patients (83.7% overall percent of correct classification, AUC = 0.61). Next, the input was narrowed to 16 genes: macrophage markers (*CD68*, *CSF1R*, *CD163*, *CSF1R*, *CSF1*, *IL10*, *CD274 (PD-L1)*, and *TNFAIP8*), T helper cells (*PDCD1/PD-1*), Tregs (*FOXP3*), apoptosis (*BCL2*, *CASP3*, and *CASP8*), NFKB pathway (*STAT3*), and metabolism (*ENO3*, *GGA3*). The overall survival of breast cancer was predicted using 16 models, namely C5, logistic regression, Bayesian network, discriminant analysis, KNN algorithm, LSVM, random trees, SVM, tree-AS, XGBoost linear, XGBoost tree, CHAID, Quest, C&R tree, random forest, and neural network (multilayer perceptron). Among all models, only random forest provided suitable modeling (input = 16 fields, overall accuracy 98.4%). The order of predictor importance was *CD274*, *FOXP3*, *ENO3*, *IL10*, *CSF1R*, *CSF1*, *BCL2*, *GGA3*, *TNFAIP8*, *CASP8*, *PDCD1*, *CASP3*, *CD163*, *TNFRSF14*, *CD68*, and *STAT3*.

Noteworthy, further analysis was performed in the breast series of the TCGA and the pan-cancer immune profiling panel. In addition to the overall survival, other survival variables were tested, including the disease-specific survival, disease-free interval, and progression-free interval. The multilayer perceptron analysis also failed to predict the survival of the patients with good performance. Additional analyses were performed. Different types of training were tested: batch, online, and mini-batch. Two types of optimization algorithms were also tested: scaled conjugate gradient, and gradient descent. The training options for the scaled conjugate gradient were the following: initial lambda (0.0000005), initial sigma (0.00005), interval center (0), and interval offset (±0.5). The training options for the gradient descent were initial learning rate (0.4), momentum (0.9), interval center (0), and the interval offset (±0.5). Of note, batch training can use both a scaled conjugate gradient and gradient descent. However, online and mini-batch are restricted to gradient descent. The training options of gradient descent in case of online and mini-batch were initial learning rate (0.4), lower boundary of learning rate (0.001), learning rate reduction, in epochs (10), momentum (0.9), interval center (0), and interval offset (±0.5). We tried improving the network performance by changing all the training parameters, but no significant improvement in performance was achieved.

References

1. Harris, N.L.; Jaffe, E.S.; Diebold, J.; Flandrin, G.; Muller-Hermelink, H.K.; Vardiman, J. Lymphoma classification—From controversy to consensus: The R.E.A.L. and WHO Classification of lymphoid neoplasms. *Ann. Oncol.* **2000**, *11* (Suppl. 1), 3–10. [CrossRef] [PubMed]
2. Campo, E.; Swerdlow, S.H.; Harris, N.L.; Pileri, S.; Stein, H.; Jaffe, E.S. The 2008 WHO classification of lymphoid neoplasms and beyond: Evolving concepts and practical applications. *Blood* **2011**, *117*, 5019–5032. [CrossRef] [PubMed]
3. Swerdlow, S.H.; Campo, E.; Pileri, S.A.; Harris, N.L.; Stein, H.; Siebert, R.; Advani, R.; Ghielmini, M.; Salles, G.A.; Zelenetz, A.D.; et al. The 2016 revision of the World Health Organization classification of lymphoid neoplasms. *Blood* **2016**, *127*, 2375–2390. [CrossRef] [PubMed]
4. Jaffe, E.S. Diagnosis and classification of lymphoma: Impact of technical advances. *Semin. Hematol.* **2019**, *56*, 30–36. [CrossRef]
5. De Leval, L.; Jaffe, E.S. Lymphoma Classification. *Cancer J.* **2020**, *26*, 176–185. [CrossRef] [PubMed]
6. Campo, E.; Jaffe, E.S.; Cook, J.R.; Quintanilla-Martinez, L.; Swerdlow, S.H.; Anderson, K.C.; Brousset, P.; Cerroni, L.; de Leval, L.; Dirnhofer, S.; et al. The International Consensus Classification of Mature Lymphoid Neoplasms: A Report from the Clinical Advisory Committee. *Blood* **2022**, *140*, 1229–1250. [CrossRef]
7. Alaggio, R.; Amador, C.; Anagnostopoulos, I.; Attygalle, A.D.; Araujo, I.B.O.; Berti, E.; Bhagat, G.; Borges, A.M.; Boyer, D.; Calaminici, M.; et al. The 5th edition of the World Health Organization Classification of Haematolymphoid Tumours: Lymphoid Neoplasms. *Leukemia* **2022**, *36*, 1720–1748. [CrossRef]
8. Chiorazzi, N.; Rai, K.R.; Ferrarini, M. Chronic lymphocytic leukemia. *N. Engl. J. Med.* **2005**, *352*, 804–815. [CrossRef]
9. Mozas, P.; Sorigué, M.; López-Guillermo, A. Follicular lymphoma: An update on diagnosis, prognosis, and management. *Med. Clin.* **2021**, *157*, 440–448. [CrossRef]
10. Randall, C.; Fedoriw, Y. Pathology and diagnosis of follicular lymphoma and related entities. *Pathology* **2020**, *52*, 30–39. [CrossRef]
11. Donzel, M.; Baseggio, L.; Fontaine, J.; Pesce, F.; Ghesquières, H.; Bachy, E.; Verney, A.; Traverse-Glehen, A. New Insights into the Biology and Diagnosis of Splenic Marginal Zone Lymphomas. *Curr. Oncol.* **2021**, *28*, 50297. [CrossRef] [PubMed]
12. Vilarrasa-Blasi, R.; Verdaguer-Dot, N.; Belver, L.; Soler-Vila, P.; Beekman, R.; Chapaprieta, V.; Kulis, M.; Queirós, A.C.; Parra, M.; Calasanz, M.J.; et al. Insights into the mechanisms underlying aberrant SOX11 oncogene expression in mantle cell lymphoma. *Leukemia* **2022**, *36*, 583–587. [CrossRef] [PubMed]
13. Navarro, A.; Beà, S.; Jares, P.; Campo, E. Molecular Pathogenesis of Mantle Cell Lymphoma. *Hematol. Oncol. Clin. N. Am.* **2020**, *34*, 795–807. [CrossRef] [PubMed]
14. Scott, D.W.; Mottok, A.; Ennishi, D.; Wright, G.W.; Farinha, P.; Ben-Neriah, S.; Kridel, R.; Barry, G.S.; Hother, C.; Abrisqueta, P.; et al. Prognostic Significance of Diffuse Large B-Cell Lymphoma Cell of Origin Determined by Digital Gene Expression in Formalin-Fixed Paraffin-Embedded Tissue Biopsies. *J. Clin. Oncol.* **2015**, *33*, 2848–2856. [CrossRef] [PubMed]
15. Robetorye, R.S.; Ramsower, C.A.; Rosenthal, A.C.; Yip, T.K.; Wendel Spiczka, A.J.; Glinsmann-Gibson, B.J.; Rimsza, L.M. Incorporation of Digital Gene Expression Profiling for Cell-of-Origin Determination (Lymph2Cx Testing) into the Routine Work-Up of Diffuse Large B-Cell Lymphoma. *J. Hematop.* **2019**, *12*, 3–10. [CrossRef]
16. Ferry, J.A. Burkitt's lymphoma: Clinicopathologic features and differential diagnosis. *Oncologist* **2006**, *11*, 375–383. [CrossRef]
17. Molyneux, E.M.; Rochford, R.; Griffin, B.; Newton, R.; Jackson, G.; Menon, G.; Harrison, C.J.; Israels, T.; Bailey, S. Burkitt's lymphoma. *Lancet* **2012**, *379*, 1234–1244. [CrossRef]
18. Dunleavy, K.; Little, R.F.; Wilson, W.H. Update on Burkitt Lymphoma. *Hematol. Oncol. Clin. N. Am.* **2016**, *30*, 1333–1343. [CrossRef] [PubMed]
19. Carreras, J.; Hamoudi, R. Artificial Neural Network Analysis of Gene Expression Data Predicted Non-Hodgkin Lymphoma Subtypes with High Accuracy. *Mach. Learn. Knowl. Extr.* **2021**, *3*, 36. [CrossRef]
20. IBM Cloud Education. IBM Cloud Learn Hub. Machine Learning. July, 2020. Available online: https://www.ibm.com/cloud/learn/machine-learning (accessed on 22 July 2022).
21. Kavlakoglu, E. AI vs. Machine Learning vs. Deep Learning vs. Neural Networks: What's the Difference? Available online: https://www.ibm.com/cloud/blog/ai-vs-machine-learning-vs-deep-learning-vs-neural-networks (accessed on 22 July 2022).
22. IBM Cloud Education. Deep Learning. 1 May 2020. Available online: https://www.ibm.com/cloud/learn/deep-learning (accessed on 22 July 2022).
23. Delua, J. Supervised vs. Unsupervised Learning: What's the Difference? IBM Analytics. 12 March 2021. Available online: https://www.ibm.com/cloud/blog/supervised-vs-unsupervised-learning (accessed on 22 July 2022).
24. Blais, A.; Mertz, D. An Introduction to Neural Networks. Pattern Learning with the Back-Propagation Algorithm. 19 August 2018. Available online: https://developer.ibm.com/articles/l-neural/ (accessed on 22 July 2022).
25. IBM Corporation. Introduction to Neural Networks. 28 February 2021. Available online: https://www.ibm.com/docs/en/spss-statistics/27.0.0?topic=networks-introduction-neural (accessed on 22 July 2022).
26. IBM Corporation. *IBM SPSS Neural Networks V27*; IBM Corporation: Armonk, NY, USA, 2020; pp. 10504–11785.

27. Carreras, J.; Hamoudi, R.; Nakamura, N. Artificial Intelligence Analysis of Gene Expression Data Predicted the Prognosis of Patients with Diffuse Large B-Cell Lymphoma. *Tokai J. Exp. Clin. Med.* **2020**, *45*, 37–48.
28. Carreras, J.; Kikuti, Y.Y.; Miyaoka, M.; Hiraiwa, S.; Tomita, S.; Ikoma, H.; Kondo, Y.; Ito, A.; Shiraiwa, S.; Hamoudi, R.; et al. A Single Gene Expression Set Derived from Artificial Intelligence Predicted the Prognosis of Several Lymphoma Subtypes; and High Immunohistochemical Expression of TNFAIP8 Associated with Poor Prognosis in Diffuse Large B-Cell Lymphoma. *AI* **2020**, *1*, 23. [CrossRef]
29. Carreras, J.; Kikuti, Y.Y.; Miyaoka, M.; Hiraiwa, S.; Tomita, S.; Ikoma, H.; Kondo, Y.; Ito, A.; Nakamura, N.; Hamoudi, R. Artificial Intelligence Analysis of the Gene Expression of Follicular Lymphoma Predicted the Overall Survival and Correlated with the Immune Microenvironment Response Signatures. *Mach. Learn. Knowl. Extr.* **2020**, *2*, 35. [CrossRef]
30. Carreras, J.; Kikuti, Y.Y.; Miyaoka, M.; Hiraiwa, S.; Tomita, S.; Ikoma, H.; Kondo, Y.; Ito, A.; Nakamura, N.; Hamoudi, R. A Combination of Multilayer Perceptron, Radial Basis Function Artificial Neural Networks and Machine Learning Image Segmentation for the Dimension Reduction and the Prognosis Assessment of Diffuse Large B-Cell Lymphoma. *AI* **2021**, *2*, 8. [CrossRef]
31. Carreras, J.; Kikuti, Y.Y.; Roncador, G.; Miyaoka, M.; Hiraiwa, S.; Tomita, S.; Ikoma, H.; Kondo, Y.; Ito, A.; Shiraiwa, S.; et al. High Expression of Caspase-8 Associated with Improved Survival in Diffuse Large B-Cell Lymphoma: Machine Learning and Artificial Neural Networks Analyses. *BioMedInformatics* **2021**, *1*, 3. [CrossRef]
32. Carreras, J.; Kikuti, Y.Y.; Miyaoka, M.; Roncador, G.; Garcia, J.F.; Hiraiwa, S.; Tomita, S.; Ikoma, H.; Kondo, Y.; Ito, A.; et al. Integrative Statistics, Machine Learning and Artificial Intelligence Neural Network Analysis Correlated CSF1R with the Prognosis of Diffuse Large B-Cell Lymphoma. *Hemato* **2021**, *2*, 11. [CrossRef]
33. Carreras, J.; Hiraiwa, S.; Kikuti, Y.Y.; Miyaoka, M.; Tomita, S.; Ikoma, H.; Ito, A.; Kondo, Y.; Roncador, G.; Garcia, J.F.; et al. Artificial Neural Networks Predicted the Overall Survival and Molecular Subtypes of Diffuse Large B-Cell Lymphoma Using a Pancancer Immune-Oncology Panel. *Cancers* **2021**, *13*, 6384. [CrossRef] [PubMed]
34. Carreras, J.; Nakamura, N.; Hamoudi, R. Artificial Intelligence Analysis of Gene Expression Predicted the Overall Survival of Mantle Cell Lymphoma and a Large Pan-Cancer Series. *Healthcare* **2022**, *10*, 155. [CrossRef]
35. Carreras, J.; Kikuti, Y.Y.; Miyaoka, M.; Hiraiwa, S.; Tomita, S.; Ikoma, H.; Kondo, Y.; Ito, A.; Hamoudi, R.; Nakamura, N. The Use of the Random Number Generator and Artificial Intelligence Analysis for Dimensionality Reduction of Follicular Lymphoma Transcriptomic Data. *BioMedInformatics* **2022**, *2*, 17. [CrossRef]
36. Subramanian, A.; Tamayo, P.; Mootha, V.K.; Mukherjee, S.; Ebert, B.L.; Gillette, M.A.; Paulovich, A.; Pomeroy, S.L.; Golub, T.R.; Lander, E.S.; et al. Gene set enrichment analysis: A knowledge-based approach for interpreting genome-wide expression profiles. *Proc. Natl. Acad. Sci. USA* **2005**, *102*, 15545–15550. [CrossRef]
37. Mootha, V.K.; Lindgren, C.M.; Eriksson, K.F.; Subramanian, A.; Sihag, S.; Lehar, J.; Puigserver, P.; Carlsson, E.; Ridderstråle, M.; Laurila, E.; et al. PGC-1alpha-responsive genes involved in oxidative phosphorylation are coordinately downregulated in human diabetes. *Nat Genet.* **2003**, *34*, 267–273. [CrossRef]
38. Cheson, B.D.; Pfistner, B.; Juweid, M.E.; Gascoyne, R.D.; Specht, L.; Horning, S.J.; Coiffier, B.; Fisher, R.I.; Hagenbeek, A.; Zucca, E.; et al. International Harmonization Project on Lymphoma. Revised response criteria for malignant lymphoma. *J. Clin. Oncol.* **2007**, *25*, 579–586. [CrossRef] [PubMed]
39. Aguirre-Gamboa, R.; Gomez-Rueda, H.; Martínez-Ledesma, E.; Martínez-Torteya, A.; Chacolla-Huaringa, R.; Rodriguez-Barrientos, A.; Tamez-Peña, J.G.; Treviño, V. SurvExpress: An online biomarker validation tool and database for cancer gene expression data using survival analysis. *PLoS ONE* **2013**, *8*, e74250. [CrossRef]
40. Ma, M.C.J.; Tadros, S.; Bouska, A.; Heavican, T.; Yang, H.; Deng, Q.; Moore, D.; Akhter, A.; Hartert, K.; Jain, N.; et al. Subtype-specific and co-occurring genetic alterations in B-cell non-Hodgkin lymphoma. *Haematologica* **2022**, *107*, 690–701. [CrossRef]
41. Herold, T.; Jurinovic, V.; Metzeler, K.H.; Boulesteix, A.L.; Bergmann, M.; Seiler, T.; Mulaw, M.; Thoene, S.; Dufour, A.; Pasalic, Z.; et al. An eight-gene expression signature for the prediction of survival and time to treatment in chronic lymphocytic leukemia. *Leukemia* **2011**, *25*, 1639–1645. [CrossRef]
42. Herold, T.; Mulaw, M.A.; Jurinovic, V.; Seiler, T.; Metzeler, K.H.; Dufour, A.; Schneider, S.; Kakadia, P.M.; Spiekermann, K.; Mansmann, U.; et al. High expression of MZB1 predicts adverse prognosis in chronic lymphocytic leukemia, follicular lymphoma and diffuse large B-cell lymphoma and is associated with a unique gene expression signature. *Leuk. Lymphoma* **2013**, *54*, 1652–1657. [CrossRef] [PubMed]
43. Lenz, G.; Wright, G.; Dave, S.S.; Xiao, W.; Powell, J.; Zhao, H.; Xu, W.; Tan, B.; Goldschmidt, N.; Iqbal, J.; et al. Lymphoma/Leukemia Molecular Profiling Project. Stromal gene signatures in large-B-cell lymphomas. *N. Engl. J. Med.* **2008**, *359*, 2313–2323. [CrossRef] [PubMed]
44. Cardesa-Salzmann, T.M.; Colomo, L.; Gutierrez, G.; Chan, W.C.; Weisenburger, D.; Climent, F.; González-Barca, E.; Mercadal, S.; Arenillas, L.; Serrano, S.; et al. High microvessel density determines a poor outcome in patients with diffuse large B-cell lymphoma treated with rituximab plus chemotherapy. *Haematologica* **2011**, *96*, 996–1001. [CrossRef]
45. Shaknovich, R.; Geng, H.; Johnson, N.A.; Tsikitas, L.; Cerchietti, L.; Greally, J.M.; Gascoyne, R.D.; Elemento, O.; Melnick, A. DNA methylation signatures define molecular subtypes of diffuse large B-cell lymphoma. *Blood* **2010**, *116*, e81–e89. [CrossRef]
46. Hummel, M.; Bentink, S.; Berger, H.; Klapper, W.; Wessendorf, S.; Barth, T.F.; Bernd, H.W.; Cogliatti, S.B.; Dierlamm, J.; Feller, A.C.; et al. Molecular Mechanisms in Malignant Lymphomas Network Project of the Deutsche Krebshilfe. A biologic definition of Burkitt's lymphoma from transcriptional and genomic profiling. *N. Engl. J. Med.* **2006**, *354*, 2419–2430. [CrossRef] [PubMed]

47. Richter, J.; Schlesner, M.; Hoffmann, S.; Kreuz, M.; Leich, E.; Burkhardt, B.; Rosolowski, M.; Ammerpohl, O.; Wagener, R.; Bernhart, S.H.; et al. ICGC MMML-Seq Project. Recurrent mutation of the ID3 gene in Burkitt lymphoma identified by integrated genome, exome and transcriptome sequencing. *Nat. Genet.* **2012**, *44*, 1316–1320. [CrossRef] [PubMed]
48. Jais, J.P.; Haioun, C.; Molina, T.J.; Rickman, D.S.; de Reynies, A.; Berger, F.; Gisselbrecht, C.; Brière, J.; Reyes, F.; Gaulard, P.; et al. Groupe d'Etude des Lymphomes de l'Adulte. The expression of 16 genes related to the cell of origin and immune response predicts survival in elderly patients with diffuse large B-cell lymphoma treated with CHOP and rituximab. *Leukemia* **2008**, *22*, 1917–1924. [CrossRef]
49. Leich, E.; Salaverria, I.; Bea, S.; Zettl, A.; Wright, G.; Moreno, V.; Gascoyne, R.D.; Chan, W.C.; Braziel, R.M.; Rimsza, L.M.; et al. Follicular lymphomas with and without translocation t(14;18) differ in gene expression profiles and genetic alterations. *Blood* **2009**, *114*, 826–834. [CrossRef] [PubMed]
50. Rosenwald, A.; Wright, G.; Wiestner, A.; Chan, W.C.; Connors, J.M.; Campo, E.; Gascoyne, R.D.; Grogan, T.M.; Muller-Hermelink, H.K.; Smeland, E.B.; et al. The proliferation gene expression signature is a quantitative integrator of oncogenic events that predicts survival in mantle cell lymphoma. *Cancer Cell* **2003**, *3*, 185–197. [CrossRef]
51. Scott, D.W.; Abrisqueta, P.; Wright, G.W.; Slack, G.W.; Mottok, A.; Villa, D.; Jares, P.; Rauert-Wunderlich, H.; Royo, C.; Clot, G.; et al. Lymphoma/Leukemia Molecular Profiling Project. New Molecular Assay for the Proliferation Signature in Mantle Cell Lymphoma Applicable to Formalin-Fixed Paraffin-Embedded Biopsies. *J. Clin. Oncol.* **2017**, *35*, 1668–1677. [CrossRef]
52. Hanamura, I.; Huang, Y.; Zhan, F.; Barlogie, B.; Shaughnessy, J. Prognostic value of cyclin D2 mRNA expression in newly diagnosed multiple myeloma treated with high-dose chemotherapy and tandem autologous stem cell transplantations. *Leukemia* **2006**, *20*, 1288–1290. [CrossRef] [PubMed]
53. Zhan, F.; Huang, Y.; Colla, S.; Stewart, J.P.; Hanamura, I.; Gupta, S.; Epstein, J.; Yaccoby, S.; Sawyer, J.; Burington, B.; et al. The molecular classification of multiple myeloma. *Blood* **2006**, *108*, 2020–2028. [CrossRef]
54. Zhan, F.; Barlogie, B.; Arzoumanian, V.; Huang, Y.; Williams, D.R.; Hollmig, K.; Pineda-Roman, M.; Tricot, G.; van Rhee, F.; Zangari, M.; et al. Gene-expression signature of benign monoclonal gammopathy evident in multiple myeloma is linked to good prognosis. *Blood* **2007**, *109*, 1692–1700. [CrossRef] [PubMed]
55. Chen, L.; Wang, S.; Zhou, Y.; Wu, X.; Entin, I.; Epstein, J.; Yaccoby, S.; Xiong, W.; Barlogie, B.; Shaughnessy, J.D., Jr.; et al. Identification of early growth response protein 1 (EGR-1) as a novel target for JUN-induced apoptosis in multiple myeloma. *Blood* **2010**, *115*, 61–70. [CrossRef]
56. Qiang, Y.W.; Ye, S.; Huang, Y.; Chen, Y.; Van Rhee, F.; Epstein, J.; Walker, B.A.; Morgan, G.J.; Davies, F.E. MAFb protein confers intrinsic resistance to proteasome inhibitors in multiple myeloma. *BMC Cancer* **2018**, *18*, 724. [CrossRef]
57. Went, M.; Sud, A.; Försti, A.; Halvarsson, B.M.; Weinhold, N.; Kimber, S.; van Duin, M.; Thorleifsson, G.; Holroyd, A.; Johnson, D.C.; et al. PRACTICAL consortium. Identification of multiple risk loci and regulatory mechanisms influencing susceptibility to multiple myeloma. *Nat. Commun.* **2018**, *9*, 3707. [CrossRef] [PubMed]
58. Wang, L.; Li, L.R.; Young, K.H. New agents and regimens for diffuse large B cell lymphoma. *J. Hematol. Oncol.* **2020**, *13*, 175. [CrossRef] [PubMed]
59. Carreras, J. Artificial Intelligence Analysis of Celiac Disease Using an Autoimmune Discovery Transcriptomic Panel Highlighted Pathogenic Genes including BTLA. *Healthcare* **2022**, *10*, 1550. [CrossRef] [PubMed]
60. Carreras, J. Artificial Intelligence Analysis of Ulcerative Colitis Using an Autoimmune Discovery Transcriptomic Panel. *Healthcare* **2022**, *10*, 1476. [CrossRef] [PubMed]
61. Lisson, C.S.; Lisson, C.G.; Mezger, M.F.; Wolf, D.; Schmidt, S.A.; Thaiss, W.M.; Tausch, E.; Beer, A.J.; Stilgenbauer, S.; Beer, M. Deep Neural Networks and Machine Learning Radiomics mode for Prediction of Relapse in Mantle Cell Lymphoma. *Cancers* **2022**, *14*, 2008. [CrossRef]
62. Sadik, M.; López-Urdaneta, J.; Ulén, J.; Enqvist, O.; Krupic, A.; Kumar, R.; Andersson, P.O.; Trägårdh, E. Artificial intelligence could alert for focal skeleton/bone marrow uptake in Hodgkin's lymphoma patients staged with FDG-PET/CT. *Sci. Rep.* **2021**, *11*, 10382. [CrossRef]
63. Wang, Y.J.; Baratto, L.; Hawk, K.E.; Theruvath, A.J.; Pribnow, A.; Thakor, A.S.; Gatidis, S.; Lu, R.; Gummidipundi, S.E.; Garcia-Diaz, J.; et al. Artificial intelligence enables whole-body positron emission tomography scans with minimal radiation exposure. *Eur. J. Nucl. Med. Mol. Imaging* **2021**, *48*, 2771–2781. [CrossRef] [PubMed]
64. Pinochet, P.; Eude, F.; Becker, S.; Shah, V.; Sibille, L.; Toledano, M.N.; Modzelewski, R.; Vera, P.; Decazes, P. Evaluation of an Automatic Classification Algorithm Using Convolutional Neural Networks in Oncological Positron Emission Tomography. *Front. Med.* **2021**, *8*, 628179. [CrossRef]
65. El Hussein, S.; Chen, P.; Medeiros, L.J.; Wistuba, I.I.; Jaffray, D.; Wu, J.; Khoury, J.D. Artificial intelligence strategy integrating morphologic and architectural biomarkers provides robust diagnostic accuracy for disease progression in chronic lymphocytic leukemia. *J. Pathol.* **2022**, *256*, 4–14. [CrossRef]
66. Swiderska-Chadaj, Z.; Hebeda, K.M.; van den Brand, M.; Litjens, G. Artificial intelligence to detect MYC translocation in slides of diffuse large B-cell lymphoma. *Virchows Arch.* **2021**, *479*, 617–621. [CrossRef]
67. Steinbuss, G.; Kriegsmann, M.; Zgorzelski, C.; Brobeil, A.; Goeppert, B.; Dietrich, S.; Mechtersheimer, G.; Kriegsmann, K. Deep Learning for the Classification of Non-Hodgkin Lymphoma on Histopathological Images. *Cancers* **2021**, *13*, 2419. [CrossRef]
68. Zhang, X.; Zhang, K.; Jiang, M.; Yang, L. Research on the classification of lymphoma pathological images based on deep residual neural network. *Technol. Health Care* **2021**, *29*, 335–344. [CrossRef]

69. Tang, G.; Fu, X.; Wang, Z.; Chen, M. A Machine Learning Tool Using Digital Microscopy (Morphogo) for the Identification of Abnormal Lymphocytes in the Bone Marrow. *Acta Cytol.* **2021**, *65*, 354–357. [CrossRef]
70. Yu, W.H.; Li, C.H.; Wang, R.C.; Yeh, C.Y.; Chuang, S.S. Machine Learning Based on Morphological Features Enables Classification of Primary Intestinal T-Cell Lymphomas. *Cancers* **2021**, *13*, 5463. [CrossRef]
71. Zhou, M.; Wu, K.; Yu, L.; Xu, M.; Yang, J.; Shen, Q.; Liu, B.; Shi, L.; Wu, S.; Dong, B.; et al. Development and Evaluation of a Leukemia Diagnosis System Using Deep Learning in Real Clinical Scenarios. *Front. Pediatr.* **2021**, *9*, 693676. [CrossRef] [PubMed]
72. Zhang, J.; Cui, W.; Guo, X.; Wang, B.; Wang, Z. Classification of digital pathological images of non-Hodgkin's lymphoma subtypes based on the fusion of transfer learning and principal component analysis. *Med. Phys.* **2020**, *47*, 4241–4253. [CrossRef]
73. Mohlman, J.S.; Leventhal, S.D.; Hansen, T.; Kohan, J.; Pascucci, V.; Salama, M.E. Improving Augmented Human Intelligence to Distinguish Burkitt Lymphoma from Diffuse Large B-Cell Lymphoma Cases. *Am. J. Clin. Pathol.* **2020**, *153*, 743–759. [CrossRef] [PubMed]
74. Li, D.; Bledsoe, J.R.; Zeng, Y.; Liu, W.; Hu, Y.; Bi, K.; Liang, A.; Li, S. A deep learning diagnostic platform for diffuse large B-cell lymphoma with high accuracy across multiple hospitals. *Nat. Commun.* **2020**, *11*, 6004. [CrossRef]
75. Miyoshi, H.; Sato, K.; Kabeya, Y.; Yonezawa, S.; Nakano, H.; Takeuchi, Y.; Ozawa, I.; Higo, S.; Yanagida, E.; Yamada, K.; et al. Deep learning shows the capability of high-level computer-aided diagnosis in malignant lymphoma. *Lab. Investig.* **2020**, *100*, 1300–1310. [CrossRef] [PubMed]
76. Zorman, M.; de la Rosa, J.L.S.; Dinevski, D. Classification of follicular lymphoma images: A holistic approach with symbol-based machine learning methods. *Wien. Klin. Wochenschr.* **2011**, *123*, 700–709. [CrossRef]
77. Zhao, M.; Mallesh, N.; Höllein, A.; Schabath, R.; Haferlach, C.; Haferlach, T.; Elsner, F.; Lüling, H.; Krawitz, P.; Kern, W. Hematologist-Level Classification of Mature B-Cell Neoplasm Using Deep Learning on Multiparameter Flow Cytometry Data. *Cytom. A* **2020**, *97*, 1073–1080. [CrossRef] [PubMed]
78. Gaidano, V.; Tenace, V.; Santoro, N.; Varvello, S.; Cignetti, A.; Prato, G.; Saglio, G.; De Rosa, G.; Geuna, M. A Clinically Applicable Approach to the Classification of B-Cell Non-Hodgkin Lymphomas with Flow Cytometry and Machine Learning. *Cancers* **2020**, *12*, 1684. [CrossRef] [PubMed]
79. Zhan, M.; Chen, Z.B.; Ding, C.C.; Qu, Q.; Wang, G.Q.; Liu, S.; Wen, F.Q. Machine learning to predict high-dose methotrexate-related neutropenia and fever in children with B-cell acute lymphoblastic leukemia. *Leuk. Lymphoma* **2021**, *62*, 2502–2513. [CrossRef] [PubMed]
80. Buciński, A.; Marszałł, M.P.; Krysiński, J.; Lemieszek, A.; Załuski, J. Contribution of artificial intelligence to the knowledge of prognostic factors in Hodgkin's lymphoma. *Eur. J. Cancer Prev.* **2010**, *19*, 308–312. [CrossRef] [PubMed]
81. Xu-Monette, Z.Y.; Zhang, H.; Zhu, F.; Tzankov, A.; Bhagat, G.; Visco, C.; Dybkaer, K.; Chiu, A.; Tam, W.; Zu, Y.; et al. A refined cell-of-origin classifier with targeted NGS and artificial intelligence shows robust predictive value in DLBCL. *Blood Adv.* **2020**, *4*, 3391–3404. [CrossRef]
82. Zhang, W.; Yang, L.; Guan, Y.Q.; Shen, K.F.; Zhang, M.L.; Cai, H.D.; Wang, J.C.; Wang, Y.; Huang, L.; Cao, Y.; et al. Novel bioinformatic classification system for genetic signatures identification in diffuse large B-cell lymphoma. *BMC Cancer* **2020**, *20*, 714. [CrossRef] [PubMed]
83. Parodi, S.; Manneschi, C.; Verda, D.; Ferrari, E.; Muselli, M. Logic Learning Machine and standard supervised methods for Hodgkin's lymphoma prognosis using gene expression data and clinical variables. *Health Inform. J.* **2018**, *24*, 54–65. [CrossRef]
84. Schmitz, R.; Wright, G.W.; Huang, D.W.; Johnson, C.A.; Phelan, J.D.; Wang, J.Q.; Roulland, S.; Kasbekar, M.; Young, R.M.; Shaffer, A.L.; et al. Genetics and Pathogenesis of Diffuse Large B-Cell Lymphoma. *N. Engl. J. Med.* **2018**, *378*, 1396–1407. [CrossRef]
85. Carreras, J.; Kikuti, Y.Y.; Hiraiwa, S.; Miyaoka, M.; Tomita, S.; Ikoma, H.; Ito, A.; Kondo, Y.; Itoh, J.; Roncador, G.; et al. High PTX3 expression is associated with a poor prognosis in diffuse large B-cell lymphoma. *Cancer Sci.* **2022**, *113*, 334–348. [CrossRef]
86. Siegel, R.L.; Miller, K.D.; Jemal, A. Cancer statistics, 2020. *CA Cancer J. Clin.* **2020**, *70*, 7–30. [CrossRef]
87. Loibl, S.; Poortmans, P.; Morrow, M.; Denkert, C.; Curigliano, G. Breast cancer. *Lancet* **2021**, *397*, 1750–1769. [CrossRef]
88. Kuchenbaecker, K.B.; Hopper, J.L.; Barnes, D.R.; Phillips, K.A.; Mooij, T.M.; Roos-Blom, M.J.; Jervis, S.; van Leeuwen, F.E.; Milne, R.L.; Andrieu, N.; et al. Risks of Breast, Ovarian, and Contralateral Breast Cancer for BRCA1 and BRCA2 Mutation Carriers. *JAMA* **2017**, *317*, 2402–2416. [CrossRef]
89. Chen, S.; Parmigiani, G. Meta-analysis of BRCA1 and BRCA2 penetrance. *J. Clin. Oncol.* **2007**, *25*, 1329–1333. [CrossRef]
90. Basurto-Hurtado, J.A.; Cruz-Albarran, I.A.; Toledano-Ayala, M.; Ibarra-Manzano, M.A.; Morales-Hernandez, L.A.; Perez-Ramirez, C.A. Diagnostic Strategies for Breast Cancer Detection: From Image Generation to Classification Strategies Using Artificial Intelligence Algorithms. *Cancers* **2022**, *14*, 3442. [CrossRef] [PubMed]
91. Wang, Z.; Li, M.; Wang, H.; Jiang, H.; Yao, Y.; Zhang, H.; Xin, J. Breast Cancer Detection Using Extreme Learning Machine Based on Feature Fusion with CNN Deep Features. *IEEE Access* **2019**, *7*, 105146–105158. [CrossRef]
92. Yap, M.H.; Pons, G.; Marti, J.; Ganau, S.; Sentis, M.; Zwiggelaar, R.; Davison, A.K.; Marti, R. Automated Breast Ultrasound Lesions Detection Using Convolutional Neural Networks. *IEEE J. Biomed. Health Inform.* **2018**, *22*, 1218–1226. [CrossRef]
93. Teare, P.; Fishman, M.; Benzaquen, O.; Toledano, E.; Elnekave, E. Malignancy Detection on Mammography Using Dual Deep Convolutional Neural Networks and Genetically Discovered False Color Input Enhancement. *J. Digit. Imaging* **2017**, *30*, 499–505. [CrossRef] [PubMed]
94. Shen, L.; Margolies, L.R.; Rothstein, J.H.; Fluder, E.; McBride, R.; Sieh, W. Deep Learning to Improve Breast Cancer Detection on Screening Mammography. *Sci. Rep.* **2019**, *9*, 12495. [CrossRef] [PubMed]

95. Gamage, T.P.B.; Malcolm, D.T.K.; Talou, G.D.M.; Mîra, A.; Doyle, A.; Nielsen, P.M.F.; Nash, M.P. An automated computational biomechanics workflow for improving breast cancer diagnosis and treatment. *Interface Focus* **2019**, *9*, 20190034. [CrossRef]
96. Bouron, C.; Mathie, C.; Seegers, V.; Morel, O.; Jézéquel, P.; Lasla, H.; Guillerminet, C.; Girault, S.; Lacombe, M.; Sher, A.; et al. Prognostic Value of Metabolic, Volumetric and Textural Parameters of Baseline [18F]FDG PET/CT in Early Triple-Negative Breast Cancer. *Cancers* **2022**, *14*, 637. [CrossRef] [PubMed]
97. Mughal, B.; Sharif, M.; Muhammad, N. Bi-model processing for early detection of breast tumor in CAD system. *Eur. Phys. J. Plus* **2017**, *132*, 266. [CrossRef]
98. Wang, S.; Rao, R.V.; Chen, P.; Zhang, Y.; Liu, A.; Wei, L. Abnormal Breast Detection in Mammogram Images by Feed-forward Neural Network Trained by Jaya Algorithm. *Fundam. Inform.* **2017**, *151*, 191–211. [CrossRef]
99. Muduli, D.; Dash, R.; Majhi, B. Automated breast cancer detection in digital mammograms: A moth flame optimization based ELM approach. *Biomed. Signal Process. Control* **2020**, *59*, 101912. [CrossRef]
100. Shiji, T.P.; Remya, S.; Lakshmanan, R.; Pratab, T.; Thomas, V. Evolutionary intelligence for breast lesion detection in ultrasound images: A wavelet modulus maxima and SVM based approach. *J. Intell. Fuzzy Syst.* **2020**, *38*, 6279–6290. [CrossRef]
101. Chakraborty, J.; Midya, A.; Rabidas, R. Computer-aided detection and diagnosis of mammographic masses using multi-resolution analysis of oriented tissue patterns. *Expert Syst. Appl.* **2018**, *99*, 168–179. [CrossRef]
102. Zahoor, S.; Shoaib, U.; Lali, I.U. Breast Cancer Mammograms Classification Using Deep Neural Network and Entropy-Controlled Whale Optimization Algorithm. *Diagnostics* **2022**, *12*, 557. [CrossRef]
103. Jagtap, A.D.; Shin, Y.; Kawaguchi, K.; Em Karniadakis, G. Deep Kronecker neural networks: A general framework for neural networks with adaptive activation functions. *Neurocomputing* **2022**, *468*, 165–180. [CrossRef]
104. Ali, M.J.; Raza, B.; Shahid, A.R. Multi-level Kronecker Convolutional Neural Network (ML-KCNN) for Glioma Segmentation from Multi-modal MRI Volumetric Data. *J. Digit. Imaging* **2021**, *34*, 905–921. [CrossRef]

onco

MDPI

Article

Regulation of Epithelial–Mesenchymal Transition Pathway and Artificial Intelligence-Based Modeling for Pathway Activity Prediction

Shihori Tanabe [1,*], Sabina Quader [2], Ryuichi Ono [3], Horacio Cabral [4], Kazuhiko Aoyagi [5], Akihiko Hirose [1], Edward J. Perkins [6], Hiroshi Yokozaki [7] and Hiroki Sasaki [8]

1 Division of Risk Assessment, Center for Biological Safety and Research, National Institute of Health Sciences, Kawasaki 210-9501, Japan
2 Innovation Center of NanoMedicine (iCONM), Kawasaki Institute of Industrial Promotion, Kawasaki 210-0821, Japan
3 Division of Cellular and Molecular Toxicology, Center for Biological Safety and Research, National Institute of Health Sciences, Kawasaki 210-9501, Japan
4 Department of Bioengineering, Graduate School of Engineering, The University of Tokyo, Tokyo 113-0033, Japan
5 Department of Clinical Genomics, National Cancer Center Research Institute, Tokyo 104-0045, Japan
6 Environmental Laboratory, US Army Engineer Research and Development Center, Vicksburg, MS 39180, USA
7 Department of Pathology, Kobe University of Graduate School of Medicine, Kobe 650-0017, Japan
8 Department of Translational Oncology, National Cancer Center Research Institute, Tokyo 104-0045, Japan
* Correspondence: stanabe@nihs.go.jp; Tel.: +81-44-270-6686

Simple Summary: Molecular network pathways are activated or inactivated under various conditions. Previously, we revealed that epithelial–mesenchymal transition (EMT) is a feature of diffuse-type gastric cancer. Here, we modeled the activation states of EMT in the development pathway using molecular pathway images and artificial intelligence (AI). The regulation of EMT in the development pathway was activated in diffuse-type gastric cancer (GC) and inactivated in intestinal-type GC. AI modeling with molecular pathway images generated a highly accurate Elastic-Net Classifier models that was validated with 10 additional activated and 10 inactivated pathway images.

Abstract: Because activity of the epithelial–mesenchymal transition (EMT) is involved in anti-cancer drug resistance, cancer malignancy, and shares some characteristics with cancer stem cells (CSCs), we used artificial intelligence (AI) modeling to identify the cancer-related activity of the EMT-related pathway in datasets of gene expression. We generated images of gene expression overlayed onto molecular pathways with Ingenuity Pathway Analysis (IPA). A dataset of 50 activated and 50 inactivated pathway images of EMT regulation in the development pathway was then modeled by the DataRobot Automated Machine Learning platform. The most accurate models were based on the Elastic-Net Classifier algorithm. The model was validated with 10 additional activated and 10 additional inactivated pathway images. The generated models had false-positive and false-negative results. These images had significant features of opposite labels, and the original data were related to Parkinson's disease. This approach reliably identified cancer phenotypes and treatments where EMT regulation in the development pathway was activated or inactivated thereby identifying conditions where therapeutics might be applied or developed. As there are a wide variety of cancer phenotypes and CSC targets that provide novel insights into the mechanism of CSCs' drug resistance and cancer metastasis, our approach holds promise for modeling and simulating cellular phenotype transition, as well as predicting molecular-induced responses.

Keywords: artificial intelligence; epithelial–mesenchymal transition; Ingenuity Pathway Analysis; machine learning; molecular pathway network

Citation: Tanabe, S.; Quader, S.; Ono, R.; Cabral, H.; Aoyagi, K.; Hirose, A.; Perkins, E.J.; Yokozaki, H.; Sasaki, H. Regulation of Epithelial–Mesenchymal Transition Pathway and Artificial Intelligence-Based Modeling for Pathway Activity Prediction. *Onco* **2023**, 3, 13–25. https://doi.org/10.3390/onco3010002

Academic Editor: Galatea Kallergi

Received: 17 November 2022
Revised: 29 December 2022
Accepted: 3 January 2023
Published: 6 January 2023

1. Introduction

Molecular network pathways are activated or inactivated under many different conditions. Previously, we found that diffuse-type gastric cancer (GC) has a feature of epithelial–mesenchymal transition (EMT) [1–3]. EMT is involved in anti-cancer drug resistance, cancer malignancy, metastasis, and cancer stem cells (CSCs) [4–7]. Experiments in anti-cancer drug-resistant cancer cell lines indicate that EMT is involved in cancer cell drug resistance [8], highlighting the significance of EMT targeting in cancer treatment [6].

Several signaling pathways involved in EMT contribute to drug resistance [6]. Tumor growth factor beta (TGFβ) signaling activates SMAD2/3, which then complexes with SMAD4 to form a trimetric SMAD complex, leading to the transcription of EMT transcription factors [9]. Wnt/β-catenin signaling activates Snail transcription to induce EMT [6,10]. Recent studies have also revealed the role of EMT in autophagy and CSCs during metastasis [11,12]. However, the relationship between the EMT pathway activation state and therapeutic responsiveness is not fully understood.

Understanding the activity state of the EMT pathway in cancer cells may be an important clue for identifying therapeutic targets in malignant cancers. To effectively predict EMT activity and potential therapeutic responsiveness, molecular pathway images were used to capture activity of EMT-related pathways of datasets in Ingenuity Pathway Analysis (IPA), followed by artificial intelligence (AI) modeling with images of gene expression activity in the pathway.

2. Materials and Methods

2.1. Data Analysis of Diffuse- and Intestinal-Type GC

We used RNA sequencing data of diffuse- and intestinal-type GC, which are publicly available in The Cancer Genome Atlas (TCGA) of the cBioPortal for Cancer Genomics database at the National Cancer Institute (NCI) Genomic Data Commons (GDC) data portal [13–17]. Publicly available data on stomach adenocarcinoma in the TCGA, Stomach Adenocarcinoma (TCGA, PanCancer Atlas), [13–16] were compared between diffuse-type GC, which is genomically stable (n = 50), and intestinal GC, which has a feature of chromosomal instability (n = 223), in TCGA Research Network publications, as previously described [1,14,18].

2.2. Network Analysis

Data on intestinal- and diffuse-type GC from the TCGA cBioPortal for Cancer Genomics were uploaded and analyzed using IPA (Qiagen, CA, USA) [19,20]. The datasets of gene expression in diseases were searched in IPA, and datasets with absolute values in z-score in the top 60 for activated state and inactivated state (total of 120) in regulation of EMT in the development pathway were extracted for AI prediction modeling and evaluation. Among 120 analyses in the activity plot of regulation of EMT in the development pathway, 50 activated and 50 inactivated analyses (total of 100) were used to generate AI models and 10 activated and 10 inactivated analyses (total of 20) were withheld for use in validating the generated model. The 100 analyses (50 activated and 50 inactivated states) found in the database of IPA and newly used to generate AI-based models are summarized in Table 1.

Table 1. Analyses in the regulation of EMT in the development pathway for AI prediction modeling.

Analysis Name	Disease State	Target Gene	Treatment	EMT
996-Breast ductal carcinoma torin 2 28190	Breast ductal carcinoma	Mtor	Torin 2	TRUE
16332-Fibrocystic breast disease neratinib 7038	Fibrocystic breast disease	Her2; egfr	Neratinib	TRUE
16885-Fibrocystic breast disease erlotinib 7651	Fibrocystic breast disease	Egfr	Erlotinib	TRUE
116-Bone osteosarcoma (OS) MK2206 2727	Bone osteosarcoma (OS)		MK2206	TRUE
1766-Breast ductal carcinoma brivanib 8512	Breast ductal carcinoma	Vegfr; fgfr	Brivanib	TRUE
47-Huntington's disease (HD) haloperidol 12804	Huntington's disease (HD)		Haloperidol	TRUE
4874-Melanoma crizotinib 22540	Melanoma	Alk and ros1	Crizotinib	TRUE
6785-Non-small cell lung carcinoma ZSTK474 24663	Non-small cell lung carcinoma	PI3K	ZSTK474	TRUE
7-Normal control differentiation medium 10230	Normal control		Differentiation medium	TRUE
13972-Prostate adenocarcinoma (PRAD) PI103 4415	Prostate adenocarcinoma (PRAD)	PI3K	PI103	TRUE
16046-Prostate adenocarcinoma (PRAD) MK2206 6720	Prostate adenocarcinoma (PRAD)	AKT	MK2206	TRUE
7063-Breast adenocarcinoma linifanib 24973	Breast adenocarcinoma	Ftk; vegf; pdgf	Linifanib	TRUE
7923-Breast adenocarcinoma PF3758309 25928	Breast adenocarcinoma	FAK4	PF3758309	TRUE
2-Breast carcinoma beta-estradiol (E2) 3915	Breast carcinoma		B-estradiol (E2)	TRUE
10974-Breast ductal carcinoma KIN001-043 1084	Breast ductal carcinoma	GSK3β	KIN001-043	TRUE
1116-Breast ductal carcinoma QL-X-138 1291	Breast ductal carcinoma	BTK; MNK	QL-X-138	TRUE
29-Colon cancer GSK525762A; trametinib 3009	Colon cancer		GSK525762A; trametinib	TRUE
35-Colon cancer active JQ1 1658	Colon cancer		Active JQ1	TRUE
13176-Colorectal adenocarcinoma BGJ398 3531	Colorectal adenocarcinoma	FGFR	BGJ398	TRUE
12948-Colorectal adenocarcinoma AZ628 3277	Colorectal adenocarcinoma	BRAF; BRAFV600E; C-RAF-1	AZ628	TRUE
12715-Colorectal adenocarcinoma AT7519 3019	Colorectal adenocarcinoma	CDK	AT7519	TRUE
6-Disease control IL-1 beta 15814	Disease control		IL-1β	TRUE
17104-Fibrocystic breast disease canertinib 7896	Fibrocystic breast disease	Egfr; her2; erbb4	Canertinib	TRUE
17239-Fibrocystic breast disease torin 1 8045	Fibrocystic breast disease	Mtor	Torin 1	TRUE
16449-Fibrocystic breast disease AZD8330 7167	Fibrocystic breast disease	MEK	AZD8330	TRUE
17590-Fibrocystic breast disease mitoxantrone 8435	Fibrocystic breast disease	Topoisomerase	Mitoxantrone	TRUE
7-Fibrosis DMSO 7394	Fibrosis		DMSO	TRUE
20894-Hepatocellular carcinoma (LIHC) chelerythrine chloride 12106	Hepatocellular carcinoma (LIHC)	PKC	Chelerythrine chloride	TRUE
59-Huntington's disease (HD) nortriptyline 12817	Huntington's disease (HD)		Nortriptyline	TRUE
2-Lung adenocarcinoma (LUAD) Transfection_HOXC6 631	Lung adenocarcinoma (LUAD)		Transfection_HOXC6	TRUE
3-Major depressive disorder differentiation medium 3130	Major depressive disorder		Differentiation medium	TRUE
5612-Melanoma AT7867 23361	Melanoma	AKT1/2/3; p70s6k/PKA	AT7867	TRUE
5173-Melanoma lapatinib 22873	Melanoma	Her2; egfr	Lapatinib	TRUE
91-Non-small cell lung carcinoma BGT226 27235	Non-small cell lung carcinoma	PI3K; mtor	BGT226	TRUE
14456-Normal control WYE125132 4953	Normal control	Mtor	WYE125132	TRUE
28175-Normal control glesatinib 20196	Normal control	C-met; tek; vegfr; ron	Glesatinib	TRUE
60-Normal control 567	Normal control			TRUE
2-Normal control culture medium 1187	Normal control		Culture medium	TRUE

Table 1. *Cont.*

Analysis Name	Disease State	Target Gene	Treatment	EMT
9914-Normal control EX527 28140	Normal control	SIRT1	EX527	TRUE
4-Normal control suberoylanilide hydroxamic acid (SAHA) 2204	Normal control		Suberoylanilide hydroxamic acid (SAHA)	TRUE
27560-Normal control BMS509744 19513	Normal control	ITK	BMS509744	TRUE
14256-Normal control AZD8055 4731	Normal control	Mtor	AZD8055	TRUE
19-Normal control no serum 3447	Normal control		No serum	TRUE
5-Parkinson's disease (PD) differentiation medium 4389	Parkinson's disease (PD)		Differentiation medium	TRUE
23661-Prostate adenocarcinoma (PRAD) AZD5438 15181	Prostate adenocarcinoma (PRAD)	CDK	AZD5438	TRUE
25661-Breast adenocarcinoma omipalisib 17403	Breast adenocarcinoma	Pi3k	Omipalisib	TRUE
90-Prostate adenocarcinoma (PRAD) monolayer culture 4346	Prostate adenocarcinoma (PRAD)		Monolayer culture	TRUE
8-Normal control lipopolysaccharide (LPS) 4907	Normal control		Lipopolysaccharide (LPS)	TRUE
2-Acute myeloid leukemia (LAML) lipopolysaccharide (LPS) 9357	Acute myeloid leukemia (LAML)		Lipopolysaccharide (LPS)	TRUE
25084-Breast adenocarcinoma CGP60474 16762	Breast adenocarcinoma	CDK1; CDK2	CGP60474	TRUE
20-Non-small cell lung carcinoma IFN gamma 13421	Non-small cell lung carcinoma		Ifnγ	FALSE
7-Normal control co-culture 3087	Normal control		Co-culture	FALSE
5-Normal control hypoxia 13911	Normal control		Hypoxia	FALSE
1-Normal control IFN alpha 4636	Normal control		Ifnα	FALSE
11-Normal control differentiation medium 10205	Normal control		Differentiation medium	FALSE
3-Normal control Infection_human betaherpesvirus 5 (HHV5) 15858	Normal control		Infection_human betaherpesvirus 5 (HHV5)	FALSE
31-Bone osteosarcoma (OS) 1,9-pyrazoloanthrone 2804	Bone osteosarcoma (OS)		1,9-pyrazoloanthrone	FALSE
57-Coronavirus disease 2019 (COVID-19) 96	Coronavirus disease 2019 (COVID-19)			FALSE
17503-Fibrocystic breast disease HG6-64-1 8339	Fibrocystic breast disease	B-RAF	HG6-64-1	FALSE
11-Genetic disease 444	Genetic disease			FALSE
4-Glioblastoma (GBM) differentiation medium 6303	Glioblastoma (GBM)		Differentiation medium	FALSE
23448-Hepatocellular carcinoma (LIHC) imatinib 14944	Hepatocellular carcinoma (LIHC)	BCR-ABL	Imatinib	FALSE
86-Huntington's disease (HD) sodium butyrate 12847	Huntington's disease (HD)		Sodium butyrate	FALSE
21-Mantle cell lymphoma DMSO 3032	Mantle cell lymphoma		DMSO	FALSE
5-Non-alcoholic steatohepatitis (NASH) none 11484	Non-alcoholic steatohepatitis (NASH)		None	FALSE
10431-Normal control RAF265 482	Normal control	C-RAF; B-RAF; B-RAFV600E	RAF265	FALSE
11-Normal control differentiation medium 4490	Normal control		Differentiation medium	FALSE
14744-Normal control dasatinib 5273	Normal control	Src family	Dasatinib	FALSE
65-Normal control IL-3 17225	Normal control		IL-3	FALSE
14639-Normal control saracatinib 5156	Normal control	Src; bcr-abl	Saracatinib	FALSE
3-Normal control DHA-5-HT 4554	Normal control		DHA-5-HT	FALSE

Table 1. *Cont.*

Analysis Name	Disease State	Target Gene	Treatment	EMT
28-Prostatic intraepithelial neoplasia (PIN) plumbagin 49	Prostatic intraepithelial neoplasia (PIN)		Plumbagin	FALSE
4-Normal control differentiation medium 3415	Normal control		Differentiation medium	FALSE
9-Huntington's disease (HD) meclizine 12851	Huntington's disease (HD)		Meclizine	FALSE
6-Normal control culture medium 593	Normal control		Culture medium	FALSE
22597-Normal control GSK429286A 13998	Normal control	ROCK1; ROCK2	GSK429286A	FALSE
8-Normal control 3-D culture; co-culture; differentiation 3017	Normal control		3D culture; co-culture; differentiation medium	FALSE
110-Normal control 109	Normal control			FALSE
26-Bone osteosarcoma (OS) nilotinib 2798	Bone osteosarcoma (OS)		Nilotinib	FALSE
26025-Breast adenocarcinoma saracatinib 17808	Breast adenocarcinoma	Src; bcr-abl	Saracatinib	FALSE
11577-Breast ductal carcinoma crizotinib 1754	Breast ductal carcinoma	Alk and ros1	Crizotinib	FALSE
17316-Fibrocystic breast disease KIN001-043 8131	Fibrocystic breast disease	GSK3β	KIN001-043	FALSE
2-Fibrosis SB525334 7389	Fibrosis		SB525334	FALSE
52-Huntington's disease (HD) meclizine 12810	Huntington's disease (HD)		Meclizine	FALSE
1-Normal control culture medium 1186	Normal control		Culture medium	FALSE
17-Normal control differentiation medium 4496	Normal control		Differentiation medium	FALSE
6-Normal control hypoxia 13912	Normal control		Hypoxia	FALSE
2-Major depressive disorder differentiation medium 3129	Major depressive disorder		Differentiation medium	FALSE
11-Disease control none 4051	Disease control		None	FALSE
10-Normal control 3-D culture; co-culture; differentiation 2995	Normal control		3D culture; co-culture; differentiation medium	FALSE
5-Normal control lipopolysaccharide (LPS) 15704	Normal control		Lipopolysaccharide (LPS)	FALSE
1-Normal control differentiation medium 1246	Normal control		Differentiation medium	FALSE
6-Normal control 151	Normal control		3d culture; none	FALSE
10-Normal control differentiation medium 4489	Normal control		Differentiation medium	FALSE
13-Normal control co-culture 3079	Normal control		Co-culture	FALSE
13051-Colorectal adenocarcinoma BMS777607 3393	Colorectal adenocarcinoma	C-MET; AXL; RON; TYRO3	BMS777607	FALSE
27-Huntington's disease (HD) meclizine 12782	Huntington's disease (HD)		Meclizine	FALSE
8-Normal control GW3965 10098	Normal control		GW3965	FALSE
11-Normal control 368	Normal control			FALSE
6-Normal control culture medium 1191	Normal control		Culture medium	FALSE

2.3. *AI Prediction Modeling*

To create a prediction model using multi-modal data including images and text descriptions of molecular networks, an enterprise AI platform (DataRobot Automated Machine Learning version 7.2; DataRobot Inc. (Boston, MA, USA) was used. For the modeling, the 100 molecular networks on the regulation of EMT in the development pathway were collected and input as image data in the DataRobot (50 images in the activated state and 50 images in the inactivated state), which automatically created and tuned prediction models using various machine-learning algorithms (e.g., eXtreme gradient-boosted trees, random forest, regularized regression such as Elastic Net, Neural Networks) [21–23]. Finally, the AI model with the highest predictive accuracy on DataRobot was identified, and various insights (such as Permutation Importance or Partial Dependence Plot) obtained from the model were reviewed. To calculate the accuracy of the model, 20 additional image data (10 images in the activated state and 10 images in the inactivated state) that were not used as training data for the AI model creation were added for validation.

2.4. Statistical Analysis

The RNA sequencing data on diffuse- and intestinal-type GC was analyzed via Student's *t*-test. The z-scores of intestinal- and diffuse-type GC samples were compared, and the difference was considered significant at $p < 0.00001$, following previous reports [1,18]. The activation z-score in each pathway was calculated in IPA to show the level of activation.

3. Results

3.1. Regulation of the EMT in Development Pathway in Diffuse- and Intestinal-Type GC

3.1.1. Gene Expression Mapping in Regulation of the EMT in the Development Pathway in Diffuse- and Intestinal-Type GC

Alterations in gene expression in diffuse- and intestinal-type GC was mapped to a canonical pathway, "Regulation of the EMT in development pathway" (Figure 1) based on the previous gene expression analysis results [1]. Red or green color indicates upregulated or downregulated genes, respectively. In the regulation of EMT in the development pathway, frizzled and adenomatous polyposis coli regulator of the WNT signaling pathway (APC) was upregulated, while SUFU negative regulator of hedgehog signaling (SUFU), pygopus family PHD finger 2 (PYGO2), and BRCA1 was downregulated in diffuse-type GC compared to intestinal-type GC. APC encodes a tumor suppressor protein that acts as an antagonist of the Wnt signaling pathway. APC is also involved in other processes, including cell migration and adhesion, transcriptional activation, and apoptosis. SUFU is associated with β-catenin binding, protein kinase binding, and transcription regulation.

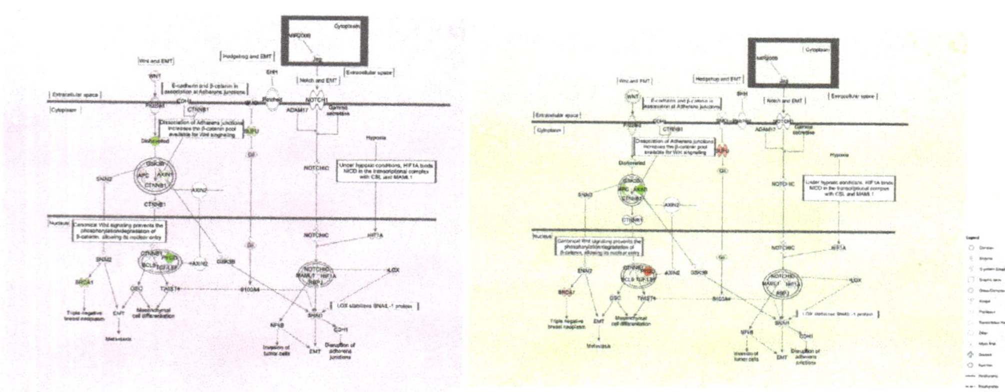

(**a**) (**b**)

Figure 1. Regulation of the epithelial–mesenchymal transition (EMT) in development pathway in diffuse- and intestinal-type gastric cancer (GC). (**a**) Gene expression alteration in diffuse-type GC in regulation of the EMT in development pathway; (**b**) Gene expression alteration in intestinal-type GC in regulation of the EMT in development pathway. Red or green color indicates upregulated or downregulated genes, respectively. The intensity of colors indicates the degree of up- or downregulation. A solid or dashed line indicates direct or indirect interaction, respectively.

3.1.2. Molecular Activity Prediction in Regulation of the EMT in Development Pathway in Diffuse- and Intestinal-Type GC

The prediction of molecular activity in the regulation of the EMT in the development pathway in diffuse- and intestinal-type GC was mapped (Figure 2). GSK3β, SNAI1, NFκB, LOX, and EMT are activated, whereas SNAI2 and E-cadherin are inactivated in diffuse-type GC compared to intestinal-type GC. Notch receptor 1 (NOTCH1) intracellular domain (NOTCHIC) was predicted to be activated in the CSL-HIF1A-MAML1-NICD complex, which consists of hypoxia-inducible factor 1 subunit alpha (HIF1A), mastermind-like transcriptional coactivator 1 (MAML1), NOTCH1, and recombination signal binding for immunoglobulin kappa J region (RBPJ) in the nucleus, and β-catenin (CTNNB1) was predicted to be activated in β-catenin-APC-AXIN-GSK3β complex in the cytoplasm in diffuse-type GC compared to intestinal-type GC.

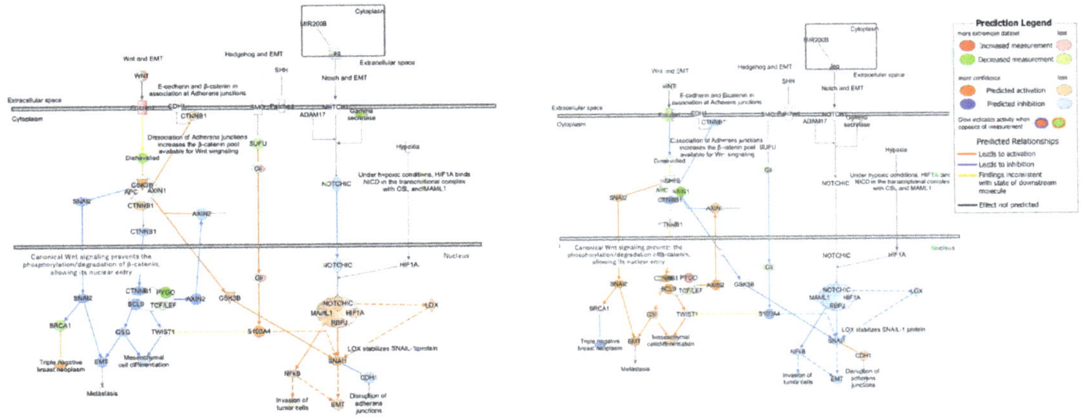

(**a**) (**b**)

Figure 2. Molecular activity prediction in regulation of the EMT in development pathway in diffuse- and intestinal-type GC. (**a**) Molecular activity prediction in diffuse-type GC; (**b**) molecular activity prediction in intestinal-type GC. Red or green color indicates upregulated or downregulated genes, respectively. The intensity of colors indicates the degree of up- or downregulation. A solid or dashed line indicates direct or indirect interaction respectively. Orange or blue color indicates predicted activation or inhibition, respectively. The intensity of colors indicates the confidence level of the prediction.

3.2. *Activity Plot of Regulation of the EMT in Development Pathway*

In total, 6216 analyses were found to be involved in the regulation of the EMT in the development pathway (as of September 2021) (Figure 3). In subsequent AI modeling analyses, samples with "NA" in the case treatment and blank in the disease state were excluded.

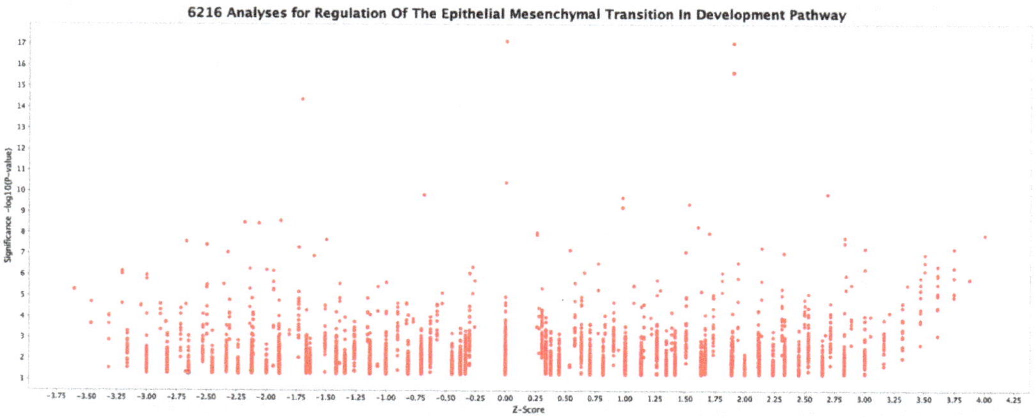

Figure 3. Activity plot of regulation of EMT in development pathway (6216 analyses, as of September 2021).

3.3. *AI Modeling and Validation of the Prediction Model*

The activation state of regulation of EMT in the development pathway was modeled by machine learning, including deep learning, using 50 activated and 50 inactivated images of the regulation of EMT in development pathway (Figure 4). DataRobot was used for machine-learning modeling and 34 models were automatically created, including an ElasticNet Classifier (L2/Binomial Deviance) model. DataRobot also highlighted the parts of the image data critical to the prediction accuracy of the model in an activation map (Figure 4).

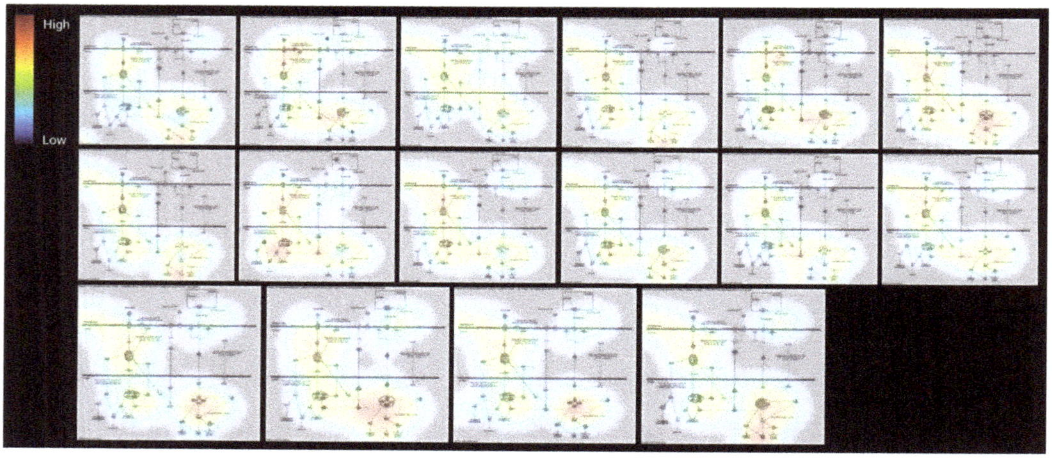

Figure 4. Activation map of AI modeling (DataRobot).

To validate the ElasticNet Classifier model, predictions were made using data on 10 activated and 10 inactivated pathway images that were not used to train the model (Table 2). The results showed that the prediction accuracy for the additional 20 images was 100% (AUC = 1.0).

Table 2. Validation of the model ElasticNet_Classifier_(L2/Binomial_Deviance).

Analysis Name	Disease State	Target Gene	Tissue	Treatment	EMT	Prediction	Label
18092-Breast adenocarcinoma CP466722 8993	breast adeno-carcinoma	ATM	Breast	Cp466722	TRUE	0.9693384	1
25525-Breast adenocarcinoma celastrol 17252	breast adeno-carcinoma	multiple targets	Breast	Celastrol	TRUE	0.99965132	1
25083-Breast adenocarcinoma CGP60474 16761	breast adeno-carcinoma	CDK1; CDK2	Breast	Cgp60474	TRUE	0.99881416	1
18267-Breast adenocarcinoma AZD8055 9187	breast adeno-carcinoma	mTOR	Breast	Azd8055	TRUE	0.99731849	1
7513-Breast adenocarcinoma OTSSP167 25473	breast adeno-carcinoma	MELK	Breast	Otssp167	TRUE	0.9991679	1
18469-Breast adenocarcinoma HG6-64-1 9411	breast adeno-carcinoma	B-RAF	Breast	Hg6-64-1	TRUE	0.99314697	1
25636-Breast adenocarcinoma HG6-64-1 17375	breast adeno-carcinoma	B-RAF	Breast	Hg6-64-1	TRUE	0.99867832	1
14-Breast carcinoma estradiol 1431	breast carcinoma		Breast	Estradiol	TRUE	0.99207239	1
895-Breast ductal carcinoma GSK1059615 27068	breast ductal carcinoma	PI3K; mTOR	Breast	Gsk1059615	TRUE	0.98180702	1
1263-Breast ductal carcinoma lapatinib 2924	breast ductal carcinoma	HER2; EGFR	Breast	Lapatinib	TRUE	0.99916824	1
9-Normal control olive pollen extract 16317	Normal control		Peripheral blood	Olive pollen extract	FALSE	0.00276633	0
37-Normal control 257	Normal control		Lung		FALSE	0.00027655	0
21926-Normal control rebastinib 13253	Normal control	BCR-ABL	Kidney	Rebastinib	FALSE	0.03588748	0
4-Normal control mock 16535	Normal control		Bone marrow	Mock	FALSE	0.00030339	0
15884-Normal control withaferin A 6539	Normal control	IKKβ	Breast	Withaferin A	FALSE	0.00271459	0

Table 2. *Cont.*

Analysis Name	Disease State	Target Gene	Tissue	Treatment	EMT	Prediction	Label
4-Normal control lipopolysaccharide (LPS) 15703	Normal control		Embryo	Lipopoly saccharide (LPS)	FALSE	0.00194256	0
10-Normal control co-culture 3076	Normal control		Peripheral blood	Co-culture	FALSE	0.00115878	0
6-Normal control actinomycin D 4750	Normal control		Fetal kidney	Actinomycin D	FALSE	0.01263976	0
2-Melanoma 35	Melanoma		Skin		FALSE	0.02006465	0
490-MYD88 deficiency lipopolysaccharide (LPS); polymyxin 12583	MYD88 deficiency		Peripheral blood	Lipopoly saccharide (LPS); polymyxin B	FALSE	0.03955118	0

3.4. Regulation of EMT in the Development Pathway in Other Diseases Than Cancer

The results of the modeling of regulation of EMT in the development pathway found one false-positive and one false-negative result in the model Elastic-Net Classifier in the process of the model generation (Figure 5). The analysis of the false-negative result was Parkinson's disease with a z-score of 3 (Figure 5a). The analysis of the false-positive result was a genetic disease with a z-score of −2.646 (Figure 5b).

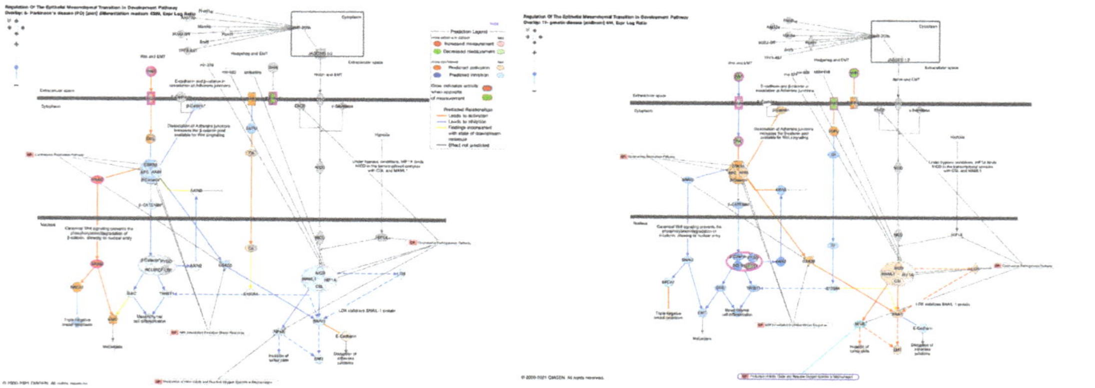

(a) (b)

Figure 5. Regulation of EMT in development pathway in diseases. (**a**) Parkinson's disease (PD) (skin) differentiation medium 4389, p value = 1.89×10^{-2}, z-score = 3; Gene identifiers marked with an asterisk (*) indicate that multiple identifiers in the dataset file map to a single gene in the Global Molecular Network. (**b**) genetic disease (midbrain) 444, p value = 4.75×10^{-2}, z-score = −2.646.

4. Discussion

Our result demonstrates that the canonical pathway of regulation of the EMT in the development pathway was activated in diffuse-type GC but not in intestinal-type GC. Specifically, the pathway mapping of gene expression revealed that Frizzled and APC were upregulated, while SUFU, PYGO2, and BRCA1 were downregulated in diffuse-type GC compared to intestinal-type GC. Frizzled proteins are a family of Wnt receptors involved in carcinogenesis [24]. It was previously shown that Frizzled-7 affected stemness and chemotherapeutic resistance in GC [25]. Accordingly, targeting inhibition of Frizzled-7 attenuated spheroid formation and stemness, as well as the resistance to cisplatin, an anti-cancer drug, in GC cells may have a therapeutic effect [25]. Besides Frizzled-7, the expression of Frizzled-10 was shown to have interesting correlation with cancer evolution. Importantly, as Frizzled-10 is not expressed in fully proliferative healthy tissue, but is highly expressed in certain cancerous tissue, it has the potential to be used as a prospective receptor molecule for targeted therapy. Intriguingly, it was found that while in GC, a decrease in cytoplasmic expression of Frizzled-10 is associated with increasing malignancy, while in colon cancer, the opposite is true; increased cytoplasmic expression of Frizzled-10 is crucial for the late stages of colon cancer progression and metastasis [24]. The co-localized expression of Frizzled family in different sub-types of cancer would confer progressive features on cancer.

APC is essential as a tumor suppressor protein in colorectal cancer and for its destruction complex functions, though its specific molecular activity has not been fully resolved [26]. The modeling or simulation of the cellular phenotype transition in EMT and diseases and predicting the molecular-induced responses in diseases would be useful for future investigation.

SUFU, PYGO2, and BRCA1 were downregulated in diffuse-type GC compared to intestinal-type GC. Previous findings have reported that SUFU, a regulator of Wnt signaling, was downregulated in GC and inhibited by miRNA-324-5p [27]. It was suggested that miRNA-324-5p induces EMT by inhibiting SUFU in GC [27]. PYGO2 was reported to be increased in human breast cancer [28]. The expression of PYGO2 was also assessed in glioma tissue samples and the results showed a positive correlation between tumor grade and PYGO2 overexpression [29]. The expression of PYGO2 was overexpressed in drug-resistant cell lines of GC and GC tissue after neoadjuvant chemotherapy [30]. It may be possible that PYGO2 has a different expression profile in diffuse-type GC compared to intestinal-type GC. BRCA1 was also downregulated in diffuse-type GC compared to intestinal-type GC. We have previously shown that the role of BRCA1 in the DNA damage response pathway was activated in intestinal-type GC compared to diffuse-type GC [18]. Accordingly, BRCA1 is rather important to intestinal-type GC.

The current study successfully generated AI-based models using 50 activated and 50 inactivated images of EMT gene regulation in the development pathway. The analyses in the database were selected based on the diseases and the treatment (Tables 1 and 2). Diseases in activated states of EMT regulation in the development pathway included bone osteosarcoma [31], breast carcinoma [32], and colon cancer [33]. AI application in gastrointestinal diseases would be a promising approach [34].

An interesting point of our current study is that the machine-learning modeling revealed that an IPA analysis of Parkinson's disease had a false-negative prediction result (Figure 5a). The color of the picture seems to be inactivated, which is in accordance with the prediction result as inactivated. Furthermore, it seems that EMT activation in the WNT pathway via SNAI2 resulted in the prediction being activated, whereas CSL-HIF1A-MAML1-NICD complex-induced EMT via SNAI1 was predicted as inactivated. In addition to Parkinson's disease, the machine-learning modeling revealed that an analysis of another unrelated genetic disease had a false-positive prediction result (Figure 5b). On the other hand, based on the analysis, GSK3β and SNAI1 were predicted as activated, while SNAI2 was inactivated (Figure 5b). The activation of GSK3β could be associated with the mediator role of GSK3β in the cross-talk of EMT signaling pathways [35].

5. Conclusions

The regulation of EMT in the development pathway was activated in diffuse-type GC and inactivated in intestinal-type GC. AI modeling with molecular pathway images generated the Elastic-Net Classifier model. The validation with 10 activated and 10 inactivated new pathway images, which were not used for the modeling, resulted in high accuracy. The modeling of the cellular phenotype transition in EMT and diseases will be studied in the near future.

Author Contributions: Conceptualization, S.T.; methodology, S.T.; formal analysis, S.T.; investigation, S.T.; writing—original draft preparation, S.T.; writing—review and editing, S.T., S.Q., R.O., H.C., K.A., A.H., E.J.P., H.Y. and H.S.; visualization, S.T.; project administration, S.T.; funding acquisition, S.T., S.Q., R.O. and A.H. All authors have read and agreed to the published version of the manuscript.

Funding: This research was funded by the Japan Agency for Medical Research and Development (AMED) Grant Number JP20ak0101093 (S.T., R.O. and A.H.), JP21mk0101216 (S.T.), JP22mk0101216 (S.T.), and Strategic International Collaborative Research Program, Grant Number JP20jm0210059 (S.T. and S.Q.), Japan Society for the Promotion of Science (JSPS) KAKENHI Grant Number 21K12133 (S.T. and R.O.).

Institutional Review Board Statement: Not applicable.

Informed Consent Statement: Not applicable.

Data Availability Statement: Not applicable.

Acknowledgments: The authors would like to acknowledge Shinpei Ijichi for assisting with the DataRobot Automated Machine Learning platform. The authors are grateful to all colleagues including members of the National Institute of Health Sciences, Japan for their support. This research was supported by the Ministry of Health, Labour, and Welfare, Japan.

Conflicts of Interest: The authors declare no conflict of interest.

References

1. Tanabe, S.; Quader, S.; Ono, R.; Cabral, H.; Aoyagi, K.; Hirose, A.; Yokozaki, H.; Sasaki, H. Molecular Network Profiling in Intestinal- and Diffuse-Type Gastric Cancer. *Cancers* **2020**, *12*, 3833. [CrossRef]
2. Landeros, N.; Santoro, P.M.; Carrasco-Avino, G.; Corvalan, A.H. Competing Endogenous RNA Networks in the Epithelial to Mesenchymal Transition in Diffuse-Type of Gastric Cancer. *Cancers* **2020**, *12*, 2741. [CrossRef]
3. Perrot-Applanat, M.; Vacher, S.; Pimpie, C.; Chemlali, W.; Derieux, S.; Pocard, M.; Bieche, I. Differential gene expression in growth factors, epithelial mesenchymal transition and chemotaxis in the diffuse type compared with the intestinal type of gastric cancer. *Oncol. Lett.* **2019**, *18*, 674–686. [CrossRef]
4. Tanabe, S.; Quader, S.; Cabral, H.; Ono, R. Interplay of EMT and CSC in Cancer and the Potential Therapeutic Strategies. *Front. Pharmacol.* **2020**, *11*, 904. [CrossRef]
5. Lambert, A.W.; Pattabiraman, D.R.; Weinberg, R.A. Emerging Biological Principles of Metastasis. *Cell* **2017**, *168*, 670–691. [CrossRef]
6. Du, B.; Shim, J.S. Targeting Epithelial-Mesenchymal Transition (EMT) to Overcome Drug Resistance in Cancer. *Molecules* **2016**, *21*, 965. [CrossRef]
7. Zhang, Y.; Weinberg, R.A. Epithelial-to-mesenchymal transition in cancer: Complexity and opportunities. *Front. Med.* **2018**, *12*, 361–373. [CrossRef]
8. Sommers, C.L.; Heckford, S.E.; Skerker, J.M.; Worland, P.; Torri, J.A.; Thompson, E.W.; Byers, S.W.; Gelmann, E.P. Loss of epithelial markers and acquisition of vimentin expression in adriamycin- and vinblastine-resistant human breast cancer cell lines. *Cancer Res.* **1992**, *52*, 5190–5197.
9. Kaimori, A.; Potter, J.; Kaimori, J.Y.; Wang, C.; Mezey, E.; Koteish, A. Transforming growth factor-beta1 induces an epithelial-to-mesenchymal transition state in mouse hepatocytes in vitro. *J. Biol. Chem.* **2007**, *282*, 22089–22101. [CrossRef]
10. Yook, J.I.; Li, X.-Y.; Ota, I.; Fearon, E.R.; Weiss, S.J. Wnt-dependent Regulation of the E-cadherin Repressor Snail*. *J. Biol. Chem.* **2005**, *280*, 11740–11748. [CrossRef]
11. Babaei, G.; Aziz, S.G.; Jaghi, N.Z.Z. EMT, cancer stem cells and autophagy; The three main axes of metastasis. *Biomed Pharm.* **2021**, *133*, 110909. [CrossRef] [PubMed]
12. Hill, C.; Wang, Y. The importance of epithelial-mesenchymal transition and autophagy in cancer drug resistance. *Cancer Drug Resist.* **2020**, *3*, 38–47. [CrossRef] [PubMed]
13. Sanchez-Vega, F.; Mina, M.; Armenia, J.; Chatila, W.K.; Luna, A.; La, K.C.; Dimitriadoy, S.; Liu, D.L.; Kantheti, H.S.; Saghafinia, S.; et al. Oncogenic Signaling Pathways in The Cancer Genome Atlas. *Cell* **2018**, *173*, 321–337.e310. [CrossRef] [PubMed]
14. Bass, A.J.; Thorsson, V.; Shmulevich, I.; Reynolds, S.M.; Miller, M.; Bernard, B.; Hinoue, T.; Laird, P.W.; Curtis, C.; Shen, H.; et al. Comprehensive molecular characterization of gastric adenocarcinoma. *Nature* **2014**, *513*, 202–209. [CrossRef]
15. Cerami, E.; Gao, J.; Dogrusoz, U.; Gross, B.E.; Sumer, S.O.; Aksoy, B.A.; Jacobsen, A.; Byrne, C.J.; Heuer, M.L.; Larsson, E.; et al. The cBio Cancer Genomics Portal: An Open Platform for Exploring Multidimensional Cancer Genomics Data. *Cancer Discov.* **2012**, *2*, 401–404. [CrossRef]
16. Gao, J.; Aksoy, B.A.; Dogrusoz, U.; Dresdner, G.; Gross, B.; Sumer, S.O.; Sun, Y.; Jacobsen, A.; Sinha, R.; Larsson, E.; et al. Integrative Analysis of Complex Cancer Genomics and Clinical Profiles Using the cBioPortal. *Sci. Signal.* **2013**, *6*, pl1. [CrossRef]

17. Grossman, R.L.; Heath, A.P.; Ferretti, V.; Varmus, H.E.; Lowy, D.R.; Kibbe, W.A.; Staudt, L.M. Toward a Shared Vision for Cancer Genomic Data. *N. Engl. J. Med.* **2016**, *375*, 1109–1112. [CrossRef]
18. Tanabe, S.; Quader, S.; Ono, R.; Cabral, H.; Aoyagi, K.; Hirose, A.; Yokozaki, H.; Sasaki, H. Cell Cycle Regulation and DNA Damage Response Networks in Diffuse- and Intestinal-Type Gastric Cancer. *Cancers* **2021**, *13*, 5786. [CrossRef]
19. Krämer, A.; Green, J.; Pollard, J., Jr.; Tugendreich, S. Causal analysis approaches in Ingenuity Pathway Analysis. *Bioinformatics* **2013**, *30*, 523–530. [CrossRef]
20. Pospisil, P.; Iyer, L.K.; Adelstein, S.J.; Kassis, A.I. A combined approach to data mining of textual and structured data to identify cancer-related targets. *BMC Bioinform.* **2006**, *7*, 354. [CrossRef]
21. Breiman, L. Random Forests. *Mach. Learn.* **2001**, *45*, 5–32. [CrossRef]
22. Friedman, J.H. Greedy function approximation: A gradient boosting machine. *Ann. Stat.* **2001**, *29*, 1189–1232. [CrossRef]
23. Liu, D.; Wang, X.; Li, L.; Jiang, Q.; Li, X.; Liu, M.; Wang, W.; Shi, E.; Zhang, C.; Wang, Y.; et al. Machine Learning-Based Model for the Prognosis of Postoperative Gastric Cancer. *Cancer Manag. Res.* **2022**, *14*, 135–155. [CrossRef]
24. Scavo, M.P.; Fucci, L.; Caldarola, L.; Mangia, A.; Azzariti, A.; Simone, G.; Gasparini, G.; Krol, S. Frizzled-10 and cancer progression: Is it a new prognostic marker? *Oncotarget* **2018**, *9*, 824–830. [CrossRef]
25. Cheng, Y.; Li, L.; Pan, S.; Jiang, H.; Jin, H. Targeting Frizzled-7 Decreases Stemness and Chemotherapeutic Resistance in Gastric Cancer Cells by Suppressing Myc Expression. *Med. Sci. Monit.* **2019**, *25*, 8637–8644. [CrossRef]
26. Nusse, R.; Clevers, H. Wnt/β-Catenin Signaling, Disease, and Emerging Therapeutic Modalities. *Cell* **2017**, *169*, 985–999. [CrossRef]
27. Peng, Y.; Zhang, X.; Lin, H.; Deng, S.; Qin, Y.; Yuan, Y.; Feng, X.; Wang, J.; Chen, W.; Hu, F.; et al. SUFU mediates EMT and Wnt/β-catenin signaling pathway activation promoted by miRNA-324-5p in human gastric cancer. *Cell Cycle* **2020**, *19*, 2720–2733. [CrossRef]
28. Chi, Y.; Wang, F.; Zhang, T.; Xu, H.; Zhang, Y.; Shan, Z.; Wu, S.; Fan, Q.; Sun, Y. miR-516a-3p inhibits breast cancer cell growth and EMT by blocking the Pygo2/Wnt signalling pathway. *J. Cell Mol. Med.* **2019**, *23*, 6295–6307. [CrossRef]
29. Wang, Z.X.; Chen, Y.Y.; Li, B.A.; Tan, G.W.; Liu, X.Y.; Shen, S.H.; Zhu, H.W.; Wang, H.D. Decreased pygopus 2 expression suppresses glioblastoma U251 cell growth. *J. Neurooncol.* **2010**, *100*, 31–41. [CrossRef]
30. Zhang, D.; Liu, Y.; Wu, Q.; Zheng, Y.; Kaweme, N.M.; Zhang, Z.; Cai, M.; Dong, Y. Pygo2 as a novel biomarker in gastric cancer for monitoring drug resistance by upregulating MDR1. *J. Cancer* **2021**, *12*, 2952–2959. [CrossRef]
31. Ye, C.; Ho, D.J.; Neri, M.; Yang, C.; Kulkarni, T.; Randhawa, R.; Henault, M.; Mostacci, N.; Farmer, P.; Renner, S.; et al. DRUG-seq for miniaturized high-throughput transcriptome profiling in drug discovery. *Nat. Commun.* **2018**, *9*, 4307. [CrossRef] [PubMed]
32. Hammerich-Hille, S.; Kaipparettu, B.A.; Tsimelzon, A.; Creighton, C.J.; Jiang, S.; Polo, J.M.; Melnick, A.; Meyer, R.; Oesterreich, S. SAFB1 mediates repression of immune regulators and apoptotic genes in breast cancer cells. *J. Biol. Chem.* **2010**, *285*, 3608–3616. [CrossRef] [PubMed]
33. Wyce, A.; Matteo, J.J.; Foley, S.W.; Felitsky, D.J.; Rajapurkar, S.R.; Zhang, X.P.; Musso, M.C.; Korenchuk, S.; Karpinich, N.C.; Keenan, K.M.; et al. MEK inhibitors overcome resistance to BET inhibition across a number of solid and hematologic cancers. *Oncogenesis* **2018**, *7*, 35. [CrossRef] [PubMed]
34. Tanabe, S.; Perkins, E.J.; Ono, R.; Sasaki, H. Artificial intelligence in gastrointestinal diseases. *Artif. Intell. Gastroenterol.* **2021**, *2*, 69–76. [CrossRef]
35. Gonzalez, D.M.; Medici, D. Signaling mechanisms of the epithelial-mesenchymal transition. *Sci. Signal.* **2014**, *7*, re8. [CrossRef]

cancers

MDPI

Article

Machine Learning Model to Stratify the Risk of Lymph Node Metastasis for Early Gastric Cancer: A Single-Center Cohort Study

Ji-Eun Na [1,2,†], Yeong-Chan Lee [3,†], Tae-Jun Kim [1,*], Hyuk Lee [1,*], Hong-Hee Won [3], Yang-Won Min [1], Byung-Hoon Min [1], Jun-Haeng Lee [1], Poong-Lyul Rhee [1] and Jae J. Kim [1]

1 Department of Medicine, Samsung Medical Center, Sungkyunkwan University School of Medicine, Seoul 06351, Korea; jieun90.na@samsung.com (J.-E.N.); yangwon.min@samsung.com (Y.-W.M.); lamsu.min@samsung.com (B.-H.M.); jh2145.lee@samsung.com (J.-H.L.); pl.rhee@samsung.com (P.-L.R.); jaej.kim@samsung.com (J.J.K.)

2 Department of Medicine, Inje University Haeundae Paik Hospital, Busan 48108, Korea

3 Department of Digital Health, Samsung Advanced Institute for Health Sciences & Technology (SAIHST), Sungkyunkwan University of Medicine, Seoul 06351, Korea; conan_8th@naver.com (Y.-C.L.); wonhh@skku.edu (H.-H.W.)

* Correspondence: taejunk91@gmail.com (T.-J.K.); leehyuk@skku.edu (H.L.); Tel.: +82-234-103-409 (T.-J.K. & H.L.); Fax: +82-234-106-983 (T.-J.K. & H.L.)

† These authors contributed equally to this work.

Simple Summary: Endoscopic resection (ER) is a treatment option for clinically T1a early gastric cancer (EGC) without suspicion of lymph node metastasis (LNM). In patients with non-curative resection after ER, additional surgery is recommended owing to the LNM risk. However, of those patients treated with additional surgery after ER, the actual rate of LNM was about 5–10%; that is, the other patients underwent unnecessary surgeries. Therefore, it is crucial to estimate LNM risk in EGC patients to determine additional management after ER. We derived a machine learning (ML) model to stratify the LNM risk in EGC patients and validate its performance. The constructed ML model, which showed good performance with an area under the receiver operating characteristic of 0.85 or higher, could stratify LNM risk into very low (<1%), low (<3%), intermediate (<7%), and high (≥7%) risk categories. These findings suggest that the ML model can stratify the LNM risk in EGC patients.

Abstract: Stratification of the risk of lymph node metastasis (LNM) in patients with non-curative resection after endoscopic resection (ER) for early gastric cancer (EGC) is crucial in determining additional treatment strategies and preventing unnecessary surgery. Hence, we developed a machine learning (ML) model and validated its performance for the stratification of LNM risk in patients with EGC. We enrolled patients who underwent primary surgery or additional surgery after ER for EGC between May 2005 and March 2021. Additionally, patients who underwent ER alone for EGC between May 2005 and March 2016 and were followed up for at least 5 years were included. The ML model was built based on a development set (70%) using logistic regression, random forest (RF), and support vector machine (SVM) analyses and assessed in a validation set (30%). In the validation set, LNM was found in 337 of 4428 patients (7.6%). Among the total patients, the area under the receiver operating characteristic (AUROC) for predicting LNM risk was 0.86 in the logistic regression, 0.85 in RF, and 0.86 in SVM analyses; in patients with initial ER, AUROC for predicting LNM risk was 0.90 in the logistic regression, 0.88 in RF, and 0.89 in SVM analyses. The ML model could stratify the LNM risk into very low (<1%), low (<3%), intermediate (<7%), and high (≥7%) risk categories, which was comparable with actual LNM rates. We demonstrate that the ML model can be used to identify LNM risk. However, this tool requires further validation in EGC patients with non-curative resection after ER for actual application.

Keywords: early gastric cancer; machine learning model; risk stratification; lymph node metastasis

Citation: Na, J.-E.; Lee, Y.-C.; Kim, T.-J.; Lee, H.; Won, H.-H.; Min, Y.-W.; Min, B.-H.; Lee, J.-H.; Rhee, P.-L.; Kim, J.J. Machine Learning Model to Stratify the Risk of Lymph Node Metastasis for Early Gastric Cancer: A Single-Center Cohort Study. *Cancers* **2022**, *14*, 1121. https://doi.org/10.3390/cancers14051121

Academic Editors: Hamid Khayyam, Ali Madani, Rahele Kafieh and Ali Hekmatnia

Received: 14 January 2022
Accepted: 20 February 2022
Published: 22 February 2022

1. Introduction

Early gastric cancer (EGC) describes a gastric tumor confined to the submucosa with or without lymph node metastasis (LNM). Endoscopic resection (ER) is recommended as a minimally invasive treatment for clinically mucosal EGC without suspicion of LNM [1–4]. In cases of non-curative resection after ER that do not satisfy the expanded criteria of curative resection, additional surgery is recommended, considering the risk of LNM [5,6]; however, LNM is found in only 5–10% of those patients after surgery [7–10]. Therefore, overtreatment is a concern. To address this, the recently revised guidelines excluded piecemeal resection and a positive lateral margin from the factors of non-curative resection after ER for which additional surgery is primarily recommended [1,4,11].

Furthermore, in Japan, patients who have non-curative resection after ER, excluding piecemeal resection and a positive lateral margin, are classified as "endoscopic curability (eCura) C-2"; patients in the eCura C-2 category are further stratified into low (2.5%), intermediate (6.7%), and high (22.7%) LNM risk categories based on the eCura scoring system [2,12,13]. In the low-risk category, there is no difference in cancer recurrence or cancer-specific mortality between patients who undergo no additional treatment and those who undergo additional surgery [14]. Hence, this LNM risk stratification system suggests that additional surgery after non-curative resection may be determined on an individual basis, considering the LNM risk, the patient's condition, and the benefits and limitations of additional surgery [11,12,14].

Another area of concern is that some patients who were confirmed non-curative resection after ER without actual LNM may be unnecessarily exposed to surgery-related risks. The rates of postoperative complications and overall mortality after gastric cancer surgery are 10–26% and 0.3–2.3%, respectively, and comorbidities, body mass index, and lymph node dissection have been reported as risk factors [15–21]. In addition, the potential for long-term health problems after gastric cancer surgery, such as reflux, gastroparesis, gallstone, and osteoporosis, must be considered [22,23]. Therefore, it is clinically significant to predict the LNM risk among EGC patients who undergo non-curative resection after ER to prevent unnecessary surgery.

To stratify the LNM risk in EGC patients, we created a machine learning (ML) model for predicting LNM risk and validated its performance.

2. Materials and Methods

2.1. Patients

We included patients who underwent surgery for EGC between May 2005 and March 2021 at Samsung Medical Center. Additionally, patients who underwent additional surgery after ER owing to complications or non-curative resection were included. Moreover, patients who underwent ER alone for EGC without surgery between May 2005 and March 2016 were included and followed up for at least 5 years. After excluding patients with missing data, a total of 14,760 patients who underwent surgery (n = 12,631) or ER alone (n = 2129) were included (Figure 1). The patients were randomly divided into the development set (70%) and validation set (30%).

2.2. Definition, Outcome, Data Sources, and Study Variables

LNM was defined based on surgical specimens of patients who underwent surgery. In patients who underwent ER alone, regional LN recurrence was determined based on computed tomography scans during follow-up.

The outcome consisted of establishing the ML model for predicting LNM risk in EGC patients and validating its performance. We divided the entire cohort into a development set (70%) for derivation of the ML model and a validation set (30%) for validation. Since the actual target participants were patients treated with ER for EGC, the performance of the ML model was evaluated for total patients and initial ER patients, respectively, using three methods in the development set and validation set. First, the area under the receiver operating characteristic (AUROC), sensitivity, and specificity of the ML model were analyzed. Second, we assessed whether the ML model could stratify the risk of LNM into very low-, low-, intermediate-, and high-risk categories. In the development set, we listed the predicted values calculated by the ML model and selected cutoffs at the points where the actual LNM rates were 1%, 3%, and 7%. An actual LNM rate <1% was allocated into the very low-, <3% into the low-, <7% into the intermediate-, and ≥7% into the high-risk categories. The 3% and 7% criteria for the low-, intermediate-, and high-risk categories were based on the previous literature [12]. Additionally, we set a very-low risk category of predicted LNM risk with <1%. This ML model for stratifying LNM risk was applied to the total patients and patients with initial ER in the validation set. Third, we evaluated the ability of the ML model to discriminate patients with negligible risk of LNM at a high-sensitivity cutoff of 100% to predict LNM. From a clinical perspective, the utility of a risk score depends on its ability to discriminate patients at low risk for LNM, i.e., it is ideal to identify patients who do not need surgery and those who need surgery.

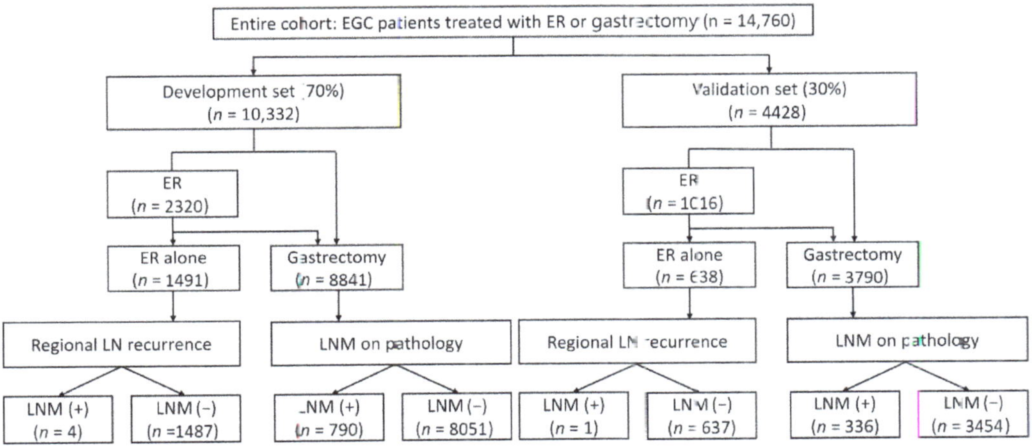

Figure 1. Diagram of patient selection.

Non-curative resection was defined as not satisfying an expanded criterion for curative resection. The expanded criteria for curative resection were en bloc resection, negative horizontal and vertical margins, absence of lymphovascular invasion, and one of the following: (a) differentiated mucosal cancer without ulcerative lesions, regardless of the tumor size; (b) differentiated mucosal cancer with ulcerative lesions that were ≤3 cm in size; (c) undifferentiated mucosal cancer without ulcerative lesions that were ≤2 cm in size; or (d) differentiated cancer invasion to the submucosa <500 µm from the muscularis mucosa that was ≤3 cm in size.

Data were collected retrospectively from the electronic medical records, including age, sex, number of tumors, tumor location (upper third, middle third, and lower third), size (mm), gross type (non-depressed and depressed), differentiation (well, moderate, signet, and poor), Lauren classification (intestinal, diffuse, and mixed), depth of invasion (lamina propria, muscularis mucosa, submucosal invasion <500 µm from the muscularis mucosa (SM1), and submucosal invasion ≥500 µm from the muscularis mucosa (SM2/3)), lymphatic invasion, venous invasion, and perineural invasion.

2.3. Establishment of the Machine Learning Model

The ML model was implemented using 3 methods to produce an optimal model based on the development set (70%): logistic regression, support vector machine (SVM), and random forest (RF) We constructed the ML model in the cohort of total patients and patients with initial ER, respectively. This design considered our actual target as EGC patients who were feasible ER. A randomized search algorithm with fivefold nested cross-validation in the development set was conducted for hyperparameter optimization of each method. The algorithm was optimized by randomly searching the given hyperparameter space 1000 times using the development set (Table S1). We selected this search algorithm rather than grid or Bayesian search algorithms because these three methods are fast enough to search all given spaces and have relatively few hyperparameters. The best hyperparameters in a model were chosen when the model had the highest AUROC. The performance of the models with the best hyperparameters was evaluated in the validation set (30%). We defined the weighted factors of 14.0 through the imbalanced rate of the classes. We confirmed the feature importance as permutating a specific variable 100 times. We publicly opened the codes and models at https://github.com/YeongChanLee/Predict-LNM (accessed on 21 February 2022).

2.4. Statistical Analysis

Baseline characteristics were compared between the development and validation sets and presented as means (standard deviation) and frequencies (%) for continuous and categorical variables, respectively. The performance of the ML model was evaluated using AUROC, sensitivity, and specificity. The sensitivity and specificity were derived using Youden's index. The risk probability was calculated for the stratification of LNM risk based on the logistic regression, RF, and SVM analyses in the development set. Predicted LNM risk was classified into very low-, low-, intermediate-, and high-risk categories according to the actual LNM rate with a cutoff <1%, <3%, and <7%. We analyzed whether the categories of predicted LNM risk correlated with the real LNM rate. As a subanalysis, the performance of the ML model was compared with the eCura system as a clinical model in cases defined as non-curative resection after ER for EGC in the validation set, using AUROC, net reclassification improvement (NRI), and specificity at a high-sensitivity cutoff of 95%. The ML model was developed using Scikit-learn 0.24.1 and Python 3.8.5. Statistical analyses were performed using R (version 3.5.1, Vienna, Austria).

3. Results

3.1. Baseline Characteristics

A total of 14,760 patients were eligible for analysis; 10,332 patients were randomly sorted into the development set and 4428 into the validation set. LNM was found in 794 of 10,332 patients (7.7%) in the development set and 337 of 4428 patients (7.6%) in the validation set. The baseline characteristics of the development and validation sets are shown in Table 1. They were comparable in most variables, including age, sex, number of tumors, size, gross type, differentiation, Lauren classification, depth of invasion, lymphatic invasion, venous invasion, and perineural invasion. However, the middle-third of the stomach was the most frequent tumor location in the development set whereas the lower-third of the stomach was the most frequent tumor location in the validation set ($p = 0.013$).

Table 1. Baseline characteristics of the development set and validation set.

Variable	Development (n = 10,332)	Validation (n = 4428)	p Value [a]
Age [†]	58 ± 11	58 ± 11	0.413
Gender			0.789
Male	6697 (65)	2881 (65)	
Female	3635 (35)	1547 (35)	
tumors	512 (5)	201 (5)	
Location			0 013
Upper	1083 (11)	483 (11)	
Middle	4773 (46)	1929 (44)	
Lower	4476 (43)	2016 (45)	
Size (mm) [†]	27 ± 18	27 ± 18	0.645
Gross type			0.823
Non-depressed	2568 (25)	1109 (25)	
Depressed	7764 (75)	3319 (75)	
Differentiation			0.999
Well	1214 (12)	523 (12)	
Moderate	4053 (39)	1741 (39)	
Signet	2315 (22)	989 (22)	
Poorly	2750 (27)	1175 (27)	
Histologic type by Lauren			0.122
Intestinal	5198 (50)	2271 (51)	
Diffuse	3867 (38)	1666 (38)	
Mixed	1267 (12)	491 (11)	
Depth of invasion			0.983
Lamina propria	2568 (25)	1114 (25)	
Muscularis mucosa	3767 (37)	1612 (37)	
SM1	1069 (10)	455 (10)	
SM2/3	2928 (28)	1247 (28)	
Lymphatic invasion, present	1571 (15)	682 (15)	0.780
Venous invasion, present	154 (2)	72 (2)	0.588
Perineural invasion, present	232 (2)	96 (2)	0.817

[†] Mean ± standard deviation presented for continuous variables. Values are expressed as n (%); unless otherwise specified. [a] p-value calculated using Student's t-test for continuous variables or Pearson's chi-square test for categorical variables for overall data. SM1: submucosal invasion <500 µm from the muscularis mucosa; SM2/3: submucosal invasion ≥500 µm from the muscularis mucosa.

3.2. Derivation of the Machine Learning Model

In the development set, LNM was found in 794 of 10,332 patients (7.7%) in the total patients, and in 42 of 2320 patients (1.8%) in patients with initial ER. The derivatated ML model showed good to excellent performance in the development set; in the total patients, logistic regression was AUROC (95% CI), 0.86 (0.85–0.88); sensitivity, 0.80; and specificity, 0.76; RF was AUROC (95% CI), 0.95 (0.94–0.95); sensitivity, 0.91; and specificity, 0.86; and SVM was AUROC (95% CI), 0.87 (0.85–0.88); sensitivity, 0.79; and specificity, 0.78. In patients with initial ER, logistic regression was AUROC (95% CI), 0.88 (0.83–0.92); sensitivity, 0.86; and specificity 0.82; RF was AUROC (95% CI), 0.95 (0.93–0.97); sensitivity, 0.93; and specificity, 0.88; and SVM was AUROC (95% CI), 0.88 (0.83–0.92); sensitivity, 0.93; and specificity, 0.73 (Figure 2).

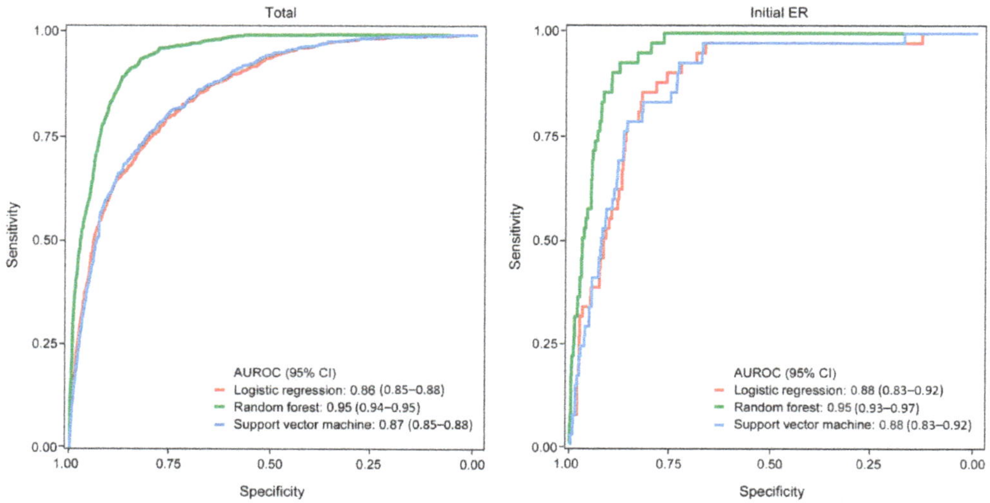

Figure 2. AUROC of the ML model for the prediction of LNM in the development set (total number = 10,332, number of patients with initial ER = 2320).

In the development set, LNM risk was predicted using the ML model (logistic regression, RF, and SVM), and the cutoff for the categories of very low, low, intermediate, and high risk was set as the value of the actual LNM rate of <1%, <3%, and <7% in the total patients and initial ER patients, respectively (Table 2). As an example, in the total patients, LNM risk was stratified using logistic regression into very low (<1%)-, low (<3%)-, intermediate (<7%)-, and high (≥7%)-risk categories, and the cutoff was determined by the actual LNM rate. Each category showed a real LNM rate of 0.2%, 1.4%, 4.1%, and 18.4% (Table 2).

Table 2. Determination of the cutoff for stratification of LNM risk based on the predictive value of the ML model and actual LNM rate in the development set. (**A**) Total patients. (**B**) Patients with initial ER.

(A) Total Patients (*n* = 10,332) and LNM (*n* = 794)				
		Logistic regression		
n of patients	*n* of LNM	Rate (%)	Risk probability	Risk category
1863	3	0.2	<1%	Very low
3105	42	1.4	≥1% to <3%	Low
1656	67	4.1	≥3% to <7%	Intermediate
3708	682	18.4	≥7%	High
		Random forest		
n of patients	*n* of LNM	Rate (%)	Risk probability	Risk category
5589	2	<0.1	<1%	Very low
1859	24	1.3	≥1% to <3%	Low
412	18	4.4	≥3% to <7%	Intermediate
2472	750	30.3	≥7%	High
		Support vector machine		
n of patients	*n* of LNM	Rate (%)	Risk probability	Risk category
2277	5	0.2	<1%	Very low
2691	35	1.3	≥1% to <3%	Low
1656	65	3.9	≥3% to <7%	Intermediate
3708	689	18.6	≥7%	High

Table 2. *Cont.*

(B) Initial ER(n = 2320) and LNM (n = 42)				
Logistic regression				
n of patients	n of LNM	Rate (%)	Risk probability	Risk category
1492	1	0.1	<1%	Very low
368	5	1.4	≥1% to <3%	Low
92	3	3.3	≥3% to <7%	Intermediate
368	33	9.0	≥7%	High
Random forest				
n of patients	n of LNM	Rate (%)	Risk probability	Risk category
1722	0	0	<1%	Very low
322	4	1.2	≥1% to <3%	Low
46	2	4.4	≥3% to <7%	Intermediate
230	36	15.7	≥7%	High
Support vector machine				
n of patients	n of LNM	Rate (%)	Risk probability	Risk category
1491	1	0.1	<1%	Very low
136	2	1.5	≥1% to <3%	Low
445	15	3.3	≥3% to <7%	Intermediate
206	24	10.4	≥7%	High

LNM, lymph node metastasis.

3.3. *Validation of the Machine Learning Model*

In the validation set, LNM was found in 337 of 4428 patients (7.6%) in the total patients, and in 24 of 1016 patients (2.4%) in patients with initial ER. In the validation set, the ML model showed a good performance in the total patients and patients with initial ER. In total patients, logistic regression was AUROC (95% CI), 0.86 (0.84–0.88); sensitivity, 0.80; and specificity, 0.75; RF was AUROC (95% CI), 0.85 (0.83–0.87); sensitivity, 0.82; and specificity, 0.72; and SVM was AUROC (95% CI), 0.86 (0.84–0.88); sensitivity, 0.69; and specificity, 0.85. In patients with initial ER, logistic regression was AUROC (95% CI), 0.90 (0.86–0.94); sensitivity, 0.92; and specificity, 0.77; RF was AUROC (95% CI), 0.88 (0.82–0.92); sensitivity, 0.92; and specificity, 0.74; and SVM was AUROC (95% CI), 0.89 (0.85–0.93); sensitivity, 0.92; and specificity, 0.78 (Figure 3).

In the validation set, logistic regression and SVM showed the possibility of stratifying the risk of LNM for total patients and patients with initial ER. The predicted LNM risk was correlated with the actual LNM rate. In the total patients, the actual LNM rate according to the very low-, low-, intermediate-, and high-risk categories was 0.1%, 1.6%, 4.3%, and 17.7% based on logistic regression and 0.1%, 1.6%, 4.2%, and 18.1% based on SVM, respectively. In patients with initial ER, the actual LNM rate according to the very low-, low-, intermediate-, and high-risk categories was 0.2%, 2.5%, 0.0%, and 11.9% based on logistic regression and 0.2%, 1.7%, 4.5%, and 13.0% based on SVM, respectively. In contrast, in the analysis using RF, the actual LNM rate was 1.3%, 5.3%, 7.4%, and 23.1% of the total patients and 0.4%, 5.0%, 10.0%, and 12.0% of patients with initial ER, which was higher than that of the predicted category of LNM risk (Table 3)

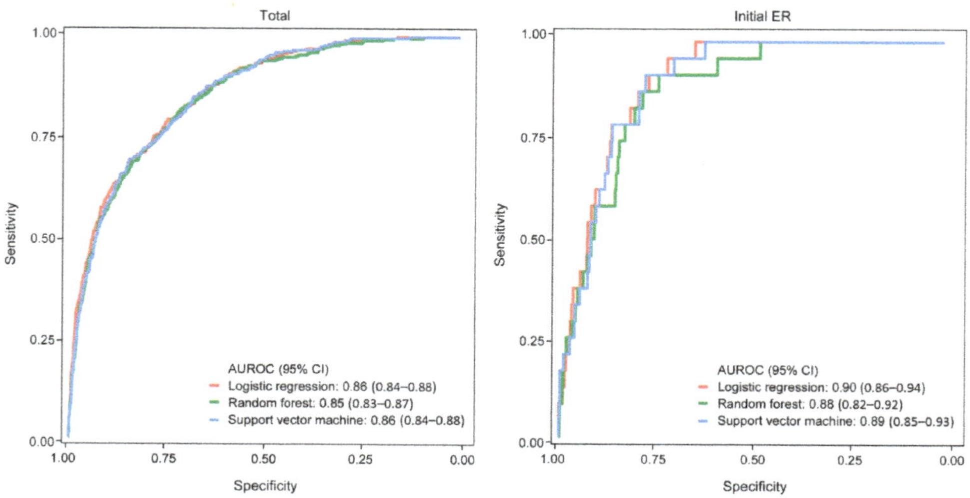

Figure 3. AUROC of the ML model for the prediction of LNM in the validation set (total number = 4428, number with initial ER = 1016).

Table 3. Risk stratification of LNM by the ML model and the actual rate in the validation set. (**A**) Total patients. (**B**) Patients with initial ER.

(A) Total Patients (*n* = 4428) and LNM (*n* = 337)				
		Logistic regression		
Risk probability	Risk category	*n* of patients	*n* of LNM	Rate (%)
<1%	Very low	801	1	0.1
≥1% to <3%	Low	1335	21	1.6
≥3% to <7%	Intermediate	708	34	4.8
≥7%	High	1584	281	17.7
		Random forest		
Risk probability	Risk category	*n* of patients	*n* of LNM	Rate (%)
<1%	Very low	2403	30	1.3
≥1% to <3%	Low	793	50	6.3
≥3% to <7%	Intermediate	176	13	7.4
≥7%	High	1056	244	23.1
		Support vector machine		
Risk probability	Risk category	*n* of patients	*n* of LNM	Rate (%)
<1%	Very low	978	1	0.1
≥1% to <3%	Low	1138	19	1.6
≥3% to <7%	Intermediate	678	30	4.2
≥7%	High	1297	287	18.1
(B) Patients with Initial ER (*n* = 1016) and LNM (*n* = 24)				
		Logistic regression		
Risk probability	Risk category	*n* of patients	*n* of LNM	Rate (%)
<1%	Very low	656	1	0.2
≥1% to <3%	Low	160	4	2.5
≥3% to <7%	Intermediate	40	0	0
≥7%	High	160	19	11.9

Table 3. *Cont.*

(B) Patients with Initial ER (n = 1016) and LNM (n = 24)				
Random forest				
Risk probability	Risk category	n of patients	n of LNM	Rate (%)
<1%	Very low	756	3	0.4
≥1% to <3%	Low	140	7	5.0
≥3% to <7%	Intermediate	20	2	10.0
≥7%	High	100	12	12.0
Support vector machine				
Risk probability	Risk category	n of patients	n of LNM	Rate (%)
<1%	Very low	655	1	0.2
≥1% to <3%	Low	59	1	1.7
≥3% to <7%	Intermediate	191	9	4.5
≥7%	High	87	13	13.0

In the total patients in the validation set, the specificities of the ML model at the high-sensitivity cutoff of 100% were 49%, 46%, and 49% in the logistic regression, RF, and SVM analyses, respectively. In patients with initial ER, the specificities of the ML model at the high-sensitivity cutoff of 100% were 71%, 57%, and 70% in the logistic regression, RF, and SVM analyses, respectively (Figure 4).

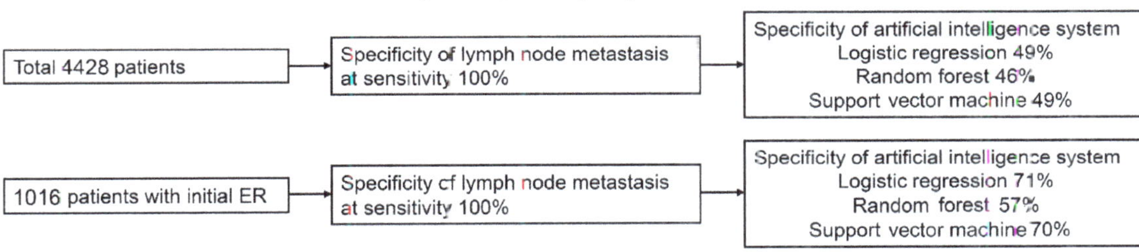

Figure 4. Identification of patients with negligible risk of lymph node metastasis at the high-sensitivity cutoff in the validation set.

In the validation set, as a subanalysis in the patients with non-curative resection after ER for EGC, LNM was found in 21 of 362 patients (5.8%). The AUROC of the ML model was 0.76, 0.73, and 0.75 in the logistic regression, RF, and SVM analyses, respectively, and the AUROC of the eCura system was 0.72. Logistic regression (NRI, 0.46) and SMV (NRI, 0.21) improved the performance compared to the eCura system. The specificities of the ML model at the high-sensitivity cutoff of 95% were 39%, 38%, and 38% in the logistic regression, RF, and SVM analyses, respectively, which were higher than the specificity of 9% for the eCura system (Figure S1).

4. Discussion

Here, we demonstrated the utility of an ML model for predicting the LNM risk in EGC patients. In the validation set, the AUROC of each ML model showed a good performance, ranging from 0.85 to 0.90. Furthermore, each ML model could stratify the LNM risk as very low, low, intermediate, and high risk, and those stratified groups showed a consistent actual LNM rate. In addition, these showed specificities of about 0.50 or higher at a matched sensitivity of 100%, indicating that it could discriminate patients with negligible risk of LNM while identifying the patients who needed surgery owing to the LNM risk with 100% sensitivity. This tool can easily be applied in clinical practice to categorize the LNM risk and identify patients with negligible LNM risk under the assumption of maximum sensitivity.

Non-curative resection after ER for EGC patients is a clinical concern. Physicians determine further strategies under careful consideration, accounting for the patient's comorbidities associated with surgical risk and individual preference, and the characteristics of the tumor and surgical procedure. Despite additional surgery owing to non-curative resection after ER, the rate of LNM is only 5–10%; hence, among the patients with non-curative resection, it is clinically significant to identify patients at low risk of LNM to prevent unnecessary surgery. The current guidelines have been revised to address these issues and recommend a more detailed strategy after non-curative resection [1,2,4,11]. In the JGCA guidelines (5th edition), among the factors of non-curative resection, piecemeal resection or a positive lateral margin is defined as eCura C-1, and other factors are described as eCura C-2. Based on these classifications, physicians can determine the appropriate therapeutic options, such as additional ER or coagulation for patients in eCura C-1. For eCura C-2, the eCura scoring system was built based on large-scale data and stratifies LNM risk as low (0–1 point), intermediate (2–4 points), or high (5–7 points) [11,12]. In patients with the low-risk category, there is no difference in cancer recurrence or cancer-specific mortality between patients who receive no additional treatment and those who undergo additional surgery [14]. Similarly, reports that investigated LNM risk in patients with early colon cancer after ER were conducted to prevent unnecessary surgery or excess treatment using the AI system and clinical guidelines [24–27]. This reflects the necessity for detailed guidance on additional strategies through the stratification of LNM risk in EGC patients with non-curative resection after ER; therefore, this study has clinical significance.

The strength of this study is that it is the first to develop an ML model to predict LNM in patients with EGC and validate its good performance. Furthermore, our study was based on a large sample size and investigated three models (logistic regression, RF, and SVM) to develop an optimal ML model. Considering that the target participants were patients who underwent ER for EGC, the performance of the ML model was verified not only for the total patients but also the patients who received ER as the initial treatment for EGC. In our study, the very low-risk group had an LNM rate of <1%. This is a stricter category than the classifications of previous reports that defined a low risk of LNM as <3%, including nomograms and the eCura system for predicting LNM in EGC patients [11,28]. In addition to the variables included in the nomogram and the eCura system, our ML model was constructed based on various variables, including the number of tumors, tumor location, Lauren classification, perineural invasion, age, sex, gross type, tumor size, differentiation, depth of invasion, lymphatic invasion, and venous invasion [12,28]. Moreover, we utilized the ability of the ML model to comprehensively interpret various factors by subdividing the data of the variables assessed in previous reports [12,28]. For example, the depth of invasion was subdivided into the lamina propria, muscularis mucosae, SM1, and SM2/3.

We evaluated the performance of the ML model using clinically relevant outcomes. In estimating LNM risk in patients with non-curative resection after ER for EGC, achieving a high sensitivity to predict LNM is essential for long-term outcomes. Furthermore, there is a need to identify patients at low risk for LNM to prevent unnecessary surgery. Our ML model showed specificities of 49% in the total patients and 71% in the patients with initial ER at the high-sensitivity cutoff of 100%. When examining only patients with non-curative resection after ER, our ML model showed specificities ranging from 38% to 39% at the high-sensitivity cutoff of 95%, which is significantly increased compared to the specificity of 9% for the eCura system. The sensitivity of 95% was set based on the highest sensitivity achieved by the eCura system. Therefore, the ML model has great clinical potential in that it had better specificity than the eCura system at a high-sensitivity cutoff, despite there being no significant difference in the value of AUROC.

This study had several limitations. First, there may be selection bias due to the exclusion of missing data and the study's retrospective nature; however, this study was designed to develop the ML model, including major factors without missing data. Second, this was a single-center study, and the results need to be validated in other institutions. In addition, it is necessary to validate the performance of the ML model in patients undergoing non-curative resection after ER for EGC. Through this additional validation, we can anticipate the improved version of the ML model by reinforcement learning and suggest that the ML model can be a valuable tool in clinical applications. Third, most of the variables included in our ML model are based on the pathology after ER. For estimation of LNM risk, several major variables, such as lymphatic invasion, vertical margin, and the depth of invasion, could not be assessed by endoscopy alone. Fourth, the comparison of long-term survival was not analyzed according to the stratification of LNM risk, as there were some cases with insufficient follow-up because the follow-up ended in March 2021.

In conclusion, the ML model showed good performance in the prediction and stratification of LNM risk in patients with EGC. Based on this finding, we suggest that the ML model has the potential to be a clinically useful tool for estimating LNM risk among patients with non-curative resection after ER.

Supplementary Materials: The following supporting information can be downloaded at: https://www.mdpi.com/article/10.3390/cancers14051121/s1. Figure S1: Performance of the ML model and eCura system for predicting LNM in patients with non-curative resection after ER. AUROC, area under the receiver operating characteristic NRI, net reclassification index. Table S1: Best hyperparameters selected from the search algorithm.

Author Contributions: Study concept and design: J.-E.N. and T.-J.K.; Acquisition, analysis, or interpretation of data: J.-E.N., Y.-C.L., T.-J.K. and H.L.; Writing and drafting of the manuscript: J.-E.N., Y.-C.L., T.-J.K. and H.L.; Critical revision of the manuscript for important intellectual content: T.-J.K., H.L., H.-E.W., Y.-W.M., B.-H.M., J.-H.L., P.-L.R. and J.J.K.; Statistical analysis: Y.-C.L. All authors have read and agreed to the published version of the manuscript.

Funding: This research received no external funding.

Institutional Review Board Statement: The study was conducted according to the guidelines of the Declaration of Helsinki and approved by the Institutional Review Board of Samsung Medical Center (2021-09-155 and 30 September 2021).

Informed Consent Statement: Informed consents were waived for this study due to the retrospective and observational design.

Data Availability Statement: The data presented in this study are available on request from the corresponding author. The data are not publicly available due to personal privacy.

Conflicts of Interest: The authors declare no conflict of interest.

References

1. Park, C.H.; Yang, D.H.; Kim, J.W.; Kim, J.H.; Kim, J.H.; Min, Y.W.; Lee, S.H.; Bae, J.H.; Chung, H.; Choi, K.D.; et al. Clinical Practice Guideline for Endoscopic Resection of Early Gastrointestinal Cancer. *Clin. Endosc.* **2020**, *53*, 142–166. [CrossRef] [PubMed]
2. Japanese Gastric Cancer Association. Japanese gastric cancer treatment guidelines 2018 (5th edition). *Gastric Cancer* **2021**, *24*, 1–21. [CrossRef] [PubMed]
3. Draganov, P.V.; Wang, A.Y.; Othman, M.O.; Fukami, N. AGA Institute Clinical Practice Update: Endoscopic Submucosal Dissection in the United States. *Clin. Gastroenterol. Hepatol.* **2019**, *17*, 16–25.e1. [CrossRef] [PubMed]
4. Pimentel-Nunes, P.; Dinis-Ribeiro, M.; Ponchon, T.; Repici, A.; Vieth, M.; De Ceglie, A.; Amato, A.; Berr, F.; Bhandari, P.; Bialek, A.; et al. Endoscopic submucosal dissection: European Society of Gastrointestinal Endoscopy (ESGE) Guideline. *Endoscopy* **2015**, *47*, 829–854. [CrossRef] [PubMed]
5. Suzuki, S.; Gotoda, T.; Hatta, W.; Oyama, T.; Kawata, N.; Takahashi, A.; Yoshifuku, Y.; Hoteya, S.; Nakagawa, M.; Hirano, M.; et al. Survival Benefit of Additional Surgery after Non-Curative Endoscopic Submucosal Dissection for Early Gastric Cancer: A Propensity Score Matching Analysis. *Ann. Surg. Oncol.* **2017**, *24*, 3353–3360. [CrossRef]
6. Li, D.; Luan, H.; Wang, S.; Zhou, Y. Survival benefits of additional surgery after non-curative endoscopic resection in patients with early gastric cancer: A meta-analysis. *Surg. Endosc.* **2019**, *33*, 711–716. [CrossRef]
7. Hoteya, S.; Iizuka, T.; Kikuchi, D.; Ogawa, O.; Mitani, T.; Matsui, A.; Furuhata, T.; Yamashita, S.; Yamada, A.; Kaise, M. Clinicopathological Outcomes of Patients with Early Gastric Cancer after Non-Curative Endoscopic Submucosal Dissection. *Digestion* **2016**, *93*, 53–58. [CrossRef]
8. Hatta, W.; Gotoda, T.; Oyama, T.; Kawata, N.; Takahashi, A.; Yoshifuku, Y.; Hoteya, S.; Nakamura, K.; Hirano, M.; Esaki, M.; et al. Is radical surgery necessary in all patients who do not meet the curative criteria for endoscopic submucosal dissection in early gastric cancer? A multi-center retrospective study in Japan. *J. Gastroenterol.* **2017**, *52*, 175–184. [CrossRef]

9. Suzuki, H.; Oda, I.; Abe, S.; Sekiguchi, M.; Nonaka, S.; Yoshinaga, S.; Saito, Y.; Fukagawa, T.; Katai, H. Clinical outcomes of early gastric cancer patients after noncurative endoscopic submucosal dissection in a large consecutive patient series. *Gastric Cancer* **2017**, *20*, 679–689. [CrossRef]
10. Yang, H.J.; Kim, S.G.; Lim, J.H.; Choi, J.; Im, J.P.; Kim, J.S.; Kim, W.H.; Jung, H.C. Predictors of lymph node metastasis in patients with non-curative endoscopic resection of early gastric cancer. *Surg. Endosc.* **2015**, *29*, 1145–1155. [CrossRef] [PubMed]
11. Hatta, W.; Gotoda, T.; Koike, T.; Masamune, A. History and future perspectives in Japanese guidelines for endoscopic resection of early gastric cancer. *Dig. Endosc.* **2020**, *32*, 180–190. [CrossRef] [PubMed]
12. Hatta, W.; Gotoda, T.; Oyama, T.; Kawata, N.; Takahashi, A.; Yoshifuku, Y.; Hoteya, S.; Nakagawa, M.; Hirano, M.; Esaki, M.; et al. A Scoring System to Stratify Curability after Endoscopic Submucosal Dissection for Early Gastric Cancer: "eCura system". *Am. J Gastroenterol.* **2017**, *112*, 874–881. [CrossRef] [PubMed]
13. Niwa, H.; Ozawa, R.; Kurahashi, Y.; Kumamoto, T.; Nakanishi, Y.; Okumura, K.; Matsuda, I.; Ishida, Y.; Hirota, S.; Shinohara, H. The eCura system as a novel indicator for the necessity of salvage surgery after non-curative ESD for gastric cancer: A case-control study. *PLoS ONE* **2018**, *13*, e0204039. [CrossRef]
14. Hatta, W.; Gotoda, T.; Oyama, T.; Kawata, N.; Takahashi, A.; Yoshifuku, Y.; Hoteya, S.; Nakagawa, M.; Hirano, M.; Esaki, M.; et al. Is the eCura system useful for selecting patients who require radical surgery after noncurative endoscopic submucosal dissection for early gastric cancer? A comparative study. *Gastric Cancer* **2018**, *21*, 481–489. [CrossRef] [PubMed]
15. Kim, W.; Song, K.Y.; Lee, H.J.; Han, S.U.; Hyung, W.J.; Cho, G.S. The impact of comorbidity on surgical outcomes in laparoscopy-assisted distal gastrectomy: A retrospective analysis of multicenter results. *Ann. Surg.* **2008**, *248*, 793–799. [CrossRef]
16. Kunisaki, C.; Makino, H.; Takagawa, R.; Sato, K.; Kawamata, M.; Kanazawa, A.; Yamamoto, N.; Nagano, Y.; Fujii, S.; Ono, H.A.; et al. Predictive factors for surgical complications of laparoscopy-assisted distal gastrectomy for gastric cancer. *Surg. Endosc.* **2009**, *23*, 2085–2093. [CrossRef] [PubMed]
17. Martin, A.N.; Das, D.; Turrentine, F.E.; Bauer, T.W.; Adams, R.B.; Zaydfudim, V.M. Morbidity and Mortality after Gastrectomy: Identification of Modifiable Risk Factors. *J. Gastrointest. Surg.* **2016**, *20*, 1554–1564. [CrossRef] [PubMed]
18. Ryu, K.W.; Kim, Y.W.; Lee, J.H.; Nam, B.H.; Kook, M.C.; Choi, I.J.; Bae, J.M. Surgical complications and the risk factors of laparoscopy-assisted distal gastrectomy in early gastric cancer. *Ann. Surg. Oncol.* **2008**, *15*, 1625–1631. [CrossRef] [PubMed]
19. Kurita, N.; Miyata, H.; Gotoh, M.; Shimada, M.; Imura, S.; Kimura, W.; Tomita, N.; Baba, H.; Kitagawa, Y.; Sugihara, K.; et al. Risk Model for Distal Gastrectomy When Treating Gastric Cancer on the Basis of Data from 33,917 Japanese Patients Collected Using a Nationwide Web-Based Data Entry System. *Ann. Surg.* **2015**, *262*, 295–303. [CrossRef]
20. Watanabe, M.; Miyata, H.; Gotoh, M.; Baba, H.; Kimura, W.; Tomita, N.; Nakagoe, T.; Shimada, M.; Kitagawa, Y.; Sugihara, K.; et al. Total gastrectomy risk model: Data from 20,011 Japanese patients in a nationwide internet-based database. *Ann. Surg.* **2014**, *260*, 1034–1039. [CrossRef] [PubMed]
21. Park, J.H.; Lee, H.J.; Oh, S.Y.; Park, S.H.; Berlth, F.; Son, Y.G.; Kim, T.H.; Huh, Y.J.; Yang, J.Y.; Lee, K.G.; et al. Prediction of Postoperative Mortality in Patients with Organ Failure after Gastric Cancer Surgery. *World J. Surg.* **2020**, *44*, 1569–1577. [CrossRef] [PubMed]
22. Shin, D.W.; Yoo, S.H.; Sunwoo, S.; Yoo, M.W. Management of long-term gastric cancer survivors in Korea. *J. Korean Med. Assoc.* **2016**, *59*, 256–265. [CrossRef]
23. Shin, D.W.; Suh, B.; Lim, H.; Suh, Y.S.; Choi, Y.J.; Jeong, S.M.; Yun, J.M.; Song, S.O.; Park, Y. Increased Risk of Osteoporotic Fracture in Postgastrectomy Gastric Cancer Survivors Compared with Matched Controls: A Nationwide Cohort Study in Korea. *Am. J. Gastroenterol.* **2019**, *114*, 1735–1743. [CrossRef] [PubMed]
24. Kudo, S.E.; Ichimasa, K.; Villard, B.; Mori, Y.; Misawa, M.; Saito, S.; Hotta, K.; Saito, Y.; Matsuda, T.; Yamada, K.; et al. Artificial Intelligence System to Determine Risk of T1 Colorectal Cancer Metastasis to Lymph Node. *Gastroenterology* **2021**, *160*, 1075–1084.e2. [CrossRef] [PubMed]
25. Labianca, R.; Nordlinger, B.; Beretta, G.D.; Mosconi, S.; Mandala, M.; Cervantes, A.; Arnold, D. ESMO Guidelines Working Group. Early colon cancer: ESMO Clinical Practice Guidelines for diagnosis, treatment and follow-up. *Ann. Oncol.* **2013**, *24* (Suppl. 6), vi64–vi72. [CrossRef] [PubMed]
26. Hashiguchi, Y.; Muro, K.; Saito, Y.; Ito, Y.; Ajioka, Y.; Hamaguchi, T.; Hasegawa, K.; Hotta, K.; Ishida, H.; Ishiguro, M.; et al. Japanese Society for Cancer of the Colon and Rectum (JSCCR) guidelines 2019 for the treatment of colorectal cancer. *Int. J. Clin. Oncol.* **2020**, *25*, 1–42. [CrossRef] [PubMed]
27. Shaukat, A.; Kaltenbach, T.; Dominitz, J.A.; Robertson, D.J.; Anderson, J.C.; Cruise, M.; Burke, C.A.; Gupta, S.; Lieberman, D.; Syngal, S.; et al. Endoscopic Recognition and Management Strategies for Malignant Colorectal Polyps: Recommendations of the US Multi-Society Task Force on Colorectal Cancer. *Gastroenterology* **2020**, *159*, 1916–1934.e2. [CrossRef] [PubMed]
28. Kim, S.M.; Min, B.H.; Ahn, J.H.; Jung, S.H.; An, J.Y.; Choi, M.G.; Sohn, T.S.; Bae, J.M.; Kim, S.; Lee, H.; et al. Nomogram to predict lymph node metastasis in patients with early gastric cancer: A useful clinical tool to reduce gastrectomy after endoscopic resection. *Endoscopy* **2020**, *52*, 435–444. [CrossRef] [PubMed]

diagnostics

MDPI

Article

Evaluation of Computer-Aided Detection (CAD) in Screening Automated Breast Ultrasound Based on Characteristics of CAD Marks and False-Positive Marks

Jeongmin Lee, Bong Joo Kang *, Sung Hun Kim and Ga Eun Park

Department of Radiology, Seoul Saint Mary's Hospital, College of Medicine, The Catholic University of Korea, Seoul 06591, Korea; jmlee328@gmail.com (J.L.); rad-ksh@catholic.ac.kr (S.H.K.); hoonhoony@naver.com (G.E.P.)
* Correspondence: lionmain@catholic.ac.kr; Tel.: +82-2-2258-6253

Abstract: The present study evaluated the effectiveness of computer-aided detection (CAD) system in screening automated breast ultrasound (ABUS) and analyzed the characteristics of CAD marks and the causes of false-positive marks. A total of 846 women who underwent ABUS for screening from January 2017 to December 2017 were included. Commercial CAD was used in all ABUS examinations, and its diagnostic performance and efficacy in shortening the reading time (RT) were evaluated. In addition, we analyzed the characteristics of CAD marks and the causes of false-positive marks. A total of 1032 CAD marks were displayed based on the patient and 534 CAD marks on the lesion. Five cases of breast cancer were diagnosed. The sensitivity, specificity, PPV, and NPV of CAD were 60.0%, 59.0%, 0.9%, and 99.6% for 846 patients. In the case of a negative study, it was less time-consuming and easier to make a decision. Among 530 false-positive marks, 459 were identified clearly for pseudo-lesions; the most common cause was marginal shadowing, followed by Cooper's ligament shadowing, peri-areolar shadowing, rib, and skin lesions. Even though CAD does not improve the performance of ABUS and a large number of false-positive marks were detected, the addition of CAD reduces RT, especially in the case of negative screening ultrasound.

Keywords: computer-aided detection; automated breast ultrasound; breast

Citation: Lee, J.; Kang, B.J.; Kim, S.H.; Park, G.E. Evaluation of Computer-Aided Detection (CAD) in Screening Automated Breast Ultrasound Based on Characteristics of CAD Marks and False-Positive Marks. *Diagnostics* **2022**, *12*, 583. https://doi.org/10.3390/diagnostics12030583

Academic Editor: Ralph A. Bundschuh

Received: 20 January 2022
Accepted: 23 February 2022
Published: 24 February 2022

1. Introduction

Mammographic screening has reduced the rate of breast cancer mortality [1]. Recent guidelines for screening of breast cancer recommend mammography starting at age 45 or 50 years [2,3]. Although the incidence of breast cancer in Asian women is still lower than in Western countries, morbidity and mortality continue to increase in Asian countries [4]. The peak age of breast cancer in Asian countries is 40–49 years, whereas in Western countries the peak is around 60 to 70 years [5]. Asian women tend to have breasts with higher density compared with Western women [6]. Further, dense breast is an independent risk factor for developing breast cancer [7].

Real-time B-mode ultrasonography has emerged as an alternative imaging technique for breast cancer screening [8]. Ultrasound elastography can quantify stiffness distribution of tissue lesions and complements conventional B-mode ultrasonography. The development of computer-aided diagnosis has improved the reliability of the system, whilst the inception of machine learning, such as deep learning, has further extended its power by facilitating automated segmentation and tumor classification [9].

Automated breast ultrasonography (ABUS) was proposed as a supplementary screening modality recently, for increased cancer detection combined with digital mammography (DM), especially in dense breasts [10–12]. In addition, ABUS has been proposed in the diagnostic setting in a few recent studies [13].

However, due to the large number of images in a single scan, the reading time (RT) of a full ABUS examination can be prolonged and cancers may be easily overlooked [14]. For this reason, computer-aided detection (CAD) software for ABUS has been developed to facilitate the radiological interpretation of ABUS examinations [15]. Few studies investigated the effect of commercially available CAD systems for ABUS on the RT and screening performance of breast radiologists [16]. However, before using the CAD system clinically, it is necessary to analyze the characteristics of CAD marks. It could be useful for radiologists to have knowledge about the characteristics of CAD marks and the causes of false-positive marks.

In this study, we evaluated the effectiveness of computer-aided detection (CAD) system in screening automated breast ultrasound (ABUS) through diagnostic performance and reading time (RT). We also investigated and analyzed the characteristics of CAD marks and the causes of false-positive marks, to distinguish between true and false marks.

2. Materials and Methods

This retrospective study was approved by the institutional review board (IRB) of our institution. The need for informed consent was waived by the ethics committee due to the retrospective design. All procedures involving human participants were in accordance with the ethical standards of IRB issued by our institution, and assessments were carried out in accordance with the tenets of the Declaration of Helsinki of 1975, and its revision in 2013.

2.1. ABUS Acquisitions

The ABUS examinations were performed with the ACUSON S2000 Automated Breast Volume Scanner system (Siemens, Erlangen, Germany). This ABUS system acquires 3D B-mode ultrasound volumes over an area of $15.4 \times 16.8 \times 6$ cm^3 volume data sets of the breast in one sweep using a mechanically driven linear array transducer (14L5). Adequate depth and focus can be obtained using predefined settings for different breast cup sizes. All ABUS examinations were performed by a single trained radiographer. To ensure coverage of the entire breast, three overlapping acquisitions including antero-posterior, medial, and lateral views were performed. The scan thickness was displayed at 1 mm intervals without overlap. A dedicated ABUS workstation was used to reconstruct the transverse slices into a 3D volume that can be read in a multiplanar hanging, with sagittal and coronal reconstructions.

2.2. CAD System

A prototype workstation was designed and developed specifically for high-throughput ABUS screening in this observer study (MeVis Medical Solutions, Bremen, Germany). In this prototype, each user action was logged with timestamps, which were subsequently used to estimate the time spent per case. The workstation was integrated with a commercially developed CAD software (QVCAD, Qview Medical Inc., Los Altos, CA, USA), which is designed to detect suspicious candidate regions in an ABUS volume highlighted with the so-called CAD marks (Figure 1).

In addition, the QVCAD software provides an "intelligent" minimum intensity projection (MinIP) of the breast tissue in a 3D ABUS volume that can be used for rapid navigation through ABUS scans for enhancement of the possible suspicious regions. The CAD-based MinIP integrated with a multiplanar hanging protocol for ABUS displays the conventional ABUS planes. By clicking on the dark spot, the 3D multiplanar hanging automatically snaps to the corresponding 3D location. The crosshair is focused on a breast lesion that is marked by the CAD software with a green circular marker. The same lesion is also enhanced and visualized as a dark spot in the MinIP. A screenshot of the CAD-aided reading environment is presented in Figure 1.

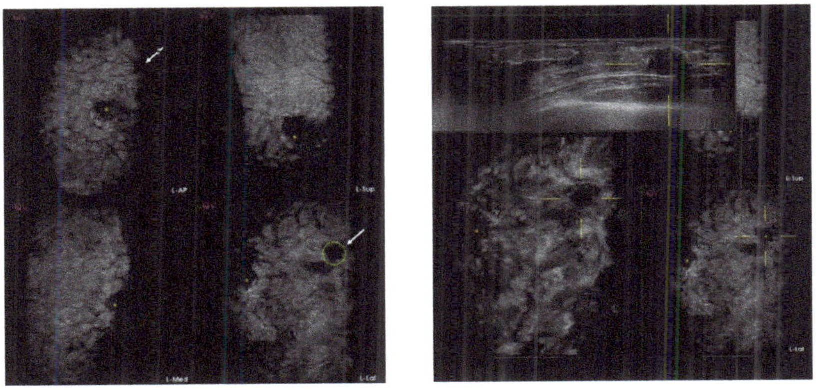

(**a**) (**b**)

Figure 1. Screening automated breast ultrasound (ABUS) of a 44-year-old woman shows a true-positive mark. (**a**) Computer-aided detection (CAD)-based minimum intensity projection (MinIP) of an ABUS scan of the antero-posterior (AP), medial, and lateral sides of the left breast. There is one dark spot (arrows) with a green circle. (**b**) The lesion showing a dark spot with a green circle laterally on the left breast confirms invasive ductal carcinoma.

The number of CAD markers displayed per ABUS volume could be adjusted by changing the values of the false-positive rate (FPR) in the configuration setting of the CAD software. According to the manual from the manufacturer, FPR was defined as the total number of false-positive CAD markers in non-cancer volumes divided by the total number of non-cancer volumes. In this study, we set the FPR to 0.2 (i.e., 1 false-positive CAD marker in non-cancer volume per 5 non-cancer volumes), which was its default setting as in previous studies [16–18].

2.3. *Study Design*

The study included a total of 846 women aged 40–49 years who underwent ABUS screening from January 2017 to December 2017. The CAD (QVCADTM) system was used in all ABUS examinations and its diagnostic performance was evaluated retrospectively.

We evaluated glandular tissue component (GTC), which was classified as minimal (<25% of the fibroglandular tissue (FGT)), mild (25–49% of the FGT), moderate (50–74% of the FGT), or marked (≥75% of the FGT) in each woman based on bilateral breast images [19].

We analyzed whether CAD addition shortened the RT. The RT was determined by the expert breast radiologists based on their subjective perception in each of the following cases: (1) CAD with ABUS = ABUS only, (2) CAD with ABUS > ABUS only, (3) CAD with ABUS < ABUS only. We defined there is a difference when RT was shortened by more than 1 min.

Furthermore, we analyzed the characteristics of CAD marks including the size of the marked lesion, lesion type (mass or non-mass), tissue composition under ultrasound, and the causes of false-positive marks. The false-positive mark was defined as the mark located on the typical benign lesion or pseudo-lesions that require no additional studies following ABUS. The number of marks per patient and per lesion and the frequency of false-positive marks were also evaluated.

Two board-certified expert breast radiologists determined the characteristics of CAD marks based on consensus. In addition, the pseudo-lesions were also evaluated by two expert breast radiologists with consensus. The characteristics of pseudo-lesions were analyzed including the number, size, and location (right or left; antero-posterior, medial or lateral; upper, mid, or lower; inner, mid, or outer).

All women with suspicious lesions were recalled and US-guided 14G core-needle biopsy was performed. Patients who were not disease-positive were followed up in 2 years with radiologic examination using mammography or ultrasonography.

3. Results

A total of 846 women participated in the study, and the median age at enrollment was 44 years (mean age ± standard deviation = 43.9 ± 3.0 years). Based on ABUS screening, five breast cancers were diagnosed pathologically over a two-year follow-up (Figure 1). The sensitivity, specificity, PPV, NPV, and accuracy of CAD for cancer detection were 60.0%, 59.0%, 0.9%, 99.6% and 59.0%, respectively, for 846 patients, while those values for 1032 CAD marks were 60.0%, 48.3%, 0.6%, 99.6%, and 48.4%, respectively.

Based on the lesion type detected, the large mass lesions were more than the non-mass lesions (60 vs. 11). Based on tissue composition under ultrasound, the number of minimal-to-mild cases in GTC was higher than moderate-to-marked cases (668 vs. 178). The rate of CAD positivity in moderate-to-marked lesions was higher than in minimal-to-mild. Table 1 summarizes the screening performance of CAD for ABUS per patient and per lesion.

Table 1. Sensitivity (SEN), specificity (SPE), positive predictive value (PPV), negative predictive value (NPV), and accuracy per patient and per computer aided detection (CAD) mark.

Total (n = 846)	Benign	Malig	SEN	SPE	PPV	NPV	Accuracy	*p*-Value
CAD								
CAD (−)	496	2	60	59	0.9	99.6	59	0.407
CAD (+)	345	3						
Mark No. 1 [#]	324	2						
Mark No. 2	20	-						
Mark No. 3	1	1						
ABUS Category								
<4	827	0	100	98.3	26.3	100	0.6	<0.0001
=>4	14	5						
Lesion type								
non-mass	10	1						
mass	56	4						
Non-mass (n = 11)								
CAD (−)	3	1	-	30	-	75	27.3	0.364
CAD (+)	7	0						
Mark No. 1	6	-						
Mark No. 2	1	-						
Mark No. 3	-	-						
Mass (n = 60)								
CAD (−)	27	1	75	48.2	9.4	96.4	50	0.616
CAD (+)	29	3						
Mark No. 1	23	2						
Mark No. 2	5	-						
Mark No. 3	1	1						
Tissue Composition								
1–2	665	3	40	79.1	1.1	99.6	78.8	0.284
3–4	176	2						
Tissue Composition (1–2, n = 668)								
CAD (−)	409	1	66.7	61.5	0.8	99.8	61.5	0.563
CAD (+)	256	2						
Mark No. 1	241	1						
Mark No. 2	14	-						
Mark No. 3	1	1						

Table 1. *Cont.*

Total (n = 846)	Benign	Malig	SEN	SPE	PPV	NPV	Accuracy	*p*-Value
Tissue Composition (3–4, n = 178)								
CAD (−)	87	1	50	49.4	1.1	98.9	49.4	1
CAD (+)	89	1						
Mark No. 1	83	1						
Mark No. 2	6	-						
Mark No. 3	-	-						
Total (n = 1032)								
CAD (−)	496	2	60	48.3	0.6	99.6	48.4	1
CAD (+)	531	3						
Mark Number 1 [€]	220	3						
Mark Number 2	79	-						
Mark Number 3	36	-						
Mark Number 4	6	-						
Mark Number 5	3	-						
Mark Number 6	1	-						

[#] Mark No. denotes the number of CAD marks per lesion. [€] Mark Number indicates the number of CAD marks per patient.

In the absence of the CAD mark, the readers determined that the reading time for CAD with ABUS was less than for ABUS only and easier to make a decision (Table 2). Table 2 summarizes the number and characteristics of CAD marks per patient.

Table 2. Characteristics of number for computer-aided detection (CAD) marks per patient and reading time (RT).

	Mark No. [#]				
	0	1	2, 3	Total (1,2,3)	*p*-Value *
Size					0.702
mean ± SD	12.2 ± 7.6	11.5 ± 6.3	18.9 ± 17.6	13 ± 9.9	
median(IQR)	10 (7, 14.5)	10 (7, 13)	15 (3, 20)	10 (7, 14)	
Mass type					0.743
non-mass	28 (87.5)	25 (80.7)	7 (87.5)	32 (82)	
mass	4 (12.5)	6 (19.4)	1 (12.5)	7 (18)	
Tissue composition					0.004
1, 2	410 (82.3)	242 (74.2)	16 (72.7)	258 (74.1)	
3, 4	88 (17.7)	84 (25.8)	6 (27.3)	90 (25.9)	
Reading time					<0.0001
CAD with ABUS = ABUS	16 (3.2)	39 (12.1)	10 (50)	49 (14.3)	
CAD with ABUS > ABUS	-	279 (86.4)	10 (50)	289 (84.3)	
CAD with ABUS < ABUS	482 (96.8)	5 (1.6)	-	5 (1.5)	

Values are expressed as numbers (percentages) for categorical variables and means (SD), median (IQR) others. * *p*-value was calculated between 0 with total (1,2,3) using Chi-square test, Fisher's exact test, or *t*-test. [#] Mark No. indicates the number of CAD marks per lesion.

Of 846 patients, 1032 CAD marks were marked in 534 lesions of 348 patients with a mean CAD mark per person of 0.8 (SD ± 1) (range 0–6) (Table 3). No CAD mark was detected in 498 patients (48.3%).

The characteristic CAD marks were determined by two reviewers by consensus as suspicious malignant lesions (0.8%, n = 4), benign lesions (13.3%. n = 71), and clear pseudo-lesions (86%, n = 459).

Among 530 false-positive marks, 459 marks were marked on the clear pseudolesions (Figures 2–4); the most common cause was marginal shadowing (209, 39.1%), followed by

Cooper's ligament shadowing (143, 26.8%), peri-areolar shadowing (64, 12%), rib (37, 6.9 %), and skin lesions (6, 1.1%).

Table 3. Number and characteristics of computer-aided detection (CAD) mark.

Characteristics of All CAD Mark (n = 1032)		
Mean and median No. of CAD marks per patient		
mean ± SD	0.8 ± 1	
median (IQR)	1 (0, 1)	
No. of CAD mark per patient	n	%
0 (498)	498	48.3
1 (1 × 223)	223	21.6
2 (2 × 79)	158	15.3
3 (3 × 36)	108	10.5
4 (4 × 6)	24	2.3
5 (5 × 3)	15	1.5
6 (6 × 1)	6	0.6
Characteristics of CAD marks per lesion (n = 534)	n	%
Suspicious	4	0.8
Benign	71	13.3
Fat	35	6.6
Benign mass	19	3.6
Cyst	9	1.7
Fibrosis/heterogenous parenchyma	8	1.5
False-positive marks for pseudolesions	459	86
Marginal shadowing	209	39.1
Cooper's ligament shadowing	143	26.8
Periareolar shadowing	64	12
Rib	37	6.9
Skin lesion	6	1.1

Values are expressed as numbers (percentages) for categorical variables and means (SD), median (IQR) others. Values are expressed as numbers (percentages) for categorical variables.

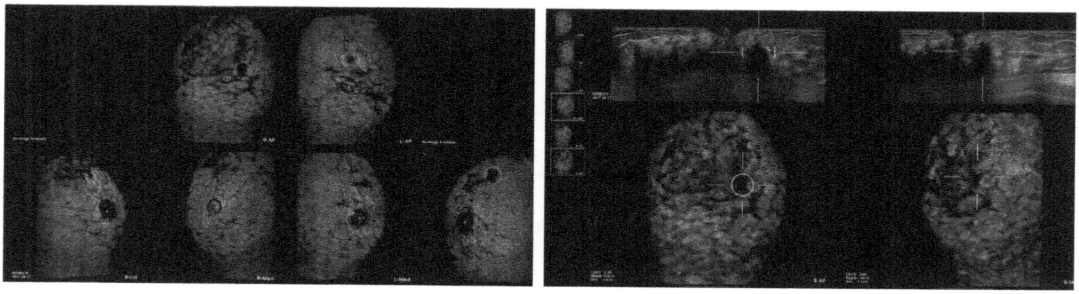

(a) (b)

Figure 2. *Cont.*

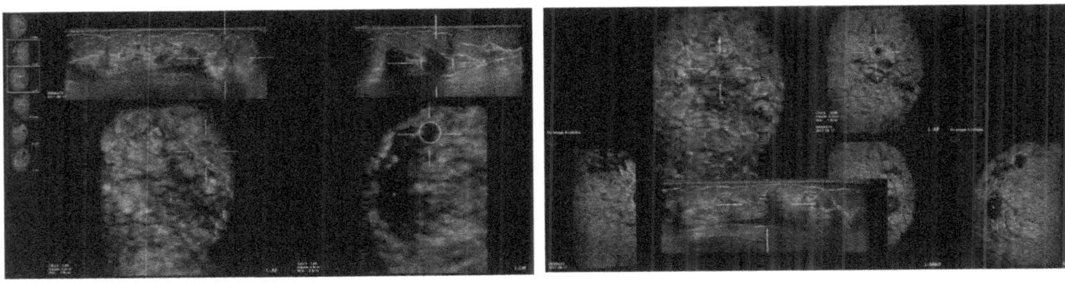

(c) (d)

Figure 2. Screening automated breast ultrasound (ABUS) of a 45-year-old woman reveals false-positive marks due to shadowing. (**a**) CAD-based minimum intensity projection (MinIP) of an ABUS scan of the AP, medial, and lateral sides of both breasts. There are three dark spots with green circles. (**b**) The lesion showing a dark spot with a green circle on AP side of the right breast confirms the pseudolesion due to periareolar shadowing in the transverse scan. (**c**,**d**) The lesion showing a dark spot with a green circle on AP side of the left breast confirms the pseudolesion due to Cooper's ligament shadowing in the transverse scan. The lesion showing a dark spot with a green circle laterally on the left breast confirms the pseudolesion due to marginal shadowing in the transverse scan.

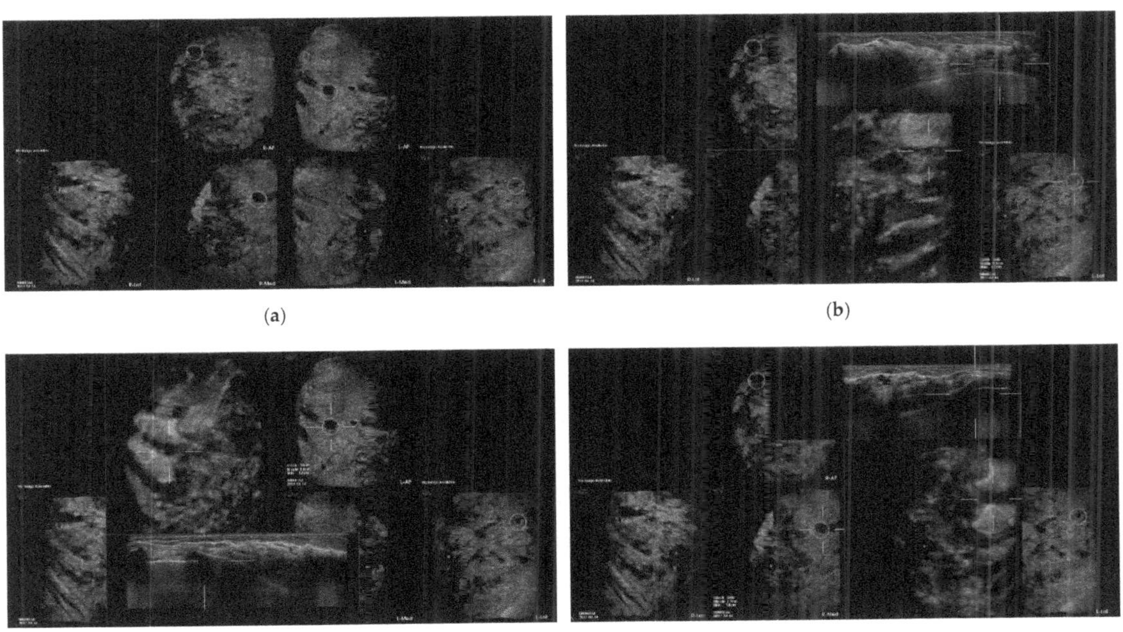

(a) (b)

(c) (d)

Figure 3. Screening automated breast ultrasound (ABUS) of a 42-year-old woman shows false-positive marks due to rib. (**a**) CAD-based minimum-intensity projection (MinIP) of an ABUS scan in the AP, medial, and lateral sides of both breasts. There are four dark spots with green circles. (**b**–**d**) The lesions showing dark spots with green circles in both AP and right medial sides of both breasts confirm pseudolesions due to ribs in the transverse scan.

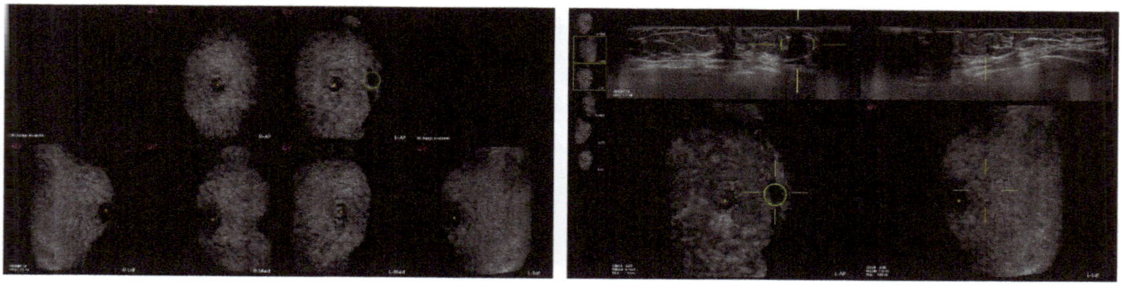

(**a**) (**b**)

Figure 4. Screening automated breast ultrasound (ABUS) of a 48-year-old woman reveals false-positive marks due to skin lesions. (**a**) CAD-based minimum intensity projection (MinIP) of an ABUS scan in the AP, medial, and lateral sides of both breasts. There is a dark spot with a green circle. (**b**) The lesion showing a dark spot with a green circle on the AP side of the left breast confirms the pseudolesion due to a skin lesion in the transverse scan.

The false-positive marks on pseudo-lesions were frequently detected in the upper portion than in the mid-to-lower portion, and in the outer portion than in the mid-to-inner portion of breast (Table 4). There were more marks in the lateral view than in AP or medial views (Table 4).

Table 4. Characteristics of false-positive marks associated with pseudolesions (n = 459).

	Mark No. #				
	All	1	2, 3	*p*-Value	*p*-Value *
Mark Location				0.337	0.002
right	262 (57.1)	251 (56.7)	11 (68.8)		
left	197 (42.9)	192 (43.3)	5 (31.3)		
Mark Location				0.806	0.026
antero-posterior	142 (29.9)	131 (29.6)	11 (34.4)		
medial	147 (31)	137 (30.9)	10 (31.3)		
lateral	186 (39.2)	175 (39.5)	11 (34.4)		
Mark Site				0.674	<0.0001
upper	377 (82.1)	362 (81.7)	15 (93.8)		
mid	30 (6.5)	30 (6.8)	-		
lower	52 (11.3)	51 (11.5)	1 (6.3)		
Mark Site				0.572	<0.0001
inner	101 (22)	96 (21.7)	5 (31.3)		
mid	139 (30.3)	134 (30.3)	5 (31.3)		
outer	219 (47.7)	213 (48.1)	6 (37.5)		
Tissue Composition				0.843	<0.0001
1, 2	305 (66.5)	294 (66.4)	11 (68.8)		
3, 4	154 (33.6)	149 (33.6)	5 (31.3)		

Values represent numbers (percentages) for categorical variables. *p*-value was calculated between MarkNo1 with MarkNo2,3 using Chi-square test. * *p*-value was calculated only in a group using Chi-square test. # Mark No. indicates the number of CAD marks per lesion.

4. Discussion

In this study, we evaluated the effectiveness of computer-aided detection (CAD) system in screening automated breast ultrasound (ABUS) through diagnostic performance and reading time (RT). A total of 846 patients displayed 1032 CAD marks and 534 CAD marks based on lesions. The sensitivity, specificity, PPV, NPV, and accuracy of CAD were 60.0%, 59.0%, 0.9%, 99.6% and 59.0% for 846 patients, respectively, while those of 1032 CAD marks were 60.0%, 48.3%, 0.6%, 99.6%, and 48.4%, respectively. The relatively higher NPV compared with other parameters indicates that the exam can be concluded with a negative study if no CAD mark is detected on ABUS. The presence of marks in multiple views did not suggest malignancy in this study. In the absence of the CAD mark, the readers determined that the reading time for CAD with ABUS was less than for ABUS only and easier to make a decision.

Several studies have reported that the performance of ABUS was comparable to that of hand-held ultrasound [20–22]. In addition, four prospective studies using ABUS demonstrated an increased cancer detection of 1.9–7.7 per 1000 examinations similar to hand-held ultrasound [10,11,14,23].

However, while the ABUS can yield standardized and structured images regardless of the experience of the operator, it takes much more time and effort to interpret the exams [24]. For this reason, the CAD system has been suggested as a supplementary method for interpreting ABUS results. However, the CAD system showed a high negative predictive value, and there were many false-positive CAD marks, which implied typical benign or pseudo-lesions that do not require further investigation. Usually, the false-positive imaging results can affect the recall rate of the screening modality. The recall rate varied from 8.8% in the J-START study to 10.7% in the American College of Radiology Imaging Network (ACRIN) study [25,26]. However, few studies reported the characteristics of the causes of false-positive marks.

In addition to the diagnostic performance of CAD on ABUS, the previous studies evaluated the RT of CAD on ABUS [27–29]. Yang et al. reported that using CAD in the concurrent-reading mode, all readers saved 32% (16 s per 50 s per volume) in RT with a higher area under the receiver operating characteristic curve values compared with non-CAD mode [28]. Jiang et al. reported that although not all studies were interpreted faster with the CAD system, on average the savings were approximately 1 min per case [29]. In our study, it was less time-consuming and easier to make a clinical decision, especially in the case of a negative study.

In this study, we investigated and analyzed the characteristics of CAD marks and the causes of false-positive marks, to distinguish between true and false marks. Among 530 false-positive marks, 459 were identified clearly for pseudo-lesions; the most common cause was marginal shadowing, followed by Cooper's ligament shadowing, peri-areolar shadowing, rib, and skin lesions, all of which were easily distinguishable radiologically. The false marks for pseudo-lesions were detected more frequently in the upper rather than in the mid-to-lower portion and in the outer rather than in the mid-to-inner portion, probably because of bulkiness and flexibility of the upper and outer portion of the breast.

ABUS is a standardized examination with multiple advantages in both screening and diagnostic settings, including increased detection of breast cancer, improved workflow, and reduced examination time. However, ABUS has disadvantages and even some limitations. Disadvantages regarding image acquisition are the inability to assess the axilla, vascularization, and lesion elasticity. The limitations of interpretation include motion- or lesion-related artifacts due to poor positioning and the lack of contact [30]. In the review article about the pros and cons of ABUS by Ioana Boca et al., marginal shadowing and Cooper's ligament shadowing were defined as artifacts due to insufficient compression [30]. Peri-areolar shadowing is defined as a nipple artifact [30]. Despite the promising detection rate with CAD software in breast cancer, radiologists should determine whether a CAD software-marked lesion is a true- or false-positive lesion, given its positive predictive value and high false-positive rate [17]. The knowledge of these artifacts improves the diagnostic performance of radiologists.

There are several limitations to this study. First, we used only image data obtained with equipment from a single vendor, with a small number of participants. In addition, this study was performed only in academic institutions by a limited number of users, board-certified expert breast radiologists, and does not represent varying clinical environments. Second, the absence of the numerical result of RT is the limitation of this study. The RT was determined by the expert breast radiologists based on their subjective perception. Finally, in our study, the expert radiologists' decision was a gold standard for suspicious lesions or pseudo-lesions. However, a large number of marks await the radiologist's rational judgment. Therefore, CAD users should be familiar with marks in various situations before using them, and the review summarizes the characteristics of CAD marks only without radiological evaluation. The knowledge of the characteristics of CAD marks and the causes of false-positive marks could improve the diagnostic performance of radiologists.

5. Conclusions

In conclusion, even though CAD addition does not improve the performance of screening ABUS and is associated with a large number of false-positive marks, CAD addition improves the negative predictive value and reduces RT, especially for negative screening ultrasound.

Author Contributions: Conceptualization, B.J.K.; methodology, J.L.; formal analysis, G.E.P. and J.L.; investigation, S.H.K. and B.J.K.; resources, B.J.K. and S.H.K.; data curation, S.H.K. and G.E.P.; writing—original draft preparation, J.L.; writing—review and editing, B.J.K.; visualization, J.L. and G.E.P.; supervision, B.J.K. and S.H.K.; project administration, B.J.K. All authors have read and agreed to the published version of the manuscript.

Funding: This research received no external funding.

Institutional Review Board Statement: The study was conducted according to the guidelines of the Declaration of Helsinki, and approved by the Institutional Review Board (or Ethics Committee) of Seoul Saint Mary's Hospital (protocol code KC16ECMI0552 and date of approval 1 September 2016).

Informed Consent Statement: Patient consent was waived due to the retrospective design of this study.

Data Availability Statement: All data generated and analyzed during this study are included in this published article. Raw data supporting the findings of this study are available from the corresponding author on request.

Conflicts of Interest: The authors declare no conflict of interest.

References

1. Elmore, J.G.; Armstrong, K.; Lehman, C.D.; Fletcher, S.W. Screening for breast cancer. *JAMA* **2005**, *293*, 1245–1256. [CrossRef]
2. Siu, A.L.; Force USPST. Screening for Breast Cancer: U.S. Preventive Services Task Force Recommendation Statement. *Ann. Intern. Med.* **2016**, *164*, 279–296. [CrossRef] [PubMed]
3. Oeffinger, K.C.; Fontham, E.T.; Etzioni, R.; Herzig, A.; Michaelson, J.S.; Shih, Y.C.; Walter, L.C.; Church, T.R.; Flowers, C.R.; LaMonte, S.J.; et al. Breast Cancer Screening for Women at Average Risk: 2015 Guideline Update from the American Cancer Society. *JAMA* **2015**, *314*, 1599–1614. [CrossRef]
4. Morimoto, T.; Nagao, T.; Okazaki, K.; Kira, M.; Nakagawa, Y.; Tangoku, A. Current status of breast cancer screening in the world. *Breast Cancer* **2009**, *16*, 2–9. [CrossRef] [PubMed]
5. Leong, S.P.; Shen, Z.Z.; Liu, T.J.; Agarwal, G.; Tajima, T.; Paik, N.S.; Sandelin, K.; Derossis, A.; Cody, H.; Foulkes, W.D. Is breast cancer the same disease in Asian and Western countries? *World J. Surg.* **2010**, *34*, 2308–2324. [CrossRef]
6. Rajaram, N.; Mariapun, S.; Eriksson, M.; Tapia, J.; Kwan, P.Y.; Ho, W.K.; Harun, F.; Rahmat, K.; Czene, K.; Taib, N.A.M.; et al. Differences in mammographic density between Asian and Caucasian populations: A comparative analysis. *Breast Cancer Res. Treat.* **2017**, *161*, 353–362. [CrossRef] [PubMed]
7. Boyd, N.F.; Martin, L.J.; Yaffe, M.J.; Minkin, S. Mammographic density and breast cancer risk: Current understanding and future prospects. *Breast Cancer Res.* **2011**, *13*, 223. [CrossRef]
8. Teh, W.; Wilson, A. The role of ultrasound in breast cancer screening. A consensus statement by the European Group for Breast Cancer Screening. *Eur. J. Cancer* **1998**, *34*, 449–450. [CrossRef]
9. Mao, Y.-J.; Lim, H.-J.; Ni, M.; Yan, W.-H.; Wong, D.W.-C.; Cheung, J.C.-W. Breast Tumour Classification Using Ultrasound Elastography with Machine Learning: A Systematic Scoping Review. *Cancers* **2022**, *14*, 367. [CrossRef]
10. Brem, R.F.; Tabár, L.; Duffy, S.W.; Inciardi, M.F.; Guingrich, J.A.; Hashimoto, B.E.; Lander, M.R.; Lapidus, R.L.; Peterson, M.K.; Rapelyea, J.A.; et al. Assessing improvement in detection of breast cancer with three-dimensional automated breast US in women with dense breast tissue: The SomoInsight Study. *Radiology* **2015**, *274*, 663–673. [CrossRef]
11. Giuliano, V.; Giuliano, C. Improved breast cancer detection in asymptomatic women using 3D-automated breast ultrasound in mammographically dense breasts. *Clin. Imaging* **2013**, *37*, 480–486. [CrossRef] [PubMed]

12. Nothacker, M.; Duda, V.; Hahn, M.; Warm, M.; Degenhardt, F.; Madjar, H.; Weinbrenner, S.; Albert, U.-S. Early detection of breast cancer: Benefits and risks of supplemental breast ultrasound in asymptomatic women with mammographically dense breast tissue. A systematic review. *BMC Cancer* **2009**, *9*, 335. [CrossRef] [PubMed]
13. Wenkel, E.; Heckmann, M.; Heinrich, M.; Schwab, S.; Uder, M.; Schulz-Wendtland, R.; Bautz, W.A.; Janka, R. *Automated Breast Ultrasound: Lesion Detection and BI-RADS™ Classification—A Pilot Study*; RöFo-Fortschritte auf dem Gebiet der Röntgenstrahlen und der Bildgebenden Verfahren; Georg Thieme Verlag KG Stuttgart: New York, NY, USA, 2008; pp. 804–808.
14. Wilczek, B.; Wilczek, H.E.; Rasouliyan, L.; Leifland, K. Adding 3D automated breast ultrasound to mammography screening in women with heterogeneously and extremely dense breasts: Report from a hospital-based, high-volume, single-center breast cancer screening program. *Eur. J. Radiol.* **2016**, *85*, 1554–1563. [CrossRef]
15. Tan, T.; Mordang, J.J.; van Zelst, J.; Grivegnée, A.; Gubern-Mérida, A.; Melendez, J.; Mann, R.M.; Zhang, W.; Platel, B.; Karssemeijer N. Computer-aided detection of breast cancers using Haar-like features in automated 3D breast ultrasound. *Med. Phys.* **2015**, *42*, 1498–1504. [CrossRef] [PubMed]
16. van Zelst, J.C.; Tan, T.; Clauser, P.; Domingo, A.; Dorrius, M.D.; Drieling, D.; Golatta, M.; Gras, F.; de Jong, M.; Pijnappel, R.; et al. Dedicated computer-aided detection software for automated 3D breast ultrasound; an efficient tool for the radiologist in supplemental screening of women with dense breasts. *Eur. Radiol.* **2018**, *28*, 2996–3006. [CrossRef]
17. Kim, Y.; Rim, J.; Kim, S.M.; La Yun, B.; Park, S.Y.; Ahn, H.S.; Kim, B.; Jang, M. False-negative results on computer-aided detection software in preoperative automated breast ultrasonography of breast cancer patients. *Ultrasonography* **2021**, *40*, 83. [CrossRef]
18. Van Zelst, J.; Tan, T.; Platel, B.; De Jong, M.; Steenbakkers, A.; Mourits, M.; Grivegnee, A.; Borelli, C.; Karssemeijer, N.; Mann, R.M.; et al. Improved cancer detection in automated breast ultrasound by radiologists using computer aided detection. *Eur. J. Radiol.* **2017**, *89*, 54–59. [CrossRef]
19. Kim, W.H.; Lee, S.H.; Chang, J.M.; Cho, N.; Moon, W.K. Background echotexture classification in breast ultrasound: Inter-observer agreement study. *Acta Radiol.* **2017**, *58*, 1427–1433. [CrossRef]
20. Jia, M.; Lin, X.; Zhou, X.; Yan, H.; Chen, Y.; Liu, P.; Bao, L.; Li, A.; Basu, P.; Qiao, Y.; et al. Diagnostic performance of automated breast ultrasound and handheld ultrasound in women with dense breasts. *Breast Cancer Res. Treat.* **2020**, *181*, 589–597. [CrossRef]
21. Zhang, X.; Chen, J.; Zhou, Y.; Mao, F.; Lin, Y.; Shen, S.; Sun, Q.; Ouyang, Z. Diagnostic value of an automated breast volume scanner compared with a hand-held ultrasound: A meta-analysis. *Gland Surg.* **2019**, *8*, 698–711. [CrossRef]
22. Vourtsis, A.; Kachulis, A. The performance of 3D ABUS versus HHUS in the visualisation and BI-RADS characterisation of breast lesions in a large cohort of 1,886 women. *Eur. Radiol.* **2018**, *28*, 592–601. [CrossRef] [PubMed]
23. Kelly, K.M.; Dean, J.; Comulada, W.S.; Lee, S.J. Breast cancer detection using automated whole breast ultrasound and mammography in radiographically dense breasts. *Eur. Radiol.* **2010**, *20*, 734–742. [CrossRef] [PubMed]
24. van Zelst, J.C.M.; Mann, R.M. Automated Three-dimensional Breast US for Screening: Technique, Artifacts, and Lesion Characterization. *Radiographics* **2018**, *38*, 663–683. [CrossRef]
25. Berg, W.A.; Bandos, A.I.; Mendelson, E.B.; Lehrer, D.; Jong, R.A.; Pisano, E.D. Ultrasound as the Primary Screening Test for Breast Cancer: Analysis from ACRIN 6666. *J. Natl. Cancer Inst.* **2016**, *108*, djv367. [CrossRef] [PubMed]
26. Ohuchi, N.; Suzuki, A.; Sobue, T.; Kawai, M.; Yamamoto, S.; Zheng, Y.-F.; Shiono, Y.N.; Saito, H.; Kuriyama, S.; Tohno, E.; et al. Sensitivity and specificity of mammography and adjunctive ultrasonography to screen for breast cancer in the Japan Strategic Anti-cancer Randomized Trial (J-START): A randomised controlled trial. *Lancet* **2016**, *387*, 341–348. [CrossRef]
27. Jiang, Y. Concurrent-Read CaD helps streamline automated Breast ultrasound (aBus) interpretation. *Breast Imaging* **2018**, *11*, 40–42.
28. Yang, S.; Gao, X.; Liu, L.; Shu, R.; Yan, J.; Zhang, G.; Xiao, Y.; Ju, Y.; Zhao, N.; Song, H. Performance and reading time of automated breast US with or without computer-aided detection. *Radiology* **2019**, *292*, 540–549. [CrossRef]
29. Jiang, Y.; Inciardi, M.F.; Edwards, A.V.; Papaioannou, J. Interpretation time using a concurrent-read computer-aided detection system for automated breast ultrasound in breast cancer screening of women with dense breast tissue. *Am. J. Roentgenol.* **2018**, *211*, 452–461. [CrossRef]
30. Ciurea, A.I.; Ciortea, C.A.; Dudea, S.M. Pros and Cons for Automated Breast Ultrasound (ABUS): A Narrative Review. *J. Pers. Med.* **2021**, *11*, 703.

cancers

MDPI

Article

Cost-Effectiveness of Artificial Intelligence Support in Computed Tomography-Based Lung Cancer Screening

Sebastian Ziegelmayer *,†, Markus Graf †, Marcus Makowski, Joshua Gawlitza † and Felix Gassert †

Institute of Diagnostic and Interventional Radiology, School of Medicine, Klinikum Rechts der Isar, Technical University Munich, Ismaninger Straße 22, 81675 Munich, Germany; markus.m.graf@tum.de (M.G.); marcus.makowski@tum.de (M.M.); joshua.gawlitza@tum.de (J.G.); felix.gassert@tum.de (F.G.)

* Correspondence: s.ziegelmayer@tum.de

† These authors contributed equally to this work.

Simple Summary: Lung cancer screening with low-dose CT (LDCT) has been shown to significantly reduce cancer-related mortality and is recommended by the United States Preventive Services Task Force (USPSTF). With pending recommendation in Europe and millions of patients enrolling in the program, deep learning algorithms could reduce the number of false positive and negative findings. Therefore, we evaluated the cost-effectiveness of using an AI algorithm for the initial screening scan using a Markov simulation. We found that AI support at initial screening is a cost-effective strategy up to a cost of USD 1240 per patient screening, given a willingness-to-pay of USD 100,000 per quality-adjusted life years (QALYs).

Abstract: Background: Lung cancer screening is already implemented in the USA and strongly recommended by European Radiological and Thoracic societies as well. Upon implementation, the total number of thoracic computed tomographies (CT) is likely to rise significantly. As shown in previous studies, modern artificial intelligence-based algorithms are on-par or even exceed radiologist's performance in lung nodule detection and classification. Therefore, the aim of this study was to evaluate the cost-effectiveness of an AI-based system in the context of baseline lung cancer screening. Methods: In this retrospective study, a decision model based on Markov simulation was developed to estimate the quality-adjusted life-years (QALYs) and lifetime costs of the diagnostic modalities. Literature research was performed to determine model input parameters. Model uncertainty and possible costs of the AI-system were assessed using deterministic and probabilistic sensitivity analysis. Results: In the base case scenario CT + AI resulted in a negative incremental cost-effectiveness ratio (ICER) as compared to CT only, showing lower costs and higher effectiveness. Threshold analysis showed that the ICER remained negative up to a threshold of USD 68 for the AI support. The willingness-to-pay of USD 100,000 was crossed at a value of USD 1240. Deterministic and probabilistic sensitivity analysis showed model robustness for varying input parameters. Conclusion: Based on our results, the use of an AI-based system in the initial low-dose CT scan of lung cancer screening is a feasible diagnostic strategy from a cost-effectiveness perspective.

Keywords: lung cancer screening; deep learning; cost-effectiveness analysis; AI-support system

Citation: Ziegelmayer, S.; Graf, M.; Makowski, M.; Gawlitza, J.; Gassert, F. Cost-Effectiveness of Artificial Intelligence Support in Computed Tomography-Based Lung Cancer Screening. *Cancers* **2022**, *14*, 1729. https://doi.org/10.3390/cancers14071729

Academic Editors: Hamid Khayyam, Ali Madani, Rahele Kafieh and Ali Hekmatnia

Received: 1 March 2022
Accepted: 23 March 2022
Published: 29 March 2022

1. Introduction

Based on the findings of the national lung screening trial (NLST), in 2014 the United States Preventive Service task force recommended the annual lung cancer screening of patients between 55 and 80 years with 20 pack years of smoking history [1,2]. In contrast to the high and further increasing incidence of lung cancer globally, the incidence of lung cancer was relatively low in the NLST. Nonetheless, the NLST was able to show a significant reduction in lung cancer related mortality due to the annual screening with low-dose computed tomography (CT). Consequently, a European Position Statement followed in 2017, strongly recommending the CT-based lung cancer screening as well [3]. This recommendation is further supported by the Dutch-Belgian lung-cancer screening trial (Nederlands-Leuvens Longkanker Screenings Onderzoek (NELSON)), which also showed a significant reduction in lung cancer mortality for high-risk patients who participated in the screening [4]. With several ongoing pilot projects in Europe, the widespread introduction of lung cancer screening seems to be only a matter of time.

Nevertheless, the benefits of lung cancer screening are limited by false negative and false positive findings, which not only result in high costs but also affect clinical outcome and quality of life [2,5,6]. Currently, low dose CT-scans in the screening setting are evaluated based on standardized systems like Lung-RADS (Lung imaging reporting and data system), which improve the diagnostic accuracy for radiologists and reduces costs by decreasing the need for further diagnostic tests [7,8]. Even after a recent revision of the reporting system, observer variability will remain a relevant limitation [9,10].

The rapid development of artificial intelligence (AI) in the medical field has shown promising results for cancer screening and recent AI-models may achieve or exceed the diagnostic performance of sub-specialized experts, for example in breast cancer screening [11]. While long-standing CAD (computer aided diagnosis/detection) systems show mixed results for lung cancer detection [12–14], novel neural networks, convolutional neural networks (CNN) in particular, seem to have a positive effect on the diagnostic performance of radiologists [15]. Ardila et al. showed that a 3D-CNN outperformed radiologists in low-dose CT screening scans when no prior scans were available, indicating a favorable benefit for screening initiation.

Among other constraints, the health economic impact of AI systems is an important factor in the decision to implement models in routine clinical practice. Despite the imminent deployment of lung cancer screening and the promising results of AI-systems, no study has been performed to evaluate the utilization of neural networks in lung cancer screening compared to the stand-alone low dose CT-scan from an economic point of view. Therefore, the aim of our study was to evaluate the cost effectiveness of an AI-system for the initial scan of annual lung cancer screening and present the first results on identifying a cost margin for a clinical integration.

2. Materials and Methods

2.1. Model Structure

A decision model including the diagnostic strategies of conventional CT and CT augmented by AI was created and used as a decision tree, as shown in Figure 1.

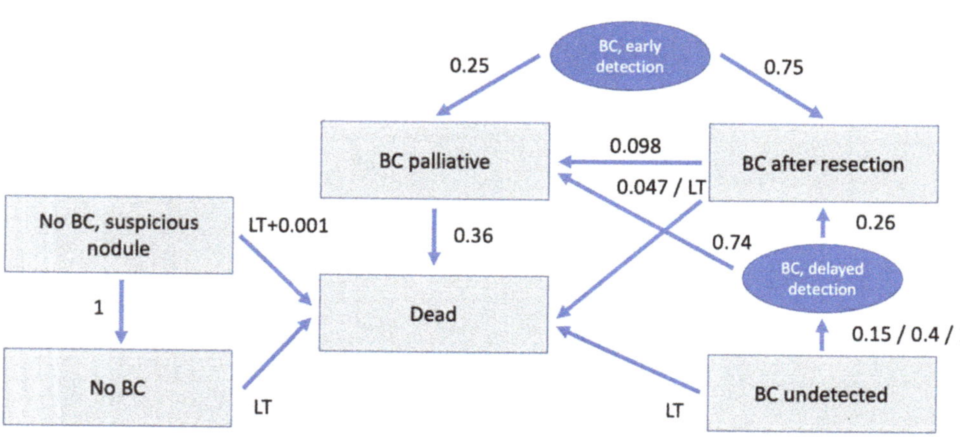

Figure 1. Markov model with possible states of disease and transition probabilities between states. BC = bronchial cancer; LT = life tables.

For calculation of costs and benefits in the different iterations a Markov transition state model was created. The model included the stages:

- No BC (patients without BC = true negative);
- No BC, Suspicious nodule (patients without BC but suspicious nodule = false positive);
- BC undetected (patients with undetected BC = false negative);
- BC after resection (patients with BC after resection);
- BC palliative (patients with BC which is unresectable/palliative);
- Dead.

Additionally, for better simulation and understanding of the model, the states "BC delayed detection" and "BC early detection" were created, which only served for transition. The Markov model reflects the different states a patient can be assigned to. Taking into account transition probabilities between the states as well as costs and effectiveness (displayed in Quality of Life) in those states during several iterations, cumulative costs and cumulative effectiveness within a defined time horizon can be calculated by adding those up throughout the iterations.

Analysis of the model was performed using a dedicated decision analysis software (TreeAge Pro Version 19.1.1, Williamstown, MA, USA).

2.2. Input Parameters

There was no requirement for an ethical approval for this analysis based on commonly available data. Model input parameters were based on current literature. Age-specific risk of death was derived from the US life tables [16]. Age at the diagnostic procedure was set to 60 years and willingness-to-pay was set to USD 100,000 per quality adjusted life year (QALY) at a discount rate of 3%, as reported previously [17,18]. The discount rate reflects the loss in economic value or effectiveness when there is a delay in realizing a benefit or incurring costs. The pre-test probability of BC was set to 2.635% for the risk group consisting of female and male smokers risk for an interval of 30 years, according to published data from Jacob et al. [19]. All input parameters and corresponding references are listed in Table 1.

Table 1. Input parameters.

Pre-test-Probability of BC	2.635	Jacob et al. [19]
Age at diagnostic procedure	60 years	US Preventive Services Task Force [1]
Assumed WTP	USD 100,000,00	Assumption
Discount rate	3.00%	Assumption
Markov model time	20 years	Assumption
	Diagnostic Test Performances	
Sensitivity for BC CT	77.9%	Ardila et al. [15]
Specificity for BC CT	87.7%	Ardila et al. [15]
Sensitivity for BC CT + AI	97.7%	Ardila et al. [15]
Specificity for BC CT + AI	98.4%	Ardila et al. [15]
	Costs (Acute)	
CT	USD 161.00	Medicare (71.250) [20]
	Costs (Long Term)	
No BC	USD 0.00	
Follow up if false positive	USD 2256.00	ten Haaf et al. [21]
Curative therapy BC/resection cost	USD 36,305.00	Cowper et al. [22]
BC undetected	USD 0	Assumption
BC after resection	USD 4283.00	ten Haaf et al. [21]
Therapy BC, palliative	USD 60,000.00	ten Haaf et al. [21]
Dead	USD 0	Assumption

Table 1. *Cont.*

	Utilities	
No BC	1	Assumption
Follow up if false positive	0.98	Gareen et al. [23]
Curative therapy BC/resection	0.79	Grutters et al. [24]
BC undetected	1	Assumption
BC after resection	0.933	Möller et al. [25]
BC palliative	0.63	Doyle et al. [26]
Dead	0	Assumption
	Transition Probabilities	
Verification of suspicious nodule as no BC	100%	Assumption
Death if no BC but suspicious nodule	0.001 (invasive diagnostics) + life tables	The National Lung Screening Trial Research Team [2]
Resection rate of BC after early detection	75%	The National Lung Screening Trial Research Team [2]
Death after curative resection	4.70%	Green et al./Toker et al. [27,28]
Recurrence after resection	9.80%	Lou et al. [29]
Detection of initially undetected BC	15% 1st, 40% 2nd, 100% 3rd year	Scholten et al. [30]
Death with undetected BC	life tables	
Resection rate of BC after delayed detection	26%	Hunbogi et al. [31]
Death with palliative care	36%	Cancer Stat Facts: Lung and Bronchus Cancer, National Cancer Institute [32]
Death without BC	life tables	

AI = artificial intelligence; BC = bronchial cancer; CT = computed tomography; QALY = quality adjusted life year; WTP = willingness-to-pay.

2.3. Diagnostic Test Performances

Sensitivity and specificity values for CT detection of BC with and without AI were derived from the literature (Table 1).

2.4. Costs

From a United States (US) healthcare perspective, costs were estimated based on Medicare data and available literature (Table 1). The long-term costs of the follow up in case of false positive was estimated at USD 2256 including the costs for a follow up CT examination and a possible bronchoscopy and biopsy [21]. The resection costs of BC were set to USD 36,305, according to Cowper et al. [22]. annual costs of palliative BC patients were estimated at USD 60,000 [21].

2.5. Utilities

Utility is measured in the additional quality-adjusted life years (QALY) which are gained through each diagnostic procedure. According to previous studies, quality of life (QOL) for curative BC patients was set to 0.79 for the first year after resection and 0.933 for the following years [24,25]. In accordance with the literature, QOL for palliative BC patients was set to 0.63 [26]. These values were then used for calculations in a Markov model specifically designed as mentioned above.

2.6. Transition Probabilities

Transition probabilities were derived from a systematic review of the recent literature and are shown in Table 1. Probability of successful resection of (early) detected BC was estimated at 75%, according to the national lung screening trial research team [2]. Risk of secondary occurrence of cancer/metastases after resection of the primary tumor was assumed to be 9.80% [29]. Annual mortality rate of curative patients was set to 4.7% and to 36.0% for palliative patients [28,32,33].

2.7. Cost-Effectiveness Analysis

The cost-effectiveness analysis was performed based on Markov simulations with a run time of 20 years (20 iterations) after initial diagnostic procedure. The discount rate was set to 3.0% and willingness-to-pay was set to USD 100,000 per QALY according to current recommendations [18].

In the base-case scenario, cost-effectiveness was determined with costs of CT + AI identical to costs of CT only, meaning costs of USD 0 for additional use of AI. Based on these results, maximum costs for AI were calculated for several willingness-to-pay thresholds. For evaluation of model uncertainty and influence of alteration of each variable on the model, a deterministic sensitivity analysis was performed. Results were visualized in a tornado diagram.

Based on the Markov model, Monte-Carlo simulations were used to perform a probabilistic sensitivity analysis with a total of 30,000 iterations. This method is used to account for the variation of input-parameters among different individuals.

3. Results

3.1. Cost-Effectiveness Analysis

Simulations of a time horizon of 20 years resulted in average cumulative costs of USD 4310.82 for CT + AI and USD 4378.44 for CT if additional diagnostic costs for the use of AI were set to USD 0 in the base case scenario. In this scenario, average cumulative effectiveness was at 13.76 QALYs for CT + AI and at 13.75 QALYs for CT. To better understand the impact of input parameters on the model, costs and effectiveness as well as distribution of the different outcomes are shown in Figure 2. Different overall costs and effectiveness derive from different distribution of the outcomes "true positive", "false negative", "true negative", and "false positive" based on different sensitivity and specificity of the two methods. The incremental cost-effectiveness ratio in the base case scenario was negative, meaning both, lower cost and higher effectiveness for CT + AI.

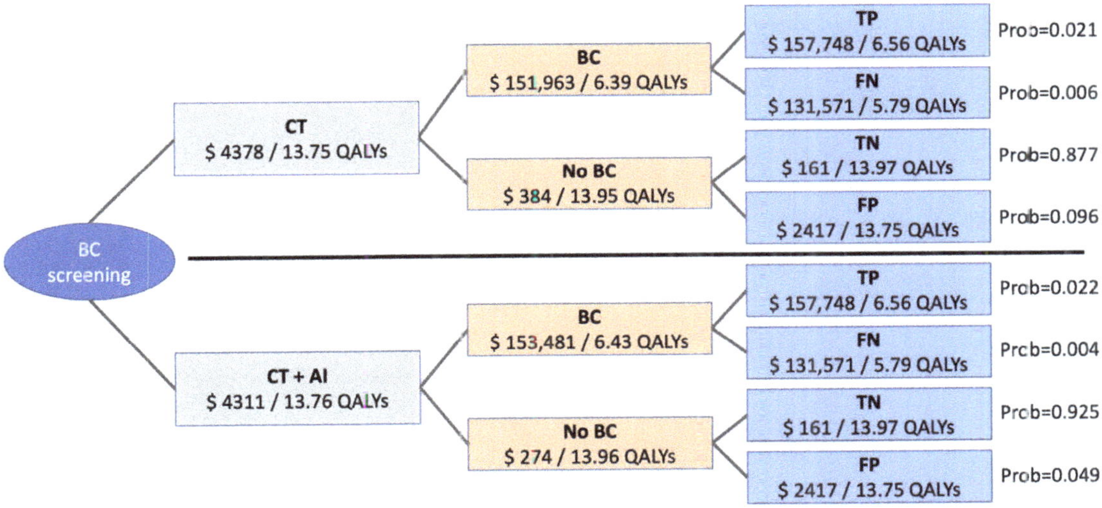

Figure 2. Roll-back of the economic model showing costs and effectiveness of the different outcomes. Distributions leading to overall costs and effectiveness are different for CT and CT + AI depending on sensitivity and specificity of the two methods and indicated as probabilities. BC = bronchial cancer; CT = computed tomography; TP = true positive; TN = true negative; FP = false positive; FN = false negative; Prob = probability.

3.2. Sensitivity Analysis

Probabilistic sensitivity analysis and Monte Carlo simulation was performed to determine the distribution of the resulting ICER-values and is visualized in Figure 3. Monte Carlo simulation reflects the difference between costs (=incremental costs) and effectiveness (=incremental effectiveness) for a certain amount of notional scenarios/iterations. All iterations with an ICER-value below the willingness-to-pay of USD 100,000 per QALY were considered cost-effective.

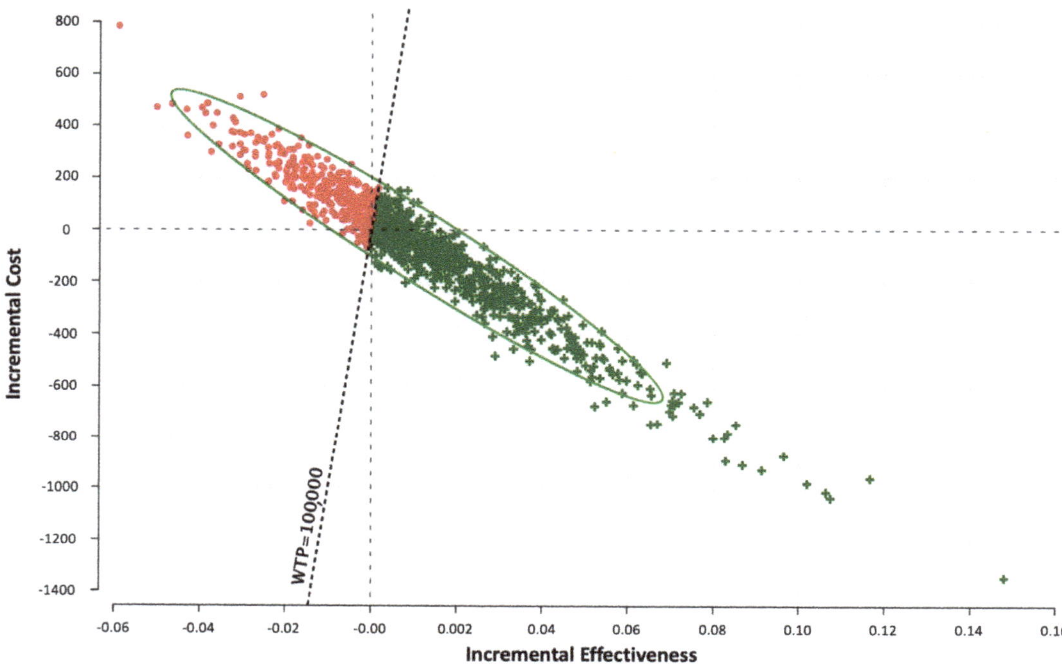

Figure 3. Probabilistic sensitivity analysis utilizing Monte-Carlo simulations (30,000 iterations). Incremental cost-effectiveness scatter plot for CT + AI vs. CT. iterations with an ICER-value below the willingness-to-pay of USD 100,000 per QALY are shown as green crosses. WTP = willingness-to-pay.

Deterministic sensitivity analysis was performed to account for variability of input parameters in the base case scenario. Results are displayed as a tornado diagram in Figure 4A.

Applying wide ranges of variation for the different input parameters, ICER stayed below USD 0/QALY for the sensitivities of the diagnostic modalities and the probabilities of resectability in early and delayed diagnosis. Although ICER turned positive when varying the specificity of CT and CT + AI, the willingness-to-pay threshold of USD 100,000/QALY was not crossed in any of the cases.

3.3. Threshold Analysis

To determine the maximum possible costs for the use of AI at a willingness-to-pay of USD 100,000/QALY, a threshold analysis was performed. As shown in Figure 5, ICER remained negative until costs of AI were raised to USD 68.

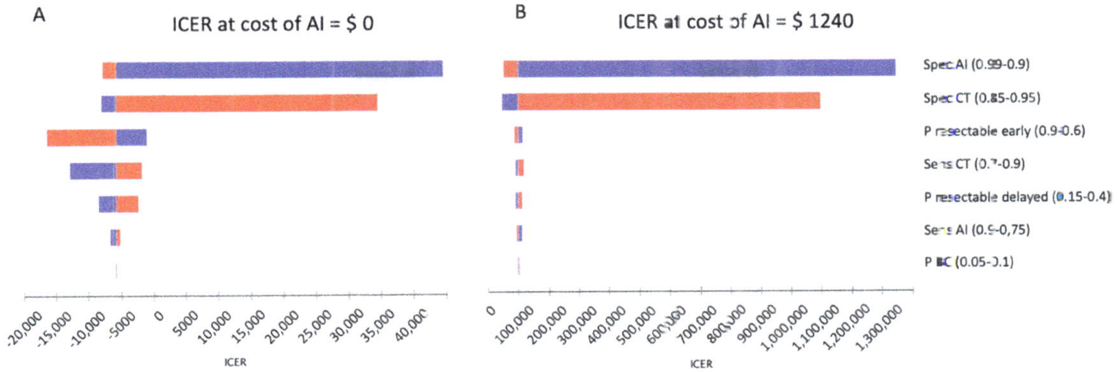

Figure 4. (**A**) Tornado diagram showing the impact of input parameters on incremental cost-effectiveness ratio (ICER) in the base case scenario. Assuming a willingness-to-pay threshold of USD 100,000 per QALY, CT + AI remained cost-effective in all cases. (**B**) Tornado diagram showing the impact of input parameters on incremental cost-effectiveness ratio (ICER) when costs of AI were set to USD 1240 with an expected value of USD 100.000 per QALY. Blue bars show changes when decreasing the value of an input parameter as compared to the base case scenario and red bars when increasing the respective value. Sens = sensitivity; Spec = specificity; CT = computed tomography; AI = artificial intelligence; P = probability.

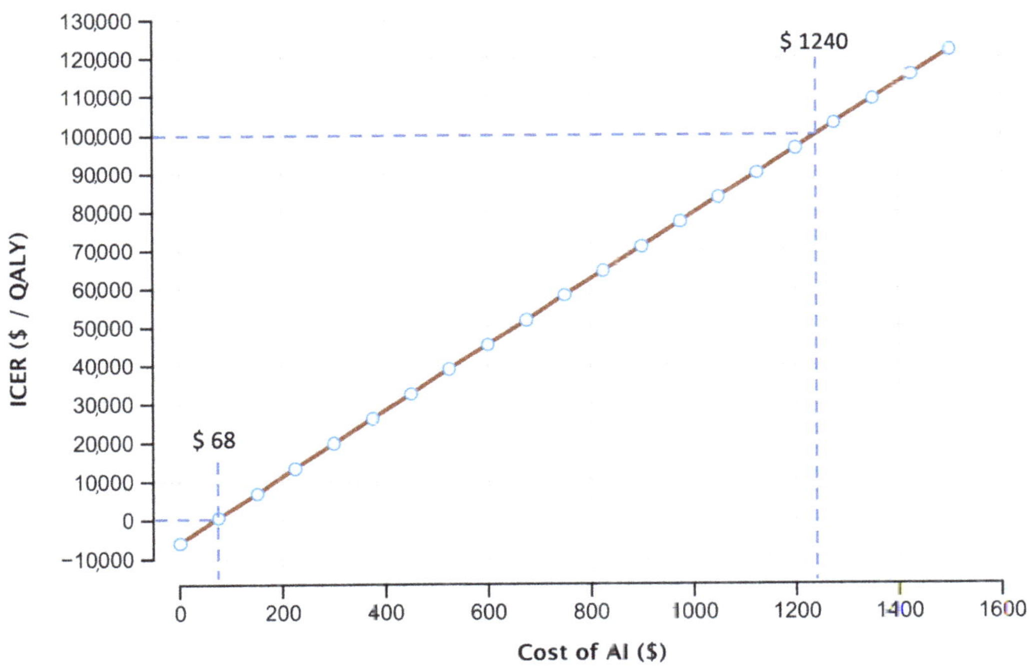

Figure 5. One-way sensitivity analysis for costs of AI (USD) and the corresponding incremental cost effectiveness ratio (ICER in USD/QALY). Thresholds indicate values at an ICER of USD 0/QALY and USD 100,000/QALY. ICER = incremental cost-effectiveness ratio; AI = artificial intelligence; QALY = quality adjusted life year.

Raising costs of AI further, the assumed willingness-to-pay threshold of USD 100,000/QALY is only crossed at a value USD 1240. Influence in different input parameters in this second base case scenario setting costs of AI to USD 1240 are shown in Figure 4B. To account for possible variation of the willingness-to-pay, Table 2 displays possible costs for AI depending on different willingness-to-pay thresholds. Due to the cost's dependency on the ICER, the cost for AI directly is further influenced by the systems performance, resulting in a higher price for a better system due to the increased ICER.

Table 2. Cost of AI at different WTP-thresholds.

WTP (USD/QALY)	0	20,000	40,000	60,000	80,000	100,000	120,000	150,000	200,000
Cost of AI (USD)	68	302	537	771	1006	1240	1475	1826	2412

4. Discussion

The widespread integration of lung cancer screening is proving to be a complex and challenging undertaking. Nevertheless, lung cancer screening is a cost-effective method to reduce lung cancer mortality. AI-models for cancer detection and classification have proved to be of benefit in lung cancer screening in several studies [15,34].

In the present study, we show that a state-of-the-art AI-model (3D-convolutional neural network according to Ardila et al.) is a cost-effective method for the baseline screening scan [15]. Despite promising results of AI in the health care sector, studies evaluating the economic impact and cost effectiveness remain sparse [35]. To our knowledge, no study has been conducted to investigate the cost-effectiveness of an AI-system in lung cancer screening. Based on the superior performance of the AI-model without prior imaging, we simulated an implementation for the initial screening scan using input parameters derived from published screening cohorts [2,15,36,37], to ensure comparability to the standard screening setting.

Our base case estimate for screening with an AI system compared to current low-dose CT screening yielded a negative ICER up to costs of USD 68 for the AI system, indicating that using an AI system in the screening setting results in lower cost and higher effectiveness up to these costs per patient scan. Furthermore, the ICER remained below the applied willingness-to-pay up to costs of USD 1240. To account for variations in input parameters, we performed a deterministic sensitivity analysis for the base case scenario and the maximum cost-effective costs (USD 1240). The specificity of the diagnostic strategy had the greatest influence for both scenarios, due to the low lung cancer rate in screening cohorts. For the base case scenario all input variations resulted in an ICER below the willingness-to-pay by a large margin, indicating robust cost-effectiveness. Adding AI support showed a reduced number of false-positives and an increased number of true negatives in our simulation. In particular, the reduction of false-positives highly impacts the value of a screening method, as not only costs in the form of unnecessary follow-up examination and possibly further, partly invasive examinations are reduced, but also patients do not have to experience the psychological distress of a possible cancer diagnosis [38]. Additionally, the false positive rates and the frequency of invasive diagnostic procedures were more frequent at the baseline CT, ranging from 7.9% to 49.3% for the false positive rate and 3.7% for additional invasive procedures [2,39], further emphasizing the benefit of AI support for the initial screening. As shown by Audelan et al., the sensitivity and specificity of AI in lung cancer screening can further be improved, consequently allowing for an additional reduction of costs and increased effectiveness [40].

Despite promising results, our study underlies several limitations. First, the cost-effectiveness was only evaluated for the initial scan in the lung cancer screening. This is due to published literature, focusing on the superiority of AI lung nodule detection and classification in initial CT of the thorax without prior imaging for comparison. According to Ardila et al., deep-learning algorithms are superior to radiologists in lung cancer screening detection, when no prior imaging is available for comparison, but is on-par as soon as previous examinations are available for the reader. Consequently, further research has to be conducted to evaluate the cost-effectiveness of AI-based computer-aided diagnosis systems in longitudinal screening, beyond the initial scan [15]. Further, our evaluation is focused on the sole AI system performance in comparison to the human reader—the radiologist. However, several studies have shown promising results for the collaboration of both, often referred to as the "Centaur model" [33]. Such systems were shown not only to be beneficial in patient care but cost-effective as well [41]. Despite dealing with different challenges compared to lung cancer, for thyroid nodule detection, AI systems outperform thyroid cancer specialized radiologists in nodule classification, but the combination of specialized radiologists with AI-support showed an even higher specificity and positive predictive value when compared to the AI system alone [42]. Therefore, further research is needed to evaluate the combination of AI models and specialized thorax radiologists in lung cancer detection and its cost-effectiveness. Lastly, cost-effectiveness analysis with decision-based models is highly dependent on the input parameters, while deterministic sensitivity analysis may incorporate parameter variation to a certain degree, and recommendations for each individual case cannot be derived from the model.

5. Conclusions

To conclude, in our study we show that screening with an AI-model in the initial screening scan is a cost-effective strategy in low-dose CT lung cancer screening with robustness to variation of input parameters. Defining thresholds for cost of AI results might help faster translate AI systems into clinical use.

Author Contributions: Conceptualization, S.Z. and F.G.; methodology, F.G. and J.G.; validation, M.G., S.Z.; formal analysis, F.G.; investigation, S.Z., M.G. and J.G.; resources, M.M.; data curation, M.G.; writing—original draft preparation, S.Z. and J.G.; writing—review and editing, M.G., F.G. and M.M.; visualization, F.G. All authors have read and agreed to the published version of the manuscript.

Funding: This research received no external funding.

Institutional Review Board Statement: Ethical review and approval were waived for this study due to this analysis is based on commonly available data.

Informed Consent Statement: Not applicable.

Data Availability Statement: The data that support the findings of this study are listed in Table 1.

Conflicts of Interest: The authors declare no conflict of interest.

References

1. Moyer Virginia, A. On behalf of the us preventive services task force screening for lung cancer: Us preventive services task force recommendation statement. *Ann. Intern. Med.* **2014**, *160*, 330–338.
2. National Lung Screening Trial Research Team. Reduced lung-cancer mortality with low-dose computed tomographic screening. *N. Engl. J. Med.* **2011**, *365*, 395–409. [CrossRef] [PubMed]
3. Oudkerk, M.; Devaraj, A.; Vliegenthart, R.; Henzler, T.; Prosch, H.; Heussel, C.P.; Bastarrika, G.; Sverzellati, N.; Mascalchi, M.; Delorme, S. European position statement on lung cancer screening. *Lancet Oncol.* **2017**, *18*, e754–e766. [PubMed]
4. De Koning, H.J.; van der Aalst, C.M.; de Jong, P.A.; Scholten, E.T.; Nackaerts, K.; Heuvelmans, M.A.; Lammers, J.-W.J.; Weenink, C.; Yousaf-Khan, U.; Horeweg, N. Reduced lung-cancer mortality with volume ct screening in a randomized trial. *N. Engl. J. Med.* **2020**, *382*, 503–513. [CrossRef] [PubMed]
5. Rasmussen, J.F.; Siersma, V.; Pedersen, J.H.; Heleno, B.; Saghir, Z.; Brodersen, J. Healthcare costs in the danish randomised controlled lung cancer ct-screening trial: A registry study. *Lung Cancer* **2014**, *83*, 347–355. [CrossRef]
6. Wiener, R.S.; Schwartz, L.M.; Woloshin, S.; Welch, H.G. Population-based risk for complications after transthoracic needle lung biopsy of a pulmonary nodule: An analysis of discharge records. *Ann. Intern. Med.* **2011**, *155*, 137–144. [CrossRef]
7. Kastner, J.; Hossain, R.; Jeudy, J.; Dako, F.; Mehta, V.; Dalal, S.; Dharaiya, E.; White, C. Lung-rads version 1.0 versus lung-rads version 1.1: Comparison of categories using nodules from the national lung screening trial. *Radiology* **2021**, *300*, 203704.
8. McKee, B.J.; Regis, S.M.; McKee, A.B.; Flacke, S.; Wald, C. Performance of acr lung-rads in a clinical ct lung screening program. *J. Am. Coll. Radiol.* **2016**, *13*, R25–R29. [CrossRef]
9. Mehta, H.J.; Mohammed, T.-L.; Jantz, M.A. The american college of radiology lung imaging reporting and data system: Potential drawbacks and need for revision. *Chest* **2017**, *151*, 539–543. [CrossRef]

10. Singh, S.; Pinsky, P.; Fineberg, N.S.; Gierada, D.S.; Garg, K.; Sun, Y.; Nath, P.H. Evaluation of reader variability in the interpretation of follow-up ct scans at lung cancer screening. *Radiology* **2011**, *259*, 263–270. [CrossRef]
11. McKinney, S.M.; Sieniek, M.; Godbole, V.; Godwin, J.; Antropova, N.; Ashrafian, H.; Back, T.; Chesus, M.; Corrado, G.S.; Darzi, A. International evaluation of an ai system for breast cancer screening. *Nature* **2020**, *577*, 89–94. [CrossRef] [PubMed]
12. Brown, M.S.; Goldin, J.G.; Rogers, S.; Kim, H.J.; Suh, R.D.; McNitt-Gray, M.F.; Shah, S.K.; Truong, D.; Brown, K.; Sayre, J.W. Computer-aided lung nodule detection in ct: Results of large-scale observer test1. *Acad. Radiol.* **2005**, *12*, 681–686. [CrossRef] [PubMed]
13. De Hoop, B.; de Boo, D.W.; Gietema, H.A.; van Hoorn, F.; Mearadji, B.; Schijf, L.; van Ginneken, B.; Prokop, M.; Schaefer-Prokop, C. Computer-aided detection of lung cancer on chest radiographs: Effect on observer performance. *Radiology* **2010**, *257*, 532–540. [CrossRef] [PubMed]
14. Jeon, K.N.; Goo, J.M.; Lee, C.H.; Lee, Y.; Choo, J.Y.; Lee, N.K.; Shim, M.-S.; Lee, I.S.; Kim, K.G.; Gierada, D.S. Computer-aided nodule detection and volumetry to reduce variability between radiologists in the interpretation of lung nodules at low-dose screening ct. *Investig. Radiol.* **2012**, *47*, 457. [CrossRef]
15. Ardila, D.; Kiraly, A.P.; Bharadwaj, S.; Choi, B.; Reicher, J.J.; Peng, L.; Tse, D.; Etemadi, M.; Ye, W.; Corrado, G. End-to-end lung cancer screening with three-dimensional deep learning on low-dose chest computed tomography. *Nat. Med.* **2019**, *25*, 954–961. [CrossRef]
16. Arias, E.; Xu, J.; Kochanek, K.D. United states life tables, 2016. *Natl. Vital Stat. Rep.* **2019**, *68*, 4.
17. Cameron, D.; Ubels, J. Norstr öm f: On what basis are medical cost-effectiveness thresholds set. *Clashing Opin. Absence Data A Syst. Rev. Glob. Health Action* **2018**, *11*, 1447828. [CrossRef]
18. Sanders, G.D.; Neumann, P.J.; Basu, A.; Brock, D.W.; Feeny, D.; Krahn, M.; Kuntz, K.M.; Meltzer, D.O.; Owens, D.K.; Prosser, L.A. Recommendations for conduct, methodological practices, and reporting of cost-effectiveness analyses: Second panel on cost-effectiveness in health and medicine. *JAMA* **2016**, *316*, 1093–1103. [CrossRef]
19. Jacob, L.; Freyn, M.; Kalder, M.; Dinas, K.; Kostev, K. Impact of tobacco smoking on the risk of developing 25 different cancers in the uk: A retrospective study of 422,010 patients followed for up to 30 years. *Oncotarget* **2018**, *9*, 17420. [CrossRef]
20. Procedure Price Lookup for Outpatient Services. Medicare.gov 71275. 2021. Available online: https://www.medicare.gov/procedure-price-lookup/cost/71275/ (accessed on 9 January 2022).
21. Ten Haaf, K.; Tammemägi, M.C.; Bondy, S.J.; van der Aalst, C.M.; Gu, S.; McGregor, S.E.; Nicholas, G.; de Koning, H.J.; Paszat, L.F. Performance and cost-effectiveness of computed tomography lung cancer screening scenarios in a population-based setting: A microsimulation modeling analysis in Ontario, Canada. *PLoS Med.* **2017**, *14*, e1002225. [CrossRef]
22. Cowper, P.A.; Feng, L.; Kosinski, A.S.; Tong, B.C.; Habib, R.H.; Putnam, J.B., Jr.; Onaitis, M.W.; Furnary, A.P.; Wright, C.D.; Jacobs, J.P. Initial and longitudinal cost of surgical resection for lung cancer. *Ann. Thorac. Surg.* **2021**, *111*, 1827–1833. [CrossRef] [PubMed]
23. Gareen, I.F.; Duan, F.; Greco, E.M.; Snyder, B.S.; Boiselle, P.M.; Park, E.R.; Fryback, D.; Gatsonis, C. Impact of lung cancer screening results on participant health-related quality of life and state anxiety in the national lung screening trial. *Cancer* **2014**, *120*, 3401–3409. [CrossRef] [PubMed]
24. Grutters, J.P.; Joore, M.A.; Wiegman, E.M.; Langendijk, J.A.; de Ruysscher, D.; Hochstenbag, M.; Botterweck, A.; Lambin, P.; Pijls-Johannesma, M. Health-related quality of life in patients surviving non-small cell lung cancer. *Thorax* **2010**, *65*, 903–907. [CrossRef]
25. Möller, A.; Sartipy, U. Long-term health-related quality of life following surgery for lung cancer. *Eur. J. Cardio-Thorac. Surg.* **2012**, *41*, 362–367. [CrossRef] [PubMed]
26. Doyle, S.; Lloyd, A.; Walker, M. Health state utility scores in advanced non-small cell lung cancer. *Lung Cancer* **2008**, *62*, 374–380. [CrossRef] [PubMed]
27. Green, A.; Hauge, J.; Iachina, M.; Jakobsen, E. The mortality after surgery in primary lung cancer: Results from the danish lung cancer registry. *Eur. J. Cardio-Thorac. Surg.* **2016**, *49*, 589–594. [CrossRef]
28. Toker, A.; Dilege, S.; Ziyade, S.; Eroglu, O.; Tanju, S.; Yilmazbayhan, D.; Kilicarslan, Z.; Kalayci, G. Causes of death within 1 year of resection for lung cancer. Early mortality after resection. *Eur. J. Cardio-Thorac. Surg.* **2004**, *25*, 515–519. [CrossRef]
29. Lou, F.; Huang, J.; Sima, C.S.; Dycoco, J.; Rusch, V.; Bach, P.B. Patterns of recurrence and second primary lung cancer in early-stage lung cancer survivors followed with routine computed tomography surveillance. *J. Thorac. Cardiovasc. Surg.* **2013**, *145*, 75–82. [CrossRef] [PubMed]
30. Scholten, E.T.; Horeweg, N.; de Koning, H.J.; Vliegenthart, R.; Oudkerk, M.; Willem, P.T.M.; de Jong, P.A. Computed tomographic characteristics of interval and post screen carcinomas in lung cancer screening. *Eur. Radiol.* **2015**, *25*, 81–88. [CrossRef]
31. Thorsteinsson, H.; Alexandersson, A.; Oskarsdottir, G.N.; Skuladottir, R.; Isaksson, H.J.; Jonsson, S.; Gudbjartsson, T. Resection rate and outcome of pulmonary resections for non–small-cell lung cancer: A nationwide study from iceland. *J. Thorac. Oncol.* **2012**, *7*, 1164–1169. [CrossRef]
32. Cancer Stat Facts: Lung and Bronchus Cancer. 2021. Available online: https://seer.cancer.gov/statfacts/html/lungb.html (accessed on 9 January 2022).
33. Goldstein, I.M.; Lawrence, J.; Miner, A.S. Human-machine collaboration in cancer and beyond: The centaur care model. *JAMA Oncol.* **2017**, *3*, 1303–1304. [CrossRef] [PubMed]

34. Liang, M.; Tang, W.; Xu, D.M.; Jirapatnakul, A.C.; Reeves, A.P.; Henschke, C.I.; Yankelevitz, D. Low-dose ct screening for lung cancer: Computer-aided detection of missed lung cancers. *Radiology* **2016**, *281*, 279–288. [CrossRef] [PubMed]
35. Wolff, J.; Pauling, J.; Keck, A.; Baumbach, J. Systematic review of economic impact studies of artificial intelligence in health care. *J. Med. Internet Res.* **2020**, *22*, e16866. [CrossRef] [PubMed]
36. Pastorino, U.; Rossi, M.; Rosato, V.; Marchianò, A.; Sverzellati, N.; Morosi, C.; Fabbri, A.; Galeone, C.; Negri, E.; Sozzi, G. Annual or biennial ct screening versus observation in heavy smokers: 5-year results of the mild trial. *Eur. J. Cancer Prev.* **2012**, *21*, 308–315. [CrossRef]
37. Van Klaveren, R.J.; Oudkerk, M.; Prokop, M.; Scholten, E.T.; Nackaerts, K.; Vernhout, R.; van Iersel, C.A.; van den Bergh, K.A.; Westeinde, S.V.; van der Aalst, C. Management of lung nodules detected by volume ct scanning. *N. Engl. J. Med.* **2009**, *361*, 2221–2229. [CrossRef]
38. Van den Bergh, K.A.; Essink-Bot, M.-L.; Borsboom, G.J.; Scholten, E.T.; Prokop, M.; de Koning, H.J.; van Klaveren, R.J. Short-term health-related quality of life consequences in a lung cancer ct screening trial (nelson). *Br. J. Cancer* **2010**, *102*, 27–34. [CrossRef]
39. Jonas, D.E.; Reuland, D.S.; Reddy, S.M.; Nagle, M.; Clark, S.D.; Weber, R.P.; Enyioha, C.; Malo, T.L.; Brenner, A.T.; Armstrong, C. Screening for lung cancer with low-dose computed tomography: Updated evidence report and systematic review for the us preventive services task force. *JAMA* **2021**, *325*, 971–987. [CrossRef]
40. Audelan, B.; Lopez, S.; Fillard, P.; Diascorn, Y.; Padovani, B.; Delingette, H. *Validation of Lung Nodule Detection a Year before Diagnosis in Nlst Dataset Based on a Deep Learning System*; European Respiratory Society: Lausanne, Switzerland, 2021.
41. Hoverman, J.R.; Klein, I.; Harrison, D.W.; Hayes, J.E.; Garey, J.S.; Harrell, R.; Sipala, M.; Houldin, S.; Jameson, M.D.; Abdullahpour, M. Opening the black box: The impact of an oncology management program consisting of level i pathways and an outbound nurse call system. *J. Oncol. Pract.* **2014**, *10*, 63–67. [CrossRef]
42. Peng, S.; Liu, Y.; Lv, W.; Liu, L.; Zhou, Q.; Yang, H.; Ren, J.; Liu, G.; Wang, X.; Zhang, X. Deep learning-based artificial intelligence model to assist thyroid nodule diagnosis and management: A multicentre diagnostic study. *Lancet Digit. Health* **2021**, *3*, e250–e259. [CrossRef]

cancers

Review

Advancements in Oncology with Artificial Intelligence—A Review Article

Nikitha Vobugari [1], Vikranth Raja [2], Udhav Sethi [3], Kejal Gandhi [1], Kishore Raja [4] and Salim R. Surani [5,*]

1 Department of Internal Medicine, Medstar Washington Hospital Center, Washington, DC 20010, USA; nikitha.vobugari@medstar.net (N.V.); kejal.d.gandhi@medstar.net (K.G.)
2 Department of Medicine, P.S.G Institute of Medical Sciences and Research, Coimbatore 641004, Tamil Nadu, India; drvikranthraja@gmail.com
3 School of Computer Science, University of Waterloo, Waterloo, ON N2L 3G1, Canada; udhav.sethi@uwaterloo.ca
4 Department of Pediatric Cardiology, University of Minnesota, Minneapolis, MN 55454, USA; drkishoreraja@gmail.com
5 Department of Pulmonary and Critical Care, Texas A&M University, College Station, TX 77843, USA
* Correspondence: surani@tamu.edu; Fax: +1-361-8507-563

Simple Summary: With the advancement of artificial intelligence, including machine learning, the field of oncology has seen promising results in cancer detection and classification, epigenetics, drug discovery, and prognostication. In this review, we describe what artificial intelligence is and its function, as well as comprehensively summarize its evolution and role in breast, colorectal, and central nervous system cancers. Understanding the origin and current accomplishments might be essential to improve the quality, accuracy, generalizability, cost-effectiveness, and reliability of artificial intelligence models that can be used in worldwide clinical practice. Students and researchers in the medical field will benefit from a deeper understanding of how to use integrative AI in oncology for innovation and research.

Abstract: Well-trained machine learning (ML) and artificial intelligence (AI) systems can provide clinicians with therapeutic assistance, potentially increasing efficiency and improving efficacy. ML has demonstrated high accuracy in oncology-related diagnostic imaging, including screening mammography interpretation, colon polyp detection, glioma classification, and grading. By utilizing ML techniques, the manual steps of detecting and segmenting lesions are greatly reduced. ML-based tumor imaging analysis is independent of the experience level of evaluating physicians, and the results are expected to be more standardized and accurate. One of the biggest challenges is its generalizability worldwide. The current detection and screening methods for colon polyps and breast cancer have a vast amount of data, so they are ideal areas for studying the global standardization of artificial intelligence. Central nervous system cancers are rare and have poor prognoses based on current management standards. ML offers the prospect of unraveling undiscovered features from routinely acquired neuroimaging for improving treatment planning, prognostication, monitoring, and response assessment of CNS tumors such as gliomas. By studying AI in such rare cancer types, standard management methods may be improved by augmenting personalized/precision medicine. This review aims to provide clinicians and medical researchers with a basic understanding of how ML works and its role in oncology, especially in breast cancer, colorectal cancer, and primary and metastatic brain cancer. Understanding AI basics, current achievements, and future challenges are crucial in advancing the use of AI in oncology.

Keywords: artificial intelligence; machine learning; deep learning; convolutional neural networks; support vector machine; breast oncology; brain tumors; colon cancer

Citation: Vobugari, N.; Raja, V.; Sethi, U.; Gandhi, K. Raja, K.; Surani, S.R. Advancements in Oncology with Artificial Intelligence—A Review Article. *Cancers* **2022**, *14*, 1349. https://doi.org/10.3390/cancers14051349

Academic Editors: Hamid Khayyam and Rahele Kafieh

Received: 21 February 2022
Accepted: 28 February 2022
Published: 6 March 2022

1. Introduction

Artificial intelligence (AI) is a field in which computers are programmed to mimic human intelligence. The abundance of data in the field of medicine makes it a good candidate for problem solving using machine learning (ML). In oncology, ML can be used to diagnose and classify tumors, detect early-stage tumors, gather genetic and histopathological data, assist in pre- and post-operative planning, and predict overall survival outcomes [1]. Deep Learning (DL), a type of ML, has proven to be effective in automating time-consuming steps such as detection and segmentation of lesions [2–4].

AI-based models have demonstrated excellent accuracy rates of cancer detection on screening mammography and breast cancer (BC) prediction based on genetics and hormonal factors [5–7]. AI plays a crucial role in early detection, classification, histopathological aspects, genetics, and molecular markers detection in colorectal cancer (CRC) [8–10]. As a result of extensive data in present-day screening and improvements in life expectancy caused by early detection of breast and colon cancer, we review the potential of AI-based diagnostics and therapeutics. Because mammograms and colonoscopies are widely used in the general population worldwide, AI can be used extensively in future studies on cancer screening to build generalizable AI systems [11]. AI has made its way into other cancer types, which we do not review here. For instance, lung cancer screening is reserved for smokers, and the United States Preventive Services Task Force (USPSTF) approved low-dose chest computed tomography (CT) scans in 2013, and prostate cancer screening has not yet been approved universally [11,12]. CNS cancers are relatively rare and have a poor prognosis. Studying AI in such rare tumors can provide a scope of precision of AI integration in improving the current standard management. In the area of central nervous system (CNS) tumors, AI and radiomics have notably enhanced detection rates and reduced several time-consuming steps in glioma grading, pre- and intraoperative planning, and postoperative follow-up [13–15].

This review article outlines how AI works in simple terminology that medical professionals can understand, how it has improved breast cancer screening, colon polyp detection, and colorectal cancer screening, as well as the implications it has in the management of CNS tumors. A literature search was conducted on PubMed, Google Scholar, arXiv, and Scopus. This is not a systematic review but a narrative review of the literature. We conclude with existing obstacles and future speculations of standardizing AI screening in oncology, as well as proposals for integrating AI basics into medical school curricula.

2. How Does Artificial Intelligence Work?

AI is a broad concept that aims to simulate human cognitive ability. ML, an approach to AI, is the study of how computer systems can learn to perform a task or predict an outcome without being explicitly programmed [16]. Mitchell et al. (1997) succinctly defines this learning process as follows: A computer program is said to learn from experience (E) with respect to some class of tasks (T) and performance measure (P), if its performance at tasks in T, as measured by P, improves with experience E. A simple example of such a task is the classification of suspicious abnormality on a screening mammogram as probable malignant or benign [17]. To learn to perform this task, a computer program would experience a dataset containing examples of correctly classified cases of benign and malignant breast lesions and come up with a model that can generalize beyond these data. Its ability to then classify previously unseen examples of breast lesions correctly would be evaluated through a quantitative measure of its performance, such as accuracy, sensitivity, and specificity.

2.1. Subtypes of Machine Learning

Algorithms for ML are typically categorized into supervised, unsupervised, or reinforcement learning. Supervised learning algorithms experience a dataset that contains a label (or correct answer) for each data point. Examples of supervised learning algorithms include support vector machine (SVM) [18,19], linear regression, logistic regression, and k-nearest neighbors [20,21]. In contrast, unsupervised algorithms such as k-means clustering [22,23], affinity propagation [24], and gaussian mixture model [25] study a dataset that does not contain labels and learn to derive structure from the given data. A reinforcement learning system trains an agent to behave in an environment by assigning it with a reward for desired behaviors or penalizing it for undesired ones. The overall objective of an ML algorithm can be interpreted as learning an approximate function of the data. This function should take as input a set of features that describe the data and output a prediction corresponding to the learning task. Classical ML algorithms are generally good at approximating linear or simple non-linear functions [13,26].

2.2. Deep Learning

DL is a type of ML that enables the learning of complex non-linear functions of the data. Most modern DL methods use neural networks as their learning model, which are loosely inspired by neuroscience [27]. The fundamental computational unit of a neural network is called a neuron. It computes a weighted sum of its inputs and then applies a non-linear operation (often called the activation function) to the sum to compute the output (See Figure 1a). Common activation functions include sigmoid, tanh, and rectified linear activation unit (ReLU) functions. A neural network comprises one or more layers of neurons, with each layer feeding on the outputs of the previous layer. Information flows forward through the network from the input, through a series of intermediate layers (called hidden layers) and finally to the output (see Figure 1b). As the number of layers and units within a layer increase, a neural network can represent functions of increasing complexity. This architecture gives neural networks the ability to learn their own complex features instead of being constrained to the hand-picked features provided as input to the model.

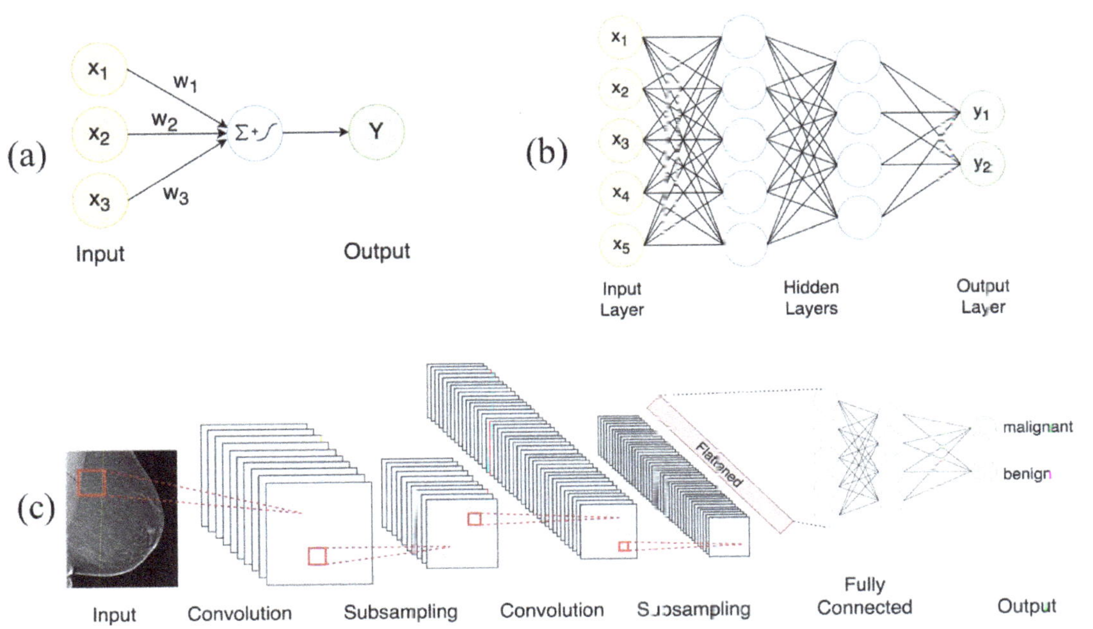

Figure 1. (**a**): Neuron, the fundamental computational unit of a neural network, computes the weighted sum of its inputs (X_1, X_2, X_3) and applies a non-linear operation to give output (Y). (**b**): An example of a feedforward neural network with two hidden layers, with five and four neurons, respectively. (**c**): An example of a convolutional neural network (CNN) applied to the classification of a screening mammogram as probable malignant or benign.

During training, the parameters of the neural network are learned in order to fit the dataset for a given task. This corresponds to minimizing some notion of a cost function, which measures the model's performance on the task. After each forward pass through the network, the cost function is used to compute the error between the predicted and expected output. An algorithm called backpropagation allows this cost information to flow backward through the neural network while adjusting the network parameters. Backpropagation computes the gradients of the cost function with respect to the network parameters, which determine the level of adjustment to be made to the parameters in each iteration [28]. These gradients are then used to update the network parameters using an optimization algorithm such as stochastic gradient descent (SGD) [29,30].

Apart from the simple feed-forward model discussed above, there are other specialized architectures of neural networks suited for specific tasks. For instance, convolutional neural networks (CNNs) have a grid-like topology and are well suited to process two or three-dimensional inputs such as images [31]. CNNs are designed to capture spatial context and learn correlations between local features, due to which they yield superior performance on image tasks, such as the classification of breast lesions in a screening mammogram as probable malignant or benign (See Figure 1c). CNN-based architectures have also been applied to biomedical segmentation applications [32]. However, CNNs face computational and memory efficiency limitations in three-dimensional (3D) segmentation tasks. More efficient methods have been proposed for the segmentation of 3D data, such as magnetic resonance imaging (MRI) volumes [33]. A recent architecture, occupancy networks for semantic segmentation (OSS-Net) [34], is built upon occupancy networks (O-Net) and contains efficient representations for 3D geometry, which allows for more accurate and faster 3D segmentation [35].

Another family of neural networks, called recurrent neural networks (RNNs), are designed to operate on sequential data. RNNs are well equipped to process sequential inputs of variable lengths for tasks such as machine translation and language modeling. Long Short Term Memory networks (LSTMs) are a special kind of RNNs capable of learning long-term dependencies between inputs [36]. Another technique called attention allows a model to selectively focus on parts of the input data as needed by enhancing specific parts of the input and diminishing others [37]. Recently, a network architecture called the Transformer has achieved state-of-the-art performance in a number of machine learning tasks [38]. Transformers discard recurrence and convolutions entirely, instead relying exclusively on attention mechanisms. Attention-based transformers have demonstrated state-of-the-art segmentation performance and may prove relevance to the field of oncology [39].

3. Breast Cancer

BC is the most prevalent cancer originally reported in National Cancer Institute Statistics, 2020 [40]. BC is a major cause of cancer-related mortality after lung cancer [41]. The death rates of BC have decreased annually from 1989 to 2017, attributed to the advancements in screening and therapies [41]. AI has shown enormous benefits in screening mammograms, BC predictive tools formulation, and drug development [5,6,42–44].

3.1. Screening Mammogram

A screening mammogram is one of the most widely performed screening tests, but these mammograms have limitations of very high false positive and false negative rates [14,42]. The AI models reduced the workload and resulted in a 69% reduction in false positive rates and a higher sensitivity rate in screening mammograms [2,42]. AI in BC screening has good accuracy rates with some methodological issues and evidence gaps [14,45].

In the context of mammography, DL algorithms such as CNNs are principally used; the mechanism of the algorithm is illustrated in Figure 1c. The performance of AI is measured by sensitivity, specificity, the area under the curve (AUC), and computation time [46]. Different DL models have been studied with various classification systems to identify abnormalities in mammograms, with overall sensitivity rates ranging from 88% to 96% [47–49]. Detection rates are augmented by the positive reinforcement of an AUC over 0.96 after biopsy confirmation [50]. A new AI model from Transpara 1.4.0 screenpoint medical BV, Nijmegen, the Netherlands, expedites interpretation and reduces workload by 20–50% by excluding mammograms with a low likelihood of cancer, allowing radiologists to concentrate on challenging cases [2,51]. The detection performance of radiologists using AI-aided systems was compared to radiologists using conventional systems. Radiologists with AI-aided systems achieved higher AUC rates, sensitivity, and classification performance [52,53].

Conventional computer-aided detection (CADe) in mammograms is hampered by high false positive and false negative rates. AI-based CAD systems have proven to reduce false positive rates by 69% and increase in sensitivity ranging from 84% to 91% [42,54]. The concept of double readers (mammogram read by two radiologists independently or together) is used in Europe to reduce false positives and false negatives. The use of AI in place of the second reader maintained a non-inferior performance and reduced the workload by 88% in a simulation study [55]. In another study, a single radiologist assessment was combined with an AI algorithm achieved higher interpretative accuracy with a specificity of 92% vs. 85.9% of a single radiologist's interpretation. However, any single AI algorithm did not outperform radiologists' accuracy rates [14]. Double readers are not a standard practice in the United States, but a prospect of cost-effective AI integration with radiologists can increase overall sensitivity. However, the acceptable miss rate threshold should be carefully considered. Another study used the breast imaging reporting and data system (BI-RADS) to incorporate radiologists' subjective thresholds while using evidence-based data to train AI. The study showed a reduction in false positives by 47.3% and a slight increase in false negatives by 26.7% [56]. AI also has the advantage of not increasing the interpretation time. AI CADe takes 20% less time than traditional CADe, but the same amount of time as radiologists [57]. Although further studies are required to assess the exact costs of AI mammography, the overall reduction in false positives could make it cost-effective [57]. DL models are being incorporated into digital breast tomosynthesis, and contrast-enhanced digital mammography datasets for volumetric assessment of breasts in three dimensions to further increase detection accuracy and reduce workload by 70% [7,58,59]. Radiomics is an approach to extract relevant quantitative properties, also known as features, from clinical, histopathological, and radiological data. It has been applied to breast imaging to further improve accuracy rates [60]. A more detailed description of radiomics is described in Section 5.2.

3.2. Genetics and Hormonal Aspects in Breast Cancer Prediction

Artificial neural networks (ANNs) achieved remarkable accuracy, measured by AUC of 0.909, 0.886, and 0.883, when assessed for their ability to predict 5-, 10-, 15-year BC-related survival rates, respectively, based on factors such as age, tumor size, axillary nodal status, histological type, mitotic count, nuclear pleomorphism, and axillary nodal status [61]. Hybrid-DL models incorporate genetics, histopathology, and radiology data, which outperform traditional models such as Gail (which calculates BC risk in the next five years based on medical and reproductive history, not takes into account BRCA gene association) and Tyrer–Cuzick models (calculates the likelihood of carrying BRCA1 or BRCA2 mutations based on personal and familial historical data) [5,6].

4. Colonic Polyps and Colorectal Cancer

CRC is the third most common cancer in the United States, with the incidence of approximately 147,950 new cases in the year 2020. AI has shown great success in screening, diagnosis, and treatment of CRC. AI is bringing about a new era for CRC screening and detection with computer-assisted techniques for adenoma detection and characterization, computer-aided drug delivery techniques, and robotic surgery. Other benefits of AI include the incorporation of ANN to effectively screen with personal health data [62].

4.1. Colorectal Cancer Screening

By detecting adenomas and preventing progression to carcinoma, screening has significantly reduced the incidence of CRC over the past decade. This has resulted in recommendations for routine screening starting at age 45 [63]. The current screening methods for CRCs include invasive procedures (colonoscopy (gold standard) and flexible sigmoidoscopy), minimally invasive procedures (capsular endoscopy), and non-invasive procedures (CT colonography or virtual colonoscopy, stool for occult blood, fecal immunochemical test, and multitarget stool DNA).

A few AI models have been tested to predict the risk of CRC and high-risk colonic polyps (CPs) from historical data and complete blood counts (CBCs). One such software, ColonFLag, predicts polyps and CRCs according to age, sex, CBC, and demographic information. Scores were compared to gold standard colonoscopy and converted to percentiles, then categories were made, such as CRC, high-risk polyps, and benign polyps [64]. Another retrospective study (MeScore, Calgary, Alberta, Canada) compared CBC results 3–6 months before colonoscopy with those from colonoscopy in two unrelated groups (Israeli and the UK). AUC for CRC diagnosis was 0.82 $\pm$ 0.01. Specificity for 50% detection was 87 $\pm$ 2% a year before diagnosis and 85 $\pm$ 2% for localized cancers [65]. Study results point to the possibility of an early and noninvasive preliminary screening that can be integrated into electronic medical records to flag high-risk patients who can then be aggressively screened to balance the risks and benefits of colonoscopy in young people. Another ANN model designed to screen a large population based only on personal health information from big data also achieved optimal results [62]. However, these models are not currently practiced and require further validation for generalizability.

4.2. Colonic Polyps Detection

Colonoscopy is the gold standard invasive testing for the detection of colonic adenoma and CRC. An adenoma is the most common precancerous lesion. Adenoma detection rate (ADR) measures a gastroenterologist's ability to detect an adenoma. ADR is inversely related to the adenoma miss rate and the risk of post-colonoscopy CRC. ADR ranges from 7% to 53%, while AMRs vary from 6% to 27% based on healthcare facilities. Several factors have been postulated to explain these differences, including quality of preprocedural bowel preparation, time of withdrawal, operator experience and training, procedure sedation, cecal intubation rate, visualization of flexures (blind spots), and use of image enhanced endoscopy and presence of flat or diminutive (less than 5 mm) and small (<10 mm but >5 mm) polyps. Studies show that endoscopists with higher ADR during screening colonoscopy are more effective in preventing subsequent CRC risk for patients [66,67].

In recent years, CADe and computer-aided diagnosis (CADx) systems have been developed to automate polyp detection during colonoscopy and further characterize them. Because of its ability to detect diminutive polyps, real-time AI-aided colonoscopy has a greater ADR than colonoscopy (OR 1.53, 95% CI 1.32–1.77; $p < 0.001$), derived from a metanalysis data [4,68,69]. An AI system, GI Genius, uses green squares to highlight suspicious lesions during a colonoscopy by generating a sound for each marker and displaying it as a video of the endoscopy. Several meta-analyses demonstrate excellent detection rates for polyp detection using AI-assisted algorithms with AUC 0.90, sensitivity 95%, and specificity 88% [8].

4.3. Colon Polyps Classification

AI-based classification of CP into cancerous vs. non-cancerous lesions on CT colonography and capsular endoscopy is a fascinating discovery. CT colonography differentiation by texture analysis based on gradient and curvature of high-order images and random forest models significantly improved the accuracy of the classification of CPs [70,71]. AI-assisted CAD model revealed an inverse correlation of CP sphericity with adenoma detection sensitivity and a direct correlation with adenoma detection accuracy. This model can effectively detect flat colonic lesions and CRCs on CT colonography [72]. Capsule endoscopy is another noninvasive diagnostic tool for gastrointestinal tract inspection, but it is a time-consuming process to process a large amount of data. Stack sparse autoencoding with image manifold constraint, a DL-based AI, is utilized to correctly identify capsular polyps from capsular endoscopic images with a rate of 98% accuracy and time effectiveness [73]. An ANN model with logistic regression showed a predictive risk of distant metastasis in CRC patients based on several clinical factors, such as pathologic stage grouping, first treatment, sex, age at diagnosis, ethnicity, marital status, and high-risk behavior variables [74]. With DL models, tumors can be segmented and delineated more accurately, and faster region-based CNNs are trained to read MRI images, enabling faster and more accurate diagnosis of CRC metastasis [75,76].

4.4. Histopathological Aspects, Genetics, and Molecular Marker Detection

Histopathological characterization is the gold standard for the classification of polyps [77]. However, one of the biggest challenges is the significant intra- and inter-observer variability. The use of DL and CNN models to automate image analysis can allow pathologists to classify CPs with an overall accuracy of 95% or more [10]. These DL models analyze whole slides and hematoxylin- and eosin-stained slides to identify four different stages, including normal mucosa, early preneoplastic lesions, adenomas, and cancer [9,10,78].

AI-based models were used to identify gene expressions, gene profiling, and non-coding micro-ribonucleotides (mi-RNAs) for diagnosis, prognosis, and targeted therapy planning [79–81]. The use of near-infrared (NIR) spectroscopy and counter propagation artificial neural networks (CP-ANNs) in the determination of mutant vs. wild B-rapidly accelerated fibrosarcoma (BRAF) gene mutations were shown to be highly accurate, specific, and sensitive [79]. Mutant BRAF is associated with a poor prognosis, and this AI model can assist in prognosticating and managing these patients aggressively. Backpropagation and learning vector quantization (LVQ) neural networks demonstrate a remarkable role in assessing the genetic profiling database from the cancer genome atlas (TCGA) in improving CRC diagnosis [81]. Several neural networks, including S-Kohonen, backpropagation, and SVM, were compared for predicting the risk of relapse after surgery. The S-Kohonen neural network was found to be the most accurate [82]. Non-coding mi-RNA plays an important role in tumorigenesis and progression of cancer by interfering with various cell signaling pathways, including, WNT/beta-catenin, phosphoinositide-3-kinase (PI3 K)/protein kinase B (Akt), epidermal growth factor receptor (EGFR), NOTCH1, mechanistic target of rapamycin (mTOR), and TP53. The identification of miRNAs through AI models aids in the diagnosis, prognosis, and targeted treatment of CRCs [80,83–86].

In the early detection of CRC, ML-based AI can help isolate circulating tumor cells in peripheral smear and analyze serum specific biomarkers, such as leucine-rich alpha-2-glycoprotein 1 (LRG1), EGFR, inter-alpha trypsin inhibitor heavy chain family member 4 (ITIH4), hemopexin (HPX), and superoxide dismutase 3 (SOD3) [87,88].

5. Central Nervous System Cancers

In the United States, primary brain tumors have an annual incidence of 14.8 per 100,000 people and have a male predominance. Despite significant advances in imaging modalities, surgical techniques, chemotherapy, radiotherapy, and radiosurgery, primary brain tumors such as glioblastoma multiforme (GBM) remains challenging to manage [89]. GBM is one of the primary intracranial neoplasms and accounts for nearly 60% of all primary brain tumors worldwide. Primary or metastatic CNS cancers are challenging to manage because of their rapid proliferation, prominent neovascularization, invasion to distant sites, and poor response to chemotherapy due to the blood–brain barrier. Clinical management includes initial observation, grading, accessing the depth of infiltration, segmentation and location of the tumor, histopathological evaluation, and identification of molecular markers. As a result, clinicians have to manually compile all the data for validation in order to formulate a treatment plan. In this regard, AI has proven to be useful in the diagnosis and management of CNS malignancies [26].

5.1. Central Nervous System Neoplasm Detection

AI has made significant advances in the diagnosis and classification of brain tumors in recent years. MRI is currently the gold standard tool for tumor detection and characterization [90]. Conventional MRI methods such as T_1 and T_2 weighted imaging and fluid-attenuated inversion recovery (FLAIR) sequences have the disadvantage of nonspecific contrast enhancement and a high likelihood of missing tumor foci infiltration. In order to enhance detection chances, perfusion MRI with dynamic susceptibility-weighted contrast material enhancement, dynamic contrast enhancement, and arterial spin labeling are also used to evaluate the neoangiogenic properties of brain tumors such as GBM. In addition to identifying tissue microstructure, diffusion-weighted imaging shows neoplastic infiltration in areas of the brain that appear normal on conventional magnetic resonance (MR) images. The use of MR spectroscopy can also be used to identify chemical metabolites such as choline, creatine, and N-acetyl aspartate, which are useful for glioma grading and identifying tumor infiltrated regions [91]. By automating these steps, AI has enhanced detection rates and efficiency of radiologists, which, in turn, has reduced the amount of time traditionally spent in diagnosing a disease. CNN-based DL can also detect millimeter-sized brain tumors and can distinguish GBMs from metastatic brain lesions [3,92]. MRI technologies provide structured anatomical information on tumors, but tumor differentiation is always based on histopathological evaluation, which is invasive, time-consuming, and expensive. It remains challenging to identify low-grade gliomas from high-grade gliomas on imaging, even with AI systems. Attention-based transformers are currently being investigated for the first time in glioma classification, and their use may offer a breakthrough [39,93].

5.2. Radiomics

A comprehensive analysis of clinical, histopathological, and radiological data combined with ML/DL image processing has paved the way for a new translational field in neuro-oncology called radiomics [60,94,95]. AI-based radiomics provides enhanced noninvasive tumor characterization by enabling histopathologic classification/grading within minutes even at surgery time, prognostication, monitoring, and treatment response evaluation [96,97]. AI algorithms are able to analyze these images at the pixel level, so they can provide information not visible to the human eye and allow for more accurate grading [3]. Radiomics involves a set of the complex multi-step processes with manual, automatic, and semi-automatic segmentations. Two main types of radiomics are described: feature-based and DL-based. Both provide more accurate and reliable results than human readers. The feature-based radiomics algorithms evaluate subsets of specific features from segmented regions and volumes of interest (VOI) into mathematical representations. This multistep process includes image pre-processing (noise reduction, spatial resampling, and intensity modification), precise tumor segmentation (manual vs. DL-based techniques), feature extraction (histogram-based, textural, and higher-order statistics features), feature selection (filter methods, wrapper approaches, and embedded techniques), and model generation and evaluation (neural networks, SVM, decision trees/ random forests, linear regression, and logistic regression models) [95,98]. DL radiomics use CNNs, in which the model learns in a cascading fashion without any prior description of features and requires a large amount of data in the learning process. The cascading technique processes data to obtain useful information, removes redundancies, and prevents overfitting [27,31,98].

5.3. Histopathological Aspects, Genetics, and Molecular Marker Detection

Traditional histopathological evaluation of cranial tumors identifies the microscopic features with areas of neovascularization, central necrosis, endothelial hyperplasia, and regions of infiltration. These are sometimes overlapping and could lead to false-positive results [99]. To overcome this complexity, digital slide scanners are now used to convert microscopic slides into image files interpreted by AI-based algorithms such as SVM and decision trees. SVMs have shown higher precision rates [98]. The AI-based algorithms analyze pathological specimens of gliomas and predict outcomes based on genetic and molecular markers, including isocitrate dehydrogenase (IDH) mutation status, 1 p/19 co-deletion status, O-6-methylguanine-DNA methyltransferase (MGMT) methylation status, epidermal growth factor receptor splice variant III (EGFRvIII), Ki-67 marker expression, prediction of p53 status in gliomas, prediction of mutations in BRAF, and catenin β-1 in craniopharyngiomas [96,98,100–103]. IDH mutation leads to the accumulation of an oncometabolite called D-2 hydroxyglutarate. This mutation is an important prognosticator in GBM. CNN-based AI has detected this biomarker from conventional MRI modalities [100]. O-6-MGMT promoter hypermethylation (encoding for DNA repair protein), which is exhibited in about 33%–57% diffuse gliomas, is a better prognostic factor owing to increased sensitivity to alkylating agents such as temozolomide [98,101,104]. AI types such as supervised machine learning combined with texture features have been found to detect this methylation status. Performing principal component analysis on the final layer of CNN indicated that features, such as nodular and heterogeneous enhancement and "masslike FLAIR edema", predicted MGMT methylation status with up to 83% accuracy [105]. EGFRvIII mutation is found in about 40% of GBM. Tumors with this mutation have been found to exhibit deep peritumoral infiltration, which is consistent with a more aggressive phenotype. EGFR mutation is also associated with increased neovascularization and cell density [106]. 1 p/19 codeletion status has been shown to have a protective effect on the prognosis. This codeletion is observed in oligodendrogliomas [102]. CNN-based AI can be employed to detect this codeletion. Ki-67 marker expression indicates tumor cell proliferation. Traditionally, this marker is detected via immunohistochemical studies on the extracted tumor sample. This method is invasive and time-consuming. Identifying this marker is essential in making a differential diagnosis and treatment plan. AI-based radiomics has been developed to detect this marker from fluorodeoxyglucose positron emission tomography (FDG PET) and MRI images [107].

5.4. *AI in Pre- and Intra-Operative Planning, Postoperative Follow-Up, and Metastasis*

5.4.1. Preoperative Assessment

Segmentation, volumetric assessment, and differentiating the tumor from healthy brain tissue and peripheral edema, quantitative measurements such as risk stratification, treatment response, and outcome prognosis are essential elements in the treatment planning of CNS tumors [108,109]. In traditional radiographic imaging, contrast-enhanced radiographic images are used to estimate tumor volume or burden; however, single-dimension imaging may not be as accurate in the volumetric assessment of nonuniform tumors, such as high-grade tumors including GBMs. Another challenge is differentiating tumor borders from surrounding edema [110]. AI algorithms such the random forest, CNN, and SVM have been applied to the tumor segments to overcome these challenges, and they have been shown to provide precise and accurate localization of the tumor. A two-step protocol with CNN and transfer learning models led to precise and accurate localization of glioma [111]. 3D-U-Net CNN on 18 -fluoroethyl-tyrosine-PET, when used for automated segmentation of gliomas, showed 88% sensitivity, 78% positive prediction, 99% negative prediction, and 99% specificity [32,112].

5.4.2. Intraoperative Modalities

High-grade tumors such as GBM have a rapid proliferation rate and invade the surrounding regions beyond the enhancing regions on the radiological images, and excision of these areas could be missed [26,113]. AI-based DL algorithms have been developed to facilitate the surgeons to remove maximum tumor regions and less of the normal healthy brain tissue simultaneously. Three-dimensional CNNs have shown promising results in aiding stereotactic radiation therapy planning. It is often difficult to differentiate among primary brain tumors, primary CNS lymphoma, and brain metastases in some situations. However, AI-based algorithms such as decision tree and multivariate logistic regression models have been developed to differentiate among these entities by using diffusion tensor imaging and dynamic susceptibility-weighted contrast-enhanced MRI [114–116].

5.4.3. Postoperative Surveillance

MRI with gadolinium contrast is the standard for determining postoperative tumor growth and tumor response [117]. CNN-based AI algorithm techniques determine accurate tumor size compared to linear methods. The ability of CNN models to differentiate the true progression from pseudo-progression and ML algorithms to differentiate radiation necrosis from tumor recurrence is revolutionary [109,110,118]. Additionally, CNN and SVM create a superior model to predict the treatment response and survival outcomes from clinical, imaging, genetic, and molecular marker data [26].

6. Precision and Personalized Medicine

AI has moved towards an era of personalized treatment in oncology with remarkable aid in oncologic drug development, clinical decision support systems, chemotherapy, immunotherapy, and radiation therapy [43]. AI algorithms have been developed to assess several factors such as oncogenetic mutation profile and drug sensitivity prediction showing overall expected prognosis, efficacy, and adverse effects with a particular treatment option in a patient with particular cancer [43,119]. In a study, an ML algorithm was designed to predict the effects of chemotherapy drugs, including gemcitabine and taxols, in correlation to patients' genetic signatures [120]. In another study, an AI-based screening system based on homologous recombination (HR) deficiency was developed to detect cancer cells with HR defects can further narrow patients who would benefit from poly ADP-ribose polymerase (PARP) inhibitors in BC patients [44]. A DL algorithm was used to identify anticancer drugs that inhibit PI3K alpha and tankyrase, promising targets for CRC treatment [121]. An ML-based drug specificity detection by examining protein–protein interactions of anticancer drug and S100A9, a calcium-binding protein, may represent a potential therapeutic target for CRC [122]. These avenues of discovery of new anticancer targeted therapy by ML models is a fascinating step towards much effective therapeutic options. ML models can also be trained to interpret screening data to predict responses to new drugs or combinational therapies [123]. An ability to synthesize and assess a large amount of chemical data also plays a role in cancer drug development by narrowing the prediction towards a specific formula; beyond the traditional experimental methods in which DL systems are currently being explored [124,125]. Learning clinical big data of cancer patients with AI can generate personalized treatment options based on DL assessed factors, including clinical, genetic, cancer-type, and stage of cancer of a patient [126]. Moreover, AI application in radiotherapy is quite distinct. AI can help radiologists plan radiation treatment regimens with automation software as effective as conventional treatment layouts in a robust, time-effective manner [127,128]. With the upcoming role of immunotherapy in managing various cancers, ML-based platforms are trained to predict the therapeutic response of immunotherapy effects in programmed cell death protein 1 (PD-1) sensitive advanced solid tumors [129,130]. AI can thus support and even surpass the capability of humans in anticancer drug development and aid in personalized treatment plans in a time-effective manner.

7. Generalizing Artificial Intelligence, Barriers, and Future Directions

A number of factors challenge the generalizability of AI systems, including possible bias, external validation of AI performance, the requirement for heterogeneous data and standardized techniques [46].

7.1. AI Performance Interpretation

In order for AI to perform in clinical practice, it must be both internally and externally validated. In internal validation, the accuracy of AI is compared to expected results when AI algorithms are tested by using previously used questions [131]. Internal validation performance tools rely on sensitivity, specificity, and AUC. The problem with interpreting AUC is that it does not consider the clinical context. For instance, different sensitivity and specificity can provide similar AUCs. In order to measure AI performance, studies should report AUC along with sensitivities and specificities at clinically relevant thresholds, this is referred to as "net benefit" [132]. As an example, high false-positive and false-negative rates continue to be a challenge in DL screening mammograms, for which balancing the net benefit would be important [42]. Thus, prior to concluding that an AI system can outperform a human reader, it is important to carefully interpret its diagnostic performance. Furthermore, the sensitivity, specificity, and accuracy of diagnostic tests are independent of real-life prevalence. As a result, robust clinical diagnostic, and predictive performance verification of AI for clinical applicability requires external validation. For external validation, a representative patient population and prospectively collected data would be necessary to train AI algorithms [131]. Moreover, internal validation poses the challenge of overestimating AI performance by familiarizing itself too much with training data, known as overfitting [131]. By separating unused training datasets, including newly recruited patients, and comparing results with those of independent investigators at different sites, it is possible to improve generalizability and minimize overfitting [131]. In a recent study, curated large mammogram screening datasets from the UK and the US revealed a promising path to generalizing AI performance [55].

7.2. Standardization of Techniques

An AI model that could be universally applicable must be taught a large amount of heterogeneous clinical data in order to become generalizable [3,54,107]. AI-based infrastructure and data storage systems are not available at all institutes, which is one of the biggest barriers [133]. There is also a lack of standardization of staining reagents, protocols, and section thicknesses of radiologic images, which can further hinder the generalizability of AI in clinical practice worldwide [1,54]. A number of automated CNN-based tools such as HistoQC, Deep Focus, and GAN-based image generators are being developed by societies such as the American College of Radiology Data Science Institute to standardize image sections [1,91]. In the field of radiomics, another challenge involves compliance with appropriate quality controls, ranging from image processing to feature extraction and from mechanics and feature extraction to algorithms for making predictions [134]. There are several emerging initiatives using DLs and CNNs to normalize or standardize images, including, "image biomarker standardization technique" [134,135]. ML algorithms are treated as a "black box" because of a lack of understanding of its inner working. This can pose a challenge when dealing with regulated healthcare data. This necessitates transparent AI algorithms and the interpretation of AI-based results to ensure no mistakes are made [26,136]. A few recently developed methods, such as saliency maps and principal component analysis, are helping interpret the workings of these algorithms [105,137].

7.3. Bias in Artificial Intelligence

Quality and quantity of data are key factors that determine the performance and objectivity of an ML system. AI can be biased in a number of ways—from assumptions made by engineers who develop AI to bias in the data used to train it. When training data are derived from a homogenous population, they may be poorly generalizable, which can potentially exacerbate racial/ethnic disparities, for example [138]. Thus, when training the AI, it is important to include diverse ethnic, age, and sex groups, as well as examples of benign and malignant tumors. Similarly, to integrate precision medicine and AI in real-world clinical settings, it is necessary to consider environmental factors, limitations of care in resource-poor locations, and co-morbidities [139]. There is also the possibility of bias introduced when radiologists' opinion is regarded as the "gold standard" rather than the actual ground truth or the absolute outcome of the case, benign or malignant [46]. As an example, several AI models in screening mammography are compared with radiologists instead of the gold standard biopsy results, introducing bias [46]. In order to overcome this problem, including interval cancers in testing sets and relying on reports from experienced radiologists might be helpful.

7.4. Ethical and Legal Perspectives

Creating future models that address the ethical issues and challenges of incorporating AI into preexisting systems requires an awareness of these issues. Few societies, such as the Department of Health and Social Care, the US Food and Drug Administration, and other global partnerships, oversee and regulate the use of AI in medicine [46,140]. The National Health Service (NHS) Trusts in the United Kingdom regulate the use of patient care data in AI in an anonymized format for research purposes [46]. In order for AI in oncology to achieve global standardization, more international organizations must be formed that can oversee future AI studies within ethical and legal boundaries to protect patient privacy.

8. Integrative Training of Computer Science and Medical Professionals

In order for AI to be effectively integrated into healthcare in general, as well as oncology, formal training of medical professionals and researchers would be critical. Numerous societies and reviews have recommended formal training, but current medical education and health informatics standards do not include mandatory AI education, and competency standards have yet to be established [141,142]. There have been efforts in the radiology community to determine students' opinions about AI applications in radiology in order to develop formal training tools. A few of these are frameworks for teaching, principles for regulating the use of AI tools, special training for evaluating AI technology, and integrating computer science, health informatics, and statistics curriculum during medical school [143–145]. Few institutes in the United States have proposed initiatives for AI in medical education, which were originally submitted by the American Medical Association. Among these initiatives are medical students working with data specialists, radiology residents working with technology base companies to develop computer-aided detection in mammography, offering a summer course by scientists or engineers to update new technologies, and involving medical students in engineering labs to create innovative ideas in health care [136]. Another framework would provide AI training for students in various fields, including medical students, health informatics students, and computer science students [142]. In order to improve patient care, medical students should become proficient in interpreting AI technologies, comparing efficiency in patient care and discussing ethical issues related to using AI tools [142]. Furthermore, medical professionals should understand the limitations and barriers of AI in clinical applications, as well as the distinction between correct and incorrect information [146,147]. In health informatics, students should be taught how to apply appropriate ML algorithms to analyze complicated medical data, integrate data analytics, and formulate questions to visualize large data sets. Students studying computer science should be trained in Python, R, and SQL programming in order to solve complex medical problems [142]. Education tools that integrate medical professionals, health informatics students, and computer science students can pave the way for further developments in the fields of medicine and oncology.

9. Conclusions

Computer systems are capable of learning tasks and predicting outcomes without being explicitly programmed through AI. DL, a subset of ML, utilizes neural networks and enables learning complex, non-linear functions from data. CNNs are well suited to process two- to three-dimensional inputs such as images, while RNNs can handle sequential inputs of variable length such as textual data. Recently developed attention-based DL systems are capable of selectively focusing on data, resulting in better accuracy in cancer detection rates. AI has shown promising results in oncology in several areas, including detection and classification, molecular characterization of tumors, cancer genetics, drug discovery, predicting treatment outcomes and survival rates, and moving the trend towards personalized medicine. In screening mammography, various DL models have demonstrated non-inferior cancer detection performance, with overall sensitivity rates of 88–96%. Radiologists with AI-assisted systems have achieved higher AUC rates and have reduced their workloads. Different real time CADe and CADx AI systems have demonstrated a higher ADR by automating polyp detection and detecting diminutive polyps during colonoscopy. The use of machines to improve cancer detection at an early stage on screening mammograms and colonoscopies has the potential to be tested for application across the globe for more efficient patient care. Several AI-based cancer detection methods have been developed for other cancer types, including lung, prostate, and cervical cancer. It is possible to pursue future objectives to implement AI worldwide in all cancer types.

CNS tumors such as GBM continue to have a poor prognosis. AI-based radiomics allows for the identification of tumors without invasive methods, by allowing for the classification and grading of tumors within minutes. Radiomics is largely used in CNS tumors identification and grading. State-of-the-art attention-based transformers are currently being studied to improve glioma classification. Analyzing histopathological, genetic, or molecular markers can be made easier with AI. With the advancement of AI, oncology has moved to a more personalized era. AI has revolutionized drug development, clinical decision support systems, chemotherapy, immunotherapy, and radiotherapy.

A better understanding of the ethical implications of the use of AI, including its performance interpretation, standardization of techniques, and the identification and correction of bias, is required for more reliable, accurate, and generalizable AI models. Global organizations must be formed to provide guidance and regulation of AI in oncology. Formal integrated training for medical, health informatics, and computer science students could drive further advances of AI in medicine and oncology.

Author Contributions: N.V., V.R. and U.S. were involved in the literature search and writing the manuscript. K.G. and K.R. contributed to the literature search. S.R.S. conceptualized the idea, was involved in writing, review, and revision of the manuscript. All authors have read and agreed to the published version of the manuscript.

Funding: This research study received no external funding.

Conflicts of Interest: All authors declare that they have no conflict of interest.

References

1. Shimizu, H.; Nakayama, K.I. Artificial intelligence in oncology. *Cancer Sci.* **2020**, *111*, 1452–1460. [CrossRef] [PubMed]
2. Rodriguez-Ruiz, A.; Lång, K.; Gubern-Merida, A.; Teuwen, J.; Broeders, M.; Gennaro, G.; Clauser, P.; Helbich, T.H.; Chevalier, M.; Mertelmeier, T.; et al. Can we reduce the workload of mammographic screening by automatic identification of normal exams with artificial intelligence? A feasibility study. *Eur. Radiol.* **2019**, *29*, 4825–4832. [CrossRef] [PubMed]
3. Aneja, S.; Chang, E.; Omuro, A. Applications of artificial intelligence in neuro-oncology. *Curr. Opin. Neurol.* **2019**, *32*, 850–856. [CrossRef] [PubMed]
4. Wang, P.; Berzin, T.M.; Glissen Brown, J.R.; Bharadwaj, S.; Becq, A.; Xiao, X.; Liu, P.; Li, L.; Song, Y.; Zhang, D.; et al. Real-time automatic detection system increases colonoscopic polyp and adenoma detection rates: A prospective randomised controlled study. *Gut* **2019**, *68*, 1813–1819. [CrossRef] [PubMed]
5. Yala, A.; Lehman, C.; Schuster, T.; Portnoi, T.; Barzilay, R. A deep learning mammography-based model for improved breast cancer risk prediction. *Radiology* **2019**, *292*, 60–66. [CrossRef]
6. Akselrod-Ballin, A.; Chorev, M.; Shoshan, Y.; Spiro, A.; Hazan, A.; Melamed, R.; Barkan, E.; Herzel, E.; Naor, S.; Karavani, E.; et al. Predicting breast cancer by applying deep learning to linked health records and mammograms. *Radiology* **2019**, *292*, 331–342. [CrossRef] [PubMed]
7. Raya-Povedano, J.L.; Romero-Martín, S.; Elías-Cabot, E.; Gubern-Mérida, A.; Rodríguez-Ruiz, A.; Álvarez-Benito, M. AI-based Strategies to Reduce Workload in Breast Cancer Screening with Mammography and Tomosynthesis: A Retrospective Evaluation. *Radiology* **2021**, *300*, 57–65. [CrossRef]
8. Lui, T.K.L.; Guo, C.G.; Leung, W.K. Accuracy of artificial intelligence on histology prediction and detection of colorectal polyps: A systematic review and meta-analysis. *Gastrointest. Endosc.* **2020**, *92*, 11–22.e6. [CrossRef]

9. Korbar, B.; Olofson, A.; Miraflor, A.; Nicka, C.; Suriawinata, M.; Torresani, L.; Suriawinata, A.; Hassanpour, S. Deep learning for classification of colorectal polyps on whole-slide images. *J. Pathol. Inform.* **2017**, *8*, 30. [CrossRef]
10. Sena, P.; Fioresi, R.; Faglioni, F.; Losi, L.; Faglioni, G.; Roncucci, L. Deep learning techniques for detecting preneoplastic and neoplastic lesions in human colorectal histological images. *Oncol. Lett.* **2019**, *18*, 6101–6107. [CrossRef]
11. Gates, T.J. Screening for cancer: Evaluating the evidence. *Am. Fam. Physician* **2001**, *63*, 513–522. [PubMed]
12. Pinsky, P.F. Lung cancer screening with low-dose CT: A world-wide view. *Transl. Lung Cancer Res.* **2018**, *7*, 234–242. [CrossRef] [PubMed]
13. Rajkomar, A.; Dean, J.; Kohane, I. Machine Learning in Medicine. *N. Engl. J. Med.* **2019**, *380*, 1347–1358. [CrossRef] [PubMed]
14. Schaffter, T.; Buist, D.S.M.; Lee, C.I.; Nikulin, Y.; Ribli, D.; Guan, Y.; Lotter, W.; Jie, Z.; Beng, H.D.; Wang, S.; et al. Evaluation of Combined Artificial Intelligence and Radiologist Assessment to Interpret Screening Mammograms. *JAMA Netw. Open* **2020**, *3*, e200265. [CrossRef] [PubMed]
15. Rasouli, P.; Moghadam, A.D.; Eslami, P.; Pasha, M.A.; Aghdaei, H.A.; Mehrvar, A.; Nezami-Asl, A.; Iravani, S.; Sadeghi, A.; Zali, M.R. The role of artificial intelligence in colon polyps detection. *Gastroenterol. Hepatol. Bed Bench* **2020**, *13*, 191–199. [PubMed]
16. Mitchell, T.; Jordan, M. Machine learning: Trends, perspectives, and prospects. *Science* **2015**, *349*, 255–260.
17. Mitchell, T. *Machine Learning*; McGraw Hill: New York, NY, USA, 1997; pp. 870–877.
18. Boser, B.E.; Guyon, I.M.; Vapnik, V.N. Training algorithm for optimal margin classifiers. In Proceedings of the Fifth Annual Workshop on Computational Learning Theory, Pittsburgh, PA, USA, 27–29 July 1992; pp. 144–152.
19. Cortes, C.; Vapnik, V. Support-vector networks. *Mach. Learn.* **1995**, *20*, 273–297. [CrossRef]
20. Fix, E.; Hodges, J.L. Discriminatory Analysis. Nonparametric Discrimination: Consistency Properties. *Int. Stat. Rev./Rev. Int. Stat.* **1989**, *57*, 238. [CrossRef]
21. Cover, T.M.; Hart, P.E. Nearest Neighbor Pattern Classification. *IEEE Trans. Inf. Theory* **1967**, *13*, 21–27. [CrossRef]
22. Hartigan, J.A.; Wong, M.A. Algorithm AS 136 A K-Means Clustering Algorithm. *J. R. Stat. Soc. Ser. B Methodol.* **2012**, *28*, 100–108. [CrossRef]
23. Macqueen, J. Some methods for classification and analysis of multivariate observations. In *Proceedings of the Fifth Berkeley Symposium on Mathematical Statistics and Probability, Berkeley, CA, USA, 21 June–18 July 1965, 27 December 1965–7 January 1966*; University of California Press: Berkeley, CA, USA, 1967; Volume 1, pp. 281–297.
24. Frey, B.J.; Dueck, D. Clustering by passing messages between data points. *Science* **2007**, *315*, 972–976. [CrossRef] [PubMed]
25. Clarke, M.R.B.; Duda, R.O.; Hart, P.E. Pattern Classification and Scene Analysis. *J. R. Stat. Soc. Ser. A* **1974**, *137*, 442. [CrossRef]
26. Daisy, P.S.; Anitha, T.S. Can artificial intelligence overtake human intelligence on the bumpy road towards glioma therapy? *Med. Oncol.* **2021**, *38*, 53. [CrossRef] [PubMed]
27. Lecun, Y.; Bengio, Y.; Hinton, G. Deep learning. *Nature* **2015**, *521*, 436–444. [CrossRef]
28. Rumelhart, D.E.; Hinton, G.E.; Williams, R.J. Learning representations by back-propagating errors. *Nature* **1986**, *323*, 533–536. [CrossRef]
29. Robbins, H.; Monro, S. A Stochastic Approximation Method. *Ann. Math. Stat.* **1951**, *22*, 400–407. [CrossRef]
30. Kiefer, J.; Wolfowitz, J. Stochastic Estimation of the Maximum of a Regression Function. *Ann. Math. Stat.* **1952**, *23*, 462–466. [CrossRef]
31. Vaz, J.M.; Balaji, S. Convolutional neural networks (CNNs): Concepts and applications in pharmacogenomics. *Mol. Divers.* **2021**, *25*, 1569–1584. [CrossRef]
32. Ronneberger, O.; Fischer, P.; Brox, T. U-net: Convolutional networks for biomedical image segmentation. In *Medical Image Computing and Computer-Assisted Intervention–MICCAI, Proceedings of the 18th International Conference, Munich, Germany, 5–9 October 2015*; Lecture Notes in Computer Science Series; Springer: Cham, Switzelrand, 2015; Volume 9351, pp. 234–241.
33. Milletari, F.; Navab, N.; Ahmadi, S.A. V-Net: Fully convolutional neural networks for volumetric medical image segmentation. In Proceedings of the 4th International Conference on 3D Vision, 3DV, Stanford, CA, USA, 25–28 October 2016; pp. 565–571.
34. Reich, C.; Prangemeier, T.; Cetin, Ö.; Koeppl, H. OSS-Net: Memory Efficient High Resolution Semantic Segmentation of 3D Medical Data. *arXiv* **2021**, arXiv:2110.10640.
35. Mescheder, L.; Oechsle, M.; Niemeyer, M.; Nowozin, S.; Geiger, A. Occupancy networks: Learning 3D reconstruction in function space. In Proceedings of the IEEE/CVF Conference on Computer Vision and Pattern Recognition (CVPR), Long Beach, CA, USA, 15–20 June 2019; pp. 4455–4465.
36. Hochreiter, S.; Urgen Schmidhuber, J. Long Shortterm Memory. *Neural Comput.* **1997**, *9*, 17351780. [CrossRef]
37. Luong, M.T.; Pham, H.; Manning, C.D. Effective approaches to attention-based neural machine translation. In Proceedings of the 2015 Conference on Empirical Methods in Natural Language Processing, Lisbon, Portugal, 17–21 September 2015; pp. 1412–1421.
38. Vaswani, A.; Shazeer, N.; Parmar, N.; Uszkoreit, J.; Jones, L.; Gomez, A.N.; Kaiser, Ł.; Polosukhin, I. Attention is all you need. In *Advances in Neural Information Processing Systems 30, Proceedings of the 31st Annual Conference on Neural Information Processing Systems (NIPS 2017), Long Beach, CA, USA, 4–9 December 2017*; Curran Associates: Red Hook, NY, USA, 2017; pp. 5999–6009.
39. Prangemeier, T.; Reich, C.; Koeppl, H. Attention-Based Transformers for Instance Segmentation of Cells in Microstructures. In Proceedings of the 2020 IEEE International Conference on Bioinformatics and Biomedicine (BIBM), Seoul, Korea, 16–19 December 2020; pp. 700–707.
40. National Cancer Institute. Cancer Statistics. Available online: https://www.cancer.gov/about-cancer/understanding/statistics (accessed on 28 January 2022).

41. DeSantis, C.E.; Ma, J.; Gaudet, M.M.; Newman, L.A.; Miller, K.D.; Goding Sauer, A.; Jemal, A.; Siegel, R.L. Breast cancer statistics, 2019. *CA Cancer J. Clin.* **2019**, *69*, 438–451. [CrossRef] [PubMed]
42. Batchu, S.; Liu, F.; Amireh, A.; Waller, J.; Umair, M. A Review of Applications of Machine Learning in Mammography and Future Challenges. *Oncology* **2021**, *99*, 483–490. [CrossRef] [PubMed]
43. Liang, G.; Fan, W.; Luo, H.; Zhu, X. The emerging roles of artificial intelligence in cancer drug development and precision therapy. *Biomed. Pharmacother.* **2020**, *128*, 110255. [CrossRef] [PubMed]
44. Gulhan, D.C.; Lee, J.J.K.; Melloni, G.E.M.; Cortés-Ciriano, I.; Park, P.J. Detecting the mutational signature of homologous recombination deficiency in clinical samples. *Nat. Genet.* **2019**, *51*, 912–919. [CrossRef] [PubMed]
45. Houssami, N.; Kirkpatrick-Jones, G.; Noguchi, N.; Lee, C.I. Artificial Intelligence (AI) for the early detection of breast cancer: A scoping review to assess AI's potential in breast screening practice. *Expert Rev. Med. Devices* **2019**, *16*, 351–362. [CrossRef]
46. Hickman, S.E.; Baxter, G.C.; Gilbert, F.J. Adoption of artificial intelligence in breast imaging: Evaluation, ethical constraints and limitations. *Br. J. Cancer* **2021**, *125*, 15–22. [CrossRef]
47. Agnes, S.A.; Anitha, J.; Pandian, S.I.A.; Peter, J.D. Classification of Mammogram Images Using Multiscale all Convolutional Neural Network (MA-CNN). *J. Med. Syst.* **2020**, *44*, 30. [CrossRef]
48. Rodriguez-Ruiz, A.; Lång, K.; Gubern-Merida, A.; Broeders, M.; Gennaro, G.; Clauser, P.; Helbich, T.H.; Chevalier, M.; Tan, T.; Mertelmeier, T.; et al. Stand-Alone Artificial Intelligence for Breast Cancer Detection in Mammography: Comparison With 101 Radiologists. *J. Natl. Cancer Inst.* **2019**, *111*, 916–922. [CrossRef]
49. Al-antari, M.A.; Al-masni, M.A.; Kim, T.S. Deep Learning Computer-Aided Diagnosis for Breast Lesion in Digital Mammogram. *Adv. Exp. Med. Biol.* **2020**, *1213*, 59–72.
50. Aboutalib, S.S.; Mohamed, A.A.; Berg, W.A.; Zuley, M.L.; Sumkin, J.H.; Wu, S. Deep learning to distinguish recalled but benign mammography images in breast cancer screening. *Clin. Cancer Res.* **2018**, *24*, 5902–5909. [CrossRef]
51. Yala, A.; Schuster, T.; Miles, R.; Barzilay, R.; Lehman, C. A deep learning model to triage screening mammograms: A simulation study. *Radiology* **2019**, *293*, 38–46. [CrossRef] [PubMed]
52. Watanabe, A.T.; Lim, V.; Vu, H.X.; Chim, R.; Weise, E.; Liu, J.; Bradley, W.G.; Comstock, C.E. Improved Cancer Detection Using Artificial Intelligence: A Retrospective Evaluation of Missed Cancers on Mammography. *J. Digit. Imaging* **2019**, *32*, 625–637. [CrossRef] [PubMed]
53. Rodríguez-Ruiz, A.; Krupinski, E.; Mordang, J.J.; Schilling, K.; Heywang-Köbrunner, S.H.; Sechopoulos, I.; Mann, R.M. Detection of breast cancer with mammography: Effect of an artificial intelligence support system. *Radiology* **2019**, *290*, 305–314. [CrossRef] [PubMed]
54. Chan, H.P.; Samala, R.K.; Hadjiiski, L.M. CAD and AI for breast cancer—Recent development and challenges. *Br. J. Radiol.* **2020**, *93*, 20190580. [CrossRef]
55. McKinney, S.M.; Sieniek, M.; Godbole, V.; Godwin, J.; Antropova, N.; Ashrafian, H.; Back, T.; Chesus, M.; Corrado, G.C.; Darzi, A.; et al. International evaluation of an AI system for breast cancer screening. *Nature* **2020**, *577*, 89–94. [CrossRef]
56. Zeng, J.; Gimenez, F.; Burnside, E.S.; Rubin, D.L.; Shachter, R. A Probabilistic Model to Support Radiologists' Classification Decisions in Mammography Practice. *Med. Decis. Mak.* **2019**, *39*, 208–216. [CrossRef]
57. Mayo, R.C.; Leung, J.W.T. Impact of artificial intelligence on women's imaging: Cost-benefit analysis. *Am. J. Roentgenol.* **2019**, *212*, 1172–1173. [CrossRef]
58. Zhang, X.; Zhang, Y.; Han, E.Y.; Jacobs, N.; Han, Q.; Wang, X.; Liu, J. Classification of whole mammogram and tomosynthesis images using deep convolutional neural networks. *IEEE Trans. Nanobiosci.* **2018**, *17*, 237–242. [CrossRef]
59. Gao, F.; Wu, T.; Li, J.; Zheng, B.; Ruan, L.; Shang, D.; Patel, B. SD-CNN: A shallow-deep CNN for improved breast cancer diagnosis. *Comput. Med. Imaging Graph.* **2018**, *70*, 53–62.
60. Tagliafico, A.S.; Piana, M.; Schenone, D.; Lai, R.; Massone, A.M.; Houssami, N. Overview of radiomics in breast cancer diagnosis and prognostication. *Breast* **2020**, *49*, 74–80. [CrossRef]
61. Lundin, M.; Lundin, J.; Burke, H.B.; Toikkanen, S.; Pylkkänen, L.; Joensuu, H. Artificial neural networks applied to survival prediction in breast cancer. *Oncology* **1999**, *57*, 281–286. [CrossRef] [PubMed]
62. Nartowt, B.J.; Hart, G.R.; Muhammad, W.; Liang, Y.; Stark, G.F.; Deng, J. Robust Machine Learning for Colorectal Cancer Risk Prediction and Stratification. *Front. Big Data* **2020**, *3*, 6. [CrossRef] [PubMed]
63. Shaukat, A.; Kahi, C.J.; Burke, C.A.; Rabeneck, L.; Sauer, B.G.; Rex, D.K. ACG Clinical Guidelines: Colorectal Cancer Screening 2021. *Am. J. Gastroenterol.* **2021**, *116*, 458–479. [CrossRef] [PubMed]
64. Hilsden, R.J.; Heitman, S.J.; Mizrahi, B.; Narod, S.A.; Goshen, R. Prediction of findings at screening colonoscopy using a machine learning algorithm based on complete blood counts (ColonFlag). *PLoS ONE* **2018**, *13*, e0207848. [CrossRef]
65. Kinar, Y.; Kalkstein, N.; Akiva, P.; Levin, B.; Half, E.E.; Goldshtein, I.; Chodick, G.; Shalev, V. Development and validation of a predictive model for detection of colorectal cancer in primary care by analysis of complete blood counts: A binational retrospective study. *J. Am. Med. Inform. Assoc.* **2016**, *23*, 879–890. [CrossRef] [PubMed]
66. Corley, D.A.; Jensen, C.D.; Marks, A.R.; Zhao, W.K.; Lee, J.K.; Doubeni, C.A.; Zauber, A.G.; de Boer, J.; Fireman, B.H.; Schottinger, J.E.; et al. Adenoma Detection Rate and Risk of Colorectal Cancer and Death. *N. Engl. J. Med.* **2014**, *370*, 1298–1306. [CrossRef]
67. Coe, S.G.; Wallace, M.B. Assessment of adenoma detection rate benchmarks in women versus men. *Gastrointest. Endosc.* **2013**, *77*, 631–635. [CrossRef]

68. Mori, Y.; Kudo, S.E.; Berzin, T.M.; Misawa, M.; Takeda, K. Computer-aided diagnosis for colonoscopy. *Endoscopy* **2017**, *49*, 813–819. [CrossRef]
69. Nazarian, S.; Glover, B.; Ashrafian, H.; Darzi, A.; Teare, J. Diagnostic accuracy of artificial intelligence and computer-aided diagnosis for the detection and characterization of colorectal polyps: Systematic review and meta-analysis. *J. Med. Internet Res.* **2021**, *23*, e27370. [CrossRef]
70. Song, B.; Zhang, G.; Lu, H.; Wang, H.; Zhu, W.; Pickhardt, P.J.; Liang, Z. Volumetric texture features from higher-order images for diagnosis of colon lesions via CT colonography. *Int. J. Comput. Assist. Radiol. Surg.* **2014**, *9*, 1021–1031. [CrossRef]
71. Grosu, S.; Wesp, P.; Graser, A.; Maurus, S.; Schulz, C.; Knösel, T.; Cyran, C.C.; Ricke, J.; Ingrisch, M.; Kazmierczak, P.M. Machine learning-based differentiation of benign and premalignant colorectal polyps detected with CT colonography in an asymptomatic screening population: A proof-of-concept study. *Radiology* **2021**, *299*, 326–335. [CrossRef] [PubMed]
72. Taylor, S.A.; Iinuma, G.; Saito, Y.; Zhang, J.; Halligan, S. CT colonography: Computer-aided detection of morphologically flat T1 colonic carcinoma. *Eur. Radiol.* **2008**, *18*, 1666–1673. [CrossRef] [PubMed]
73. Yuan, Y.; Meng, M.Q.H. Deep learning for polyp recognition in wireless capsule endoscopy images. *Med. Phys.* **2017**, *44*, 1379–1389. [CrossRef] [PubMed]
74. Biglarian, A.; Bakhshi, E.; Gohari, M.R.; Khodabakhshi, R. Artificial neural network for prediction of distant metastasis in colorectal cancer. *Asian Pac. J. Cancer Prev.* **2012**, *13*, 927–930. [CrossRef]
75. Lu, Y.; Yu, Q.; Gao, Y.; Zhou, Y.; Liu, G.; Dong, Q.; Ma, J.; Ding, L.; Yao, H.; Zhang, Z.; et al. Identification of metastatic lymph nodes in MR imaging with faster region-based convolutional neural networks. *Cancer Res.* **2018**, *78*, 5135–5143. [CrossRef]
76. Trebeschi, S.; Van Griethuysen, J.J.M.; Lambregts, D.M.J.; Lahaye, M.J.; Parmer, C.; Bakers, F.C.H.; Peters, N.H.G.M.; Beets-Tan, R.G.H.; Aerts, H.J.W.L. Deep Learning for Fully-Automated Localization and Segmentation of Rectal Cancer on Multiparametric, M.R. *Sci. Rep.* **2017**, *7*, 5301. [CrossRef]
77. Lieberman, D.A.; Rex, D.K.; Winawer, S.J.; Giardiello, F.M.; Johnson, D.A.; Levin, T.R. Guidelines for colonoscopy surveillance after screening and polypectomy: A consensus update by the us multi-society task force on colorectal cancer. *Gastroenterology* **2012**, *143*, 844–857. [CrossRef]
78. Yoon, H.; Lee, J.; Oh, J.E.; Kim, H.R.; Lee, S.; Chang, H.J.; Sohn, D.K. Tumor Identification in Colorectal Histology Images Using a Convolutional Neural Network. *J. Digit. Imaging* **2019**, *32*, 131–140. [CrossRef]
79. Zhang, X.; Yang, Y.; Wang, Y.; Fan, Q. Detection of the BRAF V600E mutation in colorectal cancer by NIR spectroscopy in conjunction with counter propagation artificial neural network. *Molecules* **2019**, *24*, 2238. [CrossRef]
80. Galamb, O.; Barták, B.K.; Kalmár, A.; Nagy, Z.B.; Szigeti, K.A.; Tulassay, Z.; Igaz, P.; Molnár, B. Diagnostic and prognostic potential of tissue and circulating long non-coding RNAs in colorectal tumors. *World J. Gastroenterol.* **2019**, *25*, 5026–5048. [CrossRef]
81. Wang, Q.; Wei, J.; Chen, Z.; Zhang, T.; Zhong, J.; Zhong, B.; Yang, P.; Li, W.; Cao, J. Establishment of multiple diagnosis models for colorectal cancer with artificial neural networks. *Oncol. Lett.* **2019**, *17*, 3314–3322. [CrossRef] [PubMed]
82. Hu, H.P.; Niu, Z.J.; Bai, Y.P.; Tan, X.H. Cancer classification based on gene expression using neural networks. *Genet. Mol. Res.* **2015**, *14*, 17605–17611. [CrossRef] [PubMed]
83. Chang, K.H.; Miller, N.; Kheirelseid, E.A.H.; Lemetre, C.; Ball, G.R.; Smith, M.J.; Regan, M.; McAnena, O.J.; Kerin, M.J. MicroRNA signature analysis in colorectal cancer: Identification of expression profiles in stage II tumors associated with aggressive disease. *Int. J. Colorectal Dis.* **2011**, *26*, 1415–1422. [CrossRef] [PubMed]
84. Amirkhah, R.; Farazmand, A.; Gupta, S.K.; Ahmadi, H.; Wolkenhauer, O.; Schmitz, U. Naïve Bayes classifier predicts functional microRNA target interactions in colorectal cancer. *Mol. Biosyst.* **2015**, *11*, 2126–2134. [CrossRef] [PubMed]
85. Herreros-Villanueva, M.; Duran-Sanchon, S.; Martín, A.C.; Pérez-Palacios, R.; Vila-Navarro, E.; Marcuello, M.; Diaz-Centeno, M.; Cubiella, J.; Diez, M.S.; Bujanda, L.; et al. Plasma MicroRNA Signature Validation for Early Detection of Colorectal Cancer. *Clin. Transl. Gastroenterol.* **2019**, *10*, e00003. [CrossRef]
86. Xuan, P.; Dong, Y.; Guo, Y.; Zhang, T.; Liu, Y. Dual convolutional neural network based method for predicting disease-related miRNAs. *Int. J. Mol. Sci.* **2018**, *19*, 3732. [CrossRef]
87. Gupta, P.; Gulzar, Z.; Hsieh, B.; Lim, A.; Watson, D.; Mei, R. Analytical validation of the CellMax platform for early detection of cancer by enumeration of rare circulating tumor cells. *J. Circ. Biomark.* **2019**, *8*, 1849454419899214. [CrossRef]
88. Ivancic, M.M.; Megna, B.W.; Sverchkov, Y.; Craven, M.; Reichelderfer, M.; Pickhardt, P.J.; Sussman, M.R.; Kennedy, G.D. Noninvasive Detection of Colorectal Carcinomas Using Serum Protein Biomarkers. *J. Surg. Res.* **2020**, *246*, 160–169. [CrossRef]
89. Hanif, F.; Muzaffar, K.; Perveen, K.; Malhi, S.M.; Simjee, S.U. Glioblastoma multiforme: A review of its epidemiology and pathogenesis through clinical presentation and treatment. *Asian Pac. J. Cancer Prev.* **2017**, *18*, 3–9.
90. Brindle, K.M.; Izquierdo-García, J.L.; Lewis, D.Y.; Mair, R.J.; Wright, A.J. Brain tumor imaging. *J. Clin. Oncol.* **2017**, *35*, 2432–2438. [CrossRef]
91. Rudie, J.D.; Rauschecker, A.M.; Bryan, R.N.; Davatzikos, C.; Mohan, S. Emerging Applications of Artificial Intelligence in Neuro-Oncology. *Radiology* **2019**, *290*, 607–618. [CrossRef] [PubMed]
92. Artzi, M.; Bressler, I.; Ben Bashat, D. Differentiation between glioblastoma, brain metastasis and subtypes using radiomics analysis. *J. Magn. Reson. Imaging* **2019**, *50*, 519–528. [CrossRef] [PubMed]

93. Wang, W.; Chen, C.; Ding, M.; Yu, H.; Zha, S.; Li, J. TransBTS: Multimodal Brain Tumor Segmentation Using Transformer. In *Medical Image Computing and Computer Assisted Intervention–Miccai 2021, Proceedings of the 24th International Conference, Strasbourg, France, 27 September–1 October 2021*; Lecture Notes in Computer Science Series; Springer: Cham, Switzerland, 2021; Volume 12901, pp. 109–119.
94. Aerts, H.J.W.L. The potential of radiomic-based phenotyping in precision medicine: A review. *JAMA Oncol.* **2016**, *2*, 1636–1642. [CrossRef] [PubMed]
95. Rizzo, S.; Botta, F.; Raimondi, S.; Origgi, D.; Fanciullo, C.; Morganti, A.G.; Bellomi, M. Radiomics: The facts and the challenges of image analysis. *Eur. Radiol. Exp.* **2018**, *2*, 36. [CrossRef]
96. Forghani, R. Precision Digital Oncology: Emerging Role of Radiomics-based Biomarkers and Artificial Intelligence for Advanced Imaging and Characterization of Brain Tumors. *Radiol. Imaging Cancer* **2020**, *2*, e190047. [CrossRef] [PubMed]
97. National Cancer Institute. Artificial Intelligence Expedites Brain Tumor Diagnosis. Available online: https://mednar.com/mednar/desktop/en/service/link/track?redirectUrl=https%3A%2F%2Fwww.cancer.gov%2Fnews-events%2Fcancer-currents-blog%2F2020%2Fartificial-intelligence-brain-tumor-diagnosis-surgery&collectionCode=MEDNAR-NCI&searchId=5ee02aa9-a656-481b-bbb7 (accessed on 28 January 2022).
98. Abdel Razek, A.A.K.; Alksas, A.; Shehata, M.; AbdelKhalek, A.; Abdel Baky, K.; El-Baz, A.; Helmy, E. Clinical applications of artificial intelligence and radiomics in neuro-oncology imaging. *Insights Imaging* **2021**, *12*, 152. [CrossRef] [PubMed]
99. Bera, K.; Schalper, K.A.; Rimm, D.L.; Velcheti, V.; Madabhushi, A. Artificial intelligence in digital pathology—New tools for diagnosis and precision oncology. *Nat. Rev. Clin. Oncol.* **2019**, *16*, 703–715. [CrossRef]
100. Wu, S.; Meng, J.; Yu, Q.; Li, P.; Fu, S. Radiomics-based machine learning methods for isocitrate dehydrogenase genotype prediction of diffuse gliomas. *J. Cancer Res. Clin. Oncol.* **2019**, *145*, 543–550. [CrossRef]
101. Korfiatis, P.; Kline, T.L.; Coufalova, L.; Lachance, D.H.; Parney, I.F.; Carter, R.E.; Buckner, J.C.; Erickson, B.J. MRI texture features as biomarkers to predict MGMT methylation status in glioblastomas. *Med. Phys.* **2016**, *43*, 2835–2844. [CrossRef]
102. Li, Y.; Liu, X.; Xu, K.; Qian, Z.; Wang, K.; Fan, X.; Li, S.; Wang, Y.; Jiang, T. MRI features can predict EGFR expression in lower grade gliomas: A voxel-based radiomic analysis. *Eur. Radiol.* **2018**, *28*, 356–362. [CrossRef]
103. Chen, X.; Wang, Y.; Yu, J.; Tong, Y.; Shi, Z.; Chen, L.; Chen, H.; Yang, Z. Noninvasive molecular diagnosis of craniopharyngioma with MRI-based radiomics approach. *BMC Neurol.* **2019**, *19*, 6. [CrossRef] [PubMed]
104. Houy, N.; Le Grand, F. Personalized oncology with artificial intelligence: The case of temozolomide. *Artif. Intell. Med.* **2019**, *99*, 101693. [CrossRef] [PubMed]
105. Chang, P.; Grinband, J.G.; Weinberg, B.D.; Bardis, M.; Bardis, M.; Cadena, G.; Su, M.Y.; Cha, S.; Filippi, C.G.; Bota, D.; et al. Deep-learning convolutional neural networks accurately classify genetic mutations in gliomas. *Am. J. Neuroradiol.* **2018**, *39*, 1201–1207. [CrossRef] [PubMed]
106. Tykocinski, E.S.; Grant, R.A.; Kapoor, G.S.; Krejza, J.; Bohman, L.E.; Gocke, T.A.; Chawla, S.; Halpern, C.H.; Lopinto, J.; Melhem, E.R.; et al. Use of magnetic perfusion-weighted imaging to determine epidermal growth factor receptor variant III expression in glioblastoma. *Neuro-Oncology* **2012**, *14*, 613–623. [CrossRef]
107. Lohmann, P.; Galldiks, N.; Kocher, M.; Heinzel, A.; Filss, C.P.; Stegmayr, C.; Mottaghy, F.M.; Fink, G.R.; Jon Shah, N.; Langen, K.J. Radiomics in neuro-oncology: Basics, workflow, and applications. *Methods* **2021**, *188*, 112–121. [CrossRef]
108. Reardon, D.A.; Galanis, E.; DeGroot, J.F.; Cloughesy, T.F.; Wefel, J.S.; Lamborn, K.R.; Lassman, A.B.; Gilbert, M.R.; Sampson, J.H. Wick, W.; et al. Clinical trial end points for high-grade glioma: The evolving landscape. *Neuro Oncol.* **2011**, *13*, 353–361. [CrossRef]
109. Peng, L.; Parekh, V.; Huang, P.; Lin, D.D.; Sheikh, K.; Baker, B.; Kirschbaum, T.; Silvestri, F.; Son, J.; Robinson, A.; et al. Distinguishing True Progression From Radionecrosis After Stereotactic Radiation Therapy for Brain Metastases with Machine Learning and Radiomics. *Int. J. Radiat. Oncol. Biol. Phys.* **2018**, *102*, 1236–1243. [CrossRef]
110. Shaver, M.M.; Kohanteb, P.A.; Chiou, C.; Bardis, M.D.; Chantaduly, C.; Bota, D.; Filippi, C.G.; Weinberg, B.; Grinband, J.; Chow, D.S.; et al. Optimizing neuro-oncology imaging: A review of deep learning approaches for glioma imaging. *Cancers* **2019**, *11*, 829. [CrossRef]
111. Cui, S.; Mao, L.; Jiang, J.; Liu, C.; Xiong, S. Automatic semantic segmentation of brain gliomas from MRI images using a deep cascaded neural network. *J. Healthc. Eng.* **2018**, *2018*, 4940593. [CrossRef]
112. Blanc-Durand, P.; Van Der Gucht, A.; Schaefer, N.; Itti, E.; Prior, J.O. Automatic lesion detection and segmentation of18F-FET PET in gliomas: A full 3D U-Net convolutional neural network study. *PLoS ONE* **2018**, *13*, e0195798. [CrossRef]
113. Hambardzumyan, D.; Bergers, G. Glioblastoma: Defining Tumor Niches. *Trends Cancer* **2015**, *1*, 252–265. [CrossRef] [PubMed]
114. Charron, O.; Lallement, A.; Jarnet, D.; Noblet, V.; Clavier, J.B.; Meyer, P. Automatic detection and segmentation of brain metastases on multimodal MR images with a deep convolutional neural network. *Comput. Biol. Med.* **2018**, *95*, 43–54. [CrossRef] [PubMed]
115. Liu, Y.; Stojadinovic, S.; Hrycushko, B.; Wardak, Z.; Lau, S.; Lu, W.; Yan, Y.; Jiang, S.B.; Zhen, X.; Timmerman, R.; et al. A deep convolutional neural network-based automatic delineation strategy for multiple brain metastases stereotactic radiosurgery. *PLoS ONE* **2017**, *12*, e0185844. [CrossRef] [PubMed]
116. Wang, S.; Kim, S.; Chawla, S.; Wolf, R.L.; Knipp, D.E.; Vossough, A.; O'Rourke, D.M.; Judy, K.D.; Poptani, H.; Melhem, E.R. Differentiation between glioblastomas, solitary brain metastases, and primary cerebral lymphomas using diffusion tensor and dynamic susceptibility contrast-enhanced MR imaging. *Am. J. Neuroradiol.* **2011**, *32*, 507–514. [CrossRef] [PubMed]
117. Liu, Y.; Xu, X.; Yin, L.; Zhang, X.; Li, L.; Lu, H. Relationship between glioblastoma heterogeneity and survival time: An MR imaging texture analysis. *Am. J. Neuroradiol.* **2017**, *38*, 1695–1701. [CrossRef]

118. Zhang, Z.; Yang, J.; Ho, A.; Jiang, W.; Logan, J.; Wang, X.; Brown, P.D.; McGovern, S.L.; Guha-Thakurta, N.; Ferguson, S.D.; et al. A predictive model for distinguishing radiation necrosis from tumour progression after gamma knife radiosurgery based on radiomic features from MR images. *Eur. Radiol.* **2018**, *28*, 2255–2263. [CrossRef]
119. Lind, A.P.; Anderson, P.C. Predicting drug activity against cancer cells by random forest models based on minimal genomic information and chemical properties. *PLoS ONE* **2019**, *14*, e0219774. [CrossRef]
120. Dorman, S.N.; Baranova, K.; Knoll, J.H.M.; Urquhart, B.L.; Mariani, G.; Carcangiu, M.L.; Rogan, P.K. Genomic signatures for paclitaxel and gemcitabine resistance in breast cancer derived by machine learning. *Mol. Oncol.* **2016**, *10*, 85–100. [CrossRef]
121. Berishvili, V.P.; Voronkov, A.E.; Radchenko, E.V.; Palyulin, V.A. Machine Learning Classification Models to Improve the Docking-based Screening: A Case of PI3K-Tankyrase Inhibitors. *Mol. Inform.* **2018**, *37*, 1800030. [CrossRef]
122. Lee, J.; Kumar, S.; Lee, S.Y.; Park, S.J.; Kim, M. Development of predictive models for identifying potential S100A9 inhibitors based on machine learning methods. *Front. Chem.* **2019**, *7*, 779. [CrossRef]
123. Sharma, A.; Rani, R. Ensembled machine learning framework for drug sensitivity prediction. *IET Syst. Biol.* **2020**, *14*, 39–46. [CrossRef] [PubMed]
124. Vamathevan, J.; Clark, D.; Czodrowski, P.; Dunham, I.; Ferran, E.; Lee, G.; Li, B.; Madabhushi, A.; Shah, P.; Spitzer, M.; et al. Applications of machine learning in drug discovery and development. *Nat. Rev. Drug Discov.* **2019**, *18*, 463–477. [CrossRef] [PubMed]
125. Baskin, I.I. The power of deep learning to ligand-based novel drug discovery. *Expert Opin. Drug Discov.* **2020**, *15*, 755–764. [CrossRef] [PubMed]
126. Printz, C. Artificial intelligence platform for oncology could assist in treatment decisions. *Cancer* **2017**, *123*, 905. [CrossRef]
127. Lou, B.; Doken, S.; Zhuang, T.; Wingerter, D.; Gidwani, M.; Mistry, N.; Ladic, L.; Kamen, A.; Abazeed, M.E. An image-based deep learning framework for individualising radiotherapy dose: A retrospective analysis of outcome prediction. *Lancet Digit. Health* **2019**, *1*, e136–e147. [CrossRef]
128. Meyer, P.; Noblet, V.; Mazzara, C.; Lallement, A. Survey on deep learning for radiotherapy. *Comput. Biol. Med.* **2018**, *98*, 126–146. [CrossRef]
129. Sun, R.; Limkin, E.J.; Vakalopoulou, M.; Dercle, L.; Champiat, S.; Han, S.R.; Verlingue, L.; Brandao, D.; Lancia, A.; Ammari, S.; et al. A radiomics approach to assess tumour-infiltrating CD8 cells and response to anti-PD-1 or anti-PD-L1 immunotherapy: An imaging biomarker, retrospective multicohort study. *Lancet Oncol.* **2018**, *19*, 1180–1191. [CrossRef]
130. Bulik-Sullivan, B.; Busby, J.; Palmer, C.D.; Davis, M.J.; Murphy, T.; Clark, A.; Busby, M.; Duke, F.; Yang, A.; Young, L.; et al. Deep learning using tumor HLA peptide mass spectrometry datasets improves neoantigen identification. *Nat. Biotechnol.* **2019**, *37*, 55–71. [CrossRef]
131. Park, S.H.; Han, K. Methodologic guide for evaluating clinical performance and effect of artificial intelligence technology for medical diagnosis and prediction. *Radiology* **2018**, *286*, 800–809. [CrossRef]
132. Halligan, S.; Altman, D.G.; Mallett, S. Disadvantages of using the area under the receiver operating characteristic curve to assess imaging tests: A discussion and proposal for an alternative approach. *Eur. Radiol.* **2015**, *25*, 932–939. [CrossRef]
133. Halling-Brown, M.D.; Warren, L.M.; Ward, D.; Lewis, E.; Mackenzie, A.; Wallis, M.G.; Wilkinson, L.S.; Given-Wilson, R.M.; McAvinchey, R.; Young, K.C. OPTIMAM mammography image database: A large-scale resource of mammography images and clinical data. *Radiol. Artif. Intell.* **2021**, *3*, e200103. [CrossRef] [PubMed]
134. Zwanenburg, A.; Leger, S.; Vallières, M.; Löck, S. Image biomarker standardisation initiative. *Radiology* **2020**, *295*, 328–338. [CrossRef] [PubMed]
135. Drozdzal, M.; Chartrand, G.; Vorontsov, E.; Shakeri, M.; Di Jorio, L.; Tang, A.; Romero, A.; Bengio, Y.; Pal, C.; Kadoury, S. Learning normalized inputs for iterative estimation in medical image segmentation. *Med. Image Anal.* **2018**, *44*, 1–13. [CrossRef] [PubMed]
136. Paranjape, K.; Schinkel, M.; Panday, R.N.; Car, J.; Nanayakkara, P. Introducing artificial intelligence training in medical education. *JMIR Med. Educ.* **2019**, *5*, e16048. [CrossRef] [PubMed]
137. Zeiler, M.D.; Fergus, R. Visualizing and understanding convolutional networks. In *Computer Vision—ECCV 2014, Proceedings of the 13th European Conference, Zurich, Switzerland, 6–12 September 2014*; Springer: Cham, Switzerland, 2014; Volume 8689, pp. 818–833.
138. Noseworthy, P.A.; Attia, Z.I.; Brewer, L.P.C.; Hayes, S.N.; Yao, X.; Kapa, S.; Friedman, P.A.; Lopez-Jimenez, F. Assessing and Mitigating Bias in Medical Artificial Intelligence: The Effects of Race and Ethnicity on a Deep Learning Model for ECG Analysis. *Circ. Arrhythmia Electrophysiol.* **2020**, *13*, e007988. [CrossRef] [PubMed]
139. Johnson, K.B.; Wei, W.Q.; Weeraratne, D.; Frisse, M.E.; Misulis, K.; Rhee, K.; Zhao, J.; Snowdon, J.L. Precision Medicine, AI, and the Future of Personalized Health Care. *Clin. Transl. Sci.* **2021**, *14*, 86–93. [CrossRef] [PubMed]
140. Van Ginneken, B.; Schaefer-Prokop, C.M.; Prokop, M. Computer-aided diagnosis: How to move from the laboratory to the clinic. *Radiology* **2011**, *261*, 719–732. [CrossRef]
141. Mantas, J.; Ammenwerth, E.; Demiris, G.; Hasman, A.; Haux, R.; Hersh, W.; Hovenga, E.; Lun, K.C.; Marin, H.; Martin-Sanchez, F.; et al. Recommendations of the international medical informatics association (IMIA) on education in biomedical and health informatics. *Methods Inf. Med.* **2010**, *49*, 105–120.
142. Hasan Sapci, A.; Aylin Sapci, H. Artificial intelligence education and tools for medical and health informatics students: Systematic review. *JMIR Med. Educ.* **2020**, *6*, e19285. [CrossRef]
143. The Royal College of Radiologists. Clinical Radiology Webinars. Available online: https://www.rcr.ac.uk/clinical-radiology/events/webinars (accessed on 28 January 2022).

144. SFR-IA Group; CERF; French Radiology Community. Artificial intelligence and medical imaging 2018: French Radiology Community white paper. *Diagn. Interv. Imaging* **2018**, *99*, 727–742. [CrossRef]
145. Tang, A.; Tam, R.; Cadrin-Chênevert, A.; Guest, W.; Chong, J.; Barfett, J.; Chepelev, L.; Cairns, R.; Mitchell, J.R.; Cicero, M.D.; et al. Canadian Association of Radiologists White Paper on Artificial Intelligence in Radiology. *Can. Assoc. Radiol. J.* **2018**, *69*, 120–135. [CrossRef] [PubMed]
146. Park, S.H.; Do, K.H.; Kim, S.; Park, J.H.; Lim, Y.S. What should medical students know about artificial intelligence in medicine? *J. Educ. Eval. Health Prof.* **2019**, *16*, 1149130. [CrossRef] [PubMed]
147. Hasan Sapci, A.; Aylin Sapci, H. Teaching hands-on informatics skills to future health informaticians: A competency framework proposal and analysis of health care informatics curricula. *JMIR Med. Inform.* **2020**, *8*, e15748. [CrossRef] [PubMed]

MDPI
St. Alban-Anlage 66
4052 Basel
Switzerland
Tel. +41 61 683 77 34
Fax +41 61 302 89 18
www.mdpi.com

MDPI Books Editorial Office
E-mail: books@mdpi.com
www.mdpi.com/books

www.ingramcontent.com/pod-product-compliance
Lightning Source LLC
Chambersburg PA
CBHW041014270926
51850CB00024B/39